Computer Graphics
and
Geometric Modeling

Springer Science+Business Media, LLC

David Salomon

Computer Graphics and Geometric Modeling

With 335 Illustrations

 Springer

David Salomon
Department of Computer Science
California State University
Northridge, CA 91330-8281
USA

Library of Congress Cataloging-in-Publication Data
Salomon, D. (David), 1938–
    Computer graphics and geometric modeling / David Salomon.
      p.   cm.
    Includes bibliographical references and index.
    ISBN 978-1-4612-7170-3   ISBN 978-1-4612-1504-2 (eBook)
    DOI 10.1007/978-1-4612-1504-2
    1. Computer graphics.   2. Mathematical models.   I. Title.
T385.S243   1999
006.6—dc21   98-33424

Printed on acid-free paper.

© 1999 Springer Science+Business Media New York
Originally published by Springer-Verlag New York, Inc. in 1999
Softcover reprint of the hardcover 1st edition 1999

All rights reserved. This work may not be translated or copied in whole or in part without the written permission of the publisher Springer Science+Business Media, LLC, except for
brief excerpts in connection with reviews or scholarly analysis. Use in connection with any form of information storage and retrieval, electronic adaptation, computer software, or by similar or dissimilar methodology now known or hereafter developed is forbidden.
The use of general descriptive names, trade names, trademarks, etc., in this publication, even if the former are not especially identified, is not to be taken as a sign that such names, as understood by the Trade Marks and Merchandise Marks Act, may accordingly be used freely by anyone.

Production managed by Jenny Wolkowicki; manufacturing supervised by Jeffrey Taub.
Camera-ready copy provided by the author.

9 8 7 6 5 4 3 2 1

ISBN 978-1-4612-7170-3

To my family, without whose help, patience, and support this book would not have been written.

Preface

Joseph-Louis Lagrange (1736–1813), one of the greatest mathematicians of the 18th century, made important contributions to the theory of numbers and to analytical and celestial mechanics. His most important work is *Mécanique Analytique* (1788), the textbook on which all subsequent work in this field is based. A contemporary reader is surprised to find no diagrams or figures of any kind in this book on mechanics. This reflects one extreme approach to graphics, namely considering it unimportant or even detracting as a teaching tool and not using it. Today, of course, this approach is unthinkable. Graphics, especially computer graphics, is commonly used in texts, advertisements, and movies to illustrate concepts, to emphasize points being discussed, and to entertain.

Our approach to graphics has been completely reversed since the days of Lagrange, and it seems that much of this change is due to the use of computers. Computer graphics today is a mature, successful, and growing field. It is used by many people for many purposes and it is enjoyed by even more people. One criterion for the maturity of a field of study is its size. When a certain discipline becomes so big that no one person can keep all of it in their head, we say that that discipline has matured (or has come of age). This is what happened to computer graphics in the last decade or so. It is now a large field consisting of many subfields such as curve and surface design, rendering methods, and computer animation. Even a person who has written a book covering the entire field cannot claim that they keep all that material in their head all the time, which is precisely the reason why textbooks are being written.

> Lagrange was born in Turin, Italy and his original name was Giuseppe Luigi LaGrangia. Here is a quote from the preface to his book:
>
> "The reader will find no figures in this work. The methods which I set forth do not require either constructions or geometrical or mechanical reasonings, but only algebraic operations, subject to a regular and uniform rule of procedure."

The material presented here has been developed during many years of teaching computer graphics. It has been revised and distilled many times, with many examples and exercises added. The text emphasizes the mathematics behind computer graphics and is intended for readers who want to understand how graphics programs work and how present-day computer graphics can generate and display realistic-looking curves, surfaces, and solid objects.

Most of the necessary mathematical background (such as vectors and matrices) is covered in the Appendix. However, some math concepts that are used only once (such as the mediation operator and points vs. vectors) are discussed right where they are introduced.

The many exercises sprinkled in the text are not a cosmetic feature. They deal with important topics, and should be worked out. Answers are provided but they should be consulted only as a last resort.

The main topics covered in this book are the following:

1. Scan conversion methods. These are used to select the best pixels for generating lines, circles, and other geometrical figures.

2. Geometric transformations and projections. This topic starts with simple concepts of translating and rotating objects, and ends with a complete picture of how an observer (or a camera) can be made to move around a three-dimensional scene, step by step, and project it on a two-dimensional screen. Perspective projections (a traditionally confusing topic) are discussed in two ways, first using the traditional approach, found in most texts, then using a novel, coordinate-free, vector-based approach. As a bonus, transformations and projections in many dimensions are discussed, as well as several interesting types of nonlinear projections.

3. Curves and surfaces. A realistic-looking object is calculated and displayed on the screen by simulating the light reflected from its surface, which is why surfaces are important in computer graphics. The key to constructing a surface, however, is to know how to draw a single curve. This is the main reason why curves are also important. The discussion is mathematical and it proceeds from curves to wire-frame surfaces, to the calculation of the normal to a surface, to the rendering of a solid surface. Quite a few methods are discussed, each illustrated by examples.

4. Several other topics are presented, such as CRT operation, antialiasing, computer animation, color and color perception, halftoning and dithering, polygonal surfaces, and compression of graphics files.

An important feature of this text is the attention to "orphans." Those are topics that most texts on computer graphics either mention briefly or completely ignore. Examples are perspective projections, curves, surfaces, quaternions, and image compression. The reader will find that this text discusses orphans in great detail, including numerous examples and exercises.

The book is intended as a textbook for a two-course sequence on computer graphics and geometric modeling for graduate and advanced undergraduate students. However, it is the author's belief that the book can also serve as a professional book, widening the horizons of professionals in other fields who are interested in a thorough, mathematical exposition of the principles, methods, and techniques used in computer graphics. The mathematical background required for a complete understanding of this material includes polynomials, vectors, matrices, determinants, and differentiation.

Preface

Historical Notes

The history of computer graphics started in the early 1950s. This is very early, considering that the history of the modern digital electronic computer itself started in the late 1940s. However, due to high hardware prices, the field was originally the domain of a few lucky individuals, and it was only in the 1970s that it started growing fast and eventually became the wide discipline that we know today. Here is a short chronology.

By 1951 the Whirlwind computer installed at MIT had two 16-inch graphics displays (actually, modified oscilloscopes). Surprisingly, there were no immediate users.

Plotters came into use as graphics output devices in 1953.

In 1955, the SAGE air defense system started its operations. It used vector-scanned monitors as its main output and light pens as its input devices.

Digital Equipment Corporation (DEC) was founded in 1957. It started making minicomputers that were later used in the early development of computer graphics.

Light pens came into wide use in 1958, the same year as the first microfilm recorder.

In 1959, a partnership of General Motors and IBM produced the first piece of drawing software, the DAC-1 (Design Augmented by Computers). Users could input the three-dimensional description of a car, view the car in perspective, and rotate it.

It was in the 1960s that the field got its first big push. In 1961, Ivan Sutherland developed Sketchpad, a drawing program, as his Ph.D. thesis at MIT. Sketchpad used a light pen as its main input device and an oscilloscope (modified to do vector scan) as its output device. The first version handled two-dimensional figures only, and was later extended to draw, transform, and project three-dimensional objects, and also to perform engineering calculations such as stress analysis. One important feature of Sketchpad was its ability to recognize constraints. The user could draw, e.g., a rough square, then instruct the software to convert it to an exact square. Another feature was the ability to deal with objects, not just individual curve segments. The user could build an object out of segments, then ask the software to scale it. For information and images related to Sketchpad, see http://www.sun.com/960710/feature3/sketchpad.html#sketch.

At about the same time, Steven Russell, another MIT student, developed the first video game, *Spacewar*. This program was written for the PDP-1 and was later used by DEC salesmen to attract customers for that minicomputer.

In 1963, the first computer-generated film, titled *Simulation of a two-giro gravity attitude control system*, was created by E. E. Zajac at Bell laboratories. Other researchers at Bell, Boeing, and Lawrence Radiation Laboratory followed soon with more films.

The first digitizer, the RAND tablet, appeared in 1964.

In the mid-1960s, interest in computer graphics was picking up. More and more companies started projects involving graphics, which gave IBM the idea of developing the first graphics terminal, the IBM 2250. At about the same time, David Evans and Ivan Sutherland cofounded their company which made, among other things, vector scan displays. Those displays are historically important since they gave a tremendous boost to computer graphics throughout the sixties.

In 1966, Sutherland developed the first three-dimensional head-mounted display (HMD). It displayed a stereoscopic pair of wire-frame images. This device was rediscovered in the 1980s and is commonly used today in virtual-reality applications.

In the late 1960s, both Sutherland and Evans were invited to develop a program in computer science at the University of Utah in Salt Lake City. Computer graphics quickly became the specialty of their department, and for years maintained its position as the primary world center for this field. Many important methods and techniques were developed at the UU computer graphics lab, among them illumination models, hidden-surface algorithms, and basic rendering techniques for polygonal surfaces. Names of UU students such as Phong Bui-Tuong, Henri Gouraud, James Blinn, and Ed Catmull are associated with many basic algorithms still in use today. A short history of computer graphics at UU can be found at URL http://www.cs.utah.edu/~riloff/cs-history.html.

Computer graphics in the 1960s was expensive since hardware was expensive. There were no personal computers or workstations. Users had to pay for mainframe time by the second or buy expensive minicomputers. Display monitors used vector scan and were black and white. The result was that only computer professionals could do computer graphics and the software was noninteractive and nonportable.

The advent of the microprocessor, in the mid-1970s, was another factor in the rapid advance of computer graphics. Personal computers appeared on the market and suddenly anyone could afford to own a computer. This encouraged the formation of small companies that developed computer animation, mostly to be used in TV commercials. Names such as Abel and Associates, Information International Inc., Digital Effects, and Systems Simulation Ltd. became well known and produced short pieces that demonstrated dazzling effects.

SIGGRAPH, the Special Interest Group on Computer Graphics (part of the ACM), held its first conference in 1973. It attracted 1200 attendees and later conferences boasted as many as 30,000 participants and hundreds of exhibitors.

The famous Utah teapot (see page 479) was developed in 1975. This is perhaps the best known three-dimensional model in computer graphics. The original teapot this model is based on can be seen at the Computer Museum in Boston.

It was during the 1970s that activity in basic computer graphics research started moving from UU first to NYIT, the New York Institute of Technology, then to Lucasfilm. Computer animation and computer painting were two topics seriously developed at those places.

The technique of (and hardware for) raster scan was developed in the 1970s by Richard Shoupe at Xerox Palo Alto Research Center (PARC). Workers in the field soon realized the advantages of raster scan and the word "pixel" entered the field of computer graphics.

Like any other mature discipline, computer graphics eventually got its first periodic publication. *Computer Graphics World* started carrying news and reviews in late 1977.

Fractals, developed by Benoit Mandelbrot in the 1960s and 1970s, were applied to computer graphics in the late 1970s by Loren Carpenter and others.

Preface

Ray tracing, a sophisticated rendering method, was developed by Turner Whitted of Bell labs and published in 1980.

Silicon Graphics Inc. (SGI) was founded in 1982 and has been building high-performance graphics computers since.

The technique of particle systems was developed in the early eighties at Lucasfilm. Morphing was developed at the same time at NYIT.

The data glove, very popular today for virtual-reality applications, was developed at Atari in 1983.

Radiosity came out of Cornell University in 1984. This is a sophisticated rendering method that simulates light reflection between surfaces by determining the exchange of energy between them.

GUI, graphical user interfaces, appeared in 1984 with the release of the first Macintosh computer.

In 1985 came the first ISO standard, the High Sierra, for CD-ROMs. The Commodore Amiga personal computer was also introduced the same year. It immediately became popular for what today are called multimedia applications.

The 1980s saw the emergence of raster-scan display monitors as the main graphics output device. This technology has benefited from experience gained with television and has resulted in the cheap, reliable color monitors of today.

The late 1980s and early 1990s also saw the developments of graphics standards such as GKS and PHIGS.

MS Windows 3.0 was first shipped in 1990 and, of course, gave a tremendous boost to the concept of GUI. More and more applications were developed to run under MS Windows.

Released in 1997, *Toy Story* is the first full-length (79 minutes, which translates to more than 114,000 animation frames at 24 frames per second) feature film that's completely computer-animated. It represents a milestone in computer graphics and marks the beginning of an era when computer graphics rendering techniques have become so sophisticated that viewers may find it impossible to tell if an image is real or if it is a mathematical model being rendered.

> Shaw's plays are the price we pay for Shaw's prefaces.
> — James Agate

Resources for Computer Graphics

As is natural to expect, the World Wide Web has many resources for computer graphics. There are also many periodicals on various aspects of graphics. Here is a list of some of the most important resources, current as of 1998.

- http://www.siggraph.org/ is the official home page of SIGGRAPH, the special interest group for graphics, one of many SIGs that are part of the ACM.

- http://www.siggraph.org/conferences/fundamentals has course notes from SIGGRAPH conferences.

- http://www.primenet.com/~grieggs/cg_faq.html by John Grieggs contains answers to frequently asked questions on graphics, as well as pointers to other resources.

- The most recent version of Richard Parent's book on computer animation is at http://www.cis.ohio-state.edu/~parent/book/outline.html.

- http://mambo.ucsc.edu/psl/cg.html is a jumping point to many sites that deal with computer graphics.

- A similar site is http://www.cs.rit.edu/~ncs/graphics.html that also has many links to CG sites.

- A very extensive site of computer-graphics-related pointers is http://ls7-www.informatik.uni-dortmund.de/html/englisch/servers.html.

- Search the Internet under "history of computer graphics" for many sites.

- It is also a good idea to search the Internet for subjects such as computer graphics, computer animation, image processing, computer vision, and computer-assisted design (CAD).

- *IEEE Computer Graphics and Applications* is a technical journal carrying research papers and news. See http://computer.org/cga.

- *Animation Magazine* is a monthly publication covering the entire animation field, computer and otherwise. Check either at http://animag.com/ or URL http://www.bcdonline.com/animag/.

- *Computer Graphics World* is a monthly publication concentrating on news (http://www.cgw.com/).

- *Digital Imaging* is a bimonthly reporting on the digital imaging industry.

A Word on Notation

It is common to represent nonscalar quantities such as points, vectors, and matrices with boldface. Below are example of the notation used here:

x, y, z, t, u, v	Italics are used for scalar quantities such as coordinates and parameters.
$\mathbf{P}, \mathbf{Q}_i, \mathbf{v}, \mathbf{M}$	Boldface is used for points, vectors, and matrices.
$\vec{\mathbf{CP}}$	An alternative notation for vectors, used when the two endpoints of the vector are known.
$\mathbf{P}(t), \mathbf{P}(u,v)$	Boldface with arguments is used for nonscalar functions such as curves and surfaces.
$\begin{pmatrix} a_{11} & a_{12} \\ a_{21} & a_{22} \end{pmatrix}$	Parentheses are used for matrices.

Preface

$\begin{vmatrix} a_{11} & a_{12} \\ a_{21} & a_{22} \end{vmatrix}$	Vertical bars are used for determinants.
$\|\mathbf{v}\|$	The absolute value (length) of vector $\mathbf{v}$.
$\mathbf{A}^T$	The transpose of matrix $\mathbf{A}$.
$x^*, \mathbf{P}^*$	The transformed values of scalars and points.
$f^u(u), \mathbf{P}^t(t), \mathbf{P}^{tt}(t)$	The derivatives (first, second,...) of scalar and vector functions.
$\dfrac{df(u)}{du}, \dfrac{d\mathbf{P}(t)}{dt}$	Alternative notation for derivatives.
$\dfrac{df^2(u)}{du^2}, \dfrac{d\mathbf{P}^2(t)}{dt^2}$	Alternative notation for higher-order derivatives.
$\dfrac{\partial f(u,v)}{\partial u}, \dfrac{\partial \mathbf{P}(u,v)}{\partial v}$	Partial derivatives.
$f(x)\|_{x_0}$ or $f(x_0)$	Value of function $f(x)$ at point x_0.
$\sum_{i=1}^{n} x_i$	The sum $x_1 + x_2 + \cdots + x_n$.
$\prod_{i=1}^{n} x_i$	The product $x_1 x_2 \ldots x_n$.

▶ **Exercise 1:** What is the meaning of $(\mathbf{P}_1, \mathbf{P}_2, \mathbf{P}_3, \mathbf{P}_4)$?

⏩ The attention symbol, shown on the left, is used to attract the reader's attention when important concepts are introduced.

The left triangle ◀ is the QED symbol, indicating the end of a proof (of which there are just a few in this book).

A book of this magnitude cannot be written without the help and encouragement of many people, but this book is an exception. The author would like to thank Nelson H. F. Beebe who went over the manuscript, made important suggestions, and pointed out many errors. J. Robert Henderson rendered valuable help with mathematical topics. Apart from this, the entire text is the product of the author who alone is responsible for any remaining mistakes, errors, and omissions.

The author welcomes any comments, suggestions and corrections. They should be e-mailed to david.salomon@csun.edu. An errata list, as well as other information, will be kept on the author's web page http://www.ecs.csun.edu/~dxs.

Fasten your seatbelts, it's going to be a bumpy night!
— Bette Davis (as Margo Channing) *All About Eve (1950)*

Contents

	Preface		vii
1.	**First Principles**		1
	1 Graphics Output	2	
	2 Bitmap Scaling	16	
	3 Bitmap Rotation	20	
	4 A Practical Drawing Program	23	
2.	**Scan-Converting Methods**		27
	1 Scan-Converting Lines	27	
	2 Midpoint Subdivision	28	
	3 DDA Methods	28	
	4 Double-Step DDA	34	
	5 Best-Fit DDA	38	
	6 Scan-Converting in Parallel	39	
	7 Scan-Converting Circles	42	
	8 Thick Curves	49	
	9 Antialiasing	50	
3.	**Transformations and Projections**		59
	1 Introduction	59	
	2 Two-Dimensional Transformations	61	
	3 Windowing	88	
	4 Clipping	90	
	5 Three-Dimensional Transformations	91	
	6 Transforming the Coordinate System	102	
	7 Projections	103	
	8 Parallel Projections	103	
	9 Perspective Projections	112	
	10 Application: Stereo Image	137	
	11 The Viewing Volume	141	
	12 Going Beyond the Third Dimension	144	
	13 Nonlinear Projections	153	

4. Curves — 173

1	Points and Vectors	174
2	Parametric Blending	180
3	Curve Representations	181
4	The Lagrange Polynomial	198
5	The Newton Polynomial	205
6	Spline Methods for Curves	206
7	Hermite Interpolation	207
8	The Cubic Spline Curve	225
9	The Quadratic Spline	247
10	Cardinal Splines	248
11	Parabolic Blending: Catmull-Rom Curves	251
12	Kochanek-Bartels Splines	258
13	Fitting a PC to Experimental Points	262
14	The Bézier Curve	266
15	Subdivision Curves	321
16	The B-Spline	328
17	The Beta Spline	389
18	Barycentric Sums Revisited	393
19	Symmetry in Curves	394
20	Conic Sections	397
21	Parametric Space of a Curve	401
22	Curvature and Torsion	402
23	The Hough Transform	410

5. Surfaces — 415

1	Input Three-Dimensional Points	416
2	Basic Concepts	417
3	Polygonal Surfaces	419
4	Delaunay Triangulation	427
5	Bilinear Surfaces	434
6	Lofted Surfaces	439
7	Coons Surfaces	443
8	The Cartesian Product	456
9	The Biquadratic Surface Patch	457
10	The Bicubic Surface Patch	459
11	Catmull-Rom Surfaces	468
12	Rectangular Bézier Surfaces	471
13	Triangular Bézier Surfaces	483
14	Converting Bézier Patches	488
15	The Gregory Patch	495
16	Gordon Surfaces	498
17	Uniform B-Spline Surfaces	499
18	Surfaces of Revolution	516
19	Sweep Surfaces	526
20	Polygonal Surfaces by Subdivision	530
21	Curves on Surfaces	533
22	Surface Normals	535

Contents

6. Rendering **537**

1	Introduction	537
2	A Simple Shading Model	538
3	Gouraud and Phong Shading	548
4	Palette Optimization	549
5	Ray Tracing	551
6	Texturing	552
7	Bump Mapping	554

7. Color **557**

1	Color and the Eye	557
2	The HLS Color Model	559
3	The HSV Color Model	559
4	The RGB Color Model	561
5	Additive and Subtractive Colors	563
6	Complementary Colors	567
7	Spectral Density	567
8	The CIE Standard	571

8. Computer Animation **575**

1	Background	575
2	Interpolating Positions	578
3	Interpolating Orientations: I	583
4	Interpolating Orientations: II	593
5	Nonuniform Interpolation	600
6	Morphing	606
7	Free-Form Deformations	607

9. Image Compression **609**

1	Introduction	610
2	Variable-Size Codes	611
3	Run-Length Encoding	612
4	Fax Compression	615
5	Cell Encoding	622
6	Quadtrees	624
7	Progressive Image Compression	630
8	FELICS	634
9	The Golomb Code	642
10	Progressive FELICS	643
11	MLP	646
12	Differential Lossless Image Compression	654
13	Wavelets	656

10. Short Topics — 661

1	Graphics Standards	661
2	Boundary Fill	668
3	Halftoning	669
4	Dithering	671
5	Fractals	680
6	A Fractal Line	681
7	Branching Rules	684
8	Iterated Function Systems (IFS)	685
9	Image Processing	688

A. Mathematical Topics — 693

1	Fourier Transforms	693
2	Forward Differences	698
3	Coordinate Systems	700
4	Vector Algebra	702
5	Matrices	709
6	Trigonometric Identities	711
7	The Greek Alphabet	715
8	Complex Numbers	716
9	Quaternions	717
10	Groups	719
11	Fields	720

References — 723

Answers to Exercises — 733

Index — 833

> To me style is just the outside of content, and content the inside of style, like the outside and the inside of the human body both go together, they can't be separated.
> — Jean-Luc Godard

Computer Graphics
and
Geometric Modeling

1
First Principles

"An image is worth a thousand words" is a well-known phrase. It reflects the truth because the human eye–brain combination is a high-capacity, sophisticated data processor. We can "digest" a huge amount of information if we receive it visually, as an image, rather than as a list of numbers. This is the reason for the success of computer graphics. However, if one image conveys a lot of information, many images are even more informative. This is why *computer animation* is so popular. A piece of animation can both teach and entertain.

> So it is not just science—reasoning about the physical world—that involves virtual reality. All reasoning, all thinking, and all external experience are forms of virtual reality.
> — David Deutsch *The Fabric Of Reality*.

Computer graphics has progressed over the years in three stages. The first stage was to develop algorithms and software to calculate and display a single image consisting of smooth, curved, realistic-looking surfaces. The second stage was to extend the basic algorithms in order to create and display an entire animation made of many *frames*, where each frame is an image. The third stage is *virtual reality*, where computer graphics goes one step beyond passively watching animation. The main features of virtual reality are:

1. Interaction. Once a virtual three-dimensional world is built, a user can walk through it at will, sometimes also "grabbing" objects and manipulating them.

2. Realistic views. With the use of a special helmet where each eye watches its own display, the user can look in one direction while moving in another. We say that this adds *degrees of freedom* to the display. The helmet also allows a stereo pair of images to be displayed, adding to the visual realism.

▶ **Exercise 1.1:** Use your crystal ball (or, in its absence, the answer provided here) to predict the next stage beyond virtual reality.

1.1 Graphics Output

The two traditional graphics output devices are the CRT (Cathode Ray Tube) and the printer (hard copy). They are both two-dimensional, so a three-dimensional image has to be *projected* before it can be output (see Section 3.7 for projections). There are many more graphics output devices currently used, but the CRT is the only one discussed here.

> For such an advanced civilization as ours to be without images that are adequate to it is as serious a defect as being without memory.
> — Werner Herzog

1.1.1 The CRT

An image can be displayed by a computer on a CRT in two different ways, *raster scan* and *vector scan* (the latter is sometimes called *random scan*). Both methods use the CRT as the output device (Figure 1.1a), but they differ in many respects. They control the electron beam in the CRT in different ways, they represent the graphics data in memory differently, and they also require different pieces of hardware to interface the computer to the CRT.

A CRT is the same kind of tube used in a common TV set. It has an electron gun (the cathode) that emits a stream of electrons (Figure 1.1a). The front surface is positively charged, so it attracts the electrons, and is coated with a phosphor compound that emits light when hit by the beam. The flash of light only lasts a fraction of a second, so, in order to get a constant display, the picture has to be refreshed about 20 times a second. (The actual refresh rate depends on the *persistence* of the compound (Figure 1.1b). For certain types of work, such as architectural drawing, long persistence is acceptable. For animation, short persistence is a must.)

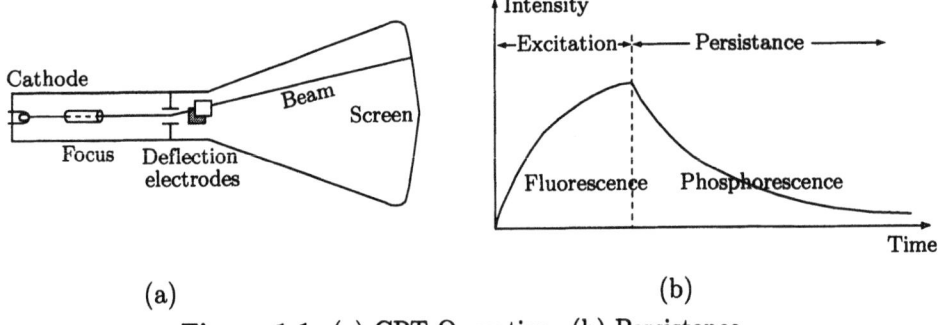

Figure 1.1: (a) CRT Operation. (b) Persistence.

The electron beam can be turned off and on very rapidly. It can also be deflected horizontally and vertically by two pairs (X and Y) of electrodes. Displaying

1 First Principles

a single point on the screen is done by turning the beam off, moving it to the part of the screen where the point should appear, and turning it on. This is done by special hardware (the CRT controller or graphics interface) that receives information from the program.

1.1.2 Standard Television

A consumer television set uses one of three international standards. The standard used in the United States is called NTSC (National Television Standards Committee), although the new digital standard (Section 1.1.3) is slowly becoming popular. NTSC specifies a television transmission of 525 lines (today, this would be $2^9 = 512$ lines but since television was developed before the advent of computers and binary numbers, the NTSC standard has nothing to do with powers of 2). In practice, though, only 480 lines are visible on the screen. Since the aspect ratio (height/width) of a television screen is $3:4$, each line has a size of $\frac{4}{3}480 = 640$ pixels. The resolution of a standard television set is thus 480×640. This may be considered, at best, medium resolution. (This is the reason why text is so hard to read on a standard television.) For more information on standard television, see [Pritchard 77].

Many computer graphics applications cannot, therefore, use a standard television set as a screen. Current high-quality CRTs start at a resolution of $1K \times 1K$ and can go much higher.

To display a complex image, the program has first to prepare it in memory. It then starts the CRT controller, which displays every element of the image and is also responsible for refreshing it. The details depend on the scan method used.

A word on color. Most color CRTs today use the *shadow mask* technique (Figure 1.2). They have three guns emitting three separate electron beams. Each beam is associated with one color but the beams themselves, of course, consist of electrons and thus have no color. The beams are adjusted such that they always converge a short distance behind the screen. By the time they reach the screen, they have diverged a bit, and they strike three different (but very close) points.

The screen is coated with dots made of three types of phosphor compounds that emit red, green, and blue light, respectively, when excited. At the plane of convergence, there is a thin, perforated metal screen: the shadow mask. When the three beams converge at a hole in the mask, they pass through, diverge, and hit three points coated with different phosphor compounds. The points glow at the three colors and the observer sees a mixture of red, green, and blue whose precise color depends on the intensities of the three beams (see discussion of spatial integration in Sections 7.4 and 10.3). When the beams are deflected a little, they hit the mask and are absorbed. After some more deflection, they converge at another hole and hit the screen at another triplet of points.

1.1.3 High-Definition Television

The NTSC standard was created in 1953, after four years of testing, for black-and-white television transmissions (NTSC stands for National Television Standards Committee). It specifies the shape of the signal sent by a TV transmitter. This is an analog signal, with amplitude that goes up and down during each scan line

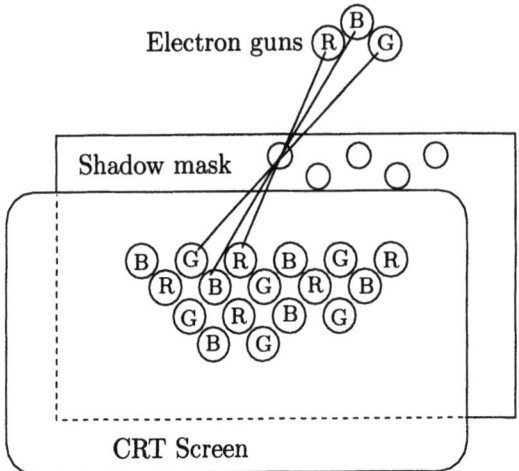

Figure 1.2: A Shadow Mask.

in response to the black and white parts of the line. Color was later added to this standard, but it had to added such that black-and-white television sets would be able to display the color signal in black and white. The result was phase modulation of the black-and-white carrier, a kludge (TV engineers call it NSCT "never the same color twice").

With the explosion of computers and digital equipment in the last two decades came the realization that a digital signal is a better, more reliable way of sending images over the air. In such a signal, the image is sent pixel by pixel, where each pixel is represented by a number specifying its color. The digital signal is still a wave, but the amplitude of the wave no longer represents the image. Rather, the wave is *modulated* to carry binary information. The term modulation means that something in the wave is modified to distinguish between the zeros and ones being sent. An FM digital signal, for example, modifies (modulates) the frequency of the wave. This type of wave uses one frequency to represent a binary zero and another to represent a one. The DTV (Digital TV) standard uses a modulation technique called 8-VSB (for *vestigial sideband*), which provides robust and reliable terrestrial transmission. The 8-VSB modulation technique allows for a broad coverage area, reduces interference with existing analog broadcasts, and is itself immune from interference.

History of DTV: The Advanced Television Systems Committee (ATSC), established in 1982, is an international organization developing technical standards for advanced video systems. Even though these standards are voluntary, they are generally adopted by the ATSC members and other manufacturers. There are currently about 80 ATSC member companies and organizations, which represent the many facets of the television, computer, telephone, and motion picture industries.

The ATSC Digital Television Standard adopted by the U.S. Federal Communications Commission (FCC) is based on a design by the Grand Alliance (a coalition of electronics manufacturers and research institutes) which was a finalist in the first round of DTV proposals under the FCC's Advisory Committee on Advanced

1 First Principles

Television Systems (ACATS). The ACATS is composed of representatives of the computer, broadcasting, telecommunications, manufacturing, cable television, and motion picture industries. Its mission is to assist in the adoption of an HDTV transmission standard and to promote the rapid implementation of HDTV in the United States.

The ACATS announced an open competition: anyone could submit a proposed HDTV standard, and the best system would be selected as the new television standard for the United States. To ensure a fast transition to HDTV, the FCC promised that every television station in the nation would be temporarily loaned an additional channel of broadcast spectrum.

The ACATS worked with the ATSC to review the proposed DTV standard, and gave its approval to final specifications for the various parts—audio, transport, format, compression, and transmission. The ATSC documented the system as a standard and ACATS adopted the Grand Alliance system in its recommendation to the FCC in late 1995.

In late 1996, corporate members of the ATSC reached an agreement on the DTV standard (Document A/53) and asked the FCC to approve it. On December 31, 1996, the FCC formally adopted every aspect of the ATSC standard except for the video formats. These video formats nevertheless remain a part of the ATSC standard and are expected to be used by broadcasters in the foreseeable future.

HTDV Specifications: The NTSC standard in use since the 1930s specifies an interlaced image composed of 525 lines where the odd numbered lines (1, 3, 5, ...) are drawn on the screen first, followed by the even numbered lines (2, 4, 6, ...). The two fields are woven together and drawn in $1/30$ of a second, allowing for 30 screen refreshes each second. In contrast, a noninterlaced picture displays the entire image at once. This *progressive scan* type of image is what's used by today's computer monitors.

The digital TVs that have been available since mid-1998 use an aspect ratio of $16/9$ and can display both the interlaced and progressive-scan images in several different resolutions—one of the best features of digital video. These formats include 525-line progressive scan (525P), 720-line progressive scan (720P), 1050-line progressive scan (1050P), and 1080-interlaced (1080I), all with square pixels.

Our present analog television sets cannot deal with the new, digital signal broadcast by TV stations, but inexpensive converters will be available (in the form of a small box that can comfortably sit on top of a TV set) to translate the digital signals to analog ones (and lose image information in the process).

The NTSC standard calls for 525 scan lines and an aspect ratio of $4/3$. This implies $\frac{4}{3} \times 525 = 700$ pixels per line, yielding a total of $525 \times 700 = 367,500$ pixels on the screen. (This is the theoretical total since only 480 lines are actually visible.) In comparison, a DTV format calling for 1080 scan lines and an aspect ratio of $16/9$ is equivalent to 1920 pixels per line, bringing the total number of pixels to $1080 \times 1920 = 2,073,600$, about 5.64 times more than the NTSC interlaced standard.

▶ **Exercise 1.2:** The NTSC aspect ratio is $4/3 = 1.33$ and that of DTV is $16/9 = 1.77$. Which one looks better?

In addition to the 1080 × 1920 DTV format, the ATSC DTV standard calls for a lower-resolution format with just 720 scan lines, implying $\frac{16}{9} \times 720 = 1280$ pixels per line. Each of these resolutions can be refreshed at one of three different rates: 60 frames/second (for live video) and 24 or 30 frames/second (for material originally produced on film). The refresh rates can be considered *temporal resolution*. The result is a total of six different formats. Table 1.3 summarizes the screen capacities and the necessary transmission rates of the six formats. With high resolution and 60 frames per second, the transmitter must be able to send 124,416,000 bits/sec (about 14.83 Mbytes/sec), which is why this format uses compression. (It uses MPEG-2. Other formats can also use this compression method.) The fact that DTV can have different spatial and temporal resolutions allows for trade-offs. Certain types of material (such as fast-moving horse or car races) may look better at high refresh rates even with low spatial resolution, while other material (such as museum-quality paintings) should ideally be watched in high spatial resolution even with low refresh rates.

Lines × pixels	Total # of pixels	Refresh rate		
		24	30	60
1080 × 1920	2,073,600	49,766,400	62,208,000	124,416,000
720 × 1280	921,600	22,118,400	27,648,000	55,296,000

Table 1.3: Resolutions and Capacities of Six DTV Formats.

Digital television (DTV) is a broad term encompassing all types of digital transmission. HDTV is a subset of DTV indicating 1080 scan lines. Another type of DTV is Standard Definition Television (SDTV) which has a picture quality slightly better than a good analog picture. (SDTV has resolution of 640×480 at 30 frames/sec and an aspect ratio of 4:3.) Since generating an SDTV picture requires fewer pixels, a broadcasting station will be able to transmit multiple channels of SDTV within its 6-MHz allowed frequency range. HDTV also incorporates Dolby Digital sound technology to bring together a complete presentation.

1.1.4 Vector Scan

In this method, the program prepares commands in a buffer in memory. Each command starts with a code (for example, 0 for point, 1 for line, 2 for circle, etc.), followed by more information. A few examples are listed below. The program then starts the CRT controller and gives it the start address of the buffer. The CRT controller fetches the commands from the buffer in memory and executes each to display a geometric object on the screen. Each command is executed by turning the beam off, moving it to where the object should start, turning it on, and moving it to draw the entire object. At the end of the buffer, the program should place a `goto` command that sends the CRT controller to the start of the buffer, to refresh the display. Examples of common commands are the following:

Point: | 0 | x1 | y1 | Line: | 1 | x1 | y1 | x2 | y2 |

1 First Principles

Circle: | 2 | x1 | y1 | r |

Goto: | 3 | address |

The main advantage of vector scan is a smooth display. Slanted lines, arcs, circles, and other objects come out smooth and perfect. However, the hardware is complex (it has to be able, for example, to sweep the beam in a perfect circle around the screen); also, a complex image means a big buffer, which may cause the display to flicker due to a low refresh rate. Today, this method is rarely used.

1.1.5 Raster Scan

Virtually all current graphics displays use this method. Its principle is that the program prepares the entire image in a buffer (called the bitmap) in memory in terms of dots. The final image consists of dots, and a number is stored in the bitmap for each dot, specifying the color of the dot. The CRT controller scans the bitmap number by number and draws a dot, called a *pixel*, on the screen for each number. If the display is bilevel (just two colors, black and white or foreground and background), then the color of a pixel is specified by one bit in the bitmap, zero for background and one for foreground. If the screen can display 2^n shades of gray, then the bitmap should contain an n-bit number for each pixel. This is sometimes referred to as a bitmap with n bitplanes. For color displays, each pixel must be represented by three numbers in the bitmap.

> Mouse droppings: [MS-DOS] n. Pixels (usually single) that are not properly restored when the mouse pointer moves away from a particular location on the screen, producing the appearance that the mouse pointer has left droppings behind.
> —Eric Raymond, *The Hacker's Dictionary*.

It is useful to imagine the screen as a rectangle with r rows and c columns. A pixel has screen coordinates (x,y), where x is in the range $[0, c-1]$ and y is in the range $[0, r-1]$. However, the bitmap is stored in memory as a one-dimensional array with index values from 0 to $r \times c - 1$. If the bitmap is stored row by row in array `bitmap`, then the index of pixel (x,y) is `yc+x`, and the pixel is accessed in the array by `bitmap[yc+x]:=`.... Unfortunately, the common operation of accessing a pixel requires a multiplication. To simplify it, the computer designer should choose a value for c that's a power of 2. The multiplication can then be replaced by a shift.

▶ **Exercise 1.3:** On some screens, the rows are numbered from the bottom up (the top screen line is row $r-1$ instead of row 0). What would be the pixel's index in array `bitmap` in such a case?

A typical bitmap in current computers represents a pixel by three bytes, each corresponding to one of the three basic colors. Each basic color can thus have 256 intensities and a pixel can have one of $2^{24} \approx 16.8$ million colors. For a $1K \times 1K$ screen resolution, this requires a bitmap of size $3 \times 2^{10} \times 2^{10} = 3 \times 2^{20} = 3$ Mbytes.

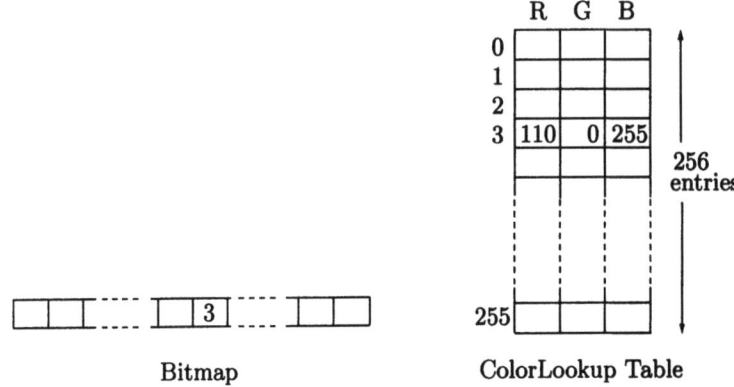

Figure 1.4: A Bitmap and a Color Lookup Table.

Doubling the resolution to $2K \times 2K$ requires a bitmap four times as big. Graphics applications are one reason why today's computers require large memories.

To save memory it is possible to use a *color lookup table* (Figure 1.4). The bitmap may consist of just 1 byte per pixel, containing a number between 0 and 255. That number is used, by the CRT controller, as a pointer to the color lookup table, which has 256 entries, each 3 bytes wide, specifying the intensities of red, green, and blue. The program first decides what colors are going to be used, and stores their values in the lookup table. Each color can be selected from a palette of $2^{24} \approx 16.8$ million colors, but, because of the size of the lookup table, a maximum of 256 colors can be displayed simultaneously. The program then scan-converts the objects to be displayed and stores pointers in the bitmap.

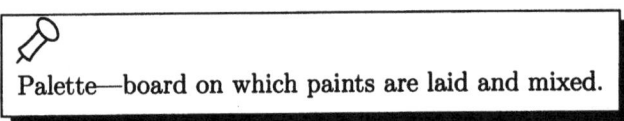

Palette—board on which paints are laid and mixed.

The CRT controller has a simple task. It scans the screen line by line and, at the same time, reads numbers from the bitmap and uses them to turn the beam on and off during each scan. The entire image is drawn as dots on scan lines (with nothing between the scan lines). In the case of a color lookup table, the controller reads a byte from the bitmap, uses it as a pointer, goes to the table, and uses the values in the 3 bytes to adjust the intensities of the three electron beams. This process is repeated for every pixel.

In practice, the CRT controller performs an interlaced scan. It first scans all the odd-numbered lines, then all the even-numbered ones, then starts the refresh. This is supposed to create a smooth display even for low refresh rates, because of the way our eyes work.

The advantages of raster scan are simple hardware, fast scan, and no flickering (the time it takes for a complete scan doesn't depend on the complexity of the image). The main disadvantage is that a picture appears on the screen as a large collection of pixels. Increasing the resolution of the screen allows for more complex images, but slanted lines and curved objects always have jagged edges. Vector scan,

1 First Principles

in contrast, does not involve any pixels and generates smooth lines and curves on the screen.

Another disadvantage is that, in addition to the bitmap, another data structure (a geometric one) is needed, to keep track of the individual image elements. Imagine a line whose pixels are turned on in the bitmap. If the line has to be erased, there is no way to tell what bits in the bitmap belong to the line. The line information (code and coordinates of endpoints) has first to be stored in the geometric data structure, then scan-converted from there. To erase the line, its information should first be found in the geometric data structure. It should be scan-converted and each pixel, in turn, erased.

Historically, vector scan was the first method used, in the 1960s and 1970s, in computer graphics. In the early 1980s, however—with the advent of high-resolution, low-cost color monitors—raster scan became the dominant display method. The reason for the popularity of raster scan is that in a complex image with many colors, a vector display requires many short vectors, thus increasing the scan time close to flickering. A raster scan takes the same time to complete a scan, regardless of the number of different pixels displayed, so the complexity of the image and the number of colors do not degrade the performance.

1.1.6 The Refresh Operation

The CRT controller saves the coordinates of the currently refreshed pixel in two internal registers, x and y. The registers are used to calculate the bitmap address, $yc+x$, of the pixel (Figure 1.5), allowing the graphics controller to read the pixel's value from the bitmap. Notice how the multiplication yc is achieved by simply placing y to the left of x and concatenating them. The value read is then converted to a voltage that's used to turn the electron beam on or off. In a color CRT, three numbers are read from the bitmap for each pixel, and each is converted to a voltage that's used to control the intensity of one beam.

While the bitmap is read, the CRT controller increments register x, to point to the next pixel. When register x overflows, it automatically returns to zero (think of an odometer going from $9\ldots99$ to $0\ldots00$) and the overflow signal is used to clock (i.e., to trigger) register y and also to move the beam to the start of the next scan line. When register y overflows, it also returns to zero and its overflow signal is used to return the beam to the start of the first scan line.

1.1.7 The Resolution of Film

The concept of resolution is easy to define in raster-scan computer graphics hardware because of the use of pixels, which are easy to count. With the advent of digital cameras featuring higher and higher resolutions, it is only natural to compare those cameras to traditional, film-based cameras and to ask, What is the resolution of film? (see also discussion of the resolution of the eye in Section 7.6).

Intuitively, we feel that film is a continuous medium, but in fact the image is recorded on film in small particles (grains) of various silver compounds. The resolution of the film is thus related to the size of these grains. One approach to quantifying the resolution is to say that the size of the grains determines the resolution, but only indirectly. This approach measures the resolution of film by

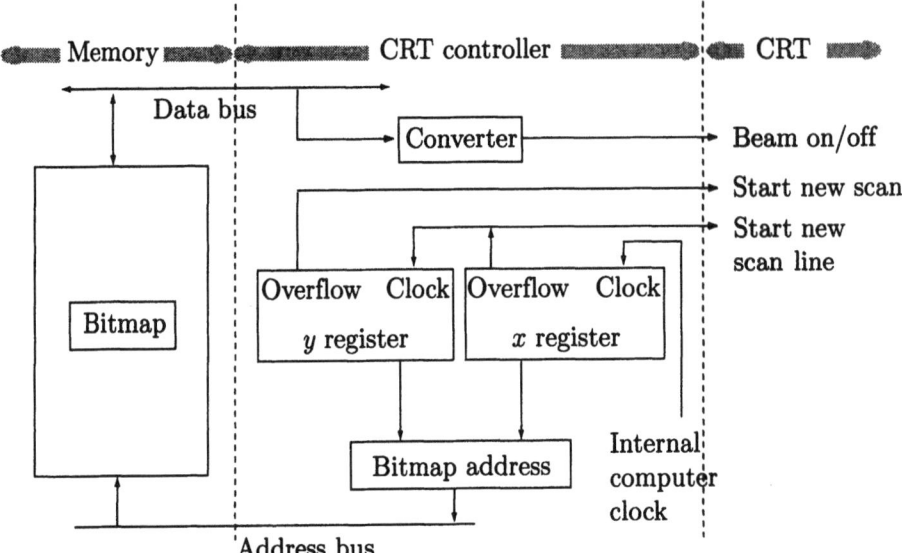

Figure 1.5: CRT Refresh.

trying to answer the following question experimentally: At what resolution can an observer no longer distinguish between an image on film and its digitized copy? Experiments with hundreds of observers with 35-mm film seem to indicate a range of resolutions from 1000×1000 to 1500×1500, depending on the observer and on film quality.

Another, more quantitative approach to this problem uses the concept of *line pairs per millimeter* (LPPM). A chart, such as the ones of Figure 1.6, is photographed and the film, after being properly developed and printed, is observed with a magnifying glass for the densest group of lines that can still be resolved. (The chart should have the same aspect ratio as the film, and should be photographed from such a distance that its outer frame will coincide with the boundary of the film.)

Suppose, for example, that a group of lines on the chart has a density of 50 lines per millimeter, with 50 gaps between them, for a total of 100 lines plus gaps per mm. If this is the highest density that can be resolved on the film (i.e., any denser groups of lines look like a gray blur), then each line and each gap can be considered a pixel, and we say that the film has an LPPM value of 100 and a resolution equivalent to 100 bit/mm. The reason for the term *line pair* is that the gaps are also considered (white) lines. A 35mm-wide film that tests at 100 LPPM thus contains the information equivalent of 3500 bits horizontally.

The patterns of Figure 1.6a consist of a half-circle of wedges. The observer has to determine the distance from the center to the point where the diverging lines can be seen individually. This distance is inversely related to the resolution (such a chart, of course, has to be calibrated before it can be used). Similar, more sophisticated charts may include, in addition, some long wedges, a series of line widths, and a range of font sizes.

1 First Principles

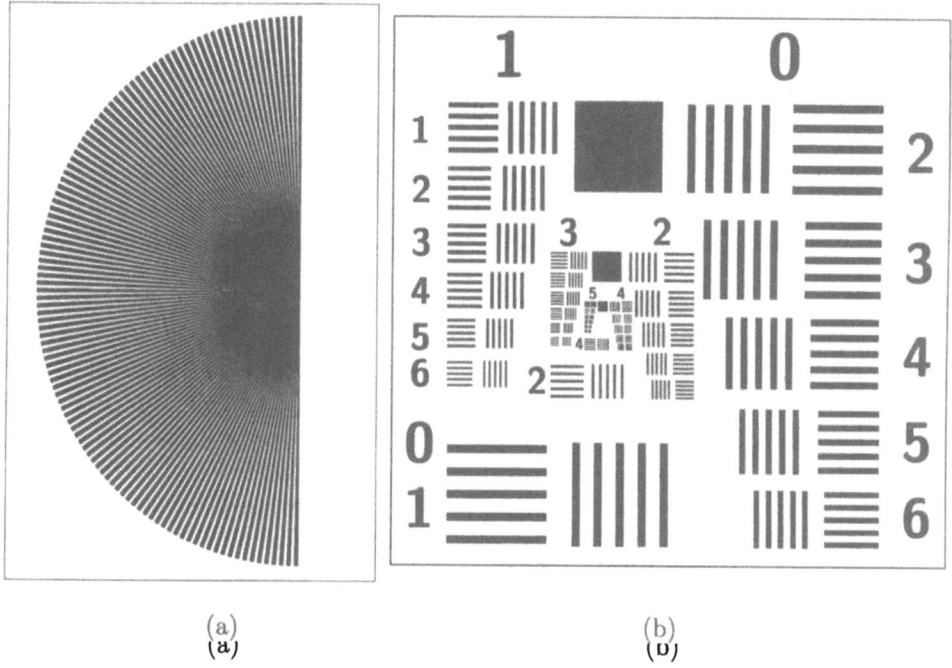

(a) (b)

Figure 1.6: Line Charts for LPPM Measurements.

Element	Group				
	0	1	2	3	4
1	10.0	22.0	40.0	80.0	160
2	11.2	22.4	44.9	89.8	—
3	12.6	25.2	50.4	101.0	—
4	14.1	28.3	56.6	113.0	—
5	15.9	31.7	63.5	127.0	—
6	17.8	35.6	71.3	143.0	—

Table 1.7: LPPM Values for Figure 1.6b.

Figure 1.6b consists of three parts, identical except for size, that are placed one inside the other. Each part contains two groups, for a total of six groups. The groups are numbered 0 through 5, and each consists of six elements (1–6). An element constitutes 10 lines, 5 horizontal and 5 vertical. Group 0 consists of the five elements numbered 2–6 on the right of the chart, and element 1 at the bottom left. Groups 2 and 4 are shaped like group 0 and are placed inside the chart. The six elements of group 1 are placed on the left side, and groups 3 and 5 are small copies of group 1 and are placed inside the chart. The widths of the lines go down from element to element, from a value w in element 1 of a group to $w/2$ in element 1 of the following group.

Given a piece of film, the chart is photographed on it and an observer should identify the smallest pair (group, element) in which all 10 lines can clearly be distinguished. The LPPM value of the film can then be found in Table 1.7 (that's limited to just groups 0–3 and the first element of group 4).

It is obvious that the *contrast* between lines and gaps in the chart affects what can be seen on the film. Dark lines are more visible than gray ones, so the measurement is done in practice with charts that have a contrast of 1000:1. At such contrast, most color films can resolve 100–125 LPPM, equivalent to 3500–4375 bits horizontal resolution for 35-mm-wide film. Slow black-and-white film may have two to three times this resolution. However, when taking real-life pictures of images with low-contrast objects, the resolution of most commercial films, as measured by LPPM, may go down to about 30 LPPM.

The next point to be mentioned has to do with image magnification. When a 35-mm-wide negative with 100 bits per mm (i.e., 3500 pixels horizontally) is enlarged to, say, 7 in. (about 178 mm, or a magnification factor of 5), the large picture will still have the same 3500 pixels horizontally, implying a resolution of only $3500/178 = 20$ bits/mm. The resolution as measured by the LPPM method has gone down by a factor of 5, but we know from experience that if the original image on the film is sharp, it can be magnified by a factor of 5 (i.e., to 7 in.) or even more without loss of image details, something that's impossible with digital images. This fact suggests that the LPPM approach to measuring film resolution is not ideal and that there is a fundamental difference between digital images and images on film.

Film type	ISO	LPPM 1.6:1	LPPM 1000:1	Total
Tri-X pan	400	50	100	8,640,000
T-Max 400	400	50	125	13,500,000
Plus-X pan	125	50	125	13,500,000
K-64	64	50	100	8,640,000
K-25	25	63	100	8,640,000
T-Max 100	100	63	200	34,560,000
Fuji Velvia	50	*	150	19,440,000
Panatomic-X	32	80	200	34,560,000
Ektar 25	25	80	200	34,560,000
Tech-Pan	25	100	320	88,473,600
Tech-Pan 120	25	100	320	368,640,000

* Not specified by Fuji.

Table 1.8: Sensitivities, LPPMs, and Total Number of Pixels for Various Commercial Films.

▶ **Exercise 1.4:** The frame size of still 35 mm film is 24×36 mm. Calculate the total number of pixels per frame assuming 100 LPPM.

Table 1.8 lists LPPM values (for two different contrasts) and total number of pixels (for a 1000:1 contrast) for various commercially available films (based on manufacturers' data sheets). The numbers vary widely because of the large number of different varieties of film emulsions available. The table raises another difficulty

1 First Principles

with the LPPM method, namely that even though film like Tech-Pan can have resolution of 320 LPPM, the actual resolution obtained may be lower because of the lens used in the camera. Most commercial lenses can't resolve 320 LPPM.

▶ **Exercise 1.5:** It is clear that film resolution, as measured by LPPM, depends on the film emulsion (its light sensitivity and grain size) and degree of contrast of the test chart. What other factors can affect the measured resolution?

No comparison of digital and film images is complete without discussing color quality. Digital images with only 8 or 16 bits per pixel suffer from a "banding" effect, where the image seem to consist of bands of different colors because the colors cannot change continuously (they are quantized). However, most observers agree that 12 bits/pixel (i.e., 2^{12} or 4096 colors) result, in images with no bands, which are indistinguishable from those of film.

Another interesting fact is that current commercial-grade CCD devices (which are widely used in digital cameras, video cameras, and amateur telescopes) provide resolution of 8–9 LPPM at best.

1.1.8 Windows and BitBlt

In modern computers, the operating system maintains the screen as a bitmap. Everything displayed on the screen, text, cursor, and graphics is saved in the bitmap as pixels and is constantly refreshed on the screen by the CRT controller.

A useful feature supported by such an operating system is *multiple windows*. At any time, the screen may display several windows (or parts thereof). They can be opened, moved, resized, and closed. The window manager, part of the operating system, is responsible for these operations.

The main task of the window manager is to save anything on the screen that gets obscured by a moving or changing window. Since all windows are rectangular, saving part of the screen means saving a rectangular part of the bitmap. This part is saved in a buffer in memory and can easily be used to restore the original screen.

Many graphics packages offer a procedure called BitBlt—an acronym that stands for "BIT Boundary bLock Transfer" and is pronounced "bitblit." This procedure can save parts of the bitmap in, and also restore parts from, memory. The operating system must have routines to provide BitBlt with fresh memory areas and to reclaim unneeded areas. This is done by a dynamic memory allocation method such as first fit, best fit, or the buddy system.

Because this is a common operation, it should be fast; today there are VLSI circuits that do BitBlt in hardware at speeds of about 100 million pixels/sec. Such a circuit can do more than just save and restore. It can also perform a logical operation while restoring the image, a feature that can be applied to create useful effects.

We call the saved pixels "source" and the current bitmap (in which the source pixels are going to be overwritten) the "destination." Instead of simply writing each source bit into the bitmap, thereby erasing a destination bit, the BitBlt circuit writes a bit that's a function of the source and destination bits. The rule is `dst:=dst op src`, where `op` is any logical operation on bits. Table 1.9 shows examples of logical operations (the rows are the source and the columns are the destinations).

$$\begin{array}{c|cc} & 0 & 1 \\ \hline 0 & 0 & 0 \\ 1 & 0 & 1 \end{array} \quad \begin{array}{c|cc} & 0 & 1 \\ \hline 0 & 0 & 1 \\ 1 & 1 & 0 \end{array} \quad \begin{array}{c|cc} & 0 & 1 \\ \hline 0 & 0 & 1 \\ 1 & 1 & 1 \end{array} \quad \begin{array}{c|cc} & 0 & 1 \\ \hline 0 & 0 & 1 \\ 1 & 0 & 0 \end{array}$$

Table 1.9: Some Logical Truth Tables.

▶ **Exercise 1.6:** How many logical operations are possible in principle?

Here are a few examples of logical operations performed by BitBlt (zero represents a white pixel and one represents a black pixel).

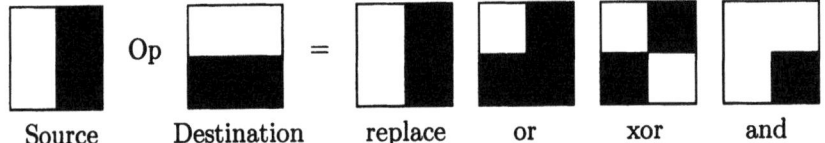

Source   Destination   replace   or   xor   and

- **dst:=src.** This is the `replace` operation. It draws the new source on the screen, regardless of what's on the screen now (i.e., regardless of the destination). It amounts to overpainting the new source on the screen (a destructive write). This is common when windows are created and moved. It can also be used to erase areas by drawing in white (the background color).

- **dst:=0.** This clears the screen.

- **dst:=not dst.** This inverts the screen.

> Bit flipping. A process of inverting bits, changing ones to zeros and vice versa. For example, in a graphics program, to invert a black-and-white bit-mapped image (to change black to white and vice versa), the program could simply flip the bits that compose the bitmap.

- **dst:=dst or src.** This adds the source to the destination (a nondestructive write). Oring a gray pattern on a white screen results in the pattern. Oring the same pattern on a black background has no effect. The paint brush in many paint programs uses this mode.

- The **and** operation **dst:=dst and src** can be used to erase destination pixels selectively.

- **dst:=dst xor src.** A pixel would be black whenever the source and destination pixels differ. This turns out to be an important mode. A common example is erasing an object from the bitmap. Imagine an image consisting of simple geometric objects. To erase an object, the coordinates of all its pixels have to be calculated, so that they can be erased. The problem is that objects may intersect; erasing all the pixels of an object will therefore leave holes in all the objects that happen to intersect it.

1 First Principles

The solution is to draw in xor mode when the object is originally drawn and also when it is erased. In this way the intersection of two objects (which is normally just one pixel or a few pixels) would be white, and after erasing one figure, it would turn back to black.

Figure 1.10 illustrates an example (see also Plate 3). In Figure 1.10a, a horizontal line is drawn in xor mode by flipping pixels from white to black. In Figure 1.10b, a vertical line is drawn, also in xor mode, so the pixel at the intersection point becomes white. In Figure 1.10c, a slanted line is drawn, so the same pixel becomes black again. In Figure 1.10d, the vertical line is erased, and in Figure 1.10e, the horizontal line is erased as well. The pixel at the intersection keeps changing its color, but the picture as a whole does not look bad. Note that both the drawing and erasing are done with a source of 1. The BitBlt operation is `dst:=dst xor 1`.

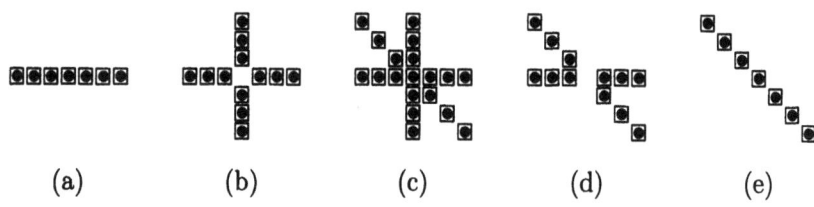

(a) (b) (c) (d) (e)

Figure 1.10: XOR at Line Intersections.

▶ **Exercise 1.7:** Is the erase operation common enough to justify xor drawing?

▶ **Exercise 1.8:** Can xor drawing be used when drawing in color (i.e., when a pixel consists of more than 1 bit)?

Xor drawing can also be used with a color lookup table. If an original entry in the bitmap is 3 (=0011) and we draw a new object with a source of 5 (=0101), then the bitmap entry is going to change to (0011 xor 0101)=0110=6. The pixel at the intersection point now points to entry 6 in the color lookup table. If we now erase the original pixels of 3, then (3 xor 6) will result in 5.

Example: The concept of *transparency* is important when adding images to a bitmap. A common example is a web page. Such a page normally has a certain background and new images added to the page obscure that background. When adding a new image to our page, we may want certain parts of the image to be transparent. Such parts should show the background instead of any image pixels. Transparency is achieved by declaring one of the colors used in the image a transparent color (a notable example is the GIF89 graphics file format). The browser displaying the page finds this declaration in the image file.

This example shows how several BitBlt operations can be used to implement transparency. We assume that a bitmap and an image are given (parts a and b, respectively, of Figure 1.11) and that the image is to added to a certain region of the bitmap (the bottom-left 4×4 part). We also assume that the image file contains information about the transparent parts. The operation proceeds in the following four steps:

1. Create a bitmap the size of the image, set all the transparent pixels in it to one and all the opaque pixels to zero. This bitmap is called a mask (Figure 1.11c).

2. Perform an exclusive-or (XOR) of the image with the proper bitmap region (in our example, the bottom-left 4 × 4 part). Store the result in the bitmap (Figure 1.11d).

3. Perform a logical AND of the bitmap and the mask and store it in the bitmap (Figure 1.11e). This operation leaves all the transparent areas unchanged and sets the opaque areas to zeros.

4. XOR the proper bitmap region and the image. This operation sets all the transparent areas back to their original values (because two XOR operations cancel each other out). The opaque areas now contain the desired image bits (Figure 1.11f) because an exclusive-or with a zero leaves a bit unchanged.

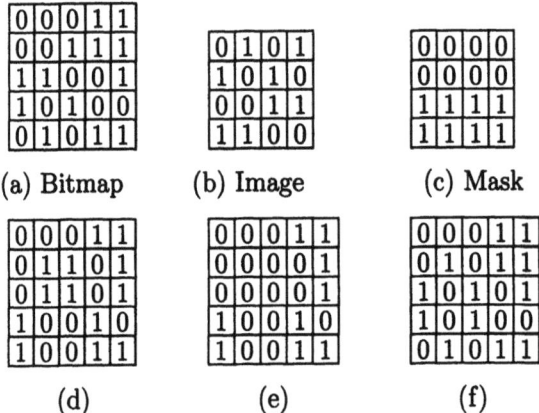

Figure 1.11: Transparency in a Bitmap.

1.2 Bitmap Scaling

An image is made of pixels, so its size is measured in pixels, not in inches or centimeters. When an image has to be displayed in a window on the screen, it sometimes has to be scaled to fit the window's size. There are two approaches to this problem:

1. To expand an image, simply copy each pixel several times. To expand an image to three times its width and twice its height, copy each pixel three times horizontally and twice vertically, creating six identical pixels. This is simple, but the result is a coarse image, visibly made of large rectangular blocks. To shrink the image to a third of its width and half its height, select every third pixel in a scan line and select every other scan line. The rest of the original image is ignored. This, again, is fast, but the resulting image may be missing important details and may be noticeably different from the original image. Below we show how this method may be used with arbitrary scale factors.

2. If high-quality results are important, the new pixels should be *calculated* from the original ones, not simply copied. To expand an image with scale factors of, say, 3×4, we need to create 3×4 = 12 new pixels for every original pixel. Let $P[i, j]$ represent the value of the pixel at row i and column j of the original image.

1 First Principles

The 12 new pixels generated from $P[i,j]$ should be computed by interpolating the values of $P[i,j]$ and its three near neighbors. The neighbor on the right of $P[i,j]$ is $P[i,j+1]$ and the two neighbors below it are $P[i+1,j]$ and $P[i+1,j+1]$. Each of the 12 new pixels is computed as a weighted average of these 4 pixels, where the weights (which add up to 1) are given by Table 1.12. The table consists of 12 subtables, each of size 2×2. They are arranged in three columns and four rows. Each entry in the table equals the product of the weights of its row and column. The four entries of each subtable add up to one. Each of the 12 new pixels is calculated as a weighted sum of the 4 pixels

$$P[i,j], \quad P[i,j+1],$$
$$P[i+1,j], \quad P[i+1,j+1],$$

where the weights are taken from the corresponding subtable.

	1	0	2/3	1/3	1/3	2/3
1:	1	0	2/3	1/3	1/3	2/3
0:	0	0	0	0	0	0
3/4:	3/4·1	3/4·0	3/4·2/3	3/4·1/3	3/4·1/3	3/4·2/3
1/4:	1/4·1	1/4·0	1/4·2/3	1/4·1/3	1/4·1/3	1/4·2/3
1/2:	1/2·1	1/2·0	1/2·2/3	1/2·1/3	1/2·1/3	1/2·2/3
1/2:	1/2·1	1/2·0	1/2·2/3	1/2·1/3	1/2·1/3	1/2·2/3
1/4:	1/4·1	1/4·0	1/4·2/3	1/4·1/3	1/4·1/3	1/4·2/3
3/4:	3/4·1	3/4·0	3/4·2/3	3/4·1/3	3/4·1/3	3/4·2/3

Table 1.12: Interpolation Weights for 3×4 Scaling.

The first of the 12 new pixels should thus be

$$1 \times P[i,j] + 0 \times P[i,j+1] + 0 \times P[i+1,j] + 0 \times P[i+1,j+1].$$

The last one is

$$\frac{1}{12}P[i,j] + \frac{2}{12}P[i,j+1] + \frac{3}{12}P[i+1,j] + \frac{6}{12}P[i+1,j+1].$$

To shrink an image by factors of $n \times m$, we have to replace a group of $n \times m$ original pixels by a single pixel. This pixel is simply computed as an average of the original $n \times m$ pixels.

An important point that should be mentioned is that the values stored in the bitmap are not always the intensities or colors of the pixels. Often the bitmap contains pointers to a color lookup table. In such a case, the interpolation shown in Table 1.12 should be performed on the pixel values in the lookup table, not on the pointers in the bitmap.

If the scale factors are not integers, then a simple, fast algorithm can be developed to determine how the new pixels depend on the original ones. We present the basic ideas assuming that an image is to be expanded by copying each original pixel several times. When we think of expanding an image, we normally imagine it to be stretched from all sides until it fits the new, larger space. To understand the algorithm, we should think of the expansion process as starting with a grid consisting of a few, large pixels and creating another grid with more, *smaller* pixels that has the same area as the original. Figure 1.13a shows an example of an original 4×4 image (whose pixels are numbered 1 through 16) and an expanded image of 10×10 pixels superimposed on it. Both are the same size, which shows us that each original pixel has to be copied $10/4 = 2.5$ times. In practice, a pixel will be copied either two or three times to obtain the required scaling. Our algorithm uses the quantity $4/10 = 0.4$ to determine how many copies of a pixel to generate. A variable diff (for *differential*) is set to zero and is incremented by 0.4 each time a pixel is copied. Each time diff crosses an integer boundary, the algorithm moves to the next original pixel and starts copying it. The process is summarized in Table 1.14. It shows how pixels 1 and 3 are copied three times each, and pixels 2 and 4, twice each. Figure 1.13b shows the final 100 pixels.

The discussion above assumes an expansion, but this method also works for shrinking, where the scale factor is less than one. Shrinking a 10×10 image to, for example, 4×4 is done by incrementing diff by $10/4 = 2.5$. The results are summarized in Table 1.15.

▶ **Exercise 1.9:** Show the original 10×10 image (with pixels numbered 1 to 100) and the final 4×4 image using a diagram similar to Figure 1.13b.

In general, diff is incremented by the inverse of the scale factor, but the main point is that both diff and the amount it is incremented by should be integers, since floating-point operations are considerably slower than the same operations on integers. We therefore declare diff an integer and increment it by the rounded value of "1000/scale factor." Each time diff crosses another multiple of 1000, we move to the next pixel in the original image and start copying it. Figure 1.16 is a pseudo-code algorithm for scaling one scan line. It copies pixels from the original bitmap P[i,j] to a scaled bitmap Q[x,y] where each scan line is N pixels wide. The values of i and x don't change in this example.

The code of Figure 1.16 can be extended to loop over scan lines. Another pair of diff and accum_diff variables is needed, since the scaling factors in the horizontal and vertical directions may be different. The loop shown in Figure 1.16 now becomes the inner loop of a double-loop algorithm whose outer loop goes over the scan lines. Once the inner loop creates the first scan line Q[1,y], the outer loop copies it as many times as necessary, using the vertical differential. It should be mentioned that this algorithm, which uses integers and a differential, is called a *DDA algorithm* (for Digital Differential Analyzer). Section 2.3 discusses DDA algorithms for scan-converting lines.

(a)

1	1	1	2	2	3	3	3	4	4
1	1	1	2	2	3	3	3	4	4
1	1	1	2	2	3	3	3	4	4
5	5	5	6	6	7	7	7	8	8
5	5	5	6	6	7	7	7	8	8
9	9	9	10	10	11	11	11	12	12
9	9	9	10	10	11	11	11	12	12
9	9	9	10	10	11	11	11	12	12
13	13	13	14	14	15	15	15	16	16
13	13	13	14	14	15	15	15	16	16

(b)

Figure 1.13: Bitmap Scaling.

`diff`:	0	0.4	0.8	1.2	1.6	2.0	2.4	2.8	3.2	3.6
Pixel:	1	1	1	2	2	3	3	3	4	4

Table 1.14: A 10/4 Expansion by Copying.

`diff`:	0	2.5	5	7.5
Pixel:	1	3	6	8

Table 1.15: A 4/10 Shrinking by Copying.

> Isn't life a series of images that change as they repeat themselves?
> — Andy Warhol

```
diff:=round(1000/scale);
accum_diff:=0;
j:=1;
for y:=1 to N do
Q[x,y]:=P[i,j];
a:=⌊accum_diff/1000⌋;
accum_diff:=accum_diff+diff;
b:=⌊accum_diff/1000⌋;
j:=j+(b-a);
endfor;
```

Figure 1.16: Scaling One Scan Line.

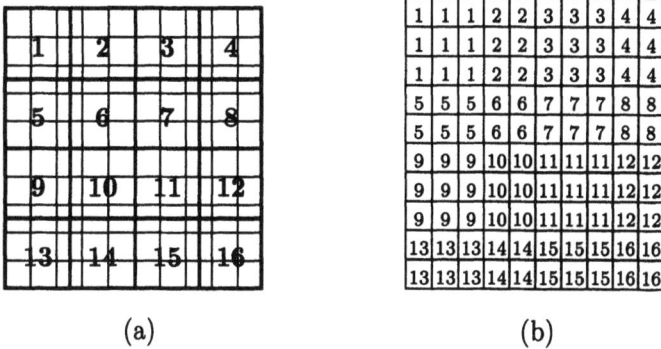

1.3 Bitmap Rotation

The problem of rotating a bitmap is common. Imagine that an image has been scan-converted and is already stored in a small bitmap. It is easy to transfer this image to a large bitmap, but what if we want to rotate it before transferring? We show three approaches to this problem. All three assume that an $n \times n$ bitmap A is given, with origin at the center, and has to be rotated clockwise θ degrees about its center into another bitmap B.

The first approach uses the two-dimensional rotation matrix $\mathbf{R}$, Equation (3.4). Each bitmap pixel (x, y) is multiplied by this matrix to produce the coordinates (x^*, y^*) of the rotated pixel. The transformation is

$$(x^*, y^*) = (x, y)\mathbf{R} = (x, y)\begin{pmatrix} \cos\theta & -\sin\theta \\ \sin\theta & \cos\theta \end{pmatrix} = (x\cos\theta + y\sin\theta, -x\sin\theta + y\cos\theta).$$

This is simple but produces unacceptable results since it is easy to show that some pixels of B may remain empty (i.e., not get any of the pixels of A mapped to them), while two pixels of A may get mapped to the same pixel of B. Figure 1.17 is a typical example. A 9×9 bitmap is rotated $45°$ and the mapping is calculated by the *Mathematica* code

```
t=Pi/4;
Brot[x_,y_]:={x Cos[t]+y Sin[t],-x Sin[t]+y Cos[t]};
Do[Print[x,",",y,N[Brot[x,y],1]], {x,-4,4},{y,-4,4}]
```

The resulting list shows that, for example, pixel $(0, 4)$ is mapped to $(3, 3)$ (the pixels numbered "1" circled), pixel $(1, 3)$ is mapped to $(3, 1)$ ("2" circled), pixel $(0, 3)$ goes to $(2, 2)$ ("3" circled), pixel $(1, 4)$ is mapped to $(4, 2)$ ("4" circled), but no pixel is mapped to $(3, 2)$ (the one with "?"). Also, the two pixels $(2, 3)$ and $(2, 4)$ are mapped to $(4, 1)$ (the gray pixels).

▶ **Exercise 1.10:** Perform the same calculation and analysis for a 3×3 bitmap.

The second approach tries to overcome these problems by looping over the pixels $B[i, j]$ of the final bitmap B and calculating the pixel of A that should map to $B[i, j]$. Since the rotation is represented by $(x^*, y^*) = (x, y)\mathbf{R}$, we use the relation $(x, y) = (x^*, y^*)\mathbf{R}^{-1}$ to find a pixel of A for every pixel of B. The matrix inverse $\mathbf{R}^{-1}$ can be calculated once, before we start the loop. This approach produces better results but is computationally intensive since the relation above requires four multiplications and two additions for each pixel being mapped.

The third approach is due to [Paeth 86]. It is based on Exercise 3.21 that shows that a two-dimensional rotation about the origin can be expressed as the sequence of three shears:

$$\begin{pmatrix} \cos\theta & -\sin\theta \\ \sin\theta & \cos\theta \end{pmatrix} = \begin{pmatrix} 1 & -\tan\frac{\theta}{2} \\ 0 & 1 \end{pmatrix}\begin{pmatrix} 1 & 0 \\ \sin\theta & 1 \end{pmatrix}\begin{pmatrix} 1 & -\tan\frac{\theta}{2} \\ 0 & 1 \end{pmatrix} = \mathbf{A} \cdot \mathbf{B} \cdot \mathbf{A}.$$

It is easy to see that the first shear transforms a pixel (x, y) to the pixel $(x, y - x\tan(\theta/2))$. This transformation preserves the column x of pixel (x, y) and changes

1 First Principles

only its row. Figure 1.18a shows how each column is moved by this shearing vertically, up or down, and how the extreme columns, where the column number x is large, are moved more than the inner columns. The center column, where $x = 0$, is not moved at all. The same thing is true for the third shear, while the second shear, which transforms (x, y) to $(x + y\sin\theta, y)$, preserves rows and changes only columns.

The following two points summarize the principles of this approach:

1. Let's assume that the image is in grayscale and let $P(x, y)$ denote the intensity of pixel (x, y). Since the quantity $\tan(\theta/2)$ is, in general, noninteger, the source pixel (x, y) is normally moved such that it covers f percent of one of the destination pixels (x, y^*) and $(1 - f)$ percent of the adjacent pixel $(x, y^* + 1)$ (Figure 1.18b, where the source pixels are dashed and the destination pixels are shown in gray). The intensity of destination pixel (x, y^*) should thus be set to the sum $fP(x, y) + (1 - f)P(x, y - 1)$. In this way, the sum of the intensities of all the destination pixels in bitmap column x will equal the sum of the intensities of all the source pixels in the same bitmap column.

```
procedure xshear(α,r,c);
 for i:=-r to r do
  skew:=i*α;
  int:=floor(shear);
  f:=frac(shear);
  PrevLeft:=0;
  for j:=-c to c do
   p:=Sbitmap[i,j];
   LeftPart:=p*f;
   Dbitmap[i,j+int]:=p-LeftPart+PrevLeft;
   PrevLeft:=LeftPart;
  endfor;
  Dbitmap[i,int]:=PrevLeft;
 endfor;
end;
```

Figure 1.19: Shearing in the x Direction.

2. Figure 1.17 shows that the source and destination bitmaps are incompatible in the sense that certain destination pixels have no corresponding source pixels and certain source pixels are moved outside the bitmap by the rotation. This incompatibility gets worse the larger the rotation angle, up to an angle of 45°. Because of the symmetry of a bitmap, our method assumes that the rotation does not exceed 45°. A rotation θ of more than 45° can be achieved by rotating in the opposite direction by $90° - \theta$. A rotation of more than 90° can be done by first swapping pixels (x, y) and $(y, n - x)$, which amounts to rotating the bitmap clockwise 90°, then using the present approach to rotate by less than 45°. We

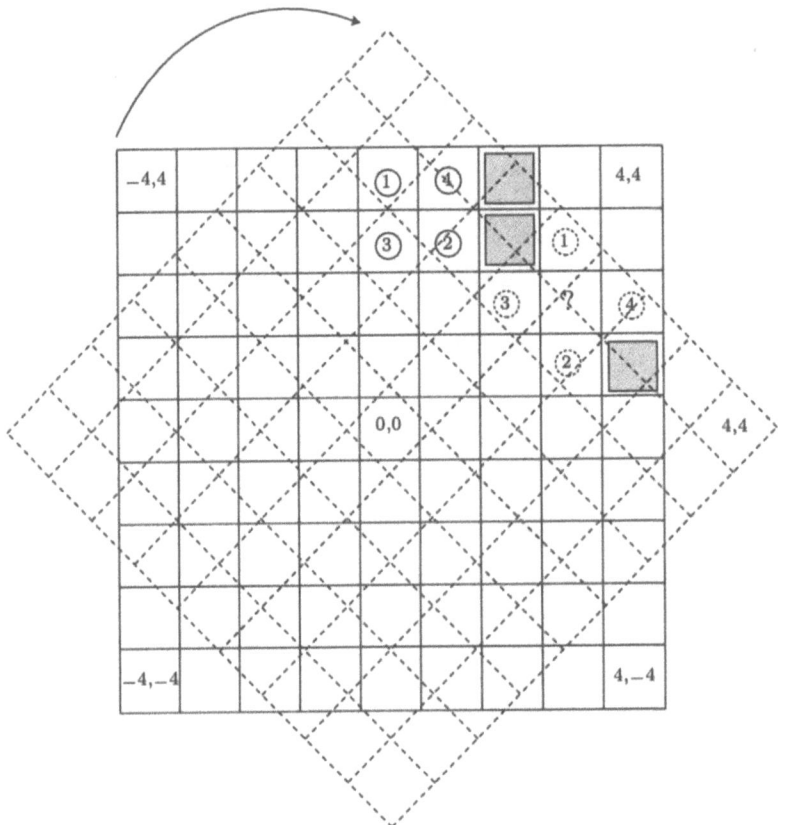

Figure 1.17: A 45° Bitmap Rotation.

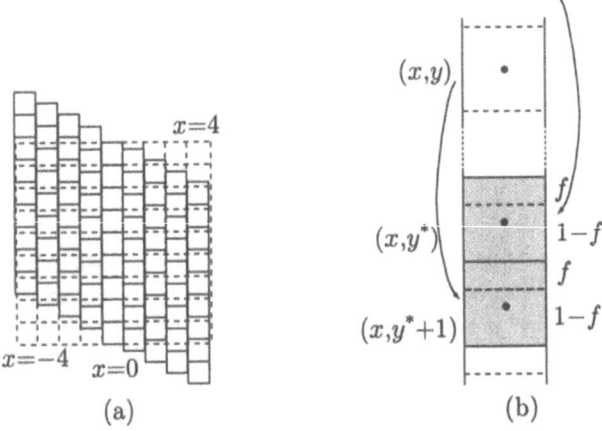

Figure 1.18: Rotation by Shearing.

1 First Principles

also assume that the rotation is done about the center of the bitmap and that the coordinates of the center pixel are $(0,0)$.

Figure 1.19 is a pseudo-code description of the routine for shearing in the x direction. The `yshear` routine for shearing in the y direction is similar. The main program inputs (or computes) the rotation angle θ, and then makes the three procedure calls

`xshear(`$-\tan(\theta/2)$`,r,c);  yshear(`$\sin\theta$`,r,c);  xshear(`$-\tan(\theta/2)$`,r,c);`

Notice that `p-LeftPart` equals `p*(1-f)` which is the right part of the contribution to the intensity of the new pixel. This part is added to `PrevLeft`, which is the left part of the contribution.

1.4 A Practical Drawing Program

Programs for artistic and technical drawing and illustrating have been popular since the mid-1980s, and this section describes their main data structures and operations. There are also painting programs, but they are not discussed here. The only background necessary for this section is an understanding of the xor operation (Section 1.1.8) and of the bitmap in general.

A typical drawing/illustration program starts with a blank screen and a menu of graphics objects that it can draw and manipulate. These normally include line, circle/oval, box (square/rectangle, optionally with rounded corners), polygon, cubic Bézier curves (Section 4.14), and text. It may also be possible to paste in a small bitmap prepared outside the program (perhaps a scanned painting or photograph). The user selects a menu option, supplies the necessary data, and the program displays the graphics object on the screen. In this way, the screen fills up with different items, each a simple graphics object, that may intersect, obscure each other, and have different colors. The user then starts editing the image by selecting various objects and deleting, moving, or modifying them. The discussion here concentrates on selecting one of possibly many graphics objects on the screen.

Figure 1.20a shows a simple drawing consisting of a line, a circle, and a rectangle. We initially assume that the different objects do not intersect. The user points the cursor (the arrow in the figure) anywhere on the line (or within a few pixels of the line) and clicks to select the line. The program has to identify the object selected and highlight it, thereby confirming the selection to the user. The problem is that the program knows only the position of the cursor on the screen (its screen coordinates). These are the coordinates of a pixel on the selected object, and the immediate task of the program is to quickly find (1) all the other pixels of this object, (2) its type (line, circle, etc.), and (3) its other data (in the case of a line, the coordinates of the two endpoints; in the case of a circle, its center and radius).

Two data structures are used by the program to perform this task. The first is the *codemap*, an array the size of the bitmap where each location contains the serial number of a graphics object on the screen. The second is a geometric data structure containing the type of each graphics object drawn so far and its color and specific data. Figure 1.20b shows how the serial numbers of the three objects drawn so far are stored in the codemap (the rest of the codemap is assumed to have been initialized to zeros). Figure 1.20c shows the geometric data structure. Notice

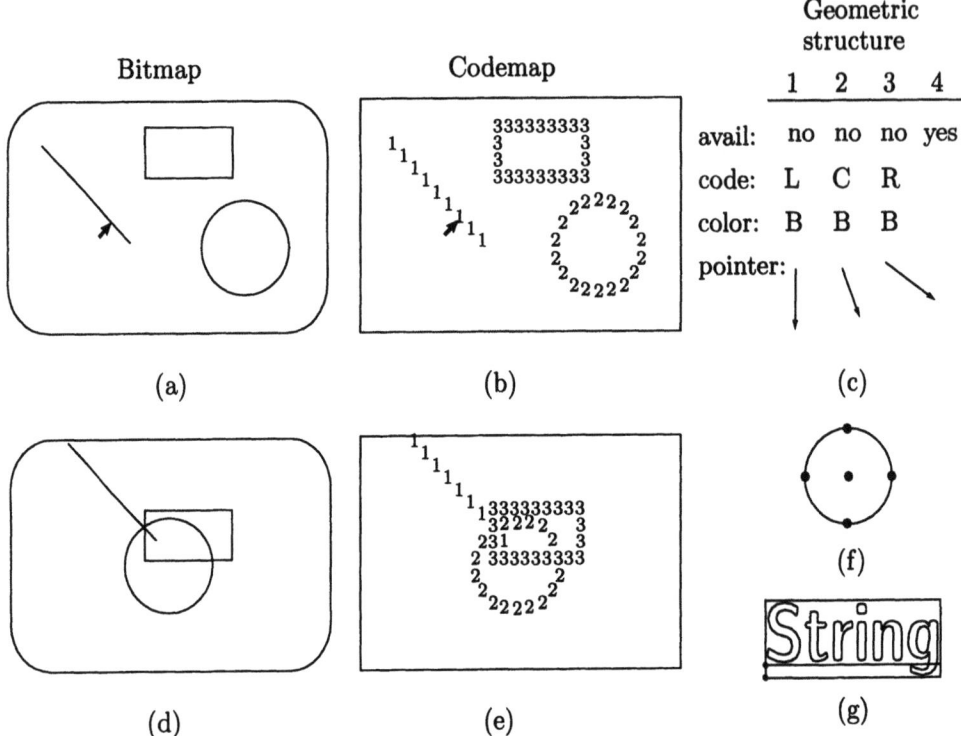

Figure 1.20: Data Structures for a Drawing Program.

that the serial numbers are implied in that structure and don't actually have to be stored there.

Once the program detects a click, it inputs the cursor coordinates and checks the corresponding location of the codemap. In our example, it finds serial number 1. (Both the bitmap and the codemap are shown as two-dimensional arrays, but, in practice, they are normally one-dimensional.) The program examines location 1 of the geometric data structure, finds that the graphics object with serial number 1 is a line, and finds the pointer to its specific data. That data consist of the coordinates of the two endpoints. The program then calculates all the pixels of the line (using the same scan-converting method that was originally used to draw the line) and highlights them. The two endpoints may be highlighted with a different color, making them more obvious to the user. This takes about the same amount of time as drawing the line in the first place. The program starts waiting for the next user response/command.

▶ **Exercise 1.11:** Rewrite the preceding paragraph for the case in which the user selects the circle.

Text presents a special problem. The drawing program gets the characters of text from the different fonts and knows nothing about their shapes. In order to select a string of text, the user normally has to click on its reference point (which is either the bottom-left corner of the text or the left point of the baseline, Figure 1.20g). Any clicks within the text string are ignored by the program (although a sophisticated program may get the width of the character from the font file, and this way be able to identify clicks anywhere on the baseline).

After selecting a graphics object, typical user responses may be to delete the object, move it, or reshape it. The user may also want to copy the object, group it with some other objects, or perform other operations, but we discuss only the three operations.

Deleting an object is done by using the same scan-converting method. Every pixel calculated by the method is erased (or is exclusive-ored with the color of the object being erased) and the program also clears the serial number of the pixel in the codemap. When done, the program uses the serial number to mark the corresponding entry of the geometric data structure "available" and deletes the storage used for the specific data.

To move a graphics object, the user grabs it at a certain pixel or anchor point and drags it to its new position. The program creates an outline of the object and moves it with the cursor. When the dragging stops, the program calculates the x and y displacements (the final cursor position minus its initial position) and moves the object to its new location pixel by pixel. Each pixel at position (x, y) is erased and drawn at position $(x + d_x, y + d_y)$. An xor may be used instead of straight erase and draw. After moving a pixel, the program moves its serial number from position (x, y) to position $(x + d_x, y + d_y)$ in the codemap. There is no need to change the geometric data structure, but the specific object information may have to be updated. (If the object is simple, such as a line or a circle, the outline of the object is the object itself. If the object is more complex, such as text or a bitmap, the outline of the object may be its bounding rectangle.)

Reshaping an object can be done by rotating, scaling, or shearing it. These transformations are discussed in Chapter 3, where it is shown that they can all be expressed by a 3×3 transformation matrix where only six of the nine elements vary. Reshaping is, thus, similar to moving. The user may enter a command or make their wish known by dragging. Once the program knows what transformation is required, it prepares a specific transformation matrix **T**. Each pixel **P** is then moved from its original position to position **P·T**, a step that's followed by moving its serial number in the codemap in the same way. The geometric data structure may have to be updated since reshaping may change the type of an object. A circle may become an oval; a square may turn into a rectangle or a parallelogram. Notice that rotating and scaling a bitmap causes special problems and is discussed in Sections 1.3 and 1.2.

We now turn to the case where graphics objects may intersect. Figure 1.20d shows three intersecting objects. We assume that the line was drawn first, followed by the circle, and then by the rectangle. A glance at Figure 1.20e shows that the codes at the intersection points are those of the latest object that passes through the point. If the user clicks on an intersection point, the latest object through it will be selected. This does not seem bad, until we consider the case where that object is moved. If the rectangle of Figure 1.20d is moved, its two intersection points would be set to zero. After many such moves, the codemap may become meaningless. We outline two solutions to this problem:

1. When an object with serial number n is drawn, the program checks each codemap location before it sets it to n. If a location is nonzero, its value is saved in a linked list before the location is set to n. We now have to have a linked list associated with each codemap location. We visualize the codemap as an array of structures, where each structure consists of two fields—a code and a pointer to a list. Initially, all the codes are set to zero and all the pointers are set to null. When the first code is stored in a location, the pointer remains null. When another code is stored in the location, the original code is saved in a list node and the pointer is set to point to that node. When a code is removed from a location, the pointer is checked. If it is not null, it is followed and the most recent code is restored. Modern computers can perform these operations in a fraction of a second. When this method is used, clicking on an intersection point will select the most recent object that passes through it.

2. When an object with serial number n is drawn, the program exclusive-or's n with the codemap locations. The value stored in a codemap location is, thus, the xor of the serial numbers of all the objects that pass through it. When an object with serial number m is deleted or moved, the program performs an xor of a codemap location with m. No lists are necessary, so this method is simple and fast. The following examples show the values stored in a hypothetical codemap location when objects 1, 5, 7, 1, and 5 are added then deleted (we assume 4-bit serial numbers)

$$0000 \operatorname{xor} 0001 = 0001, \quad 0001 \operatorname{xor} 0101 = 0100, \quad 0100 \operatorname{xor} 0111 = 0011,$$
$$0011 \operatorname{xor} 0001 = 0010, \quad 0010 \operatorname{xor} 0101 = 0111.$$

The downside is that a codemap location corresponding to an intersection point does not contain any of the serial numbers of the intersecting objects. In our example, the contents of the hypothetical location varies from 4 to 3 to 2, ending with 7. Thus, the program should flag all codemap locations where objects intersect, and should ignore any user clicks at those points.

▸ **Exercise 1.12:** What is a good way to implement such flags?

Pro cut shrimp cage
An anagram of "Computer Graphics"

2
Scan-Converting Methods

In a raster-scan system, two steps are necessary in order to display a geometric figure: (1) an algorithm should be executed to select the best pixels (the ones closest to the ideal figure) and (2) the selected pixels should be turned on.

Step 2 is simple. It only requires setting bits in the bitmap (or using xor). Most current compilers have a built-in function to do this. All that the program has to say is "`putpixel(row,col);`" or "`plot(r,c,color);`" or something similar. Step 1, however, is more complex. The *scan-converting* algorithm has to be fast and it must depend on the shape of the figure. This chapter discusses scan-converting algorithms for straight lines and for circles.

2.1 Scan-Converting Lines

The simplest geometric figure, after the point, is the straight line. Its explicit equation is $y = ax + b$, where a is the slope and b, the y-intercept (see Section 4.3.1 for other ways to represent lines). In practice, the coordinates of the two endpoints (x_1, y_1) and (x_2, y_2) are given, instead of a and b, but it is easy to express a and b in terms of the endpoints. The slope a is simply $(y_2 - y_1)/(x_2 - x_1)$ or $\Delta y/\Delta x$. The value of b is obtained from $y_1 = ax_1 + b$, which implies

$$b = y_1 - ax_1 = y_1 - \frac{y_2 - y_1}{x_2 - x_1}x_1 = \frac{y_1(x_2 - x_1) - x_1(y_2 - y_1)}{x_2 - x_1} = \frac{y_1 x_2 - x_1 y_2}{x_2 - x_1}.$$

Our first algorithm uses (x_1, y_1) and (x_2, y_2) to calculate a and b, then executes the loop of Figure 2.1. However, this loop is very slow because it uses multiplications and also because it works with real quantities that eventually have to be rounded

```
var a, b, x, y, x1, x2, y1, y2: real;
a:=(y2-y1)/(x2-x1);
b:=y1-a*x1;
x:=x1;
repeat
y:=a*x+b;
point(round(x),round(y));
x:=x+0.01;
until x>x2;
```

Figure 2.1: Scan Convert $y = ax + b$.

to integers. A better algorithm should use just additions/subtractions and just integers.

▶ **Exercise 2.1:** The loop of Figure 2.1 has another disadvantage; what is it?

2.2 Midpoint Subdivision

This is a completely different approach to the problem of scan-converting lines. It involves a procedure Midpoint that divides the line segment in two, calculates the midpoint of the segment, plots it, and calls itself recursively twice, to do the same thing for the two halves. The following is a listing in the C language:

```
void Midpoint(int a1,int b1, int a2, int b2)
{
int midx,midy;
midx=(a1+a2)/2; midy=(b1+b2)/2;
putpixel(midx,midy,Color); /* Turbo C */
if(abs(a1-midx)>1 || abs(b1-midy)>1) Midpoint(a1,b1,midx,midy);
if(abs(midx-a2)>1 || abs(midy-b2)>1) Midpoint(midx,midy,a2,b2);
}
```

The main program needs only input the two endpoints; then, execute the command Midpoint(x1,y1,x2,y2);. In practice, the procedure may be written as a nonrecursive one, manipulating the stack explicitly. One attractive feature of this method is the simple arithmetic. Only two divisions are required and even they can be replaced by a shift.

2.3 DDA Methods

Better algorithms should be arithmetically simple and should use just integers. There is a large class of such methods, called DDA (for Digital Differential Analyzer) methods. Section 1.2 discusses one such method, for bitmap scaling. A few DDA methods for scan-converting lines are described below.

> The original Differential Analyzer was an analog computer built, in 1927, by the American scientist and engineer Vannevar Bush (1890–1974). It was based on the use of mechanical integrators that could be interconnected in any desired manner.

2.3.1 Simple DDA

This algorithm uses the relationship $a(x+1)+b = (ax+b)+a = y+a$, which implies that if we increment x by 1 and would like to stay on the line, we should increment y by a. The following code assumes that the two endpoints $(x1, y1)$ and $(x2, y2)$ are given. This algorithm does not work for vertical lines, where $a = \infty$.

```
var x, x1, x2: integer; a, b, y: real;
calculate a and b;
x:=x1; y:=a*x+b;
repeat
point(x,round(y));
x:=x+1; y:=y+a;
until x>x2;
```

This is still inefficient because y is still a real quantity, but it has another, more important drawback. Imagine a very steep line, where $a \gg 1$. Since our loop increments x by 1, successive pixels will have y coordinates that differ greatly and the result may be a fragmented line, made of a few disconnected pixels. To correct this problem, we have to check the slope, and if it is greater than 45°, increment y, not x, in steps of 1. From $y = ax + b$, we get $y + 1 = ax + b + 1 = ax + 1 + b = a(x+1/a)+b$. Therefore, when y is incremented by 1, x should be incremented by $1/a$. The algorithm becomes

```
var Δx,Δy,L:integer; x,y,a,G,H:real;
Δx:=x2-x1; Δy:=y2-y1;
x:=x1; y:=y1;
if Δx>Δy then G:=1; H:=a;
else G:=1/a; H:=1 endif;
for L:=1 to max(Δx,Δy)+1 do
point(round(x),round(y));
x:=x+G; y:=y+H;
endfor;
```

Ideally, the number of pixels generated when a line is scan-converted should equal the length of the line. This is because the length is measured in screen units, which are pixels. A line of length L displayed by fewer than L pixels would be fragmented. If the same line is generated by more than L pixels, it would look brighter than other lines.

The total number of pixels drawn by simple DDA is $\max(\Delta x, \Delta y)$. For lines that are close to horizontal or close to vertical, $\max(\Delta x, \Delta y) \approx$ length, which is the ideal case. For a 45° line, $\Delta x = \Delta y$, so Δx pixels are drawn. The length of such a line equals $\sqrt{\Delta x^2 + \Delta y^2} = \sqrt{2\Delta x^2} = \sqrt{2}\Delta x \approx 1.41\Delta x$ which implies that $\Delta x \approx 0.71$ length. For such a line, we, therefore, get only about 71% of the ideal number of pixels. Consequently, such lines look dim.

▶ **Exercise 2.2:** Given the two endpoints $(x1, y1) = (1,2)$ and $(x2, y2) = (4,6)$, execute the simple DDA algorithm manually and show the pixels generated. Also, calculate the length of the line.

2.3.2 A Variation

The following pseudo-code is a variation of the simple DDA. The quantity length is not the length of the line but is related to it. Because of the way it is defined, one of the quantities x_incr, y_incr equals ±1, which simplifies the algorithm.

```
procedure SimpleDDA(x1,y1,x2,y2: integer);
begin Δx,Δy,length,i:integer; x,y,x_incr,y_incr:real;
Δx:=x2-x1; Δy:=y2-y1;
length:=max(abs(Δx), abs(Δy));
x_incr:=Δx/length; y_incr:=Δy/length;
x:=x1; y:=y1;
for i:=1 to length+1 do
point(round(x),round(y));
x:=x+x_incr; y:=y+y_incr;
endfor;
end;
```

Notice that the other quantity (x_incr or y_incr) is still real.

2.3.3 Symmetrical DDA

The simpler DDA methods are based on a loop, where during each iteration, the (x, y) coordinates of the previous pixel are incremented by $\epsilon \Delta x$ and $\epsilon \Delta y$, respectively, for some quantity ϵ. A typical procedure is

```
procedure SymmDDA(x1,y1,x2,y2: integer);
calculate eps;
xIncr:=eps*Δx;
yIncr:=eps*Δy;
x:=x1+.5; y:=y1+.5;
repeat
Plot(trunc(x),trunc(y));
x:=x+xIncr; y:=y+yIncr;
until x=x2 or y=y2;
end;
```

(See the end of this section for an explanation of x:=x1+.5; y:=y1+.5; and for the use of trunc instead of round.)

In the symmetrical DDA method, we set $\epsilon = 2^{-n}$, where n is defined by

$$2^{n-1} \leq \max(|\Delta x|, |\Delta y|) < 2^n.$$

This sets ϵ to 1 over the (approximate) length of the line. It also sets the x and y increments to values less than 1.

As an example, consider the line from $(0,0)$ to $(7,5)$. Its equation is $y = (5/7)x$. For this line, we have $\Delta x = 7$ and $\Delta y = 5$, leading to $2^2 = 4 \leq \max(7,5) < 8 = 2^3$. Therefore, n is set to 3, implying $\epsilon = 2^{-3} = 1/8$. The (x, y) coordinates of the previous pixel are incremented by $\epsilon \Delta x = 7/8$ and $\epsilon \Delta y = 5/8$, respectively. The nine steps of the loop are summarized in Table 2.2. The last column of the table

2 Scan-Converting Methods

(+.−) compares the ideal y coordinate of the point (which equals 5/7 times the x coordinate) to the y coordinate actually displayed. A "+" is shown if the point displayed is above the ideal point, a "−" is shown in the opposite case, and a period is shown in the ideal case.

Point No.	Start (x,y)		Truncated to	Ideal y	+.−
1	$(.5,.5)$	$=(.5,.5)$	$(0,0)$	0	.
2	$(.5+\frac{7}{8},.5+\frac{5}{8})$	$=(11/8,9/8)$	$(1,1)$	5/7	+
3	$(\frac{11}{8}+\frac{7}{8},\frac{9}{8}+\frac{5}{8})$	$=(18/8,14/8)$	$(2,1)$	10/7	−
4	$(\frac{18}{8}+\frac{7}{8},\frac{14}{8}+\frac{5}{8})$	$=(25/8,19/8)$	$(3,2)$	15/7	−
5	$(\frac{25}{8}+\frac{7}{8},\frac{19}{8}+\frac{5}{8})$	$=(32/8,25/8)$	$(4,3)$	20/7	+
6	$(\frac{32}{8}+\frac{7}{8},\frac{24}{8}+\frac{5}{8})$	$=(39/8,29/8)$	$(4,3)$	20/7	+
7	$(\frac{39}{8}+\frac{7}{8},\frac{29}{8}+\frac{5}{8})$	$=(46/8,34/8)$	$(5,4)$	25/7	+
8	$(\frac{46}{8}+\frac{7}{8},\frac{34}{8}+\frac{5}{8})$	$=(54/8,39/8)$	$(6,4)$	30/7	−
9	$(\frac{54}{8}+\frac{7}{8},\frac{39}{8}+\frac{5}{8})$	$=(61/8,44/8)$	$(7,5)$	35/7	.

Table 2.2: Symmetrical DDA Example.

Note that point 6 is identical to point 5 and should not be displayed. If we ignore point 6, we end up with a line where the first and last points are smack on the ideal line, points 3, 4, and 8 are below the line, and points 2, 5, and 7 are above it. This is the reason for the name *Symmetrical DDA*.

How can we avoid plotting point 6? The ideal solution is to use special hardware where both x and y are stored in registers that have two parts. The left part of each register holds the integer value of the variable, and the right part, the fractional value. The x and y increments, which are less than one, are added to the fractional parts. Whenever a fractional part overflows, the overflow signal increments the corresponding integer part. If neither fractional register overflows, the point is not plotted.

This method creates the *truncated* value in the integer parts, rather than the rounded values, since this is faster. Since we still need the rounded, not truncated, values, we initialize both fractional parts to 0.5 rather than to zero. Notice that `trunc(x+0.5)` equals `round(x)`.

2.3.4 Quadrantal DDA

This is a more sophisticated DDA method. Its principle is to increment, in each step, either x or y but not both. From step to step, we move along the line either horizontally or vertically, but not diagonally (Figure 2.3).

The key to the algorithm is to realize that the ratio

$$\frac{\text{number of } y \text{ increments}}{\text{number of } x \text{ increments}}$$

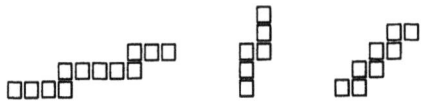

Figure 2.3: Quadrantal DDA.

should equal the slope $\Delta y/\Delta x$. The implementation uses an auxiliary variable, Err, that is initially set to zero. In each step, if Err>0, we increment x by 1 and decrement Err by Δy. If Err<0, we increment y by 1 and increment Err by Δx. Thus, if Err is positive, we move along the line horizontally and decrement Err. After a number of steps, Err becomes negative and we start moving vertically, incrementing Err.

If $\Delta x > \Delta y$, we end up with more x than y increments, which is appropriate since $\Delta x > \Delta y$ implies a slope that's less than 45°. If $\Delta x < \Delta y$, there will be more steps in the y direction. The quantity Err, thus, oscillates around zero all the time. The method is called *quadrantal* because the details of the steps depend on the direction of the line. If the direction is in the range 0°–90°, we increment both x and y by 1. If the direction is in the range 90°–180°, we increment x by -1 and y by 1; similarly, for the other two ranges. Hence, the algorithm must start by determining the range of the direction, and this is done by comparing Δx and Δy and looking at their signs. The pseudo-code below shows the loop for the first quadrant. It assumes that the two endpoints $(x1, y1)$ and $(x2, y2)$ of the line segment are given.

```
var x1,y1,x2,y2,Δx,Δy: integer;
Δx:=x2-x1; Δy:=y2-y1;
Err:=0;
repeat
plot(x1,y1);
if Err>0 then
  x1:=x1+1;
  Err:=Err-Δy
else
  y1:=y1+1;
  Err:=Err+Δx
endif;
until x1>=x2 and y1>=y2;
```

The quadrantal DDA method features the following:

- No multiplication, division, or real numbers are used.

- There are no diagonal moves, just horizontal and vertical ones. The line is therefore made of segments with a one-pixel overlap.

- If Err happens to be zero in some loop iteration, it means that the current pixel is located right on the ideal line. Our program is written such that for Err=0, it executes the **else** part and thus increments $y1$. When the current pixel is positioned

2 Scan-Converting Methods

on the ideal line, we normally don't care whether the next pixel is drawn in the x or in the y direction. There is one exception, however—the second pixel! The first pixel is drawn at point $(x1, y1)$, and since initially Err is set to zero, the second pixel will always be drawn at $(x1, y1+1)$. This may be annoying if the line is close to horizontal, since it produces the pattern ▯▯▯▯▯▯▯ instead of ▯▯▯▯▯▯▯. The solution is to set Err initially to

$$\text{Err} = \begin{cases} +\Delta x/2, & \text{if } \Delta x > \Delta y, \\ -\Delta y/2, & \text{otherwise,} \end{cases}$$

instead of to zero. This produces better looking lines, but notice that Err is no longer an integer.

The number of pixels in a quadrantal line segment is easily calculated. For a line close to horizontal or close to vertical, the number equals the length of the line (one pixel per x or per y value, as in the simple DDA method). For a line slanted at 45°, quadrantal DDA produces two pixels per x value (except the two extreme values), so the line looks as in Figure 2.3. The length of the line is $\sqrt{2}\Delta x$, so the ratio (number of pixels/line length) equals $2\Delta x/\sqrt{2}\Delta x = \sqrt{2} \approx 1.41$. Thus, there is an excess of 41%. Such lines are still too bright and also slow to calculate.

▶ **Exercise 2.3:** Given the two endpoints $(1, 1)$ and $(5, 5)$, calculate the pixels for the straight line between them obtained by simple DDA and by quadrantal DDA.

2.3.5 Octantal DDA

The main idea in this method is to move along the line either horizontally or diagonally, but not vertically (the precise rule depends on the slope). The main feature of such lines is that they are made of segments that do not overlap (Figure 2.4), so they look smoother than similar quadrantal lines.

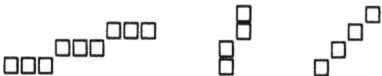

Figure 2.4: Octantal DDA.

Eight sets of rules are necessary. If the direction of the line is in the range 0°–45° (the first octant, where $0 \leq \text{slope} \leq 1$), only horizontal and diagonal moves are allowed. If the direction is in the second octant (45°–90°), only vertical and diagonal moves are allowed, not horizontal. In the third octant, the moves are the same as in the second one, but with negative x increments. The precise rules for four of the eight octants are summarized as follows:

In octant 1:

1. If Err<=0, move horizontally ($+x$) and update Err:=Err+Δy.
2. If Err>0, move diagonally ($+x, +y$) and update Err:=Err+$\Delta y - \Delta x$.

In octant 2:

1. If Err<0, move diagonally ($+x, +y$) and update Err:=Err+$\Delta y - \Delta x$.

2. If `Err>=0`, move vertically ($+y$) and update `Err:=Err`$-\Delta x$.

In octant 7:

1. If `Err<0`, move diagonally ($+x, -y$) and update `Err:=Err`$-\Delta y - \Delta x$.
2. If `Err>=0`, move vertically ($-y$) and update `Err:=Err`$-\Delta x$.

In octant 8:

1. If `Err<=0`, move horizontally ($+x$) and update `Err:=Err`$-\Delta y$.
2. If `Err>0`, move diagonally ($+x, -y$) and update `Err:=Err`$-\Delta y - \Delta x$.

The rules for octants 3–6 are similar.

Note that the program can be made more compact (although a bit slower) by using the symmetry between certain octants. The only difference between octants 2 and 7, for example, is in the y coordinate. If the endpoints are $(1,1)$ and $(2,-7)$ (which corresponds to octant 7), then the program can calculate the line for points $(1,1)$ and $(2,7)$ (octant 2), and transform the line to octant 7 by reversing the y coordinates of all the pixels.

▶ **Exercise 2.4:** Do it.

The main advantage of the octantal over the quadrantal DDA is the number of pixels per line, which is closer to the length of the line. Lines drawn using this method look finer and more precise. For lines close to horizontal or vertical, the number of pixels is the same as in the quadrantal DDA method. For a 45° line, however, there is an improvement. The line looks as in Figure 2.4, one pixel per x value. Hence, the number of pixels is Δx and the ratio (# of pixels/line length) equals $\Delta x/\sqrt{2}\Delta x = 1/\sqrt{2} \approx 0.71 = 1 - 0.29$. There is, thus, a pixel shortage of 29% (compared to an excess of 41% in quadrantal DDA).

▶ **Exercise 2.5:** Intuitively it seems that the 45° line shown in Figure 2.4 is the best representation of such a line. It looks straight and precise. Any additional pixels would make it too thick. It seems that the octantal DDA method produces ideal 45° lines, so how can we say that they have a shortage of pixels?

2.4 Double-Step DDA

This is an extension of the quadrantal and octantal methods, where in each iteration, two pixels, instead of just one, are calculated and displayed. The method is due to [Wu and Rokne 87]. It is based on a loop, where, in each iteration, a simple decision is made based on the sign of a discriminator D. The method is fast, since only half the number of iterations is required. We discuss the details of the method for lines in the first quadrant (i.e., with slopes between 0° and 45°), but it can easily be extended to the other three quadrants.

Figure 2.5 shows that if pixel P was selected in some iteration, then the following iteration must select "its" two pixels according to patterns 1–4. No other patterns are possible for a line in the first quadrant. In order to select one of four patterns, we need to ask three questions, and this translates to three `if` statements in each iteration. Since other DDA methods execute only one `if` per iteration, our method must be improved, otherwise it will turn out to be slow, even though it does two pixels per iteration.

The key to improving the method is to realize that lines with higher slopes (close to 45°) do not use pattern 1 and lines close to horizontal do not use pattern

2 Scan-Converting Methods

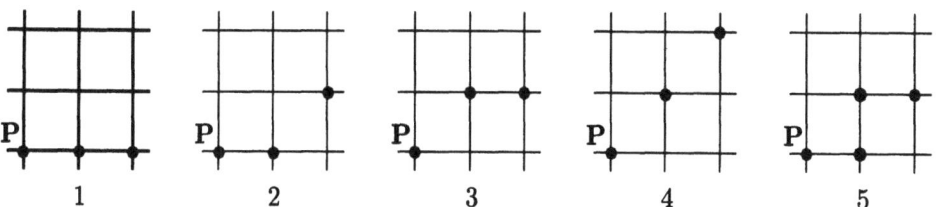

Figure 2.5: The Five Patterns for Quadrant-1 Lines.

4. Hence, our method should start by checking the slope.

There are two approaches to improving the double-step method:

1. We reduce the number of patterns to three by selecting pattern 5 (with the two gray pixels) whenever we need patterns 2 or 3. The algorithm checks the slope, as shown below. If the slope is small enough, only patterns 1 and 5 will be used. The algorithm goes into a loop, where, in each iteration, it uses a simple test to select either pattern 1 or 5. If the slope is high, the algorithm goes into a similar loop, where it selects either pattern 4 or 5 in each iteration.

This results in a fast algorithm that has another advantage. Since pattern 5 has two gray pixels, the resulting line is somewhat antialiased. (The important topic of antialiasing is discussed in Section 2.9.) The following pseudo-code procedure summarizes this approach (but does not show how D is calculated):

```
procedure dbstep(x1,y1, x2,y2);
dx=x2-x1; dy=y2-y1;
x:=x1; y:=y1;
if dy/dx>.5 (high slope)
then
 while x≤x2 do
  if D<0 then pattern(5) else pattern(4);
 endwhile;
else        (low slope)
 while x≤x2 do
  if D<0 then pattern(1) else pattern(5);
 endwhile;
endif;
procedure pattern(patt: integer);
case patt of
 1: pixel(x+1,y); pixel(x+2,y);
 2: pixel(x+1,y); pixel(x+2,y+1); y:=y+1;
 3: pixel(x+1,y+1); pixel(x+2,y+1); y:=y+1;
 4: pixel(x+1,y+1); pixel(x+2,y+2); y:=y+2;
end (case)
x:=x+2;
end;
```

2. We stay with the four patterns and divide the algorithm into two loops, as in method 1. This time, each loop has to have two if statements to select one of three patterns. This is somewhat slow but can be used in a monochromatic display, where no gray can be shown.

Both methods are discussed below, but note, again, that they are restricted to lines in the first quadrant.

To restrict our choices, let's first distinguish higher slopes (where we select patterns 2, 3, and 4, but never 1) from lower ones (where pattern 1, 2, or 3 should be used, but never pattern 4). Figure 2.6 shows the pixel that was selected at iteration $i - 1$. We name this pixel P and denote its coordinates (x_{i-1}, y_{i-1}). The next iteration (iteration i) should select one of the three pixels L (pattern 1), K (patterns 2 or 3), or J (pattern 4). Three important dimensions in the figure are a—the distance of the line from pixel P, b—its distance from L, and c—its distance from J.

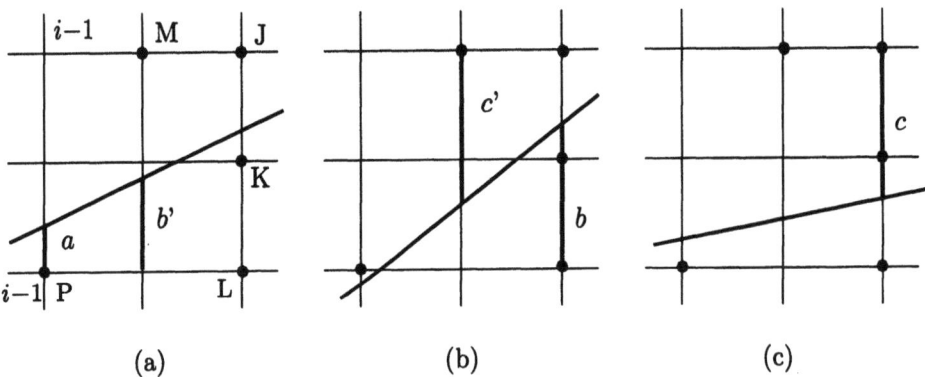

Figure 2.6: Details of Calculations.

A look at the diagram should convince the reader that a should satisfy $a \leq 0.5$, since, otherwise, iteration $i - 1$ would have selected the pixel above P. Note that the line may go below pixel P, but even in this case, it should satisfy $|a| \leq 0.5$.

Pattern 4 is only selected for lines with higher slopes. The smallest slope for which pattern 4 (pixel J) will be selected satisfies $a = 0.5$ and $b = 1.5$ (the line is farthest from P and is midway between K and J, see Figure 2.6a). This corresponds to a slope of 0.5.

Pattern 1 is only selected for lines with lower slopes. The highest slope for which this pattern could possibly be selected corresponds to $a = -0.5$ and $b = 0.5$ (the line is below pixel P and midway between L and K). This, again, corresponds to a slope of 0.5.

We therefore say that slope 0.5, which corresponds to a slope angle of 22.5°, separates patterns 1 and 4, so they cannot occur in the same line. Figure 2.6a shows that the condition for the line to be closer to pixel K (patterns 2 or 3) than to J (pattern 4) is $(b - c)/2 < 0.5$, which corresponds to $b - c < 1$ or $b - c - 1 < 0$. Similarly, Figure 2.6c shows that the line will be closer to pixel L (pattern 1) than

2 Scan-Converting Methods

to K (patterns 2 or 3) if it satisfies $(c - b)/2 > 0.5$, which corresponds to $c - b > 1$ or $b - c + 1 < 0$.

The algorithm should therefore first look at the slope and do the following:

1. If the slope is < 0.5, there should be a loop that checks the sign of $b - c + 1$. If it is negative, pattern 1 should be selected; otherwise, patterns 2 or 3.

2. Similarly, if the slope is > 0.5, there should be loop where the sign of $b - c - 1$ should be checked in each iteration. If it is negative, patterns 2 or 3 should be selected; otherwise, pattern 4.

What is needed now is a way to compute the sign of both $b-c+1$ and $b-c-1$ by using just integer quantities and simple operations. We first develop a simple test for the sign of $b - c + 1$. We first observe that our line, which goes from point (x_1, y_1) to (x_2, y_2), can be translated such that it starts at the origin and this will not change the pattern of pixels. We therefore assume that our line has the equation $y = (\Delta y / \Delta x)x$ (the y-intercept is zero), without loss of generality.

We next observe that the height of the line at $x = x_{i-1} + 2$ is $(\Delta y/\Delta x)(x_{i-1} + 2)$. The same height can also be written as the sum $y_{i-1} + b$, so we get $b = (\Delta y/\Delta x)(x_{i-1}+2) - y_{i-1}$. To find a similar expression for c, we subtract the height of the line at $x = x_{i-1} + 2$ from the height of pixel J (which is $y_{i-1} + 2$). The result is $c = y_{i-1} + 2 - (\Delta y/\Delta x)(x_{i-1} + 2)$.

The quantity $b - c + 1$ can now be written

$$b - c + 1 = \frac{\Delta y}{\Delta x}(x_{i-1} + 2) - y_{i-1} - \left[y_{i-1} + 2 - \frac{\Delta y}{\Delta x}(x_{i-1} + 2)\right] + 1$$

$$= 2\frac{\Delta y}{\Delta x}(x_{i-1} + 2) - 2y_{i-1} - 1.$$

We now define a new quantity D_i (the discriminator for the sign of $b - c + 1$) by means of Δx and Δy as

$$D_i = \Delta x(b - c + 1) = 2\Delta y(x_{i-1} + 2) - 2\Delta x\, y_{i-1} - \Delta x.$$

Note that since the line is in the first quadrant, Δx is never negative (this assumption should be changed when extending the algorithm to other quadrants). This means that the sign of D_i is the same as that of $b - c + 1$ and we can use D_i in our loop to determine which pattern to use.

Before starting the loop, we should set $D_1 = 4\Delta y - \Delta x$. In each iteration, a new value D_{i+1} should be calculated from the previous value D_i. From $D_{i+1} = 2\Delta y(x_i + 2) - 2\Delta x\, y_i - \Delta x$, we get $D_{i+1} - D_i = 2\Delta y(x_i - x_{i-1}) - 2\Delta x(y_i - y_{i-1})$. Since $x_i - x_{i-1} = 2$ and since $y_i - y_{i-1}$ equals either 1 (for patterns 2 and 3) or 0 (for pattern 1), we can write

$$D_{i+1} = \begin{cases} D_i + 4\Delta y, & \text{if } D_i < 0 \text{ (pattern 1)}, \\ D_i + 4\Delta y - 2\Delta x, & \text{otherwise (patterns 2 and 3)}. \end{cases}$$

The discriminator for the sign of $b - c - 1$ is derived in a similar way, and the result is $D_1 = 4\Delta y - 3\Delta x$ and

$$D_{i+1} = \begin{cases} D_i + 4\Delta y - 2\Delta x, & \text{if } D_i < 0 \text{ (patterns 2 and 3)}, \\ D_i + 4(\Delta y - \Delta x), & \text{otherwise (pattern 4)}. \end{cases}$$

This completes the algorithms for approach 1, where patterns 1, 4, and 5 are used. In order to implement approach 2, where patterns 1–4 are used, we need another discriminator D_t. This is calculated by a process similar to the one used to obtain D_i, except that point M is used in Figure 2.6 instead of pixel P as the base for the calculations. Its distance c' from the line is shown in Figure 2.6b. By the definition of a discriminator, we get

$$D_t = \begin{cases} \Delta x(b' - c' + 1), & \text{if slope} < 0.5, \\ \Delta x(b' - c' - 1), & \text{if slope} \geq 0.5, \end{cases} \quad (2.1)$$

where Figure 2.6 provides the relations $b' = b - \Delta y/\Delta x$ and $c' = c + \Delta y/\Delta x$, which imply $D_t = D_i - 2\Delta y$. Also, from Figure 2.6 we see that when $b' - c' + 1 < 0$, pattern 2 should be used, otherwise pattern 3. Combining this knowledge with $D_t = D_i - 2\Delta y$ and with Equation (2.1), we find that the relation $b' - c' + 1 < 0$ is equivalent to

$$D_i < \begin{cases} 2\Delta y, & \text{if slope} < 0.5, \\ 2(\Delta y - \Delta x), & \text{if slope} \geq 0.5. \end{cases}$$

The last point to discuss is the termination of the loop. Since each iteration increments x by 2, we cannot simply check to see whether x is greater than x_2, as this may result in one extra pixel. The solution is to realize that the extra pixel will result only if the line is made of an odd number of pixels. Since our lines are in the first quadrant, the number of pixels equals Δx. We therefore change the loop if Δx is odd, and terminate it when $x = x_2 - 1$. The last pixel is then drawn separately, outside the loop, at (x_2, y_2).

2.5 Best-Fit DDA

This is an interesting, fast method based on Euclid's algorithm. It is shown here for the first octant only (not including slopes of 0° and 45°, but drawing lines for these slopes is trivial), but it is easy to generalize to the other octants since it produces a string of bits indicating the moves from pixel to pixel. For the first octant, a 0 bit means a horizontal move and a bit of 1 means a diagonal move. For the second octant, a 0 bit means a vertical move and a 1 bit means a diagonal move. Section 2.3.5 discusses the symmetry between certain octants. The only difference between octants 2 and 7, for example, is in the y coordinate.

Since the algorithm is based on strings, we use the string notation rev(str) for the reverse of the string in variable str, we use str1+str2 to denote the concatenation of the two strings, we use substring(str,a,b) for the substring from position a to position b of str, and we use len(str) for the length of string str.

2 Scan-Converting Methods

(String operations are discussed in any introductory text on computer programming.) Table 2.7 is a pseudo-code listing of the algorithm

The algorithm calls for several string reversals. This operation is slow when done by software. On a graphics computer, however, a special machine instruction may be added that reverses a register, making it possible to reverse a string in just one clock cycle.

Example: A line from $(1,2)$ to $(18,12)$. The slope is $(12-2)/(18-1) = 10/17$ (i.e., in the first octant). Table 2.8 summarizes the steps. The final result is the string 10101011010110101. Note that it consists of 10 ones and 7 zeros. It indicates 17 steps—10 of them horizontal and 7 diagonal—following the first pixel at $(1,2)$.

▶ **Exercise 2.6:** Apply this algorithm by hand to the lines from $(4,4)$ to $(12,6)$, and from $(4,4)$ to $(14,7)$.

DDA methods generate different types of lines, but they all obey the following rule: In a line that's close to horizontal, steps in the y direction (or diagonal steps) are isolated (never two consecutive steps) and steps in the x direction occur in runs of the same size (except, perhaps, the first and last runs). In a line that's close to vertical, the situation is the reverse.

2.6 Scan-Converting in Parallel

Parallel computers are becoming more and more common, and researchers are constantly looking for parallel versions of important algorithms. It turns out that scan-converting algorithms can be modified to run on parallel computers. This section shows how to generalize scan-converting methods for lines so they can run on an MIMD (Multi Instruction, Multi Data) computer. Such a computer consists of several processors, all running simultaneously, communicating with each other either through shared memory or by means of message passing.

The main principle in developing a parallel algorithm is to divide (or partition) the problem into subproblems that are **identical**. In this way, only one program has to be written, and the individual processors execute identical copies. The problem of scan-converting lines can be divided into subproblems in several ways as follows:

1. Assuming that the MIMD computer has p processors, divide the interval Δx (or Δy, whichever is greater) into p equal subintervals, and assign each subinterval to a processor. Each processor calculates the best pixels in its subinterval, and the total time is thus reduced by a factor of p. The main problem is the error variable. Its start value in each subinterval should equal its last value in the preceding subinterval, which is initially unknown. Assigning `Err:=0` in every subinterval leads to less than ideal results.

2. Consider the *bounding rectangle* of the line segment (Figure 2.9a). It contains $\Delta x \times \Delta y$ pixels. We assume that the MIMD computer has at least $\Delta x \times \Delta y$ processors arranged in a two-dimensional grid where the processor in position (i,j) becomes "responsible" for the pixel at relative position (i,j) in the rectangle. The processor calculates the distance d of its pixel from the line. If d is less than a preset value, the processor turns the pixel on (or, if XOR is used, flips it). This method is fast, but requires many processors.

```
procedure bestfit(x1,y1,x2,y2: integer);
var str1,str2: 0..1;
    x,y,i: integer;
    done: boolean;
begin
 done:=false;
 str1:='0'; str2:='1';
 y:=y2-y1;
 x:=(x2-x1)-y;
 repeat
  case
   x>y: str2:=rev(str2)+str1; x:=x-y;
   x=y: done:=true;
   x<y: str1:=rev(str1)+str2; y:=y-x;
  end; (* case *)
 until done;
 str1:=rev(str2)+str1;
 (* or str1:=rev(str1)+str2 *)
 duplicate str1 x times;
 x:=x1; y:=y1; pixel(x,y);
 for i:=1 to len(str1) do
  begin (* actual drawing *)
  x:=x+1;
  if substring(str1,i,i)='1' then y:=y+1
  pixel(x,y);
 end; (* for *)
end.
```

Table 2.7: The Best-Fit DDA Method.

x	y	str1	str2
17	10	0	1
7	10	0	1
7	3	01	1
4	3	01	101
1	3	01	10101
1	2	1010101	10101
1	1	101010110101	10101

Table 2.8: Best-Fit DDA Example.

2 Scan-Converting Methods

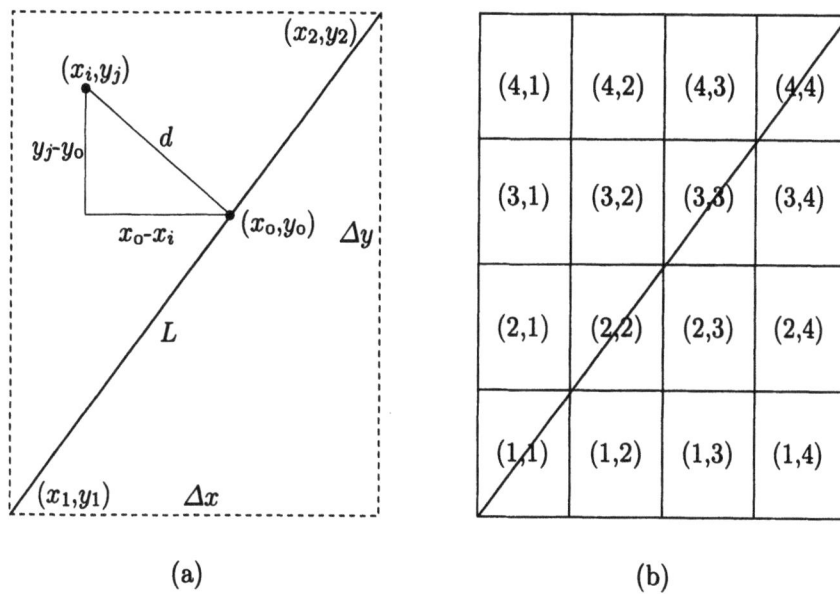

Figure 2.9: (a) A Bounding Rectangle. (b) Divided Among 16 Processors.

Figure 2.9 shows the details of the calculation. The line segment L goes from (x_1, y_1) to (x_2, y_2). We draw a perpendicular d from point (x_i, y_j) to the line. The point where it intersects the line is denoted (x_0, y_0). The two triangles $(L, \Delta y, \Delta x)$ and $(d, y_j - y_0, x_0 - x_i)$ are similar. We can, thus, multiply corresponding sides to get

$$d \times L = \Delta y(x_0 - x_i) + \Delta x(y_j - y_0),$$

yielding

$$d = -\frac{\Delta y}{L} x_i + \frac{\Delta x}{L} y_j + \frac{x_0 \Delta y - y_0 \Delta x}{L} = A x_i + B y_j + C.$$

The three constants A, B, and C are the same for all pixels (see Exercise 2.7). They are calculated—normally by processor 0—and sent to all the other processors. When a processor receives these values, it calculates the distance d of "its" pixel from the line and decides whether or not to turn it on.

▶ **Exercise 2.7:** Show why $C = (x_0 \Delta y - y_0 \Delta x)/L$ is a constant that does not depend on x_0 or y_0.

3. If the MIMD computer does not have enough processors, each processor can be assigned a row or column in the bounding rectangle. The processor then calculates the intersection of the line with "its" row or column. If the line is close to horizontal, there should be one pixel per column, so each processor should be assigned a column x_i. The processor then solves the equation

$$y = a x_i + b = \frac{y_2 - y_1}{x_2 - x_1} x_i + \frac{y_1 x_2 - x_1 y_2}{x_2 - x_1}$$

for y and sets pixel (x_i, y).

If the line is close to vertical, there should be one pixel per row and each processor should be assigned a row y_j. The processor then solves the equation

$$y_j = ax + b = \frac{y_2 - y_1}{x_2 - x_1}x + \frac{y_1 x_2 - x_1 y_2}{x_2 - x_1}$$

for x and sets pixel (x, y_j).

4. If the number of processors is very limited, each may be assigned more than just a row or a column of pixels. As an example, let's assume that we have an MIMD computer with 16 processors, organized in a two-dimensional grid and numbered from processor $(1,1)$ to processor $(4,4)$. If the bounding box contains $n \times n$ pixels, we can divide them evenly among the 16 processors and assign each processor a square area of $m \times m$ pixels, where $m = n/4$. Processor number (i,j) would, in this case, be "responsible" for the $m \times m$ square of pixels whose bottom-left corner is at position $(i \times \Delta x/m, j \times \Delta y/m)$ (Figure 2.9b).

2.7 Scan-Converting Circles

Because of the high symmetry of the circle, it is possible to scan-convert it in many ways. Only a few of these methods are discussed in this section, but many more can be found in the charming article by [Blinn 87].

2.7.1 Obvious Methods

Obvious methods for scan-converting a circle are easy to derive and to implement but are slow and inefficient.

The first obvious method is based on the Cartesian equation of a circle $x^2 + y^2 = R^2$, which yields $y = \sqrt{R^2 - x^2}$. This expression is used in the loop below to create one-quarter of the circle, which is then duplicated to complete the circle.

```
for x:=0 to R step eps do
y:=sqrt(R*R-x*x);
plot(x,y); plot(-x,y);
plot(x,-y); plot(-x,-y);
end;
```

The method is slow, but a more important disadvantage is that the pixels are not uniformly distributed over the quarter circle. This is a result of the equal x increments of the loop (Figure 2.10).

The next obvious method solves this problem by using the parametric equation $x = R\cos\theta$, $y = R\sin\theta$, which expresses the circle in terms of polar coordinates.

```
for theta:=0 to pi/2 step eps do
x:=R*cos(theta); y:=R*sin(theta);
plot(x,y); plot(-x,y);
plot(x,-y); plot(-x,-y);
end;
```

This method is still very inefficient because of the use of trigonometric functions and also because some pixels may be set multiple times.

2 Scan-Converting Methods

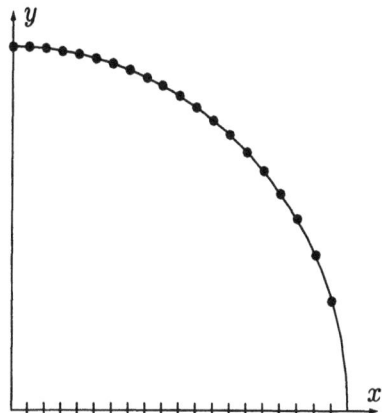

Figure 2.10: Equal Increments of x.

2.7.2 A Circle in Polar Coordinates

This is a fast algorithm, even though it uses real numbers. It is presented here for a complete circle (the loop goes from 0° to 360°), but it can easily be modified to calculate just one-quarter or one octant and use symmetry to complete the circle.

The circle equation in polar coordinates is $x = R\cos\theta$, $y = R\sin\theta$. A computer program calculates the circle pixel by pixel, by iterating and varying a variable k from 0 to $n-1$ (where n, the number of steps, is specified by the user). As a result, it is convenient to write the above equations as $x_k = R\cos(\theta_k)$, $y_k = R\sin(\theta_k)$, where $\theta_k = 2\pi k/n$.

The main point of this method is to calculate trigonometric functions just once. The user has to input a small value $\Delta\theta$, which the program uses as an angle increment. In each step, it increments $\theta_{k+1} = \theta_k + \Delta\theta$. We use the trigonometric identities for the sum of angles (Section A.6):

$$\cos(\alpha + \beta) = \cos\alpha\cos\beta - \sin\alpha\sin\beta, \quad \sin(\alpha + \beta) = \sin\alpha\cos\beta + \cos\alpha\sin\beta;$$

from this we get

$$\begin{aligned} x_{k+1} &= R\cos(\theta_{k+1}) \\ &= R\cos(\theta_k + \Delta\theta) \\ &= R\left[\cos(\theta_k)\cos(\Delta\theta) - \sin(\theta_k)\sin(\Delta\theta)\right] \\ &= x_k\cos(\Delta\theta) - y_k\sin(\Delta\theta). \end{aligned}$$

Similarly, $y_{k+1} = x_k\sin(\Delta\theta) + y_k\cos(\Delta\theta)$.

The program needs to calculate $\sin(\Delta\theta)$ and $\cos(\Delta\theta)$ only once, then loop n times. A pseudo-code algorithm appears below. (Note the quantities a and b. They are added to every pixel, which creates a circle centered at point (a, b) rather than at the origin.)

```
input(n,delta,a,b,R);
xk:=R; yk:=0;
dcos:=Cos(delta); dsin:=Sin(delta);
for k:=0 to n-1 do
xn:=xk*dcos-yk*dsin;
yn:=xk*dsin+yk*dcos;
xk:=xn; yk:=yn;
pixel(round(xn)+a,round(yn)+b);
end;
```

▶ **Exercise 2.8:** Select $\Delta\theta = 5°$, $a = b = 0$ and use this method to calculate the 18 equally spaced points of the first quadrant of a circle of radius 1.

2.7.3 Bresenham-Michener Circle Method

This is a DDA method based on a loop that starts at point $(0, R)$ and ends at point $(R/\sqrt{2}, R/\sqrt{2})$ to create one octant of the circle. Each pixel calculated is used to calculate seven more pixels, in the remaining seven octants, to create the complete circle (see also Exercise 3.15). To move from $(0, R)$ to $(R/\sqrt{2}, R/\sqrt{2})$, we need to increment x and decrement y. The loop in Figure 2.11 is set such that the x coordinate is incremented in every step, but the y coordinate is only decremented in certain steps (i.e., conditionally). The results are good (i.e., the pixels are fairly uniformly distributed) because in this octant the circle is close to horizontal.

In each step (except the first), the algorithm examines two points, **S** and **T**, that differ only in their y coordinates, and it selects the one that's closer to the true circle. The algorithm maintains a variable d_i (calculated using just additions, subtractions, and shifts) that is updated every step. The sign of d_i is used as an indicator, telling the program whether to decrement y at the step or not. The general form of the loop is as in Figure 2.11.

```
x:=0; y:=R;
while x<y do
 Plot(x,y);
 .
 .
 .
 if d>0 then
  .
  .
  y:=y-1;
 else
  ...
 endif;
 x:=x+1;
endwhile;
```

Figure 2.11: Main Loop of Bresenham's Circle Algorithm.

2 Scan-Converting Methods

If a point $\mathbf{P}_{i-1} = (x_{i-1}, y_{i-1})$ has been selected in a certain step, then the next step should increment x from x_{i-1} to $x_{i-1}+1$ and either set $y_i = y_{i-1}$ or decrement $y_i = y_{i-1} - 1$. The next step should, therefore, select either point $\mathbf{S}_i = (x_{i-1}+1, y_{i-1})$ or $\mathbf{T}_i = (x_{i-1}+1, y_{i-1}-1)$. The two quantities D_S and D_T are defined based on the distances from the points to the circle (Figure 2.11):

$$D_S = (x_{i-1}+1)^2 + y_{i-1}^2 - R^2,$$
$$D_T = R^2 - [(x_{i-1}+1)^2 + (y_{i-1}-1)^2],$$
$$d_i = D_S - D_T = 2(x_{i-1}+1)^2 + y_{i-1}^2 + (y_{i-1}-1)^2 - 2R^2.$$

Note that these quantities are based on the distances of points $\mathbf{S}$ and $\mathbf{T}$ from the true circle. They are not the distances themselves, since this would require a square root calculation. If the circle passes closer to point $\mathbf{S}$, then $D_S < D_T$ and d_i is negative. If the circle passes closer to $\mathbf{T}$, then d_i is positive. Hence, the sign of d_i indicates which point to select in iteration i.

The only remaining detail is the recalculation of d_i in each iteration. It turns out that calculating d_{i+1} is very simple if it is done in terms of d_i. This, in fact, is the main advantage of the method. We start with

$$d_{i+1} = 2(x_i+1)^2 + y_i^2 + (y_i-1)^2 - 2R^2.$$

We already know that x_i always equals $x_{i-1}+1$, but the value of y_i depends on the choice of point. If $d_i > 0$ (point $\mathbf{T}$ selected), then $y_i = y_{i-1} - 1$ and

$$\begin{aligned} d_{i+1} &= 2(x_{i-1}+2)^2 + (y_{i-1}-1)^2 + (y_{i-1}-2)^2 - 2R^2 \\ &= 2[(x_{i-1}+1)^2 + 2x_{i-1}+3] + [y_{i-1}^2 - 2y_{i-1}+1] \\ &\quad + (y_{i-1}-1)^2 - 2y_{i-1} + 3 - 2R^2 \\ &= d_i + 4x_{i-1} + 2 \cdot 3 - 2y_{i-1} + 1 - 2y_{i-1} + 3 \\ &= d_i + 4(x_{i-1} - y_{i-1}) + 10. \end{aligned}$$

If, however, $d_i < 0$ (point $\mathbf{S}$ selected), then $y_i = y_{i-1}$ and

$$\begin{aligned} d_{i+1} &= [2(x_{i-1}+2)^2] + y_{i-1}^2 + (y_{i-1}-1)^2 - 2R^2 \\ &= [2(x_{i-1}+1)^2 + 4x_{i-1} + 6] + y_{i-1}^2 + (y_{i-1}-1)^2 - 2R^2 \\ &= d_i + 4x_{i-1} + 6. \end{aligned}$$

Hence, updating the value of d_i in either case is simple and does not require any arithmetic operations beyond addition, subtraction, and shifting. Notice also that d_{i+1} depends on (x_{i-1}, y_{i-1}) and not on (x_i, y_i). The program should therefore update the value of d **before** updating the values of x and y.

The initial value of d_i, namely d_1, is easily found by substituting $(x_{i-1}, y_{i-1}) = (0, R)$. This gives

$$\begin{aligned} d_1 &= 2(0+1)^2 + R^2 + (R-1)^2 - 2R^2 \\ &= 2 + R^2 + R^2 - 2R + 1 - 2R^2 \\ &= 3 - 2R. \end{aligned}$$

```
procedure Bresenham(R);              procedure Plot8(x,y);
x:=0; y:=R; d:=3-2*R;                Plot(x,y);
while x<y do                         Plot(-x,-y);
Plot8(x,y);                          Plot(-x,y);
if d>0 then                          Plot(x,-y);
  d:=d+4*(x-y)+10;                   Plot(y,x);
  y:=y-1;                            Plot(-y,-x);
else                                 Plot(-y,x);
  d:=d+4*x+6;                        Plot(y,-x);
endif;                               end; {Plot8}
x:=x+1;
endwhile;
if x=y then Plot8(x,y)
end; {Bresenham}
```

Figure 2.12: Bresenham's Circle Algorithm.

The final program is shown in Figure 2.12.

2.7.4 Circle and Ellipse

The parametric representation of a circle of radius R centered on the origin is

$$\mathbf{C}(t) = (R\cos(2\pi t), R\sin(2\pi t)), \quad \text{where } 0 \leq t \leq 1.$$

Note that this expression can also be used to calculate the vertices of an n-sided regular polygon. A circle with a center at point (m, n) is given by the expression $(m + R\cos(2\pi t), n + R\sin(2\pi t))$.

It is possible to represent the circle by

$$\mathbf{C}(t) = (x(t), y(t)) = \left(R\left(\frac{1-t}{1+t}\right), R\left(\frac{2\sqrt{t}}{1+t}\right)\right), \quad 0 \leq t \leq \infty. \quad (2.2)$$

This is a circle of radius R since it is easy to see that $x^2(t) + y^2(t) = R^2$.

When t varies in the range $[0, 1]$, Equation (2.2) creates the first quadrant of the circle, since $C(0) = (R, 0)$ and $C(1) = (0, R)$. When t varies in the range $[1, \infty]$, it describes the second quadrant since $C(1) = (0, R)$ and $C(\infty) = (-R, 0)$. This is easy to see when one realizes that $x(1/t) = -x(t)$ and $y(1/t) = y(t)$. Hence, if a value t_0 in the range $[0, 1]$ creates a point (x, y), then the value $1/t_0$ (which is in the range $[1, \infty]$) will create the point $(-x, y)$.

An ellipse is the locus of all the points for which the sum of the distances to two fixed points, called the foci, is constant. An ellipse centered on the origin with foci at points $(-c, 0)$ and $(c, 0)$ is called *canonical*. Its implicit representation is $(x/a)^2 + (y/b)^2 = 1$, where $2a$ and $2b$ are the major and minor axes, respectively.

2 Scan-Converting Methods

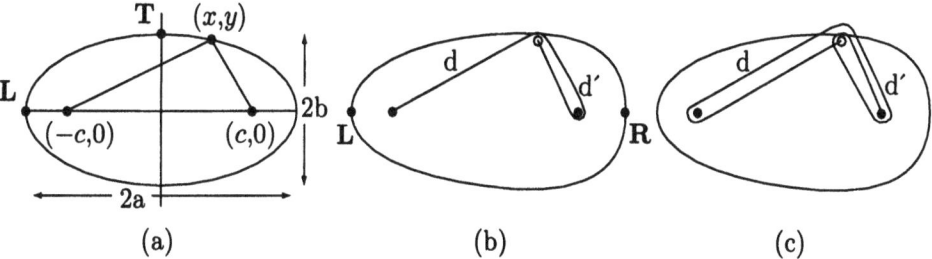

Figure 2.13: Ellipse and Oval.

▶ **Exercise 2.9:** Show that $a^2 = b^2 + c^2$.

Ovals: In much the same way that an ellipse can be considered an extension of the circle, the shape of an egg can be considered a generalization of the ellipse. The shape of an egg isn't well defined, but the great 19th-century physicist James Maxwell discovered, at the tender age of 15, how to create an entire family of oval shapes that resemble an egg [May 62]. Figure 2.13b shows how to attach a string to the left focus and to the pen (the hollow circle) such that the distance d of the pen from the left focus equals **twice** its distance d' from the right focus. Moving the pen traces an oval shape reminiscent of an egg. Figure 2.13c shows a similar oval created by attaching a string to the right focus and to the pen in a way that keeps the sum $2d + 3d'$ constant. The entire family of oval shapes can be created by keeping the sum $ad + bd'$ constant for various integer values of a and b.

▶ **Exercise 2.10:** Express the maximum and minimum values of d in terms of the length S of the string and the distance F between the foci.

The ellipse can be represented parametrically similar to a circle:

$$\mathbf{E}(t) = (a\cos(2\pi t), b\sin(2\pi t)), \quad 0 \le t \le 1,$$

or

$$\mathbf{E}(t) = \left(a\left(\frac{1-t}{1+t}\right), b\left(\frac{2\sqrt{t}}{1+t}\right)\right), \quad 0 \le t \le \infty.$$

The eccentricity of the ellipse measures how much it deviates from a circle. It is defined as $e = c/a$. For a circle, $e = 0$. When $e = 1$, the ellipse reduces to a line from $(-c, 0)$ to $(c, 0)$. The eccentricity of the Earth's orbit around the Sun is $\approx 1/60$.

When an area is scaled (expanded or shrunk), the determinant of the scaling matrix (page 63) equals the scaling factor. This can be used to determine the area πab of the ellipse. The equations of the circle and the ellipse are $x^2 + y^2 = R^2$ and $(x/a)^2 + (y/b)^2 = 1$. Therefore, if point (x, y) is on the circle, it can be transformed to the ellipse by the scaling transformation

$$\begin{pmatrix} a/R & 0 \\ 0 & b/R \end{pmatrix}.$$

The determinant of this matrix equals ab/R^2, so the area of the ellipse equals the circle area times ab/R^2 or $\pi R^2 \times ab/R^2 = \pi ab$.

Superellipse

This interesting and little known curve was developed by the French mathematician Gabriel Lamé (1795–1870) in 1818.

The parametric representation of the ellipse is $(a\cos\theta, b\sin\theta)$ where $0 \le \theta \le 2\pi$. The superellipse has the generalized representation $(a\cos^n\theta, b\sin^n\theta)$, where θ varies in the same range and n is any non-negative real number. The implicit representation is $|x/a|^n + |y/b|^n = 1$.

It is easy to see that the extreme values of x are $-a$ and a, and the extreme values of y are $-b$ and b. The curve, thus, lies inside the rectangle of width $2a$ and height $2b$, centered on the origin. The "corner" points of the superellipse are $(\pm\sigma a, \pm\sigma b)$ where $\sigma = 2^{-1/n}$ is called the *superness* of the curve.

The following values of n are especially interesting (Figure 2.14):

1. $n = 0$ (superness $= 0$). The superellipse becomes a rectangle. If $n = 0$ and $a = b$, it becomes a square.
2. $0 < n < 1$. The superellipse becomes a rounded-corner rectangle.
3. $n = 1$ (superness $= 0.5$). The superellipse becomes an ellipse. If $n = 1$ and $a = b$, it becomes a circle.
4. $n = 2$ (superness ≈ 0.7071). The superellipse becomes diamond shaped.
5. $n > 2$ (superness > 0.7071). The superellipse becomes a pinched diamond. When $n \to \infty$ (superness $\to 1$), it approaches the shape of a plus sign.

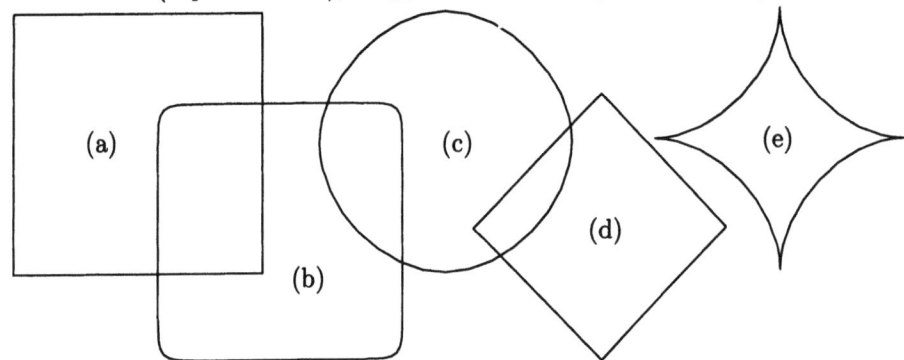

Figure 2.14: Five Superellipses.

Check also [Lamé 98] for an interactive JAVA applet. The superellipse was popularized by the multi-artist Piet Hein, who designed one of the Stockholm city squares as a superellipse with $n = 2.5$ (or superness of $2^{-0.4} \approx 0.7578$). Regardless of its shape, the superellipse passes through the same points when $\theta = 0, \pi/2, \pi, 3\pi/2$. When $a = b$, the superellipse is reduced to the supercircle $(a\cos^n\theta, a\sin^n\theta)$. It can be generalized to the three-dimensional superellipsoid

$$(a\cos^n\theta\cos^m\phi, b\cos^n\theta\sin^m\phi, c\sin^n\theta), \quad \text{where } -\pi/2 \le \theta \le \pi/2 \text{ and } -\pi \le \phi \le \pi.$$

This solid can take many shapes, but all are bounded in the box whose dimensions are $2a \times 2b \times 2c$.

2 Scan-Converting Methods

> A circle no doubt has a certain appealing simplicity at first glance, but one look at an ellipse should have convinced even the most mystical of astronomers that the perfect simplicity of the circle is akin to the vacant smile of complete idiocy. Compared to what an ellipse can tell us, a circle has little to say. Possibly our own search for cosmic simplicities in the physical universe is of this circular kind—a projection of our uncomplicated mentality on an infinitely intricate external world.
>
> — Eric Temple Bell, *Mathematics: Queen and Servant of Science*.

2.8 Thick Curves

So far, we have implicitly assumed that our curves are one-pixel wide. How can we scan-convert **thick** lines and circles? Two simple methods are briefly described:

1. Replicating pixels. The user specifies the width w of the line. Any desired scan-converting method is used to determine the best pixels. If the line is close to horizontal, each pixel selected is replicated by turning on several pixels above it and below it (Figure 2.15a), to obtain a column w pixels high. If the line is close to vertical, several pixels to its left and right are turned on instead.

This is a simple, fast method, but it produces lines shaped like parallelograms instead of rectangles (Figure 2.15b). Such lines don't connect very well and don't always have the right width (if the slope is 45°, the width is $w/\sqrt{2}$; see Figure 2.15b). When drawing a curve with this method, the program should constantly check the slope to determine whether horizontal or vertical replication is needed. For lines that are not very thick, however, this method may be satisfactory.

2. Using a drawing pen. The user specifies the shape of a pen, which can be a square, a rectangle, a circle, or anything else. The program uses any scan-converting method to determine the best pixels, and draws the shape of the pen (its footprint) centered on every pixel selected (Figure 2.15c). This is slow, since successive copies of the pen may highly overlap, but the results are usually acceptable.

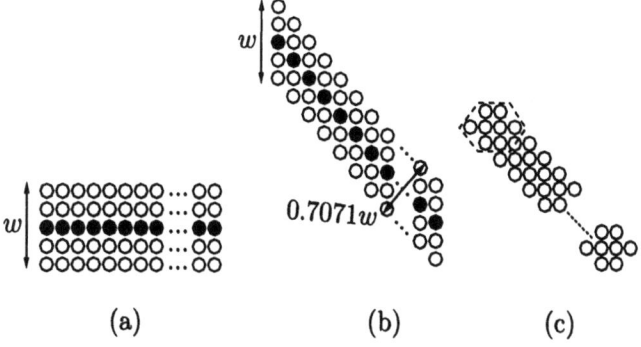

Figure 2.15: Thick Lines.

2.8.1 General Considerations

When a new scan-converting algorithm is developed or when an existing algorithm is selected for practical use, the following three points should be considered:

1. Is the algorithm symmetric? A symmetric scan-conversion algorithm calculates the same set of pixels when run from $\mathbf{P}_1$ to $\mathbf{P}_2$ as when run from $\mathbf{P}_2$ to $\mathbf{P}_1$. Most scan-conversion algorithms are not symmetric, which can lead to problems when paths are being drawn and erased. In such a case, it is important not only to use the same algorithm to draw and to erase a path but also to remember which endpoint is $\mathbf{P}_1$ and which is $\mathbf{P}_2$.

▶ **Exercise 2.11:** Is the quadrantal DDA method symmetric?

2. Does the algorithm calculate any pixels more than once? This may not be important in modern computers, where the bitmap is located in RAM and the output device is a CRT. It was, however, important in some old types of displays, such as the (now obsolete) Tektronix storage tubes, where setting a pixel multiple times made it brighter.

▶ **Exercise 2.12:** Does the symmetrical DDA method satisfy point 2?

3. Are points generated in order of minimal adjacent distance? This is important when the output device is a pen plotter. In such a device, it is essential to minimize pen travel in order to save time and to end up with high-quality output. An algorithm that sends the pen all over the place would not only be very slow on a plotter, but may result in poor plotting quality. This feature of the algorithm may even be significant when the bitmap is stored in RAM, since parts of the RAM may be loaded into a cache memory, and setting pixels all over the bitmap may cause unnecessary loading of memory pages into cache.

▶ **Exercise 2.13:** The Bresenham-Michener circle method (Section 2.7.3) calculates pixels in the second octant and uses each pixel to calculate and plot seven more pixels. Suggest a way to modify it to satisfy point 3.

2.9 Antialiasing

All scan-converting methods result in jagged lines. It turns out that the human eye is very sensitive to jagged edges and even high resolution does not eliminate this problem.

Better looking lines can be created by using an *antialiasing* method. Several such methods exist and all are based on the observation that a pixel, rather than being a perfect dot, occupies a small area on the screen. Similarly, a line displayed on the screen or printed on paper has some finite width and is not infinitely thin. For most pixels, only part of the pixel area is inside the line. An antialiasing method calculates this part and sets the pixel's intensity (or shade of gray) accordingly. Figure 2.16a shows some pixels drawn this way.

Even a cursory look at this diagram shows that it is not enough to just set the intensities of pixels according to how close they are to the mathematical line. The line generated by Figure 2.16a consists of dark and bright areas and looks worse than an aliased line. A good antialiasing method should generate lines with every pixel having the same average brightness. Figure 2.16b shows an example where

2 Scan-Converting Methods

for every pixel on the line, a "mate" pixel, either above or below it, is drawn such that the sum of intensities of the pair is constant. If a pixel is centered on the line, its intensity is high and that of its mate is low. If a pixel is slightly off the line, its intensity is low and that of its mate is adjusted to bring the sum of the two intensities to the same constant. In Figure 2.16c, each pixel has two mates.

> Jaggies: /jag'eez/ n. The "stairstep" effect observable when an edge (especially a linear edge of very shallow or steep slope) is rendered on a pixel device (as opposed to a vector display).
>
> —Eric Raymond, *The Hacker's Dictionary*

Regardless of the method used, the hardware should be capable of displaying several colors, intensities, or shades of gray. This, of course, is very common today. Antialiasing cannot be done on a monochromatic (bilevel) output device.

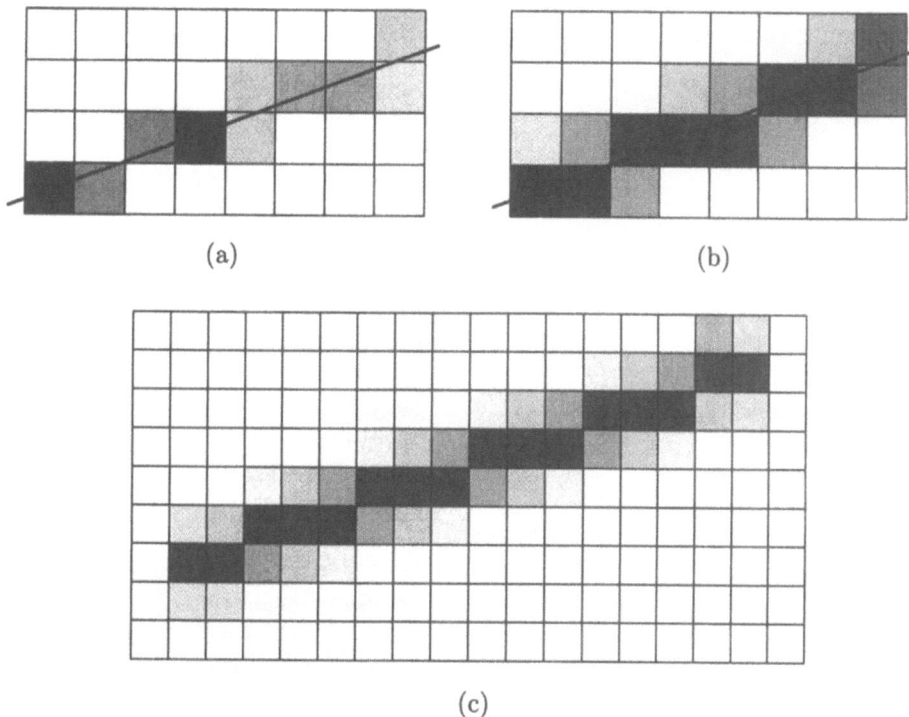

Figure 2.16: Gray Pixels for Antialiasing.

There are two main approaches to antialiasing. One approach modifies existing scan-conversion algorithms to calculate different intensities for the pixels *while* the image is being generated. The other approach uses the original algorithms to create the entire image first (aliased), then does the antialiasing by scanning the bitmap (or a copy of it) and performing some operation on the values stored there.

We illustrate the first approach by modifying a simple DDA line algorithm. Figure 2.17 shows the original and modified algorithms for the first octant. The original algorithm draws one pixel at each iteration, whereas the modified algorithm draws two. A main pixel is drawn at the position calculated by the iteration and an auxiliary pixel is drawn right above it. Assuming that the hardware can generate $n+1$ intensities (or shades of gray) labeled 0 through n, we want to calculate an intensity proportional to the value of the Err variable. A look at the algorithm shows that Err varies in the range $[-\Delta x, +\Delta y]$. The expression

$$\begin{cases} \lfloor n \times \text{Err}/\Delta y \rfloor, & \text{Err} > 0, \\ 0, & \text{Err} = 0, \\ -\lfloor n \times \text{Err}/\Delta x \rfloor, & \text{Err} < 0 \end{cases}$$

is therefore proportional to Err and is in the range $[0, n]$. If we assign it to the integer variable Ip, we get a valid intensity level.

The principle is to draw the main pixel with intensity n-Ip and the auxiliary pixel with intensity Ip. The sum of intensities assigned to the two pixels in each iteration is therefore n.

```
                                    x,y,n,Int,Δx,Δy: integer;
                                    Err: real;
                                    Δx:=x2-x1; Δy:=y2-y1;
                                    Err:=-(Δx div 2);
                                    x:=x1; y:=y1;
                                    Pixel(x,y,n);
Δx:=x2-x1; Δy:=y2-y1;               while x<x2 do begin
Err:=-(Δx div 2);                    Err:=Err+Δy;
x:=x1; y:=y1;                        if Err>=0 then
Pixel(x,y);                          begin
while x<x2 do                          y:=y+1;
begin                                  Err:=Err-Δx
 Err:=Err+Δy;                        end;
 if Err>=0 then                      x:=x+1;
 begin                               if Err>0 then Ip:=n*Err/Δy;
   y:=y+1;                           if Err=0 then Ip:=0;
   Err:=Err-Δx                       if Err<0 then Ip:=-n*Err/Δx;
 end;                                Pixel(x,y,n-Ip);
 x:=x+1;                             Pixel(x,y+1,Ip);
 Pixel(x,y);                        end (* while *);
end (* while *);
```

     Original   Modified

Figure 2.17: First Octant DDA Algorithm.

2 Scan-Converting Methods

▶ **Exercise 2.14:** Given the two endpoints $(1,1)$ and $(2,7)$ and the value $n = 15$ (i.e., 16 colors), calculate the values of `Err` for which the main pixel and the auxiliary pixel have identical intensities.

The same principle can be used to modify any scan-conversion method. The method can be improved and adjusted depending on the kind of figure to be antialiased (a thin line, a thick line, a polygon, etc.) and on the hardware used (circular pixels, square pixels, color CRT, or just shades of gray). One important improvement has to do with the case of half the maximum intensity. If two adjacent pixels are drawn in half the maximum intensity, the eye will perceive them as dimmer than one pixel at full intensity. The simple intensity calculation above can be improved by increasing the intensity of the two pixels in a group when `Ip` is close to $n/2$. We need a function of `Ip` that will reach a maximum at `Ip` $= n/2 \to 2\text{Ip}/n = 1$ and will be zero when `Ip` $= 0$ and `Ip` $= 1$. Such a function is $\alpha(1 - |1 - 2\text{Ip}/n|)$. The intensity of each of the two pixels drawn at each iteration should therefore be multiplied by the factor

$$1 + \alpha\left(1 - \left|1 - \frac{2\text{Ip}}{n}\right|\right),$$

where α depends on the kind of figure and on the hardware. The value of α should be adjusted experimentally until the best result is obtained.

2.9.1 Pitteway-Watkinson Algorithm

This method is based on the first approach. It has been developed to antialias the edges of a filled polygon. It provides a simple way of calculating the area of a pixel that's inside the polygon. Figure 2.18a shows a polygon edge with slope $a = 3/10$. A careful look shows that as we move from pixel 1 to 2 to 3, the areas under the edge grow (the small areas marked "shade" should be considered part of pixels 3, 6, and 9). When moving diagonally, from 3 to 4, or from 6 to 7, the area shrinks. (We assume that each pixel covers an area of size 1×1 and that the first pixel is half covered.)

The rule at the heart of this algorithm is: When moving horizontally, the pixel area occupied by the polygon (i.e., that part of the pixel that's inside the polygon) grows by the slope a. When moving diagonally, the area shrinks by $1 - a$. Figure 2.18b shows how the area added to a pixel can be divided into two triangles, each with a base of size 1 and a height of size a.

Assuming that the hardware supports intensity levels 0 through I, we define a quantity i as the integer nearest $a \times I$ and either increment the intensity by i, or decrement it by $1 - i$, as we move from pixel to pixel along the edge of the filled polygon.

The algorithm as described above works only for polygon edges with slopes $0° \leq a \leq 45°$ (the first octant) but can easily be extended to slopes in other octants.

The second approach to antialiasing proceeds as follows: (1) scan-conversion algorithms are used to create all the objects in the image (lines, circles, filled polygons, etc.) in a buffer in memory and (2) an algorithm scans all the pixels in the buffer and creates an antialiased image in the bitmap by changing intensities or adding/deleting pixels.

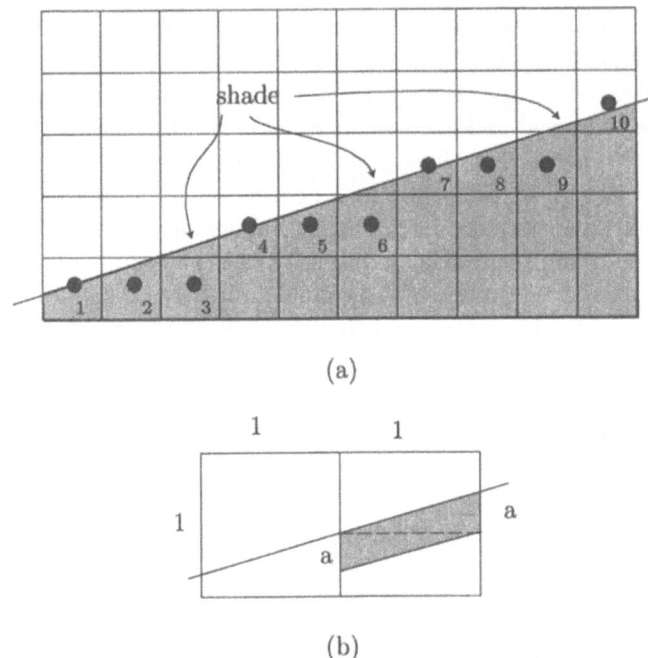

Figure 2.18: The Pitteway-Watkinson Algorithm.

The two similar methods of *supersampling* and *filtering* use this approach are described in Sections 2.9.2 and 2.9.3, respectively.

> The Image is more than an idea. It is a vortex or cluster of fused ideas and is endowed with energy.
> — Ezra Pound.

2.9.2 Supersampling

This method is based on the observation that aliasing (i.e., the problem of jagged edges) decreases with higher resolution. The scan-converting algorithm is executed, when supersampling is used, for a resolution higher than that of the hardware. If the hardware resolution is, say, 256×256 with 16 colors (i.e., 4 bit planes), then the original (black and white) algorithm is executed in a buffer that's four times wider and taller (i.e., of size $1K \times 1K$). Each pixel in the bitmap now corresponds to 16 black and white pixels in the buffer and its (antialiased) value is set to be proportional to the number of black pixels in this group of 16.

The algorithm scans the buffer group by group. For each group, the number of black pixels is computed. It can be between 0 and 16, but the pixel in the bitmap can only take the values 0 through 15. We, therefore, have to set the bitmap pixel

2 Scan-Converting Methods

to a value slightly lower than the sum, for example,

$$\text{round}\left[\frac{15}{16} \times \sum(\text{black pixels})\right].$$

Hence, a group with 15 black pixels (the dashed box in Figure 2.19a) yields intensity 14, but in a group with 4 black pixels, the rounding results in intensity 4.

If the hardware supports many colors and high resolution, the algorithm has to use a large buffer and it becomes computationally intensive. With a bitmap resolution of $1K \times 1K$ and 8-bit pixels, the buffer must have a size of $8K \times 8K = 64M$ bits and it is divided into 1M (about a million) groups of 64 bits each. This is an example of a graphics algorithm that can benefit from parallel execution.

Figure 2.16c is an example of a line antialiased by this method. Supersampling works well with the jagged edges of filled polygons, but not with thin lines. Figure 2.19b shows an example of such a line. Each group sums to 4, so each results in a pixel of intensity 4. The final line will therefore still be aliased and will also be dim.

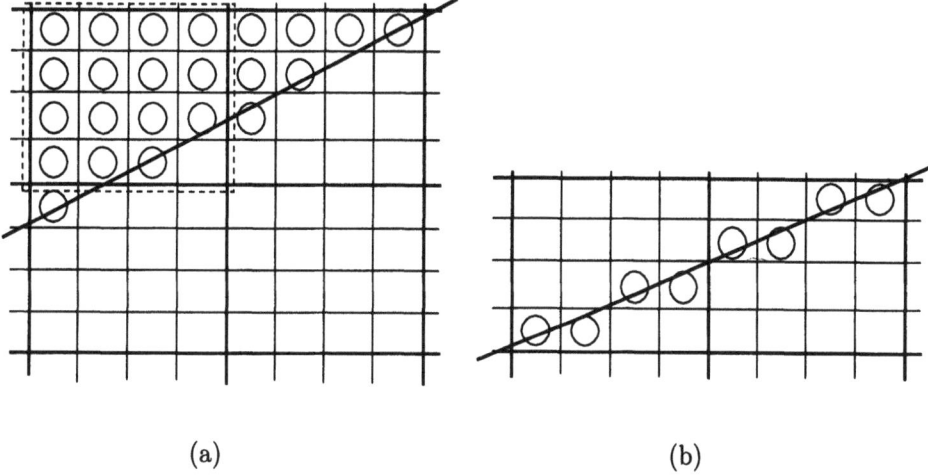

Figure 2.19: Antialiasing.

2.9.3 Filtering

The principle of this method is averaging. The original scan-conversion algorithms are applied to generate all the objects in the hardware resolution, in either black and white or shades of gray. The pixels are stored in a buffer. The algorithm then scans the buffer in small overlapping groups, each centered on a pixel (Figure 2.20a). The pixel values in each group are averaged and a value is assigned to the pixel in the bitmap, depending on the average. Different weights are assigned to the pixels in the group, depending on their distances from the center. Figure 2.20b shows typical weights for a 3×3 group. Note that the weights add up to 1.

A pixel at position (i, j) in the bitmap is now assigned the weighted average of itself and its eight nearest neighbors as follows:

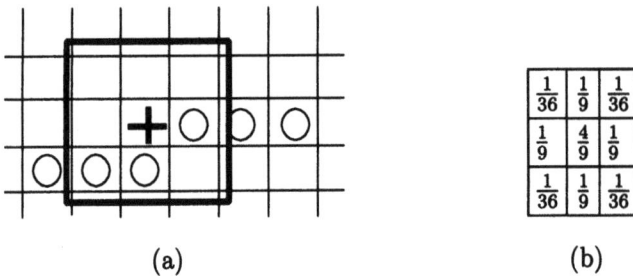

(a) (b)

Figure 2.20: Filtering.

$$P(i,j) = \frac{1}{36}I(i-1,j+1) + \frac{1}{9}I(i,j+1) + \frac{1}{36}I(i+1,j+1)$$
$$+ \frac{1}{9}I(i-1,j) + \frac{4}{9}I(i,j) + \frac{1}{9}I(i+1,j)$$
$$+ \frac{1}{36}I(i-1,j-1) + \frac{1}{9}I(i,j-1) + \frac{1}{36}I(i+1,j-1).$$

Note that special treatment should be given to the pixels at the boundary of the bitmap, since they have fewer than eight neighbors. Alternatively, two more rows (one at the top and one at the bottom) can be added to the bitmap, together with two additional columns. They can be set to zero (or to other values, if special effects are needed), so that every pixel will have eight neighbors.

In the case of black-and-white pixels, this weighted sum has a value in the range $[0, 1]$ and has to be scaled to whatever range $[0, n]$ is supported by the hardware. In the case of multilevel pixels, a different scaling is necessary. For high-resolution hardware, larger filter groups, such as 5×5 or 7×7, produce better results.

It is interesting to note that supersampling is a special case of filtering. The 4×4 supersampling example described earlier corresponds to a 4×4 filtering matrix where all the weights equal $1/16$.

> Identifying tanks and soldiers in pictures beamed back from a KH-11 Keyhole satellite is often a matter of counting dots on a computer monitor. "With 6-in. resolution you get a pixel for each shoulder and one for the head," says John Pike, space intelligence expert at the Federation of American Scientists. "That's hardly enough even to differentiate between military and civilian."
>
> — *TIME* Magazine, March 4, 1991, p. 32.

2.9.4 Convolution

Both supersampling and filtering are special cases of the general technique of *convolution*. Many practical image processing problems start with two functions that have to be combined. Normally, one function represents some kind of a signal (for

2 Scan-Converting Methods

example, a voltage that represents color, or discrete numbers representing pixel intensities) and the other one, a weight. The signal has to be combined with the weight over a certain range. For continuous functions $f(x)$ and $g(x)$, the convolution is defined as

$$C(x) = \int_{-\infty}^{\infty} f(t)g(x-t)\,dt.$$

For the discrete functions normally used in computer graphics, the convolution is defined as:

$$C(i,j) = \sum_{x=0}^{x_{max}-1} \sum_{y=0}^{y_{max}-1} I(x,y)f(i-x, j-y).$$

In our case, $I(x,y)$ is the intensity of the pixel at position (x,y), and $f(u,w)$ is the filter. The 3×3 filter of Figure 2.20b can be written as a convolution if it is defined as

$$f(u,w) = \begin{cases} 4/9, & u = w = 0, \\ 1/9, & u = 0 \text{ and } |w| = 1, \\ 1/9, & w = 0 \text{ and } |u| = 1, \\ 1/36, & |u| = |w| = 1, \\ 0, & |u| > 1 \text{ or } |w| > 1. \end{cases}$$

The last line in the definition takes care of cases where either $i - x$ or $j - y$ are negative.

(See also Section A.1.)

> Passing through to Fifth Avenue Rodney began to scan the numbers on the nearest houses.
> Horatio Alger, Jr., *Cast Upon the Breakers*

> In a rapid glance she scanned his figure, beaming with freshness and health.
> Lev Tolstoy, *Anna Karenina*

3
Transformations and Projections

Anyone doing graphics on a computer realizes very quickly that transformations are an integral part of the process of developing an image. If an image has two identical parts (for example, two hands), only one need be constructed. The other one is obtained by copying the first, then moving, reflecting, and rotating it to bring it to the right position. Often, we want to zoom on a small part of an image so more detail can be seen. Sometimes, it is useful to zoom out so a large image can be seen in its entirety on the screen, even though no details can then be discerned. Operations such as moving an image, rotating, reflecting, and scaling it are called *geometric transformations*. They are discussed in the first part of this chapter. A general reference for transformations is [Mortenson 95]. The second part of the chapter discusses projections. A three-dimensional object has to be projected down to two dimensions before it can be printed or displayed on a two-dimensional graphics output device.

3.1 Introduction

A geometric transformation is a function f whose domain and range are points. The notation $\mathbf{P}^* = f(\mathbf{P})$ implies that applying f to a point $\mathbf{P}$ yields a transformed point $\mathbf{P}^*$. Since our transformations are called *geometric*, they should have geometric interpretation, so not every function f can serve. Years of study and experience have shown that in order to be meaningful as a geometric transformation, a function must satisfy two conditions: it has to be *onto* and *one-to-one*.

- A general function f maps its domain D into its range R. If every point in R has a corresponding point in D, then the function maps its domain *onto* its range. An example is $f(x) = \lfloor x \rfloor$, which maps the real numbers onto the integers. Every integer has a real number (in fact, infinitely many real numbers) that map to it.

Another example is $g(x) = 1/x$, a mapping from the real numbers into the real numbers. This mapping is not onto since no real number maps to zero. Requiring a transformation to be onto makes sense since it guarantees that there will not be any special points $\mathbf{P}^*$ that cannot be reached by the transformation.

- An arbitrary function may map two distinct points x and y into the same point. Function $f(x)$ above maps the two distinct numbers 9.2 and 9.9 into the integer 9. A *one-to-one* function satisfies $x \neq y \to f(x) \neq f(y)$. Function $g(x)$ above is one-to-one. Requiring a transformation to be one-to-one makes sense since it means that a given point $\mathbf{P}^*$ is the transformed image of one point only, making it possible to reconstruct the inverse transformation.

▻ **Definition**: A geometric transformation is a function that is both onto and one-to-one, and whose range and domain are points.

▸ **Exercise 3.1**: Do either of the two real functions $f_1(x,y) = (x^2, y)$ and $f_2(x,y) = (x^3, y)$ satisfy the definition above?

There are two ways to look at geometric transformations. We can interpret them as either moving points to new locations or as moving the entire coordinate system while leaving the points alone. The latter interpretation is discussed in Section 3.6, but the reader should realize that whatever interpretation is used, the movement caused by a geometric transformation is *instantaneous*. We should not think of a point as moving along a path from its original location to a new location, but rather as being grabbed and immediately planted in its new location.

▻ Combining transformations is an important operation that is discussed in detail in Section 3.2.2. The intention of the present paragraph is to make it clear that such a combination (sometimes called a *product*) amounts to a *composition* of functions. If functions f and g represent two transformations, then the composition $g \circ f$ represents the product of the two transformations. Such a composition is also denoted $\mathbf{P}^* = g(f(\mathbf{P}))$. It can be shown that combining transformations is associative, i.e., $g \circ (f \circ h) = (g \circ f) \circ h$. This fact, together with a few other basic properties of transformations, makes it possible to identify *groups* of transformations. The mathematical concept of a group is discussed in Section A.10, and it is clear that a set of transformations forms a group if it includes the identity transformation, if it is closed, and if every transformation in the set has an inverse that is also included in the set.

An example of a group of transformations is the set of two-dimensional rotations about the origin through angles of 0° and 180°. This two-element set is a group since a zero-degree rotation is an identity transformation and since a 180° rotation is the inverse of itself.

▸ **Exercise 3.2**: Is the operation of combining transformations commutative?

Another important example of a group of transformations is the set of *linear transformations* that map a point $\mathbf{P} = (x, y, z)$ to a point $\mathbf{P}^* = (x^*, y^*, z^*)$, where

$$\begin{aligned} x^* &= a_{11}x + a_{12}y + a_{13}z + a_{14}, \\ y^* &= a_{21}x + a_{22}y + a_{23}z + a_{24}, \\ z^* &= a_{31}x + a_{32}y + a_{33}z + a_{34}. \end{aligned} \quad (3.1)$$

3 Transformations and Projections

▶ Such transformations are called *affine* and are defined more precisely on pages 75 and 179.

A little thinking shows that the coefficients a_{i4} of Equation (3.1) represent quantities that are added to the transformed coordinates (x^*, y^*, z^*), thereby simply *translating* $\mathbf{P}^*$ along the coordinate axes. This is why the detailed discussion below starts by temporarily ignoring these coefficients, which leads to the simple system of equations

$$\begin{aligned} x^* &= a_{11}x + a_{12}y + a_{13}z, \\ y^* &= a_{21}x + a_{22}y + a_{23}z, \\ z^* &= a_{31}x + a_{32}y + a_{33}z. \end{aligned} \quad (3.2)$$

If the 3×3 coefficient matrix of this system of equations is nonsingular or, equivalently, if the determinant of the coefficient matrix is nonzero (see Section A.5 for a refresher on matrices and determinants), then the system is easy to invert and can be expressed in the form

$$\begin{aligned} x &= b_{11}x^* + b_{12}y^* + b_{13}z^*, \\ y &= b_{21}x^* + b_{22}y^* + b_{23}z^*, \\ z &= b_{31}x^* + b_{32}y^* + b_{33}z^*, \end{aligned} \quad (3.3)$$

where the b_{ij}'s are expressed in terms of the a_{ij}'s. It is now easy to see that, for example, the two-dimensional line $Ax + By + C = 0$ is transformed by Equation (3.3) to the two-dimensional line

$$(Ab_{11} + Bb_{21})x^* + (Ab_{12} + Bb_{22})y^* + C = 0.$$

▶ **Exercise 3.3:** Show that Equation (3.3) maps the general second-degree curve

$$Ax^2 + Bxy + Cy^2 + Dx + Ey + F = 0$$

to another second-degree curve.

In general, an affine transformation maps any curve of degree n to another curve of the same degree.

3.2 Two-Dimensional Transformations

In practical work, a complete two-dimensional image is built on the screen object by object and is then edited before it is considered satisfactory. One aspect of editing is to *transform* objects. Typical transformations are moving or sliding (translation), reflecting or flipping (mirror image), zooming (scaling), rotating, and shearing.

The transformation can be applied to every pixel of the object. Alternatively, it can be applied only to some key points that define the object (such as the four corners of a rectangle), following which the object is reconstructed from the transformed key points.

The same principle applies to a three-dimensional image. It is made up of individual three-dimensional objects that can be transformed individually, following which the entire image should be projected on the two-dimensional screen (or

> As soon as we use words like "image," we are already thinking of how one shape corresponds to the other—of how you might move one shape to bring it into coincidence with the other. Bilateral symmetry means that if you reflect the left half in a mirror, then you obtain the right half. Reflection is a mathematical concept, but it is not a shape, a number, or a formula. It is a *transformation*—that is, a rule for moving things around.
> —Ian Stewart, *Nature's Numbers*, 1995.

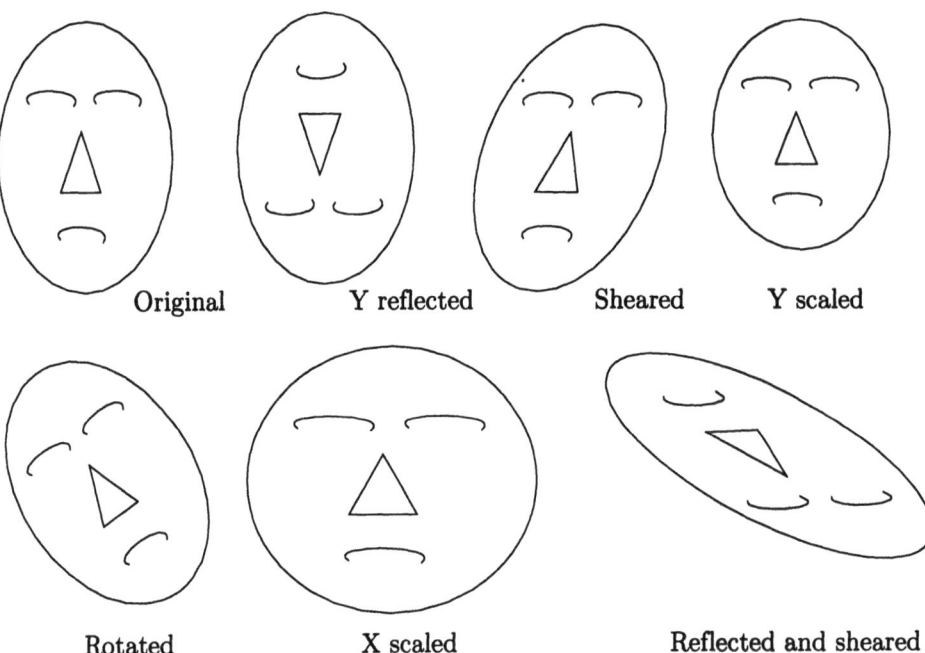

Figure 3.1: Two-Dimensional Transformations.

other output device). We first take a look at the mathematics of two-dimensional transformations.

We use the notation $\mathbf{P} = (x,y)$ for a point and $\mathbf{P}^* = (x^*, y^*)$ for the transformed point. The simplest linear transformation is $x^* = ax + cy$ and $y^* = bx + dy$, in which each of the new coordinates is a linear combination of the two old ones. This transformation can be written $\mathbf{P}^* = \mathbf{PT}$, where $\mathbf{T}$ is the 2×2 matrix $\begin{pmatrix} a & b \\ c & d \end{pmatrix}$ (see Section A.5 for matrices).

To understand the functions of the four matrix elements, we start by setting $b = c = 0$. The transformation becomes $x^* = ax$, $y^* = dy$. Such a transformation is called *scaling*. If applied to all the points of an object, all the x dimensions are scaled by a factor of a and all the y dimensions are scaled by a factor of d. Note

3 Transformations and Projections

that a and d can also be less than 1, causing shrinking of the object. If any of a and d equal -1, the transformation is a *reflection*. Any other negative values cause both scaling and reflection.

Note that scaling an object by factors of a and d changes its area by a factor of $a \times d$ and that this factor is also the value of the determinant of the scaling matrix $\begin{pmatrix} a & 0 \\ 0 & d \end{pmatrix}$.

Below are examples of scaling and reflection. In **A**, the y coordinates are scaled by a factor of 2. In **B**, the x coordinates are reflected. In **C**, the x dimensions are shrunk to 0.001 their original values. In **D**, the figure is shrunk to a vertical line.

$$\mathbf{A} = \begin{pmatrix} 1 & 0 \\ 0 & 2 \end{pmatrix}; \quad \mathbf{B} = \begin{pmatrix} -1 & 0 \\ 0 & 1 \end{pmatrix}; \quad \mathbf{C} = \begin{pmatrix} 0.001 & 0 \\ 0 & 1 \end{pmatrix}; \quad \mathbf{D} = \begin{pmatrix} 0 & 0 \\ 0 & 1 \end{pmatrix}.$$

▶ **Exercise 3.4:** What scaling transformation changes a circle to an ellipse?

The next step is to set $a = 1$ and $d = 1$ (no scaling or reflection) and explore the effect of b and c. The transformation becomes $x^* = x + cy$, $y^* = bx + y$. We first try the matrix $\begin{pmatrix} 1 & 1 \\ 0 & 1 \end{pmatrix}$ and transform the four points at $(1,0)$, $(3,0)$, $(1,1)$, and $(3,1)$. They are transformed to $(1,1)$, $(3,3)$, $(1,2)$, and $(3,4)$. When we plot the original and the transformed points (Figure 3.2a) it becomes obvious that the original rectangle has been sheared vertically and was transformed into a parallelogram. A similar effect occurs when we try the matrix $\begin{pmatrix} 1 & 0 \\ 1 & 1 \end{pmatrix}$. The quantities b and c are therefore responsible for *shearing*. Figure 3.2b shows the connection between shearing and the operation of scissors. This is the reason for the name shearing.

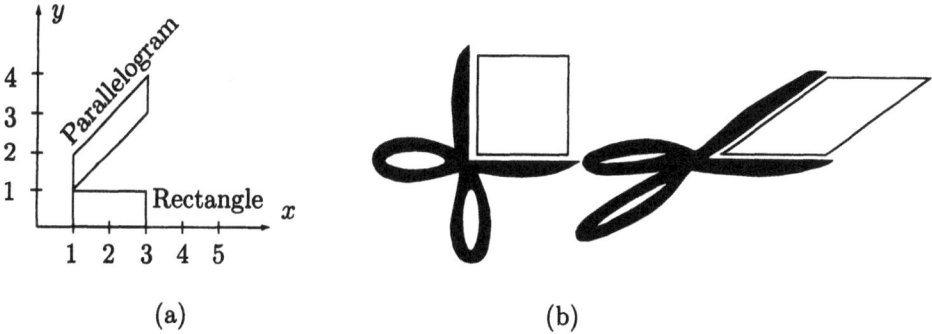

Figure 3.2: Scissors and Shearing.

▶ **Exercise 3.5:** Apply the shearing transformation $\begin{pmatrix} 1 & -1 \\ 0 & 1 \end{pmatrix}$ to the four points $(1,0)$, $(3,0)$, $(1,1)$ and $(3,1)$. What are the transformed points? What geometrical figure do they represent?

The next important transformation is *rotation*. Figure 3.3 shows a point **P** rotated clockwise about the origin through an angle θ to become $\mathbf{P}^*$. Simple trigonometry gives $x = R \cos \alpha$ and $y = R \sin \alpha$. From this, we get the expressions for x^* and y^*

$$x^* = R\cos(\alpha - \theta) = R\cos\alpha\cos\theta + R\sin\alpha\sin\theta = x\cos\theta + y\sin\theta,$$
$$y^* = R\sin(\alpha - \theta) = -R\cos\alpha\sin\theta + R\sin\alpha\cos\theta = -x\sin\theta + y\cos\theta.$$

3.2 Two-Dimensional Transformations

Hence, the clockwise rotation matrix in two dimensions is

$$\begin{pmatrix} \cos\theta & -\sin\theta \\ \sin\theta & \cos\theta \end{pmatrix}, \text{ which also equals the product } \begin{pmatrix} \cos\theta & 0 \\ 0 & \cos\theta \end{pmatrix}\begin{pmatrix} 1 & -\tan\theta \\ \tan\theta & 1 \end{pmatrix}. \quad (3.4)$$

This shows that any rotation in two dimensions is a combination of scaling (and, perhaps, reflection) and shearing; an unexpected result (that's true for all angles satisfying $\tan\theta \neq \infty$).

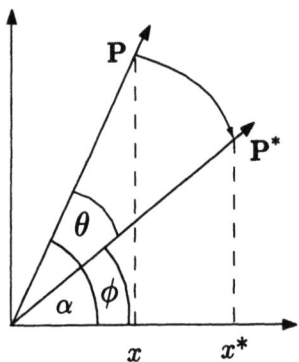

Figure 3.3: Clockwise Rotation.

▸ **Exercise 3.6:** Show how a 45° rotation is achieved by scaling followed by shearing.

▸ **Exercise 3.7:** Derive the rotation matrix in two dimensions by using the polar coordinates (r, θ) of points instead of the Cartesian coordinates (x, y).

A rotation matrix has the following property: When any row is multiplied by itself, the result is 1, and when a row is multiplied by another row, the result is 0. The same thing is true for columns. Such a matrix is called *orthonormal*.

Matrix $\mathbf{T}_1$ below rotates counterclockwise. Matrix $\mathbf{T}_2$ reflects about the line $y = x$, and matrix $\mathbf{T}_3$ reflects about the line $y = -x$. Note the determinants of these matrices. In general, a determinant of $+1$ indicates pure rotation, whereas a determinant of -1 indicates pure reflection. (As a reminder, $\det\begin{pmatrix} a & b \\ c & d \end{pmatrix} = ad - bc$. See also Section A.5.)

$$\mathbf{T}_1 = \begin{pmatrix} \cos\theta & \sin\theta \\ -\sin\theta & \cos\theta \end{pmatrix}; \quad \mathbf{T}_2 = \begin{pmatrix} 0 & 1 \\ 1 & 0 \end{pmatrix}; \quad \mathbf{T}_3 = \begin{pmatrix} 0 & -1 \\ -1 & 0 \end{pmatrix}. \quad (3.5)$$

▸ **Exercise 3.8:** Prove that a y reflection (i.e., reflection about the x axis) followed by a reflection through the line $y = -x$ is a pure rotation.

▸ **Exercise 3.9:** Prove that the transformation matrix

$$\begin{pmatrix} \dfrac{1-t^2}{1+t^2} & \dfrac{2t}{1+t^2} \\ \dfrac{-2t}{1+t^2} & \dfrac{1-t^2}{1+t^2} \end{pmatrix}$$

produces pure rotation.

3 Transformations and Projections

▶ **Exercise 3.10:** For what values of A does the following matrix represent pure rotation and for what values does it represent pure reflection?

$$\begin{pmatrix} a/A & b/A & 0 \\ -b/A & a/A & 0 \\ 0 & 0 & 1 \end{pmatrix}.$$

■ **A 90° Rotation:** In the case of a 90° clockwise rotation, the rotation matrix is

$$\begin{pmatrix} \cos(90) & -\sin(90) \\ \sin(90) & \cos(90) \end{pmatrix} = \begin{pmatrix} 0 & -1 \\ 1 & 0 \end{pmatrix}. \tag{3.6}$$

A point $\mathbf{P} = (x, y)$ is thus transformed to the point $(y, -x)$. For a counterclockwise 90° rotation, (x, y) is transformed to $(-y, x)$. This is called the *negate and exchange* rule.

> Representations rotated not always by one hundred and eighty degrees, but sometimes by ninety or forty-five, completely subvert habitual perceptions of space; the outline of Europe, for instance, a shape familiar to anyone who has been even only to junior school, when swung around ninety degrees to the right, with the west at the top, begins to look like Denmark.
>
> —Georges Perec, *Life, A User's Manual*.

3.2.1 Homogeneous Coordinates

Unfortunately, our simple 2×2 transformation matrix cannot generate all the necessary transformations! In particular, it cannot generate *translation*. This is easy to see by arguing that any object containing the origin will, after any of the transformations above, still contain the origin (i.e., the result of $(0, 0)\mathbf{T}$ is $(0, 0)$ for any matrix $\mathbf{T}$).

Translations can be expressed by $x^* = x + m$, $y^* = y + n$, and one way to implement them is to generalize our transformations to $\mathbf{P}^* = \mathbf{PT} + (m, n)$, where $\mathbf{T}$ is the familiar 2×2 transformation matrix. A more elegant approach, however, is to stay with $\mathbf{P}^* = \mathbf{PT}$ and to extend $\mathbf{T}$ to the 3×3 matrix

$$\mathbf{T} = \begin{pmatrix} a & b & 0 \\ c & d & 0 \\ m & n & 1 \end{pmatrix}. \tag{3.7}$$

This approach is called *homogeneous coordinates* and is commonly used in projective geometry. It makes it possible to unify all the two-dimensional transformations within one 3×3 matrix with six elements. The problem is that a two-dimensional point cannot be multiplied by a 3×3 matrix. This is solved by representing our points in homogeneous coordinates, which is done by extending the point (x, y) to the triplet $(x, y, 1)$. The rules for using homogeneous coordinates are the following:

The Golden Ratio

Start with a straight segment of length l and divide it into two parts a and b such that $a + b = l$ and $l/a = a/b$.

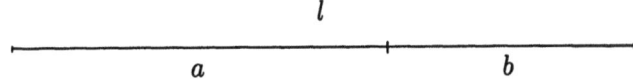

The ratio a/b is a constant called the *Golden Ratio* and is denoted ϕ. It is one of the important mathematical constants, such as π and e, and was already known to the ancient Greeks. It is believed that geometric figures can be made more pleasing to the human eye if they involve this ratio. One example is the golden rectangle, whose sides are x and $x\phi$ long. Many classical buildings and paintings involve this ratio. [Huntley 70] is a lively introduction to the Golden Ratio. It illustrates properties such as

$$\phi = \sqrt{1 + \sqrt{1 + \sqrt{1 + \sqrt{1 + \cdots}}}} \quad \text{and} \quad \phi = 1 + \cfrac{1}{1 + \cfrac{1}{1 + \cfrac{1}{\cdots}}}.$$

The value of ϕ is easy to calculate. The basic ratio $l/a = a/b = \phi$ implies $(a+b)/a = a/b = \phi$, which, in turn, means $1 + b/a = \phi$ or $1 + 1/\phi = \phi$, an equation that can be written $\phi^2 - \phi - 1 = 0$. This equation is easy to solve, yielding $\phi = (1 + \sqrt{5})/2 \approx 1.618\ldots$.

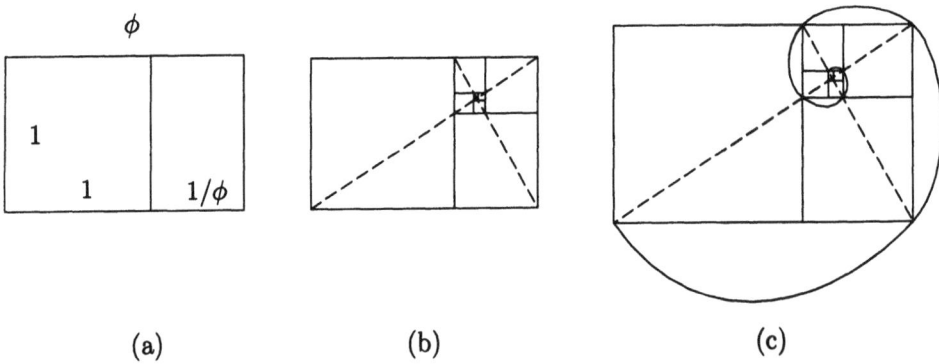

(a) (b) (c)

Figure 3.4: The Golden Ratio.

The equation $\phi = 1 + 1/\phi$ illustrates another unusual property of ϕ. Imagine the golden rectangle with sides $1 \times \phi$ (Figure 3.4a). Such a rectangle can be divided into a 1×1 square and a smaller golden rectangle of dimensions $1 \times 1/\phi$. The smaller rectangle can now be divided into a $1/\phi \times 1/\phi$ square, and an even smaller golden rectangle (Figure 3.4b). When this process continues, the rectangles converge to a point. Figure 3.4c shows how a logarithmic spiral can be drawn through corresponding corners of the rectangles.

3 Transformations and Projections

1. To transform a point (x, y) to homogeneous coordinates, simply add a third component of 1. Hence, $(x, y) \Rightarrow (x, y, 1)$.

2. To transform the triplet (a, b, c) from homogeneous coordinates back into a pair (x, y), divide by the third component. Hence, $(a, b, c) \Rightarrow (a/c, b/c)$.

This means that a point (x, y) has an infinite number of representations in homogeneous coordinates. Any triplet (ax, ay, a) is a valid representation of the point provided that $a \neq 0$. This suggests a way to intuitively understand homogeneous coordinates. We can consider the triplet (ax, ay, a) a point in three-dimensional space. When a varies from 0 to ∞, the point travels along a straight ray from the origin to infinity. The direction of the ray is determined by x and y, but not by a. Therefore, each two-dimensional point (x, y) corresponds to a ray in three-dimensional space. To find the "real" location of the point, we look at the $z = 1$ plane. All points on this plane have coordinates $(x, y, 1)$ so we only have to strip off the 1 in order to see where the point is located. Section 3.5 shows that homogeneous coordinates can also be used to transform three-dimensional points.

▶ **Exercise 3.11:** Write the transformation matrix that performs (1) a y-reflection, (2) a translation by -1 in the x and y directions, and (3) a 180° rotation about the origin. Apply this compound transformation to the four corners $(1, 1)$, $(1, -1)$, $(-1, 1)$, and $(-1, -1)$ of a square centered on the origin. What are the transformed corners?

⏵ Matrix (3.7) is the general transformation matrix in two dimensions. It produces the most general linear transformation $x^* = ax + cy + m$, $y^* = bx + dy + n$ and it shows that this transformation depends on just six numbers.

We can gain a deeper understanding of homogeneous coordinates when we add two more numbers to matrix (3.7), writing it as

$$\begin{pmatrix} a & b & p \\ c & d & q \\ m & n & 1 \end{pmatrix}. \tag{3.8}$$

A general point (x, y) is now transformed to

$$(x, y, 1) \begin{pmatrix} a & b & p \\ c & d & q \\ m & n & 1 \end{pmatrix} = (ax + cy + m, bx + dy + n, px + qy + 1).$$

Applying rule 2 shows that the transformed point (x^*, y^*) is given by

$$x^* = \frac{ax + cy + m}{px + qy + 1}, \quad y^* = \frac{bx + dy + n}{px + qy + 1}.$$

To understand what this means, we apply this result to the four points $(2, 1)$, $(6, 1)$, $(2, 5)$, and $(6, 5)$ that constitute the four corners of a square (Figure 3.5a). Using the simple transformation

$$\begin{pmatrix} 1 & 0 & 1 \\ 0 & 1 & 1 \\ 0 & 0 & 1 \end{pmatrix}$$

(i.e., no scaling, rotation, shearing, or translation and $p = q = 1$), the points are transformed to
$$\mathbf{P}_1 = (2, 1) \to (2, 1, 4) \to (1/2, 1/4),$$
$$\mathbf{P}_2 = (6, 1) \to (6, 1, 8) \to (3/4, 1/8),$$
$$\mathbf{P}_3 = (2, 5) \to (2, 5, 8) \to (1/4, 5/8),$$
$$\mathbf{P}_4 = (6, 5) \to (6, 5, 12) \to (1/2, 5/12).$$

The transformed points (Figure 3.5b) also seem to form a square, but one that's viewed from a different direction and seen in perspective. This suggests that our transformation (using just p and q, without scaling, reflection, rotation, or shearing) has moved the square from its original position in the xy plane to another plane. Such transformations are called *projections* and are useful when dealing with objects in three-dimensional space.

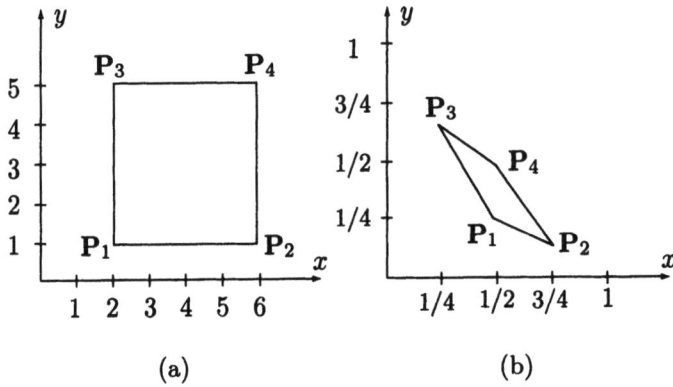

Figure 3.5: A Two-Dimensional Projection of a Square.

3.2.2 Combining Transformations

Matrix notation is useful when working with transformations, since it makes it easy to combine transformations. To combine transformations **A**, **B**, and **C**, we write the three transformation matrices and multiply them. An example is an x-reflection, followed by a y-scaling, followed by a 45° rotation

$$\begin{pmatrix} -1 & 0 \\ 0 & 1 \end{pmatrix} \begin{pmatrix} 1 & 0 \\ 0 & 2 \end{pmatrix} \begin{pmatrix} 0.707 & -0.707 \\ 0.707 & 0.707 \end{pmatrix} = \begin{pmatrix} -0.707 & 0.707 \\ 1.414 & 1.414 \end{pmatrix}.$$

In general, matrix multiplication is noncommutative, reflecting the fact that geometric transformations are also noncommutative. It is easy to convince yourself that, for example, a rotation about the origin followed by a translation is not the same as a translation followed by a rotation about the origin.

Note that all the transformations discussed above are performed about the origin. Figure 3.6a shows an object rotated 40° clockwise. It is easy to see that the center of rotation is the origin. If we want, for example, to rotate an object about a point **P**, we have to translate both the object and the point such that **P** goes to the

3 Transformations and Projections

origin (Figure 3.6b), then rotate the object, and finally translate back (Figure 3.6c). Similarly, to reflect an object through an arbitrary line, we have to (1) translate the line (and the object) until it passes through the origin, (2) rotate the line (and the object) until it coincides with one of the coordinate axes, (3) reflect through that axis, (4) rotate back, and (5) translate back.

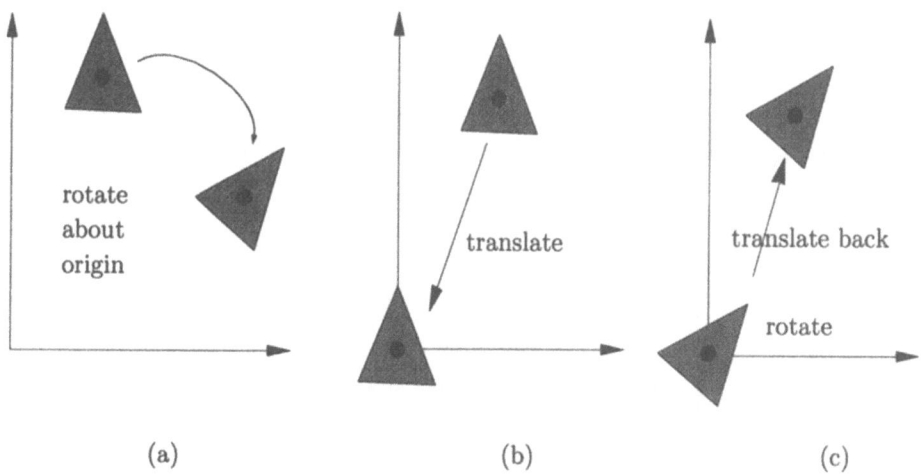

(a)   (b)   (c)

Figure 3.6: Rotation About a Point.

(Transformations are usually done about the origin. See Exercise 3.55 for an example on how this affects scaling in three dimensions.)

▶ **Exercise 3.12:** Derive the rotation matrix for a two-dimensional rotation about a point (x_0, y_0) using just trigonometry (i.e., without using translation).

Example: Reflection through the line $y = x + 1$. This line has a slope of 1 (i.e., it makes an angle of 45° with the x axis) and it intercepts the y axis at $y = 1$. We first translate down one unit, then rotate clockwise by 45°, then reflect through the x axis, rotate back, and translate back. The result is (α stands for both $\sin 45°$ and $\cos 45°$)

$$\mathbf{T} = \begin{pmatrix} 1 & 0 & 0 \\ 0 & 1 & 0 \\ 0 & -1 & 1 \end{pmatrix} \begin{pmatrix} \alpha & -\alpha & 0 \\ \alpha & \alpha & 0 \\ 0 & 0 & 1 \end{pmatrix} \begin{pmatrix} 1 & 0 & 0 \\ 0 & -1 & 0 \\ 0 & 0 & 1 \end{pmatrix} \begin{pmatrix} \alpha & \alpha & 0 \\ -\alpha & \alpha & 0 \\ 0 & 0 & 1 \end{pmatrix} \begin{pmatrix} 1 & 0 & 0 \\ 0 & 1 & 0 \\ 0 & 1 & 1 \end{pmatrix}$$

$$= \begin{pmatrix} 0 & 2\alpha^2 & 1 \\ 2\alpha^2 & 0 & 0 \\ -2\alpha^2 & 1 & 1 \end{pmatrix} = \begin{pmatrix} 0 & 1 & 0 \\ 1 & 0 & 0 \\ -1 & 1 & 1 \end{pmatrix}.$$

(This is because $2\alpha^2 = \sin^2 45° + \cos^2 45° = 1$.) Note that $\det \mathbf{T} = -1$, indicating pure reflection.

▶ **Exercise 3.13:** Prove that the result in the example is correct.

Example: Reflection about an arbitrary line. Given the line $y = ax + b$, it is possible to reflect a point about this line by transforming the line to the x axis, reflecting about that axis, and transforming the line back. Since a is the slope (i.e., the tangent of the angle α between the line and the x axis) and b is the y-intercept, the individual transformations needed are (1) a translation of $-b$ units in the y direction, (2) a clockwise rotation of α degrees about the origin, (3) a reflection about the x axis, (4) a counterclockwise rotation, and (5) a reverse translation. The combined transformation matrix is thus

$$\mathbf{T}_{\text{reflect}} = \begin{pmatrix} 1 & 0 & 0 \\ 0 & 1 & 0 \\ 0 & -b & 1 \end{pmatrix} \begin{pmatrix} \cos\alpha & -\sin\alpha & 0 \\ \sin\alpha & \cos\alpha & 0 \\ 0 & 0 & 1 \end{pmatrix} \begin{pmatrix} 1 & 0 & 0 \\ 0 & -1 & 0 \\ 0 & 0 & 1 \end{pmatrix}$$
$$\times \begin{pmatrix} \cos\alpha & \sin\alpha & 0 \\ -\sin\alpha & \cos\alpha & 0 \\ 0 & 0 & 1 \end{pmatrix} \begin{pmatrix} 1 & 0 & 0 \\ 0 & 1 & 0 \\ 0 & b & 1 \end{pmatrix} \quad (3.9)$$
$$= \begin{pmatrix} \cos(2\alpha) & \sin(2\alpha) & 0 \\ \sin(2\alpha) & -\cos(2\alpha) & 0 \\ -b\sin(2\alpha) & 2b\cos^2\alpha & 1 \end{pmatrix}.$$

The determinant of this transformation matrix equals -1, as should be for pure reflection. For the two special cases $a = b = 0$ and $a = 45°$ and $b = 0$, Equation (3.9) reduces to

$$\begin{pmatrix} 1 & 0 & 0 \\ 0 & -1 & 0 \\ 0 & 0 & 1 \end{pmatrix} \quad \text{and} \quad \begin{pmatrix} 0 & 1 & 0 \\ 1 & 0 & 0 \\ 0 & 0 & 1 \end{pmatrix}, \quad \text{respectively.}$$

One feature that makes Equation (3.9) less than general is the way the sine and cosine are obtained from the tangent of a known angle. Given that the slope a equals $\tan\alpha$, we can calculate

$$a = \tan\alpha = \frac{\sin\alpha}{\cos\alpha} = \frac{\sin\alpha}{\sqrt{1-\sin^2\alpha}},$$

which yields $\sin^2\alpha = a^2/(1+a^2)$ or

$$\sin\alpha = \pm\frac{a}{\sqrt{1+a^2}} \quad \text{and} \quad \cos\alpha = \pm\frac{1}{\sqrt{1+a^2}}.$$

The signs depend on the angle (rather, on the quadrant the angle happens to be in) and cannot be determined in a general way.

▶ **Exercise 3.14:** Calculate the numerical value of matrix $\mathbf{T}_{\text{reflect}}$ for the case $\alpha = 30°$ and $b = 1$.

3 Transformations and Projections

▸ **Exercise 3.15:** Bresenham's circle algorithm takes advantage of the symmetry of the circle by calculating the pixels for only one octant and duplicating each pixel seven times. Since the circle has very high symmetry, is it possible to improve that algorithm even more by doing half an octant and duplicating each pixel 15 times? (When the answer to this becomes clear, it will also become clear why this exercise appears here and not in Section 2.7.3.)

Another feature that makes Equation (3.9) less than general is the use of the explicit representation $y = ax + b$. This representation is limited since it cannot express vertical lines (for which a would be infinite). When reflecting about an arbitrary line, it is better to use the more general implicit representation of a straight line $ax + by + c = 0$. The slope of this line is $-a/b$ and substituting $b = 0$ yields a vertical line.

Given such a line, we start with a point $\mathbf{P} = (x, y)$ and its reflection $\mathbf{P}^* = (x^*, y^*)$ about the line. It is clear that the segment $\mathbf{PP}^*$ must be perpendicular to the line (Section 4.3.1), so its equation is $bx - ay + d = 0$. Since both $\mathbf{P}$ and $\mathbf{P}^*$ are on such a line, they satisfy $bx - ay + d = 0$ and $bx^* - ay^* + d = 0$. Subtracting yields

$$b(x - x^*) = a(y - y^*). \tag{3.10}$$

We assume that $\mathbf{P}^*$ is the reflection of $\mathbf{P}$ about the line $ax + by + c = 0$, so the midpoint of segment $\mathbf{PP}^*$, which is the point $\big((x + x^*)/2, (y + y^*)/2\big)$, must be on this line and must therefore satisfy

$$a\frac{x + x^*}{2} + b\frac{y + y^*}{2} + c = 0. \tag{3.11}$$

Equations (3.10) and (3.11) can easily be solved for x^* and y^*. The solutions are the following:

$$\begin{aligned}\mathbf{P}^* = (x^*, y^*) &= \left(x - \frac{2a(ax + by + c)}{a^2 + b^2}, y - \frac{2b(ax + by + c)}{a^2 + b^2}\right) \\ &= \left(\frac{(b^2 - a^2)x - 2aby - 2ac}{a^2 + b^2}, \frac{-2abx + (a^2 - b^2)y - 2bc}{a^2 + b^2}\right).\end{aligned} \tag{3.12}$$

Equation (3.12) is easy to verify intuitively for vertical and for horizontal lines. When b is zero, the line becomes the vertical line $x = -c/a$ and Equation (3.12) reduces to

$$\mathbf{P}^* = (x^*, y^*) = \left(x - \frac{2a(ax + c)}{a^2}, y\right) = \left(-x - \frac{2c}{a}, y\right).$$

When $a = 0$, the line is the horizontal $y = -c/b$, and the same equation reduces to

$$\mathbf{P}^* = (x^*, y^*) = \left(x, y - \frac{2b(by + c)}{b^2}\right) = \left(x, -y - \frac{2c}{b}\right).$$

The transformation matrix for reflection about an arbitrary line $ax+by+c=0$ is directly obtained from Equation (3.12):

$$\mathbf{T} = \begin{pmatrix} b^2-a^2 & -2ab & 0 \\ -2ab & a^2-b^2 & 0 \\ -2ac & -2bc & \frac{1}{a^2+b^2} \end{pmatrix}. \quad (3.13)$$

Its determinant is

$$\det \mathbf{T} = \frac{(b^2-a^2)(a^2-b^2) - 4a^2b^2}{a^2+b^2} = -\frac{a^4+2a^2b^2+b^4}{a^2+b^2} = -(a^2+b^2),$$

which equals -1 (pure reflection) for lines expressed in the standard form (Section 4.3.1) where $a^2+b^2=1$.

▸ **Exercise 3.16:** Use Equation (3.12) to obtain the transformation rule for reflection about a line that passes through the origin.

We turn now to the product of two reflections about the two arbitrary lines L_1: $ax+by+c=0$ and L_2: $dx+ey+f=0$ (Figure 3.7a). This product can be calculated from Equation (3.13) as the matrix product

$$\begin{pmatrix} b^2-a^2 & -2ab & 0 \\ -2ab & a^2-b^2 & 0 \\ -2ac & -2bc & \frac{1}{a^2+b^2} \end{pmatrix} \begin{pmatrix} e^2-d^2 & -2de & 0 \\ -2de & d^2-e^2 & 0 \\ -2df & -2ef & \frac{1}{d^2+e^2} \end{pmatrix},$$

but this product is complex and hard to interpret geometrically. In order to simplify it, we assume, without loss of generality, that both lines pass through the origin and that the first is also horizontal (Figure 3.7b). The first assumption means that the lines intersect at the origin and that $c=f=0$. The second assumption means that the first line is identical to the x axis, so $a=0$ and $b=1$. Also, $f=0$ implies $dx+ey=0$ or $y=-(d/e)x$. The quantity $-d/e$ is the slope (i.e., $\tan\theta$) of the second line, so we conclude that

$$-\frac{d}{e} = -\tan\theta = -\frac{\sin\theta}{\cos\theta}, \quad \text{implying } d^2+e^2=1.$$

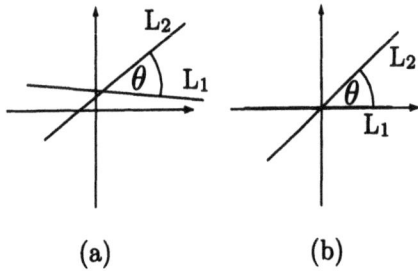

(a)   (b)

Figure 3.7: Reflections About Two Intersecting Lines.

3 Transformations and Projections

Under these assumptions, the matrix product above becomes

$$\begin{pmatrix} 1 & 0 & 0 \\ 0 & -1 & 0 \\ 0 & 0 & 1 \end{pmatrix} \begin{pmatrix} e^2 - d^2 & -2de & 0 \\ -2de & d^2 - e^2 & 0 \\ 0 & 0 & 1 \end{pmatrix}$$

$$= \begin{pmatrix} e^2 - d^2 & -2de & 0 \\ 2de & e^2 - d^2 & 0 \\ 0 & 0 & 1 \end{pmatrix}$$

$$= \begin{pmatrix} \cos(2\theta) & -\sin(2\theta) & 0 \\ \sin(2\theta) & \cos(2\theta) & 0 \\ 0 & 0 & 1 \end{pmatrix}, \qquad (3.14)$$

leading to the important conclusion that the product of two reflections about arbitrary lines is a rotation through an angle 2θ about the intersection point of the lines, where θ is the angle between the lines. It can be shown that the opposite is also true; any rotation is the product of two reflections about two intersecting lines.

The above discussion assumes that both lines pass through the origin. In the special case where $\theta = 0$, such lines would be identical so reflecting a point $\mathbf{P}$ about them would move it back to itself. However, matrix (3.14) reduces, for $\theta = 0$, to the identity matrix, so it valid even for identical lines.

In the special case where the lines are parallel, their intersection point is at infinity and a rotation about a center at infinity is a translation.

▶ **Exercise 3.17:** Given the two parallel lines $y = 0$ and $y = c$, calculate the double reflection of a point (x, y) about them.

▶ **Exercise 3.18:** Consider the shearing transformation $\mathbf{T}_a$ of Equation (3.15), followed by the 90° rotation $\mathbf{T}_b$. What is the combined transformation and what kind of transformation is it?

$$\mathbf{T}_a = \begin{pmatrix} 0 & 1 & 0 \\ 2 & 0 & 0 \\ 0 & 0 & 1 \end{pmatrix}, \quad \mathbf{T}_b \begin{pmatrix} \cos 90° & -\sin 90° & 0 \\ \sin 90° & \cos 90° & 0 \\ 0 & 0 & 1 \end{pmatrix}. \qquad (3.15)$$

▶ **Exercise 3.19:** Given the two rotations

$$\mathbf{T}_1 = \begin{pmatrix} \cos \theta_1 & -\sin \theta_1 & 0 \\ \sin \theta_1 & \cos \theta_1 & 0 \\ 0 & 0 & 1 \end{pmatrix} \quad \text{and} \quad \mathbf{T}_2 = \begin{pmatrix} \cos \theta_2 & -\sin \theta_2 & 0 \\ \sin \theta_2 & \cos \theta_2 & 0 \\ 0 & 0 & 1 \end{pmatrix},$$

calculate the combined transformation $\mathbf{T}_1\mathbf{T}_2$. Is it identical to a rotation through $(\theta_1 + \theta_2)$?

▶ **Exercise 3.20:** Given the two shearing transformations

$$\mathbf{T}_1 = \begin{pmatrix} 1 & b & 0 \\ 0 & 1 & 0 \\ 0 & 0 & 1 \end{pmatrix} \quad \text{and} \quad \mathbf{T}_2 = \begin{pmatrix} 1 & 0 & 0 \\ c & 1 & 0 \\ 0 & 0 & 1 \end{pmatrix},$$

calculate the combined transformation $\mathbf{T}_1\mathbf{T}_2$. Is it identical to a shearing by factors b and c?

▶ **Exercise 3.21:** Prove that three successive shearings about the x, y, and x axes is equivalent to a rotation about the origin.

▶ **Exercise 3.22:** Matrix $\begin{pmatrix} a & 0 \\ 0 & d \end{pmatrix}$ scales an object by factors a and d along the x and y axes, respectively. If we want to scale the object by the same factors, but in the i and j directions (see Figure 3.8, where i and j are perpendicular and form an angle θ with the x and y axes, respectively), we need to (1) rotate the object θ degrees clockwise, (2) scale along the x and y axes using matrix $\begin{pmatrix} a & 0 \\ 0 & d \end{pmatrix}$, and (3) rotate back. Write the three transformation matrices and their product. Discuss the case $a = d$ (uniform scaling).

▶ **Exercise 3.23:** A similar exercise with shearing. Matrix $\begin{pmatrix} 1 & b \\ c & 1 \end{pmatrix}$ shears an object by factors c and b along the x and y axes, respectively. Calculate the matrix that shears the object by the same factors, but in the i and j directions (Figure 3.8).

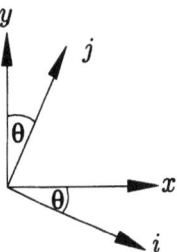

Figure 3.8: Scaling Along Rotated Axes.

▶ **Exercise 3.24:** Discuss scaling relative to a point (x_0, y_0) and show that the result is identical to the product of a translation followed by scaling followed by a reverse translation.

Using Equation (Ans.2) in the Answers to Exercises, it is easy to explore the effect of two consecutive scaling transformations, with scaling factors of k_1 and k_2 and about points $\mathbf{P}_1 = (x_1, y_1)$ and $\mathbf{P}_2 = (x_2, y_2)$, respectively. We simply multiply the two transformation matrices

$$\begin{pmatrix} k_1 & 0 & 0 \\ 0 & k_1 & 0 \\ x_1(1-k_1) & y_1(1-k_1) & 1 \end{pmatrix} \begin{pmatrix} k_2 & 0 & 0 \\ 0 & k_2 & 0 \\ x_2(1-k_2) & y_2(1-k_2) & 1 \end{pmatrix}$$
$$= \begin{pmatrix} k_1 k_2 & 0 & 0 \\ 0 & k_1 k_2 & 0 \\ k_2(1-k_1)x_1 + (1-k_2)x_2 & k_2(1-k_1)y_1 + (1-k_2)y_2 & 1 \end{pmatrix}. \quad (3.16)$$

The result is similar to Equation (Ans.2) except for the bottom row. It seems that the product of two scalings is a third scaling with a factor $k_1 k_2$, but about what

3 Transformations and Projections

point? To write Equation (3.16) in the form of Equation (Ans.2), we write

$$k_2(1-k_1)x_1 + (1-k_2)x_2 = x_c(1-k_1k_2),$$
$$k_2(1-k_1)y_1 + (1-k_2)y_2 = y_c(1-k_1k_2)$$

and solve for (x_c, y_c), obtaining

$$x_c = \frac{k_2(1-k_1)x_1 + (1-k_2)x_2}{1-k_1k_2},$$
$$y_c = \frac{k_2(1-k_1)y_1 + (1-k_2)y_2}{1-k_1k_2}.$$

The center of the double scaling is, thus, point

$$\mathbf{P}_c = \frac{k_2(1-k_1)}{1-k_1k_2}\mathbf{P}_1 + \frac{1-k_2}{1-k_1k_2}\mathbf{P}_2.$$

This is a point on the straight segment connecting $\mathbf{P}_1$ and $\mathbf{P}_2$.

In the special case $\mathbf{P}_1 = \mathbf{P}_2$, it is easy to see that the center of the double scaling is $\mathbf{P}_c = \mathbf{P}_1 = \mathbf{P}_2$.

▶ **Exercise 3.25:** What is the result of two consecutive scalings with the same scaling factors but about two different points?

▶ **Exercise 3.26:** Show that all the points with coordinates (t^2, t), where $0 \leq t \leq 1$, lie, after being transformed by

$$\begin{pmatrix} -1 & 0 & 1 \\ 0 & 2 & 0 \\ 1 & 0 & 1 \end{pmatrix},$$

on the perimeter of the unit circle $x^2 + y^2 = 1$ (Hint: See Figure 4.5b.)

It is easy to see that the transformations discussed above can change lengths and angles. Scaling changes the lengths of objects. Rotation and shearing change angles. One thing that's preserved, though, is parallel lines. A pair of parallel lines will remain parallel after any scaling, reflection, rotation, shearing, and translation. A transformation that preserves parallelism (and also maps finite points to finite points) is called *affine* (see page 179 for a more formal definition of affine transformations).

3.2.3 Fast Rotations

Rotation involves the calculation of the transcendental sine and cosine functions, which is time-consuming. If many rotations are needed, it is preferable to precalculate the trigonometric functions for many angles and store them in a table. This section shows how to do this using integers only, a method that results in much faster rotations than using floating-point numbers.

The method is illustrated for the first quadrant (rotation angles of 0° to 90°) in increments of 1°. Notice that rotations in other quadrants can be achieved by a first-quadrant rotation followed by a reflection. The following *Mathematica* code generates 91 sine values, from $\sin 0° = 0$ to $\sin 90° = 1$, multiplies each by $2^{14} = 16,384$, rounds them, and stores them in a table as 16-bit integers, ranging from 0 to 16,384.

```
d2r=Pi/180;
Table[Round[N[16384*Sin[i*d2r]]], {i,0,90}]
```

θ	$\sin\theta$	θ	$\sin\theta$	θ	$\sin\theta$	θ	$\sin\theta$	θ	$\sin\theta$
0	0	1	286	2	572	3	857	4	1143
5	1428	6	1713	7	1997	8	2280	9	2563
10	2845	11	3126	12	3406	13	3686	14	3964
15	4240	16	4516	17	4790	18	5063	19	5334
20	5604	21	5872	22	6138	23	6402	24	6664
25	6924	26	7182	27	7438	28	7692	29	7943
30	8192	31	8438	32	8682	33	8923	34	9162
35	9397	36	9630	37	9860	38	10087	39	10311
40	10531	41	10749	42	10963	43	11174	44	11381
45	11585	46	11786	47	11982	48	12176	49	12365
50	12551	51	12733	52	12911	53	13085	54	13255
55	13421	56	13583	57	13741	58	13894	59	14044
60	14189	61	14330	62	14466	63	14598	64	14726
65	14849	66	14968	67	15082	68	15191	69	15296
70	15396	71	15491	72	15582	73	15668	74	15749
75	15826	76	15897	77	15964	78	16026	79	16083
80	16135	81	16182	82	16225	83	16262	84	16294
85	16322	86	16344	87	16362	88	16374	89	16382
90	16384								

Table 3.9: Sine Values as 16-Bit Integers.

The 91 values are shown in Table 3.9, but notice that they are only approximations of the true sine values (even floating-point sine values are, in general, just approximations, but usually better than our integers). This means that many successive rotations of a point may place it farther and farther away from its true position. When we perform many successive rotations of an object which consists of many points, placing points away from where they should be generally results in a deformation of the object.

We assume that the points are represented by coordinates which are 16-bit integers. Calculating the rotated coordinates (x^*, y^*) of a point (x, y) is now done by, for example,

$$x^* = \text{rshift}(x \times \text{Table}(90 - \theta), 14) - \text{rshift}(y \times \text{Table}(\theta), 14),$$
$$y^* = \text{rshift}(x \times \text{Table}(\theta), 14) + \text{rshift}(y \times \text{Table}(90 - \theta), 14).$$

3 Transformations and Projections

Notice how the needed cosine values are obtained from the end of the table. This method works since the table has 91 entries. Multiplying a 16-bit integer coordinate by a 16-bit integer sine value yields a 32-bit product. The right shift effectively divides the product by $2^{14} = 16{,}384$, which is needed since our integer sine values have been premultiplied by this scale factor.

▸ **Exercise 3.27:** Use this method to calculate the results of rotating point $(1,2)$ by $60°$ and by $80°$. In each case, compare the results to those obtained when a built-in sine and cosine functions are used.

3.2.4 CORDIC Rotations

We all use calculators to compute values of common functions, but have you ever wondered how a calculator finds the value of, say, $\tan 72.81°$ so fast? Most calculators (as well as the Intel 80x87 family of coprocessors) use CORDIC (COordinate Rotation, DIgital Computer), a unified method for computing many elementary functions. CORDIC was originally proposed by [Volder 59] and was extended by [Walther 71]. The original references are hard to find but are included in [Swartzlander 90]. Here, we show how CORDIC can be used to implement fast rotations.

Consider a rotation about the origin where the rotation angle θ satisfies $0° \leq \theta < 90°$ (it is in the first quadrant). The special case $\theta = 90°$ can be implemented by the negate and exchange rule (Equation (3.6)). Rotations in other quadrants can be achieved by a first-quadrant rotation, followed by a reflection.

The rotation is expressed by (see Equation (3.4))

$$(x^*, y^*) = (x, y) \begin{pmatrix} \cos\theta & -\sin\theta \\ \sin\theta & \cos\theta \end{pmatrix} \quad (3.17)$$

Since $\theta < 90°$, we know that $\cos\theta \neq 0$, so we can factor out $\cos\theta$, giving

$$(x^*, y^*) = \cos\theta \, (x, y) \begin{pmatrix} 1 & -\tan\theta \\ \tan\theta & 1 \end{pmatrix}$$

We now express θ as the sum $\sum_{i=0}^{m} \theta_i$, where angles θ_i are defined by the relation $\tan\theta_i = 2^{-i}$ or $\theta_i \stackrel{\text{def}}{=} \arctan(2^{-i})$. The first 16 θ_i, for $i = 0, 1, \ldots, 15$ are listed in Table 3.10.

In order to express any angle θ as the sum of these particular θ_i, some θ_i will have to be subtracted. Consider, for example, $\theta = 58°$. We start with $\theta_0 = 45°$. Since $\theta_0 < \theta$, we add θ_1. The sum $\theta_0 + \theta_1 = 45 + 26.5651 = 71.5651$ is greater than θ, so we subtract θ_2. The new sum, 57.5289, is less than θ, so we add θ_3, and so on.

▸ **Exercise 3.28:** We want to be able to express any angle θ in the range $[0°, 90°)$ by adding and subtracting a number of consecutive θ_i, from θ_0 to some θ_m, without skipping any θ_i in between. Is that possible?

It is easy to write a program that decides which of the θ_i's should be added and which should be subtracted. We thus end up with

$$\theta = \sum_{i=0}^{m} d_i \theta_i = \sum_{i=0}^{m} d_i \arctan(2^{-i}), \quad \text{where} \quad d_i = \pm 1.$$

3.2 Two-Dimensional Transformations

i	θ_i (degrees)	θ_i (radians)	K_i
0	45.	0.785398	0.70710678118654746
1	26.5651	0.463648	0.63245553203367577
2	14.0362	0.244979	0.61357199107789628
3	7.12502	0.124355	0.60883391251775243
4	3.57633	0.0624188	0.60764825625616825
5	1.78991	0.0312398	0.60735177014129604
6	0.895174	0.0156237	0.60727764409352614
7	0.447614	0.00781234	0.60725911229889284
8	0.223811	0.00390623	0.60725447933256249
9	0.111906	0.00195312	0.60725332108987529
10	0.0559529	0.000976562	0.60725303152913446
11	0.0279765	0.000488281	0.60725295913894495
12	0.0139882	0.000244141	0.60725294104139727
13	0.00699411	0.00012207	0.60725293651701029
14	0.00349706	0.0000610352	0.60725293538591352
15	0.00174853	0.0000305176	0.60725293510313938

Table 3.10: The First 16 θ_i's and Scale Factors.

Once the number m of necessary d_i and their values have been determined, we rotate (x, y) to (x^*, y^*) in a loop where each iteration rotates a point (x_i, y_i) through an angle $d_i\theta_i$ to a point (x_{i+1}, y_{i+1}). A general iteration can be expressed:

$$\begin{aligned}(x_{i+1}, y_{i+1}) &= \cos(d_i\theta_i)\,(x_i, y_i)\begin{pmatrix} 1 & -\tan\theta_i \\ \tan\theta_i & 1 \end{pmatrix} \\ &= \cos(d_i\theta_i)\,(x_i, y_i)\begin{pmatrix} 1 & -d_i 2^{-i} \\ d_i 2^{-i} & 1 \end{pmatrix} \\ &= \cos(d_i\theta_i)\,(x_i + y_i d_i 2^{-i}, y_i - x_i d_i 2^{-i}). \end{aligned} \quad (3.18)$$

We interpret the result (x_{i+1}, y_{i+1}) of an iteration as the vector from the origin to point (x_{i+1}, y_{i+1}). Equation (3.18) shows that this vector is the product of two terms. The second term, $(x_i + y_i d_i 2^{-i}, y_i - x_i d_i 2^{-i})$, determines the direction of the vector, while the first term, $\cos(d_i\theta_i)$, affects only the magnitude of the vector. The second term is easy to calculate since it involves just shifts. We know that d_i is just a sign and that a product of the form $x_i 2^{-i}$ can be computed by shifting x_i i positions to the right. The problem is to calculate the first term, $\cos(d_i\theta_i)$, and to multiply the two terms.

This is why CORDIC proceeds by first performing all the iterations

$$(x_{i+1}, y_{i+1}) \leftarrow (x_i + y_i d_i 2^{-i}, y_i - x_i d_i 2^{-i}),$$

using just right shifts and additions/subtractions; the cosine terms are ignored. The result is a vector which points in the right direction but is too long (Figure 3.12).

3 Transformations and Projections

To bring this vector to its right size, it should be multiplied by the scale factor

$$K_m = \prod_{i=0}^{m} \cos\theta_i.$$

(Notice that $\cos(d_i\theta_i) = \cos\theta_i$ since cosine is an even function.) This is discouraging since it seems that m multiplications are needed just to calculate the scale factor K_m. However, the first 16 scale factors are listed in Table 3.10 and even a quick glance shows that they converge to the number $0.60725\ldots$. Reference [Vachss 87] shows that K_m can be obtained simply by using the first m bits of this number and ignoring the rest.

Using the identity $\sin^2\theta + \cos^2\theta = 1$ and the definition $\tan\theta_i = 2^{-i}$, we get

$$\cos\theta_i = \frac{1}{\sqrt{1+tan^2\theta_i}} = \frac{1}{\sqrt{1+2^{-2i}}},$$

which is why the scale factors of Table 3.10 were so easily calculated to a high precision by the code
`N[Table[Product[(2^(-2i)+1)^(-1/2),{i,0,n}],{n,0,16}],17]//TableForm`.

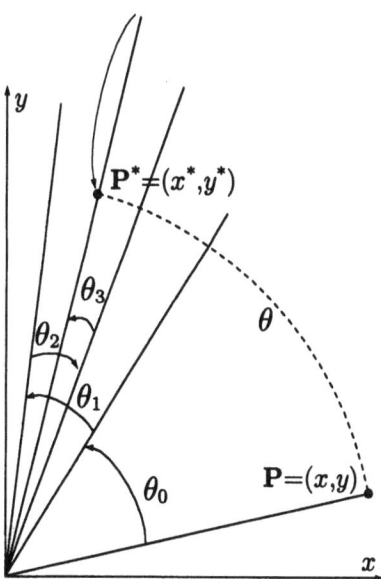

Figure 3.11: CORDIC Rotation.

▶ **Exercise 3.29:** Suggest another way to calculate K_m.

Any practical CORDIC implementation (see [Jarvis 90] for a C program) should have the following two features:

1. Since CORDIC uses only shifts and additions/subtractions, it should use fixed-point, instead of floating-point, arithmetic. This is fast, since shifting and adding fixed-point numbers can be done using integer operations. Notice that all the

numbers involved in the computations are less than unity except perhaps the original coordinates (x, y). A software package for graphics using this method should thus use normalized coordinates (fixed-point numbers in the interval $[0, 1]$) throughout, and perform all the calculations on these small numbers. Each iteration results in a pair (x_{i+1}, y_{i+1}) that's slightly larger than its predecessor, but the last iteration results in a pair that can be larger than (x, y) by a factor of at most $1/0.60725\ldots = 1.64676\ldots$. This pair is then scaled down when multiplied by K_m. The final step is to scale the final coordinates up.

All this suggests a 32-bit fixed-point format where the leftmost bit is reserved, as usual, for the sign, the next two bits are the integer part, and the remaining 29 bits are the fractional part (29 bits are equivalent to 9 decimal digits). The largest number that can be represented by this format is $11.11\ldots1_2 = 3.999\ldots$ and the smallest one is $110\ldots0_2 = -4$. It's a good idea to reserve two bits for the integer part because (1) even though all the numbers involved are 1 or less, some intermediate results may be greater than 1 and (2) this makes it possible to represent the important constants π, e, and ϕ (the Golden Ratio).

2. Earlier we said "It is easy to write a program that decides which of the θ_i's should be added and which should be subtracted." The practical way to do this is to initialize a variable z to θ and try to drive z to zero during the iterations. In iteration i the program should calculate both $z + \theta_i$ and $z - \theta_i$, select the value that's closer to zero, use it to decide whether to add or subtract θ_i, then update z. If $z - \theta_i$ is closer to zero, then θ_i should be added; otherwise, θ_i should be subtracted. An example is $\theta = 58°$. We initialize z to 58. In iteration 0, it is clear that $58 - 45 = 13$ is closer to zero than $58 + 45$. The program therefore adds θ_0 and updates z to 13. In iteration 1, the program finds that $13 - 26.5651 = -13.5651$ is closer to zero than $13 + 26.5651$, so it adds θ_1 and updates z to -13.5651. In iteration 2, the program discovers that $-13.5651 + 14.0362 = 0.4711$ is closer to zero than $-13.5651 - 14.0362$, so it subtracts θ_2 and updates z to 0.4711.

Finally, we realize that there is really no need to compare $z + \theta_i$ and $z - \theta_i$ in iteration i. We simply start by selecting $d_0 = +1$ and update z by subtracting $z \leftarrow z - \theta_0$, $z \leftarrow z - \theta_1$, etc., until we get a negative value in z. We then change d_i to -1 (the new sign of z) and update z by $z \leftarrow z - d_i\theta_i$ (which now amounts to adding θ_i to z). This is summarized by the *Mathematica* code of Figure 3.12. (But note that the Sign function in *Mathematica* returns $+1$, 0, or -1, while we need a result of $+1$ or -1. The code as shown is simple but not completely general.)

```
t=Table[ArcTan[2.^{-i}], {i,0,15}];  (* arctans in radians *)
d=1; x=2.1; y=0.34; z=46. Degree;
Do[{Print[i,", ",x,", ",y,", ",z,", ",d],
  xn=x+y d 2^{-i}, yn=y-x d 2^{-i},
  zn=z-d t[[i+1]], d=Sign[zn], x=xn, y=yn, z=zn}, {i,0,14}]
Print[0.60725x,", ",0.60725y]
```

Figure 3.12: *Mathematica* Code for CORDIC Rotations.

▶ **Exercise 3.30:** Instead of using the complex CORDIC method, wouldn't it be

3 Transformations and Projections

simpler to perform a rotation by a direct use of Equation (3.17)? After all, this only requires the calculation of one sine and one cosine values?

3.2.5 Similarities

A *similarity* is a transformation that scales all distances by a fixed factor. It is easy to show that it is produced by the special transformation matrix

$$\begin{pmatrix} a & c & 0 \\ -c & a & 0 \\ m & n & 1 \end{pmatrix}.$$

Proof: Translations preserve distances, so we can ignore the translation part of the matrix above and restrict ourselves to the matrix $\begin{pmatrix} a & c \\ -c & a \end{pmatrix}$. It transforms a point $\mathbf{P} = (x, y)$ to the point $\mathbf{P}^* = (x^*, y^*) = (ax - cy, cx + ay)$. Given the two transformations $\mathbf{P}_1 \to \mathbf{P}_1^*$ and $\mathbf{P}_2 \to \mathbf{P}_2^*$, it is straightforward to show that

$$\begin{aligned}
\texttt{distance}^2(\mathbf{P}_1^*\mathbf{P}_2^*) &= \left((\Delta x^*)^2 + (\Delta y^*)^2\right) \\
&= [(ax_2 - cy_2) - (ax_1 - cy_1)]^2 + [(cx_2 + ay_2) - (cx_1 + ay_1)]^2 \\
&= (a\Delta x - c\Delta y)^2 + (c\Delta x + a\Delta y)^2 \\
&= a^2\Delta x^2 - 2a\Delta xc\Delta y + c^2\Delta y^2 + c^2\Delta x^2 + 2c\Delta xa\Delta y + a^2\Delta y^2 \\
&= (a^2 + c^2)(\Delta x^2 + \Delta y^2) \\
&= (a^2 + c^2)\texttt{distance}^2(\mathbf{P}_1\mathbf{P}_2),
\end{aligned}$$

implying that $\texttt{distance}(\mathbf{P}_1^*\mathbf{P}_2^*) = \sqrt{a^2 + c^2}\,\texttt{distance}(\mathbf{P}_1\mathbf{P}_2)$ (end of proof).

In general, a similarity is a transformation of the form $\mathbf{P}^* = (x^*, y^*) = (ax - cy + m, \pm(cx + ay) + n)$, where the ratio of expansion (or shrinking) is $k = \sqrt{a^2 + c^2}$. If k is positive, the similarity is called *direct*; if k is negative, the similarity is *opposite*.

▶ **Exercise 3.31:** What if $k = 0$?

Using the ratio k, we can write a similarity (ignoring the translation part) as the product

$$\begin{pmatrix} a & c & 0 \\ -c & a & 0 \\ 0 & 0 & 1 \end{pmatrix} \begin{pmatrix} k & 0 & 0 \\ 0 & k & 0 \\ 0 & 0 & 1 \end{pmatrix} \begin{pmatrix} a/k & c/k & 0 \\ -c/k & a/k & 0 \\ 0 & 0 & 1 \end{pmatrix},$$

which shows that a similarity is a combination of scaling/reflection (by a factor k) and a rotation. (The definition of k implies that $(a/k)^2 + (c/k)^2 = 1$, so we can consider c/k and a/k the sine and cosine of the rotation angle, respectively.)

3.2.6 A 180° Rotation

Another interesting example of combining transformations is a 180° rotation about a fixed point $\mathbf{P} = (P_x, P_y)$. This combination is called *halfturn*. It is performed, as usual, by translating $\mathbf{P}$ to the origin, rotating about the origin, and translating

3.2 Two-Dimensional Transformations

back. The transformation matrix is (notice that $\cos(180°) = -1$)

$$\mathbf{T} = \begin{pmatrix} 1 & 0 & 0 \\ 0 & 1 & 0 \\ -P_x & -P_y & 1 \end{pmatrix} \begin{pmatrix} -1 & 0 & 0 \\ 0 & -1 & 0 \\ 0 & 0 & 1 \end{pmatrix} \begin{pmatrix} 1 & 0 & 0 \\ 0 & 1 & 0 \\ P_x & P_y & 1 \end{pmatrix} = \begin{pmatrix} -1 & 0 & 0 \\ 0 & -1 & 0 \\ 2P_x & 2P_y & 1 \end{pmatrix}.$$

A general point (x,y) is transformed by a halfturn to

$$(x, y, 1) \begin{pmatrix} -1 & 0 & 0 \\ 0 & -1 & 0 \\ 2P_x & 2P_y & 1 \end{pmatrix} = (-x + 2P_x, -y + 2P_y, 1) \qquad (3.19)$$

(Figure 3.13a), but it's more interesting to explore the effect of two consecutive halfturns, about points **P** and **Q**. The second halfturn transforms point $(-x + 2P_x, -y + 2P_y, 1)$ to

$$(-x + 2P_x, -y + 2P_y, 1) \begin{pmatrix} -1 & 0 & 0 \\ 0 & -1 & 0 \\ 2Q_x & 2Q_y & 1 \end{pmatrix} = (x - 2P_x + 2Q_x, y - 2P_y + 2Q_y, 1).$$

(3.20)

If $\mathbf{P} = \mathbf{Q}$, then the result of the second halfturn is (x, y), showing how two identical 180° rotations return a point to its original location. If **P** and **Q** are different, the result is a *translation* of the original point (x, y) by factors $-2P_x + 2Q_x$ and $-2P_y + 2Q_y$ (Figure 3.13b).

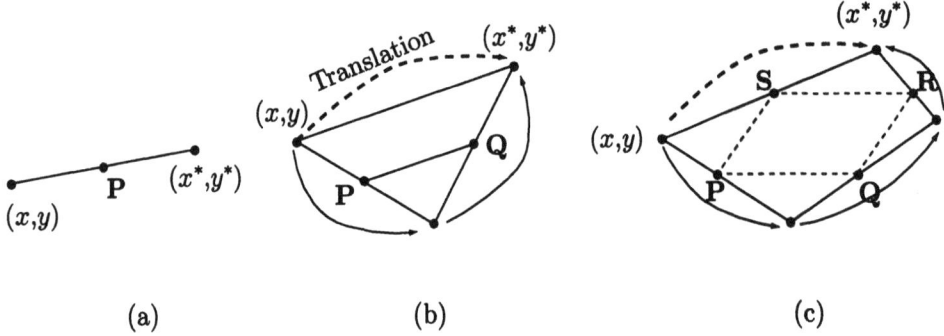

Figure 3.13: Halfturns.

▶ **Exercise 3.32:** What is the result of three consecutive halfturns about the distinct points **P**, **Q**, and **R**?

3.2.7 Glide Reflection

This transformation is a special combination of three reflections. Imagine the two vertical parallel lines $x = L$ and $x = M$ and the horizontal line $y = N$ (Figure 3.14a). Reflecting a point $\mathbf{P} = (x, y)$ about the line $x = L$ is done by translating the line

3 Transformations and Projections

to the y axis, reflecting about that axis, and translating back. The transformation matrix is

$$\begin{pmatrix} 1 & 0 & 0 \\ 0 & 1 & 0 \\ -L & 0 & 1 \end{pmatrix} \begin{pmatrix} -1 & 0 & 0 \\ 0 & 1 & 0 \\ 0 & 0 & 1 \end{pmatrix} \begin{pmatrix} 1 & 0 & 0 \\ 0 & 1 & 0 \\ L & 0 & 1 \end{pmatrix} = \begin{pmatrix} -1 & 0 & 0 \\ 0 & 1 & 0 \\ 2L & 0 & 1 \end{pmatrix},$$

and the transformed point is

$$(x, y, 1) \begin{pmatrix} -1 & 0 & 0 \\ 0 & 1 & 0 \\ 2L & 0 & 1 \end{pmatrix} = (-x + 2L, y, 1).$$

Reflecting this point about the line $x = M$ results in

$$(-x + 2L, y, 1) \begin{pmatrix} -1 & 0 & 0 \\ 0 & 1 & 0 \\ 2M & 0 & 1 \end{pmatrix} = (x - 2L + 2M, y, 1)$$

(a translation), and reflecting this about the horizontal line $y = N$ yields

$$(x - 2L + 2M, y, 1) \begin{pmatrix} 1 & 0 & 0 \\ 0 & -1 & 0 \\ 0 & 2N & 1 \end{pmatrix} = (x - 2L + 2M, -y + 2N, 1).$$

This particular glide reflection is therefore a translation in x and a reflection in y. A general glide reflection is the product of three reflections, the first two about parallel lines L and M and the third about a line N perpendicular to them (Figure 3.14b).

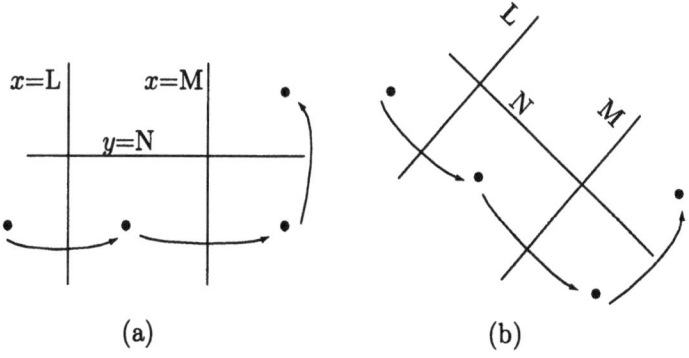

Figure 3.14: Glide Reflection.

3.2.8 Improper Rotations

A rotation followed by a reflection about one of the coordinate axes is called an *improper rotation*. The transformation matrices for the two possible improper rotations in two dimensions (Figure 3.15) are

$$\begin{pmatrix} \cos\theta & -\sin\theta \\ \sin\theta & \cos\theta \end{pmatrix} \begin{pmatrix} 1 & 0 \\ 0 & -1 \end{pmatrix} = \begin{pmatrix} \cos\theta & \sin\theta \\ \sin\theta & -\cos\theta \end{pmatrix},$$

$$\begin{pmatrix} \cos\theta & -\sin\theta \\ \sin\theta & \cos\theta \end{pmatrix} \begin{pmatrix} -1 & 0 \\ 0 & 1 \end{pmatrix} = \begin{pmatrix} -\cos\theta & -\sin\theta \\ -\sin\theta & \cos\theta \end{pmatrix},$$

and the transformation rules therefore are

$$x^* = x\cos\theta + y\sin\theta, \quad y^* = x\sin\theta - y\cos\theta,$$
$$x^* = -x\cos\theta - y\sin\theta, \quad y^* = -x\sin\theta + y\cos\theta.$$

Notice that the determinant of an improper rotation matrix equals -1, like that of a pure reflection.

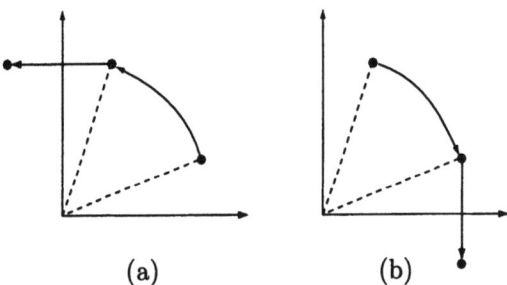

Figure 3.15: Improper Rotations.

An improper rotation differs from a rotation in one important aspect. When we rotate an object through a small angle and repeat this transformation, the object seems to move smoothly along a circle. Each time we repeat an improper rotation, however, the object "jumps" from one side of the coordinate plane to the other. The total effect is very different from that of a smooth circular movement.

3.2.9 Decomposing Transformations

Sometimes, a certain transformation A may be equivalent to the combined effects of several different transformations B, C, and D. We say that A can be *decomposed* into B, C, and D. Mathematically, this is equivalent to saying that the original transformation matrix T_A equals the product $T_B T_C T_D$. We have already seen that a rotation in two dimensions can be decomposed into a scaling followed by a shearing; here are other examples.

It may come as a surprise that the general two-dimensional transformation matrix, Equation (3.7), can be written as a product of shearing, scaling, rotation,

3 Transformations and Projections

and translation:

$$
\begin{pmatrix} a & b & 0 \\ c & d & 0 \\ m & n & 1 \end{pmatrix} = \begin{pmatrix} 1 & 0 & 0 \\ (ac+bd)/A^2 & 1 & 0 \\ 0 & 0 & 1 \end{pmatrix} \begin{pmatrix} A & 0 & 0 \\ 0 & (ad-bc)/A & 0 \\ 0 & 0 & 1 \end{pmatrix}
$$
$$
\times \begin{pmatrix} a/A & b/A & 0 \\ -b/A & a/A & 0 \\ 0 & 0 & 1 \end{pmatrix} \begin{pmatrix} 1 & 0 & 0 \\ 0 & 1 & 0 \\ m & n & 1 \end{pmatrix},
$$
(3.21)

where $A = \sqrt{a^2+b^2}$. The third matrix produces rotation since $(a/A)^2+(b/A)^2 = 1$.

Even something as simple as shearing in one direction can be written as the product of a unit shearing and two scalings:

$$
\begin{pmatrix} 1 & 0 & 0 \\ c & 1 & 0 \\ 0 & 0 & 1 \end{pmatrix} = \begin{pmatrix} 1/c & 0 & 0 \\ 0 & 1 & 0 \\ 0 & 0 & 1 \end{pmatrix} \begin{pmatrix} 1 & 0 & 0 \\ 1 & 1 & 0 \\ 0 & 0 & 1 \end{pmatrix} \begin{pmatrix} c & 0 & 0 \\ 0 & 1 & 0 \\ 0 & 0 & 1 \end{pmatrix}.
$$

Even the simple transformation of a unit shearing can be decomposed into a product that involves a scaling and two rotations. Note that the Golden Ratio ϕ is involved,

$$
\begin{pmatrix} 1 & 0 & 0 \\ 1 & 1 & 0 \\ 0 & 0 & 1 \end{pmatrix} = \begin{pmatrix} \cos\alpha & -\sin\alpha & 0 \\ \sin\alpha & \cos\alpha & 0 \\ 0 & 0 & 1 \end{pmatrix} \begin{pmatrix} \phi & 0 & 0 \\ 0 & 1/\phi & 0 \\ 0 & 0 & 1 \end{pmatrix} \begin{pmatrix} \cos\beta & \sin\beta & 0 \\ -\sin\beta & \cos\beta & 0 \\ 0 & 0 & 1 \end{pmatrix},
$$

where $\alpha = \tan^{-1}\phi \approx 58.28°$ and $\beta = \tan^{-1}(1/\phi) \approx 31.72°$.

(This is indeed a surprising result. It means that a clockwise rotation of 58.28°, followed by a scaling of ϕ in the x direction and $1/\phi$ in the y direction, followed by a counterclockwise rotation of 31.72° is equivalent to a unit shear in the x direction. This is illustrated by Figure 3.16.)

> Geometry has two great treasures: one the Theorem of Pythagoras; the other, the division of a line into extreme and mean ratio. The first we may compare to a measure of gold; the second we may name a precious jewel.
> — Johannes Kepler.

▶ **Exercise 3.33:** Given the transformation

$$x^* = 3x - 2y + 1, \quad y^* = 4x + 5y - 6,$$

calculate the transformation matrix and decompose it into a product of four matrices as shown in Equation (3.21).

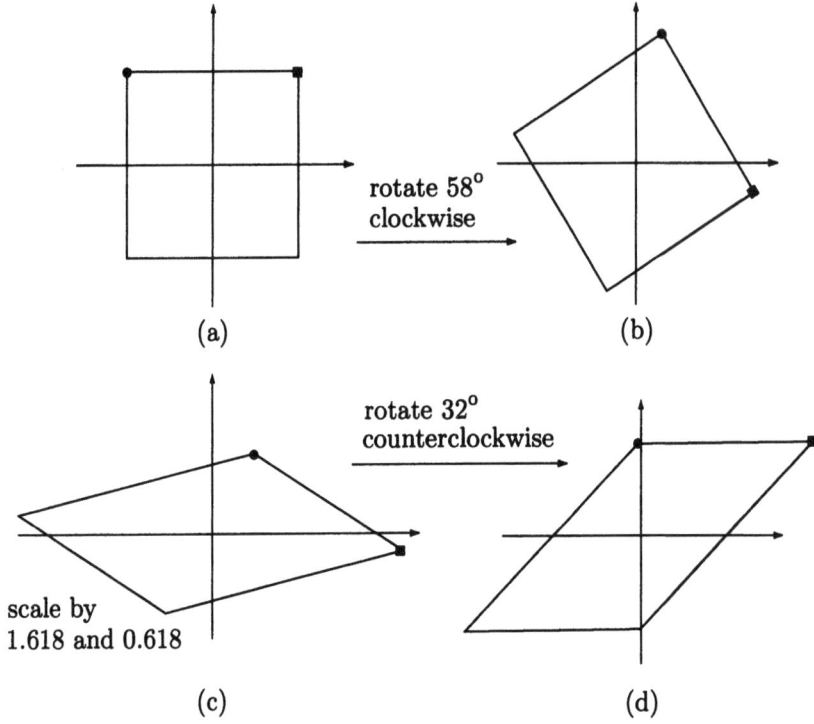

Figure 3.16: Shearing Decomposed into Rotation and Scaling.

3.2.10 Reconstructing Transformations

Given a sequence of two-dimensional transformations, we normally write the 3 × 3 matrix for each, then multiply the matrices. The result is another 3 × 3 matrix that's used to transform all the points of an object. An interesting question is: Given the points of an object before and after a transformation, can we reconstruct the transformation matrix from them?

The answer is yes! The general two-dimensional transformation matrix depends on six numbers, so all we need are six equations involving transformed points. Since each point consists of two numbers, three points are enough to reconstruct the transformation matrix. Given three points both before ($\mathbf{P}_1, \mathbf{P}_2, \mathbf{P}_3$) and after ($\mathbf{P}_1^*, \mathbf{P}_2^*, \mathbf{P}_3^*$), a transformation we can write the three equations $\mathbf{P}_1^* = \mathbf{P}_1\mathbf{T}$, $\mathbf{P}_2^* = \mathbf{P}_2\mathbf{T}$, and $\mathbf{P}_3^* = \mathbf{P}_3\mathbf{T}$ and solve for the six elements of $\mathbf{T}$.

Example: The three points $(1, 1)$, $(1, 0)$, and $(0, 1)$ are transformed to $(3, 4)$, $(2, -1)$, and $(0, 2)$, respectively. We write the general transformation $(x^*, y^*) = (ax + cy + m, bx + dy + n)$ for the three sets

$$(3, 4) = (a + c + m, b + d + n),$$
$$(2, -1) = (a + m, b + n),$$
$$(0, 2) = (c + m, d + n),$$

3 Transformations and Projections

and this is easily solved to yield $a = 3$, $b = 2$, $c = 1$, $d = 5$, $m = -1$, and $n = -3$. The transformation matrix is thus

$$\mathbf{T} = \begin{pmatrix} 3 & 2 & 0 \\ 1 & 5 & 0 \\ -1 & -3 & 1 \end{pmatrix}.$$

▶ **Exercise 3.34:** Inverse transformations. From $\mathbf{P}^* = \mathbf{PT}$, we get $\mathbf{P}^*\mathbf{T}^{-1} = \mathbf{PTT}^{-1}$ or $\mathbf{P} = \mathbf{P}^*\mathbf{T}^{-1}$. We can therefore reconstruct an original point $\mathbf{P}$ from the transformed one, $\mathbf{P}^*$, if we know the inverse of the transformation matrix $\mathbf{T}$. In general, the inverse of the 3×3 matrix

$$\mathbf{T} = \begin{pmatrix} a & b & 0 \\ c & d & 0 \\ m & n & 1 \end{pmatrix}$$

is

$$\mathbf{T}^{-1} = \frac{1}{ad - bc} \begin{pmatrix} d & -b & 0 \\ -c & a & 0 \\ cn - dm & bm - an & 1 \end{pmatrix}. \tag{3.22}$$

Calculate the inverses of the transformation matrices for scaling, shearing, rotation, and translation, and discuss their properties.

▶ **Exercise 3.35:** Given that the four points

$$\mathbf{P}_1 = (0,0), \quad \mathbf{P}_2 = (0,1), \quad \mathbf{P}_3 = (1,1), \quad \text{and } \mathbf{P}_4 = (1,0)$$

are transformed to

$$\mathbf{P}_1^* = (0,0), \quad \mathbf{P}_2^* = (2,3), \quad \mathbf{P}_3^* = (8,4), \quad \text{and } \mathbf{P}_4^* = (6,1),$$

reconstruct the transformation matrix.

3.2.11 A Note

All the expressions derived so far for transformations are based on the basic relation $\mathbf{P}^* = \mathbf{PT}$. Some authors prefer the equivalent relation $\mathbf{P}^* = \mathbf{TP}$, which changes the mathematics somewhat. If we want the coordinates of the transformed point to be the same as before, i.e., $x^* = ax + cy + m$, $y^* = bx + dy + n$, we have to write the relation $\mathbf{P}^* = \mathbf{TP}$

$$\begin{pmatrix} x^* \\ y^* \\ 1 \end{pmatrix} = \begin{pmatrix} a & c & m \\ b & d & n \\ 0 & 0 & 1 \end{pmatrix} \begin{pmatrix} x \\ y \\ 1 \end{pmatrix}.$$

The first difference is that both $\mathbf{P}$ and $\mathbf{P}^*$ are columns instead of rows. This is because of the rules of matrix multiplication. The second difference is that the new transformation matrix **T** is the transpose of the original one. Hence, rotation, for example, is achieved by the matrices

$$\begin{pmatrix} \cos\theta & \sin\theta & 0 \\ -\sin\theta & \cos\theta & 0 \\ 0 & 0 & 1 \end{pmatrix}$$ for a clockwise rotation, $$\begin{pmatrix} \cos\theta & -\sin\theta & 0 \\ \sin\theta & \cos\theta & 0 \\ 0 & 0 & 1 \end{pmatrix}$$ for a counterclockwise rotation.

and

Similarly, translation is done by

$$\begin{pmatrix} 1 & 0 & m \\ 0 & 1 & n \\ 0 & 0 & 1 \end{pmatrix}$$

instead of

$$\begin{pmatrix} 1 & 0 & 0 \\ 0 & 1 & 0 \\ m & n & 1 \end{pmatrix}.$$

3.3 Windowing

Given a large image, where only part of it can be displayed, we can think of the screen as a *window* into the image (this is not the same as MS Windows!). This introduces the problem of coordinate transformation. We call the original, large image *the world*. A point with world coordinates (x, y) may end up being displayed, say, on the left of the screen, where its coordinates will be $(0, Y)$ for some Y. Thus the point has to be translated. Furthermore, we may want to display part of the world on only part of the screen. Such a part is a rectangle called a *viewport*. The image being displayed has therefore to be *scaled* and then translated again, in the screen, into the viewport. Fortunately, transforming coordinates from world to screen to viewport is simple.

We assume that the window occupies the rectangle $(W_{xl}, W_{xr}, W_{yb}, W_{yt})$ in the world. We also assume that the viewport occupies the rectangle $(V_{xl}, V_{xr}, V_{yb}, V_{yt})$ on the screen. To transform a point (X_w, Y_w) in world coordinates to the point (X_s, Y_s) in screen coordinates, we proceed in three steps:

1. We calculate the distances between the point and the bottom-left corner of the window. They are $X_w - W_{xl}$ and $Y_w - W_{yb}$.
2. We scale these distances by the relative sizes of the viewport and the window.

$$\frac{V_{xr} - V_{xl}}{W_{xr} - W_{xl}}(X_w - W_{xl}), \qquad \frac{V_{yt} - V_{yb}}{W_{yt} - W_{yb}}(Y_w - W_{yb}).$$

3. We add the coordinates of the viewport's left-bottom corner, V_{xl}, V_{yb}

$$X_s = \frac{V_{xr} - V_{xl}}{W_{xr} - W_{xl}}(X_w - W_{xl}) + V_{xl} = aX_w + m,$$
$$Y_s = \frac{V_{xr} - V_{xl}}{W_{xr} - W_{xl}}(Y_w - W_{yb}) + V_{yb} = cY_w + n.$$

3 Transformations and Projections

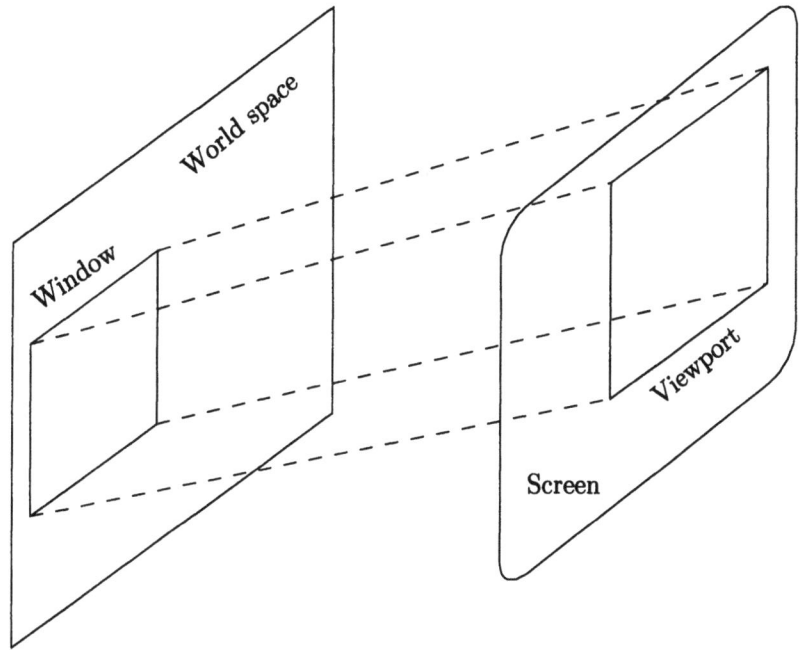

Figure 3.17: Windowing.

We only need to calculate the four quantities a, c, m, and n once. They are then used to transform all the points (X_w, Y_w). Windowing is therefore an example of scaling and translation where the transformation is

$$(X_s, Y_s, 1) = (X_w, Y_w, 1) \begin{pmatrix} a & 0 & 0 \\ 0 & c & 0 \\ m & n & 1 \end{pmatrix}.$$

If an element of the world image is a line, we can transform the two endpoints, then clip the line in the viewport (see Section 3.4 for clipping).

3.3.1 Summary

The general two-dimensional affine transformation is given by $x^* = ax + cy + m$, $y^* = bx + dy + n$. This section shows the values or constraints that should be assigned to the four coefficients a, b, c, and d in order to get certain types transformations (we ignore translations).

- A general affine transformation is obtained when $ad - bc \neq 0$. For $ad - bc = +1$, the transformation is rotation, and for $ad - bc = -1$, it is reflection.

- The case $ad - bc = 0$ corresponds to a singular transformation.

- The identity transformation is obtained when $a = d = 1$ and $b = c = 0$.

- An isometry is obtained by $a^2 + b^2 = c^2 + d^2 = 1$ and $ac + bd = 0$. An isometry is a transformation that preserves distances. If **P** and **Q** are two points on an object, then the distance d between them is preserved, meaning that the distance between $\mathbf{P}^*$ and $\mathbf{Q}^*$ is the same d. Rotations, reflections, and translations are isometries.

- A similarity is obtained for $a^2 + b^2 = c^2 + d^2$ and $ac + bd = 0$. A similarity is a transformation that preserves the ratios of lengths. A typical similarity is scaling, but it may be combined with rotation, reflection, and translation.

- An *equiareal* transformation (preserving areas) is obtained when $|ad - bc| = 1$.

- A shearing in the x direction is caused by $a = d = 1$ and $b = 0$. Similarly, a shearing in the y direction corresponds to $a = d = 1$ and $c = 0$.

- A uniform scaling is $a = d > 0$ and $b = c = 0$ (the identity is a special case of scaling).

- A uniform reflection is $a = d < 0$ and $b = c = 0$.

- A rotation is the result of $a = d = \cos\theta$ and $b = -c = \sin\theta$.

3.4 Clipping

When working with a viewport, parts of the image may have to be clipped. Imagine a drawing program where the user has developed a large image consisting of lines, circles, and other geometric figures. In addition to the bitmap, the program maintains another data structure with the geometric information (the endpoints of lines, radius and center of circles, etc.). The user now specifies a viewport (perhaps by means of the mouse) and asks the program to copy the contents of the viewport onto a file; thus, it becomes a new image. There are two ways to create such a copy:

1. Copy the bitmap pixels that are inside the viewport. The new, copied picture becomes a (small) bitmap. It can be printed or pasted into another bitmap, but the software has lost the geometric information of the image. It no longer knows what the original lines and circles were.

2. Use the geometric information to figure out what lines and circles appear in the viewport and clip those items to the boundaries of the viewport. This is why clipping is important.

3.4.1 Cohen-Sutherland Line Clipping

This is an efficient method for clipping lines. The main idea is to classify lines according to the relation between their endpoints and the boundaries of the viewport. It is easy to see that if both endpoints of a line are above the viewport, the entire line is outside the viewport and should not be drawn. The same is true if the two endpoints are below, to the left of, or to the right of the viewport. If both endpoints are inside the viewport, the entire line should be displayed (no clipping). The algorithm proceeds as follows:

1. Each of the two endpoints of a line segment is compared to the boundaries of the viewport and is assigned a 4-bit number according to the result (Figure 3.18). Thus, if an endpoint is above and to the left of the viewport, it is assigned the number 1001.

3 Transformations and Projections

2. The two numbers assigned to the endpoints are logically "anded." If the result is nonzero, the two endpoints are on the same side of the viewport, meaning the entire line is outside the viewport and should be ignored (case 1 in Figure 3.18). If the result is zero, there are three subcases:

2.1. Both numbers are "0000," meaning that both endpoints are inside the viewport. There is no need to clip and the entire line should be displayed (case 2 in Figure 3.18).

2.2. Only one number is "0000." One endpoint is inside the viewport and the other outside it. One clipping point should be calculated (point "a" in Figure 3.18).

2.3. None of the two numbers is "0000." Both endpoints are, in this case, outside the viewport and two clipping points ("b" and "c" in Figure 3.18) should be calculated.

Bit #	If set then point is
3	above window
2	below window
1	right of window
0	left of window

3210	3210	3210
1001	1000	1010
0001	0000	0010
0101	0100	0110

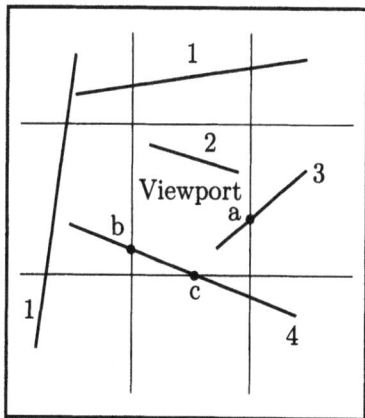

Figure 3.18: The Cohen-Sutherland Clipping Algorithm.

To calculate a clipping point, the program has to calculate the intersection of its line segment with all four boundaries of the viewport.

3.5 Three-Dimensional Transformations

Three-dimensional transformations are important because a three-dimensional object may have a complex, unfamiliar shape and may have to be scaled, reflected, and rotated in order to comprehend its shape when seen on a flat screen.

3.5.1 Three-Dimensional Coordinate Systems

Figure 3.19a illustrates the two types, left-handed and right handed, of three-dimensional coordinate systems. The rule for a right-handed system is: if you align your right thumb with the positive x axis and your right index finger with the positive y axis, then your right middle finger will point in the direction of positive z. The rule for a left-handed system uses the left hand in a similar manner. It is also possible to define a left-handed coordinate system as the mirror image of a right-handed one.

The difference between left-handed and right-handed coordinate systems becomes important when a three-dimensional object is projected on a two-dimensional

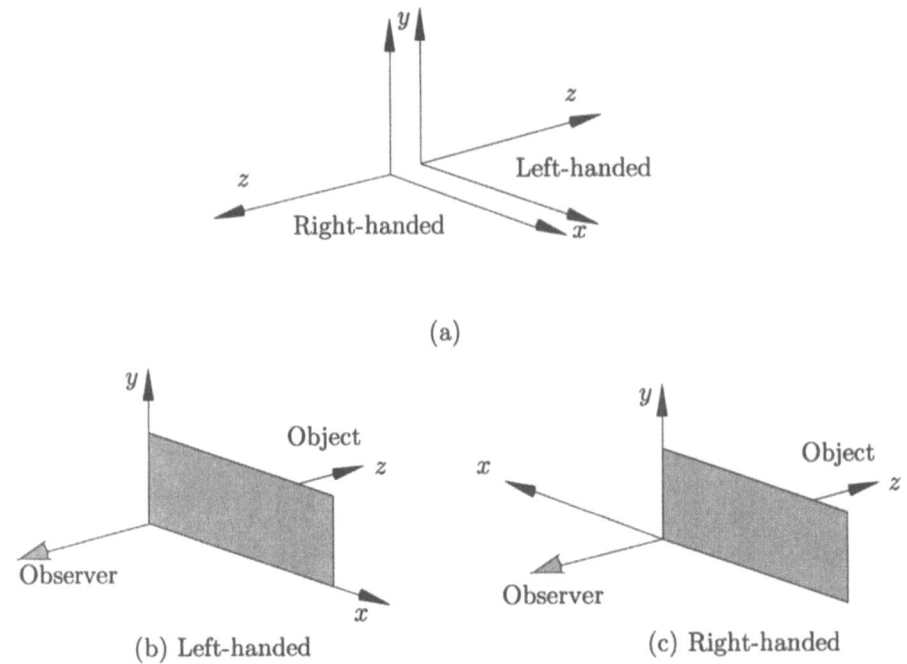

Figure 3.19: Three-Dimensional Coordinate Systems.

screen (Section 3.7). We assume that the screen is positioned at the xy plane with its origin (i.e., its bottom-left corner) at the origin of the three-dimensional system. We also assume that the object to be projected is located on the positive side of the z axis and the observer is located on the negative side, looking at the projection of the image on the screen. Figure 3.19b shows that in a left-handed three-dimensional coordinate system, the directions of the positive x and y axes on the screen coincide with those of the three-dimensional x and y axes. In a right-handed system (Figure 3.19c), though, the two-dimensional x axis (on the screen) and the three-dimensional x axis point in opposite directions.

> **Principle: Express co-ordinate ideas in similar form.**
> This principle, that of parallel construction, requires that expressions of similar content and function should be outwardly similar. The likeness of form enables the reader to recognize more readily the likeness of content and function. Familiar instances from the Bible are the Ten Commandments, the Beatitudes, and the petitions of the Lord's Prayer.
> —W. Strunk and E. B. White, *The Elements of Style*.

3 Transformations and Projections

3.5.2 Basic Principles

We develop three-dimensional transformations using the same method as with two-dimensional transformations. A three-dimensional point $\mathbf{P} = (x, y, z, 1)$ is transformed to a point $\mathbf{P}^* = (x^*, y^*, z^*, 1)$ by multiplying it by a 4×4 matrix

$$\mathbf{T} = \begin{pmatrix} a & b & c & p \\ d & e & f & q \\ h & i & j & r \\ l & m & n & s \end{pmatrix}. \tag{3.23}$$

The last column of $\mathbf{T}$ is not $(0,0,0,1)$ because it is used for projections (page 116). As a result, the product $\mathbf{PT}$ is the quartet (X, Y, Z, H), where H equals $xp + yq + zr + s$ and is generally not 1. The three coordinates (x^*, y^*, z^*) of $\mathbf{P}^*$ are obtained by dividing (X, Y, Z) by H. Hence, $(x^*, y^*, z^*) = (X/H, Y/H, Z/H)$.

The top left 3×3 part of $\mathbf{T}$ is responsible for scaling and reflection (a, e, and j), shearing (b, c, f, and d, h, i), and rotation (all nine). The three quantities l, m, and n are responsible for translation, and the only new parameters are those in the last column (p, q, r, s).

To understand the meaning of s, we examine the matrix $\mathbf{T} = \begin{pmatrix} 1 & & & \\ & 1 & & \\ & & 1 & \\ & & & s \end{pmatrix}$. Multiplying $\mathbf{P}$ by $\mathbf{T}$ transforms $(x, y, z, 1)$ into (x, y, z, s), so the new point has coordinates $(x/s, y/s, z/s)$. The parameter s is therefore responsible for global scaling (by a factor of $1/s$). It is identical to transforming by $\begin{pmatrix} 1/s & & & \\ & 1/s & & \\ & & 1/s & \\ & & & 1 \end{pmatrix}$.

Rotation in three dimensions is more difficult. One approach is to write three rotation matrices that rotate about the three axes:

$$\begin{pmatrix} \cos\theta & -\sin\theta & 0 & 0 \\ \sin\theta & \cos\theta & 0 & 0 \\ 0 & 0 & 1 & 0 \\ 0 & 0 & 0 & 1 \end{pmatrix}, \begin{pmatrix} \cos\theta & 0 & -\sin\theta & 0 \\ 0 & 1 & 0 & 0 \\ \sin\theta & 0 & \cos\theta & 0 \\ 0 & 0 & 0 & 1 \end{pmatrix}, \begin{pmatrix} 1 & 0 & 0 & 0 \\ 0 & \cos\theta & -\sin\theta & 0 \\ 0 & \sin\theta & \cos\theta & 0 \\ 0 & 0 & 0 & 1 \end{pmatrix}. \tag{3.24}$$

The matrices of Equation (3.24) are sometimes written in the slightly different form:

$$\begin{pmatrix} \cos\theta & \sin\theta & 0 & 0 \\ -\sin\theta & \cos\theta & 0 & 0 \\ 0 & 0 & 1 & 0 \\ 0 & 0 & 0 & 1 \end{pmatrix}, \begin{pmatrix} \cos\theta & 0 & -\sin\theta & 0 \\ 0 & 1 & 0 & 0 \\ \sin\theta & 0 & \cos\theta & 0 \\ 0 & 0 & 0 & 1 \end{pmatrix}, \begin{pmatrix} 1 & 0 & 0 & 0 \\ 0 & \cos\theta & \sin\theta & 0 \\ 0 & -\sin\theta & \cos\theta & 0 \\ 0 & 0 & 0 & 1 \end{pmatrix} \tag{3.25}$$

(where the first and third ones have their direction of rotation reversed). This is suitable for a right-handed coordinate system since each of the matrices of (3.25) describes a clockwise rotation about one of the axes when viewed in the positive direction of that axis. The first matrix, for example, describes a rotation about the z axis that's clockwise (i.e., from positive x to positive y) when viewed in the direction of positive z.

The rotation matrices of Eqs. (3.24) and (3.25) are simple but not very useful since, in practice, we rarely know how to break a general rotation into three rotations about the coordinate axes. There are some cases, however, where rotations about the coordinate axes are common. One such case is discussed in Section 3.8.2; two more are presented below.

Case 1: Rotations about the coordinate axes are common in the motion of a submarine or an airplane. These vehicles have three degrees of freedom and have three natural, mutually perpendicular axes of rotation that are traditionally given the names *roll*, *pitch*, and *yaw* (Figure 3.20). Roll is a rotation about the direction of motion of the vehicle. An airplane rolls when it banks by dipping one wing and lifting the other. Pitch is an up or down rotation about an axis that goes through the wings. An airplane uses its elevators for this. Yaw is a left-right rotation about a vertical axis, accomplished by the rudder. These terms originated with sailors since a ship can yaw and also has limited roll and pitch capabilities.

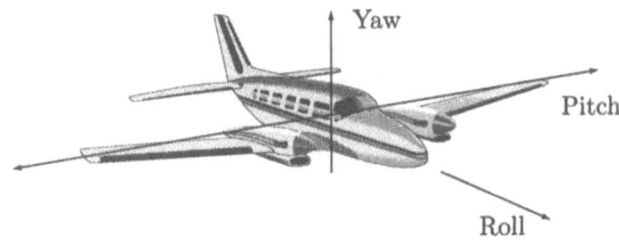

Figure 3.20: Roll, Pitch, and Yaw.

Case 2: Another example of an application where rotations about the three coordinate axes are common is L-systems. This is a system of formal notation, developed by the biologist Aristid Lindenmayer (hence the "L") in 1968 as a tool to describe the morphology of plants [Lindenmayer 68]. In the 1970s, this notation was adopted by computer scientists and used to define formal languages. Since 1984, it has also been used to describe and draw many types of fractals. Today, L-systems are used to generate tilings, geometric art, and even music.

The main idea of L-systems is to define a complex object by (1) defining an initial simple object, called the *axiom*, and (2) writing rules that show how to replace parts of the axiom. The rules are written in terms of *turtle moves*, a concept originally introduced by the LOGO programming language [Abelson and DiSessa 82].

Imagine a computer-controlled turtle moving on a plane. At any time, the state of the turtle is completely defined by specifying its position (x, y) and heading α. The turtle obeys the following commands:

F: Move a distance d forward (where d is a user-controlled parameter) and draw a straight-line segment on the plane's surface. This moves the turtle from point (x, y) to point $(x + d\cos\alpha, y + d\sin\alpha)$.

f: Move as above but without drawing anything.

+: Turn right (clockwise). The turtle changes its heading from α to $\alpha + \delta$, where δ is a parameter. The current position does not change and nothing is drawn.

3 Transformations and Projections

−: Turn left (i.e., counterclockwise). The turtle changes its heading from α to $\alpha-\delta$.

[Prusinkiewicz 89] is a basic reference for L-systems. The application that interests us here is the extension of the concept of turtle movements to three dimensions. In two dimensions the heading of the turtle is a single number. In three dimensions it is a three-dimensional vector **H** (see Figure 3.21). The "F" and "f" commands are identical to the two-dimensional ones; they move the turtle d units in the direction of **H** (without changing **H**). When it comes to changing the heading, however, the two commands + and − are not enough since orientations in three dimensions are complex. It turns out that two vectors **H** and **U** (for "heading" and "up") are required in order to completely specify the orientation of a three-dimensional object. To understand this, imagine a turtle looking in direction **H**. Now rotate the turtle about **H**. This does not move the turtle, nor does it change its heading **H**, but it changes what the turtle sees.

The "up" vector **U** (see also Section 8.1) points in the direction from the bottom part (legs and belly) to the top part (shell) of the turtle and is perpendicular to the heading **H**. Rotating the turtle about **H** amounts to rotating **U** about **H**. It is also useful to add a third vector that we arbitrarily call **L** (for "left") that's perpendicular to both **H** and **U** such that **U**, **H**, and **L** form a right-handed local coordinate system for the turtle. They correspond to yaw, roll, and pitch (Figure 3.20), respectively. **L** is defined by $\mathbf{L} = \mathbf{U} \times \mathbf{H}$.

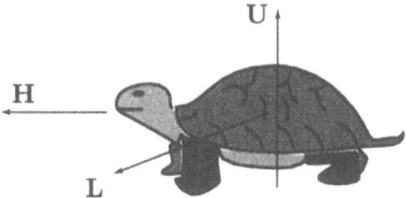

Figure 3.21: Turtle Orientation in Three Dimensions.

The orientation of the turtle can now be changed systematically by the six simple commands $h+$, $h-$, $u+$, $u-$, $l+$, and $l-$. The first two commands rotate the turtle $\pm\theta$ degrees about its heading (i.e., about the **H** vector); they change **U** and **L** but not **H**. Similarly, commands $u+$ and $u-$ rotate the turtle about **U** and commands $l+$ and $l-$ rotate it about **L**. These commands are implemented by constructing a matrix **B** whose columns are **H**, **U**, and **L** and multiplying it by one of the three rotation matrices

$$\mathbf{R}_H = \begin{pmatrix} 1 & 0 & 0 \\ 0 & \cos\theta & -\sin\theta \\ 0 & \sin\theta & \cos\theta \end{pmatrix}, \quad \mathbf{R}_U = \begin{pmatrix} \cos\theta & 0 & -\sin\theta \\ 0 & 1 & 0 \\ \sin\theta & 0 & \cos\theta \end{pmatrix},$$

$$\mathbf{R}_L = \begin{pmatrix} \cos\theta & -\sin\theta & 0 \\ \sin\theta & \cos\theta & 0 \\ 0 & 0 & 1 \end{pmatrix}.$$

Thus, the commands $h+$, $u+$, and $l+$ are executed by $\mathbf{B}\cdot\mathbf{R}_H$, $\mathbf{B}\cdot\mathbf{R}_U$, and $\mathbf{B}\cdot\mathbf{R}_L$, respectively.

▶ **Exercise 3.36:** How should the commands $h-$, $u-$, and $l-$ be executed?

> The following true story is an example of Aristid's modesty. At one of the American conferences somebody asked him what the L in "L-systems" stands for. Aristid's answer was "Languages."
>
> —Grzegorz Rozenberg.

The direction of a three-dimensional rotation generated by the matrices of (3.24) can be found by the following rule: Write down the sequence "x, y, z" and erase the symbol that corresponds to the axis of rotation. Two symbols are left and are denoted l and r. Draw the coordinate axes such that the positive direction of l will be up and the positive direction of r will be to the right. The rotation will then be counterclockwise, from positive r to positive l to negative r to negative l (Figure 3.22 and see also Exercise 3.43).

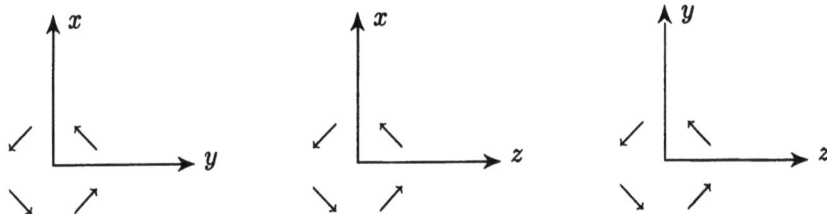

Figure 3.22: Direction of Three-Dimensional Rotations.

Example: A rotation about the z axis produced by the leftmost matrix of (3.24). After erasing z, the two symbols left are x and y. We draw the coordinate axes such that positive x is up and positive y is to the right. The matrix produces counterclockwise rotation. To achieve clockwise rotation, either use a negative angle or the inverse of the rotation matrix. Inverting an arbitrary matrix may involve a bit of work (page 710), but inverting our rotation matrices involves just changing the signs of the sine functions.

Example: A compound transformation consisting of (1) a translation by l, m, and n units along the three coordinate axes, (2) a rotation of θ degrees about the x axis, (3) a rotation of ϕ degrees about the y axis, and (4) the reverse translation.

3 Transformations and Projections

The four transformation matrices are

$$\mathbf{T}_r = \begin{pmatrix} 1 & 0 & 0 & 0 \\ 0 & 1 & 0 & 0 \\ 0 & 0 & 1 & 0 \\ l & m & n & 1 \end{pmatrix}, \quad \mathbf{T}_{rr} = \begin{pmatrix} 1 & 0 & 0 & 0 \\ 0 & 1 & 0 & 0 \\ 0 & 0 & 1 & 0 \\ -l & -m & -n & 1 \end{pmatrix},$$

$$\mathbf{R}_x = \begin{pmatrix} 1 & 0 & 0 & 0 \\ 0 & \cos\theta & \sin\theta & 0 \\ 0 & -\sin\theta & \cos\theta & 0 \\ 0 & 0 & 0 & 1 \end{pmatrix}, \quad \mathbf{R}_y = \begin{pmatrix} \cos\phi & 0 & -\sin\phi & 0 \\ 0 & 1 & 0 & 0 \\ \sin\phi & 0 & \cos\phi & 0 \\ 0 & 0 & 0 & 1 \end{pmatrix}.$$

Their product equals the 4×4 matrix

$$\mathbf{T} = \mathbf{T}_r \mathbf{R}_x \mathbf{R}_y \mathbf{T}_{rr}$$

$$= \begin{pmatrix} \cos\phi & 0 & -\sin\phi & 0 \\ \sin\phi\sin\theta & \cos\theta & \cos\phi\sin\theta & 0 \\ \cos\theta\sin\phi & -\sin\theta & \cos\phi\cos\theta & 0 \\ \begin{matrix}-l+l\cos\phi \\ +m\cos(\phi-\theta)/2 \\ -m\cos(\phi+\theta)/2 \\ +n\sin(\phi-\theta)/2 \\ +n\sin(\phi+\theta)/2\end{matrix} & \begin{matrix}-m \\ +m\cos\theta \\ -n\sin\theta \end{matrix} & \begin{matrix}[-2n+n\cos(\phi-\theta) \\ +n\cos(\phi+\theta) \\ -2l\sin\phi \\ -m\sin(\phi-\theta) \\ +m\sin(\phi+\theta)]/2 \end{matrix} & 1 \end{pmatrix}.$$

Substituting the values $\theta = 30°$, $\phi = 45°$, and $l = m = n = -1$, we get the 4 × 4 matrix

$$\mathbf{T} = \begin{pmatrix} 0.7071 & 0 & -0.7071 & 0 \\ 0.3540 & 0.866 & 0.3540 & 0 \\ 0.6124 & -0.50 & 0.6124 & 0 \\ -0.673 & 0.634 & 0.7410 & 1 \end{pmatrix}.$$

A point at $(1, 2, 3)$, for example, is transformed by $\mathbf{T}$ to the point

$$(1, 2, 3, 1)\mathbf{T} = (2.5793, 0.866, 2.5791, 1).$$

▶ **Exercise 3.37:** Perform the same operations for the compound transformation $\mathbf{T}_r \mathbf{R}_x \mathbf{T}_{rr}$.

3.5.3 General Rotations

In practice, we normally don't know how to express an arbitrary rotation as a product of rotations about the coordinate axes, so we have to derive this important transformation explicitly. In general, a point $\mathbf{P}$ is to be rotated through an angle θ about a line that passes through the origin. (If the line does not go through the origin, it has to be translated first. It is also obvious now that a general rotation in three dimensions is fully defined by four numbers—one is the rotation angle and the other three are the components of the rotation axis.) The point ends up at $\mathbf{P}^*$. We connect $\mathbf{P}$ to the origin and call the resulting vector $\mathbf{r}$. Rotating point $\mathbf{P}$ to $\mathbf{P}^*$ is identical to rotating the vector $\mathbf{r}$ to $\mathbf{r}^*$. We denote by $\mathbf{u}$ the unit vector in the direction of rotation. Figure 3.23a shows that the component OC of **r** along **u** is left unchanged, but the component CP is rotated to CP*. The distance OC is seen from the diagram to be $(\mathbf{r}\bullet\mathbf{u})$, so the vector $\vec{OC}$ can be written $(\mathbf{r}\bullet\mathbf{u})\mathbf{u}$. From $\mathbf{r} = \vec{OC} + \vec{CP}$, we get $\vec{CP} = \mathbf{r} - (\mathbf{r}\bullet\mathbf{u})\mathbf{u}$ or, in terms of magnitudes, $|\vec{CP}| = |\mathbf{r} - (\mathbf{r}\bullet\mathbf{u})\mathbf{u}|$. It can also be seen from the diagram that $|\vec{CP}| = |\mathbf{r}|\sin\phi$. Since **u** is a unit vector, we can write $|\mathbf{u}\times\mathbf{r}| = |\mathbf{r}|\sin\phi$. We thus get $|\vec{CP}| = |\mathbf{r} - (\mathbf{r}\bullet\mathbf{u})\mathbf{u}| = |\mathbf{u}\times\mathbf{r}|$.

Figure 3.23b shows the situation when looking from the origin in the **u** direction. Note that the vector $\vec{CQ}$ is perpendicular to both **u** and **r**, so it is in the direction of $\mathbf{u}\times\mathbf{r}$.

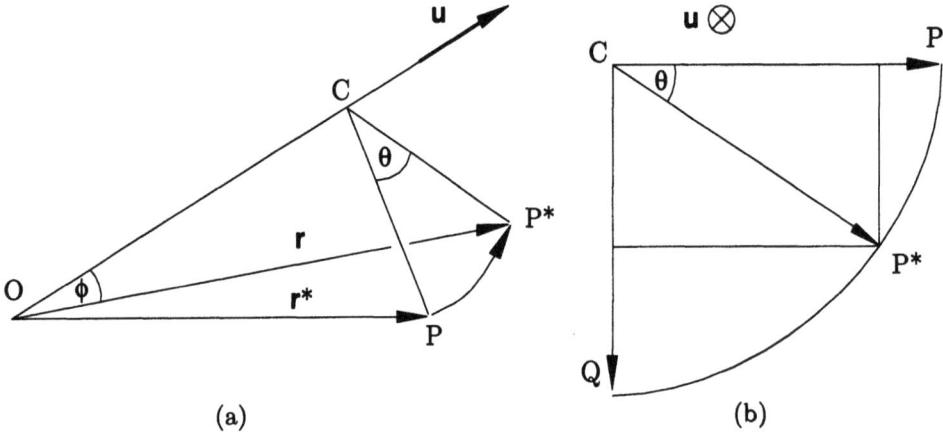

Figure 3.23: A General Rotation.

The next step is to resolve CP* into its components. From Figure 3.23b, we get

$$\vec{CP^*} = \cos\theta[\mathbf{r} - (\mathbf{r}\bullet\mathbf{u})\mathbf{u}] + \sin\theta[\mathbf{r} - (\mathbf{r}\bullet\mathbf{u})\mathbf{u}] = \cos\theta[\mathbf{r} - (\mathbf{r}\bullet\mathbf{u})\mathbf{u}] + \sin\theta(\mathbf{u}\times\mathbf{r}),$$

which can be used to express $\mathbf{r}^*$:

$$\mathbf{r}^* = \vec{OC} + \vec{CP^*} = (\mathbf{r}\bullet\mathbf{u})\mathbf{u} + \cos\theta[\mathbf{r} - (\mathbf{r}\bullet\mathbf{u})\mathbf{u}] + \sin\theta(\mathbf{u}\times\mathbf{r}). \qquad (3.26)$$

Using Equations (A.9) and (A.11) (page 703), we can rewrite this as $\mathbf{r}^* = (\mathbf{u}\mathbf{u}^T)\mathbf{r} + \cos\theta\mathbf{r} - \cos\theta(\mathbf{u}\mathbf{u}^T)\mathbf{r} + \sin\theta\mathbf{U}\mathbf{r}$, where

$$\mathbf{U} = \begin{pmatrix} 0 & -u_z & u_y \\ u_z & 0 & -u_x \\ -u_y & u_x & 0 \end{pmatrix}.$$

▶ The result can now be summarized as $\mathbf{r}^* = \mathbf{M}\mathbf{r}$, where

$$\mathbf{M} = \mathbf{u}\mathbf{u}^T + \cos\theta(\mathbf{I} - \mathbf{u}\mathbf{u}^T) + \sin\theta\mathbf{U} \qquad (3.27)$$

3 Transformations and Projections

$$= \begin{pmatrix} u_x^2 + \cos\theta(1-u_x^2) & u_x u_y(1-\cos\theta) - u_z\sin\theta & u_x u_z(1-\cos\theta) + u_y\sin\theta \\ u_x u_y(1-\cos\theta) + u_z\sin\theta & u_y^2 + \cos\theta(1-u_y^2) & u_y u_z(1-\cos\theta) - u_x\sin\theta \\ u_x u_z(1-\cos\theta) - u_y\sin\theta & u_y u_z(1-\cos\theta) + u_x\sin\theta & u_z^2 + \cos\theta(1-u_z^2) \end{pmatrix}.$$

Direction Cosines: If $\mathbf{v} = (v_x, v_y, v_z)$ is a three-dimensional vector, its *direction cosines* are defined as

$$N_1 = \frac{v_x}{|\mathbf{v}|}, \quad N_2 = \frac{v_y}{|\mathbf{v}|}, \quad N_3 = \frac{v_z}{|\mathbf{v}|}.$$

These are the cosines of the angles between the direction of $\mathbf{v}$ and the three coordinate axes. It is easy to verify that $N_1^2 + N_2^2 + N_3^2 = 1$. If $\mathbf{u} = (u_x, u_y, u_z)$ is a unit vector, then $|\mathbf{u}| = 1$ and u_x, u_y, and u_z are the direction cosines of $\mathbf{u}$.

It can be shown that a rotation through an angle $-\theta$ is performed by the transpose $\mathbf{M}^T$. Consider the two successive and opposite rotations $\mathbf{r}^* = \mathbf{Mr}$ and $\mathbf{r}' = \mathbf{M}^T \mathbf{r}^*$. On one hand, they can be expressed as the product $\mathbf{r}' = \mathbf{M}^T \mathbf{r}^* = \mathbf{M}^T \mathbf{M r}$. On the other hand, they rotate in opposite directions, so they return all points to their original positions; therefore $\mathbf{r}'$ must be equal to $\mathbf{r}$. We end up with $\mathbf{r} = \mathbf{M}^T \mathbf{M r}$ or $\mathbf{M M}^T = \mathbf{I}$, where $\mathbf{I}$ is the identity matrix. The transpose $\mathbf{M}^T$ thus equals the inverse, $\mathbf{M}^{-1}$ of $\mathbf{M}$, which shows that a rotation matrix $\mathbf{M}$ is orthogonal (Section A.5).

Example: a rotation about the z axis. $\mathbf{u} = (0, 0, 1)$ so

$$\mathbf{u}\mathbf{u}^T = \begin{pmatrix} 0 & 0 & 0 \\ 0 & 0 & 0 \\ 0 & 0 & 1 \end{pmatrix} \text{ and } \mathbf{U} = \begin{pmatrix} 0 & -1 & 0 \\ 1 & 0 & 0 \\ 0 & 0 & 0 \end{pmatrix}, \text{ hence } \mathbf{M} = \begin{pmatrix} \cos\theta & -\sin\theta & 0 \\ \sin\theta & \cos\theta & 0 \\ 0 & 0 & 1 \end{pmatrix},$$

which is the familiar rotation matrix about the z axis.

3.5.4 Quaternions

Section A.9 is a general introduction to quaternions and should be reviewed before reading ahead. What makes quaternions useful to us is the fact that they can elegantly express arbitrary rotations. If we want to rotate a point $\mathbf{P}$ by an angle θ about a direction $\mathbf{v}$, we should first prepare the quaternion $\mathbf{q} = [\cos(\theta/2), \sin(\theta/2)\mathbf{u}]$, where $\mathbf{u} = \mathbf{v}/|\mathbf{v}|$ is a unit vector in the direction of $\mathbf{v}$. The rotation can then be expressed as the triple product $\mathbf{q} \cdot [0, \mathbf{P}] \cdot \mathbf{q}^{-1}$. Note that our $\mathbf{q}$ is a unit quaternion, since $\sin^2(\theta/2) + \cos^2(\theta/2) = 1$.

▶ **Exercise 3.38:** Prove that the triple product $\mathbf{q} \cdot [0, \mathbf{P}] \cdot \mathbf{q}^{-1}$ really performs a rotation of $\mathbf{P}$ about $\mathbf{v}$ (or $\mathbf{u}$). (Hint: perform the multiplications and show that they produce Equation (3.26).)

As an example of quaternion rotation, consider a 90° rotation of point $\mathbf{P} = (0, 1, 1)$ about the y axis. The quaternion involved is $\mathbf{q} = [\cos 45°, \sin 45°(0, 1, 0)]$. It is a unit quaternion, so its inverse is $\mathbf{q}^{-1} = [\cos 45°, -\sin 45°(0, 1, 0)]$. The rotated point is thus

$$\mathbf{q}[0,\mathbf{P}]\mathbf{q}^{-1}$$
$$= [-\sin 45°, (\sin 45, \cos 45°, \cos 45°)] [0, (0,1,1)] [\cos 45°, -\sin 45°(0,1,0)]$$
$$= [0, (1,1,0)].$$

The quaternion resulting from the triple product always has a zero scalar. We ignore the scalar and find that the point has been moved, by the rotation, from the $x = 0$ plane to the $z = 0$ plane.

Figure 3.24 illustrates this particular rotation about the y axis and also makes it easy to understand the rule for the direction of the quaternion rotation $\mathbf{q}[0,\mathbf{P}]\mathbf{q}^{-1}$. The rule is:

▶ Let $\mathbf{q} = [s,\mathbf{v}]$ be a rotation quaternion in a right-handed three-dimensional coordinate system. To an observer looking in the direction of $\mathbf{v}$, the triple product $\mathbf{q}[0,\mathbf{P}]\mathbf{q}^{-1}$ rotates point $\mathbf{P}$ clockwise. For a negative rotation angle, the rotation is counterclockwise. In a left-handed coordinate system (Figure 3.24b), the direction of rotation is the opposite.

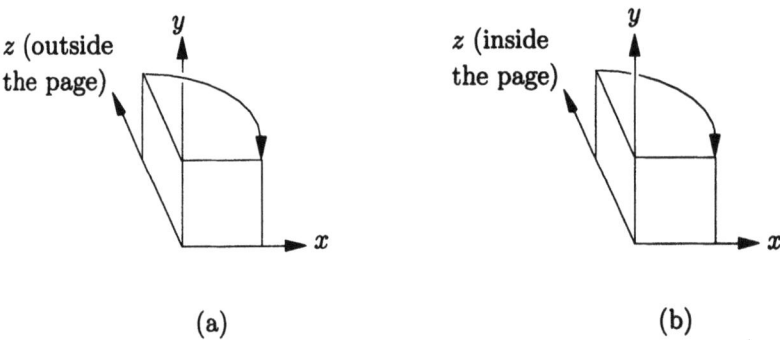

Figure 3.24: Rotation in a Right-Handed (a) and in a Left-Handed (b) Coordinate Systems.

3.5.5 Concatenating Rotations

Sometimes, we have to perform two consecutive rotations on an object. This turns out to be easy and numerically stable with a quaternion representation.

If $\mathbf{q}_1$ and $\mathbf{q}_2$ are unit quaternions representing the two rotations, then associativity of quaternion multiplication implies that the combined rotation of $\mathbf{q}_1$ followed by $\mathbf{q}_2$ is represented by the quaternion $\mathbf{q}_2 \cdot \mathbf{q}_1$. The proof is

$$\mathbf{q}_2 \cdot (\mathbf{q}_1 \cdot \mathbf{P} \cdot \mathbf{q}_1^{-1}) \cdot \mathbf{q}_2^{-1} = (\mathbf{q}_2 \cdot \mathbf{q}_1) \cdot \mathbf{P} \cdot (\mathbf{q}_1^{-1} \cdot \mathbf{q}_2^{-1}) = (\mathbf{q}_2 \cdot \mathbf{q}_1) \cdot \mathbf{P} \cdot (\mathbf{q}_2 \cdot \mathbf{q}_1)^{-1}.$$

Quaternion multiplication involves fewer operations than matrix multiplication, so combining rotations by means of quaternions is faster. Performing fewer multiplications also implies better numerical accuracy.

3 Transformations and Projections

In general, we use 4×4 transformation matrices to express three-dimensional transformations, so we would like to be able to express the rotation $\mathbf{P}^* = \mathbf{q}[0, \mathbf{P}]\mathbf{q}^{-1}$ as $\mathbf{P}^* = \mathbf{PM}$ where $\mathbf{M}$ is a 4×4 matrix. Given the two quaternions $\mathbf{q}_1 = w_1 + x_1\mathbf{i} + y_1\mathbf{j} + z_1\mathbf{k} = (w_1, x_1, y_1, z_1)$ and $\mathbf{q}_2 = w_2 + x_2\mathbf{i} + y_2\mathbf{j} + z_2\mathbf{k} = (w_2, x_2, y_2, z_2)$, their product is

$$\mathbf{q}_1 \cdot \mathbf{q}_2 = (w_1 w_2 - x_1 x_2 - y_1 y_2 - z_1 z_2) + (w_1 x_2 + x_1 w_2 + y_1 z_2 - z_1 y_2)\mathbf{i}$$
$$+ (w_1 y_2 - x_1 z_2 + y_1 w_2 + z_1 x_2)\mathbf{j} + (w_1 z_2 + x_1 y_2 - y_1 x_2 + z_1 w_2)\mathbf{k}.$$

The first step is to realize that each term in this product depends linearly on the coefficients of $\mathbf{q}_1$. This product can therefore be expressed as

$$\mathbf{q}_1 \cdot \mathbf{q}_2 = \mathbf{q}_2 \cdot \mathbf{L}(\mathbf{q}_1) = (x_2, y_2, z_2, w_2) \begin{pmatrix} w_1 & z_1 & -y_1 & -x_1 \\ -z_1 & w_1 & x_1 & -y_1 \\ y_1 & -x_1 & w_1 & -z_1 \\ x_1 & y_1 & z_1 & w_1 \end{pmatrix}.$$

When $\mathbf{L}(\mathbf{q}_1)$ multiplies the row vector $\mathbf{q}_2$, the result is a row vector representation for $\mathbf{q}_1 \cdot \mathbf{q}_2$. Each term also depends linearly on the coefficients of $\mathbf{q}_2$, so the same product can also be expressed as

$$\mathbf{q}_1 \cdot \mathbf{q}_2 = \mathbf{q}_1 \cdot \mathbf{R}(\mathbf{q}_2) = (x_1, y_1, z_1, w_1) \begin{pmatrix} w_2 & -z_2 & y_2 & -x_2 \\ z_2 & w_2 & -x_2 & -y_2 \\ -y_2 & x_2 & w_2 & -z_2 \\ x_2 & y_2 & z_2 & w_2 \end{pmatrix}.$$

When $\mathbf{R}(\mathbf{q}_2)$ multiplies the row vector $\mathbf{q}_1$, the result is also a row vector representation for $\mathbf{q}_1 \cdot \mathbf{q}_2$.

We can now write the triple product $\mathbf{q} \cdot [0, \mathbf{P}] \cdot \mathbf{q}^{-1}$ in terms of the matrices $\mathbf{L}(\mathbf{q})$ and $\mathbf{R}(\mathbf{q})$:

$$\mathbf{q}[0, \mathbf{P}]\mathbf{q}^{-1} = \mathbf{q}([0, \mathbf{P}] \cdot \mathbf{q}^{-1}) = \mathbf{q}([0, \mathbf{P}]\mathbf{R}(\mathbf{q}^{-1}))$$
$$= ([0, \mathbf{P}]\mathbf{R}(\mathbf{q}^{-1}))\mathbf{L}(\mathbf{q}) = [0, \mathbf{P}](\mathbf{R}(\mathbf{q}^{-1})\mathbf{L}(\mathbf{q}))$$
$$= [0, \mathbf{P}]\mathbf{M},$$

where matrix $\mathbf{M}$ is

$$\mathbf{M} = \mathbf{R}(\mathbf{q}^{-1}) \cdot \mathbf{L}(\mathbf{q})$$
$$= \begin{pmatrix} w & z & -y & x \\ -z & w & x & y \\ y & -x & w & z \\ -x & -y & -z & w \end{pmatrix} \begin{pmatrix} w & z & -y & -x \\ -z & w & x & -y \\ y & -x & w & -z \\ x & y & z & w \end{pmatrix}$$
$$= \begin{pmatrix} w^2+x^2-y^2-z^2 & 2xy+2wz & 2xz-2wy & 0 \\ 2xy-2wz & w^2-x^2+y^2-z^2 & 2yz+2wx & 0 \\ 2xz+2wy & 2yz-2wx & w^2-x^2-y^2+z^2 & 0 \\ 0 & 0 & 0 & w^2+x^2+y^2+z^2 \end{pmatrix}.$$

Since ours are unit quaternions, they satisfy $w^2 + x^2 + y^2 + z^2 = 1$, so we can write the final result

$$\mathbf{M} = \begin{pmatrix} 1 - 2y^2 - 2z^2 & 2xy + 2wz & 2xz - 2wy & 0 \\ 2xy - 2wz & 1 - 2x^2 - 2z^2 & 2yz - 2wx & 0 \\ 2xz + 2wy & 2yz - 2wx & 1 - 2x^2 - 2y^2 & 0 \\ 0 & 0 & 0 & 1 \end{pmatrix}. \qquad (3.28)$$

In a left-handed coordinate system, the same rotation is expressed by the triple product $\mathbf{q}^{-1}[0, \mathbf{P}]\mathbf{q}$ or, equivalently, by $\mathbf{P}^* = \mathbf{P} \cdot \mathbf{M}^T$, where $\mathbf{M}^T$ is the transpose of $\mathbf{M}$.

3.6 Transforming the Coordinate System

Our discussion so far has assumed that points are transformed in a static coordinate system. It is also possible (and sometimes useful) to transform the coordinate system instead of the points. To understand the main idea, let's consider the simple example of translation. Suppose that a point $\mathbf{P}$ is transformed to a point $\mathbf{P}^*$ by translating it m and n units in the x and y directions, respectively. How can the transformation be reversed? We consider two ways:

1. Suppose that the original transformation was $\mathbf{P}^* = \mathbf{PT}$, where

$$\mathbf{T} = \begin{pmatrix} 1 & 0 & 0 \\ 0 & 1 & 0 \\ m & n & 1 \end{pmatrix}.$$

It is clear that the transformation matrix

$$\mathbf{S} = \begin{pmatrix} 1 & 0 & 0 \\ 0 & 1 & 0 \\ -m & -n & 1 \end{pmatrix}$$

will transform $\mathbf{P}^*$ back to $\mathbf{P}$. However, it is trivial to show, by using Equation (3.22), that $\mathbf{S}$ is the inverse of $\mathbf{T}$.

2. The transformation can be reversed by translating the coordinate system in the reverse directions (i.e., by $-m$ and $-n$ units) by using an (unknown) transformation matrix $\mathbf{M}$.

Since the two methods produce the same result, we conclude that $\mathbf{M} = \mathbf{S} = \mathbf{T}^{-1}$. Transforming the coordinate axes is therefore done by a matrix that's the inverse of transforming a point. This is true for any affine transformations, not just translation.

> It was his own room. There was no doubt about that. But it had undergone a surprising transformation.
> —Charles Dickens, *A Christmas Carol.*

3 Transformations and Projections

3.7 Projections

In general, a projection is a transformation from n dimensions to $n-1$ dimensions. Projections are classified into linear and nonlinear. The two main types of linear projections are *parallel* and *perspective*. There are several different varieties of each (Figure 3.25) and the whole topic is usually treated in Projective Geometry texts. After a short discussion of parallel projections, we discuss perspective projections in detail, since they result in an easy-to-interpret two-dimensional image of a three-dimensional object. Finally, Section 3.13 discusses nonlinear projections.

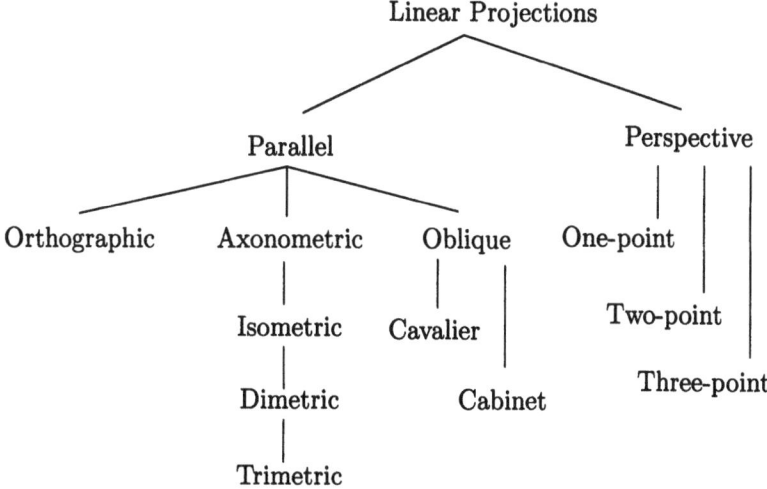

Figure 3.25: Hierarchy of Projections.

In the discussion that follows, we assume a three-dimensional object (or objects) that should be projected, a projection plane on which the object is projected, and an observer viewing the projection. In order for the projection to make sense, we assume that the object and the observer are located on two different sides of the projection plane. We also assume that the projection plane is perpendicular to the line of sight of the observer, since this is how we normally view objects displayed on a screen or printed on paper.

3.8 Parallel Projections

Like any other projections used in practice, parallel projections project a three-dimensional object onto a two-dimensional *projection plane* such as a screen, paper, or film.

The principle of all parallel projections is to select a direction **v**, and construct a ray that starts at a general point **P** on the object and goes in the direction **v**. The point **P*** where this ray intercepts the projection plane becomes the projection of **P**. The process is repeated for all the points on the object, creating a set of parallel rays, which is the reason for the name *parallel projections*. Figure 3.26 illustrates this principle. In Figure 3.26a, the rays are perpendicular to the projection plane, and in

Figure 3.26b, they strike at a different angle. This is why the latter method is called *oblique projection* (Section 3.8.3). Figure 3.26c shows a different interpretation of the principle of parallel projections. We can imagine that all the rays originate at a *center of projection* located at infinity; thus, they are parallel. This interpretation is useful since it unifies parallel and perspective projections. With this interpretation in mind, we state the general rule of projections:

➤ *Rule*: Given a projection plane and an object on one side of it, select a center of projection on the other side of the projection plane and construct rays that go from this center, through the projection plane to points on the object. If a ray going to point **P** on the object penetrates the projection plane at point **P***, then **P*** is the projection of **P**. If the center of projection is at infinity, the projection is parallel. If the center of projection is at the observer, the projection is perspective (Section 3.9).

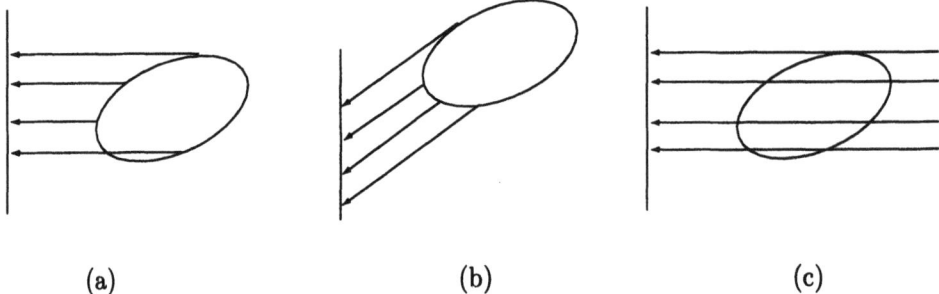

(a)   (b)   (c)

Figure 3.26: Parallel Projections.

There are three types of parallel projections: orthographic, axonometric, and oblique.

3.8.1 Orthographic Projections

This is the simplest of the three parallel projections. The idea is to project an object "flat" on one of the three coordinate planes. In the special case where the observer is located on the z axis, this type of projection is done by setting the z coordinates of all the points of the object to zero. Figure 3.26a illustrates how this is done. The result is not very attractive, but it can be useful for certain applications, most notably technical drawing. The transform of point $\mathbf{P} = (x, y, z)$ on the object is point $\mathbf{P}^* = (x, y)$ on the projection plane. Matrix $\mathbf{T}_z$ is the formal representation of this projection. Matrices $\mathbf{T}_x$ and $\mathbf{T}_y$ can be used to orthographically project points on the yz and the xz planes, respectively.

$$\mathbf{T}_x = \begin{pmatrix} 0 & 0 & 0 & 0 \\ 0 & 1 & 0 & 0 \\ 0 & 0 & 1 & 0 \\ 0 & 0 & 0 & 1 \end{pmatrix}, \mathbf{T}_y = \begin{pmatrix} 1 & 0 & 0 & 0 \\ 0 & 0 & 0 & 0 \\ 0 & 0 & 1 & 0 \\ 0 & 0 & 0 & 1 \end{pmatrix}, \mathbf{T}_z = \begin{pmatrix} 1 & 0 & 0 & 0 \\ 0 & 1 & 0 & 0 \\ 0 & 0 & 0 & 0 \\ 0 & 0 & 0 & 1 \end{pmatrix}. \quad (3.29)$$

It is clear that one orthographic projection of a three-dimensional object rarely gives enough information about the object. It is therefore common to group three such projections, together with one axonometric or perspective projection, on one page (or one screen), as illustrated by Figure 3.27, in order to convey more information. If the object is complex, it is possible to group six orthographic projections together, so as to show flat projections of the object from above, below, from the left, the right, its front, and its back. For more complex objects, sectional views may be necessary. Such a view is obtained by passing an imaginary plane through the object and drawing a projection of the plane.

Notice that the object shown in Figure 3.27 has two properties that make it especially easy to project, namely it is a cube and its edges are aligned with the coordinate axes. In general, if the main edges of the object are not aligned with the coordinate axes, its orthographic projections along the axes may be confusing, and it is preferable to rotate the object, if this is permissible, and align it before being projected. If the object is not a cube and does not resemble a cube, the best option is to select on the object three axes that are judged the "main" ones and align them with the coordinate axes. If the object is so complex that it is impossible to find three such axes, then the designer should consider drawing six projections, drawing some sectional views of the object, or using a nonorthographic projection.

▸ **Exercise 3.39:** Try to interpret the three orthographic projections of Figure 3.28.

The main advantage of orthographic projections has to do with measuring dimensions. The projection of a segment of length l on the object is a segment of length l on the projection plane. This helps in manufacturing an object directly from a drawing and is the main reason orthographic projections are used in technical drawing.

> He did not know how to meet her charge. He wanted to say that literature was above politics. But they were friends of many years' standing and their careers had been parallel, first at the University and then as teachers: he could not risk a grandiose phrase with her.
> —James Joyce, *The Dead*.

3.8.2 Axonometric Projections

An orthographic projection of an object normally shows the details of one of its main faces. It may be detailed and it may show the true shape of that face with all true dimensions, but it shows little or nothing of the rest of the object. Axonometric projections show more of the object in each projection but at the price of having wrong dimensions and angles. Such a projection typically shows three or more faces of the object, but some of the dimensions are shrunk. This makes it hard to figure out the dimensions of object details just by measuring them on the projection. An axonometric projection shows the true shape of a face of the object (with true dimensions) only if the face happens to be parallel to the projection plane. Otherwise, the shape of the face is distorted and its dimensions are shrunk. The axonometric projection is still parallel, so a group of parallel lines on the object will appear parallel in the projection.

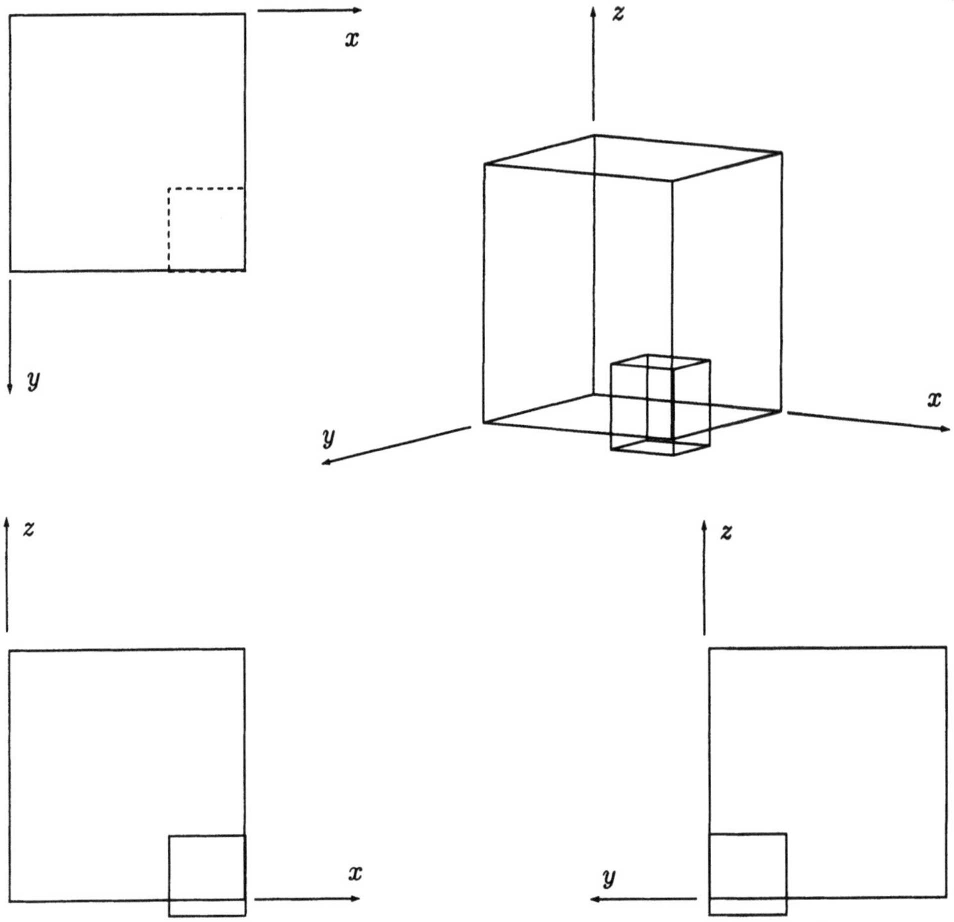

3.8 Parallel Projections

Figure 3.27: Three Orthographic Projections.

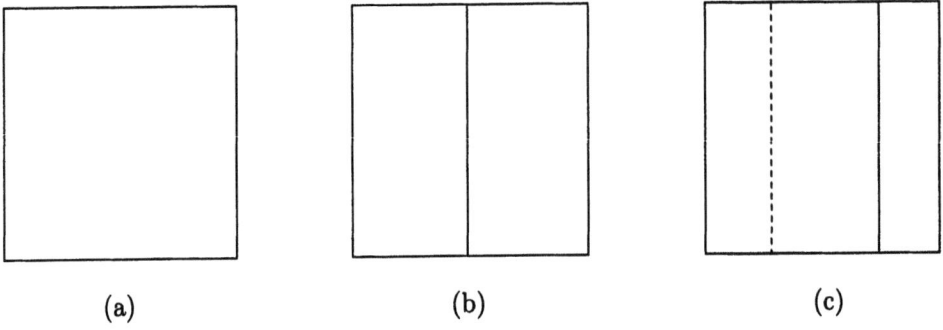

Figure 3.28: Three Orthographic Projections for Exercise 3.39.

3 Transformations and Projections

To construct an axonometric projection, the object may have to be rotated first, to bring the desired faces toward the projection plane. It is then projected on that plane in parallel. We assume that the projection plane is the xy plane, so the projection is done by clearing the z coordinates of all the points or, equivalently, by multiplying each point, after rotating it, by matrix $\mathbf{T}_z$ of Equation (3.29). Assuming that we first rotate the object ϕ degrees about the y axis, then θ degrees about the x axis, the combined rotation/projection matrix is (see (3.25))

$$\mathbf{T} = \begin{pmatrix} \cos\phi & 0 & -\sin\phi & 0 \\ 0 & 1 & 0 & 0 \\ \sin\phi & 0 & \cos\phi & 0 \\ 0 & 0 & 0 & 1 \end{pmatrix} \begin{pmatrix} 1 & 0 & 0 & 0 \\ 0 & \cos\theta & \sin\theta & 0 \\ 0 & -\sin\theta & \cos\theta & 0 \\ 0 & 0 & 0 & 1 \end{pmatrix} \begin{pmatrix} 1 & 0 & 0 & 0 \\ 0 & 1 & 0 & 0 \\ 0 & 0 & 0 & 0 \\ 0 & 0 & 0 & 1 \end{pmatrix}$$

$$= \begin{pmatrix} \cos\phi & \sin\phi\sin\theta & 0 & 0 \\ 0 & \cos\theta & 0 & 0 \\ \sin\phi & -\cos\phi\sin\theta & 0 & 0 \\ 0 & 0 & 0 & 1 \end{pmatrix}.$$

(3.30)

To find how various dimensions have shrunk, we start with the vector $(1, 0, 0, 1)$. This is a unit vector in the direction of the x axis, expressed in homogeneous coordinates. Multiplying it by $\mathbf{T}$ gives another vector that we denote $(x_1, x_2, 0, 1)$. Its magnitude is $s_x = \sqrt{x_1^2 + x_2^2}$, and since the original vector had magnitude 1, the quantity s_x expresses the ratio of magnitudes or the factor by which all dimensions in the x direction have shrunk after the transformation/projection $\mathbf{T}$. In a similar way, selecting unit vectors $(0, 1, 0, 1)$ and $(0, 0, 1, 1)$ in the y and z directions, and multiplying them by $\mathbf{T}$ produces vectors $(y_1, y_2, 0, 1)$ and $(z_1, z_2, 0, 1)$ and shrinking factors $s_y = \sqrt{y_1^2 + y_2^2}$ and $s_z = \sqrt{z_1^2 + z_2^2}$ in the y and z directions, respectively. Figure 3.29a shows a unit cube rotated such that its three sides, which used to be parallel to the coordinate axes, seem to have different lengths. Such an axonometric projection is called *trimetric*.

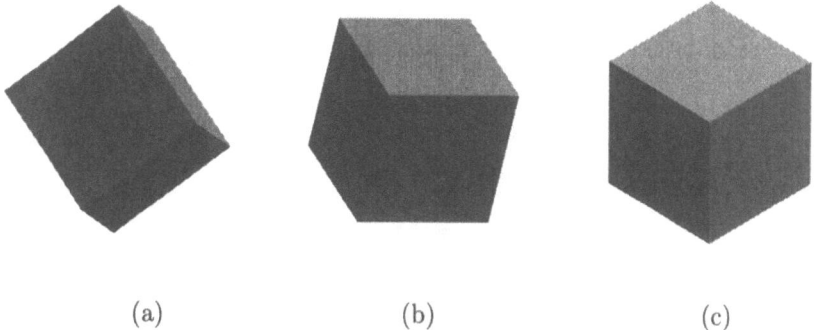

(a) (b) (c)

Figure 3.29: Three Axonometric Projections.

Figure 3.29b shows the same unit cube rotated such that two of its three sides seem to have the same length while the third side looks shorter. Such an axonometric projection is called *dimetric*. Similarly, Figure 3.29c shows the same unit cube rotated such that all its sides seem to have the same length. This type of axonometric projection is called *isometric*.

Matrix **T** of Equation (3.30) can be used to calculate the special rotations that produce a dimetric projection. Consider the product of a unit vector in the x direction and **T**:

$$(1,0,0,1)\mathbf{T} = (1,0,0,1)\begin{pmatrix} \cos\phi & \sin\phi\sin\theta & 0 & 0 \\ 0 & \cos\theta & 0 & 0 \\ \sin\phi & -\cos\phi\sin\theta & 0 & 0 \\ 0 & 0 & 0 & 1 \end{pmatrix} \quad (3.31)$$

$$= (\cos\phi, \sin\phi\sin\theta, 0, 1).$$

Any vector in the x direction thus shrinks, after being rotated by matrix **T**, by a factor s_x given by Equation (3.32). The same equation also gives the shrink factors s_y and s_z of any vector in the y and z directions.

$$\begin{aligned} s_x &= \sqrt{\cos^2\phi + \sin^2\phi\sin^2\theta}, \\ s_y &= \sqrt{\cos^2\theta}, \\ s_z &= \sqrt{\sin^2\phi + \cos^2\phi\sin^2\theta}. \end{aligned} \quad (3.32)$$

If we want a dimetric projection where equal-size segments in the x and y directions will have equal sizes after the projection, we set $s_x = s_y$ or, equivalently,

$$\cos^2\phi + \sin^2\phi\sin^2\theta = \cos^2\theta,$$

which produces the relation

$$\sin^2\phi = \frac{\sin^2\theta}{1-\sin^2\theta}. \quad (3.33)$$

Equation (3.33) together with the expression for s_z^2 yields

$$\begin{aligned} s_z^2 &= \sin^2\phi + \cos^2\phi\sin^2\theta = \sin^2\phi + (1-\sin^2\phi)\sin^2\theta \\ &= \sin^2\phi(1-\sin^2\theta) + \sin^2\theta \\ &= \frac{\sin^2\theta}{1-\sin^2\theta}(1-\sin^2\theta) + \sin^2\theta, \end{aligned}$$

or $2\sin^4\theta - (2+s_z^2)\sin^2\theta + s_z^2 = 0$, a quadratic equation in $\sin^2\theta$ whose solutions are $\sin^2\theta = s_z^2/2$ and $\sin^2\theta = 1$. The second solution cannot be used in Equation (3.33) and has to be discarded. The first solution produces

$$\theta = \sin^{-1}\left(\pm\frac{s_z}{\sqrt{2}}\right) \quad \text{and} \quad \phi = \sin^{-1}\left(\pm\frac{s_z}{\sqrt{2-s_z^2}}\right). \quad (3.34)$$

3 Transformations and Projections

Since the sine function has values in the range $[-1, 1]$, the argument of $\sin^{-1}$ must be in this range. The expression $s_z/\sqrt{2}$ is in this range when $-1.414 \leq s_z \leq +1.414$, and the expression $s_z/\sqrt{2-s_z^2}$ is in this range when $-1 \leq s_z \leq +1$. Since s_z is a shrinking factor, it is non-negative, so the above discussion shows that it must be in the range $[0, 1]$. Also, since Equation (3.34) contains a "$\pm$", any value of s_z produces four solutions.

Example: Given $s_z = 1/2$, we calculate θ and ϕ:

$$\theta = \sin^{-1}\left(\pm \frac{0.5}{\sqrt{2}}\right) = \sin^{-1}(\pm 0.35355) = \pm 20.7°,$$

$$\phi = \sin^{-1}\left(\pm \frac{0.5}{\sqrt{2-0.5^2}}\right) = \sin^{-1}(\pm 0.378) = \pm 22.2°.$$

▶ **Exercise 3.40:** Repeat the example for $s_z = 0.625$.

▶ **Exercise 3.41:** Calculate θ and ϕ for $s_x = s_z$ (equal shrink factors in the x and z directions).

The condition for an isometric projection (Figure 3.29c) is $s_x = s_y = s_z$. We already know that $s_x = s_y$ results in Equation (3.33). Similarly, it is easy to see that $s_y = s_z$ results in $\cos^2\theta = \sin^2\phi + \cos^2\phi \sin^2\theta$, which can be written

$$\sin^2\phi = \frac{1 - 2\sin^2\theta}{1 - \sin^2\theta}. \tag{3.35}$$

Equations (3.33) and (3.35) result in $\sin^2\theta = 1 - 2\sin^2\theta$ or $\sin^2\theta = 1/3$, yielding $\theta = \pm 35.26°$. Rotation angle ϕ can now be calculated from Equation (3.33):

$$\sin^2\phi = \frac{1/3}{1 - 1/3} = 1/2, \quad \text{yielding } \phi = \pm 45°.$$

The shrink factors can be calculated from, for example, $s_y = \cos^2\theta = \sqrt{2/3} \approx 0.8165$.

We conclude that the isometric projection is the most useful but also the most restrictive of the three axonometric projections. Given a diagram with the isometric projection of an object, we can measure distances on the diagram and divide them by 0.8165 to obtain actual dimensions on the object. However, the diagram must show the object (whose main edges are assumed to be originally aligned with the coordinate axes) after being rotated by $\pm 45°$ about the y axis and by $\pm 35.26°$ about the x axis. If these rotations don't show enough object features, a less restrictive projection, such as dimetric or trimetric, must be used.

3.8.3 Oblique Projections

As has been mentioned earlier, an oblique projection is a special case of a parallel projection (i.e., with a center of projection at infinity) where the projecting rays are not perpendicular to the projection plane. In the same way that axonometric projections show more object details than orthographic projections but make it harder to figure out dimensions on the object, oblique projections generally show

> What recurrent impressions of the same were possible by hypothesis?
> Retreating, at the terminus of the Great Northern Railway, Amiens Street, with constant uniform acceleration, along parallel lines meeting at infinity, if produced: along parallel lines, reproduced from infinity, with constant uniform retardation, at the terminus of the Great Northern Railway, Amiens street, returning.
> —James Joyce, *Ulysses.*

more object details than axonometric projections but distort angles and dimensions even more. In an oblique projection, only those faces of the object that are parallel to the projection plane are projected with their true dimensions. Other faces are distorted.

Two oblique projections, *cavalier* and *cabinet*, are used in practice. In the former, the projecting rays strike the projection plane at an angle of 45° (Figure 3.30a,b). In the latter, the rays hit the projection plane at an angle of 63.43° (Figure 3.30c, d).

Because of the special 45° angle the three shrink factors of a cavalier projection are equal, as shown below. In a cabinet projection, the shrink factors in the two directions perpendicular to the plane of projection equal 1/2.

Figure 3.31a illustrates the geometry of oblique projections and can be used to derive their transformation matrix. In the diagram, we assume that the projection plane is $z = 0$ (the xy plane) and that all the projecting rays hit this plane at an angle θ. Two projecting rays are shown, one projecting the special point $\mathbf{P} = (0, 0, 1)$ to a point $(a, b, 0)$ and the other projecting $\mathbf{Q} = (0, 0, z)$, a general point on the z axis, to a point $(A, B, 0)$. The origin $(0, 0, 0)$ is projected onto itself, so the projection of the unit segment from the origin to $\mathbf{P}$ is the segment of size s from the origin to $(a, b, 0)$. The value s is, thus, the shrink factor of the oblique projection. The three quantities a, b, and s can also be related by $a = s\cos\phi$ and $b = s\sin\phi$, where ϕ is measured on the projection plane. The shrink factor s can also be related to the projection angle θ by $\tan\theta = 1/s$ or $s = \cot\theta$.

We now consider the projecting ray from $\mathbf{Q}$ to $(A, B, 0)$. Since $\mathbf{Q}$ is at a distance z from the origin, the distance on the projection plane between the origin and point $(A, B, 0)$ is sz. We thus have the relations $A = sz\cos\phi$ and $B = sz\sin\phi$. The next step is to consider the projection of a general point (x, y, z). Since all the projecting rays are parallel, it should not be hard to see that moving a point from $(0, 0, z)$ to $(x, 0, z)$ moves its projection from $(A, B, 0)$ to $(x+A, B, 0)$. Similarly, moving a point from $(0, 0, z)$ to $(0, y, z)$ moves its projection from $(A, B, 0)$ to $(A, y+B, 0)$. A general point located at (x, y, z) should thus be projected to a point at $(x + A, y + B, 0)$. Thus, the rule of oblique projections is

$$(x, y, z) \longrightarrow (x + sz\cos\phi, y + sz\sin\phi, 0), \tag{3.36}$$

3 Transformations and Projections

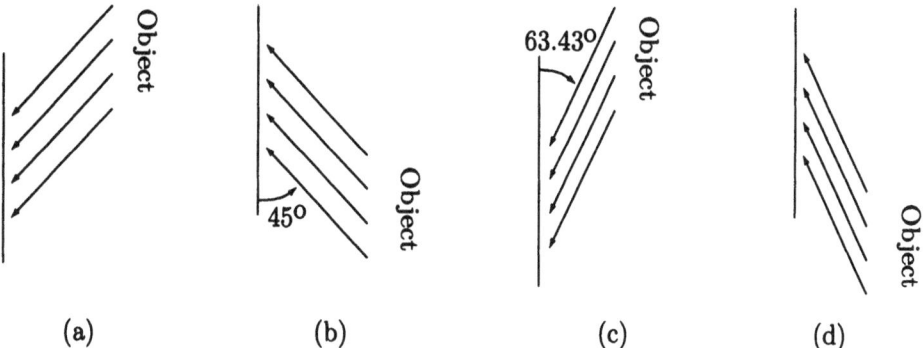

Figure 3.30: Oblique Projections.

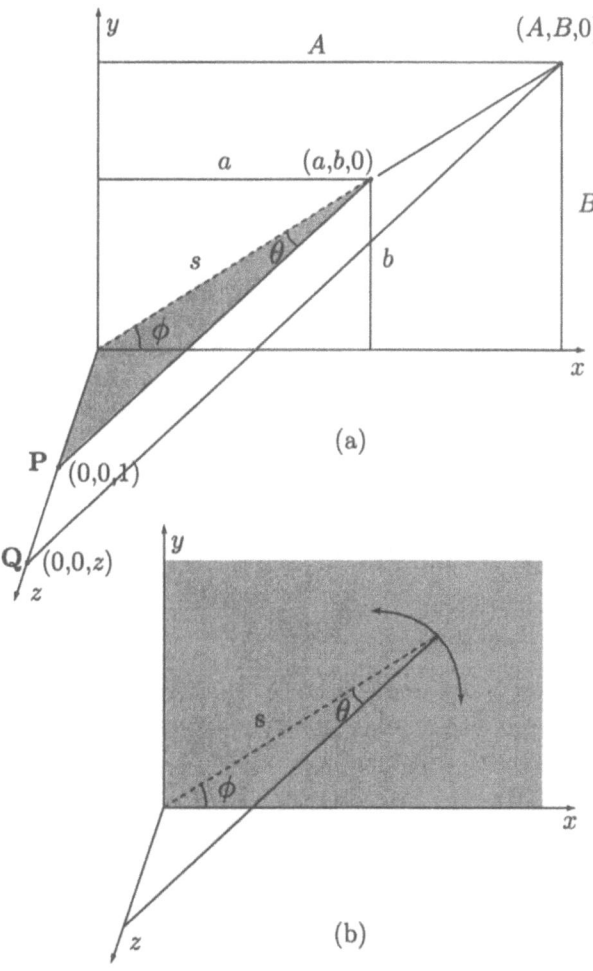

Figure 3.31: Oblique Projections.

which can be written in terms of a transformation matrix

$$\mathbf{P}^* = \mathbf{PT} = (x, y, z) \begin{pmatrix} 1 & 0 & 0 & 0 \\ 0 & 1 & 0 & 0 \\ s\cos\phi & s\sin\phi & 0 & 0 \\ 0 & 0 & 0 & 1 \end{pmatrix}. \quad (3.37)$$

It is now easy to examine the three special cases:

1. A cavalier projection. It is defined as the case where the projection angle is $45°$, which implies $s = \cot(45°) = 1$. All edges and segments thus have shrink factors of 1.

2. A projection angle of $90°$. A value $\theta = 90°$ implies a shrink factor $s = \cot(90°) = 0$. Matrix $\mathbf{T}$ of Equation (3.37) reduces to matrix $\mathbf{T}_z$ of Equation (3.29), showing how the oblique projection reduces in this case to an orthographic projection.

3. A cabinet projection. It is defined as the case where the projection angle is $63.43°$, which implies $s = \cot(63.43°) = 1/2$. All edges and segments perpendicular to the projection plane have shrink factors of $1/2$.

Figure 3.31b shows how ϕ and θ are independent. For a given projection angle θ, it is possible to assign ϕ any value by rotating the triangle in the figure. In practice, this means that an object can be projected several times, with different values of ϕ but with the same projection angle θ. Such projections may give a lot of visual information about the object while having the same shrink factors.

> I will sette as I doe often in woorke use, a paire of paralleles, or [twin] lines of one lengthe, thus =, bicause noe 2. thynges, can be moare equalle.
> — Robert Recorde, 1557.

3.9 Perspective Projections

Perspective projection is important because this is how we observe objects in real life. This type of projection is obtained from the general rule of projection (page 104), when the observer is selected as the center of projection. We know from our everyday experience that the perspective projection has the following three properties:

a. The farther away an object is, the smaller it appears to an observer (this is not true for parallel projections).

b. All lines on the object parallel to the line of sight of the observer appear to converge to a *vanishing point*.

c. The amount of perspective seen by an observer depends on the distance between the observer and the object. A nearby observer sees more perspective, which means a bigger difference in size between the front and the back of the object being viewed.

3 Transformations and Projections

Perspective projections have traditionally been confusing, so they are introduced here gradually, step by step, with examples. The steps are as follows:

1. State the rule for perspective projection.
2. Derive the expressions of a general transformed point $\mathbf{P}^*$.
3. Prove that our rule is the right one by showing that it satisfies the three properties a–c.
4. Write the 4×4 transformation matrix that generates perspective projections.
5. Discuss the meaning of the term *vanishing points*.
6. Generalize the results by showing how to move the observer to any point.
7. Consider how to transform the depth (the z coordinate) of a point. This is discussed in Section 3.9.9.

Step 1: To simplify the mathematics, we assume that the observer is positioned on the negative side of the z axis, at a distance k from the origin, looking in the direction of positive z (Figure 3.32). Notice that both the position $(0,0,-k)$ and the direction $(0,0,1)$ of the observer must be given. We further assume that the plane of projection (the screen or paper on which the two-dimensional picture is to be drawn) is the xy plane. Let $\mathbf{P} = (x,y,z)$ be a three-dimensional point on the object and let $\mathbf{P}^* = (x^*, y^*)$ be the transformed point on the projection plane. It is important to understand that the observer and the projection plane constitute a single unit. When the observer is moved or is transformed in any way, the projection plane has to be transformed in the same way as if it were rigidly attached to the observer.

The rule for perspective projection states that in order to obtain $\mathbf{P}^*$, we need to draw a straight line from $\mathbf{P}$ to the center of projection, which is the observer. The point where this line pierces the projection plane is $\mathbf{P}^*$. This is a special case of the general rule of projections (page 104).

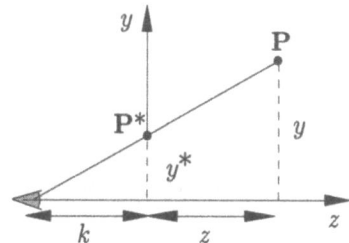

Figure 3.32: Principle of Perspective Projection.

Step 2:. Figure 3.32 makes it clear that

$$\frac{x^*}{k} = \frac{x}{z+k} \quad \text{and} \quad \frac{y^*}{k} = \frac{y}{z+k},$$

therefore

$$x^* = \frac{x}{(z/k)+1} \quad \text{and} \quad y^* = \frac{y}{(z/k)+1}. \tag{3.38}$$

> **Brunelleschi and the Invention of Perspective Projection**
>
> The perspective projection allows the artist and graphics designer to translate three-dimensional objects into realistic-looking two-dimensional images. Today it is generally believed that the first person to experiment with perspective and to discuss its principles was Filippo Brunelleschi (1377–1446), the Florentine architect, sculptor, and engineer (although some experts credit Paolo Uccello or Leon Battista Alberti with the original idea of perspective).
>
> Brunelleschi is best known today for his ingenious design of the great dome that crowns the Santa Maria del Fiore Cathedral in Florence, but history may judge that his greatest achievement was the perspective projection. For more on Brunelleschi's life and work, see [Battisti 81].
>
> Tommaso Cassai (1401–1427?), better known as Masaccio, was the first of the great Renaissance painters to use perspective in his paintings, thereby inaugurating the modern era in painting. Especially celebrated is his fresco *The Holy Trinity*, where the barrel vaulted ceiling uses perspective in an incredibly complex way.
>
> Leonardo da Vinci, who certainly knew about perspective, developed his own projection, now known as aerial, or atmospheric, perspective. This method of adding depth to a two-dimensional painting is based on the perception that contrasts of color and shade appear greater in nearby objects than in those far away, and that warm colors appear to advance while cool colors appear to recede. Aerial perspective is also used in East Asian art, where zones of mist are sometimes used to separate near and far parts of the scene.
>
> Another great artist interested in perspective was the German painter Albrecht Dürer (1471–1528).

(Notice that in a right-handed coordinate system; the sign of x should be reversed, as discussed in Section 3.5.1.)

Step 3:. The three properties of perspective projection mentioned earlier can now be proved.

Property a: When the object is distant (located away from the observer, which corresponds to large z values), Equation (3.38) produces small values for x^* and y^*. Specifically

$$\lim_{z \to \infty} x^* = 0, \qquad \lim_{z \to \infty} y^* = 0.$$

The object therefore appears smaller to the observer (it shrinks toward the origin).

Property b: Imagine a line parallel to the z axis (i.e., the line of sight). All points on this line have the same x and y coordinates and differ only by their z coordinates. When such a line is extended to infinity, its projection on the screen approaches $(0,0)$ regardless of the x and y coordinates of the points on the line. This shows that every line parallel to the z axis has a projection, on the screen, that converges to the origin. The origin (the center of the screen) is, therefore, the *vanishing point* of such lines.

3 Transformations and Projections

▶ **Exercise 3.42:** Almost every group of parallel lines on the object (not just lines parallel to the z axis) will converge to a vanishing point. What groups of parallel lines will still appear to the observer as parallel?

Property c: Changing the distance $d = k + z$ between the observer and the object should affect the amount of perspective. This is shown by selecting two points $\mathbf{P}_1 = (x_1, y_1, z_1)$ and $\mathbf{P}_2 = (x_1, y_1, z_2)$ on our object (with the same x and y coordinates) and considering their projections $\mathbf{P}_1^* = (x_1^*, y_1^*)$, $\mathbf{P}_2^* = (x_2^*, y_2^*)$. Specifically, the ratio x_1^*/x_2^* should be calculated:

$$\frac{x_1^*}{x_2^*} = \frac{x_1}{(z_1/k)+1} \bigg/ \frac{x_1}{(z_2/k)+1} = \frac{(z_2/k)+1}{(z_1/k)+1} = \frac{z_2+k}{z_1+k} = \frac{d_2}{d_1}.$$

When we move the object or the observer away from each other, both d_1 and d_2 increase, bringing the ratio $x_1^*/x_2^* = d_2/d_1$ closer to 1. The two projected points get closer on the screen, which means that the object is projected on the screen with less perspective.

Figure 3.33 shows the difference between increasing z (moving the object away from the screen) and increasing k (moving the observer away from the screen). In either case, d gets larger, which implies less perspective. However, in the former case, we get a smaller projection on the screen (in the limit, the projection degenerates to a point), whereas in the latter case, we get less perspective (in the limit, the projection becomes parallel).

Note that in the latter case, the size of the picture on the screen does not change significantly. However, since the observer is located far away, they will see the entire screen small and, as a result, a small projected image.

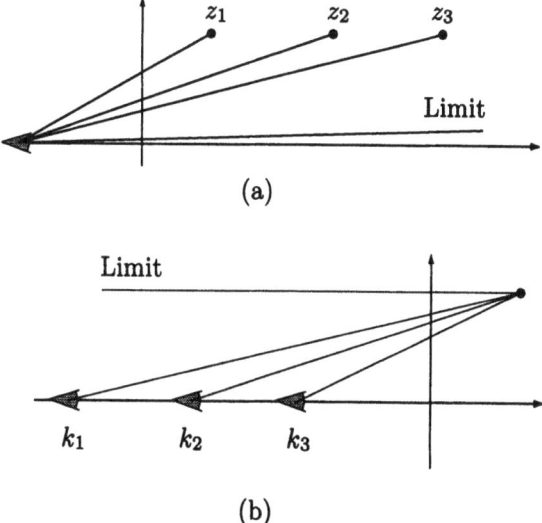

Figure 3.33: (a) Increasing z. (b) Increasing k.

Step 4:. The 4×4 transformation matrix that generates the transformations of Equation (3.38) is

$$\mathbf{T}_p = \begin{pmatrix} 1 & 0 & 0 & 0 \\ 0 & 1 & 0 & 0 \\ 0 & 0 & 0 & r \\ 0 & 0 & 0 & 1 \end{pmatrix}. \quad (3.39)$$

This is easy to verify by multiplying $(x, y, z, 1) \times \mathbf{T}_p$ and substituting $r = 1/k$. Similarly, the transformation matrices

$$\mathbf{T}_x = \begin{pmatrix} 0 & 0 & 0 & p \\ 0 & 1 & 0 & 0 \\ 0 & 0 & 1 & 0 \\ 0 & 0 & 0 & 1 \end{pmatrix}, \quad \mathbf{T}_y = \begin{pmatrix} 1 & 0 & 0 & 0 \\ 0 & 0 & 0 & q \\ 0 & 0 & 1 & 0 \\ 0 & 0 & 0 & 1 \end{pmatrix},$$

generate perspective projections for cases where the observer is located on the negative side of the x (or y) axis, at a distance of $1/p$ (or $1/q$) from the origin, looking in the direction of positive x (or y). The plane of projection is assumed to be the yz (or xz) plane. The parameters p (or q) have meanings similar to r above.

Step 5:. Imagine a cube centered on the origin. It will appear to the observer as in Figure 3.34a. This is the simple case of one vanishing point. If the same cube is now rotated 45°, it will appear as in Figure 3.34b with *two* vanishing points. The picture seen by the observer has changed considerably by just changing the orientation of the object. The observer, the coordinate axes, and the shape of the object have not changed. The question is when and why do we get two vanishing points. To understand the answer, let's imagine a sphere observed in real life. It has no perspective at all. True, it looks bigger when it gets nearer, but there are no vanishing points and no converging lines. The same is true for a shapeless smooth blob. Its front looks bigger than its back, but there is no perspective since there are no converging lines. The reason for the lack of perspective is that such objects feature curved surfaces with no straight lines.

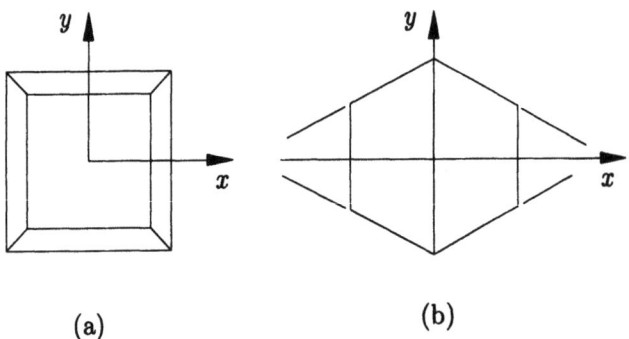

Figure 3.34: One-Point and Two-Point Perspective Projections.

Perspective can only be observed if the surface of the object has straight lines on it. A group of straight, parallel lines on the surface of an object will almost

3 Transformations and Projections

always converge to a vanishing point and create the effect of perspective. In fact, the only case where such a group will not converge is when the lines are parallel to the projection plane. Thus, in the case of the cube in Figure 3.34a, before the rotation only lines parallel to the z axis converged. All the other lines on the surface of the cube were parallel to the x or y axes and therefore parallel to the projection plane. After the rotation, lines parallel to the y axis remained parallel to the projection plane. However, the other two groups of lines now converge, to two different vanishing points.

Clearly, the rotation does not have to be 45°. Even the smallest rotation will produce such an effect. A small rotation (Figure 3.35a) will not change the position of the z-parallel lines by much; therefore, the original vanishing point will move just a little. The x-parallel lines will no longer be parallel but will be close to parallel. As a result, their vanishing point (which originally was at infinity) will not move by much and will, therefore, be far away.

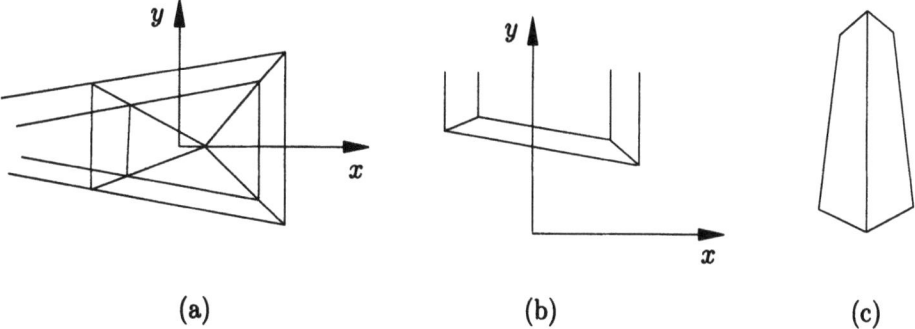

(a) (b) (c)

Figure 3.35: Perspective Projection.

Figure 3.35b shows the same rotated cube after it has been raised in the y direction. The y-parallel lines are still parallel to the projection plane, so their projections do not converge. The observer, however, now sees only part of the cube. To see all of it again, the observer has either to move up with the cube or to look up at it. Looking up means rotating the line of sight of the observer (about the x axis). Since the projection plane is perpendicular to the line of sight, rotating that line causes a rotation of the projection plane. Once the observer rotates, however, the y-parallel lines are no longer parallel to the projection plane and their projections therefore converge. The result is three vanishing points (Figure 3.35c). This is what we see when we stand in front of a skyscraper, looking up.

Step 6:. The next point to consider is the generalization of the mathematics. Up until now we have assumed that the observer is located at point $(0, 0, -k)$, looking toward the origin. We also assumed that the projection plane is at the xy plane, with its top and bottom oriented in the positive and negative y directions, respectively. This is called the *standard position* of the observer. In general, the observer may be located at any point, looking in any direction, with the screen perpendicular to the line of sight of the observer and at a distance of k units. In the general case, the direction of the top of the screen should also be specified. We outline four approaches to the general case:

Approach 1: The first approach uses elementary trigonometry to derive the projection matrix. This is discussed in Section 3.9.1.

Approach 2: This approach assumes that the observer starts at the standard position and is supposed to be transformed to a general location by means of translations and rotations. The idea is to leave the observer at the standard position and instead perform the reverse transformations on the object. This approach is outlined in Section 3.9.2.

Approach 3: Here, we use translations and/or rotations to bring the observer from its general location to the standard position. The same transformations are also applied to all the points of the object. Each point is then projected using the standard projection matrix, Equation (3.39). This is discussed in Section 3.9.3.

Approach 4: The fourth approach uses vector analysis to derive the equations of the projected point. Sections 3.9.5 and 3.9.8 discuss two such methods.

> Reality is a question of perspective; the further you get from the past, the more concrete and plausible it seems—but as you approach the present, it inevitably seems incredible.
> — Salman Rushdie (1981).

3.9.1 General Perspective: I

To illustrate the first approach, we work out a simple example. We rotate the observer θ degrees counterclockwise about the y axis from positive z to positive x (Figure 3.36). The observer ends up at point

$$(0,0,-k) \begin{pmatrix} \cos\theta & 0 & -\sin\theta \\ 0 & 1 & 0 \\ \sin\theta & 0 & \cos\theta \end{pmatrix} = (-k\sin\theta, 0, -k\cos\theta) = (-k\alpha, 0, -k\beta), \quad (3.40)$$

where $\alpha = \sin\theta$ and $\beta = \cos\theta$ (notice that $\alpha^2 + \beta^2 = 1$). We select a general point $\mathbf{P} = (l, m, n)$ on the object being projected and we are looking for its projection $\mathbf{P}^*$ on the new projection plane. Notice that the new projection plane is still perpendicular to the line of sight of the observer and is still at a distance of k units. It is no longer identical to the xy plane, but it still contains the origin.

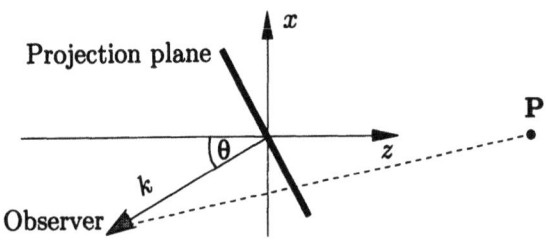

Figure 3.36: Observer Rotated About the y Axis.

3 Transformations and Projections

▶ **Exercise 3.43:** The previous paragraph talks about rotating the observer counterclockwise, but Equation (3.40) looks like Equation (3.4), which generates clockwise rotation. What's the explanation?

■ The first task is to find the equation of the line segment from the observer to point **P**. We use the parametric equation $\mathbf{P}(t) = (\mathbf{P}_2-\mathbf{P}_1)t+\mathbf{P}_1$ (Equation (A.12), page 704). When applied to the observer (that we denote $\mathbf{P}_1$) and to point **P** (that we denote $\mathbf{P}_2$), it yields

$$\mathbf{P}(t) = (l+k\alpha, m, n+k\beta)t + (-k\alpha, 0, -k\beta)$$
$$= ((l+k\alpha)t - k\alpha, mt, (n+k\beta)t - k\beta)$$
$$= (P_x(t), P_y(t), P_z(t)).$$

■ Next, we find the equation of the projection plane. Vector $(-k\alpha, 0, -k\beta)$ is perpendicular to the plane (it is the *normal* to the plane), so it is perpendicular to any general vector (x, y, z) on the plane. This is why their dot product (Section A.4) is zero. From $(-k\alpha, 0, -k\beta) \bullet (x, y, z) = 0$, we get the plane equation $\alpha x = -\beta z$.

▶ **Exercise 3.44:** Why doesn't this equation involve y?

■ The next task is to find the intersection point of the line and the projection plane. This is obtained at the value t_0 that satisfies $\alpha P_x(t_0) = -\beta P_z(t_0)$ or

$$\alpha\big((l+k\alpha)t_0 - k\alpha\big) = -\beta\big((n+k\beta)t_0 - k\beta\big).$$

The solution is

$$t_0 = \frac{k(\alpha^2 + \beta^2)}{\alpha l + \beta n + k(\alpha^2 + \beta^2)} = \frac{k}{\alpha l + \beta n + k}.$$

The intersection point is $\mathbf{P}(t_0)$.

■ The next task is to find the three coordinates of the projected point $\mathbf{P}^* = \mathbf{P}(t_0)$. The x coordinate is

$$x^* = P_x(t_0) = (l+k\alpha)t_0 - k\alpha = (l+k\alpha)\frac{k}{\alpha l + \beta n + k} - k\alpha = \frac{lk\beta^2 - nk\alpha\beta}{\alpha l + \beta n + k}.$$

The y coordinate is

$$y^* = P_y(t_0) = mt_0 = \frac{mk}{\alpha l + \beta n + k}$$

and the z coordinate is

$$z^* = P_z(t_0) = (n+k\beta)t_0 - k\beta = (n+k\beta)\frac{k}{\alpha l + \beta n + k} - k\beta = \frac{-lk\alpha\beta + nk\alpha^2}{\alpha l + \beta n + k}.$$

From $(x^*, y^*, z^*) = (X/H, Y/H, Z/H)$, we get

$$X = lk\beta^2 - nk\alpha\beta,$$
$$Y = mk,$$
$$Z = -lk\alpha\beta + nk\alpha^2,$$
$$H = \alpha l + \beta n + k.$$

■ Using the four expressions above and keeping in mind that (l, m, n) are the coordinates of point **P**, it is easy to figure out the transformation matrix that projects **P** to **P***:

$$(l, m, n, 1)\mathbf{T} = (X, Y, Z, H) \quad \text{implies} \quad \mathbf{T} = \begin{pmatrix} k\beta^2 & 0 & -k\alpha\beta & \alpha \\ 0 & k & 0 & 0 \\ -k\alpha\beta & 0 & k\alpha^2 & \beta \\ 0 & 0 & 0 & k \end{pmatrix}. \quad (3.41)$$

▷ **Exercise 3.45:** Calculate the values of matrix (3.41) for the three special cases $\theta = 0°$, $45°$, and $90°$.

▷ **Exercise 3.46:** Given the point $\mathbf{P} = (\beta l, m, -\alpha l)$, calculate its projection. Explain the result!

▷ **Exercise 3.47:** Imagine rotating the observer, who is now at $(-k\alpha, 0, -k\beta)$, a second time, by an angle ϕ about the x axis. The new position of the observer is

$$(-k\alpha, 0, -k\beta)\begin{pmatrix} 1 & 0 & 0 \\ 0 & \cos\phi & -\sin\phi \\ 0 & \sin\phi & \cos\phi \end{pmatrix}$$
$$= (-k\sin\theta, -k\cos\theta\sin\phi, -k\cos\theta\cos\phi)$$
$$= (k\alpha, -k\beta\gamma, -k\beta\delta),$$

where $\gamma = \sin\phi$ and $\delta = \cos\phi$. Derive the projection matrix for this case, using steps similar to the ones above.

▷ **Exercise 3.48:** After two rotations, the observer may be located at any point in space. This still does not represent the most general case because there is another constraint. What is it?

It is important to realize that matrix (3.41) isn't as useful as it seems at first. It generates the coordinates of projected points, but those coordinates are on the plane $\alpha x = -\beta z$. In practice, we want to display the projected points on the screen, which is two-dimensional, so we have to go through another step. We have to define two local axes on $\alpha x = -\beta z$, then figure out the coordinates of the projected points relative to those axes. This is why the approaches outlined in Sections 3.9.2 and 3.9.3 are preferable. They project points onto the xy plane, where they effectively have just two coordinates. Before looking at these approaches, however, here is a short summary of the current section.

3 Transformations and Projections

Summary: The method of this section proceeds in the following steps:

1. Find the equation of the line segment connecting a general point **P** to the observer.
2. Find the equation of the projection plane.
3. Find the intersection point of the line and the plane.
4. Convert the coordinates of the intersection point to screen coordinates.

It is possible to use the approach of this section in order to figure out the projection matrix for the general case where the observer may be located at any point **B**, looking in an arbitrary given direction **D**. This, however, is messy because in addition to **B**, **D**, and k, another vector is needed, to define the "up" direction of the projection plane. In this section, we started with the "up" direction in the positive y direction. After the two rotations, that direction changes, but it is fully determined by the rotations and does not need to be explicitly specified. Another drawback of this approach is that points are projected on a three-dimensional plane, so they have three dimensions. We want to display the projected image on the two-dimensional screen, so we need points with two dimensions. The following two sections show how to project points on the xy plane, which effectively makes them two-dimensional.

> Perspective, as its inventor remarked, is a beautiful thing. What horrors of damp huts, where human beings languish, may not become picturesque through aerial distance! What hymning of cancerous vices may we not languish over as sublimest art in the safe remoteness of a strange language and artificial phrase! Yet we keep a repugnance to rheumatism and other painful effects when presented in our personal experience.
>
> — George Eliot, *Daniel Deronda*.

3.9.2 General Perspective: II

This approach is ideal for cases where the observer starts at the standard position and we want to transform it (them?) by means of translations and/or rotations (but no scaling or shearing) to a new location, so it can observe the object from a different direction. This approach is useful, for example, in computer animation and is used in Section 8.4.3. Suppose that we have to perform a transformation **A** (where $\mathbf{A} = \mathbf{T}_1 \cdot \mathbf{T}_2 \cdots \mathbf{T}_n$ is a product of translations and rotations) on the observer, to transform it to a new location. The idea is to leave the observer alone and instead apply the inverse transformation $\mathbf{A}^{-1}$ on the object. Section A.5 shows that the inverse of **A** is $\mathbf{B} = \mathbf{A}^{-1} = \mathbf{T}_n^{-1} \cdots \mathbf{T}_2^{-1} \cdot \mathbf{T}_1^{-1}$ (where $\mathbf{T}_i^{-1}$ is the reverse of transformation $\mathbf{T}_i$).

The point is that transforming the observer with **A** or transforming the object with **B** will bring them to the *same relative position*. Once the object has been transformed, we can use matrix $\mathbf{T}_p$ (Equation (3.39)) to do the perspective projection since the observer is still located at the standard position. In practice, there is, of course, no need to actually transform the object. All that we have to do is calculate matrix **B** and multiply each point of the object by $\mathbf{B} \cdot \mathbf{T}_p$.

3.9 Perspective Projections

Example: an observer at the standard position and an object close to the origin (Figure 3.37). Suppose that we want to translate the observer to the origin, rotate it 45° counterclockwise, then translate it $-k$ units in both the x and z directions (Figure 3.37a,b,c). The transformation matrices are

$$\mathbf{T}_1 = \begin{pmatrix} 1 & 0 & 0 & 0 \\ 0 & 1 & 0 & 0 \\ 0 & 0 & 1 & 0 \\ 0 & 0 & k & 1 \end{pmatrix}, \quad \mathbf{T}_2 = \begin{pmatrix} \cos 45° & 0 & -\sin 45° & 0 \\ 0 & 1 & 0 & 0 \\ \sin 45° & 0 & \cos 45° & 0 \\ 0 & 0 & 0 & 1 \end{pmatrix},$$

$$\mathbf{T}_3 = \begin{pmatrix} 1 & 0 & 0 & 0 \\ 0 & 1 & 0 & 0 \\ 0 & 0 & 1 & 0 \\ -k & 0 & -k & 1 \end{pmatrix}.$$

The reverse transformations, performed in reverse order are (Figure 3.37d,e,f)

$$\mathbf{B} = \begin{pmatrix} 1 & 0 & 0 & 0 \\ 0 & 1 & 0 & 0 \\ 0 & 0 & 1 & 0 \\ k & 0 & k & 1 \end{pmatrix} \begin{pmatrix} \cos 45° & 0 & \sin 45° & 0 \\ 0 & 1 & 0 & 0 \\ -\sin 45° & 0 & \cos 45° & 0 \\ 0 & 0 & 0 & 1 \end{pmatrix} \begin{pmatrix} 1 & 0 & 0 & 0 \\ 0 & 1 & 0 & 0 \\ 0 & 0 & 1 & 0 \\ 0 & 0 & -k & 1 \end{pmatrix}$$

$$= \begin{pmatrix} \sin 45° & 0 & \sin 45° & 0 \\ 0 & 1 & 0 & 0 \\ -\sin 45° & 0 & \sin 45° & 0 \\ 0 & 0 & -k + 2k \sin 45° & 1 \end{pmatrix}.$$

Any point $\mathbf{P} = (x, y, z, 1)$ on the object can be projected to a point $\mathbf{P}^*$ on the screen by

$$\mathbf{P}^* = \mathbf{PBT}_p = (x, y, z, 1) \begin{pmatrix} a & 0 & 0 & a/k \\ 0 & 1 & 0 & 0 \\ -a & 0 & 0 & a/k \\ 0 & 0 & 0 & 2a \end{pmatrix}$$

$$= (a(x - z), y, 0, a(2k + x + z)/k),$$

resulting in

$$x^* = \frac{k(x - z)}{2k + x + z}, \quad y^* = \frac{yk}{a(2k + x + z)},$$

where $a = \sin 45°$. A comparison of parts (c) and (f) in Figure 3.37 shows how the observer and the object end up in the same relative positions.

Since transforming the observer involves just translations and rotations, it is possible to transform the observer from the standard position to any point by means of (1) a translation to the origin, (2) a general rotation about the origin, and (3) another translation from the origin to its final location. The two translations are easy to express, and Section 3.9.3 discusses how to derive the rotation matrix that would leave the observer looking in a given direction $\mathbf{D}$.

3 Transformations and Projections

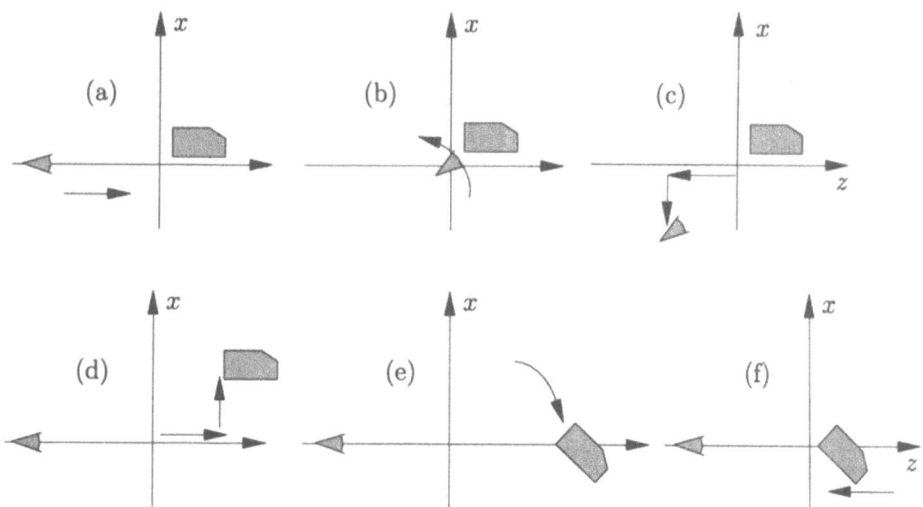

Figure 3.37: Transforming Observer and Object.

It should again be emphasized that the observer and the projection plane constitute a single unit and should be transformed together. Even though this approach transforms the object and not the observer, it is still important to make sure that the object remains on the other side of the projection plane from the observer after all the transformations.

On the other side of the screen, it all looks so easy.
— Jeff Bridges (as Kevin Flynn) in *Tron* (1982).

3.9.3 General Perspective: III

This section illustrates the third approach to general perspective projections. We assume that the observer is located at an arbitrary point **B** and is looking in a given direction **D** (Figure 3.38a). The idea is to translate **B** to the origin (Figure 3.38b), rotate such that direction **D** will coincide with $(0, 0, 1)$ (i.e., looking in the positive z direction, Figure 3.38c), then translate from the origin to point $(0, 0, -k)$ (Figure 3.38d). These three transformations bring the observer to the standard position. The same transformations are then applied to every point **P** of the image, thereby bringing the observer and the image to the same relative positions they had before the transformations. One way to understand this approach is to imagine that the observer and all the image points are transformed as one solid object, such that the observer ends up at the standard position. Another way to look at this approach is to imagine that we transform the coordinate axes (Section 3.6) while the observer and the image are left in their places.

3.9 Perspective Projections

Since the observer is located at the standard position, matrix $\mathbf{T}_p$ (Equation (3.39)) can now be used to project image points. This approach has the advantage that all the image points are projected on the xy plane, so they effectively become two-dimensional.

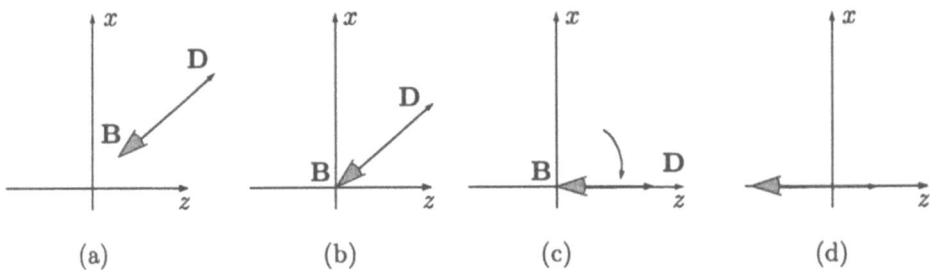

Figure 3.38: Transforming the Observer.

This approach is developed below for the general case but is illustrated first for a case where the coordinates of $\mathbf{B}$ and the components of $\mathbf{D}$ are known numbers. We assume that an observer is located at $\mathbf{B} = (1, 1, 1)$ and is looking in direction $\mathbf{D} = (1, 0, 1)$ (i.e., between the directions of positive x and positive z). Matrix $\mathbf{T}_1$ below translates from $(1, 1, 1)$ to the origin. Matrix $\mathbf{T}_2$ rotates by 45° from the positive x to the positive z direction. Matrix $\mathbf{T}_3$ translates from the origin to point $(0, 0, -k)$. The result is (we denote $s = \cos 45° = \sin 45° = 1/\sqrt{2}$),

$$\mathbf{T} = \mathbf{T}_1 \mathbf{T}_2 \mathbf{T}_3 \mathbf{T}_p$$

$$= \begin{pmatrix} 1 & 0 & 0 & 0 \\ 0 & 1 & 0 & 0 \\ 0 & 0 & 1 & 0 \\ -1 & -1 & -1 & 1 \end{pmatrix} \begin{pmatrix} s & 0 & s & 0 \\ 0 & 1 & 0 & 0 \\ -s & 0 & s & 0 \\ 0 & 0 & 0 & 1 \end{pmatrix} \begin{pmatrix} 1 & 0 & 0 & 0 \\ 0 & 1 & 0 & 0 \\ 0 & 0 & 1 & 0 \\ 0 & 0 & -k & 1 \end{pmatrix} \begin{pmatrix} 1 & 0 & 0 & 0 \\ 0 & 1 & 0 & 0 \\ 0 & 0 & 0 & r \\ 0 & 0 & 0 & 1 \end{pmatrix}$$

$$= \begin{pmatrix} s & 0 & s & 0 \\ 0 & 1 & 0 & 0 \\ -s & 0 & s & 0 \\ 0 & -1 & -2s-k & 1 \end{pmatrix} \begin{pmatrix} 1 & 0 & 0 & 0 \\ 0 & 1 & 0 & 0 \\ 0 & 0 & 0 & r \\ 0 & 0 & 0 & 1 \end{pmatrix}$$

$$= \begin{pmatrix} s & 0 & 0 & sr \\ 0 & 1 & 0 & 0 \\ -s & 0 & 0 & sr \\ 0 & -1 & 0 & 1-kr-2rs \end{pmatrix}$$

$$= \begin{pmatrix} s & 0 & 0 & sr \\ 0 & 1 & 0 & 0 \\ -s & 0 & 0 & sr \\ 0 & -1 & 0 & -2rs \end{pmatrix}. \tag{3.42}$$

(Recall that $k = 1/r$.) The projection of any point $\mathbf{P} = (x, y, z)$ is calculated by $\mathbf{P}^* = \mathbf{P}\mathbf{T}$.

3 Transformations and Projections

Example 1: Point $\mathbf{P} = (1,1,1)$ is projected to $\mathbf{P}^* = (0,0,0)$ since

$$(1,1,1,1)\begin{pmatrix} s & 0 & 0 & sr \\ 0 & 1 & 0 & 0 \\ -s & 0 & 0 & sr \\ 0 & -1 & 0 & -2rs \end{pmatrix} = (0,0,0,0).$$

Example 2: Point $\mathbf{P} = (2k, 0, 2k)$ is projected to $\mathbf{P}^* = (0, -1/\sqrt{2}(2-r), 0)$ since

$$(2k, 0, 2k, 1)\begin{pmatrix} s & 0 & 0 & sr \\ 0 & 1 & 0 & 0 \\ -s & 0 & 0 & sr \\ 0 & -1 & 0 & -2rs \end{pmatrix} = (0, -1, 0, 2s(2-r)).$$

▸ **Exercise 3.49:** The product

$$(0,0,0,1)\begin{pmatrix} s & 0 & 0 & sr \\ 0 & 1 & 0 & 0 \\ -s & 0 & 0 & sr \\ 0 & -1 & 0 & -2rs \end{pmatrix}$$

equals $(0, -1, 0, -2sr)$, which suggests that the origin $(0,0,0)$ is projected on the screen at point $\mathbf{P}^* = (0, k/\sqrt{2}, 0)$. This, however, does not make sense since point $(0,0,0)$ was originally "behind" the observer and should remain behind it after all the transformations. What's the explanation?

Mighty is geometry; joined with art, resistless.
— Euripides.

Here is another example. An observer is located at $\mathbf{B} = (-k\sin\theta, 0, -k\cos\theta) = (-k\alpha, 0, -k\beta)$ and is looking in direction $\mathbf{D} = (\alpha, 0, \beta)$ (i.e., toward the origin). Matrices $\mathbf{T}_1$, $\mathbf{T}_2$, $\mathbf{T}_3$, and $\mathbf{T}_p$ below are similar to the ones from the previous example. The result is

$$\mathbf{T} = \mathbf{T}_1 \mathbf{T}_2 \mathbf{T}_3 \mathbf{T}_p$$

$$= \begin{pmatrix} 1 & 0 & 0 & 0 \\ 0 & 1 & 0 & 0 \\ 0 & 0 & 1 & 0 \\ k\alpha & 0 & k\beta & 1 \end{pmatrix}\begin{pmatrix} \beta & 0 & \alpha & 0 \\ 0 & 1 & 0 & 0 \\ -\alpha & 0 & \beta & 0 \\ 0 & 0 & 0 & 1 \end{pmatrix}\begin{pmatrix} 1 & 0 & 0 & 0 \\ 0 & 1 & 0 & 0 \\ 0 & 0 & 1 & 0 \\ 0 & 0 & -k & 1 \end{pmatrix}\begin{pmatrix} 1 & 0 & 0 & 0 \\ 0 & 1 & 0 & 0 \\ 0 & 0 & 0 & r \\ 0 & 0 & 0 & 1 \end{pmatrix}$$

$$= \begin{pmatrix} \beta & 0 & \alpha & 0 \\ 0 & 1 & 0 & 0 \\ -\alpha & 0 & \beta & 0 \\ 0 & 0 & 0 & 1 \end{pmatrix}\begin{pmatrix} 1 & 0 & 0 & 0 \\ 0 & 1 & 0 & 0 \\ 0 & 0 & 0 & r \\ 0 & 0 & 0 & 1 \end{pmatrix}$$

$$= \begin{pmatrix} \beta & 0 & 0 & \alpha r \\ 0 & 1 & 0 & 0 \\ -\alpha & 0 & 0 & \beta r \\ 0 & 0 & 0 & 1 \end{pmatrix}. \tag{3.43}$$

It is easy to see that for $\theta = 0$ (where $\alpha = 0$ and $\beta = 1$), matrix (3.43) reduces to matrix (3.39).

▶ **Exercise 3.50:** Assuming an observer positioned as in the above example, calculate the projection of point $\mathbf{P} = (\beta l, m, -\alpha l)$.

▶ **Exercise 3.51:** Projection matrices (3.43) and (3.41) correspond to the same geometry, so one would think that they should be identical. Why are they different?

We now develop this approach for the general case where an observer is located at an arbitrary point $\mathbf{B} = (a, b, c)$ looking in an arbitrary direction $\mathbf{D} = (d, e, f)$, where vector $\mathbf{D}$ is assumed to be normalized, i.e., $d^2 + e^2 + f^2 = 1$. Translating the observer to the origin is done, as usual, by matrix $\mathbf{T}_1$:

$$\mathbf{T}_1 = \begin{pmatrix} 1 & 0 & 0 & 0 \\ 0 & 1 & 0 & 0 \\ 0 & 0 & 1 & 0 \\ -a & -b & -c & 1 \end{pmatrix}.$$

The main task is to rotate vector $\mathbf{D}$ so it coincides with the positive z direction. The rotation should be done about an axis that's perpendicular to both $\mathbf{D}$ and the z axis. A general vector in this direction is obtained by the cross-product

$$\mathbf{D} \times (0, 0, 1) = (d, e, f) \times (0, 0, 1) = (e, -d, 0).$$

Normalizing this vector yields

$$\mathbf{u} = \frac{(e, -d, 0)}{\sqrt{e^2 + d^2}} = \left(\frac{e}{\sqrt{1 - f^2}}, \frac{-d}{\sqrt{1 - f^2}}, 0 \right).$$

Vector $\mathbf{u}$ is a unit vector in the direction of rotation. The rotation angle θ is the angle between vectors $\mathbf{D}$ and $z = (0, 0, 1)$. Since both are unit vectors, we get $\cos \theta = \mathbf{D} \bullet (0, 0, 1) = f$ and $\sin \theta = \sqrt{1 - \cos^2 \theta} = \sqrt{1 - f^2}$.

The rotation matrix is obtained from Equation (3.27)

$$\mathbf{T}_2 = \begin{pmatrix} \frac{e^2 + f - f^3 - e^2 f}{1 - f^2} & \frac{-ed}{1+f} & d & 0 \\ \frac{-ed}{1+f} & \frac{d^2 + f - f^3 - d^2 f}{1 - f^2} & e & 0 \\ -d & -e & f & 0 \\ 0 & 0 & 0 & 1 \end{pmatrix}.$$

The other two tasks are to translate the observer from the origin to point $(0, 0, -k)$ by means of $\mathbf{T}_3$ and to use matrix $\mathbf{T}_p$ to project from the standard

3 Transformations and Projections

position,

$$T_3 = \begin{pmatrix} 1 & 0 & 0 & 0 \\ 0 & 1 & 0 & 0 \\ 0 & 0 & 1 & 0 \\ 0 & 0 & -k & 1 \end{pmatrix}, \quad T_p = \begin{pmatrix} 1 & 0 & 0 & 0 \\ 0 & 1 & 0 & 0 \\ 0 & 0 & 0 & r \\ 0 & 0 & 0 & 1 \end{pmatrix}.$$

The result is the matrix product

$$\mathbf{T}_g = \mathbf{T}_1 \mathbf{T}_2 \mathbf{T}_3 \mathbf{T}_p \tag{3.44}$$

$$= \begin{pmatrix} \frac{e^2+f+f^2}{1+f} & \frac{-de}{1+f} & 0 & dr \\ \frac{-de}{1+f} & \frac{d^2+f+f^2}{1+f} & 0 & er \\ -d & -e & 0 & fr \\ \frac{cd+bde-ae^2-af+cdf-af^2}{1+f} & \frac{-bd^2+ce+ade-bf+cef-bf^2}{1+f} & 0 & -(ad+be+cf)r \end{pmatrix}.$$

For the special case of an observer located at $\mathbf{B} = (-k\sin\theta, 0, -k\cos\theta) = (-k\alpha, 0, -k\beta)$ and looking in direction $\mathbf{D} = (\alpha, 0, \beta)$ this reduces to matrix (3.43).

▸ **Exercise 3.52:** Calculate matrix (3.44) twice. First for the case where $\mathbf{D} = (0,0,1)$ (observer looking in the positive z direction), then for $\mathbf{D} = (0,0,1)$ and $\mathbf{B} = (0,0,-k)$ (the standard position).

▸ **Exercise 3.53:** Assuming an observer at point $\mathbf{B} = (0,1,0)$ looking in direction $\mathbf{D} = (0, 1/\sqrt{2}, 1/\sqrt{2})$, calculate the projection of point $\mathbf{P} = (0, 1, 10)$.

Matrix $\mathbf{T}_g$ of Equation (3.44) contains the expression $1+f$ in the denominators of certain elements, which may cause undefined values when $f = -1$. Since we assume that vector $\mathbf{D}$ is normalized, $d^2 + e^2 + f^2$ must be equal to 1, so $f = -1$ implies $d = e = 0$, which, in turn, implies $\mathbf{D} = (0, 0, -1)$, i.e., an observer looking in the negative z direction. It turns out that $\mathbf{T}_g$ can be used even in this case. When $d = e = 0$, we can write

$$T_g[1,1] = \frac{e^2+f+f^2}{1+f} = \frac{f(1+f)}{1+f} = f = -1,$$

$$T_g[2,2] = \frac{d^2+f+f^2}{1+f} = \frac{f(1+f)}{1+f} = f = -1.$$

$$T_g[4,1] = \frac{cd+bde-ae^2-af+cdf-af^2}{1+f} = -af\frac{1+f}{1+f} = a,$$

$$T_g[4,2] = \frac{-bd^2+ce+ade-bf+cef-bf^2}{1+f} = -bf\frac{1+f}{1+f} = b.$$

Matrix elements $T_g[1,2] = T_g[2,1] = -de/(1+f)$ have the indefinite form 0/0, but we set them to zero. Matrix $\mathbf{T}_g$ becomes

$$\mathbf{T}_g = \begin{pmatrix} -1 & 0 & 0 & 0 \\ 0 & -1 & 0 & 0 \\ 0 & 0 & 0 & -r \\ a & b & 0 & cr \end{pmatrix}.$$

This is a matrix that transforms a point $\mathbf{P} = (x, y, z, 1)$ to point

$$\mathbf{P}^* = \left(\frac{-x+a}{(c-z)r}, \frac{-y+b}{(c-z)r}, 0\right).$$

Following are two quick tests of this matrix:

1. When the observer $\mathbf{B}$ is located at the standard location $(0, 0, -k)$, the above matrix transforms an arbitrary point $\mathbf{P} = (x, y, z)$ to the point

$$\mathbf{P}^* = \left(\frac{x}{(k+z)r}, \frac{y}{(k+z)r}, 0\right) = \left(\frac{x}{1+z/k}, \frac{y}{1+z/k}, 0\right),$$

which is the familiar Equation (3.38).

2. When the observer $\mathbf{B}$ is located at $(1, 1, 1)$, point $(x, y, z) = (1, 1, -1)$ is transformed to

$$\left(\frac{1-1}{(1+1)r}, \frac{1-1}{(1+1)r}, 0\right) = (0, 0, 0).$$

The reader should visualize this situation with the help of a diagram to see why the result is correct.

3.9.4 n-Point Perspective

The purpose of this section is to clear up the meaning of the term *n-point perspective* and to show the difference between it and the number of vanishing points actually seen by an observer. We start by discussing the number of vanishing points an object may have.

Figure 3.39a shows a cube head-on. It is easy to see that it features one vanishing point. Figure 3.39b shows the same cube with two vanishing points, after it has been rotated. In Figure 3.39c, the same cube is seen having three vanishing points because we assume that the observer is located very close to it (it's like being very close to a skyscraper and looking up). The same cube features one, two, or three vanishing points depending on how an observer looks at it. In general, an object may have *any number* of vanishing points. It turns out that the concept of vanishing points has to do with the shape of the object and the direction it is being observed from. A sphere, for example, is curved and smooth. There are no straight lines on its surface and, therefore, no vanishing points. The same thing is true for any smooth blob.

In order to feature vanishing points, an object must have groups of straight lines on it and the lines in each group must be parallel. The lines may be created by the intersection of two planes on the object, as in the case of a cube, or they may be painted or scribed on the surface of the object. Any such group of lines creates a vanishing point, except if the lines are perpendicular to the direction of observation. This is why rotating the cube has added a vanishing point. There is a group of parallel lines on the cube that were perpendicular to the direction of observation before the rotation but not after the rotation. This group appears to converge to a new vanishing point after the rotation. The conclusion is that an object may have any number of vanishing points depending on its shape, on lines

3 Transformations and Projections

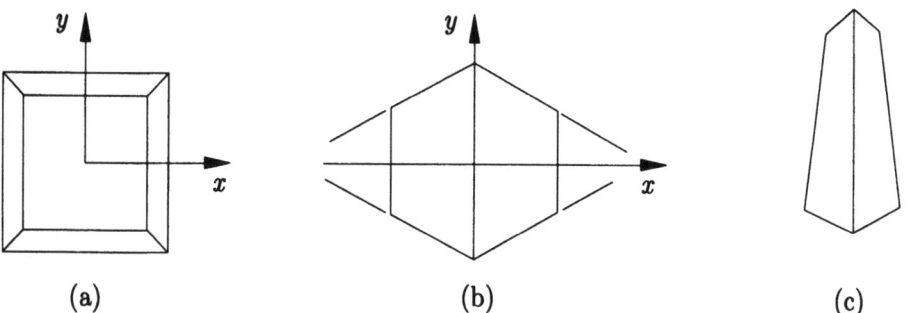

Figure 3.39: n-Point Perspectives.

that happen to be on its surface, and on the direction it is viewed from. Figure 3.40 shows a familiar object, a circular staircase, that features many vanishing points.

▶ **Exercise 3.54:** How many vanishing points?

The well-known drawing *Relativity* by M. C. Escher features three vanishing points (see discussion in [Ernst 76]).

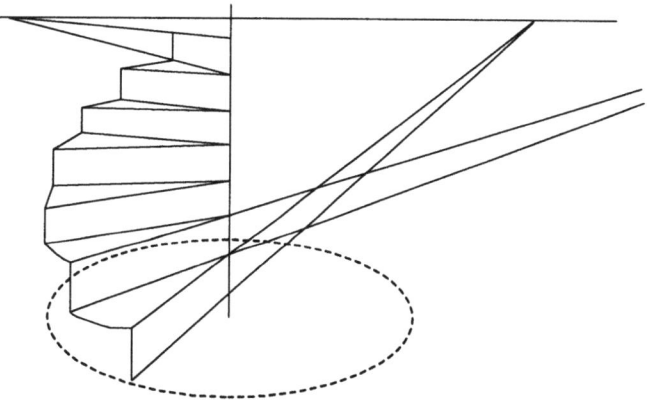

Figure 3.40: Many Vanishing Points.

Now for the term *n-point perspective*. Matrix (3.43) has two nonzero elements αr and βr in its fourth column. A similar situation is true for matrix (3.42). The case where two of the three quantities p, q, and r are nonzero is called a *two-point perspective*. It has nothing to do with the number of vanishing points actually seen. Instead, it has to do with the orientation of the projection plane relative to the coordinate system. A two-point perspective means that the projection plane intercepts two of the three coordinate axes. Which two axes depends on which two of the three numbers are nonzero. A three-point perspective (as in Equation (Ans.3)) means that the projection plane intercepts all three coordinate axes. There are no zero-point or four-point perspectives.

In summary, an object may appear to have any number of vanishing points, depending on its shape, orientation, and lines on its surface. The term *n-point perspective*, however, indicates how many of the quantities p, q, and r in the current transformation matrix are nonzero. It also tells how many of the three coordinate axes are intercepted by the projection plane.

▶ **Exercise 3.55:** Consider the simple transformation of *scaling*. The transformation matrix is

$$\begin{pmatrix} T_1 & 0 & 0 & 0 \\ 0 & T_2 & 0 & 0 \\ 0 & 0 & T_3 & 0 \\ 0 & 0 & 0 & 1 \end{pmatrix}.$$

When combined with perspective projection, it yields

$$\begin{pmatrix} T_1 & 0 & 0 & 0 \\ 0 & T_2 & 0 & 0 \\ 0 & 0 & T_3 & 0 \\ 0 & 0 & 0 & 1 \end{pmatrix} \begin{pmatrix} 1 & 0 & 0 & 0 \\ 0 & 1 & 0 & 0 \\ 0 & 0 & 0 & r \\ 0 & 0 & 0 & 1 \end{pmatrix} = \begin{pmatrix} T_1 & 0 & 0 & 0 \\ 0 & T_2 & 0 & 0 \\ 0 & 0 & 0 & T_3 r \\ 0 & 0 & 0 & 1 \end{pmatrix}.$$

Hence, a point $(x, y, z, 1)$ is transformed to $(T_1 x, T_2 y, 0, T_3 rz + 1)$, which implies

$$x^* = \frac{T_1 x}{T_3 rz + 1}, \qquad y^* = \frac{T_2 y}{T_3 rz + 1}.$$

In the special case of uniform scaling $T_1 = T_2 = T_3 = T$, we get $x^* = x/(rz + 1/T)$, $y^* = y/(rz + 1/T)$. The problem is that when T gets large (large magnification), $1/T$ becomes small, resulting in

$$x^* \approx \frac{x}{rz} = \frac{xk}{z}, \qquad y^* \approx \frac{y}{rz} = \frac{yk}{z}.$$

We don't seem to get the expected magnification. What's the explanation?

3.9.5 General Perspective: IV

This approach uses vector analysis to calculate perspective projection in the general case. Readers unfamiliar with vector analysis should first read Section A.4. Figure 3.41a shows an observer at point **B**, looking in an arbitrary direction **a**. The screen is, of course, perpendicular to the line of sight **a** and we assume that it is located at a distance of |**a**| from the observer. The center of the screen is at point **C**. Note that vector **a** gives both the direction of view of the observer and the distance between the observer and the screen.

We select an arbitrary point **P** on the object and connect it with the observer. The intersection of line **BP** and the screen is, of course, the projected point **P***. Vector **b** indicates the position of the observer. Vector **c** indicates the direction **CP*** on the screen. Vector **d** is the position vector of point **P***. Vector **e** connects **B** to **P**. Vector **p** points from the origin to point **P**. Vector **d** − **b** connects point **B** to point **P***.

3 Transformations and Projections

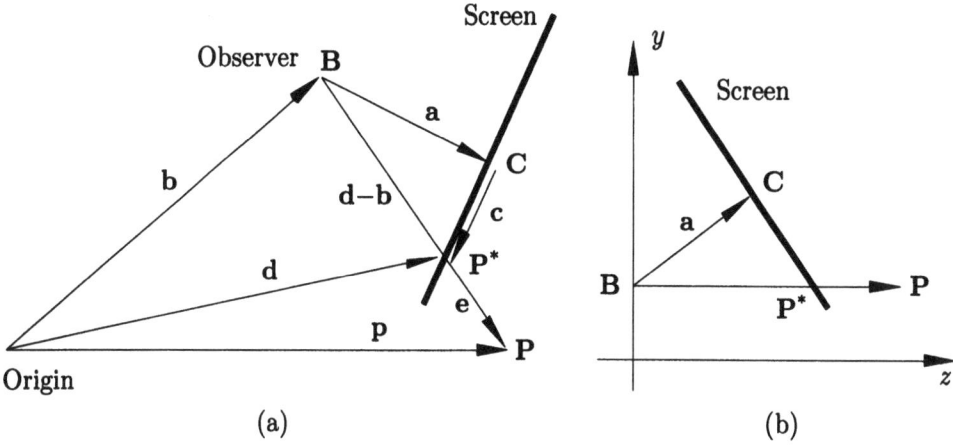

Figure 3.41: (a) General Perspective with Vectors. (b) Example.

Vector **p** is the sum $\mathbf{p} = \mathbf{b} + \mathbf{e}$, which implies $\mathbf{e} = \mathbf{p} - \mathbf{b}$. From $\mathbf{d} = \mathbf{b} + \mathbf{a} + \mathbf{c}$, we get $\mathbf{c} = \mathbf{d} - \mathbf{b} - \mathbf{a}$. Vector $\mathbf{d} - \mathbf{b}$ is in the direction of **e**, so we can write $\mathbf{d} - \mathbf{b} = \alpha \mathbf{e} = \alpha(\mathbf{p} - \mathbf{b})$, where α is a real number. This implies $\mathbf{c} = \alpha(\mathbf{p} - \mathbf{b}) - \mathbf{a}$ or

$$\mathbf{d} = \mathbf{b} + \mathbf{a} + \mathbf{c} = \mathbf{b} + \alpha(\mathbf{p} - \mathbf{b}). \tag{3.45}$$

Since the line of sight is perpendicular to the screen, we get $\mathbf{a} \bullet \mathbf{c} = 0$, which implies $\mathbf{a} \bullet [\alpha(\mathbf{p} - \mathbf{b}) - \mathbf{a}] = 0$, or $\alpha \mathbf{a} \bullet (\mathbf{p} - \mathbf{b}) = \mathbf{a} \bullet \mathbf{a}$, or

$$\alpha = \frac{|\mathbf{a}|^2}{\mathbf{a} \bullet (\mathbf{p} - \mathbf{b})}. \tag{3.46}$$

Before we continue with the analysis, the following cases should be discussed:

1. α is positive. This is the normal case. It means that the observer and point **P** are on different sides of the screen and it is meaningful to project the point on the screen.

2. α is zero. This implies a vector **a** of magnitude zero, or an observer positioned at the screen. Either the observer or the screen should be moved before anything can be meaningfully displayed.

3. α is negative. This implies that **P** and the observer are on the same side of the screen, so **P** should not be projected.

4. α is undefined. This occurs when $\mathbf{a} \bullet (\mathbf{p} - \mathbf{b}) = 0$, meaning that **a** is perpendicular to $\mathbf{p} - \mathbf{b}$ and, therefore, to **e**. Vector **e** (the line of sight) is therefore parallel to the screen, so point **P** should not be projected.

After calculating α we can proceed in one of two ways: (1) we can use Equation (3.45) to calculate vector **d**, which points directly to **P*** on the screen, or (2) we can calculate the screen coordinates of vector **c**. In the latter case, we consider the center of the screen (point **C**) our origin and we define two unit vectors **u** and **w** to serve as local axes on the screen. The screen coordinates of **c** are, in this case, the projections $\mathbf{u} \bullet \mathbf{c}$ and $\mathbf{w} \bullet \mathbf{c}$ of **c** on these axes.

To calculate **u** and **w**, we recall that they should be on the screen (and therefore perpendicular to **a**) and also perpendicular to each other. We can therefore write $\mathbf{a} \bullet \mathbf{u} = \mathbf{a} \bullet \mathbf{w} = \mathbf{u} \bullet \mathbf{w} = 0$. It also makes sense to require that **u** be in the xy plane (which will cause **w** to point in the z direction as much as possible). Solving these equations results in

$$\mathbf{u} = (a_y, -a_x, 0) \quad \text{and} \quad \mathbf{w} = (a_x a_z, a_y a_z, -a_x^2 - a_y^2). \tag{3.47}$$

To convert **u** and **w** to unit vectors, they should further be divided by their absolute values.

Note that **u** and **w** are undefined if **a** points in the z direction (if $\mathbf{a} = (0, 0, a_z)$ then $\mathbf{u} = \mathbf{w} = (0, 0, 0)$, undefined direction). However, in this case the screen is parallel to the xy plane, so we can simply define the local coordinate axes as $\mathbf{u} = (1, 0, 0) = \mathbf{i}$ and $\mathbf{w} = (0, 1, 0) = \mathbf{j}$.

3.9.6 Example 1

To keep this example simple, all the points involved are on the yz plane, i.e., their x coordinates are zero. This example is illustrated in Figure 3.41b.

We assume an observer at $\mathbf{B} = (0, 1, 0)$, looking in direction $(0, 1, 1)$ (i.e., 45° in the yz plane). Vector **a** must point in this direction and we assume $\mathbf{a} = (0, 2, 2)$ (i.e., the center of the screen is at a distance of $|\mathbf{a}| = \sqrt{2^2 + 2^2} = \sqrt{8}$ units from the observer). We further assume that the point **P** to be projected is at $(0, 1, 10)$. The center of the screen (point **C**) is easily seen to be at $\mathbf{b} + \mathbf{a} = (0, 1, 0) + (0, 2, 2) = (0, 3, 2)$.

The first step is to calculate α

$$\alpha = \frac{|\mathbf{a}|^2}{\mathbf{a} \bullet (\mathbf{p} - \mathbf{b})} = \frac{8}{(0, 2, 2) \bullet (0 - 0, 1 - 1, 10 - 0)} = \frac{2}{5}.$$

The next step is to calculate $\mathbf{d} = \mathbf{b} + \alpha(\mathbf{p} - \mathbf{b}) = (0, 1, 0) + (2/5)(0, 0, 10) = (0, 1, 4)$. The projected point $\mathbf{P}^*$ is therefore at $(0, 1, 4)$. (See the diagram to convince yourself that the precise value of the z coordinate of **P** is unimportant in this case.)

Next, we calculate the local coordinates of this point on the screen. Vector **c** is first calculated by $\mathbf{c} = \alpha(\mathbf{p} - \mathbf{b}) - \mathbf{a} = (2/5)(0, 0, 10) - (0, 2, 2) = (0, -2, 2)$. The local axes on the screen are calculated next from Equation (3.47). They are $\mathbf{u} = (2, 0, 0)$ and $\mathbf{w} = (0, 4, -4)$. To convert them into unit vectors, we divide each by its magnitude, producing $\mathbf{u} = (1, 0, 0)$ and $\mathbf{w} = (0, 1/\sqrt{2}, -1/\sqrt{2})$. (Note that **u** is in the x direction and **w** is in the yz plane.)

The screen coordinates of **c** are thus $\mathbf{u} \bullet \mathbf{c} = (1, 0, 0) \bullet (0, -2, 2) = 0$ and $\mathbf{w} \bullet \mathbf{c} = (0, 1/\sqrt{2}, -1/\sqrt{2}) \bullet (0, -2, 2) = -\sqrt{8}$. The projected point is therefore $\sqrt{8}$ units away from the center of the screen **C**. Note that this equals the absolute value of vector **c**. (This should be compared with Exercise 3.53.)

As an added bonus, let's calculate the equation of the screen. Let (x, y, z) be a general point on the screen. The vector from the center (point **C**) to (x, y, z) is $(x - 0, y - 3, z - 2)$. This vector must be perpendicular to the normal to the screen

3 Transformations and Projections

(vector **a**), resulting in

$$0 = \mathbf{a} \bullet (x, y-3, z-2) = (0,2,2) \bullet (x, y-3, z-2), \quad \text{or} \quad y+z=5.$$

This equation relates the y and z coordinates of all the points on the screen. Any point with coordinates $(x, y, 5-y)$ is therefore on the screen regardless of the value of x. Note that the projected point $\mathbf{P}^*$ also satisfies this relation.

▶ **Exercise 3.56:** Generalize the previous example for the case of a general point $\mathbf{P} = (x, y, z)$.

3.9.7 Example 2

This is, again, a simple example, illustrated in Figure 3.42. The screen is centered on the origin at a 45° angle and the observer is at point $(-k/\sqrt{2}, 0, -k/\sqrt{2})$, a distance of k units from the screen. To simplify the notation, we introduce the quantity $\psi = k/\sqrt{2}$. From Figure 3.41a it is clear that $\mathbf{a} = (\psi, 0, \psi)$ and $\mathbf{b} = -\mathbf{a} = (-\psi, 0, -\psi)$. The center of the screen is, as always, at $\mathbf{a}+\mathbf{b}$, which is point $(0,0,0)$.

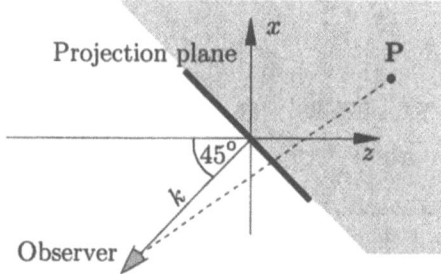

Figure 3.42: Observer Rotated About the y Axis.

The first step is to calculate α:

$$\alpha = \frac{|\mathbf{a}|^2}{\mathbf{a} \bullet (\mathbf{P} - \mathbf{b})}$$
$$= \frac{2\psi^2}{(\psi, 0, \psi) \bullet (x+\psi, y, z+\psi)}$$
$$= \frac{2\psi}{x+z+2\psi}.$$

(Try to convince yourself that α is positive in the gray area above and to the right of the screen because $x+z+2\psi$ is positive in this area.)

The next step is to calculate vector $\mathbf{d}$:

$$\mathbf{d} = \mathbf{b} + \alpha(\mathbf{P} - \mathbf{b})$$
$$= (-\psi, 0, -\psi) + \frac{2\psi}{x+z+2\psi}(x+\psi, y, z+\psi)$$
$$= \frac{\psi}{x+z+2\psi}(x-z, y, z-x).$$

Notice that $\mathbf{P} = (0,0,0)$ is transformed to $\mathbf{P}^* = (0,0,0)$. Also, every point $\mathbf{P} = (x, 0, -x)$ is transformed to $\mathbf{P}^* = (0,0,0)$.

Since the screen is centered at the origin, we have $\mathbf{c} = \alpha(\mathbf{P}-\mathbf{b})-\mathbf{a} = \alpha(\mathbf{P}-\mathbf{b})+\mathbf{b} = \mathbf{d}$. The next step is to calculate the local screen vectors $\mathbf{u}$ and $\mathbf{w}$ from Equation (3.47). This is straightforward and results in $\mathbf{u} = (0, -\psi, 0)$ and $\mathbf{w} = (\psi, 0, -\psi)$. After normalization, these become $\mathbf{u} = (0, -1, 0)$ and $\mathbf{w} = (1/\sqrt{2}, 0, -1/\sqrt{2})$. Notice that $\mathbf{u}$ is the y axis and $\mathbf{w}$ is in the xz plane.

The screen equation is obtained from $\mathbf{a} \bullet (x, y, z) = 0$, which implies $\psi(x+z) = 0$ or $x = -z$. The last step is to calculate the transformation matrix. From

$$x^* = \frac{X}{H} = \frac{\psi(x-z)}{x+z+2\psi}, \quad y^* = \frac{Y}{H} = \frac{\psi y}{x+z+2\psi}, \quad z^* = \frac{Z}{H} = \frac{\psi(z-x)}{x+z+2\psi},$$

we get

$$(X, Y, Z, H) = (x, y, z, 1) \begin{pmatrix} \psi & 0 & -\psi & 1 \\ 0 & \psi & 0 & 0 \\ -\psi & 0 & \psi & 1 \\ 0 & 0 & 0 & 2\psi \end{pmatrix}.$$

(Notice the two 1's in the last column. They indicate that the projection plane intercepts the x and z axes, but not the y axis. This is a two-point perspective.)

3.9.8 General Perspective: V

This approach to the problem of perspective projection also uses vector analysis, but we assume that the following is given (Figure 3.43):

1. The position of the observer (vector $\mathbf{b}$).
2. The direction and distance from the observer to the projection plane (vector $\mathbf{a}$).
3. An "up" vector $\mathbf{Z}$. This will determine the direction of the local screen vector $\mathbf{w}$.
4. Two viewing half-angles h and v. This approach is handy when we want to limit the projected image to certain viewing angles.

We proceed in the following simple steps:

1. Calculate vector $\mathbf{U}$ as perpendicular to both $\mathbf{a}$ and $\mathbf{Z}$. $\mathbf{U} = \mathbf{a} \times \mathbf{Z}$.
2. Calculate vector $\mathbf{W}$ as perpendicular to both $\mathbf{U}$ and $\mathbf{a}$. $\mathbf{W} = \mathbf{U} \times \mathbf{a}$. Vector $\mathbf{W}$ is in the $Z\mathbf{a}$ plane and is perpendicular to $\mathbf{a}$. It will serve to calculate vector $\mathbf{w}$ on the screen in step 4.
3. Calculate $\mathbf{C} = \mathbf{b} + \mathbf{a}$. This is the center of the screen.
4. Calculate the screen vectors $\mathbf{u}$ and $\mathbf{w}$. They are in the directions of $\mathbf{U}$ and $\mathbf{W}$, respectively, but their sizes are determined by the viewing angles

$$\mathbf{u} = \frac{\mathbf{U}}{|\mathbf{U}|}|\mathbf{a}| \tan h, \quad \mathbf{w} = \frac{\mathbf{W}}{|\mathbf{W}|}|\mathbf{a}| \tan v.$$

5. Calculate $\alpha = \frac{|\mathbf{a}|^2}{\mathbf{a}\bullet(\mathbf{P}-\mathbf{b})}$ and vectors $\mathbf{d} = \mathbf{b} + \alpha(\mathbf{P}-\mathbf{b})$ and $\mathbf{c} = \alpha(\mathbf{P}-\mathbf{b}) - \mathbf{a}$ in the usual way.

3 Transformations and Projections

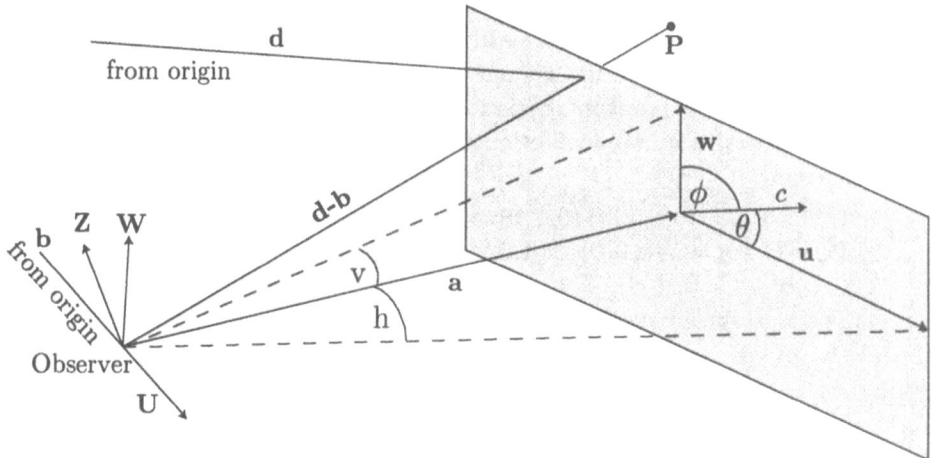

Figure 3.43: A Viewing Geometry.

6. Now that c is known, we use it to calculate the two scale factors c_x and c_y:

$$c_x = \frac{|\mathbf{c}|\cos\theta}{|\mathbf{u}|} = \frac{1}{|\mathbf{u}|^2}(\mathbf{c}\bullet\mathbf{u}), \qquad c_y = \frac{|\mathbf{c}|\cos\phi}{|\mathbf{w}|} = \frac{1}{|\mathbf{w}|^2}(\mathbf{c}\bullet\mathbf{w}).$$

These are numbers in the range $[-1, 1]$. Any point $\mathbf{P} = (x, y, z)$ for which either c_x or c_y are greater than 1 or less than -1 is therefore outside the screen and should not be displayed.

The range of values of c_x and c_y assumes that the origin of the screen is at its center. The actual screen coordinates (s_x, s_y) of a pixel depend on the dimensions of the screen (measured in pixels). They are given by

$$s_x = \text{(half the screen width)} \times c_x, \quad s_y = \text{(half the screen height)} \times c_y.$$

If the origin is at the bottom-left corner, then

$$s_x = \text{(half the screen width)} + \text{(half the screen width)} \times c_x,$$
$$s_y = \text{(half the screen height)} + \text{(half the screen height)} \times c_y.$$

If it is at the top-left corner,

$$s_x = \text{(half the screen width)} + \text{(half the screen width)} \times c_x,$$
$$s_y = \text{(half the screen height)} - \text{(half the screen height)} \times c_y.$$

Example: We apply the method above to the standard case depicted in Figure 3.32, where the screen is part of the xy plane and is centered on the origin and the observer is located k units from the origin on the negative z axis. Assuming that the two half-angles h and v are given, we need to calculate scale factors c_x

and c_y that will make it possible to determine for any given point **P** whether its projection on the xy plane is inside or outside the screen.

It is clear that $\mathbf{b} = (0,0,-k)$ and $\mathbf{a} = (0,0,k) = -\mathbf{b}$. We also select the positive y direction as our "up" direction, so $\mathbf{Z} = (0,1,0)$. To express the final results in a general way, we denote $m = \tan h$ and $n = \tan v$. The calculation is straightforward.

1. $\mathbf{U} = \mathbf{a} \times \mathbf{Z} = (0,0,k) \times (0,1,0) = (-k,0,0)$.
2. $\mathbf{W} = \mathbf{U} \times \mathbf{a} = (-k,0,0) \times (0,0,k) = (0,k^2,0)$.
3. $\mathbf{C} = \mathbf{b} + \mathbf{a} = (0,0,0)$. The center of the screen is at the origin.
4. The local screen axes are

$$\mathbf{u} = \frac{\mathbf{U}}{|\mathbf{U}|}|\mathbf{a}|\tan h = (-km, 0, 0), \quad \mathbf{w} = \frac{\mathbf{W}}{|\mathbf{W}|}|\mathbf{a}|\tan v = (0, kn, 0).$$

5. The three quantities α, **d**, and **c** are calculated next:

$$\alpha = \frac{|\mathbf{a}|^2}{\mathbf{a} \bullet (\mathbf{P} - \mathbf{b})} = \frac{k^2}{(0,0,k) \bullet (x,y,z+k)} = \frac{k}{z+k},$$

$$\mathbf{d} = \mathbf{b} + \alpha(\mathbf{P} - \mathbf{b}) = (0,0,k) + \frac{k}{z+k}(x,y,z+k) = \frac{k}{z+k}(x,y,0),$$

$$\mathbf{c} = \alpha(\mathbf{P} - \mathbf{b}) - \mathbf{a} = \alpha(\mathbf{P} - \mathbf{b}) + \mathbf{b} = \mathbf{d}.$$

6. The scale factors c_x and c_y can now be calculated

$$c_x = \frac{\mathbf{c} \bullet \mathbf{u}}{|\mathbf{u}|^2} = \frac{\frac{k}{z+k}(-xkm)}{k^2m^2} = \frac{-x}{m(z+k)},$$

$$c_y = \frac{\mathbf{c} \bullet \mathbf{w}}{|\mathbf{w}|^2} = \frac{\frac{k}{z+k}(ykn)}{k^2n^2} = \frac{y}{n(z+k)}.$$
(3.48)

As a simple application of these results, let's select $h = v = 45°$, which implies $m = n = 1$. Let's also assume screen dimensions of 100×100 pixels, a local origin at the center of the screen, and $k = 1$. For point $\mathbf{P} = (1,2,1)$, we get the scale factors

$$c_x = \frac{-x}{m(z+k)} = \frac{-1}{1+1} = -0.5, \quad c_y = \frac{y}{n(z+k)} = \frac{2}{1+1} = 1.$$

The screen coordinates are thus $s_x = 50 \times (-0.5) = -25$ and $s_y = 1 \times 50 = 50$ (the top of the screen). However, any point with coordinates $(1, y, 1)$ where $y > 2$ would produce a scale factor $c_y > 1$, implying that its projection is outside the screen.

▶ **Exercise 3.57:** Why is Equation (3.48) asymmetric with respect to x and y (i.e., why $-x$ and not $-y$)?

3 Transformations and Projections

3.9.9 Perspective Depth

The perspective projection converts a three-dimensional point to a two-dimensional one. It completely erases any information about the depth (the z coordinate) of the original point. However, certain algorithms for hidden surface removal need precisely such information. We therefore need to generalize our perspective projection to create a third coordinate z^*, with information about the original z coordinate of the projected point. The most obvious definition is $z^* = z$, but this definition does not preserve straight lines.

Imagine two three-dimensional points $\mathbf{P}_1 = (x_1, y_1, z_1)$ and $\mathbf{P}_2 = (x_2, y_2, z_2)$ projected to the points

$$\mathbf{P}_1^* = \left(\frac{x_1 k}{k+z_1}, \frac{y_1 k}{k+z_1}, z_1\right) \quad \text{and} \quad \mathbf{P}_2^* = \left(\frac{x_2 k}{k+z_2}, \frac{y_2 k}{k+z_2}, z_2\right).$$

Note that the two projected points are not necessarily on the projection plane. We say that they are located in the *image space*.

The line $\mathbf{P}(t) = \mathbf{P}_1 + (\mathbf{P}_2 - \mathbf{P}_1)t$ (Equation (A.12), page 704) connects the two original points. The line $\mathbf{P}^*(u) = \mathbf{P}_1^* + (\mathbf{P}_2^* - \mathbf{P}_1^*)u$ connects the two projected ones. It can be shown that an arbitrary point $\mathbf{P}(t_0)$ on this line is projected to a point that's not on $\mathbf{P}^*(u)$.

This is why the perspective depth projection is not defined simply as $z^* = z$, but in a more complex way, as $z^* = z/(k+z)$. This definition preserves depth information, since it satisfies $z_1 > z_2 \Rightarrow z_1^* > z_2^*$. It also preserves straight lines.

▸ **Exercise 3.58:** Prove the above claim.

3.10 Application: Stereo Image

Stereo (from the Greek στερεοσ), —solid, three-dimensional.

The ideal graphics output device should be three-dimensional. Unfortunately, only few such devices are available today and they are expensive and cumbersome. This is why stereo pictures, displayed on a two-dimensional screen or printed on paper, are so interesting. The reason we see real-life objects in three dimensions is that our eyes are separated (by about 70–75 mm) and hence look at the same object from slightly different directions (Figure 3.44a). They see slightly different images, which are "fused" by the brain to create the three-dimensional image.

The principle of stereo images is therefore to display two, slightly different images of the same object and to make sure that each eye sees just one image. This is normally achieved by either (1) displaying the two images in two different colors and watching them through special glasses that allow one color to reach just the left eye and the other, just the right eye or (2) holding a piece of cardboard between

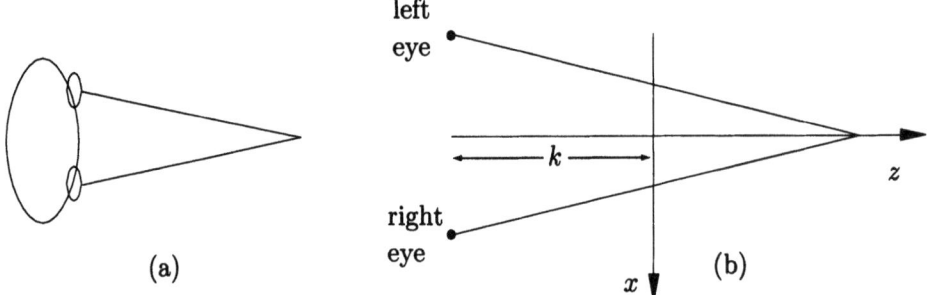

Figure 3.44: Principle of Stereo Images.

the eyes and the display, to partially block the view such that each eye sees only one image.

The simplest way to calculate the two stereo images is to use translation and perspective projection. This is, therefore, a useful application of the concepts described earlier. Figure 3.44b shows each eye as an observer. The left eye is located at $(-e, 0, -k)$ and the right one at $(e, 0, -k)$. To create the image seen by the left eye, we first have to translate it to the origin, then follow with a standard perspective projection. The transformations are

$$\begin{pmatrix} 1 & 0 & 0 & 0 \\ 0 & 1 & 0 & 0 \\ 0 & 0 & 1 & 0 \\ e & 0 & 0 & 1 \end{pmatrix} \begin{pmatrix} 1 & 0 & 0 & 0 \\ 0 & 1 & 0 & 0 \\ 0 & 0 & 0 & r \\ 0 & 0 & 0 & 1 \end{pmatrix} = \begin{pmatrix} 1 & 0 & 0 & 0 \\ 0 & 1 & 0 & 0 \\ 0 & 0 & 0 & r \\ e & 0 & 0 & 1 \end{pmatrix} = \mathbf{T}_{left}.$$

The transformation for the right eye is similarly

$$\begin{pmatrix} 1 & 0 & 0 & 0 \\ 0 & 1 & 0 & 0 \\ 0 & 0 & 0 & r \\ -e & 0 & 0 & 1 \end{pmatrix} = \mathbf{T}_{right}.$$

The stereo pair is created by transforming each point $\mathbf{P}$ on the original image twice, to the two points $\mathbf{P}_{left} = \mathbf{P}\,\mathbf{T}_{left}$ and $\mathbf{P}_{right} = \mathbf{P}\,\mathbf{T}_{right}$. The value selected for e depends on how the picture is to be viewed. For the dual-color method described earlier, $2e$ should equal the distance between real eyes (about 70–75 mm). This is a small value, so there is not much difference between $\mathbf{P}_{right}$ and $\mathbf{P}_{left}$. The two images highly overlap.

For a general point $\mathbf{P} = (x, y, z)$, the projections for both eyes are

$$\mathbf{P}_{left} = (x, y, z, 1)\mathbf{T}_{left} = (x + e, y, 0, zr + 1) \to \left(\frac{x+e}{zr+1}, \frac{y}{zr+1} \right),$$

$$\mathbf{P}_{right} = (x, y, z, 1)\mathbf{T}_{right} = (x - e, y, 0, zr + 1) \to \left(\frac{x-e}{zr+1}, \frac{y}{zr+1} \right).$$

3 Transformations and Projections

This means that the smaller z is (i.e., the closer the point is to the observer), the greater the difference between what the two eyes see. A good way to visualize this is to imagine an object sliding past the observer. The front of the object slides faster than the back, an effect known as *parallax*.

As an example, consider the two points $\mathbf{P} = (5,0,1)$ and $\mathbf{Q} = (5,0,2)$. They differ in their z coordinate only. Assuming that $e = 2$ and $r = 3$, their projections are

$$\mathbf{P}_{left} = \left(\frac{5+2}{3+1}, 0\right) = \left(\frac{7}{4}, 0\right), \quad \mathbf{P}_{right} = \left(\frac{5-2}{3+1}, 0\right) = \left(\frac{3}{4}, 0\right),$$

$$\mathbf{Q}_{left} = \left(\frac{5+2}{2 \cdot 3 + 1}, 0\right) = \left(\frac{7}{7}, 0\right), \quad \mathbf{Q}_{right} = \left(\frac{5-2}{2 \cdot 3 + 1}, 0\right) = \left(\frac{3}{7}, 0\right).$$

The difference between $\mathbf{P}_{left}$ and $\mathbf{P}_{right}$ is $7/4 - 3/4 = 1$, whereas the difference between $\mathbf{Q}_{left}$ and $\mathbf{Q}_{right}$ is only $7/7 - 3/7 = 4/7$.

Figure 3.45 is an example of a stereo pair of a polyline connecting the eight corners of a cube. It was produced by the *Mathematica* code listed in Figure 3.46.

A more sophisticated approach to generating a stereo image is shown in Figure 3.47a. The two eyes are located at $(e, 0, -k)$ and $(-e, 0, -k)$ and they view the general point $\mathbf{P} = (x, y, z)$ from different directions. Point $\mathbf{P}$ is projected twice on the projection plane, at points $\mathbf{P}_L$ and $\mathbf{P}_R$, using the general rule for perspective projections. Assuming that the distance between the eyes is $2e$, Figure 3.47c,d shows how to calculate the x coordinates of points $\mathbf{P}_L$ and $\mathbf{P}_R$, respectively. Using similar triangles, Figure 3.47c yields

$$\frac{x-e}{k+z} = \frac{x_L - w}{k} \quad \text{or} \quad x_L = \frac{x-e}{1+z/k} + e = \frac{x + ez/k}{1 + z/k},$$

and, similarly, from Figure 3.47d we get

$$\frac{x+e}{k+z} = \frac{x_R + w}{k} \quad \text{or} \quad x_R = \frac{x+e}{1+z/k} - e = \frac{x - ez/k}{1 + z/k}.$$

Since both eyes are at $y = 0$, the y^* coordinates of both $\mathbf{P}_L$ and $\mathbf{P}_R$ are given by

$$y^* = \frac{y}{1 + z/k}.$$

We thus obtain the transformation matrices $\mathbf{T}_L$ and $\mathbf{T}_R$ that transform $\mathbf{P}$ to $\mathbf{P}_L$ and $\mathbf{P}_R$,

$$\mathbf{T}_L = \begin{pmatrix} 1 & 0 & 0 & 0 \\ 0 & 1 & 0 & 0 \\ e/k & 0 & 0 & 1/k \\ 0 & 0 & 0 & 1 \end{pmatrix}, \quad \mathbf{T}_R = \begin{pmatrix} 1 & 0 & 0 & 0 \\ 0 & 1 & 0 & 0 \\ -e/k & 0 & 0 & 1/k \\ 0 & 0 & 0 & 1 \end{pmatrix}. \quad (3.49)$$

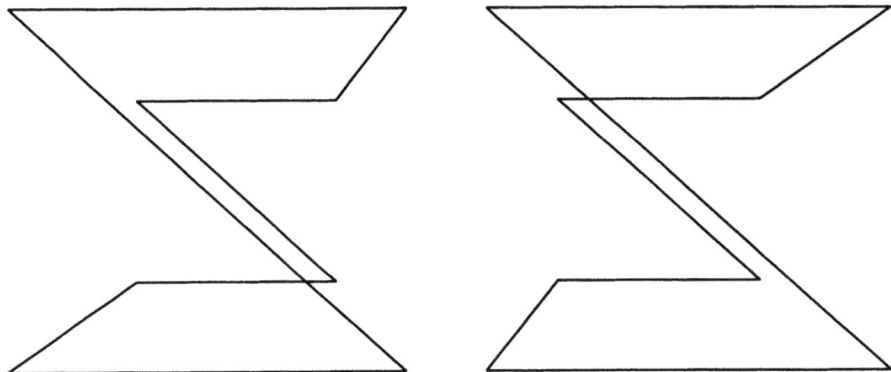

Figure 3.45: Example of a Stereo Image Pair.

```
Tl={{1,0,0,0},{0,1,0,0},{0,0,0,r},{e,0,0,1}};
Tr={{1,0,0,0},{0,1,0,0},{0,0,0,r},{-e,0,0,1}};
pt={{1,1,1,1},{-1,1,1,1},{1,-1,1,1},{-1,-1,1,1},
    {1,1,-1,1},{-1,1,-1,1},{1,-1,-1,1},
    {-1,-1,-1,1},{1,1,1,1}};
e=.1; r=3;
qt=Table[0, {i,9}, {j,4}];
Do[qt[[i]]=pt[[i]].Tr, {i,1,9}];  (* use Tl for other image *)
Do[qt[[i,1]]=qt[[i,1]]/qt[[i,4]], {i,1,9}];
Do[qt[[i,2]]=qt[[i,2]]/qt[[i,4]], {i,1,9}];
ListPlot[Table[{qt[[i,1]],qt[[i,2]]}, {i,1,9}],
 PlotJoined->True, Axes->False]
```

Figure 3.46: *Mathematica* Code for a Stereo Image Pair.

Figure 3.47b shows how to select reasonable values for e and k. In real life, the distance between the eyes is about 75 mm (about 3 in.). Normal reading distance is about 20 in. Using the values 3 and 20, we get $\tan\theta/2 = 1.5/20$, yielding $\theta/2 = 4.29°$ or $\theta = 8.58°$. This is the average stereo angle between the eyes. To get a stereo pair that will look natural and will be free of distortions, we should select values for e and k that should maintain this angle. A natural value for k is 4 in. since this is the focal length of the lenses used by most commercial stereoscopes. If we reduce k from 20 to 4 (a factor of 5), we should reduce e from 3 to $3/5 = 0.6$ to maintain the same stereo angle.

A stereo pair is therefore calculated by substituting $e = 0.6$, $k = 4$ in Equation (3.49) and computing $\mathbf{P}_L = \mathbf{T}_L \cdot \mathbf{P}$ and $\mathbf{P}_R = \mathbf{T}_R \cdot \mathbf{P}$ for every point $\mathbf{P}$ of the object.

▶ **Exercise 3.59:** What would be good values for e and k assuming a distance of 2.5 in. between the eyes?

The discussion above also shows how to build a stereoscope to view a stereo pair. In a piece of cardboard, cut two circular holes with a diameter of about

3 Transformations and Projections

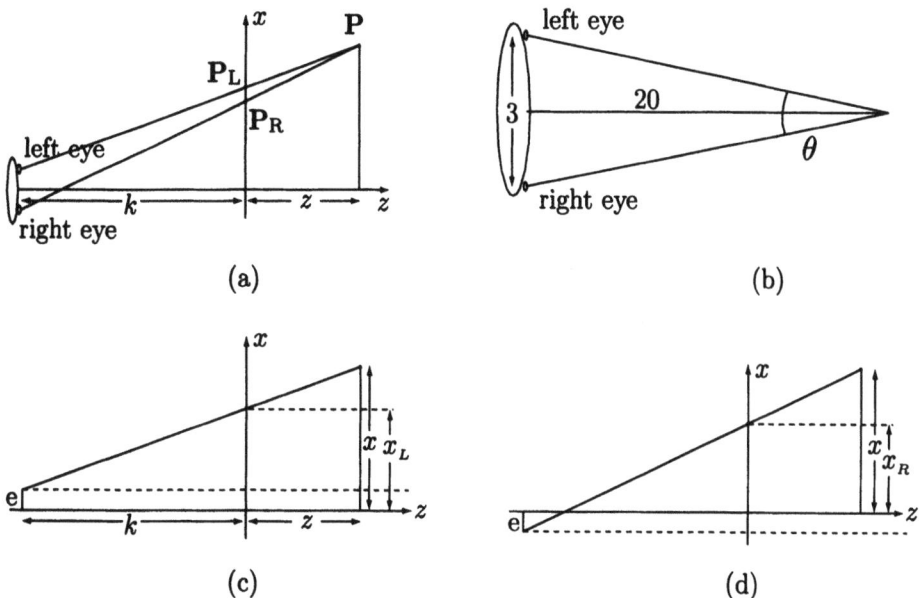

Figure 3.47: Perspective Projection of a Stereo Pair.

1.5 in. each and with about 3 in. separation between their centers. Place a lens with a focal length of 4 in. in each hole. Look at a stereo pair that's 4 in. away through the lenses, using another piece of cardboard to make sure each eye sees only one image. More sophisticated devices are available from several sources, such as StereoGraphics Corp. URL http://www.stereographics.com/ and Edmund Scientific http://www.edsci.com/.

3.11 The Viewing Volume

In order to display realistic images, we have to limit the items being displayed to those that would actually be seen by an observer located at $(0, 0, -k)$ and looking at the image projected on the screen. There are three cases to consider:

1. The observer and the object being projected should be located on different sides of the projection plane. Any parts of the object located on the same side as the observer should not be projected. Such parts should be identified and ignored. If the software does not do that, such parts would be projected in a wrong way, upside down and back to front. As an example, consider points P_1 and P_2 in Figure 3.48a. The former is on the other side of the screen from the observer and is therefore projected correctly. The latter is on the observer's side of the screen and is projected on the negative side of the x axis. Including points such as P_2 in the projection creates a confusing effect.

2. Those parts of the scene that are very far away will normally not be seen by an actual observer and should not be displayed on the screen. It is customary to define a plane located at $z = K$ (the far plane) to identify those parts of the image with z coordinates $> K$ and to clip them off.

3. The screen and the far plane now define a truncated pyramid, called the

3.11 The Viewing Volume

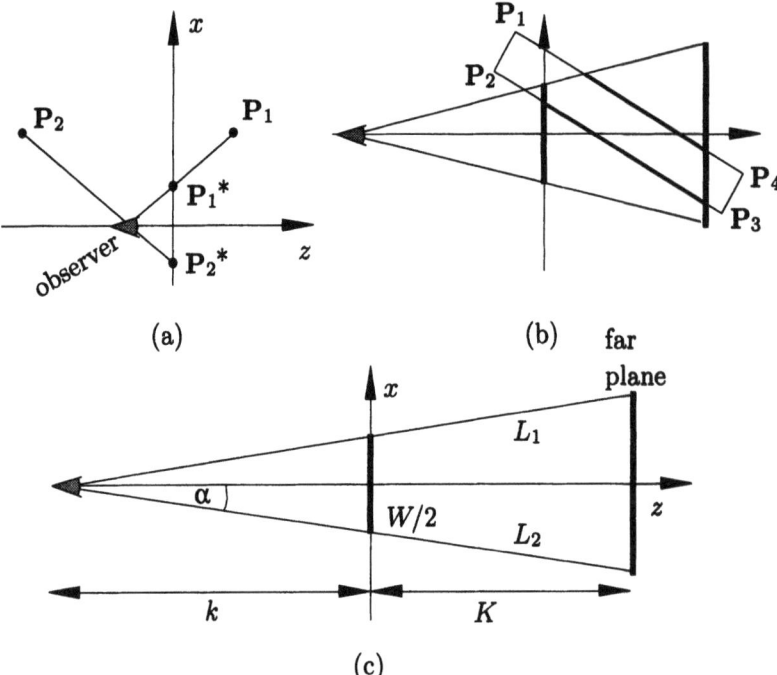

Figure 3.48: The Viewing Volume in Three Dimensions.

viewing volume. Those parts of the image that are outside it are not visible to the observer and should not be displayed.

Imagine a picture made up of points connected with straight lines. Before displaying the picture, the software should determine which points are outside the viewing volume. Those points should not be displayed, but should not be ignored either. Figure 3.48b shows four points connected to form a rectangle. Notice how some of the lines connecting the points should not be displayed and others should be *clipped*. In general, only those parts of the image that are inside the viewing volume should be displayed.

It is easy to determine if a point $\mathbf{P} = (x, y, z)$ is inside the viewing volume. We assume that the screen is a square, W units on a side. Figure 3.48c shows two of the four lines that bound the pyramid. It is easy to see that $\tan \alpha = (W/2)/k = W/(2k)$. This is also the slope of line L_1. The x-intercept of the line is $W/2$, so the line's equation is $x = W/2(z/k+1)$. The equation of L_2 is, similarly, $x = -W/2(z/k+1)$. Since the diagram is symmetric with respect to x and y, we conclude that point $\mathbf{P}$ is located inside the pyramid if its coordinates satisfy $|x|, |y| \leq W/2(z/k+1)$.

▶ **Exercise 3.60:** Assume that the distance k of the observer from the screen equals the size W of the screen. What will be the width of the field of view of the observer?

Let's assume that two points, $\mathbf{P}$ and $\mathbf{Q}$, are part of the total image and are to be connected with a straight line. The first step is to determine, for each point, whether it is located inside or outside the viewing volume (if a point is located on the edge of the viewing volume, it is considered to be inside). In the second step,

3 Transformations and Projections

three cases should be distinguished:

1. Both points are inside the viewing volume. The line connecting them is completely inside the volume and should be fully displayed. This is because the viewing volume is convex (it is a convex polyhedron).

2. One point is inside and the other is outside the viewing volume. The line connecting them intercepts the volume at exactly one point. (This, again, is a result of the convexity of the viewing volume.) The interception point should be calculated and the line should be clipped.

3. Both points are outside. The line connecting them is either completely outside (and should thus be ignored) or it intercepts the viewing volume at two points. Both interception points should be calculated and the line segment connecting them should be displayed. (There is also the degenerate case where both interception points are identical; the line just touches the viewing volume. In such a case the line can be ignored, or just one pixel displayed.)

3.11.1 Application: Flight Simulation

People have been fascinated by flight since the dawn of history. It is therefore not surprising that simple, inexpensive flight simulators for personal computers appeared as soon as CRTs became a common output device. A flight simulator, even a simple one, is a complex program since it has to simulate the behavior of an airplane and display both the interior (instruments) and exterior (the view from the cockpit) in real time. This section is concerned with displaying the view from the cockpit and we show that this task is an application of the important concept of *viewing volume*.

Figure 3.49a shows part of a typical World War II bomber. It is obvious that the field of view of the pilots in the cockpit is restricted. They see a lot of sky and part of the airplane, but only distant parts of the ground in front and on the sides. The bombardier, however, has almost a 180° field of view and can see all the way from 6 o'clock (the ground below their feet) to 12 o'clock (straight up).

Figure 3.49b is a schematic diagram showing the viewing volume of the pilot (ignoring the curvature of the Earth). We assume that the flight simulator has to display the pilot's view on a screen placed k units in front of the pilot. It is obvious that the view depends on the precise shape of the aircraft (this determines the orientation of lines L_1 and L_2). Most of the screen in the figure is a projection of the sky and only a small part shows a projection of the ground in front of the aircraft. It is also trivial to use similar triangles to obtain the basic perspective expression (Equation (3.38))

$$\frac{z+k}{y} = \frac{k}{y^*} \quad \text{or} \quad y^* = \frac{ky}{z+k} = \frac{y}{z/k+1}.$$

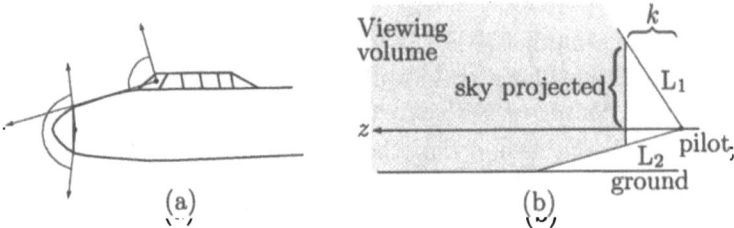

Figure 3.49: a. Two Fields of View. b. A Viewing Volume.

3.12 Going Beyond the Third Dimension

We seem to live in a three-dimensional world. There is plenty of evidence for this. A point (i.e., a location in space) can be fully specified by three numbers. Three lines can be placed in space perpendicular to each other, but no fourth line can be added that will be perpendicular to all three. Still, people think of the fourth dimension, are fascinated by it, and try to "understand" it, visualize it, or at least associate it with something familiar from the real world. Geometers have been working with multidimensional geometries for a long time; modern physics deals with multi-dimensional quantities, and science-fiction writers explore the implications of living in non-three-dimensional worlds. A notable example of the latter is *Flatland*, a study of a two-dimensional world occupied by flat living beings [Abbott 80], but the interested reader should also try [L'Engle 62].

▶ **Exercise 3.61:** Relativity theory deals with space-time. Is time the long-sought-after fourth dimension?

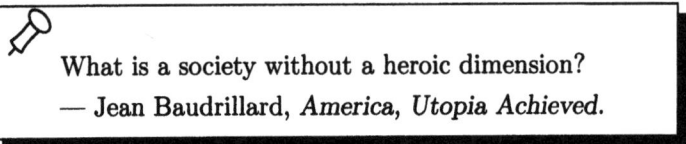

What is a society without a heroic dimension?
— Jean Baudrillard, *America, Utopia Achieved.*

Computer graphics uses mathematics to create lifelike three-dimensional objects, transform them in a variety of ways, and project them on a two-dimensional output device. Because of the use of mathematics, it is possible to extend these concepts to any number of dimensions. This section discusses the following three topics:

1. How simple geometrical figures such as a cube, a tetrahedron, or a pyramid can be constructed in any number of dimensions.

2. How transformations, such as rotation and translation, can be performed in any number of dimensions.

3. How an n-dimensional object can be projected to an object in $n-1$ dimensions.

What this section does not explain is how an n-dimensional object looks and how to visualize such an object from its projections. After reading this material, an advanced reader should be able to construct, say, a four-dimensional cube, rotate it, and display its projection in three dimensions (using virtual reality) and two dimensions. The author believes, however, that no amount of reading will teach

3 Transformations and Projections

the reader how to interpret the projections in order to figure out what such a cube "really looks like" in four dimensions. Dealing with multidimensional objects is a good way to learn geometry and to practice the mathematics of transformations and projections, but it does not seem to add to our understanding of the real world around us.

We start with a multidimensional cube. This is a good example since a cube is a simple, symmetric object that's defined by a small number of points and is bounded by straight edges. Our familiarity with three-dimensional cubes makes it easy for us to agree with the following statement: A two-dimensional cube is a square, a one-dimensional cube is a straight segment, and a zero-dimensional cube is a point. These "cubes" are shown in Figure 3.50, and examining them helps us understand how cubes of higher dimensions can be constructed. The rule for constructing a higher-dimensional cube is: To construct an n-dimensional cube start with an $(n-1)$-dimensional cube, duplicate it, and drag the copy away from the original while tracing out perpendicular lines from each point of the original cube to its copy (a process known as "extrusion"). Constructing, for example, a three-dimensional cube with this rule is a convincing experience since this cube is made of two copies of a two-dimensional cube and those copies are connected by lines that are perpendicular to both copies.

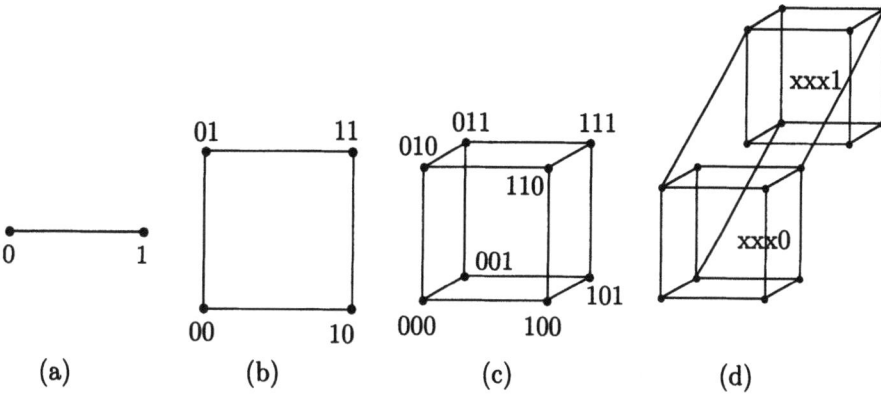

Figure 3.50: Cubes of Various Dimensions.

A four-dimensional cube is thus constructed by duplicating a three-dimensional cube, dragging the copy away, and tracing out perpendicular lines from each of the original eight points to its copy. The result is an object consisting of 16 points. This object can be drawn (Figure 3.50d), but it is clear that there is something wrong with the drawing since the lines connecting the two three-dimensional copies (only some are shown) are not perpendicular to both copies. This is where our geometric imagination fails and we have to rely on our mathematics. We simply extend our methods and principles without attempting to visualize the resulting structures.

In the case of a cube, it is especially easy to extend the mathematics beyond three dimensions. Figure 3.50 shows that a cube of n dimensions consists of 2^n points, each of which is connected to n other points. The total number of edges thus seems to be $n2^n$, but since this counts each edge twice, the right number is $n2^n/2$. Constructing an n-dimensional cube should be done by selecting the right 2^n points and connecting each to n other points.

> Circles to square and cubes to double
> Would give a man excessive trouble.
> — Matthew Prior.

Consider an n-dimensional unit cube with one corner at the origin. To figure out the coordinates of the 2^n points comprising it, we start with small values of n. For $n = 0$, the cube consists of a single point so this point should be the origin. When using this cube to construct a one-dimensional cube, we duplicate the point and drag it one unit along the (only) coordinate axis. It makes sense to number the two points 0 (the point at the origin) and 1 (its copy at distance one). A two-dimensional cube is constructed by duplicating points 0 and 1 and tracing out unit segments between the copies and the originals. We now have to distinguish between the original points and their copies, and this is done by adding one bit to their numbers. The original points may be assigned numbers 00 and 01, while the copies may be called 10 and 11 (if we decide to add the new bit on the right, these numbers will become 00, 10, and 01, 11, respectively).

When this is carried out to higher and higher dimensions, it becomes clear that each point in an n-dimensional cube is assigned an n-bit number. Since there are 2^n n-bit numbers, it is obvious that every n-bit number corresponds to a point in that cube. These facts lead to the conclusion that the number assigned to a point can be interpreted as its coordinates. For example, in a four-dimensional cube, a point that's assigned the number 0100 has coordinates $(0, 1, 0, 0)$.

The next question is how to connect the points. Point 0100 in a four-dimensional cube should be connected to four other points in the cube (its four nearest neighbors) but which ones are they? To answer this question, we look again at the process of numbering the points. When this process is examined closely, we realize that each bit in the number of a point corresponds to one of the dimensions of the cube. In the three-dimensional cube of Figure 3.50c, for example, the leftmost bit of the number of each point corresponds to the left–right dimension, the middle bit corresponds to the up–down dimension, and the rightmost bit corresponds to the front–back dimension.

The conclusion is that the n points that are the nearest neighbors of a given point **P** in an n-dimensional cube are the ones whose numbers differ from **P**'s number by one bit. The four nearest neighbors of point $(0, 1, 0, 0)$ in a four-dimensional cube are thus points $(1, 1, 0, 0)$, $(0, 0, 0, 0)$, $(0, 1, 1, 0)$, and $(0, 1, 0, 1)$. Once this is understood, it is possible to develop software that calculates the points and their connections for an arbitrary number n of dimensions.

The next example of a multidimensional geometric object is the tetrahedron. A three-dimensional tetrahedron is a pyramid with a triangular base. Each of its four faces is a triangle. This suggests that the two-dimensional tetrahedron is an equilateral triangle. Each of the three edges of such a triangle is a straight segment,

3 Transformations and Projections

which, in turn, suggests that a one-dimensional tetrahedron is a straight segment. A little thinking leads to the following rule for constructing multidimensional tetrahedrons: An n-dimensional tetrahedron is constructed by selecting the midpoint of the $(n-1)$-dimensional tetrahedron and dragging it into the nth dimension. Figure 3.51 illustrates the first two steps of this process.

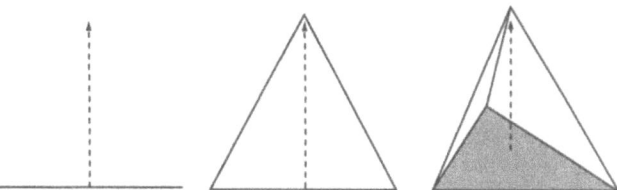

Figure 3.51: Tetrahedrons of Various Dimensions.

▶ **Exercise 3.62:** Show the rules for constructing a multidimensional pyramid and octahedron.

In the above discussion, we have used familiar concepts from one, two, and three dimensions to draw conclusions about vertices and edges of four-dimensional objects. This method is general and can be used to derive more properties of those objects. Here are two examples:

1. A straight segment (a one-dimensional object) is bounded by two points (zero-dimensional objects). A square (a two-dimensional object) is bounded by four segments (one-dimensional objects). A cube (a three-dimensional object) is bounded by eight squares (two-dimensional objects). We conclude that a four-dimensional cube must be bounded by 16 $(= 2^4)$ three-dimensional cubes and an n-dimensional cube must be bounded by 2^n $(n-1)$-dimensional cubes.

2. When the size of a straight segment is doubled, its length grows by a factor of 2^1. When the size of each side of a square is doubled, its area grows by a factor of 2^2. When the size of every side of a cube is doubled, its volume grows by a factor of 2^3. This is why length, area, and volume are one, two, and three-dimensional attributes, respectively. By extension, the "hypervolume" of the four-dimensional cube grows by a factor of 2^4 when every side of the cube is doubled in size.

▶ **Exercise 3.63:** What exactly is the hypervolume of the four-dimensional cube?

> I regret that it has been necessary for me in this lecture to administer such a large dose of four-dimensional geometry. I do not apologize, because I am really not responsible for the fact that nature in its most fundamental aspect is four-dimensional. Things are what they are.
> — Alfred North Whitehead.

We now turn to the problem of transforming a multidimensional object. We demonstrate that this can also be solved by extending the rules and notation used in two- and three-dimensional transformations and by giving up any attempts to visualize higher-dimensional objects either before or after the transformations.

3.12 Going Beyond the Third Dimension

We start with rotation. In two dimensions, there is only one, familiar rotation matrix, namely

$$\mathbf{R}_2 = \begin{pmatrix} \cos\theta & -\sin\theta \\ \sin\theta & \cos\theta \end{pmatrix}.$$

In three dimensions (we limit ourselves to rotations about the coordinate axes), there are three rotation matrices:

$$\mathbf{R}_{3,1} = \begin{pmatrix} 1 & 0 & 0 \\ 0 & \cos\theta & -\sin\theta \\ 0 & \sin\theta & \cos\theta \end{pmatrix}, \quad \mathbf{R}_{3,2} = \begin{pmatrix} \cos\theta & 0 & -\sin\theta \\ 0 & 1 & 0 \\ \sin\theta & 0 & \cos\theta \end{pmatrix},$$

$$\mathbf{R}_{3,3} = \begin{pmatrix} \cos\theta & -\sin\theta & 0 \\ \sin\theta & \cos\theta & 0 \\ 0 & 0 & 1 \end{pmatrix}.$$

These matrices can be interpreted by noticing that $\mathbf{R}_2$ has four elements, all of which are used in the rotation. They are all set to either the sine or the cosine of the rotation angle. In contrast, each of the three matrices $\mathbf{R}_{3,i}$ corresponds to a rotation about coordinate axis i and has a special structure. Its ith row and ith column are zero except for element (i,i), which is set to one. This guarantees that the ith coordinate of a point $\mathbf{P} = (x,y,z)$ being rotated will be preserved. This also leaves exactly four available entries in the matrix and they are set to the sine and cosine of the rotation angle, as in the two-dimensional case.

Once this is understood, it is easy to see how rotation matrices for higher dimensions can be constructed. A two-dimensional rotation matrix does not preserve any of the coordinates of the rotated point, while a three-dimensional rotation matrix preserves one coordinate of a rotated point. We can extend this concept and require a four-dimensional rotation matrix to preserve two of the four coordinates of a point (and change the other two). Such a matrix should therefore have rows i and j and columns i and j set to 0 (where i and j are any integers in the range $[1,4]$), except that the diagonal elements (i,i) and (j,j) should be set to 1's. This leaves exactly four empty positions and they should be set to the sine and cosine of the rotation angle, as in the two-dimensional case. To figure out the total number of four-dimensional rotation matrices, we recall that two objects can be selected from four objects in $\binom{4}{2} = 6$ ways, so there must be six such matrices. Two of them, corresponding to $ij = 12$ and $ij = 24$, respectively, are

$$\mathbf{R}_{4,12} = \begin{pmatrix} 1 & 0 & 0 & 0 \\ 0 & 1 & 0 & 0 \\ 0 & 0 & \cos\theta & -\sin\theta \\ 0 & 0 & \sin\theta & \cos\theta \end{pmatrix}, \quad \mathbf{R}_{4,24} = \begin{pmatrix} \cos\theta & 0 & -\sin\theta & 0 \\ 0 & 1 & 0 & 0 \\ \sin\theta & 0 & \cos\theta & 0 \\ 0 & 0 & 0 & 1 \end{pmatrix}.$$

In general, a rotation matrix in n dimensions should preserve $n-2$ of the n coordinates of a point and change the remaining two. After setting $n-2$ of its rows and columns to 0 (and setting the corresponding $n-2$ diagonal elements to 1's), the remaining four elements are set to the sine and cosine of the rotation angle, as in

3 Transformations and Projections

the two-dimensional case. The number of such matrices is $\binom{n}{2}$, the number of ways to select two objects out of n.

Another useful interpretation of multidimensional rotations is to think of $\mathbf{R}_2$ as rotating points on the xy plane and of $\mathbf{R}_{3,1}$, $\mathbf{R}_{3,2}$, and $\mathbf{R}_{3,3}$ as rotating points on the yz, xz, and xy planes, respectively. When this interpretation is carried over to higher dimensions, we can say that, for example, $\mathbf{R}_{4,12}$ rotates points on the zw plane (i.e., the plane formed by the third and fourth coordinates). In four dimensions, the four coordinate axes x, y, z, and w form six planes, and in n dimensions, there are $\binom{n}{2}$ coordinate planes.

▶ **Exercise 3.64:** What are the six coordinate planes in four dimensions?

The next transformation we consider is scaling/reflection. In two and three dimensions, these transformations are produced by matrices with scaling/reflection factors along the diagonal and with all nondiagonal elements set to zero. The generalization to n dimensions is immediate. Shearing in n dimensions is produced by a matrix with 1's along the diagonal and with the nondiagonal elements set to certain values. An $n \times n$ matrix has $n^2 - n$ nondiagonal elements, so even for small values of n, such as 3 or 4, it is practically impossible to predict the effects of shearing on any but the simplest objects.

▶ **Exercise 3.65:** Show how to translate an n-dimensional object along any of the n coordinate axes.

The next concept that should be extended to higher dimensions is projections. We discuss several techniques to project a four-dimensional object in three and two dimensions. Once these techniques are grasped, their generalization to n dimensions is straightforward. The first technique is parallel projection.

This type of projection is discussed in detail for the three-dimensional case in Section 3.8, where it is shown that when the observer is located at $(0, 0, -k)$, an orthographic projection is done by simply deleting the z coordinate of all the projected points. To extend this to four dimensions, we assume that the observer is located at $(0, 0, 0, -k)$ and we delete the fourth coordinate of all the points of the object. When this is applied to the four-dimensional cube, the result is two identical three-dimensional cubes. In practice, we may want to project from four to two dimensions and this can be done by deleting two of the four coordinates.

▶ **Exercise 3.66:** What is the result of applying this technique to parallel-projecting a three-dimensional cube onto one dimension.

Figure 3.52 shows 3 two-dimensional parallel projections of a four-dimensional unit cube. The cube is shown projected on three different planes and it is clear that the projections "hint" at a cubic shape, but, at least in this author's opinion, they do not add to our "understanding" of the shape of the four-dimensional cube. Figure 3.53 is a listing of basic *Mathematica* code that generates such projections.

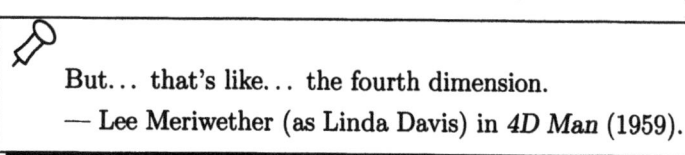

But... that's like... the fourth dimension.
— Lee Meriwether (as Linda Davis) in *4D Man* (1959).

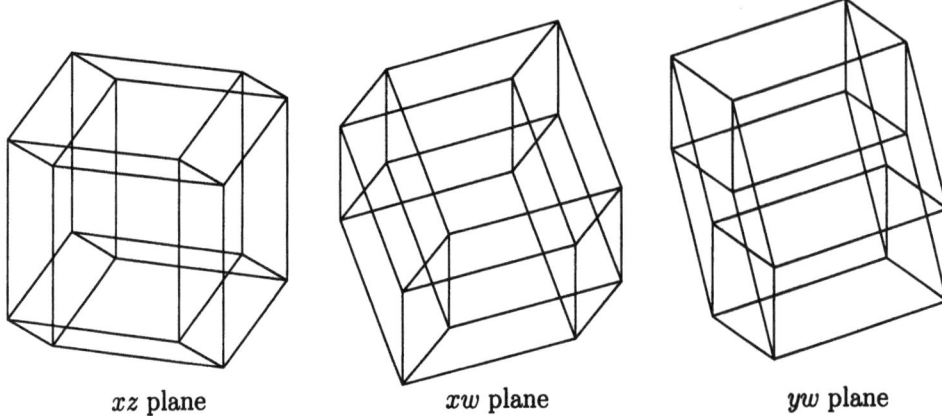

xz plane   *xw* plane   *yw* plane

Figure 3.52: Parallel Projections of a Four-Dimensional Cube.

```
Clear[a,h,i,j,k,l,p,t,fi,prm]; (* rotating a hypercube *)
h={ (* The 16 corner points *)
{{{{0,0,0,0},{0,1,0,0}},{{1,0,0,0},{1,1,0,0}}},
 {{{0,0,1,0},{0,1,1,0}},{{1,0,1,0},{1,1,1,0}}}},
{{{{0,0,0,1},{0,1,0,1}},{{1,0,0,1},{1,1,0,1}}},
 {{{0,0,1,1},{0,1,1,1}},{{1,0,1,1},{1,1,1,1}}}}};
t=75 Degree; a=2; (* "a" specifies what coordinate to delete *)
r12={{1,0,0,0},{0,1,0,0},{0,0,Cos[t],-Sin[t]},{0,0,Sin[t],Cos[t]}};
r24={{Cos[t],0,-Sin[t],0},{0,1,0,0},{Sin[t],0,Cos[t],0},{0,0,0,1}};
r14={{1,0,0,0},{0,Cos[t],-Sin[t],0},{0,Sin[t],Cos[t],0},{0,0,0,1}};
p=h; p=p.r14; (* rotate points in list 'p' *)
fi[i_]:=If[i==1,2,1]
(* Each call to "prm" generates four lines with 3D endpoints *)
prm[i_,j_,k_,l_]:={
 Line[{Delete[p[[i,j,k,l]],a],Delete[p[[fi[i],j,k,l]],a]}],
 Line[{Delete[p[[i,j,k,l]],a],Delete[p[[i,fi[j],k,l]],a]}],
 Line[{Delete[p[[i,j,k,l]],a],Delete[p[[i,j,fi[k],l]],a]}],
 Line[{Delete[p[[i,j,k,l]],a],Delete[p[[i,j,k,fi[l]]],a]}]}
(* "prm" is invoked 16 times, for the 16 corners of the hypercube *)

Show[Graphics3D[
Flatten[{prm[1,1,1,1],prm[2,1,1,1],prm[1,2,1,1],prm[1,1,2,1],
 prm[1,1,1,2],prm[2,2,1,1],prm[1,2,2,1],prm[1,1,2,2],
 prm[2,1,1,2],prm[2,1,2,1],prm[1,2,1,2],prm[2,2,2,1],
 prm[2,2,1,2],prm[2,1,2,2],prm[1,2,2,2],prm[2,2,2,2]}]],Boxed->False]
```

Figure 3.53: *Mathematica* Code for Generating Hypercube Projections.

3 Transformations and Projections

The second technique discussed here is slicing or contouring. A slice (or a contour) of a three-dimensional object is generated by specifying a plane and deleting all the object parts that are not located on this plane. A typical example is a slice along the z axis at, for example, $z = 1$. It shows an outline of the object consisting of all the object points with $z = 1$. Slicing a three-dimensional cube along one of the coordinate axes produces a square, but slicing the same cube along a slanted plane produces a triangle. This explains why Figure 3.54, which shows nine parallel slices of a four-dimensional cube, consists of pyramids, truncated pyramids, and truncated cubes.

The third technique discussed here is perspective projection of an n-dimensional object. We start with the perspective projection from three to two dimensions, given by Equation (3.38), duplicated here:

$$x^* = \frac{x}{(z/k)+1}, \quad y^* = \frac{y}{(z/k)+1}. \tag{3.38}$$

We first consider a simple case. Figure 3.55 shows how a two-dimensional object can be projected down to one dimension. We assume an observer located on the negative side of the y axis, at point $(0, -k)$ and we project two-dimensional points that have positive y coordinates on the x axis (this is the projection "plane"). We use the rule of perspective projection (page 113), which says: Connect the projected point $\mathbf{P}$ to the observer with a straight segment and place the projected point $\mathbf{P^*}$ where this segment intercepts the projection plane—in our case the x axis. Using similar triangles, we get

$$\frac{x}{y+k} = \frac{x^*}{k}, \quad \text{from which we derive} \quad x^* = \frac{x}{1+y/k}. \tag{3.50}$$

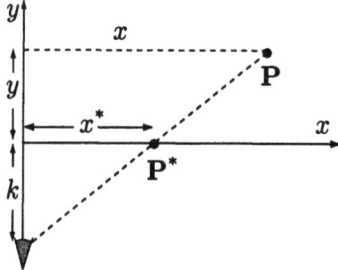

Figure 3.55: Perspective Projection from Two Dimensions to One.

Equations (3.38) and (3.50) show how to extend perspective projection to higher dimensions. To project a four-dimensional point $\mathbf{P} = (x, y, z, w)$ to three dimensions, we assume an observer located at $(0, 0, 0, -k)$ and generalize Eqs. (3.38) and (3.50) to

$$x^* = \frac{x}{(w/k)+1}, \quad y^* = \frac{y}{(w/k)+1}, \quad z^* = \frac{z}{(w/k)+1}. \tag{3.51}$$

Figure 3.54: Nine Slices of a Four-Dimensional Cube.

3 Transformations and Projections

The resulting three-dimensional point **P*** can be further projected to a point **P**** in two dimensions by applying Equation (3.38). The extension of Equation (3.51) to n dimensions is now obvious. We assume an n-dimensional point $\mathbf{P} = (d_1, d_2, \ldots, d_n)$ and an observer located at the special point $(0, 0, \ldots, 0, -k)$. We denote the $(n-1)$-dimensional projected point $\mathbf{P}^* = (d_1^*, d_2^*, \ldots, d_{n-1}^*)$ and it is given by

$$d_1^* = \frac{d_1}{(d_n/k)+1}, \quad d_2^* = \frac{d_2}{(d_n/k)+1}, \ldots, d_{n-1}^* = \frac{d_{n-1}}{(d_n/k)+1}. \qquad (3.52)$$

Figure 3.56 is the familiar one-point perspective projection of a three-dimensional cube. The front face looks bigger than the back face, but each is a square. When this is compared to the one-point perspective projection of a four-dimensional cube, shown next to it, the similarity becomes obvious. What we see is a large front "face" and, centered inside it, a small back "face" of the hypercube (recall that each face is a three-dimensional cube).

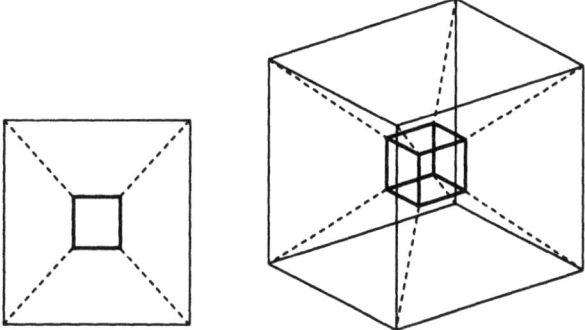

Figure 3.56: One-Point Perspective Projections of a Cube and a Hypercube.

See [Banchoff 96] for a lively discussion of multidimensional objects and their transformations and projections.

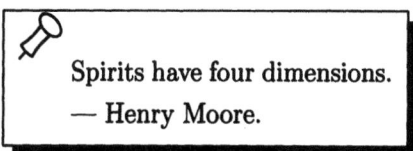

Spirits have four dimensions.
— Henry Moore.

3.13 Nonlinear Projections

In addition to the parallel and perspective projections, other projections may be developed that are useful in special situations or that create ornamental or artistic effects. It seems that the number of such projections is vast and is limited only by the imagination of the mathematicians coming up with them. This section discusses some of the more common nonlinear projections, including the false perspective, the 180° fisheye projection, the 360° panoramic projection, the telescopic and microscopic projections, projections of a sphere, and circle inversion. These projections create unusual effects and are mathematically simple and easy to understand. However, since they are nonlinear, they generally cannot be represented by means of transformation matrices.

3.13.1 False Perspective

Equation (3.38) is the main expression for perspective projection:

$$x^* = \frac{x}{1 + (z/k)}, \quad y^* = \frac{y}{1 + (z/k)}.$$

It shows that the coordinates of the projected point $\mathbf{P}^*$ are calculated by dividing by the z coordinate (the depth) of the original point $\mathbf{P}$. The idea in false (or pseudo) perspective is to artificially introduce perspective (or an effect similar to perspective) into a two-dimensional image, thereby making it appear three-dimensional. Since points in such an image have just x and y coordinates, it is natural to change Equation (3.38) to

$$x^* = \frac{x}{1 + f(x,y)}, \quad y^* = \frac{y}{1 + f(x,y)}, \quad (3.53)$$

where $f(x,y)$ is a function to be selected according to the desired effect. For example, the function

$$f(x,y) = -\frac{1}{2}e^{-ax^2 - by^2},$$

where a and b are real constants, gets the value -0.5 for $x = y = 0$ (the origin) and approaches zero for very large x or y coordinates (positive or negative). Points (x,y) near the origin are therefore projected to $(2x, 2y)$, while points on the edges of the image are hardly affected by this projection. This has the effect of magnifying the center of the image, thereby making it appear closer. Other functions may create different effects.

> By introducing unusual vanishing points and forcing elements of a composition to obey them, Escher was able to render scenes in which the "up/down" and "left/right" orientations of its elements shift, depending on how the viewer's eye takes it in. In his perspective study for "High and Low," the artist has placed five vanishing points: top left and right, bottom left and right, and center. The result is that in the bottom half of the composition the viewer is looking up, but in the top half he or she is looking down. To emphasize what he has accomplished, Escher has made the top and bottom halves depictions of the same composition.
> — B. Sidney Smith.

3.13.2 The 180° Fisheye Projection

This type of projection is named after the fisheye camera lenses, but the same principle is used in peepholes installed in doors. The idea is to take the half-sphere of space (with infinite radius) that's in front of an observer and project it into a flat unit circle. This is done in a two-step process. In the first step, illustrated in

3 Transformations and Projections

Figure 3.57a, all the points in the half-sphere are projected into an infinitely large circle on the xy plane. In the second step, all the points of this circle are moved closer to the center and end up on the unit circle located in front of the observer (who is represented by point V in Figure 3.57b).

The first step uses either parallel or perspective projection to project points onto a plane. Figure 3.57a shows an observer V located at point $(0,0,1)$ and two three-dimensional points $\mathbf{P}$ and $\mathbf{Q}$. Point $\mathbf{P}$ is projected on the xy plane in perspective and point $\mathbf{Q}$ is projected in parallel. These two types of projections have already been described in detail, so no further discussion of this step is necessary.

In the second step, each point on the xy plane is moved toward the origin by halving its angle of view θ. Figure 3.57b shows a point $\mathbf{P}$ on the xy plane where the angle between the z axis and line VP is θ. The point is moved closer to the origin and becomes $\mathbf{P}^*$ with a view angle of $\theta/2$. Since both $\mathbf{P}$ and $\mathbf{P}^*$ are on the xy plane, we can consider this transformation scaling in two dimensions. The transformed point $\mathbf{P}^*$ equals $s\mathbf{P}$, where the scale factor s is less than one (i.e., shrinking). However, it is easy to see intuitively that points located away from the origin will be scaled more than points closer to the origin. The scale factor s is thus not constant; it depends on $\mathbf{P}$, which makes this type of projection *nonlinear*.

Figure 3.57b shows that $\tan\theta = \mathbf{P}/(\text{VO}) = \mathbf{P}$, implying $\theta = \arctan|\mathbf{P}|$. Similarly, the transformed point satisfies $|\mathbf{P}^*| = \tan(\theta/2)$, which yields the scaling factor

$$s = \frac{|\mathbf{P}^*|}{|\mathbf{P}|} = \frac{\tan(\theta/2)}{|\mathbf{P}|} = \frac{\tan\big((\arctan|\mathbf{P}|)/2\big)}{|\mathbf{P}|}.$$

Notice that points that are the farthest on the xy plane have an angle of view close to 90°. Their projections will therefore have an angle close to 45°. A view angle of 45° means that the distance of such a projected point from the origin equals the distance VO of the observer from the origin, which is 1. The result is that all the points on the xy plane are moved by the fisheye projection onto the unit circle located in the xy plane and centered on the origin. Figure 3.58 shows the fisheye projection of two simple images. The distortions introduced by this nonlinear projection are noticeable.

A well-known example of the fisheye projection is the drawing *Hand with Reflecting Sphere* by M. C. Escher [Ernst 76].

> A story of particular facts is a mirror which obscures and distorts that which should be beautiful; poetry is a mirror which makes beautiful that which it distorts.
> — Percy Bysshe Shelley, *A Defence of Poetry.*

3.13.3 The 360° Panoramic Projection

A visitor to an unusually beautiful spot sometimes wishes they could see the view behind them as well as in front of them at the same time. This is the effect provided by the *panoramic projection*. To understand the principle behind this type

3.13 Nonlinear Projections

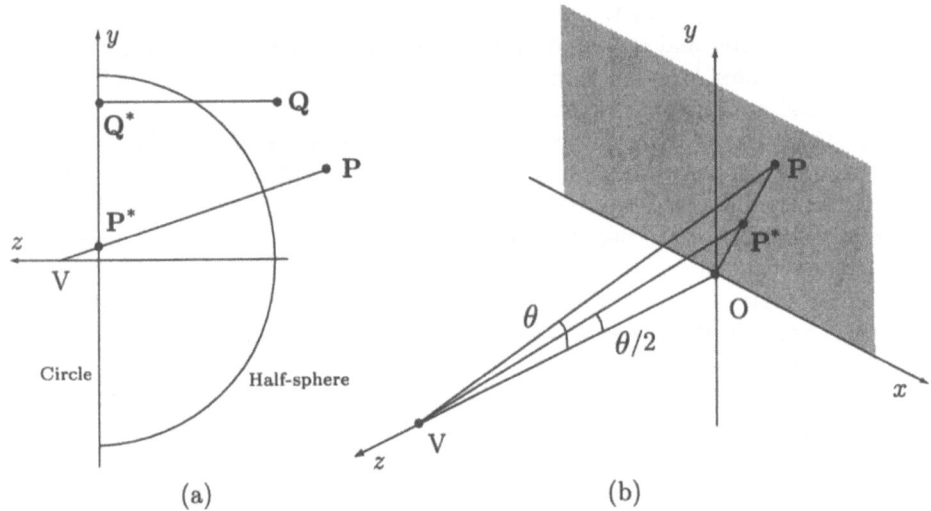

Figure 3.57: 180° Fisheye Projection.

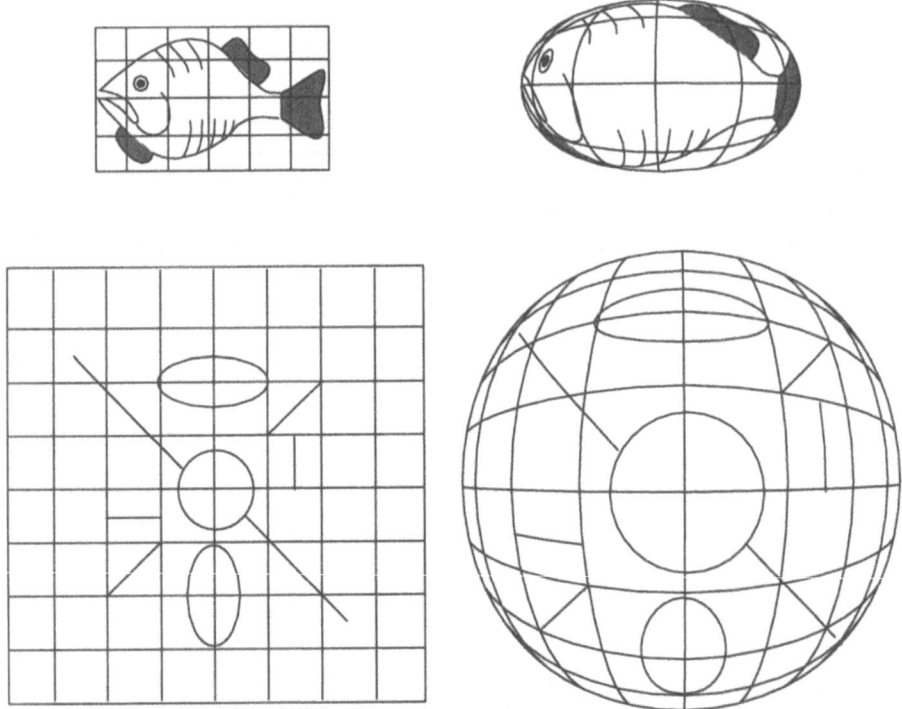

Figure 3.58: Examples of the Fisheye Projection.

3 Transformations and Projections

of projection, imagine a rectangle made of transparent material being rolled into a cylinder and placed around an observer (Figure 3.59a). The observer is assumed to be at the origin, which is also the center of the cylinder, looking at the view outside through the transparent cylinder. The observer now starts turning around. We imagine that everything that the observer sees is magically fused into the cylinder material. After the observer has turned a full circle, the entire cylinder is full of images. The cylinder is now unrolled and is hung flat, as a rectangular picture on a wall. The image shown in this picture is a 360° panorama of the view seen by the observer.

The Mesdag Panorama

The Mesdag Panorama is a painting depicting a 360° panoramic view of the surroundings of Scheveningen, a fishing port northwest of The Hague, as seen by the painter in 1881.

The painting is huge, measuring 120×14 meters (390×45 feet) for an area of about 17,000 square feet. It is folded into a cylinder and several observers can enter from below and stand at the center, turning, watching, and admiring.

The Mesdag panorama was painted by the 19th-century Dutch painter Hendrik Willem Mesdag (1831–1915), with the help of S. Mesdag-van Houten, Theophile de Bock, G.H. Breitner, B.J. Blommers, and A. Nijberck.

Similar panoramas were exhibited throughout Europe and America during the 19th century (they were sometimes called *cycloramas*). The Mesdag panorama is one of the last panorama paintings still in existence. It can be viewed at the Museum Panorama Mesdag in The Hague, The Netherlands.

See [Mesdag Documentation Society 98] for more information.

Figure 3.59a shows a cylinder centered about the origin. It is easy to see how a general point **P** is projected to a point **P*** on the cylinder. Figure 3.59b shows the cylinder unrolled. Point **P** is located in the same place in space, but its projection has moved with the cylinder.

Figure 3.59c shows the geometry of the problem. We assume that the dimensions of the original rectangle are $2Y \times 2Z$. When rolled into a cylinder of radius R, the perimeter of the cylinder satisfies $2\pi R = 2Y$, so $R = Y/\pi$. Consider a general point $\mathbf{P} = (x, y, z)$ viewed by the observer. When the cylinder is eventually unrolled, **P** will be projected to a point $\mathbf{P}^* = (x^*, y^*, z^*)$ and our problem is to determine the coordinates of **P*** as functions of x, y, z, Y, and Z.

The x^* coordinate is simple. The figure shows that all the points on the unrolled cylinder have the same x coordinate. We can set it to R or, even simpler, to zero. The y^* coordinate should equal the length of the arc subtended by θ, which is $R\theta$. Angle θ depends on the x and y coordinates of **P** but not on its z coordinate. The relation is $(x, y) = D(\cos\theta, \sin\theta)$ where D is the distance (projected on the xy

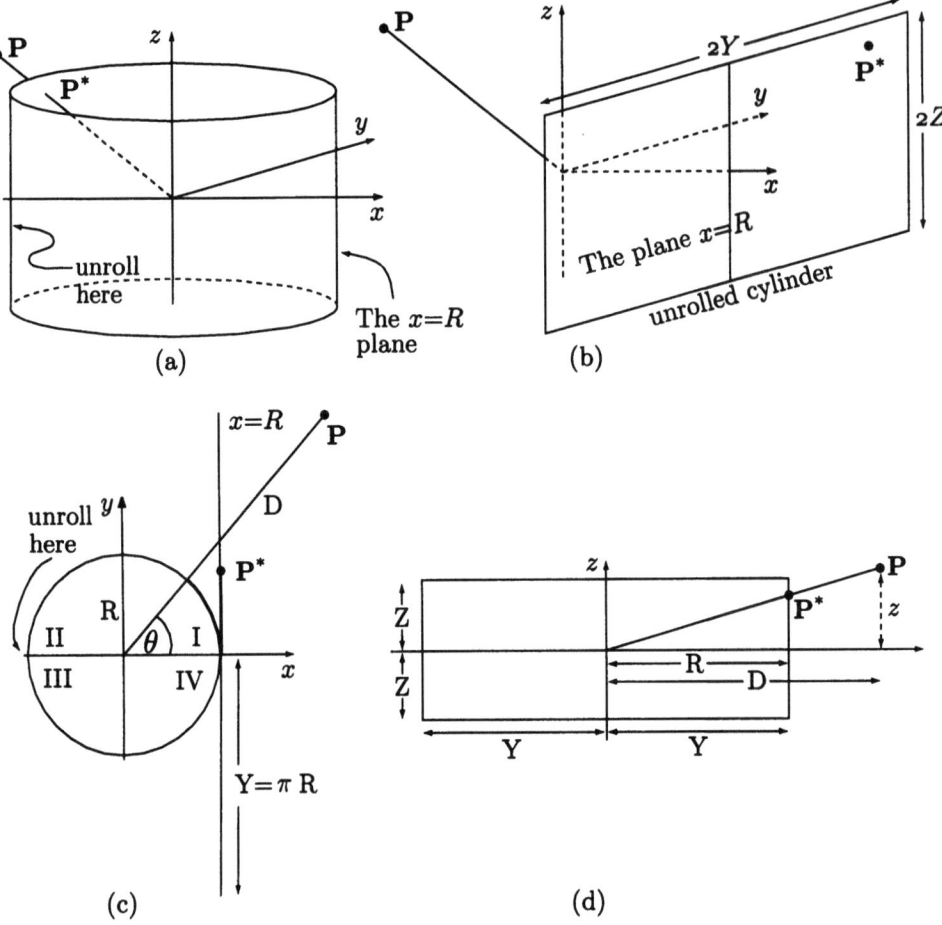

Figure 3.59: 360° Panoramic Projection.

plane) of **P** from the origin. It equals $\sqrt{x^2+y^2}$. From this we get

$$\frac{(x,y)}{\sqrt{x^2+y^2}} = (\cos\theta, \sin\theta),$$

or

$$\theta = \arcsin\frac{y}{\sqrt{x^2+y^2}} = \arccos\frac{x}{\sqrt{x^2+y^2}} = \arctan\left(\frac{y}{x}\right).$$

Notice that the signs of x and y determine the quadrant number. If θ is in quadrants III or IV, then y^* should be negative.

The z^* coordinate is calculated by perspective projection. Figure 3.59d shows how this is done with similar triangles:

$$\frac{z}{D} = \frac{z^*}{R} \to z^* = \frac{zR}{D} = \frac{zY}{\pi\sqrt{x^2+y^2}}.$$

3 Transformations and Projections

▸ **Exercise 3.67:** It seems that the projected point $\mathbf{P}^*$ is given by

$$(x^*, y^*, z^*) = \left(0, \pm R\theta, \frac{zY}{\pi\sqrt{x^2+y^2}}\right),$$

so its coordinates depend on x, y, z, and Y, but not on Z. What's the explanation?

The panoramic projection leads naturally to the concept of *curved perspective*. This comes up when we consider the panoramic projection of a straight line. Figure 3.60a shows a cylinder and a line A in space. Several projection lines are shown going from A to the center of the cylinder. These lines are contained in a plane L and we know from elementary geometry that the intersection of a cylinder and a plane is, in general, an ellipse (Figure 3.60b). The projection of A on the cylinder is therefore an elliptical arc. When the cylinder is unrolled, this arc turns into a sinusoidal curve (Figure 3.60c).

▸ **Exercise 3.68:** Prove this claim!

This means that the panoramic projection turns straight lines into curved ones, resulting in curved perspective. Two special cases should be considered. One is when the plane is perpendicular to the cylinder (corresponding to an angle $\theta = 0°$ in Figure Ans.13 in the Answers to Exercises Chapter) and the other, when it is parallel to the axis of the cylinder (corresponding to an angle $\theta = 90°$ in Figure Ans.13). In the former case, the intersection is a circle and the sinusoidal curve has zero amplitude (i.e., it degenerates into a straight line). In the latter case, the intersection is an infinite ellipse and the sinusoidal curve has infinite amplitude; it degenerates into three lines.

Figure 3.60d shows an observer positioned at the center of a cylinder and looking to the north. Three horizontal infinitely long lines are shown. The projections of lines 1 and 2 are ellipses and become the sinusoidals shown in Figure 3.60e. The projection of line 2 is a half-circle (not shown) that becomes a straight line when the cylinder is unrolled. This shows how horizontal straight lines are projected by curved perspective into either horizontal or curved lines. Vertical lines are always projected into vertical straight lines.

An intuitive way to understand (and accept) curved perspective is to print the curved projection of a familiar scene on paper, roll the sheet of paper into a cylinder, and look at the scene inside (this may be simple if the projection incorporates less than 360°). Once seen this way, all the curved lines on the paper should look straight. This method also provides a simple test of any software used to calculate the projection.

Commercial software for creating cylinder-shaped panoramas already exists. Two examples are the Apple QuickTime VR *Authoring Studio* and *PhotoVista* from Live Picture Inc. A qualitative discussion of curved perspective can be found in [Ernst 76], pp. 102–103. [Skov 92] provides a detailed derivation. The well-known drawing *High and Low* by M. C. Escher is an example of curved perspective. The following quote, from [Ernst 76], suggests a way to generalize curved perspective:

"Perhaps it has already struck you that the cylinder perspective used by Escher, leading to curved lines in place of the straight lines prescribed

3.13 Nonlinear Projections

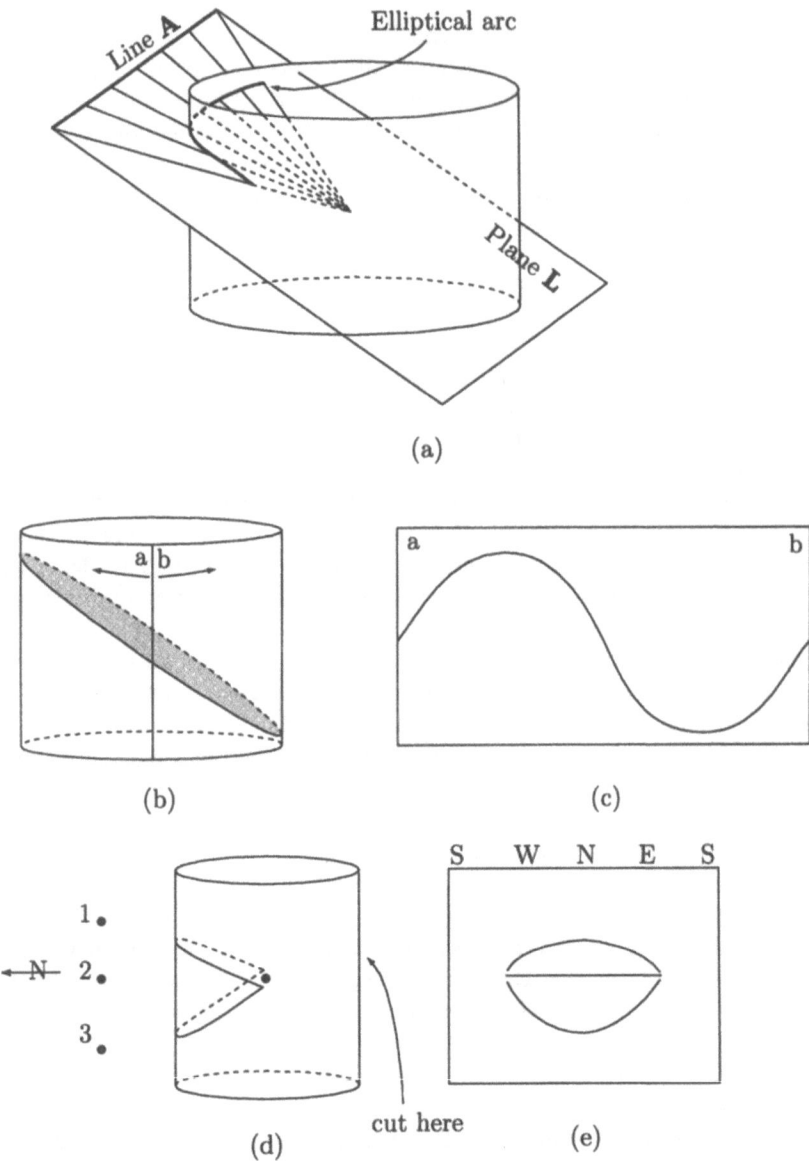

Figure 3.60: Line, Plane, and Cylinder.

3 Transformations and Projections

by traditional perspective, could be developed even further. Why not a spherical picture around the eye of the viewer instead of a cylindrical one? A fish-eye objective produces scenes as they would appear on a spherical picture. Escher certainly did give some thought to this, but he did not put the idea into practice, and therefore we will not pursue this further."

The idea raised by Ernst (but not pursued by Escher) is to imagine a transparent sphere placed around the observer, where everything seen by the observer through the sphere is fused onto the sphere's material. The sphere is then "unrolled" flat, resulting in a spherical perspective. The trouble with this idea is that a sphere cannot be unrolled into a flat surface without introducing further distortions (see Section 3.13.7).

> What you see on these screens up here is a fantasy; a computer enhanced hallucination!
> — John Wood (as Stephen Falken) in *WarGames* (1983).

Viewers who are willing to accept a highly distorted projection (and this is not a trivial sacrifice since we develop panoramic projections in order to see *more* of the image, not less) might be interested in *spherical panoramic projection*, a generalization of curved perspective that's described below.

Imagine a transparent sphere of radius R centered on the origin, where an observer is located, looking through the sphere in the z direction. The sphere is now truncated by selecting a value θ in the range $[0, \pi/2]$ and removing the parts of the sphere above and below latitude θ. The remaining part is shaped like a barrel (Figure 3.61a). The barrel is now cut behind the observer and is unrolled into a flat, two-dimensional figure resembling a Band-Aid (Figure 3.61c) that's called a *band*. The image seen by the observer through the barrel is displayed on this band, in contrast with curved perspective, where the projected image is displayed on a rectangle.

At its center, the band has a width of $2\pi R$ (the circumference of the sphere), whereas its width at the top and bottom equals $2\pi R \cos\theta$. The height of the band is $2R\theta$. Truncating the sphere into a barrel makes it possible to control the amount of distortion in the final projected image. Small values of θ result in a narrow band, its shape close to a rectangle. Only a small part of the scene around the observer is displayed on this band but with little distortion. When θ is set close to $\pi/2$, the band becomes taller and its shape approaches a circle. It includes more of the scene (only those parts directly above and below the observer are omitted) but with more distortions, especially at the top and bottom.

As in curved perspective, horizontal lines are projected on the band as sinusoidals, but we now show that even vertical lines, which in curved perspective were projected straight, now become curved. Figure 3.61b shows the barrel from above (i.e., looking in the y direction). A long vertical line (parallel to the y axis) is shown and we assume that a general point on this line is projected to a point **v** on the barrel. After the barrel is unrolled, the y coordinate of point **v** varies in the range $[-R\theta, +R\theta]$. The x coordinate depends on the y coordinate and equals the radius of the barrel at height y times the angle ϕ. The radius of the bar-

3.13 Nonlinear Projections

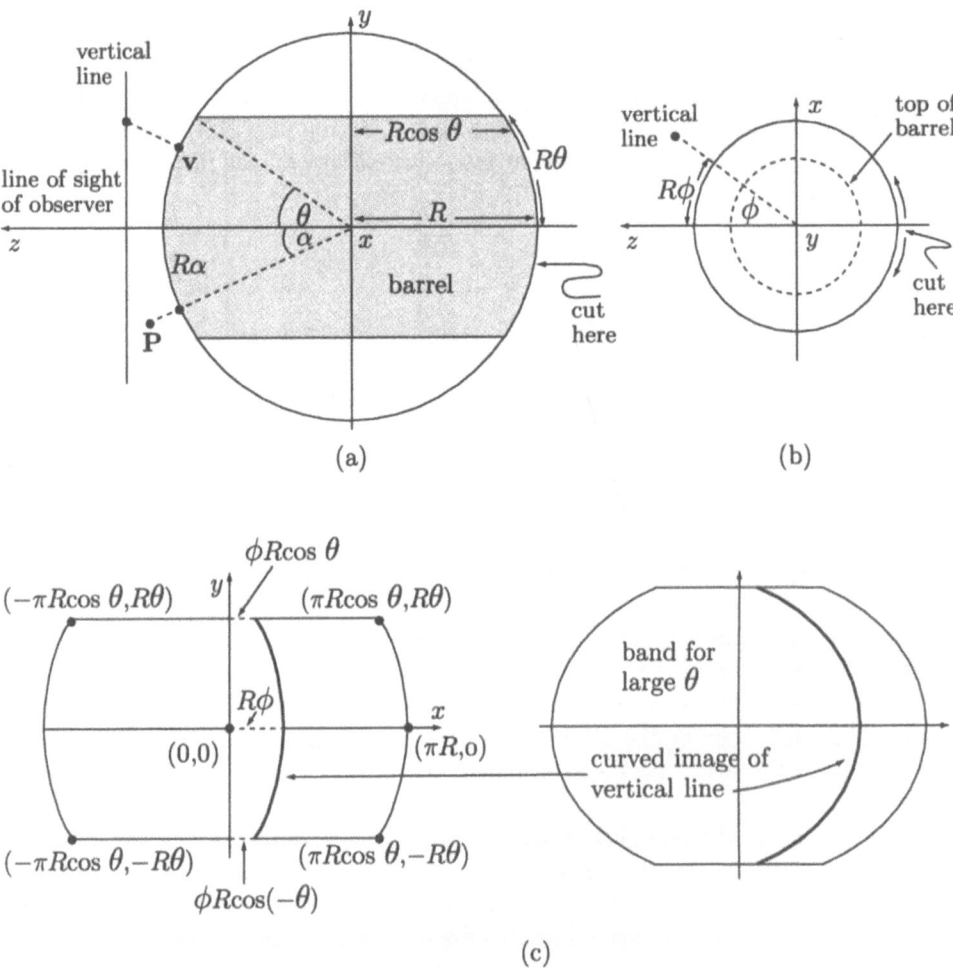

Figure 3.61: Spherical Panoramic Perspective.

> For me it remains an open question whether [this work] pertains to the realm of mathematics or to that of art.
> — M. C. Escher.

3 Transformations and Projections

rel at height y is easily seen to be $R\cos(y/R)$, so point **v** is located on the band at position $(\phi R\cos(y/R), y)$, where $-R\theta \leq y \leq +R\theta$. This position varies from $(\phi R\cos(-\theta), -R\theta)$ to $(\phi R, 0)$ to $(\phi R\cos(\theta), R\theta)$ when y varies from $-R\theta$ to 0 to $R\theta$. The projection of the vertical line on the band is, thus, the thick curve shown in Figure 3.61c. It is easy to see that the closer θ is to $\pi/2$, the smaller $\cos\theta$ is and the more curved the projection.

Given an arbitrary point $\mathbf{P} = (x, y, z)$, it is relatively easy to calculate the xy coordinates of its projection on the band. Figure 3.61b shows the situation on the xz plane and makes it clear that the x coordinate of the projected point on the band is the arc $R\phi$. Since $\tan\phi = x/z$, we get the x coordinate as $R\arctan(x/z)$. Similarly, Figure 3.61a shows that the y coordinate of the projected point on the band is the arc $R\alpha$ or $R\arctan(y/z)$. The projected point thus has band coordinates $(R\phi, R\alpha)$ or $(R\arctan(x/z), R\arctan(y/z))$. Both ϕ and α can vary in the range $[-\pi, +\pi]$, so the projected x coordinate varies in the range $[-\pi R, +\pi R]$. The projected y coordinate varies in the same range, but it is clear from the figure that any point $\mathbf{P}$ for which $|\alpha| > |\theta|$ is projected outside the barrel (i.e., on one of the sphere parts that have been removed) and should consequently be rejected.

The *IPIX Wizard* program from IPIX Inc. can create a spherical panorama from two scanned fisheye photographs.

To some people, spherical perspective may seem less interesting (and perhaps also less useful) than cylindrical perspective, as the following quote, from a virtual-reality professional, suggests.

> Our market is not craving [sphere-shaped panoramas] right now. You can convey a sense of place without looking at the sky or floor.
> —David Palermo, Apple Computer (1998).

3.13.4 Panoramic Cameras

The dictionary definition of *panorama* is "a picture taken in three-dimensional space and presented on a continuous surface encircling the viewer."

Still cameras today use a large variety of lenses, from extreme wide angle to powerful telephoto, but even the best wide-angle lenses cannot capture an image that spans more than 180°. Most fisheye lenses can capture 180° images, but the result is highly distorted, especially along the edges. Photographers like to be able to stand at a given point and capture an image of everything visible to them from all sides, which is one reason why a panoramic camera is a handy thing to have (if you need other reasons, think of the military...).

There are currently three types of cameras that can capture panoramic images: a rotating camera, a swing-lens camera, and a camera with a parabolic panoramic lens system. The first two produce undistorted images. Here is a description of all three (followed by a note on pinhole cameras).

A rotating camera, as its name implies, works by rotating on its base, transferring the image to the film while moving the film in the opposite direction, so the film stays stationary relative to the ground. Examples of this type are the Swiss-made

RoundShot (http://www.roundshot.ch/) and the Hulcherama camera, invented and built by Charles A. Hulcher (http://www.hulchercamera.com/). Following is some information on the latter type.

The Hulcherama is a slit-scanning panoramic camera that works by rotating on its base. An electronically controlled motor is responsible for uniform rotation (the rate of rotation may be varied from 1 s to 144 s per revolution). During the rotation, the image passes through the lens, then through an adjustable narrow slit onto the film (Figure 3.62a). The slit masks out most of the image but lets a narrow portion pass through, which is how any optical distortion is minimized. As the camera rotates in one direction, the film moves past the slit in the opposite direction. The camera rotation and film movement are synchronized so that the film is stationary relative to the image being photographed. As the camera makes a complete revolution, 8.9 in. of film pass behind the slit, creating a 360° panoramic image with a height of 2.25 in. The aspect ratio is thus a pleasing $2.25:8.9 \approx 1:4$. It is possible to let the camera rotate more than one revolution (possibly changing the image each time) and a roll of 120-format film is long enough for three revolutions (the Hulcherama uses standard 120 or 220 roll film).

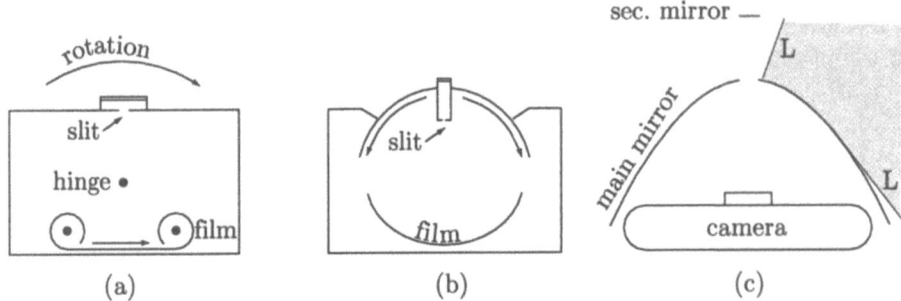

Figure 3.62: Panoramic Cameras.

A swing-lens camera (Figure 3.62b) has a lens that rotates during an exposure, thereby "painting" the image on the film through a narrow, vertical, constant-width slit. In order to keep the same distance between the film and the lens, the film has to be curved. An advantage of this type of camera is that the lens only has to cover the vertical dimension of the film and the width of the slit, so it does not have to be complex. A disadvantage is the field of view, which is less than 180°; a complete 360° panorama is created by taking several shots and combining them using special equipment (for a film camera) or special software (for a digital camera). Examples of this type are the Widelux (now discontinued) and the Noblex.

The Noblex 135-S takes a 136°-wide image and uses standard 35-mm film. The Noblex-150 provides a 146° angle-of-view, uses 120 film, and produces six 5-in.-wide images on a roll. It can take multiple exposures on the same film.

A panoramic lens system (Figure 3.62c) is somewhat similar to a reflecting telescope. Its main part is a *convex* parabolic mirror (in contrast to the mirrors

3 Transformations and Projections

used in telescopes, which are concave) that captures the entire (or almost the entire) half-sphere of image above it and sends it up, where it is reflected by a small, flat mirror and sent down, through a hole in the main mirror, to a camera. There are no moving parts, no rotating parts, no need for multiple images, and no need to stitch multiple photos together. The price for all this (aside from the price of the system) is image distortion. This kind of system can be used with any camera, digital or film.

Since the mirror captures everything above it and on all sides, the only way for the photographer to stay out of the picture is to crawl under the camera. A panoramic lens system is therefore used while mounted on a tripod and operated from below.

An example of this type is the Portal S1 panoramic lens system, made by the BeHere company of Cupertino, California (http://www.behere.com/). It is 12.5 in. in diameter, 13 in. tall, and weighs less than 10 lbs. It has a 35-mm Nikon mount, so any Nikon-compatible camera body, digital or film, can be used with the Portal S1. The depth of field of the Portal is from 1 in. to infinity (there is no need to focus the camera). Its lateral field of view is, of course, 360°, but its vertical field of view is limited to the gray area in the diagram and it equals 100° (the angle between the two lines marked L). When anything outside this area is reflected in the main mirror, it cannot reach the secondary mirror.

If a film camera is used, the film can later be scanned, then processed with special software provided by the manufacturer. This software flattens the donut-shaped image and can also perform other processing such as evening out the lighting, correcting brightness and contrast, and slightly sharpening the edges. The image can then be saved in one of the popular panoramic formats such as QuickTime VR.

▶ **Exercise 3.69:** Explain why the image produced by a panoramic-lens system is shaped like a donut.

Note: The pinhole used to be the first camera of many a poor youngster. This is simply a box with a small hole in front and film or light-sensitive paper loaded in back. The shutter can be as simple as a piece of tape that's removed to expose the film, then reapplied manually, or it can be a bought, cable-operated shutter assembly. If the hole is small enough, the resulting image is sharp; if the film is wide, this primitive device can produce wide-angle images.

See also http://laplaza.org/~pinhole/ for more examples of panoramic cameras. URL http://www.shortcourses.com/ has information on panoramic cameras and on creating panoramic images.

> I've finally figured out what's wrong with photography. It's a one-eyed man looking through a little 'ole. Now, how much reality can there be in that?
> — David Hockney.

3.13.5 Telescopic Projection

Because it transforms a three-dimensional point **P** to another three-dimensional point **P***, the telescopic projection is really a transformation, not a projection.

Nevertheless, it can be useful since it simulates the way a telescope works. It brings objects closer to the observer and it does this nonlinearly. The farther away an object is, the more it is moved toward the observer. This transformation is based on the *thin lens equation* from optics. Figure 3.63a shows such a lens, centered on the origin. A point **P** at a distance z from the origin is also shown, with its transformed point **P*** at a distance z^*, and an observer at a distance k on the other side of the origin. The thin lens equation is

$$\frac{1}{k} = \frac{1}{z} + \frac{1}{z^*} \quad \text{or} \quad z^* = \frac{kz}{z-k}. \tag{3.54}$$

Equation (3.54) shows that when z is close to zero, z^* is also close to zero (there is not much difference between a point and its projection), but when z approaches infinity, z^* approaches k. A point at infinity is brought to a distance of k from the origin and a point close to the origin is not moved much.

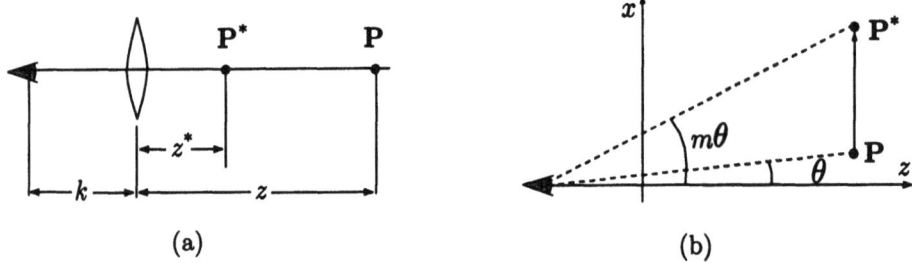

Figure 3.63: Telescopic and Microscopic Projections.

▶ **Exercise 3.70:** What should be the distance z of a point in order for it to be moved to a distance $z^* = k/2$ by the telescopic transformation?

After a three-dimensional scene has been telescoped point by point, we can use perspective projection to display it in two dimensions.

3.13.6 Microscopic Projection

A sample observed through a microscope is normally thin. We can therefore assume that points that go through a microscopic projection have the same (or similar) z coordinates. In contrast to a telescope, which brings points closer to the observer, a microscope "opens up" the points. Figure 3.63b shows how this is done by moving points away from the z axis. If the view angle of a point **P** is θ, then the microscope places its projection **P*** such that its view angle is $m\theta$, where m is the magnification power of the microscope. The projection rule is thus

$$\frac{x}{z+k} = \tan\theta \quad \text{and} \quad \frac{x^*}{z+k} = \tan(m\theta). \tag{3.55}$$

Calculating x^* therefore involves the two steps $\theta = \arctan(x/(z+k))$ and $x^* = (z+$

3 Transformations and Projections

$k) \tan(m\theta)$. For small angles, $\tan\theta$ is close to θ, so we can write as an approximation

$$\frac{x^*}{z+k} = m\frac{x}{z+k} \quad \text{or} \quad x^* = mx.$$

This is a linear scaling transformation where both x and y are scaled by a factor of m, while z is left unchanged. The transformation matrix is

$$\begin{pmatrix} m & 0 & 0 & 0 \\ 0 & m & 0 & 0 \\ 0 & 0 & 1 & 0 \\ 0 & 0 & 0 & 1 \end{pmatrix}.$$

3.13.7 Sphere Projections

Even though most of us believe that the Earth is a sphere (or close to a sphere, but see [Flat Earth Society 98] for a conflicting opinion) we prefer to see it mapped on a flat surface. Mapping a three-dimensional point located on the surface of a sphere to a two-dimensional point is a projection. There are different ways to project a sphere on a flat surface, but none is isometric, i.e., none preserves distances and angles. Three simple projections, cylindrical, Mercator and stereographic, are discussed.

Figure 3.64a shows how a sphere of radius r can be mapped into a cylinder of the same radius. Simply wrap the cylinder such that it touches the sphere along the equator and project every point on the sphere to the cylinder along the line segment that goes from the point to the cylinder and that's normal to the cylinder. The cylinder is then unrolled into a flat rectangle of dimensions $2r \times 2\pi r = 4\pi r^2$. Notice that this equals the surface area of the sphere. As we move away, north or south, from the equator, this projection shrinks the north–south distances and stretches the east–west distances. This projection is very similar to the panoramic projection.

The Mercator projection is similar, only now the sphere is wrapped by a cylinder of infinite height. A point **P** on the surface of the sphere is projected onto the cylinder along the line segment that goes from the center of the sphere to **P** and onto the cylinder. This projection is named after Gerhardus Mercator (1512–1594), a Flemish mathematician and geographer, who used it in his world map of 1569.

It is easy to see that the distance between the projected latitudes increases as we move away from the Equator (latitude 0°) and that the poles are mapped to infinity. In practice, the height of the cylinder has to be finite, so latitudes up to about 80° only can be mapped.

Figure 3.64b shows that the length of latitude θ on the sphere is $2\pi R \cos\theta$. When projected on the cylinder, this length is stretched to $2\pi R$. The stretch factor is thus $1/\cos\theta$, so this is a nonlinear projection. The lines of longitude, however, are equally spaced on the cylinder. At latitude 60°, the stretch factor is $1/\cos 60° = 2$, while at 80°, it is $1/\cos 80° \approx 5.76$.

The Mercator projection does not seem to distort shapes. The stretching of northern areas shows many details that are not visible on other maps. Canada, for

3.13 Nonlinear Projections

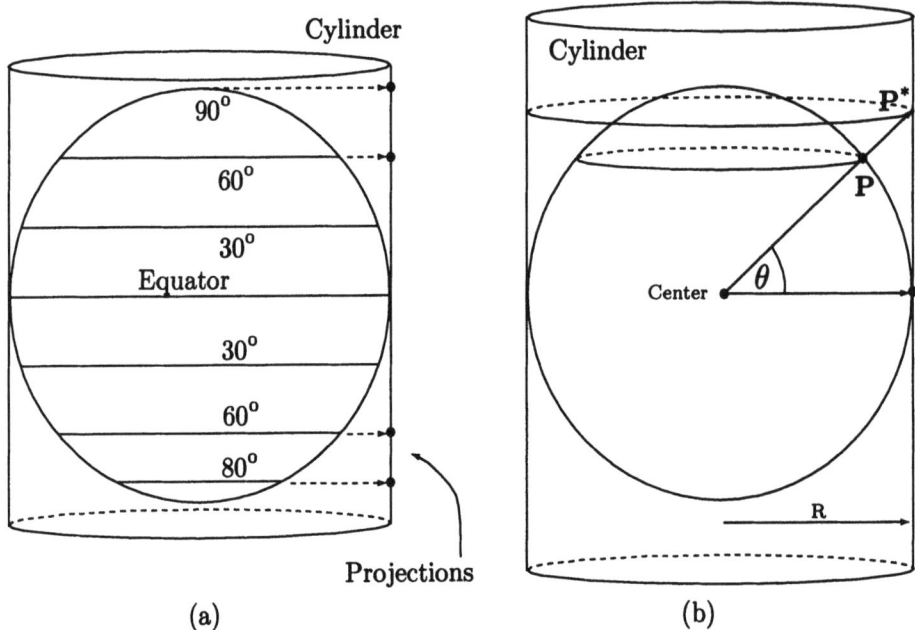

Figure 3.64: Projecting a Sphere on a Cylinder.

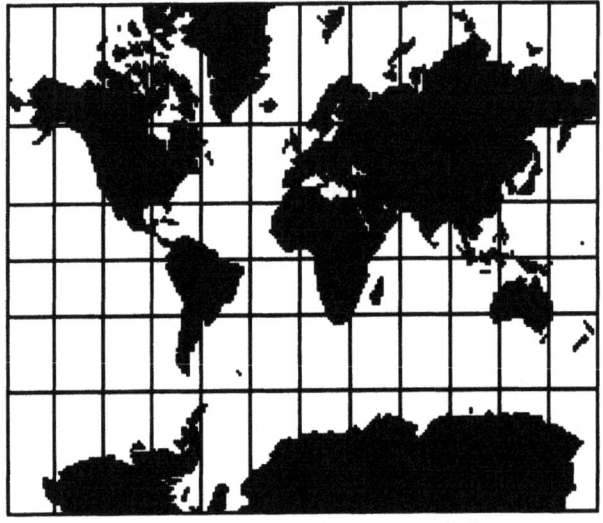

Figure 3.65: A Mercator Projection.

3 Transformations and Projections

> What in water did Bloom, waterlover, drawer of water, watercarrier, returning to the range, admire?
> Its universality: its democratic equality and constancy to its nature in seeking its own level: its vastness in the ocean of Mercator's projection: its unplumbed profundity in the Sundam trench of the Pacific exceeding 8000 fathoms: the restlessness of its waves and surface particles visiting in turn all points of its seaboard.
> —James Joyce, *Ulysses*.

example, is twice as large as the United States and the United States shows larger than Brazil on the Mercator map of the Earth. Greenland appears much larger on this map than on a globe of a comparable size (Figure 3.65).

The principle of the *stereographic projection* (Figure 3.66) is to have the projection plane tangent to the south pole of the sphere and project a point **P** on the sphere by extending the segment from the north pole to **P** until it reaches the projection plane. The north pole itself does not have any projection (it projects to infinity) and the main property of this projection is that it preserves angles.

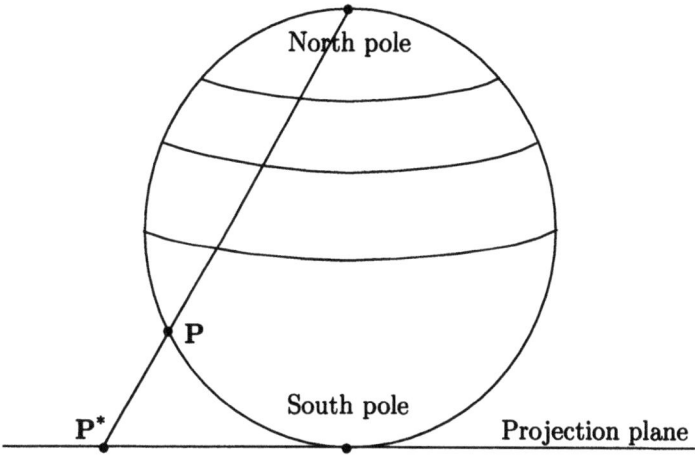

Figure 3.66: Stereographic Projection of a Sphere.

> Those who forget good and evil and seek only to know the facts are more likely to achieve good than those who view the world through the distorting medium of their own desires.
> — Bertrand Russell, *A Free Man's Worship and Other Essays*.

3.13.8 Circle Inversion

This projection transforms a two-dimensional image to another two-dimensional image. Figure 3.67 shows the unit circle about the origin and a general point **P** with polar coordinates (r,θ). The projected point is defined to be $\mathbf{P}^* = (1/r, \theta)$. Since the point and its projection have the same angle θ, they are located on the same straight line through the origin. If $r > 1$, then **P** is outside the unit circle and $\mathbf{P}^*$ is inside it since $1/r < 1$. This projection therefore inverts points with respect to the unit circle. It is easy to see that points on the circle are projected to themselves and that this projection is undefined for the origin where $r = 0$ (although we can say that the origin is projected to the *point at infinity*). Since **P** is moved to $\mathbf{P}^*$ along the line that connects **P** to the origin, we can think of this projection as scaling and express it as

$$\mathbf{P}^* = (x^*, y^*) = \frac{(x,y)}{|x^2+y^2|} = \frac{\mathbf{P}}{|x^2+y^2|} = s\mathbf{P}.$$

Notice that the scale factor s depends on **P**, showing that this type of projection is nonlinear.

This projection has a number of interesting features. Among them are the following:

1. Any circle that intersects the unit circle at right angles is projected to itself.
2. The angle between two projected lines is preserved. Circle inversion is thus a *conformal* projection.
3. Circles that do not pass through the origin are projected into circles (that do not pass through the origin and generally have a different radius).
4. Similarly, lines that do not pass through the origin are projected into circles that *do* pass through the origin.
5. A circle centered on the origin is projected to another circle similarly centered.

Here is a proof of feature 4. Figure 3.68 shows a line L that does not pass through the origin. Consequently, there must be a perpendicular to L from the origin. The point where this perpendicular meets L is denoted **P** and its projection is denoted $\mathbf{P}^*$. We now select another arbitrary point **Q** on L and denote its projection $\mathbf{Q}^*$. It is obvious that $OP \cdot OP^* = 1$ and $OQ \cdot OQ^* = 1$, so we conclude that $OP/OQ^* = OQ/OP^*$. This shows that triangles OPQ and OP^*Q^* are similar (notice that they have a common angle), which, in turn, implies that angles OPQ and OQ^*P^* are equal. Since the former is a right angle, the latter must also be 90°. However, point **Q** is an arbitrary point on L, so angle OQ^*P^* equals 90° for any point **Q** on L, showing that the projection $\mathbf{Q}^*$ lies on a circle that passes through the origin O and has a diameter OP^*. The projection of **P** is $\mathbf{P}^*$ and the projection of the origin is the point (or points) at infinity. Line L of Figure 3.68 passes inside the unit circle. For lines outside this circle, the diagram looks different but the proof is identical.

▶ **Exercise 3.71:** Use similar arguments to prove feature 3.

3 Transformations and Projections

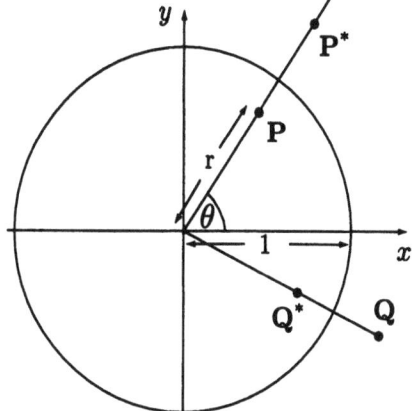

Figure 3.67: Circle Inversion.

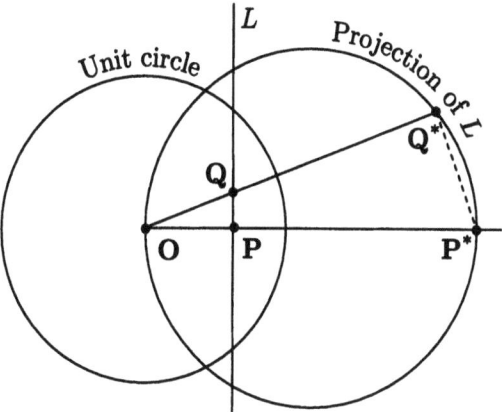

Figure 3.68: Circular Inversion of a Line.

▶ **Exercise 3.72:** The discussion so far has assumed inversion with respect to the unit circle. Given a circle C of radius R about the origin, show how to project a point **P** with respect to it.

> He felt not quite canny, as though the world were suffering a dreamlike distortion of perspective which he might arrest by shutting it all out for a few minutes and then looking at it afresh.
>
> Thomas Mann, *Death in Venice*

4
Curves

Curves are important in computer graphics for two reasons:

1. The main task of computer graphics is to generate realistic-looking images. Most real objects are seen because they reflect light, so the computer has to simulate the way light is reflected from a surface. This requires a mathematical model of the surface of an object and such a model is normally based on a model for a single curve. An understanding of computer graphics should therefore start with curves, continue with surfaces, and include a model of light reflection from a surface.

2. Some important graphics problems are two-dimensional and can be solved with curves, without the need for surfaces and light reflection. Examples are computer art (drawing and painting), technical drawing, and computer-aided manufacturing (CAM). Clothes designers, for example, can benefit from a program that can draw general curves and that can later automatically cut material along a given curve. In fact, the fair cubic splines (Section 4.8.7) were originally developed to design and cut insoles and other shoe parts.

A lot of research effort has gone into curves in the last 30 years because of these reasons. Many sophisticated curve methods are known today—some are specialized and others are general purpose. All of them, however, make extensive use of points, vectors, or both. This why this chapter starts with a discussion of points and vectors (check also Section A.4). It then introduces the concept of parametric representation of curves, and it continues with the following examples of practical methods for curves:

1. Lagrange interpolation. This type of curve uses a high-degree polynomial to interpolate data points.

2. Newton interpolation. This results in the same polynomial as the Lagrange method, but the approach is more practical.

3. Hermite interpolation. It constructs a single curve segment that is defined by means of two endpoints and two tangent vectors.

4. Cubic splines. This is a complete curve made up of Hermite segments with C^2 continuity (Section 4.3.6). Three types of those curves, namely special, general, and fair are discussed.

5. Quadratic splines. This type of curve is similar to the cubic spline but is simpler.

6. Cardinal splines. This curve is made of segments where each segment is defined by means of four points but only goes from the second point to the third one. A new parameter is added to these splines to control the *tension* of the curve.

7. Catmull-Rom curves (also called Catmull-Rom splines or Overhauser splines). This curve is also made of segments, each a blend of two parabolas. It can also be defined as a special case of the cardinal spline.

8. Kochanek-Bartels splines. This is an extension of the cardinal spline where three parameters are added to control tension, continuity, and bias. These splines are especially useful in computer animation.

9. Cubic and quadratic blending. Section 4.14.6 shows how simple cubic or quadratic expressions (based on Bernstein polynomials) can be used to blend three or four points.

10. Bézier curves. This is an approximating curve based on control points that's made up of segments (usually cubic or quadratic). This is the method of choice of many commercial graphics programs, which is why it is discussed here in great detail.

11. B-splines. This approximating curve features local control and any desired continuity. There are several varieties of this curve, among them an interpolating B-spline, where the points are data points, and a rational version. The rational B-spline is the most general type of curve in use today and many other types of curves can be obtained from it as special cases.

12. Beta-splines. They are a modification of B-splines where tension parameters are added for greater control of curve shape.

13. Conic sections. These are familiar and are useful in many practical situations.

> When producers want to know what the public wants, they graph it as curves. When they want to tell the public what to get, they say it in curves.
> — Marshall McLuhan.

4.1 Points and Vectors

This is a short discussion of certain properties of points and vectors, presenting results that will be needed later. Everything that follows applies to two-dimensional, as well as to three-dimensional points. The examples, however, are given in terms of two-dimensional points.

Points and vectors are similar but not identical. A point has no dimensions; it represents a location in space. A vector is a mathematical entity that has two attributes: direction and magnitude. It is possible to associate a point **P** with the

4 Curves

vector **v** connecting it to the origin (Figure 4.1c). This association is useful, but, generally the point and the vector are different.

Example: If the coordinates of a point are $(3,4)$, then the point is located 3 units away from the y axis and 4 units away from the x axis. If, on the other hand, a vector is represented in terms of the same pair $(3,4)$, then the direction of the vector is $4/3$ (it moves 3 units in the x direction for every 4 units in the y direction) and its magnitude is $\sqrt{3^2 + 4^2} = 5$.

Both points and vectors are written as pairs. However, a vector pair represents direction and magnitude, whereas a point pair indicates a position relative to the coordinate axes. Consider the two vectors **w** in Figure 4.1b. They look different but are identical, since they have the same direction and magnitude. One goes from point **a** to point **b** and can be written $\mathbf{w} = \mathbf{b} - \mathbf{a}$. The other can be written as $\mathbf{w} = \mathbf{d} - \mathbf{c}$. Points **a** and **c** are different since they occupy different positions relative to the coordinate axes. Points **b** and **d** are also different. However, the differences $\mathbf{b} - \mathbf{a}$ and $\mathbf{d} - \mathbf{c}$ are identical and so are the vectors.

Section A.4 discusses operations on vectors. Here, we study operations on points. Let $\mathbf{P}_0 = (x_0, y_0)$ and $\mathbf{P}_1 = (x_1, y_1)$ be two points. The *difference* $\mathbf{P}_1 - \mathbf{P}_0 = (x_1 - x_0, y_1 - y_0) = (\Delta x, \Delta y)$ is well defined. It is the vector (the direction and distance) from $\mathbf{P}_0$ to $\mathbf{P}_1$ (Figure 4.1a). Note that it is a pair, not a single number.

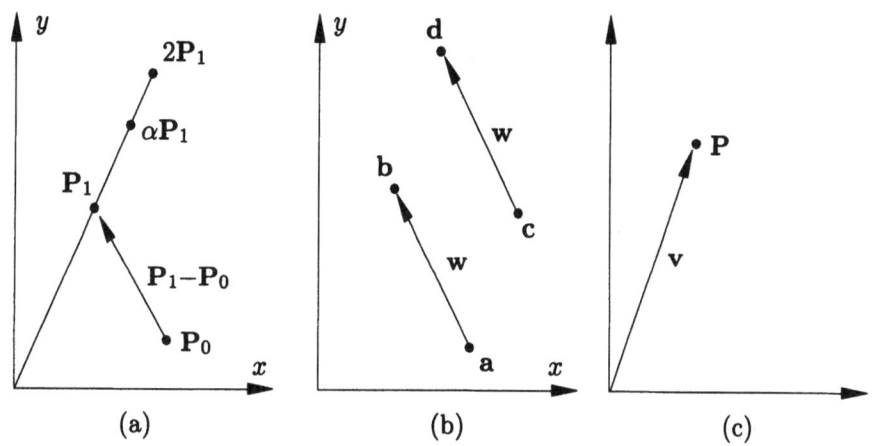

Figure 4.1: Operations on Points.

Example: The two points $\mathbf{P}_0 = (5, 4)$ and $\mathbf{P}_1 = (2, 6)$ are subtracted to produce the pair $\mathbf{P}_1 - \mathbf{P}_0 = (-3, 2)$. The new pair, however, is a vector, not a point, since it represents a direction and a distance. To get from $\mathbf{P}_0$ to $\mathbf{P}_1$, we need to move -3 units in the x direction and 2 units in the y direction. Similarly, $\mathbf{P}_0 - \mathbf{P}_1$ is the direction from $\mathbf{P}_1$ to $\mathbf{P}_0$. The distance between the points is $\sqrt{(-3)^2 + 2^2}$. These properties do not depend on the particular coordinate axes used. If, for example, we translate the origin—or, equivalently, translate the points—m units in the x direction and n units in the y direction, the points will have new coordinates, but the difference will not change. The difference of two points is thus a vector.

▶ The same property (the difference of points being independent of the coordinate axes) holds after rotation, scaling, shearing, and reflection: the so-called *affine transformations* (or mappings). This is why the operation of subtracting two points is affinely invariant.

The product $\alpha\mathbf{P}$, where α is a real number, is a point on the line connecting $\mathbf{P}$ with the origin (Figure 4.1a). This operation is also affinely invariant. As a corollary, we conclude that any three *collinear* points have the form $\mathbf{P}_0$, $\mathbf{P}_1 = \alpha\mathbf{P}_0$, and $\mathbf{P}_2 = \beta\mathbf{P}_0$ or, equivalently, satisfy the relation

$$\mathbf{P}_0 = \frac{1}{2\alpha}\mathbf{P}_1 + \frac{1}{2\beta}\mathbf{P}_2 = a\mathbf{P}_1 + b\mathbf{P}_2$$

for some real numbers a and b. Three collinear points are therefore not independent since any of them can be expressed in terms of the other two.

▶ **Exercise 4.1:** Given the three points $\mathbf{P}_0 = (12, -3)$, $\mathbf{P}_1 = (24, -6)$, and $\mathbf{P}_2 = (18, -4.5)$, are they collinear?

▶ **Exercise 4.2:** What can we say about four collinear points?

The sum of points, however, is not well defined (but see note below). One might think that adding two points is like adding vectors. The lines connecting the points with the origin are added, to produce a sum vector. In fact, as Figure 4.2a shows, this operation depends on the coordinate axes. Moving the origin (or moving the points) will move the sum of the vectors, thus changing the sum of the points. This is why the sum of points is, in general, undefined.

The sum of vectors, however, is well defined since it represents a sum of directions and magnitudes. The sum of a point and a vector is similarly well defined (Figure 4.2b).

▶ **Exercise 4.3:** The sum of a point and a vector is well defined, but is it a point or a vector?

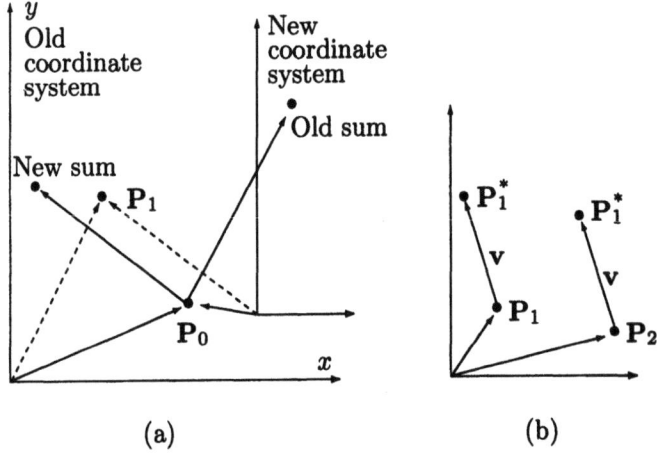

Figure 4.2: (a) Adding Points. (b) Adding a Point and a Vector.

4 Curves

Note: There is, however, one important special case where the sum of points is defined. This is the so-called *barycentric sum*. If we multiply each point by a weight and if the weights add up to 1, then the sum of the weighted points is affinely invariant, i.e., it is a valid point. Here is the (simple) proof:

If $\sum_{i=0}^{n} w_i = 1$, then

$$\sum_{i=0}^{n} w_i \mathbf{P}_i$$
$$= \mathbf{P}_0 + \sum_{i=1}^{n} w_i \mathbf{P}_i - (1 - w_0)\mathbf{P}_0$$
$$= \mathbf{P}_0 + w_1 \mathbf{P}_1 + w_2 \mathbf{P}_2 + \cdots + w_n \mathbf{P}_n - (w_1 + \cdots + w_n)\mathbf{P}_0$$
$$= \mathbf{P}_0 + w_1(\mathbf{P}_1 - \mathbf{P}_0) + w_2(\mathbf{P}_2 - \mathbf{P}_0) + \cdots + w_n(\mathbf{P}_n - \mathbf{P}_0)$$
$$= \mathbf{P}_0 + \sum_{i=1}^{n} w_i(\mathbf{P}_i - \mathbf{P}_0). \tag{4.1}$$

This is the sum of the point $\mathbf{P}_0$ and the vector $\sum_{i=1}^{n} w_i(\mathbf{P}_i - \mathbf{P}_0)$, and we already know that the sum of a point and a vector is a point. ◄

A special case is the barycentric sum of two points $(1-t)\mathbf{P}_0 + t\mathbf{P}_1$. This is a point on the line from $\mathbf{P}_0$ to $\mathbf{P}_1$. In fact, the entire straight segment from $\mathbf{P}_0$ to $\mathbf{P}_1$ is obtained when t is varied from 0 to 1 (Figure 4.3a).

Proof: Let $\mathbf{P}(t) = (1-t)\mathbf{P}_0 + t\mathbf{P}_1$. Clearly, $\mathbf{P}(0) = \mathbf{P}_0$ and $\mathbf{P}(1) = \mathbf{P}_1$. Also, since $\mathbf{P}(t) = t(\mathbf{P}_1 - \mathbf{P}_0) + \mathbf{P}_0$, $\mathbf{P}(t)$ is a linear function of t, meaning a straight line in t. The tangent vector is the derivative $\frac{d\mathbf{P}}{dt}$ and it equals $\mathbf{P}_1 - \mathbf{P}_0$, the direction from $\mathbf{P}_0$ to $\mathbf{P}_1$. Notice that this is a vector, not a slope. When $t = 1/2$, we get $\mathbf{P}(0.5) = \frac{1}{2}\mathbf{P}_1 + \frac{1}{2}\mathbf{P}_0$, the midpoint between $\mathbf{P}_0$ and $\mathbf{P}_1$ ◄

The concept of barycentric weights is so useful that the two numbers $1-t$ and t are called the *barycentric coordinates* of point $\mathbf{P}(t)$ with respect to $\mathbf{P}_0$ and $\mathbf{P}_1$.

> The word *barycentric* is derived from *barycenter*, meaning "center of gravity," because such weights are used to calculate the center of gravity of an object. Barycentric weights have many uses in geometry in general and in curve and surface design in particular.

Another useful example is the barycentric coordinates of a two-dimensional point with respect to the three corners of a triangle. Imagine a triangle with corners $\mathbf{P}_0$, $\mathbf{P}_1$, and $\mathbf{P}_2$ (Figure 4.3b). Any point $\mathbf{P}$ inside the triangle can be expressed as the weighted combination

$$\mathbf{P} = u\mathbf{P}_0 + v\mathbf{P}_1 + w\mathbf{P}_2, \quad \text{where} \quad u + v + w = 1. \tag{4.2}$$

The proof is that Equation (4.2) can be written explicitly as three equations in the three unknowns u, v, w:

$$P_x = uP_{0x} + vP_{1x} + wP_{2x},$$
$$P_y = uP_{0y} + vP_{1y} + wP_{2y}, \quad (4.3)$$
$$1 = u + v + w.$$

The solutions are unique provided that the three equations are independent. ◀

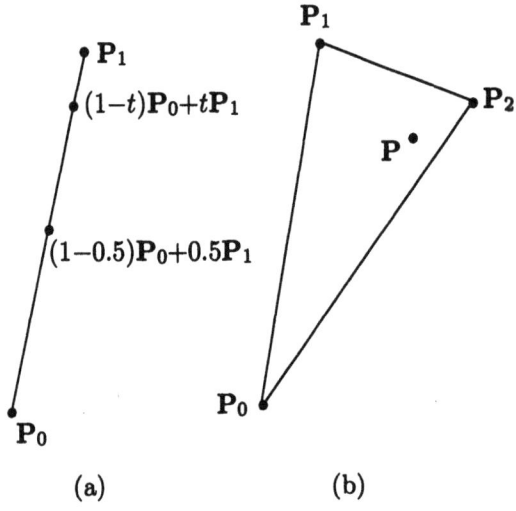

Figure 4.3: Line and Triangle.

▸ **Exercise 4.4:** Show that Equation (4.3) consists of three independent equations if the three points P_0, P_1, and P_2 are independent.

▸ **Exercise 4.5:** Show that the barycentric coordinates of point P_0 with respect to P_0, P_1, and P_2 are $(1,0,0)$. Also discuss the barycentric coordinates of points outside the triangle.

Example: Let $P_0 = (1,1)$, $P_1 = (2,3)$, $P_2 = (5,1)$, and $P = (2,2)$. Equation (4.3) becomes

$$(2,2) = u(1,1) + v(2,3) + w(5,1); \quad u + v + w = 1,$$

or

$$2 = u + 2v + 5w,$$
$$2 = u + 3v + w, \quad \text{which yield} \quad \begin{cases} u = 3/8, \\ v = 1/2, \\ w = 1/8. \end{cases}$$
$$1 = u + v + w,$$

▸ **Exercise 4.6:** For a given triangle, calculate the (x, y, z) coordinates of the point with barycentric coordinates $(1/3, 1/3, 1/3)$. This point is called the *centroid* and is, in a sense, the center of the triangle. (Imagine cutting the triangle out of a piece of cardboard. If you try to support it at the centroid, it will balance.) See also Section 5.3.8 for more triangle centers.

Note: This material is useful for the triangular Bézier surface patches described in Section 5.13.

▶ The barycentric combination is the most fundamental operation on points; so much so that it is used to define affine transformations (see also page 75). The definition is: A transformation of points in space is affine if it leaves barycentric combinations invariant. Hence, if $\mathbf{P} = \sum w_i \mathbf{P}_i$ and $\sum w_i = 1$, and if $\mathbf{T}$ is an affine transformation, then $\mathbf{TP} = \sum w_i \mathbf{TP}_i$. All the transformations discussed earlier—scaling, shearing, etc.—are affine.

Note: The difference of two points is a vector. We can consider such a difference a weighted sum where the weights add up to zero (they are $+1$ and -1). It turns out that a weighted sum of points where the weights add up to zero is a vector.

Proof: Let
$$\mathbf{Q} = \sum_{i=1}^{n} w_i \mathbf{P}_i, \quad \text{where} \quad \sum w_i = 0,$$
and let $\mathbf{P}$ be a point. The sum $\mathbf{R} = \mathbf{Q} + \mathbf{P}$ is barycentric (since its coefficients add up to 1) and is therefore a point. The difference $\mathbf{R} - \mathbf{P} = \mathbf{Q}$ is a difference of points and, hence, is a vector ◀

Note: Multiplying a point by a number produces a point, so if $\mathbf{P}$ is a point, then $-\mathbf{P}$ is also a point. It is located on the line connecting $\mathbf{P}$ with the origin, on the other side of the origin from $\mathbf{P}$. Once this is understood, we notice that the sum of points $\mathbf{P} + \mathbf{Q}$ can be written as the difference of points $\mathbf{P} - (-\mathbf{Q})$. This difference is, of course, the vector from point $-\mathbf{Q}$ to point $\mathbf{P}$ (Figure 4.4), so we conclude that the sum $\mathbf{P} + \mathbf{Q}$ of two points is well defined but is not very useful since it tells us something about the relative positions of $\mathbf{P}$ and $-\mathbf{Q}$, not $\mathbf{P}$ and $\mathbf{Q}$. Assuming that Figure 4.4 depicts the points $\mathbf{Q} = (-5, -1)$ and $\mathbf{P} = (4, 3)$, the sum $\mathbf{P} + \mathbf{Q}$ equals $(-5, -1) + (4, 3) = (-1, 2)$. This shows that in order to get from point $-\mathbf{Q}$ to point $\mathbf{P}$, we need to move one negative step in the x direction for every two steps in the y direction.

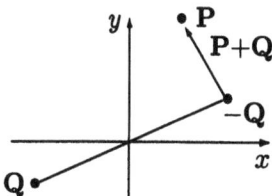

Figure 4.4: Adding Two Points.

▶ Exercise 4.7: Let $\mathbf{P}$ and $\mathbf{Q}$ be points and let $\mathbf{v}$ and $\mathbf{w}$ be vectors. What is the sum $\mathbf{P} - \mathbf{Q} + \mathbf{v} + \mathbf{w}$?

Summary: The following operations have been discussed here:
$$\text{point} - \text{point} = \text{vector},$$
$$\text{scalar} \times \text{point} = \text{point},$$

$$\text{vector} \pm \text{vector} = \text{vector},$$
$$\text{scalar} \times \text{vector} = \text{vector},$$
$$\text{point} + \text{vector} = \text{point}.$$

The operation point + point is left undefined (since it is not useful). A barycentric sum of points is a point, and a weighted sum of points where the weights add up to zero is a vector. The distance between the two points $\mathbf{P}_0$ and $\mathbf{P}_1$ is $\sqrt{(x_0 - x_1)^2 + (y_0 - y_1)^2}$.

Section A.4 discusses more operations on vectors.

4.2 Parametric Blending

This is a family of techniques that make it possible to adjust the value of some quantity in small steps, without any discontinuities. Blending can be thought of as averaging or interpolating. Examples are the following (see also Section 4.14.6):

1. Numbers. The average of the two numbers 15 and 18 is $(15 + 18)/2 = 16.5$. This can also be written as $0.5 \times 15 + 0.5 \times 18$, which can be interpreted as the *blend*, or the weighted sum, of the two numbers, where each is assigned a weight of 0.5. If 15 should be assigned a greater weight than 18, then a blend such as $0.9 \times 15 + 0.1 \times 18$ can be calculated. An example of a practical algorithm that blends numbers is Gouraud shading, in which reflection intensities are interpolated. This method smooths out the shading of a polygonal surface and is discussed in detail in Section 6.3.

2. Points. If $\mathbf{P}_1$ and $\mathbf{P}_2$ are points, then the expression $\alpha \mathbf{P}_1 + \beta \mathbf{P}_2$ is a blend of the two points, in which α and β are the weights (or the coefficients). If $\alpha + \beta = 1$, then the blend is a point on the line connecting $\mathbf{P}_1$ and $\mathbf{P}_2$.

3. Rotations. A rotation is described by means of the rotation angle (one number) and the axis of rotation (three numbers). These four numbers can be described by a quaternion (Section 3.5.4) and two quaternions can also be blended, resulting in a smooth sequence of rotations that proceeds in small, equal steps from an initial rotation to a final one. This is useful in computer animation and is discussed in Section 8.4.2.

4. Curve construction. Given a number of points, a curve can be created as a weighted sum of the points. It has the form $\sum w_i(t) \mathbf{P}_i$, where the weights $w_i(t)$ are barycentric. Such a curve is a *blend* of the points. For each value of t, the blend is different, but we have to make sure that the sum of the weights is always one. It is possible to blend vectors, in addition to points, as part of the curve, and the weights of these vectors don't have to satisfy any particular requirement. Most of the curve methods described in this chapter generate a curve as a blend of points, vectors, or both.

5. Surfaces. Using the same principle, points, vectors, and curves can be blended to form a surface patch.

6. Images. Section 10.9.1 shows how image processing can be done by blending a general image with a special one.

4.3 Curve Representations

As mentioned in the Preface, one of our main aims is to understand how a solid, realistic-looking curved surface can be specified by the user, calculated by software, and displayed pixel by pixel. The first step toward this goal is an understanding of curves. Once we have an algorithm to calculate and display a general curve, we may try to generalize it to a surface.

In practice, curves (and surfaces) are specified by the user in terms of points and are developed in an interactive process. The user starts by entering the coordinates of points, either by scanning a rough image of the desired shape and digitizing certain points on the image, or by drawing a rough shape on the screen and selecting certain points with the mouse or a similar pointing device. After the curve has been drawn, the user may want to change its shape by moving, adding, or deleting points. Such points can be used in two different ways:

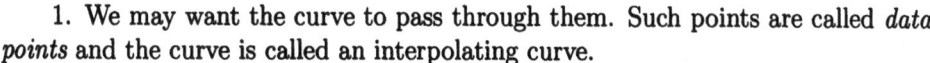

1. We may want the curve to pass through them. Such points are called *data points* and the curve is called an interpolating curve.

2. We may want the points to control the shape of the curve by exerting a "pull" on it. A point may pull part of the curve toward it, allowing the user to change the shape of the curve by moving the point. Generally, however, the curve does not pass through the point. Such points are called *control points* and the curve is called an approximating curve.

We start with the concepts of explicit, implicit, and parametric curves. We then show how polynomials can be used to fit a curve through given data points and how polynomials can be generalized to splines (parametric curves made of segments). The discussion applies to both plane (two-dimensional) and space (three-dimensional) curves. Any algorithm for plane parametric curves can easily be extended to three dimensions and the only difference is that plane curves have both a slope *and* a tangent vector, whereas space curves have a tangent vector only (see discussion below).

An expression such as $y = f(x)$ describes a curve. This is called the *explicit* representation of the curve. It is not very general since it cannot represent vertical lines and is also single-valued. For each value of x, only a single value of y can be calculated. The *implicit* representation of a curve has the form $F(x,y) = 0$ and can represent multivalued curves (more than one y value for an x value). These curve representations can be used only when the function is known. In practical work—where complex curves (in the shape of a car or of a toaster, e.g.) are needed—the function is normally unknown, which is why a different approach is needed.

The curve representation used in practice is called the *parametric representation*. A two-dimensional parametric curve has the form $\mathbf{P}(t) = \big(f(t), g(t)\big)$ or $\mathbf{P}(t) = \big(x(t), y(t)\big)$. The functions f and g give the (x, y) coordinates of any point on the curve, and the points are obtained when the parameter t is varied over a certain interval $[a, b]$, usually $[0, 1]$.

A simple example of a two-dimensional parametric curve is $\mathbf{P}(t) = (2t - 1, t^2)$. When t is varied from 0 to 1, the curve proceeds from the initial point $\mathbf{P}(0) = (-1, 0)$ to the final point $\mathbf{P}(1) = (1, 1)$. The x coordinate is linear in t and the y coordinate behaves as t^2.

Note that the first derivative $\frac{d\mathbf{P}(t)}{dt}$, which we denote $\mathbf{P}^t(t)$ or $(P_x^t(t), P_y^t(t))$, is the tangent vector to the curve at any point. The tangent is a vector and, therefore, possesses a direction (which is the direction of the curve at the point) and a magnitude (which indicates the speed of the curve). The tangent, however, is not the slope of the curve. The tangent is a pair of numbers, whereas the slope is a single number. The slope equals $\tan\theta$, where θ is the angle between the tangent vector and the x axis. The slope of a two-dimensional parametric curve is obtained by

$$\frac{dy}{dx} = \frac{\frac{dy}{dt}}{\frac{dx}{dt}} = \frac{P_y^t(t)}{P_x^t(t)}.$$

An example is the curve $\mathbf{P}(t) = (x(t), y(t)) = (1 + t, t^2)$. Its tangent vector is $\mathbf{P}^t(t) = (t, 2t)$ and the slope is $2t/t = 2$. The slope is constant, which suggests that the curve is a straight line. This is also easy to see from the tangent vector. The direction of this vector is always the same since it can be described by saying "for every step in the x direction, move two steps in the y direction."

Circle Representations

Because of its high symmetry, the circle can be represented in many ways. The following representations all assume a center at the origin. For a circle centered at point (m, n), just add the pair (m, n) to every pixel before plotting it.

1. Cartesian implicit representation. $x^2 + y^2 = R^2$, where $-\pi \leq x \leq \pi$. For each value of x, there are two values of y.
2. Polar representation. $\mathbf{P}(R, \theta) = (R\cos\theta, R\sin\theta)$, where $0 \leq \theta \leq 2\pi$.
3. Parametric representations. Four different representations are shown, but notice that the first of them is identical to the polar representation above.
   3.1. $\mathbf{P}(t) = R(\cos t, \sin t)$, where $0 \leq t \leq 2\pi$.
   3.2. Substituting $t = \tan(u/2)$ yields $\mathbf{P}(t) = R[(1 - t^2)/(1 + t^2), 2t/(1 + t^2)]$. When $0 \leq t \leq 1$, this generates the first quadrant from $(R, 0)$ to $(0, R)$ (see also Figure 4.5b).
   3.3. $\mathbf{P}(t) = R(t, \pm\sqrt{1 - t^2})$. When $0 \leq t \leq 1$ this generates the first quadrant from $(0, R)$ to $(R, 0)$ and, simultaneously, the third quadrant from $(0, -R)$ to $(-R, 0)$.
   3.4. $\mathbf{P}(t) = (0.441, -0.441)t^3 + (-1.485, -0.162)t^2 + (0.044, 1.603)t + (1, 0)$. When $0 \leq t \leq 1$, this generates the first quadrant from $(1, 0)$ to $(0, 1)$.

(See also Section A.2 and Equations (2.2) and (Ans.32).)

▶ **Exercise 4.8:** Show how an expression such as in 3.4 is derived.

▶ **Exercise 4.9:** Figure 4.5a shows a polygon inscribed in a circle. It is clear that adding sides to the polygon brings it closer to the circle. Calculate the difference $R - d$ as a function of n, the number of polygon sides.

The particle paradigm: Better insight into the behavior of parametric functions can be gained by thinking of the curve $\mathbf{P}(t) = (x(t), y(t))$ as a path traced out by a particle. The parameter t can then be interpreted as time and the first two derivatives $\mathbf{P}^t(t)$ and $\mathbf{P}^{tt}(t)$ can be interpreted as the velocity and acceleration of

4 Curves

the particle, respectively. It turns out that different parametric representations of the same curve may have different "speeds." The particle represented by Example 3.1, for example, "moves" along the circle at speed $\mathbf{P}^t(t) = (-\sin t, \cos t)$, which is constant since $|\mathbf{P}^t(t)| = \sqrt{\sin^2 t + \cos^2 t} = 1$. The particle of circle representation 3.2, on the other hand, moves at the variable velocity

$$\mathbf{P}^t(t) = \left(\frac{-4t}{(1+t^2)^2}, \frac{2(1-t^2)}{(1+t^2)^2}\right).$$

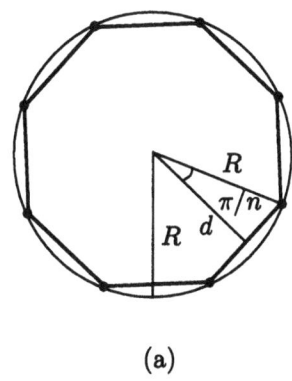
(a)

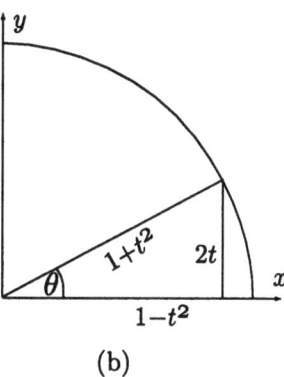
(b)

Figure 4.5: (a) A Polygon Inscribed in a Circle.
(b) A Parametric Representation.

▶ **Exercise 4.10:** Show that this velocity varies with t.

▶ **Exercise 4.11:** What three-dimensional curve is described by the parametric function $(\cos t, \sin t, t)$? (Hint: see Section 5.6.1).

See also page 517 for the parametric representations of the sphere, the ellipsoid, and of the torus as a small circle rotating around a larger circle. Section A.6.1 discusses great circles on a sphere.

We conclude this discussion with the immortal words of Galileo (see [Knuth 84] page 101).

> The area of a circle is a mean proportional between any two regular and similar polygons of which one circumscribes it and the other is isoperimetric with it. In addition, the area of the circle is less than that of any circumscribed polygon and greater than that of any isoperimetric polygon. And further, of these circumscribed polygons, the one that has the greater number of sides has a smaller area than the one that has a lesser number; but, on the other hand, the isoperimetric polygon that has the greater number of sides is the larger.
> [Galileo, 1638]

4.3.1 Straight Lines

The straight line is the simplest curve, so it is the first to be discussed. It can be represented in a number of ways.

Explicit Representation. This is the well-known expression $y = ax + b$, where a is the slope and b is the y intercept. This is a simple, intuitive representation that has one disadvantage—it cannot handle vertical lines (for which the slope is infinite).

Given the line $y = ax + b$, any line perpendicular to it should have a slope of $-1/a$, so its equation is $y = -x/a + c$, where c, the y intercept, can be any real number.

Given two lines $y = ax + b$ and $y = cx + d$, their intersection point (x_0, y_0) is located on both lines so it satisfies $ax_0+b = cx_0+d$, which implies $x_0 = (d-b)/(a-c)$ and
$$y_0 = ax_0 + b = a(d-b)/(a-c) + b = (ad - bc)/(a - c).$$

Such a point exists if $a \neq c$, i.e., if the lines have different slopes (if they have the same slope, they are either parallel or identical and they don't have a single intersection point). See also Section 4.3.3.

Given the slope a of a line and one point (x_0, y_0) on it, the line equation is $y - y_0 = a(x - x_0)$ or $y = ax + y_0 - ax_0$.

> Time can be thought of as a line (theoretically, of infinite length) on which is located, as a continuously moving point, the present moment. Anything ahead of the present moment is in the future, and anything behind it is in the past.
> —Randolph Quirk et al., *A Comprehensive Grammar of the English Language.*

Implicit Representation. This has the form $ax + by + c = 0$ and its advantage is that vertical lines can also be represented. Notice that this representation can be written
$$y = -\frac{a}{b}x - \frac{c}{b},$$
showing that the slope is $-a/b$ and the y intercept is $-c/b$. The three coefficients a, b, and c are therefore not independent. If we change them such that the ratios $-a/b$ and $-c/b$ are maintained, we end up with the same line. A simple example is the lines $3x + 2y + 4 = 0$ and $6x + 4y + 8 = 0$ that are obviously identical. The expression $\alpha ax + \alpha by + \alpha c = 0$ thus describes the same line regardless of the value of α. In order to standardize our notation, we select an α such that $(\alpha a)^2 + (\alpha b)^2 = 1$. When the line $ax + by + c = 0$ is given in *standard form* (or canonical form), we can therefore assume that $a^2 + b^2 = 1$.

A natural problem at this point is to find the equation of the line perpendicular to $ax + by + c = 0$. Since the slope of the original line is $-a/b$, the slope of the perpendicular should be $+b/a$. Any line perpendicular to $ax + by + c = 0$ can therefore be expressed as $bx - ay + d = 0$.

4 Curves

The equation of the line that passes through two given points (x_0, y_0) and (x_1, y_1) is

$$\frac{y - y_0}{x - x_0} = \frac{y_1 - y_0}{x_1 - x_0},$$

which can be written $(y_1 - y_0)x + (x_0 - x_1)y + (x_1 y_0 - y_1 x_0) = 0$. This has the implicit form $ax + by + c = 0$.

A variation is the *intercept representation* of a line $x/a + y/b = 1$. It is easy to see that this line intercepts the coordinate axes at points $x = a$ and $y = b$, respectively.

The implicit representation can also be used to divide the entire two-dimensional plane into two parts. The function $L(x, y) = ax + by + c$ divides the plane into the two regions: $L(x, y) > 0$ and $L(x, y) < 0$. The boundary between them is the line $L(x, y) = 0$.

▶ *Parametric Representation.* The parametric representation of the line segment from $\mathbf{P}_0$ to $\mathbf{P}_1$ is

$$\mathbf{P}(t) = (1 - t)\mathbf{P}_0 + t\mathbf{P}_1 = \mathbf{P}_0 + (\mathbf{P}_1 - \mathbf{P}_0)t = \mathbf{P}_0 + t\mathbf{d}. \quad (4.4)$$

Note that the weights $1 - t$ and t are barycentric. The tangent vector is $\frac{d\mathbf{P}}{dt} = \mathbf{P}_1 - \mathbf{P}_0 = \mathbf{d}$, the direction from $\mathbf{P}_0$ to $\mathbf{P}_1$.

See also Section 4.23 for the normal parametric representation of a straight line.

Note how the straight line is obtained in the figure as a linear, barycentric combination of the two vectors $\mathbf{P}_0$ and $\mathbf{P}_1$, with coefficients $(1-t)$ and t. We can think of this combination as a vector that pivots from $\mathbf{P}_0$ to $\mathbf{P}_1$ and at the same time changes its magnitude, so its tip always touches the line.

The expression $\mathbf{P}_0 + t\mathbf{d}$ is also useful. It describes the line as the sum of the point $\mathbf{P}_0$ and the vector $t\mathbf{d}$, a vector pointing from $\mathbf{P}_0$ to $\mathbf{P}_1$, whose magnitude depends on t. This representation is useful in cases where the direction of the line and one point on it are known. Notice that t varies in the range $[-\infty, +\infty]$.

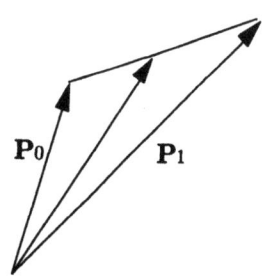

4.3.2 Distance of a Point From a Line

Given a line in parametric form $\mathbf{L}(t) = \mathbf{P}_0 + t\mathbf{v}$ (where $\mathbf{v}$ is a vector in the direction of the line) and a point $\mathbf{P}$, what is the distance between them? Assume that $\mathbf{Q}$ is the point on $\mathbf{L}(t)$ that's the closest to $\mathbf{P}$. Point $\mathbf{Q}$ can be expressed as $\mathbf{Q} = \mathbf{P}(t_0) = \mathbf{P}_0 + t_0 \mathbf{v}$ for some t_0. The vector from $\mathbf{Q}$ to $\mathbf{P}$ is $\mathbf{P} - \mathbf{Q}$. Since $\mathbf{Q}$ is the closest point to $\mathbf{P}$, this vector should be perpendicular to the line. We thus end up with the condition $(\mathbf{P} - \mathbf{Q}) \bullet \mathbf{v} = 0$ or $(\mathbf{P} - \mathbf{P}_0 - t_0 \mathbf{v}) \bullet \mathbf{v} = 0$, which is satisfied by

$$t_0 = \frac{(\mathbf{P} - \mathbf{P}_0) \bullet \mathbf{v}}{\mathbf{v} \bullet \mathbf{v}}.$$

Substituting this value of t_0 in the line equation gives

$$\mathbf{Q} = \mathbf{P}_0 + \frac{(\mathbf{P} - \mathbf{P}_0) \bullet \mathbf{v}}{\mathbf{v} \bullet \mathbf{v}} \mathbf{v}. \tag{4.5}$$

The distance between $\mathbf{Q}$ and $\mathbf{P}$ is the magnitude of vector $\mathbf{P} - \mathbf{Q}$.

This method always works since vector $\mathbf{v}$ cannot be zero (otherwise there would be no line).

In the two-dimensional case, the line can be represented explicitly as $y = ax+b$ and the problem can be easily solved using just elementary trigonometry. Figure 4.6 shows a general point $\mathbf{P} = (P_x, P_y)$ at a distance d from a line $y = ax + b$. It is easy to see that the vertical distance e between the line and $\mathbf{P}$ is $|P_y - aP_x - b|$. We also know from trigonometry that

$$1 = \sin^2 \alpha + \cos^2 \alpha = \tan^2 \alpha \cos^2 \alpha + \cos^2 \alpha = \cos^2 \alpha (1 + \tan^2 \alpha),$$

implying

$$\cos^2 \alpha = \frac{1}{1 + \tan^2 \alpha}.$$

We thus get

$$d = e \cos \alpha = e\sqrt{\cos^2 \alpha} = \frac{e}{\sqrt{1 + \tan^2 \alpha}} = \frac{|P_y - aP_x - b|}{\sqrt{1 + a^2}}. \tag{4.6}$$

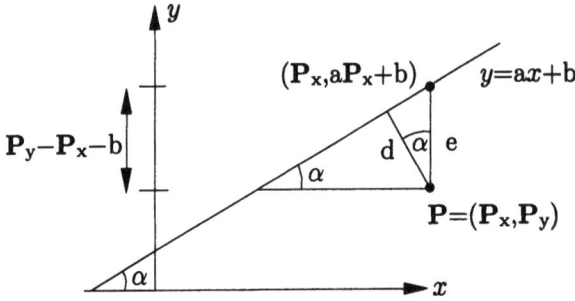

Figure 4.6: Distance Between $\mathbf{P}$ and $y = ax + b$.

▶ **Exercise 4.12:** Many math problems can be solved in more than one way and this problem is a good example. It is easy to solve by approaching it from different directions. Show some approaches to the solution.

> A man who boasts about never changing his views is a man who's decided always to travel in a straight line—the kind of idiot who believes in absolutes.
> — Honoré de Balzac, *Père Goriot*.

4 Curves

4.3.3 Intersection of Lines

Here is a simple, fast algorithm for finding the intersection point(s) of two line segments. Assuming that the two segments $\mathbf{P}_1 + \alpha(\mathbf{P}_2 - \mathbf{P}_1)$ and $\mathbf{P}_3 + \beta(\mathbf{P}_4 - \mathbf{P}_3)$ are given (Equation (4.4)), their intersection point satisfies

$$\mathbf{P}_1 + \alpha(\mathbf{P}_2 - \mathbf{P}_1) = \mathbf{P}_3 + \beta(\mathbf{P}_4 - \mathbf{P}_3),$$

or

$$\alpha(\mathbf{P}_2 - \mathbf{P}_1) - \beta(\mathbf{P}_4 - \mathbf{P}_3) + (\mathbf{P}_1 - \mathbf{P}_3) = 0.$$

This can also be written $\alpha\mathbf{A} + \beta\mathbf{B} + \mathbf{C} = 0$, where $\mathbf{A} = \mathbf{P}_2 - \mathbf{P}_1$, $\mathbf{B} = \mathbf{P}_3 - \mathbf{P}_4$, and $\mathbf{C} = \mathbf{P}_1 - \mathbf{P}_3$. The solutions are

$$\alpha = \frac{B_y C_x - B_x C_y}{A_y B_x - A_x B_y}, \qquad \beta = \frac{A_x C_y - A_y C_x}{A_y B_x - A_x B_y}.$$

The calculation of $\mathbf{A}$, $\mathbf{B}$, and $\mathbf{C}$ requires six subtractions. The calculation of α and β requires three subtractions, six multiplications (since the denominators are identical), and two divisions.

Example 1: To calculate the intersection of the line segment from $\mathbf{P}_1 = (-1, 1)$ to $\mathbf{P}_2 = (1, -1)$ with the line segment from $\mathbf{P}_3 = (-1, -1)$ to $\mathbf{P}_4 = (1, 1)$, we first calculate

$$\mathbf{A} = \mathbf{P}_2 - \mathbf{P}_1 = (2, -2), \quad \mathbf{B} = \mathbf{P}_3 - \mathbf{P}_4 = (-2, -2), \quad \mathbf{C} = \mathbf{P}_1 - \mathbf{P}_3 = (0, 2).$$

Then calculate

$$\alpha = \frac{0 + 4}{4 + 4} = \frac{1}{2}, \qquad \beta = \frac{4 - 0}{4 + 4} = \frac{1}{2}.$$

The lines intersect at their midpoints.

Example 2: The line segment from $\mathbf{P}_1 = (0, 0)$ to $\mathbf{P}_2 = (1, 0)$ and the line segment from $\mathbf{P}_3 = (2, 0)$ to $\mathbf{P}_4 = (2, 1)$ don't intersect. However, the calculation shows the values of α and β necessary for them to intersect,

$$\mathbf{A} = \mathbf{P}_2 - \mathbf{P}_1 = (1, 0), \quad \mathbf{B} = \mathbf{P}_3 - \mathbf{P}_4 = (0, -1), \quad \mathbf{C} = \mathbf{P}_1 - \mathbf{P}_3 = (-2, 0),$$

yields

$$\alpha = \frac{2 - 0}{0 + 1} = 2, \qquad \beta = \frac{0 - 0}{0 + 1} = 0.$$

The lines would intersect at $\alpha = 2$ (i.e., if we extend the first segment to twice its length beyond $\mathbf{P}_2$) and $\beta = 0$ (i.e., point $\mathbf{P}_3$).

▶ **Exercise 4.13**: How can we identify overlapping lines (i.e., the case of infinitely many intersection points) and parallel lines (no intersection points)?

> The description of right lines and circles, upon which geometry is founded, belongs to mechanics. Geometry does not teach us to draw these lines, but requires them to be drawn.
> —Isaac Newton, 1687.

4.3.4 Parametric Curves

Generally, it is impossible to tell much about the behavior of a parametric curve $\mathbf{P}(t) = (x(t), y(t))$ by examining the two components $x(t)$ and $y(t)$ independently. Each of the two functions may have features that do not exist in the combination. The reverse is also true—the combined curve may have features that do not exist in any of the two components.

Here is an example of two smooth curves whose combination is a parametric plane curve with a cusp (a sharp corner). The following two curves are polynomials in t:

$$x(t) = -18t^2 + 18t + 2, \quad y(t) = -16t^3 + 24t^2 - 12t + 5, \quad \text{where} \quad 0 \leq t \leq 1.$$

They are smooth, since their derivatives $x'(t) = -36t + 18$ and $y'(t) = -48t^2 + 48t - 12$ are continuous in the range $0 \leq t \leq 1$. However, the combined curve

$$\mathbf{P}(t) = (0, -16)t^3 + (-18, 24)t^2 + (18, -12)t + (2, 5)$$

features a sharp corner (a cusp or a kink), since its tangent vector

$$\mathbf{P}^t(t) = 3(0, -16)t^2 + 2(-18, 24)t + (18, -12)$$

satisfies $\mathbf{P}^t(0.5) = (0, 0)$.

▸ **Exercise 4.14:** Find two curves $x(t)$ and $y(t)$, each with a cusp, such that the combined curve $\mathbf{P}(t) = (x(t), y(t))$ is smooth.

▸ **Exercise 4.15:** Find three curves $x(t)$, $y(t)$, and $z(t)$, each a cubic polynomial, such that the combined curve $\mathbf{P}(t) = (x(t), y(t), z(t))$ is not a cubic polynomial.

A word about the notation used here. We have used the letter $\mathbf{P}$ to denote both points and curves. In later chapters, the same letter is used to denote surfaces. The reason is that a surface is sometimes represented in terms of curves, and curves are many times expressed in terms of points. It is always easy to tell what a particular $\mathbf{P}$ stands for by counting the number of free parameters. Something like $\mathbf{P}(u, w)$ denotes a surface since it depends on two variable parameters, whereas $\mathbf{P}(0, w)$ is a curve and $\mathbf{P}(u_0, 1)$ (for a fixed u_0) is a point.

One important feature of curves is *independence of the coordinate axes*. We don't want the curve to change shape when the coordinate axes (or the points

Parametric curves used in computer graphics are normally based on polynomials, since they are simple functions that are easy to calculate and are flexible enough to create many different shapes. However, in principle, any functions can be used to create a parametric curve. Here is an example that uses the smooth sine and cosine curves to create the nonsmooth parametric curve shown on the right. It is defined by the simple expression

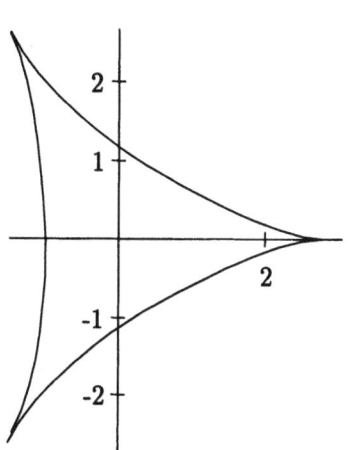

$$\mathbf{P}(t) = (2\cos(t) + \cos(2t), 2\sin(t) - \sin(2t)),$$

where $0 \le t \le 2\pi$. This curve has cusps at $t = 0$, $t = 0.261799$, and $t = 0.523599$. Another example of a parametric curve that's not a simple polynomial is the circular Bézier curve, Equation (4.141).

defining the curve) are rigidly moved or rotated. Here is an example of how such a thing can happen. Consider the parametric curve

$$\mathbf{P}(t) = (1-t)^3 \mathbf{P}_0 + t^3 \mathbf{P}_1 = \left((1-t)^3 x_0 + t^3 x_1, (1-t)^3 y_0 + t^3 y_1\right).$$

It is easy to see that $\mathbf{P}(0) = \mathbf{P}_0$ and $\mathbf{P}(1) = \mathbf{P}_1$ (the curve passes through the two points). What kind of a curve is $\mathbf{P}(t)$? The tangent vector of our curve is

$$\left(\frac{dx}{dt}, \frac{dy}{dt}\right) = \left(-3(1-t)^2 x_0 + 3t^2 x_1, -3(1-t)^2 y_0 + 3t^2 y_1\right).$$

To calculate the slope, we have to select actual points. We start with the two points $\mathbf{P}_0 = (0,0)$ and $\mathbf{P}_1 = (5,6)$. The slope of the curve is

$$\frac{dy}{dx} = \frac{dy}{dt} \bigg/ \frac{dx}{dt} = \frac{-3(1-t)^2 0 + 3t^2 \times 6}{-3(1-t)^2 0 + 3t^2 \times 5} = \frac{6}{5} = \text{constant},$$

so the curve is a straight line.

Next, we translate both points by the same amount $(0,-1)$, so the new points are $\mathbf{P}_0 = (0,-1)$ and $\mathbf{P}_1 = (5,5)$. The new slope is

$$\frac{3(1-t)^2 + 15t^2}{15t^2} = \frac{1}{5}\left(\frac{1}{t} - 1\right) + 1.$$

It is no longer constant and the curve is therefore no longer a straight line. The curve has changed its shape just because its endpoints have been moved!

It turns out that a curve of the form

$$P(t) = \sum_{i=0}^{n} w_i P_i, \quad \text{where} \quad \sum_{i=0}^{n} w_i = 1,$$

is independent of the particular coordinate axes used. This is arguably the most important property of barycentric weights.

It is easy to extend the concept of parametric curves to three dimensions (space curves). There are two small differences: (1) $P(t)$ should be of the form $(x(t), y(t), z(t))$ and (2) the slope of a three-dimensional curve is not defined. It has, however, a tangent vector dP/dt.

▸ **Exercise 4.16:** Prove that the parametric curve

$$P(t) = P + 2\alpha(Q - P)t + (1 - 2\alpha)(Q - P)t^2, \quad 0 \leq t \leq 1 \qquad (4.7)$$

(where α is any real number) is a straight line, even though it is a polynomial of degree 2 in t. Note that the curve goes from point P to point Q.

4.3.5 Uniform and Nonuniform Parametric Curves

So far, we have assumed that the parameter t of a parametric curve $P(t) = (x(t), y(t))$ varies in the range $[0, 1]$. It is also possible to vary t in other ranges, and such curves may be useful in special situations. This idea arises naturally when we try to fit a curve to a given set of data points. One question that should be answered in such a case is what value should the parameter t have at each point. It turns out that this is both a problem and an opportunity. A practical, interactive algorithm for curve design should make it possible to treat the values of t at the data points as parameters, and thus to produce an entire family of curves, all of whose members pass through the given data points (but behave differently between points). This gives the designer an extra tool that can be used to construct the right curve.

The two approaches to this problem are (1) increment t by one for each point and (2) increment t by different values. The former approach yields a *uniform* parametric curve, while the latter results in a *nonuniform* parametric curve. Uniform parametric curves are normally easy to calculate and they produce good results when the points are roughly equally spaced. However, when the spacing of the points is very different, a uniform curve may look strange and unnatural, even though it passes through all the data points. This is when a nonuniform parametric curve should be used.

In such a curve it is common to increase the value of t at point P_i by the distance $|P_i - P_{i-1}|$. Notice that this distance is the chord length from point P_{i-1} to point P_i. If this convention is used, then t starts at zero and is assigned the accumulated chord length at every data point. If the curve does not oscillate much between data points, the chord length is a good approximation to the arc length of the curve, so t is assigned, in such a case, values that are close to the arc length. A curve $P(s)$ where the parameter is the arc length s has a tangent vector $P^s(s)$ of magnitude one (it's a unit vector). If we express such a curve as $P(s) = (x(s), y(s))$,

4 Curves

then $(x^s(s), y^s(s))$ is a unit vector, which implies that $|x^s(s)| \leq 1$ and $|y^s(s)| \leq 1$. This, in turn, means that the slopes of both curves $x(s)$ and $y(s)$ are bounded between -1 and $+1$, so the two curves are never too steep and are generally "well behaved."

4.3.6 Curve Continuity

In practice, a complete curve is normally made up of segments, so it is important to look at the way segments can connect. There are two types of continuities: *geometric* and *parametric*. If two consecutive segments meet at a point, the total curve is said to have G^0 geometric continuity. It may look as in Figure 4.7a. If, in addition, the directions of the tangent vectors of the two segments are the same at the point, the curve has G^1 geometric continuity at the point. The two segments thus connect smoothly (Figure 4.7b). In general, a curve has geometric continuity G^n at a join point if every pair of the first n derivatives of the two segments have the same direction at the point. If the same derivatives also have identical *magnitudes* at the point, then the curve is said to have C^n parametric continuity at the point.

A G^2 or a C^2 continuity at a point means that the two segments connect smoothly at the point, but it also guarantees that they don't twist much when they are close to the point (Figure 4.7c). The curve is therefore *tight* in the vicinity of the point.

A C^k continuity is more restrictive than G^k, so a curve that has C^k continuity at a join point also has G^k continuity at the point, but there is an exception. Imagine two segments connecting at a point, where both have tangent vectors of $(0, 0, 0)$ at the point. The vectors are identical, so the curve has C^1 continuity at the point. However, Exercise 4.51 (page 230) shows that the two segments may move in different directions at the point, in which case the curve won't have G^1 continuity.

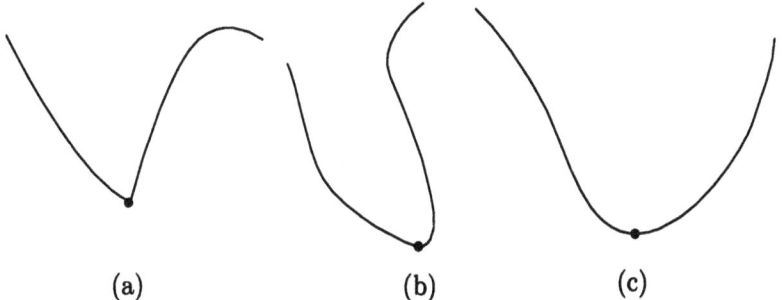

(a)   (b)   (c)

Figure 4.7: (a) G^0 Continuity (a Sharp Angle). (b) G^1 Continuity (a Smooth Connection). (c) G^2 Continuity (a Tight Curve).

The reason for having two types of continuities has to do with parameter substitution (see box). Given a curve segment $\mathbf{P}(t)$ where $0 \leq t \leq 1$, we can substitute $T = t^2$. The new segment $\mathbf{Q}(T) = \mathbf{Q}(t^2)$, where $0 \leq T \leq 1$, is identical in shape to $\mathbf{P}(t)$. The two identical curves must, of course, have the same tangents. However,

Parameter Substitution

Instead of naming the parameter t, we can give it a different name. Moreover, we can use a function of t as the parameter. It can be shown that if $g(t)$ is a function that increases monotonically with t (i.e., if $t_2 > t_1$ implies $g(t_2) > g(t_1)$), then the curve $\mathbf{P}(g(t))$ will have the same shape as $\mathbf{P}(t)$ (although $g(t)$ will normally have to vary in a different range than t).

For two-dimensional curves, the substitution does not affect the slope since

$$\frac{\frac{dy(g)}{dg}\big/\frac{dg(t)}{dt}}{\frac{dx(g)}{dg}\big/\frac{dg(t)}{dt}} = \frac{\frac{dy(t)}{dt}}{\frac{dx(t)}{dt}} = \frac{dy(t)}{dx(t)}.$$

their calculated tangent vectors have different magnitudes since

$$\frac{d\mathbf{Q}(t^2)}{dt} = 2t\frac{d\mathbf{Q}(t)}{dt} = 2t\frac{d\mathbf{P}(t)}{dt}.$$

This is why we separate the direction and the magnitude of the tangent vectors when considering curve continuities. If the directions of the tangent vectors are equal, they produce a smooth join and we call this case G^1 continuity (which is normally all that is required in practice).

Example: Consider the two straight segments $\mathbf{P}(t) = (8t, 6t)$ and $\mathbf{Q}(t) = (4(t+2), 3(t+2))$. The first goes from $(0,0)$ to $(8,6)$ and the second, from $(8,6)$ to $(12,9)$. Their tangent vectors are $\mathbf{P}^t(t) = (8,6)$ and $\mathbf{Q}^t(t) = (4,3)$. The segments connect smoothly at $(8,6)$ (in fact, they look like one straight segment), but their tangent vectors are different at that point! The total curve thus has G^1 continuity at point $(8,6)$, but not C^1 continuity.

It is interesting to note, however, that the unit tangent vectors **are** equal at the joint. The magnitude of $\mathbf{P}^t(t)$ is $\sqrt{8^2 + 6^2} = 10$ and that of $\mathbf{Q}^t(t) = \sqrt{4^2 + 3^2} = 5$. The two unit tangent vectors are therefore equal to $(8/10, 6/10) = (4/5, 3/5)$. The unit tangent vector thus provides a better measure of the tangent than the tangent vector itself. Another natural vector that's associated with every point of a smooth curve is the curvature, discussed in Section 4.22.

A curve whose tangent vector and curvature vector (Section 4.22.6) are everywhere continuous is said to have G^2 (second-order geometric) continuity.

You can do anything you like with me except paint me, Hughie dear. I have to draw the line somewhere. But that's just what you *can't* do—draw a line, I mean. I like you in every way, as you well know, except as a painter. You would have been a good painter if you had never painted—did I invent that?

— L. P. Hartley *The Hireling*.

4 Curves

4.3.7 PC Curves

Parametric curves used in computer graphics are based on polynomials. A polynomial of degree one has the form $\mathbf{P}_1(t) = \mathbf{A}t + \mathbf{B}$ and is, therefore, a straight line so it can only be used in special cases. A polynomial of degree 2 (quadratic) has the form $\mathbf{P}_2(t) = \mathbf{A}t^2 + \mathbf{B}t + \mathbf{C}$ and is a conic section (Section 4.20), so it can only have a few different shapes. A polynomial of degree 3 (cubic) has the form $\mathbf{P}_3(t) = \mathbf{A}t^3 + \mathbf{B}t^2 + \mathbf{C}t + \mathbf{D}$ and is the simplest one that can have complex shapes and can also be a space curve. (The complexity of this polynomial is limited, though. It can have at most one loop, and, if it does not have a loop, it can have at most two inflection points). Polynomials of higher degrees are sometimes needed, but, in general, they wiggle too much and are hard to control. They also have more coefficients, so they require more input data to calculate all the coefficients. As a result, a complete curve is constructed in practice of segments, each a parametric cubic polynomial (also called a PC). Such a curve is a piecewise polynomial curve, sometimes also called a *spline* (see definition on page 226).

Figure 4.8 shows seven data points and two curves that fit them. The dashed curve is a polynomial of degree 6; the solid curve is a spline. It is easy to see that the polynomial oscillates, whereas the spline curve is tight and is, therefore, more pleasing to the eye.

▶ **Exercise 4.17:** Prove that a quadratic polynomial must be a plane curve.

▶ **Exercise 4.18:** Why does a high-degree polynomial wiggle?

Question: The word "quad" comes from Latin for "four," so why is a degree-2 polynomial called quadratic? While we are at it, why is a degree-3 polynomial called cubic?

Answer: A square of side length n has four sides (it is quadratic), but its area is n^2 and this is associated with a degree-2 polynomial, which has terms up to x^2. Similarly, a cube of side length n has volume n^3, which is why "cubic" has become associated with a degree-3 polynomial.

A single PC segment is determined by means of points (data or control) or tangent vectors. Sometimes, continuity considerations are also used to constrain the curve. Whatever the input data are, the segment always has the same form and requires the calculation of four coefficients. Four equations are therefore needed, which calls for four given quantities. We express a PC segment using the notation

$$\mathbf{P}(t) = (t^3, t^2, t, 1) \begin{pmatrix} m_{11} & m_{12} & m_{13} & m_{14} \\ m_{21} & m_{22} & m_{23} & m_{24} \\ m_{31} & m_{32} & m_{33} & m_{34} \\ m_{41} & m_{42} & m_{43} & m_{44} \end{pmatrix} \begin{pmatrix} \mathbf{G}_1 \\ \mathbf{G}_2 \\ \mathbf{G}_3 \\ \mathbf{G}_4 \end{pmatrix} = \mathbf{T}(t) \cdot \mathbf{M} \cdot \mathbf{G},$$

where $\mathbf{M}$ is the basis matrix that depends on the method used, and $\mathbf{G}$ is the geometry vector, consisting of the four given quantities (points or tangent vectors).

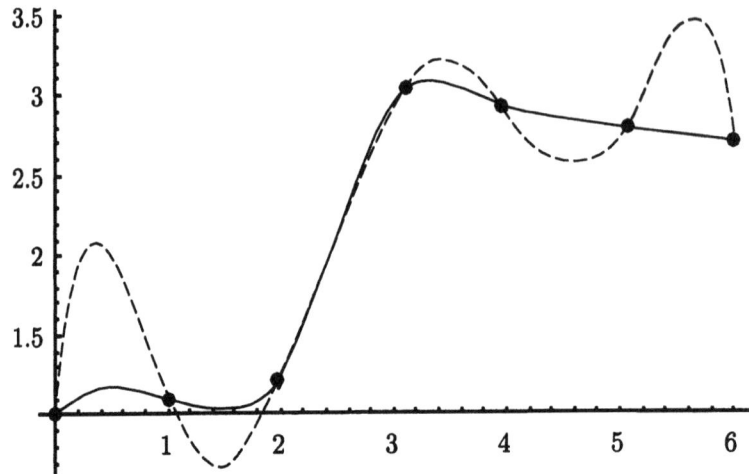

Figure 4.8: Polynomial and Spline Fit.

```
Clear[points];
points={{0,1},{1,1.1},{2,1.2},{3,3},{4,2.9},{5,2.8},{6,2.7}};
InterpolatingPolynomial[points,x];
Interpolation[points,InterpolationOrder->3];
Show[ListPlot[points,Prolog->AbsolutePointSize[5]],
  Plot[%%,{x,0,6},PlotStyle->Dashing[{0.05,0.05}]],
  Plot[%[x],{x,0,6}]]
```

Mathematica Code for Figure 4.8.

The segment can also be written as the weighted sum

$$\begin{aligned}\mathbf{P}(t) &= (t^3 m_{11} + t^2 m_{21} + t m_{31} + m_{41})\mathbf{G}_1 + (t^3 m_{12} + t^2 m_{22} + t m_{32} + m_{42})\mathbf{G}_2 \\ &\quad + (t^3 m_{13} + t^2 m_{23} + t m_{33} + m_{43})\mathbf{G}_3 + (t^3 m_{14} + t^2 m_{24} + t m_{34} + m_{44})\mathbf{G}_4 \\ &= B_1(t)\mathbf{G}_1 + B_2(t)\mathbf{G}_2 + B_3(t)\mathbf{G}_3 + B_4(t)\mathbf{G}_4 = \mathbf{B} \cdot \mathbf{G},\end{aligned}$$

where matrix $\mathbf{B}$ equals the product $\mathbf{T} \cdot \mathbf{M}$ and the $B_i(t)$ are the weights. They are also called the *blending functions*, since they blend the four given quantities. If the quantities being blended are points, the four weights should be barycentric, which implies that the sum of the elements of $\mathbf{M}$, in such a case, should equal 1 (since the 16 elements of $\mathbf{M}$ are the elements of the $B_i(t)$'s).

A PC segment can also be written in the form

$$\mathbf{P}(t) = \mathbf{A}t^3 + \mathbf{B}t^2 + \mathbf{C}t + \mathbf{D} = (t^3, t^2, t, 1)\begin{pmatrix} A_x & A_y & A_z \\ B_x & B_y & B_z \\ C_x & C_y & C_z \\ D_x & D_y & D_z \end{pmatrix} = \mathbf{T}(t) \cdot \mathbf{C}.$$

4 Curves

Its first derivative is

$$\frac{d\mathbf{P}(t)}{dt} = \frac{d\mathbf{T}(t)}{dt} \cdot \mathbf{C} = (3t^2, 2t, 1, 0)\mathbf{C}$$

and it is the *tangent vector* of the curve. This vector points in the direction of the tangent to the curve, but its magnitude is also important. It describes the *speed* of the curve.

In physics, if the function $x(t)$ describes the position of an object at time t, then $dx(t)/dt$ describes its velocity, and $d^2x(t)/dt^2$ gives its acceleration. This is also true for curves, but the speed in question is not the speed of drawing the curve on the screen! Rather, it is the distance covered on the curve when t is incremented in equal steps (see the particle paradigm of Section 4.3).

This concept is important in computer animation. Imagine a camera moving along the curve while t is incremented in equal steps. The speed of the camera at a point is given by the magnitude of the tangent vector at that point. If we want the camera to move at a constant speed, all tangent vectors must have the same magnitude. This implies that the second derivatives must have magnitude 0, corresponding to zero acceleration.

Before we start looking at actual PC curves, we illustrate the concepts above by means of two simple examples, a polyline and four points.

4.3.8 Polyline

Given a set of data points the problem is to find the straight segments that connect successive pairs of points. The complete curve is called a *polyline* and is made of straight segments.

A straight line from point $\mathbf{P}_i$ to point $\mathbf{P}_j$ is given by $\mathbf{P}(t) = (1-t)\mathbf{P}_i + t\mathbf{P}_j$ (Equation (4.4)). It can be written

$$\mathbf{P}_{ij}(t) = (1-t, t)\begin{pmatrix}\mathbf{P}_i \\ \mathbf{P}_j\end{pmatrix} = (1-t, t)\begin{pmatrix}1 & 0 \\ 0 & 1\end{pmatrix}\begin{pmatrix}\mathbf{P}_i \\ \mathbf{P}_j\end{pmatrix}.$$

The basis matrix in this case is the 2×2 identity matrix and the geometry vector is $(\mathbf{P}_i, \mathbf{P}_j)^T$, where the superscript T indicates transpose.

4.3.9 Four Points

Four points $\mathbf{P}_1$, $\mathbf{P}_2$, $\mathbf{P}_3$, and $\mathbf{P}_4$ are given. We are looking for a PC curve that passes through these points and has the form

$$\mathbf{P}(t) = \mathbf{a}t^3 + \mathbf{b}t^2 + \mathbf{c}t + \mathbf{d} = (t^3, t^2, t, 1)(\mathbf{a}, \mathbf{b}, \mathbf{c}, \mathbf{d})^T = \mathbf{T}(t)\mathbf{A}, \qquad (4.8)$$

where each of the four coefficients $\mathbf{a}$, $\mathbf{b}$, $\mathbf{c}$, and $\mathbf{d}$ is a pair (or a triplet), $\mathbf{T}(t)$ is the row vector $(t^3, t^2, t, 1)$, and $\mathbf{A}$ is the column vector $(\mathbf{a}, \mathbf{b}, \mathbf{c}, \mathbf{d})^T$. Calculating the curve therefore involves finding the values of the four unknowns $\mathbf{a}$, $\mathbf{b}$, $\mathbf{c}$, and $\mathbf{d}$.

Since the four points can be located anywhere, we cannot assume anything about their positions and we make the general assumption that $\mathbf{P}_1$ and $\mathbf{P}_4$ are the two endpoints $\mathbf{P}(0)$ and $\mathbf{P}(1)$ of the curve, and that $\mathbf{P}_2$ and $\mathbf{P}_3$ are the two (equally spaced) interior points $\mathbf{P}(1/3)$ and $\mathbf{P}(2/3)$. We therefore write $\mathbf{P}(0) = \mathbf{P}_1$, $\mathbf{P}(1/3) = \mathbf{P}_2$, $\mathbf{P}(2/3) = \mathbf{P}_3$, and $\mathbf{P}(1) = \mathbf{P}_4$, or, explicitly,

$$\begin{aligned}\mathbf{a}(0)^3 + \mathbf{b}(0)^2 + \mathbf{c}(0) + \mathbf{d} &= \mathbf{P}_1,\\ \mathbf{a}(1/3)^3 + \mathbf{b}(1/3)^2 + \mathbf{c}(1/3) + \mathbf{d} &= \mathbf{P}_2,\\ \mathbf{a}(2/3)^3 + \mathbf{b}(2/3)^2 + \mathbf{c}(2/3) + \mathbf{d} &= \mathbf{P}_3,\\ \mathbf{a}(1)^3 + \mathbf{b}(1)^2 + \mathbf{c}(1) + \mathbf{d} &= \mathbf{P}_4.\end{aligned}$$

The solutions of this system of equations are

$$\begin{aligned}\mathbf{a} &= -(9/2)\mathbf{P}_1 + (27/2)\mathbf{P}_2 - (27/2)\mathbf{P}_3 + (9/2)\mathbf{P}_4,\\ \mathbf{b} &= 9\mathbf{P}_1 - (45/2)\mathbf{P}_2 + 18\mathbf{P}_3 - (9/2)\mathbf{P}_4,\\ \mathbf{c} &= -(11/2)\mathbf{P}_1 + 9\mathbf{P}_2 - (9/2)\mathbf{P}_3 + \mathbf{P}_4,\\ \mathbf{d} &= \mathbf{P}_1.\end{aligned}$$

Substituting these solutions into Equation (4.8) gives

$$\begin{aligned}\mathbf{P}(t) =& \bigl(-(9/2)\mathbf{P}_1 + (27/2)\mathbf{P}_2 - (27/2)\mathbf{P}_3 + (9/2)\mathbf{P}_4\bigr)t^3 \\ & + \bigl(9\mathbf{P}_1 - (45/2)\mathbf{P}_2 + 18\mathbf{P}_3 - (9/2)\mathbf{P}_4\bigr)t^2 \\ & + \bigl(-(11/2)\mathbf{P}_1 + 9\mathbf{P}_2 - (9/2)\mathbf{P}_3 + \mathbf{P}_4\bigr)t + \mathbf{P}_1.\end{aligned}$$

After rearranging, this becomes

$$\begin{aligned}\mathbf{P}(t) =& (-4.5t^3 + 9t^2 - 5.5t + 1)\mathbf{P}_1 + (13.5t^3 - 22.5t^2 + 9t)\mathbf{P}_2 \\ & + (-13.5t^3 + 18t^2 - 4.5t)\mathbf{P}_3 + (4.5t^3 - 4.5t^2 + t)\mathbf{P}_4 \\ =& G_1(t)\mathbf{P}_1 + G_2(t)\mathbf{P}_2 + G_3(t)\mathbf{P}_3 + G_4(t)\mathbf{P}_4 \\ =& \mathbf{G}(t)\mathbf{P},\end{aligned} \qquad (4.9)$$

where

$$\begin{aligned}G_1(t) &= (-4.5t^3 + 9t^2 - 5.5t + 1), & G_3(t) &= (-13.5t^3 + 18t^2 - 4.5t),\\ G_2(t) &= (13.5t^3 - 22.5t^2 + 9t), & G_4(t) &= (4.5t^3 - 4.5t^2 + t),\end{aligned} \qquad (4.10)$$

$\mathbf{P}$ is the column $(\mathbf{P}_1, \mathbf{P}_2, \mathbf{P}_3, \mathbf{P}_4)^T$ and $\mathbf{G}(t)$ is the matrix $(G_1(t), G_2(t), G_3(t), G_4(t))$ (see also Exercise 4.26 for a different approach to this polynomial).

The functions $G_i(t)$ are called *blending functions* since they create any point on the curve as a blend of the four given points. Note that they are barycentric (they should be, since they blend points). We can also write

$$G_1(t) = (t^3, t^2, t, 1)(-4.5, 9, -5.5, 1)^T$$

4 Curves

and similarly for $G_2(t)$, $G_3(t)$, and $G_4(t)$. In matrix notation, this becomes

$$\mathbf{G}(t) = (t^3, t^2, t, 1) \begin{pmatrix} -4.5 & 13.5 & -13.5 & 4.5 \\ 9.0 & -22.5 & 18 & -4.5 \\ -5.5 & 9.0 & -4.5 & 1.0 \\ 1.0 & 0 & 0 & 0 \end{pmatrix} = \mathbf{T}(t)\,\mathbf{N}. \qquad (4.11)$$

The curve can now be written $\mathbf{P}(t) = \mathbf{G}(t)\mathbf{P} = \mathbf{T}(t)\,\mathbf{N}\mathbf{P}$. Matrix $\mathbf{N}$ is called the basis matrix and $\mathbf{P}$ is the geometry vector. Since $\mathbf{P}(t)$ equals the product $\mathbf{T}(t)\,\mathbf{A}$ (Equation (4.8)), we get the result $\mathbf{A} = \mathbf{N}\mathbf{P}$.

To calculate the curve, we only need to calculate the four quantities $\mathbf{a}$, $\mathbf{b}$, $\mathbf{c}$, and $\mathbf{d}$ (that constitute vector $\mathbf{A}$), and write Equation (4.8) using the numerical values of $\mathbf{a}$, $\mathbf{b}$, $\mathbf{c}$, and $\mathbf{d}$.

Example: (This example is in two dimensions, each of the four points $\mathbf{P}_i$ and the four coefficients $\mathbf{a}, \mathbf{b}, \mathbf{c}, \mathbf{d}$ is a pair. For three-dimensional curves the method is the same except that triplets are used instead of pairs.) Given the four two-dimensional points $\mathbf{P}_1 = (0,0)$, $\mathbf{P}_2 = (1,0)$, $\mathbf{P}_3 = (1,1)$, and $\mathbf{P}_4 = (0,1)$, we set up the equation

$$\begin{pmatrix} \mathbf{a} \\ \mathbf{b} \\ \mathbf{c} \\ \mathbf{d} \end{pmatrix} = \mathbf{A} = \mathbf{N}\mathbf{P} = \begin{pmatrix} -4.5 & 13.5 & -13.5 & 4.5 \\ 9.0 & -22.5 & 18 & -4.5 \\ -5.5 & 9.0 & -4.5 & 1.0 \\ 1.0 & 0 & 0 & 0 \end{pmatrix} \begin{pmatrix} (0,0) \\ (1,0) \\ (1,1) \\ (0,1) \end{pmatrix}.$$

The solutions are

$$\mathbf{a} = -4.5(0,0) + 13.5(1,0) - 13.5(1,1) + 4.5(0,1) = (0,-9),$$
$$\mathbf{b} = 19(0,0) - 22.5(1,0) + 18(1,1) - 4.5(0,1) = (-4.5, 13.5),$$
$$\mathbf{c} = -5.5(0,0) + 9(1,0) - 4.5(1,1) + 1(0,1) = (4.5, -3.5),$$
$$\mathbf{d} = 1(0,0) - 0(1,0) + 0(1,1) - 0(0,1) = (0,0).$$

So the curve $\mathbf{P}(t)$ that passes through the given points is

$$\mathbf{P}(t) = \mathbf{T}(t)\,\mathbf{A} = (0,-9)t^3 + (-4.5, 13.5)t^2 + (4.5, -3.5)t.$$

It is now easy to calculate and verify that $\mathbf{P}(0) = (0,0) = \mathbf{P}_1$, and

$$\mathbf{P}(1/3) = (0,-9)(1/27) + (-4.5, 13.5)(1/9) + (4.5, -3.5)(1/3) = (1,0) = \mathbf{P}_2,$$
$$\mathbf{P}(1) = (0,-9)1^3 + (-4.5, 13.5)1^2 + (4.5, -3.5)1 = (0,1) = \mathbf{P}_4.$$

▶ **Exercise 4.19:** Calculate $\mathbf{P}(2/3)$ and verify that it equals $\mathbf{P}_3$.

▶ **Exercise 4.20:** Imagine the circular arc of radius 1 in the first quadrant (a quarter circle). Write the coordinates of the four points that are equally spaced on this arc. Use the coordinates to calculate a PC approximating this arc. Calculate point $\mathbf{P}(1/2)$. How far does it deviate from the midpoint of the true quarter circle?

▸ **Exercise 4.21:** Calculate the PC that passes through the four points $\mathbf{P}_1, \ldots, \mathbf{P}_4$ assuming that only the three relative coordinates $\Delta_1 = \mathbf{P}_2 - \mathbf{P}_1$, $\Delta_2 = \mathbf{P}_3 - \mathbf{P}_2$, and $\Delta_3 = \mathbf{P}_4 - \mathbf{P}_3$ are given. Show a numeric example.

The main advantage of this method is its simplicity. Given the four points, it is easy to calculate the PC that passes through them. This, however, is also the reason for the main disadvantage of the method. It produces only *one* PC that passes through four given points. If that PC does not have the required shape, too bad. This simple curve method is not interactive.

Even though this method is not very useful for curve drawing, it may be useful for interpolation. Given two points $\mathbf{P}_1$ and $\mathbf{P}_2$, we know that the point midway between them is their average, $(\mathbf{P}_1+\mathbf{P}_2)/2$. A natural question is: Given four points $\mathbf{P}_1$ through $\mathbf{P}_4$, what point is located midway between them? We can answer this question by calculating the average, $(\mathbf{P}_1+\mathbf{P}_2+\mathbf{P}_3+\mathbf{P}_4)/4$, but this weighted sum assigns the same weight to each of the four points. If we want to assign more weight to the interior points $\mathbf{P}_2$ and $\mathbf{P}_3$, we can calculate the PC that passes through the points and compute $\mathbf{P}(0.5)$. The result is

$$\mathbf{P}(0.5) = -0.625\mathbf{P}_1 + 0.5625\mathbf{P}_2 + 0.5625\mathbf{P}_3 - 0.625\mathbf{P}_4.$$

This is a weighted sum that assigns more weight to the interior points. Notice that the weights are barycentric. This method can be extended to a two-dimensional grid of points (Section 5.10.1).

> A precisian professor had the habit of saying: "...quartic polynomial $ax^4 + bx^3 + cx^2 + dx + e$, where e need not be the base of the natural logarithms."
> —J. E. Littlewood (1885–1977) *A Mathematician's Miscellany.*

▸ **Exercise 4.22:** The preceding method makes sense if the four points are (approximately) equally spaced along the curve. If they are not, the following may be done: Instead of using $1/3$ and $2/3$ as the intermediate values, the user may specify values α and β such that $\mathbf{P}_2 = \mathbf{P}(\alpha)$ and $\mathbf{P}_3 = \mathbf{P}(\beta)$. Generalize Equation (4.11) such that it depends on α and β.

4.4 The Lagrange Polynomial

Given the $n+1$ data points $\mathbf{P}_0 = (x_0, y_0)$, $\mathbf{P}_1 = (x_1, y_1), \ldots, \mathbf{P}_n = (x_n, y_n)$, the problem is to find a function $y = f(x)$ that will pass through all of them. We first try an expression of the form $y = \sum_{i=0}^{n} y_i L_i^n(x)$. This is a weighted sum of the individual y_i coordinates where the weights depend on the x_i coordinates. This sum will pass through the points if

$$L_i^n(x) = \begin{cases} 1, & x = x_i, \\ 0, & \text{otherwise.} \end{cases}$$

A good mathematician can easily guess that such functions are given by

$$L_i^n(x) = \frac{\Pi_{j \neq i}(x - x_j)}{\Pi_{j \neq i}(x_i - x_j)} = \frac{(x-x_0)(x-x_1)\cdots(x-x_{i-1})(x-x_{i+1})(x-x_n)}{(x_i-x_0)\cdots(x_i-x_{i-1})(x_i-x_{i+1})\cdots(x_i-x_n)}.$$

4 Curves

(Note that $(x - x_i)$ is missing from the numerator and $(x_i - x_i)$ is missing from the denominator.) The function $y = \sum_{i=0}^{n} y_i L_i^n(x)$ is called the Lagrange polynomial (and is denoted LP) because (1) it was originally developed by Lagrange and (2) it is a polynomial of degree n.

Definition: A polynomial of degree n in x is the function

$$P_n(x) = \sum_{i=0}^{n} a_i x^i = a_0 + a_1 x + a_2 x^2 + \cdots + a_n x^n,$$

where a_i are the coefficients of the polynomial (in our case, they are real numbers).

> **Definition**: Polynomial is a function in which the output is the sum of terms that are the products of constant values and the input raised to some integer power. The polynomial of a polynomial is another polynomial. From a time complexity point of view, polynomials are well-behaved.

Calculating a polynomial involves additions, multiplications, and exponentiations. There are two methods, however, that greatly simplify this calculation. They are the following:

1. Horner's rule. A degree-3 polynomial is written

$$P(x) = ((a_3 x + a_2)x + a_1)x + a_0.$$

It's easy to see that only additions and multiplications are left and exponentiations are no longer needed.

2. Forward differences. This is one of Newton's many contributions to mathematics and it is described in detail in Section A.2. Only the first step requires multiplications. All other steps are done with additions only.

These methods make polynomials very desirable to use. In practice, however, we have to use them in parametric form (see Section 4.6) since any explicit function $y = f(x)$ is limited in shape (note that the LP, for example, cannot be calculated if two of the $n + 1$ given data points have the same x coordinate).

The LP has two properties that make it impractical for interactive curve design: (1) it is of a high degree and (2) it is unique.

1. Writing $P_n(x) = 0$ creates an equation of degree n in x. It has n solutions (some may be complex numbers), so, when plotted as a curve, it intercepts the x axis n times. For large n, such a curve would be too loose and it may also oscillate a lot. In practice, we normally prefer tight curves.

2. It is easy to prove that the LP is unique (see below). In principle, there are infinitely many curves that pass through any set of given points and the one we are looking for may not be the LP. Any useful mathematical method for curve design should make it easy for the designer to change the shape of the curve by changing the values of parameters.

> *Theorem*: There is only one polynomial of degree n that passes through any given set of $n+1$ points.
>
> *Proof*: A root of the polynomial $P_n(x)$ is a value x_r such that $P_n(x_r) = 0$. A polynomial $P_n(x)$ can have at most n distinct roots (unless it is the zero polynomial). Suppose that there is another polynomial $Q_n(x)$ that passes through the same $n+1$ data points. At the points, we thus have $P_n(x_i) = Q_n(x_i) = y_i$ or $(P_n - Q_n)(x_i) = 0$. The difference $(P_n - Q_n)$ is a polynomial whose degree must be $\leq n$, so it cannot have more than n distinct roots. On the other hand, this difference is 0 at the $n+1$ data points, so it has $n+1$ roots. We conclude that it must be the zero polynomial, which implies that $P_n(x)$ and $Q_n(x)$ are identical. ◄

As a result of the two properties, we conclude that a useful curve design method should use polynomials of a low degree and should depend on parameters that affect the shape of the curve. Designing a curve is an interactive process where the designer selects values for the parameters, plots the curve, examines it, and repeats the process until the shape of the curve is satisfactory.

To get a curve that's a low-degree polynomial, we can use composite LPs. We divide our data points into groups $\mathbf{P}_0 \ldots \mathbf{P}_3$, $\mathbf{P}_3 \ldots \mathbf{P}_6$ of four points each and fit a degree-3 LP through each group. Since the groups overlap, the polynomial segments will join and will make one continuous curve. However, the different segments may not join smoothly and there is still no control over the shape of the curve. To achieve this, we use *spline curves* (Section 4.6).

▶ **Exercise 4.23:** Calculate the LP between the two points $\mathbf{P}_0 = (x_0, y_0)$ and $\mathbf{P}_1 = (x_1, y_1)$. What kind of a curve is it?

The LP can also be expressed in parametric form. Given the $n+1$ data points $\mathbf{P}_0, \mathbf{P}_1, \ldots, \mathbf{P}_n$, we want to find a polynomial $\mathbf{P}(t)$ that passes through all of them, such that $\mathbf{P}(t_0) = \mathbf{P}_0$, $\mathbf{P}(t_1) = \mathbf{P}_1, \ldots, \mathbf{P}(t_n) = \mathbf{P}_n$, where $t_0 = 0$, $t_n = 1$, and $t_1, \ldots, t_{n-1}$ are certain values between 0 and 1 (the t_i are called *knot* values). The LP gets the form $\mathbf{P}(t) = \sum_{i=0}^n \mathbf{P}_i L_i^n(t)$. This is a weighted sum of the individual points where the weights (or basis functions) are given by

$$L_i^n(t) = \frac{\Pi_{j \neq i}(t - t_j)}{\Pi_{j \neq i}(t_i - t_j)}. \tag{4.12}$$

Note that $\sum_{i=0}^n L_i^n(t) = 1$.

▶ **Exercise 4.24:** Calculate the parametric LP between the two general points $\mathbf{P}_0$ and $\mathbf{P}_1$.

▶ **Exercise 4.25:** Calculate the parametric LP between the three points $\mathbf{P}_0 = (0,0)$, $\mathbf{P}_1 = (0,1)$, and $\mathbf{P}_2 = (1,1)$.

4 Curves

▶ **Exercise 4.26:** Calculate the parametric LP between the four equally spaced points P_1, P_2, P_3, and P_4 and show that it is identical to the interpolating PC given by Equation (4.9).

The parametric LP is also mentioned on page 498, in connection with Gordon surfaces.

The LP has another disadvantage. If the resulting curve is not satisfactory, the user may want to fine-tune it by adding one more point. However, all the basis functions $L_i^n(t)$ will have to be recalculated in such a case, since they also depend on the points, not only on the knot values. This disadvantage makes the LP slow to use in practice, which is why the Newton polynomial (Section 4.5) is sometimes used instead.

4.4.1 The Quadratic Lagrange Polynomial

Equation (4.12) can easily be used to obtain the Lagrange polynomial for three points P_0, P_1, and P_2. The weights in this case are

$$L_0^2(t) = \frac{\prod_{j\neq 0}^2 (t-t_j)}{\prod_{j\neq 0}^2 (t_0-t_j)} = \frac{(t-t_1)(t-t_2)}{(t_0-t_1)(t_0-t_2)},$$

$$L_1^2(t) = \frac{\prod_{j\neq 1}^2 (t-t_j)}{\prod_{j\neq 1}^2 (t_1-t_j)} = \frac{(t-t_0)(t-t_2)}{(t_1-t_0)(t_1-t_2)}, \qquad (4.13)$$

$$L_2^2(t) = \frac{\prod_{j\neq 2}^2 (t-t_j)}{\prod_{j\neq 2}^2 (t_2-t_j)} = \frac{(t-t_0)(t-t_1)}{(t_2-t_0)(t_2-t_1)},$$

and the polynomial $P_2(t) = \sum_{i=0}^2 P_i L_i^2(t)$ is easy to calculate once the values of t_0, t_1, and t_2 have been determined.

The *Uniform Quadratic Lagrange Polynomial* is obtained when $t_0 = 0$, $t_1 = 1$, and $t_2 = 2$. Equation (4.13) yields

$$\begin{aligned}\mathbf{P}_{2u}(t) &= \frac{t^2-3t+2}{2}\mathbf{P}_0 - (t^2-2t)\mathbf{P}_1 + \frac{t^2-t}{2}\mathbf{P}_2 \\ &= (t^2, t, 1)\begin{pmatrix} 1/2 & -1 & 1/2 \\ -3/2 & 2 & -1/2 \\ 1 & 0 & 0 \end{pmatrix}\begin{pmatrix} \mathbf{P}_0 \\ \mathbf{P}_1 \\ \mathbf{P}_2 \end{pmatrix}.\end{aligned} \qquad (4.14)$$

The *Nonuniform Quadratic Lagrange Polynomial* is obtained when $t_0 = 0$, $t_1 = t_0 + \Delta_0 = \Delta_0$, and $t_2 = t_1 + \Delta_1 = \Delta_0 + \Delta_1$ for some positive Δ_0 and Δ_1. Equation (4.13) gives

$$L_0^2(t) = \frac{(t-\Delta_0)(t-\Delta_1)}{(-\Delta_0)(-\Delta_1)}, \quad L_1^2(t) = \frac{(t-0)(t-\Delta_1)}{\Delta_0(\Delta_0-\Delta_1)}, \quad L_2^2(t) = \frac{(t-0)(t-\Delta_0)}{\Delta_1(\Delta_1-\Delta_0)},$$

and the nonuniform polynomial is

$$\mathbf{P}_{2nu}(t) = (t^2, t, 1) \begin{pmatrix} \frac{1}{\Delta_0(\Delta_0+\Delta_1)} & -\frac{1}{\Delta_0\Delta_1} & \frac{1}{(\Delta_0+\Delta_1)\Delta_1} \\ \frac{-1}{\Delta_0+\Delta_1} - \frac{1}{\Delta_0} & \frac{1}{\Delta_0}+\frac{1}{\Delta_1} & -\frac{1}{\Delta_1}+\frac{1}{\Delta_0+\Delta_1} \\ 1 & 0 & 0 \end{pmatrix} \begin{pmatrix} \mathbf{P}_0 \\ \mathbf{P}_1 \\ \mathbf{P}_2 \end{pmatrix}.$$

(4.15)

For $\Delta_0 = \Delta_1 = 1$, Equation (4.15) reduces to the uniform polynomial, Equation (4.14). For $\Delta_0 = \Delta_1 = 1/2$, the parameter t varies in the "standard" range $[0, 1]$ and Equation (4.15) becomes

$$\mathbf{P}_{2std}(t) = (t^2, t, 1) \begin{pmatrix} 2 & -4 & 2 \\ -3 & 4 & -1 \\ 1 & 0 & 0 \end{pmatrix} \begin{pmatrix} \mathbf{P}_0 \\ \mathbf{P}_1 \\ \mathbf{P}_2 \end{pmatrix}. \qquad (4.16)$$

In most cases, Δ_0 and Δ_1 should be set to the chord lengths $|\mathbf{P}_1-\mathbf{P}_0|$ and $|\mathbf{P}_2-\mathbf{P}_1|$, respectively.

Example: The three points $\mathbf{P}_0 = (1,0)$, $\mathbf{P}_1 = (1.3,.5)$, and $\mathbf{P}_2 = (4,0)$ are given. The uniform LP is obtained when $\Delta_0 = \Delta_1 = 1$ and it equals

$$\mathbf{P}_{2u}(t) = \bigl(1 - 0.9t + 1.2t^2, 0.5(2-t)t\bigr).$$

Many nonuniform polynomials are possible. We select the one that's obtained when the Δ values are the chord lengths between the points. In our case, they are $\Delta_0 = |\mathbf{P}_1 - \mathbf{P}_0| \approx 0.583$ and $\Delta_1 = |\mathbf{P}_2 - \mathbf{P}_1| \approx 2.75$. This polynomial is

$$\mathbf{P}_{2nu}(t) = (1 + 0.433t + 0.14t^2, 1.04t - 0.312t^2).$$

These uniform and nonuniform polynomials are shown in Figure 4.9. The figure illustrates how the nonuniform curve based on the chord lengths between the points features smaller curvature overall. Such a curve is generally considered a better interpolation of the three points.

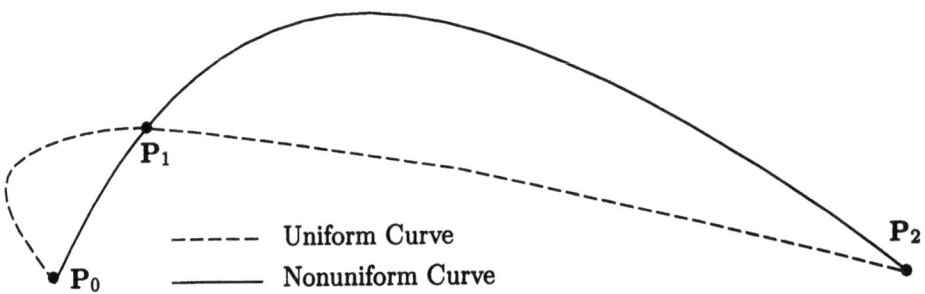

Figure 4.9: Three-Point Lagrange Polynomials.

4 Curves

Figure 4.10 shows three examples of nonuniform Lagrange polynomials that pass through the three points $\mathbf{P}_0 = (1,1)$, $\mathbf{P}_1 = (2,2)$, and $\mathbf{P}_2 = (4,0)$. The value of Δ_0 is 1.414, the chord length between $\mathbf{P}_0$ and $\mathbf{P}_1$. The chord length between $\mathbf{P}_1$ and $\mathbf{P}_2$ is 5.66 and Δ_1 is first assigned this value, then half this value, and finally twice it. The three resulting curves illustrate how the Lagrange polynomial can be shaped by changing the Δ_i parameters. The three polynomials in this case are

$$(1 + 0.354231t + 0.249634t^2, 1 + 1.76716t - 0.749608t^2),$$
$$(1 + 0.70738t - 0.000117766t^2, 1 + 1.1783t - 0.333159t^2),$$
$$(1 + 0.777945t - 0.0500221t^2, 1 + 0.919208t - 0.149925t^2).$$

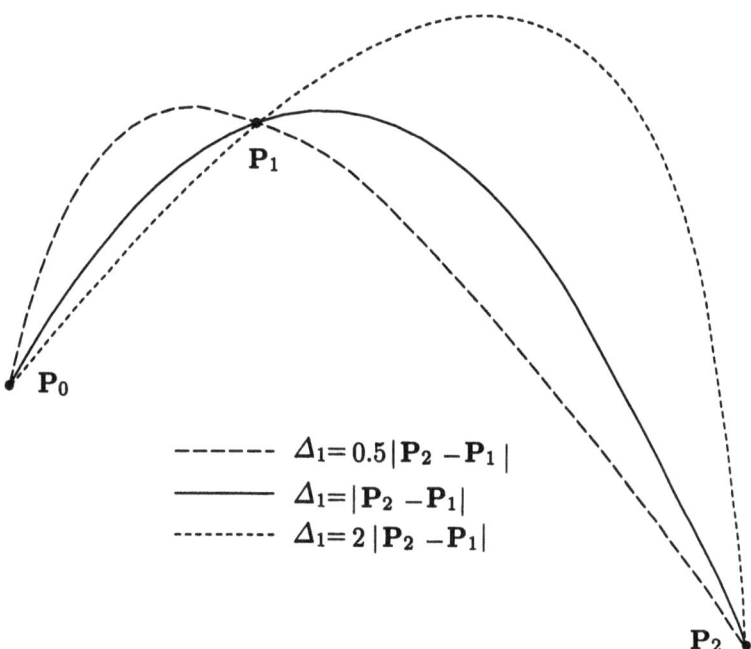

Figure 4.10: Three-Point Nonuniform Lagrange Polynomials.

4.4.2 The Cubic Lagrange Polynomial

Equation (4.12) is now applied to the cubic Lagrange polynomial that interpolates the four points $\mathbf{P}_0$, $\mathbf{P}_1$, $\mathbf{P}_2$, and $\mathbf{P}_3$. The weights in this case are

$$L_0^3(t) = \frac{\prod_{j \neq 0}^3 (t - t_j)}{\prod_{j \neq 0}^3 (t_0 - t_j)} = \frac{(t - t_1)(t - t_2)(t - t_3)}{(t_0 - t_1)(t_0 - t_2)(t_0 - t_3)},$$

$$L_1^3(t) = \frac{\prod_{j \neq 1}^3 (t - t_j)}{\prod_{j \neq 1}^3 (t_1 - t_j)} = \frac{(t - t_0)(t - t_2)(t - t_3)}{(t_1 - t_0)(t_1 - t_2)(t_1 - t_3)},$$

$$L_2^3(t) = \frac{\prod_{j\neq 2}^3(t-t_j)}{\prod_{j\neq 2}^3(t_2-t_j)} = \frac{(t-t_0)(t-t_1)(t-t_3)}{(t_2-t_0)(t_2-t_1)(t_2-t_3)},$$

$$L_3^3(t) = \frac{\prod_{j\neq 3}^3(t-t_j)}{\prod_{j\neq 3}^3(t_3-t_j)} = \frac{(t-t_0)(t-t_1)(t-t_2)}{(t_3-t_0)(t_3-t_1)(t_3-t_2)}, \tag{4.17}$$

and the polynomial $\mathbf{P}_3(t) = \sum_{i=0}^3 \mathbf{P}_i L_i^3(t)$ is easy to calculate once the values of t_0, t_1, t_2, and t_3 have been determined.

The Nonuniform Cubic Lagrange Polynomial is obtained when $t_0 = 0$, $t_1 = t_0 + \Delta_0 = \Delta_0$, $t_2 = t_1 + \Delta_1 = \Delta_0 + \Delta_1$, and $t_3 = t_2 + \Delta_2 = \Delta_0 + \Delta_1 + \Delta_2$ for positive Δ_i. The expression for the polynomial is

$$\mathbf{P}_{3nu}(t) = (t^3, t^2, t, 1)\mathbf{Q}\begin{pmatrix}\mathbf{P}_0\\\mathbf{P}_1\\\mathbf{P}_2\\\mathbf{P}_3\end{pmatrix}, \tag{4.18}$$

where $\mathbf{Q}$ is the matrix

$$\mathbf{Q} = \begin{pmatrix} \frac{1}{(-\Delta_0)(-\Delta_0-\Delta_1)(-\Delta_0-\Delta_1-\Delta_2)} & \frac{1}{\Delta_0(-\Delta_1)(-\Delta_1-\Delta_2)} & \frac{1}{(\Delta_0+\Delta_1)\Delta_1(-\Delta_2)} & \frac{1}{(\Delta_0+\Delta_1+\Delta_2)(\Delta_1+\Delta_2)\Delta_2} \\ -\frac{3\Delta_0+2\Delta_1+\Delta_2}{(-\Delta_0)(-\Delta_0-\Delta_1)(-\Delta_0-\Delta_1-\Delta_2)} & -\frac{2\Delta_0+2\Delta_1+\Delta_2}{\Delta_0(-\Delta_1)(-\Delta_1-\Delta_2)} & -\frac{2\Delta_0+\Delta_1+\Delta_2}{(\Delta_0+\Delta_1)\Delta_1(-\Delta_2)} & -\frac{2\Delta_0+\Delta_1}{(\Delta_0+\Delta_1+\Delta_2)(\Delta_1+\Delta_2)\Delta_2} \\ \frac{\Delta_0(\Delta_0+\Delta_1)+(\Delta_0+\Delta_1)(\Delta_0+\Delta_1+\Delta_2)+(\Delta_0+\Delta_1+\Delta_2)\Delta_0}{(-\Delta_0)(-\Delta_0-\Delta_1)(-\Delta_0-\Delta_1-\Delta_2)} & \frac{(\Delta_0+\Delta_1)(\Delta_0+\Delta_1+\Delta_2)}{\Delta_0(-\Delta_1)(-\Delta_1-\Delta_2)} & \frac{\Delta_0(\Delta_0+\Delta_1+\Delta_2)}{(\Delta_0+\Delta_1)\Delta_1(-\Delta_2)} & \frac{\Delta_0(\Delta_0+\Delta_1)}{(\Delta_0+\Delta_1+\Delta_2)(\Delta_1+\Delta_2)\Delta_2} \\ -\frac{\Delta_0(\Delta_0+\Delta_1)(\Delta_0+\Delta_1+\Delta_2)}{(-\Delta_0)(-\Delta_0-\Delta_1)(-\Delta_0-\Delta_1-\Delta_2)} & 0 & 0 & 0 \end{pmatrix}.$$

The Uniform Cubic Lagrange Polynomial. We construct the "standard" case, where t varies from 0 to 1. This implies $t_0 = 0$, $t_1 = 1/3$, $t_2 = 2/3$, and $t_3 = 1$. Equation (4.18) reduces to

$$\mathbf{P}_{3u}(t) = (t^3, t^2, t, 1)\begin{pmatrix} -9/2 & 27/2 & -27/2 & 9/2 \\ 9 & -45/2 & 18 & -9/2 \\ -11/2 & 9 & -9/2 & 1 \\ 1 & 0 & 0 & 0 \end{pmatrix}\begin{pmatrix}\mathbf{P}_0\\\mathbf{P}_1\\\mathbf{P}_2\\\mathbf{P}_3\end{pmatrix}. \tag{4.19}$$

4.5 The Newton Polynomial

We again assume that $n+1$ data points $\mathbf{P}_0, \mathbf{P}_1, \ldots, \mathbf{P}_n$ are given and are assigned knot values

$$t_0 = 0 < t_1 < \cdots < t_{n-1} < t_n = 1.$$

We are looking for a curve expressed by the degree-n parametric polynomial

$$\mathbf{P}(t) = \sum_{i=0}^{n} N_i(t) \mathbf{A}_i,$$

where the basis functions $N_i(t)$ depend on the knot values only and not on the data points. Only the (unknown) coefficients $\mathbf{A}_i$ depend on the points. This definition (originally proposed by Newton) is useful because it produces coefficients such that $\mathbf{A}_i$ depends only on points $\mathbf{P}_0, \ldots, \mathbf{P}_i$. If the user decides to add a point $\mathbf{P}_{n+1}$, only one coefficient, $\mathbf{A}_{n+1}$, and one basis function, $N_{n+1}(t)$, need be calculated.

The definition of the basis functions is

$$N_0(t) = 1 \quad \text{and} \quad N_i(t) = (t - t_0)(t - t_1) \cdots (t - t_{i-1}), \quad \text{for} \quad i = 1, \ldots, n.$$

To calculate the unknown coefficients, we write the equations

$$\mathbf{P}_0 = \mathbf{P}(t_0) = \mathbf{A}_0,$$
$$\mathbf{P}_1 = \mathbf{P}(t_1) = \mathbf{A}_0 + \mathbf{A}_1(t_1 - t_0),$$
$$\mathbf{P}_2 = \mathbf{P}(t_2) = \mathbf{A}_0 + \mathbf{A}_1(t_2 - t_0) + \mathbf{A}_2(t_2 - t_0)(t_2 - t_1),$$
$$\vdots$$
$$\mathbf{P}_n = \mathbf{P}(t_n) = \mathbf{A}_0 + \cdots.$$

These equations don't have to solved simultaneously. Each can easily be solved after all its predecessors have been solved. The solutions are

$$\mathbf{A}_0 = \mathbf{P}_0,$$
$$\mathbf{A}_1 = \frac{\mathbf{P}_1 - \mathbf{P}_0}{t_1 - t_0},$$
$$\mathbf{A}_2 = \frac{\mathbf{P}_2 - \mathbf{P}_0 - \dfrac{(\mathbf{P}_1 - \mathbf{P}_0)(t_2 - t_0)}{t_1 - t_0}}{(t_2 - t_0)(t_2 - t_1)} = \frac{\dfrac{\mathbf{P}_2 - \mathbf{P}_1}{t_2 - t_1} - \dfrac{\mathbf{P}_1 - \mathbf{P}_0}{t_1 - t_0}}{t_2 - t_0}.$$

This obviously gets very complicated quickly, so we use the concept of *divided differences* to express all the solutions in compact notation. The divided difference of the knots $t_i t_k$ is denoted $[t_i t_k]$ and is defined as

$$[t_i t_k] \stackrel{\text{def}}{=} \frac{\mathbf{P}_i - \mathbf{P}_k}{t_i - t_k}.$$

The solutions can now be expressed by

$$\mathbf{A}_0 = \mathbf{P}_0,$$
$$\mathbf{A}_1 = \frac{\mathbf{P}_1 - \mathbf{P}_0}{t_1 - t_0} = [t_1 t_0],$$
$$\mathbf{A}_2 = [t_2 t_1 t_0] = \frac{[t_2 t_1] - [t_1 t_0]}{t_2 - t_0},$$
$$\mathbf{A}_3 = [t_3 t_2 t_1 t_0] = \frac{[t_3 t_2 t_1] - [t_2 t_1 t_0]}{t_3 - t_0},$$
$$\vdots$$
$$\mathbf{A}_n = [t_n \ldots t_1 t_0] = \frac{[t_n \ldots t_1] - [t_{n-1} \ldots t_0]}{t_n - t_0}.$$

▶ **Exercise 4.27:** Given the same points and knot values as in Exercise 4.25, calculate the Newton polynomial that passes through the points.

▶ **Exercise 4.28:** The tangent vector to a curve $\mathbf{P}(t)$ is the derivative $\frac{d\mathbf{P}(t)}{dt}$, which we denote by $\mathbf{P}^t(t)$. Calculate the tangent vectors to the curve of Exercises 4.25 and 4.27 at the three points. Also calculate the slopes of the curve at the points.

4.6 Spline Methods for Curves

Spline methods are common in computer graphics. Normally, such a method is originally developed for curves (both plane and space) and is later extended to surfaces. This chapter discusses several types of splines.

> Spline approximation contains the delicious paradox of Prokofieff's Classical Symphony: it seems as though it might have been written several centuries ago, but of course it could not have been.
> —Philip J. Davis, 1964.

A parametric curve has the form $\mathbf{P}(t) = (x(t), y(t), z(t))$, where the parameter t varies in a certain range, usually $[0, 1]$. The simplest functions x, y, and z are polynomials. They are simple to calculate and can represent complex curves. A quadratic (degree-2) polynomial is usually not general enough to represent complex curves.

▶ **Exercise 4.29:** a. Why is the last sentence true? b. Prove that a quadratic polynomial cannot have an inflection point.

As a result, the simplest polynomial that is useful in curve design is the cubic. A curve based on such polynomials is called a *parametric cubic* (PC) curve and its general form is given by Equation (4.8), which is called the *algebraic representation* of the curve. Depending on the method used, the user may specify four points on the curve, or two endpoints and their tangent vectors, or anything else. The user's specifications are then used to derive the *geometric representation* of the PC.

4 Curves

▸ **Exercise 4.30:** Show how to find the distance between a PC and a given point $\mathbf{P}_0 = (x_0, y_0)$. You can limit the discussion to two dimensions.

4.7 Hermite Interpolation

This method calculates a PC that's a single curve segment connecting two given points. Since a PC has four coefficients, two points are not enough and more data are needed. The Hermite interpolation method requires that the tangent vectors at the start and at the end of the curve be given also. They are the extra data needed to completely calculate a Hermite segment. It should be emphasized that Hermite interpolation does not produce a complete curve, only a single segment. A practical curve method must be able to generate a long, complex curve connecting many data points and may do this using the principles developed here. An example is the cubic spline method (Section 4.8).

Each point on the Hermite segment is expressed as a combination of the two endpoints and the two tangents. The four quantities are blended by suitable blending functions (Figure 4.11a, where parts b,c are shown for comparison and are discussed elsewhere). The method is called *Hermite interpolation*, after Charles Hermite, who derived the blending functions. This method is interactive. If the curve turns out to have a wrong shape, the user can change it by modifying the tangent vectors.

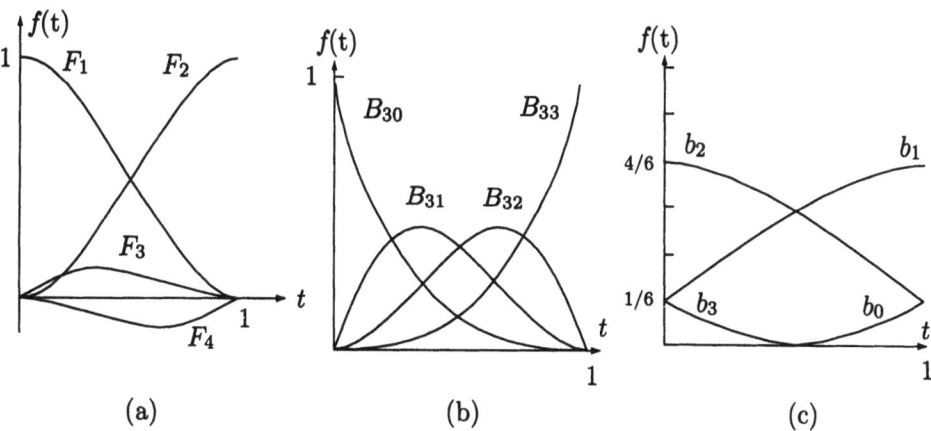

Figure 4.11: Weight Functions of (a) Hermite, (b) Bernstein, and (c) B-Spline Interpolations.

▸ **Exercise 4.31:** In the case of a four-point PC, we can change the shape of the curve by changing the position of points. Why then is the four-point method considered noninteractive?

The result of several such changes is a set of curves, all passing through the same endpoints but with different directions at the points. Figure 4.12 illustrates the behavior of the Hermite curve when the start tangent vector is varied. In Figure 4.12a, the directions of the initial tangents are varied, whereas in Figure 4.12b,

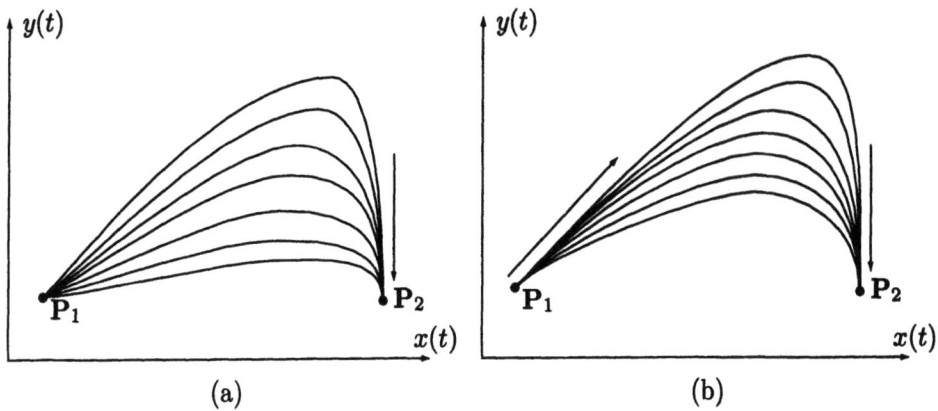

Figure 4.12: Hermite Tangent Vectors. (a) Different Directions, (b) Different Magnitudes.

they all point in a 45° direction and their magnitudes increase. It is easy to see that the larger the magnitude, the longer the curve continues in the initial direction.

The opposite is also true. As the start tangent gets shorter, the curve spends less "time" going in the direction of that vector. It turns toward the endpoint "early." In the limit, when both tangents have zero magnitude, the curve reduces to the straight line from $\mathbf{P}_1$ to $\mathbf{P}_2$ (see also Exercise 4.37).

The reason the magnitudes, and not just the directions, of the tangents affect the shape of the curve is that the three-dimensional Hermite segment is a PC and calculating a PC involves 4 coefficients, each a triplet, for a total of 12 unknown numbers. The two endpoints supply six known quantities and the two tangents should supply the remaining six. However, if we consider only the direction of a vector and not its magnitude, then the vectors $(1, 0.5, 0.3)$, $(2, 1, 0.6)$, and $(4, 2, 1.2)$ are all equal. In such a case, only two of the three vector components are independent and two vectors supply only four independent quantities.

▶ **Exercise 4.32:** Prove this claim!

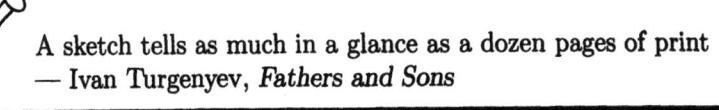

A sketch tells as much in a glance as a dozen pages of print
— Ivan Turgenyev, *Fathers and Sons*

The tangent vector to a curve $\mathbf{P}(t)$ is the derivative $d\mathbf{P}(t)/dt$, which we denote $\mathbf{P}^t(t)$. Equation (4.8) is the *algebraic representation* of a PC. It shows that the derivative is

$$\mathbf{P}^t(t) = 3\mathbf{a}t^2 + 2\mathbf{b}t + \mathbf{c}. \qquad (4.20)$$

We denote the two given points by $\mathbf{P}_1$ and $\mathbf{P}_2$ and the two given tangents by $\mathbf{P}_1^t$ and $\mathbf{P}_2^t$. The four quantities are now used to calculate the *geometric representation* of the PC by writing equations that relate the four unknown coefficients $(\mathbf{a}, \mathbf{b}, \mathbf{c}, \mathbf{d})$

4 Curves

to the four known ones, $\mathbf{P}_1$, $\mathbf{P}_2$, $\mathbf{P}_1^t$, and $\mathbf{P}_2^t$. The equations are $\mathbf{P}(0) = \mathbf{P}_1$, $\mathbf{P}(1) = \mathbf{P}_2$, $\mathbf{P}^t(0) = \mathbf{P}_1^t$, and $\mathbf{P}^t(1) = \mathbf{P}_2^t$. Their explicit forms are

$$\begin{aligned} \mathbf{a}\cdot 0^3 + \mathbf{b}\cdot 0^2 + \mathbf{c}\cdot 0 + \mathbf{d} &= \mathbf{P}_1, \\ \mathbf{a}\cdot 1^3 + \mathbf{b}\cdot 1^2 + \mathbf{c}\cdot 1 + \mathbf{d} &= \mathbf{P}_2, \\ 3\mathbf{a}\cdot 0^2 + 2\mathbf{b}\cdot 0 + \mathbf{c} &= \mathbf{P}_1^t, \\ 3\mathbf{a}\cdot 1^2 + 2\mathbf{b}\cdot 1 + \mathbf{c} &= \mathbf{P}_2^t. \end{aligned} \quad (4.21)$$

This is easy to solve and the solutions are

$$\begin{aligned} \mathbf{a} &= 2\mathbf{P}_1 - 2\mathbf{P}_2 + \mathbf{P}_1^t + \mathbf{P}_2^t, \\ \mathbf{b} &= -3\mathbf{P}_1 + 3\mathbf{P}_2 - 2\mathbf{P}_1^t - \mathbf{P}_2^t, \\ \mathbf{c} &= \mathbf{P}_1^t, \\ \mathbf{d} &= \mathbf{P}_1. \end{aligned} \quad (4.22)$$

Substituting these solutions into Equation (4.8) gives

$$\mathbf{P}(t) = (2\mathbf{P}_1 - 2\mathbf{P}_2 + \mathbf{P}_1^t + \mathbf{P}_2^t)t^3 + (-3\mathbf{P}_1 + 3\mathbf{P}_2 - 2\mathbf{P}_1^t - \mathbf{P}_2^t)t^2 + \mathbf{P}_1^t t + \mathbf{P}_1, \quad (4.23)$$

which, after rearranging, becomes

$$\begin{aligned} \mathbf{P}(t) &= (2t^3 - 3t^2 + 1)\mathbf{P}_1 + (-2t^3 + 3t^2)\mathbf{P}_2 + (t^3 - 2t^2 + t)\mathbf{P}_1^t + (t^3 - t^2)\mathbf{P}_2^t \\ &= F_1(t)\mathbf{P}_1 + F_2(t)\mathbf{P}_2 + F_3(t)\mathbf{P}_1^t + F_4(t)\mathbf{P}_2^t \\ &= (F_1(t), F_2(t), F_3(t), F_4(t))(\mathbf{P}_1, \mathbf{P}_2, \mathbf{P}_1^t, \mathbf{P}_2^t)^T \\ &= \mathbf{F}(t)\mathbf{B}, \end{aligned} \quad (4.24)$$

where

$$\begin{aligned} F_1(t) &= (2t^3 - 3t^2 + 1), \quad F_2(t) = (-2t^3 + 3t^2), \\ F_3(t) &= (t^3 - 2t^2 + t), \quad F_4(t) = (t^3 - t^2), \end{aligned} \quad (4.25)$$

$\mathbf{B}$ is the column $(\mathbf{P}_1, \mathbf{P}_2, \mathbf{P}_1^t, \mathbf{P}_2^t)^T$, and $\mathbf{F}(t)$ is the row $(F_1(t), F_2(t), F_3(t), F_4(t))$. Equations (4.23) and (4.24) are the geometric representation of the Hermite PC segment.

Functions $F_i(t)$ are called the *Hermite blending functions* since they create any point on the curve as a blend of the four given quantities. They are shown in Figure 4.11a. Note that $F_1(t) + F_2(t) \equiv 1$. These two functions blend points, not tangent vectors, and should thus be barycentric. We can also write $F_1(t) = (t^3, t^2, t, 1)(2, -3, 0, 1)^T$ and similarly for $F_2(t)$, $F_3(t)$, and $F_4(t)$. In matrix notation this becomes

$$\mathbf{F}(t) = (t^3, t^2, t, 1)\begin{pmatrix} 2 & -2 & 1 & 1 \\ -3 & 3 & -2 & -1 \\ 0 & 0 & 1 & 0 \\ 1 & 0 & 0 & 0 \end{pmatrix} = \mathbf{T}(t)\,\mathbf{H}.$$

The curve can now be written

$$\mathbf{P}(t) = \mathbf{F}(t)\mathbf{B} = \mathbf{T}(t)\,\mathbf{H}\,\mathbf{B} = (t^3, t^2, t, 1) \begin{pmatrix} 2 & -2 & 1 & 1 \\ -3 & 3 & -2 & -1 \\ 0 & 0 & 1 & 0 \\ 1 & 0 & 0 & 0 \end{pmatrix} \begin{pmatrix} \mathbf{P}_1 \\ \mathbf{P}_2 \\ \mathbf{P}_1^t \\ \mathbf{P}_2^t \end{pmatrix}. \quad (4.26)$$

Since, from Equation (4.8), $\mathbf{P}(t) = \mathbf{T}(t)\mathbf{A}$, we get $\mathbf{A} = \mathbf{H}\mathbf{B}$. Matrix $\mathbf{H}$ is called the Hermite basis matrix.

Following is *Mathematica* code to display a single Hermite curve segment.

```
Clear[T,H,B]; (* Hermite Interpolation *)
T={t^3,t^2,t,1};
H={{2,-2,1,1},{-3,3,-2,-1},{0,0,1,0},{1,0,0,0}};
B={{0,0},{2,1},{1,1},{1,0}};
ParametricPlot[T.H.B,{t,0,1},PlotRange->All]
```

▶ **Exercise 4.33:** Express the midpoint $\mathbf{P}(0.5)$ of a Hermite segment in terms of the two endpoints and two tangent vectors. Draw a diagram to illustrate the geometric interpretation of the result.

4.7.1 Hermite Derivatives

The concept of blending can be applied to calculating the derivatives of a curve, not just to the curve itself. One way to calculate $\mathbf{P}^t(t)$ is to differentiate $\mathbf{T}(t) = (t^3, t^2, t, 1)$. The result is

$$\mathbf{P}^t(t) = \mathbf{T}^t(t)\mathbf{H}\mathbf{B} = (3t^2, 2t, 1, 0)\mathbf{H}\mathbf{B}.$$

A more general method is to use the relation $\mathbf{P}(t) = \mathbf{F}(t)\mathbf{B}$, which implies

$$\mathbf{P}^t(t) = \mathbf{F}^t(t)\mathbf{B} = (F_1^t(t), F_2^t(t), F_3^t(t), F_4^t(t))\mathbf{B}.$$

The individual derivatives $F_i^t(t)$ can be obtained from Equation (4.24). The results can be expressed as

$$\mathbf{P}^t(t) = (t^3, t^2, t, 1) \begin{pmatrix} 0 & 0 & 0 & 0 \\ 6 & -6 & 3 & 3 \\ -6 & 6 & -4 & -2 \\ 0 & 0 & 1 & 0 \end{pmatrix} \begin{pmatrix} \mathbf{P}_1 \\ \mathbf{P}_2 \\ \mathbf{P}_1^t \\ \mathbf{P}_2^t \end{pmatrix} = \mathbf{T}(t)\mathbf{H}_t\mathbf{B}. \quad (4.27)$$

Similarly, the second derivatives of the Hermite segment can be expressed as

$$\mathbf{P}^{tt}(t) = (t^3, t^2, t, 1) \begin{pmatrix} 0 & 0 & 0 & 0 \\ 0 & 0 & 0 & 0 \\ 12 & -12 & 6 & 6 \\ -6 & 6 & -4 & -2 \end{pmatrix} \begin{pmatrix} \mathbf{P}_1 \\ \mathbf{P}_2 \\ \mathbf{P}_1^t \\ \mathbf{P}_2^t \end{pmatrix} = \mathbf{T}(t)\mathbf{H}_{tt}\mathbf{B}. \quad (4.28)$$

These expressions make it easy to calculate the first and second derivatives at any point on a Hermite segment. Similar expressions can be derived for any other curves that are based on blending geometrical quantities.

▶ **Exercise 4.34:** What is H_{ttt}?

A Two-Dimensional Example: The two two-dimensional points $\mathbf{P}_1 = (0,0)$ and $\mathbf{P}_2 = (1,0)$ and the two tangents $\mathbf{P}_1^t = (1,1)$ and $\mathbf{P}_2^t = (0,-1)$ are given. The segment should, therefore, start at the origin, going in a 45° direction, and end at point $(1,0)$, going straight down. The calculation of $\mathbf{P}(t)$ is straightforward:

$$\mathbf{P}(t) = \mathbf{T}(t)\,\mathbf{A} = \mathbf{T}(t)\,\mathbf{H}\,\mathbf{B}$$

$$= (t^3, t^2, t, 1)\begin{pmatrix} 2 & -2 & 1 & 1 \\ -3 & 3 & -2 & -1 \\ 0 & 0 & 1 & 0 \\ 1 & 0 & 0 & 0 \end{pmatrix}\begin{pmatrix} (0,0) \\ (1,0) \\ (1,1) \\ (0,-1) \end{pmatrix}$$

$$= (t^3, t^2, t, 1)\begin{pmatrix} 2(0,0) - 2(1,0) + 1(1,1) + 1(0,-1) \\ -3(0,0) + 3(1,0) - 2(1,1) - 1(0,-1) \\ 0(0,0) + 0(1,0) + 1(1,1) + 0(0,-1) \\ 1(0,0) + 0(1,0) + 0(1,1) + 0(0,-1) \end{pmatrix}$$

$$= (t^3, t^2, t, 1)\begin{pmatrix} (-1,0) \\ (1,-1) \\ (1,1) \\ (0,0) \end{pmatrix}$$

$$= (-1,0)t^3 + (1,-1)t^2 + (1,1)t. \tag{4.29}$$

▶ **Exercise 4.35:** Use Equation (4.29) to show that the segment really passes through points $(0,0)$ and $(1,0)$. Calculate the tangent vectors and use them to show that the segment really starts and ends in the right directions.

▶ **Exercise 4.36:** Repeat the above example with $\mathbf{P}_1^t = (2,2)$. The new curve segment should go through the same points, in the same directions. However, it should continue longer in the original 45° direction, since the size of the new tangent is $\sqrt{2^2+2^2} = 2\sqrt{2}$, longer than the previous one, which is $\sqrt{1^2+1^2} = \sqrt{2}$.

▶ **Exercise 4.37:** Calculate the Hermite curve for two given points $\mathbf{P}_1$ and $\mathbf{P}_2$ assuming that the tangent vectors at the two points are zero (indeterminate). What kind of a curve is this?

▶ **Exercise 4.38:** Use the Hermite method to calculate PC segments for the cases where the known quantities are as follows:

1. The three tangent vectors at the start, middle, and end of the segment.

2. The two interior points $\mathbf{P}(1/3)$ and $\mathbf{P}(2/3)$, and the two extreme tangent vectors $\mathbf{P}^t(0)$ and $\mathbf{P}^t(1)$.

3. The two extreme points $\mathbf{P}(0)$ and $\mathbf{P}(1)$, and the two interior tangent vectors $\mathbf{P}^t(1/3)$ and $\mathbf{P}^t(2/3)$ (this is similar to case 2, so the reader should consider it a voluntary exercise).

A Three-Dimensional Example: Given the two points $\mathbf{P}_1 = (0,0,0)$ and $\mathbf{P}_2 = (1,1,1)$ and the two tangent vectors $\mathbf{P}_1^t = (1,0,0)$ and $\mathbf{P}_2^t = (0,1,0)$, the curve segment is

$$\mathbf{P}(t) = (t^3, t^2, t, 1) \begin{pmatrix} 2 & -2 & 1 & 1 \\ -3 & 3 & -2 & -1 \\ 0 & 0 & 1 & 0 \\ 1 & 0 & 0 & 0 \end{pmatrix} \begin{pmatrix} (0,0,0) \\ (1,1,1) \\ (1,0,0) \\ (0,1,0) \end{pmatrix}$$
$$= (-t^3 + t^2 + t, -t^3 + 2t^2, -2t^3 + 3t^2). \qquad (4.30)$$

4.7.2 PC Conic Approximations

Hermite interpolation can be used to approximate a conic section (see Section 4.20 for more on conics). Given three points $\mathbf{P}_0$, $\mathbf{P}_1$, and $\mathbf{P}_2$ and a scalar α, we can use the 4-tuple

$$(\mathbf{P}_0, \mathbf{P}_2, 4\alpha(\mathbf{P}_1 - \mathbf{P}_0), 4\alpha(\mathbf{P}_2 - \mathbf{P}_1)), \quad \text{where } 0 \leq \alpha \leq 1, \qquad (4.31)$$

as two points and two vectors, to create a segment that approximates a conic section. We obtain an ellipse when $0 \leq \alpha < 0.5$, a parabola when $\alpha = 0.5$, and a hyperbola when $0.5 < \alpha \leq 1$ (see below for a circle).

The tangent vectors at the two ends are $\mathbf{P}^t(0) = 4\alpha(\mathbf{P}_1 - \mathbf{P}_0)$ and $\mathbf{P}^t(1) = 4\alpha(\mathbf{P}_2 - \mathbf{P}_1)$ (note their directions). The tangent vector halfway is $\mathbf{P}^t(0.5) = (1.5 - \alpha)(\mathbf{P}_2 - \mathbf{P}_0)$. It is parallel to the vector $\mathbf{P}_2 - \mathbf{P}_0$.

The case of the parabola is especially useful and is explicitly shown here. Substituting $\alpha = 0.5$ in Equation (4.31) and using Equation (4.26) yields the Hermite segment

$$\mathbf{P}(t) = (t^3, t^2, t, 1) \begin{pmatrix} 2 & -2 & 1 & 1 \\ -3 & 3 & -2 & -1 \\ 0 & 0 & 1 & 0 \\ 1 & 0 & 0 & 0 \end{pmatrix} \begin{pmatrix} \mathbf{P}_0 \\ \mathbf{P}_2 \\ 2(\mathbf{P}_1 - \mathbf{P}_0) \\ 2(\mathbf{P}_2 - \mathbf{P}_1) \end{pmatrix}$$
$$= (1-t)^2 \mathbf{P}_0 + 2t(1-t)\mathbf{P}_1 + t^2 \mathbf{P}_2.$$

This is the parabola produced in Exercise 4.62.

▶ **Exercise 4.39:** We know that any three points $\mathbf{P}_0$, $\mathbf{P}_1$, and $\mathbf{P}_2$ define a unique parabola (i.e., a triangle defines a parabola). Use Hermite interpolation to calculate the parabola from $\mathbf{P}_0$ to $\mathbf{P}_2$ whose start and end tangents go in the directions from $\mathbf{P}_0$ to $\mathbf{P}_1$ and from $\mathbf{P}_1$ to $\mathbf{P}_2$, respectively.

Hermite interpolation provides a simple way to draw approximate circles and circular arcs. Figure 4.13a shows how this method is used to construct a circular arc of unit radius about the origin. We assume that an arc spanning an angle 2θ is needed and we place its two endpoints $\mathbf{P}_1$ and $\mathbf{P}_2$ at locations $(\cos\theta, -\sin\theta)$ and $(\cos\theta, \sin\theta)$, respectively. This arc is symmetric about the x axis, but we later show how to rotate it to have a general arc. Since a circle is always perpendicular to its radius, we select as our start and end tangents two vectors that are perpendicular to $\mathbf{P}_1$ and $\mathbf{P}_2$. They are $\mathbf{P}_1^t = a(\sin\theta, \cos\theta)$ and $\mathbf{P}_2^t = a(-\sin\theta, \cos\theta)$, where a

4 Curves

is a parameter to be determined. The curve segment defined by these points and vectors is, as usual,

$$\mathbf{P}(t) = (t^3, t^2, t, 1) \begin{pmatrix} 2 & -2 & 1 & 1 \\ -3 & 3 & -2 & -1 \\ 0 & 0 & 1 & 0 \\ 1 & 0 & 0 & 0 \end{pmatrix} \begin{pmatrix} (\cos\theta, -\sin\theta) \\ (\cos\theta, \sin\theta) \\ a(\sin\theta, \cos\theta) \\ a(-\sin\theta, \cos\theta) \end{pmatrix} \quad (4.32)$$

$$= (2t^3 - 3t^2 + 1)(\cos\theta, -\sin\theta) + (-2t^3 + 3t^2)(\cos\theta, \sin\theta)$$
$$+ (t^3 - 2t^2 + t)a(\sin\theta, \cos\theta) + (t^3 - t^2)a(-\sin\theta, \cos\theta).$$

We need an equation in order to determine a and we get it by requiring that the curve segment passes through the circular arc at its center, i.e., $\mathbf{P}(0.5) = (1, 0)$. This produces the equation

$$(1, 0) = \mathbf{P}(0.5) = \left(\frac{2}{8} - \frac{3}{4} + 1\right)(\cos\theta, -\sin\theta) + \left(-\frac{2}{8} + \frac{3}{4}\right)(\cos\theta, \sin\theta)$$
$$+ \left(\frac{1}{8} - \frac{2}{4} + \frac{1}{2}\right) a(\sin\theta, \cos\theta) + \left(\frac{1}{8} - \frac{1}{4}\right) a(-\sin\theta, \bar{\cos}\theta)$$
$$= \frac{1}{8}(8\cos\theta + 2a\sin\theta, 0),$$

whose solution is

$$a = \frac{4(1 - \cos\theta)}{\sin\theta}.$$

The curve can now be written

$$\mathbf{P}(t) = (t^3, t^2, t, 1) \begin{pmatrix} 2 & -2 & 1 & 1 \\ -3 & 3 & -2 & -1 \\ 0 & 0 & 1 & 0 \\ 1 & 0 & 0 & 0 \end{pmatrix} \begin{pmatrix} (\cos\theta, -\sin\theta) \\ (\cos\theta, \sin\theta) \\ \left(4(1-\cos\theta), \frac{4(1-\cos\theta)}{\tan\theta}\right) \\ \left(-4(1-\cos\theta), \frac{4(1-\cos\theta)}{\tan\theta}\right) \end{pmatrix}.$$

This curve provides an excellent approximation to a circular arc, even for angles θ as large as 90°.

▶ **Exercise 4.40:** Write Equation (4.32) for $\theta = 90°$; calculate $\mathbf{P}(0.25)$ and the deviation of the curve from a true circle at this point.

In general, an arc with a unit radius is not symmetric about the x axis but may look as in Figure 4.13b, where $\mathbf{P}_1$ and $\mathbf{P}_2$ are any points at a distance of one unit from the origin. All that's necessary to calculate the arc from Equation (4.32) is the value of θ (where 2θ is the angle between $\mathbf{P}_1$ and $\mathbf{P}_2$) and this can be calculated numerically from the two points using the relations

$$\theta = (\theta_1 - \theta_2)/2, \quad \cos\theta_1 = \mathbf{P}_1 \bullet (1, 0), \quad \cos\theta_2 = \mathbf{P}_2 \bullet (1, 0),$$
$$\cos(2\theta) = \cos(\theta_1 - \theta_2) = \cos\theta_1\cos\theta_2 + \sin\theta_1\sin\theta_2,$$
$$\cos\theta = \pm\sqrt{(1 + \cos^2\theta)/2}, \quad \sin\theta = \sqrt{1 - \cos^2\theta}.$$

214   4.7 Hermite Interpolation

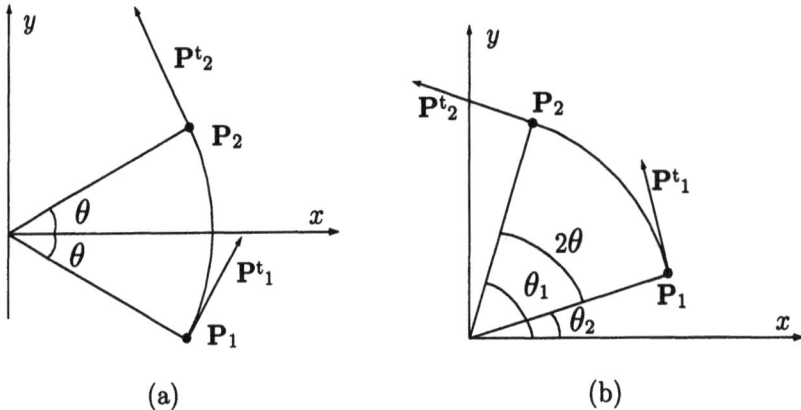

Figure 4.13: Hermite Segment and a Circular Arc.

4.7.3 Degree-5 Hermite Interpolation

It is possible to extend the basic idea of Hermite interpolation to polynomials of higher degree. Naturally, more data are needed in order to calculate such a polynomial, and these data are supplied by the user in the form of higher-order derivatives of the curve. If the user specifies the endpoints, the two extreme tangent vectors, and the two extreme second derivatives, the software can use these six pieces of data to calculate the six coefficients of a fifth-degree polynomial that interpolates the two points. In general, if the two endpoints and the first k pairs of derivatives at the extreme points are known (a total of $2k+2$ items), they can be used to calculate an interpolating polynomial of degree $2k+1$. These higher-degree polynomials are not as useful as the cubic, but the fifth-degree polynomial is shown here, as a demonstration of Hermite interpolation.

Given two endpoints $\mathbf{P}_1$ and $\mathbf{P}_2$, the values of two tangent vectors $\mathbf{P}_1^t$ and $\mathbf{P}_2^t$, and of two second derivatives $\mathbf{P}_1^{tt}$ and $\mathbf{P}_2^{tt}$, we can calculate the polynomial

$$\mathbf{P}(t) = \mathbf{a}t^5 + \mathbf{b}t^4 + \mathbf{c}t^3 + \mathbf{d}t^2 + \mathbf{e}t + \mathbf{f}$$

by writing the six equations

$$\mathbf{P}(0) = \mathbf{a}t^5 + \mathbf{b}t^4 + \mathbf{c}t^3 + \mathbf{d}t^2 + \mathbf{e}t + \mathbf{f}|_0 = \mathbf{f} = \mathbf{P}_1,$$
$$\mathbf{P}(1) = \mathbf{a}t^5 + \mathbf{b}t^4 + \mathbf{c}t^3 + \mathbf{d}t^2 + \mathbf{e}t + \mathbf{f}|_1 = \mathbf{a} + \mathbf{b} + \mathbf{c} + \mathbf{d} + \mathbf{e} + \mathbf{f} = \mathbf{P}_2,$$
$$\mathbf{P}^t(0) = 5\mathbf{a}t^4 + 4\mathbf{b}t^3 + 3\mathbf{c}t^2 + 2\mathbf{d}t + \mathbf{e}|_0 = \mathbf{e} = \mathbf{P}_1^t,$$
$$\mathbf{P}^t(1) = 5\mathbf{a}t^4 + 4\mathbf{b}t^3 + 3\mathbf{c}t^2 + 2\mathbf{d}t + \mathbf{e}|_1 = 5\mathbf{a} + 4\mathbf{b} + 3\mathbf{c} + 2\mathbf{d} + \mathbf{e} = \mathbf{P}_2^t,$$
$$\mathbf{P}^{tt}(0) = 20\mathbf{a}t^3 + 12\mathbf{b}t^2 + 6\mathbf{c}t + 2\mathbf{d}|_0 = 2\mathbf{d} = \mathbf{P}_1^{tt},$$
$$\mathbf{P}^{tt}(1) = 20\mathbf{a}t^3 + 12\mathbf{b}t^2 + 6\mathbf{c}t + 2\mathbf{d}|_1 = 20\mathbf{a} + 12\mathbf{b} + 6\mathbf{c} + 2\mathbf{d} = \mathbf{P}_2^{tt}.$$

After solving for the six unknown coefficients, the fifth-degree Hermite interpolating

4 Curves

polynomial is obtained:

$$\begin{aligned}
\mathbf{P}(t) &= F_1(t)\mathbf{P}_1 + F_2(t)\mathbf{P}_2 + F_3(t)\mathbf{P}_1^t + F_4(t)\mathbf{P}_2^t + F_5(t)\mathbf{P}_1^{tt} + F_6(t)\mathbf{P}_2^{tt} \\
&= (-6t^5 + 15t^4 - 10t^3 + 1)\mathbf{P}_1 + (6t^5 - 15t^4 + 10t^3)\mathbf{P}_2 \\
&\quad + (-3t^5 + 8t^4 - 6t^3 + t)\mathbf{P}_1^t + (-3t^5 + 7t^4 - 4t^3)\mathbf{P}_2^t \\
&\quad + (-(1/2)t^5 + (3/2)t^4 - (3/2)t^3 + (1/2)t^2)\mathbf{P}_1^{tt} + ((1/2)t^5 - t^4 + (1/2)t^3)\mathbf{P}_2^{tt} \\
&= (t^5, t^4, t^3, t^2, t, 1) \begin{pmatrix} -6 & 6 & -3 & -3 & -1/2 & 1/2 \\ 15 & -15 & 8 & 7 & 3/2 & -1 \\ -10 & 10 & -6 & -4 & -3/2 & 1/2 \\ 0 & 0 & 0 & 0 & 1/2 & 0 \\ 0 & 0 & 1 & 0 & 0 & 0 \\ 1 & 0 & 0 & 0 & 0 & 0 \end{pmatrix} \begin{pmatrix} \mathbf{P}_1 \\ \mathbf{P}_2 \\ \mathbf{P}_1^t \\ \mathbf{P}_2^t \\ \mathbf{P}_1^{tt} \\ \mathbf{P}_2^{tt} \end{pmatrix}.
\end{aligned}$$

4.7.4 Editing the Hermite Segment

The Hermite method is interactive. The points normally cannot be moved, but the tangent vectors can be modified. Even if their directions cannot be changed, their magnitudes normally are not fixed by the user and can be changed to modify the shape of the curve.

The following simple experiment sheds light on this point. We start with the Hermite segment that is defined by the two endpoints $\mathbf{P}_1 = (0,0)$ and $\mathbf{P}_2 = (2,1)$ and by the two tangent vectors $\mathbf{P}^t(0) = (1,1)$ and $\mathbf{P}^t(1) = (1,0)$. The curve starts in the 45° direction and ends in a horizontal direction. The curve is easy to calculate. Its expression is

$$\mathbf{P}(t) = (t^3, t^2, t, 1) \begin{pmatrix} 2 & -2 & 1 & 1 \\ -3 & 3 & -2 & -1 \\ 0 & 0 & 1 & 0 \\ 1 & 0 & 0 & 0 \end{pmatrix} \begin{pmatrix} (0,0) \\ (2,1) \\ (1,1) \\ (1,0) \end{pmatrix} = -(2,1)t^3 + (3,1)t^2 + (1,1)t. \tag{4.33}$$

Suppose that the user decides that the curve should be raised a bit, but should still start and end at the same points, going in the same directions at the endpoints. Since the designer cannot move the points and cannot change the directions of the tangent vectors, the only way to edit the curve is to change the magnitudes of the tangents.

To keep the same directions, the new tangent vectors should have the form (a, a) and $(b, 0)$, where a and b are two new parameters that have to be calculated. To raise the curve, we go through the following steps:

1. Calculate the midpoint of the curve. This is $\mathbf{P}(0.5) = (1, 5/8)$.
2. Decide by how much to raise it. Let's say we decide to raise it to $(1, 1)$.
3. Calculate a new curve $\mathbf{Q}(t)$, based on the tangents (a, a) and $(b, 0)$.
4. Require that the new curve pass through $(1, 1)$ and calculate a and b from this requirement.

The general form of the new curve is

$$\mathbf{Q}(t) = (t^3, t^2, t, 1) \begin{pmatrix} 2 & -2 & 1 & 1 \\ -3 & 3 & -2 & -1 \\ 0 & 0 & 1 & 0 \\ 1 & 0 & 0 & 0 \end{pmatrix} \begin{pmatrix} (0,0) \\ (2,1) \\ (a,a) \\ (b,0) \end{pmatrix}$$

$$= (a+b-4, a-2)t^3 + (-2a-b+6, 3-2a)t^2 + (a,a)t. \quad (4.34)$$

The requirement $\mathbf{Q}(0.5) = (1,1)$ can now be written

$$(a+b-4, a-2)/8 + (-2a-b+6, 3-2a)/4 + (a,a)/2 = (1,1),$$

which yields the two equations $a+b-4+2(-2a-b+6)+4a = 8$ and $a-2+2(3-2a)+4a = 8$. The solutions are $a = b = 4$ and the curve thus has the form

$$\mathbf{Q}(t) = (4,2)t^3 - (6,5)t^2 + (4,4)t. \quad (4.35)$$

A simple check verifies that this curve really starts at $(0,0)$, ends at $(2,1)$, has the extreme tangents $(4,4)$ and $(4,0)$, and passes midway through $(1,1)$.

Raising the midpoint from $(1, 5/8)$ to $(1,1)$ has completely changed the curve (Equations (4.33) and (4.35) are different). It now starts going in the same 45° direction, then starts going up, reaches point $(1,1)$, starts going down, and still has "time" to arrive at point $(2,1)$ moving horizontally. An interesting question is: How much can we raise the midpoint? If we raise it from $(1, 5/8)$ to, say, $(1, 100)$, would the curve be able to change directions, climb up, pass through the new midpoint, dive down, and still approach $(2,1)$ moving horizontally?

To check this, let's assume that we raise the midpoint from $(1, 5/8)$ to $(1, 5/8 + \alpha)$, where α is a real number. The curve is constrained by $\mathbf{Q}(0.5) = (1, 5/8 + \alpha)$, which yields the equation

$$(a+b-4, a-2)/8 + (-2a-b+6, 3-2a)/4 + (a,a)/2 = (1, 5/8 + \alpha).$$

The solutions are $a = b = 1 + 8\alpha$. This means that α can vary without limit. When α is positive, the curve is pulled up. Negative values of α push the curve down. The value $\alpha = -1/8$ is special. It implies $a = b = 0$ and results in the curve $\mathbf{Q}(t) = (6t^2 - 4t^3, 3t^2 - 2t^3)$. The parameter substitution $u = 3t^2 - 2t^3$ yields $\mathbf{Q}(u) = (2u, u)$. This curve is the straight line from $(0,0)$ to $(2,1)$. Its midpoint is $(1, 1/2)$.

▶ **Exercise 4.41:** Values $\alpha < -1/8$ result in negative a and b. Can they still be used in Equation (4.34)?

▶ **Exercise 4.42:** How can we coerce the curve of Equation (4.34) to have point $(1,0)$ as its midpoint?

Note: Raising the curve is done by increasing the size of the tangent vectors. This forces the curve to continue longer in the initial and final directions. This is also the reason why too much raising causes undesirable effects. Figure 4.14 shows

4 Curves

the original curve ($\alpha = 0$) and the effects of increasing α. For $\alpha = 0.4$, the curve is raised and still has a reasonable shape. However, for larger values of α, the curve gets tight, develops a *cusp* (a kink), then starts looping on itself. It is easy to see that when $\alpha = 5/8$, the tangent vector becomes indefinite at the midpoint ($t = 0.5$).

Proof: The tangent vector of the curve of Equation (4.34) is

$$\mathbf{Q}^t(t) = 3(a+b-4, a-2)t^2 + 2(-2a-b+6, 3-2a)t + (a,a).$$

From $a = b = 1 + 8\alpha$, we get

$$\mathbf{Q}^t(t) = (48\alpha - 6, 24\alpha - 3)t^2 + (6 - 48\alpha, 2 - 32\alpha)t + (1 + 8\alpha, 1 + 8\alpha).$$

For $\alpha = 5/8$, this reduces to $\mathbf{Q}^t(t) = (24, 12)t^2 - (24, 18)t + (6, 6)$, so $\mathbf{Q}^t(0.5) = (0,0)$. ◂

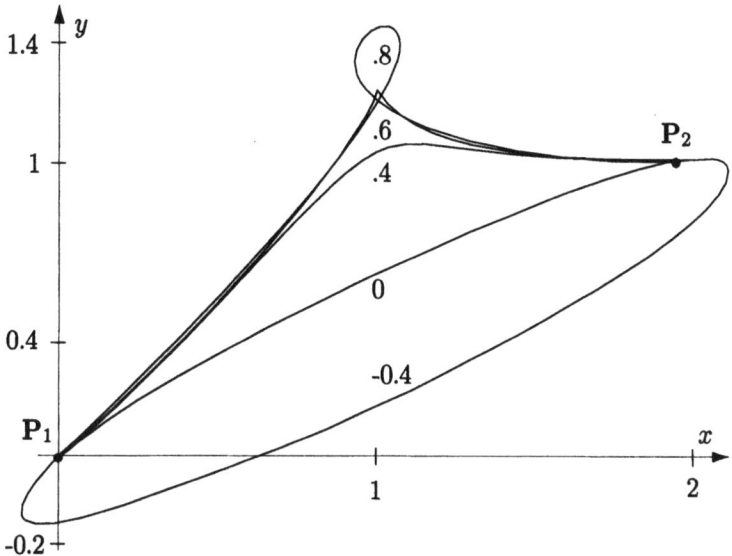

Figure 4.14: Effects of Changing α.

▶ **Exercise 4.43**: Given the two endpoints $\mathbf{P}_1 = (0,0)$ and $\mathbf{P}_2 = (1,0)$ and the two tangent vectors $\mathbf{P}_1^t = \alpha(\cos\theta, \sin\theta)$ and $\mathbf{P}_1^t = \alpha(\cos\theta, -\sin\theta)$ (Figure 4.15), calculate the value of α for which the Hermite segment from $\mathbf{P}_1$ to $\mathbf{P}_2$ has a cusp.

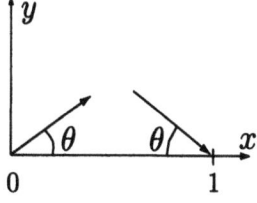

Figure 4.15: Tangents for Exercise 4.43.

The following problem may sometimes occur in practice. Given two endpoints $\mathbf{P}_1$ and $\mathbf{P}_2$, two *unit* tangent vectors $\mathbf{T1}$ and $\mathbf{T2}$, and a third point $\mathbf{P}_3$, find scale factors α and β such that the Hermite segment $\mathbf{P}(t)$ defined by points $\mathbf{P}_1$ and $\mathbf{P}_2$ and tangents $\alpha\mathbf{T1}$ and $\beta\mathbf{T2}$, respectively, will pass through $\mathbf{P}_3$. Also find the value t_0 for which $\mathbf{P}(t_0) = \mathbf{P}_3$.

We start with Equation (4.24), which, in our case, becomes

$$\mathbf{P}_3 = F_1(t_0)\mathbf{P}_1 + F_2(t_0)\mathbf{P}_2 + F_3(t_0)\alpha\mathbf{T1} + F_4(t_0)\beta\mathbf{T2},$$

where the $F_i(t)$ are given by Equation (4.25). Since $F_1(t) + F_2(t) \equiv 1$ we can write

$$\mathbf{P}_3 - \mathbf{P}_1 = F_2(t_0)(\mathbf{P}_2 - \mathbf{P}_1) + \alpha F_3(t_0)\mathbf{T1} + \beta F_4(t_0)\mathbf{T2}.$$

This can now be written as the three scalar equations

$$\begin{aligned} x_3 - x_1 &= F_2(t_0)(x_2 - x_1) + \alpha F_3(t_0)T1_x + \beta F_4(t_0)T2_x, \\ y_3 - y_1 &= F_2(t_0)(y_2 - y_1) + \alpha F_3(t_0)T1_y + \beta F_4(t_0)T2_y, \\ z_3 - z_1 &= F_2(t_0)(z_2 - z_1) + \alpha F_3(t_0)T1_z + \beta F_4(t_0)T2_z. \end{aligned} \quad (4.36)$$

This is a system of three equations in the three unknowns α, β, and t_0. In principle, it should have one solution, but solving it is awkward since t_0 is included in the $F_i(t_0)$ functions, which are degree-3 polynomials in t_0. The first step is to isolate the two products $\alpha F_3(t_0)$ and $\beta F_4(t_0)$ in the first two equations. This yields

$$\begin{pmatrix} \alpha F_3(t_0) \\ \beta F_4(t_0) \end{pmatrix} = \begin{pmatrix} T1_x & T2_x \\ T1_y & T2_y \end{pmatrix}^{-1} \left[\begin{pmatrix} x_3 - x_1 \\ y_3 - y_1 \end{pmatrix} - \begin{pmatrix} x_2 - x_1 \\ y_2 - y_1 \end{pmatrix} F_2(t_0) \right].$$

This result is used in step two to eliminate $\alpha F_3(t_0)$ and $\beta F_4(t_0)$ from the third equation:

$$\begin{aligned} z_3 - z_1 &= F_2(t_0)(z_2 - z_1) + (T1_z, T2_z)\begin{pmatrix} \alpha F_3(t_0) \\ \beta F_4(t_0) \end{pmatrix} \\ &= F_2(t_0)(z_2 - z_1) \\ &\quad + (T1_z, T2_z)\begin{pmatrix} T1_x & T2_x \\ T1_y & T2_y \end{pmatrix}^{-1} \left[\begin{pmatrix} x_3 - x_1 \\ y_3 - y_1 \end{pmatrix} - \begin{pmatrix} x_2 - x_1 \\ y_2 - y_1 \end{pmatrix} F_2(t_0) \right]. \end{aligned}$$

We now have an equation with the single unknown t_0. Step three is to simplify the result above by using the value $F_2(t_0) = -2t_0^3 + 3t_0^2$:

$$\begin{vmatrix} x_2 - x_1 & y_2 - y_1 & z_2 - z_1 \\ T1_x & T1_y & T1_z \\ T2_x & T2_y & T2_z \end{vmatrix} (-2t_0^3 + 3t_0^2) = \begin{vmatrix} x_3 - x_1 & y_3 - y_1 & z_3 - z_1 \\ T1_x & T1_y & T1_z \\ T2_x & T2_y & T2_z \end{vmatrix}. \quad (4.37)$$

(See Equation (A.17) for the matrix inverse.) Step four is to solve Equation (4.37) for t_0. Once t_0 is known, α and β can be computed from the other equations.

4 Curves

Equation (4.37), however, is cubic in t_0, so it may have to be solved numerically and it may have between zero and three real solutions t_0. Any acceptable solution t_0 must be a real number in the range $[0,1]$ and must result in positive α and β.

This, of course, is a slow, tedious approach and should only be used as a last resort, when nothing else works.

4.7.5 Truncating and Segmenting

Surfaces and solid objects are constructed of curves. When surfaces are joined, clipped, or intersected, there is sometimes a need to truncate curves. In general, the problem of truncating a curve starts with a parametric curve $\mathbf{P}(t)$ and the two values t_i and t_j. A new curve $\mathbf{Q}(T)$ needs be calculated, which is identical to the segment $\mathbf{P}(t_i) \to \mathbf{P}(t_j)$ (Figure 4.16a) when T varies from 0 to 1. The discussion in this section is limited to Hermite segments. The endpoints of the new curve are $\mathbf{Q}(0) = \mathbf{P}(t_i)$ and $\mathbf{Q}(1) = \mathbf{P}(t_j)$. To understand how the two extreme tangent vectors of $\mathbf{Q}(T)$ are calculated, we first need to discuss reparametrization of parametric curves.

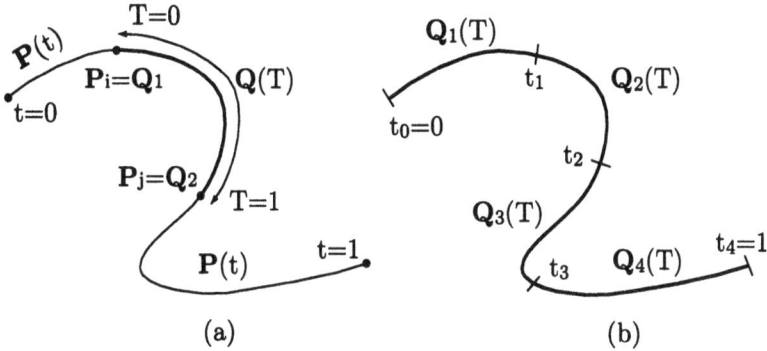

Figure 4.16: Truncating and Segmenting.

Reparametrization is the case where a new parameter $T(t)$ is substituted for the original parameter t. Notice that $T(t)$ is a function of t. One example of reparametrization is reversing the direction of a curve. It is easy to see that when t goes from 0 to 1, the simple function $T = 1 - t$ goes from 1 to 0. The two curves $\mathbf{P}(t)$ and $\mathbf{P}(1-t)$ have the same shape and location but move in opposite directions. Another example of reparametrization is a curve $\mathbf{P}(t)$ with a parameter $0 \le t \le 1$ being transformed to a curve $\mathbf{Q}(T)$ with a parameter $a \le T \le b$ (Section 4.8.6 has an example). The simplest relation between T and t is linear, i.e., $T = at + b$. We can examine this relation from two different points:

1. At two different points i and j along the curve, the parameters are related by $T_i = at_i + b$ and $T_j = at_j + b$, respectively. Subtracting yields $T_j - T_i = a(t_j - t_i)$, so $a = (T_j - T_i)/(t_j - t_i)$.
2. $T = at + b$ gives $dT = a\,dt$.

These two points can be combined to

$$\frac{dt}{dT} = \frac{1}{a} = \frac{t_j - t_i}{T_j - T_i}. \tag{4.38}$$

Equation (4.38) is used to calculate the extreme tangent vectors of our new curve $\mathbf{Q}(T)$. Since it goes from point $\mathbf{P}(t_i)$ (where $T = 0$) to point $\mathbf{P}(t_j)$ (where $T = 1$), we have $T_j - T_i = 1$. The tangent vectors of $\mathbf{Q}(T)$ are therefore

$$\mathbf{Q}^T(T) = \frac{d\mathbf{Q}(T)}{dT} = \frac{d\mathbf{P}(t)}{dt}\frac{dt}{dT} = \mathbf{P}^t(t) \cdot (t_j - t_i).$$

The two extreme tangents are $\mathbf{Q}^T(0) = (t_j - t_i)\mathbf{P}^t(t_i)$ and $\mathbf{Q}^T(1) = (t_j - t_i)\mathbf{P}^t(t_j)$. The new curve can now be calculated by

$$\mathbf{Q}(T) = (T^3, T^2, T, 1)\mathbf{H} \begin{pmatrix} \mathbf{P}(t_i) \\ \mathbf{P}(t_j) \\ (t_j - t_i)\mathbf{P}^t(t_i) \\ (t_j - t_i)\mathbf{P}^t(t_j) \end{pmatrix}, \tag{4.39}$$

where $\mathbf{H}$ is the Hermite matrix, Equation (4.26).

▶ **Exercise 4.44:** Calculate the PC segment $\mathbf{Q}(T)$ that's the truncation of $\mathbf{P}(t) = (-1, 0)t^3 + (1, -1)t^2 + (1, 1)t$ (Equation (4.29)) from $t_i = 0.25$ to $t_j = 0.75$.

Segmenting a curve is the problem of calculating several truncations. We assume that we are given values $0 = t_0 < t_1 < t_2 < \cdots < t_n = 1$, and we want to break a given curve $\mathbf{P}(t)$ into n segments such that segment i will go from point $\mathbf{P}(t_{i-1})$ to point $\mathbf{P}(t_i)$ (Figure 4.16b). Equation (4.39) gives segment i as

$$\mathbf{Q}_i(T) = (T^3, T^2, T, 1)\mathbf{H} \begin{pmatrix} \mathbf{P}(t_{i-1}) \\ \mathbf{P}(t_i) \\ (t_i - t_{i-1})\mathbf{P}^t(t_{i-1}) \\ (t_i - t_{i-1})\mathbf{P}^t(t_i) \end{pmatrix}.$$

4.7.6 Special and Degenerate Hermite Segments

The following special cases result in Hermite curve segments that are either especially simple (degenerate) or especially interesting

- The case $\mathbf{P}_1 = \mathbf{P}_2$ and $\mathbf{P}_1^t = \mathbf{P}_2^t = 0$. Equation (4.23) yields $\mathbf{P}(t) = \mathbf{P}_1$; the curve degenerates to a point.

- The case $\mathbf{P}_1^t = \mathbf{P}_2^t = \mathbf{P}_2 - \mathbf{P}_1$. The two tangents point in the same direction, from $\mathbf{P}_1$ to $\mathbf{P}_2$. Equation (4.23) yields

$$\mathbf{P}(t) = \big(2\mathbf{P}_1 - 2\mathbf{P}_2 + 2(\mathbf{P}_2 - \mathbf{P}_1)\big)t^3 + \big(-3\mathbf{P}_1 + 3\mathbf{P}_2 - 3(\mathbf{P}_2 - \mathbf{P}_1)\big)t^2$$
$$+ (\mathbf{P}_2 - \mathbf{P}_1)t + \mathbf{P}_1$$
$$= (\mathbf{P}_2 - \mathbf{P}_1)t + \mathbf{P}_1. \tag{4.40}$$

The curve degenerates to a straight line.

4 Curves

- The case $\mathbf{P}_1 = \mathbf{P}_2$. Equation (4.23) yields $\mathbf{P}(t) = (\mathbf{P}_1^t + \mathbf{P}_2^t)t^3 + (-2\mathbf{P}_1^t - \mathbf{P}_2^t)t^2 + \mathbf{P}_1^t t + \mathbf{P}_1$. It is easy to see that this curve satisfies $\mathbf{P}(0) = \mathbf{P}(1)$. It is closed (although it is not a circle).

- The case $\mathbf{P}_1^t = \mathbf{P}_2^t = (x_2 - x_1, y_2 - y_1, 0)$. Equation (4.23) yields

$$\begin{aligned}\mathbf{P}(t) &= \big(2\mathbf{P}_1 - 2\mathbf{P}_2 + 2(x_2 - x_1, y_2 - y_1, 0)\big)t^3 \\ &+ \big(-3\mathbf{P}_1 + 3\mathbf{P}_2 - 3(x_2 - x_1, y_2 - y_1, 0)\big)t^2 \\ &+ (x_2 - x_1, y_2 - y_1, 0)t + (x_1, y_1, z_1) \\ &= \big(x_1 + (x_2 - x_1)t, y_1 + (y_2 - y_1)t, z_1 + (z_2 - z_1)(3t^2 - 2t^3)\big).\end{aligned}$$

This is linear in x and y, so it is planar.

4.7.7 Special and Degenerate Curves

Parametric curves in general, not just Hermite segments, exhibit special behavior when their derivatives satisfy certain conditions. Here are four examples:

1. If the first derivative $\mathbf{P}^t(t)$ of a curve $\mathbf{P}(t)$ is zero for all values of t, then $\mathbf{P}(t)$ degenerates to the point $\mathbf{P}(0)$.
2. If $\mathbf{P}^t(t) \neq 0$ and $\mathbf{P}^t(t) \times \mathbf{P}^{tt}(t) = 0$, then $\mathbf{P}(t)$ is a straight line.
3. If $\mathbf{P}^t(t) \times \mathbf{P}^{tt}(t) \neq 0$ and $|\mathbf{P}^t(t)\, \mathbf{P}^{tt}(t)\, \mathbf{P}^{ttt}(t)| = 0$, then $\mathbf{P}(t)$ is a plane curve. (The notation $|\mathbf{a}\ \mathbf{b}\ \mathbf{c}|$ refers to the determinant whose three columns are $\mathbf{a}$, $\mathbf{b}$, and $\mathbf{c}$.)
4. Finally, if both $\mathbf{P}^t(t) \times \mathbf{P}^{tt}(t)$ and $|\mathbf{P}^t(t)\, \mathbf{P}^{tt}(t)\, \mathbf{P}^{ttt}(t)|$ are nonzero, the curve $\mathbf{P}(t)$ is nonplanar (i.e., it is a space curve).

4.7.8 Hermite Straight Segments

Equation (4.40) shows that the Hermite segment can sometimes degenerate into a straight line. This section describes variations on Hermite straight segments. Specifically, we look in detail at the case where the two extreme tangent vectors point in the same direction, from $\mathbf{P}_1$ to $\mathbf{P}_2$, but have different magnitudes. We denote them by $\mathbf{P}_1^t = \alpha(\mathbf{P}_2 - \mathbf{P}_1)$ and $\mathbf{P}_2^t = \beta(\mathbf{P}_2 - \mathbf{P}_1)$, where α and β can be any real numbers. Equation (4.40) is obtained in the special case $\alpha = \beta = 1$.

The Hermite segment is expressed as $\mathbf{P}(t) = \mathbf{F}(t)\mathbf{B}$, where the four $F_i(t)$ functions are given by Equation (4.25), and $\mathbf{B}$ is the geometry vector, which, in our case, has the form

$$\mathbf{B} = \big(\mathbf{P}_1, \mathbf{P}_2, \alpha(\mathbf{P}_2 - \mathbf{P}_1), \beta(\mathbf{P}_2 - \mathbf{P}_1)\big)^T.$$

This can be written (since $F_1(t) + F_2(t) \equiv 1$)

$$\begin{aligned}\mathbf{P}(t) &= F_1(t)\mathbf{P}_1 + F_2(t)\mathbf{P}_2 + F_3(t)\alpha(\mathbf{P}_2 - \mathbf{P}_1) + F_4(t)\beta(\mathbf{P}_2 - \mathbf{P}_1) \\ &= \mathbf{P}_1 + (F_2(t) + \alpha F_3(t) + \beta F_4(t))(\mathbf{P}_2 - \mathbf{P}_1) \\ &= \mathbf{P}_1 + \big((1 - 2t^3 + 3t^2) + \alpha(t^3 - 2t^2 + t) + \beta(t^3 - t^2)\big)(\mathbf{P}_2 - \mathbf{P}_1) \\ &= \mathbf{P}_1 + \big((\alpha + \beta - 2)t^3 - (2\alpha + \beta - 3)t^2 + \alpha t\big)(\mathbf{P}_2 - \mathbf{P}_1). \end{aligned} \qquad (4.41)$$

This has the form $\mathbf{P}(t) = \mathbf{P}_1 + G(t)(\mathbf{P}_2 - \mathbf{P}_1)$, which shows that all the points of $\mathbf{P}(t)$ lie on the straight line that passes through $\mathbf{P}_1$ and has the tangent vector $(\mathbf{P}_2 - \mathbf{P}_1)$. The precise form of $\mathbf{P}(t)$ depends on the values and signs of α and β. The rest of this section analyzes several cases in detail. The remaining cases can be analyzed similarly. See also Exercise 4.66.

Case 1 is when $\alpha = \beta = 1$, which leads to Equation (4.40), a straight segment from $\mathbf{P}_1$ to $\mathbf{P}_2$.

Case 2 is when $\alpha = \beta = 0$. Equation (4.41) reduces in this case to

$$\mathbf{P}(t) = \mathbf{P}_1 + (-2t^3 + 3t^2)(\mathbf{P}_2 - \mathbf{P}_1), \tag{4.42}$$

or $\mathbf{P}(T) = \mathbf{P}_1 + T(\mathbf{P}_2 - \mathbf{P}_1)$, where $T = -2t^3 + 3t^2$. This also is a straight segment from $\mathbf{P}_1$ to $\mathbf{P}_2$ but moving at a variable speed. It accelerates up to point $\mathbf{P}(0.5)$, then decelerates.

▶ **Exercise 4.45:** Prove this claim!

Case 3 is when $\alpha = \beta = -1$. Equation (4.41) in this case becomes

$$\mathbf{P}(t) = \mathbf{P}_1 + (-4t^3 + 6t^2 - t)(\mathbf{P}_2 - \mathbf{P}_1), \tag{4.43}$$

which is the curve shown in Figure 4.17a. It consists of three straight segments, but we can also think of it as a straight line that goes from $\mathbf{P}_1$ *backward* to a certain point $\mathbf{P}(i)$, then reverses direction, passes points $\mathbf{P}_1$ and $\mathbf{P}_2$, stops at point $\mathbf{P}(j)$, reverses direction again, and ends at $\mathbf{P}_2$. We can calculate i and j by calculating the tangent of Equation (4.43) and equating it to zero. The tangent vector is $\mathbf{P}^t(t) = (-12t^2 + 12t - 1)(\mathbf{P}_2 - \mathbf{P}_1)$ and the roots of the quadratic equation $-12t^2 + 12t - 1 = 0$ are (approximately) 0.083 and 0.92.

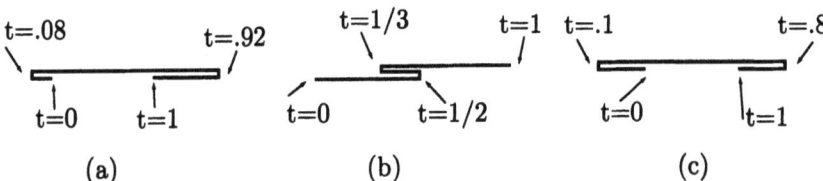

Figure 4.17: Straight Hermite Segments.

Case 4 is when $\alpha > 0$, $\beta > 0$. As an example, we try the values $\alpha = 2$ and $\beta = 4$. Equation (4.41) in this case becomes

$$\mathbf{P}(t) = \mathbf{P}_1 + (4t^3 - 5t^2 + 2t)(\mathbf{P}_2 - \mathbf{P}_1). \tag{4.44}$$

This curve also consists of three straight segments (Figure 4.17b), but it looks different. It goes forward from $\mathbf{P}_1$ to a certain point $\mathbf{P}(i)$, then reverses direction, goes to point $\mathbf{P}(j)$, reverses direction again, and continues to $\mathbf{P}_2$. We can calculate i and j by calculating the tangent of Equation (4.44) and equating it to zero. The

4 Curves

tangent vector is $\mathbf{P}^t(t) = (12t^2 - 10t + 2)(\mathbf{P}_2 - \mathbf{P}_1)$ and the roots of the quadratic equation $12t^2 - 10t + 2 = 0$ are $1/3$ and $1/2$.

Case 5 is when $\alpha < 0$, $\beta < 0$. As an example, we try the values $\alpha = -2$ and $\beta = -4$. Equation (4.41) in this case becomes

$$\mathbf{P}(t) = \mathbf{P}_1 + (-8t^3 + 11t^2 - 2t)(\mathbf{P}_2 - \mathbf{P}_1). \tag{4.45}$$

This curve again consists of three straight segments as in case 3, but points i and j are different (Figure 4.17c). The tangent of Equation (4.45) is $\mathbf{P}^t(t) = (-24t^2 + 22t - 2)(\mathbf{P}_2 - \mathbf{P}_1)$, and the roots of the quadratic equation $-24t^2 + 22t - 2 = 0$ are (approximately) 0.1 and 0.8.

Table 4.18 summarizes the nine possible cases of Equation (4.41).

Case	1	2	3	4	5	6	7	8	9
α	1	0	-1	>0	<0	>0	<0	≤ 0	≥ 0
β	1	0	-1	>0	<0	≤ 0	≥ 0	>0	<0

Table 4.18: Nine Cases of Straight Hermite Segments.

4.7.9 A Variant Hermite Segment

The Hermite method assumes that two points and two tangents are given. These constitute four known quantities, so four equations should be set and solved for four unknowns. A variation on this theme assumes that two points and just one tangent are given. These constitute only three quantities, so only three equations can be set and only three unknowns solved and calculated. The idea is to calculate this variant curve as a quadratic (degree-2) polynomial, which has only three coefficients. We denote the points as usual $\mathbf{P}_1$ and $\mathbf{P}_2$ and we denote the tangent vector (which is assumed to be the start tangent, but can also be the end tangent) $\mathbf{P}_1^t$. The quadratic polynomial is denoted by $\mathbf{P}(t) = \mathbf{a}t^2 + \mathbf{b}t + \mathbf{c}$, its tangent vector is $\mathbf{P}^t(t) = 2\mathbf{a}t + \mathbf{b}$, and we can immediately set up the three equations $\mathbf{P}(0) = \mathbf{P}_1$, $\mathbf{P}(1) = \mathbf{P}_2$, and $\mathbf{P}^t(0) = \mathbf{P}_1^t$ whose explicit forms are

$$\begin{aligned}
\mathbf{a} \cdot 0^2 + \mathbf{b} \cdot 0 + \mathbf{c} &= \mathbf{P}_1, \\
\mathbf{a} \cdot 1^2 + \mathbf{b} \cdot 1 + \mathbf{c} &= \mathbf{P}_2, \\
2\mathbf{a} \cdot 0 + \mathbf{b} &= \mathbf{P}_1^t.
\end{aligned} \tag{4.46}$$

The solutions are $\mathbf{c} = \mathbf{P}_1$, $\mathbf{b} = \mathbf{P}_1^t$, and $\mathbf{a} = \mathbf{P}_2 - \mathbf{b} - \mathbf{c} = \mathbf{P}_2 - \mathbf{P}_1 - \mathbf{P}_1^t$.

The quadratic polynomial is thus

$$\begin{aligned}
\mathbf{P}(t) &= (\mathbf{P}_2 - \mathbf{P}_1 - \mathbf{P}_1^t)t^2 + \mathbf{P}_1^t t + \mathbf{P}_1 \\
&= (-t^2 + 1)\mathbf{P}_1 + t^2 \mathbf{P}_2 + (-t^2 + t)\mathbf{P}_1^t \\
&= (t^2, t, 1) \begin{pmatrix} -1 & 1 & -1 \\ 0 & 0 & 1 \\ 1 & 0 & 0 \end{pmatrix} \begin{pmatrix} \mathbf{P}_1 \\ \mathbf{P}_2 \\ \mathbf{P}_1^t \end{pmatrix}.
\end{aligned} \tag{4.47}$$

Its tangent vector is $\mathbf{P}^t(t) = 2\mathbf{a}t + \mathbf{b} = 2(\mathbf{P}_2 - \mathbf{P}_1 - \mathbf{P}_1^t)t + \mathbf{P}_1^t$, which implies that the end tangent is

$$\mathbf{P}^t(1) = 2(\mathbf{P}_2 - \mathbf{P}_1) - \mathbf{P}_1^t. \tag{4.48}$$

Figure 4.19 shows the simple geometric interpretation of this.

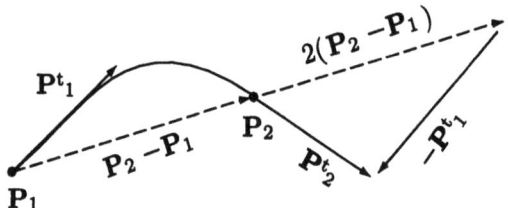

Figure 4.19: The Geometric Interpretation of the End Tangent.

▶ **Exercise 4.46:** Obtain the nonuniform version of this quadratic polynomial assuming that the parameter t varies from zero to some positive number Δ.

▶ **Exercise 4.47:** Calculate a quadratic parametric polynomial $\mathbf{P}(t) = \mathbf{a}t^2 + \mathbf{b}t + \mathbf{c}$ assuming that only the two extreme tangent vectors $\mathbf{P}^t(0)$ and $\mathbf{P}^t(1)$ are given.

▶ **Exercise 4.48:** Use your curve design skills to obtain the cubic polynomial equation of the curve segment $\mathbf{P}(t)$ defined by the following three conditions: (1) The two endpoints $\mathbf{P}_1$ and $\mathbf{P}_2$ are given, (2) the end tangent $\mathbf{P}_2^t$ is given, and (3) the start second derivative $\mathbf{P}^{tt}(0)$ is zero.

4.7.10 Nonuniform Hermite Segments

This the case where the parameter t varies in the interval $[0, \Delta]$, where Δ can be any real positive number. The derivation of the nonuniform Hermite segment is similar to the uniform case. Equation (4.21) becomes

$$\mathbf{a} \cdot 0^3 + \mathbf{b} \cdot 0^2 + \mathbf{c} \cdot 0 + \mathbf{d} = \mathbf{P}_1,$$
$$\mathbf{a}\Delta^3 + \mathbf{b}\Delta^2 + \mathbf{c}\Delta + \mathbf{d} = \mathbf{P}_2,$$
$$3\mathbf{a} \cdot 0^2 + 2\mathbf{b} \cdot 0 + \mathbf{c} = \mathbf{P}_1^t,$$
$$3\mathbf{a}\Delta^2 + 2\mathbf{b}\Delta + \mathbf{c} = \mathbf{P}_2^t,$$

with solutions

$$\mathbf{a} = \frac{2(\mathbf{P}_1 - \mathbf{P}_2)}{\Delta^3} + \frac{\mathbf{P}_1^t + \mathbf{P}_2^t}{\Delta^2},$$
$$\mathbf{b} = \frac{3(\mathbf{P}_2 - \mathbf{P}_1)}{\Delta^2} - \frac{2\mathbf{P}_1^t}{\Delta} - \frac{\mathbf{P}_2^t}{\Delta},$$
$$\mathbf{c} = \mathbf{P}_1^t,$$
$$\mathbf{d} = \mathbf{P}_1.$$

4 Curves

The curve segment can now be expressed, similar to Equation (4.26), in the form

$$\mathbf{P}(t) = (t^3, t^2, t, 1) \begin{pmatrix} \frac{2}{\Delta^3} & \frac{-2}{\Delta^3} & \frac{1}{\Delta^2} & \frac{1}{\Delta^2} \\ \frac{-3}{\Delta^2} & \frac{3}{\Delta^2} & \frac{-2}{\Delta} & \frac{-1}{\Delta} \\ 0 & 0 & 1 & 0 \\ 1 & 0 & 0 & 0 \end{pmatrix} \begin{pmatrix} \mathbf{P}_1 \\ \mathbf{P}_2 \\ \mathbf{P}_1^t \\ \mathbf{P}_2^t \end{pmatrix} = \mathbf{T}(t)\mathbf{H}_{nu}\mathbf{B}. \quad (4.49)$$

It is easy to verify that matrix $\mathbf{H}_{nu}$ reduces to $\mathbf{H}$ for $\Delta = 1$. Figure 4.20 shows a typical nonuniform Hermite segment drawn three times for $\Delta = 0.5, 1$, and 2. A careful examination of the three curves shows that increasing the value of Δ causes the curve segment to continue longer in its initial and final directions; it has the same effect as increasing the magnitude of the tangent vectors in the case of the uniform Hermite segment. Once this is grasped, the reader should not be surprised to learn that the nonuniform curve of Equation (4.49) can also be expressed as

$$\mathbf{P}(t) = (t^3, t^2, t, 1) \begin{pmatrix} 2 & -2 & 1 & 1 \\ -3 & 3 & -2 & -1 \\ 0 & 0 & 1 & 0 \\ 1 & 0 & 0 & 0 \end{pmatrix} \begin{pmatrix} \mathbf{P}_1 \\ \mathbf{P}_2 \\ \Delta\mathbf{P}_1^t \\ \Delta\mathbf{P}_2^t \end{pmatrix}. \quad (4.50)$$

This shows that the nonuniform Hermite curve segment isn't very useful since it is a special case of the uniform curve. Any nonuniform Hermite curve can also be obtained as a uniform Hermite curve by adjusting the magnitudes of the tangent vectors.

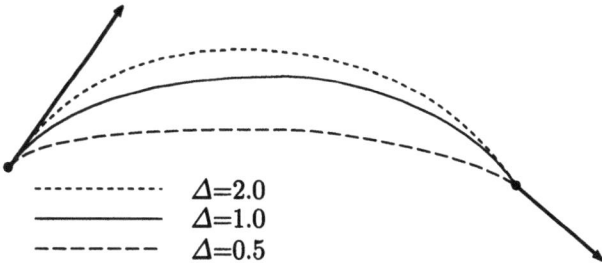

Figure 4.20: A Nonuniform Hermite Segment.

4.8 The Cubic Spline Curve

The Hermite curve is just a single segment connecting two points. Its main use is as a key to developing more sophisticated curve methods. Imagine that n two-dimensional data points are given and are numbered $\mathbf{P}_1$ through $\mathbf{P}_n$. There are infinitely many curves that pass through all the points in order of their numbers (Figure 4.21b), but the eye tends to trace *one* imaginary smooth curve through the points, especially if the points are arranged in a familiar pattern. It is therefore useful to have a computer program that will do the same thing. Since the computer does not recognize familiar patterns the way humans do, such a method should be interactive. This way the user can help the program select the right curve, or can edit and modify the first result if it is not satisfactory.

It is possible to extend the idea of a PC to fit a curve through many data points. The LP (Section 4.4) does that, but it has important disadvantages. We therefore use a different method. We try to fit a curve through n given points by smoothly connecting $n-1$ individual segments, each a PC that can easily be calculated and displayed. Such a curve is called a *cubic spline*. The spline method has the advantage of being interactive; the user can change the shape of the curve interactively by changing the tangent vectors at the beginning and the end of the curve.

Definition: A spline is a set of polynomials of degree k that are smoothly connected at certain data points. At each data point, two polynomials should connect, and their first derivatives (tangent vectors) should have the same values. The definition also requires that all their derivatives up to the $(k-1)$st be the same at the point.

A cubic spline is a set of PCs where the individual curves meet at the interior points, their first derivatives have the same values at the points, and the same is true for their second derivatives.

Given the n data points $\mathbf{P}_1, \mathbf{P}_2, \ldots, \mathbf{P}_n$, we look for $n-1$ parametric cubics $\mathbf{P}_1(t), \mathbf{P}_2(t), \ldots, \mathbf{P}_{n-1}(t)$ such that $\mathbf{P}_k(t)$ is the polynomial segment from point $\mathbf{P}_k$ to point $\mathbf{P}_{k+1}$ (Figure 4.21a). The PCs will have to be smoothly connected at the $n-2$ interior points $\mathbf{P}_2, \mathbf{P}_3, \ldots, \mathbf{P}_{n-1}$, which means that their first derivatives will have to match at every interior point. The definition of a spline requires that their second derivatives match too. This requirement is important since (1) it provides us with the necessary equations and (2) it results in a smooth curve in the sense that once the curve is drawn, the eye can no longer detect the positions of the original data points.

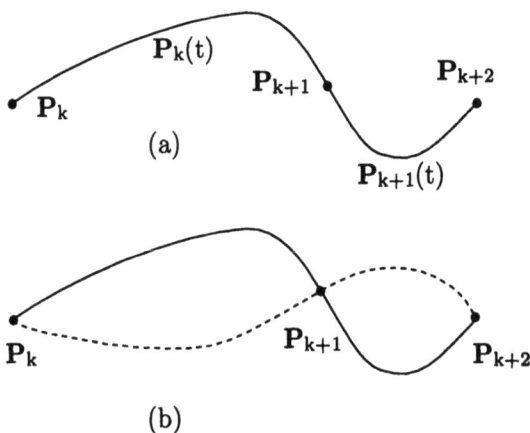

Figure 4.21: Two Different Curves.

The principle is to divide the set of n points into $n-1$ overlapping sets of

4 Curves

two points each and to fit a PC to each set. The sets are $(\mathbf{P}_1, \mathbf{P}_2)$, $(\mathbf{P}_2, \mathbf{P}_3), \ldots,$ $(\mathbf{P}_{n-1}, \mathbf{P}_n)$. To calculate the $n-1$ PC segments, we use the Hermite geometric representation of a PC (Equations (4.23) and (4.24)). In addition to the two points, this representation requires the two tangent vectors at the points. These vectors are unknown and have to be calculated. Since there are n such vectors (one per point), n equations are needed.

The equations (see below) are derived from the requirement that the second derivatives of the individual PC segments match at every interior point. However, there are only $n-2$ interior points, so we can only have $n-2$ equations, enough to solve for only $n-2$ unknowns.

The key to solving this problem is to ask the user to provide the program with the values of two tangent vectors (usually the first and last ones). Now, the equations can easily be solved, yielding the remaining $n-2$ tangents. This seems a strange way to solve equations, but it has the advantage of being interactive. If the resulting curve looks wrong, the user can repeat the calculation with two new tangent vectors. The details are shown later, but here is a summary of the steps involved:

1. The n data points are input into the program.
2. The user provides values (guesses or estimates) for two tangent vectors.
3. The program sets up $n-2$ equations, with the remaining $n-2$ tangent vectors as the unknowns, and solves them.
4. The program loops $n-1$ times. In each iteration, it uses two adjacent points and their tangent vectors to calculate one PC segment.

We start with three general points, $\mathbf{P}_k$, $\mathbf{P}_{k+1}$, and $\mathbf{P}_{k+2}$, of which $\mathbf{P}_{k+1}$ must be an interior point and the other two can be either interior or endpoints. The range of values of k is, therefore, $[1, n-2]$. The PC segment from $\mathbf{P}_k$ to $\mathbf{P}_{k+1}$ is denoted $\mathbf{P}_k(t)$. We therefore have $\mathbf{P}_k(0) = \mathbf{P}_k$ and $\mathbf{P}_k(1) = \mathbf{P}_{k+1}$. The tangent vectors of $\mathbf{P}_k(t)$ at the endpoints are still unknown and are denoted by $\mathbf{P}_k^t$ and $\mathbf{P}_{k+1}^t$. The first step is to express segment $\mathbf{P}_k(t)$ geometrically—in terms of the two endpoints and the two tangents—which has been done by Equation (4.23). Applied to our segment, it becomes

$$\mathbf{P}_k(t) = \mathbf{P}_k + \mathbf{P}_k^t t + \left[3(\mathbf{P}_{k+1} - \mathbf{P}_k) - 2\mathbf{P}_k^t - \mathbf{P}_{k+1}^t\right] t^2 \\ + \left[2(\mathbf{P}_k - \mathbf{P}_{k+1}) + \mathbf{P}_k^t + \mathbf{P}_{k+1}^t\right] t^3. \quad (4.51)$$

When the same equation is applied to the next segment $\mathbf{P}_{k+1}(t)$ (from $\mathbf{P}_{k+1}$ to $\mathbf{P}_{k+2}$), it becomes

$$\mathbf{P}_{k+1}(t) = \mathbf{P}_{k+1} + \mathbf{P}_{k+1}^t t + \left[3(\mathbf{P}_{k+2} - \mathbf{P}_{k+1}) - 2\mathbf{P}_{k+1}^t - \mathbf{P}_{k+2}^t\right] t^2 \\ + \left[2(\mathbf{P}_{k+1} - \mathbf{P}_{k+2}) + \mathbf{P}_{k+1}^t + \mathbf{P}_{k+2}^t\right] t^3. \quad (4.52)$$

▶ **Exercise 4.49:** Where do we use the assumption that the first derivatives of segments $\mathbf{P}_k(t)$ and $\mathbf{P}_{k+1}(t)$ are equal at the interior point $\mathbf{P}_{k+1}$?

Next, we use the requirement that the second derivatives of the two PCs be equal at the interior points. The second derivative $\mathbf{P}^{tt}(t)$ of $\mathbf{P}(t)$ is obtained by differentiating Equation (4.20):

$$\mathbf{P}^{tt}(t) = 6\mathbf{a}t + 2\mathbf{b}. \tag{4.53}$$

Equality of the second derivatives at the interior point $\mathbf{P}_{k+1}$ implies

$$\mathbf{P}^{tt}_k(1) = \mathbf{P}^{tt}_{k+1}(0) \quad \text{or} \quad 6\mathbf{a}_k \times 1 + 2\mathbf{b}_k = 6\mathbf{a}_{k+1} \times 0 + 2\mathbf{b}_{k+1}. \tag{4.54}$$

Using the values of **a** and **b** from Equation (4.22), we get

$$6\left[2(\mathbf{P}_k - \mathbf{P}_{k+1}) + \mathbf{P}^t_k + \mathbf{P}^t_{k+1}\right] + 2\left[3(\mathbf{P}_{k+1} - \mathbf{P}_k) - 2\mathbf{P}^t_k - \mathbf{P}^t_{k+1}\right]$$
$$= 2\left[3(\mathbf{P}_{k+2} - \mathbf{P}_{k+1}) - 2\mathbf{P}^t_{k+1} - \mathbf{P}^t_{k+2}\right], \tag{4.55}$$

which, after simple algebraic manipulations, becomes

$$\mathbf{P}^t_k + 4\mathbf{P}^t_{k+1} + \mathbf{P}^t_{k+2} = 3(\mathbf{P}_{k+2} - \mathbf{P}_k). \tag{4.56}$$

The three quantities on the left side of Equation (4.56) are unknown. The two quantities on the right side are known.

Equation (4.56) can be written $n-2$ times for all the interior points $\mathbf{P}_{k+1} = \mathbf{P}_2, \mathbf{P}_3, \ldots, \mathbf{P}_{n-1}$ to obtain a system of $n-2$ linear algebraic equations expressed in matrix form as

$$n-2\left\{\underbrace{\begin{pmatrix} 1 & 4 & 1 & 0 & \cdots & 0 \\ 0 & 1 & 4 & 1 & \cdots & 0 \\ & & \ddots & \ddots & \ddots & \vdots \\ 0 & \cdots & \cdots & 1 & 4 & 1 \end{pmatrix}}_{n} \begin{pmatrix} \mathbf{P}^t_1 \\ \mathbf{P}^t_2 \\ \vdots \\ \mathbf{P}^t_n \end{pmatrix} = \begin{pmatrix} 3(\mathbf{P}_3 - \mathbf{P}_1) \\ 3(\mathbf{P}_4 - \mathbf{P}_2) \\ \vdots \\ 3(\mathbf{P}_n - \mathbf{P}_{n-2}) \end{pmatrix}\right. . \tag{4.57}$$

Equation (4.57) is really $n-2$ equations in the n unknowns $\mathbf{P}^t_1, \mathbf{P}^t_2, \ldots, \mathbf{P}^t_n$. A good approach to the solution is to let the user specify the values of the two extreme tangents $\mathbf{P}^t_1$ and $\mathbf{P}^t_n$. Once these values are substituted in Equation (4.57), it's easy to solve it and get values for the remaining $n-2$ tangents, $\mathbf{P}^t_2, \ldots, \mathbf{P}^t_{n-1}$. The n tangent vectors are now used to calculate the original coefficients **a**, **b**, **c**, and **d** of each segment by means of Equations (4.22), (4.23), or (4.26), which should be written and solved $n-1$ times, once for each segment of the spline.

The reader should notice that the matrix of coefficients of Equation (4.57) is tridiagonal and, therefore, diagonally dominant and thus nonsingular (Section A.5). This means that the equation can always be solved and that it has a unique solution.

This approach to solving Equation (4.57) is called the *clamped* end condition. Its advantage is that the user can change the shape of the curve by entering new values for $\mathbf{P}^t_1$ and $\mathbf{P}^t_n$ and recalculating. This allows for interactive design, where each step brings the curve closer to the desired shape. Figure 4.21b is an example of two cubic splines that pass through the same points and differ in $\mathbf{P}^t_1$ and $\mathbf{P}^t_n$ only. It illustrates how the shape of the entire curve can be radically changed by modifying the two extreme tangents.

4 Curves

It is possible to let the user specify any two tangent vectors, not just the two extreme ones. However, controlling the two extreme tangents makes it easier to edit the curve in practical situations.

The downside of the cubic spline is the following:

1. There is no local control. Modifying the extreme tangent vectors changes Equation (4.57) and results in a different set of n tangent vectors. The entire curve is modified!

2. Equation (4.57) is a system of n equations that, for large values of n, may be too slow to solve.

> Picnic Blues (anagram of Cubic Spline.)

4.8.1 Example

Given the four points $\mathbf{P}_1 = (0,0)$, $\mathbf{P}_2 = (1,0)$, $\mathbf{P}_3 = (1,1)$, and $\mathbf{P}_4 = (0,1)$, we are looking for three PC segments $\mathbf{P}_1(t)$, $\mathbf{P}_2(t)$, and $\mathbf{P}_3(t)$ that will connect smoothly at the two interior points $\mathbf{P}_2$ and $\mathbf{P}_3$ and will constitute the spline. We further select an initial direction $\mathbf{P}_1^t = (1,-1)$ and a final direction $\mathbf{P}_4^t = (-1,-1)$. Figure 4.22 shows the points, the two extreme tangent vectors, and the resulting curve.

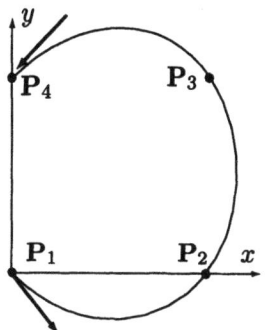

Figure 4.22: A Cubic Spline Example.

We first write Equation (4.57) for our special case ($n=4$):

$$\begin{pmatrix} 1 & 4 & 1 & 0 \\ 0 & 1 & 4 & 1 \end{pmatrix} \begin{pmatrix} (1,-1) \\ \mathbf{P}_2^t \\ \mathbf{P}_3^t \\ (-1,-1) \end{pmatrix} = \begin{pmatrix} 3[(1,1)-(0,0)] \\ 3[(0,1)-(1,0)] \end{pmatrix} = \begin{pmatrix} (3,3) \\ (-3,3) \end{pmatrix}.$$

or
$$(1,-1) + 4\mathbf{P}_2^t + \mathbf{P}_3^t = (3,3)$$
$$\mathbf{P}_2^t + 4\mathbf{P}_3^t + (-1,-1) = (-3,3).$$

This is a system of two equations in two unknowns. It is easy to solve and the solutions are $\mathbf{P}_2^t = (\frac{2}{3}, \frac{4}{5})$ and $\mathbf{P}_3^t = (-\frac{2}{3}, \frac{4}{5})$.

We now write Equation (4.26) three times, for the three spline segments. For the first segment, Equation (4.26) becomes

$$\mathbf{P}_1(t) = (t^3, t^2, t, 1) \begin{pmatrix} 2 & -2 & 1 & 1 \\ -3 & 3 & -2 & -1 \\ 0 & 0 & 1 & 0 \\ 1 & 0 & 0 & 0 \end{pmatrix} \begin{pmatrix} (0,0) \\ (1,0) \\ (1,-1) \\ (\frac{2}{3}, \frac{4}{5}) \end{pmatrix}$$

$$= (-\tfrac{1}{3}, -\tfrac{1}{5})t^3 + (\tfrac{1}{3}, \tfrac{6}{5})t^2 + (1, -1)t.$$

The second segment is calculated in a similar way:

$$\mathbf{P}_2(t) = (t^3, t^2, t, 1) \begin{pmatrix} 2 & -2 & 1 & 1 \\ -3 & 3 & -2 & -1 \\ 0 & 0 & 1 & 0 \\ 1 & 0 & 0 & 0 \end{pmatrix} \begin{pmatrix} (1,0) \\ (1,1) \\ (\frac{2}{3}, \frac{4}{5}) \\ (-\frac{2}{3}, \frac{4}{5}) \end{pmatrix}$$

$$= (0, -\tfrac{2}{5})t^3 + (-\tfrac{2}{3}, \tfrac{3}{5})t^2 + (\tfrac{2}{3}, \tfrac{4}{5})t + (1, 0).$$

Finally, we write, for the third segment,

$$\mathbf{P}_3(t) = (t^3, t^2, t, 1) \begin{pmatrix} 2 & -2 & 1 & 1 \\ -3 & 3 & -2 & -1 \\ 0 & 0 & 1 & 0 \\ 1 & 0 & 0 & 0 \end{pmatrix} \begin{pmatrix} (1,1) \\ (0,1) \\ (-\frac{2}{3}, \frac{4}{5}) \\ (-1,-1) \end{pmatrix}$$

$$= (\tfrac{1}{3}, -\tfrac{1}{5})t^3 - (\tfrac{2}{3}, \tfrac{3}{5})t^2 + (-\tfrac{2}{3}, \tfrac{4}{5})t + (1, 1),$$

which completes the example.

▸ **Exercise 4.50:** Check to make sure that the three polynomial segments really connect at the two interior points. What are the tangent vectors at the points?

▸ **Exercise 4.51:** Redo the example above with an indefinite initial direction $\mathbf{P}_1^t = (0,0)$. What does it mean for a curve to start going in an indefinite direction?

4.8.2 Relaxed Cubic Splines

The original approach to the cubic spline curve was to let the user specify the two extreme tangent vectors. This approach is known as the *clamped* end condition. It is possible to use different end conditions, and the one described in this section is based on the simple idea of setting the two extreme second derivatives of the curve, $\mathbf{P}_1^{tt}(0)$ and $\mathbf{P}_{n-1}^{tt}(1)$, to zero. If we think of the second derivative as the acceleration of the curve (see the particle paradigm of Section 4.3), then this end condition implies constant speeds and, hence, small curvatures at both ends of the curve. This is why this end condition is called *relaxed*.

It is easy to calculate the relaxed cubic spline. The second derivative of the parametric cubic $\mathbf{P}(t)$ is $\mathbf{P}^{tt}(t) = 6a t + 2 \mathbf{b}$ (Equation (4.53)). The end condition $\mathbf{P}_1^{tt}(0) = 0$ implies $2\mathbf{b}_1 = 0$ or, from Equation (4.22)

$$-3\mathbf{P}_1 + 3\mathbf{P}_2 - 2\mathbf{P}_1^t - \mathbf{P}_2^t = 0, \quad \text{which yields} \quad \mathbf{P}_1^t = \tfrac{3}{2}(\mathbf{P}_2 - \mathbf{P}_1) - \tfrac{1}{2}\mathbf{P}_2^t. \quad (4.58)$$

4 Curves

The other end condition, $\mathbf{P}^{tt}_{n-1}(1) = 0$, implies $6a_{n-1} + 2b_{n-1} = 0$ or, from Equation (4.22)

$$6\left(2\mathbf{P}_{n-1} - 2\mathbf{P}_n + \mathbf{P}^t_{n-1} + \mathbf{P}^t_n\right) + 2\left(-3\mathbf{P}_{n-1} + 3\mathbf{P}_n - 2\mathbf{P}^t_{n-1} - \mathbf{P}^t_n\right) = 0,$$

or

$$\mathbf{P}^t_n = \tfrac{3}{2}(\mathbf{P}_n - \mathbf{P}_{n-1}) - \tfrac{1}{2}\mathbf{P}^t_{n-1}. \tag{4.59}$$

Substituting Eqs. (4.58) and (4.59) in Equation (4.57) results in

$$n-2\left\{\underbrace{\begin{pmatrix} 1 & 4 & 1 & 0 & \cdots & 0 \\ 0 & 1 & 4 & 1 & \cdots & 0 \\ & & \ddots & \ddots & & \vdots \\ 0 & \cdots & 1 & 4 & 1 & 0 \\ 0 & \cdots & & 1 & 4 & 1 \end{pmatrix}}_{n} \begin{pmatrix} \tfrac{3}{2}(\mathbf{P}_2 - \mathbf{P}_1) - \tfrac{1}{2}\mathbf{P}^t_2 \\ \mathbf{P}^t_2 \\ \vdots \\ \mathbf{P}^t_{n-1} \\ \tfrac{3}{2}(\mathbf{P}_n - \mathbf{P}_{n-1}) - \tfrac{1}{2}\mathbf{P}^t_{n-1} \end{pmatrix}\right. \tag{4.60}$$

$$= \begin{pmatrix} 3(\mathbf{P}_3 - \mathbf{P}_1) \\ 3(\mathbf{P}_4 - \mathbf{P}_2) \\ \vdots \\ 3(\mathbf{P}_{n-1} - \mathbf{P}_{n-3}) \\ 3(\mathbf{P}_n - \mathbf{P}_{n-2}) \end{pmatrix}.$$

This is a system of $n-2$ equations in the $n-2$ unknowns $\mathbf{P}^t_2, \mathbf{P}^t_3, \ldots, \mathbf{P}^t_{n-1}$. Calculating the relaxed cubic spline is done in the following steps:

1. Set up Equation (4.60) and solve it to obtain the $n-2$ interior tangent vectors.

2. Use $\mathbf{P}^t_2$ to calculate $\mathbf{P}^t_1$ from Equation (4.58). Similarly, use $\mathbf{P}^t_{n-1}$ to calculate $\mathbf{P}^t_n$ from Equation (4.59).

3. Now that the values of all n tangent vectors are known, write and solve Equation (4.23) or (4.26) $n-1$ times, each time calculating one spline segment.

The clamped cubic spline is interactive. The curve can be modified by changing the two extreme tangent vectors. The relaxed cubic spline, on the other hand, is not interactive. The only way to edit or modify it is to move the points or add points. The points, however, are data points that may be dictated by the problem on hand or that may be given by a user, so it may not always be possible to move them.

Example: We use the same four points $\mathbf{P}_1 = (0,0)$, $\mathbf{P}_2 = (1,0)$, $\mathbf{P}_3 = (1,1)$, and $\mathbf{P}_4 = (0,1)$ of Section 4.8.1. The first step is to set up Equation (4.60) and solve it to obtain the two interior tangent vectors $\mathbf{P}^t_2$ and $\mathbf{P}^t_3$.

$$\begin{pmatrix} 1 & 4 & 1 & 0 \\ 0 & 1 & 4 & 1 \end{pmatrix} \begin{pmatrix} (\tfrac{3}{2},0) - \tfrac{1}{2}\mathbf{P}^t_2 \\ \mathbf{P}^t_2 \\ \mathbf{P}^t_3 \\ (-\tfrac{3}{2},0) - \tfrac{1}{2}\mathbf{P}^t_3 \end{pmatrix} = \begin{pmatrix} (3,3) \\ (-3,3) \end{pmatrix}.$$

The solutions are
$$\mathbf{P}_2^t = \left(\frac{3}{5}, \frac{2}{3}\right), \quad \mathbf{P}_3^t = \left(-\frac{3}{5}, \frac{2}{3}\right).$$

The second step is to calculate $\mathbf{P}_1^t$ and $\mathbf{P}_4^t$:

$$\mathbf{P}_1^t = \frac{3}{2}(\mathbf{P}_2 - \mathbf{P}_1) - \frac{1}{2}\mathbf{P}_2^t = \left(\frac{3}{2}, 0\right) - \frac{1}{2}\left(\frac{3}{5}, \frac{2}{3}\right) = \left(\frac{6}{5}, -\frac{1}{3}\right),$$

$$\mathbf{P}_4^t = \frac{3}{2}(\mathbf{P}_4 - \mathbf{P}_3) - \frac{1}{2}\mathbf{P}_3^t = \left(-\frac{3}{2}, 0\right) - \frac{1}{2}\left(-\frac{3}{5}, \frac{2}{3}\right) = \left(-\frac{6}{5}, -\frac{1}{3}\right).$$

Now that the values of all four tangent vectors are known, the last step is to write and solve Equation (4.23) or (4.26) three times to calculate each of the three segments of our example curve.

For the first segment, Equation (4.26) becomes

$$\mathbf{P}_1(t) = (t^3, t^2, t, 1) \begin{pmatrix} 2 & -2 & 1 & 1 \\ -3 & 3 & -2 & -1 \\ 0 & 0 & 1 & 0 \\ 1 & 0 & 0 & 0 \end{pmatrix} \begin{pmatrix} (0,0) \\ (1,0) \\ (\frac{6}{5}, -\frac{1}{3}) \\ (\frac{3}{5}, \frac{2}{3}) \end{pmatrix}$$

$$= (-\tfrac{1}{5}, \tfrac{1}{3})t^3 + (\tfrac{6}{5}, -\tfrac{1}{3})t.$$

For the second segment, Equation (4.26) becomes

$$\mathbf{P}_2(t) = (t^3, t^2, t, 1) \begin{pmatrix} 2 & -2 & 1 & 1 \\ -3 & 3 & -2 & -1 \\ 0 & 0 & 1 & 0 \\ 1 & 0 & 0 & 0 \end{pmatrix} \begin{pmatrix} (1,0) \\ (1,1) \\ (\frac{3}{5}, \frac{2}{3}) \\ (-\frac{3}{5}, \frac{2}{3}) \end{pmatrix}$$

$$= (0, -\tfrac{2}{3})t^3 + (-\tfrac{3}{5}, 1)t^2 + (\tfrac{3}{5}, \tfrac{2}{3})t + (1, 0).$$

▶ **Exercise 4.52:** Do the third PC segment.

4.8.3 Cyclic Cubic Splines

The *cyclic* end condition is ideal for a closed cubic spline (Section 4.8.5) and also for a periodic cubic spline (Section 4.8.4). The condition is that the tangent vectors be equal at the two extremes of the curve (i.e., $\mathbf{P}_1^t = \mathbf{P}_n^t$) and the same for the second derivatives $\mathbf{P}_1^{tt} = \mathbf{P}_n^{tt}$. Notice that the curve doesn't have to be closed, i.e., a segment from $\mathbf{P}_n$ to $\mathbf{P}_1$ is not required.

Applying Equation (4.20) to the first condition yields

$$\mathbf{P}_1^t(0) = \mathbf{P}_{n-1}^t(1)$$

or

$$3\mathbf{a}_1 t^2 + 2\mathbf{b}_1 t + \mathbf{c}_1 \big|_{t=0} = 3\mathbf{a}_{n-1} t^2 + 2\mathbf{b}_{n-1} t + \mathbf{c}_{n-1} \big|_{t=1}$$

or

$$\mathbf{c}_1 = 3\mathbf{a}_{n-1} + 2\mathbf{b}_{n-1} + \mathbf{c}_{n-1}. \tag{4.61}$$

4 Curves

Applying Equation (4.53) to the second condition yields

$$\mathbf{P}_1^{tt}(0) = \mathbf{P}_{n-1}^{tt}(1)$$

or

$$6\mathbf{a}_1 t + 2\mathbf{b}_1|_{t=0} = 6\mathbf{a}_{n-1} t + 2\mathbf{b}_{n-1}|_{t=1}$$

or

$$2\mathbf{b}_1 = 6\mathbf{a}_{n-1} + 2\mathbf{b}_{n-1}. \tag{4.62}$$

Subtracting Equations (4.61) and (4.62) yields $\mathbf{c}_1 - 2\mathbf{b}_1 = -3\mathbf{a}_{n-1} + \mathbf{c}_{n-1}$ or, from Equation (4.22),

$$\mathbf{P}_1^t - 2[-3\mathbf{P}_1 + 3\mathbf{P}_2 - 2\mathbf{P}_1^t - \mathbf{P}_2^t] = -3[2\mathbf{P}_{n-1} - 2\mathbf{P}_n + \mathbf{P}_{n-1}^t + \mathbf{P}_n^t] + \mathbf{P}_{n-1}^t.$$

This can be written

$$\mathbf{P}_1^t + 4\mathbf{P}_1^t + 3\mathbf{P}_n^t = 6(\mathbf{P}_2 - \mathbf{P}_1 + \mathbf{P}_n - \mathbf{P}_{n-1}) - (\mathbf{P}_2^t + \mathbf{P}_{n-1}^t).$$

Using the end condition $\mathbf{P}_1^t = \mathbf{P}_n^t$, we get

$$\mathbf{P}_1^t = \mathbf{P}_n^t = \tfrac{3}{4}(\mathbf{P}_2 - \mathbf{P}_1 + \mathbf{P}_n - \mathbf{P}_{n-1}) - \tfrac{1}{4}(\mathbf{P}_2^t + \mathbf{P}_{n-1}^t). \tag{4.63}$$

Substituting Equation (4.63) in Equation (4.57) results in

$$n-2 \left\{ \underbrace{\begin{pmatrix} 1 & 4 & 1 & 0 & \cdots & 0 \\ 0 & 1 & 4 & 1 & \cdots & 0 \\ & & \ddots & \ddots & & \vdots \\ 0 & \cdots & 1 & 4 & 1 & 0 \\ 0 & \cdots & \cdots & 1 & 4 & 1 \end{pmatrix}}_{n} \begin{pmatrix} \tfrac{3}{4}(\mathbf{P}_2 - \mathbf{P}_1 + \mathbf{P}_n - \mathbf{P}_{n-1}) - \\ \tfrac{1}{4}(\mathbf{P}_2^t + \mathbf{P}_{n-1}^t) \\ \mathbf{P}_2^t \\ \vdots \\ \mathbf{P}_{n-1}^t \\ \tfrac{3}{4}(\mathbf{P}_2 - \mathbf{P}_1 + \mathbf{P}_n - \mathbf{P}_{n-1}) - \\ \tfrac{1}{4}(\mathbf{P}_2^t + \mathbf{P}_{n-1}^t) \end{pmatrix} \right. \tag{4.64}$$

$$= \begin{pmatrix} 3(\mathbf{P}_3 - \mathbf{P}_1) \\ 3(\mathbf{P}_4 - \mathbf{P}_2) \\ \vdots \\ 3(\mathbf{P}_{n-1} - \mathbf{P}_{n-3}) \\ 3(\mathbf{P}_n - \mathbf{P}_{n-2}) \end{pmatrix},$$

which is a system of $n-2$ equations in the $n-2$ unknowns $\mathbf{P}_2^t, \mathbf{P}_3^t, \ldots, \mathbf{P}_{n-1}^t$. Notice that in the case of a closed curve, these equations are somehow simplified since $\mathbf{P}_1 = \mathbf{P}_n$. Calculating the cyclic cubic spline is done in the following steps:

1. Set up Equation (4.64) and solve it to obtain the $n-2$ interior tangent vectors.

2. Use $\mathbf{P}_2^t$ and $\mathbf{P}_{n-1}^t$ to calculate $\mathbf{P}_1^t$ and $\mathbf{P}_n^t$ from Equation (4.63).

3. Now that the values of all n tangent vectors are known, write and solve Equation (4.23) or (4.26) $n-1$ times, each time calculating one spline segment.

4.8 The Cubic Spline Curve

Example: We select the five points $\mathbf{P}_1 = \mathbf{P}_5 = (0,-1)$, $\mathbf{P}_2 = (1,0)$, $\mathbf{P}_3 = (0,1)$, and $\mathbf{P}_4 = (-1,0)$ and calculate the cubic spline for these points with the cyclic end condition. Notice that the curve is closed since $\mathbf{P}_1 = \mathbf{P}_5$. Also, since the points are symmetric about the origin, we can expect the resulting four PC segments to be similar. We start with Equation (4.64),

$$\begin{pmatrix} 1 & 4 & 1 & 0 & 0 \\ 0 & 1 & 4 & 1 & 0 \\ 0 & 0 & 1 & 4 & 1 \end{pmatrix} \begin{pmatrix} \frac{3}{4}(\mathbf{P}_2 - \mathbf{P}_1 + \mathbf{P}_5 - \mathbf{P}_4) - \frac{1}{4}(\mathbf{P}_2^t + \mathbf{P}_4^t) \\ \mathbf{P}_2^t \\ \mathbf{P}_3^t \\ \mathbf{P}_4^t \\ \frac{3}{4}(\mathbf{P}_2 - \mathbf{P}_1 + \mathbf{P}_5 - \mathbf{P}_4) - \frac{1}{4}(\mathbf{P}_2^t + \mathbf{P}_4^t) \end{pmatrix} = \begin{pmatrix} 3(\mathbf{P}_3 - \mathbf{P}_1) \\ 3(\mathbf{P}_4 - \mathbf{P}_2) \\ 3(\mathbf{P}_5 - \mathbf{P}_3) \end{pmatrix},$$

which is solved to yield $\mathbf{P}_2^t = (0, 3/2)$, $\mathbf{P}_3^t = (-3/2, 0)$, and $\mathbf{P}_4^t = (0, -3/2)$. These values are used to solve Equation (4.63):

$$\mathbf{P}_1^t = \mathbf{P}_5^t = \tfrac{3}{4}(\mathbf{P}_2 - \mathbf{P}_1 + \mathbf{P}_5 - \mathbf{P}_4) - \tfrac{1}{4}(\mathbf{P}_2^t + \mathbf{P}_4^t),$$

which gives $\mathbf{P}_1^t = \mathbf{P}_5^t = (3/2, 0)$. The four segments can now be calculated in the usual way. For the first segment, Equation (4.26) becomes

$$\mathbf{P}_1(t) = (t^3, t^2, t, 1) \begin{pmatrix} 2 & -2 & 1 & 1 \\ -3 & 3 & -2 & -1 \\ 0 & 0 & 1 & 0 \\ 1 & 0 & 0 & 0 \end{pmatrix} \begin{pmatrix} (0,-1) \\ (1,0) \\ (\tfrac{3}{2},0) \\ (0,\tfrac{3}{2}) \end{pmatrix}$$
$$= -(\tfrac{1}{2}, \tfrac{1}{2})t^3 + (0, \tfrac{3}{2})t^2 + (\tfrac{3}{2}, 0)t + (0, -1).$$

For the second segment, Equation (4.26) becomes

$$\mathbf{P}_2(t) = (t^3, t^2, t, 1) \begin{pmatrix} 2 & -2 & 1 & 1 \\ -3 & 3 & -2 & -1 \\ 0 & 0 & 1 & 0 \\ 1 & 0 & 0 & 0 \end{pmatrix} \begin{pmatrix} (1,0) \\ (0,1) \\ (0,\tfrac{3}{2}) \\ (-\tfrac{3}{2},0) \end{pmatrix}$$
$$= (\tfrac{1}{2}, -\tfrac{1}{2})t^3 + (-\tfrac{3}{2}, 0)t^2 + (0, \tfrac{3}{2})t + (1, 0).$$

▶ **Exercise 4.53:** Do the third and fourth PC segments.

Notice how the symmetry of the problem causes the coefficients of $\mathbf{P}_1(t)$ and $\mathbf{P}_3(t)$ to have opposite signs, and the same for the coefficients of $\mathbf{P}_2(t)$ and $\mathbf{P}_4(t)$.

It is also possible to have an *anticyclic* end condition for the cubic spline. It requires that the two extreme tangent vectors have the same magnitudes but opposite directions $\mathbf{P}_1^t = -\mathbf{P}_n^t$ and the same condition for the second derivatives $\mathbf{P}_1^{tt} = -\mathbf{P}_n^{tt}$. Such an end condition makes sense for curves such as the cross section of a vase.

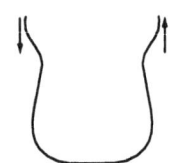

Following steps similar to the ones for the cyclic case, we get for the anticyclic end condition

$$\mathbf{P}_1^t = -\mathbf{P}_n^t = \frac{3}{4}(\mathbf{P}_2 - \mathbf{P}_1 - \mathbf{P}_n + \mathbf{P}_{n-1}) - \frac{1}{4}(\mathbf{P}_2^t - \mathbf{P}_{n-1}^t). \tag{4.65}$$

4 Curves

▶ **Exercise 4.54:** Given the three points $\mathbf{P}_1 = (-1,0)$, $\mathbf{P}_2 = (0,1)$, and $\mathbf{P}_3 = (1,0)$, calculate the anticyclic cubic spline for them and compare it to the clamped cubic spline for the same points.

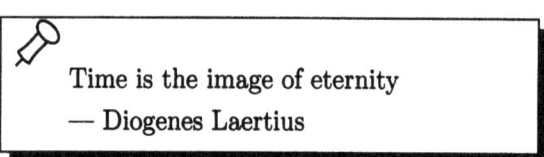

Time is the image of eternity
— Diogenes Laertius

4.8.4 Periodic Cubic Splines

A periodic function $f(x)$ is one that repeats itself. If p is the period of the function, then $f(x + p) = f(x)$ for any x. A two-dimensional cubic spline is periodic if it has the same extreme tangent vectors (i.e., if it starts and ends going in the same direction) and if its two extreme points $\mathbf{P}(0)$ and $\mathbf{P}(1)$ have the same y coordinate. If the curve satisfies these conditions, then we can place consecutive copies of it side by side and the result will look like one curve repeating itself.

The case of a three-dimensional periodic cubic spline is less clear. It seems that the two extreme points can be any points (they don't have to have the same y or z coordinates or any other relationship), so the condition for periodicity is that the curve will have the same start and end tangents, i.e., it will be cyclic.

Example: Exercise 4.11 shows that the parametric expression $(\cos t, \sin t, t)$ describes a helix (see also Section 5.6.1 for a double helix). Modifying this expression to $\mathbf{P}(t) = (0.05t + \cos t, \sin t, .1t)$ creates a helix that moves in the x direction as it climbs up in the z direction. Figure 4.23 shows its behavior. This curve starts at $\mathbf{P}(0) = (1, 0, 0)$ and ends at $\mathbf{P}(10\pi) = (0.5\pi + 1, 0, 0.5\pi)$. There is no special relation between the start and end points, but the curve is periodic since both its start and end tangents equal $\mathbf{P}^t(0) = \mathbf{P}^t(10\pi) = (0.05, 1, 0.1)$. We can construct another period of this curve by copying it, moving the copy parallel to itself, and placing it such that the start point of the copy is at the end point of the original curve.

Notice that it is possible to make the start and end points even more unrelated by, for example, tilting the helix also in the y direction as it climbs up in the z direction. This kind of effect achieved by an expression such as

$$\mathbf{P}(t) = (0.05t + \cos t, -0.05t^2 + \sin t, 0.1t).$$

4.8.5 Closed Cubic Splines

This is the case where a new curve segment should be added, from $\mathbf{P}_n$ to $\mathbf{P}_1$, closing the curve. In such a curve, every point is interior, so Equation (4.57) becomes a system of n equations in the same n unknowns. No user input is needed, which means that the only way to edit such a curve is to move/add/delete points. It is convenient to denote $\mathbf{P}_{n+1}$ by $\mathbf{P}_1$ and $\mathbf{P}_{n+2}$ by $\mathbf{P}_2$. Equation (4.57) becomes

4.8 The Cubic Spline Curve

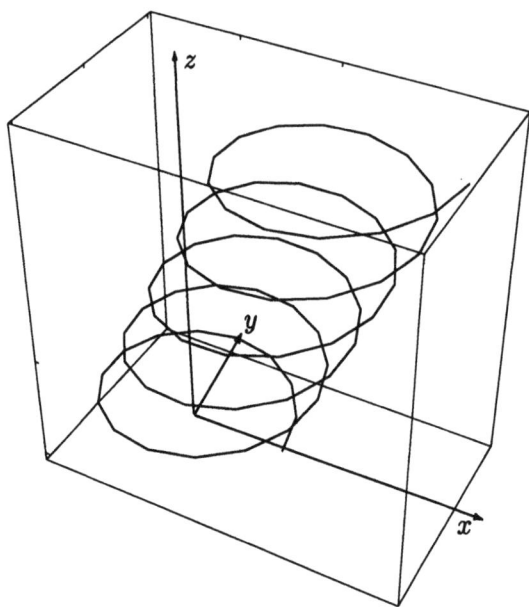

Figure 4.23: The Helix as a Periodic Curve; Created by
`ParametricPlot3D[{.05t+Cos[t],Sin[t],.1t}, {t,0,10Pi}]`

$$n\left\{\underbrace{\begin{pmatrix} 1 & 4 & 1 & \cdots & 0 & \cdots & 0 \\ 0 & 1 & 4 & 1 & \cdots & \cdots & 0 \\ & & \ddots & \ddots & & & \vdots \\ 0 & \cdots & \cdots & \cdots & 1 & 4 & 1 \\ 1 & \cdots & \cdots & \cdots & 0 & 1 & 4 \\ 4 & 1 & 0 & \cdots & 0 & 0 & 1 \end{pmatrix}}_{n} \begin{pmatrix} \mathbf{P}_1^t \\ \mathbf{P}_2^t \\ \vdots \\ \mathbf{P}_{n-1}^t \\ \mathbf{P}_n^t \end{pmatrix} = \begin{pmatrix} 3(\mathbf{P}_3 - \mathbf{P}_1) \\ 3(\mathbf{P}_4 - \mathbf{P}_2) \\ \vdots \\ 3(\mathbf{P}_{n+1} - \mathbf{P}_{n-1}) \\ 3(\mathbf{P}_{n+2} - \mathbf{P}_n) \end{pmatrix}. \quad (4.66)$$

Example: Given the four points of Section 4.8.1, $\mathbf{P}_1 = (0,0)$, $\mathbf{P}_2 = (1,0)$, $\mathbf{P}_3 = (1,1)$, and $\mathbf{P}_4 = (0,1)$, we are looking for four PC segments $\mathbf{P}_1(t)$, $\mathbf{P}_2(t)$, $\mathbf{P}_3(t)$, and $\mathbf{P}_4(t)$ that would connect smoothly at the four points. Equation (4.66) becomes

$$\begin{pmatrix} 1 & 4 & 1 & 0 \\ 0 & 1 & 4 & 1 \\ 1 & 0 & 1 & 4 \\ 4 & 1 & 0 & 1 \end{pmatrix} \begin{pmatrix} \mathbf{P}_1^t \\ \mathbf{P}_2^t \\ \mathbf{P}_3^t \\ \mathbf{P}_4^t \end{pmatrix} = \begin{pmatrix} 3(\mathbf{P}_3 - \mathbf{P}_1) \\ 3(\mathbf{P}_4 - \mathbf{P}_2) \\ 3(\mathbf{P}_1 - \mathbf{P}_3) \\ 3(\mathbf{P}_2 - \mathbf{P}_4) \end{pmatrix}. \quad (4.67)$$

Its solutions are $\mathbf{P}_1^t = (3/4, -3/4)$, $\mathbf{P}_2^t = (3/4, 3/4)$, $\mathbf{P}_3^t = (-3/4, 3/4)$, and $\mathbf{P}_4^t = (-3/4, -3/4)$, and the four spline segments are

$$\mathbf{P}_1(t) = (-1/2, 0)t^3 + (3/4, 3/4)t^2 + (3/4, -3/4)t,$$
$$\mathbf{P}_2(t) = (0, -1/2)t^3 + (-3/4, 3/4)t^2 + (3/4, 3/4)t + (1, 0),$$

4 Curves

$$\mathbf{P}_3(t) = (1/2, 0)t^3 + (-3/4, -3/4)t^2 + (-3/4, 3/4)t + (1, 1),$$
$$\mathbf{P}_4(t) = (0, 1/2)t^3 + (3/4, -3/4)t^2 + (-3/4, -3/4)t + (0, 1).$$

4.8.6 Nonuniform Cubic Splines

All the different types of cubic splines discussed earlier assume that the parameter t varies in the range $[0, 1]$ in every segment. These types of cubic spline are therefore *uniform* or *normalized*. The *nonuniform cubic spline* is obtained by adding another parameter t_k to every spline segment and letting t vary in the range $[0, t_k]$. Since there are $n - 1$ spline segments connecting the n data points, this adds $n - 1$ parameters to the curve, which makes it easier to fine-tune the shape of the curve. The nonuniform cubic splines are especially useful in cases where the data points are unequally spaced. In regions where the points are closely spaced, the normalized cubic spline tends to develop loops and overshoots. In regions where the points are widely spaced, it tends to "cut corners," i.e., to be too tight. Careful selection of the t_k parameters can overcome these tendencies.

The calculation of the nonuniform cubic spline is based on that of the normalized one. We simply rewrite some of the basic equations substituting t_k for 1 as the final value of t. We start with Equation (4.21) that becomes, for the first spline segment,

$$\mathbf{a} \cdot 0^3 + \mathbf{b} \cdot 0^2 + \mathbf{c} \cdot 0 + \mathbf{d} = \mathbf{P}_1,$$
$$\mathbf{a}(t_1)^3 + \mathbf{b}(t_1)^2 + \mathbf{c}(t_1) + \mathbf{d} = \mathbf{P}_2,$$
$$3\mathbf{a} \cdot 0^2 + 2\mathbf{b} \cdot 0 + \mathbf{c} = \mathbf{P}_1^t,$$
$$3\mathbf{a}(t_1)^2 + 2\mathbf{b}(t_1) + \mathbf{c} = \mathbf{P}_2^t,$$

with solutions

$$\mathbf{a} = \frac{2(\mathbf{P}_1 - \mathbf{P}_2)}{t_1^3} + \frac{\mathbf{P}_1^t}{t_1^2} + \frac{\mathbf{P}_2^t}{t_1^2},$$
$$\mathbf{b} = \frac{3(\mathbf{P}_2 - \mathbf{P}_1)}{t_1^2} - \frac{2\mathbf{P}_1^t}{t_1} - \frac{\mathbf{P}_2^t}{t_1}, \qquad (4.68)$$
$$\mathbf{c} = \mathbf{P}_1^t,$$
$$\mathbf{d} = \mathbf{P}_1.$$

Equation (4.23) now becomes

$$\mathbf{P}(t) = \left[\frac{2(\mathbf{P}_1 - \mathbf{P}_2)}{t_1^3} + \frac{\mathbf{P}_1^t}{t_1^2} + \frac{\mathbf{P}_2^t}{t_1^2}\right] t^3 + \left[\frac{3(\mathbf{P}_2 - \mathbf{P}_1)}{t_1^2} - \frac{2\mathbf{P}_1^t}{t_1} - \frac{\mathbf{P}_2^t}{t_1}\right] t^2 + \mathbf{P}_1^t t + \mathbf{P}_1. \qquad (4.69)$$

Equation (4.54) becomes

$$\mathbf{P}_k^{tt}(t_k) = \mathbf{P}_{k+1}^{tt}(0) \qquad \text{or} \qquad 6\mathbf{a}_k \times t_k + 2\mathbf{b}_k = 6\mathbf{a}_{k+1} \times 0 + 2\mathbf{b}_{k+1}, \qquad (4.70)$$

and Equation (4.55) is now

$$2\left[\frac{3(\mathbf{P}_{k+1}-\mathbf{P}_k)}{t_k^2} - \frac{2\mathbf{P}_k^t}{t_k} - \frac{\mathbf{P}_{k+1}^t}{t_k}\right] + 6t_k\left[\frac{2(\mathbf{P}_k-\mathbf{P}_{k+1})}{t_k^3} + \frac{\mathbf{P}_k^t}{t_k^2} + \frac{\mathbf{P}_{k+1}^t}{t_k^2}\right]$$
$$= 2\left[\frac{3(\mathbf{P}_{k+2}-\mathbf{P}_{k+1})}{t_{k+1}^2} - \frac{2\mathbf{P}_{k+1}^t}{t_{k+1}} - \frac{\mathbf{P}_{k+2}^t}{t_{k+1}}\right]. \tag{4.71}$$

Equation (4.56) now becomes

$$t_{k+1}\mathbf{P}_k^t + 2(t_k + t_{k+1})\mathbf{P}_{k+1}^t + t_k\mathbf{P}_{k+2}^t$$
$$= \frac{3}{t_k t_{k+1}}\left[t_k^2(\mathbf{P}_{k+2}-\mathbf{P}_{k+1}) + t_{k+1}^2(\mathbf{P}_{k+1}-\mathbf{P}_k)\right]. \tag{4.72}$$

This produces the new version of Equation (4.57):

$$n-2\left\{\begin{pmatrix} t_2 & 2(t_1+t_2) & t_1 & 0 & 0 & \cdots & 0 \\ 0 & t_3 & 2(t_2+t_3) & t_2 & 0 & \cdots & 0 \\ & & \ddots & & \ddots & & \vdots \\ 0 & 0 & \cdots & \cdots & t_{n-1} & 2(t_{n-1}+t_{n-2}) & t_{n-2} \end{pmatrix}\begin{pmatrix} \mathbf{P}_1^t \\ \mathbf{P}_2^t \\ \vdots \\ \mathbf{P}_n^t \end{pmatrix}\right.$$
$$\underbrace{}_{n}$$
$$= \begin{pmatrix} \frac{3}{t_1 t_2}\left[t_1^2(\mathbf{P}_3-\mathbf{P}_2) + t_2^2(\mathbf{P}_2-\mathbf{P}_1)\right] \\ \frac{3}{t_2 t_3}\left[t_2^2(\mathbf{P}_4-\mathbf{P}_3) + t_3^2(\mathbf{P}_3-\mathbf{P}_2)\right] \\ \vdots \\ \frac{3}{t_{n-2}t_{n-1}}\left[t_{n-2}^2(\mathbf{P}_n-\mathbf{P}_{n-1}) + t_{n-1}^2(\mathbf{P}_{n-1}-\mathbf{P}_{n-2})\right] \end{pmatrix}. \tag{4.73}$$

This is, again, a system of $n-2$ equations in the n unknowns $\mathbf{P}_1^t, \mathbf{P}_2^t, \ldots, \mathbf{P}_n^t$. After the user supplies values for the two extreme tangent vectors $\mathbf{P}_1^t$ and $\mathbf{P}_n^t$, this system can be solved, yielding the values of the remaining $n-2$ tangent vectors. Each of the $n-1$ spline segments can now be calculated by means of Equation (4.68) that is written for the first segment in compact form as follows:

$$\begin{pmatrix} \mathbf{a} \\ \mathbf{b} \\ \mathbf{c} \\ \mathbf{d} \end{pmatrix} = \begin{pmatrix} 2/t_1^3 & -2/t_1^3 & 1/t_1^2 & 1/t_1^2 \\ -3/t_1^2 & 3/t_1^2 & -2/t_1 & -1/t_1 \\ 0 & 0 & 1 & 0 \\ 1 & 0 & 0 & 0 \end{pmatrix}\begin{pmatrix} \mathbf{P}_1 \\ \mathbf{P}_2 \\ \mathbf{P}_1^t \\ \mathbf{P}_2^t \end{pmatrix}. \tag{4.74}$$

Notice how each of Equations (4.68) through (4.74) reduces to the original equation when all the t_i are set to 1. The nonuniform cubic spline can now be calculated in the following steps:

1. The user inputs values for the two extreme tangent vectors and for the $n-1$ parameters t_k. The software sets up and solves Equation (4.73) to calculate the remaining tangent vectors.

2. The software sets up and solves Equation (4.74) $n-1$ times, once for each of the spline segments.

4 Curves

3. Each segment $\mathbf{P}_k(t)$ is plotted by varying t from 0 to t_k.

Before looking at an example, it is useful to try to understand the advantage of having the extra parameters t_k. Equation (4.68) shows that a large value of t_k for spline segment $\mathbf{P}_k(t)$ means small **a** and **b** coefficients (since t_k appears in the denominators) and, hence, a small second derivative $\mathbf{P}_k^{tt}(t) = 6\mathbf{a}_k + 2\mathbf{b}_k$ for that segment. Since the second derivative can be interpreted as the acceleration of the curve, we can predict that a large t_k implies small overall acceleration for segment k. Thus, most of the segment will be close to a straight line. This is also easy to see when we substitute small **a** and **b** in $\mathbf{P}_k(t) = \mathbf{a}t^3 + \mathbf{b}t^2 + \mathbf{c}t + \mathbf{d}$. The dominant part of the segment becomes $\mathbf{c}t + \mathbf{d}$, which brings it close to linear. If the start and end directions of the segment are very different, the entire segment cannot be a straight line, so, in order to minimize its overall second derivative, the segment will end up consisting of two or three parts, each close to a straight line, with short, highly curved corner segments connecting them (Figure 4.24). Such a geometry has a small overall second derivative. This knowledge is useful when designing curves, which is why the nonuniform cubic spline should not be dismissed as impractical. It may be the best method for certain curves.

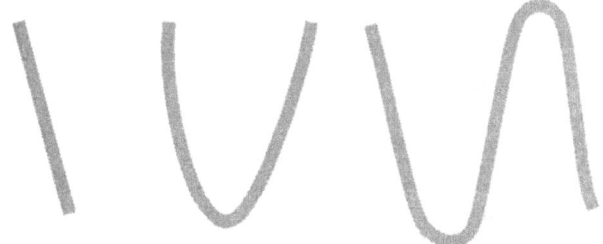

Figure 4.24: Curves with Small Overall Second Derivative.

Example: The four points of Section 4.8.1 are used in this example. They are $\mathbf{P}_1 = (0,0)$, $\mathbf{P}_2 = (1,0)$, $\mathbf{P}_3 = (1,1)$, and $\mathbf{P}_4 = (0,1)$. We also select the same initial and final directions $\mathbf{P}_1^t = (1,-1)$ and $\mathbf{P}_4^t = (-1,-1)$. We decide to use $t_k = 2$ for each of the three spline segments to illustrate how large t_k values create a curve very different from the one of Section 4.8.1. Equation (4.73) becomes

$$\begin{pmatrix} t_2 & 2(t_1+t_2) & t_1 & 0 \\ 0 & t_3 & 2(t_2+t_3) & t_2 \end{pmatrix} \begin{pmatrix} (1,-1) \\ \mathbf{P}_2^t \\ \mathbf{P}_3^t \\ (-1,-1) \end{pmatrix}$$

$$= \begin{pmatrix} \frac{3}{t_1 t_2}[t_1^2(\mathbf{P}_3 - \mathbf{P}_2) + t_2^2(\mathbf{P}_2 - \mathbf{P}_1)] \\ \frac{3}{t_2 t_3}[t_2^2(\mathbf{P}_4 - \mathbf{P}_3) + t_3^2(\mathbf{P}_3 - \mathbf{P}_2)] \end{pmatrix}.$$

For $t_1 = t_2 = t_3 = 2$, this yields $\mathbf{P}_2^t = (1/6, 1/2)$ and $\mathbf{P}_3^t = (-1/6, 1/2)$. Equa-

4.8 The Cubic Spline Curve

tion (4.74) is now written and solved three times:

Segment 1 $\begin{pmatrix} a \\ b \\ c \\ d \end{pmatrix} = \begin{pmatrix} 2/t_1^3 & -2/t_1^3 & 1/t_1^2 & 1/t_1^2 \\ -3/t_1^2 & 3/t_1^2 & -2/t_1 & -1/t_1 \\ 0 & 0 & 1 & 0 \\ 1 & 0 & 0 & 0 \end{pmatrix} \begin{pmatrix} (0,0) \\ (1,0) \\ (1,-1) \\ (1/6,1/2) \end{pmatrix}.$

Segment 2 $\begin{pmatrix} a \\ b \\ c \\ d \end{pmatrix} = \begin{pmatrix} 2/t_2^3 & -2/t_2^3 & 1/t_2^2 & 1/t_2^2 \\ -3/t_2^2 & 3/t_2^2 & -2/t_2 & -1/t_2 \\ 0 & 0 & 1 & 0 \\ 1 & 0 & 0 & 0 \end{pmatrix} \begin{pmatrix} (1,0) \\ (1,1) \\ (1/6,1/2) \\ (-1/6,1/2) \end{pmatrix}.$

Segment 3 $\begin{pmatrix} a \\ b \\ c \\ d \end{pmatrix} = \begin{pmatrix} 2/t_3^3 & -2/t_3^3 & 1/t_3^2 & 1/t_3^2 \\ -3/t_3^2 & 3/t_3^2 & -2/t_3 & -1/t_3 \\ 0 & 0 & 1 & 0 \\ 1 & 0 & 0 & 0 \end{pmatrix} \begin{pmatrix} (1,1) \\ (0,1) \\ (-1/6,1/2) \\ (-1,-1) \end{pmatrix}.$

This yields the coefficients for the three spline segments:

$$\mathbf{P}_1(t) = (1/24, -1/8)t^3 + (-1/3, 3/4)t^2 + (1, -1)t,$$
$$\mathbf{P}_2(t) = (0, 0)t^3 + (-1/12, 0)t^2 + (1/6, 1/2)t + (1, 0),$$
$$\mathbf{P}_3(t) = -(1/24, 1/8)t^3 + (-1/12, 0)t^2 + (-1/6, 1/2)t + (1, 1).$$

The result is shown in Figure 4.25. It should be compared with the uniform curve of Figure 4.22 that's based on the same four points. (Remember that t varies from 0 to 2 in each of the segments above.)

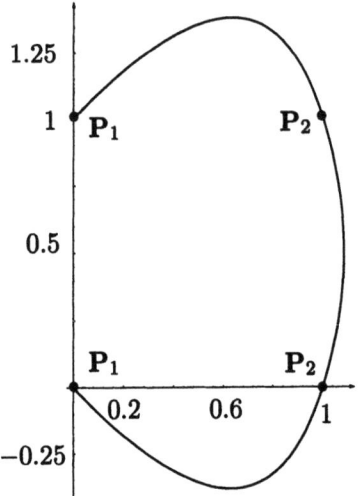

Figure 4.25: A Nonuniform Cubic Spline Example.

4 Curves

4.8.7 Fair Cubic Splines

The term "fair curve" refers to a spline curve where each segment is close to a circular arc. Such a curve does not have very flat or very curved segments (e.g., segments with loops) and is generally considered pleasing to the eye. It is useful in artistic applications and in font design, where the aim is to get nice looking curves rather than a precise fit to a given set of points. The approach presented here is based on [Manning 74] and is illustrated by Figure 4.26. In Figure 4.26a we see four Hermite segments connecting the same two endpoints, all with 45° tangent vectors. The difference between them is the magnitude of the vectors. If we denote the distance between the points l, then Figure 4.26a(i)–(iii) shows curves whose tangent vectors have magnitudes smaller than, equal to, and greater than l, respectively. In Figure 4.26a(iv), the left tangent has magnitude $> l$ and the right one $< l$, resulting in a curve that's "pulled" to the right. Of these four, curve (ii) can be considered "fair," since it is close to a special circular arc, namely the arc that passes through the endpoints of the curve *at the same angle* as the tangent vectors.

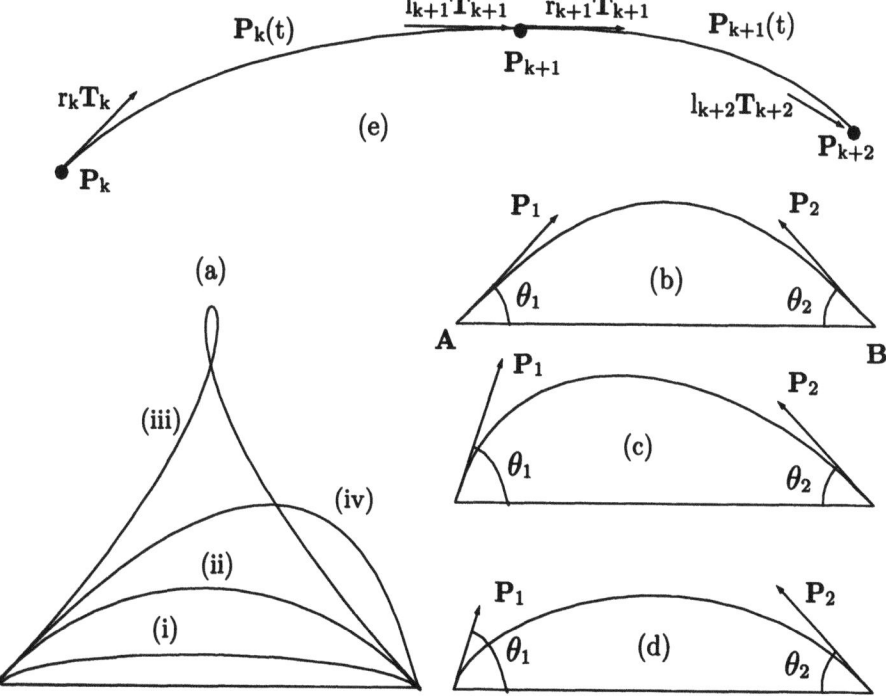

Figure 4.26: Varying Tangent Vector Magnitudes.

▶ **Exercise 4.55:** Two given points P_1 and P_2 define a straight segment $P_1 \to P_2$. Your task is to construct a "fair" Hermite segment with P_1 and P_2 as its endpoints. Here is what you need to show. Imagine the circular arc that passes through P_1 and P_2 and makes angles θ with the line $P_1 \to P_2$. Show that if the center point

$\mathbf{P}(0.5)$ of the Hermite segment is located on this circular arc, then the magnitudes of the tangent vectors satisfy

$$|\mathbf{P}_1^t| = |\mathbf{P}_2^t| = \frac{2|\mathbf{P}_2 - \mathbf{P}_1|}{1 + \cos\theta}. \qquad (4.75)$$

Figure 4.26b shows a Hermite segment with tangent vectors whose magnitudes satisfy Equation (4.75). The curve is therefore close to the special circular arc mentioned earlier. In Figure 4.26c, the left tangent vector $\mathbf{P}_1$ has been rotated (but has the same magnitude), thereby increasing angle θ_1 and pulling the curve to the left. Figure 4.26d shows how the curve can (approximately) be returned to its former shape by shortening $\mathbf{P}_1$. This suggests a way to build a complete cubic spline where every segment is fair. At every internal point $\mathbf{P}_{k+1}$, the "incoming" and "outgoing" tangent vectors should go in the same direction but should have *different magnitudes*. Figure 4.26e shows the terminology used. The incoming tangent vector at interior point $\mathbf{P}_{k+1}$ is $\mathbf{P}_k^t(1) = l_{k+1}\mathbf{T}_{k+1}$, where all $\mathbf{T}_i$ are unit vectors and l_{k+1} is a scalar (the magnitude of the tangent). Similarly, the outgoing tangent vector at $\mathbf{P}_{k+1}$ is $\mathbf{P}_{k+1}^t(0) = r_{k+1}\mathbf{T}_{k+1}$, the same unit vector, but with a different magnitude r_{k+1}.

Equation (4.75) can be considered the definition of a fair Hermite segment in the special case $\theta_1 = \theta_2$. The previous paragraph suggests a way to extend it to cases where $\theta_1 \neq \theta_2$. We need to express the magnitudes of the tangents $\mathbf{P}_1^t$ and $\mathbf{P}_2^t$ in terms of the endpoints and the angles θ_1 and θ_2 such that $\theta_1 > \theta_2$ will result in $|\mathbf{P}_1^t| < |\mathbf{P}_2^t|$. One way to achieve this effect is to define

$$r_1 = |\mathbf{P}_1^t| = \frac{2|\mathbf{P}_2 - \mathbf{P}_1|}{1 + \alpha\cos\theta_2 + (1-\alpha)\cos\theta_1}, \qquad (4.76)$$

$$l_2 = |\mathbf{P}_2^t| = \frac{2|\mathbf{P}_2 - \mathbf{P}_1|}{1 + \alpha\cos\theta_1 + (1-\alpha)\cos\theta_2}, \qquad (4.77)$$

where α is a parameter in the range $[0,1]$ to be determined by the user. This, of course, is not the only way to define a fair curve, but this definition has two useful properties:

1. For $\theta_1 = \theta_2$, Equations (4.76) and (4.77) reduce to Equation (4.75).

2. If $\theta_1 > \theta_2$, then $\cos\theta_1 < \cos\theta_2$ (for fair curves, we can assume angles between 0° and 90°). In order to achieve $|\mathbf{P}_1^t| < |\mathbf{P}_2^t|$, we need a situation where $(1-\alpha)\cos\theta_1 < \alpha\cos\theta_2$. This is satisfied when $1 - \alpha \leq \alpha$, i.e., when $0.5 \leq \alpha \leq 1$. [Manning 74] suggests $\alpha = 2/3$, but it seems that α should be left as a user-defined parameter, especially for closed fair curves, which are discussed later, where there is no other parameter for the user to adjust.

The condition for slope continuity at the $n-2$ interior points is thus written $h_{k+1}\mathbf{P}_k^t(1) = \mathbf{P}_{k+1}^t(0)$, where h_{k+1} is the ratio of the tangent vectors' magnitudes. We denote by $\mathbf{T}_k$ a unit tangent vector, so we can write $h_{k+1}l_{k+1}\mathbf{T}_{k+1} = r_{k+1}\mathbf{T}_{k+1}$ or $h_{k+1} = r_{k+1}/l_{k+1}$ for $k = 1, 2, \ldots, n-2$. The two tangents $\mathbf{P}_k^t(1)$ and $\mathbf{P}_{k+1}^t(0)$ go in the same direction, so h_{k+1} must be positive.

4 Curves

The quantities $l_2, l_3, \ldots, l_n$ and $r_1, r_2, \ldots, r_{n-1}$ make a total of $2n-2$ unknowns that have to be calculated. The equations to calculate them are obtained by the requirement that the curvatures of the spline segments be equal at the $n-2$ interior points. When Equation (4.220) is used for the curvature, this requirement becomes

$$\frac{d^2\mathbf{P}_k(1)}{ds^2} = \frac{d^2\mathbf{P}_{k+1}(0)}{ds^2}$$

or

$$\frac{\mathbf{P}_k^{tt}(1)}{\mathbf{P}_k^t(1) \bullet \mathbf{P}_k^t(1)} - \frac{\mathbf{P}_k^t(1) \bullet \mathbf{P}_k^{tt}(1)}{\left(\mathbf{P}_k^t(1) \bullet \mathbf{P}_k^t(1)\right)^2}\mathbf{P}_k^t(1)$$
$$= \frac{\mathbf{P}_{k+1}^{tt}(0)}{\mathbf{P}_{k+1}^t(0) \bullet \mathbf{P}_{k+1}^t(0)} - \frac{\mathbf{P}_{k+1}^t(0) \bullet \mathbf{P}_{k+1}^{tt}(0)}{\left(\mathbf{P}_{k+1}^t(0) \bullet \mathbf{P}_{k+1}^t(0)\right)^2}\mathbf{P}_{k+1}^t(0).$$

This is simplified by substituting $h_{k+1}\mathbf{P}_k^t(1) = \mathbf{P}_{k+1}^t(0)$ and multiplying by $(\mathbf{P}_k^t(1) \bullet \mathbf{P}_k^t(1))$, yielding

$$\mathbf{P}_k^{tt}(1) - \frac{\mathbf{P}_k^t(1) \bullet \mathbf{P}_k^{tt}(1)}{\mathbf{P}_k^t(1) \bullet \mathbf{P}_k^t(1)}\mathbf{P}_k^t(1) = \frac{\mathbf{P}_{k+1}^{tt}(0)}{h_{k+1}^2} - \frac{h_{k+1}^2\left(\mathbf{P}_k^t(1) \bullet \mathbf{P}_{k+1}^{tt}(0)\right)}{h_{k+1}^4\left(\mathbf{P}_k^t(1) \bullet \mathbf{P}_k^t(1)\right)}\mathbf{P}_k^t(1)$$

or

$$h_{k+1}^2\mathbf{P}_k^{tt}(1) - \mathbf{P}_{k+1}^{tt}(0) = \frac{h_{k+1}^2\left(\mathbf{P}_k^t(1) \bullet \mathbf{P}_k^{tt}(1)\right) - \left(\mathbf{P}_k^t(1) \bullet \mathbf{P}_{k+1}^{tt}(0)\right)}{\mathbf{P}_k^t(1) \bullet \mathbf{P}_k^t(1)}\mathbf{P}_k^t(1). \quad (4.78)$$

Equation (4.78) can be written

$$h_{k+1}^2\mathbf{P}_k^{tt}(1) - \mathbf{P}_{k+1}^{tt}(0) = M_{k+1}\mathbf{P}_k^t(1), \quad (4.79)$$

where the quantity M_{k+1} that is defined by

$$M_{k+1} \stackrel{\text{def}}{=} \frac{h_{k+1}^2\left(\mathbf{P}_k^t(1) \bullet \mathbf{P}_k^{tt}(1)\right) - \left(\mathbf{P}_k^t(1) \bullet \mathbf{P}_{k+1}^{tt}(0)\right)}{\mathbf{P}_k^t(1) \bullet \mathbf{P}_k^t(1)},$$

is a scalar combining all the scalar quantities from the right-hand side of Equation (4.78).

> Fair: To draw and adjust (the lines of a ship's hull being designed) to produce regular surfaces of the correct form.
> — A dictionary definition.

The next step is to replace the two second derivatives on the left side of Equation (4.79) with expressions containing the unknowns l_k, r_k, and $\mathbf{T}_k$ (and, perhaps, some known quantities, such as the points $\mathbf{P}_k$). This will provide equations whose solutions will yield the tangent vectors $\mathbf{P}_k^t$ at all the points. We start with the second derivative of a PC $\mathbf{P}^{tt}(t) = 6\mathbf{a}t + \mathbf{b}$ (Equation (4.53)), where $\mathbf{a}$ and $\mathbf{b}$ are given by Equation (4.22). The two second derivatives used in Equation (4.79) are (see also Figure 4.26e) as follows:

1. From segment $\mathbf{P}_k(t)$,

$$\begin{aligned}
\mathbf{P}_k^{tt}(1) &= 6\mathbf{a}_k \times 1 + \mathbf{b}_k \\
&= 6[2(\mathbf{P}_k - \mathbf{P}_{k+1}) + \mathbf{P}_k^t + \mathbf{P}_{k+1}^t] + 2[3(\mathbf{P}_{k+1} - \mathbf{P}_k) - 2\mathbf{P}_k^t - \mathbf{P}_{k+1}^t] \\
&= -6(\mathbf{P}_{k+1} - \mathbf{P}_k) + 2\mathbf{P}_k^t + 4\mathbf{P}_{k+1}^t \\
&= -6(\mathbf{P}_{k+1} - \mathbf{P}_k) + 2r_k\mathbf{T}_k + 4l_{k+1}\mathbf{T}_{k+1}.
\end{aligned} \quad (4.80)$$

2. From segment $\mathbf{P}_{k+1}(t)$,

$$\begin{aligned}
\mathbf{P}_{k+1}^{tt}(0) &= 6\mathbf{a}_{k+1} \times 0 + \mathbf{b}_{k+1} \\
&= 2[3(\mathbf{P}_{k+2} - \mathbf{P}_{k+1}) - 2\mathbf{P}_{k+1}^t - \mathbf{P}_{k+2}^t] \\
&= 6(\mathbf{P}_{k+2} - \mathbf{P}_{k+1}) - 4r_{k+1}\mathbf{T}_{k+1} - 2l_{k+2}\mathbf{T}_{k+2}.
\end{aligned} \quad (4.81)$$

Substituting Eqs. (4.80) and (4.81) into Equation (4.79) and taking into account that $h_k = r_k/l_k$ for $k = 1, 2, \ldots, n-2$, we get

$$\begin{aligned}
\frac{r_{k+1}^2}{l_{k+1}^2} &- 6(\mathbf{P}_{k+1} - \mathbf{P}_k) + 2r_k\mathbf{T}_k + 4l_{k+1}\mathbf{T}_{k+1} \\
&- [6(\mathbf{P}_{k+2} - \mathbf{P}_{k+1}) - 4r_{k+1}\mathbf{T}_{k+1} - 2l_{k+2}\mathbf{T}_{k+2}] \\
&= M_{k+1}\mathbf{P}_k^t(1).
\end{aligned}$$

Multiplying by l_k^2 and simplifying yields

$$\begin{aligned}
r_{k+1}^2[-6(\mathbf{P}_{k+1} - \mathbf{P}_k) &+ 2r_k\mathbf{T}_k + 4l_{k+1}\mathbf{T}_{k+1}] \\
- l_{k+1}^2[6(\mathbf{P}_{k+2} - \mathbf{P}_{k+1}) &- 4r_{k+1}\mathbf{T}_{k+1} - 2l_{k+2}\mathbf{T}_{k+2}] \\
= l_{k+1}^2 M_{k+1}\mathbf{P}_k^t(1) &= l_{k+1}^3 M_{k+1}\mathbf{T}_{k+1}
\end{aligned}$$

or

$$\begin{aligned}
-6r_{k+1}^2(\mathbf{P}_{k+1} - \mathbf{P}_k) &+ 2r_{k+1}^2 r_k\mathbf{T}_k + 4r_{k+1}^2 l_{k+1}\mathbf{T}_{k+1} \\
-6l_{k+1}^2(\mathbf{P}_{k+2} - \mathbf{P}_{k+1}) &+ 4l_{k+1}^2 r_{k+1}\mathbf{T}_{k+1} + 2l_{k+1}^2 l_{k+2}\mathbf{T}_{k+2} \\
= l_{k+1}^3 M_{k+1}\mathbf{T}_{k+1}.
\end{aligned}$$

Dividing by 2 and changing signs results in

$$\begin{aligned}
3r_{k+1}^2(\mathbf{P}_{k+1} - \mathbf{P}_k) &- 3l_{k+1}^2(\mathbf{P}_{k+2} - \mathbf{P}_{k+1}) - r_{k+1}^2 r_k\mathbf{T}_k - l_{k+1}^2 l_{k+2}\mathbf{T}_{k+2} \\
&= -\frac{1}{2}l_{k+1}^3 M_{k+1}\mathbf{T}_{k+1} - 2r_{k+1}^2 l_{k+1}\mathbf{T}_{k+1} + 2l_{k+1}^2 r_{k+1}\mathbf{T}_{k+1} \stackrel{\text{def}}{=} L_{k+1}\mathbf{T}_{k+1},
\end{aligned} \quad (4.82)$$

where L_{k+1} is defined as everything that multiplies $\mathbf{T}_{k+1}$ on the right-hand side of Equation (4.82).

Equation (4.82) is a vector equation that can be written $n-2$ times, for $k = 1, 2, \ldots, n-2$. It therefore yields $2(n-2)$ or $3(n-2)$ equations, depending on

4 Curves

whether the original data points $\mathbf{P}_k$ are two- or three-dimensional. However, more equations are needed. To derive them, we turn our attention to Figure 4.27, which shows the relation between a unit tangent vector $\mathbf{T}$ and the angle θ between it and the line connecting the two endpoints

$$\cos\theta_k = \mathbf{T}_k \bullet \frac{\mathbf{P}_{k+1} - \mathbf{P}_k}{|\mathbf{P}_{k+1} - \mathbf{P}_k|}.$$

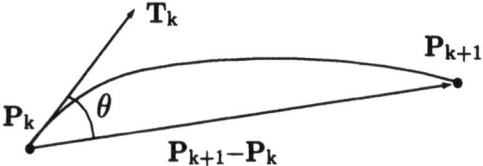

Figure 4.27: Relation Between θ and $\mathbf{T}$.

For segment $\mathbf{P}_k(t)$, we get from Equation (4.77)

$$l_{k+1} = |\mathbf{P}_{k+1}^t| = \frac{2|\mathbf{P}_{k+1} - \mathbf{P}_k|}{1 + \alpha \mathbf{T}_k \bullet \frac{\mathbf{P}_{k+1}-\mathbf{P}_k}{|\mathbf{P}_{k+1}-\mathbf{P}_k|} + (1-\alpha)\mathbf{T}_{k+1} \bullet \frac{\mathbf{P}_{k+1}-\mathbf{P}_k}{|\mathbf{P}_{k+1}-\mathbf{P}_k|}}$$
$$= \frac{2|\mathbf{P}_{k+1} - \mathbf{P}_k|}{1 + [\alpha \mathbf{T}_k + (1-\alpha)\mathbf{T}_{k+1}] \bullet \frac{\mathbf{P}_{k+1}-\mathbf{P}_k}{|\mathbf{P}_{k+1}-\mathbf{P}_k|}}. \tag{4.83}$$

From Equation (4.76)

$$r_k = |\mathbf{P}_k^t| = \frac{2|\mathbf{P}_{k+1} - \mathbf{P}_k|}{1 + \alpha \mathbf{T}_{k+1} \bullet \frac{\mathbf{P}_{k+1}-\mathbf{P}_k}{|\mathbf{P}_{k+1}-\mathbf{P}_k|} + (1-\alpha)\mathbf{T}_k \bullet \frac{\mathbf{P}_{k+1}-\mathbf{P}_k}{|\mathbf{P}_{k+1}-\mathbf{P}_k|}}$$
$$= \frac{2|\mathbf{P}_{k+1} - \mathbf{P}_k|}{1 + [\alpha \mathbf{T}_{k+1} + (1-\alpha)\mathbf{T}_k] \bullet \frac{\mathbf{P}_{k+1}-\mathbf{P}_k}{|\mathbf{P}_{k+1}-\mathbf{P}_k|}}. \tag{4.84}$$

Equations (4.83) and (4.84) are scalar. They can be written for each of the $n-1$ spline segments (i.e., for $k = 1, 2, \ldots, n-1$), providing $2n - 2$ additional equations. The total number of equations is thus $(2n-4) + (2n-2) = 4n-6$ for the two-dimensional case and $(3n-6) + (2n-2) = 5n-8$ for the three-dimensional case. The unknowns are

$$l_2, l_3, \ldots, l_n, r_1, r_2, \ldots, r_{n-1}, L_2, L_3, \ldots, L_{n-1} \text{ and } \mathbf{T}_1, \mathbf{T}_2, \ldots, \mathbf{T}_n,$$

a total of $(n-1) + (n-1) + (n-2) + n = 4n - 4$ unknowns. It is important to realize that in the two-dimensional case, each unit vector $\mathbf{T}_k$ is equivalent to only one unknown (since it can be written $(\cos\theta, \sin\theta)$ for some angle θ). In the three-dimensional case, each $\mathbf{T}_k$ is equivalent to two unknowns, so the number of unknowns in that case is $5n - 4$. We thus end up with $4n - 6$ equations and $4n - 4$

unknowns (in the two-dimensional case) or $5n - 8$ equations and $5n - 4$ unknowns (in the three-dimensional case). We seem to be two or four equations short, but we already know from past experience with cubic splines that this apparent problem can be turned to our advantage. The solution, of course, is to ask the user to supply the values of the two extreme unit tangents $\mathbf{T}_1$ and $\mathbf{T}_n$. This provides the two or four necessary values to bring the number of unknowns down to the number of equations. This, together with the free parameter α, turns fair cubic splines into an interactive method.

The discussion so far has assumed an open curve, but closed curves are also useful in practical problems. In fact, the original fair cubic splines were developed by [Manning 74] to help design insoles for shoes; a useful, practical example of a closed curve. A closed curve does not have endpoints; every point can be considered interior. Equation (4.82) can thus be written n times, providing $2n$ or $3n$ equations. A closed curve also requires n segments, instead of $n-1$. Each of Equations (4.83) and (4.84) can thus be written n times, once for each segment. The total number of equations is therefore $4n$ or $5n$. The unknowns are

$$l_1, l_2, \ldots, l_n, r_1, r_2, \ldots, r_n, L_1, L_2, \ldots, L_n \text{ and } \mathbf{T}_1, \mathbf{T}_2, \ldots, \mathbf{T}_n,$$

a total of $4n$ or $5n$ unknowns. A closed curve can thus be computed based on the n data points, without any additional user input. If it is unsatisfactory, it can be edited by moving the points or changing the value of the parameter α.

One drawback of this method is that the equations are not linear. This makes it complicated to solve them and it also means that there may be either no solutions or several different solutions. A simple, iterative algorithm for solving the equations is the following:

1. Guess reasonable initial values for the unit tangents $\mathbf{T}_k$. A good idea is to set $\mathbf{T}_k$ in the direction $\mathbf{P}_{k+1} - \mathbf{P}_k$.

2. Using these values, solve Equations (4.83) and (4.84) to calculate initial values for all the l_k and r_k unknowns.

3. Substitute the current values of $\mathbf{T}_k$, l_k, and r_k in the left-hand side of Equation (4.82) to obtain better values for the products $L_k \mathbf{T}_k$. Once such a product is known, $\mathbf{T}_k$ can be calculated since its magnitude is 1.

4. Repeat steps 2 and 3 until the process converges (i.e., until none of l_k, r_k, and $\mathbf{T}_k$ changes between consecutive iterations by more than a preset threshold).

As has been mentioned earlier, this process is not guaranteed to converge, or it may converge to one of several possible solutions. Another difficulty has to do with the constants L_k. We never need their magnitudes, but their signs are important, since they give the sense of direction at the interior points. One way to handle the L_k is to keep the angle between the two vectors $(\mathbf{P}_{k+2}-\mathbf{P}_{k+1})$ and $(\mathbf{P}_{k+1}-\mathbf{P}_k)$ small for every interior point $\mathbf{P}_{k+1}$ (see Figure 4.26e). This will produce individual spline segments, none of which turns too much, resulting in tangents $\mathbf{T}_k$, $\mathbf{T}_{k+1}$, and $\mathbf{T}_{k+2}$ that don't differ much in direction. Such a situation corresponds to $L_k > 0$ for every k. An alternative is to keep the sign of each of the new products $L_k \mathbf{T}_k$ obtained in step 3 identical to the sign of $\mathbf{T}_k$ used earlier in that step. This guarantees that none of the new $\mathbf{T}_k$ will change much in direction during an iteration.

4 Curves

4.9 The Quadratic Spline

The cubic spline curve is useful in certain practical applications, which raises the question of splines of different degrees based on the same concepts. It turns out that splines of degrees higher than 3 are not useful since they are more computationally intensive and tend to have many undesirable inflection points (i.e., they tend to "wiggle" excessively). Splines of degree 1 are, of course, just polylines, but quadratic splines can be useful in some situations. Such a spline is easy to derive and to calculate. Each spline segment is a quadratic polynomial, i.e., a parabolic arc, so it results in fewer oscillations in the curve. On the other hand, quadratic spline segments can connect with at most C^1 continuity since their second derivative is constant. A quadratic spline curve thus may not be as smooth as a cubic spline that passes through the same points.

The quadratic spline curve is derived in this section based on the variant Hermite segment of Section 4.7.9. Each segment $\mathbf{P}_i(t)$ is thus a quadratic polynomial defined by its two endpoints $\mathbf{P}_i$ and $\mathbf{P}_{i+1}$ and by its start tangent vector $\mathbf{P}_i^t$. Equation (4.48) shows that the end tangent of such a segment is $\mathbf{P}_i^t(1) = 2(\mathbf{P}_{i+1} - \mathbf{P}_i) - \mathbf{P}_i^t$. The first two spline segments are

$$\mathbf{P}_1(t) = (\mathbf{P}_2 - \mathbf{P}_1 - \mathbf{P}_1^t)t^2 + \mathbf{P}_1^t t + \mathbf{P}_1,$$
$$\mathbf{P}_2(t) = (\mathbf{P}_3 - \mathbf{P}_2 - \mathbf{P}_2^t)t^2 + \mathbf{P}_2^t t + \mathbf{P}_2.$$

At their joint point $\mathbf{P}_2$, they have the tangent vectors $\mathbf{P}_1^t(1) = 2(\mathbf{P}_2 - \mathbf{P}_1) - \mathbf{P}_1^t$ and $\mathbf{P}_2^t(0) = \mathbf{P}_2^t$. In order to achieve C^1 continuity we should have $\mathbf{P}_1^t(1) = \mathbf{P}_2^t(0)$ or $2(\mathbf{P}_2 - \mathbf{P}_1) - \mathbf{P}_1^t = \mathbf{P}_2^t$. This equation can be written $\mathbf{P}_1^t + \mathbf{P}_2^t = 2(\mathbf{P}_2 - \mathbf{P}_1)$, and when duplicated $n - 1$ times, for the $n - 1$ interior points, the result is

$$n-1\left\{\underbrace{\begin{pmatrix} 1 & 1 & 0 & 0 & \cdots & 0 & 0 \\ 0 & 1 & 1 & 0 & \cdots & 0 & 0 \\ & & \ddots & \ddots & & \vdots \\ 0 & 0 & 0 & 0 & \cdots & 1 & 1 \end{pmatrix}}_{n} \begin{pmatrix} \mathbf{P}_1^t \\ \mathbf{P}_2^t \\ \vdots \\ \mathbf{P}_n^t \end{pmatrix} = 2 \begin{pmatrix} \mathbf{P}_2 - \mathbf{P}_1 \\ \mathbf{P}_3 - \mathbf{P}_2 \\ \vdots \\ \mathbf{P}_n - \mathbf{P}_{n-1} \end{pmatrix}. \quad (4.85)$$

As with the cubic spline, there are more unknowns than equations (n unknowns and $n - 1$ equations), and the standard technique is to ask the user to provide a value for one of the unknown tangent vectors, normally $\mathbf{P}_1^t$.

Example: We use the four points as in the example of Section 4.8.1, namely $\mathbf{P}_1 = (0,0)$, $\mathbf{P}_2 = (1,0)$, $\mathbf{P}_3 = (1,1)$, and $\mathbf{P}_4 = (0,1)$. We also select the same start tangent $\mathbf{P}_1^t = (1,-1)$. Equation (4.85) becomes

$$\begin{pmatrix} 1 & 1 & 0 & 0 \\ 0 & 1 & 1 & 0 \\ 0 & 0 & 1 & 1 \end{pmatrix} \begin{pmatrix} \mathbf{P}_1^t \\ \mathbf{P}_2^t \\ \mathbf{P}_3^t \\ \mathbf{P}_4^t \end{pmatrix} = 2 \begin{pmatrix} \mathbf{P}_2 - \mathbf{P}_1 \\ \mathbf{P}_3 - \mathbf{P}_2 \\ \mathbf{P}_4 - \mathbf{P}_3 \end{pmatrix} = \begin{pmatrix} (2,0) \\ (0,2) \\ (-2,0) \end{pmatrix},$$

with solutions $\mathbf{P}_2^t = (1,1)$, $\mathbf{P}_3^t = (-1,1)$, and $\mathbf{P}_4^t = (-1,-1)$. The three spline segments are thus

$$\begin{aligned}\mathbf{P}_1(t) &= (\mathbf{P}_2 - \mathbf{P}_1 - \mathbf{P}_1^t)t^2 + \mathbf{P}_1^t t + \mathbf{P}_1 = (t, t^2 - t),\\ \mathbf{P}_2(t) &= (\mathbf{P}_3 - \mathbf{P}_2 - \mathbf{P}_2^t)t^2 + \mathbf{P}_2^t t + \mathbf{P}_2 = (-t^2 + t + 1, t),\\ \mathbf{P}_3(t) &= (\mathbf{P}_4 - \mathbf{P}_3 - \mathbf{P}_3^t)t^2 + \mathbf{P}_3^t t + \mathbf{P}_3 = (-t + 1, -t^2 + t + 1).\end{aligned}$$

Their tangent vectors are $\mathbf{P}_1^t(t) = (1, 2t-1)$, $\mathbf{P}_2^t(t) = (-2t+1, 1)$, and $\mathbf{P}_3^t(t) = (-1, -2t+1)$. It is easy to see that $\mathbf{P}_1^t(1) = \mathbf{P}_2^t(0) = (1,1)$ and $\mathbf{P}_2^t(1) = \mathbf{P}_3^t(0) = (-1, 1)$. Also, the end tangent of the entire curve is $\mathbf{P}_3^t(1) = (-1, -1)$, the same as for the cubic case.

4.10 Cardinal Splines

This is another application of Hermite blending. Cardinal splines overcome the main disadvantages of cubic splines, namely the lack of local control and the need to solve a system of linear equations that may be large (its size depends on the number of data points). Cardinal splines also include an additional parameter T that controls the *tension* of the curve. The price for all this is the loss of second-order continuity. Strictly speaking, this loss means that the cardinal spline isn't really a spline (see the definition of splines on page 226), but its definition, its mathematics, and its behavior are so similar to those of other splines that the name *cardinal spline* has stuck.

Perhaps the best way to visualize a spline under tension is to think of the data points as nails driven into a board and of the spline as a rubber band strung above or below the nails. To add tension, simply pull the rubber band, which tightens it, bringing each segment closer to a straight line.

The cardinal spline for n given points is calculated and drawn in overlapping segments, each depending on four points only. Each point participates in only four curve segments, so moving one point affects only those segments and not the entire curve. The curve therefore features *local control*. The individual segments connect smoothly; their first derivatives are equal at the connection points (the curve thus has first-order continuity). However, the second derivatives of the segments are generally different at the connection points.

We first organize the points in overlapping groups of four consecutive points each. The groups are

$[\mathbf{P}_1, \mathbf{P}_2, \mathbf{P}_3, \mathbf{P}_4]$, $[\mathbf{P}_2, \mathbf{P}_3, \mathbf{P}_4, \mathbf{P}_5]$, $[\mathbf{P}_3, \mathbf{P}_4, \mathbf{P}_5, \mathbf{P}_6]$, ..., $[\mathbf{P}_{n-3}, \mathbf{P}_{n-2}, \mathbf{P}_{n-1}, \mathbf{P}_n]$.

We then use Hermite interpolation (i.e., two points and two tangent vectors) to calculate a curve segment $\mathbf{P}(t)$ for each group. Denoting the four points of a group by $\mathbf{P}_1$, $\mathbf{P}_2$, $\mathbf{P}_3$, and $\mathbf{P}_4$, we select the two interior points $\mathbf{P}_2$ and $\mathbf{P}_3$ as the start and end points of the segment and define the two necessary tangent vectors as $s(\mathbf{P}_3 - \mathbf{P}_1)$ and $s(\mathbf{P}_4 - \mathbf{P}_2)$, where s is a real number. Thus, segment $\mathbf{P}(t)$ goes from $\mathbf{P}_2$ to $\mathbf{P}_3$ and its two extreme tangent vectors are proportional to the vectors $\mathbf{P}_3 - \mathbf{P}_1$ and $\mathbf{P}_4 - \mathbf{P}_2$ (Figure 4.28a). The proportionality constant s is related to the tension parameter T (see below).

4 Curves

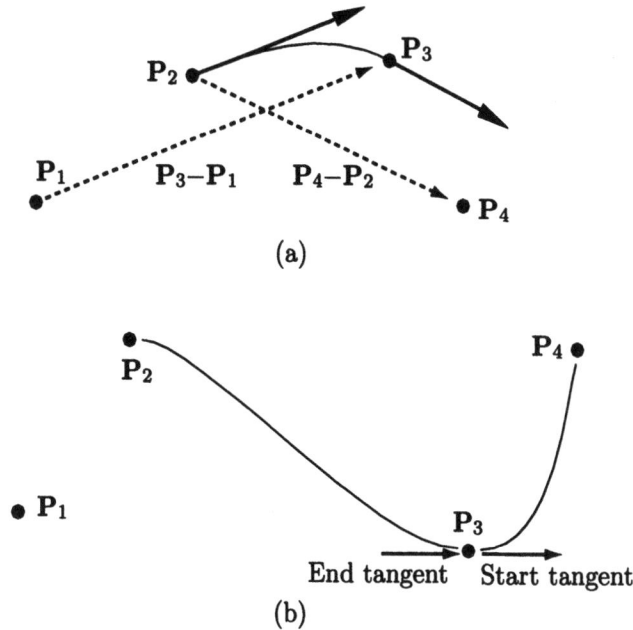

Figure 4.28: Tangent Vectors in a Cardinal Spline.

This choice of the tangent vectors guarantees that the individual segments of the cardinal spline will connect smoothly. Figure 4.28b shows that the end tangent of the segment for group $[\mathbf{P}_1, \mathbf{P}_2, \mathbf{P}_3, \mathbf{P}_4]$ is identical to the start tangent of the next group, $[\mathbf{P}_2, \mathbf{P}_3, \mathbf{P}_4, \mathbf{P}_5]$.

Segment $\mathbf{P}(t)$ is therefore defined by

$$\mathbf{P}(0) = \mathbf{P}_2, \quad \mathbf{P}(1) = \mathbf{P}_3,$$
$$\mathbf{P}^t(0) = s(\mathbf{P}_3 - \mathbf{P}_1), \quad \mathbf{P}^t(1) = s(\mathbf{P}_4 - \mathbf{P}_2) \tag{4.86}$$

and is easily calculated by applying Hermite interpolation (Equation (4.26)) to the four quantities of Equation (4.86)

$$\begin{aligned}\mathbf{P}(t) &= (t^3, t^2, t, 1)\begin{pmatrix} 2 & -2 & 1 & 1 \\ -3 & 3 & -2 & -1 \\ 0 & 0 & 1 & 0 \\ 1 & 0 & 0 & 0 \end{pmatrix}\begin{pmatrix} \mathbf{P}_2 \\ \mathbf{P}_3 \\ s(\mathbf{P}_3 - \mathbf{P}_1) \\ s(\mathbf{P}_4 - \mathbf{P}_2) \end{pmatrix} \\ &= (t^3, t^2, t, 1)\begin{pmatrix} -s & 2-s & s-2 & s \\ 2s & s-3 & 3-2s & -s \\ -s & 0 & s & 0 \\ 0 & 1 & 0 & 0 \end{pmatrix}\begin{pmatrix} \mathbf{P}_1 \\ \mathbf{P}_2 \\ \mathbf{P}_3 \\ \mathbf{P}_4 \end{pmatrix}.\end{aligned} \tag{4.87}$$

Tension in the cardinal spline can now be controlled by changing the lengths of the tangent vectors by means of the parameter s. A long tangent vector (obtained by a large s) causes the curve to continue longer in the direction of the tangent. A short tangent has the opposite effect; the curve moves a short distance in the direction of the tangent, then quickly changes direction and moves toward the end point. A zero-length tangent (corresponding to $s = 0$) produces a straight line between the endpoints (infinite tension). In principle, the parameter s can be varied from 0 to ∞. In practice, we use only values in the range $[0, 1]$. However, since $s = 0$ produces maximum tension, we cannot intuitively think of s as the tension parameter and we need to define another parameter, T.

The tension parameter T is defined as $s = (1-T)/2$, which implies $T = 1 - 2s$. The value $T = 0$ results in $s = 1/2$. The curve, in this case, is defined as having tension zero and is called the *Catmull-Rom spline* (Section 4.11). Increasing T from 0 to 1 decreases s from 1/2 to 0, thereby reducing the magnitude of the tangent vectors down to 0. This produces curves with more tension. Exercise 4.37 tells us that when the tangent vectors have magnitude zero, the Hermite curve segment is a straight line, so the entire cardinal spline curve becomes a polyline, the curve with maximum tension. Decreasing T from 0 to -1 increases s from 1/2 to 1. The result is a curve with more slack at the data points.

To illustrate this behavior mathematically, we rewrite Equation (4.87) to show its dependence on s:

$$\mathbf{P}(t) = s(-t^3 + 2t^2 - t)\mathbf{P}_1 + s(-t^3 + t^2)\mathbf{P}_2 + (2t^3 - 3t^2 + 1)\mathbf{P}_2 \\ + s(t^3 - 2t^2 + t)\mathbf{P}_3 + (-2t^3 + 3t^2)\mathbf{P}_3 + s(t^3 - t^2)\mathbf{P}_4. \quad (4.88)$$

For $s = 0$, Equation (4.88) becomes $(2t^3 - 3t^2 + 1)\mathbf{P}_2 + (-2t^3 + 3t^2)\mathbf{P}_3$, which can be simplified to $(3t^2 - 2t^3)(\mathbf{P}_3 - \mathbf{P}_2) + \mathbf{P}_2$. Substituting $u = 3t^2 - 2t^3$ reduces this to $u(\mathbf{P}_3 - \mathbf{P}_2) + \mathbf{P}_2$, which is the straight line from $\mathbf{P}_2$ to $\mathbf{P}_3$.

For large s, we use Equation (4.88) to calculate the mid-curve value $\mathbf{P}(0.5)$:

$$\mathbf{P}(0.5) = \frac{s}{8}[(\mathbf{P}_3 - \mathbf{P}_1) + (\mathbf{P}_2 - \mathbf{P}_4)] + 0.5(\mathbf{P}_2 + \mathbf{P}_3) \\ = \frac{s}{8}\left[\mathbf{P}^t(0) - \mathbf{P}^t(1)\right] + 0.5(\mathbf{P}_2 + \mathbf{P}_3).$$

This, of course, is identical to Equation (Ans.10). The left term is the difference of the two tangent vectors, multiplied by $s/8$. As s grows, this term grows without limit. The right term is the midpoint between $\mathbf{P}_2$ and $\mathbf{P}_3$. Adding the two terms produces a point that's far away (for large s) from the midpoint, showing that the curve moves a long distance away from the start point $\mathbf{P}_2$ before changing direction and starting toward the end point $\mathbf{P}_3$. Large values of s therefore feature low tension.

The tension of the curve can thus be increased by setting s close to 0 (or, equivalently, setting T close to 1); it can be decreased by increasing s (or, equivalently, decreasing T toward 0).

▸ **Exercise 4.56:** What happens when $T > 1$?

4 Curves

For $T = 0$, we get $s = 0.5$ and Equation (4.87) reduces to

$$\mathbf{P}(t) = (t^3, t^2, t, 1) \begin{pmatrix} -0.5 & 1.5 & -1.5 & 0.5 \\ 1 & -2.5 & 2 & -0.5 \\ -0.5 & 0 & 0.5 & 0 \\ 0 & 1 & 0 & 0 \end{pmatrix} \begin{pmatrix} \mathbf{P}_1 \\ \mathbf{P}_2 \\ \mathbf{P}_3 \\ \mathbf{P}_4 \end{pmatrix}. \quad (4.89)$$

This is the Catmull-Rom spline discussed in Section 4.11.

Example: Given the four points $(1,0)$, $(3,1)$, $(6,2)$, and $(2,3)$, we use Equation (4.87) to calculate the cardinal spline segment from $(3,1)$ to $(6,2)$:

$$\mathbf{P}(t) = (t^3, t^2, t, 1) \begin{pmatrix} -s & 2-s & s-2 & s \\ 2s & s-3 & 3-2s & -s \\ -s & 0 & s & 0 \\ 0 & 1 & 0 & 0 \end{pmatrix} \begin{pmatrix} (1,0) \\ (3,1) \\ (6,2) \\ (2,3) \end{pmatrix}$$
$$= t^3(4s-6, 4s-2) + t^2(-9s+9, -6s+3) + t(5s, 2s) + (3,1).$$

For high tension (i.e., $T = 1$ or $s = 0$), this reduces to the straight line

$$\mathbf{P}(t) = (-6, -2)t^3 + (9, 3)t^2 + (3, 1) = (3, 1)(-2t^3 + 3t^2) + (3, 1) = (3, 1)u + (3, 1).$$

For $T = 0$ (or $s = 1/2$), this cardinal spline reduces to the Catmull-Rom curve

$$\mathbf{P}(t) = (-4, 0)t^3 + (4.5, 0)t^2 + (2.5, 1)t + (3, 1). \quad (4.90).$$

4.11 Parabolic Blending: Catmull-Rom Curves

The Catmull-Rom curve (or the Catmull-Rom spline) is the special case of a cardinal spline with tension $T = 0$. In this section, we describe an approach to the Catmull-Rom spline where each spline segment is derived as the blend of two parabolas.

This approach to the Catmull-Rom curve proceeds in the following steps:

1. Organize the points in overlapping groups of three consecutive points each. The groups are

$$[\mathbf{P}_1, \mathbf{P}_2, \mathbf{P}_3], \quad [\mathbf{P}_2, \mathbf{P}_3, \mathbf{P}_4], \quad [\mathbf{P}_3, \mathbf{P}_4, \mathbf{P}_5], \quad \cdots \quad [\mathbf{P}_{n-2}, \mathbf{P}_{n-1}, \mathbf{P}_n].$$

2. Fit two parabolas, one through the first three points, $\mathbf{P}_1$, $\mathbf{P}_2$, and $\mathbf{P}_3$, and the other through the overlapping group, $\mathbf{P}_2$, $\mathbf{P}_3$, and $\mathbf{P}_4$.

3. Calculate the first curve segment from $\mathbf{P}_2$ to $\mathbf{P}_3$ as a linear blend of the two parabolas, using the two barycentric weights $1 - t$ and t.

4. Fit a third parabola, through points $\mathbf{P}_3$, $\mathbf{P}_4$, and $\mathbf{P}_5$ and calculate the second curve segment, from $\mathbf{P}_3$ to $\mathbf{P}_4$, as a linear blend of the second and third parabolas.

5. Repeat until the last segment, from $\mathbf{P}_{n-2}$ to $\mathbf{P}_{n-1}$, is calculated as a linear blend of the $(n-3)$rd and the $(n-2)$nd parabolas.

Parabola and Hyperbola

The parabola is defined by means of a fixed point, **F**, called the focus, and a line, called the directrix. The definition is: The parabola is the locus of all points **P** whose distance from the focus equals their distance from the directrix. In the special (canonical) case where $\mathbf{F} = (a, 0)$ and the directrix is the vertical line $x = -a$, the distances are $\sqrt{(x-a)^2 + y^2}$ and $(x+a)$. Equating them gives the implicit representation of the parabola $y^2 = 2ax$. The parametric representation is $(at^2, 2at)$, where $-\infty \le t \le \infty$ (see also Exercises 4.62 and 4.39).

Note that the parabola extends to infinity, so any practical implementation has to be limited to just part of it.

The hyperbola is defined as the locus of all points for which the difference of the distances from two fixed points (the foci) is constant. For the canonical hyperbola, the foci are at $(-c, 0)$ and $(c, 0)$. Its implicit representation is $(x/a)^2 - (y/b)^2 = 1$ and its parametric representation is $(a\sec(t), b\tan(t))$. One wing of the hyperbola is described when t varies in $[-\pi/2, \pi/2]$ and the other when t is in the range $[\pi/2, 3\pi/2]$. Also, $c^2 = a^2 + b^2$. Contrast this with the ellipse!

The main part of a reflecting telescope is a mirror. The rays of light coming from a star can be considered parallel for all practical purposes since stars are so distant. It turns out that in order to reflect these parallel rays into a sharp focus, the mirror has to be a parabola. The proof is straightforward and requires no mathematics.

Parallel rays that have left the star together arrive at points A, B, C, and D on line E simultaneously. Without the mirror, they would also have arrived at points A″, B″, C″, and D″ simultaneously but, because of the mirror, they arrive at points A′, B′, C′, and D′ and get reflected. In order for them to arrive simultaneously at one point (i.e., to meet at the focus), the mirror has to satisfy A′A″=A′F (and similar conditions for all other points). Adding distance AA′ to both sides of this equation yields AA′+A′F=AA′+A′A″, but distance AA′+A′A″ equals AA″, which is the same as BB″, CC″, etc. We therefore get AA′+A′F=constant. The sum of the distances of point A′ from line E and from point F is constant and that is the definition of a parabola, so point A′ must lie on a parabola.

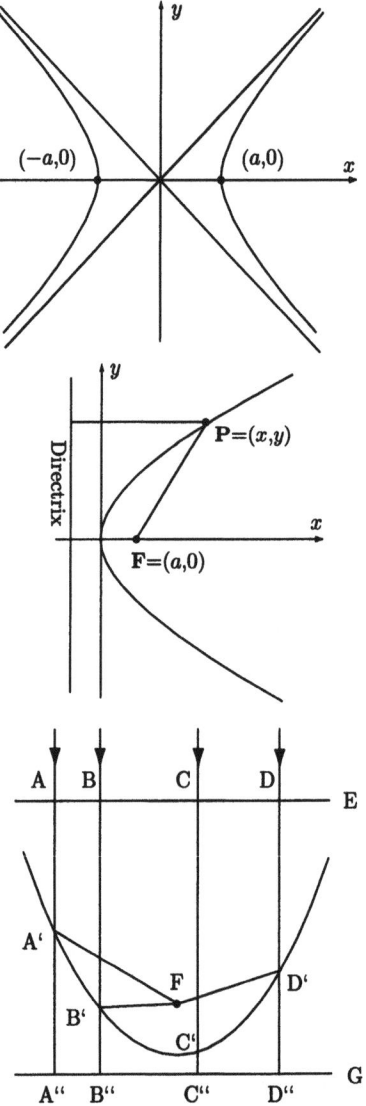

4 Curves

Each parabola is defined by three points (which, of course, are on the same plane) and is thus flat. However, the two parabolas that make up the segment are not generally on the same plane, so their blend is not necessarily flat and can twist in space.

The two original parabolas are denoted by $\mathbf{Q}(u) = (u^2, u, 1)\mathbf{H}_{123}$ and $\mathbf{R}(w) = (w^2, w, 1)\mathbf{H}_{234}$, where $\mathbf{H}_{123}$ and $\mathbf{H}_{234}$ are column vectors, each depending on the three points involved. They will have to be calculated. The expression for the blended segment is $\mathbf{P}(t) = (1-t)\mathbf{Q}(u) + t\mathbf{R}(w)$. Since this expression depends on t only, we have to express parameters u and w in terms of t. We try the linear expressions $u = at + b$, $w = ct + d$.

To calculate a, b, c, and d, we write the end conditions for the two parabolas and for the curve segment (Figure 4.29a):

$$\mathbf{Q}(0) = \mathbf{P}_1, \quad \mathbf{Q}(0.5) = \mathbf{P}_2, \quad \mathbf{Q}(1) = \mathbf{P}_3,$$
$$\mathbf{R}(0) = \mathbf{P}_2, \quad \mathbf{R}(0.5) = \mathbf{P}_3, \quad \mathbf{R}(1) = \mathbf{P}_4,$$
$$\mathbf{P}(0) = \mathbf{P}_2, \quad\quad\quad\quad\quad\quad\quad \mathbf{P}(1) = \mathbf{P}_3.$$

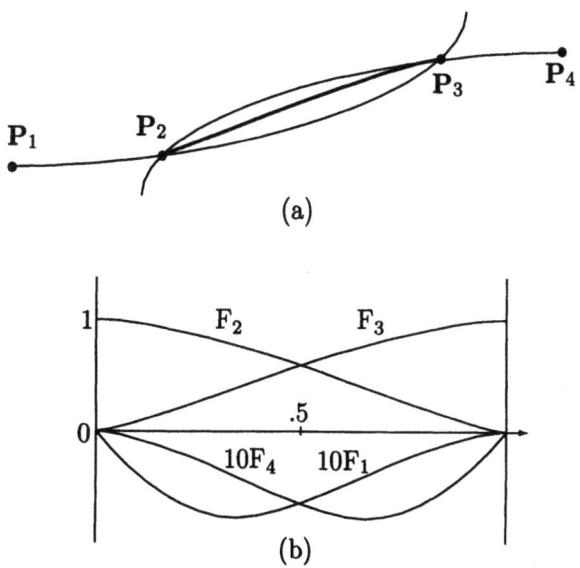

Figure 4.29: Parabolic Blending: (a) Two Parabolas. (b) The Blend Functions.

For point $\mathbf{P}_2$, we get (1) $u = 0.5$ and $t = 0$, implying $b = 0.5$, and (2) $w = 0$ and $t = 0$, implying $d = 0$. For point $\mathbf{P}_3$, we similarly get (1) $u = 1$ and $t = 1$, implying $a + b = 1 \Rightarrow a = 0.5$, and (2) $w = 0.5$ and $t = 1$, implying $c = 0.5$. This results in $u = (1+t)/2$ and $w = t/2$.

4.11 Parabolic Blending: Catmull-Rom Curves

$\mathbf{Q}(\alpha) = \mathbf{P}_2$ and $\mathbf{R}(\beta) = \mathbf{P}_3$ (where $0 \leq \alpha, \beta \leq 1$ and their values depend on the placement of the points) and derive the expression for the curve from there.

Here is a summary of the results: The three parameters are now related by $u = (1-\alpha)t + \alpha$ and $w = \beta t$. The two parabolas are given by

$$\begin{pmatrix} \mathbf{P}_1 \\ \mathbf{P}_2 \\ \mathbf{P}_3 \end{pmatrix} = \begin{pmatrix} 0 & 0 & 1 \\ \alpha^2 & \alpha & 1 \\ 1 & 1 & 1 \end{pmatrix} \mathbf{H}_{123},$$

implying

$$\mathbf{H}_{123} = \mathbf{M}^{-1} \begin{pmatrix} \mathbf{P}_1 \\ \mathbf{P}_2 \\ \mathbf{P}_3 \end{pmatrix} = \begin{pmatrix} \frac{1}{\alpha} & \frac{-1}{\alpha(1-\alpha)} & \frac{1}{1-\alpha} \\ \frac{-(1+\alpha)}{\alpha} & \frac{1}{\alpha(1-\alpha)} & \frac{-\alpha}{1-\alpha} \\ 1 & 0 & 0 \end{pmatrix} \begin{pmatrix} \mathbf{P}_1 \\ \mathbf{P}_2 \\ \mathbf{P}_3 \end{pmatrix},$$

and

$$\begin{pmatrix} \mathbf{P}_2 \\ \mathbf{P}_3 \\ \mathbf{P}_4 \end{pmatrix} = \begin{pmatrix} 0 & 0 & 1 \\ \beta^2 & \beta & 1 \\ 1 & 1 & 1 \end{pmatrix} \mathbf{H}_{234},$$

implying

$$\mathbf{H}_{234} = \mathbf{M}^{-1} \begin{pmatrix} \mathbf{P}_2 \\ \mathbf{P}_3 \\ \mathbf{P}_4 \end{pmatrix} = \begin{pmatrix} \frac{1}{\beta} & \frac{-1}{\beta(1-\beta)} & \frac{1}{1-\beta} \\ \frac{-(1+\beta)}{\beta} & \frac{1}{\beta(1-\beta)} & \frac{-\beta}{1-\beta} \\ 1 & 0 & 0 \end{pmatrix} \begin{pmatrix} \mathbf{P}_2 \\ \mathbf{P}_3 \\ \mathbf{P}_4 \end{pmatrix}.$$

The final expression of the curve is

$$\mathbf{P}(t) = (t^3, t^2, t, 1) \begin{pmatrix} \frac{-(1-\alpha)^2}{\alpha} & \frac{(1-\alpha)+\alpha\beta}{\alpha} & \frac{-(1-\alpha)-\alpha\beta}{1-\beta} & \frac{\beta^2}{1-\beta} \\ \frac{2(1-\alpha)^2}{\alpha} & \frac{-2(1-\alpha)-\alpha\beta}{\alpha} & \frac{2(1-\alpha)-\beta(1-2\alpha)}{1-\beta} & \frac{-\beta^2}{1-\beta} \\ \frac{-(1-\alpha)^2}{\alpha} & \frac{(1-2\alpha)}{\alpha} & \alpha & 0 \\ 0 & 1 & 0 & 0 \end{pmatrix} \begin{pmatrix} \mathbf{P}_1 \\ \mathbf{P}_2 \\ \mathbf{P}_3 \\ \mathbf{P}_4 \end{pmatrix}.$$

4.11.2 Bessel's Algorithm

The cardinal spline and the Catmull-Rom curve are based on the particular way the two extreme tangent vectors of each four-point segment are defined. Equation (4.86) defines $\mathbf{P}^t(0) = s(\mathbf{P}_3 - \mathbf{P}_1)$ and $\mathbf{P}^t(1) = s(\mathbf{P}_4 - \mathbf{P}_2)$. So far, these definitions, which seem arbitrary, have been used because they make sense. They can, however, be explained (or justified) by a simple method called *Bessel's algorithm*. The idea is to calculate a quadratic interpolating polynomial $\mathbf{Q}_s(t)$ for the first three points $\mathbf{P}_0$, $\mathbf{P}_1$, and $\mathbf{P}_2$ and define $\mathbf{P}^t(0)$ as the tangent vector of $\mathbf{Q}_s(t)$ at point $\mathbf{P}_1$ (Figure 4.30). Similarly, another quadratic interpolating polynomial $\mathbf{Q}_e(t)$ is calculated for the last

4 Curves

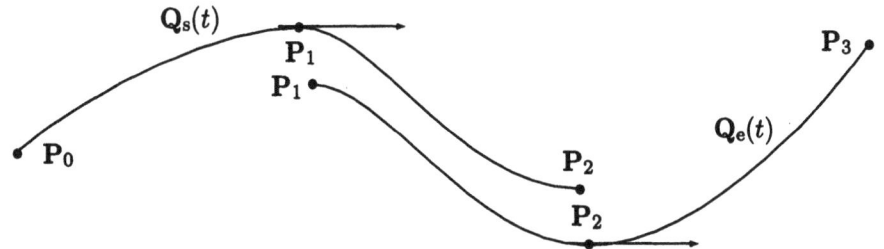

Figure 4.30: Bessel's Algorithm.

three points $\mathbf{P}_1$, $\mathbf{P}_2$, and $\mathbf{P}_3$, and $\mathbf{P}^t(1)$ is defined as the tangent vector of $\mathbf{Q}_e(t)$ at point $\mathbf{P}_2$.

Friedrich Wilhelm Bessel (1784–1846): German astronomer and mathematician. Best known for making the first accurate measurement of the distance to a star

The uniform quadratic Lagrange polynomial (Equation (4.14)) is used as our interpolating polynomial:

$$\mathbf{Q}_s(t) = \frac{t^2 - 3t + 2}{2}\mathbf{P}_0 - (t^2 - 2t)\mathbf{P}_1 + \frac{t^2 - t}{2}\mathbf{P}_2$$

$$= (t^2, t, 1) \begin{pmatrix} 1/2 & -1 & 1/2 \\ -3/2 & 2 & -1/2 \\ 1 & 0 & 0 \end{pmatrix} \begin{pmatrix} \mathbf{P}_0 \\ \mathbf{P}_1 \\ \mathbf{P}_2 \end{pmatrix}.$$

The parameter t varies in the range $[0, 2]$, so $\mathbf{Q}_s(1)$ gives the middle point. The tangent vector of $\mathbf{Q}_s(t)$ is

$$\mathbf{Q}_s^t(t) = \frac{2t - 3}{2}\mathbf{P}_0 - (2t - 2)\mathbf{P}_1 + \frac{2t - 1}{2}\mathbf{P}_2$$

$$= (2t, 1, 0) \begin{pmatrix} 1/2 & -1 & 1/2 \\ -3/2 & 2 & -1/2 \\ 1 & 0 & 0 \end{pmatrix} \begin{pmatrix} \mathbf{P}_0 \\ \mathbf{P}_1 \\ \mathbf{P}_2 \end{pmatrix}.$$

Thus, $\mathbf{Q}_s^t(1) = (\mathbf{P}_2 - \mathbf{P}_0)/2$. Similarly,

$$\mathbf{Q}_e(t) = \frac{t^2 - 3t + 2}{2}\mathbf{P}_1 - (t^2 - 2t)\mathbf{P}_2 + \frac{t^2 - t}{2}\mathbf{P}_3$$

$$= (t^2, t, 1) \begin{pmatrix} 1/2 & -1 & 1/2 \\ -3/2 & 2 & -1/2 \\ 1 & 0 & 0 \end{pmatrix} \begin{pmatrix} \mathbf{P}_1 \\ \mathbf{P}_2 \\ \mathbf{P}_3 \end{pmatrix},$$

which yields $\mathbf{Q}_e^t(1) = (\mathbf{P}_3 - \mathbf{P}_1)/2$.

It is also possible to use the nonuniform quadratic Lagrange polynomial (Equation (4.15)). If we select

$$\mathbf{Q}_s(t) = (t^2, t, 1) \begin{pmatrix} \frac{1}{\Delta_0(\Delta_0 + \Delta_1)} & -\frac{1}{\Delta_0 \Delta_1} & \frac{1}{(\Delta_0 + \Delta_1)\Delta_1} \\ \frac{-1}{\Delta_0 + \Delta_1} - \frac{1}{\Delta_0} & \frac{1}{\Delta_0} + \frac{1}{\Delta_1} & -\frac{1}{\Delta_1} + \frac{1}{\Delta_0 + \Delta_1} \\ 1 & 0 & 0 \end{pmatrix} \begin{pmatrix} \mathbf{P}_0 \\ \mathbf{P}_1 \\ \mathbf{P}_2 \end{pmatrix},$$
(4.93)

then the tangent vector at point $\mathbf{P}_1$ becomes

$$\begin{aligned}\mathbf{Q}_s^t(\Delta_0) &= -\frac{\Delta_1}{\Delta_0(\Delta_0 + \Delta_1)}\mathbf{P}_0 + \frac{\Delta_1 - \Delta_0}{\Delta_0 \Delta_1}\mathbf{P}_1 + \frac{\Delta_0}{(\Delta_0 + \Delta_1)\Delta_1}\mathbf{P}_2 \\ &= \left(\frac{\Delta_1}{\Delta_0 + \Delta_1}\right)\left(\frac{\mathbf{P}_1 - \mathbf{P}_0}{\Delta_0}\right) + \left(\frac{\Delta_0}{\Delta_0 + \Delta_1}\right)\left(\frac{\mathbf{P}_2 - \mathbf{P}_1}{\Delta_1}\right).\end{aligned}$$
(4.94)

It is easy to see that Equation (4.94) reduces to $(\mathbf{P}_2 - \mathbf{P}_0)/2$ when $\Delta_0 = \Delta_1 = 1$.

▶ **Exercise 4.58:** Use Equation (4.15) to represent $\mathbf{Q}_e(t)$ and calculate the tangent vector $\mathbf{Q}_e^t(\Delta_1)$.

4.12 Kochanek-Bartels Splines

This is an extension of the cardinal spline where, in addition to the tension parameter T, two new parameters, c and b, are introduced to control the *continuity* and *bias* of the curve. The main reference is [Kochanek and Bartels 84].

In a cardinal spline, tension is controlled by changing the magnitudes of the tangent vectors. However, the two tangent vectors at a data point—the one of the "arriving" segment, and the one of the "departing" segment—are identical. In a Kochanek-Bartels spline, in contrast, the two tangents at each data point are different and it is this difference that allows control of the *continuity* and *bias*.

We assume that $\mathbf{P}_k$ is a data point where two spline segments $\mathbf{P}_{k-1}(t)$ and $\mathbf{P}_k(t)$ meet (Figure 4.31f). Point $\mathbf{P}_k$ is the last endpoint of segment $\mathbf{P}_{k-1}(t)$, and the first endpoint of segment $\mathbf{P}_k(t)$. We denote the two tangent vectors $\mathbf{P}_a \stackrel{\text{def}}{=} \mathbf{P}_{k-1}^t(1)$ and $\mathbf{P}_d \stackrel{\text{def}}{=} \mathbf{P}_k^t(0)$ (a stands for "arriving" and d stands for "departing"). In a cardinal spline the two tangents $\mathbf{P}_a$ and $\mathbf{P}_d$ are identical and are proportional to the vector $\mathbf{P}_{k+1} - \mathbf{P}_{k-1}$ (the chord surrounding $\mathbf{P}_k$). This guarantees a smooth connection of the two segments.

Continuity: This refers to a situation where the two segments do not connect smoothly at point $\mathbf{P}_k$. Such a situation is sometimes useful in computer animation, where the curve defines an animation path that should not be completely smooth, but should feature jumps and jerks. This effect is achieved by making $\mathbf{P}_a$ and $\mathbf{P}_d$ different. They are, somewhat arbitrarily, defined as

$$\mathbf{P}_a = \frac{1-c}{2}(\mathbf{P}_k - \mathbf{P}_{k-1}) + \frac{1+c}{2}(\mathbf{P}_{k+1} - \mathbf{P}_k),$$
$$\mathbf{P}_d = \frac{1+c}{2}(\mathbf{P}_k - \mathbf{P}_{k-1}) + \frac{1-c}{2}(\mathbf{P}_{k+1} - \mathbf{P}_k),$$

4 Curves

where c is the *continuity parameter*. A continuous curve (i.e., $\mathbf{P}_a = \mathbf{P}_d$) is obtained for $c = 0$. For $c \neq 0$, the two tangents are different and the curve has a sharp corner (a kink or a cusp) at point $\mathbf{P}_k$, and this corner becomes more pronounced for large values of c. The case $c = -1$ implies $\mathbf{P}_a = \mathbf{P}_k - \mathbf{P}_{k-1}$ (the "arriving" chord) and $\mathbf{P}_d = \mathbf{P}_{k+1} - \mathbf{P}_k$ (the "departing" chord). The case $c = 1$ produces tangent vectors in the opposite directions: $\mathbf{P}_a = \mathbf{P}_{k+1} - \mathbf{P}_k$ and $\mathbf{P}_d = \mathbf{P}_k - \mathbf{P}_{k-1}$.

It is interesting to note that increasing the tension or decreasing the continuity produce curves that may look similar. However, even when they have identical shapes, the curves are different since they "move" at different speeds. Imagine moving along a curve segment, varying t in equal steps. In a high-tension curve, the speed is nonuniform; the distance between consecutive pixels starts small, gets bigger, then gets smaller toward the end. In a low-continuity curve, the same path may be traversed but at a uniform speed (the distance between consecutive pixels is fixed). This fact is illustrated in Figure 4.31a,b and is important in animation, where the designer may want either uniform or accelerated motion along a path.

The reason for this behavior is the size of the tangent vectors. In a high-tension curve, the tangent vector starts with magnitude zero as the curve leaves point $\mathbf{P}_{k-1}$; its magnitude grows as the curve approaches its midpoint, then shrinks back to zero as the curve approaches point $\mathbf{P}_k$. Since the tangent vector is the first derivative, its magnitude corresponds to the "speed" of the curve. In a low-continuity curve, on the other hand, the magnitude of the tangent vector changes from $(\mathbf{P}_k - \mathbf{P}_{k-2})/2$ at point $\mathbf{P}_{k-1}$ to $(\mathbf{P}_{k+1} - \mathbf{P}_{k-1})/2$ at point $\mathbf{P}_k$. The magnitude thus depends on the four points and is normally nonzero (it is zero only if $\mathbf{P}_k = \mathbf{P}_{k-2}$ or $\mathbf{P}_{k+1} = \mathbf{P}_{k-1}$).

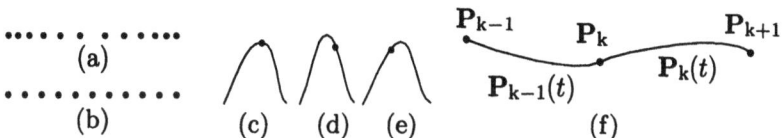

Figure 4.31: Effects of (a) High Tension; (b) Low Continuity; (c) $b = 0$; (d) $b = -1$; (e) $b = 1$; (f) Two Segments.

Bias: In a cardinal spline with zero tension, both tangent vectors at point $\mathbf{P}_k$ have the values

$$(\mathbf{P}_{k+1} - \mathbf{P}_{k-1})/2 = \big((\mathbf{P}_k - \mathbf{P}_{k-1}) + (\mathbf{P}_{k+1} - \mathbf{P}_k)\big)/2.$$

A good interpretation of this value is: The direction of the curve at point $\mathbf{P}_k$ is the average of the two chords connecting at $\mathbf{P}_k$.

Bias is an additional parameter b (widely misunderstood), being introduced to control the direction of the curve at $\mathbf{P}_k$. With bias, the arriving and departing tangents are defined (again arbitrarily) as

$$\mathbf{P}_a = \mathbf{P}_d = \frac{1+b}{2}(\mathbf{P}_k - \mathbf{P}_{k-1}) + \frac{1-b}{2}(\mathbf{P}_{k+1} - \mathbf{P}_k).$$

The vectors are identical, so the two curve segments $\mathbf{P}_a(t)$ and $\mathbf{P}_d(t)$ are still connected smoothly at $\mathbf{P}_k$. Setting $b = 1$ changes both tangents to $\mathbf{P}_k - \mathbf{P}_{k-1}$, the chord on the left of $\mathbf{P}_k$. The other extreme value, $b = -1$, changes them to the chord on the right of $\mathbf{P}_k$. Both cases are illustrated in Figure 4.31d,e, whereas Figure 4.31c shows the effect of zero bias ($b = 0$).

Bias is used in computer animation to obtain the effect of overshooting a point ($b = 1$) or undershooting it ($b = -1$).

The final result is that the tangent vector that departs point $\mathbf{P}_k$ is defined by

$$\mathbf{P}_d = \mathbf{P}_k^t(0) = \tfrac{1}{2}(1-T)(1+b)(1-c)(\mathbf{P}_k - \mathbf{P}_{k-1}) + \tfrac{1}{2}(1-T)(1-b)(1+c)(\mathbf{P}_{k+1} - \mathbf{P}_k).$$

Similarly, the tangent vector arriving at point $\mathbf{P}_{k+1}$ is defined by

$$\mathbf{P}_a = \mathbf{P}_k^t(1) = \tfrac{1}{2}(1-T)(1+b)(1+c)(\mathbf{P}_{k+1} - \mathbf{P}_k) + \tfrac{1}{2}(1-T)(1-b)(1-c)(\mathbf{P}_{k+2} - \mathbf{P}_{k+1}),$$

and the entire Kochanek-Bartels curve segment from $\mathbf{P}_k$ to $\mathbf{P}_{k+1}$ is constructed by

$$\mathbf{P}(t) = (t^3, t^2, t, 1)\mathbf{H} \begin{pmatrix} \mathbf{P}_k \\ \mathbf{P}_{k+1} \\ \mathbf{P}_d \\ \mathbf{P}_a \end{pmatrix},$$

where $\mathbf{H}$ is the Hermite matrix of Equation (4.26).

Note that the second derivatives are generally not continuous at the data points.

Example: The five points $\mathbf{P}_1 = (1,0)$, $\mathbf{P}_2 = (3,1)$, $\mathbf{P}_3 = (4,2)$, $\mathbf{P}_4 = (5,1)$, and $\mathbf{P}_5 = (7,0)$ are given. We calculate the two Kochanek-Bartels segments $\mathbf{P}_1(t)$ and $\mathbf{P}_2(t)$ that go from $\mathbf{P}_2$ to $\mathbf{P}_3$ and from $\mathbf{P}_3$ to $\mathbf{P}_4$, respectively. The three parameters are set to $T = -0.3$, $c = 0$, and $b = 1$ to illustrate overshooting.

For segment $\mathbf{P}_1(t)$, the tangent vector leaving the start point $\mathbf{P}_2$ is given by

$$\begin{aligned}\mathbf{P}_1^t(0) &= \tfrac{1}{2}(1-T)(1+b)(1-c)(\mathbf{P}_2 - \mathbf{P}_1) + \tfrac{1}{2}(1-T)(1-b)(1+c)(\mathbf{P}_3 - \mathbf{P}_2) \\ &= \tfrac{1}{2}(1+0.3)(1+1)(1-0)[(3,1) - (1,0)] \\ &= (2.6, 1.3).\end{aligned}$$

The tangent vector arriving at $\mathbf{P}_2$ is given by

$$\begin{aligned}\mathbf{P}_1^t(1) &= \tfrac{1}{2}(1-T)(1+b)(1+c)(\mathbf{P}_3 - \mathbf{P}_2) + \tfrac{1}{2}(1-T)(1-b)(1-c)(\mathbf{P}_4 - \mathbf{P}_3) \\ &= \tfrac{1}{2}(1+0.3)(1+1)(1+0)[(4,2) - (3,1)] \\ &= (1.3, 1.3).\end{aligned}$$

The segment is thus

$$\begin{aligned}\mathbf{P}_1(t) &= (t^3, t^2, t, 1) \begin{pmatrix} 2 & -2 & 1 & 1 \\ -3 & 3 & -2 & -1 \\ 0 & 0 & 1 & 0 \\ 1 & 0 & 0 & 0 \end{pmatrix} \begin{pmatrix} (3,1) \\ (4,2) \\ (2.6, 1.3) \\ (1.3, 1.3) \end{pmatrix} \\ &= (1.9, 0.6)t^3 + (-3.5, -0.9)t^2 + (2.6, 1.3)t + (3, 1).\end{aligned}$$

4 Curves

To illustrate the overshoot effect created by setting $b = 1$, we first calculate the tangent vector of the segment

$$\mathbf{P}_1^t(t) = (5.7, 1.8)t^2 + (-7, -1.8)t + (2.6, 1.3).$$

In order to have a maximum, this vector must be equal to $(x, 0)$ for some positive x and for some t in the range $[0, 1]$. To find this value, we try to solve the equations $5.7t^2 - 7t + 2.6 = x$ and $1.8t^2 - 1.8t + 1.3 = 0$. These are quadratic equations in t, which can be written $At^2 + Bt + C$. They have solutions only if the discriminant $B^2 - 4AC$ is non-negative. However, it is easy to see that the second discriminant, $1.8^2 - 4 \times 1.8 \times 1.3$, is negative. The first segment therefore climbs from $(3, 1)$ to $(4, 2)$ without ever being horizontal.

▶ **Exercise 4.59:** Calculate the second segment and show how it overshoots $\mathbf{P}_3$.

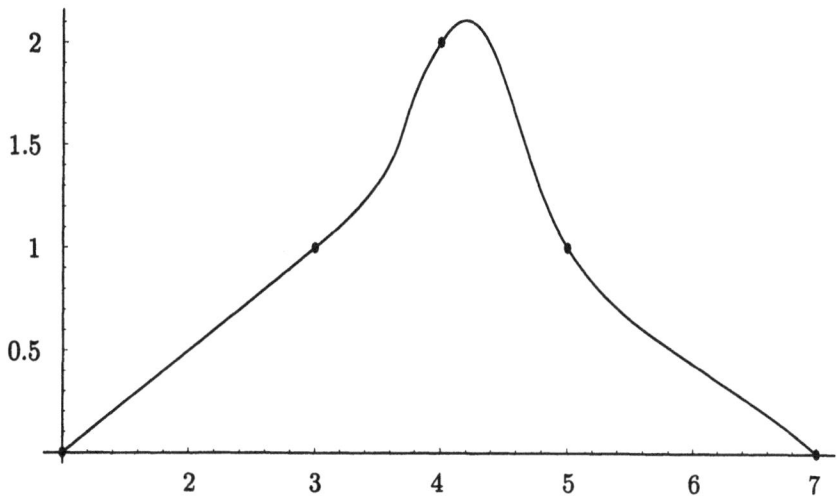

Figure 4.32: A Kochanek-Bartels Curve.

Figure 4.32 shows the entire curve. The two extreme points $\mathbf{P}_1$ and $\mathbf{P}_5$ have been duplicated, making it possible to calculate four curve segments. The diagram was created by *Mathematica* code similar to

```
T={t^3,t^2,t,1};
H={{2,-2,1,1},{-3,3,-2,-1},{0,0,1,0},{1,0,0,0}};
P={{1,0},{3,1},{4,2},{5,1},{7,0}};
Tens=-0.3; b=1; c=0;
Pd=0.5(1-Tens)(1+b)(1-c)(P[[2]]-P[[1]])
 +0.5(1-Tens)(1-b)(1+c)(P[[3]]-P[[2]]);
Pa=0.5(1-Tens)(1+b)(1+c)(P[[3]]-P[[2]])
 +0.5(1-Tens)(1-b)(1-c)(P[[4]]-P[[3]]);
ParametricPlot[T.H.{P[[2]],P[[3]],Pd,Pa},{t,0,1},PlotRange->All];
```

4.13 Fitting a PC to Experimental Points

The spline methods discussed so far use data points. The curve methods of Sections 4.14 and 4.16 use control points. In the case of data points, the curve has to pass through all of them. Control points exert a pull on the curve, so each of them pulls the curve toward itself (Section 4.3). The method described here, due to [Plass and Stone 83], uses *experimental points* ("epoints" for short). Such points are typically obtained as a result of a science experiment, but can also be input by scanning an image. Given n epoints $\mathbf{P}_1, \mathbf{P}_2,\ldots, \mathbf{P}_n$, the problem is to calculate a PC curve that will pass close to all the points but will not necessarily pass *through* them. A user-controlled tolerance parameter controls the closeness of the fit.

Since a PC is fully defined by means of just four coefficients, it cannot have a very complex shape, so it may not be able to follow a set of epoints that meander all over the place. In such a case, the curve will have to be calculated as a spline where each segment is a PC and the segments fit together, either smoothly or with corner joints. In this section, we show how to calculate one such PC, so we assume that the n epoints don't describe a complex curve. To distinguish between simple and complex curves quantitatively, we connect the n epoints with $n-1$ straight segments, resulting in an open polygon. Experience shows that the set of points is simple and will allow a PC to follow it if (1) all the angles between consecutive segments are in the range $[135°, 180°]$, (2) the curve has at most one loop, and (3) if it does not have a loop, it can have at most two inflection points.

We denote our single PC segment by

$$\begin{aligned}\mathbf{P}(t) &= \bigl(u(t), w(t)\bigr)\\ &= \mathbf{a}_3 t^3 + \mathbf{a}_2 t^2 + \mathbf{a}_1 t + \mathbf{a}_0\\ &= (a_{3x}, a_{3y})t^3 + (a_{2x}, a_{2y})t^2 + (a_{1x}, a_{1y})t + (a_{0x}, a_{0y}),\end{aligned} \qquad (4.95)$$

where the four vector quantities **a**, **b**, **c**, and **d** have to be calculated. Together, they constitute eight numbers, so we can say that a PC segment has eight degrees of freedom. To understand the method, let's imagine that we have somehow found a PC segment $\mathbf{P}(t)$ that passes close to all n epoints. We can use this PC to find the n values of the parameter t where the curve passes closest to each of the n epoints. Denoting these values by $t_1, t_2,\ldots, t_n$, we use them to label the epoints $\mathbf{P}_{t_1}, \mathbf{P}_{t_2},\ldots, \mathbf{P}_{t_n}$. Now imagine the opposite situation where we still don't know the PC segment, but we already have the epoints somehow labeled correctly. Using the coordinates of the n epoints and the n values of t, we could, in such a case, calculate a curve using the least-squares fitting technique.

The idea is to start with an initial set of estimated t values, use least squares to calculate a PC segment from this set, use this PC to calculate a better set of t values, and repeat until the curve obtained is close enough to all the epoints. Convergence is not guaranteed, but experience shows that epoints that satisfy the three conditions stated earlier normally result in a reasonably shaped curve in just a few iterations.

The initial set of estimated t values is based on the lengths of the polygon's edges. Denoting the polygon vertices (i.e., the epoints) by $\mathbf{P}_i = (x_i, y_i)$, we define

4 Curves

a quantity s_k as the sum of the polygon's edges from $\mathbf{P}_1$ to $\mathbf{P}_k$:

$$s_1 = 0, \quad s_k = \sum_{i=1}^{k-1} |\mathbf{P}_{i+1}-\mathbf{P}_i| = \sum_{i=1}^{k-1} \sqrt{(x_{i+1}-x_i)^2 + (y_{i+1}-y_i)^2}, \quad k=2,3,\ldots,n.$$

The initial value of t_k is now defined as the ratio s_k/s_n, resulting in $t_1 = 0$, $t_n = 1$, and, in general, $0 \leq t_i \leq 1$.

▸ **Exercise 4.60:** Given the eight epoints $\mathbf{P}_1 = (2,5)$, $\mathbf{P}_2 = (2,8)$, $\mathbf{P}_3 = (5,11)$, $\mathbf{P}_4 = (8,8)$, $\mathbf{P}_5 = (11,4)$, $\mathbf{P}_6 = (14,8)$, $\mathbf{P}_7 = (13,8)$, and $\mathbf{P}_8 = (11,10)$, draw them in the xy plane, draw the open polygon connecting them, indicate the "bad" polygon vertices, and calculate the quantities s_k and t_k.

Once a set of t_i values is available, a PC curve segment can be calculated by least squares. The principle is to compute values for the four coefficients $\mathbf{a}_i$ that will minimize the expression

$$S(\mathbf{a}_0,\mathbf{a}_1,\mathbf{a}_2,\mathbf{a}_3) = \sum_{j=1}^{n}(\mathbf{P}(t_j)-\mathbf{P}_j)^2 = \sum_{j=1}^{n}\left(\sum_{i=0}^{3}\mathbf{a}_i t_j^i - \mathbf{P}_j\right)^2.$$

We consider this expression a function S of the four coefficients $\mathbf{a}_i$ and minimize it by (1) writing the four partial derivatives of S,

$$\frac{\partial S(\mathbf{a}_0,\mathbf{a}_1,\mathbf{a}_2,\mathbf{a}_3)}{\partial \mathbf{a}_k} = \sum_{j=1}^{n} 2 \sum_{i=0}^{3}\left(\mathbf{a}_i t_j^i - \mathbf{P}_j\right)t_j^k, \quad 1 \leq k \leq 4,$$

(2) equating each to zero,

$$\sum_{i=0}^{3}\left(\sum_{j=1}^{n} t_j^i t_j^k\right) \mathbf{a}_i = \sum_{j=1}^{n} \mathbf{P}_j t_j^k, \quad 1 \leq k \leq 4,$$

which can also be written

$$\begin{aligned}\mathbf{a}_0(t_1^0 t_1^k + t_2^0 t_2^k + \cdots + t_n^0 t_n^k) + \mathbf{a}_1(t_1^1 t_1^k + t_2^1 t_2^k + \cdots + t_n^1 t_n^k) \\ + \mathbf{a}_2(t_1^2 t_1^k + t_2^2 t_2^k + \cdots + t_n^2 t_n^k) + \mathbf{a}_3(t_1^3 t_1^k + t_2^3 t_2^k + \cdots + t_n^3 t_n^k) \\ = \mathbf{P}_1 t_1^k + \mathbf{P}_2 t_2^k + \cdots + \mathbf{P}_n t_n^k, \quad 1 \leq k \leq 4,\end{aligned} \quad (4.96)$$

and then (3) solving the resulting system of four linear equations in the four unknowns $\mathbf{a}_i$.

Having produced values for the four coefficients $\mathbf{a}_i$, we use the resulting PC to calculate a better set of t values. For each epoint, we find the value of t that produces the point nearest it on the PC and assign that t value to the epoint. Mathematically, this amounts to finding the minimum distance between an epoint $\mathbf{P}_j = (x_j,y_j)$ and the curve $\mathbf{P}(t) = (u(t),w(t))$. Since the distance involves a square root, we use the square of the distance (a similar method is used in the Bresenham-Michener circle method, Section 2.7.3).

Our problem is, therefore, to minimize the function

$$D(t) = |\mathbf{P}(t) - \mathbf{P}_j| = \bigl(u(t) - x_j\bigr)^2 + \bigl(w(t) - y_j\bigr)^2,$$

and we do this by differentiating it with respect to t, equating the derivative to zero, and solving for t. Thus,

$$2(u(t) - x_j)u^t(t) + 2(w(t) - y_j)w^t(t) = 0. \qquad (4.97)$$

Since $u(t)$ and $w(t)$ are cubic polynomials in t, their derivatives are quadratic polynomials. The left side of Equation (4.97) is thus a degree-5 polynomial in t, so a numerical solution is required. We use the Newton-Raphson method, a general, fast, iterative method for finding roots of functions. Given a function $f(t)$, the method requires an initial value of t (a guess or an estimate) and updates this value by the iteration

$$t \leftarrow t - \frac{f(t)}{f'(t)}.$$

If the initial value is close to a root, convergence is fast but is not guaranteed. In our case, function $f(t)$ is given by Equation (4.97), and we always have an estimate for t. Our Newton-Raphson iteration thus becomes

$$t \leftarrow t - \frac{2(u(t) - x_j)u^t(t) + 2(w(t) - y_j)w^t(t)}{u^t(t)^2 + w^t(t)^2 + (u(t) - x_j)u^{tt}(t) + (w(t) - y_j)w^{tt}(t)}.$$

Experience shows that one iteration is enough to produce a new t value much better than its predecessor.

The new t values may be located outside the interval $[0, 1]$, so one last step is needed, where all n new t values are linearly scaled to bring them back into the right interval, if necessary. Here is how it's done.

If the new t_1 is positive, it should not be scaled or changed in any way. This is the algorithm's way of telling us that a better fit would be achieved if the curve did not start at the first epoint. However, if t_1 becomes negative (e.g., if $t_1 = -\alpha$), it should be incremented by α to bring it back to zero, and all the other t_i's should be incremented by quantities that get smaller with i until they reach zero for $i = n$ (i.e., t_n should not be changed). Similarly, if the new t_n is less than 1, it should not be scaled, but if it exceeds 1 (by a quantity β), it should be decremented by β and all the other t_i's should be decremented by quantities that get smaller with i until they reach zero for $i = 1$.

Once this is grasped, it is easy to guess how a general value t_i should be scaled. It should be incremented by α multiplied by some weight and decremented by β multiplied by another weight, such that the weights add up to 1. If the new t_1 is positive, α should be set to 0. Similarly, if the new t_n is less than 1, β should be set to 0. The result is

$$t_i \leftarrow t_i + \alpha \frac{n-i}{n-1} - \beta \frac{i-1}{n-1}. \qquad (4.98)$$

4 Curves

▶ **Exercise 4.61:** Given the eight new t values, -0.1, 0.1, 0.2, 0.3, 0.4, 0.6, 0.8, and 1.2, use Equation (4.98) to scale them.

These are the steps of the iteration. The loop continues until none of the t values changes significantly or, alternatively, until the maximum distance between an epoint and the curve falls below a preset threshold. If this does not happen after a certain, fixed number of iterations, the loop stops and displays an error message (curve does not converge to epoints). Figure 4.33 is an example of a spline fitting a set of 20 epoints. It is easy to see how the fit improves even after a small number of iterations.

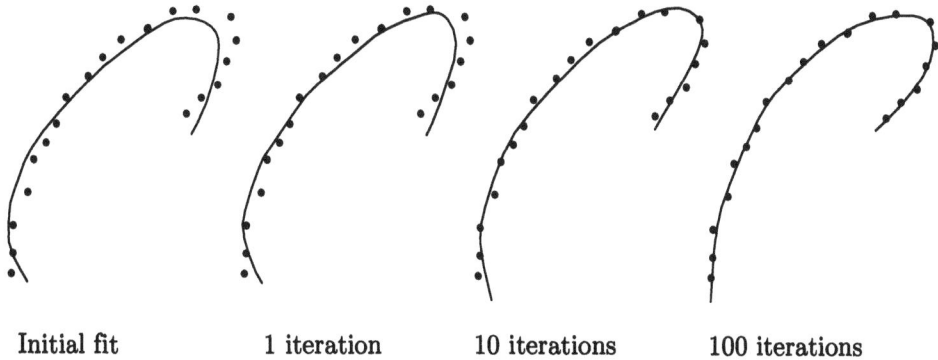

Initial fit   1 iteration   10 iterations   100 iterations

Figure 4.33: Spline Fit to Many Epoints.

We next discuss how to add constraints to the PC segment that's being calculated. When the initial t values are calculated by $t_k = s_k/s_n$, the first value, t_1, becomes zero, and the last value, t_n, is set to 1. The PC segment thus starts at the first epoint $\mathbf{P}_1$ and ends at $\mathbf{P}_n$. When the t values are updated in an iteration, both t_1 and t_n may get new values. If t_1 goes below zero, it is scaled back to zero. However, if it becomes positive (e.g., $t_1 = 0.05$), it is not changed. This means that point $\mathbf{P}(0.05)$ on the curve would be closest to $\mathbf{P}_1$. The start of the curve [point $\mathbf{P}(0)$] would, in this case, be located "before" $\mathbf{P}_1$. A similar situation may happen at the end of the curve, where $\mathbf{P}(1)$ may move "past" the last epoint $\mathbf{P}_n$.

Fitting a PC segment to a set of epoints in this way generally means that the curve may be "longer" than the set. Sometimes, we want the curve to start and end at the two extreme epoints, so we have to "constrain" it. Another aspect of constraining arises when we are given a complex set of epoints, where more than one PC segment is needed to fit all the points. In such a case, we have to consider the problem of joining individual segments. A segment may therefore have to be constrained by specifying its start and/or end tangent vectors.

The point to understand is that each added constraint reduces the quality of the fit. The reason is that a PC depends on four vector coefficients, which constitute eight scalar quantities (it has eight degrees of freedom). Adding a constraint means fixing one or more of those quantities, thereby reducing the number of degrees of freedom, and thus leading to a worse fit. The number of constraints should, therefore, be kept small (no more than one or two).

Adding constraints is done by generalizing the cubic polynomials $u(t)$ and $w(t)$. Instead of writing them in the form

$$\bigl(u(t), w(t)\bigr) = \mathbf{a}_3 t^3 + \mathbf{a}_2 t^2 + \mathbf{a}_1 t + \mathbf{a}_0,$$

we express them as

$$\bigl(u(t), w(t)\bigr) = \mathbf{a}_1 F_1(t) + \mathbf{a}_2 F_2(t) + \mathbf{a}_3 F_3(t) + \mathbf{a}_4 F_4(t),$$

where the $F_i(t)$ are any four linearly independent cubic polynomials. Both $u(t)$ and $w(t)$ remain cubic polynomials, but certain choices of the $F_i(t)$'s may make it easy to constrain the endpoints or the extreme tangents of the PC segment.

One such choice is the set of four *Hermite blending functions* of Equation (4.25), duplicated here:

$$\begin{aligned}F_1(t) &= 2t^3 - 3t^2 + 1, & F_2(t) &= -2t^3 + 3t^2, \\ F_3(t) &= t^3 - 2t^2 + t, & F_4(t) &= t^3 - t^2.\end{aligned} \quad (4.25)$$

We know from Equation (4.24) that if a PC segment $\mathbf{P}(t)$ is expressed as the weighted sum

$$\mathbf{P}(t) = \mathbf{a}_1 F_1(t) + \mathbf{a}_2 F_2(t) + \mathbf{a}_3 F_3(t) + \mathbf{a}_4 F_4(t), \quad (4.99)$$

then $\mathbf{a}_1$ and $\mathbf{a}_2$ are the endpoints of the segment, and $\mathbf{a}_3$ and $\mathbf{a}_4$ are its two extreme tangents. We can now add constraints by using Equation (4.99) instead of Equation (4.95) and preassigning values to some of the four $\mathbf{a}_i$ coefficients. For example, if we want the initial tangent vector to be in the "up" direction $(0,1)$, we assign $\mathbf{a}_3$ the value $(0,1)$ and end up with Equation (4.96) becoming a system of three equations in the three unknowns $\mathbf{a}_1$, $\mathbf{a}_2$, and $\mathbf{a}_4$. It is now obvious that the more constraints [i.e., the more coefficients are assigned values and eliminated from Equation (4.96)], the fewer are the possibilities for fitting the PC segment to the epoints. There is, therefore, a trade-off between a good fit and more constraints.

4.14 The Bézier Curve

The Bézier curve is extremely important in practical work, so it is discussed here in detail. Two approaches to the design of a Bézier curve are described, one using Bernstein polynomials and the other using the mediation operator.

The Bézier curve is a parametric curve $\mathbf{P}(t)$ that is a polynomial function of the parameter t. The degree of the polynomial depends on the number of points used to define the curve. The method uses *control points* and produces an approximating curve. The curve does not pass through the interior points but is attracted by them (however, see Exercise 4.66 for an exception). It is as if the points exert a pull on the curve. Each point influences the direction of the curve by pulling it toward itself, and that influence is strongest when the curve gets nearest the point. Figure 4.34 shows some examples of cubic Bézier curves. Such a curve is defined by four points and is expressed as a cubic polynomial. Notice that one has a cusp and another one has a loop. Interactive curve design is done by changing the positions of the points, adding more points, or deleting some.

4 Curves

Historical Notes

Pierre Etienne Bézier (pronounced "Bez-yea") was an applied mathematician with the French car manufacturer Reñault. In the early 1960s, encouraged by his employer, he began searching for ways to automate the process of designing cars. His methods have been the basis of the modern field of Computer Aided Geometric Design (CAGD), a field with practical applications in many areas.

It is interesting to note that Paul de Faget de Casteljau—an applied mathematician with Citroën—was the first, in 1959, to develop the various Bézier methods but—because of the secretiveness of his employer—never published it (except for two internal technical memos that were discovered in 1975). This is why the entire field is named after the second person, Bézier, who developed it.

Bézier and de Casteljau did their work while working for car manufacturers. It is little known that Steven Anson Coons of MIT did most of his work on surfaces (around 1967) while a consultant for Ford. Another mathematician, William J. Gordon, has generalized the Coons surfaces, in 1969, as part of his work for General Motors research labs. It seems that car manufacturers have been very innovative in the CAGD area.

How does one go about deriving such a curve? We describe two approaches to the design—a weighted sum and a linear interpolation—and prove that they are identical.

4.14.1 Pascal Triangle and the Binomial Theorem

The Pascal triangle and the binomial theorem are related since they both use the same numbers. The Pascal triangle is an infinite triangular matrix that's built from the edges inside

$$
\begin{array}{ccccccc}
 & & & 1 & & & \\
 & & 1 & & 1 & & \\
 & & 1 & 2 & 1 & & \\
 & 1 & 3 & & 3 & 1 & \\
 1 & & 4 & 6 & 4 & & 1 \\
1 & 5 & 10 & & 10 & 5 & 1 \\
\cdots & & \cdots & & \cdots & &
\end{array}
$$

We first fill the left and right edges with ones, then compute each interior element as the sum of the two elements directly above it. As can be expected, it is not hard to obtain an explicit expression for the general element of the Pascal triangle. We first number the rows from 0 starting at the top, and the columns from 0 starting on the left. A general element is denoted $\binom{i}{j}$. We then observe that the top two rows (corresponding to $i = 0, 1$) consist of 1's and that every other row can be obtained

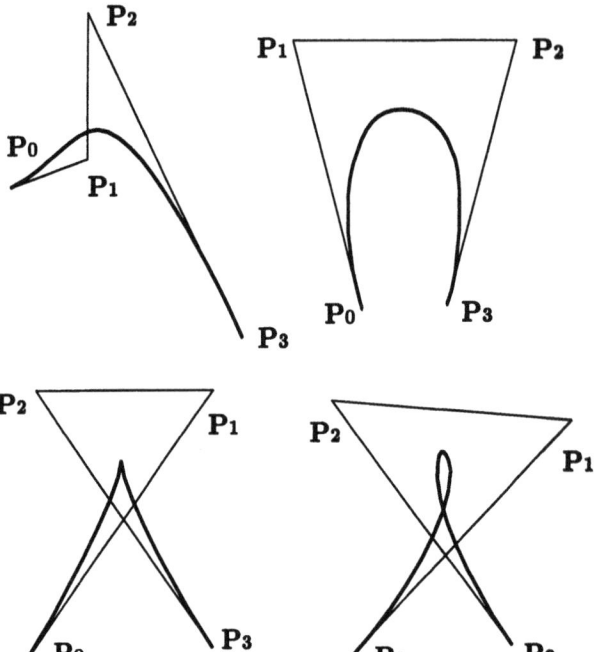

Figure 4.34: Some Cubic Bézier Curves and Their Control Points.

as the sum of its predecessor and a shifted version of its predecessor. For example,

$$
\begin{array}{r}
1\ 3\ 3\ 1 \\
+\quad 1\ 3\ 3\ 1 \\
\hline
1\ 4\ 6\ 4\ 1
\end{array}
$$

This shows that the elements of the triangle satisfy

$$\binom{i}{0} = \binom{i}{1} = 1, \quad i = 0, 1, \ldots,$$

$$\binom{i}{j} = \binom{i-1}{j-1} + \binom{i-1}{j}, \quad i = 2, 3, \ldots, \quad j = 1, \ldots, (i-1).$$

From this, it is easy to get the explicit expression

$$
\begin{aligned}
\binom{i}{j} &= \binom{i-1}{j-1} + \binom{i-1}{j} \\
&= \frac{(i-1)!}{(j-1)!(i-j)!} + \frac{(i-1)!}{j!(i-1-j)!} \\
&= \frac{j(i-1)!}{j!(i-j)!} + \frac{(i-j)(i-1)!}{j!(i-j)!} \\
&= \frac{i!}{j!(i-j)!}.
\end{aligned}
$$

4 Curves

The general element of the Pascal triangle is thus the well-known *binomial coefficient*

$$\binom{i}{j} = \frac{i!}{j!(i-j)!}.$$

The binomial coefficient is one of Newton's many contributions to mathematics. His binomial theorem states that

$$(a+b)^n = \sum_{i=0}^{n} \binom{n}{i} a^i b^{n-i}. \tag{4.100}$$

This equation can be written in a symmetric way by denoting $j = n - i$. The result is

$$(a+b)^n = \sum_{\substack{i+j=n \\ i,j \geq 0}} \frac{(i+j)!}{i!j!} a^i b^j, \tag{4.101}$$

from which we can easily guess the *trinomial theorem*

$$(a+b+c)^n = \sum_{\substack{i+j+k=n \\ i,j,k \geq 0}} \frac{(i+j+k)!}{i!j!k!} a^i b^j c^k. \tag{4.102}$$

4.14.2 The Bernstein Form of the Bézier Curve

The first approach to the Bézier curve expresses it as a weighted sum of the points (with, of course, barycentric weights). Each control point is multiplied by a weight and the products are then added. Let's denote the control points by $\mathbf{P}_0, \mathbf{P}_1, \ldots, \mathbf{P}_n$ (n is therefore defined as 1 less than the number of points) and denote the weights by B_i. The weighted sum is

$$\mathbf{P}(t) = \sum_{i=0}^{n} \mathbf{P}_i B_i, \quad 0 \leq t \leq 1.$$

The result, $\mathbf{P}(t)$, depends on the parameter t. Since the points are given by the user, it is obvious that the weights must depend on t. We thus denote them $B_i(t)$. How should $B_i(t)$ behave as a function of t?

We first look at $B_0(t)$, the weight associated with the first point. We want the first point to affect the curve mostly at the beginning, i.e., when t is close to 0. As t grows toward 1 (i.e., as the curve moves away from point $\mathbf{P}_0$), we want $B_0(t)$ to go down to 0. When $B_0(t) = 0$, the first point no longer influences the shape of the curve.

Next, we turn to $B_1(t)$. This weight function should start small, should have a maximum when the curve approaches the second point $\mathbf{P}_1$, and should then start dropping until it reaches zero. A natural question is: When does the curve reach the second point (for what value of t)? The answer is: It depends on the number of points. For three points (the case $n = 2$), the Bézier curve passes closest to the second point (the interior point) when $t = 0.5$. For four points, the curve is nearest the second point when $t = 1/3$. It is now clear that the weight functions must also depend on n and we denote them by $B_{n,i}(t)$. Hence, $B_{3,1}(t)$ should start from 0, have a maximum at $t = 1/3$, and go down to 0 from there. Figure 4.35 shows the desired behavior of $B_{n,i}(t)$ for $n = 2, 3, 4$. The five different weights of $B_{4,i}(t)$ have their maxima at $t = 0$, $1/4$, $1/2$, $3/4$, and 1 (see also Figure 4.11b).

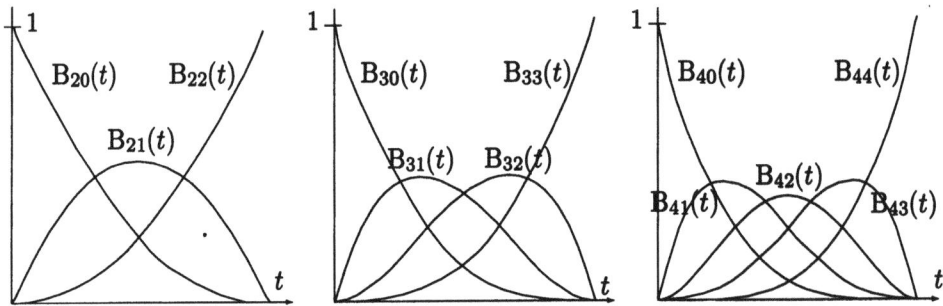

Figure 4.35: The Bernstein Polynomials for $n = 2, 3, 4$.

The functions eventually used by Bézier (and by de Casteljau) were derived by the Russian mathematician Sergeï Natanovich Bernshteĭn in 1912, as part of his work on approximation theory (see Chapter 6 of [Davis 63]). They are known as the Bernstein polynomials and are defined by

$$B_{n,i}(t) = \binom{n}{i} t^i (1-t)^{n-i}, \quad \text{where} \quad \binom{n}{i} = \frac{n!}{i!(n-i)!} \quad (4.103)$$

are the binomial coefficients. They feature the desired behavior and have a few more useful properties, which are discussed below. (In calculating the curve, we assume that the quantity 0^0, which is normally undefined, equals 1.)

The Bézier curve is now defined as

$$\mathbf{P}(t) = \sum_{i=0}^{n} \mathbf{P}_i B_{n,i}(t), \quad \text{where } B_{n,i}(t) = \binom{n}{i} t^i (1-t)^{n-i} \text{ and } 0 \leq t \leq 1. \quad (4.104)$$

Each control point (which is a pair or a triplet of coordinates) is multiplied by its weight, which is in the range $[0, 1]$. The weights act as *blending functions* that blend together the contributions of the different points.

Following is *Mathematica* code to calculate and plot the Bernstein polynomials and the Bézier curve:

```
(* Just the base functions bern. Note how "pwr" handles 0^0 *)
Clear[pwr,bern,n,i,t]
pwr[x_,y_]:=If[x==0 && y==0, 1, x^y];
bern[n_,i_,t_]:=Binomial[n,i]pwr[t,i]pwr[1-t,n-i]
  (* t^i \[Times] (1-t)^(n-i) *)
```

4 Curves

```
Plot[Evaluate[Table[bern[5,i,t], {i,0,5}]], {t,0,1},
 DefaultFont->{"cmr10", 10}]
Clear[i,t,pnts,pwr,bern,bzCurve,g1,g2]; (* Cubic Bezier curve *)
(* either read points from file
pnts=ReadList["DataPoints",{Number,Number}]; *)
(* or enter them explicitly *)
pnts={{0,0},{.7,1},{.3,1},{1,0}}; (* 4 points for a cubic curve *)
pwr[x_,y_]:=If[x==0 && y==0, 1, x^y];
bern[n_,i_,t_]:=Binomial[n,i]pwr[t,i]pwr[1-t,n-i]
bzCurve[t_]:=Sum[pnts[[i+1]]bern[3,i,t], {i,0,3}]
g1=ListPlot[pnts, Prolog->AbsolutePointSize[4], PlotRange->All,
 AspectRatio->Automatic, DisplayFunction->Identity]
g2=ParametricPlot[bzCurve[t], {t,0,1}, DisplayFunction->Identity]
Show[g1,g2, DisplayFunction->$DisplayFunction]
```

When Bézier started—in the early 1960s—searching for such functions, he set the following requirements:

1. The functions should be such that the curve passes through the first and last control points.

2. The tangent to the curve at the start point should be $\mathbf{P}_1 - \mathbf{P}_0$, i.e., the curve should start at point $\mathbf{P}_0$ going toward $\mathbf{P}_1$. The same property should hold at the last point.

3. The same requirement is generalized for higher derivatives of the curve at the two extreme endpoints. Hence, $\mathbf{P}^{tt}(0)$ should depend only on the first point $\mathbf{P}_0$ and its two neighbors $\mathbf{P}_1$ and $\mathbf{P}_2$. In general, $\mathbf{P}^{(k)}(0)$ should only depend on $\mathbf{P}_0$ and its k neighbors $\mathbf{P}_1, \ldots, \mathbf{P}_k$. This feature provides complete control over the continuity at the joints between separate Bézier curves (Section 4.14.5).

4. The weight functions should be symmetric with respect to t and $(1-t)$. This means that a reversal of the sequence of control points would not affect the shape of the curve.

5. The weights should be barycentric, to guarantee that the shape of the curve is independent of the coordinate system.

The definition just provided, using Bernstein polynomials as the weights, satisfies all these requirements.

Examples: In the special case $n = 2$ (three control points), the weights are

$$B_{2,0}(t) = \binom{2}{0}t^0(1-t)^{2-0} = 1 \cdot 1 \cdot (1-t)^2 = (1-t)^2,$$
$$B_{2,1}(t) = \binom{2}{1}t^1(1-t)^{2-1} = 2t(1-t),$$
$$B_{2,2}(t) = \binom{2}{2}t^2(1-t)^{2-2} = t^2,$$

and the curve is

$$\mathbf{P}(t) = (1-t)^2 \mathbf{P}_0 + 2t(1-t)\mathbf{P}_1 + t^2 \mathbf{P}_2$$
$$= \left((1-t)^2, 2t(1-t), t^2\right)(\mathbf{P}_0, \mathbf{P}_1, \mathbf{P}_2)^T$$

$$= (t^2, t, 1) \begin{pmatrix} 1 & -2 & 1 \\ -2 & 2 & 0 \\ 1 & 0 & 0 \end{pmatrix} \begin{pmatrix} \mathbf{P}_0 \\ \mathbf{P}_1 \\ \mathbf{P}_2 \end{pmatrix}. \qquad (4.105)$$

This is the *quadratic* Bézier curve.

▶ **Exercise 4.62:** Given three points $\mathbf{P}_1$, $\mathbf{P}_2$, and $\mathbf{P}_3$, calculate the parabola that goes from $\mathbf{P}_1$ to $\mathbf{P}_3$ and whose start and end tangent vectors point in directions $\mathbf{P}_2 - \mathbf{P}_1$ and $\mathbf{P}_3 - \mathbf{P}_2$, respectively.

In the special case $n = 3$, the four weight functions are

$$B_{3,0}(t) = \binom{3}{0} t^0 (1-t)^{3-0} = 1 \cdot 1 \cdot (1-t)^3 = (1-t)^3,$$
$$B_{3,1}(t) = \binom{3}{1} t^1 (1-t)^{3-1} = 3t(1-t)^2,$$
$$B_{3,2}(t) = \binom{3}{2} t^2 (1-t)^{3-2} = 3t^2(1-t),$$
$$B_{3,3}(t) = \binom{3}{3} t^3 (1-t)^{3-3} = 1 \cdot t^3 (1-t)^0 = t^3,$$

and the curve is thus

$$\mathbf{P}(t) = (1-t)^3 \mathbf{P}_0 + 3t(1-t)^2 \mathbf{P}_1 + 3t^2(1-t) \mathbf{P}_2 + t^3 \mathbf{P}_3 \qquad (4.106)$$
$$= \left[(1-t)^3, 3t(1-t)^2, 3t^2(1-t), t^3 \right] [\mathbf{P}_0, \mathbf{P}_1, \mathbf{P}_2, \mathbf{P}_3]^T$$
$$= \left[(1 - 3t + 3t^2 - t^3), (3t - 6t^2 + 3t^3), (3t^2 - 3t^3), t^3 \right] [\mathbf{P}_0, \mathbf{P}_1, \mathbf{P}_2, \mathbf{P}_3]^T$$
$$= (t^3, t^2, t, 1) \begin{pmatrix} -1 & 3 & -3 & 1 \\ 3 & -6 & 3 & 0 \\ -3 & 3 & 0 & 0 \\ 1 & 0 & 0 & 0 \end{pmatrix} \begin{pmatrix} \mathbf{P}_0 \\ \mathbf{P}_1 \\ \mathbf{P}_2 \\ \mathbf{P}_3 \end{pmatrix}. \qquad (4.107)$$

⮕ It is clear that $\mathbf{P}(t)$ is a cubic polynomial in t. It is the *cubic* Bézier curve.

In general, the Bézier curve is a polynomial of degree n.

▶ **Exercise 4.63:** The cubic curve of Equation (4.107) is drawn when the parameter t varies in the range $[0, 1]$. Show how to substitute t with a new parameter u such that the curve will be drawn when $-1 \leq u \leq +1$.

▶ **Exercise 4.64:** Calculate the Bernstein polynomials for $n = 4$.

It can be proved by induction that the general, $n + 1$-point Bézier curve can be represented by

$$\mathbf{P}(t) = (t^n, t^{n-1}, \ldots, t, 1) \mathbf{N} \begin{pmatrix} \mathbf{P}_1 \\ \mathbf{P}_2 \\ \vdots \\ \mathbf{P}_{n-1} \\ \mathbf{P}_n \end{pmatrix} = \mathbf{T}(t) \cdot \mathbf{N} \cdot \mathbf{P}, \qquad (4.108)$$

4 Curves

where

$$N = \begin{pmatrix} \binom{n}{0}\binom{n}{n}(-1)^n & \binom{n}{1}\binom{n-1}{n-1}(-1)^{n-1} & \cdots & \binom{n}{n}\binom{n-n}{n-n}(-1)^0 \\ \binom{n}{0}\binom{n}{n-1}(-1)^{n-1} & \binom{n}{1}\binom{n-1}{n-2}(-1)^{n-2} & \cdots & 0 \\ \vdots & \vdots & \cdots & 0 \\ \binom{n}{0}\binom{n}{1}(-1)^1 & \binom{n}{1}\binom{n-1}{0}(-1)^0 & \cdots & 0 \\ \binom{n}{0}\binom{n}{0}(-1)^0 & 0 & \cdots & 0 \end{pmatrix}.$$

Matrix N is symmetric and its below-diagonal elements are all zeros. Its determinant therefore equals (up to a sign) the product of the diagonal elements, which are all nonzero. A nonzero determinant implies a nonsingular matrix. Thus, matrix N always has an inverse. N can also be written as the product AB, where

$$A = \begin{pmatrix} \binom{n}{n}(-1)^n & \binom{n}{1}\binom{n-1}{n-1}(-1)^{n-1} & \cdots & \binom{n}{n}\binom{n-n}{n-n}(-1)^0 \\ \binom{n}{n-1}(-1)^{n-1} & \binom{n}{1}\binom{n-1}{n-2}(-1)^{n-2} & \cdots & 0 \\ \vdots & \vdots & \cdots & 0 \\ \binom{n}{1}(-1)^1 & \binom{n}{1}\binom{n-1}{0}(-1)^0 & \cdots & 0 \\ \binom{n}{0}(-1)^0 & 0 & \cdots & 0 \end{pmatrix}$$

and

$$B = \begin{pmatrix} \binom{n}{0} & 0 & \cdots & 0 \\ 0 & \binom{n}{1} & \cdots & 0 \\ \vdots & & \ddots & \vdots \\ 0 & 0 & \cdots & \binom{n}{n} \end{pmatrix}.$$

Figure 4.36 shows the Bézier N matrices for $n = 1, 2, \ldots, 7$.

▶ **Exercise 4.65:** Calculate the Bézier curve for the case $n = 1$ (two control points). What kind of a curve is it?

▶ **Exercise 4.66:** Generally, the Bézier curve passes through the first and last control points, but not through the intermediate points. Consider the case of three points P_0, P_1, and P_2 on a straight line. Intuitively, it seems that the curve will be a straight line and thus would pass through the interior point P_1. Is that so?

The Bézier curve can also be represented in a very compact and elegant way as $P(t) = (1 - t + tE)^n P_0$, where E is the shift operator defined by $EP_i = P_{i+1}$ (i.e., applying E to point P_i produces point P_{i+1}).

Proof: Define an operator E by $EP_i = P_{i+1}$. This implies $EP_0 = P_1$, $E^2 P_0 = P_2$, and $E^i P_0 = P_i$

The Bézier curve can now be written

$$P(t) = \sum_{i=0}^{n} \binom{n}{i} t^i (1-t)^{n-i} P_i = \sum_{i=0}^{n} \binom{n}{i} t^i (1-t)^{n-i} E^i P_0$$

$$= \sum_{i=0}^{n} \binom{n}{i} (tE)^i (1-t)^{n-i} P_0 = (tE + (1-t))^n P_0,$$

$$\mathbf{B}_1 = \begin{pmatrix} -1 & 1 \\ 1 & 0 \end{pmatrix},$$

$$\mathbf{B}_2 = \begin{pmatrix} 1 & -2 & 1 \\ -2 & 2 & 0 \\ 1 & 0 & 0 \end{pmatrix},$$

$$\mathbf{B}_3 = \begin{pmatrix} -1 & 3 & -3 & 1 \\ 3 & -6 & 3 & 0 \\ -3 & 3 & 0 & 0 \\ 1 & 0 & 0 & 0 \end{pmatrix},$$

$$\mathbf{B}_4 = \begin{pmatrix} 1 & -4 & 6 & -4 & 1 \\ -4 & 12 & -12 & 4 & 0 \\ 6 & -12 & 6 & 0 & 0 \\ -4 & 4 & 0 & 0 & 0 \\ 1 & 0 & 0 & 0 & 0 \end{pmatrix},$$

$$\mathbf{B}_5 = \begin{pmatrix} -1 & 5 & -10 & 10 & -5 & 1 \\ 5 & -20 & 30 & -20 & 5 & 0 \\ -10 & 30 & -30 & 10 & 0 & 0 \\ 10 & -20 & 10 & 0 & 0 & 0 \\ -5 & 5 & 0 & 0 & 0 & 0 \\ 1 & 0 & 0 & 0 & 0 & 0 \end{pmatrix},$$

$$\mathbf{B}_6 = \begin{pmatrix} 1 & -6 & 15 & -20 & 15 & -6 & 1 \\ -6 & 30 & -60 & 60 & -30 & 6 & 0 \\ 15 & -60 & 90 & -60 & 15 & 0 & 0 \\ -20 & 60 & -60 & 20 & 0 & 0 & 0 \\ 15 & -30 & 15 & 0 & 0 & 0 & 0 \\ -6 & 6 & 0 & 0 & 0 & 0 & 0 \\ 1 & 0 & 0 & 0 & 0 & 0 & 0 \end{pmatrix},$$

$$\mathbf{B}_7 = \begin{pmatrix} -1 & 7 & -21 & 35 & -35 & 21 & -7 & 1 \\ 7 & -42 & 105 & -140 & 105 & -42 & 7 & 0 \\ -21 & 105 & -210 & 210 & -105 & 21 & 0 & 0 \\ 35 & -140 & 210 & -140 & 35 & 0 & 0 & 0 \\ -35 & 105 & -105 & 35 & 0 & 0 & 0 & 0 \\ 21 & -42 & 21 & 0 & 0 & 0 & 0 & 0 \\ -7 & 7 & 0 & 0 & 0 & 0 & 0 & 0 \\ 1 & 0 & 0 & 0 & 0 & 0 & 0 & 0 \end{pmatrix}.$$

Figure 4.36: The First Seven Bézier Basis Matrices.

4 Curves

where the last step is an application of the binomial theorem, Equation (4.100). ◄

Examples: For $n = 1$, we get

$$\mathbf{P}(t) = (1 - t + tE)\mathbf{P}_0 = \mathbf{P}_0(1-t) + \mathbf{P}_1 t.$$

For $n = 2$, we get

$$\begin{aligned}\mathbf{P}(t) &= (1 - t + tE)^2 \mathbf{P}_0 \\ &= (1 - t + tE - t + t^2 - t^2 E + tE - t^2 E + t^2 E^2)\mathbf{P}_0 \\ &= \mathbf{P}_0(1 - 2t + t^2) + \mathbf{P}_1(2t - 2t^2) + \mathbf{P}_2 t^2 \\ &= \mathbf{P}_0(1+t)^2 + \mathbf{P}_1 2t(1-t) + \mathbf{P}_2 t^2.\end{aligned}$$

Given $n + 1$ control points $\mathbf{P}_0$ through $\mathbf{P}_n$, we can represent the Bézier curve for the points by $\mathbf{P}_n^{(n)}(t)$, where the quantity $\mathbf{P}_i^{(j)}(t)$ is defined recursively by

$$\mathbf{P}_i^{(j)}(t) = \begin{cases} (1-t)\mathbf{P}_{i-1}^{(j-1)}(t) + t\mathbf{P}_i^{(j-1)}(t), & \text{for } j > 0, \\ \mathbf{P}_i, & \text{for } j = 0. \end{cases} \quad (4.109)$$

The following examples show how the definition above is used to generate the quantities $\mathbf{P}_i^{(j)}(t)$ and why $\mathbf{P}_n^{(n)}(t)$ is the degree-n curve:

$$\begin{aligned}\mathbf{P}_0^{(0)}(t) &= \mathbf{P}_0, \quad \mathbf{P}_1^{(0)}(t) = \mathbf{P}_1, \quad \mathbf{P}_2^{(0)}(t) = \mathbf{P}_2, \ldots, \mathbf{P}_n^{(0)}(t) = \mathbf{P}_n, \\ \mathbf{P}_1^{(1)}(t) &= (1-t)\mathbf{P}_0^{(0)}(t) + t\mathbf{P}_1^{(0)}(t) = (1-t)\mathbf{P}_0 + t\mathbf{P}_1, \\ \mathbf{P}_2^{(2)}(t) &= (1-t)\mathbf{P}_1^{(1)}(t) + t\mathbf{P}_2^{(1)}(t) \\ &= (1-t)\big((1-t)\mathbf{P}_0 + t\mathbf{P}_1\big) + t\big((1-t)\mathbf{P}_1 + t\mathbf{P}_2\big) \\ &= (1-t)^2 \mathbf{P}_0 + 2t(1-t)\mathbf{P}_1 + t^2 \mathbf{P}_2, \\ \mathbf{P}_3^{(3)}(t) &= (1-t)\mathbf{P}_2^{(2)}(t) + t\mathbf{P}_3^{(2)}(t) \\ &= (1-t)\big((1-t)\mathbf{P}_1^{(1)}(t) + t\mathbf{P}_2^{(1)}(t)\big) + t\big((1-t)\mathbf{P}_2^{(1)}(t) + t\mathbf{P}_3^{(1)}(t)\big) \\ &= (1-t)^2 \mathbf{P}_1^{(1)}(t) + 2t(1-t)\mathbf{P}_2^{(1)}(t) + t^2 \mathbf{P}_3^{(1)}(t) \\ &= (1-t)^2\big((1-t)\mathbf{P}_0 + t\mathbf{P}_1\big) + 2t(1-t)\big((1-t)\mathbf{P}_1 + t\mathbf{P}_2\big) \\ &\quad + t^2\big((1-t)\mathbf{P}_2 + t\mathbf{P}_3\big) \\ &= (1-t)^3 \mathbf{P}_0 + 3t(1-t)^2 \mathbf{P}_1 + 3t^2(1-t)\mathbf{P}_2 + t^3 \mathbf{P}_3,\end{aligned}$$

4.14.3 Fast Calculation of the Curve

Calculating the Bézier curve is straightforward but slow. However, with a little thinking, it can be speeded up considerably, which makes this curve very useful in practice. Three methods are discussed in this section.

Method 1: We notice the following points:

- The calculation involves the binomials $\binom{n}{i}$ for $i = 0, 1, \ldots, n$, which, in turn, require the factorials $0!, 1!, \ldots, n!$. The factorials can be precalculated once (each one from its predecessor) and stored in a table. They can then be used to calculate all the necessary binomials and those can also be stored in a table.

- The calculation involves terms of the form t^i for $i = 0, 1, \ldots, n$ and for many t values in the range $[0, 1]$. These can also be precalculated and stored in a two-dimensional table where they can be accessed later, using t and i as indexes. This has the advantage that the values of $(1-t)^{n-i}$ can be read from the same table (using $1-t$ and $n-i$ as row and column indexes).

The calculation now reduces to a sum where each term is a product of four quantities—one control point and three numbers from tables. Instead of

$$\sum_{i=0}^{n} \binom{n}{i} t^i (1-t)^{n-i} \mathbf{P}_i,$$

we need to compute

$$\sum_{i=0}^{n} \text{Table}_1[i, n] \cdot \text{Table}_2[t, i] \cdot \text{Table}_2[1-t, n-i] \cdot \mathbf{P}_i.$$

Method 2: Once n is known, each of the $n+1$ Bernstein polynomials $B_{n,i}(t)$, $i = 0, 1, \ldots, n$, can be precalculated for all the necessary values of t and stored in a table. The curve can now be calculated as the sum

$$\sum_{i=0}^{n} \text{Table}[t, i] \mathbf{P}_i,$$

showing that each pixel requires $n+1$ table lookups, $n+1$ multiplications, and n additions.

Method 3: The idea of *forward differences* (Section A.2) is used, together with the Taylor series representation, to speed up the calculation significantly. The Bézier curve, which we denote by $\mathbf{B}(t)$, is drawn pixel by pixel in a loop where t is incremented from 0 to 1 in fixed, small steps of Δt. The idea of forward differences is to find a quantity $\mathbf{dB}$ such that $\mathbf{B}(t + \Delta t) = \mathbf{B}(t) + \mathbf{dB}$ for any value of t. If such a $\mathbf{dB}$ can be found, then it is enough to calculate $\mathbf{B}(0)$ (which, as we know, is simply $\mathbf{P}_0$) and use forward differences to calculate

$$\mathbf{B}(0 + \Delta t) = \mathbf{B}(0) + \mathbf{dB},$$
$$\mathbf{B}(2\Delta t) = \mathbf{B}(\Delta t) + \mathbf{dB} = \mathbf{B}(0) + 2\mathbf{dB},$$

and, in general,

$$\mathbf{B}(i\Delta t) = \mathbf{B}((i-1)\Delta t) + \mathbf{dB} = \mathbf{B}(0) + i\,\mathbf{dB}.$$

The point is that $\mathbf{dB}$ should not depend on t. If $\mathbf{dB}$ turns out to depend on t, then as we advance t from 0 to 1, we would have to use different values of

4 Curves

dB, slowing down the calculations. The fastest way to calculate the curve is to precalculate **dB** before the loop starts and to repeatedly add this precalculated value to **B**(t) inside the loop.

We calculate **dB** by using the *Taylor series* representation of the Bézier curve. In general, the Taylor series representation of a function $f(t)$ at a point $f(t + \Delta t)$ is the infinite sum

$$f(t + \Delta t) = f(t) + f'(t)\Delta t + \frac{f''(t)\Delta^2 t}{2!} + \frac{f'''(t)\Delta^3 t}{3!} + \cdots.$$

In order to avoid dealing with an infinite sum, we limit our discussion to cubic Bézier curves. These are the most common Bézier curves and are used by many popular graphics applications. They are defined by four control points and are given by Equations (4.106) and (4.107):

$$\mathbf{B}(t) = (1-t)^3 \mathbf{P}_0 + 3t(1-t)^2 \mathbf{P}_1 + 3t^2(1-t)\mathbf{P}_2 + t^3 \mathbf{P}_3$$

$$= (t^3, t^2, t, 1) \begin{pmatrix} -1 & 3 & -3 & 1 \\ 3 & -6 & 3 & 0 \\ -3 & 3 & 0 & 0 \\ 1 & 0 & 0 & 0 \end{pmatrix} \begin{pmatrix} \mathbf{P}_0 \\ \mathbf{P}_1 \\ \mathbf{P}_2 \\ \mathbf{P}_3 \end{pmatrix}.$$

These curves are thus cubic polynomials in t, meaning that only their first three derivatives are nonzero. In order to simplify the calculation of their derivatives, we need to express these curves in the form $\mathbf{B}(t) = \mathbf{a}t^3 + \mathbf{b}t^2 + \mathbf{c}t + \mathbf{d}$ (Equation (4.8)). This is done by

$$\mathbf{B}(t) = (1-t)^3 \mathbf{P}_0 + 3t(1-t)^2 \mathbf{P}_1 + 3t^2(1-t)\mathbf{P}_2 + t^3 \mathbf{P}_3$$
$$= \bigl(3(\mathbf{P}_1 - \mathbf{P}_2) - \mathbf{P}_0 + \mathbf{P}_3\bigr)t^3 + \bigl(3(\mathbf{P}_0 + \mathbf{P}_2) - 6\mathbf{P}_1\bigr)t^2 + 3(\mathbf{P}_1 - \mathbf{P}_0)t + \mathbf{P}_0$$
$$= \mathbf{a}t^3 + \mathbf{b}t^2 + \mathbf{c}t + \mathbf{d},$$

so $\mathbf{a} = 3(\mathbf{P}_1 - \mathbf{P}_2) - \mathbf{P}_0 + \mathbf{P}_3$, $\mathbf{b} = 3(\mathbf{P}_0 + \mathbf{P}_2) - 6\mathbf{P}_1$, $\mathbf{c} = 3(\mathbf{P}_1 - \mathbf{P}_0)$, and $\mathbf{d} = \mathbf{P}_0$. These relations can also be expressed in matrix notation:

$$\begin{pmatrix} \mathbf{a} \\ \mathbf{b} \\ \mathbf{c} \\ \mathbf{d} \end{pmatrix} = \begin{pmatrix} -1 & 3 & -3 & 1 \\ 3 & -6 & 3 & 0 \\ -3 & 3 & 0 & 0 \\ 1 & 0 & 0 & 0 \end{pmatrix} \begin{pmatrix} \mathbf{P}_0 \\ \mathbf{P}_1 \\ \mathbf{P}_2 \\ \mathbf{P}_3 \end{pmatrix}.$$

The curve is now easy to differentiate:

$$\mathbf{B}^t(t) = 3\mathbf{a}t^2 + 2\mathbf{b}t + \mathbf{c}, \quad \mathbf{B}^{tt}(t) = 6\mathbf{a}t + 2\mathbf{b}, \quad \mathbf{B}^{ttt}(t) = 6\mathbf{a};$$

and the Taylor series representation gives

$$\mathbf{dB} = \mathbf{B}(t + \Delta t) - \mathbf{B}(t)$$
$$= \mathbf{B}^t(t)\Delta t + \frac{\mathbf{B}^{tt}(t)\Delta^2 t}{2} + \frac{\mathbf{B}^{ttt}(t)\Delta^3 t}{6}$$
$$= 3\mathbf{a}\,t^2 \Delta t + 2\mathbf{b}\,t\Delta t + \mathbf{c}\Delta t + 3\mathbf{a}\,t\Delta^2 t + \mathbf{b}\Delta^2 t + \mathbf{a}\Delta^3 t.$$

This seems like a failure since the value obtained for **dB** is a function of t (it should be denoted by **dB**(t) instead of just **dB**) and is also slow to calculate. However, the original cubic curve **B**(t) is a degree-3 polynomial in t, whereas **dB**(t) is only a degree-2 polynomial. This suggests a way out of our problem. We can try to express **dB**(t) by means of the Taylor series, similar to what we did with the original curve **B**(t). This should give us a forward difference **ddB**(t) that's a polynomial of degree 1 in t. The quantity **ddB**(t) can, in turn, be represented by another Taylor series to produce a forward difference **dddB** that's a degree-0 polynomial in t, i.e., a constant. Once we do that, we will end up with an algorithm of the form

```
precalculate certain quantities;
B = P0;
for t:=0 to 1 step Δt do
PlotPixel(B);
B:=B+dB; dB:=dB+ddB; ddB:=ddB+dddB;
endfor;
```

The quantity **ddB**(t) is obtained by

$$\mathbf{dB}(t + \Delta t) = \mathbf{dB}(t) + \mathbf{ddB}(t) = \mathbf{dB}(t) + \mathbf{dB}^t(t)\Delta t + \frac{\mathbf{dB}(t)^{tt}\Delta^2 t}{2},$$

yielding

$$\mathbf{ddB}(t) = \mathbf{dB}^t(t)\Delta t + \frac{\mathbf{dB}(t)^{tt}\Delta^2 t}{2}$$
$$= (6\mathbf{a}\,t\Delta t + 2\mathbf{b}\Delta t + 3\mathbf{a}\Delta^2 t)\Delta t + \frac{6\mathbf{a}\Delta t\Delta^2 t}{2}$$
$$= 6\mathbf{a}\,t\Delta^2 t + 3\mathbf{b}\Delta^2 t + 6\mathbf{a}\Delta^3 t.$$

Finally, the constant **dddB** is similarly obtained by $\mathbf{ddB}(t + \Delta t) = \mathbf{ddB}(t) + \mathbf{dddB} = \mathbf{ddB}(t) + \mathbf{ddB}^t(t)\Delta t$, yielding $\mathbf{dddB} = \mathbf{ddB}^t(t)\Delta t = 6\mathbf{a}\Delta^3 t$.

The four quantities involved in the calculation of the curve are thus

$$\mathbf{B}(t) = \mathbf{a}t^3 + \mathbf{b}t^2 + \mathbf{c}t + \mathbf{d},$$
$$\mathbf{dB}(t) = 3\mathbf{a}\,t^2\Delta t + 2\mathbf{b}\,t\Delta t + \mathbf{c}\Delta t + 3\mathbf{a}\,t\Delta^2 t + \mathbf{b}\Delta^2 t + \mathbf{a}\Delta^3 t,$$
$$\mathbf{ddB}(t) = 6\mathbf{a}\,t\Delta^2 t + 3\mathbf{b}\Delta^2 t + 6\mathbf{a}\Delta^3 t,$$
$$\mathbf{dddB} = 6\mathbf{a}\Delta^3 t.$$

They all have to be calculated at $t = 0$, as functions of the four control points $\mathbf{P}_i$, before the loop starts:

$$\mathbf{B}(0) = \mathbf{d} = \mathbf{P}_0,$$
$$\mathbf{dB}(0) = \mathbf{c}\Delta t + \mathbf{b}\Delta^2 t + \mathbf{a}\Delta^3 t$$
$$= 3\Delta t(\mathbf{P}_1 - \mathbf{P}_0) + \Delta^2 t\big(3(\mathbf{P}_0 + \mathbf{P}_2) - 6\mathbf{P}_1\big)$$
$$+ \Delta^3 t\big(3(\mathbf{P}_1 - \mathbf{P}_2) - \mathbf{P}_0 + \mathbf{P}_3\big)$$

4 Curves

$$= 3\Delta t(\mathbf{P}_1 - \mathbf{P}_0) + 3\Delta^2 t(\mathbf{P}_0 - 2\mathbf{P}_1 + \mathbf{P}_2)$$
$$+ \Delta^3 t\big(3(\mathbf{P}_1 - \mathbf{P}_2) - \mathbf{P}_0 + \mathbf{P}_3\big),$$

$$\mathbf{ddB}(0) = 2\mathbf{b}\Delta^2 t + 6\mathbf{a}\Delta^3 t$$
$$= 2\Delta^2 t\big(3(\mathbf{P}_0 + \mathbf{P}_2) - 6\mathbf{P}_1\big) + 6\Delta^3 t\big(3(\mathbf{P}_1 - \mathbf{P}_2) - \mathbf{P}_0 + \mathbf{P}_3\big)$$
$$= 6\Delta^2 t(\mathbf{P}_0 - 2\mathbf{P}_1 + \mathbf{P}_2) + 6\Delta^3 t\big(3(\mathbf{P}_1 - \mathbf{P}_2) - \mathbf{P}_0 + \mathbf{P}_3\big),$$

$$\mathbf{dddB} = 6\mathbf{a}\Delta^3 t = 6\Delta^3 t\big(3(\mathbf{P}_1 - \mathbf{P}_2) - \mathbf{P}_0 + \mathbf{P}_3\big).$$

The above relations can be expressed in matrix notation as follows:

$$\begin{pmatrix} \mathbf{dddB} \\ \mathbf{ddB}(0) \\ \mathbf{dB}(0) \\ \mathbf{B}(0) \end{pmatrix} = \begin{pmatrix} 6 & 0 & 0 & 0 \\ 6 & 2 & 0 & 0 \\ 1 & 1 & 1 & 0 \\ 0 & 0 & 0 & 1 \end{pmatrix} \begin{pmatrix} \Delta^3 t & 0 & 0 & 0 \\ 0 & \Delta^2 t & 0 & 0 \\ 0 & 0 & \Delta t & 0 \\ 0 & 0 & 0 & 1 \end{pmatrix} \begin{pmatrix} \mathbf{a} \\ \mathbf{b} \\ \mathbf{c} \\ \mathbf{d} \end{pmatrix}$$

$$= \begin{pmatrix} 6 & 0 & 0 & 0 \\ 6 & 2 & 0 & 0 \\ 1 & 1 & 1 & 0 \\ 0 & 0 & 0 & 1 \end{pmatrix} \begin{pmatrix} \Delta^3 t & 0 & 0 & 0 \\ 0 & \Delta^2 t & 0 & 0 \\ 0 & 0 & \Delta t & 0 \\ 0 & 0 & 0 & 1 \end{pmatrix} \begin{pmatrix} -1 & 3 & -3 & 1 \\ 3 & -6 & 3 & 0 \\ -3 & 3 & 0 & 0 \\ 1 & 0 & 0 & 0 \end{pmatrix} \begin{pmatrix} \mathbf{P}_0 \\ \mathbf{P}_1 \\ \mathbf{P}_2 \\ \mathbf{P}_3 \end{pmatrix}$$

$$= \begin{pmatrix} -6\Delta^3 t & 18\Delta^3 t & -18\Delta^3 t & 6\Delta^3 t \\ 6\Delta^2 t - 6\Delta^3 t & -12\Delta^2 t + 18\Delta^3 t & 6\Delta^2 t - 18\Delta^3 t & 6\Delta^3 t \\ 3\Delta^2 t - \Delta^3 t - 3\Delta t & -6\Delta^2 t + 3\Delta^3 t + 3\Delta t & 3\Delta^2 t - 3\Delta^3 t & \Delta^3 t \\ 1 & 0 & 0 & 0 \end{pmatrix} \begin{pmatrix} \mathbf{P}_0 \\ \mathbf{P}_1 \\ \mathbf{P}_2 \\ \mathbf{P}_3 \end{pmatrix}$$

$$= \mathbf{Q} \begin{pmatrix} \mathbf{P}_0 \\ \mathbf{P}_1 \\ \mathbf{P}_2 \\ \mathbf{P}_3 \end{pmatrix},$$

where $\mathbf{Q}$ is a 4×4 matrix that can be calculated once Δt is known.

A detailed examination of the above expressions shows that the following quantities have to be precalculated: $3\Delta t$, $3\Delta^2 t$, $\Delta^3 t$, $6\Delta^2 t$, $6\Delta^3 t$, $\mathbf{P}_0 - 2\mathbf{P}_1 + \mathbf{P}_2$, and $3(\mathbf{P}_1 - \mathbf{P}_2) - \mathbf{P}_0 + \mathbf{P}_3$. We thus end up with the simple, fast algorithm shown in Figure 4.37. For readers who are interested in a quick test, Figure 4.38 is the corresponding *Mathematica* code.

Each point on the curve (i.e., each pixel in the loop) is calculated by three additions and three assignments only. There are no multiplications and no table lookups. This is indeed a very fast algorithm!

4.14.4 Properties of the Weights and the Curve

1. The weights add up to 1 (they are barycentric). This is easily proved using Newton's binomial theorem $(a+b)^n = \sum_{i=0}^{n} \binom{n}{i} a^i b^{n-i}$:

$$1 = \big(t + (1-t)\big)^n = \sum_{i=0}^{n} \binom{n}{i} t^i (1-t)^{n-i} = \sum_{i=0}^{n} B_{n,i}(t). \tag{4.110}$$

```
Q1:=3Δt;
Q2:=Q1×Δt;   // 3Δ²t
Q3:=Δ³t;
Q4:=2Q2;     // 6Δ²t
Q5:=6Q3;     // 6Δ³t
Q6:=P₀ - 2P₁ + P₂;
Q7:=3(P₁ - P₂) - P₀ + P₃;
B:=P₀;
dB:=(P₁ - P₀)Q1+Q6×Q2+Q7×Q3;
ddB:=Q6×Q4+Q7×Q5;
dddB:=Q7×Q5;
for t:=0 to 1 step Δt do
Pixel(B);
B:=B+dB; dB:=dB+ddB; ddB:=ddB+dddB;
endfor;
```

Figure 4.37: A Fast Bézier Curve Algorithm.

```
n=3; Clear[q1,q2,q3,q4,q5,Q6,Q7,B,dB,ddB,dddB,p0,p1,p2,p3,tabl];
p0={0,1}; p1={5,.5}; p2={0,.5}; p3={0,1};  (* Four points *)
dt=.01; q1=3dt; q2=3dt^2; q3=dt^3; q4=2q2; q5=6q3;
Q6=p0-2p1+p2; Q7=3(p1-p2)-p0+p3;
B=p0; dB=(p1-p0) q1+Q6 q2+Q7 q3; (* space indicates *)
ddB=Q6 q4+Q7 q5; dddB=Q7 q5;     (* multiplication *)
tabl={};
Do[{tabl=Append[tabl,B], B=B+dB, dB=dB+ddB, ddB=ddB+dddB},
                                          {t,0,1,dt}];
ListPlot[tabl];
```

Figure 4.38: *Mathematica* Code for Figure 4.37.

2. The curve passes through the two endpoints $\mathbf{P}_0$ and $\mathbf{P}_n$. We assume that $0^0 = 1$ and observe that

$$B_{n,0}(0) = \binom{n}{0} 0^0 (1-0)^{n-0} = 1 \cdot 1 \cdot 1^n = 1,$$

which implies

$$\mathbf{P}(0) = \sum_{i=0}^{n} \mathbf{P}_i B_{n,i}(0) = \mathbf{P}_0 B_{n,0}(0) = \mathbf{P}_0.$$

Also, since

$$B_{n,n}(1) = \binom{n}{n} 1^n (1-1)^{(n-n)} = 1 \cdot 1 \cdot 0^0 = 1,$$

we get

$$\mathbf{P}(1) = \sum_{i=0}^{n} \mathbf{P}_i B_{n,i}(1) = \mathbf{P}_n B_{n,n}(1) = \mathbf{P}_n.$$

4 Curves

3. Another interesting property of the Bézier curve is its symmetry with respect to the numbering of the control points. If we number the points $\mathbf{P}_n, \mathbf{P}_{n-1}, \ldots, \mathbf{P}_0$, we get the same curve, only it goes from right (point $\mathbf{P}_0$) to left (point $\mathbf{P}_n$). The Bernstein polynomials satisfy the identity $B_{n,j}(t) = B_{n,n-j}(1-t)$, which can be proved directly and which can be used to prove the symmetry

$$\sum_{j=0}^{n} \mathbf{P}_j B_{n,j}(t) = \sum_{j=0}^{n} \mathbf{P}_{n-j} B_{n,j}(1-t).$$

4. The first derivative (the tangent vector) of the curve is straightforward to derive:

$$\mathbf{P}^t(t) = \sum_{i=0}^{n} \mathbf{P}_i B'_{n,i}(t)$$

$$= \sum_{0}^{n} \mathbf{P}_i \binom{n}{i} \left[i t^{i-1} (1-t)^{n-i} + t^i (n-i)(1-t)^{n-i-1}(-1) \right]$$

$$= \sum_{0}^{n} \mathbf{P}_i \binom{n}{i} i t^{i-1} (1-t)^{n-i} - \sum_{0}^{n-1} \mathbf{P}_i \binom{n}{i} t^i (n-i)(1-t)^{n-1-i}$$

(using the identity $n\binom{n-1}{i-1} = i\binom{n}{i}$, we get)

$$= n \sum_{1}^{n} \mathbf{P}_i \binom{n-1}{i-1} t^{i-1} (1-t)^{(n-1)-(i-1)} - n \sum_{0}^{n-1} \mathbf{P}_i \binom{n-1}{i} t^i (1-t)^{n-1-i}$$

(but $\binom{n-1}{i-1} t^{i-1} (1-t)^{(n-1)-(i-1)} = B_{n-1,i-1}(t)$, so)

$$= n \sum_{0}^{n-1} \mathbf{P}_{i+1} B_{n-1,i}(t) - n \sum_{0}^{n-1} \mathbf{P}_i B_{n-1,i}(t)$$

$$= n \sum_{0}^{n-1} [\mathbf{P}_{i+1} - \mathbf{P}_i] B_{n-1,i}(t)$$

$$= n \sum_{0}^{n-1} \Delta \mathbf{P}_i B_{n-1,i}(t), \quad \text{where} \quad \Delta \mathbf{P}_i = \mathbf{P}_{i+1} - \mathbf{P}_i. \tag{4.111}$$

Note that the tangent vector is another Bézier curve, based on the n control points $\Delta\mathbf{P}_i$ ($\Delta\mathbf{P}_i$ is the difference of two points, hence it is a vector, but since it is represented by a pair or a triplet, we can conveniently think of it as a point). As a result, the second derivative is obviously another Bézier curve based on the $n-1$ control points $\Delta^2 \mathbf{P}_i = \Delta\mathbf{P}_{i+1} - \Delta\mathbf{P}_i = \mathbf{P}_{i+2} - 2\mathbf{P}_{i+1} + \mathbf{P}_i$.

5. The weight functions $B_{n,i}(t)$ have a maximum at $t = i/n$. To see this, we first differentiate the weights

$$B'_{n,i}(t) = \binom{n}{i} \left[i t^{i-1} (1-t)^{n-i} + t^i (n-i)(1-t)^{n-i-1}(-1) \right]$$
$$= \binom{n}{i} i t^{i-1} (1-t)^{n-i} - \binom{n}{i} t^i (n-i)(1-t)^{n-1-i},$$

then equate the derivative to zero $\binom{n}{i}i\,t^{i-1}(1-t)^{n-i} - \binom{n}{i}t^i(n-i)(1-t)^{n-1-i} = 0$. Dividing by $t^{i-1}(1-t)^{n-i-1}$ yields $i(1-t) - t(n-i) = 0$ or $t = i/n$.

6. The two derivatives $\mathbf{P}^t(0)$ and $\mathbf{P}^t(1)$ are easy to derive from Equation (4.111) and are used to reshape the curve. They are $\mathbf{P}^t(0) = n(\mathbf{P}_1 - \mathbf{P}_0)$ and $\mathbf{P}^t(1) = n(\mathbf{P}_n - \mathbf{P}_{n-1})$. Since n is always positive, we get that $\mathbf{P}^t(0)$, the initial tangent of the curve, points in the direction from $\mathbf{P}_0$ to $\mathbf{P}_1$. This initial tangent can easily be controlled by moving point $\mathbf{P}_1$; similarly for the final tangent.

▶ 7. The Bézier curve features global control. This means that moving one control point $\mathbf{P}_i$ changes the entire curve. Most of the change, however, occurs at the vicinity of $\mathbf{P}_i$. This feature stems from the fact that the weight functions $B_{n,i}(t)$ are nonzero for all values of t (except $t = 0$ and $t = 1$) and thus any change in a control point $\mathbf{P}_i$ affects the contribution of the term $\mathbf{P}_i B_{n,i}(t)$ for all values of t.

8. The concept of the *convex hull* of a set of points is introduced in Section 5.3.5. Here, we show a connection between the Bézier curve and the convex hull. Let $\mathbf{P}_1$, $\mathbf{P}_2, \ldots, \mathbf{P}_n$ be a given set of points and let a point $\mathbf{P}$ be constructed as a barycentric sum of these points with non-negative weights, i.e.,

$$\mathbf{P} = \sum_{i=1}^{n} a_i \mathbf{P}_i, \quad \text{where} \quad \sum a_i = 1 \text{ and } a_i \geq 0. \tag{4.112}$$

It can be shown that the set of all points $\mathbf{P}$ satisfying Equation (4.112) lie in the convex hull of the given set of points. The Bézier curve, Equation (4.104), satisfies Equation (4.112) for all values of t, so all its points lie in the convex hull of the set of control points. Thus, the curve is said to have the *convex hull property*. The importance of this property is that it makes the Bézier curve more predictable. A designer specifying a set of control points needs just a little experience to visualize the shape of the curve, since the convex hull property guarantees that the curve will not "stray" far from the control points.

4.14.5 Connecting Bézier Curves

Since the Bézier curve is a polynomial of degree n, it is slow to calculate for large values of n. It is preferable to connect several Bézier segments, each defined by a few points—typically four to six—into one smooth curve. The condition for smooth connection of two such segments is simple to derive. We assume that the control points are divided into two sets $\mathbf{P}_0, \mathbf{P}_1, \ldots, \mathbf{P}_n$ and $\mathbf{Q}_0, \mathbf{Q}_1, \ldots, \mathbf{Q}_m$, where the last point $\mathbf{P}_n$ of the first set is identical to the first point $\mathbf{Q}_0$ of the second set. Since the tangent vectors satisfy

$$\mathbf{Q}^t(0) = m(\mathbf{Q}_1 - \mathbf{Q}_0) \quad \text{and} \quad \mathbf{P}^t(1) = n(\mathbf{P}_n - \mathbf{P}_{n-1}),$$

the condition for a smooth connection is $m\mathbf{Q}_1 - m\mathbf{Q}_0 = n\mathbf{P}_n - n\mathbf{P}_{n-1}$. Since $\mathbf{Q}_0 = \mathbf{P}_n$, the condition becomes

$$\mathbf{P}_n = \frac{m}{m+n}\mathbf{Q}_1 + \frac{n}{m+n}\mathbf{P}_{n-1}. \tag{4.113}$$

4 Curves

Hence, the condition for smooth linking is that the three points $\mathbf{P}_{n-1}$, $\mathbf{P}_n$, and $\mathbf{Q}_1$ be *collinear* (see Section 4.1). In the special case where $n = m$, Equation (4.113) reduces to $\mathbf{P}_n = 0.5\mathbf{Q}_1 + 0.5\mathbf{P}_{n-1}$ or $2\mathbf{P}_n = \mathbf{Q}_1 + \mathbf{P}_{n-1}$, which means that $\mathbf{P}_n$ should be the midpoint between $\mathbf{Q}_1$ and $\mathbf{P}_{n-1}$.

Breaking large curves into segments has the additional advantage of easy control. The Bézier curve offers only global control, but if it is constructed of separate segments, a change in the control points in one segment will not affect the behavior of the other segments. Figure 4.39 is an example of two Bézier segments connected smoothly.

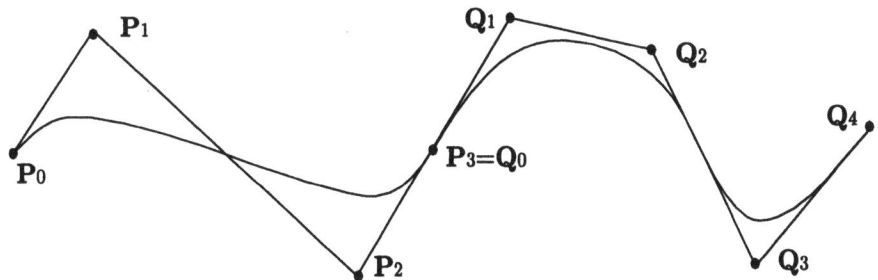

Figure 4.39: Connecting Bézier Segments.

4.14.6 Quadratic and Cubic Blending

We start with the linear blend $\mathbf{P}(t) = (1-t)\mathbf{P}_1 + t\mathbf{P}_2$ (Equation (4.4)). This means that if we select, for example, $t = 0.7$, then $\mathbf{P}(t)$ will be a blend of 30% of $\mathbf{P}_1$ and 70% of $\mathbf{P}_2$.

It is possible to blend points in nonlinear ways. An intuitive way to get, for example, quadratic blending is to square the two weights of the linear blend. However, the result, which is $\mathbf{P}(t) = (1-t)^2\mathbf{P}_1 + t^2\mathbf{P}_2$, depends on the particular coordinate axes used, since the two coefficients $(1-t)^2$ and t^2 are not barycentric. It turns out that the sum $(1-t)^2 + 2t(1-t) + t^2$ equals 1. As a result, we can use quadratic blending to blend three points, but not two.

Similarly, if we try a cubic blend by simply writing $\mathbf{P}(t) = (1-t)^3\mathbf{P}_1 + t^3\mathbf{P}_2$, we get the same problem. Cubic blending can be achieved by adding four terms with weights t^3, $3t^2(1-t)$, $3t(1-t)^2$, and $(1-t)^3$.

We therefore conclude that Bézier methods can be used for blending. The Bézier curve is a result of blending several points with the Bernstein polynomials, which add up to unity. Quadratic and cubic blending are special cases of the Bézier blending (or the Bézier interpolation).

4.14.7 The Bézier Curve as a Linear Interpolation

The original form of the Bézier curve, as developed by de Casteljau in 1959, uses an approach entirely different from that of Bézier. Specifically, it uses the two concepts of linear interpolation and the *mediation operator*.

Definition. The mediation operator $t[\mathbf{P}_0, \mathbf{P}_1]$ between two points $\mathbf{P}_0$ and $\mathbf{P}_1$ is defined as

$$t[\mathbf{P}_0, \mathbf{P}_1] = t\mathbf{P}_1 + (1-t)\mathbf{P}_0 = t(\mathbf{P}_1 - \mathbf{P}_0) + \mathbf{P}_0, \quad \text{where} \quad 0 \le t \le 1.$$

The actual definition is recursive. The mediation operator can be applied to any number of points according to

$$t[\mathbf{P}_0, \ldots, \mathbf{P}_n] = t[\,t[\mathbf{P}_0, \ldots, \mathbf{P}_{n-1}], t[\mathbf{P}_1, \ldots, \mathbf{P}_n]\,],$$

$$\vdots$$

$$t[\mathbf{P}_0, \mathbf{P}_1, \mathbf{P}_2, \mathbf{P}_3] = t[\,t[\mathbf{P}_0, \mathbf{P}_1, \mathbf{P}_2], t[\mathbf{P}_1, \mathbf{P}_2, \mathbf{P}_3]\,],$$
$$t[\mathbf{P}_0, \mathbf{P}_1, \mathbf{P}_2] = t[\,t[\mathbf{P}_0, \mathbf{P}_1], t[\mathbf{P}_1, \mathbf{P}_2]\,],$$
$$t[\mathbf{P}_0, \mathbf{P}_1] = t\mathbf{P}_1 + (1-t)\mathbf{P}_0 = t(\mathbf{P}_1 - \mathbf{P}_0) + \mathbf{P}_0, \quad \text{where} \quad 0 \le t \le 1.$$

This operator creates curves that interpolate between the points. It has the advantages of being a simple mathematical function (and thus fast to calculate) and of producing interpolation curves whose shape can easily be predicted. We now look at cases involving more and more points.

Case 1. Two points. Given the two points $\mathbf{P}_0$ and $\mathbf{P}_1$, we denote the straight segment connecting them $\mathbf{L}_{01}$. It is easy to see that $\mathbf{L}_{01} = t[\mathbf{P}_0, \mathbf{P}_1]$, since the mediation operator is a linear function of t and since $0[\mathbf{P}_0, \mathbf{P}_1] = \mathbf{P}_0$ and $1[\mathbf{P}_0, \mathbf{P}_1] = \mathbf{P}_1$. Notice that values of t below 0 or above 1 correspond to those parts of the line that don't lie between the two points. Such values may be of interest in certain cases but not in the present context. The interpolation curve between the two points is denoted by $\mathbf{P}_1(t)$ and is simply selected as the line $\mathbf{L}_{01}$ connecting the points. Hence, $\mathbf{P}_1(t) = \mathbf{L}_{01} = t[\mathbf{P}_0, \mathbf{P}_1]$. Notice that a straight line is also a polynomial of degree 1.

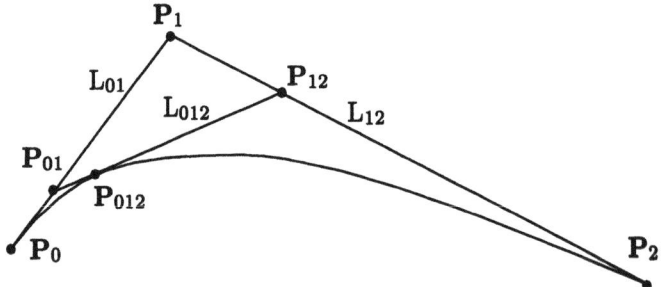

Figure 4.40: Repeated Linear Interpolation.

Case 2. Three points. Given the three points $\mathbf{P}_0$, $\mathbf{P}_1$, and $\mathbf{P}_2$ (Figure 4.40), the mediation operator can be used to build an interpolation curve between them by going through the following steps:

4 Curves

- Construct the two lines $\mathbf{L}_{01} = t[\mathbf{P}_0, \mathbf{P}_1]$, $\mathbf{L}_{12} = t[\mathbf{P}_1, \mathbf{P}_2]$.

- For some $0 \leq t_0 \leq 1$, consider the two points $\mathbf{P}_{01} = t_0[\mathbf{P}_0, \mathbf{P}_1]$ and $\mathbf{P}_{12} = t_0[\mathbf{P}_1, \mathbf{P}_2]$. Connect the points with a line $\mathbf{L}_{012}$. The expression for the line is, of course, $t[\mathbf{P}_{01}, \mathbf{P}_{12}]$ and it equals

$$\mathbf{L}_{012} = t[\mathbf{P}_{01}, \mathbf{P}_{12}] = t[\, t[\mathbf{P}_0, \mathbf{P}_1], t[\mathbf{P}_1, \mathbf{P}_2]\,] = t[\mathbf{P}_0, \mathbf{P}_1, \mathbf{P}_2].$$

- For the same t_0, select point $\mathbf{P}_{012} = t_0[\mathbf{P}_0, \mathbf{P}_1, \mathbf{P}_2]$ on $\mathbf{L}_{012}$. The point can be expressed as

$$\mathbf{P}_{012} = t_0[\mathbf{P}_0, \mathbf{P}_1, \mathbf{P}_2] = t_0[\mathbf{P}_{01}, \mathbf{P}_{12}] = t_0[\, t_0[\mathbf{P}_0, \mathbf{P}_1], t_0[\mathbf{P}_1, \mathbf{P}_2]\,].$$

Now, release t_0 and let it vary from 0 to 1. Point $\mathbf{P}_{012}$ slides along the line $\mathbf{L}_{012}$, whose endpoints will, in turn, slide along $\mathbf{L}_{01}$ and $\mathbf{L}_{12}$. The curve described by point $\mathbf{P}_{012}$ as it is sliding is the interpolation curve for $\mathbf{P}_0$, $\mathbf{P}_1$, $\mathbf{P}_2$ that we are seeking. We denote it by $\mathbf{P}_2(t)$ and its expression is easy to calculate, using the definition of $t[\mathbf{P}_i, \mathbf{P}_j]$:

$$\begin{aligned}
\mathbf{P}_2(t) &= t[\mathbf{P}_0, \mathbf{P}_1, \mathbf{P}_2] \\
&= t[\, t[\mathbf{P}_0, \mathbf{P}_1], t[\mathbf{P}_1, \mathbf{P}_2]\,] \\
&= t[t\mathbf{P}_1 + (1-t)\mathbf{P}_0, t\mathbf{P}_2 + (1-t)\mathbf{P}_1] \\
&= t[t\mathbf{P}_2 + (1-t)\mathbf{P}_1] + (1-t)[t\mathbf{P}_1 + (1-t)\mathbf{P}_0] \\
&= \mathbf{P}_0(1-t)^2 + 2\mathbf{P}_1 t(1-t) + \mathbf{P}_2 t^2.
\end{aligned}$$

$\mathbf{P}_2(t)$ is therefore the Bézier curve for three points.

Case 3. Four points. Given the four points $\mathbf{P}_0$, $\mathbf{P}_1$, $\mathbf{P}_2$, and $\mathbf{P}_3$, we follow similar steps:

- Construct the three lines $\mathbf{L}_{01} = t[\mathbf{P}_0, \mathbf{P}_1]$, $\mathbf{L}_{12} = t[\mathbf{P}_1, \mathbf{P}_2]$, and $\mathbf{L}_{23} = t[\mathbf{P}_2, \mathbf{P}_3]$.

- Select three points, $\mathbf{P}_{01} = t_0[\mathbf{P}_0, \mathbf{P}_1]$, $\mathbf{P}_{12} = t_0[\mathbf{P}_1, \mathbf{P}_2]$, $\mathbf{P}_{23} = t_0[\mathbf{P}_2, \mathbf{P}_3]$, and construct lines $\mathbf{L}_{012} = t[\mathbf{P}_0, \mathbf{P}_1, \mathbf{P}_2] = t[\mathbf{P}_{01}, \mathbf{P}_{12}]$, and $\mathbf{L}_{123} = t[\mathbf{P}_1, \mathbf{P}_2, \mathbf{P}_3] = t[\mathbf{P}_{12}, \mathbf{P}_{23}]$.

- Select two points, $\mathbf{P}_{012}$ on segment $\mathbf{L}_{012}$ and $\mathbf{P}_{123}$ on segment $\mathbf{L}_{123}$. Construct a new segment $\mathbf{L}_{0123}$ as the mediation $t[\mathbf{P}_0, \mathbf{P}_1, \mathbf{P}_2, \mathbf{P}_3] = t[\mathbf{P}_{012}, \mathbf{P}_{123}]$.

- Select a general point $\mathbf{P}_{0123}$ on $\mathbf{L}_{0123}$.

When t varies from 0 to 1, point $\mathbf{P}_{0123}$ slides along $\mathbf{L}_{0123}$, whose endpoints, in turn, slide along $\mathbf{L}_{012}$ and $\mathbf{L}_{123}$, which also slide. The entire structure, which resembles a *scaffolding* (Figure Ans.20), slides along the original three lines. The interpolation curve for the four original points is denoted by $\mathbf{P}_3(t)$ and its expression is not hard to calculate, using the expression for $\mathbf{P}_2(t) = t[\mathbf{P}_0, \mathbf{P}_1, \mathbf{P}_2]$:

$$\begin{aligned}
\mathbf{P}_3(t) = t[\mathbf{P}_0, \mathbf{P}_1, \mathbf{P}_2, \mathbf{P}_3] &= t[\,t[\mathbf{P}_0, \mathbf{P}_1, \mathbf{P}_2], t[\mathbf{P}_1, \mathbf{P}_2, \mathbf{P}_3]\,]\\
&= t[t^2\mathbf{P}_3 + 2t(1-t)\mathbf{P}_2 + (1-t)^2\mathbf{P}_1]\\
&\quad + (1-t)[t^2\mathbf{P}_2 + 2t(1-t)\mathbf{P}_1 + (1-t)^2\mathbf{P}_0]\\
&= t^3\mathbf{P}_3 + 3t^2(1-t)\mathbf{P}_2 + 3t(1-t)^2\mathbf{P}_1 + (1-t)^3\mathbf{P}_0.
\end{aligned}$$

$\mathbf{P}_3(t)$ is therefore the Bézier curve for four points.

Case 4. In the general case, $n+1$ points, $\mathbf{P}_0, \mathbf{P}_1, \ldots, \mathbf{P}_n$ (where $n > 0$), are given. The interpolation curve is, similarly, $t[\mathbf{P}_0, \mathbf{P}_1, \ldots, \mathbf{P}_n] = t[\mathbf{P}_{01\ldots n-1}, \mathbf{P}_{12\ldots n}]$. It can be proved by induction that its value is the degree-n polynomial

$$\mathbf{P}_n(t) = \sum_{i=0}^{n} \mathbf{P}_i B_{n,i}(t), \quad \text{where} \quad B_{n,i}(t) = \binom{n}{i} t^i (1-t)^{n-i},$$

which is the Bézier curve for $n+1$ points. The two approaches to curve construction, the one using Bernstein polynomials and the one using scaffolding, are thus equivalent.

Table 4.41 summarizes the process of scaffolding in the general case. The process takes n steps. In the first step, n new points are constructed between the original $n+1$ control points. In the second step, $n-1$ new points are constructed, between the n points of step 1 and so on, up to step n, where one point is constructed. The total number of points constructed during the entire process is therefore

$$n + (n-1) + (n-2) + \cdots + 2 + 1 = n(n+1)/2.$$

Step	Points constructed	# of points
1	$\mathbf{P}_{01}\,\mathbf{P}_{12}\,\mathbf{P}_{23}\ldots\mathbf{P}_{n-1,n}$	n
2	$\mathbf{P}_{012}\,\mathbf{P}_{123}\,\mathbf{P}_{234}\ldots\mathbf{P}_{n-2,n-1,n}$	$n-1$
3	$\mathbf{P}_{0123}\,\mathbf{P}_{1234}\,\mathbf{P}_{2345}\ldots\mathbf{P}_{n-3,n-2,n-1,n}$	$n-2$
$\vdots$	$\vdots$	$\vdots$
n	$\mathbf{P}_{0123\ldots n}$	$n-2$

Table 4.41: The n Steps of Scaffolding.

4.14.8 Nonsmooth Bézier Curves

The Bézier curve may have cusps (kinks or sharp corners) at points where it has to loop on itself. At such points, the curve has no definite tangent vector, so if we try to calculate the tangent, we end up with the indefinite direction $(0,0)$.

Example: Figure 4.42 shows three cubic Bézier curves. All three are generated by control points $\mathbf{P}_0 = (0,0)$ and $\mathbf{P}_3 = (1,0)$. The other two (interior) control points are as follows:

4 Curves

1. $\mathbf{P}_1 = (0.7, 1)$ and $\mathbf{P}_2 = (0.3, 1)$. These produce the smooth curve (dot-dashed) of Figure 4.42a.

2. $\mathbf{P}_1 = (1, 1)$ and $\mathbf{P}_2 = (0, 1)$. Opening up the points produces the cusp of Figure 4.42b (solid curve).

3. $\mathbf{P}_1 = (1.5, 1)$ and $\mathbf{P}_2 = (-0.5, 1)$ (points not shown). Opening up the points even more produces a loop (the dashed curve of Figure 4.42c).

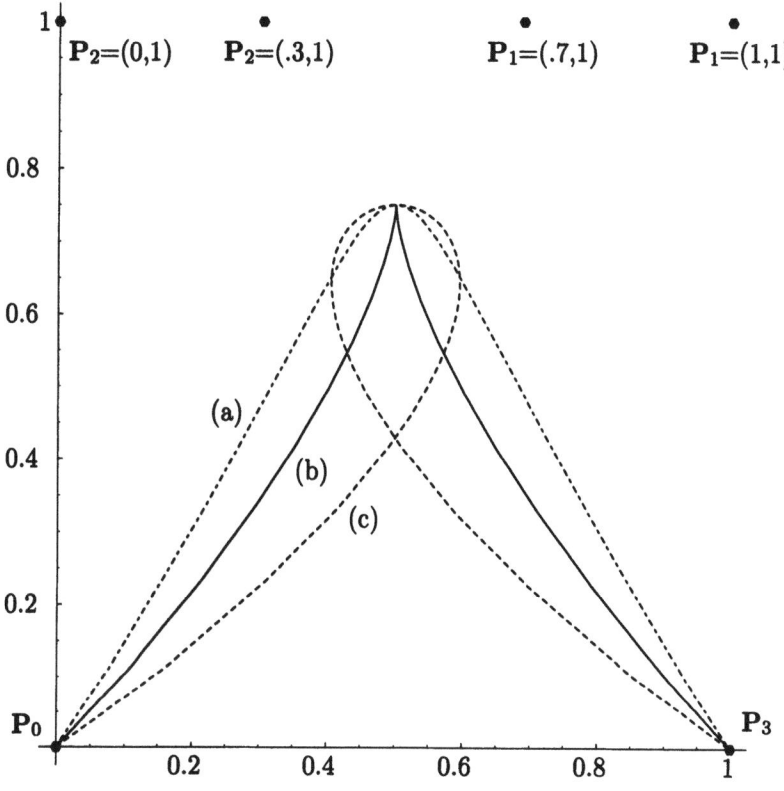

Figure 4.42: Three Bézier Curves.

▶ **Exercise 4.67:** Calculate the curve of case 2 and show that it has a cusp at its midpoint.

(See also Exercise 4.66 for another nonsmooth Bézier curve.)

4.14.9 Subdividing a Parametric Curve

Parametric curves are defined by means of points (data or control) and sometimes also vectors. Typical ways to edit such a curve are to move points around and to add points. Intuitively, it is clear that adding points allows for finer control of the shape of the curve. On the other hand, adding points means more calculations to compute and display the curve.

It therefore seems that the best method is to start with a few points, and if these are not enough and we still do not obtain the right shape of the curve, add one point (or a few points) at a time until the desired shape is achieved.

4.14 The Bézier Curve

This section discusses an approach whereby this is achieved by subdividing a parametric curve into two segments. Together, the two segments have the same shape as the original curve, but they are defined by more entities (points or vectors), making it possible to fine-tune the curve. This approach is later applied to the Bézier curve.

> The control of large numbers is possible, and like unto that of small numbers, if we subdivide them.
> — Sun Tze

We limit our discussion to cubic curves, but the method illustrated here applies to curves of any degree. Let

$$\mathbf{P}(t) = (t^3, t^2, t, 1)\mathbf{M}\begin{pmatrix}\mathbf{P}_0\\\mathbf{P}_1\\\mathbf{P}_2\\\mathbf{P}_3\end{pmatrix} \quad (4.114)$$

be any cubic parametric curve defined by four nonscalar entities (points or vectors) where the parameter t varies from 0 to 1. We now construct the two halves $\mathbf{P}_1(t)$ and $\mathbf{P}_2(t)$ of this curve given by the parameter ranges $[0, 0.5]$ and $[0.5, 1]$ (Section 4.14.10 shows how the unequal ranges $[0, \alpha]$ and $[\alpha, 1]$ can be used instead).

Each half is given by an expression similar to Equation (4.114) and is based on a new set of entities $\mathbf{Q}_i$ computed from the original set $\mathbf{P}_i$. To construct the first half of the curve, we define a new parameter $u = 2t$. When t varies in the range $[0, 0.5]$, u varies from 0 to 1. The first half of the curve is obtained from Equation (4.114) by substituting $t = u/2$:

$$\begin{aligned}\mathbf{P}_1(t) &= (u^3/8, u^2/4, u/2, 1)\mathbf{M}\begin{pmatrix}\mathbf{P}_0\\\mathbf{P}_1\\\mathbf{P}_2\\\mathbf{P}_3\end{pmatrix}\\ &= (u^3, u^2, u, 1)\begin{pmatrix}\frac{1}{8} & 0 & 0 & 0\\ 0 & \frac{1}{4} & 0 & 0\\ 0 & 0 & \frac{1}{2} & 0\\ 0 & 0 & 0 & 1\end{pmatrix}\mathbf{M}\begin{pmatrix}\mathbf{P}_0\\\mathbf{P}_1\\\mathbf{P}_2\\\mathbf{P}_3\end{pmatrix}\\ &= (u^3, u^2, u, 1)\mathbf{L}\mathbf{M}\begin{pmatrix}\mathbf{P}_0\\\mathbf{P}_1\\\mathbf{P}_2\\\mathbf{P}_3\end{pmatrix}\\ &= (u^3, u^2, u, 1)\mathbf{M}\begin{pmatrix}\mathbf{Q}_0\\\mathbf{Q}_1\\\mathbf{Q}_2\\\mathbf{Q}_3\end{pmatrix}.\end{aligned} \quad (4.115)$$

The last line of Equation (4.115) expresses $\mathbf{P}_1(t)$ in terms of new entities $\mathbf{Q}_i$. This

4 Curves

expression shows that these entities can be calculated from the equation

$$\mathbf{M}\begin{pmatrix} \mathbf{Q}_0 \\ \mathbf{Q}_1 \\ \mathbf{Q}_2 \\ \mathbf{Q}_3 \end{pmatrix} = \mathbf{L}\mathbf{M}\begin{pmatrix} \mathbf{P}_0 \\ \mathbf{P}_1 \\ \mathbf{P}_2 \\ \mathbf{P}_3 \end{pmatrix}, \text{ whose solution is } \begin{pmatrix} \mathbf{Q}_0 \\ \mathbf{Q}_1 \\ \mathbf{Q}_2 \\ \mathbf{Q}_3 \end{pmatrix} = \mathbf{M}^{-1}\mathbf{L}\mathbf{M}\begin{pmatrix} \mathbf{P}_0 \\ \mathbf{P}_1 \\ \mathbf{P}_2 \\ \mathbf{P}_3 \end{pmatrix}. \quad (4.116)$$

The second half, $\mathbf{P}_2(t)$, of the curve is calculated similarly. We first define a new parameter $u = 2t - 1$. When t varies in the range $[0.5, 1]$, u varies from 0 to 1. The second half of the curve is obtained from Equation (4.114) by substituting $t = (u+1)/2$:

$$\mathbf{P}_2(t) = ((u+1)^3/8, (u+1)^2/4, (u+1)/2, 1)\mathbf{M}\begin{pmatrix} \mathbf{P}_0 \\ \mathbf{P}_1 \\ \mathbf{P}_2 \\ \mathbf{P}_3 \end{pmatrix}$$

$$= (u^3, u^2, u, 1)\begin{pmatrix} \frac{1}{8} & 0 & 0 & 0 \\ \frac{3}{8} & \frac{1}{4} & 0 & 0 \\ \frac{3}{8} & \frac{2}{4} & \frac{1}{2} & 0 \\ \frac{1}{8} & \frac{1}{4} & \frac{1}{2} & 1 \end{pmatrix}\mathbf{M}\begin{pmatrix} \mathbf{P}_0 \\ \mathbf{P}_1 \\ \mathbf{P}_2 \\ \mathbf{P}_3 \end{pmatrix}$$

$$= (u^3, u^2, u, 1)\mathbf{R}\mathbf{M}\begin{pmatrix} \mathbf{P}_0 \\ \mathbf{P}_1 \\ \mathbf{P}_2 \\ \mathbf{P}_3 \end{pmatrix}$$

$$= (u^3, u^2, u, 1)\mathbf{M}\begin{pmatrix} \mathbf{Q}_4 \\ \mathbf{Q}_5 \\ \mathbf{Q}_6 \\ \mathbf{Q}_7 \end{pmatrix}. \quad (4.117)$$

The new entities $\mathbf{Q}_i$ are calculated for this second half by

$$\begin{pmatrix} \mathbf{Q}_4 \\ \mathbf{Q}_5 \\ \mathbf{Q}_6 \\ \mathbf{Q}_7 \end{pmatrix} = \mathbf{M}^{-1}\mathbf{R}\mathbf{M}\begin{pmatrix} \mathbf{P}_0 \\ \mathbf{P}_1 \\ \mathbf{P}_2 \\ \mathbf{P}_3 \end{pmatrix}. \quad (4.118)$$

Given matrix $\mathbf{M}$ and four entities $\mathbf{P}_i$, the eight new entities $\mathbf{Q}_i$ can be calculated from Equations (4.116) and (4.118). The generalization of this method to higher-degree curves is straightforward. This method is now applied to the cubic Bézier curve as an example. Matrix $\mathbf{M}$ and its inverse are

$$\mathbf{M} = \begin{pmatrix} -1 & 3 & -3 & 1 \\ 3 & -6 & 3 & 0 \\ -3 & 3 & 0 & 0 \\ 1 & 0 & 0 & 0 \end{pmatrix}, \quad \mathbf{M}^{-1} = \begin{pmatrix} 0 & 0 & 0 & 1 \\ 0 & 0 & \frac{1}{3} & 1 \\ 0 & \frac{1}{3} & \frac{2}{3} & 1 \\ 1 & 1 & 1 & 1 \end{pmatrix}.$$

The matrix products of Equations (4.116) and (4.118) now become

$$\mathbf{M}^{-1}\mathbf{LM} = \begin{pmatrix} 1 & 0 & 0 & 0 \\ \frac{1}{2} & \frac{1}{2} & 0 & 0 \\ \frac{1}{4} & \frac{2}{4} & \frac{1}{4} & 0 \\ \frac{1}{8} & \frac{3}{8} & \frac{3}{8} & \frac{1}{8} \end{pmatrix}, \quad \mathbf{M}^{-1}\mathbf{RM} = \begin{pmatrix} \frac{1}{8} & \frac{3}{8} & \frac{3}{8} & \frac{1}{8} \\ 0 & \frac{1}{4} & \frac{2}{4} & \frac{1}{4} \\ 0 & 0 & \frac{1}{2} & \frac{1}{2} \\ 0 & 0 & 0 & 1 \end{pmatrix}. \quad (4.119)$$

The eight new entities (that are control points in this case) are

$$Q_0 = P_0,$$
$$Q_1 = \frac{1}{2}P_0 + \frac{1}{2}P_1 = \frac{1}{2}(P_0 + P_1),$$
$$Q_2 = \frac{1}{4}P_0 + \frac{2}{4}P_1 + \frac{1}{4}P_2 = \frac{1}{2}\left(\frac{1}{2}(P_0 + P_1) + \frac{1}{2}(P_1 + P_2)\right),$$
$$Q_3 = \frac{1}{8}P_0 + \frac{3}{8}P_1 + \frac{3}{8}P_2 + \frac{1}{8}P_3$$
$$= \frac{1}{2}\left(\frac{1}{2}\left(\frac{1}{2}(P_0 + P_1) + \frac{1}{2}(P_1 + P_2)\right) + \frac{1}{2}\left(\frac{1}{2}(P_1 + P_2) + \frac{1}{2}(P_2 + P_3)\right)\right),$$
$$Q_4 = \frac{1}{8}P_0 + \frac{3}{8}P_1 + \frac{3}{8}P_2 + \frac{1}{8}P_3$$
$$= \frac{1}{2}\left(\frac{1}{2}\left(\frac{1}{2}(P_0 + P_1) + \frac{1}{2}(P_1 + P_2)\right) + \frac{1}{2}\left(\frac{1}{2}(P_1 + P_2) + \frac{1}{2}(P_2 + P_3)\right)\right),$$
$$Q_5 = \frac{1}{4}P_1 + \frac{2}{4}P_2 + \frac{1}{4}P_3 = \frac{1}{2}\left(\frac{1}{2}(P_1 + P_2) + \frac{1}{2}(P_2 + P_3)\right),$$
$$Q_6 = \frac{1}{2}P_1 + \frac{1}{2}P_2 = \frac{1}{2}(P_1 + P_2),$$
$$Q_7 = P_3.$$

Section 4.14.10 shows a different approach, using the mediation operator, to the problem of subdividing a curve. That approach is applied to the Bézier curve.

4.14.10 Subdividing the Bézier Curve

Bézier methods are interactive. It is possible to control the shape of the curve by moving the control points and by smoothly connecting individual segments. Imagine a situation where the points are moved and maneuvered about, but the curve "refuses" to get the right shape. This may happen when there are not enough points. There are two ways to increase the number of points. One is to add a point to a segment while increasing its degree. This is called *degree elevation* and is discussed in Section 4.14.11.

An alternative way is to subdivide a segment into two segments such that there is no change in the shape of the curve. If the segment is of degree k (i.e., based on $k+1$ control points), this is done by adding $2k-1$ new control points and deleting $k-1$ of the original points, bringing the number of points to $(k+1) + (2k-1) - (k-1) = 2k+1$. Each new segment is based on $k+1$ points and they share one

4 Curves

of the new points. With more points, it is now possible to manipulate the control points of the two segments in order to fine-tune the shape of the segments.

The new points being added consist of some of the ones constructed in the last k steps of the scaffolding process. For the case $k = 2$ (quadratic curve segments), the three points $\mathbf{P}_{01}$, $\mathbf{P}_{12}$, and $\mathbf{P}_{012}$ are added and the single point $\mathbf{P}_1$ is deleted (Figure 4.40). The two new segments consist of points $\mathbf{P}_0$, $\mathbf{P}_{01}$, and $\mathbf{P}_{012}$, and $\mathbf{P}_{012}$, $\mathbf{P}_{12}$, and $\mathbf{P}_2$. For the case $k = 3$ (cubic segments), the five points $\mathbf{P}_{01}$, $\mathbf{P}_{23}$, $\mathbf{P}_{012}$, $\mathbf{P}_{123}$, and $\mathbf{P}_{0123}$ are added and the two points $\mathbf{P}_1$ and $\mathbf{P}_2$ are deleted (Figure Ans.20). The two new segments consist of points $\mathbf{P}_0$, $\mathbf{P}_{01}$, $\mathbf{P}_{012}$, and $\mathbf{P}_{0123}$ and $\mathbf{P}_{0123}$, $\mathbf{P}_{123}$, $\mathbf{P}_{23}$, and $\mathbf{P}_3$.

Using the mediation operator to express the new points in the scaffolding in terms of the original control points produces, for the quadratic case

$$\mathbf{P}_{01} = \alpha(\mathbf{P}_0+\mathbf{P}_1), \mathbf{P}_{12} = \alpha(\mathbf{P}_1+\mathbf{P}_2), \mathbf{P}_{23} = \alpha(\mathbf{P}_2+\mathbf{P}_3), \mathbf{P}_{012} = \alpha^2(\mathbf{P}_0+2\mathbf{P}_1+\mathbf{P}_2),$$

where α is any value in the range $[0,1]$. We can therefore write

$$\begin{pmatrix} \mathbf{P}_0 \\ \mathbf{P}_{01} \\ \mathbf{P}_{012} \end{pmatrix} = \frac{1}{\alpha^2} \begin{pmatrix} \alpha^2 & 0 & 0 \\ \alpha & \alpha & 0 \\ 1 & 2 & 1 \end{pmatrix} \begin{pmatrix} \mathbf{P}_0 \\ \mathbf{P}_1 \\ \mathbf{P}_2 \end{pmatrix}, \quad \begin{pmatrix} \mathbf{P}_{012} \\ \mathbf{P}_{12} \\ \mathbf{P}_2 \end{pmatrix} = \frac{1}{\alpha^2} \begin{pmatrix} 1 & 2 & 1 \\ 0 & \alpha & \alpha \\ 0 & 0 & \alpha^2 \end{pmatrix} \begin{pmatrix} \mathbf{P}_0 \\ \mathbf{P}_1 \\ \mathbf{P}_2 \end{pmatrix}$$

for the left and right segments, respectively.

▶ **Exercise 4.68:** Use the mediation operator to calculate the scaffolding for the cubic case (four control points). Use $\alpha = 1/2$ and write the results in terms of matrices, as above.

In the general case where an $n+1$-point Bézier curve is subdivided, the $n-1$ points being deleted are $\mathbf{P}_1, \mathbf{P}_2, \ldots, \mathbf{P}_{n-1}$ (the original $n-1$ interior control points). The $2n-1$ points added are the first and last points constructed in each scaffolding step (except the last step, where only one point is constructed). Table 4.41 shows that these are points $\mathbf{P}_{01}$, $\mathbf{P}_{n-1,n}$ (from step 1), $\mathbf{P}_{012}$, $\mathbf{P}_{n-2,n-1,n}$ (from step 2), $\mathbf{P}_{0123}$, $\mathbf{P}_{n-3,n-2,n-1,n}$ (from step 3), up to $\mathbf{P}_{0123\ldots n}$ from step n.

The $2n-1$ points being added are therefore

$$\mathbf{P}_{01}, \mathbf{P}_{012}, \mathbf{P}_{0123}, \ldots, \mathbf{P}_{0123\ldots n}, \mathbf{P}_{123\ldots n}, \mathbf{P}_{23\ldots n}, \ldots, \mathbf{P}_{n-1,n}.$$

4.14.11 Degree Elevation

In this section, we approach the problem of degree elevation of the Bézier curve in a general way. We start with a Bézier curve $\mathbf{P}_n(t)$ of degree n (i.e., defined by $n+1$ control points). We then show how to select a set of $n+2$ control points such that the Bézier curve $\mathbf{P}_{n+1}(t)$ (which has degree $n+1$) defined by them will have the same shape as the original curve.

The advantage of degree elevation is that the new curve is based on more control points and is therefore easier to edit by moving the points. Its shape can be better fine-tuned than that of the original curve. The Bézier curve, however, features global control, which means that moving any point affects the shape of the entire curve. Degree elevation should therefore be used with caution.

Our approach is to create the new degree-$n+1$ curve as the combination $\mathbf{P}_{n+1}(t) = t\mathbf{P}_n(t) + (1-t)\mathbf{P}_n(t)$. We use the notation

$$\mathbf{P}_n(t) = \sum_{i=0}^{n} \binom{n}{i} t^i (1-t)^{n-i} \mathbf{P}_i \stackrel{\text{def}}{=} \langle \mathbf{P}_0, \mathbf{P}_1, \ldots, \mathbf{P}_n \rangle.$$

(The angle bracket notation implies that each point should be multiplied by the corresponding Bernstein polynomial and the products summed.)

The first step is to express $t\mathbf{P}_n(t)$ in the new notation

$$t\mathbf{P}_n(t) = \sum_{i=0}^{n} \binom{n}{i} t^{i+1}(1-t)^{n-i} \mathbf{P}_i = \sum_{k=1}^{m} \binom{m-1}{k-1} t^k (1-t)^{m-k} \mathbf{P}_{k-1}$$

$$= \sum_{k=0}^{m} \binom{m}{k} t^k (1-t)^{m-k} \frac{k}{m} \mathbf{P}_{k-1} = \left\langle 0, \frac{\mathbf{P}_0}{n+1}, \frac{2\mathbf{P}_1}{n+1}, \ldots, \frac{n\mathbf{P}_{n-1}}{n+1}, \mathbf{P}_n \right\rangle.$$

Here, we first use the substitutions $k = i+1$ and $m = n+1$, and then the identity

$$\binom{m-1}{k-1} = \frac{k}{m}\binom{m}{k}.$$

The next step is to similarly express $(1-t)\mathbf{P}_n(t)$ in the new notation:

$$(1-t)\mathbf{P}_n(t) = \left\langle \mathbf{P}_0, \frac{n\mathbf{P}_1}{n+1}, \frac{(n-1)\mathbf{P}_2}{n+1}, \ldots, \frac{\mathbf{P}_n}{n+1}, 0 \right\rangle.$$

Adding the two expressions produces

$$\mathbf{P}_{n+1}(t) = (1-t)\mathbf{P}_n(t) + t\mathbf{P}_n(t)$$

$$= \left\langle 0, \frac{\mathbf{P}_0}{n+1}, \frac{2\mathbf{P}_1}{n+1}, \ldots, \frac{n\mathbf{P}_{n-1}}{n+1}, \mathbf{P}_n \right\rangle$$

$$+ \left\langle \mathbf{P}_0, \frac{n\mathbf{P}_1}{n+1}, \frac{(n-1)\mathbf{P}_2}{n+1}, \ldots, \frac{\mathbf{P}_n}{n+1}, 0 \right\rangle$$

$$= \left\langle \mathbf{P}_0, \frac{\mathbf{P}_0 + n\mathbf{P}_1}{n+1}, \frac{2\mathbf{P}_1 + (n-1)\mathbf{P}_2}{n+1}, \ldots, \frac{n\mathbf{P}_{n-1} + \mathbf{P}_n}{n+1}, \mathbf{P}_n \right\rangle, \quad (4.120)$$

which shows the $n+2$ control points defining the new, elevated Bézier curve.

If the new control points are denoted by $\mathbf{Q}_i$, then the expression above can be summarized by the following notation:

$$\mathbf{Q}_0 = \mathbf{P}_0,$$
$$\mathbf{Q}_i = a_i \mathbf{P}_{i-1} + (1 - a_i)\mathbf{P}_i, \quad \text{where} \quad a_i = \frac{i}{n+1}, \quad i = 1, 2, \ldots, n,$$
$$\mathbf{Q}_{n+1} = \mathbf{P}_n.$$

4 Curves

▸ **Exercise 4.69:** Why will $\mathbf{P}_{n+1}(t)$ have the same shape as the original curve $\mathbf{P}_n(t)$?

▸ **Exercise 4.70:** Given the quadratic Bézier curve defined by the three control points $\mathbf{P}_0$, $\mathbf{P}_1$, and $\mathbf{P}_2$, elevate its degree twice and show the five new control points.

▸ **Exercise 4.71:** Given the four control points $\mathbf{P}_0 = (0,0)$, $\mathbf{P}_1 = (1,2)$, $\mathbf{P}_2 = (3,2)$, and $\mathbf{P}_3 = (2,0)$, elevate the degree of the Bézier curve defined by them.

4.14.12 Reparametrizing the Curve

The parameter t varies normally in the range $[0, 1]$. It is, however, easy to reparametrize the Bézier curve such that its parameter varies in an arbitrary range $[a, b]$, where a and b are real and $a \leq b$. The new curve is denoted by $\mathbf{P}_{ab}(t)$ and is simply the original curve with a different parameter:

$$\mathbf{P}_{ab}(t) = \mathbf{P}\left(\frac{t-a}{b-a}\right).$$

The two functions $\mathbf{P}_{ab}(t)$ and $\mathbf{P}(t)$ produce the same curve when t varies from a to b in the former and from 0 to 1 in the latter. Notice that the new curve has tangent vector

$$\mathbf{P}_{ab}^t(t) = \frac{1}{b-a}\mathbf{P}^t\left(\frac{t-a}{b-a}\right).$$

Reparametrization can also be used to answer the question: Given a Bézier curve $\mathbf{P}(t)$ where $0 \leq t \leq 1$, how can we calculate a curve $\mathbf{Q}(t)$ that's defined on an arbitrary part of $\mathbf{P}(t)$? More specifically, if $\mathbf{P}(t)$ is defined by control points $\mathbf{P}_i$ and if we select an interval $[a, b]$, how can we calculate control points $\mathbf{Q}_i$ such that the curve $\mathbf{Q}(t)$ based on them will go from $\mathbf{P}(a)$ to $\mathbf{P}(b)$ [i.e., $\mathbf{Q}(0) = \mathbf{P}(a)$ and $\mathbf{Q}(1) = \mathbf{P}(b)$] and will be identical to $\mathbf{P}(t)$ in that interval? As an example, if $[a, b] = [0, 0.5]$, then $\mathbf{Q}(t)$ will be identical to the first half of $\mathbf{P}(t)$. The point is that the interval $[a, b]$ does not have to be inside $[0, 1]$. We may select, for example, $[a, b] = [0.9, 1.5]$ and end up with a curve $\mathbf{Q}(t)$ that will go from $\mathbf{P}(0.9)$ to $\mathbf{P}(1.5)$ as t varies from 0 to 1. Even though the Bézier curve was originally designed with $0 \leq t \leq 1$ in mind, it can still be calculated for t values outside this range. If we like its shape in the range $[0.2, 1.1]$, we may want to calculate new control points $\mathbf{Q}_i$ and obtain a new curve $\mathbf{Q}(t)$ that has this shape when *its* parameter varies in the standard range $[0, 1]$.

Our approach is to define the new curve $\mathbf{Q}(t)$ as $\mathbf{P}([b-a]t + a)$ and express the control points $\mathbf{Q}_i$ of $\mathbf{Q}(t)$ in terms of the control points $\mathbf{P}_i$ and a and b. We illustrate this technique with the cubic Bézier curve. This curve is given by Equation (4.107) and we can thus write

$$\mathbf{Q}(t) = \mathbf{P}([b-a]t + a)$$

$$= \left(([b-a]t+a)^3, ([b-a]t+a)^2, ([b-a]t+a), 1\right) \begin{pmatrix} -1 & 3 & -3 & 1 \\ 3 & -6 & 3 & 0 \\ -3 & 3 & 0 & 0 \\ 1 & 0 & 0 & 0 \end{pmatrix} \begin{pmatrix} \mathbf{P}_0 \\ \mathbf{P}_1 \\ \mathbf{P}_2 \\ \mathbf{P}_3 \end{pmatrix}$$

$$= (t^3, t^2, t, 1) \begin{pmatrix} (b-a)^3 & 0 & 0 & 0 \\ 3a(b-a)^2 & (b-a)^2 & 0 & 0 \\ 3a^2(b-a) & 2a(b-a) & b-a & 0 \\ a^3 & a^2 & a & 1 \end{pmatrix} \begin{pmatrix} -1 & 3 & -3 & 1 \\ 3 & -6 & 3 & 0 \\ -3 & 3 & 0 & 0 \\ 1 & 0 & 0 & 0 \end{pmatrix} \begin{pmatrix} \mathbf{P}_0 \\ \mathbf{P}_1 \\ \mathbf{P}_2 \\ \mathbf{P}_3 \end{pmatrix}$$

$$= \mathbf{T}(t) \cdot \mathbf{A} \cdot \mathbf{M} \cdot \mathbf{P}$$
$$= \mathbf{T}(t) \cdot \mathbf{M} \cdot \mathbf{M}^{-1} \cdot \mathbf{A} \cdot \mathbf{M} \cdot \mathbf{P}$$
$$= \mathbf{T}(t) \cdot \mathbf{M} \cdot (\mathbf{M}^{-1} \cdot \mathbf{A} \cdot \mathbf{M}) \cdot \mathbf{P}$$
$$= \mathbf{T}(t) \cdot \mathbf{M} \cdot \mathbf{B} \cdot \mathbf{P}$$
$$= \mathbf{T}(t) \cdot \mathbf{M} \cdot \mathbf{Q},$$

where

$$\mathbf{B} = \mathbf{M}^{-1} \cdot \mathbf{A} \cdot \mathbf{M}$$
$$= \begin{pmatrix} (1-a)^3 & 3(a-1)^2 a & 3(1-a)a^2 & a^3 \\ (a-1)^2(1-b) & (a-1)(-2a-b+3ab) & a(a+2b-3ab) & a^2 b \\ (1-a)(-1+b)^2 & (b-1)(-a-2b+3ab) & b(2a+b-3ab) & ab^2 \\ (1-b)^3 & 3(b-1)^2 b & 3(1-b)b^2 & b^3 \end{pmatrix}.$$
(4.121)

The four new control points $\mathbf{Q}_i$, $i = 0, 1, 2, 3$ are thus obtained by selecting specific values for a and b, calculating matrix $\mathbf{B}$, and multiplying it by the column $\mathbf{P} = (\mathbf{P}_0, \mathbf{P}_1, \mathbf{P}_2, \mathbf{P}_3)^T$.

▶ **Exercise 4.72:** Show that the new curve $\mathbf{Q}(t)$ is independent of the particular coordinate system used.

Example: We select values $b = 2$ and $a = 1$. The new curve $\mathbf{Q}(t)$ will be identical to the part of $\mathbf{P}(t)$ from $\mathbf{P}(1)$ to $\mathbf{P}(2)$ (normally, of course, we don't calculate this part, but this example assumes that we are interested in it). Matrix $\mathbf{B}$ becomes, in this case

$$\mathbf{B} = \begin{pmatrix} 0 & 0 & 0 & 1 \\ 0 & 0 & -1 & 2 \\ 0 & 1 & -4 & 4 \\ -1 & 6 & -12 & 8 \end{pmatrix}$$

(it should now be easy to verify that each row sums up to 1) and the new control points are

$$\begin{pmatrix} \mathbf{Q}_0 \\ \mathbf{Q}_1 \\ \mathbf{Q}_2 \\ \mathbf{Q}_3 \end{pmatrix} = \mathbf{B} \begin{pmatrix} \mathbf{P}_0 \\ \mathbf{P}_1 \\ \mathbf{P}_2 \\ \mathbf{P}_3 \end{pmatrix} = \begin{pmatrix} \mathbf{P}_3 \\ -\mathbf{P}_2 + 2\mathbf{P}_3 \\ \mathbf{P}_1 - 4\mathbf{P}_2 + 4\mathbf{P}_3 \\ -\mathbf{P}_0 + 6\mathbf{P}_1 - 12\mathbf{P}_2 + 8\mathbf{P}_3 \end{pmatrix}.$$

To understand the geometrical meaning of these points, we define three auxiliary

4 Curves

points $\mathbf{R}_i$:

$$\mathbf{R}_1 = \mathbf{P}_1 + (\mathbf{P}_1 - \mathbf{P}_0),$$
$$\mathbf{R}_2 = \mathbf{P}_2 + (\mathbf{P}_2 - \mathbf{P}_1),$$
$$\mathbf{R}_3 = \mathbf{R}_2 + (\mathbf{R}_2 - \mathbf{R}_1) = \mathbf{P}_0 - 4\mathbf{P}_1 + 4\mathbf{P}_2,$$

and write the $\mathbf{Q}_i$'s in the form

$$\mathbf{Q}_0 = \mathbf{P}_3,$$
$$\mathbf{Q}_1 = \mathbf{P}_3 + (\mathbf{P}_3 - \mathbf{P}_2),$$
$$\mathbf{Q}_2 = \mathbf{Q}_1 + (\mathbf{Q}_1 - \mathbf{R}_2) = \mathbf{P}_1 - 4\mathbf{P}_2 + 4\mathbf{P}_3,$$
$$\mathbf{Q}_3 = \mathbf{Q}_2 + (\mathbf{Q}_2 - \mathbf{R}_3) = -\mathbf{P}_0 + 6\mathbf{P}_1 - 12\mathbf{P}_2 + 8\mathbf{P}_3.$$

Figure 4.43 illustrates how the four new points $\mathbf{Q}_i$ are obtained from the four original points $\mathbf{P}_i$.

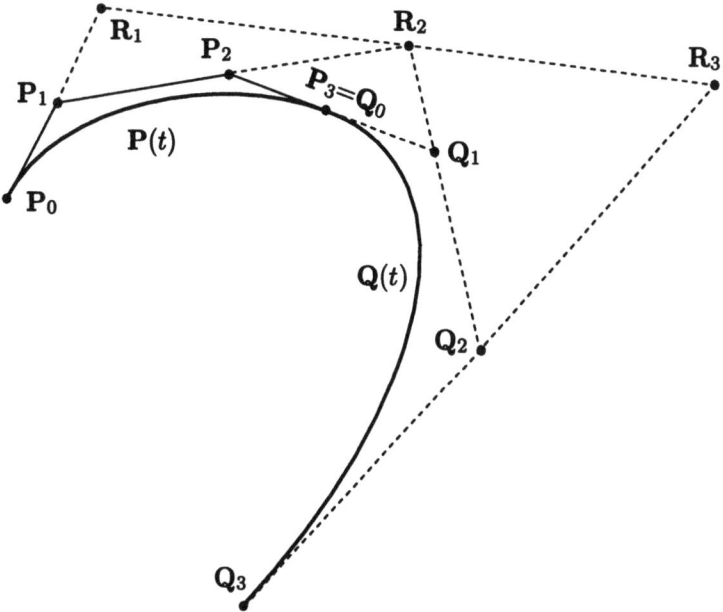

Figure 4.43: Control Points for the Case $[a, b] = [1, 2]$.

Example: We select $b = 2$ and $a = 0$. The new curve $\mathbf{Q}(t)$ will be identical to $\mathbf{P}(t)$ from $\mathbf{P}(0)$ to $\mathbf{P}(2)$. Matrix $\mathbf{B}$ becomes

$$\mathbf{B} = \begin{pmatrix} 1 & 0 & 0 & 0 \\ -1 & 2 & 0 & 0 \\ 1 & -4 & 4 & 0 \\ -1 & 6 & -12 & 8 \end{pmatrix},$$

and the new control points $\mathbf{V}_i$ are

$$\begin{pmatrix} \mathbf{V}_0 \\ \mathbf{V}_1 \\ \mathbf{V}_2 \\ \mathbf{V}_3 \end{pmatrix} = \mathbf{B} \begin{pmatrix} \mathbf{P}_0 \\ \mathbf{P}_1 \\ \mathbf{P}_2 \\ \mathbf{P}_3 \end{pmatrix} = \begin{pmatrix} \mathbf{P}_0 \\ -\mathbf{P}_0 + 2\mathbf{P}_1 \\ \mathbf{P}_0 - 4\mathbf{P}_1 + 4\mathbf{P}_2 \\ -\mathbf{P}_0 + 6\mathbf{P}_1 - 12\mathbf{P}_2 + 8\mathbf{P}_3 \end{pmatrix},$$

and it is easy to see that they satisfy $\mathbf{V}_0 = \mathbf{P}_0$, $\mathbf{V}_1 = \mathbf{R}_1$, $\mathbf{V}_2 = \mathbf{R}_3$, and $\mathbf{V}_3 = \mathbf{Q}_3$.

▸ **Exercise 4.73:** (1) Calculate matrix $\mathbf{B}$ for $a = 1$ and $b = a+x$ (where x is positive); (2) calculate the four new control points $\mathbf{Q}_i$ as functions of the $\mathbf{P}_i$'s and of b; and (3) recalculate them for $x = 0.75$.

▸ **Exercise 4.74:** Calculate matrix $\mathbf{B}$ and the four new control points $\mathbf{Q}_i$ for $a = 0$ and $b = 0.5$ (the first half of the curve).

4.14.13 Length of the Bézier Curve

The length $L(\mathbf{P})$ of the Bézier curve $\mathbf{P}(t)$ can be calculated by evaluating the integral

$$L(\mathbf{P}) = \int_0^1 |\mathbf{P}^t(t)| \, dt$$

(Section A.4.5), but this is a tedious operation. It turns out that the length can be calculated approximately (but to any desired accuracy) from the lengths of the *control polygon* and the *chord* of the curve.

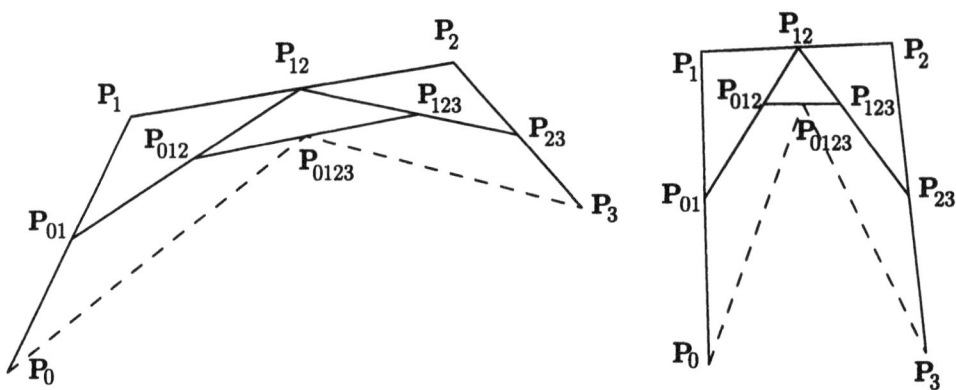

Figure 4.44: Two Subdivided Curves.

Figure 4.44 shows the control polygons of two cubic Bézier curves (themselves not shown) that have been subdivided. The original control polygon of each curve consists of the three straight segments connecting points $\mathbf{P}_0$, $\mathbf{P}_1$, $\mathbf{P}_2$, and $\mathbf{P}_3$. It is clear that the length of this polygon can be used as a (rough) approximation of the length of the curve and also that the control polygon is longer than the curve. It is also clear that the length of the control polygon after one subdivision in the

4 Curves

middle (i.e., the length of the five segments connecting points P_0, P_{01}, P_{012}, P_{123}, P_{23}, and P_3) is a better approximation and that the length will always be longer than that of the curve. We denote the length of the control polygon after k midway subdivisions $L_1^{(k)}(P)$.

The *chord* of the original curve is simply the straight segment from P_0 to P_3. The chord is shorter than the curve and is clearly not a good approximation of the length of the curve (especially for the curve on the right). However, after one midway subdivision, the chord length (the two dashed segments connecting points P_0, P_{0123}, and P_3) becomes a better approximation and it is easy to see intuitively that after k subdivisions, the chord length (which we denote by $L_0^{(k)}(P)$) becomes a better approximation.

▸ **Exercise 4.75:** What is $L_0^{(2)}(P)$?

It therefore makes sense to use **both** $L_1^{(k)}(P)$ and $L_0^{(k)}(P)$ to get a good approximation of the curve length. The former expression should be assigned more weight than the latter. The result discussed here is due to [Gravesen 93]. It states that the length of the Bézier curve $P(t)$ of order n (i.e., based on $n+1$ control points) is given by

$$L(P) = \frac{n-1}{n+1} L_1^{(k)}(P) + \frac{2}{n+1} L_0^{(k)}(P) \qquad (4.122)$$

to within 16^{-k}. A segment is supposed to be divided in the middle.

Equation (4.122) is a barycentric weighted sum of $L_1^{(k)}(P)$ and $L_0^{(k)}(P)$, where the former has a large weight (the expression $(n-1)/(n+1)$ approaches 1 for large n) and the latter has a small weight (the value $2/(n+1)$ approaches 0 for large n). The meaning of the value 16^{-k} is that the difference between the true length and Equation (4.122) decreases by a factor of 16 after each subdivision. In practice, only about three to four subdivisions are required to get the length of the curve to a high accuracy, sufficient for most practical purposes.

Notice that $L_1^{(k)}(P) \geq L(P) \geq L_0^{(k)}(P)$. An equal sign applies only if the curve is a straight line, in which case both the control polygon and the chord coincide with the curve.

4.14.14 Speed of the Bézier Curve

Speed is normally measured in units of length per units of time. The speed discussed here, however, is measured in pixels per unit of t. The problem is that incrementing t in equal steps of size Δ moves us unequal distances on the curve. This happens commonly with curves and is not specific to the Bézier curve. It causes two problems:

1. In regions of low speed, where incrementing t by a small unit Δ moves us just a small distance along the curve, the values $P(t)$ and $P(t+\Delta)$ may be so close that they may refer to the same pixel. Plotting the same pixel twice slows down the curve plotting algorithm.

2. In regions of high speed, the distance between $P(t)$ and $P(t+\Delta)$ may be more than one pixel, causing the final curve to look fragmented.

4.14 The Bézier Curve

This section discusses the speed of the Bézier curve and how it is affected by the relative positions of the control points.

The curves in Figure 4.45 were constructed by varying t in 30 small steps. The 30 pixels are not uniformly distributed along the curve. This property is a result of the shape of the weight functions and it is easy to verify just by watching the pixels drawn on the screen. The curve of Figure 4.45c is simple. It is close to a straight line (its curvature is small) and it is based on four control points that are roughly equidistant. In this curve, the pixels initially move fast; toward the middle of the curve they slow down; close to the end they speed up again.

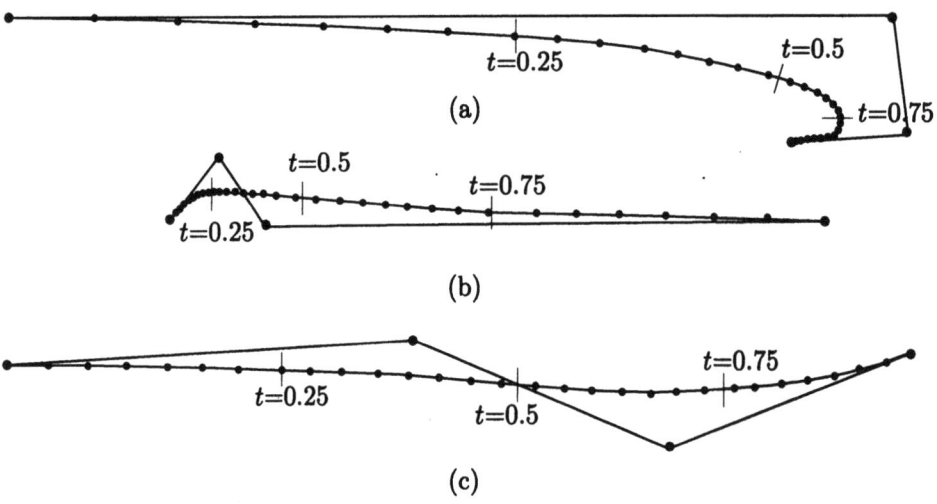

Figure 4.45: Speed of Bézier Curves.

The explanation of this behavior is simple. At the start, when t is close to zero, the shape of the curve is influenced mostly by $B_{n,0}(t)$, since the other weights are close to zero. This function, however, has a large negative slope in this region, so every small change in t changes its value (and, as a result, the value of the curve) substantially. The pixels drawn for, say, $t = 0.01$ and $t = 0.02$ will be quite separated. Toward the end, when t is close to 1, a similar situation happens with $B_{n,n}(t)$. In the middle, however, the curve is influenced by weight functions that don't slope as much, so small changes in t produce small changes in the curve. Therefore, the pixels drawn for, say, $t = 0.5$ and $t = 0.51$ are not separated as much as $\mathbf{P}(0.01)$ and $\mathbf{P}(0.02)$.

The curve of Figure 4.45b is also close to a straight line, but its four control points are not equidistant. It is easy to see how the pixels bunch together when the curve travels in the region where the first three points are located. Once out of this region, the curve "picks up speed."

Figure 4.45a is similar. The pixels are again bunched together in the vicinity of the last three points, but these points are not on a straight line, a feature that gives the curves large curvature in their area. This example shows that we can expect

4 Curves

the curve to slow down in regions with high curvature, because the control points must be close together in order to create high curvature.

To understand why the curve slows down when control points are close together, let's imagine an extreme case where $\mathbf{P}_0 = \mathbf{P}_1 = \mathbf{P}_2$. The expression for the curve in such a case is

$$\mathbf{P}(t) = \mathbf{P}_0\bigl(B_{30}(t) + B_{31}(t) + B_{32}(t)\bigr) + \mathbf{P}_3 B_{33}(t).$$

It is easy to see that the parameter t must get very close to 1 before point $\mathbf{P}_3$ would have much influence on the curve (before $B_{33}(t)$ would become larger than the sum $B_{30}(t) + B_{31}(t) + B_{32}(t)$). This is why the curve spends most of its "time" in the vicinity of the triple point $\mathbf{P}_0$, then rushes toward $\mathbf{P}_3$ when t gets close to 1.

4.14.15 Constant Speed

Sometimes it is important to move along a Bézier curve at constant speed. A practical example is computer animation, where the (imaginary) camera has to be moved along a curve and stopped to take a snapshot at $n+1$ equally spaced positions (Section 8.2). The method discussed here is based on approximating the curve by a polyline, then finding the values t_i of the parameter t that advance equal distances on the polyline and use them to move along the curve. To construct the polyline, the algorithm selects points on the curve and connects them with straight segments. In regions where the curve is close to a straight line (i.e., has low curvature), these points can be well separated. In regions where the curvature is high, the points must, of course, be close together (Figure 4.46a) to guarantee good approximation. The points are selected by applying the subdivision method of Section 4.14.10. A subdivision divides a curve into two curves that connect at a point and this point becomes a vertex of the polyline. Our algorithm thus proceeds in the following steps:

1. The first and last points of the curve are placed in the (initially empty) list of polyline points.

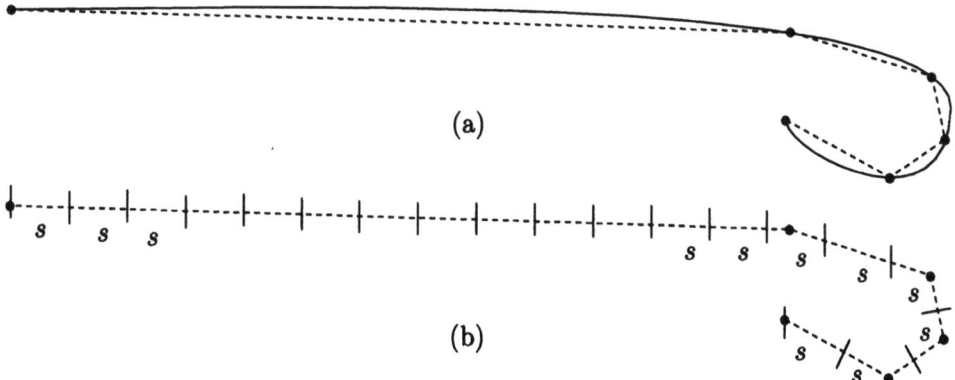

Figure 4.46: Unequally Spaced Points.

2. The curve is checked to see if it deviates from a straight line sufficiently to justify being subdivided. If yes, it is subdivided, the common point of subdivision is added to the list of polyline points, and each of the resulting two curves is recursively checked to see if it should be further subdivided. This step provides *adaptive subdivision* of the curve, i.e., only high-curvature areas are further subdivided.

3. The polyline created by the list of points provides a close approximation to the curve. The length L of this polyline is the sum of the lengths of the individual segments, so it is easy to calculate. Our original problem was to move along the curve and stop at $n+1$ equally spaced points. Now that we have a polyline of length L closely following the curve, we divide it into n chunks of size $s = L/n$ each (Figure 4.46b).

4. The $n+1$ parameter values t_i that divide the polyline into chunks of size s are calculated. These values are later used to move along the curve and stop at $n+1$ points. The better the polyline approximates the curve, the more equally spaced these points will be.

Subdividing the Bézier curve is time-consuming, so the minimum number of subdivisions should be used. At the same time, each subdivision improves the approximation of the polyline to the curve. The test used in step 2 is therefore crucial to the performance of the algorithm. This test is based on Equation (4.122) (Section 4.14.13) that defines the relation between the length of a Bézier curve and the lengths of its control polygon and its chord. The control polygon is normally longer than the curve; the chord is normally shorter. The two quantities have the same length only when the curve is a straight line.

The conclusion is that the closer the lengths of the control polygon and the chord, the closer the curve is to a straight line. The curve should thus be recursively subdivided if the test

```
if(ctrl_polygon - chord>=eps)
```
or, alternatively,
```
if(ctrl_polygon > (1+eps)*chord)
```
is satisfied, where eps is a small, user-controlled tolerance parameter (notice that the control polygon cannot be shorter than the chord, so the difference
```
ctrl_polygon - chord
```
is never negative).

Figure 4.47 is a pseudo-code for step 4. We assume that we already have a k-segment polyline based on the $k+1$ points $\mathbf{P}_0, \mathbf{P}_1, \ldots, \mathbf{P}_k$ obtained by the subdivisions. The algorithm starts by measuring the total length L of the polyline and calculating $s = L/n$, where n is an input parameter. The main loop iterates over the segments and measures n chunks of length s. For a general segment from $\mathbf{P}_{i-1}$ to $\mathbf{P}_i$, variable st measures the distance from the end of the last chunk to the end of the previous segment. A piece of size $s - st$ is still needed to complete the current chunk. This piece may require just part of the current segment, or the entire segment and part (or all) of the next one. Variable t is incremented from 0 to 1. Each time a chunk of length s is identified, t is set to the correct value at the end of this chunk. At the end of an iteration, it is always set to its value at point $\mathbf{P}_i$ (the end of the segment).

4 Curves

```
t=0;
TotSegLen=0; // total length of segments visited so far
L=0; // total length of polyline
for i=1 to k do L=L+|P_i - P_{i-1}|; endfor;
st=0; s=L/n; // size of a chunk
AddTable(0); // add initial value
for i=1 to k do // loop over k segments
  SegLen=|P_i - P_{i-1}|;
  TotSegLen=TotSegLen+SegLen;
  if(s-st≤SegLen)
  then // a chunk ends at this segment
   t=t+(s-st)/L;
   AddTable(t);
   while SegLen>s do // more chunks in
    t=t+s/L;        // this segment
    AddTable(t);
    SegLen=SegLen-s;
   endwhile;
   st=SegLen;
  else // entire segment is part of chunk
   st=st+SegLen;
  endif;
  t=t+TotSegLen/L;
endfor;
AddTable(1); // add final value
```

Figure 4.47: Measuring n Chunks on a Polyline.

4.14.16 Converting Cubic Curves

The fact that the Bézier curve has the convex hull property makes it useful to convert other types of curves to a Bézier curve. The discussion below shows how to do this for the cubic case. Let

$$\mathbf{Q}(t) = (t^3, t^2, t, 1)\mathbf{M} \begin{pmatrix} \mathbf{Q}_0 \\ \mathbf{Q}_1 \\ \mathbf{Q}_2 \\ \mathbf{Q}_3 \end{pmatrix}$$

be any cubic parametric curve where the $\mathbf{Q}_i$'s may be points, tangent vectors, or any other nonscalar quantities. The cubic Bézier curve is given by Equation (4.107),

$$\mathbf{P}(t) = (t^3, t^2, t, 1)\mathbf{B} \begin{pmatrix} \mathbf{P}_0 \\ \mathbf{P}_1 \\ \mathbf{P}_2 \\ \mathbf{P}_3 \end{pmatrix},$$

where **B** is the basis matrix

$$\mathbf{B} = \begin{pmatrix} -1 & 3 & -3 & 1 \\ 3 & -6 & 3 & 0 \\ -3 & 3 & 0 & 0 \\ 1 & 0 & 0 & 0 \end{pmatrix}.$$

For the curves to be equal, the following must be true:

$$\mathbf{B} \begin{pmatrix} \mathbf{P}_0 \\ \mathbf{P}_1 \\ \mathbf{P}_2 \\ \mathbf{P}_3 \end{pmatrix} = \mathbf{M} \begin{pmatrix} \mathbf{Q}_0 \\ \mathbf{Q}_1 \\ \mathbf{Q}_2 \\ \mathbf{Q}_3 \end{pmatrix}.$$

The solution is thus

$$\begin{pmatrix} \mathbf{P}_0 \\ \mathbf{P}_1 \\ \mathbf{P}_2 \\ \mathbf{P}_3 \end{pmatrix} = \mathbf{B}^{-1}\mathbf{M} \begin{pmatrix} \mathbf{Q}_0 \\ \mathbf{Q}_1 \\ \mathbf{Q}_2 \\ \mathbf{Q}_3 \end{pmatrix},$$

and it always exists since we know that **B** is nonsingular.

Similarly, it is possible to convert the Bézier curve into any other cubic form, provided **M** is nonsingular. The following discussion shows the relationship between the Bézier curve and the Hermite curve segment. A similar relationship between the Bézier curve and the Catmull-Rom curve is shown in Section 4.14.18.

Any set of four given control points $\mathbf{P}_0$, $\mathbf{P}_1$, $\mathbf{P}_2$, and $\mathbf{P}_3$ determines a unique (cubic) Bézier curve. It is interesting to note that there is a Hermite curve that has an identical shape. It is determined by the 4-tuple

$$(\mathbf{P}_0, \mathbf{P}_3, 3(\mathbf{P}_1 - \mathbf{P}_0), 3(\mathbf{P}_3 - \mathbf{P}_2)). \tag{4.123}$$

▶ **Exercise 4.76:** Prove this claim!

The opposite is also true. Given two points $\mathbf{P}_0$ and $\mathbf{P}_1$ and two tangent vectors $\mathbf{P}_0^t$ and $\mathbf{P}_1^t$ they define a Hermite segment. An identical Bézier segment is determined by the 4-tuple

$$\left(\mathbf{P}_0, (\mathbf{P}_0 + \tfrac{1}{3}\mathbf{P}_0^t), (\mathbf{P}_1 - \tfrac{1}{3}\mathbf{P}_1^t), \mathbf{P}_1\right). \tag{4.124}$$

4.14.17 Cubic Bézier Segments with Tension

Adding a tension parameter to a cubic Bézier segment is done in a way similar to the Cardinal spline (Section 4.10). We use Hermite interpolation (Equation (4.26)) to calculate a PC segment that starts at point $\mathbf{P}_0$ and ends at point $\mathbf{P}_3$ and whose extreme tangent vectors are $s(\mathbf{P}_1 - \mathbf{P}_0)$ and $s(\mathbf{P}_3 - \mathbf{P}_2)$ (see Equation (4.123).) Notice that the lengths of these vectors are controlled by the tension parameter s.

4 Curves

Substituting these values in Equation (4.26), we manipulate it so that it ends up looking like a cubic Bézier segment, Equation (4.107):

$$\mathbf{P}(t) = (t^3, t^2, t, 1) \begin{pmatrix} 2 & -2 & 1 & 1 \\ -3 & 3 & -2 & -1 \\ 0 & 0 & 1 & 0 \\ 1 & 0 & 0 & 0 \end{pmatrix} \begin{pmatrix} \mathbf{P}_0 \\ \mathbf{P}_3 \\ s(\mathbf{P}_1 - \mathbf{P}_0) \\ s(\mathbf{P}_3 - \mathbf{P}_2) \end{pmatrix}$$

$$= (t^3, t^2, t, 1) \begin{pmatrix} 2-s & s & -s & s-2 \\ 2s-3 & -2s & s & 3-s \\ -s & s & 0 & 0 \\ 1 & 0 & 0 & 0 \end{pmatrix} \begin{pmatrix} \mathbf{P}_0 \\ \mathbf{P}_1 \\ \mathbf{P}_2 \\ \mathbf{P}_3 \end{pmatrix}. \quad (4.125)$$

A quick check verifies that Equation (4.125) reduces to the cubic Bézier segment, Equation (4.107), for $s = 3$. This value is therefore considered the "neutral" or "standard" value of the tension parameter s. Since s controls the length of the tangent vectors, small values of s should produce the effects of higher tension and, in the extreme, the value $s = 0$ should result in indefinite tangent vectors and in the curve segment becoming a straight line. To show this, we rewrite Equation (4.125) for $s = 0$:

$$\mathbf{P}(t) = (t^3, t^2, t, 1) \begin{pmatrix} 2 & 0 & 0 & -2 \\ -3 & 0 & 0 & 3 \\ 0 & 0 & 0 & 0 \\ 1 & 0 & 0 & 0 \end{pmatrix} \begin{pmatrix} \mathbf{P}_0 \\ \mathbf{P}_1 \\ \mathbf{P}_2 \\ \mathbf{P}_3 \end{pmatrix}$$

$$= (2t^3 - 3t^2 + 1)\mathbf{P}_0 + (-2t^3 + 3t^2)\mathbf{P}_3.$$

Substituting $T = 3t^2 - 2t^3$ for t changes the above expression to the form $\mathbf{P}(T) = (\mathbf{P}_3 - \mathbf{P}_0)T + \mathbf{P}_0$, that is, a straight line from $\mathbf{P}(0) = \mathbf{P}_0$ to $\mathbf{P}(1) = \mathbf{P}_3$.

The tangent vector of Equation (4.125) is

$$\mathbf{P}^t(t) = (3t^2, 2t, 1, 0) \begin{pmatrix} 2-s & s & -s & s-2 \\ 2s-3 & -2s & s & 3-s \\ -s & s & 0 & 0 \\ 1 & 0 & 0 & 0 \end{pmatrix} \begin{pmatrix} \mathbf{P}_0 \\ \mathbf{P}_1 \\ \mathbf{P}_2 \\ \mathbf{P}_3 \end{pmatrix} \quad (4.126)$$

$$= \left(3t^2(2-s) + 2t(2s-3) - s\right)\mathbf{P}_0 + \left(3st^2 - 4st + s\right)\mathbf{P}_1$$
$$+ \left(-3st^2 + 2st\right)\mathbf{P}_2 + \left(3t^2(s-2) + 2t(3-s)\right)\mathbf{P}_3.$$

The extreme tangents are $\mathbf{P}^t(0) = s(\mathbf{P}_1 - \mathbf{P}_0)$ and $\mathbf{P}^t(1) = s(\mathbf{P}_3 - \mathbf{P}_2)$. Substituting $s = 0$ in Equation (4.126) yields the tangent vector for the case of infinite tension:

$$\mathbf{P}^t(t) = 6(t^2 - t)\mathbf{P}_0 - 6(t^2 - t)\mathbf{P}_3 = 6(t - t^2)(\mathbf{P}_3 - \mathbf{P}_0). \quad (4.127)$$

▶ **Exercise 4.77:** Since the spline segment is a straight line in this case, its tangent vector should always point in the same direction. Use Equation (4.127) to show that this is so.

See also Section 4.16.7 for a discussion of cubic B-spline with tension.

> We interrupt this program to increase dramatic tension.
> — Joe Leahy (as the Announcer) in *Freakazoid!* (1995).

4.14.18 An Interpolating Bézier Curve: I

Any set of four control points P_1, P_2, P_3, and P_4 determines a unique Catmull-Rom segment that's a cubic polynomial going from point P_2 to point P_3. It turns out that such a segment can also be written as a four-point Bézier curve from P_2 to P_3. All we have to do is find two points, X and Y, located between P_2 and P_3, such that the Bézier curve based on P_2, X, Y, and P_3 will be identical to the Catmull-Rom segment. This turns out to be easy. We start with the expressions for a Catmull-Rom segment defined by P_1, P_2, P_3, and P_4, and for a four-point Bézier curve defined by P_2, X, Y, and P_3 (Equations (4.89) and (4.107)):

$$(t^3, t^2, t, 1) \begin{pmatrix} -0.5 & 1.5 & -1.5 & 0.5 \\ 1 & -2.5 & 2 & -0.5 \\ -0.5 & 0 & 0.5 & 0 \\ 0 & 1 & 0 & 0 \end{pmatrix} \begin{pmatrix} P_1 \\ P_2 \\ P_3 \\ P_4 \end{pmatrix},$$

$$(t^3, t^2, t, 1) \begin{pmatrix} -1 & 3 & -3 & 1 \\ 3 & -6 & 3 & 0 \\ -3 & 3 & 0 & 0 \\ 1 & 0 & 0 & 0 \end{pmatrix} \begin{pmatrix} P_2 \\ X \\ Y \\ P_3 \end{pmatrix}.$$

These have to be equal for each power of t, which yields the four equations

$$\begin{aligned} -0.5P_1 + 1.5P_2 - 1.5P_3 + 0.5P_4 &= -P_2 + 3X - 3Y + P_3, \\ P_1 - 2.5P_2 + 2.0P_3 - 0.5P_4 &= 3P_2 - 6X + 3Y, \\ -0.5P_1 + 0.5P_3 &= -3P_2 + 3X, \\ P_2 &= P_2. \end{aligned}$$

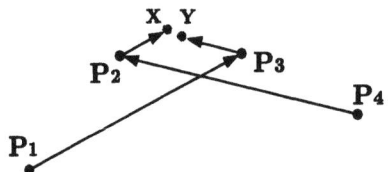

Figure 4.48: Calculating Points X and Y.

These are easily solved to produce

$$X = P_2 + \frac{1}{6}(P_3 - P_1), \quad Y = P_3 - \frac{1}{6}(P_4 - P_2). \qquad (4.128)$$

4 Curves

The difference $(\mathbf{P}_3 - \mathbf{P}_1)$ is the vector from $\mathbf{P}_1$ to $\mathbf{P}_3$. Point $\mathbf{X}$ is thus obtained by adding 1/6 of this vector to point $\mathbf{P}_2$ (Figure 4.48). Similarly, $\mathbf{Y}$ is obtained by subtracting 1/6 of the difference $(\mathbf{P}_4 - \mathbf{P}_2)$ from point $\mathbf{P}_3$.

This simple result suggests a novel approach to the problem of interactive curve design, an approach that combines the useful features of both cubic splines and Bézier curves. A cubic spline passes through the (data) points but is not highly interactive. It can be edited only by modifying the two extreme tangent vectors. A Bézier curve does not pass through the (control) points, but it is easy to manipulate and edit by moving the points. The new approach works as follows:

1. The user is asked to input n points, through which the final curve will pass.
2. The program divides the points into overlapping groups of four points and computes two auxiliary points $\mathbf{X}$ and $\mathbf{Y}$, as shown above, for each group.
3. A Bézier segment is then drawn from the second to the third point of each group, using points $\mathbf{X}$ and $\mathbf{Y}$ as its other two control points. Note that points $\mathbf{Y}$ and $\mathbf{P}_3$ of a group are on a straight line with point $\mathbf{X}$ of the next group. This guarantees that the individual segments will connect smoothly.
4. It is also possible to draw a Bézier segment from $\mathbf{P}_1$ to $\mathbf{P}_2$ (and, similarly, from $\mathbf{P}_{n-1}$ to $\mathbf{P}_n$). This segment uses the two auxiliary control points $\mathbf{X} = \mathbf{P}_1 + \frac{1}{6}(\mathbf{P}_2 - \mathbf{P}_1)$ and $\mathbf{Y} = \mathbf{P}_2 - \frac{1}{6}(\mathbf{P}_3 - \mathbf{P}_1)$.

Users find it easy to specify such a curve, since they don't have to worry about the positions of the control points. The curve is made of $n - 1$ segments and the two auxiliary control points of each segment are calculated automatically.

Such a curve is usually pleasing to the eye and rarely needs to be edited. However, if it is not satisfactory, it can be modified by moving the auxiliary control points. There are $2(n-1)$ of them, which allows for flexible control. A good program should display the auxiliary points and should make it easy for the user to grab and move any of them.

The well-known drawing program Adobe Illustrator uses a similar approach. The user specifies points with the mouse. At each point $\mathbf{P}_i$, the user presses the mouse button to fix $\mathbf{P}_i$, then drags the mouse before releasing the button, which defines two symmetrical points, $\mathbf{X}$ (following $\mathbf{P}_i$) and $\mathbf{Y}$ (preceding it). Releasing the button is a signal to the program to draw the segment from $\mathbf{P}_{i-1}$ to $\mathbf{P}_i$ (Figure 4.49).

Example: We apply this method to the six points $\mathbf{P}_0 = (1/2, 0)$, $\mathbf{P}_1 = (1/2, 1/2)$, $\mathbf{P}_2 = (0, 1)$, $\mathbf{P}_3 = (1, 3/2)$, $\mathbf{P}_4 = (3/2, 1)$, and $\mathbf{P}_5 = (1, 1/2)$. The six points yield three curve segments and the main step is to calculate the two intermediate points for each of the three segments. This is trivial:

$$\mathbf{X}_1 = \mathbf{P}_1 + (\mathbf{P}_2 - \mathbf{P}_0)/6 = (5/12, 2/3), \quad \mathbf{Y}1 = \mathbf{P}_2 - (\mathbf{P}_3 - \mathbf{P}_1)/6 = (-1/12, 5/6),$$
$$\mathbf{X}_2 = \mathbf{P}_2 + (\mathbf{P}_3 - \mathbf{P}_1)/6 = (1/12, 7/6), \quad \mathbf{Y}2 = \mathbf{P}_3 - (\mathbf{P}_4 - \mathbf{P}_2)/6 = (3/4, 3/2),$$
$$\mathbf{X}_3 = \mathbf{P}_3 + (\mathbf{P}_4 - \mathbf{P}_2)/6 = (5/4, 3/2), \quad \mathbf{Y}3 = \mathbf{P}_4 - (\mathbf{P}_5 - \mathbf{P}_3)/6 = (3/2, 7/6).$$

Once the points are available, the three segments can easily be calculated. Each is a cubic Bézier segment based on a group of four points. The groups are

$$[\mathbf{P}_1, \mathbf{X}_1, \mathbf{Y}_1, \mathbf{P}_2], \quad [\mathbf{P}_2, \mathbf{X}_2, \mathbf{Y}_2, \mathbf{P}_3], \quad [\mathbf{P}_3, \mathbf{X}_3, \mathbf{Y}_3, \mathbf{P}_4],$$

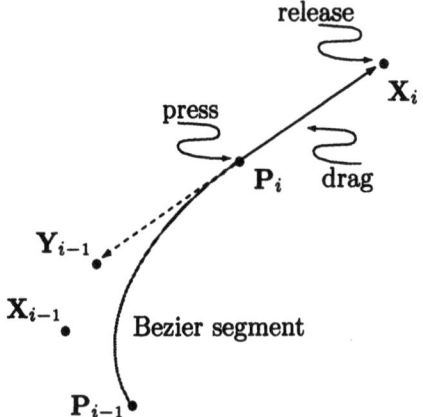

Figure 4.49: Construction of $\mathbf{X}_i$ and $\mathbf{Y}_i$ by Click and Drag.

and the three curve segments are

$$\begin{aligned}
\mathbf{P}_1(t) &= (1-t)^3\mathbf{P}_1 + 3t(1-t)^2\mathbf{X}_1 + 3t^2(1-t)\mathbf{Y}_1 + t^3\mathbf{P}_2 \\
&= \big((2-t-5t^2+4t^3)/4, (1+t)/2\big), \\
\mathbf{P}_2(t) &= (1-t)^3\mathbf{P}_2 + 3t(1-t)^2\mathbf{X}_2 + 3t^2(1-t)\mathbf{Y}_2 + t^3\mathbf{P}_3 \\
&= \big((t+7t^2-4t^3)/4, (2+t+t^2-t^3)/2\big), \\
\mathbf{P}_3(t) &= (1-t)^3\mathbf{P}_3 + 3t(1-t)^2\mathbf{X}_3 + 3t^2(1-t)\mathbf{Y}_3 + t^3\mathbf{P}_4 \\
&= \big((4+3t-t^3)/4, (3-2t^2+t^3)/2\big).
\end{aligned}$$

The 12 points and the 3 segments are shown in Figure 4.50 (where the segments have been separated intentionally), as well as the code for the entire example.

4.14.19 An Interpolating Bézier Curve: II

We start, as usual, with $n+1$ control points $\mathbf{P}_0, \ldots, \mathbf{P}_n$. Two auxiliary points $\mathbf{X}_i$ and $\mathbf{Y}_{i+1}$ are automatically calculated by the software between each pair $\mathbf{P}_i$, $\mathbf{P}_{i+1}$ of control points. After all the $\mathbf{X}_i$ and $\mathbf{Y}_i$ points have been computed, the curve is drawn as a sequence of four-point Bézier segments, each based on a group of four points $\mathbf{P}_i$, $\mathbf{X}_i$, $\mathbf{Y}_{i+1}$, and $\mathbf{P}_{i+1}$. The auxiliary points are computed as follows.

A new point, $\mathbf{Q}_i$, is defined by the relation $\mathbf{Q}_i - \mathbf{P}_i = \mathbf{P}_i - \mathbf{P}_{i-1}$. It will be recalled that the difference of two points is a vector, so $\mathbf{Q}_i = 2\mathbf{P}_i - \mathbf{P}_{i-1}$ is at the same distance from $\mathbf{P}_i$ as $\mathbf{P}_i$ is from $\mathbf{P}_{i-1}$ (Figure 4.51). Also, the direction from $\mathbf{P}_i$ to $\mathbf{Q}_i$ is the same as that from $\mathbf{P}_{i-1}$ to $\mathbf{P}_i$. The first auxiliary point $\mathbf{X}_i$ is now calculated midway between $\mathbf{Q}_i$ and $\mathbf{P}_{i+1}$, i.e.,

$$\mathbf{X}_i = \frac{\mathbf{Q}_i + \mathbf{P}_{i+1}}{2} = \frac{2\mathbf{P}_i - \mathbf{P}_{i-1} + \mathbf{P}_{i+1}}{2} = \mathbf{P}_i + \frac{1}{2}(\mathbf{P}_{i+1} - \mathbf{P}_{i-1}). \quad (4.129)$$

The second auxiliary point $\mathbf{Y}_i$ is now calculated by $\mathbf{Y}_i - \mathbf{P}_i = \mathbf{P}_i - \mathbf{X}_i$, i.e., $\mathbf{Y}_i$ is symmetric to $\mathbf{P}_i$ with respect to $\mathbf{X}_i$. Given the definition of $\mathbf{X}_i$ from Equa-

4 Curves

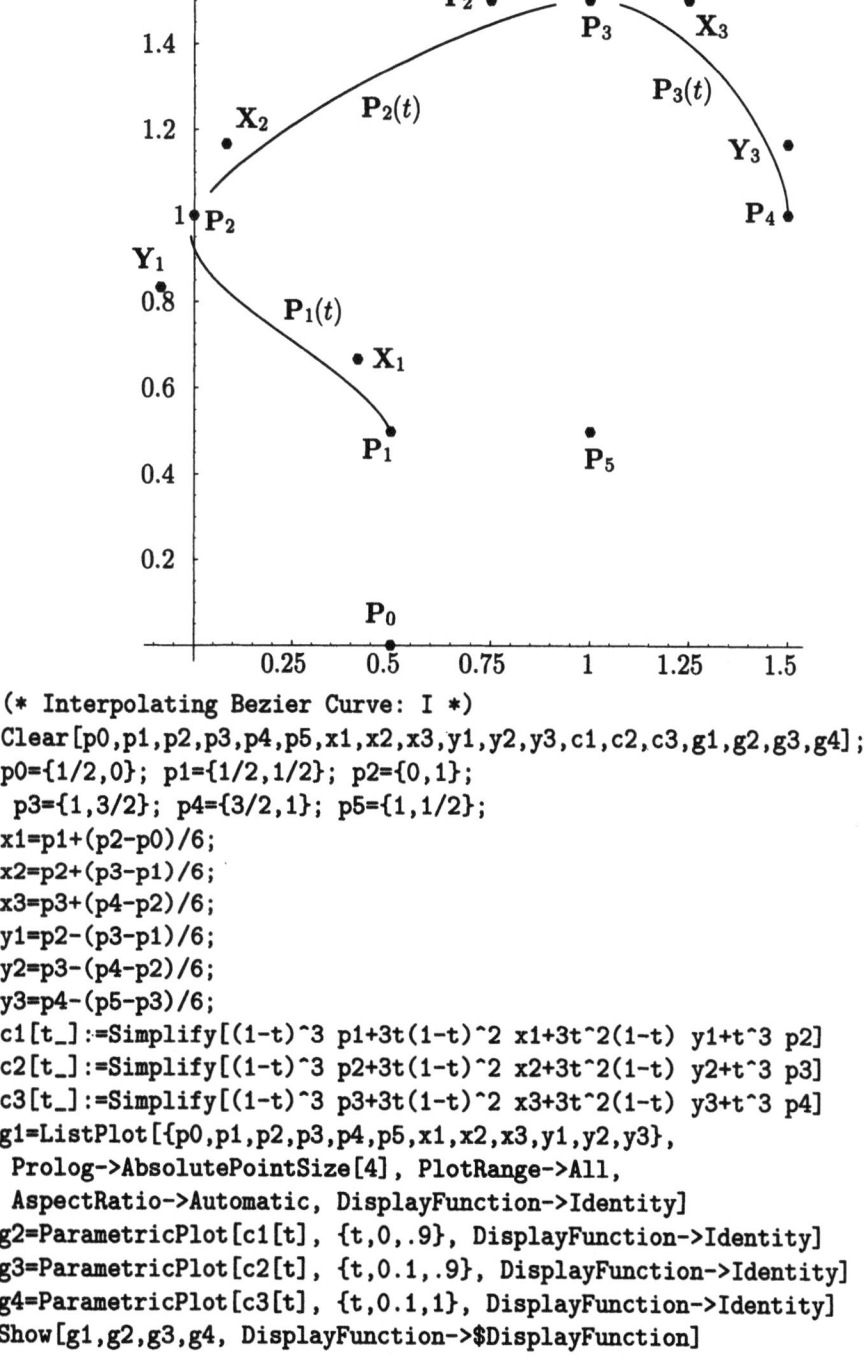

```
(* Interpolating Bezier Curve: I *)
Clear[p0,p1,p2,p3,p4,p5,x1,x2,x3,y1,y2,y3,c1,c2,c3,g1,g2,g3,g4];
p0={1/2,0}; p1={1/2,1/2}; p2={0,1};
 p3={1,3/2}; p4={3/2,1}; p5={1,1/2};
x1=p1+(p2-p0)/6;
x2=p2+(p3-p1)/6;
x3=p3+(p4-p2)/6;
y1=p2-(p3-p1)/6;
y2=p3-(p4-p2)/6;
y3=p4-(p5-p3)/6;
c1[t_]:=Simplify[(1-t)^3 p1+3t(1-t)^2 x1+3t^2(1-t) y1+t^3 p2]
c2[t_]:=Simplify[(1-t)^3 p2+3t(1-t)^2 x2+3t^2(1-t) y2+t^3 p3]
c3[t_]:=Simplify[(1-t)^3 p3+3t(1-t)^2 x3+3t^2(1-t) y3+t^3 p4]
g1=ListPlot[{p0,p1,p2,p3,p4,p5,x1,x2,x3,y1,y2,y3},
 Prolog->AbsolutePointSize[4], PlotRange->All,
 AspectRatio->Automatic, DisplayFunction->Identity]
g2=ParametricPlot[c1[t], {t,0,.9}, DisplayFunction->Identity]
g3=ParametricPlot[c2[t], {t,0.1,.9}, DisplayFunction->Identity]
g4=ParametricPlot[c3[t], {t,0.1,1}, DisplayFunction->Identity]
Show[g1,g2,g3,g4, DisplayFunction->$DisplayFunction]
```

Figure 4.50: An Interpolating Bézier Curve: I.

tion (4.129), we get

$$Y_i = 2P_i - X_i = P_i - \frac{1}{2}(P_{i+1} - P_{i-1}). \quad (4.130)$$

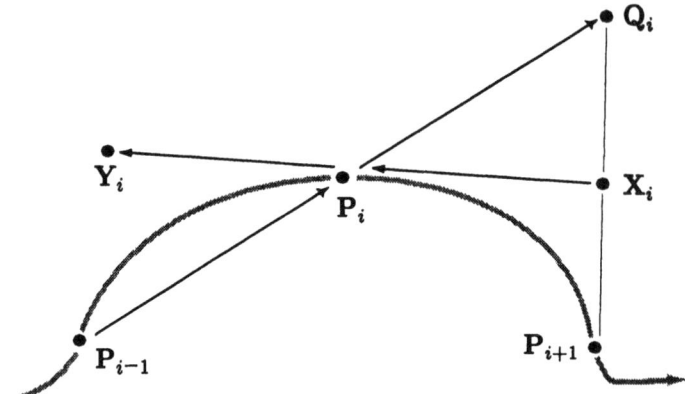

Figure 4.51: Construction of X_i and Y_{i-1}.

The final sequence of points is

$$P_0, X_0, Y_1, P_1, X_1, Y_2, P_2, X_2, Y_3, P_3, \ldots, P_{n-1}, X_{n-1}, Y_n, P_n,$$

a total of $3n + 1$ points. Notice that X_0 and Y_n cannot be calculated by Equations (4.129) and (4.130). They have to be input by the user, and they serve to establish the start and end directions of the curve.

As mentioned earlier, the curve is made up of four-point segments based on the groups

$$(P_0, X_0, Y_1, P_1), \quad (P_1, X_1, Y_2, P_2), \ldots, (P_{n-1}, X_{n-1}, Y_n, P_n).$$

Notice that this method is similar to that of Section 4.14.18, the main difference being that this method requires the user to input the values of X_0 and Y_n, whereas in Section 4.14.18, the software calculates all the auxiliary points but produces a curve from P_1 to P_{n-1} instead of from P_0 to P_n. Another difference is the factors of 1/2 and 1/6.

Experience indicates that this method produces satisfactory curves in most cases. In cases where the curve is not satisfactory, a variant simply draws the Bézier curve that's based on all $3n + 1$ points. This curve does not, of course, pass through the original control points, so it is not an interpolating curve, but it may, nevertheless, be useful in certain applications, such as computer animation (Section 8.2).

Example: We apply this method to the six points $P_0 = (1/2, 0)$, $P_1 = (1/2, 1/2)$, $P_2 = (0, 1)$, $P_3 = (1, 3/2)$, $P_4 = (3/2, 1)$, and $P_5 = (1, 1/2)$. The

4 Curves

six points yield five curve segments. The first step is to calculate the two intermediate points for each of the five segments. This is straightforward, but notice that the choice of $\mathbf{X}_0$ and $\mathbf{Y}_5$ is arbitrary:

$$\mathbf{X}_0 = \mathbf{P}_1 - \mathbf{P}_0 = (0, 1/2), \quad \mathbf{Y}_5 = \mathbf{P}_4 - \mathbf{P}_5 = (1/2, 1/2),$$
$$\mathbf{X}_1 = \mathbf{P}_1 + (\mathbf{P}_2 - \mathbf{P}_0)/2 = (1/4, 1), \quad \mathbf{Y}_1 = \mathbf{P}_1 - (\mathbf{P}_2 - \mathbf{P}_0)/2 = (3/4, 0),$$
$$\mathbf{X}_2 = \mathbf{P}_2 + (\mathbf{P}_3 - \mathbf{P}_1)/2 = (1/4, 3/2), \quad \mathbf{Y}_2 = \mathbf{P}_2 - (\mathbf{P}_3 - \mathbf{P}_1)/2 = (-1/4, 1/2),$$
$$\mathbf{X}_3 = \mathbf{P}_3 + (\mathbf{P}_4 - \mathbf{P}_2)/2 = (7/4, 3/2), \quad \mathbf{Y}_3 = \mathbf{P}_3 - (\mathbf{P}_4 - \mathbf{P}_2)/2 = (1/4, 3/2),$$
$$\mathbf{X}_4 = \mathbf{P}_4 + (\mathbf{P}_5 - \mathbf{P}_3)/2 = (3/2, 1/2), \quad \mathbf{Y}_4 = \mathbf{P}_4 - (\mathbf{P}_5 - \mathbf{P}_3)/2 = (3/2, 3/2),$$

Once the points are available, the five segments can be calculated easily. Each is a cubic Bézier segment based on a group of four points. The groups are

$$[\mathbf{P}_0, \mathbf{X}_0, \mathbf{Y}_1, \mathbf{P}_1], \quad [\mathbf{P}_1, \mathbf{X}_1, \mathbf{Y}_2, \mathbf{P}_2], \quad [\mathbf{P}_2, \mathbf{X}_2, \mathbf{Y}_3, \mathbf{P}_3],$$
$$[\mathbf{P}_3, \mathbf{X}_3, \mathbf{Y}_4, \mathbf{P}_4], \quad [\mathbf{P}_4, \mathbf{X}_4, \mathbf{Y}_5, \mathbf{P}_5],$$

and the five curve segments are

$$\mathbf{P}_1(t) = (1-t)^3 \mathbf{P}_0 + 3t(1-t)^2 \mathbf{X}_0 + 3t^2(1-t)\mathbf{Y}_1 + t^3 \mathbf{P}_1$$
$$= \big((2 - 6t + 15t^2 - 9t^3)/4, t(3 - 6t + 4t^2)/2\big),$$
$$\mathbf{P}_2(t) = (1-t)^3 \mathbf{P}_1 + 3t(1-t)^2 \mathbf{X}_1 + 3t^2(1-t)\mathbf{Y}_2 + t^3 \mathbf{P}_2$$
$$= \big((2 - 3t - 3t^2 + 4t^3)/4, (1 + 3t - 6t^2 + 4t^3)/2\big),$$
$$\mathbf{P}_3(t) = (1-t)^3 \mathbf{P}_2 + 3t(1-t)^2 \mathbf{X}_2 + 3t^2(1-t)\mathbf{Y}_3 + t^3 \mathbf{P}_3$$
$$= \big(t(3 - 3t + 4t^2)/4, (2 + 3t - 3t^2 + t^3)/2\big),$$
$$\mathbf{P}_4(t) = (1-t)^3 \mathbf{P}_3 + 3t(1-t)^2 \mathbf{X}_3 + 3t^2(1-t)\mathbf{Y}_4 + t^3 \mathbf{P}_4$$
$$= \big(1 + 9t/4 - 3t^2 + 5t^3/4, (3 - t^3)/2\big),$$
$$\mathbf{P}_5(t) = (1-t)^3 \mathbf{P}_4 + 3t(1-t)^2 \mathbf{X}_4 + 3t^2(1-t)\mathbf{Y}_5 + t^3 \mathbf{P}_5$$
$$= \big((3 - 6t^2 + 5t^3)/2, (2 - 3t + 3t^2 - t^3)/2\big).$$

The 12 points and the five segments are shown in Figure 4.52 (where the segments have been separated intentionally), as well as the code for the entire example.

4.14.20 An Interpolating Bézier Curve: III

The approach outlined here calculates an interpolating Bézier curve by solving equations. Given a set of $n+1$ data points $\mathbf{Q}_0, \mathbf{Q}_1, \ldots, \mathbf{Q}_n$, we select $n+1$ values t_i such that $\mathbf{P}(t_i) = \mathbf{Q}_i$. We require that whenever t reaches one of the values t_i, the curve will pass through a point $\mathbf{Q}_i$. The values t_i don't have to be equally spaced, a feature that provides control over the "speed" of the curve. All that's needed to calculate the curve is to calculate a set of $n+1$ control points $\mathbf{P}_i$. This is done by setting and solving the set of $n+1$ linear equations $\mathbf{P}(t_0) = \mathbf{Q}_0, \mathbf{P}(t_1) = \mathbf{Q}_1, \ldots, \mathbf{P}(t_n) = \mathbf{Q}_n$

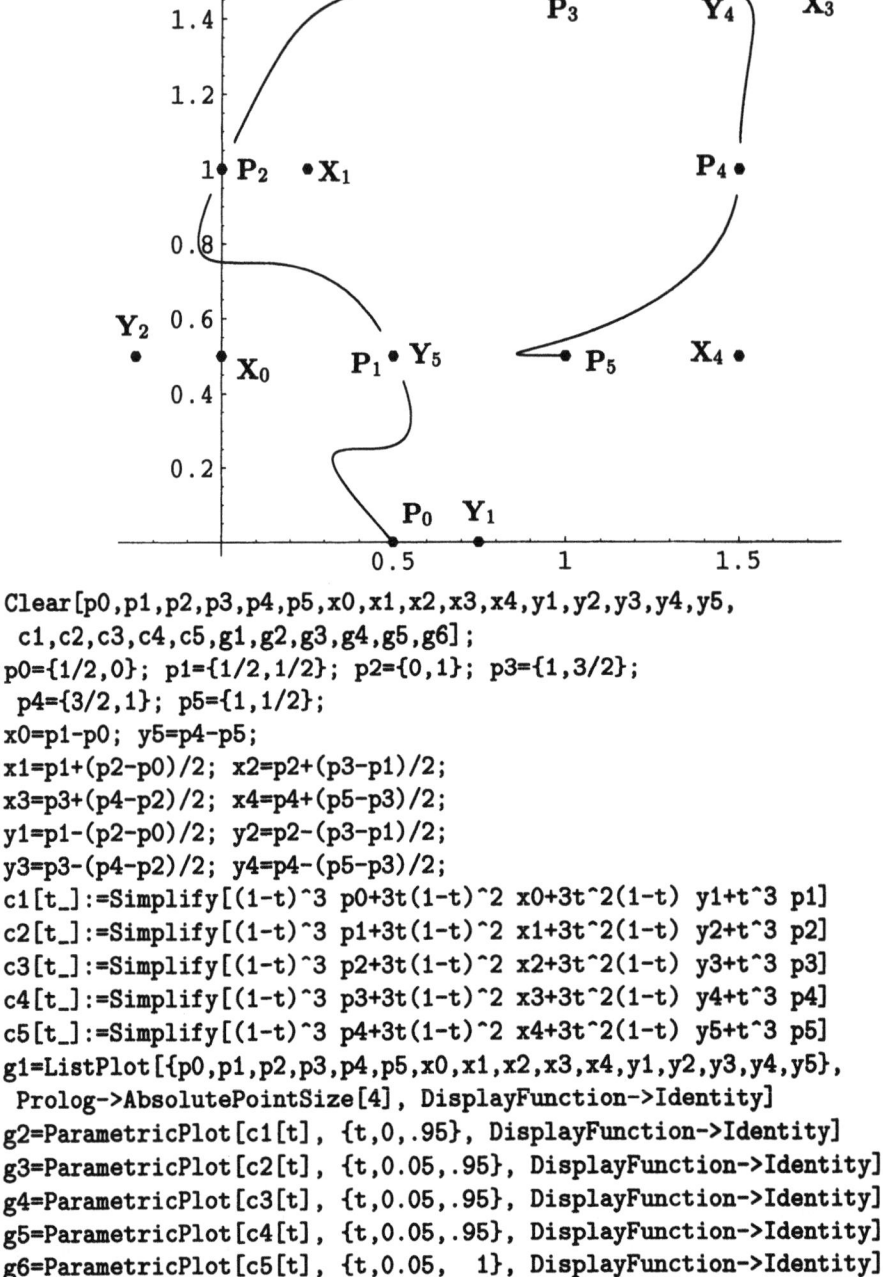

```
Clear[p0,p1,p2,p3,p4,p5,x0,x1,x2,x3,x4,y1,y2,y3,y4,y5,
  c1,c2,c3,c4,c5,g1,g2,g3,g4,g5,g6];
p0={1/2,0}; p1={1/2,1/2}; p2={0,1}; p3={1,3/2};
 p4={3/2,1}; p5={1,1/2};
x0=p1-p0; y5=p4-p5;
x1=p1+(p2-p0)/2; x2=p2+(p3-p1)/2;
x3=p3+(p4-p2)/2; x4=p4+(p5-p3)/2;
y1=p1-(p2-p0)/2; y2=p2-(p3-p1)/2;
y3=p3-(p4-p2)/2; y4=p4-(p5-p3)/2;
c1[t_]:=Simplify[(1-t)^3 p0+3t(1-t)^2 x0+3t^2(1-t) y1+t^3 p1]
c2[t_]:=Simplify[(1-t)^3 p1+3t(1-t)^2 x1+3t^2(1-t) y2+t^3 p2]
c3[t_]:=Simplify[(1-t)^3 p2+3t(1-t)^2 x2+3t^2(1-t) y3+t^3 p3]
c4[t_]:=Simplify[(1-t)^3 p3+3t(1-t)^2 x3+3t^2(1-t) y4+t^3 p4]
c5[t_]:=Simplify[(1-t)^3 p4+3t(1-t)^2 x4+3t^2(1-t) y5+t^3 p5]
g1=ListPlot[{p0,p1,p2,p3,p4,p5,x0,x1,x2,x3,x4,y1,y2,y3,y4,y5},
  Prolog->AbsolutePointSize[4], DisplayFunction->Identity]
g2=ParametricPlot[c1[t], {t,0,.95}, DisplayFunction->Identity]
g3=ParametricPlot[c2[t], {t,0.05,.95}, DisplayFunction->Identity]
g4=ParametricPlot[c3[t], {t,0.05,.95}, DisplayFunction->Identity]
g5=ParametricPlot[c4[t], {t,0.05,.95}, DisplayFunction->Identity]
g6=ParametricPlot[c5[t], {t,0.05,  1}, DisplayFunction->Identity]
Show[g1,g2,g3,g4,g5,g6, DisplayFunction->$DisplayFunction]
```

Figure 4.52: An Interpolating Bézier Curve: II.

4 Curves

that's expressed in matrix notation as follows:

$$\begin{pmatrix} B_{n,0}(t_0) & B_{n,1}(t_0) & \cdots & B_{n,n}(t_0) \\ B_{n,0}(t_1) & B_{n,1}(t_1) & \cdots & B_{n,n}(t_1) \\ \vdots & \vdots & \ddots & \vdots \\ B_{n,0}(t_n) & B_{n,1}(t_n) & \cdots & B_{n,n}(t_n) \end{pmatrix} \begin{pmatrix} \mathbf{P}_0 \\ \mathbf{P}_1 \\ \vdots \\ \mathbf{P}_n \end{pmatrix} = \begin{pmatrix} \mathbf{Q}_0 \\ \mathbf{Q}_1 \\ \vdots \\ \mathbf{Q}_n \end{pmatrix}. \quad (4.131)$$

This set can be expressed as $\mathbf{MP} = \mathbf{Q}$ and it is easily solved numerically by $\mathbf{P} = \mathbf{M}^{-1}\mathbf{Q}$. If we select $t_0 = 0$, the first row of Equation (4.131) yields $\mathbf{P}_0 = \mathbf{Q}_0$. Similarly, if we select $t_n = 1$, the last row of Equation (4.131) yields $\mathbf{P}_n = \mathbf{Q}_n$. This decreases the number of equations from $n+1$ to $n-1$.

The disadvantage of this approach is that any changes in the t_i's require a recalculation of $\mathbf{M}$ and, consequently, of $\mathbf{M}^{-1}$.

If controlling the speed of the curve is not important, we can select the $n+1$ equally spaced values $t_i = i/n$. Equation (4.131) can now be written

$$\begin{pmatrix} B_{n,0}(0/n) & B_{n,1}(0/n) & \cdots & B_{n,n}(0/n) \\ B_{n,0}(1/n) & B_{n,1}(1/n) & \cdots & B_{n,n}(1/n) \\ \vdots & \vdots & \ddots & \vdots \\ B_{n,0}(n/n) & B_{n,1}(n/n) & \cdots & B_{n,n}(n/n) \end{pmatrix} \begin{pmatrix} \mathbf{P}_0 \\ \mathbf{P}_1 \\ \vdots \\ \mathbf{P}_n \end{pmatrix} = \begin{pmatrix} \mathbf{Q}_0 \\ \mathbf{Q}_1 \\ \vdots \\ \mathbf{Q}_n \end{pmatrix}. \quad (4.132)$$

Now, if the data points $\mathbf{Q}_i$ are moved, matrix $\mathbf{M}$ (or, rather, $\mathbf{M}^{-1}$) doesn't have to be recalculated. If we number the rows and columns of $\mathbf{M}$ by 0 through n, then a general element of $\mathbf{M}$ equals

$$M_{ij} = B_{n,j}(i/n) = \binom{n}{j}(i/n)^j(1 - i/n)^{n-j} = \frac{n!(n-i)^{n-j}i^j}{j!(n-j)!n^n}.$$

Such elements can be calculated, if desired, as exact rational integers, instead of (approximate) floating-point numbers.

Example: We use Equation (4.132) to compute the interpolating Bézier curve that passes through the four points $\mathbf{Q}_0 = (0,0)$, $\mathbf{Q}_1 = (1,1)$, $\mathbf{Q}_2 = (2,1)$, and $\mathbf{Q}_3 = (3,0)$. Since the curve has to pass through the first and last point, we get $\mathbf{P}_0 = \mathbf{Q}_0 = (0,0)$ and $\mathbf{P}_3 = \mathbf{Q}_3 = (3,0)$. Since the four given points are equally spaced, it makes sense to assume that $\mathbf{P}(1/3) = \mathbf{Q}_1$ and $\mathbf{P}(2/3) = \mathbf{Q}_2$. We, thus, end up with the two equations

$$3(1/3)(1-1/3)^2\mathbf{P}_1 + 3(1/3)^2(1-1/3)\mathbf{P}_2 + (1/3)^3(3,0) = (1,1),$$
$$3(2/3)(1-2/3)^2\mathbf{P}_1 + 3(2/3)^2(1-2/3)\mathbf{P}_2 + (2/3)^3(3,0) = (2,1)$$

that are solved to yield $\mathbf{P}_1 = (1, 3/2)$ and $\mathbf{P}_2 = (2, 3/2)$. The curve is thus

$$\mathbf{P}(t) = (1-t)^3(0,0) + 3t(1-t)^2(1,3/2) + 3t^2(1-t)(2,3/2) + t^3(3,0).$$

▶ **Exercise 4.78**: Plot the curve and the eight points.

4.14.21 An Interpolating Bézier Curve: IV

Traditionally, the word *font* refers to a set of characters of type that share the same size and style, such as Times Roman 12 point. Imagine the task of a font designer about to design the next character of a new font. The designer has a rough idea of the shape of the character and needs to create a curve that will fit the outline of the character. A natural solution to the problem is to strategically place data points along the outline of the character and connect them with spline segments. The method described here uses four-point Bézier segments, where each segment goes from a data point $\mathbf{P}_k$ to the next point $\mathbf{P}_{k+1}$, using two intermediate control points that the software calculates automatically. In this way, the designer does not have to know about control points. Their (the designer's) job is to place data points strategically along the desired curve. Figure 4.53 is an example of the letter "A" in font Times. It is clear that large parts of the letter are made of straight (or close to straight) segments which require few data points. Only regions of high curvature need many points for their definition.

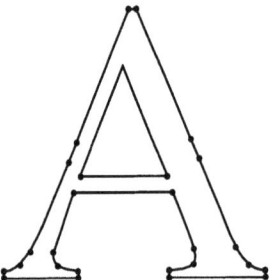

Figure 4.53: Data points For the Letter "A".

The method outlined here is due to John Hobby [Hobby 86] and is used in the Metafont software (see page 131 of [Knuth 86] for more details) to design the outlines of fonts of type. This method constructs an interpolating Bézier curve and does it by combining features borrowed from Hermite interpolation and cubic splines. The advantage of this method is that the designer can help the software in three ways as follows:

1. The designer may specify the two intermediate control points for any segment. This overrides the points calculated by the software and is usually done by the designer while looking at the first version of the curve and attempting to fine-tune its shape.

2. The designer may specify the direction of the curve at certain points. Often, it is clear to the designer that the curve should go, for example, horizontally from left to right, when it passes through data point $\mathbf{P}_k$ and such information can be very helpful to the software.

3. Another feature that can help the designer get the right curve is the ability to specify the *tension* of the curve individually for each segment. The next paragraph is copied from Section 4.10.

4 Curves

Perhaps the best way to visualize a spline under tension is to think of the data points as nails driven into a board, and of the spline as a rubber band strung above or below the nails. To add tension, simply pull the rubber band, which tightens it, bringing each segment closer to a straight line.

> Overhead the sky was half crystalline, half misty, and the night around was chill and vibrant with rich tension.
> — F. Scott Fitzgerald, *This Side of Paradise*

We start with a single Hermite segment connecting points $\mathbf{P}_1$ and $\mathbf{P}_2$. The two extreme tangent vectors are usually denoted $\mathbf{P}_1^t$ and $\mathbf{P}_2^t$. In this section, they are expressed as

$$\mathbf{P}_1^t = (f(\theta,\phi)/\tau_1)\mathbf{T}_1 \quad \text{and} \quad \mathbf{P}_2^t = (g(\theta,\phi)/\tau_2)\mathbf{T}_2,$$

where $\mathbf{T}_1$ and $\mathbf{T}_2$ are unit vectors in the directions of the tangents, and f and g are functions that determine the magnitudes of the tangents. These functions depends on θ and ϕ, which are the angles between the line $\mathbf{P}_1 \to \mathbf{P}_2$ and the two tangents. The quantities τ_1 and τ_2 are the tension parameters. The bigger they are, the shorter the tangent vectors become and the closer the curve gets to a straight line. The default value of the tension parameters is 1, but the user can specify values τ_k at any data point $\mathbf{P}_k$ along the curve.

We select functions f and g in a manner similar to that of Section 4.8.7, Equations (4.76) and (4.77):

$$g(\theta,\phi) = |\mathbf{P}_1^t| = \frac{2|\mathbf{P}_2 - \mathbf{P}_1|}{1 + \alpha\cos\phi + (1-\alpha)\cos\theta}, \qquad (4.133)$$
$$f(\theta,\phi) = |\mathbf{P}_2^t| = \frac{2|\mathbf{P}_2 - \mathbf{P}_1|}{1 + \alpha\cos\theta + (1-\alpha)\cos\phi},$$

where α is a user-defined parameter in the range $[0.5, 1]$.

Instead of constructing the Hermite segment out of the two points and two tangents, we construct it as a four-point Bézier curve, according to Equation (4.124) (Section 4.14.16):

$$\begin{aligned}&\left(\mathbf{P}_1, \left(\mathbf{P}_1 + \frac{1}{3}\mathbf{P}_1^t\right), \left(\mathbf{P}_2 - \frac{1}{3}\mathbf{P}_2^t\right), \mathbf{P}_2\right) \\ &= \left(\mathbf{P}_1, \left(\mathbf{P}_1 + \frac{f(\theta,\phi)}{3\tau_1}\mathbf{T}_1\right), \left(\mathbf{P}_2 - \frac{g(\theta,\phi)}{3\tau_2}\mathbf{T}_2\right), \mathbf{P}_2\right).\end{aligned} \qquad (4.134)$$

In this way, each segment has two interior control points, making it possible for the user to explicitly specify any control points.

The software has to set up equations that are easy to solve (i.e., linear). The equations are based on the requirements that the first and second derivatives are continuous at the $n-2$ interior points. The unknowns are the various θ and ϕ angles. Each interior point has two such angles, and each of the two extreme points has one angle. The total number of unknowns is, therefore, $1+2(n-2)+1 = 2n-2$.

Once all the angles are known, all the f and g functions and all the unit tangents $\mathbf{T}_k$ can be calculated. Using these and the tension parameters, all the interior control points

$$\mathbf{P}_k + \frac{f(\theta_k, \phi_{k+1})}{3\tau_k}\mathbf{T}_k \quad \text{and} \quad \mathbf{P}_{k+1} - \frac{g(\theta_k, \phi_{k+1})}{3\tau_{k+1}}\mathbf{T}_{k+1}, \qquad (4.135)$$

can be calculated. (Equation (4.135) shows how changing the tension parameters is equivalent to sliding the two interior control points along the lines that connect them to $\mathbf{P}_k$ and $\mathbf{P}_{k+1}$, respectively.) The last step is to calculate and draw all the $n-1$ Bézier segments that constitute the curve. Any control points and directions supplied by the user help the calculations since they reduce the number of unknowns. Any tension parameters supplied by the user should be included in the calculations.

The requirement that the first derivatives be equal at the interior points results in the $n-2$ equations

$$\theta_{k+1} + \phi_{k+1} = -\psi_{k+1}, \quad \text{for} \quad k = 1, 2, \ldots, n-2, \qquad (4.136)$$

where ψ_{k+1} is the angle between vectors $\mathbf{P}_{k+2} - \mathbf{P}_{k+1}$ and $\mathbf{P}_{k+1} - \mathbf{P}_k$. The requirement that the second derivatives be equal at the interior points leads to the so-called *mock curvature* that is obtained by the following:

1. Calculating the second derivative $\mathbf{P}_k^{tt}(t)$ of a general segment $\mathbf{P}_k(t)$ (where the functions f and g are given by Equation (4.133)). This is easier to do if the Hermite form of the segment is used, rather than the Bézier form. The second derivative of the PC segment $\mathbf{P}_k(t)$ is $\mathbf{P}_k^{tt}(t) = 6\mathbf{a}_k t + \mathbf{b}_k$ (Equation (4.53)), where $\mathbf{a}_k$ and $\mathbf{b}_k$ are given by Equation (4.22):

$$\mathbf{a}_k = 2\mathbf{P}_k - 2\mathbf{P}_{k+1} + \mathbf{P}_k^t + \mathbf{P}_{k+1}^t, \quad \mathbf{b}_k = -3\mathbf{P}_k + 3\mathbf{P}_{k+1} - 2\mathbf{P}_k^t - \mathbf{P}_{k+1}^t.$$

2. Calculating the Taylor expansion of $\mathbf{P}_k^{tt}(1)$ and $\mathbf{P}_{k+1}^{tt}(0)$ about $\theta = \phi = 0$ and retaining the linear parts.

This process results in the $n-2$ linear equations for the unknown angles

$$\frac{\tau_{k+1}^2}{|\mathbf{P}_{k+1} - \mathbf{P}_k|}\left(\frac{\theta_k + \phi_{k+1}}{\tau_k} - 3\phi_{k+1}\right) = \frac{\tau_{k+1}^2}{|\mathbf{P}_{k+2} - \mathbf{P}_{k+1}|}\left(\frac{\theta_{k+1} + \phi_{k+2}}{\tau_{k+2}} - 3\theta_{k+1}\right), \qquad (4.137)$$

where $k = 1, 2, \ldots, n-2$. The total number of Equations (4.136) and (4.137) is $2n - 4$, again two short of the number of unknowns. The user should, therefore, supply the values of two unknowns, typically the angles of the two extreme tangents. For a closed curve, every data point is interior, so the number of equations is $2n$, the same as the number of unknowns.

4.14.22 Circular Bézier Curves

The Bézier curve is a polynomial $\mathbf{P}(t)$ whose parameter t varies in the interval $[0, 1]$ and whose value is a point (in two or three dimensions). We say that the *domain* of this polynomial is the interval $[0, 1]$ and the *range* is the two- or three-dimensional Euclidean space (i.e., all the pairs or triplets of real numbers). This section, which

4 Curves

is based on [Alfeld et al. 95], describes the *circular Bézier curve*, a polynomial whose domain is a circular arc, not an interval.

We start with a unit circle C centered on the origin. An arc A of length less than π and with vertices at points v_1 and v_2 is selected such that $0 < \theta_2 - \theta_1 < \pi$ (Figure 4.54). Assume that v is a point on the arc, then v can be written as the weighted combination

$$v = b_1 v_1 + b_2 v_2, \qquad (4.138)$$

where b_1 and b_2 are the *circular barycentric coordinates* of v with respect to arc A. It is easy to see that these coordinates have the following properties:

1. The coordinates of v_1 are $(b_1, b_2) = (1, 0)$ and those of v_2 are $(b_1, b_2) = (0, 1)$.
2. Any other point v on the arc has two positive coordinates b_1 and b_2.
3. The sum $b_1 + b_2$ equals 1 for the two endpoints but is greater than 1 for any other point. Notice that point $v = 0.5b_1 + 0.5b_2$ is located midway between v_1 and v_2, but on the *straight segment* connecting them, not on the arc. In order for v to be on the arc, the sum $b_1 + b_2$ should be greater than 1.
4. The coordinates of v are invariant under rotation since they depend only on the relative positions of v, v_1, and v_2. This justifies the name *barycentric*.

Since v, v_1, and v_2 are located on the circumference of a unit circle, each can be expressed by means of one angle θ. Figure 4.54a shows that $v = (\cos\theta, \sin\theta)$, $v_1 = (\cos\theta_1, \sin\theta_1)$, and $v_2 = (\cos\theta_2, \sin\theta_2)$. Substituting this in Equation (4.138) results in

$$(\cos\theta, \sin\theta) = b_1(\cos\theta_1, \sin\theta_1) + b_2(\cos\theta_2, \sin\theta_2).$$

This is a system of two equations whose solutions are

$$\begin{aligned} b_1(\theta) &= \frac{\sin\theta_2 \cos\theta - \cos\theta_2 \sin\theta}{\sin\theta_2 \cos\theta_1 - \cos\theta_2 \sin\theta_1} = \frac{\sin(\theta_2 - \theta)}{\sin(\theta_2 - \theta_1)}, \\ b_2(\theta) &= \frac{\sin\theta \cos\theta_1 - \cos\theta \sin\theta_1}{\sin\theta_2 \cos\theta_1 - \cos\theta_2 \sin\theta_1} = \frac{\sin(\theta - \theta_1)}{\sin(\theta_2 - \theta_1)}. \end{aligned} \qquad (4.139)$$

The circular barycentric coordinates are thus expressed as linear combinations of $\sin\theta$ and $\cos\theta$, and also as ratios of arc lengths. They can also be expressed as ratios of areas of triangles:

$$b_1 = \frac{\text{area}(0, v, v_2)}{\text{area}(0, v_1, v_2)}, \quad b_2 = \frac{\text{area}(0, v_1, v)}{\text{area}(0, v_1, v_2)}.$$

The next step is to define the *circular Bernstein polynomials* of degree n,

$$B_{ni}(\theta) = \binom{n}{i} b_1(\theta)^{n-i} b_2(\theta)^i, \qquad i = 0, 1, \ldots, n, \qquad (4.140)$$

(compare with Equation (4.103)) and the *circular Bézier curve*,

$$\mathbf{P}(\theta) = \sum_{i=0}^{n} c_i (\cos\theta, \sin\theta) B_{ni}(\theta), \qquad \theta_1 \leq \theta \leq \theta_2, \qquad (4.141)$$

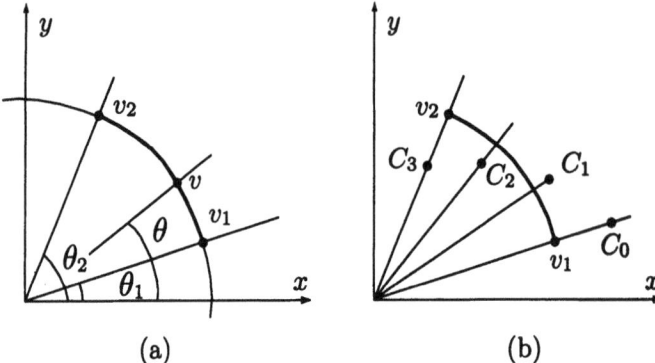

Figure 4.54: Barycentric Circular Coordinates.

where the c_i's are any real numbers. Notice that this curve is a *trigonometric polynomial*, i.e., a polynomial where each monomial consists of powers of sines and cosines instead of powers of x. Comparing this definition with the linear Bézier curve (Equation (4.104)) suggests how to define the control points of the circular curve. We divide the range $[\theta_1, \theta_2]$ into n equal intervals by defining

$$\phi_i = \theta_1 + i(\theta_2 - \theta_1)/n, \qquad i = 0, 1, \ldots, n, \qquad (4.142)$$

and define the control points

$$C_i \stackrel{\text{def}}{=} c_i(\cos \phi_i, \sin \phi_i).$$

Points C_i have a simple geometric meaning. The pair $(\cos \phi_i, \sin \phi_i)$ is a point on the circumference of the unit circle C. Multiplying it by c_i yields a point C_i on the line connecting the origin to $(\cos \phi_i, \sin \phi_i)$ (Figure 4.54b). The control points are thus equally spaced in arc length around the arc A.

From Equation (4.142) we get $\phi_0 = \theta_1$ and $\phi_n = \theta_2$. From this, it is easy to see that

$$\mathbf{P}(\theta_1) = \sum_{i=0}^{n} c_i(\cos \theta_1, \sin \theta_1) B_{ni}(\theta_1) = c_0(\cos \phi_0, \sin \phi_0) = C_0,$$

$$\mathbf{P}(\theta_2) = \sum_{i=0}^{n} c_i(\cos \theta_2, \sin \theta_2) B_{ni}(\theta_2) = c_n(\cos \phi_n, \sin \phi_n) = C_n.$$

The circular curve thus starts at C_0 and ends at C_n, reinforcing the interpretation of the C_i's as control points. There is a difference, however, between the way the control points are used in the linear and in the circular Bézier curves. In the linear curve, the user selects the control points, which may be any points, and the curve is calculated as the weighted sum of the points. In the circular curve, the user selects the constants c_i and the two values θ_1 and θ_2. Each point $\mathbf{P}(\theta_0)$ on the curve becomes the weighted sum of points $c_i(\cos \theta_0, \sin \theta_0)$, not of the control

4 Curves

points. The control points are spread evenly along the interval $[\theta_1, \theta_2]$ but are not used in calculating the curve.

Example: We select $n = 3$, $\theta_1 = 0°$, $\theta_2 = 90° = \pi/2$, and constants $c_0 = 0$, $c_1 = 0.1$, $c_2 = 0.2$, and $c_3 = 2$. We notice that $\sin(\theta_2 - \theta_1) = 1$, so the circular Bernstein polynomials for this case are

$$B_{3i}(\theta) = \binom{3}{i} b_1(\theta)^{3-i} b_2(\theta)^i = \binom{3}{i} \sin^{3-i}(\pi/2 - \theta) \sin^i(\theta)$$

and the circular curve is

$$\mathbf{P}(\theta) = (\cos\theta, \sin\theta) \left[0 \binom{3}{0} \sin^3(\pi/2 - \theta) \sin^0(\theta) + 0.1 \binom{3}{1} \sin^2(\pi/2 - \theta) \sin^1(\theta) \right.$$

$$\left. + 0.2 \binom{3}{2} \sin^1(\pi/2 - \theta) \sin^2(\theta) + 2 \binom{3}{3} \sin^0(\pi/2 - \theta) \sin^3(\theta) \right]$$

$$= (\cos\theta, \sin\theta) \left[0.3 \sin^2(\pi/2 - \theta) \sin(\theta) + 0.6 \sin(\pi/2 - \theta) \sin^2(\theta) + 2 \sin^3(\theta) \right].$$

It goes from $C_0 = c_0(\cos 0, \sin 0) = (0, 0)$ to $C_3 = c_3(\cos(\pi/2), \sin(\pi/2)) = (0, 2)$ (Figure 4.55).

▸ **Exercise 4.79:** What are the four control points in this case?

▸ **Exercise 4.80:** Calculate the four control points of the cubic circular curve defined by $\theta_1 = 0$, $\theta_2 = 90° = \pi/2$, $c_0 = 2$, $c_1 = 1.2$, $c_2 = 1.6$, and $c_3 = 1$.

4.14.23 Circles and Bézier Curves

The equation of a circle is $x^2 + y^2 = r^2$ or $y = \pm\sqrt{r^2 - x^2}$. This is not a polynomial and, in fact, Exercise 4.81 proves that a polynomial cannot represent a circle. Applying Bézier methods to circles can be done in two ways: (1) using rational Bézier curves (Section 4.14.24) or (2) producing an approximation to the circle. This section discusses the latter.

We start with a three-point example. We select the three points $\mathbf{P}_0 = (1, 0)$, $\mathbf{P}_1 = (k, k)$, and $\mathbf{P}_2 = (0, 1)$ and attempt to find the value of k such that the quadratic Bézier curve defined by the points will best approximate a quarter circle of radius 1 (Figure 4.56). The curve is given, of course, by

$$\mathbf{P}(t) = (1-t)^2(1, 0) + 2t(1-t)(k, k) + t^2(0, 1)$$
$$= \left(1 + 2t(k-1) + t^2(1-2k), 2kt + t^2(1-2k)\right) \quad (4.143)$$
$$= \left(P_x(t), P_y(t)\right),$$

and it equals the circle at its start and end points. We need a constraint that will produce an equation whose solution will yield the value of k. A reasonable constraint is to require that the curve be identical to the circle at its midpoint. This can be expressed as $\mathbf{P}(0.5) = (1/\sqrt{2}, 1/\sqrt{2})$ and it produces the equation

$$\mathbf{P}(0.5) = \frac{1}{4}(1, 0) + \frac{1}{2}(k, k) + \frac{1}{4}(0, 1) = \left(\frac{1}{\sqrt{2}}, \frac{1}{\sqrt{2}}\right),$$

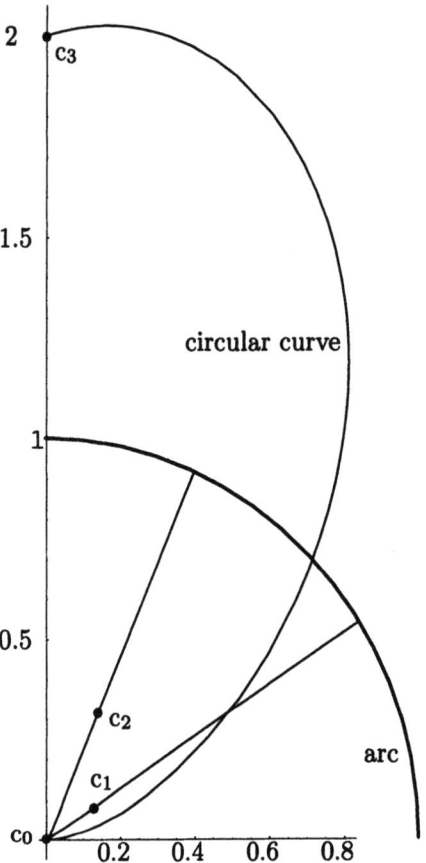

Figure 4.55: A Circular Bézier Curve.

whose solution is
$$k = \frac{2\sqrt{2} - 1}{2} \approx 0.914.$$
We also note that the tangent vector of Equation (4.143) is

$$\mathbf{P}^t(t) = \big(2(k-1) + 2t(1-2k), 2k + 2t(1-2k)\big). \tag{4.144}$$

How much does this curve deviate from a true circle of radius 1? To answer this, we first notice that the distance of a point $\mathbf{P}(t)$ from the origin is

$$D(t) = \sqrt{P_x^2(t) + P_y^2(t)} = \sqrt{\big(1 + 2t(k-1) + t^2(1-2k)\big)^2 + \big(2kt + t^2(1-2k)\big)^2}.$$

To find the maximum distance, we differentiate $D(t)$:

$$\frac{dD(t)}{dt} = \frac{2P_x(t) \cdot P_x^t(t) + 2P_y(t) \cdot P_y^t(t)}{\frac{1}{2}\sqrt{P_x^2(t) + P_y^2(t)}}$$

4 Curves

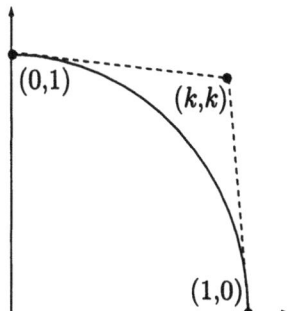

Figure 4.56: A Quadratic Bézier Curve
Approximating a Quarter Circle.

and set the result equal to 0. This yields $P_x(t) \cdot P_x^t(t) + P_y(t) \cdot P_y^t(t) = \mathbf{P}(t) \cdot \mathbf{P}^t(t) = 0$. Using Equations (4.143) and (4.144), we get the equation

$$2(k-1) + 2\big(1 + 2(k-1)^2\big)t - 6(1-2k)^2 t^2 + 4(1-2k)^2 t^3 = 0,$$

which has two roots in the interval $[0, 1]$, namely $t_1 \approx 0.33179$ and $t_2 \approx 0.66821$, close to the expected values of $1/3$ and $2/3$. Simple computation shows the maximum distance of $\mathbf{P}(t)$ from the origin to be $D(t_1) = D(t_2) = 0.995685$. The maximum deviation of this from a circle of radius one is thus 0.432%, very impressive!

▸ **Exercise 4.81:** Prove that the Bézier curve cannot be a circle.

▸ **Exercise 4.82:** Consider the quarter circle from $\mathbf{P}_0 = (1, 0)$ to $\mathbf{P}_3 = (0, 1)$. Select two points $\mathbf{P}_1$ and $\mathbf{P}_2$ such that the Bézier curve defined by the four points would be the closest possible to a circle.

▸ **Exercise 4.83:** Do the same for the oval (elliptic) arc from $(1, 0)$ to $(0, 1)$.

▸ **Exercise 4.84:** Calculate the cubic Bézier curve that approximates the circular arc of Figure 4.57 spanning an angle of 2θ. The calculation should be based on the requirement that the curve and the arc have the same endpoints and the same extreme tangent vectors.

Example: We approximate a sine wave by smoothly joining eight cubic Bézier segments (Figure 4.58). The first segment requires four control points and each of the remaining seven segments requires three additional points. The total number of points is, thus, 25. They are numbered $\mathbf{P}_0$ through $\mathbf{P}_{24}$, but because of the high symmetry of the sine wave, only the first seven points, $\mathbf{P}_0$ through $\mathbf{P}_6$, need be calculated. The rest can be obtained from these by simple translations and reflections. We require that the following three points be on the sine curve, making it easy to find their coordinates:

$$\mathbf{P}_0 = (0, 0), \ \mathbf{P}_3 = \left(\frac{\pi}{4}, \sin\left(\frac{\pi}{4}\right)\right) \approx (0.785, 0.7071), \ \mathbf{P}_6 = \left(\frac{\pi}{2}, \sin\left(\frac{\pi}{2}\right)\right) \approx (1.57, 1).$$

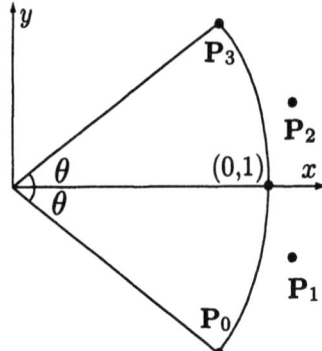

Figure 4.57: A Cubic Bézier Curve Approximating an Arc.

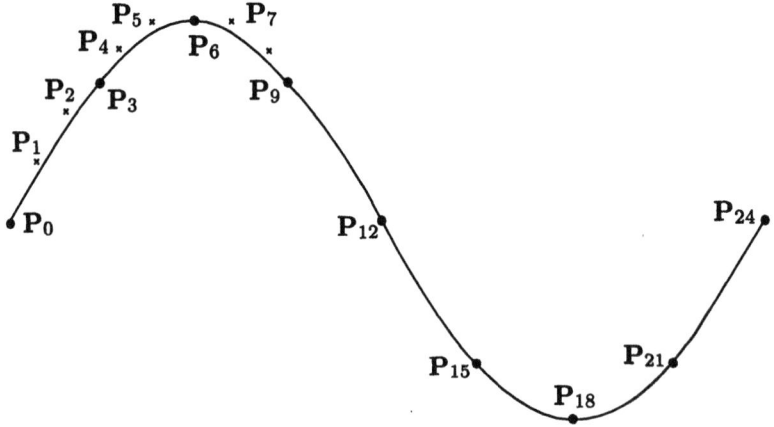

Figure 4.58: A Sine Curve Approximated by Eight Cubic Bézier Segments.

The expression for segment i (where $i = 0, 3, 6, 9, 12, 15, 18$, and 21) is

$$\mathbf{P}_i(t) = (1-t)^3\mathbf{P}_i + 3t(1-t)^2\mathbf{P}_{i+1} + 3t^2(1-t)\mathbf{P}_{i+2} + t^3\mathbf{P}_{i+3},$$

and its tangent vector is

$$\mathbf{P}_i^t(t) = -3(1-t)^2\mathbf{P}_i + (3-9t)(1-t)\mathbf{P}_{i+1} + 3t(2-3t)\mathbf{P}_{i+2} + 3t^2\mathbf{P}_{i+3}.$$

To calculate point $\mathbf{P}_1$, we require that the initial tangent $\mathbf{P}_0^t(0)$ of the first curve segment matches the initial slope of the sine wave, which is 45°. We can therefore write $\mathbf{P}_0^t(0) = (a, a)$ for any positive a and we select $a = 0.7071$ since this produces a normalized tangent vector. The result is

$$(.7071, .7071) = \mathbf{P}_0^t(0) = -3\mathbf{P}_0 + 3\mathbf{P}_1 \text{ or } \mathbf{P}_1 = (.7071, .7071)/3 = (.2357, .2357).$$

To calculate points $\mathbf{P}_2$ and $\mathbf{P}_4$, we again require that the final tangent vector $\mathbf{P}_0^t(1)$ of the first segment match the slope of the sine wave at $x_3 = \pi/4$. That

4 Curves

slope is 0.7071, so we select $(1, 0.7071)$ as the tangent vector, then normalize it to $(0.816, 0.577)$. We end up with

$$(.816, .577) = \mathbf{P}_0^t(1) = -3\mathbf{P}_2 + 3\mathbf{P}_3 \text{ or } \mathbf{P}_2 = \mathbf{P}_3 - (.816, .577)/3 = (.513, .5151).$$

By symmetry we also get $\mathbf{P}_4 = \mathbf{P}_3 + (0.816, .577)/3 = (1.057, 0.899)$.

Only point $\mathbf{P}_5$ remains to be calculated. Again, we require that the final tangent vector $\mathbf{P}_3^t(1)$ of the second segment (segment 3) match the slope of the sine wave at $\mathbf{P}_6$, which is 0. The normalized tangent vector is thus $(1, 0)$, which produces the equation

$$(1, 0) = \mathbf{P}_3^t(1) = 3\mathbf{P}_6 - 3\mathbf{P}_5, \text{ or } \mathbf{P}_5 = \mathbf{P}_6 - (1, 0)/3 = (1.237, 1).$$

Points $\mathbf{P}_7$ through $\mathbf{P}_{24}$ can be obtained from the first seven points by translation and reflection. Alternatively, the first four cubic segments can be calculated and each pixel can be used to calculate one more pixel by translation and reflection.

4.14.24 Rational Bézier Curves

It is possible to generalize the definition of the Bézier curve (Equation (4.104)) to

$$\mathbf{P}(t) = \frac{\sum_{i=0}^{n} w_i \mathbf{P}_i B_{n,i}(t)}{\sum_{i=0}^{n} w_i B_{n,i}(t)}, \qquad 0 \le t \le 1.$$

The weight functions are now ratios of polynomials and they also depend on additional weights w_i. This representation seems unnecessarily complicated (and, for most practical applications, it is), but it allows for more shapes, such as precise representation of conic sections (Section 4.20).

Section 4.16.8 shows that the Bézier curve is a special case of the B-spline curve. As a result, many current software systems use the rational B-spline (Section 4.16.18) when rational curves are required. Such a system can produce the rational Bézier curve as a special case, so there is no need to discuss this type of curve in detail here.

4.15 Subdivision Curves

The Bézier curve can be constructed as either a weighted sum of control points or by the process of scaffolding. These are two approaches that seem very different but yield the same result. A third approach to curve (and surface) design, using the process of *refinement* (also known as *subdivision*), is described in this section. This is a general approach that can produce Bézier curves, B-spline curves, and other types of curves. Its main advantage is that it can easily be extended to surfaces. The idea is to start with a given set of control points $\mathbf{P}_i$, perform a computation that results in a new set of points $\mathbf{P}_i^1$, and repeat the process, producing more and more sets of points $\mathbf{P}_i^k$. Table 4.59 shows the notation used.

$$\begin{array}{cccc}\mathbf{P}_0, & \mathbf{P}_1, & \ldots, & \mathbf{P}_n \\ \mathbf{P}_0^1, & \mathbf{P}_1^1, & \ldots, & \mathbf{P}_{n_1}^1 \\ \mathbf{P}_0^2, & \mathbf{P}_1^2, & \ldots, & \mathbf{P}_{n_2}^2 \\ & & \vdots & \\ \mathbf{P}_0^k, & \mathbf{P}_1^k, & \ldots, & \mathbf{P}_{n_k}^k \end{array}$$

Table 4.59: Refining Control Points.

Each point $\mathbf{P}_j^k$ is calculated as a weighted sum of the points $\mathbf{P}_i^{k-1}$ of the previous iteration. Thus,

$$\mathbf{P}_j^k = \sum_{i=0}^{n_{k-1}} a_{ijk} \mathbf{P}_i^{k-1} = (a_{0jk}, a_{1jk}, \ldots, a_{n_{k-1},jk}) \begin{pmatrix} \mathbf{P}_0^{k-1} \\ \mathbf{P}_1^{k-1} \\ \vdots \\ \mathbf{P}_{n_{k-1}}^{k-1} \end{pmatrix},$$

where a_{ijk} are real coefficients. Notice that each iteration produces a different number $n_k + 1$ of points. If n_k gets smaller with k, then the number of points gets smaller and smaller until a single point is left. An example is the de Casteljau scaffolding construction, a process that produces one point of the Bézier curve. At the other extreme, n_k may get larger with k, producing more points in each iteration. We then stop after a few iterations and draw the curve by drawing straight segments between the points of the last iteration. An example of this case is the Chaikin algorithm, described later.

Each iteration can be completely described by its coefficient matrix

$$\begin{pmatrix} \mathbf{P}_0^k \\ \mathbf{P}_1^k \\ \vdots \\ \mathbf{P}_{n_k}^k \end{pmatrix} = \begin{pmatrix} a_{00k} & a_{10k} & \cdots & a_{n_{k-1},0k} \\ a_{01k} & a_{11k} & \cdots & a_{n_{k-1},1k} \\ \vdots & \vdots & & \vdots \\ a_{0,n_k,k} & a_{1,n_k,k} & \cdots & a_{n_{k-1},n_k,k} \end{pmatrix} \begin{pmatrix} \mathbf{P}_0^{k-1} \\ \mathbf{P}_1^{k-1} \\ \vdots \\ \mathbf{P}_{n_{k-1}}^{k-1} \end{pmatrix} \quad (4.145)$$

$$= \begin{pmatrix} \mathbf{P}_0^k \\ \mathbf{P}_1^k \\ \vdots \\ \mathbf{P}_{n_k}^k \end{pmatrix} \mathbf{M}_k \begin{pmatrix} \mathbf{P}_0^{k-1} \\ \mathbf{P}_1^{k-1} \\ \vdots \\ \mathbf{P}_{n_{k-1}}^{k-1} \end{pmatrix},$$

where $\mathbf{M}_k$ has $n_k + 1$ rows and $n_{k-1} + 1$ columns. Since the number of iterations may be large, the number of coefficients a_{ijk} may be huge. In practice, this number is significantly reduced in three ways: (1) using a rule of calculation where most of the coefficients are zero, (2) using coefficients a_{ij} that are independent of k, and (3) using coefficients a_{ik} that are independent of j. Case 2 is called *uniform refinement* and case 3 is called *stationary refinement*.

4 Curves

Example 1: This is the de Casteljau scaffolding construction expressed as a refinement process. The rule of calculation is

$$\mathbf{P}_j^{k+1} = 0.5(\mathbf{P}_j^k + \mathbf{P}_{j+1}^k), \tag{4.146}$$

which implies that the a_i coefficients are independent of j and k (this is a stationary uniform refinement method) and are zero except for the two coefficients a_j and a_{j+1}. Since $\mathbf{P}_j^k$ depends on $\mathbf{P}_j^{k-1}$ and $\mathbf{P}_{j+1}^{k-1}$, the largest value for j is $n_k - 1$. This means that each iteration reduces the number of points by 1 (Figure 4.60a). We start with the $n+1$ points $\mathbf{P}_0, \mathbf{P}_1, \ldots, \mathbf{P}_n$. The first iteration produces n points, the second one $n-1$ points, and so on, until iteration n produces one point. That point is on the Bézier curve $\mathbf{P}(t)$ defined by the $n+1$ original control points. In fact, that point is $\mathbf{P}(0.5)$. If we generalize Equation (4.146) to $\mathbf{P}_j^{k+1} = \alpha(\mathbf{P}_j^k + \mathbf{P}_{j+1}^k)$, then the final point is $\mathbf{P}(\alpha)$. Matrix $\mathbf{M}_k$ of Equation (4.145) is

$$\mathbf{M}_k = \begin{pmatrix} 0.5 & 0.5 & 0 & 0 & \cdots & 0 \\ 0 & 0.5 & 0.5 & 0 & \cdots & 0 \\ 0 & 0 & 0.5 & 0.5 & \cdots & 0 \\ \vdots & \vdots & \vdots & \ddots & \ddots & \vdots \\ 0 & 0 & 0 & \cdots & 0.5 & 0.5 \end{pmatrix}.$$

It is independent of k and is of order $k \times (k+1)$.

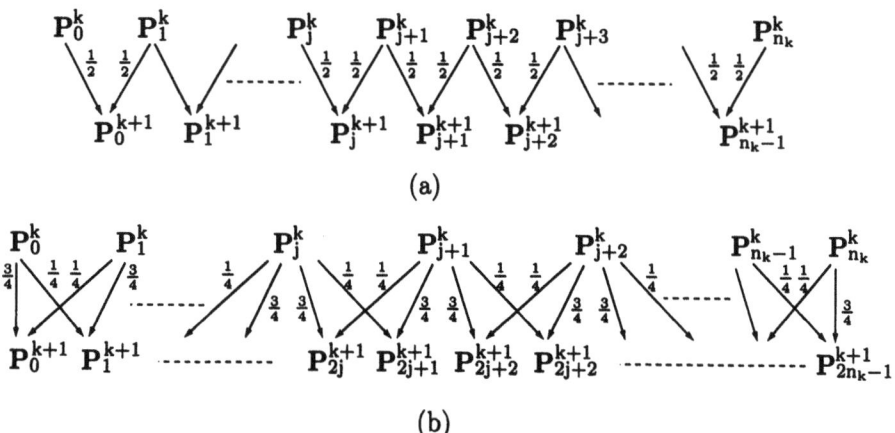

Figure 4.60: (a) De Casteljau Refinement. (b) Chaikin's Method.

Example 2: We start with the $n+1$ control points $\mathbf{P}_0, \mathbf{P}_1, \ldots, \mathbf{P}_n$ and apply the rule of refinement (Figure 4.60b)

$$\begin{aligned} \mathbf{P}_{2j}^{k+1} &= \frac{3}{4}\mathbf{P}_j^k + \frac{1}{4}\mathbf{P}_{j+1}^k, \\ \mathbf{P}_{2j+1}^{k+1} &= \frac{1}{4}\mathbf{P}_j^k + \frac{3}{4}\mathbf{P}_{j+1}^k. \end{aligned} \tag{4.147}$$

$$P_0^1 = \tfrac{3}{4}P_0 + \tfrac{1}{4}P_1, \qquad P_1^1 = \tfrac{1}{4}P_0 + \tfrac{3}{4}P_1,$$
$$P_2^1 = \tfrac{3}{4}P_1 + \tfrac{1}{4}P_2, \qquad P_3^1 = \tfrac{1}{4}P_1 + \tfrac{3}{4}P_2,$$
$$P_4^1 = \tfrac{3}{4}P_2 + \tfrac{1}{4}P_3, \qquad P_5^1 = \tfrac{1}{4}P_2 + \tfrac{3}{4}P_3,$$
$$\vdots \qquad\qquad\qquad \vdots$$
$$P_{2n-2}^1 = \tfrac{3}{4}P_{n-1} + \tfrac{1}{4}P_n, \quad P_{2n-1}^1 = \tfrac{1}{4}P_{n-1} + \tfrac{3}{4}P_n.$$

Table 4.61: First Iteration of Chaikin's Algorithm.

The first iteration starts with the original $n+1$ points and produces the $2n$ points P_i^1 shown in Table 4.61:
Each subsequent iteration doubles the number of points and brings the points closer to the curve. After k iterations (where k depends on the required precision), the curve is drawn by drawing straight segments between the points produced in the last iteration. This method is due to George Chaikin [Chaikin 74] and has a simple geometric interpretation (illustrated in Figure 4.62). Part (a) of the figure shows a control polygon made of five points. The rule of refinement is: Take a segment P_iP_{i+1} of the control polygon and place two new points Q_i and R_i at distances from P_i of 1/4 and 3/4 the segment's size, respectively (Figure 4.62b). The new points are thus defined by

$$Q_i = \frac{3}{4}P_i + \frac{1}{4}P_{i+1}, \quad R_i = \frac{1}{4}P_i + \frac{3}{4}P_{i+1}.$$

This is repeated for all the polygon segments. If we start with $n+1$ control points defining a control polygon with n sides, we end up with $2n$ new points Q_i and R_i. They should now be connected to form a new control polygon with $2n-1$ sides. As this process is repeated (Figure 4.62c), the control polygons get closer to the smooth curve shown in Figure 4.62d. This figure also shows that the midpoint of any segment of the control polygon is a point on the Chaikin curve. In fact, the midpoint of any segment generated at any stage of the refinement is a point on the Chaikin curve.

▶ **Exercise 4.85:** Is this curve a Bézier curve?

This algorithm works for closed curves too. The only necessary change is to connect the last point P_n to the first one P_0 and calculate the two auxiliary points Q_n and R_n. This can be done in a natural way if we copy P_0 and name the duplicate P_{n+1}. Figure 4.63 shows three instances in the construction of such a curve. Again, we see that the midpoint of any segment of the control polygon is a point on the closed Chaikin curve.

To identify the kind of curve that Chaikin's algorithm produces, let's consider the control polygon defined by the three points P_0, $P_1 = B$, and P_2 (Figure 4.64). Let A and C be the midpoints of segments P_0P_1 and P_1P_2, respectively, and let point P be the midpoint of points $M_{ab} = (A+B)/2$ and $M_{bc} = (B+C)/2$. This point has the following properties:

1. It is located on the Bézier curve defined by points A, B, and C because it's been constructed using the de Casteljau scaffolding process.

4 Curves

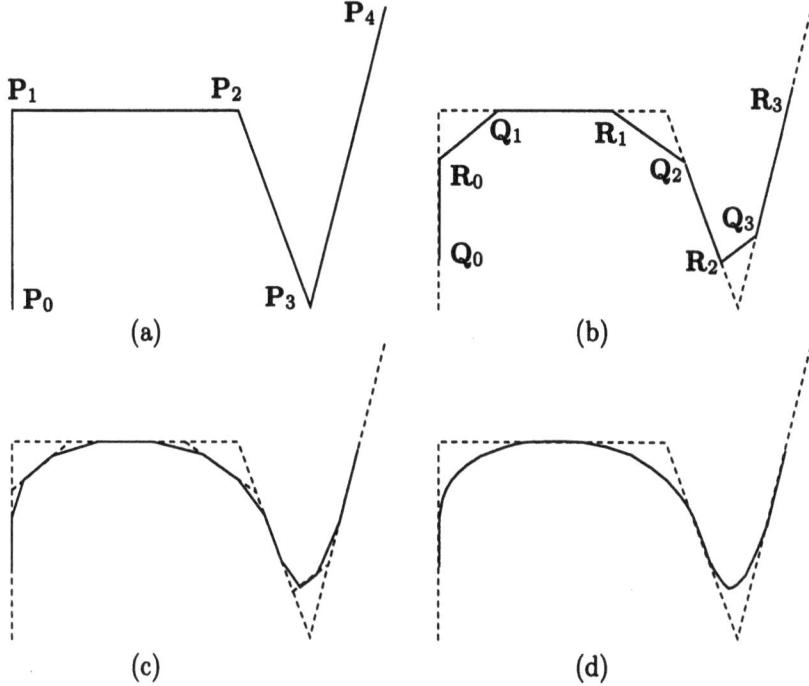

Figure 4.62: Chaikin's Algorithm for a Control Polygon.

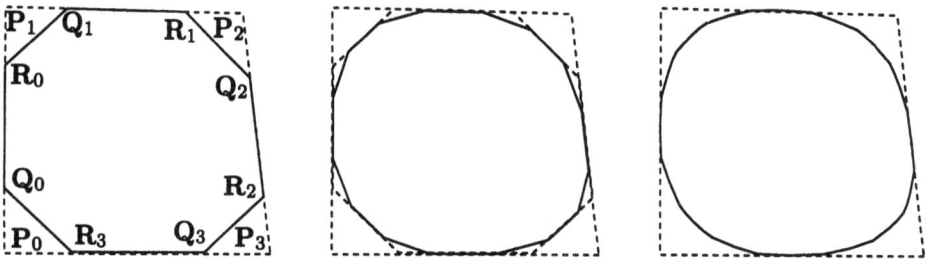

Figure 4.63: Chaikin's Algorithm for a Closed Curve.

2. It is located on the Chaikin curve defined by points P_0, P_1, and P_2. This is because points M_{ab} and M_{bc} are the points constructed by the first step of Chaikin's algorithm and we already know that the midpoint of any Chaikin segment is a point on the Chaikin curve.

▶ **Exercise 4.86:** Show that points M_{ab} and M_{bc} are the points constructed by the first step of Chaikin's algorithm.

The second refinement step produces the two midpoints, P_{01} and P_{11} (Figure 4.64) using the recursive procedures

$$B \leftarrow (A+B)/2, \quad C \leftarrow P, \quad P_{01} \leftarrow (A+2B+C)/4,$$
$$A \leftarrow P, \quad B \leftarrow (B+C)/2, \quad P_{11} \leftarrow (A+2B+C)/4.$$

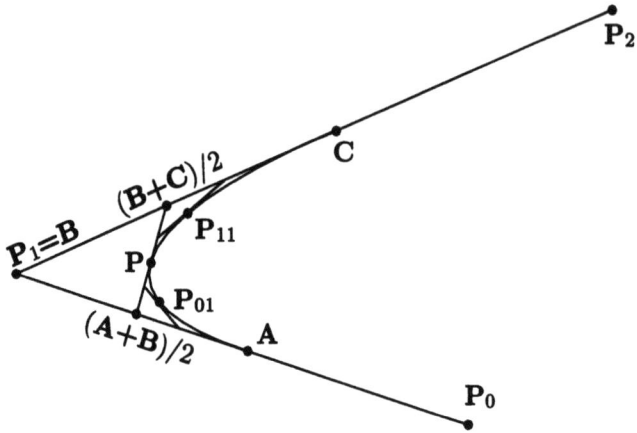

Figure 4.64: Points on the Chaikin Curve.

An argument similar to the one above shows that these two points are also on the quadratic Bézier curve defined by **A**, **B**, and **C** as well as on the Chaikin curve defined by $\mathbf{P}_0$, $\mathbf{P}_1$, and $\mathbf{P}_2$. Applying this argument to all the points generated by the refinement steps shows that they are located on both curves, which proves that the Chaikin curve defined by $\mathbf{P}_0$, $\mathbf{P}_1$, and $\mathbf{P}_2$ is identical to the quadratic Bézier curve defined by **A**, **B**, and **C**. This Bézier curve is

$$\mathbf{P}(t) = (1-t)^2 \mathbf{A} + 2t(1-t)\mathbf{B} + t^2 \mathbf{C},$$

and it is easy to express in terms of the original control points $\mathbf{P}_i$:

$$\begin{aligned}
\mathbf{P}(t) &= (t^2, t, 1) \begin{pmatrix} 1 & -2 & 1 \\ -2 & 2 & 0 \\ 1 & 0 & 0 \end{pmatrix} \begin{pmatrix} \mathbf{A} \\ \mathbf{B} \\ \mathbf{C} \end{pmatrix} \\
&= (t^2, t, 1) \begin{pmatrix} 1 & -2 & 1 \\ -2 & 2 & 0 \\ 1 & 0 & 0 \end{pmatrix} \begin{pmatrix} (\mathbf{P}_0 + \mathbf{P}_1)/2 \\ \mathbf{P}_1 \\ (\mathbf{P}_1 + \mathbf{P}_2)/2 \end{pmatrix} \\
&= (t^2, t, 1) \begin{pmatrix} 1 & -2 & 1 \\ -2 & 2 & 0 \\ 1 & 0 & 0 \end{pmatrix} \begin{pmatrix} 1/2 & 1/2 & 0 \\ 0 & 1 & 0 \\ 0 & 1/2 & 1/2 \end{pmatrix} \begin{pmatrix} \mathbf{P}_0 \\ \mathbf{P}_1 \\ \mathbf{P}_2 \end{pmatrix} \\
&= \frac{1}{2}(t^2, t, 1) \begin{pmatrix} 1 & -2 & 1 \\ -2 & 2 & 0 \\ 1 & 1 & 0 \end{pmatrix} \begin{pmatrix} \mathbf{P}_0 \\ \mathbf{P}_1 \\ \mathbf{P}_2 \end{pmatrix}.
\end{aligned} \quad (4.148)$$

The result is the quadratic B-spline curve segment, Equation (4.154).

Our conclusion is that the curve produced by Chaikin's algorithm is not a new type of curve but a quadratic B-spline. This fact lets us see this curve in a new light and it also shows a relation between the quadratic Bézier and B-spline curves.

4 Curves

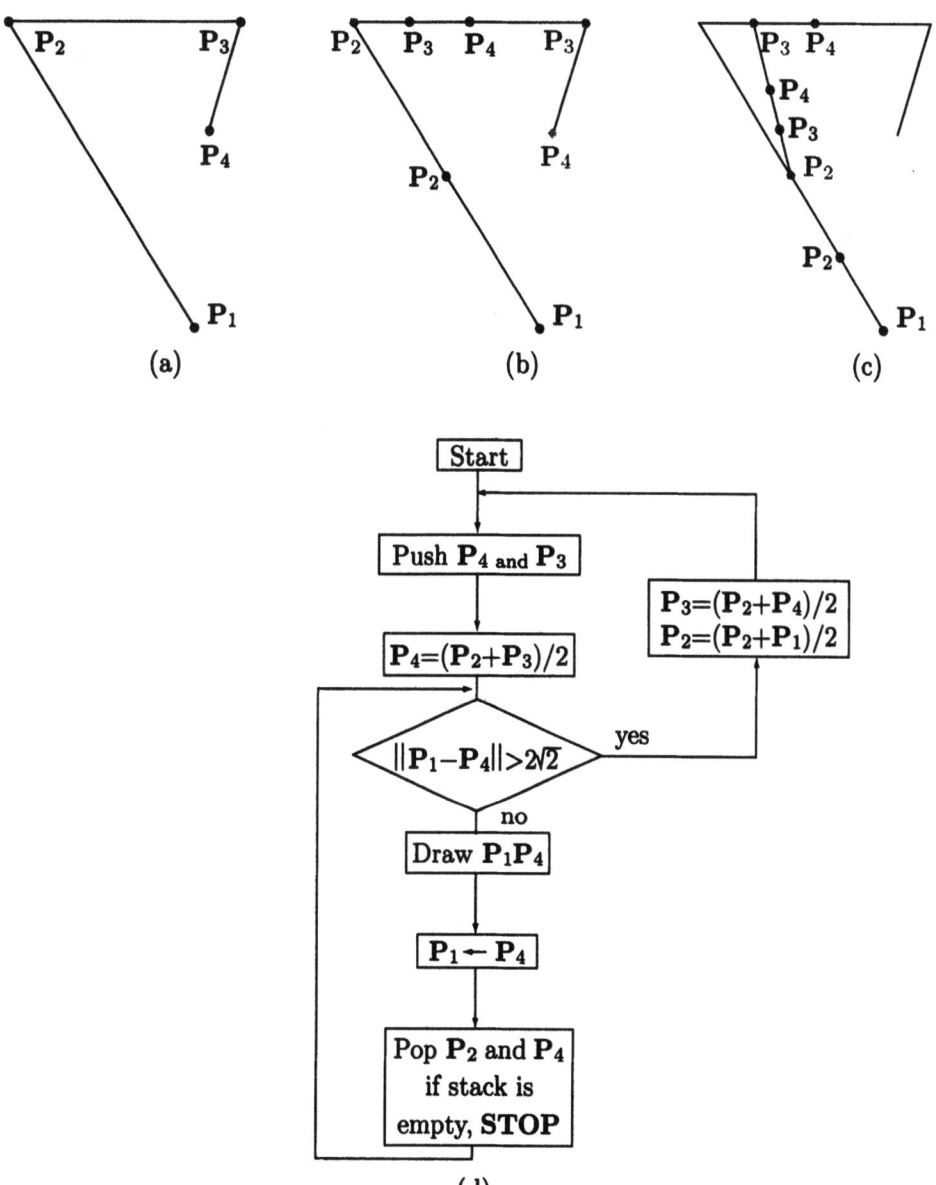

Figure 4.65: The Original Chaikin Algorithm.

The Original Chaikin Algorithm

The description of Chaikin's algorithm provided here differs from that originally proposed by George Chaikin. Here is the original description of the method, as it appears in [Chaikin 74]. We start with four points $\mathbf{P}_1$ through $\mathbf{P}_4$ (Figure 4.65a). Points $\mathbf{P}_4$ and $\mathbf{P}_3$ are pushed into a stack and a new $\mathbf{P}_4 = (\mathbf{P}_2+\mathbf{P}_3)/2$ is calculated. Points $\mathbf{P}_1$ and $\mathbf{P}_4$ are now compared. If their distance is greater than or equal to three pixels, then points $\mathbf{P}_2$ and $\mathbf{P}_3$ are recomputed according to

$$\mathbf{P}_3 = (\mathbf{P}_2 + \mathbf{P}_4)/2, \quad \mathbf{P}_2 = (\mathbf{P}_2 + \mathbf{P}_1)/2$$

(Figure 4.65b,c), points $\mathbf{P}_4$ and $\mathbf{P}_3$ are pushed into the stack, point $\mathbf{P}_4$ is recalculated, and the distance between $\mathbf{P}_1$ and $\mathbf{P}_4$ is checked. This repeats until the distance gets smaller than three pixels, in which case the short segment $\mathbf{P}_1\mathbf{P}_4$ is drawn, point $\mathbf{P}_4$ is renamed $\mathbf{P}_1$, the stack is popped twice and the resulting points are named $\mathbf{P}_2$ and $\mathbf{P}_4$, and the distance $\mathbf{P}_1\mathbf{P}_4$ is checked again. The process terminates when the stack is empty. Figure 4.65d is a flowchart of this algorithm.

▶ **Exercise 4.87:** Why compare the distance to three and not to two pixels?

▶ **Exercise 4.88:** (Easy). State this relation.

4.16 The B-Spline

These curves were first proposed in the 1940s but were seriously developed only in the 1970s, by several researchers, most notably R. Riesenfeld. They have been generalized since and much is currently known about them. The letter "B" stands for Basis, so the full name of this curve is the basis spline.

The B-spline curve overcomes the main disadvantages of the Bézier curve, namely (1) the degree of the Bézier curve depends on the number of control points, (2) it offers only global control, and (3) individual segments are easy to connect with C^1 continuity, but C^2 is hard to obtain. The B-spline curve features local control and any desired continuity. To obtain C^n continuity, the individual spline segments should be polynomials of degree n. The B-spline curve is an approximating curve, so it is based on control points. However, in addition to the control points, the user has to specify the values of certain quantities called "knots." They are real numbers that offer additional control over the shape of the curve. The basic approach taken in Section 4.16.1 ignores the knots, but they are introduced in Section 4.16.12 and their effect on the curve is explored.

There are several types of B-splines. In the *uniform* (also called periodic) B-spline (Section 4.16.1), the knot values are uniformly spaced and all the weight functions have the same shape and are shifted with respect to each other. In the *nonuniform* B-spline (Section 4.16.15), the knots are specified by the user and the weight functions are generally different. There is also an *open uniform* B-spline (Section 4.16.14), where the knots are not uniform but are specified in a simple way. In a *rational* B-spline (Section 4.16.18), the weight functions are in the form

4 Curves

of a ratio of two polynomials. In a *nonrational* B-spline, they are polynomials in t. The B-spline is an approximating curve based on control points, but there is also an *interpolating* version that passes through the points (Section 4.16.11). Section 4.16.7 shows how tension can be added to the B-spline.

4.16.1 A Basic Approach

Our approach is to first use basic assumptions to calculate the expressions for the quadratic and cubic uniform B-splines directly and without mentioning knots, and then to show how to extend the calculations to uniform B-splines of any order. Following this, we discuss a different, recursive formulation of the weight functions of the uniform, open uniform, and nonuniform B-splines.

4.16.2 The Quadratic Uniform B-Spline

We start with the quadratic uniform B-spline. We assume that $n+1$ control points, $\mathbf{P}_0, \mathbf{P}_1, \ldots, \mathbf{P}_n$, are given and we want to calculate a spline curve where each segment $\mathbf{P}_i(t)$ is a quadratic parametric polynomial based on three points, $\mathbf{P}_{i-1}$, $\mathbf{P}_i$, and $\mathbf{P}_{i+1}$. We require that the segments connect with C^1 continuity (only cubic and higher-degree polynomial segments can have C^2 or higher continuities) and that the entire curve has local control. To achieve all this, we have to give up something and we elect to give up the requirement that a segment pass through its first and last control points. We denote the start and end points of segment $\mathbf{P}_i(t)$, $\mathbf{K}_i$ and $\mathbf{K}_{i+1}$, respectively and we call them *joint points*, or just *joints*. These points will have to be calculated. Figure 4.66a shows two quadratic segments $\mathbf{P}_1(t)$ and $\mathbf{P}_2(t)$ defined by the four control points $\mathbf{P}_0$, $\mathbf{P}_1$, $\mathbf{P}_2$, and $\mathbf{P}_3$. The first segment goes from joint $\mathbf{K}_1$ to joint $\mathbf{K}_2$ and the second one goes from joint $\mathbf{K}_2$ to joint $\mathbf{K}_3$, where the joints are drawn tentatively and will have to be calculated and redrawn. Note that each segment is defined by three control points, so its control polygon has two edges. The first spline segment is defined by $\mathbf{P}_0$, $\mathbf{P}_1$, and $\mathbf{P}_2$ only, so any changes in $\mathbf{P}_3$ will not affect it. This is how local control is achieved.

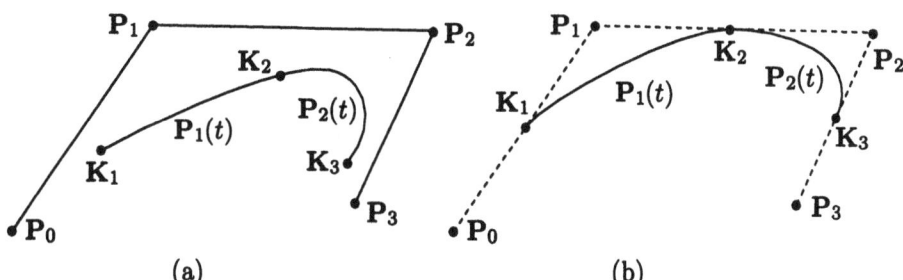

Figure 4.66: The Quadratic Uniform B-Spline.

We use the usual notation for the two segments

$$\mathbf{P}_i(t) = (t^2, t, 1)\mathbf{M} \begin{pmatrix} \mathbf{P}_{i-1} \\ \mathbf{P}_i \\ \mathbf{P}_{i+1} \end{pmatrix}, \quad i = 1, 2, \tag{4.149}$$

where **M** is the 3×3 basis matrix whose nine elements have to be calculated. We define three functions $a(t)$, $b(t)$, and $c(t)$ by:

$$(t^2, t, 1)\mathbf{M} = (t^2, t, 1)\begin{pmatrix} a_2 & b_2 & c_2 \\ a_1 & b_1 & c_1 \\ a_0 & b_0 & c_0 \end{pmatrix}$$
$$= (a_2 t^2 + a_1 t + a_0, b_2 t^2 + b_1 t + b_0, c_2 t^2 + c_1 t + c_0)$$
$$= (a(t), b(t), c(t)). \qquad (4.150)$$

The nine elements of **M** are calculated from the following three requirements:

1. The two segments should meet at a common joint point and their tangent vectors should be equal at that point. This is expressed as

$$\mathbf{P}_1(1) = \mathbf{P}_2(0), \quad \mathbf{P}_1^t(1) = \mathbf{P}_2^t(0) \qquad (4.151)$$

and produces the explicit equations

$$a(1)\mathbf{P}_0 + b(1)\mathbf{P}_1 + c(1)\mathbf{P}_2 = a(0)\mathbf{P}_1 + b(0)\mathbf{P}_2 + c(0)\mathbf{P}_3,$$
$$a'(1)\mathbf{P}_0 + b'(1)\mathbf{P}_1 + c'(1)\mathbf{P}_2 = a'(0)\mathbf{P}_1 + b'(0)\mathbf{P}_2 + c'(0)\mathbf{P}_3.$$

Since the control points $\mathbf{P}_i$ can be any, we can rewrite the two equations above in the form

$$\begin{aligned} a(1) &= 0, & a'(1) &= 0, & \text{for } \mathbf{P}_0, \\ b(1) &= a(0), & b'(1) &= a'(0), & \text{for } \mathbf{P}_1, \\ c(1) &= b(0), & c'(1) &= b'(0), & \text{for } \mathbf{P}_2, \\ 0 &= c(0), & 0 &= c'(0), & \text{for } \mathbf{P}_3. \end{aligned}$$

Using the notation of Equation (4.150), this can be written

$$\begin{aligned} a_2 + a_1 + a_0 &= 0, & 2a_2 + a_1 &= 0, \\ b_2 + b_1 + b_0 &= a_0, & 2b_2 + b_1 &= 0, \\ c_2 + c_1 + c_0 &= b_0, & 2c_2 + c_1 &= 0, \\ 0 &= c_0, & 0 &= c_1. \end{aligned} \qquad (4.152)$$

This requirement produces eight equations for the nine unknown matrix elements.

2. The entire curve should be independent of the particular coordinate system used, which implies that the weight functions of each segment should be barycentric, i.e., $a(t) + b(t) + c(t) = 1$. This condition can be written explicitly as

$$a_2 + b_2 + c_2 = 0, \quad a_1 + b_1 + c_1 = 0, \quad a_0 + b_0 + c_0 = 1, \qquad (4.153)$$

and they add three more equations.

4 Curves

We now have 11 equations for the 9 unknowns, but it is easy to show that only 9 of the 11 are independent. The sum of the first two of Equations (4.153) equals the sum of the three equations in the right column of Equation (4.152). Taking this into account, the equations can be solved uniquely, yielding

$$a_2 = 1/2, \quad a_1 = -1, \quad a_0 = 1/2,$$
$$b_2 = -1, \quad b_1 = 1, \quad b_0 = 1/2,$$
$$c_2 = 1/2, \quad c_1 = 0, \quad c_0 = 0.$$

The general quadratic B-spline segment, Equation (4.149), can now be written as

$$\mathbf{P}_i(t) = \frac{1}{2}(t^2, t, 1) \begin{pmatrix} 1 & -2 & 1 \\ -2 & 2 & 0 \\ 1 & 1 & 0 \end{pmatrix} \begin{pmatrix} \mathbf{P}_{i-1} \\ \mathbf{P}_i \\ \mathbf{P}_{i+1} \end{pmatrix} \quad (4.154)$$
$$= \frac{1}{2}(t^2 - 2t + 1)\mathbf{P}_{i-1} + \frac{1}{2}(-2t^2 + 2t + 1)\mathbf{P}_i + \frac{t^2}{2}\mathbf{P}_{i+1}, \quad i = 1, 2.$$

We are now in a position to calculate the start and end points, $\mathbf{K}_i$ and $\mathbf{K}_{i+1}$ of segment i. They are

$$\mathbf{K}_i = \mathbf{P}_i(0) = \frac{1}{2}(\mathbf{P}_{i-1} + \mathbf{P}_i), \quad \mathbf{K}_{i+1} = \mathbf{P}_i(1) = \frac{1}{2}(\mathbf{P}_i + \mathbf{P}_{i+1}).$$

The quadratic spline segment thus starts at the middle of the straight segment $\mathbf{P}_{i-1}\mathbf{P}_i$ and ends at the middle of the straight segment $\mathbf{P}_i\mathbf{P}_{i+1}$, as shown in Figure 4.66b.

The tangent vector of the general quadratic B-spline segment is easily obtained from Equation (4.154):

$$\mathbf{P}_i^t(t) = \frac{1}{2}(2t, 1, 0) \begin{pmatrix} 1 & -2 & 1 \\ -2 & 2 & 0 \\ 1 & 1 & 0 \end{pmatrix} \begin{pmatrix} \mathbf{P}_{i-1} \\ \mathbf{P}_i \\ \mathbf{P}_{i+1} \end{pmatrix} = (t-1)\mathbf{P}_{i-1} + (-2t+1)\mathbf{P}_i + t\mathbf{P}_{i+1}.$$
(4.155)

The tangent vectors at both ends of the segment are thus

$$\mathbf{P}^t(0) = \mathbf{P}_i - \mathbf{P}_{i-1}, \quad \mathbf{P}^t(1) = \mathbf{P}_{i+1} - \mathbf{P}_i,$$

i.e., each points in the direction of one of the edges of the control polygon of the spline segment.

Since a quadratic spline segment is a polynomial of degree 2, we require continuity of the first derivative only. It is easy to show that the second derivative of our segment is $\mathbf{P}_{i-1} - 2\mathbf{P}_i + \mathbf{P}_{i+1}$. It is constant for a segment but is different for different segments.

Equation (4.148) shows a relation between the quadratic B-spline and Bézier curves. A similar relation between the cubic curves is illustrated in Section 4.16.8.

Example: Given the four control points $\mathbf{P}_0 = (1,0)$, $\mathbf{P}_1 = (1,1)$, $\mathbf{P}_2 = (2,1)$, and $\mathbf{P}_3 = (2,0)$ (Figure 4.67), the first quadratic spline segment is obtained from Equation (4.154):

$$\begin{aligned}\mathbf{P}_1(t) &= \frac{1}{2}(t^2, t, 1)\begin{pmatrix} 1 & -2 & 1 \\ -2 & 2 & 0 \\ 1 & 1 & 0 \end{pmatrix}\begin{pmatrix} \mathbf{P}_0 \\ \mathbf{P}_1 \\ \mathbf{P}_2 \end{pmatrix} \\ &= \frac{1}{2}(t^2 - 2t + 1)(1,0) + \frac{1}{2}(-2t^2 + 2t + 1)(1,1) + \frac{t^2}{2}(2,1) \\ &= (t^2/2 + 1, -t^2/2 + t + 1/2).\end{aligned}$$

It starts at joint $\mathbf{K}_1 = \mathbf{P}_1(0) = (1, \frac{1}{2})$ and ends at joint $\mathbf{K}_2 = \mathbf{P}_1(1) = (\frac{3}{2}, 1)$.

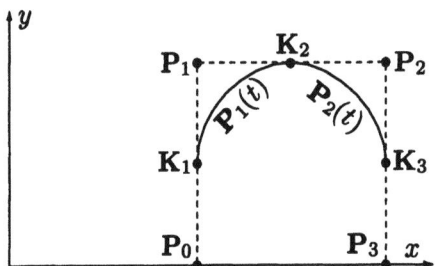

Figure 4.67: A Quadratic Uniform B-Spline Example.

The tangent vector of this segment is obtained from Equation (4.155):

$$\begin{aligned}\mathbf{P}_1^t(t) &= \frac{1}{2}(2t, 1, 0)\begin{pmatrix} 1 & -2 & 1 \\ -2 & 2 & 0 \\ 1 & 1 & 0 \end{pmatrix}\begin{pmatrix} \mathbf{P}_0 \\ \mathbf{P}_1 \\ \mathbf{P}_2 \end{pmatrix} \\ &= (t-1)(1,0) + (-2t+1)(1,1) + t(2,1) \\ &= (t, 1-t).\end{aligned}$$

The first segment thus starts going in direction $\mathbf{P}_1^t(0) = (0,1)$ (straight up) and ends going in direction $\mathbf{P}_1^t(1) = (1,0)$ (to the right).

▶ **Exercise 4.89:** Calculate the second segment, its tangent vector, and joint point $\mathbf{K}_3$.

Closed Quadratic B-Splines: Closed curves are sometimes called for and a closed B-spline curve is easy to calculate. Given the usual $n+1$ control points, we extend them cyclically to obtain the $n+3$ points

$$\mathbf{P}_n, \mathbf{P}_0, \mathbf{P}_1, \mathbf{P}_2, \ldots, \mathbf{P}_{n-1}, \mathbf{P}_n, \mathbf{P}_0$$

4 Curves

and compute the curve by applying Equation (4.154) to the $n+1$ geometry vectors

$$\begin{pmatrix} \mathbf{P}_n \\ \mathbf{P}_0 \\ \mathbf{P}_1 \end{pmatrix} \begin{pmatrix} \mathbf{P}_0 \\ \mathbf{P}_1 \\ \mathbf{P}_2 \end{pmatrix} \begin{pmatrix} \mathbf{P}_1 \\ \mathbf{P}_2 \\ \mathbf{P}_3 \end{pmatrix} \cdots \begin{pmatrix} \mathbf{P}_{n-2} \\ \mathbf{P}_{n-1} \\ \mathbf{P}_n \end{pmatrix} \begin{pmatrix} \mathbf{P}_{n-1} \\ \mathbf{P}_n \\ \mathbf{P}_0 \end{pmatrix}.$$

Example: Given the four control points $\mathbf{P}_0 = (1,0)$, $\mathbf{P}_1 = (1,1)$, $\mathbf{P}_2 = (2,1)$, and $\mathbf{P}_3 = (2,0)$ of the previous example, it is easy to close the curve by calculating the two additional segments

$$\begin{aligned} \mathbf{P}_0(t) &= \frac{1}{2}(t^2, t, 1) \begin{pmatrix} 1 & -2 & 1 \\ -2 & 2 & 0 \\ 1 & 1 & 0 \end{pmatrix} \begin{pmatrix} \mathbf{P}_3 \\ \mathbf{P}_0 \\ \mathbf{P}_1 \end{pmatrix} \\ &= \frac{1}{2}(t^2 - 2t + 1)(2,0) + \frac{1}{2}(-2t^2 + 2t + 1)(1,0) + \frac{t^2}{2}(1,1) \\ &= (t^2/2 - t + 3/2, t^2/2). \\ \mathbf{P}_3(t) &= \frac{1}{2}(t^2, t, 1) \begin{pmatrix} 1 & -2 & 1 \\ -2 & 2 & 0 \\ 1 & 1 & 0 \end{pmatrix} \begin{pmatrix} \mathbf{P}_2 \\ \mathbf{P}_3 \\ \mathbf{P}_0 \end{pmatrix} \\ &= \frac{1}{2}(t^2 - 2t + 1)(2,1) + \frac{1}{2}(-2t^2 + 2t + 1)(2,0) + \frac{t^2}{2}(1,0) \\ &= (-t^2/2 + 2, t^2/2 - t + 1/2). \end{aligned}$$

The four segments connect the four joint points $(1, 1/2)$, $(3/2, 1)$, $(2, 1/2)$, $(3/2, 0)$ and back to $(1, 1/2)$.

4.16.3 Quadratic Uniform B-Spline by Subdivision

The uniform B-spline for a group of $n+1$ control points can be constructed as a set of short segments, each a quadratic polynomial based on three control points. This section shows how Chaikin's algorithm (Section 4.15) can be applied to the construction of such a curve. We divide the original $n+1$ control points into $n-1$ overlapping groups of three points each, and use each group to calculate four new points. The groups are

$$\mathbf{P}_0\mathbf{P}_1\mathbf{P}_2, \quad \mathbf{P}_1\mathbf{P}_2\mathbf{P}_3, \ldots, \mathbf{P}_{n-2}\mathbf{P}_{n-1}\mathbf{P}_n.$$

Subdividing the first group is done by

$$\begin{pmatrix} \mathbf{P}_0^1 \\ \mathbf{P}_1^1 \\ \mathbf{P}_2^1 \\ \mathbf{P}_3^1 \end{pmatrix} = \frac{1}{4} \begin{pmatrix} 3 & 1 & 0 \\ 1 & 3 & 0 \\ 0 & 3 & 1 \\ 0 & 1 & 3 \end{pmatrix} \begin{pmatrix} \mathbf{P}_0 \\ \mathbf{P}_1 \\ \mathbf{P}_2 \end{pmatrix} = \begin{pmatrix} \frac{3}{4}\mathbf{P}_0 + \frac{1}{4}\mathbf{P}_1 \\ \frac{1}{4}\mathbf{P}_0 + \frac{3}{4}\mathbf{P}_1 \\ \frac{3}{4}\mathbf{P}_1 + \frac{1}{4}\mathbf{P}_2 \\ \frac{1}{4}\mathbf{P}_1 + \frac{3}{4}\mathbf{P}_2 \end{pmatrix},$$

and it yields the four new points $\mathbf{P}_0^1$, $\mathbf{P}_1^1$, $\mathbf{P}_2^1$, and $\mathbf{P}_3^1$. Subdividing the second group is done similarly and yields the four points $\mathbf{P}_2^1$, $\mathbf{P}_3^1$, $\mathbf{P}_4^1$, and $\mathbf{P}_5^1$, of which only the last two are new. Each subsequent group thus yields two new points when subdivided. The process is then repeated on the $2n$ segments defined by the $2n+2$ new points $\mathbf{P}_i^1$, yielding $4n+4$ points $\mathbf{P}_i^2$. These points, in turn, define $4n+2$ segments. When the number of points is large enough, the curve can be drawn by connecting each pair of adjacent points with a straight segment.

It can be shown (see page 326) that the curve obtained this way is the quadratic uniform B-spline, Equation (4.154).

4.16.4 The Cubic Uniform B-Spline

This curve is again defined by $n+1$ control points and is made of spline segments $\mathbf{P}_i(t)$, each a PC defined by four control points $\mathbf{P}_{i-1}$, $\mathbf{P}_i$, $\mathbf{P}_{i+1}$, and $\mathbf{P}_{i+2}$. The general form of segment i is thus

$$\mathbf{P}_i(t) = (t^3, t^2, t, 1)\mathbf{M} \begin{pmatrix} \mathbf{P}_{i-1} \\ \mathbf{P}_i \\ \mathbf{P}_{i+1} \\ \mathbf{P}_{i+2} \end{pmatrix}, \qquad (4.156)$$

where $\mathbf{M}$ is a 4×4 matrix whose 16 elements have to be calculated by translating the constraints on the curve into 16 equations and solving them. The constraints are (1) two segments should meet with C^2 continuity and (2) the entire curve should be independent of the particular coordinate system used. As in the quadratic case, we give up the requirement that a segment $\mathbf{P}_i(t)$ starts and ends at control points, and we denote its extreme points by $\mathbf{K}_i$ and $\mathbf{K}_{i+1}$. These joints can be calculated as soon as the expression for the segment becomes known. Figure 4.68a shows a tentative design for two cubic segments.

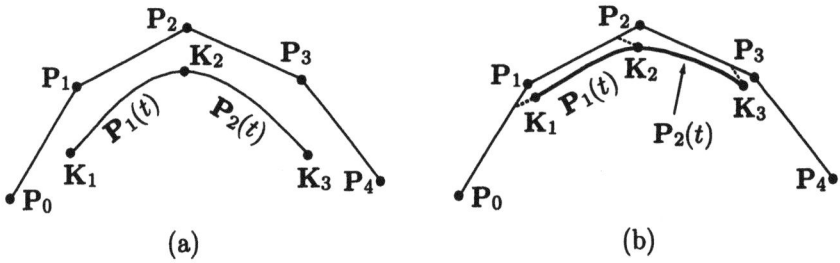

Figure 4.68: The Cubic Uniform B-Spline.

We start the calculation by writing

$$\begin{aligned}(t^3,t^2,t,1)\mathbf{M} &= (t^3,t^2,t,1)\begin{pmatrix} a_3 & b_3 & c_3 & d_3 \\ a_2 & b_2 & c_2 & d_2 \\ a_1 & b_1 & c_1 & d_1 \\ a_0 & b_0 & c_0 & d_0 \end{pmatrix} \\ &= (a_3t^3 + a_2t^2 + a_1t + a_0, b_3t^3 + b_2t^2 + b_1t + b_0, \\ &\quad c_3t^3 + c_2t^2 + c_1t + c_0, d_3t^3 + d_2t^2 + d_1t + d_0) \\ &= (a(t), b(t), c(t), d(t)).\end{aligned}$$

4 Curves

The first three constraints are expressed by

$$\mathbf{P}_1(1) = \mathbf{P}_2(0), \quad \mathbf{P}_1^t(1) = \mathbf{P}_2^t(0), \quad \mathbf{P}_1^{tt}(1) = \mathbf{P}_2^{tt}(0),$$

or, explicitly

$$a(1)\mathbf{P}_0 + b(1)\mathbf{P}_1 + c(1)\mathbf{P}_2 + d(1)\mathbf{P}_3 = a(0)\mathbf{P}_1 + b(0)\mathbf{P}_2 + c(0)\mathbf{P}_3 + d(0)\mathbf{P}_4,$$
$$a'(1)\mathbf{P}_0 + b'(1)\mathbf{P}_1 + c'(1)\mathbf{P}_2 + d'(1)\mathbf{P}_3 = a'(0)\mathbf{P}_1 + b'(0)\mathbf{P}_2 + c'(0)\mathbf{P}_3 + d'(0)\mathbf{P}_4,$$
$$a''(1)\mathbf{P}_0 + b''(1)\mathbf{P}_1 + c''(1)\mathbf{P}_2 + d''(1)\mathbf{P}_3 = a''(0)\mathbf{P}_1 + b''(0)\mathbf{P}_2 + c''(0)\mathbf{P}_3 + d''(0)\mathbf{P}_4.$$

Using the definitions of $a(t)$ and its relatives, this can be written as

$$\begin{aligned}
a_3 + a_2 + a_1 + a_0 &= 0, & 3a_3 + 2a_2 + a_1 &= 0, & 6a_3 + 2a_2 &= 0, \\
b_3 + b_2 + b_1 + b_0 &= a_0, & 3b_3 + 2b_2 + b_1 &= a_1, & 6b_3 + 2b_2 &= 2a_2, \\
c_3 + c_2 + c_1 + c_0 &= b_0, & 3c_3 + 2c_2 + c_1 &= b_1, & 6c_3 + 2c_2 &= 2b_2, \\
d_3 + d_2 + d_1 + d_0 &= c_0, & 3d_3 + 2d_2 + d_1 &= c_1, & 6d_3 + 2d_2 &= 2c_2, \\
0 &= d_0, & 0 &= d_1, & 0 &= 2d_2.
\end{aligned} \quad (4.157)$$

These are 15 equations for the 16 unknowns.

We already know from the quadratic case that the weight functions of each segment should be barycentric, i.e., $a(t) + b(t) + c(t) + d(t) = 1$. This condition can be written explicitly as

$$\begin{aligned}
a_3 + b_3 + c_3 + d_3 &= 0, & a_2 + b_2 + c_2 + d_2 &= 0, \\
a_1 + b_1 + c_1 + d_1 &= 0, & a_0 + b_0 + c_0 + d_0 &= 1,
\end{aligned} \quad (4.158)$$

and they add four more equations. We now have 19 equations, but only 16 of them are independent, since the first 3 equations of Equation (4.158) can be obtained by summing the first 4 equations of the left column of Equation (4.157). The system of equations can therefore be uniquely solved and the solutions are

$$\begin{aligned}
a_3 &= -1/6, & a_2 &= 1/2, & a_1 &= -1/2, & a_0 &= 1/6, \\
b_3 &= 1/2, & b_2 &= -1, & b_1 &= 0, & b_0 &= 2/3, \\
c_3 &= -1/2, & c_2 &= 1/2, & c_1 &= 1/2, & c_0 &= 1/6, \\
d_3 &= 1/6, & d_2 &= 0, & d_1 &= 0, & d_0 &= 0,
\end{aligned}$$

The cubic B-spline segment can now be expressed by

$$\begin{aligned}
\mathbf{P}_i(t) &= \frac{1}{6}(t^3, t^2, t, 1)\begin{pmatrix} -1 & 3 & -3 & 1 \\ 3 & -6 & 3 & 0 \\ -3 & 0 & 3 & 0 \\ 1 & 4 & 1 & 0 \end{pmatrix}\begin{pmatrix} \mathbf{P}_{i-1} \\ \mathbf{P}_i \\ \mathbf{P}_{i+1} \\ \mathbf{P}_{i+2} \end{pmatrix} \\
&= \frac{1}{6}(-t^3 + 3t^2 - 3t + 1)\mathbf{P}_{i-1} + \frac{1}{6}(3t^3 - 6t^2 + 4)\mathbf{P}_i \\
&\quad + \frac{1}{6}(-3t^3 + 3t^2 + 3t + 1)\mathbf{P}_{i+1} + \frac{t^3}{6}\mathbf{P}_{i+2}
\end{aligned} \quad (4.159)$$

The two extreme points are thus

$$\mathbf{K}_i = \mathbf{P}_i(0) = \frac{1}{6}(\mathbf{P}_{i-1} + 4\mathbf{P}_i + \mathbf{P}_{i+1}), \quad \mathbf{K}_{i+1} = \mathbf{P}_i(1) = \frac{1}{6}(\mathbf{P}_i + 4\mathbf{P}_{i+1} + \mathbf{P}_{i+2}).$$

In order to interpret them geometrically, we write them as

$$\begin{aligned}\mathbf{K}_i &= \left(\frac{1}{6}\mathbf{P}_{i-1} + \frac{5}{6}\mathbf{P}_i\right) + \frac{1}{6}(\mathbf{P}_{i+1} - \mathbf{P}_i), \\ \mathbf{K}_{i+1} &= \left(\frac{1}{6}\mathbf{P}_i + \frac{5}{6}\mathbf{P}_{i+1}\right) + \frac{1}{6}(\mathbf{P}_{i+2} - \mathbf{P}_{i+1}).\end{aligned} \quad (4.160)$$

Point $\mathbf{K}_i$ is the sum of the point $(\frac{1}{6}\mathbf{P}_{i-1} + \frac{5}{6}\mathbf{P}_i)$ and one-sixth of the vector $(\mathbf{P}_{i+1} - \mathbf{P}_i)$. Point $\mathbf{K}_{i+1}$ has a similar interpretation. Both are shown in Figure 4.68b.

▶ **Exercise 4.90:** Show another way to interpret $\mathbf{P}_i(0)$ and $\mathbf{P}_i(1)$ geometrically.

Users, especially those used to Bézier curves, find it counterintuitive that the B-spline curve does not start and end at its terminal control points. This "inconvenient" feature can be modified—and the curve made to start and end at its extreme points—by adding two *phantom* endpoints, $\mathbf{P}_{-1}$ and $\mathbf{P}_{n+1}$, at both ends of the curve, and placing those points at locations that would force the curve to start at $\mathbf{P}_0$ and end at $\mathbf{P}_n$. The calculation is simple. The first segment starts at $\frac{1}{6}[\mathbf{P}_{-1} + 4\mathbf{P}_0 + \mathbf{P}_1]$. This value will equal $\mathbf{P}_0$ if we select $\mathbf{P}_{-1} = 2\mathbf{P}_0 - \mathbf{P}_1$. Similarly, the last segment ends at $\frac{1}{6}[\mathbf{P}_{n-1} + 4\mathbf{P}_n + \mathbf{P}_{n+1}]$ and this value will equal $\mathbf{P}_n$ if we select $\mathbf{P}_{n+1} = 2\mathbf{P}_n - \mathbf{P}_{n-1}$.

Adding phantom points adds two segments to the curve, but this has the advantage that the tangents at the start and the end of the curve have known directions. The former is in the direction from $\mathbf{P}_0$ to $\mathbf{P}_1$ and the latter is from $\mathbf{P}_{n-1}$ to $\mathbf{P}_n$ (same as the end tangents of a Bézier curve). The tangent vector at the start of the first segment is $\frac{1}{2}\mathbf{P}_{-1} + \frac{1}{2}\mathbf{P}_1 = \mathbf{P}_1 - \mathbf{P}_0$, and similarly for the end tangent of the last segment.

The tangent vector of the general cubic B-spline segment is

$$\mathbf{P}_i^t(t) = \frac{1}{6}(-3t^2 + 6t - 3)\mathbf{P}_{i-1} + \frac{1}{6}(9t^2 - 12t)\mathbf{P}_i + \frac{1}{6}(-9t^2 + 6t + 3)\mathbf{P}_{i+1} + \frac{t^2}{2}\mathbf{P}_{i+2}.$$

The extreme tangent vectors are thus

$$\mathbf{P}_i^t(0) = \frac{1}{2}(\mathbf{P}_{i+1} - \mathbf{P}_{i-1}), \quad \mathbf{P}_i^t(1) = \frac{1}{2}(\mathbf{P}_{i+2} - \mathbf{P}_i). \quad (4.161)$$

They have simple geometric interpretations.

The second derivative of the cubic segment is

$$\mathbf{P}_i^{tt}(t) = \frac{1}{6}(-6t + 6)\mathbf{P}_{i-1} + \frac{1}{6}(18t - 12)\mathbf{P}_i + \frac{1}{6}(-18t + 6)\mathbf{P}_{i+1} + t\mathbf{P}_{i+2},$$

and it's easy to see that $\mathbf{P}_i^{tt}(1) = \mathbf{P}_{i+1}^{tt}(0) = \mathbf{P}_i - 2\mathbf{P}_{i+1} + \mathbf{P}_{i+2}$, which proves the C^2 continuity of this curve.

4 Curves

Example: We select the five points $\mathbf{P}_0 = (0,0)$, $\mathbf{P}_1 = (0,1)$, $\mathbf{P}_2 = (1,1)$, $\mathbf{P}_3 = (2,1)$, and $\mathbf{P}_4 = (2,0)$. They have simple, integer coordinates to simplify the computations. We use these points to calculate two cubic B-spline segments. The first one is given by Equation (4.159):

$$\mathbf{P}_1(t) = \frac{1}{6}(-t^3 + 3t^2 - 3t + 1)(0,0) + \frac{1}{6}(3t^3 - 6t^2 + 4)(0,1)$$
$$+ \frac{1}{6}(-3t^3 + 3t^2 + 3t + 1)(1,1) + \frac{t^3}{6}(2,1)$$
$$= (-t^3/6 + t^2/2 + t/2 + 1/6, t^3/6 - t^2/2 + t/2 + 5/6).$$

It starts at joint $\mathbf{K}_1 = \mathbf{P}_1(0) = (1/6, 5/6)$ and ends at joint $\mathbf{K}_2 = \mathbf{P}_1(1) = (1,1)$. Notice that these joint points can be verified from Equation (4.160). The tangent vector of this segment is

$$\mathbf{P}_1^t(t) = \frac{1}{6}(-3t^2 + 6t - 3)(0,0) + \frac{1}{6}(9t^2 - 12t)(0,1)$$
$$+ \frac{1}{6}(-9t^2 + 6t + 3)(1,1) + \frac{t^2}{2}(2,1)$$
$$= (-t^2/2 + t + 1/2, t^2/2 - t + 1/2).$$

The two extreme tangents are $\mathbf{P}_1^t(0) = (1/2, 1/2)$ and $\mathbf{P}_1^t(1) = (1,0)$. These can also be verified by Equation (4.161).

▶ **Exercise 4.91**: Calculate the second spline segment $\mathbf{P}_2(t)$, its tangent vector, and joint $\mathbf{K}_3$.

▶ **Exercise 4.92**: Use the five control points of the above example to calculate the three segments and four joints of the *quadratic* uniform B-spline defined by the points.

Exercise 4.92 shows that the same $n+1$ control points can be used to construct a quadratic or a cubic B-spline curve (or a B-spline curve of any order). This is in contrast to the Bézier curve whose order is determined by the number of control points. This is also the reason why both n and the degree of the polynomials that make up the spline segments are needed to identify a B-spline. In practice, we use n and k (the *order*) to identify a B-spline. The order is simply the degree plus 1. Thus, a B-spline defined by five control points $\mathbf{P}_0$ through $\mathbf{P}_4$ can be of order 2 (linear, with four segments), order 3 (quadratic, with three segments), order 4 (cubic, with two segments), or order 5, with one segment.

Figure 4.69a,b,c how a Bézier curve, a cubic B-spline, and a quadratic B-spline, respectively, are attracted to their control polygons.

Collinear Points: Segment two of Exercise 4.92 depends on points $\mathbf{P}_1$, $\mathbf{P}_2$, and $\mathbf{P}_3$ that are located on the line $y = 1$. This is why this segment is horizontal (and thus straight). The B-spline can, thus, connect curved and straight segments with any desired continuity. All that's necessary is to have enough collinear control points for one segment to make it straight. In the case of a quadratic B-spline, three collinear points will result in a straight segment that will connect to its neighbors

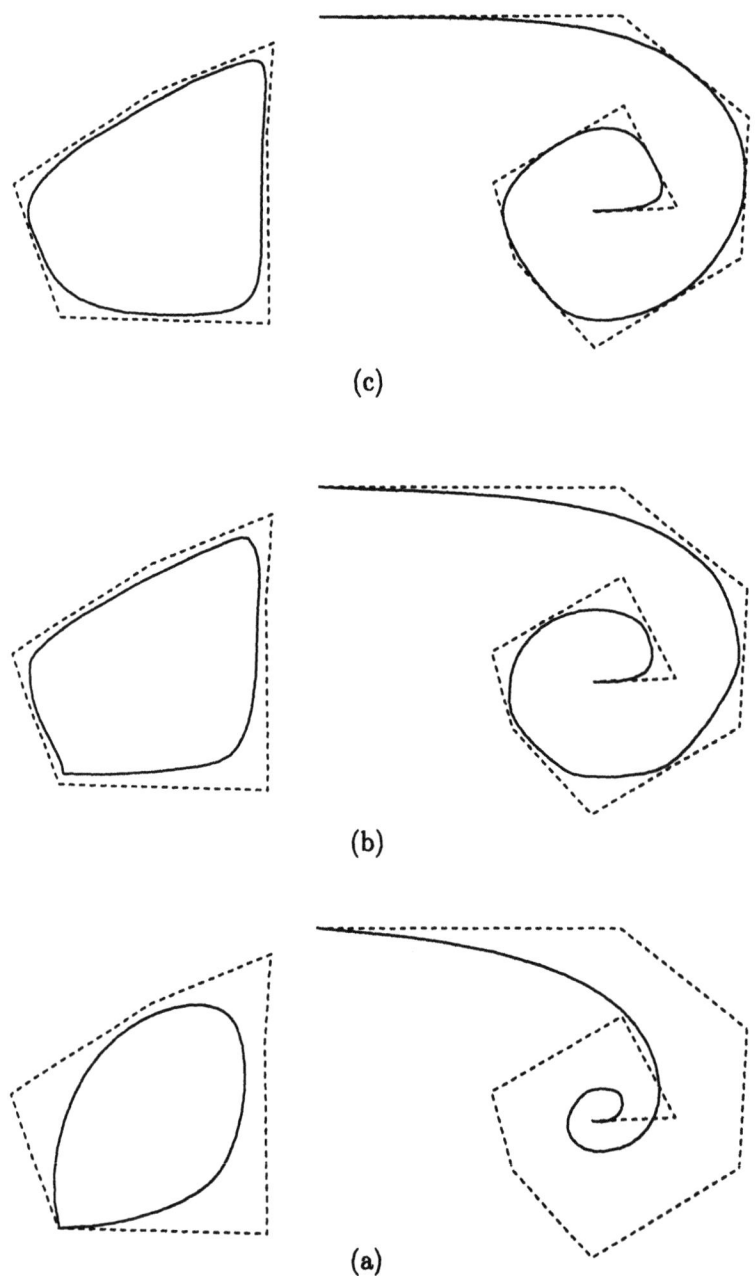

Figure 4.69: A Comparison of (a) Bézier, (b) Cubic B-Spline, and (c) Quadratic B-Spline Curves.

4 Curves

(curved or straight) with C^1 continuity. In the case of a cubic B-spline, four collinear points will result in a straight segment that will connect to its neighbors (curved or straight) with C^2 continuity, and similarly for higher-degree uniform B-splines.

A Closed Cubic B-Spline Curve: Closing a cubic B-spline is similar to closing a quadratic curve. Given the usual $n+1$ control points, we extend them cyclically to obtain the $n+4$ points

$$\mathbf{P}_n, \mathbf{P}_0, \mathbf{P}_1, \mathbf{P}_2, \ldots, \mathbf{P}_{n-1}, \mathbf{P}_n, \mathbf{P}_0, \mathbf{P}_1,$$

and compute the curve by applying Equation (4.159) to the $n+1$ geometry vectors

$$\begin{pmatrix} \mathbf{P}_n \\ \mathbf{P}_0 \\ \mathbf{P}_1 \\ \mathbf{P}_2 \end{pmatrix} \begin{pmatrix} \mathbf{P}_0 \\ \mathbf{P}_1 \\ \mathbf{P}_2 \\ \mathbf{P}_3 \end{pmatrix} \begin{pmatrix} \mathbf{P}_1 \\ \mathbf{P}_2 \\ \mathbf{P}_3 \\ \mathbf{P}_4 \end{pmatrix} \cdots \begin{pmatrix} \mathbf{P}_{n-2} \\ \mathbf{P}_{n-1} \\ \mathbf{P}_n \\ \mathbf{P}_0 \end{pmatrix} \begin{pmatrix} \mathbf{P}_{n-1} \\ \mathbf{P}_n \\ \mathbf{P}_0 \\ \mathbf{P}_1 \end{pmatrix}.$$

4.16.5 Multiple Control Points

This is the case in which two or more control points have the same coordinates. We use the uniform cubic B-spline (Equation (4.159)) as an example, but higher-degree uniform B-splines exhibit similar behavior.

We start with a double control point. Consider the cubic segment $\mathbf{P}_1(t)$ defined by the four control points $\mathbf{P}_0$, $\mathbf{P}_1 = \mathbf{P}_2$, and $\mathbf{P}_3$. Its expression is

$$\mathbf{P}_1(t) = \frac{1}{6}(-t^3 + 3t^2 - 3t + 1)\mathbf{P}_0 + \frac{1}{6}(-3t^2 + 3t + 5)\mathbf{P}_1 + \frac{t^3}{6}\mathbf{P}_3,$$

which implies $\quad \mathbf{P}_1(0) = \frac{1}{6}\mathbf{P}_0 + \frac{5}{6}\mathbf{P}_1, \quad \mathbf{P}_1(1) = \frac{5}{6}\mathbf{P}_1 + \frac{1}{6}\mathbf{P}_3.$

This segment thus starts and ends at the same points (and also has the same extreme tangent vectors) as the general cubic segment. The difference is that it is strongly attracted to the double point.

Next, we consider a triple point. The five control points $\mathbf{P}_0$, $\mathbf{P}_1 = \mathbf{P}_2 = \mathbf{P}_3$, and $\mathbf{P}_4$ define the two cubic segments

$$\mathbf{P}_1(t) = \frac{1}{6}(-t^3 + 3t^2 - 3t + 1)\mathbf{P}_0 + \frac{1}{6}(t^3 - 3t^2 + 3t + 5)\mathbf{P}_1$$
$$= (1-u)\mathbf{P}_0 + u\mathbf{P}_1, \quad \text{for } u = (t^3 - 3t^2 + 3t + 5)/6,$$
$$\mathbf{P}_2(t) = \frac{1}{6}(-t^3 + 6)\mathbf{P}_1 + \frac{t^3}{6}\mathbf{P}_4$$
$$= (1-w)\mathbf{P}_1 + w\mathbf{P}_4, \quad \text{for } w = t^3/6.$$

The parameter substitutions above show that these segments are straight (Figure 4.70). The extreme points of the two segments are

$$\mathbf{P}_1(0) = \frac{1}{6}\mathbf{P}_0 + \frac{5}{6}\mathbf{P}_1, \quad \mathbf{P}_1(1) = \mathbf{P}_1,$$
$$\mathbf{P}_2(0) = \mathbf{P}_1, \quad \mathbf{P}_2(1) = \frac{5}{6}\mathbf{P}_1 + \frac{1}{6}\mathbf{P}_4,$$

showing that the segments meet at the triple control point.

In general, a cubic segment is attracted to a double control point and passes through a triple control point. A degree-4 segment is attracted to double and triple control points and passes through quadruple points, and similarly for higher-degree uniform segments.

The tangent vectors of the two cubic segments are

$$\mathbf{P}_1^t(t) = \frac{1}{6}(-3t^2 + 6t - 3)\mathbf{P}_0 + \frac{1}{6}(3t^2 - 6t + 3)\mathbf{P}_1,$$

$$\mathbf{P}_2^t(t) = -\frac{t^2}{2}\mathbf{P}_1 + \frac{t^2}{2}\mathbf{P}_4,$$

yielding the extreme directions

$$\mathbf{P}_1^t(0) = \frac{1}{2}(\mathbf{P}_1 - \mathbf{P}_0), \quad \mathbf{P}_1^t(1) = 0 \cdot \mathbf{P}_0 + 0 \cdot \mathbf{P}_1 = (0,0),$$

$$\mathbf{P}_2^t(0) = (0,0), \quad \mathbf{P}_2^t(1) = \frac{1}{2}(\mathbf{P}_4 - \mathbf{P}_1).$$

The first segment thus starts in the direction from $\mathbf{P}_0$ to the triple point $\mathbf{P}_1$. The second segment ends going in the direction from $\mathbf{P}_1$ to $\mathbf{P}_4$. However, at the triple point, both tangents are indefinite, suggesting a cusp. It turns out that the two segments are straight lines (Figure 4.70).

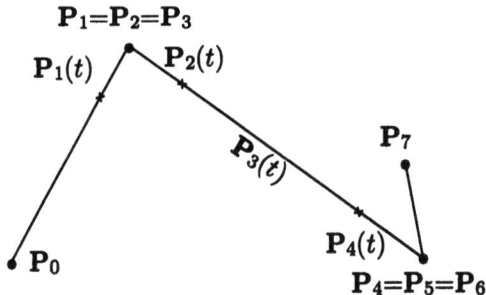

Figure 4.70: A Triple Point.

▶ **Exercise 4.93:** Given the eight control points $\mathbf{P}_0$, $\mathbf{P}_1 = \mathbf{P}_2 = \mathbf{P}_3$, $\mathbf{P}_4 = \mathbf{P}_5 = \mathbf{P}_6$, and $\mathbf{P}_7$, calculate the two cubic segments $\mathbf{P}_3(t)$ and $\mathbf{P}_4(t)$ and their start and end points (Figure 4.70).

▶ **Exercise 4.94:** Show that a cubic B-spline segment passes through its first control point if it is a triple point.

As a corollary, we deduce that a uniform cubic B-spline curve where every control point is triple is a polyline.

Example: We consider the case where both terminal points are triple and there are two other points in between. The total number of control points is eight and

4 Curves

they satisfy $\mathbf{P}_0 = \mathbf{P}_1 = \mathbf{P}_2$ and $\mathbf{P}_5 = \mathbf{P}_6 = \mathbf{P}_7$. The five cubic spline segments are

$$\mathbf{P}_1(t) = \frac{1}{6}(-t^3 + 6)\mathbf{P}_0 + \frac{t^3}{6}\mathbf{P}_3,$$

$$\mathbf{P}_2(t) = \frac{1}{6}(2t^3 - 3t^2 - 3t + 5)\mathbf{P}_0 + \frac{1}{6}(-3t^3 + 3t^2 + 3t + 1)\mathbf{P}_3 + \frac{t^3}{6}\mathbf{P}_4,$$

$$\mathbf{P}_3(t) = \frac{1}{6}(-t^3 + 3t^2 - 3t + 1)\mathbf{P}_0 + \frac{1}{6}(3t^3 - 6t^2 + 4)\mathbf{P}_3$$
$$\quad + \frac{1}{6}(-3t^3 + 3t^2 + 3t + 1)\mathbf{P}_4 + \frac{t^3}{6}\mathbf{P}_5, \quad (4.162)$$

$$\mathbf{P}_4(t) = \frac{1}{6}(-t^3 + 3t^2 - 3t + 1)\mathbf{P}_3 + \frac{1}{6}(3t^3 - 6t^2 + 4)\mathbf{P}_4$$
$$\quad + \frac{1}{6}(-2t^3 + 3t^2 + 3t + 1)\mathbf{P}_5,$$

$$\mathbf{P}_5(t) = \frac{1}{6}(-t^3 + 3t^2 - 3t + 1)\mathbf{P}_4 + \frac{1}{6}(t^3 - 3t^2 + 3t + 5)\mathbf{P}_5.$$

It is easy to see that they satisfy $\mathbf{P}_1(0) = \mathbf{P}_0$ and $\mathbf{P}_5(1) = \mathbf{P}_5$ and that they meet at the four points

$$\frac{5}{6}\mathbf{P}_0 + \frac{1}{6}\mathbf{P}_3, \quad \frac{1}{6}\mathbf{P}_0 + \frac{4}{6}\mathbf{P}_3 + \frac{1}{6}\mathbf{P}_4, \quad \frac{1}{6}\mathbf{P}_3 + \frac{4}{6}\mathbf{P}_4 + \frac{1}{6}\mathbf{P}_5, \quad \frac{1}{6}\mathbf{P}_4 + \frac{5}{6}\mathbf{P}_5.$$

If we want to keep the two extreme points as triples, we can edit this curve only by moving the two interior points $\mathbf{P}_3$ and $\mathbf{P}_4$. Moving $\mathbf{P}_4$ affects the last four segments, and moving $\mathbf{P}_3$ affects the first four segments. This type of curve is therefore similar to a Bézier curve in that it starts and ends at its extreme control points and it features only limited local control.

▶ **Exercise 4.95:** Given the eight control points $\mathbf{P}_0 = \mathbf{P}_1 = \mathbf{P}_2 = (1,0)$, $\mathbf{P}_3 = (2,1)$, $\mathbf{P}_4 = (4,0)$, and $\mathbf{P}_5 = \mathbf{P}_6 = \mathbf{P}_7 = (4,1)$, use Equation (4.162) to calculate the cubic uniform B-spline curve defined by these points and compare it to the Bézier curve defined by the points.

4.16.6 The Cubic B-Spline as a Circle

The uniform B-spline, like the Bézier curve, cannot represent a precise circle. However, the cubic B-spline can provide an excellent approximation to a circle or a circular arc using just a few control points. The following discussion shows how to place those points in order to obtain a unit circle centered on the origin. Figure 4.71b shows m equidistant control points $\mathbf{P}_i$ placed on a circle of radius R, where R has to be determined. The coordinates of those points are

$$\mathbf{P}_i = (R\cos\theta_i, R\sin\theta_i) = \left(R\cos\frac{2\pi i}{m}, R\sin\frac{2\pi i}{m}\right) \quad \text{for} \quad i = 0, 1, \ldots, m-1.$$

We divide the control points, as usual, into overlapping groups of four points each and calculate a cubic B-spline segment $\mathbf{P}_i(t)$ for each group. We require that the two terminal points $\mathbf{P}_i(0)$ and $\mathbf{P}_i(1)$ be at a distance of one unit from the origin.

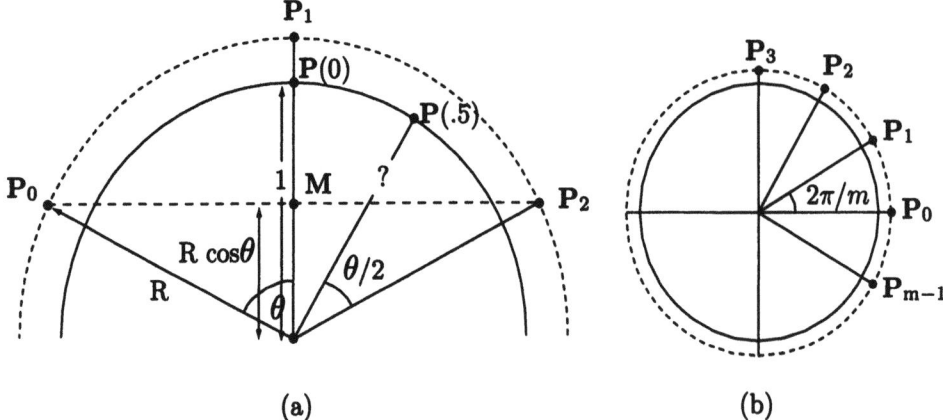

Figure 4.71: A Cubic B-Spline and a Circle.

Exercise 4.90 shows that the start point $\mathbf{P}_1(0)$ of the first segment of this curve satisfies (Figure 4.71a)

$$\mathbf{P}(0) = \frac{1}{3}\frac{\mathbf{P}_0 + \mathbf{P}_2}{2} + \frac{2}{3}\mathbf{P}_1 = \frac{1}{3}\mathbf{M} + \frac{2}{3}\mathbf{P}_1.$$

The distance of $\mathbf{P}(0)$ from the origin is therefore

$$\frac{1}{3}R\cos\theta + \frac{2}{3}R,$$

and the same is true for the end point $\mathbf{P}(1)$. On the other hand, we require that this distance equals one unit, so the result is

$$\frac{1}{3}R(\cos\theta + 2) = 1 \quad \text{or} \quad R = \frac{3}{2 + \cos\theta}. \tag{4.163}$$

Our control points should therefore have coordinates

$$\mathbf{P}_i = \left(\frac{3\cos\frac{2\pi i}{m}}{2 + \cos\frac{2\pi}{m}}, \frac{3\sin\frac{2\pi i}{m}}{2 + \cos\frac{2\pi}{m}} \right) \quad \text{for} \quad i = 0, 1, \ldots, m-1.$$

To estimate the number of control points necessary for a good approximation, we first estimate the error of this representation. Since the curve is identical to a circle at the control points, we assume that the worst approximation is obtained midway between control points, i.e., at points $\mathbf{P}_i(0.5)$. Figure 4.71a shows one such point whose distance from the origin is labeled "?." The midpoint of a cubic

4 Curves

segment, however, is easily calculated from Equation (4.159) to be

$$\mathbf{P}_1(0.5) = \frac{1}{6}\left(\frac{1}{8}\mathbf{P}_0 + \frac{23}{8}\mathbf{P}_1 + \frac{23}{8}\mathbf{P}_2 + \frac{1}{8}\mathbf{P}_3\right)$$

$$= \left(\frac{(1+\cos\theta)(11+\cos\theta)}{8(2+\cos\theta)}, \frac{\sin\theta(11+\cos\theta)}{8(2+\cos\theta)}\right),$$

where $\theta = 2\pi i/m$. The deviation from a true circle is thus

$$1 - \sqrt{\mathbf{P}_1^2(0.5)} = \frac{(1-\cos\frac{\pi}{m})^2(2-\cos\frac{\pi}{m})}{2(2+\cos\frac{2\pi}{m})}.$$

Even for $m = 4$, the deviation is only 2.77%. For $m = 5$, it is 0.94%, and for $m = 6$, it is 0.41%. The B-spline can therefore provide an excellent, fast approximation to a circle.

Example: We calculate the four segments for the case $m = 4$. The value of R is

$$R = \frac{3}{2+\cos\frac{2\pi}{4}} = 3/2,$$

so the control points are

$$\mathbf{P}_0 = (R\cos 0, R\sin 0) = (3/2, 0),$$
$$\mathbf{P}_1 = (R\cos\frac{\pi}{2}, R\sin\frac{\pi}{2}) = (0, 3/2),$$
$$\mathbf{P}_2 = (R\cos\pi, R\sin\pi) = (-3/2, 0),$$
$$\mathbf{P}_3 = (R\cos\frac{3\pi}{2}, R\sin\frac{3\pi}{2}) = (0, -3/2).$$

Equation (4.159) is used to obtain the first segment:

$$\mathbf{P}_1(t) = \frac{1}{6}(t^3, t^2, t, 1)\begin{pmatrix} -1 & 3 & -3 & 1 \\ 3 & -6 & 3 & 0 \\ -3 & 0 & 3 & 0 \\ 1 & 4 & 1 & 0 \end{pmatrix}\begin{pmatrix} (3/2, 0) \\ (0, 3/2) \\ (-3/2, 0) \\ (0, -3/2) \end{pmatrix}$$

$$= \frac{1}{4}(2t^3 - 6t, 2t^3 - 6t^2 + 4).$$

This segment goes from $(0,1)$ to $(-1,0)$ and its midpoint is at $(-22/32, 22/32) = (-0.6875, 0.6875)$. The true circle is at $(-0.7071, 0.7071)$, so the difference is ≈ 0.02. Normally, a cubic B-spline curve based on four control points has two segments but our curve is closed, so it is made of four segments.

▸ **Exercise 4.96**: Calculate the remaining three segments.

Example: Approximating a circular arc. We restrict our discussion to arcs on the unit circle centered on the origin. To specify such an arc, the user should input the coordinates of the two endpoints **S** and **E** (both at a distance of one unit from the origin) and the software should use them to calculate the coordinates of the four control points C_0, C_1, C_2, and C_3 that produce the best approximation for the arc $\mathbf{C}(t)$. Figure 4.72a shows how $\mathbf{S}$ and $\mathbf{E}$ become the endpoints $\mathbf{C}(0)$ and $\mathbf{C}(1)$ of the arc. It also shows that $\cos\theta = \mathbf{E}\bullet\mathbf{S}$. Equation (4.163) gives the distance R of the four control points from the origin and shows how to compute the two interior points

$$\mathbf{C}_1 = R\mathbf{S} = \frac{3}{2+\mathbf{E}\bullet\mathbf{S}}\mathbf{S}, \quad \mathbf{C}_2 = R\mathbf{E} = \frac{3}{2+\mathbf{E}\bullet\mathbf{S}}\mathbf{E}.$$

Control point $\mathbf{C}_0$ is found by rotating $\mathbf{C}_1$ clockwise θ degrees and the control point $\mathbf{C}_3$ is found by rotating $\mathbf{C}_2$ θ degrees counterclockwise. The rotation matrices are obtained from Equation (3.4), bearing in mind that $\cos\theta = \mathbf{E}\bullet\mathbf{S}$ and $\sin\theta = \sqrt{1-(\mathbf{E}\bullet\mathbf{S})^2}$:

$$\mathbf{C}_0 = \mathbf{C}_1\begin{pmatrix} \mathbf{E}\bullet\mathbf{S} & \sqrt{1-(\mathbf{E}\bullet\mathbf{S})^2} \\ -\sqrt{1-(\mathbf{E}\bullet\mathbf{S})^2} & \mathbf{E}\bullet\mathbf{S} \end{pmatrix},$$

$$\mathbf{C}_3 = \mathbf{C}_2\begin{pmatrix} \mathbf{E}\bullet\mathbf{S} & -\sqrt{1-(\mathbf{E}\bullet\mathbf{S})^2} \\ \sqrt{1-(\mathbf{E}\bullet\mathbf{S})^2} & \mathbf{E}\bullet\mathbf{S} \end{pmatrix}.$$

Once the four control points are known, the cubic B-spline segment can be calculated.

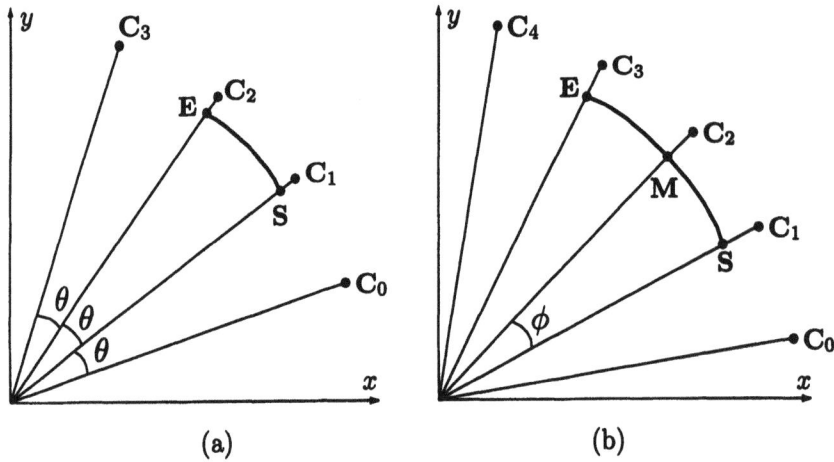

Figure 4.72: Cubic B-Splines and Arcs.

Approximating long arcs may require more than one spline segment and this can also be handled by our method. The user should again input the coordinates of the two endpoints $\mathbf{S}$ and $\mathbf{E}$ (both at a distance of one unit from the origin) and the software should use them to calculate the coordinates of five control points C_0 through C_4 (Figure 4.72b). The first step is to compute the midpoint $\mathbf{M}$ of $\mathbf{S}$ and $\mathbf{E}$. Once $\mathbf{M}$ is known, the three interior control points C_1, C_2, and C_3 can easily be calculated. The two exterior points C_0 and C_4 are found by rotating C_1 and C_3,

4 Curves

respectively. Once the five control points are known, two cubic spline segments can be calculated and, together, they constitute the arc.

▸ **Exercise 4.97:** How is M calculated?

The discussion on page 388 shows how rational B-splines can be used to generate a circle precisely.

4.16.7 Cubic B-Splines with Tension

Adding a tension parameter to the uniform cubic B-spline is done in a way similar to the Cardinal spline (Section 4.10). We use Hermite interpolation (Equation (4.26)) to calculate a PC segment that starts and ends at the same points as a cubic B-spline and whose extreme tangent vectors point in the same directions as those of the cubic B-spline, but have lengths controlled by a tension parameter s. Substituting $\frac{1}{6}\mathbf{P}_0 + \frac{4}{6}\mathbf{P}_1 + \frac{1}{6}\mathbf{P}_2$ and $\frac{1}{6}\mathbf{P}_1 + \frac{4}{6}\mathbf{P}_2 + \frac{1}{6}\mathbf{P}_3$ for the terminal points and $s(\mathbf{P}_2 - \mathbf{P}_0)$ and $s(\mathbf{P}_3 - \mathbf{P}_1)$ for the extreme tangents, we write Equation (4.26) and manipulate it such that it ends up looking like a uniform cubic B-spline segment, Equation (4.159).

$$\mathbf{P}(t) = (t^3, t^2, t, 1) \begin{pmatrix} 2 & -2 & 1 & 1 \\ -3 & 3 & -2 & -1 \\ 0 & 0 & 1 & 0 \\ 1 & 0 & 0 & 0 \end{pmatrix} \begin{pmatrix} \frac{1}{6}\mathbf{P}_0 + \frac{4}{6}\mathbf{P}_1 + \frac{1}{6}\mathbf{P}_2 \\ \frac{1}{6}\mathbf{P}_1 + \frac{4}{6}\mathbf{P}_2 + \frac{1}{6}\mathbf{P}_3 \\ s(\mathbf{P}_2 - \mathbf{P}_0) \\ s(\mathbf{P}_3 - \mathbf{P}_1) \end{pmatrix}$$

$$= \frac{1}{6}\Big[\big(t^3(2-s) + t^2(2s-3) - st + 1\big)\mathbf{P}_0 + \big(t^3(6-s) + t^2(s-9) + 4\big)\mathbf{P}_1$$

$$+ \big(t^3(s-6) + t^2(9-2s) + st + 1\big)\mathbf{P}_2 + \big(t^3(s-2) + t^2(3-s)\big)\mathbf{P}_3\Big]$$

$$= \frac{1}{6}(t^3, t^2, t, 1) \begin{pmatrix} 2-s & 6-s & s-6 & s-2 \\ 2s-3 & s-9 & 9-2s & 3-s \\ -s & 0 & s & 0 \\ 1 & 4 & 1 & 0 \end{pmatrix} \begin{pmatrix} \mathbf{P}_0 \\ \mathbf{P}_1 \\ \mathbf{P}_2 \\ \mathbf{P}_3 \end{pmatrix}. \quad (4.164)$$

A quick check verifies that Equation (4.164) reduces to the uniform cubic B-spline segment, Equation (4.159), for $s = 3$. This value is therefore considered the "neutral" or "standard" value of the tension parameter s. Since s controls the length of the tangent vectors, small values of s should produce the effects of higher tension and, in the extreme, the value $s = 0$ should result in indefinite tangent vectors and in the spline segment becoming a straight line. To show this, we rewrite Equation (4.164) for $s = 0$:

$$\mathbf{P}(t) = \frac{1}{6}(t^3, t^2, t, 1) \begin{pmatrix} 2 & 6 & -6 & -2 \\ -3 & -9 & 9 & 3 \\ 0 & 0 & 0 & 0 \\ 1 & 4 & 1 & 0 \end{pmatrix} \begin{pmatrix} \mathbf{P}_0 \\ \mathbf{P}_1 \\ \mathbf{P}_2 \\ \mathbf{P}_3 \end{pmatrix}$$

$$= \frac{1}{6}(2t^3 - 3t^2 + 1)\mathbf{P}_0 + (\frac{1}{6}6t^3 - 9t^2 + 4)\mathbf{P}_1$$

$$+ \frac{1}{6}(-6t^3 + 9t^2 + 1)\mathbf{P}_2 + \frac{1}{6}(-2t^3 + 3t^2)\mathbf{P}_3.$$

Substituting $T = 3t^2 - 2t^3$ for the parameter t changes the above expression to the form

$$\mathbf{P}(T) = \frac{1}{6}(-\mathbf{P}_0 - 3\mathbf{P}_1 + 3\mathbf{P}_2 + \mathbf{P}_3)T + \frac{1}{6}(\mathbf{P}_0 + 4\mathbf{P}_1 + \mathbf{P}_2),$$

which is a straight line from $\mathbf{P}(0) = \frac{1}{6}(\mathbf{P}_0 + 4\mathbf{P}_1 + \mathbf{P}_2)$ to $\mathbf{P}(1) = \frac{1}{6}(\mathbf{P}_1 + 4\mathbf{P}_2 + \mathbf{P}_3)$.

The tangent vector of Equation (4.164) is

$$\mathbf{P}^t(t) = \frac{1}{6}(3t^2, 2t, 1, 0) \begin{pmatrix} 2-s & 6-s & s-6 & s-2 \\ 2s-3 & s-9 & 9-2s & 3-s \\ -s & 0 & s & 0 \\ 1 & 4 & 1 & 0 \end{pmatrix} \begin{pmatrix} \mathbf{P}_0 \\ \mathbf{P}_1 \\ \mathbf{P}_2 \\ \mathbf{P}_3 \end{pmatrix}$$

$$= \frac{1}{6}\Big[\big(3t^2(2-s) + 2t(2s-3) - s\big)\mathbf{P}_0 + \big(3t^2(6-s) + 2t(s-9)\big)\mathbf{P}_1$$
$$+ \big(3t^2(s-6) + 2t(9-2s) + s\big)\mathbf{P}_2 + \big(3t^2(s-2) + 2t(3-s)\big)\mathbf{P}_3\Big]. \tag{4.165}$$

The extreme tangents are

$$\mathbf{P}^t(0) = \frac{s}{6}(\mathbf{P}_2 - \mathbf{P}_0) \quad \text{and} \quad \mathbf{P}^t(1) = \frac{s}{6}(\mathbf{P}_3 - \mathbf{P}_1).$$

Substituting $s = 0$ in Equation (4.165) yields the tangent vector for the case of infinite tension:

$$\mathbf{P}^t(t) = \frac{1}{6}\Big[6(t^2 - t)\mathbf{P}_0 + 18(t^2 - t)\mathbf{P}_1 - 18(t^2 - t)\mathbf{P}_2 - 6(t^2 - t)\mathbf{P}_3\Big] \tag{4.166}$$
$$= (t^2 - t)(\mathbf{P}_0 + 3\mathbf{P}_1 - 3\mathbf{P}_2 - \mathbf{P}_3).$$

▶ **Exercise 4.98**: Since the spline segment is a straight line in this case, its tangent vector should always point in the same direction. Use Equation (4.166) to show that this is so.

Figure 4.73 illustrates the effect of tension on a cubic B-spline. Three curves are shown, corresponding to s values of 0, 3, and 5.

See also Section 4.14.17 for a discussion of cubic Bézier curves with tension.

Sex alleviates tension and love causes it.
— Woody Allen (as Andrew) in *A Midsummer Night's Sex Comedy* (1982)

4.16.8 Cubic B-Spline and Bézier Curves

Given a cubic B-spline segment $\mathbf{P}(t)$ based on the four control points $\mathbf{P}_0$, $\mathbf{P}_1$, $\mathbf{P}_2$, and $\mathbf{P}_3$, it is easy to find four control points $\mathbf{Q}_0$, $\mathbf{Q}_1$, $\mathbf{Q}_2$, and $\mathbf{Q}_3$ such that the Bézier curve $\mathbf{Q}(t)$ defined by them will have the same shape as $\mathbf{P}(t)$. This is done by equating the matrices of Equation (4.159) that define $\mathbf{P}(t)$ to those of

4 Curves

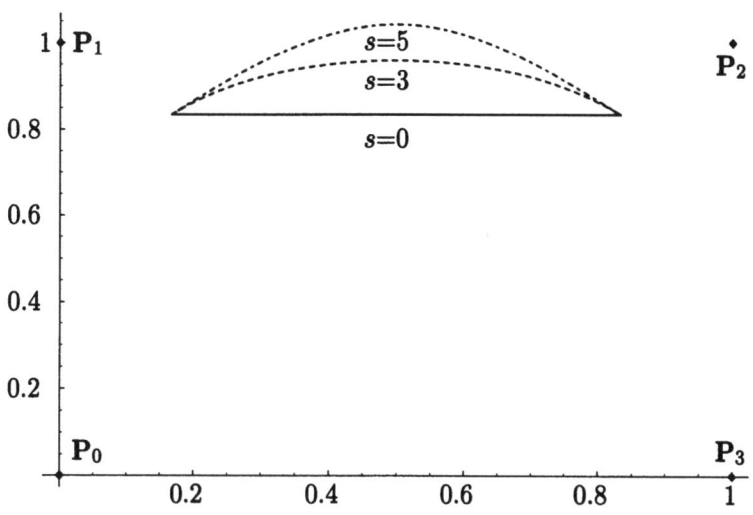

Figure 4.73: Cubic B-Splines with Tension.

```
(* Cubic B-spline with tension *)
Clear[t,s,pnts,stnp,tensMat,bsplineTensn,g1,g2,g3,g4];
pnts={{0,0},{0,1},{1,1},{1,0}};
stnp=Transpose[pnts];
tensMat={{2-s,6-s,s-6,s-2},{2s-3,s-9,9-2s,3-s},{-s,0,s,0},{1,4,1,0}};
bsplineTensn[t_]:=Module[{tmpstruc}, tmpstruc={t^3,t^2,t,1}.tensMat;
{tmpstruc.stnp[[1]],tmpstruc.stnp[[2]]}/6];
g1=ListPlot[pnts, Prolog->AbsolutePointSize[3],
 DisplayFunction->Identity];
s=0;
g2=ParametricPlot[bsplineTensn[t], {t,0,1},
 Compiled->False, DisplayFunction->Identity];
s=3;
g3=ParametricPlot[bsplineTensn[t], {t,0,1},
 Compiled->False, DisplayFunction->Identity,
 PlotStyle->AbsoluteDashing[{2,2}]];
s=5;
g4=ParametricPlot[bsplineTensn[t], {t,0,1},
 Compiled->False, DisplayFunction->Identity,
 PlotStyle->AbsoluteDashing[{1,2,2,2}]];
Show[g1,g2,g3,g4, DisplayFunction->$DisplayFunction]
```

Mathematica Code for Figure 4.73.

Equation (4.107) that define $\mathbf{Q}(t)$:

$$\begin{pmatrix} -1 & 3 & -3 & 1 \\ 3 & -6 & 3 & 0 \\ -3 & 0 & 3 & 0 \\ 1 & 4 & 1 & 0 \end{pmatrix} \begin{pmatrix} \mathbf{P}_0 \\ \mathbf{P}_1 \\ \mathbf{P}_2 \\ \mathbf{P}_3 \end{pmatrix} = \begin{pmatrix} -1 & 3 & -3 & 1 \\ 3 & -6 & 3 & 0 \\ -3 & 3 & 0 & 0 \\ 1 & 0 & 0 & 0 \end{pmatrix} \begin{pmatrix} \mathbf{Q}_0 \\ \mathbf{Q}_1 \\ \mathbf{Q}_2 \\ \mathbf{Q}_3 \end{pmatrix}.$$

The solutions are

$$\mathbf{Q}_0 = \frac{1}{6}(\mathbf{P}_0 + 4\mathbf{P}_1 + \mathbf{P}_2),$$

$$\mathbf{Q}_1 = \frac{1}{6}(4\mathbf{P}_1 + 2\mathbf{P}_2),$$

$$\mathbf{Q}_2 = \frac{1}{6}(2\mathbf{P}_1 + 4\mathbf{P}_2),$$

$$\mathbf{Q}_3 = \frac{1}{6}(\mathbf{P}_1 + 4\mathbf{P}_2 + \mathbf{P}_3).$$

Equation (4.148) shows a similar relation between the quadratic B-spline and Bézier curves.

4.16.9 Cubic Uniform B-Spline by Subdivision

This section uses an approach similar to that of Section 4.16.3. We show how Chaikin's algorithm (Section 4.15) can be applied to the construction of a cubic uniform B-spline for a group of $n+1$ control points $\mathbf{P}_i$. The points are divided into overlapping groups of four points each, and each group is used to calculate, by subdivision, a PC that becomes a segment of the entire curve. These cubic segments have C^2 continuity. Since subdivision is a recursive process, we denote the control points obtained after the kth subdivision by $\mathbf{P}_i^k$. Thus, it makes sense to denote the original control points by $\mathbf{P}_i^0$. They are divided into the groups

$$\mathbf{P}_0^0\mathbf{P}_1^0\mathbf{P}_2^0\mathbf{P}_3^0, \quad \mathbf{P}_1^0\mathbf{P}_2^0\mathbf{P}_3^0\mathbf{P}_4^0, \ldots, \mathbf{P}_{n-3}^0\mathbf{P}_{n-2}^0\mathbf{P}_{n-1}^0\mathbf{P}_n^0.$$

Subdividing the first group is done by

$$\begin{pmatrix} \mathbf{P}_0^1 \\ \mathbf{P}_1^1 \\ \mathbf{P}_2^1 \\ \mathbf{P}_3^1 \\ \mathbf{P}_4^1 \end{pmatrix} = \frac{1}{8} \begin{pmatrix} 4 & 4 & 0 & 0 \\ 1 & 6 & 1 & 0 \\ 0 & 4 & 4 & 0 \\ 0 & 1 & 6 & 1 \\ 0 & 0 & 4 & 4 \end{pmatrix} \begin{pmatrix} \mathbf{P}_0^0 \\ \mathbf{P}_1^0 \\ \mathbf{P}_2^0 \\ \mathbf{P}_3^0 \end{pmatrix} = \begin{pmatrix} \frac{1}{2}(\mathbf{P}_0^0 + \mathbf{P}_1^0) \\ \frac{1}{8}(\mathbf{P}_0^0 + 6\mathbf{P}_1^0 + \mathbf{P}_2^0) \\ \frac{1}{2}(\mathbf{P}_1^0 + \mathbf{P}_2^0) \\ \frac{1}{8}(\mathbf{P}_1^0 + 6\mathbf{P}_2^0 + \mathbf{P}_3^0) \\ \frac{1}{2}(\mathbf{P}_2^0 + \mathbf{P}_3^0) \end{pmatrix},$$

and it yields the five new points $\mathbf{P}_0^1$, $\mathbf{P}_1^1$, $\mathbf{P}_2^1$, $\mathbf{P}_3^1$, and $\mathbf{P}_4^1$. This process is repeated on each of the groups $\mathbf{P}_0^1\mathbf{P}_1^1\mathbf{P}_2^1\mathbf{P}_3^1$ and $\mathbf{P}_1^1\mathbf{P}_2^1\mathbf{P}_3^1\mathbf{P}_4^1$ to yield new control points $\mathbf{P}_i^2$, and so on, until the desired resolution is obtained.

4 Curves

After k subdivision steps, we end up with many control points $\mathbf{P}_i^k$, the first five of which are

$$\mathbf{P}_0^k = \frac{1}{2}(\mathbf{P}_0^{k-1} + \mathbf{P}_1^{k-1}),$$

$$\mathbf{P}_1^k = \frac{1}{8}(\mathbf{P}_0^{k-1} + 6\mathbf{P}_1^{k-1} + \mathbf{P}_2^{k-1}),$$

$$\mathbf{P}_2^k = \frac{1}{2}(\mathbf{P}_1^{k-1} + \mathbf{P}_2^{k-1}),$$

$$\mathbf{P}_3^k = \frac{1}{8}(\mathbf{P}_1^{k-1} + 6\mathbf{P}_2^{k-1} + \mathbf{P}_3^{k-1}),$$

$$\mathbf{P}_4^k = \frac{1}{2}(\mathbf{P}_2^{k-1} + \mathbf{P}_3^{k-1}).$$

For even values of i, the relations above can be written as

$$\begin{pmatrix} \mathbf{P}_i^k \\ \mathbf{P}_{i+1}^k \\ \mathbf{P}_{i+2}^k \end{pmatrix} = \frac{1}{8} \begin{pmatrix} 4 & 4 & 0 \\ 1 & 6 & 1 \\ 0 & 4 & 4 \end{pmatrix} \begin{pmatrix} \mathbf{P}_i^{k-1} \\ \mathbf{P}_{i+1}^{k-1} \\ \mathbf{P}_{i+2}^{k-1} \end{pmatrix} = \mathbf{A} \begin{pmatrix} \mathbf{P}_i^{k-1} \\ \mathbf{P}_{i+1}^{k-1} \\ \mathbf{P}_{i+2}^{k-1} \end{pmatrix},$$

and this can be used to calculate the limit of any triplet $(\mathbf{P}_i^k, \mathbf{P}_{i+1}^k, \mathbf{P}_{i+2}^k)$ for large k without having to go through tedious recursive calculations. We notice that

$$\begin{pmatrix} \mathbf{P}_i^k \\ \mathbf{P}_{i+1}^k \\ \mathbf{P}_{i+2}^k \end{pmatrix} = \mathbf{A}^2 \begin{pmatrix} \mathbf{P}_i^{k-2} \\ \mathbf{P}_{i+1}^{k-2} \\ \mathbf{P}_{i+2}^{k-2} \end{pmatrix} = \mathbf{A}^3 \begin{pmatrix} \mathbf{P}_i^{k-3} \\ \mathbf{P}_{i+1}^{k-3} \\ \mathbf{P}_{i+2}^{k-3} \end{pmatrix},$$

and so on up to

$$\begin{pmatrix} \mathbf{P}_i^k \\ \mathbf{P}_{i+1}^k \\ \mathbf{P}_{i+2}^k \end{pmatrix} = \mathbf{A}^{k-1} \begin{pmatrix} \mathbf{P}_i^1 \\ \mathbf{P}_{i+1}^1 \\ \mathbf{P}_{i+2}^1 \end{pmatrix} = \mathbf{A}^k \begin{pmatrix} \mathbf{P}_i^0 \\ \mathbf{P}_{i+1}^0 \\ \mathbf{P}_{i+2}^0 \end{pmatrix}.$$

Our problem is therefore to calculate the limit of $\mathbf{A}^k$ as k approaches infinity, and this can easily be done using the following theorem (see any text on matrices for the proof and for more information on eigenvalues and eigenvectors):

Theorem: Given an $n \times n$ matrix $\mathbf{A}$ for which there exist n linearly independent eigenvectors, $\mathbf{Q}^{-1}\mathbf{A}\mathbf{Q} = \Lambda$, where $\mathbf{Q}$ is the matrix whose columns are the n eigenvectors and Λ is the diagonal matrix whose diagonal elements are the eigenvalues of $\mathbf{A}$.

An immediate corollary is $\mathbf{A} = \mathbf{Q}\Lambda\mathbf{Q}^{-1}$, which implies

$$\mathbf{A}^2 = \mathbf{Q}\Lambda\mathbf{Q}^{-1}\mathbf{Q}\Lambda\mathbf{Q}^{-1} = \mathbf{Q}\Lambda^2\mathbf{Q}^{-1}$$

and, in general $\mathbf{A}^k = \mathbf{Q}\Lambda^k\mathbf{Q}^{-1}$. This theorem allows us to write our matrix $\mathbf{A}$ (after using appropriate software to calculate its eigenvalues and a set of linearly independent eigenvectors) as

$$A = \begin{pmatrix} 1 & -1 & 1 \\ -1/2 & 0 & 1 \\ 1 & 1 & 1 \end{pmatrix} \begin{pmatrix} 1/4 & 0 & 0 \\ 0 & 1/2 & 0 \\ 0 & 0 & 1 \end{pmatrix} \begin{pmatrix} 1/3 & -2/3 & 1/3 \\ -1/2 & 0 & 1/2 \\ 1/6 & 2/3 & 1/6 \end{pmatrix}.$$

Since matrix Λ is diagonal, we have

$$\lim_{k\to\infty} \Lambda^k = \lim_{k\to\infty} \begin{pmatrix} (1/4)^k & 0 & 0 \\ 0 & (1/2)^k & 0 \\ 0 & 0 & 1^k \end{pmatrix} = \begin{pmatrix} 0 & 0 & 0 \\ 0 & 0 & 0 \\ 0 & 0 & 1 \end{pmatrix}.$$

The limit of $\mathbf{A}^k$ is therefore

$$\begin{pmatrix} 1 & -1 & 1 \\ -1/2 & 0 & 1 \\ 1 & 1 & 1 \end{pmatrix} \begin{pmatrix} 0 & 0 & 0 \\ 0 & 0 & 0 \\ 0 & 0 & 1 \end{pmatrix} \begin{pmatrix} 1/3 & -2/3 & 1/3 \\ -1/2 & 0 & 1/2 \\ 1/6 & 2/3 & 1/6 \end{pmatrix} = \frac{1}{6}\begin{pmatrix} 1 & 4 & 1 \\ 1 & 4 & 1 \\ 1 & 4 & 1 \end{pmatrix},$$

so we end up with the limits

$$\lim_{k\to\infty} \begin{pmatrix} \mathbf{P}_i^k \\ \mathbf{P}_{i+1}^k \\ \mathbf{P}_{i+2}^k \end{pmatrix} = \frac{1}{6}(1,4,1) \begin{pmatrix} \mathbf{P}_i^k \\ \mathbf{P}_{i+1}^k \\ \mathbf{P}_{i+2}^k \end{pmatrix} = \frac{1}{6}(\mathbf{P}_i^k + 4\mathbf{P}_{i+1}^k + \mathbf{P}_{i+2}^k),$$

where k can be any non-negative integer.

The above limits imply the following: If after k subdivision steps, we consider three adjacent control points $\mathbf{P}_i^k$, $\mathbf{P}_{i+1}^k$, and $\mathbf{P}_{i+2}^k$, then all three will converge, after many more subdivisions, to the same limit. This limit must, therefore, be a point on the curve. Here are some examples.

Example 1: We start with $k = 0$, $i = 0$, and obtain

$$\lim_{k\to\infty} \begin{pmatrix} \mathbf{P}_0^0 \\ \mathbf{P}_1^0 \\ \mathbf{P}_2^0 \end{pmatrix} = \frac{1}{6}(1,4,1) \begin{pmatrix} \mathbf{P}_0^0 \\ \mathbf{P}_1^0 \\ \mathbf{P}_2^0 \end{pmatrix} = \frac{1}{6}(\mathbf{P}_0^0 + 4\mathbf{P}_1^0 + \mathbf{P}_2^0),$$

which is point $\mathbf{P}(0)$, as can be seen from Equation (4.159).

Example 2: The values $k = 0$, $i = 1$ result in

$$\lim_{k\to\infty} \begin{pmatrix} \mathbf{P}_1^0 \\ \mathbf{P}_2^0 \\ \mathbf{P}_3^0 \end{pmatrix} = \frac{1}{6}(1,4,1) \begin{pmatrix} \mathbf{P}_1^0 \\ \mathbf{P}_2^0 \\ \mathbf{P}_3^0 \end{pmatrix} = \frac{1}{6}(\mathbf{P}_1^0 + 4\mathbf{P}_2^0 + \mathbf{P}_3^0),$$

which is point $\mathbf{P}(1)$, as can be seen from the same equation.

4 Curves

Example 3: We assume $k = 1$, $i = 1$ (i.e., after one subdivision) and get

$$\lim_{k \to \infty} \begin{pmatrix} \mathbf{P}_1^1 \\ \mathbf{P}_2^1 \\ \mathbf{P}_3^1 \end{pmatrix} = \frac{1}{6}(1,4,1) \begin{pmatrix} \mathbf{P}_1^1 \\ \mathbf{P}_2^1 \\ \mathbf{P}_3^1 \end{pmatrix} = \frac{1}{6}(\mathbf{P}_1^1 + 4\mathbf{P}_2^1 + \mathbf{P}_3^1)$$

$$= \frac{1}{6}\left(\frac{1}{8}(\mathbf{P}_0^0 + 6\mathbf{P}_1^0 + \mathbf{P}_2^0) + \frac{4}{2}(\mathbf{P}_1^0 + \mathbf{P}_2^0) + \frac{1}{8}(\mathbf{P}_1^0 + 6\mathbf{P}_2^0 + \mathbf{P}_3^0)\right)$$

$$= \frac{1}{48}(\mathbf{P}_0^0 + 23\mathbf{P}_1^0 + 23\mathbf{P}_2^0 + \mathbf{P}_3^0).$$

Equation (4.159) tells us that this is point $\mathbf{P}(1/2)$.

4.16.10 Higher-Degree Uniform B-Splines

The methods of Sections 4.16.2 and 4.16.4 can be used to compute uniform B-splines of higher degrees. It can be shown (see, e.g., [Yamaguchi 88], p. 329) that the degree-n uniform B-spline segment is given by

$$\mathbf{P}_i(t) = (t^n, \ldots, t^2, t, 1)\mathbf{M} \begin{pmatrix} \mathbf{P}_{i-1} \\ \mathbf{P}_i \\ \mathbf{P}_{i+1} \\ \vdots \\ \mathbf{P}_{i+n-1} \end{pmatrix},$$

where the elements m_{ij} of the basis matrix $\mathbf{M}$ are given by

$$m_{ij} = \frac{1}{n!}\binom{n}{i}\sum_{k=j}^{n}(n-k)^i(-1)^{k-j}\binom{n+1}{k-j}.$$

Figure 4.74 shows a few examples of these matrices.

4.16.11 Interpolating B-Splines

The B-spline is an approximating curve. Its shape is determined by the control points $\mathbf{P}_i$, but the curve itself does not pass through those points. Instead, it passes through the joints $\mathbf{K}_i$. In our notation so far, we have assumed that the cubic uniform B-spline is based on $n + 1$ control points and passes through $n - 1$ joint points. The number of control points for the cubic curve is, thus, always two more than the number of joints.

> One person's constant is another person's variable.
> — Susan Gerhart.

This section solves the opposite problem. We show how to use the B-spline method to calculate an interpolating cubic spline curve that passes through a set of $n + 1$ given data points $\mathbf{K}_0, \mathbf{K}_1, \ldots, \mathbf{K}_n$. The curve must consist of n segments and

$$\mathbf{M}_1 = \frac{1}{1!}\begin{pmatrix} -1 & 1 \\ 1 & 0 \end{pmatrix}$$

$$\mathbf{M}_2 = \frac{1}{2!}\begin{pmatrix} 1 & -2 & 1 \\ -2 & 2 & 0 \\ 1 & 1 & 0 \end{pmatrix}$$

$$\mathbf{M}_3 = \frac{1}{3!}\begin{pmatrix} -1 & 3 & -3 & 1 \\ 3 & -6 & 3 & 0 \\ -3 & 0 & 3 & 0 \\ 1 & 4 & 1 & 0 \end{pmatrix}$$

$$\mathbf{M}_4 = \frac{1}{4!}\begin{pmatrix} 1 & -4 & 6 & -4 & 1 \\ -4 & 12 & -12 & 4 & 0 \\ 6 & -6 & -6 & 6 & 0 \\ -4 & -12 & 12 & 4 & 0 \\ 1 & 11 & 11 & 1 & 0 \end{pmatrix}$$

$$\mathbf{M}_5 = \frac{1}{5!}\begin{pmatrix} -1 & 5 & -10 & 10 & -5 & 1 \\ 5 & -20 & 30 & -20 & 5 & 0 \\ -10 & 20 & 0 & -20 & 10 & 0 \\ 10 & 20 & -60 & 20 & 10 & 0 \\ -5 & -50 & 0 & 50 & 5 & 0 \\ 1 & 26 & 66 & 26 & 1 & 0 \end{pmatrix}$$

$$\mathbf{M}_6 = \frac{1}{6!}\begin{pmatrix} 1 & -6 & 15 & -20 & 15 & -6 & 1 \\ -6 & 30 & -60 & 60 & -30 & 6 & 0 \\ 15 & -45 & 30 & 30 & -45 & 15 & 0 \\ -20 & -20 & 160 & -160 & 20 & 20 & 0 \\ 15 & 135 & -150 & -150 & 135 & 15 & 0 \\ -6 & -150 & -240 & 240 & 150 & 6 & 0 \\ 1 & 57 & 302 & 302 & 57 & 1 & 0 \end{pmatrix}$$

Figure 4.74: Some Basis Matrices for Uniform B-Splines.

4 Curves

the idea is to use the $\mathbf{K}_i$ points to calculate a new set of points $\mathbf{P}_i$, then use the new points as the control points of a cubic uniform B-spline curve. To get n cubic segments, we need $n+3$ points and we denote them by $\mathbf{P}_{-1}$ through $\mathbf{P}_{n+1}$.

Using $\mathbf{P}_i$ as our control points, Equation (4.159) shows that the general segment $\mathbf{P}_i(t)$ terminates at $\mathbf{P}_i(1) = \frac{1}{6}[\mathbf{P}_{i-2} + 4\mathbf{P}_{i-1} + \mathbf{P}_i]$. We require that the segment ends at point $\mathbf{K}_{i-1}$, which gives us the equation $\frac{1}{6}[\mathbf{P}_{i-2} + 4\mathbf{P}_{i-1} + \mathbf{P}_i] = \mathbf{K}_{i-1}$. When this equation is repeated for $0 \leq i \leq n$, we get a system of $n+1$ equations. However, there are $n+3$ unknowns ($\mathbf{P}_{-1}$ through $\mathbf{P}_{n+1}$), so we need two more equations.

The needed equations are obtained by considering the tangent vectors of the interpolating curve at its two ends. We denote the tangent at the start by $\mathbf{T}_1$. It is given by $\mathbf{T}_1 = \frac{1}{2}(\mathbf{P}_1 - \mathbf{P}_{-1})$, so it points in the direction from $\mathbf{P}_{-1}$ to $\mathbf{P}_1$; similarly for the end tangent $\mathbf{T}_n = \frac{1}{2}(\mathbf{P}_{n+1} - \mathbf{P}_{n-1})$. The result is

$$n+3\left\{\frac{1}{6}\begin{pmatrix} -3 & 0 & 3 & 0 & \ldots & 0 & 0 & 0 \\ 1 & 4 & 1 & 0 & \ldots & 0 & 0 & 0 \\ 0 & 1 & 4 & 1 & \ldots & 0 & 0 & 0 \\ \vdots & & & & & & & \vdots \\ 0 & 0 & 0 & 0 & \ldots & 4 & 1 & 0 \\ 0 & 0 & 0 & 0 & \ldots & 1 & 4 & 1 \\ 0 & 0 & 0 & 0 & \ldots & -3 & 0 & 3 \end{pmatrix}\begin{pmatrix} \mathbf{P}_{-1} \\ \mathbf{P}_0 \\ \mathbf{P}_1 \\ \vdots \\ \mathbf{P}_{n-1} \\ \mathbf{P}_n \\ \mathbf{P}_{n+1} \end{pmatrix} = \begin{pmatrix} \mathbf{T}_1 \\ \mathbf{K}_0 \\ \mathbf{K}_1 \\ \vdots \\ \mathbf{K}_{n-1} \\ \mathbf{K}_n \\ \mathbf{T}_n \end{pmatrix}\right.. \quad (4.167)$$

The user specifies the values of the two extreme tangents $\mathbf{T}_1$ and $\mathbf{T}_n$, the equations are solved, and the $\mathbf{P}_i$ points are then used in the usual way to calculate a cubic uniform B-spline that passes through the original points $\mathbf{K}_i$.

Notice that the coefficient matrix of Equation (4.167) is not diagonally dominant because of the four ± 3's. We can, however, modify it slightly by writing the system of equations in the form

$$n+3\left\{\frac{1}{6}\begin{pmatrix} -3/2 & 0 & 3/2 & 0 & \ldots & 0 & 0 & 0 \\ 1 & 4 & 1 & 0 & \ldots & 0 & 0 & 0 \\ 0 & 1 & 4 & 1 & \ldots & 0 & 0 & 0 \\ \vdots & & & & & & & \vdots \\ 0 & 0 & 0 & 0 & \ldots & 4 & 1 & 0 \\ 0 & 0 & 0 & 0 & \ldots & 1 & 4 & 1 \\ 0 & 0 & 0 & 0 & \ldots & -3/2 & 0 & 3/2 \end{pmatrix}\begin{pmatrix} \mathbf{P}_{-1} \\ \mathbf{P}_0 \\ \mathbf{P}_1 \\ \vdots \\ \mathbf{P}_{n-1} \\ \mathbf{P}_n \\ \mathbf{P}_{n+1} \end{pmatrix} = \begin{pmatrix} \mathbf{T}_1/2 \\ \mathbf{K}_0 \\ \mathbf{K}_1 \\ \vdots \\ \mathbf{K}_{n-1} \\ \mathbf{K}_n \\ \mathbf{T}_n/2 \end{pmatrix}\right..$$

(4.168)

The coefficient matrix of Equation (4.168) is columnwise diagonally dominant and is therefore nonsingular. Thus, this system of equations has a unique solution, but this system is mathematically identical to Equation (4.167), so that system of equations also has a unique solution.

Example: This is the opposite of the example on page 337. We start with $\mathbf{K}_0 = (1/6, 5/6)$, $\mathbf{K}_1 = (1,1)$, $\mathbf{K}_2 = (11/6, 5/6)$, and the two extreme tangents

$\mathbf{T}_1 = (1/2, 1/2)$ and $\mathbf{T}_2 = (1/2, -1/2)$, and set up the 5×5 system of equations

$$\frac{1}{6}\begin{pmatrix} -3 & 0 & 3 & 0 & 0 \\ 1 & 4 & 1 & 0 & 0 \\ 0 & 1 & 4 & 1 & 0 \\ 0 & 0 & 1 & 4 & 1 \\ 0 & 0 & -3 & 0 & 3 \end{pmatrix} \begin{pmatrix} \mathbf{P}_{-1} \\ \mathbf{P}_0 \\ \mathbf{P}_1 \\ \mathbf{P}_2 \\ \mathbf{P}_3 \end{pmatrix} = \begin{pmatrix} (1/2, 1/2) \\ (1/6, 5/6) \\ (1, 1) \\ (11/6, 5/6) \\ (1/2, -1/2) \end{pmatrix}.$$

This is easy to solve and the solutions are $\mathbf{P}_{-1} = (0,0)$, $\mathbf{P}_0 = (0,1)$, $\mathbf{P}_1 = (1,1)$, $\mathbf{P}_2 = (2,1)$, and $\mathbf{P}_3 = (2,0)$, identical to the original control points of the above-mentioned example.

4.16.12 A Knot Vector-Based Approach

This approach to the uniform B-spline curve assumes that the curve is a weighted sum, $\mathbf{P}(t) = \sum_{i=0}^{n} \mathbf{P}_i B_{n,i}(t)$, of the control points and it calculates the weight functions. The method is similar to the one used in deriving the Bézier curve (Section 4.14.2). The cubic uniform B-spline is used here as an example, but the same method can be applied to B-splines of any order. We assume that five control points are given (so that five weight functions, $B_{4,0}(t)$ through $B_{4,4}(t)$, will be needed) and that the curve will consist of two cubic segments. In this approach, we assume that each spline segment is traced when the parameter t varies over one unit, from an integer value u to the next integer $u+1$. The u values are called the *knots* of the B-spline. Since they are the integers $0, 1, 2, \ldots$, they are uniformly distributed, hence the name *uniform* B-spline. To trace out a two-segment spline curve, t should therefore vary in the range $[0, 2]$.

The guiding principle is that each weight function should be a PC, should have a maximum at the vicinity of "its" control point, and should drop to zero when away from the point. A general weight function should therefore have the bell shape shown in Figure 4.75a. To actually calculate such a function, we write it as the union of four parts, $b_0(t)$, $b_1(t)$, $b_2(t)$, and $b_3(t)$, each a simple PC, and each defined over one unit of t. Figure 4.75b shows how each weight $B_{4,i}(t)$ is defined over a range of five knots and is zero elsewhere

The following considerations are used to set up equations to calculate the $b_i(t)$ functions:

1. They should be barycentric.
2. They should provide C^2 continuity at the three points where they join.
3. $b_0(t)$ and its first two derivatives should be zero at the start point $b_0(0)$.
4. $b_3(t)$ and its first two derivatives should be zero at the end point $b_3(1)$.

We define $b_i(t) = A_i t^3 + B_i t^2 + C_i t + D_i$. The above conditions yield the following equations:

1. The single equation $B_{4,0}(0) + B_{4,1}(0) + B_{4,2}(0) + B_{4,3}(0) = 1$. This is a special case of condition 1. We see later that the $b_i(t)$ functions resulting from our equations are, in fact, barycentric.

4 Curves

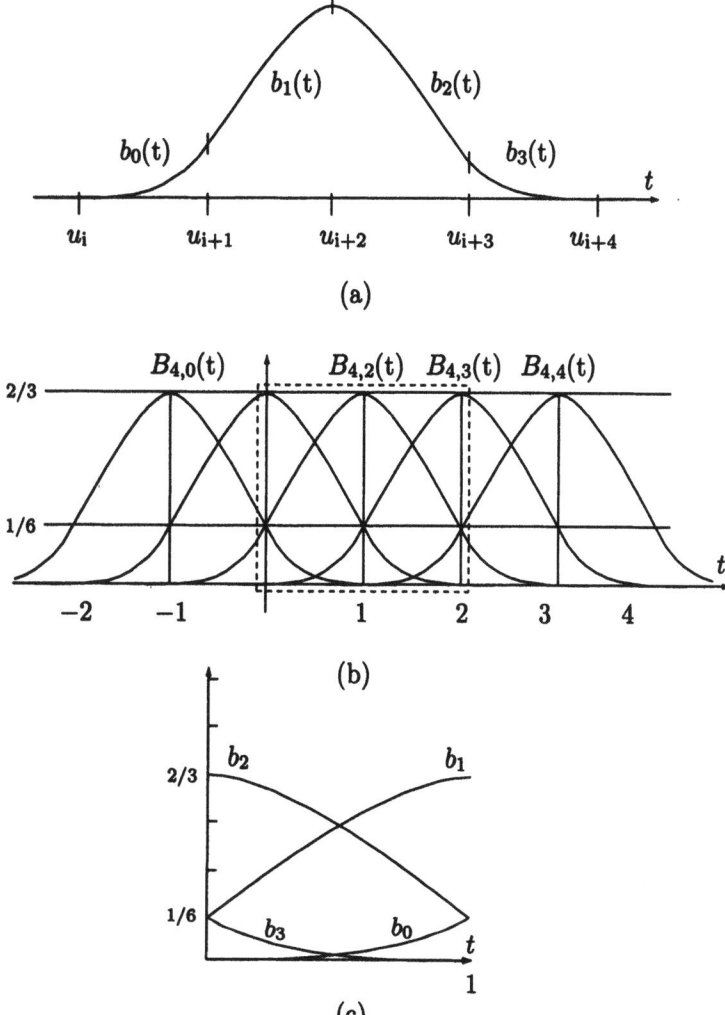

Figure 4.75: Weight Functions of the Cubic Uniform B-Spline.

2. Condition 2 yields the nine equations

$$
\begin{array}{lll}
b_0(1) = b_1(0), & b_0^t(1) = b_1^t(0), & b_0^{tt}(1) = b_1^{tt}(0), \\
b_1(1) = b_2(0), & b_1^t(1) = b_2^t(0), & b_1^{tt}(1) = b_2^{tt}(0), \\
b_2(1) = b_3(0), & b_2^t(1) = b_3^t(0), & b_2^{tt}(1) = b_3^{tt}(0).
\end{array} \quad (4.169)
$$

The first two derivatives of $b_i(t)$ are

$$\frac{db_i(t)}{dt} = b_i^t(t) = 3A_it^2 + 2B_it + C_i, \qquad \frac{d^2b_i(t)}{dt^2} = b_i^{tt}(t) = 6A_it + 2B_i,$$

so the nine equations above can be written explicitly as

$$A_0 + B_0 + C_0 + D_0 = D_1, \quad 3A_0 + 2B_0 + C_0 = C_1, \quad 6A_0 + 2B_0 = 2B_1,$$
$$A_1 + B_1 + C_1 + D_1 = D_2, \quad 3A_1 + 2B_1 + C_1 = C_2, \quad 6A_1 + 2B_1 = 2B_2,$$
$$A_2 + B_2 + C_2 + D_2 = D_3, \quad 3A_2 + 2B_2 + C_2 = C_3, \quad 6A_2 + 2B_2 = 2B_3.$$

3. Condition 3 yields the three equations

$$D_0 = 0, \quad C_0 = 0, \quad 2B_0 = 0.$$

4. Condition 4 yields the three equations

$$A_3 + B_3 + C_3 + D_3 = 0, \quad 3A_3 + 2B_3 + C_3 = 0, \quad 6A_3 + 2B_3 = 0.$$

We thus end up with 16 equations that are easy to solve. Their solutions are

$$b_0(t) = \frac{1}{6}t^3, \quad b_1(t) = \frac{1}{6}(1 + 3t + 3t^2 - 3t^3),$$
$$b_2(t) = \frac{1}{6}(4 - 6t^2 + 3t^3), \quad b_3(t) = \frac{1}{6}(1 - 3t + 3t^2 - t^3). \quad (4.170)$$

The proof that the $b_i(t)$ functions are barycentric is now trivial. Figures 4.75c and 4.11c show the shapes of the four weights.

Now that the weight functions are known, the entire curve can be expressed as the weighted sum $\mathbf{P}(t) = \sum_{i=0}^{n} \mathbf{P}_i B_{4,i}(t)$, where the weights all look the same and are shifted with respect to each other by using different ranges for t. Each weight $B_{4,i}(t)$ is nonzero only in the (open) range (u_{i-3}, u_{i+1}) (Figure 4.75b).

Each curve segment $\mathbf{P}_i(t)$ can now be expressed as the barycentric sum of the four weighted points $\mathbf{P}_{i-3}, \ldots, \mathbf{P}_i$ (or, alternatively, as a linear combination of the $B_{4,i}(t)$ functions), $\mathbf{P}_i(t) = \sum_{j=-3}^{0} \mathbf{P}_{i+j} B_{4,i+j}(t)$, where $u_i \leq t < u_{i+1}$. The next (crucial) step is to realize that in the range $u_i \leq t < u_{i+1}$, only component b_3 of B_{i-3} is nonzero and similarly for the other three weights (see the dashed box of Figure 4.75b). The segment can therefore be written

$$\mathbf{P}_i(t) = \sum_{j=3}^{0} \mathbf{P}_{i-j} b_j(t)$$
$$= \frac{1}{6}\mathbf{P}_{i-3}(-t^3 + 3t^2 - 3t + 1) + \frac{1}{6}\mathbf{P}_{i-2}(3t^3 - 6t^2 + 4)$$
$$+ \frac{1}{6}\mathbf{P}_{i-1}(-3t^3 + 3t^2 + 3t + 1) + \frac{1}{6}\mathbf{P}_i t^3 \quad (4.171)$$
$$= \frac{1}{6}(t^3, t^2, t, 1)\begin{pmatrix} -1 & 3 & -3 & 1 \\ 3 & -6 & 3 & 0 \\ -3 & 0 & 3 & 0 \\ 1 & 4 & 1 & 0 \end{pmatrix}\begin{pmatrix} \mathbf{P}_{i-3} \\ \mathbf{P}_{i-2} \\ \mathbf{P}_{i-1} \\ \mathbf{P}_i \end{pmatrix},$$

4 Curves

an expression identical (except for the choice of index i) to Equation (4.159). This approach to deriving the weight functions can be generalized for the nonuniform B-spline.

The dashed box of Figure 4.75b illustrates how the $B_{4,i}(t)$ weight functions blend the five control points in the two spline segments. The first weight, $B_{4,0}(t)$, goes down from 1/6 to 0 when t varies from 0 to 1. The first control point $\mathbf{P}_0$ thus starts by contributing 1/6 of its value to the curve, then decreases its contribution until it disappears at $t = 1$. This is why $\mathbf{P}_0$ does not contribute to the second segment. The second weight, $B_{4,1}(t)$, starts at 2/3 (when $t = 0$), goes down to 1/6 for $t = 1$ and all the way to 0 when t reaches 2. This is how the second control point $\mathbf{P}_1$ participates in the blend that generates the first two spline segments. Notice how the weight functions have their maxima at integer values of t, how only three weights are nonzero at these values, and how there are four nonzero weights for any other values of t.

Figure 4.76a shows the weight functions for the linear uniform B-spline. Each has the form of a hat, going from 0 to 1 and back to 0. They also have their maxima at integer values of t. The weight functions of the quadratic B-spline are shown in Figure 4.76b. Notice how each varies from 0 to 3/4, how they meet at a height of 1/2, and how their maxima are at half-integer values of t. The first weight, $B_{3,0}(t)$, drops from 1/2 to 0 for the first spline segment (i.e., when t varies in the range $[0, 1]$) and remains zero for the second and subsequent segments. The second weight, $B_{3,1}(t)$, climbs from 1/2 to 1, then drops back to 1/2 for the first segment. For the second segment, this weight goes down from 1/2 to 0. These diagrams provide a deeper understanding of the way the control points are blended by the uniform B-spline.

The general B-spline weight functions are normally denoted $N_{ik}(t)$ and can be defined recursively. Before delving into this topic, however, we show how the uniform B-spline curve itself can be defined recursively, similar to the recursive definition of the Bézier curve (Equation (4.109)). Given a set of $n+1$ control points $\mathbf{P}_0$ through $\mathbf{P}_n$ and a *uniform knot vector* $(t_0, t_1, \ldots, t_{n+k})$ (a set of equally spaced $n + k + 1$ nondecreasing real numbers), the B-spline of order k is defined as

$$\mathbf{P}(t) = \mathbf{P}_l^{(k-1)}(t), \quad \text{where} \quad t_l \leq t < t_{l+1} \qquad (4.172)$$

and where the quantities $\mathbf{P}_i^{(j)}(t)$ are defined recursively by

$$\mathbf{P}_i^{(j)}(t) = \begin{cases} \mathbf{P}_i, & \text{for } j = 0, \\ (1 - T_{ij})\mathbf{P}_{i-1}^{(j-1)}(t) + T_{ij}\mathbf{P}_i^{(j-1)}(t), & \text{for } j > 0, \end{cases}$$

and

$$T_{ij} = \frac{t - t_i}{t_{i+k-j} - t_i}.$$

Figure 4.77 is a pyramid that illustrates how the quantities $\mathbf{P}_i^{(k-1)}(t)$ are constructed recursively. Each $\mathbf{P}_i^{(j)}(t)$ in the figure is constructed as a barycentric sum of the two quantities immediately to its left. Equation (4.172) is the *geometric* definition of the uniform B-spline.

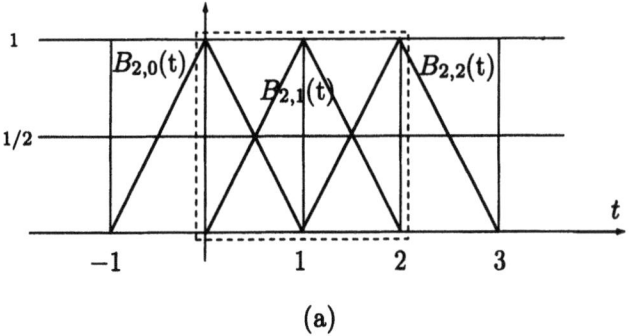

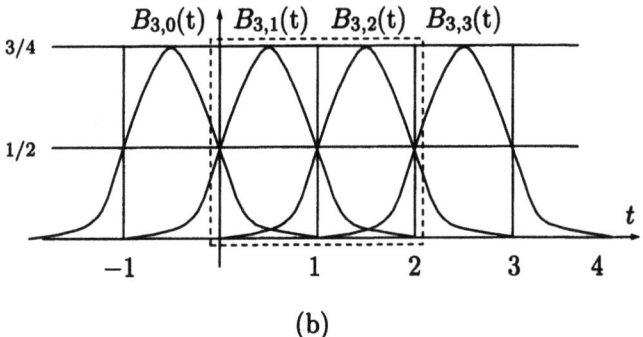

Figure 4.76: Weight Functions of the Linear and the Quadratic B-Splines.

We now turn to the *algebraic* (or analytical) definition of the general (uniform and nonuniform) B-spline curve. It is defined as the weighted sum

$$\mathbf{P}(t) = \sum_{i=0}^{n} \mathbf{P}_i N_{ik}(t),$$

where the weight functions $N_{ik}(t)$ are defined recursively by

$$N_{i1}(t) = \begin{cases} 1, & \text{if } t \in [t_i, t_{i+1}), \\ 0, & \text{otherwise} \end{cases} \qquad (4.173)$$

(note how the interval starts at t_i but does not reach t_{i+1}. Such an interval is closed on the left and open on the right) and

$$N_{ik}(t) = \frac{t - t_i}{t_{i+k-1} - t_i} N_{i,k-1}(t) + \frac{t_{i+k} - t}{t_{i+k} - t_{i+1}} N_{i+1,k-1}(t), \quad \text{where } 0 \le i \le n. \qquad (4.174)$$

The weights $N_{ik}(t)$ may be tedious to calculate in the general case, where the knots t_i can be any knot, but are easy to calculate in the special case where the knot vector is the uniform sequence $(0, 1, \ldots, n+k)$, i.e., when $t_i = i$. Here are examples for the first few values of k.

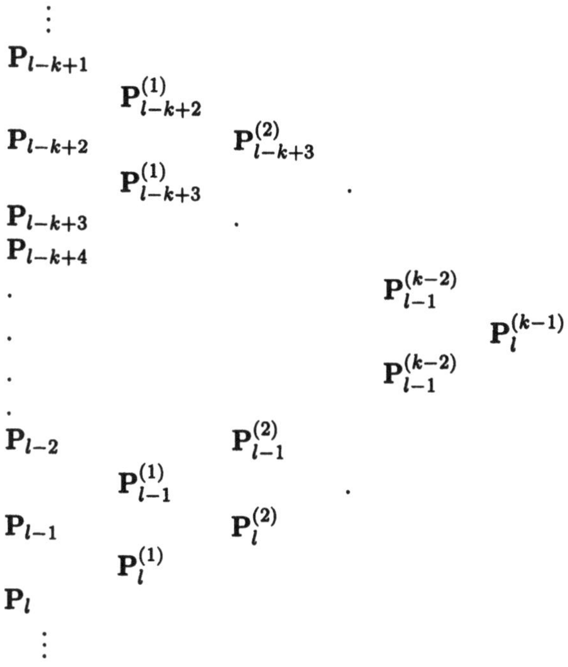

4 Curves

$$\vdots$$

$\mathbf{P}_{l-k+1}$

$\quad\quad \mathbf{P}^{(1)}_{l-k+2}$

$\mathbf{P}_{l-k+2} \quad\quad\quad\quad \mathbf{P}^{(2)}_{l-k+3}$

$\quad\quad \mathbf{P}^{(1)}_{l-k+3}$

$\mathbf{P}_{l-k+3}$

$\mathbf{P}_{l-k+4}$

$\cdot$

$\cdot \quad\quad\quad\quad\quad\quad\quad\quad\quad\quad \mathbf{P}^{(k-2)}_{l-1}$

$\cdot \quad\quad\quad\quad\quad\quad\quad\quad\quad\quad\quad\quad\quad \mathbf{P}^{(k-1)}_{l}$

$\cdot \quad\quad\quad\quad\quad\quad\quad\quad\quad \mathbf{P}^{(k-2)}_{l-1}$

$\cdot$

$\mathbf{P}_{l-2} \quad\quad\quad\quad \mathbf{P}^{(2)}_{l-1}$

$\quad\quad \mathbf{P}^{(1)}_{l-1}$

$\mathbf{P}_{l-1} \quad\quad\quad\quad \mathbf{P}^{(2)}_{l}$

$\quad\quad \mathbf{P}^{(1)}_{l}$

$\mathbf{P}_{l}$

$\vdots$

Figure 4.77: Recursive Construction of $\mathbf{P}^{(k-1)}_l(t)$.

For $k = 1$, the weight functions are defined by

$$N_{i1}(t) = \begin{cases} 1, & \text{if } t \in [i, i+1), \\ 0, & \text{otherwise.} \end{cases} \quad (4.175)$$

This results in the "step" functions shown in Figure 4.78. Notice how each step is closed on the left and open on the right and how $N_{i1}(t)$ is nonzero only in the interval $[i, i+1)$ (this interval is its *support*). It is also clear that each of them is a shifted version of its predecessor, so we can express any of them as a shifted version of the first one and write $N_{i1}(t) = N_{01}(t - i)$.

For $k = 2$, the weight functions can be calculated for any i from Equation (4.174):

$$N_{02}(t) = \frac{t - t_0}{t_1 - t_0} N_{01}(t) + \frac{t_2 - t}{t_2 - t_1} N_{11}(t)$$
$$= t N_{01}(t) + (2 - t) N_{11}(t)$$
$$= \begin{cases} t, & \text{when } 0 \leq t < 1, \\ 2 - t, & \text{when } 1 \leq t < 2, \\ 0, & \text{otherwise,} \end{cases}$$
$$N_{12}(t) = \frac{t - t_1}{t_2 - t_1} N_{11}(t) + \frac{t_3 - t}{t_3 - t_2} N_{21}(t)$$
$$= (t - 1) N_{11}(t) + (3 - t) N_{21}(t)$$

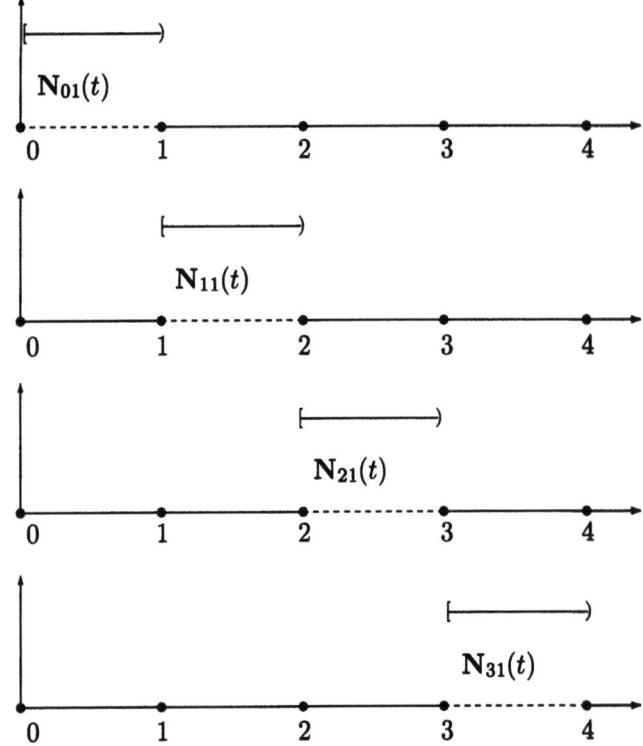

Figure 4.78: Uniform B-Spline Weight Functions for $k = 1$.

$$= \begin{cases} t-1, & \text{when } 1 \le t < 2, \\ 3-t, & \text{when } 2 \le t < 3, \\ 0, & \text{otherwise,} \end{cases}$$

$$N_{22}(t) = \frac{t-t_2}{t_3-t_2} N_{21}(t) + \frac{t_4-t}{t_4-t_3} N_{31}(t)$$

$$= (t-2)N_{21}(t) + (4-t)N_{31}(t)$$

$$= \begin{cases} t-2, & \text{when } 2 \le t < 3, \\ 4-t, & \text{when } 3 \le t < 4, \\ 0, & \text{otherwise.} \end{cases}$$

The hat-shaped functions are shown in Figure 4.79. Notice how $N_{i2}(t)$ spans the interval $[i, i+2)$. It is also obvious that each of them is a shifted version of its predecessor, so we can express any of them as a shifted version of the first one and write $N_{i2}(t) = N_{02}(t-i)$.

For $k = 3$, the calculations are similar:

$$N_{03}(t) = \frac{t-t_0}{t_2-t_0} N_{02}(t) + \frac{t_3-t}{t_3-t_1} N_{12}(t)$$

$$= \frac{t}{2} N_{02}(t) + \frac{3-t}{2} N_{12}(t)$$

4 Curves

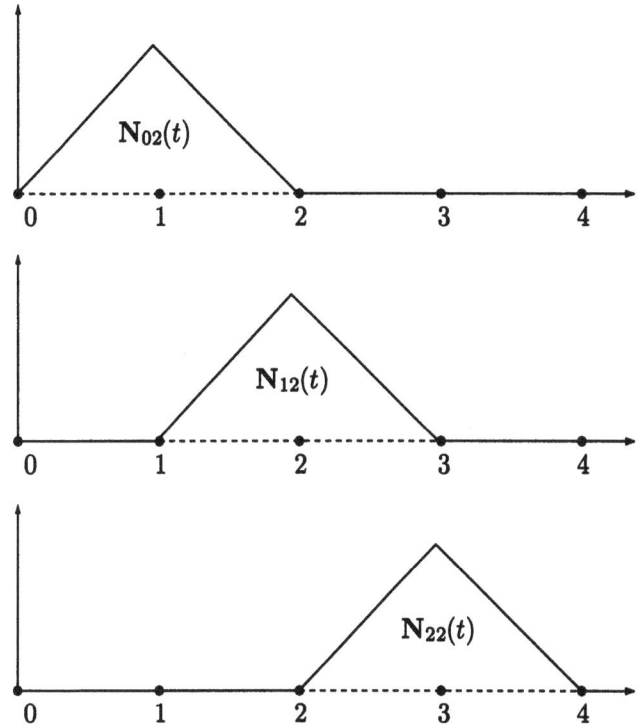

Figure 4.79: Uniform B-Spline Weight Functions for $k = 2$.

$$= \begin{cases} t^2/2, & \text{when } 0 \leq t < 1, \\ \frac{t^2}{2}(2-t) + \frac{3-t}{2}(t-1), & \text{when } 1 \leq t < 2, \\ (3-t)^2/2, & \text{when } 2 \leq t < 3, \\ 0, & \text{otherwise,} \end{cases}$$

$$= \begin{cases} t^2/2, & \text{when } 0 \leq t < 1, \\ (-2t^2 + 6t - 3)/2, & \text{when } 1 \leq t < 2, \\ (3-t)^2/2, & \text{when } 2 \leq t < 3, \\ 0, & \text{otherwise,} \end{cases}$$

$$N_{13}(t) = \frac{t-t_1}{t_3-t_1} N_{12}(t) + \frac{t_4-t}{t_4-t_2} N_{22}(t)$$

$$= \frac{t-1}{2} N_{12}(t) + \frac{4-t}{2} N_{22}(t)$$

$$= \begin{cases} (t-1)^2/2, & \text{when } 1 \leq t < 2, \\ (-2t^2 + 10t - 11)/2, & \text{when } 2 \leq t < 3, \\ (4-t)^2/2, & \text{when } 3 \leq t < 4, \\ 0, & \text{otherwise.} \end{cases}$$

Each of these curves (Figure 4.80) is a spline whose three segments are quadratic polynomials (i.e., parabolic arcs) smoothly joined at the knots. Notice again that the support of $N_{i3}(t)$ is the interval $[i, i+3)$ and that they are shifted versions of

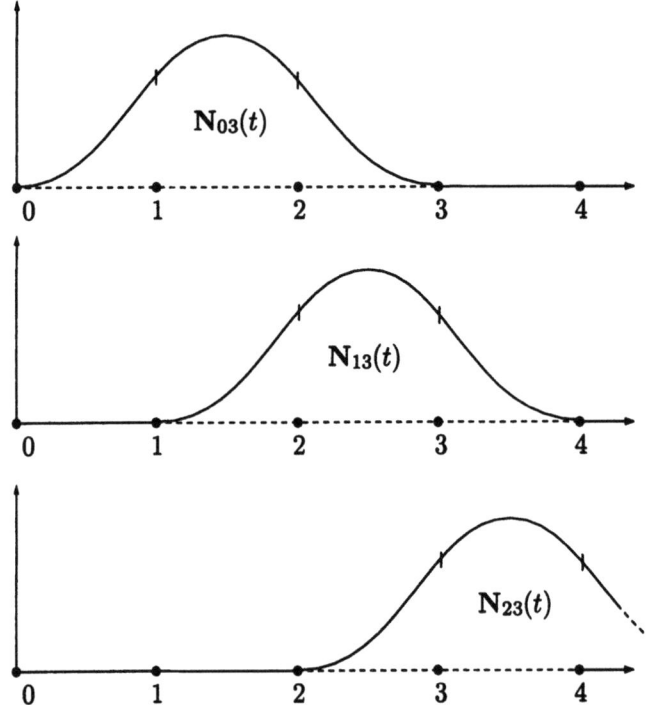

Figure 4.80: Uniform B-Spline Weight Functions for $k = 3$.

each other, allowing us to write $N_{i3}(t) = N_{03}(t-i)$.

▶ **Exercise 4.99:** How can we show that the various $N_{i3}(t)$ are shifted versions of each other?

In general, the support of $N_{ik}(t)$ is the interval $[i, i+k)$ and $N_{ik}(t) = N_{0k}(t-i)$. Figure 4.81 shows how a general weight function $N_{ik}(t)$ is constructed recursively. Each $N_{ij}(t)$ function in this triangle is constructed as a barycentric sum of the two functions immediately to its left.

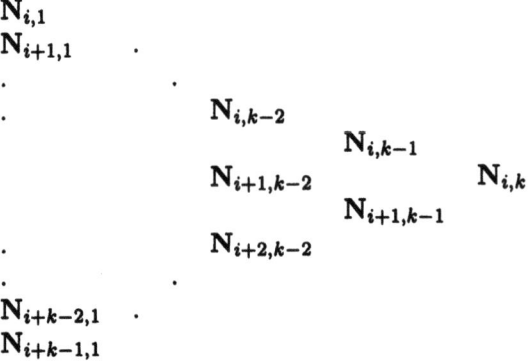

Figure 4.81: Recursive Construction of $N_{i,k}(t)$.

4 Curves

The geometric and algebraic definitions of the B-spline look different but it can be shown that they are identical. The proof of this is called the Cox-DeBoor (or DeBoor-Cox) formula [DeBoor 72].

4.16.13 Recursive Definitions of the B-Spline

The order k of the B-spline curve is an integer in the range $[2, n+1]$ (it is possible to have $k = 1$, but the curve degenerates in this case to just a plot of the control points). Each blending function $N_{ik}(t)$ has support over k intervals $[t_i, t_{i+k-1})$ and is zero outside its support. The knot vector $(t_0, t_1, \ldots, t_{n+k})$ consists of $n+k+1$ nondecreasing real numbers t_i. These values define $n+k$ subintervals $[t_i, t_{i+1})$. The two extreme values t_0 and t_n are selected based on the values of n and k. Any terms of the form $0/0$ or $x/0$ in the calculation of the blending functions are assumed to be zero. Editing the B-spline curve can be done by (1) adding, moving, or deleting control points without changing the order k, (2) changing the order k without changing the control points, and (3) increasing the size of the knot vector. The knot vector contains $n+k+1$ values, so increasing its size means that either n or k should be increased. Here are a few more properties of the curve:

1. Plotting the B-spline curve is done by varying the parameter t over the range of knot values $[t_{k-1}, t_{n+1})$.

2. Each segment of the curve (between two consecutive knot values) depends on k control points. This is why the curve has a local control and it also means that the maximum value of k is the number of control points $n+1$.

3. Any control point participates in at most k segments.

4. The curve lies inside the convex hull defined by at most k control points. This means that the curve passes close to the control points, a feature that makes it easy for a designer to place these points in order to get the right curve shape.

5. The blending functions $N_{ik}(t)$ are barycentric for any t in the interval $[t_{k-1}, t_{n+1})$. They are also non-negative and, except for $k = 1$, each has one maximum.

6. The curve and its first $k-1$ derivatives are continuous over the entire range (except that nonuniform B-splines can have discontinuities, see Figure 4.84d).

7. The entire curve can be affinely transformed by transforming the control points, then redrawing the curve from the new points.

One important difference between the B-spline and the Bézier curve is the use of a *knot vector*. This feature (which has already been mentioned) consists of a nondecreasing sequence of real numbers called *knots*. The knot vector adds flexibility to the curve and provides better control of its shape, but its use requires experience. There are three common ways to select the values in the knot vector, namely uniform, open uniform, and nonuniform. In a uniform B-spline the knot values are equally spaced. An example is $(-2, -1.5, -0.5, 0, 0.5, 1, 1.5)$, but more typical examples are a vector with normalized values between 0 and 1 $(0, 0.2, 0.4, 0.6, 0.8, 1)$ or a vector with integer values $(0, 1, 2, 3, 4, 5, 6)$. Figure 4.82 lists *Mathematica* code to calculate, print, and plot the weight functions for any set of knots.

```
(* B-spline weight functions printed and plotted *)
Clear[bspl,knt,i,k,n,t,p]
bspl[i_,k_,t_]:=If[knt[[i+k]]==knt[[i+1]],0,   (* 0<=i<=n *)
 bspl[i,k-1,t] (t-knt[[i+1]])/(knt[[i+k]]-knt[[i+1]])] \
 +If[knt[[i+1+k]]==knt[[i+2]],0,
 bspl[i+1,k-1,t] (knt[[i+1+k]]-t)/(knt[[i+1+k]]-knt[[i+2]])];
bspl[i_,1,t_]:=If[knt[[i+1]]<=t<knt[[i+2]], 1, 0];
n=4; k=3; (* Note: 0<=k<=n *)
(* knt=Table[i, {i,0,n+k}]; *) (* knots for the uniform case *)
knt={0,0,0,1,2,3,3,3}; (* knots for the NONuniform case *)
(* Show the weight functions *)
Do[Print["N(",i,",",k,",",t,")=",Simplify[bspl[i,k,t]]], {i,0,n}]
(* Plot them. Plots are separated using .97 instead of 1 *)
Do[p[i+1]=Plot[bspl[i,k,t], {t,k-.97,n+.97},
 DisplayFunction->Identity], {i,0,n}]
Show[Table[p[i+1], {i,0,n}], Ticks->None,
 DisplayFunction->$DisplayFunction]
```

Figure 4.82: *Mathematica* Code for the B-Spline Weight Functions.

4.16.14 Open Uniform B-Splines

This type of B-spline is obtained when the knot vector is uniform except at its two ends, where knot values are repeated k times. Examples are the following:

For $n = 3$ and $k = 2$, there are $n + k + 1 = 6$ knots, e.g., $(0, 0, 1, 2, 3, 3)$.
For $n = 4$ and $k = 4$, there are $n + k + 1 = 9$ knots, e.g., $(0, 0, 0, 0, 1, 2, 2, 2, 2)$.
For $n = 3$ and $k = 2$, there are $n + k + 1 = 6$ knots, e.g., $(0, 0, 0.33, 0.67, 1, 1)$.
For $n = 4$ and $k = 4$, there are $n + k + 1 = 9$ knots, e.g., $(0, 0, 0, 0, 0.5, 1, 1, 1, 1)$.

(Notice how the last two examples are normalized.) In general, given values for n and k, we can generate an integer open knot vector by setting

$$t_i = \begin{cases} 0, & \text{for } 0 \leq i < k, \\ i - k + 1, & \text{for } k \leq i \leq n, \\ n - k + 2, & \text{for } n < i \leq n + k, \end{cases} \quad \text{for} \quad 0 \leq i \leq n + k. \quad (4.176)$$

An open uniform B-spline curve starts at $\mathbf{P}_0$ and ends at $\mathbf{P}_n$. This feature makes it easy to generate closed curves of this type. The two extreme tangents of this curve go in the directions from $\mathbf{P}_0$ to $\mathbf{P}_1$ and from $\mathbf{P}_{n-1}$ to $\mathbf{P}_n$, respectively. This is why open uniform B-spline curves are similar to Bézier curves. In fact, when $k = n + 1$ (i.e., when the degree of the polynomials is n), these curves have knot vectors of the form $(0, 0, \ldots, 0, 1, 1, \ldots, 1)$ and they reduce to Bézier curves.

Example 1: Five control points $\mathbf{P}_0$ through $\mathbf{P}_4$ are given. This is a case where $n = 4$. We select order 3 (i.e., segments that are polynomials of degree 2) and use

4 Curves

Equation (4.176) to construct the knot sequence $(0,0,0,1,2,3,3,3)$. The parameter t varies from $t_{k-1} = t_2 = 0$ to $t_{n+1} = t_5 = 3$, so our curve will consist of three segments. Each of the blending functions $N_{i3}(t)$ (where $0 \le i \le n$) is nonzero over three subintervals of t and is calculated from Eqs. (4.173) and (4.174). The result is

$$N_{03}(t) = (1-t)^2, \qquad 0 \le t < 1,$$

$$N_{13}(t) = \frac{1}{2}\begin{cases} -3t^2 + 4t, & 0 \le t < 1, \\ (2-t)^2, & 1 \le t < 2, \end{cases}$$

$$N_{23}(t) = \frac{1}{2}\begin{cases} t^2, & 0 \le t < 1, \\ -2t^2 + 6t - 3, & 1 \le t < 2, \\ (3-t)^2, & 2 \le t < 3, \end{cases}$$

$$N_{33}(t) = \frac{1}{2}\begin{cases} (t-1)^2, & 1 \le t < 2, \\ -3t^2 + 14t - 15, & 2 \le t < 3, \end{cases}$$

$$N_{43}(t) = (t-2)^2, \qquad 2 \le t < 3,$$

so the three spline segments are

$$\mathbf{P}_1(t) = (1-t)^2\mathbf{P}_0 + \tfrac{1}{2}t(4-3t)\mathbf{P}_1 + \tfrac{1}{2}t^2\mathbf{P}_2, \qquad 0 \le t < 1,$$
$$\mathbf{P}_2(t) = \tfrac{1}{2}(2-t)^2\mathbf{P}_1 + \tfrac{1}{2}[t(2-t)+(t-1)(3-t)]\mathbf{P}_2 + \tfrac{1}{2}(t-1)^2\mathbf{P}_3, \qquad 1 \le t < 2,$$
$$\mathbf{P}_3(t) = \tfrac{1}{2}(3-t)^2\mathbf{P}_2 + \tfrac{1}{2}(3-t)(3t-5)\mathbf{P}_3 + (t-2)^2\mathbf{P}_4, \qquad 2 \le t < 3.$$

It is now easy to calculate where each segment starts and ends:

$$\mathbf{P}_1(0) = \mathbf{P}_0, \quad \mathbf{P}_1(1) = (\mathbf{P}_1 + \mathbf{P}_2)/2,$$
$$\mathbf{P}_2(1) = (\mathbf{P}_1 + \mathbf{P}_2)/2, \quad \mathbf{P}_2(2) = (\mathbf{P}_2 + \mathbf{P}_3)/2,$$
$$\mathbf{P}_3(2) = (\mathbf{P}_2 + \mathbf{P}_3)/2, \quad \mathbf{P}_3(3) = \mathbf{P}_4,$$

Figure 4.83 shows a typical example of the three segments (with artificial gaps between them).

▸ **Exercise 4.100:** Show that the three spline segments provide C^1 continuity at the two interior points $\mathbf{P}_1(1) = \mathbf{P}_2(1)$ and $\mathbf{P}_2(2) = \mathbf{P}_3(2)$.

Example 2: Again, five control points, but this time we select $k = n + 1 = 5$. The curve will thus consist of degree-4 polynomial segments. Since such a segment requires five points (it has five coefficients, so five equations are needed), we will end up with just one segment. Equation (4.176) is again used to construct the knot vector $(0,0,0,0,0,1,1,1,1,1)$. The parameter t varies from $t_{k-1} = t_4 = 0$ to $t_{n+1} = t_5 = 1$, showing again that the curve will consist of one segment. Since $k = n + 1$, this should be a Bézier curve.

The calculation of the blending functions $N_{i5}(t)$ (where $0 \le i \le n$) is shown below in detail. We start with the nine functions $N_{i1}(t)$ that are calculated from Equation (4.173):

$N_{01} = 1$ when $t_0 \le t < t_1$, $N_{11} = 1$ when $t_1 \le t < t_2, \ldots, N_{81} = 1$ when $t_8 \le t < t_9$.

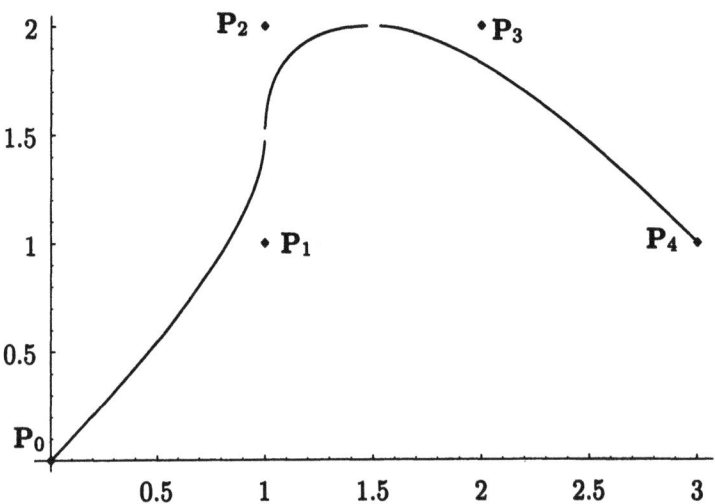

Figure 4.83: An Open Uniform B-Spline.

Since $t_0 = t_1 = t_2 = t_3 = t_4 = 0$ and $t_5 = t_6 = t_7 = t_8 = t_9 = 1$, we conclude that

$$N_{41} = 1 \quad \text{when} \quad t \in [t_4, t_5) = [0, 1),$$

and the other eight functions $N_{i1}(t)$ are zero. The next step is to calculate the eight functions $N_{i2}(t)$ from Equation (4.174):

$$N_{02}(t) = \frac{t - t_0}{t_1 - t_0} N_{01} + \frac{t_2 - t}{t_2 - t_1} N_{11} = 0,$$

$$N_{12}(t) = \frac{t - t_1}{t_2 - t_1} N_{11} + \frac{t_3 - t}{t_3 - t_2} N_{21} = 0,$$

$$N_{22}(t) = \frac{t - t_2}{t_3 - t_2} N_{21} + \frac{t_4 - t}{t_4 - t_3} N_{31} = 0,$$

$$N_{32}(t) = \frac{t - t_3}{t_4 - t_3} N_{31} + \frac{t_5 - t}{t_5 - t_4} N_{41} = 0 + (1 - t),$$

$$N_{42}(t) = \frac{t - t_4}{t_5 - t_4} N_{41} + \frac{t_6 - t}{t_6 - t_5} N_{51} = t + 0,$$

$$N_{52}(t) = \frac{t - t_5}{t_6 - t_5} N_{51} + \frac{t_7 - t}{t_7 - t_6} N_{61} = 0,$$

$$N_{62}(t) = \frac{t - t_6}{t_7 - t_6} N_{61} + \frac{t_8 - t}{t_8 - t_7} N_{71} = 0,$$

$$N_{72}(t) = \frac{t - t_7}{t_8 - t_7} N_{71} + \frac{t_9 - t}{t_9 - t_8} N_{81} = 0.$$

Only $N_{32}(t)$ and $N_{42}(t)$ are nonzero. The seven functions $N_{i3}(t)$ are calculated

4 Curves

similarly:

$$N_{03}(t) = \frac{t-t_0}{t_2-t_0}N_{02} + \frac{t_3-t}{t_3-t_1}N_{12} = 0,$$

$$N_{13}(t) = \frac{t-t_1}{t_3-t_1}N_{12} + \frac{t_4-t}{t_4-t_2}N_{22} = 0,$$

$$N_{23}(t) = \frac{t-t_2}{t_4-t_2}N_{22} + \frac{t_5-t}{t_5-t_3}N_{32} = 0 + (1-t)^2,$$

$$N_{33}(t) = \frac{t-t_3}{t_5-t_3}N_{32} + \frac{t_6-t}{t_6-t_4}N_{42} = t(1-t) + (1-t)t,$$

$$N_{43}(t) = \frac{t-t_4}{t_6-t_4}N_{42} + \frac{t_7-t}{t_7-t_5}N_{52} = t^2 + 0,$$

$$N_{53}(t) = \frac{t-t_5}{t_7-t_5}N_{52} + \frac{t_8-t}{t_8-t_6}N_{62} = 0,$$

$$N_{63}(t) = \frac{t-t_6}{t_8-t_6}N_{62} + \frac{t_9-t}{t_9-t_7}N_{72} = 0.$$

Three of the seven functions are nonzero. The six functions $N_{i4}(t)$ are

$$N_{04}(t) = \frac{t-t_0}{t_3-t_0}N_{03} + \frac{t_4-t}{t_4-t_1}N_{13} = 0,$$

$$N_{14}(t) = \frac{t-t_1}{t_4-t_1}N_{13} + \frac{t_5-t}{t_5-t_2}N_{23} = 0 + (1-t)^3,$$

$$N_{24}(t) = \frac{t-t_2}{t_5-t_2}N_{23} + \frac{t_6-t}{t_6-t_3}N_{33} = t(1-t)^2 + 2t(1-t)^2,$$

$$N_{34}(t) = \frac{t-t_3}{t_6-t_3}N_{33} + \frac{t_7-t}{t_7-t_4}N_{43} = 2t^2(1-t) + (1-t)t^2,$$

$$N_{44}(t) = \frac{t-t_4}{t_7-t_4}N_{43} + \frac{t_8-t}{t_8-t_5}N_{53} = t^3,$$

$$N_{54}(t) = \frac{t-t_5}{t_8-t_5}N_{53} + \frac{t_9-t}{t_9-t_6}N_{63} = 0.$$

Four of them are nonzero. The last step is the calculation of the five functions $N_{i5}(t)$:

$$N_{05}(t) = \frac{t-t_0}{t_4-t_0}N_{04} + \frac{t_5-t}{t_5-t_1}N_{14} = (1-t)^4,$$

$$N_{15}(t) = \frac{t-t_1}{t_5-t_1}N_{14} + \frac{t_6-t}{t_6-t_2}N_{24} = t(1-t)^3 + 3t(1-t)^3,$$

$$N_{25}(t) = \frac{t-t_2}{t_6-t_2}N_{24} + \frac{t_7-t}{t_7-t_3}N_{34} = 3t^2(1-t)^2 + 3t^2(1-t)^2,$$

$$N_{35}(t) = \frac{t-t_3}{t_7-t_3}N_{34} + \frac{t_8-t}{t_8-t_4}N_{44} = 3t^3(1-t) + (1-t)t^3,$$

$$N_{45}(t) = \frac{t-t_4}{t_8-t_4}N_{44} + \frac{t_9-t}{t_9-t_5}N_{54} = t^4.$$

All five are nonzero and they should look familiar (they are the Bernstein polynomials for $n = 4$). The curve consists of the single segment

$$\mathbf{P}(t) = \sum_{i=0}^{4} N_{i5}(t) \mathbf{P}_i$$
$$= (1-t)^4 \mathbf{P}_0 + 4t(1-t)^3 \mathbf{P}_1 + 6t^2(1-t)^2 \mathbf{P}_2 + 4t^3(1-t)\mathbf{P}_3 + t^4 \mathbf{P}_4,$$

which is the Bézier curve defined by the five points. The B-spline method is thus more general than the Bézier curve since it contains the latter as a special case.

➧ It is the multiplicity of knot values that causes the open B-spline to start and end at its extreme control points. This is easy to understand when we realize that every subinterval $[t_i, t_{i+1})$ of knots corresponds to one segment $\mathbf{P}_i(t)$ of the B-spline. When $t_i = t_{i+1}$, that segment reduces to a point. The result is that each repeat of a knot value decreases the continuity at a joint point by 1. Consider, for example, the open B-spline of order $k = 4$. The individual spline segments are degree-3 (cubic) polynomials that have C^2 continuity at their joint points. If knot t_i has multiplicity 2 (i.e., $t_i = t_{i+1}$), then segment $\mathbf{P}_i(t)$ reduces to a point and segments $\mathbf{P}_{i-1}(t)$ and $\mathbf{P}_{i+1}(t)$ meet at a joint point with C^1 continuity. If knot t_i has multiplicity 3 ($t_i = t_{i+1} = t_{i+2}$), then segments $\mathbf{P}_i(t)$ and $\mathbf{P}_{i+1}(t)$ reduce to points and segments $\mathbf{P}_{i-1}(t)$ and $\mathbf{P}_{i+2}(t)$ meet at a joint point (which in this case is a control point) with C^0 continuity. If the first knot has multiplicity 4 ($t_0 = t_1 = t_2 = t_3$), then segments $\mathbf{P}_0(t)$, $\mathbf{P}_1(t)$, and $\mathbf{P}_2(t)$ reduce to points and segment $\mathbf{P}_3(t)$ starts at that point with no continuity.

4.16.15 Nonuniform B-Splines

This type of B-spline is more general than the uniform or open B-splines, although it is not the most general. It is obtained when the knot values are not equally spaced. The only requirement is that the knots be nondecreasing. Adjusting the knot values (as well as having multiple values) is a feature that helps fine-tune the shape of the curve. Multiple knots can be used to pull the curve in a certain direction and to create a cusp or even a discontinuity at a join point. Since nonuniform B-splines can get complex, we limit the discussion in this section to order-4 (i.e., degree-3) nonuniform B-splines. This is not a serious limitation, as this type is the most commonly used and it makes it easier to understand the properties and behavior of the nonuniform B-spline.

In the case of order-4 nonuniform B-splines, the knot vector contains values from t_0 to t_{n+4} (there are four more knots than control points), so the minimum number of knots is eight (since the minimum number of control points is four) and the parameter t varies, in this case, from $t_{k-1} = t_3$ to $t_{n+1} = t_4$. Spline segment $\mathbf{P}_i(t)$ depends on control points $\mathbf{P}_{i-3}$, $\mathbf{P}_{i-2}$, $\mathbf{P}_{i-1}$, and $\mathbf{P}_i$ and its expression is

$$\mathbf{P}_i(t) = N_{i-3,4}(t)\mathbf{P}_{i-3} + N_{i-2,4}(t)\mathbf{P}_{i-2} + N_{i-1,4}(t)\mathbf{P}_{i-1} + N_{i,4}(t)\mathbf{P}_i,$$

where $3 \leq i \leq n$ and $t_i \leq t \leq t_{i+1}$. There are $n-2$ segments denoted $\mathbf{P}_3(t)$ through $\mathbf{P}_n(t)$. When $n = 3$ (four control points), the curve consists of just one segment.

4 Curves

When knot t_i has multiplicity 2 (i.e., $t_i = t_{i+1}$), segment $\mathbf{P}_i(t)$ reduces to a point. As has been mentioned earlier, it is this feature that makes the nonuniform B-spline so flexible, powerful, and, thus, practical.

The weight functions are defined recursively by Equations (4.173) and (4.174) but go up to N_{i4} only:

$$N_{i1}(t) = \begin{cases} 1, & \text{if } t \in [t_i, t_{i+1}), \\ 0, & \text{otherwise,} \end{cases}$$

$$N_{i2}(t) = \frac{t - t_i}{t_{i+1} - t_i} N_{i,1}(t) + \frac{t_{i+2} - t}{t_{i+2} - t_{i+1}} N_{i+1,1}(t),$$

$$N_{i3}(t) = \frac{t - t_i}{t_{i+2} - t_i} N_{i,2}(t) + \frac{t_{i+3} - t}{t_{i+3} - t_{i+1}} N_{i+1,2}(t), \qquad (4.177)$$

$$N_{i4}(t) = \frac{t - t_i}{t_{i+3} - t_i} N_{i,3}(t) + \frac{t_{i+4} - t}{t_{i+4} - t_{i+1}} N_{i+1,3}(t).$$

The first set, $N_{i1}(t)$, are horizontal segments. The second set, $N_{i2}(t)$, are straight lines. The third set are quadratic polynomials and the fourth set, $N_{i4}(t)$, are cubic polynomials. Each cubic segment is defined by four control points and lies in the convex hull defined by the points. Thus, segment $\mathbf{P}_i(t)$ is defined by points $\mathbf{P}_{i-3}$, $\mathbf{P}_{i-2}$, $\mathbf{P}_{i-1}$, and $\mathbf{P}_i$, while segment $\mathbf{P}_{i+1}(t)$ is defined by points $\mathbf{P}_{i-2}$, $\mathbf{P}_{i-1}$, $\mathbf{P}_i$, and $\mathbf{P}_{i+1}$.

Figure 4.84 illustrates the effect of knot multiplicities using $n = 7$ (i.e., eight points) as an example. The knot vector should contain $n + k + 1 = 7 + 4 + 1 = 12$ values and t should vary from $t_{k-1} = t_3$ to $t_{n+1} = t_8$, a total of five subintervals. The four parts of the figure show cubic B-spline curves constructed with the knot vectors

$$(-3, -2, -1, 0, 1, 2, 3, 4, 5, 6, 7, 8), \quad (-3, -2, -1, 0, 1, 1, 2, 3, 4, 5, 6, 7),$$
$$(-3, -2, -1, 0, 1, 1, 1, 2, 3, 4, 5, 6), \quad (-3, -2, -1, 0, 1, 1, 1, 1, 2, 3, 4, 5),$$

respectively. Notice that only six knots, t_3 through t_8, are really important. The rest are distinct and uniform but less important since only some of them are used in calculating the blending functions.

In Figure 4.84a, all knots have multiplicity 1, each segment is defined by four points, and adjacent segments share three points. The first segment, $\mathbf{P}_3(t)$, is defined by points $\mathbf{P}_0$, $\mathbf{P}_1$, $\mathbf{P}_2$, and $\mathbf{P}_3$, while the last segment, $\mathbf{P}_7(t)$, is defined by points $\mathbf{P}_4$, $\mathbf{P}_5$, $\mathbf{P}_6$, and $\mathbf{P}_7$. The five segments join with C^2 continuity. In Figure 4.84b, we set $t_4 = t_5$, thereby reducing segment $\mathbf{P}_4(t)$ to zero length, causing segments $\mathbf{P}_3(t)$ and $\mathbf{P}_5(t)$ to meet at join $t_4 = t_5$. However, these segments share just two control points, $\mathbf{P}_2$ and $\mathbf{P}_3$, so they have less "in common" and, consequently, join with only C^1 continuity. In Figure 4.84c, we set $t_4 = t_5 = t_6$, thus reducing segments $\mathbf{P}_4(t)$ and $\mathbf{P}_5(t)$ to zero length and causing segments $\mathbf{P}_3(t)$ and $\mathbf{P}_6(t)$ to meet. These segments share just one control point, namely $\mathbf{P}_3$, so they meet at this point, with C^0 continuity. In Figure 4.84d, we set $t_4 = t_5 = t_6 = t_7$, so now we have three zero-length segments, namely $\mathbf{P}_4(t)$, $\mathbf{P}_5(t)$, and $\mathbf{P}_6(t)$. Segments

$\mathbf{P}_3(t)$ and $\mathbf{P}_7(t)$ now have to meet, but they don't have any common control points. The result is a *discontinuity* (a break) in the curve between points $\mathbf{P}_3$ and $\mathbf{P}_4$.

Example: This long example is divided into two parts.

Part a. In this part, we calculate the blending functions and spline segments of the curve of Figure 4.84a, where the knot vector is the uniform sequence

$$(-3, -2, -1, 0, 1, 2, 3, 4, 5, 6, 7, 8).$$

The calculations are done bearing in mind that t varies from $t_3 = 0$ to $t_8 = 5$. We need to calculate all the functions $N_{i4}(t)$ that are nonzero in the five subintervals $[0, 1)$, $[1, 2)$, $[2, 3)$, $[3, 4)$, and $[4, 5)$. Four blending functions are used to construct each of the five spline segments, so segment $\mathbf{P}_3(t)$ is defined by functions $N_{04}(t)$ through $N_{34}(t)$, segment $\mathbf{P}_4(t)$ is defined by functions $N_{14}(t)$ through $N_{44}(t)$, and segment $\mathbf{P}_7(t)$ is defined by functions $N_{44}(t)$ through $N_{74}(t)$. The first step is to calculate N_{i1}:

$$N_{31} = 1 \text{ for } t \in [0, 1), \quad N_{41} = 1 \text{ for } t \in [1, 2),$$
$$N_{51} = 1 \text{ for } t \in [2, 3), \quad N_{61} = 1 \text{ for } t \in [3, 4), \quad N_{71} = 1 \text{ for } t \in [4, 5),$$

and $N_{01}, N_{11}, N_{21}, N_{81}, N_{91}, N_{10,1}$, and $N_{11,1}$ are zero in the range $0 \le t < 5$.

Step 2 is to calculate functions N_{i2} that are nonzero for $0 \le t < 5$:

$$N_{02}(t) = \frac{t - t_0}{t_1 - t_0} N_{01} + \frac{t_2 - t}{t_2 - t_1} N_{11} = 0,$$

$$N_{12}(t) = \frac{t - t_1}{t_2 - t_1} N_{11} + \frac{t_3 - t}{t_3 - t_2} N_{21} = 0,$$

$$N_{22}(t) = \frac{t - t_2}{t_3 - t_2} N_{21} + \frac{t_4 - t}{t_4 - t_3} N_{31} = (1 - t) \quad \text{for} \quad t \in [0, 1),$$

$$N_{32}(t) = \frac{t - t_3}{t_4 - t_3} N_{31} + \frac{t_5 - t}{t_5 - t_4} N_{41} = \begin{cases} t & \text{for } t \in [0, 1), \\ 2 - t & \text{for } t \in [1, 2), \end{cases}$$

$$N_{42}(t) = \frac{t - t_4}{t_5 - t_4} N_{41} + \frac{t_6 - t}{t_6 - t_5} N_{51} = \begin{cases} t - 1 & \text{for } t \in [1, 2), \\ 3 - t & \text{for } t \in [2, 3), \end{cases}$$

$$N_{52}(t) = \frac{t - t_5}{t_6 - t_5} N_{51} + \frac{t_7 - t}{t_7 - t_6} N_{61} = \begin{cases} t - 2 & \text{for } t \in [2, 3), \\ 4 - t & \text{for } t \in [3, 4), \end{cases}$$

$$N_{62}(t) = \frac{t - t_6}{t_7 - t_6} N_{61} + \frac{t_8 - t}{t_8 - t_7} N_{71} = \begin{cases} t - 3 & \text{for } t \in [3, 4), \\ 5 - t & \text{for } t \in [4, 5), \end{cases}$$

$$N_{72}(t) = \frac{t - t_7}{t_8 - t_7} N_{71} + \frac{t_9 - t}{t_9 - t_8} N_{81} = t - 3 \quad \text{for } t \in [4, 5).$$

This step terminates at $N_{72}(t)$ since $N_{82}(t)$ and its successors are zero for $0 \le t < 5$.

Step 3 requires the calculation of several functions N_{i3}:

$$N_{03}(t) = \frac{t - t_0}{t_2 - t_0} N_{02} + \frac{t_3 - t}{t_3 - t_1} N_{12} = 0,$$

4 Curves

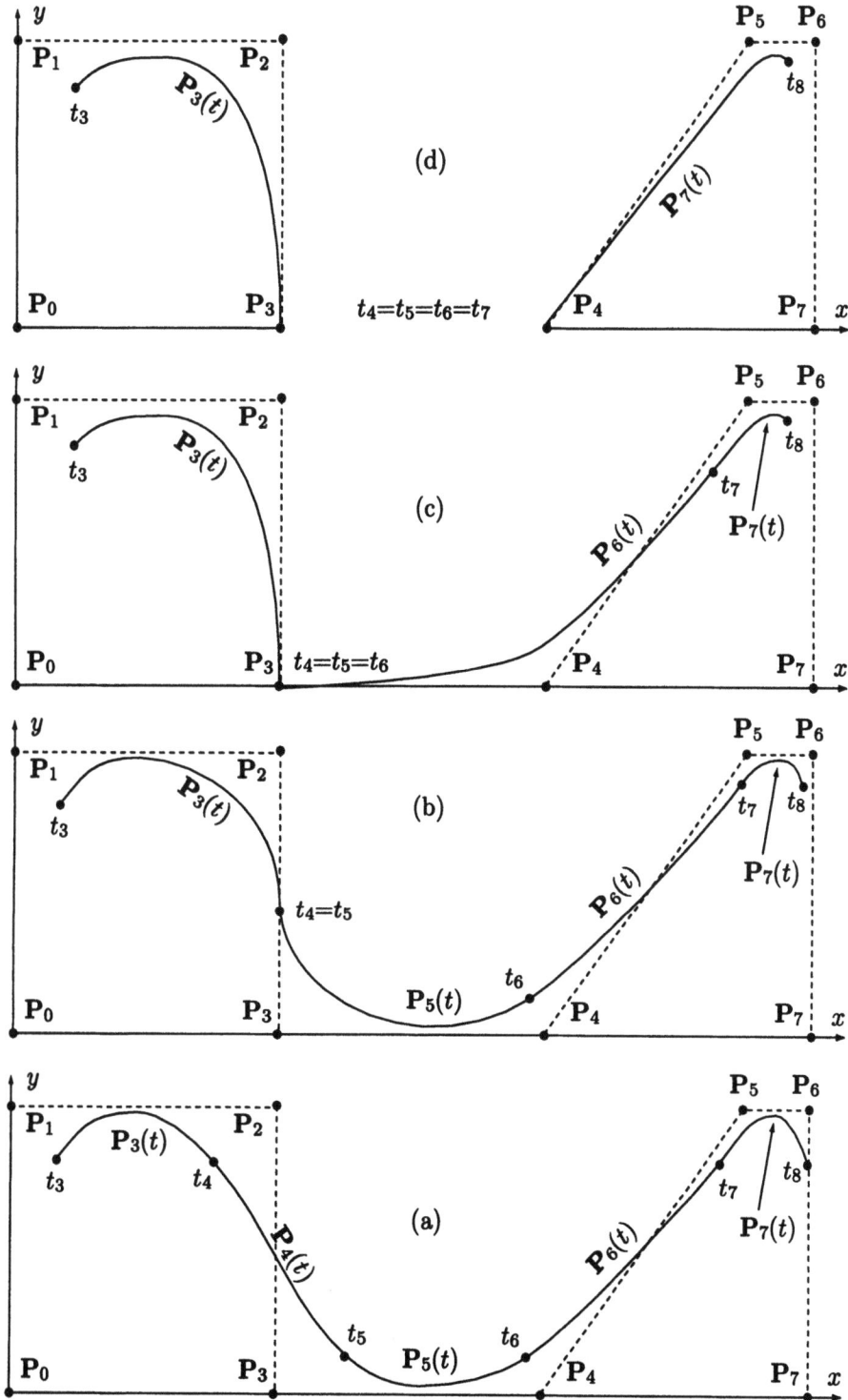

Figure 4.84: An Eight-Point Nonuniform B-Spline Curve with Multiple Knots.

4.16 The B-Spline

$$N_{13}(t) = \frac{t-t_1}{t_3-t_1}N_{12} + \frac{t_4-t}{t_4-t_2}N_{22} = \frac{1}{2}(1-t)^2 \qquad \text{for } t \in [0,1),$$

$$N_{23}(t) = \frac{t-t_2}{t_4-t_2}N_{22} + \frac{t_5-t}{t_5-t_3}N_{32} = \frac{1}{2}\begin{cases}(-2t^2+2t+1) & \text{for } t \in [0,1), \\ (2-t)^2 & \text{for } t \in [1,2),\end{cases}$$

$$N_{33}(t) = \frac{t-t_3}{t_5-t_3}N_{32} + \frac{t_6-t}{t_6-t_4}N_{42} = \frac{1}{2}\begin{cases}t^2 & \text{for } t \in [0,1), \\ (-2t^2+6t-3) & \text{for } t \in [1,2), \\ (3-t)^2 & \text{for } t \in [2,3),\end{cases}$$

$$N_{43}(t) = \frac{t-t_4}{t_6-t_4}N_{42} + \frac{t_7-t}{t_7-t_5}N_{52} = \frac{1}{2}\begin{cases}(t-1)^2 & \text{for } t \in [1,2), \\ (-2t^2+10t-11) & \text{for } t \in [2,3), \\ (4-t)^2 & \text{for } t \in [3,4),\end{cases}$$

$$N_{53}(t) = \frac{t-t_5}{t_7-t_5}N_{52} + \frac{t_8-t}{t_8-t_6}N_{62} = \frac{1}{2}\begin{cases}(t-2)^2 & \text{for } t \in [2,3), \\ (-2t^2+14t-23) & \text{for } t \in [3,4), \\ (5-t)^2 & \text{for } t \in [4,5),\end{cases}$$

$$N_{63}(t) = \frac{t-t_6}{t_8-t_6}N_{62} + \frac{t_9-t}{t_9-t_7}N_{72} = \frac{1}{2}\begin{cases}(t-3)^2 & \text{for } t \in [3,4), \\ (-2t^2+18t-39) & \text{for } t \in [4,5),\end{cases}$$

$$N_{73}(t) = \frac{t-t_7}{t_9-t_7}N_{72} + \frac{t_{10}-t}{t_{10}-t_8}N_{82} = \frac{1}{2}(t-4)^2 \qquad \text{for } t \in [4,5).$$

We stop at N_{73} since N_{83} and its successors are zero for $0 \le t < 5$.

The last step involves the calculation of eight functions N_{i4}:

$$N_{04}(t) = \frac{t-t_0}{t_3-t_0}N_{03} + \frac{t_4-t}{t_4-t_1}N_{13} = \frac{1}{6}(1-t)^3 \qquad \text{for } t \in [0,1),$$

$$N_{14}(t) = \frac{t-t_1}{t_4-t_1}N_{13} + \frac{t_5-t}{t_5-t_2}N_{23} = \frac{1}{6}\begin{cases}(3t^3-6t^2+4) & \text{for } t \in [0,1), \\ (2-t)^3 & \text{for } t \in [1,2),\end{cases}$$

$$N_{24}(t) = \frac{t-t_2}{t_5-t_2}N_{23} + \frac{t_6-t}{t_6-t_3}N_{33} = \frac{1}{6}\begin{cases}(-3t^3+3t^2+3t+1) & \text{for } t \in [0,1), \\ (3t^3-15t^2+21t-5) & \text{for } t \in [1,2), \\ (3-t)^3 & \text{for } t \in [2,3),\end{cases}$$

$$N_{34}(t) = \frac{t-t_3}{t_6-t_3}N_{33} + \frac{t_7-t}{t_7-t_4}N_{43} = \frac{1}{6}\begin{cases}t^3 & \text{for } t \in [0,1), \\ (-3t^3+12t^2-12t+4) & \text{for } t \in [1,2), \\ (3t^3-24t^2+60t-44) & \text{for } t \in [2,3), \\ (4-t)^3 & \text{for } t \in [3,4),\end{cases}$$

$$N_{44}(t) = \frac{t-t_4}{t_7-t_4}N_{43} + \frac{t_8-t}{t_8-t_5}N_{53} = \frac{1}{6}\begin{cases}(t-1)^3 & \text{for } t \in [1,2), \\ (-3t^3+21t^2-45t+31) & \text{for } t \in [2,3), \\ (3t^3-33t^2+117t-131) & \text{for } t \in [3,4), \\ (5-t)^3 & \text{for } t \in [4,5),\end{cases}$$

$$N_{54}(t) = \frac{t-t_5}{t_8-t_5}N_{53} + \frac{t_9-t}{t_9-t_6}N_{63} = \frac{1}{6}\begin{cases}(t-2)^3 & \text{for } t \in [2,3), \\ (-3t^3+30t^2-96t+100) & \text{for } t \in [3,4), \\ (3t^3-42t^2+192t-284) & \text{for } t \in [4,5),\end{cases}$$

$$N_{64}(t) = \frac{t-t_6}{t_9-t_6}N_{63} + \frac{t_{10}-t}{t_{10}-t_7}N_{73} = \frac{1}{6}\begin{cases}(t-3)^3 & \text{for } t \in [3,4), \\ (-3t^3+39t^2-165t+229) & \text{for } t \in [4,5),\end{cases}$$

4 Curves

$$N_{74}(t) = \frac{t-t_7}{t_{10}-t_7}N_{73} + \frac{t_{11}-t}{t_{11}-t_8}N_{83} = \frac{1}{6}(t-4)^3 \qquad \text{for } t \in [4,5).$$

A careful study of this last group shows that N_{84} and its successors are zero for $0 \le t < 5$.

The last group of blending functions can now be used to construct the five spline segments:

$$\begin{aligned}
\mathbf{P}_3(t) &= N_{04}(t)\mathbf{P}_0 + N_{14}(t)\mathbf{P}_1 + N_{24}(t)\mathbf{P}_2 + N_{34}(t)\mathbf{P}_3 & t \in [0,1) \\
&= \frac{1}{6}\big[(1-t)^3\mathbf{P}_0 + (3t^3 - 6t^2 + 4)\mathbf{P}_1 \\
&\quad + (-3t^3 + 3t^2 + 3t + 1)\mathbf{P}_2 + t^3\mathbf{P}_3\big], \\
\mathbf{P}_4(t) &= N_{14}(t)\mathbf{P}_1 + N_{24}(t)\mathbf{P}_2 + N_{34}(t)\mathbf{P}_3 + N_{44}(t)\mathbf{P}_4 & t \in [1,2) \\
&= \frac{1}{6}\big[(2-t)^3\mathbf{P}_1 + (3t^3 - 15t^2 + 21t - 5)\mathbf{P}_2 \\
&\quad + (-3t^3 + 12t^2 - 12t + 4)\mathbf{P}_3 + (t-1)^3\mathbf{P}_4\big], \\
\mathbf{P}_5(t) &= N_{24}(t)\mathbf{P}_2 + N_{34}(t)\mathbf{P}_3 + N_{44}(t)\mathbf{P}_4 + N_{54}(t)\mathbf{P}_5 & t \in [2,3) \\
&= \frac{1}{6}\big[(3-t)^3\mathbf{P}_2 + (3t^3 - 24t^2 + 60t - 44)\mathbf{P}_3 \\
&\quad + (-3t^3 + 21t^2 - 45t + 31)\mathbf{P}_4 + (t-2)^3\mathbf{P}_5\big], \\
\mathbf{P}_6(t) &= N_{34}(t)\mathbf{P}_3 + N_{44}(t)\mathbf{P}_4 + N_{54}(t)\mathbf{P}_5 + N_{64}(t)\mathbf{P}_6 & t \in [3,4) \\
&= \frac{1}{6}\big[(4-t)^3\mathbf{P}_3 + (3t^3 - 33t^2 + 117t - 131)\mathbf{P}_4 \\
&\quad + (-3t^3 + 30t^2 - 96t + 100)\mathbf{P}_5 + (t-3)^3\mathbf{P}_6\big], \\
\mathbf{P}_7(t) &= N_{44}(t)\mathbf{P}_4 + N_{54}(t)\mathbf{P}_5 + N_{64}(t)\mathbf{P}_6 + N_{74}(t)\mathbf{P}_7 & t \in [4,5) \\
&= \frac{1}{6}\big[(5-t)^3\mathbf{P}_4 + (3t^3 - 42t^2 + 192t - 284)\mathbf{P}_5 \\
&\quad + (-3t^3 + 39t^2 - 165t + 229)\mathbf{P}_6 + (t-4)^3\mathbf{P}_7\big].
\end{aligned}$$

A direct check verifies that each segment has barycentric weights. The entire curve starts at $\mathbf{P}_3(0) = (\mathbf{P}_0 + 4\mathbf{P}_1 + \mathbf{P}_2)/6$ and ends at $\mathbf{P}_7(5) = (\mathbf{P}_5 + 4\mathbf{P}_6 + \mathbf{P}_7)/6$. The four joint points between the segments are

$$\mathbf{P}_3(1) = \mathbf{P}_4(1) = (\mathbf{P}_1 + 4\mathbf{P}_2 + \mathbf{P}_3)/6, \quad \mathbf{P}_4(2) = \mathbf{P}_5(2) = (\mathbf{P}_2 + 4\mathbf{P}_3 + \mathbf{P}_4)/6,$$
$$\mathbf{P}_5(3) = \mathbf{P}_6(3) = (\mathbf{P}_3 + 4\mathbf{P}_4 + \mathbf{P}_5)/6, \quad \mathbf{P}_6(4) = \mathbf{P}_7(4) = (\mathbf{P}_4 + 4\mathbf{P}_5 + \mathbf{P}_6)/6.$$

The coordinates of the control points of Figure 4.84a are $\mathbf{P}_0 = (0,0)$, $\mathbf{P}_1 = (0,1)$, $\mathbf{P}_2 = (1,1)$, $\mathbf{P}_3 = (1,0)$, $\mathbf{P}_4 = (2,0)$, $\mathbf{P}_5 = (2.75, 1)$, $\mathbf{P}_6 = (3,1)$, and $\mathbf{P}_7 = (3,0)$. The curve therefore starts at $(1/6, 5/6)$, ends at $(2.96, 5/6)$, and passes through the joins $(5/6, 5/6)$, $(7/6, 1/6)$, $(1.96, 1/6)$, and $(2.67, 5/6)$.

Part b: To continue the example, we now calculate the blending functions and spline segments of the curve of Figure 4.84b where the knot vector is the nonuniform $(-3, -2, -1, 0, 1, 1, 2, 3, 4, 5, 6, 7)$. Notice that we now have $t_4 = t_5 = 1$, resulting in different blending functions and different spline segments.

It is important to realize that t varies in this case from $t_3 = 0$ to $t_8 = 4$. The five intervals of t for the five spline segments are $[0, 1)$, $[1, 1)$, $[1, 2)$, $[2, 3)$, and $[3, 4)$. The second segment $\mathbf{P}_4(t)$ has now been reduced to a single point.

The first step is to calculate N_{i1}:

$$N_{31} = 1 \text{ for } t \in [0, 1), N_{41} = 1 \text{ for } t \in [1, 1),$$
$$N_{51} = 1 \text{ for } t \in [1, 2), N_{61} = 1 \text{ for } t \in [2, 3), N_{71} = 1 \text{ for } t \in [3, 4),$$

and N_{01}, N_{11}, N_{21}, N_{81}, N_{91}, $N_{10,1}$, and $N_{11,1}$ are zero in the range $0 \le t < 4$.

Step 2 is to calculate functions N_{i2} that are nonzero for $0 \le t < 4$:

$$N_{02}(t) = \frac{t - t_0}{t_1 - t_0} N_{01} + \frac{t_2 - t}{t_2 - t_1} N_{11} = 0,$$

$$N_{12}(t) = \frac{t - t_1}{t_2 - t_1} N_{11} + \frac{t_3 - t}{t_3 - t_2} N_{21} = 0,$$

$$N_{22}(t) = \frac{t - t_2}{t_3 - t_2} N_{21} + \frac{t_4 - t}{t_4 - t_3} N_{31} = (1 - t) \quad \text{for } t \in [0, 1),$$

$$N_{32}(t) = \frac{t - t_3}{t_4 - t_3} N_{31} + \frac{t_5 - t}{t_5 - t_4} N_{41} = t \quad \text{for } t \in [0, 1),$$

$$N_{42}(t) = \frac{t - t_4}{t_5 - t_4} N_{41} + \frac{t_6 - t}{t_6 - t_5} N_{51} = 2 - t \quad \text{for } t \in [1, 2),$$

$$N_{52}(t) = \frac{t - t_5}{t_6 - t_5} N_{51} + \frac{t_7 - t}{t_7 - t_6} N_{61} = \begin{cases} t - 1 & \text{for } t \in [1, 2), \\ 3 - t & \text{for } t \in [2, 3), \end{cases}$$

$$N_{62}(t) = \frac{t - t_6}{t_7 - t_6} N_{61} + \frac{t_8 - t}{t_8 - t_7} N_{71} = \begin{cases} t - 2 & \text{for } t \in [2, 3), \\ 4 - t & \text{for } t \in [3, 4), \end{cases}$$

$$N_{72}(t) = \frac{t - t_7}{t_8 - t_7} N_{71} + \frac{t_9 - t}{t_9 - t_8} N_{81} = t - 3 \quad \text{for } t \in [3, 4).$$

This step terminates at $N_{72}(t)$ since $N_{82}(t)$ and its successors are zero for $0 \le t < 4$.

Step 3 requires the calculation of several functions N_{i3}:

$$N_{03}(t) = \frac{t - t_0}{t_2 - t_0} N_{02} + \frac{t_3 - t}{t_3 - t_1} N_{12} = 0,$$

$$N_{13}(t) = \frac{t - t_1}{t_3 - t_1} N_{12} + \frac{t_4 - t}{t_4 - t_2} N_{22} = \frac{1}{2}(1 - t)^2 \quad \text{for } t \in [0, 1),$$

$$N_{23}(t) = \frac{t - t_2}{t_4 - t_2} N_{22} + \frac{t_5 - t}{t_5 - t_3} N_{32} = \frac{1}{2}(-3t^2 + 2t + 1) \quad \text{for } t \in [0, 1),$$

$$N_{33}(t) = \frac{t - t_3}{t_5 - t_3} N_{32} + \frac{t_6 - t}{t_6 - t_4} N_{42} = \begin{cases} t^2 & \text{for } t \in [0, 1), \\ (2 - t)^2 & \text{for } t \in [1, 2), \end{cases}$$

$$N_{43}(t) = \frac{t - t_4}{t_6 - t_4} N_{42} + \frac{t_7 - t}{t_7 - t_5} N_{52} = \frac{1}{2} \begin{cases} (-3t^2 + 10t - 7) & \text{for } t \in [1, 2), \\ (3 - t)^2 & \text{for } t \in [2, 3), \end{cases}$$

$$N_{53}(t) = \frac{t - t_5}{t_7 - t_5} N_{52} + \frac{t_8 - t}{t_8 - t_6} N_{62} = \frac{1}{2} \begin{cases} (t - 1)^2 & \text{for } t \in [1, 2), \\ (-2t^2 + 10t - 11) & \text{for } t \in [2, 3), \\ (4 - t)^2 & \text{for } t \in [3, 4), \end{cases}$$

4 Curves

$$N_{63}(t) = \frac{t-t_6}{t_8-t_6}N_{62} + \frac{t_9-t}{t_9-t_7}N_{72} = \frac{1}{2}\begin{cases}(t-2)^2 & \text{for } t \in [2,3),\\ (-2t^2+14t-23) & \text{for } t \in [3,4),\end{cases}$$

$$N_{73}(t) = \frac{t-t_7}{t_9-t_7}N_{72} + \frac{t_{10}-t}{t_{10}-t_8}N_{82} = \frac{1}{2}(t-3)^2 \qquad \text{for } t \in [3,4).$$

Here we stop at N_{73} since N_{83} and its successors are zero for $0 \leq t < 4$.

The last step involves the calculation of eight functions N_{i4}:

$$N_{04}(t) = \frac{t-t_0}{t_3-t_0}N_{03} + \frac{t_4-t}{t_4-t_1}N_{13} = \frac{1}{6}(1-t)^3 \qquad \text{for } t \in [0,1),$$

$$N_{14}(t) = \frac{t-t_1}{t_4-t_1}N_{13} + \frac{t_5-t}{t_5-t_2}N_{23} = \frac{1}{12}(11t^3 - 15t^2 - 3t + 7) \qquad \text{for } t \in [0,1),$$

$$N_{24}(t) = \frac{t-t_2}{t_5-t_2}N_{23} + \frac{t_6-t}{t_6-t_3}N_{33} = \begin{cases}\frac{1}{4}(-5t^3+3t^2+3t+1) & \text{for } t \in [0,1),\\ \frac{1}{2}(2-t)^3 & \text{for } t \in [1,2),\end{cases}$$

$$N_{34}(t) = \frac{t-t_3}{t_6-t_3}N_{33} + \frac{t_7-t}{t_7-t_4}N_{43} = \begin{cases}\frac{1}{2}t^3 & \text{for } t \in [0,1),\\ \frac{1}{4}(5t^3-27t^2+45t-21) & \text{for } t \in [1,2),\\ \frac{1}{4}(3-t)^3 & \text{for } t \in [2,3),\end{cases}$$

$$N_{44}(t) = \frac{t-t_4}{t_7-t_4}N_{43} + \frac{t_8-t}{t_8-t_5}N_{53} = \begin{cases}\frac{1}{12}(-11t^3+51t^2-69t+29) & \text{for } t \in [1,2),\\ \frac{1}{12}(7t^3-57t^2+147t-115) & \text{for } t \in [2,3),\\ \frac{1}{6}(4-t)^3 & \text{for } t \in [3,4),\end{cases}$$

$$N_{54}(t) = \frac{t-t_5}{t_8-t_5}N_{53} + \frac{t_9-t}{t_9-t_6}N_{63} = \frac{1}{6}\begin{cases}(t-1)^3 & \text{for } t \in [1,2),\\ (-3t^3+21t^2-45t+31) & \text{for } t \in [2,3),\\ (3t^3-33t^2+117t-131) & \text{for } t \in [3,4),\end{cases}$$

$$N_{64}(t) = \frac{t-t_6}{t_9-t_6}N_{63} + \frac{t_{10}-t}{t_{10}-t_7}N_{73} = \frac{1}{6}\begin{cases}(t-2)^3 & \text{for } t \in [2,3),\\ (-3t^3+30t^2-96t+100) & \text{for } t \in [3,4),\end{cases}$$

$$N_{74}(t) = \frac{t-t_7}{t_{10}-t_7}N_{73} + \frac{t_{11}-t}{t_{11}-t_8}N_{83} = \frac{1}{6}(t-3)^3 \qquad \text{for } t \in [3,4).$$

This group of blending functions can now be used to construct the five spline segments

$$\mathbf{P}_3(t) = N_{04}(t)\mathbf{P}_0 + N_{14}(t)\mathbf{P}_1 + N_{24}(t)\mathbf{P}_2 + N_{34}(t)\mathbf{P}_3 \qquad t \in [0,1)$$
$$= \frac{1}{6}(1-t)^3\mathbf{P}_0 + \frac{1}{12}(11t^3-15t^2-3t+7)\mathbf{P}_1$$
$$+ \frac{1}{4}(-5t^3+3t^2+3t+1)\mathbf{P}_2 + \frac{1}{2}t^3\mathbf{P}_3,$$

$$\mathbf{P}_4(t) = N_{14}(1)\mathbf{P}_1 + N_{24}(1)\mathbf{P}_2 + N_{34}(1)\mathbf{P}_3 + N_{44}(1)\mathbf{P}_4 \qquad t \in [1,1)$$
$$= 0\mathbf{P}_1 + \frac{1}{2}\mathbf{P}_2 + \frac{1}{2}\mathbf{P}_3 + 0\mathbf{P}_4 = (\mathbf{P}_2+\mathbf{P}_3)/2 \quad \text{(a point)},$$

$$\mathbf{P}_5(t) = N_{24}(t)\mathbf{P}_2 + N_{34}(t)\mathbf{P}_3 + N_{44}(t)\mathbf{P}_4 + N_{54}(t)\mathbf{P}_5 \qquad t \in [1,2)$$
$$= \frac{1}{2}(2-t)^3\mathbf{P}_2 + \frac{1}{4}(5t^3-27t^2+45t-21)\mathbf{P}_3$$
$$+ \frac{1}{12}(-11t^3+51t^2-69t+29)\mathbf{P}_4 + \frac{1}{6}(t-1)^3\mathbf{P}_5,$$

$$\mathbf{P}_6(t) = N_{34}(t)\mathbf{P}_3 + N_{44}(t)\mathbf{P}_4 + N_{54}(t)\mathbf{P}_5 + N_{64}(t)\mathbf{P}_6 \qquad t \in [2,3)$$
$$= \frac{1}{4}(3-t)^3\mathbf{P}_3 + \frac{1}{12}(7t^3 - 57t^2 + 147t - 115)\mathbf{P}_4$$
$$+ \frac{1}{6}(-3t^3 + 21t^2 - 45t + 31)\mathbf{P}_5 + \frac{1}{6}(t-2)^3\mathbf{P}_6,$$
$$\mathbf{P}_7(t) = N_{44}(t)\mathbf{P}_4 + N_{54}(t)\mathbf{P}_5 + N_{64}(t)\mathbf{P}_6 + N_{74}(t)\mathbf{P}_7 \qquad t \in [3,4)$$
$$= \frac{1}{6}\big[(4-t)^3\mathbf{P}_4 + (3t^3 - 33t^2 + 117t - 131)\mathbf{P}_5$$
$$+ (-3t^3 + 30t^2 - 96t + 100)\mathbf{P}_6 + (t-3)^3\mathbf{P}_7\big].$$

A direct check verifies that each segment has barycentric weights. The entire curve starts at $\mathbf{P}_3(0) = (2\mathbf{P}_0 + 7\mathbf{P}_1 + 3\mathbf{P}_2)/12$ and ends at $\mathbf{P}_7(4) = (\mathbf{P}_5 + 4\mathbf{P}_6 + \mathbf{P}_7)/6$. The three joint points between the segments are

$$\mathbf{P}_3(1) = \mathbf{P}_5(1) = (\mathbf{P}_2 + \mathbf{P}_3)/2, \quad \mathbf{P}_5(2) = \mathbf{P}_6(2) = (3\mathbf{P}_3 + 7\mathbf{P}_4 + 2\mathbf{P}_5)/12,$$
$$\mathbf{P}_6(3) = \mathbf{P}_7(3) = (\mathbf{P}_4 + 4\mathbf{P}_5 + \mathbf{P}_6)/6.$$

(End of example.)

▶ **Exercise 4.101:** Calculate the blending functions and spline segments for the curves of Figure 4.84c,d.

This example illustrates the power and flexibility of the nonuniform B-spline. Other curve methods make it possible to control the shape of a curve by moving control points, by subdividing the curve and adding points, and by repeating certain points. The nonuniform B-spline method can use all these operations but can also fine-tune the curve by changing the values of knots and by using multiple knots.

4.16.16 Matrix Form of the Nonuniform B-Spline

The Cox-DeBoor recursive formula, Equations (4.173) and (4.174), is general and can be used to calculate the blending functions of the uniform, open, and nonuniform B-splines. However, it is complex and slow to calculate. Explicit, matrix-based expressions for the B-spline are simpler and faster to use. Such expressions have been derived for the uniform quadratic B-spline in Section 4.16.2 (Equation (4.154)) and for the uniform cubic B-spline in Section 4.16.4 (Equation (4.159)). Similar expressions are derived in this section for the linear, quadratic, and cubic *nonuniform* B-splines. We temporarily use the notation u instead of t for the parameter and u_i instead of t_i for the knots.

For the linear case, where $k = 2$, the Cox-DeBoor formula becomes

$$N_{i2} = \frac{u - u_i}{u_{i+1} - u_i} N_{i1}(u) + \frac{u_{i+2} - u}{u_{i+2} - u_{i+2}} N_{i+1,1}(u)$$
$$= \begin{cases} \dfrac{u - u_i}{u_{i+1} - u_i} & \text{for } u \in [u_i, u_{i+1}), \\ \dfrac{u_{i+2} - u}{u_{i+2} - u_{i+1}} & \text{for } u \in [u_{i+1}, u_{i+2}), \\ 0 & \text{otherwise.} \end{cases} \qquad (4.178)$$

4 Curves

For $i = 0$, this becomes

$$N_{02} = \begin{cases} \dfrac{u - u_0}{u_1 - u_0} & \text{for } u \in [u_0, u_1), \\ \dfrac{u_2 - u}{u_2 - u_1} & \text{for } u \in [u_1, u_2), \\ 0 & \text{otherwise.} \end{cases} \quad (4.179)$$

The other blending function N_{12} is easily obtained from Equation (4.179) by incrementing all the indices.

Blending function N_{02} is zero over the subinterval $[u_2, u_3)$ and blending function N_{12} is zero over $[u_0, u_1)$. It is, therefore, only over the interval $[u_1, u_2)$ that both these functions are nonzero, so the parameter u should vary from u_1 to u_2. Over this interval, we have

$$N_{02}(u) = \frac{u_2 - u}{u_2 - u_1}, \quad N_{12}(u) = \frac{u_3 - u}{u_3 - u_2}. \quad (4.180)$$

To derive the expression for the linear spline, we denote $\Delta = u_2 - u_1$ and define the parameter t by

$$t = \frac{u - u_1}{\Delta} = \frac{u - u_1}{u_2 - u_1}.$$

Notice that $u = u_1 \to t = 0$ and $u = u_2 \to t = 1$. Also, $u - u_1 = t\Delta$ and $u - u_2 = \Delta(t-1)$. Substituting this in Equation (4.180) yields the matrix expression for the linear nonuniform B-spline

$$\mathbf{P}(t) = (t, 1) \begin{pmatrix} -1 & 1 \\ 1 & 0 \end{pmatrix} \begin{pmatrix} \mathbf{P}_0 \\ \mathbf{P}_1 \end{pmatrix}. \quad (4.181)$$

When t varies from 0 to 1, this becomes the straight line from $\mathbf{P}_0$ to $\mathbf{P}_1$. The nonuniform linear B-spline does not depend on Δ, so it is identical to the uniform linear B-spline.

When you get an 8 on the midterm, there ain't a curve in the world that can save you.

— Unknown

Applying the Cox-DeBoor formula to Equation (4.179), we get the first quadratic blending function N_{03}:

$$N_{03}(u) = \begin{cases} \dfrac{u - u_0}{u_2 - u_0} \cdot \dfrac{u - u_0}{u_1 - u_0} & \text{for } u \in [u_0, u_1), \\ \dfrac{u - u_0}{u_2 - u_0} \cdot \dfrac{u_2 - u}{u_2 - u_1} + \dfrac{u_3 - u}{u_3 - u_1} \cdot \dfrac{u - u_1}{u_2 - u_1} & \text{for } u \in [u_1, u_2), \\ \dfrac{u_3 - u}{u_3 - u_1} \cdot \dfrac{u_3 - u}{u_3 - u_2} & \text{for } u \in [u_2, u_3), \\ 0 & \text{otherwise.} \end{cases} \quad (4.182)$$

Functions N_{13} and N_{23} are obtained from Equation (4.182) by incrementing all the indices. When this is done, we observe that each of the three blending functions N_{i3} is zero over different intervals and it is only over subinterval $[u_2, u_3)$ that all three are nonzero, and their values are

$$N_{03}(u) = \frac{u_3 - u}{u_3 - u_1} \cdot \frac{u_3 - u}{u_3 - u_2},$$
$$N_{13}(u) = \frac{u - u_1}{u_3 - u_1} \cdot \frac{u_3 - u}{u_3 - u_2} + \frac{u_4 - u}{u_4 - u_2} \cdot \frac{u - u_2}{u_3 - u_2},$$
$$N_{23}(u) = \frac{u - u_2}{u_4 - u_2} \cdot \frac{u - u_2}{u_3 - u_2}. \qquad (4.183)$$

Since the knot vector is nonuniform, the differences between consecutive knots may be different and we denote them

$$\Delta_1 = u_2 - u_1, \quad \Delta_2 = u_3 - u_2, \quad \Delta_3 = u_4 - u_3.$$

We also define $t = (u - u_2)/\Delta_2$, which implies

$$u - u_1 = t\Delta_2 + \Delta_1,$$
$$u - u_2 = t\Delta_2,$$
$$u - u_3 = (t - 1)\Delta_2,$$
$$u - u_4 = t\Delta_2 - (\Delta_2 + \Delta_3). \qquad (4.184)$$

Equations (4.183) and (4.184) yield the matrix form of the nonuniform quadratic B-spline

$$\mathbf{P}(t) = (t^2, t, 1) \begin{pmatrix} a & -a - b & b \\ -2a & 2a & 0 \\ a & 1 - a & 0 \end{pmatrix} \begin{pmatrix} \mathbf{P}_0 \\ \mathbf{P}_1 \\ \mathbf{P}_2 \end{pmatrix}, \qquad (4.185)$$

where

$$a = \frac{\Delta_2}{\Delta_1 + \Delta_2}, \quad b = \frac{\Delta_2}{\Delta_2 + \Delta_3},$$

and t varies from 0 to 1 (note that $u = u_2 \to t = 0$ and $u = u_3 \to t = 1$).

> B-splines were known to and studied by Nikolai Lobachevsky whose major contribution to mathematics is perhaps the so-called non-Euclidean (hyperbolic) geometry in the late eighteenth century. The modern version described here was developed, in the late 1970s, by C. DeBoor, M. Cox and L. Mansfield. Note that their algorithm is a generalization of de Casteljau's scaffolding method.

The last step is to get the matrix form of the nonuniform cubic B-spline. We apply the Cox-DeBoor formula to Equation (4.182) to get the first of the four

4 Curves

blending functions N_{i4}:

$$N_{04}(u) = \begin{cases} \dfrac{u-u_0}{u_3-u_0} \cdot \dfrac{u-u_0}{u_2-u_0} \cdot \dfrac{u-u_0}{u_1-u_0} & \text{for } u \in [u_0, u_1), \\[6pt] \dfrac{u-u_0}{u_3-u_0} \cdot \dfrac{u-u_0}{u_2-u_0} \cdot \dfrac{u_2-u}{u_2-u_1} \\ + \dfrac{u-u_0}{u_3-u_0} \cdot \dfrac{u_3-u}{u_3-u_1} \cdot \dfrac{u-u_1}{u_2-u_1} \\ + \dfrac{u_4-u}{u_4-u_1} \cdot \dfrac{u-u_1}{u_3-u_1} \cdot \dfrac{u-u_1}{u_2-u_1} & \text{for } u \in [u_1, u_2), \\[6pt] \dfrac{u-u_0}{u_3-u_0} \cdot \dfrac{u_3-u}{u_3-u_1} \cdot \dfrac{u_3-u}{u_3-u_2} \\ + \dfrac{u_4-u}{u_4-u_1} \cdot \dfrac{u-u_1}{u_3-u_1} \cdot \dfrac{u_3-u}{u_3-u_2} \\ + \dfrac{u_4-u}{u_4-u_1} \cdot \dfrac{u_4-u}{u_4-u_2} \cdot \dfrac{u-u_2}{u_3-u_2} & \text{for } u \in [u_2, u_3), \\[6pt] \dfrac{u_4-u}{u_4-u_1} \cdot \dfrac{u_4-u}{u_4-u_2} \cdot \dfrac{u_4-u}{u_4-u_3} & \text{for } u \in [u_3, u_4), \\[6pt] 0 & \text{otherwise.} \end{cases} \quad (4.186)$$

The remaining three blending functions N_{14}, N_{24}, and N_{34} are obtained from Equation (4.186) by incrementing all the indices. When this is done we observe, as before, that each of the four blending functions N_{i4} is zero over different intervals and it is only over subinterval $[u_3, u_4)$ that all four are nonzero. Their values are

$$\begin{aligned} N_{04}(u) &= \frac{u_4-u}{u_4-u_1} \cdot \frac{u_4-u}{u_4-u_2} \cdot \frac{u_4-u}{u_4-u_3}, \\ N_{14}(u) &= \frac{u-u_1}{u_4-u_1} \cdot \frac{u_4-u}{u_4-u_2} \cdot \frac{u_4-u}{u_4-u_3} + \frac{u_5-u}{u_5-u_2} \cdot \frac{u-u_2}{u_4-u_2} \cdot \frac{u_4-u}{u_4-u_3} \\ &\quad + \frac{u_5-u}{u_5-u_2} \cdot \frac{u_5-u}{u_5-u_3} \cdot \frac{u-u_3}{u_4-u_3}, \\ N_{24}(u) &= \frac{u-u_2}{u_5-u_2} \cdot \frac{u-u_2}{u_4-u_2} \cdot \frac{u_4-u}{u_4-u_3} + \frac{u-u_2}{u_5-u_2} \cdot \frac{u_5-u}{u_5-u_3} \cdot \frac{u-u_3}{u_4-u_3} \\ &\quad + \frac{u_6-u}{u_6-u_3} \cdot \frac{u-u_3}{u_5-u_3} \cdot \frac{u-u_3}{u_4-u_3}, \\ N_{34}(u) &= \frac{u-u_3}{u_6-u_3} \cdot \frac{u-u_3}{u_5-u_3} \cdot \frac{u-u_3}{u_4-u_3}. \end{aligned} \quad (4.187)$$

Since the knot vector is nonuniform, the differences between consecutive knots may be different and we denote them by

$$\Delta_1 = u_2 - u_1, \quad \Delta_2 = u_3 - u_2, \quad \Delta_3 = u_4 - u_3,$$
$$\Delta_4 = u_5 - u_4, \quad \Delta_5 = u_6 - u_5, \quad t = (u - u_3)/\Delta_3.$$

This implies

$$u - u_1 = t\Delta_3 + (\Delta_1 + \Delta_2),$$
$$u - u_2 = t\Delta_3 + \Delta_2,$$

$$u - u_3 = t\Delta_3,$$
$$u - u_4 = (t-1)\Delta_3, \quad (4.188)$$
$$u - u_5 = t\Delta_3 - (\Delta_3 + \Delta_4),$$
$$u - u_6 = t\Delta_3 - (\Delta_3 + \Delta_4 + \Delta_5).$$

Equations (4.187) and (4.188) yield the matrix form of the nonuniform cubic B-spline:

$$\mathbf{P}(t) = (t^3, t^2, t, 1) \begin{pmatrix} -a & a+b+c & -b-c-d & d \\ 3a & -3a-3b & 3b & 0 \\ -3a & 3a-3e & 3e & 0 \\ a & 1-a-f & f & 0 \end{pmatrix} \begin{pmatrix} \mathbf{P}_0 \\ \mathbf{P}_1 \\ \mathbf{P}_2 \\ \mathbf{P}_3 \end{pmatrix}, \quad (4.189)$$

where

$$a = \frac{\Delta_3^2}{(\Delta_1+\Delta_2+\Delta_3)(\Delta_2+\Delta_3)}, \qquad d = \frac{\Delta_3^2}{(\Delta_3+\Delta_4+\Delta_5)(\Delta_4+\Delta_5)},$$
$$b = \frac{\Delta_3^2}{(\Delta_2+\Delta_3+\Delta_4)(\Delta_2+\Delta_3)}, \qquad e = \frac{\Delta_2\Delta_3}{(\Delta_2+\Delta_3+\Delta_4)(\Delta_2+\Delta_3)},$$
$$c = \frac{\Delta_3^2}{(\Delta_2+\Delta_3+\Delta_4)(\Delta_3+\Delta_4)}, \qquad f = \frac{\Delta_2^2}{(\Delta_2+\Delta_3+\Delta_4)(\Delta_2+\Delta_3)}.$$

The quantities Δ_i are defined as differences of knot values $u_{i+1} - u_i$ and a good choice for those differences is the chord lengths between points. However, a cubic spline segment requires five Δ_i's, but there are only three chords between the four points defining it. In general, a B-spline curve is defined by $n+1$ points, having n chords between them, but $n+2$ differences Δ_i are required. A standard technique is to select

$$\Delta_1 = \Delta_2 = |\mathbf{P}_1 - \mathbf{P}_0|, \qquad \Delta_{n+1} = \Delta_{n+2} = |\mathbf{P}_n - \mathbf{P}_{n-1}|,$$

and $\Delta_i = |\mathbf{P}_{i-1} - \mathbf{P}_{i-2}|$ for $i = 3, 4, \ldots, n$.

The last topic discussed in this section is the relation between the quadratic uniform and quadratic nonuniform B-splines. Given three control points $\mathbf{Q}_0$, $\mathbf{Q}_1$, and $\mathbf{Q}_2$, the uniform quadratic B-spline $\mathbf{Q}(t)$ defined by them is given by Equation (4.154):

$$\mathbf{Q}(t) = \frac{1}{2}(t^2, t, 1) \begin{pmatrix} 1 & -2 & 1 \\ -2 & 2 & 0 \\ 1 & 1 & 0 \end{pmatrix} \begin{pmatrix} \mathbf{Q}_0 \\ \mathbf{Q}_1 \\ \mathbf{Q}_2 \end{pmatrix}. \quad (4.154)$$

The nonuniform quadratic B-spline defined by three control points $\mathbf{P}_0$, $\mathbf{P}_1$, and $\mathbf{P}_2$ is given by Equation (4.185). If we require the two curves to be identical for any value of the parameter t, we can write

$$\frac{1}{2}\begin{pmatrix} 1 & -2 & 1 \\ -2 & 2 & 0 \\ 1 & 1 & 0 \end{pmatrix}\begin{pmatrix} \mathbf{Q}_0 \\ \mathbf{Q}_1 \\ \mathbf{Q}_2 \end{pmatrix} = \begin{pmatrix} a & -a-b & b \\ -2a & 2a & 0 \\ a & 1-a & 0 \end{pmatrix}\begin{pmatrix} \mathbf{P}_0 \\ \mathbf{P}_1 \\ \mathbf{P}_2 \end{pmatrix}.$$

4 Curves

This is a system of three equations where we assume that the unknowns are the Q_i's. The solutions are

$$Q_0 = 2aP_0 + (1-2a)P_1, \quad Q_1 = P_1, \quad Q_2 = (1-2b)P_1 + 2bP_2.$$

To see the geometrical interpretation of these relations, we write

$$Q_0 = 2aP_0 + (1-2a)P_1 = 2aP_0 + 2(1-a)P_1 - P_1 = 2P(0) - P_1 = 2Q(0) - Q_1,$$

which implies $Q_0 - Q(0) = Q(0) - Q_1$. The distance between Q_0 and $Q(0)$ equals the distance between $Q(0)$ and Q_1, and a similar relation among Q_1, $Q(1)$, and Q_2.

The conclusion is that a group of three points P_0, P_1, and P_2 defining a single quadratic nonuniform B-spline segment $P(t)$ can be replaced by a group of three points Q_0, Q_1, and Q_2 defining a single quadratic *uniform* B-spline segment $Q(t)$ identical to $P(t)$. However, given a set of $n+1$ control points P_i for a nonuniform B-spline curve, they cannot, in general, be replaced by a set of $n+1$ points Q_i that produce an identical uniform B-spline curve.

4.16.17 Subdividing the B-spline Curve

The B-spline curve is easy to manipulate by moving the control points and changing the knots. Still, if the curve is based on too few points, it may "refuse" to get the right shape, no matter what. More control points can be added, in such a case, by subdividing the curve, similar to subdividing the Bézier curve (Section 4.14.10). The method described here is called the Oslo algorithm and the discussion follows [Cohen et al. 80] and [Prautzsch 84].

(Control points can also be added by raising the degree of the B-spline curve, similar to the degree elevation of the Bézier curve, Section 4.14.11. This operation is discussed in [Cohen et al. 85].)

The idea behind subdividing a curve is that there are many (even infinitely many) sets of control points that produce the same curve. Normally, we are interested in the smallest number of control points that will produce a given curve, but if we cannot get the right shape with the original $n+1$ control points, we need to find a set of $n+2$ points that will produce *the same curve*, then move the new points around, attempting to bring the curve to the desired shape.

Assuming that we have a set of $n+1$ control points P_i and a knot vector $(t_0, t_1, \ldots, t_{n+k})$, we start the subdivision process by inserting several new knots, thereby obtaining a new knot vector $(u_0, u_1, \ldots, u_{m+k})$ where $m > n$. The new, subdivided curve is based on the $m+1$ control points Q_j defined by the Oslo algorithm as

$$Q_j = \sum_{i=0}^{n} a_{ij}^k P_i, \quad \text{where} \quad 0 \leq i \leq n \quad \text{and} \quad 0 \leq j \leq m,$$

where the coefficients a_{ij}^k are defined recursively by a relation similar to the Cox-

4.16 The B-Spline

DeBoor formula

$$a_{ij}^1 = \begin{cases} 1, & t_i \le u_j < t_{i+1}, \\ 0, & \text{otherwise}, \end{cases} \quad (4.190)$$

$$a_{ij}^k = \frac{u_{j+k-1} - t_i}{t_{i+k-1} - t_i} a_{ij}^{k-1} + \frac{t_{i+k} - u_{j+k-1}}{t_{i+k} - t_{i+1}} a_{i+1,j}^{k-1}. \quad (4.191)$$

This relation guarantees that $\sum_i^n a_{ij}^k = 1$.

If the original knot vector is uniform, inserting a single knot will convert it to a nonuniform vector. However, an open knot vector can sometimes remain open after inserting new knots, as the following example shows. Suppose that we have the open vector $(0, 0, 0, 1, 2, 2, 2)$, where t varies from 0 to 2. This corresponds to a two-segment curve and we want to subdivide both segments. We first multiply each knot by 2, obtaining the vector $(0, 0, 0, 2, 4, 4, 4)$ that produces the same curve when $0 \le t < 4$. Next, we insert knots 1 and 3 to obtain the knot vector $(0, 0, 0, 1, 2, 3, 4, 4, 4)$. This vector is still open and it corresponds to the four segments $[0, 1)$, $[1, 2)$, $[2, 3)$, and $[3, 4)$.

Example: We assume four control points and quadratic segments (i.e., $k = 3$). We already know that each segment is defined by three points, so two segments are needed for this curve. The knot vector is assumed to be uniform and it goes from $t_0 = 0$ to $t_{n+k} = t_6 = 6$. The parameter t varies from $t_{k-1} = t_2 = 2$ to $t_{n+1} = t_4 = 4$; two subintervals. This again shows that the curve consists of two spline segments, the first for the subinterval $[t_2, t_3)$ and the second for $[t_3, t_4)$. We decide to subdivide the first segment. This segment is defined by points $\mathbf{P}_0$, $\mathbf{P}_1$, and $\mathbf{P}_2$ (notice that $n = 2$ for this subdivision), so the subdivision process should produce four points, $\mathbf{Q}_0$, $\mathbf{Q}_1$, $\mathbf{Q}_2$, and $\mathbf{Q}_3$ (this implies $m = 3$), such that the two quadratic segments defined by them will have the same shape as the segment being subdivided.

To perform the subdivision, we need to insert a new knot between $t_2 = 2$ and $t_3 = 3$. We (somewhat arbitrarily) select its value to be 2.5. The new knot vector is

$$(u_0, u_1, u_2, u_3, u_4, u_5, u_6, u_7) = (0, 1, 2, 2.5, 3, 4, 5, 6),$$

and it is nonuniform. The calculation of the a_{ij}^k coefficients is done by varying i from 0 to $n = 2$ and varying j from 0 to $m = 3$. It requires three steps, for $k = 1, 2, 3$ (notice that this k is not the same as the order of the B-spline).

Step 1: We use Equation (4.190). A direct comparison of the t_i and u_i knots shows that the only nonzero a_{ij}^1 coefficients are a_{00}^1, a_{11}^1, a_{22}^1, and a_{23}^1. Each has a value of 1.

Step 2: We calculate a_{ij}^2 for $j = 0, 1, 2, 3$ from Equation (4.191). For each value of j, we stop when we get coefficients that add up to 1. The nonzero coefficients are

$$a_{00}^2 = \frac{u_1 - t_0}{t_1 - t_0} a_{00}^1 + \frac{t_2 - u_1}{t_2 - t_1} a_{10}^1 = \frac{1 - 0}{1 - 0} \cdot 1 = 1,$$

$$a_{11}^2 = \frac{u_2 - t_1}{t_2 - t_1} a_{11}^1 + \frac{t_3 - u_2}{t_3 - t_2} a_{21}^1 = \frac{2 - 1}{2 - 1} \cdot 1 = 1,$$

4 Curves

$$a_{12}^2 = \frac{u_3 - t_1}{t_2 - t_1} a_{12}^1 + \frac{t_3 - u_3}{t_3 - t_2} a_{22}^1 = \frac{3 - 2.5}{3 - 2} \cdot 1 = 1/2,$$

$$a_{22}^2 = \frac{u_3 - t_2}{t_3 - t_2} a_{22}^1 + \frac{t_4 - u_3}{t_4 - t_3} a_{32}^1 = \frac{2.5 - 2}{3 - 2} \cdot 1 = 1/2,$$

$$a_{23}^2 = \frac{u_4 - t_2}{t_3 - t_2} a_{23}^1 + \frac{t_4 - u_4}{t_4 - t_3} a_{33}^1 = \frac{3 - 2}{3 - 2} \cdot 1 = 1.$$

Step 3: The coefficients of step 2 are used to calculate a_{ij}^3:

$$a_{00}^3 = \frac{u_2 - t_0}{t_2 - t_0} a_{00}^2 + \frac{t_3 - u_2}{t_3 - t_1} a_{10}^2 = \frac{2 - 0}{2 - 0} \cdot 1 = 1,$$

$$a_{01}^3 = \frac{u_3 - t_0}{t_2 - t_0} a_{01}^2 + \frac{t_3 - u_3}{t_3 - t_1} a_{11}^2 = \frac{3 - 2.5}{3 - 1} \cdot 1 = 1/4,$$

$$a_{11}^3 = \frac{u_3 - t_1}{t_3 - t_1} a_{11}^2 + \frac{t_4 - u_3}{t_4 - t_2} a_{21}^2 = \frac{2.5 - 1}{3 - 1} \cdot 1 = 3/4,$$

$$a_{12}^3 = \frac{u_4 - t_1}{t_3 - t_1} a_{12}^2 + \frac{t_4 - u_4}{t_4 - t_2} a_{22}^2 = \frac{3 - 1}{3 - 1} \cdot \frac{1}{2} + \frac{4 - 3}{4 - 2} \cdot \frac{1}{2} = 3/4,$$

$$a_{22}^3 = \frac{u_4 - t_2}{t_4 - t_2} a_{22}^2 + \frac{t_5 - u_4}{t_5 - t_3} a_{32}^2 = \frac{3 - 2}{4 - 2} \cdot \frac{1}{2} = 1/4,$$

$$a_{23}^3 = \frac{u_5 - t_2}{t_4 - t_2} a_{23}^2 + \frac{t_5 - u_5}{t_5 - t_3} a_{33}^2 = \frac{4 - 2}{4 - 2} \cdot 1 = 1.$$

The four new control points can now be calculated:

$$\mathbf{Q}_0 = \sum_{i=0}^{3} a_{i0}^3 \mathbf{P}_i = a_{00}^3 \mathbf{P}_0 = \mathbf{P}_0,$$

$$\mathbf{Q}_1 = \sum_{i=0}^{3} a_{i1}^3 \mathbf{P}_i = a_{01}^3 \mathbf{P}_0 + a_{11}^3 \mathbf{P}_1 = \frac{1}{4}\mathbf{P}_0 + \frac{3}{4}\mathbf{P}_1,$$

$$\mathbf{Q}_2 = \sum_{i=0}^{3} a_{i2}^3 \mathbf{P}_i = a_{12}^3 \mathbf{P}_1 + a_{22}^3 \mathbf{P}_2 = \frac{3}{4}\mathbf{P}_1 + \frac{1}{4}\mathbf{P}_2,$$

$$\mathbf{Q}_3 = \sum_{i=0}^{3} a_{i3}^3 \mathbf{P}_i = a_{23}^3 \mathbf{P}_2 = \mathbf{P}_2,$$

The two quadratic B-spline segments defined by $\mathbf{Q}_0\mathbf{Q}_1\mathbf{Q}_2$ and $\mathbf{Q}_2\mathbf{Q}_3\mathbf{Q}_4$ have the same shape as the original segment defined by $\mathbf{P}_0\mathbf{P}_1\mathbf{P}_2$, but they are easier to modify since they are based on four points.

4.16.18 Nonuniform Rational B-Splines (NURBS)

The use of a knot vector is one reason why the B-spline curve is more general than the Bézier and other curve methods. The $n+k+1$ knots can be used as parameters and can be varied by the user/designer to get the required shape of the curve. The rational B-spline, described in this section, uses an additional set of $n+1$ parameters

w_i, called *weights*, to add even greater flexibility to the curve. In addition to this feature, the rational B-spline has several more important advantages as follows:

1. It makes it possible to create curves that are true conic sections. We already know, from Exercise 4.81, that a polynomial cannot represent a circle. More generally, it cannot represent general conic sections. Sections 4.14.23 and 4.16.6 show how the Bézier and B-spline curves can represent approximate circles. If precise circles or conic sections are needed, then rational curves are the natural choice.

2. It is invariant under perspective projections. We know that curves that are barycentric sums are invariant under affine transformations. If we want to rotate, scale, shear, or translate such a curve, we can apply the transformation to the control points and use the new points to draw the transformed curve. There is no need to apply the transformation to every pixel on the curve. However, if we want to project a curve in perspective on a two-dimensional output device, we have to individually project every pixel on the curve. With a rational curve, we can perspective-project the control points and use the projected, two-dimensional points to calculate the projected curve.

3. It reduces to the nonrational B-spline when all the weights are set to $w_i = 1$. This means that a software package for rational B-splines can be used to generate nonrational ones (uniform, open, and nonuniform). This implies that the nonuniform rational B-spline (NURBS for short) is the most general type of curve. It can take many shapes and can easily be reduced to simpler forms. Because of this, NURBS is, today, the de facto standard for curve design and is also used by the IGES standard (Section 10.1). Two excellent references to NURBS are [Piegl 96] and [Farin 98].

Perhaps the best way to introduce rational B-splines (and rational curves in general) is by means of homogeneous coordinates. This method, introduced in Section 3.2.1, starts by adding an extra (homogeneous) dimension to points, so a two-dimensional point becomes a triplet (x, y, w) and a three-dimensional point becomes a 4-tuple (x, y, z, w). After transforming or manipulating the point, it is projected back to its original number of dimensions by dividing its coordinates by w. Given four-dimensional control points $\mathbf{Q}_i = (x_i, y_i, z_i, w_i)$, where we assume for convenience that the w_i coordinates are non-negative, we can define a (nonrational) B-spline curve as

$$\mathbf{P}_{\mathrm{nr}}(t) = \sum_{i=0}^{n} \mathbf{Q}_i N_{ik}(t).$$

From this we get the rational B-spline $\mathbf{P}_{\mathrm{r}}(t)$ by isolating that part of $\mathbf{P}_{\mathrm{nr}}(t)$ that depends on the fourth coordinates w_i and dividing by this part.

$$\mathbf{P}_{\mathrm{r}}(t) = \frac{\sum_{i=0}^{n} \mathbf{P}_i w_i N_{ik}(t)}{\sum_{i=0}^{n} w_i N_{ik}(t)} = \sum_{i=0}^{n} \mathbf{P}_i R_{ik}(t), \qquad (4.192)$$

where $\mathbf{P}_i = (x_i, y_i, z_i)$ are three-dimensional control points and $R_{ik}(t)$ are the new,

4 Curves

rational blending functions defined by

$$R_{ik}(t) = \frac{w_i N_{ik}(t)}{\sum_{i=0}^{n} w_i N_{ik}(t)}. \qquad (4.193)$$

This type of curve has most of the properties of the nonrational B-spline. The following should be mentioned in particular:

1. The new blending functions $R_{ik}(t)$ are non-negative and barycentric.
2. The curve reduces to the nonrational curve when all $w_i = 1$ (this is a direct consequence of Equation (4.193)).
3. Since the rational curve is the four-dimensional generalization of the nonrational B-spline, the algorithms for curve subdivision and degree elevation of the B-spline can be used for the rational version. They simply have to be executed on the four-dimensional control points (x_i, y_i, z_i, w_i).

So much for the definition of the rational B-spline. The main question is how to assign values to the weights in order to modify the shape of the curve in a predictable way. The rest of this section describes two approaches to this. The first one is to set all weights $w_i = 1$, then change the value of one of them and see how it affects the blending functions. The second approach is to derive specific sets of weights that will produce B-spline curves that are conic sections. The first approach is illustrated by a detailed example.

Example: This is an extension of the open B-spline example on page 364. We assume $n = 4$ (five control points), select order $k = 3$ (quadratic polynomial segments), and the knot vector $(0, 0, 0, 1, 2, 3, 3, 3)$. The parameter t varies from $t_{k-1} = t_2 = 0$ to $t_{n+1} = t_5 = 3$, so our curve consists of three segments. The nonrational blending functions $N_{i3}(t)$ are

$$N_{03}(t) = (1-t)^2, \qquad 0 \le t < 1,$$

$$N_{13}(t) = \frac{1}{2} \begin{cases} t(4-3t), & 0 \le t < 1, \\ (2-t)^2, & 1 \le t < 2, \end{cases}$$

$$N_{23}(t) = \frac{1}{2} \begin{cases} t^2, & 0 \le t < 1, \\ (-2t^2 + 6t - 3), & 1 \le t < 2, \\ (3-t)^2, & 2 \le t < 3, \end{cases}$$

$$N_{33}(t) = \frac{1}{2} \begin{cases} (t-1)^2, & 1 \le t < 2, \\ (-3t^2 + 14t - 15), & 2 \le t < 3, \end{cases}$$

$$N_{43}(t) = (t-2)^2, \qquad 2 \le t < 3.$$

Before we can calculate the rational blending functions, we have to select values for the five weights. We choose $(1, 1, w_2, 1, 1)$, where w_2 will later be assigned several different values. The result is

$$R_{03}(t) = \frac{w_0 N_{03}(t)}{\sum_{i=0}^{4} w_i N_{i3}(t)} = \frac{(1-t)^2}{(1-t)^2 + t(4-3t)/2 + w_2 t^2/2}, \qquad t \in [0,1),$$

$$R_{13}(t) = \frac{w_1 N_{13}(t)}{\sum_{i=0}^{4} w_i N_{i3}(t)} = \begin{cases} \frac{t(4-3t)/2}{(1-t)^2+t(4-3t)/2+w_2t^2/2}, & t \in [0,1) \\ \frac{(2-t)^2/2}{(2-t)^2/2+w_2(-2t^2+6t-3)/2+(t-1)^2/2}, & t \in [1,2), \end{cases}$$

$$R_{23}(t) = \frac{w_2 N_{23}(t)}{\sum_{i=0}^{4} w_i N_{i3}(t)} = \begin{cases} \frac{w_2 t^2/2}{(1-t)^2+t(4-3t)/2+w_2t^2/2}, & t \in [0,1) \\ \frac{w_2(-2t^2+6t-3)/2}{(2-t)^2/2+w_2(-2t^2+6t-3)/2+(t-1)^2/2}, & t \in [1,2) \\ \frac{w_2(3-t)^2/2}{w_2(3-t)^2/2+(-3t^2+14t-15)/2+(t-2)^2}, & t \in [2,3), \end{cases}$$

$$R_{33}(t) = \frac{w_3 N_{33}(t)}{\sum_{i=0}^{4} w_i N_{i3}(t)} = \begin{cases} \frac{(t-1)^2/2}{(2-t)^2/2+w_2(-2t^2+6t-3)/2+(t-1)^2/2}, & t \in [1,2) \\ \frac{(-3t^2+14t-15)/2}{w_2(3-t)^2/2+(-3t^2+14t-15)/2+(t-2)^2}, & t \in [2,3), \end{cases}$$

$$R_{43}(t) = \frac{w_4 N_{43}(t)}{\sum_{i=0}^{4} w_i N_{i3}(t)} = \frac{(t-2)^2}{w_2(3-t)^2/2+(-3t^2+14t-15)/2+(t-2)^2} \quad t \in [2,3).$$

We next calculate the three spline segments for the four cases $w_2 = 0, 0.5, 1$, and 5.

For $w_2 = 0$, the three segments are

$$\mathbf{P}_1(t) = \frac{(1-t)^2}{1-t^2/2}\mathbf{P}_0 + \frac{(4-3t)t}{2-t^2}\mathbf{P}_1 + 0\mathbf{P}_2,$$

$$\mathbf{P}_2(t) = \frac{(2-t)^2}{5-6t+2t^2}\mathbf{P}_1 + 0\mathbf{P}_2 + \frac{(t-1)^2}{5-6t+2t^2}\mathbf{P}_3,$$

$$\mathbf{P}_3(t) = 0\mathbf{P}_2 + \frac{15-14t+3t^2}{7-6t+t^2}\mathbf{P}_3 + \frac{2(-2+t)^2}{-7+6t-t^2}\mathbf{P}_4.$$

For $w_2 = 0.5$, they are

$$\mathbf{P}_1(t) = \frac{(1-t)^2}{1-0.25t^2}\mathbf{P}_0 + \frac{(4-3t)t}{2-0.5t^2}\mathbf{P}_1 + \frac{0.25t^2}{1-0.25t^2}\mathbf{P}_2,$$

$$\mathbf{P}_2(t) = \frac{(2-t)^2}{3.5-3t+t^2}\mathbf{P}_1 + \frac{0.25(-3+6t-2t^2)}{1.75-1.5t+0.5t^2}\mathbf{P}_2 + \frac{(t-1)^2}{3.5-0.5t^2}\mathbf{P}_3,$$

$$\mathbf{P}_3(t) = \frac{0.25(3-t)^2}{-1.25+1.5t-0.25t^2}\mathbf{P}_2 + \frac{-15+14t-3t^2}{-2.5+3.5t-0.5t^2}\mathbf{P}_3 + \frac{(t-2)^2}{-1.25+1.5t-0.25t^2}\mathbf{P}_4.$$

For $w_2 = 1$, we get

$$\mathbf{P}_1(t) = (1-t)^2\mathbf{P}_0 + \frac{(4-3t)t}{2}\mathbf{P}_1 + \frac{t^2}{2}\mathbf{P}_2,$$

$$\mathbf{P}_2(t) = \frac{(2-t)^2}{2}\mathbf{P}_1 + \frac{-3+6t-2t^2}{2}\mathbf{P}_2 + \frac{(t-1)^2}{2+6t-3t^2}\mathbf{P}_3,$$

$$\mathbf{P}_3(t) = \frac{(3-t)^2}{2}\mathbf{P}_2 + \frac{-15+14t-3t^2}{2}\mathbf{P}_3 + (t-2)^2\mathbf{P}_4.$$

4 Curves

Finally for $w_2 = 5$, the segments are

$$\mathbf{P}_1(t) = \frac{(1-t)^2}{1+2t^2}\mathbf{P}_0 + \frac{(4-3t)t}{2+4t^2}\mathbf{P}_1 + \frac{5t^2}{2+4t^2}\mathbf{P}_2,$$

$$\mathbf{P}_2(t) = \frac{(2-t)^2}{-10+24t-8t^2}\mathbf{P}_1 + \frac{5(-3+6t-2t^2)}{-10+24t-8t^2}\mathbf{P}_2 + \frac{(t-1)^2}{-10+54t-23t^2}\mathbf{P}_3,$$

$$\mathbf{P}_3(t) = \frac{5(3-t)^2}{38-24t+4t^2}\mathbf{P}_2 + \frac{-15+14t-3t^2}{38-24t+4t^2}\mathbf{P}_3 + \frac{(t-2)^2}{19-12t+2t^2}\mathbf{P}_4.$$

They are plotted in Figure 4.85 for control points $\mathbf{P}_0 = (0,0)$, $\mathbf{P}_1 = (0,1)$, $\mathbf{P}_2 = (1,0)$, $\mathbf{P}_3 = (2,1)$, and $\mathbf{P}_4 = (2,0)$. It is easy to see how weight w_2 affects the shape of the curve by controlling the amount of "pull" that point $\mathbf{P}_2$ exerts on the curve. For $w_2 = 0$, point $\mathbf{P}_2$ has no effect. The curve is defined by the four remaining points and it is identical to the control polygon of these points. As w_2 grows toward 5, the curve gets more and more attracted to $\mathbf{P}_2$.

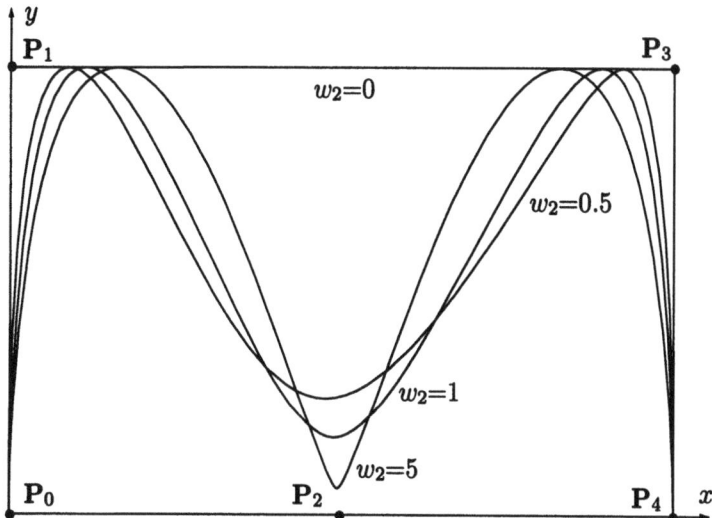

Figure 4.85: Effects of Varying Weight w_2.

Now for the second approach. We are looking for specific sets of weights that will generate conic sections. Since the conics are described by quadratic equations and each is fully defined by means of three points, it makes sense to try rational B-splines of order $k = 3$ defined by three points (i.e., $n = 2$). The conic is easier to design if the B-spline curve starts and ends at control points, so it makes sense to use an open B-spline. Since we have selected $k = n+1$, we know (from Section 4.16.14) that the open B-spline will be a Bézier curve. The knot vector for our curve is calculated by Equation (4.176) to be $(0,0,0,0,1,1,1,1)$. To simplify our task, we try the simple set of weights $(1, w_1, 1)$. Our problem is to find out for what values, if any, of w_1 we get precise conics.

There is no need to use the Cox-DeBoor recursive formula (Equation (4.174)) to calculate the blending functions since they are the quadratic Bernstein polynomials.

The curve itself can now easily be written

$$\mathbf{P}(t) = \frac{N_{03}(t)\mathbf{P}_0 + w_1 N_{13}(t)\mathbf{P}_1 + N_{23}(t)\mathbf{P}_2}{N_{03}(t) + w_1 N_{13}(t) + N_{23}(t)}$$

$$= \frac{(1-t)^2 \mathbf{P}_0 + 2w_1 t(1-t)\mathbf{P}_1 + t^2 \mathbf{P}_2}{(1-t)^2 + 2w_1 t(1-t) + t^2}. \quad (4.194)$$

▶ **Exercise 4.102:** Show that in the special case where $w_1 = 0$, the curve of Equation (4.194) reduces to the straight line between $\mathbf{P}_0$ and $\mathbf{P}_2$.

The midpoint $\mathbf{S}$ of the curve of Equation (4.194) is given by

$$\mathbf{S} = \mathbf{P}(0.5) = \frac{(\mathbf{P}_0 + \mathbf{P}_2)/2}{1 + w_1} + \frac{w_1 \mathbf{P}_1}{1 + w_1} = \frac{1}{1 + w_1}\mathbf{M} + \frac{w_1}{1 + w_1}\mathbf{P}_1 = (1-u)\mathbf{M} + u\mathbf{P}_1, \quad (4.195)$$

where $\mathbf{M} = (\mathbf{P}_0 + \mathbf{P}_2)/2$ is the midpoint of $\mathbf{P}_0$ and $\mathbf{P}_2$ and $u \stackrel{\text{def}}{=} w_1/(1+w_1)$. Point $\mathbf{S}$, which is called the *shoulder point* of the curve, thus moves along a straight line from $\mathbf{M}$ to $\mathbf{P}_1$ when w_1 varies from 0 to ∞ (or, equivalently, when u varies from 0 to 1). Equation (4.195) also yields the relation

$$w_1 = \frac{\mathbf{M} - \mathbf{S}}{\mathbf{S} - \mathbf{P}_1}, \quad (4.196)$$

which shows that w_1 is the ratio of two distances.

It can be shown (see, e.g., [Lee 86]) that the single weight w_1 determines the type of conic generated by Equation (4.194). Values in the range $(0, 1)$ generate an elliptic curve (with a circle as a special case). The value $w_1 = 1$ produces a parabolic curve, and values $w_1 > 1$ result in a hyperbolic curve. Figure 4.86 shows examples of these types of conics (notice that $\mathbf{S}$ is not necessarily the maximum point on these curves).

A circle is formed when the three control points form an isosceles triangle. If we denote the base angle of this triangle θ, it can be shown that a circular arc spanning 2θ degrees is obtained when $w_1 = \cos\theta$. The most common cases are $\theta = 60°$ and $\theta = 90°$. In the latter case (Figure 4.87b), a complete circle can easily be formed by using the symmetry of a circle and duplicating every point four times. In the former case (Figure 4.87a), a complete circle can be obtained by specifying six control points and calculating three spline segments.

As an example, let's use the three points $\mathbf{P}_0 = (0, -1)R$, $\mathbf{P}_1 = (-1.732, -1)R$, and $\mathbf{P}_2 = (-0.866, 0.5)R$ of Figure 4.87a. Substituting these points in Equation (4.194) and setting w_1 to $\cos 60° = 0.5$ yields the 60° circular arc that goes from $\mathbf{P}_0$ to $\mathbf{P}_2$:

$$\mathbf{P}(t) = \frac{(1-t)^2 \mathbf{P}_0 + 2w_1 t(1-t)\mathbf{P}_1 + t^2 \mathbf{P}_2}{(1-t)^2 + 2w_1 t(1-t) + t^2}$$

$$= R\frac{(1-t)^2(0,-1) + t(1-t)(-1.732,-1) + t^2(-0.866, 0.5)}{(1-t)^2 + t(1-t) + t^2}$$

$$= R\frac{(0.866t^2 - 1.732t, 0.5t^2 + t - 1)}{(1-t)^2 + t(1-t) + t^2}.$$

4 Curves

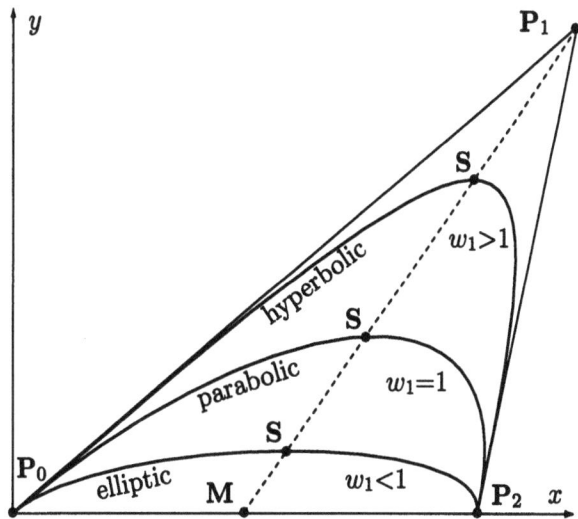

Figure 4.86: Conics Generated by Varying w_1.

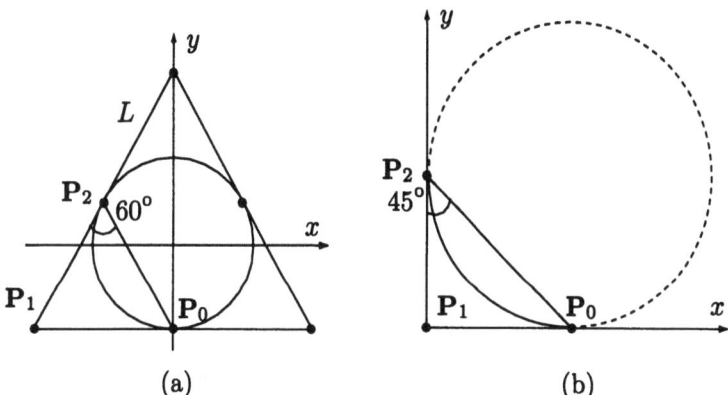

Figure 4.87: Control Points for Circles.

▶ **Exercise 4.103:** Show how to figure out the coordinates of the three points from Figure 4.87a.

▶ **Exercise 4.104:** Given the three points $\mathbf{P}_0 = (1,0)R$, $\mathbf{P}_1 = (0,0)$, and $\mathbf{P}_2 = (0,1)R$ of Figure 4.87b, calculate the quadratic rational B-spline segment defined by the points whose shape is a circular arc spanning $90°$.

4.17 The Beta Spline

This method (see [Bartels et al. 87] and [Barsky 88] for more information) is a generalization of the cubic uniform nonrational B-spline. Two parameters are added in order to control the *bias* and *tension* of the curve. The Beta spline curve is a weighted sum of control points $\mathbf{P}(t) = \sum_{i=0}^{n} \mathbf{P}_i B_{n,i}(t)$, and the derivation of the weight functions is similar to that of the B-spline. A general weight function $B_{n,i}(t)$ is expressed (as in Section 4.16.12) as the union of four parts $b_0(t)$, $b_1(t)$, $b_2(t)$, and $b_3(t)$, each a PC defined over one unit of t. We make three requirements and use them to set up 16 equations. The first requirement is that the individual curve segments meet at the joint points (remember, the joints $\mathbf{K}_i$ are points where the segments meet; they are not the control points). This can be expressed as

$$\mathbf{P}_{i-1}(1) = \mathbf{P}_i(0). \tag{4.197}$$

The second requirement is that the tangent vectors have the same direction at the joints (G^1 continuity, see Figure 4.88). This is expressed as

$$\beta_1 \mathbf{P}^t_{i-1}(1) = \mathbf{P}^t_i(0), \tag{4.198}$$

where β_1 is an input parameter specified by the user. The geometrical interpretation of β_1 is the difference in speed between one segment approaching joint $\mathbf{K}_i$ and the next segment leaving it. With $\beta_1 = 1$, the speeds are the same. Values $\beta_1 > 1$ will give the curve extra speed after the joint, so it will have less tendency to change direction. The effect is to get more tension to the right of each joint point. Values $0 \le \beta_1 < 1$ will have the opposite effect and will pull the curve closer to the tangent on the left of a joint. This form of curve reshaping is called *bias* and β_1 is the bias parameter (compare to the bias parameter of the Kochanek-Bartels spline, Section 4.12).

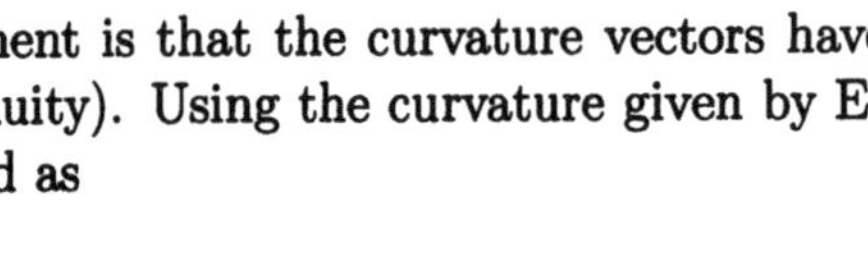

Figure 4.88: G^1 Continuity in Beta Splines.

The third requirement is that the curvature vectors have the same direction at the joints (G^2 continuity). Using the curvature given by Equation (4.217), this requirement is expressed as

$$\frac{\mathbf{P}^t_{i-1}(1) \times \mathbf{P}^{tt}_{i-1}(1) \times \mathbf{P}^t_{i-1}(1)}{|\mathbf{P}^t_{i-1}(1)|^4} = \frac{\mathbf{P}^t_i(0) \times \mathbf{P}^{tt}_i(0) \times \mathbf{P}^t_i(0)}{|\mathbf{P}^t_i(0)|^4}.$$

Using Equation (4.198), we substitute $\mathbf{P}^t_i(0)$ to get

$$\frac{\mathbf{P}^t_{i-1}(1) \times \mathbf{P}^{tt}_{i-1}(1) \times \mathbf{P}^t_{i-1}(1)}{|\mathbf{P}^t_{i-1}(1)|^4} = \frac{\beta_1 \mathbf{P}^t_{i-1}(1) \times \mathbf{P}^{tt}_{i-1}(0) \times \beta_1 \mathbf{P}^t_{i-1}(1)}{|\beta_1 \mathbf{P}^t_{i-1}(1)|^4}$$

$$= \frac{1}{\beta_1^2} \frac{\mathbf{P}^t_{i-1}(1) \times \mathbf{P}^{tt}_{i-1}(0) \times \mathbf{P}^t_{i-1}(1)}{|\mathbf{P}^t_{i-1}(1)|^4}.$$

4 Curves

The curvatures will thus be equal if

$$\beta_1^2 \mathbf{P}^{tt}_{i-1}(1) = \mathbf{P}^{tt}_i(0). \tag{4.199}$$

Notice, however, that the second derivative $\mathbf{P}^{tt}_i(0)$ may have an additional component in the direction of the tangent $\mathbf{P}^{t}_{i-1}(1)$. This component does not participate in the equations since the cross-product of a vector with itself is zero. Equation (4.199) is therefore not general. A more general condition for curvature equality is (see Figure 4.89)

$$\beta_1^2 \mathbf{P}^{tt}_{i-1}(1) + \beta_2 \mathbf{P}^{t}_{i-1}(1) = \mathbf{P}^{tt}_i(0), \quad \text{for } \beta_1 > 0. \tag{4.200}$$

The physical meaning of Equation (4.200) is: The second derivative (the "acceleration") $\mathbf{P}^{tt}_i(0)$ may have a component in the direction of the tangent (the "velocity") $\mathbf{P}^{t}_{i-1}(1)$. This component is not seen in the above equations since it does not affect the curvature. (It is like applying a force in the direction of motion. The force changes the acceleration but not the direction, so it does not affect the curvature.) This component has a value of $\beta_2 \mathbf{P}^{t}_{i-1}(1)$ and is therefore 0 for $\beta_2 = 0$.

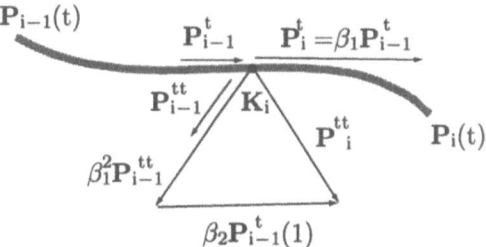

Figure 4.89: Curvature in Beta Splines.

The β_2 parameter affects the tension of the curve in the following way. Increasing β_2 moves the joint point $\mathbf{K}_i$ between segments $\mathbf{P}_{i-1}(t)$ and $\mathbf{P}_i(t)$ closer to control point $\mathbf{P}_{i-2}$ and this is true for all the joints. As a result, the curve passes nearer the control polygon, and it is this behavior that creates the effect (illusion?) of added tension.

The three requirements can be translated into the following 15 equations:

$$\begin{aligned}
0 &= b_0(0), & 0 &= b'_0(0), & 0 &= b''_0(0), \\
b_0(1) &= b_1(0), & \beta_1 b'_0(1) &= b'_1(0), & \beta_1^2 b''_0(1) + \beta_2 b'_0(1) &= b''_1(0), \\
b_1(1) &= b_2(0), & \beta_1 b'_1(1) &= b'_2(0), & \beta_1^2 b''_1(1) + \beta_2 b'_1(1) &= b''_2(0), \\
b_2(1) &= b_3(0), & \beta_1 b'_2(1) &= b'_3(0), & \beta_1^2 b''_2(1) + \beta_2 b'_2(1) &= b''_3(0), \\
b_3(1) &= 0, & \beta_1 b'_3(1) &= 0, & \beta_1^2 b''_3(1) + \beta_2 b'_3(1) &= 0
\end{aligned}$$

(compare with Equation (4.169)) and we get one more equation by the barycentric requirement $b_0(0) + b_1(0) + b_2(0) + b_3(0) = 1$.

Using appropriate software, the 16 equations can easily be solved symbolically. The solutions are

$$b_0(t) = \frac{1}{\delta}[2t^3],$$

$$b_1(t) = \frac{1}{\delta}\left[-(2\beta_2 + 2\beta_1^2 + 2\beta_1 + 2)t^3 + (3\beta_2 + 6\beta_1^2)t^2 + (6\beta_1)t + 2\right],$$

$$b_2(t) = \frac{1}{\delta}\left[(2\beta_2 + 2\beta_1^3 + 2\beta_1^2 + 2\beta_1)t^3 - (3\beta_2 + 6\beta_1^3 + 6\beta_1^2)t^2 \right. \quad (4.201)$$
$$\left. + (6\beta_1^3 - 6\beta_1)t + (\beta_2 + 4\beta_1^2 + 4\beta_1)\right],$$

$$b_3(t) = \frac{1}{\delta}\left[-(2\beta_1^3)t^3 + (6\beta_1^3)t^2 - (6\beta_1^3)t + (2\beta_1^3)\right],$$

where $\delta = \beta_2 + 2\beta_1^3 + 4\beta_1^2 + 4\beta_1 + 2 \neq 0$.

These weight functions reduce to the cubic uniform B-spline weight functions (Equation (4.170)) when $\beta_1 = 1$ and $\beta_2 = 0$. To gain more insight into the behavior of the weight functions, we consider several special cases.

- To show how β_2 acts as tension, we notice that, at the limit, where $\beta_2 \to \infty$, Equations (4.201) converge to

$$b_0(t) = 0, \quad b_1(t) = (-2t^3 + 3t^2), \quad b_2(t) = 1 - (-2t^3 + 3t^2), \quad b_3(t) = 0,$$

regardless of the value of β_1. Substituting $u = -2t^3 + 3t^2$, we get

$$b_0(u) = 0, \quad b_1(u) = t, \quad b_2(u) = 1 - t, \quad b_3(u) = 0,$$

so a curve segment has the form $\mathbf{P}_i(t) = \mathbf{P}_{i-2}(1-t) + \mathbf{P}_{i-1}t$. Each curve segment becomes, in the limit, a straight segment between control points $\mathbf{P}_{i-2}$ and $\mathbf{P}_{i-1}$.

- In a similar way, we can explore the meaning of β_1. In the limit where $\beta_1 \to \infty$, Equations (4.201) approach

$$b_0(t) = 0, \quad b_1(t) = 0, \quad b_2(t) = (t^3 - 3t^2 + 3t), \quad b_3(t) = 1 - (t^3 - 3t^2 + 3t),$$

regardless of the value of β_2. The substitution $t = (u^3 - 3u^2 + 3u)$ shows that segment $\mathbf{P}_i(t)$ becomes the straight line from $\mathbf{P}_{i-3}$ to $\mathbf{P}_{i-2}$.

- It is also easy to show that, in the limit where $\beta_1 \to 0$ and $\beta_2 \to 0$, Equations (4.201) become

$$b_0(t) = t^3, \quad b_1(t) = 1 - t^3, \quad b_2(t) = 0, \quad b_3(t) = 0,$$

that is, a straight line from $\mathbf{P}_{i-1}$ to $\mathbf{P}_i$.

In the type of β-spline described here, each parameter has the same value for all segments. The curve is called the uniformly shaped β-spline. An extension of

4 Curves

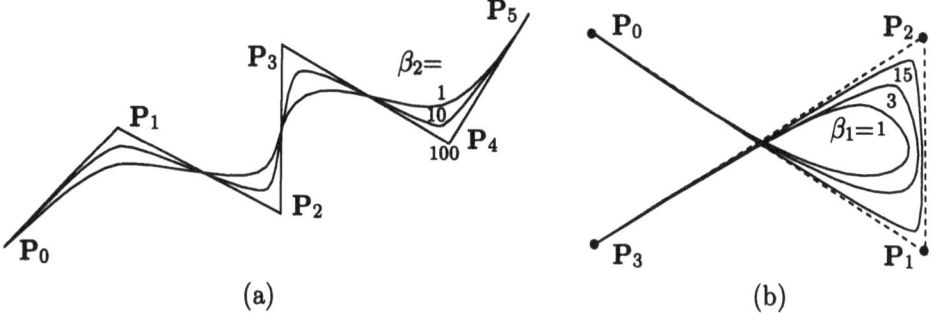

Figure 4.90: Tension in Beta Splines.

this curve that allows different values of the parameters in each segment is called the continuously shaped β-spline ([Bartels et al. 87], p. 321).

Figure 4.90a shows the effects of increasing β_2 from 0 (no tension) to 10 (moderate tension) to 100 (high tension). Figure 4.90b shows the effects of increasing β_1 from 1 to 3 to 15. The curve tends to continue longer in the direction of $\mathbf{P}_0\mathbf{P}_1$.

4.18 Barycentric Sums Revisited

The concept of barycentric combinations has been mentioned many times. Section 4.1 shows that the barycentric sum of weighted points is affinely invariant, i.e., it is a valid point. Section 4.2 discusses curves as barycentric combinations of points. Page 190 mentions that curves of the form

$$\mathbf{P}(t) = \sum_{i=0}^{n} w_i \mathbf{P}_i, \quad \text{where} \quad \sum_{i=0}^{n} w_i = 1$$

are independent of the particular coordinate axes used. This section presents a more general approach to this topic. We now know that curves are sometimes constructed as the sum

$$\begin{aligned}\mathbf{P}(t) &= \sum w_i \mathbf{P}_i + \sum u_i \mathbf{v}_i \\ &= (t^n, t^{n-1}, \ldots, t, 1)\mathbf{Q}\,(\mathbf{P}_1, \mathbf{P}_2, \ldots, \mathbf{v}_1, \mathbf{v}_2, \ldots)^T,\end{aligned} \quad (4.202)$$

where $\mathbf{P}_i$ are points and $\mathbf{v}_i$ are vectors. It turns out that such a curve is independent of the particular coordinate system used if the weights w_i are barycentric (there is no similar requirement for the u_i weights). Notice that the points can be data points, control points, or any other points. The vectors can be tangents, second derivatives or any other vectors, but the statement above is always true. It is sometimes called the *isotropic principle*.

To prove the isotropic principle, we first explore the conditions under which a rigid transformation $\mathbf{G}$ of the points $\mathbf{P}_i$ and vectors $\mathbf{v}_i$ is identical to a rigid transformation of the curve $\mathbf{P}(t)$. Mathematically, the former transformation is expressed $\sum w_i \mathbf{P}_i \mathbf{G} + \sum u_i \mathbf{v}_i \mathbf{G}$ and the latter transformation is just $\mathbf{P}(t)\mathbf{G}$. A rigid transformation can be expressed (Section 8.3) as a combination of a translation $\mathbf{T}$

and a rotation $\mathbf{R}$. The case of rotation is easily explored. We know that points and vectors are rotated by $\mathbf{P}^* = \mathbf{PR}$ and $\mathbf{v}^* = \mathbf{vR}$. If the rigid transformation is rotation, we end up with

$$\sum w_i \mathbf{P}_i^* + \sum u_i \mathbf{v}_i^* = \sum w_i \mathbf{P}_i \mathbf{R} + \sum u_i \mathbf{v}_i \mathbf{R} = (\sum w_i \mathbf{P}_i + \sum u_i \mathbf{v}_i) \mathbf{R} = \mathbf{P}(t) \mathbf{R}.$$

The weighted sum of the rotated points and vectors is always identical to the rotated curve.

Translation, however, is different. We know that points are translated by $\mathbf{P}^* = \mathbf{P} + \mathbf{T}$, but it is important to remember that a vector has just two attributes, namely direction and magnitude. Translating a vector parallel to itself does not change it. We thus end up with

$$\sum w_i \mathbf{P}_i^* + \sum u_i \mathbf{v}_i^*$$
$$= \sum w_i (\mathbf{P}_i + \mathbf{T}) + \sum u_i \mathbf{v}_i$$
$$= (\sum w_i \mathbf{P}_i + \sum u_i \mathbf{v}_i) + (\sum w_i) \mathbf{T} = \mathbf{P}(t) + (\sum w_i) \mathbf{T}.$$

This is identical to the transformed curve $\mathbf{P}^*(t) = \mathbf{P}(t) + \mathbf{T}$ only if $\sum w_i = 1$.

This proof of the isotropic principle can be extended to scaling. If $\mathbf{S}$ is a scaling transformation, then both points and vectors transform in the same way, namely $\mathbf{P}^* = \mathbf{PS}$ and $\mathbf{v}^* = \mathbf{vS}$. We end up with

$$\sum w_i \mathbf{P}_i^* + \sum u_i \mathbf{v}_i^* = \sum w_i \mathbf{P}_i \mathbf{S} + \sum u_i \mathbf{v}_i \mathbf{S} = (\sum w_i \mathbf{P}_i + \sum u_i \mathbf{v}_i) \mathbf{S} = \mathbf{P}(t) \mathbf{S}.$$

This means, for example, that relabeling the coordinate axes from, say, inches to centimeters will not change the shape of a curve if the curve satisfies the isotropic principle.

In general, it can be shown that curves (and also surfaces) defined by means of barycentric sums remain invariant under any affine transformations. They also remain invariant under parallel projections but not under perspective ones. It is also possible to show that if $\sum w_i = 1$, then matrix $\mathbf{Q}$ of Equation (4.202) has the following property: Each of its columns that corresponds to a point has a sum of 0, except the last one, which sums to 1. Examples are Equations (4.14) and (4.26), but notice that in the latter, only the first two columns correspond to points (see also the Bézier $\mathbf{N}$ matrices, Equation (4.108)).

4.19 Symmetry in Curves

When a curve $\mathbf{P}(t)$ is defined by means of points $\mathbf{P}_i$ it should not change its shape when the order of the points is reversed. This is a symmetry property that's satisfied by all well-defined curves. We use a cubic polynomial to illustrate this property and to show precisely when such a curve is well defined. Other types of curve have to satisfy different conditions in order to be well defined (i.e., to have the symmetry property). We know from experience that a cubic polynomial defined by four points

4 Curves

can always be expressed by

$$\mathbf{P}(t) = (t^3, t^2, t, 1)\mathbf{M} \begin{pmatrix} \mathbf{P}_1 \\ \mathbf{P}_2 \\ \mathbf{P}_3 \\ \mathbf{P}_4 \end{pmatrix},$$

where $\mathbf{M}$ is a 4×4 basis matrix

$$\mathbf{M} = \begin{pmatrix} m_{11} & m_{12} & m_{13} & m_{14} \\ m_{21} & m_{22} & m_{23} & m_{24} \\ m_{31} & m_{32} & m_{33} & m_{34} \\ m_{41} & m_{42} & m_{43} & m_{44} \end{pmatrix}.$$

Examples of such matrices are Eqs. (4.19) and (4.26). A similar curve $\mathbf{Q}(u)$ that's defined by the same basis matrix but the *reverse* sequence of points has the form

$$\mathbf{Q}(u) = (u^3, u^2, u, 1)\mathbf{M} \begin{pmatrix} \mathbf{P}_4 \\ \mathbf{P}_3 \\ \mathbf{P}_2 \\ \mathbf{P}_1 \end{pmatrix}.$$

Our task is to find the conditions that the 16 elements of $\mathbf{M}$ have to satisfy in order to have $\mathbf{P}(t) = \mathbf{Q}(u)$ when both t and u vary from 0 to 1.

It turns out that all we have to do is substitute $u = 1 - t$ to get

$$\mathbf{Q}(t) = \left((1-t)^3, (1-t)^2, 1-t, 1\right)\mathbf{M} \begin{pmatrix} \mathbf{P}_4 \\ \mathbf{P}_3 \\ \mathbf{P}_2 \\ \mathbf{P}_1 \end{pmatrix}$$

$$= (t^3, t^2, t, 1) \begin{pmatrix} -1 & 0 & 0 & 0 \\ 3 & 1 & 0 & 0 \\ -3 & -2 & -1 & 0 \\ 1 & 1 & 1 & 1 \end{pmatrix} \mathbf{M} \begin{pmatrix} 0 & 0 & 0 & 1 \\ 0 & 0 & 1 & 0 \\ 0 & 1 & 0 & 0 \\ 1 & 0 & 0 & 0 \end{pmatrix} \begin{pmatrix} \mathbf{P}_1 \\ \mathbf{P}_2 \\ \mathbf{P}_3 \\ \mathbf{P}_4 \end{pmatrix}$$

$$= (t^3, t^2, t, 1)\mathbf{N} \begin{pmatrix} \mathbf{P}_1 \\ \mathbf{P}_2 \\ \mathbf{P}_3 \\ \mathbf{P}_4 \end{pmatrix},$$

where $\mathbf{N}$ is the 4×4 matrix

$$\mathbf{N} = \begin{pmatrix} -m_{14} & -m_{13} & -m_{12} & -m_{11} \\ 3m_{14} + m_{24} & 3m_{13} + m_{23} & 3m_{12} + m_{22} & 3m_{11} + m_{21} \\ -3m_{14} - 2m_{24} - m_{34} & -3m_{13} - 2m_{23} - m_{33} & -3m_{12} - 2m_{22} - m_{32} & -3m_{11} - 2m_{21} - m_{31} \\ m_{14} + m_{24} + m_{34} + m_{44} & m_{13} + m_{23} + m_{33} + m_{43} & m_{12} + m_{22} + m_{32} + m_{42} & m_{11} + m_{21} + m_{31} + m_{41} \end{pmatrix}.$$

It is easy to see that the condition for $\mathbf{Q}(t) = \mathbf{P}(t)$ is $\mathbf{N} = \mathbf{M}$. This can be written explicitly as a set of 16 relations (only 14 of which are independent) that the elements of matrix $\mathbf{M}$ have to satisfy:

$$\begin{aligned}
-m_{14} &= m_{11}, & -m_{13} &= m_{12}, & -m_{12} &= m_{13}, & -m_{11} &= m_{14}, \\
3m_{14}+m_{24} &= m_{21}, & 3m_{13}+m_{23} &= m_{22}, & 3m_{12}+m_{22} &= m_{23}, & 3m_{11}+m_{21} &= m_{24}, \\
-3m_{14} - 2m_{24} - m_{34} &= m_{31}, & -3m_{13} - 2m_{23} - m_{33} &= m_{32}, \\
-3m_{12} - 2m_{22} - m_{32} &= m_{33}, & -3m_{11} - 2m_{21} - m_{31} &= m_{34}, \\
m_{14} + m_{24} + m_{34} + m_{44} &= m_{41}, & m_{13} + m_{23} + m_{33} + m_{43} &= m_{42}, \\
m_{12} + m_{22} + m_{32} + m_{42} &= m_{43}, & m_{11} + m_{21} + m_{31} + m_{41} &= m_{44}.
\end{aligned} \quad (4.203)$$

A direct check shows that the Lagrange and Bézier basis matrices of Equations (4.19) and (4.107) satisfy this condition, but the Hermite basis matrix of Equation (4.26) does not.

The reason the Hermite basis matrix does not satisfy conditions (4.203) is that those conditions were developed for a curve that depends only on points. The Hermite curve, it should be remembered, is based on both points and tangent vectors, so there is no reason for it to satisfy the same symmetry conditions. However, it can be shown that this curve is symmetric. One point that should be mentioned with regard to the Hermite curve has to do with the directions of the tangent vectors. When traversing the Hermite curve in reverse, the directions of the two tangents have to be reversed, as illustrated by Figure 4.91. With this in mind, it is easy to show that the Hermite curve is symmetric. The reversed curve is denoted $\mathbf{Q}(u) = \mathbf{a}u^3 + \mathbf{b}u^2 + \mathbf{c}u + \mathbf{d}$ and it satisfies

$$\mathbf{Q}(0) = \mathbf{P}_2, \quad \mathbf{Q}(1) = \mathbf{P}_1, \quad \mathbf{Q}^u(0) = -\mathbf{P}_2^t, \quad \mathbf{Q}^u(1) = -\mathbf{P}_1^t.$$

This is written explicitly as the four equations

$$\begin{aligned}
\mathbf{a} \cdot 0^3 + \mathbf{b} \cdot 0^2 + \mathbf{c} \cdot 0 + \mathbf{d} &= \mathbf{P}_2, \\
\mathbf{a} \cdot 1^3 + \mathbf{b} \cdot 1^2 + \mathbf{c} \cdot 1 + \mathbf{d} &= \mathbf{P}_1, \\
3\mathbf{a} \cdot 0^2 + 2\mathbf{b} \cdot 0 + \mathbf{c} &= -\mathbf{P}_2^t, \\
3\mathbf{a} \cdot 1^2 + 2\mathbf{b} \cdot 1 + \mathbf{c} &= -\mathbf{P}_1^t,
\end{aligned}$$

whose solutions are

$$\begin{aligned}
\mathbf{a} &= 2\mathbf{P}_2 - 2\mathbf{P}_1 - \mathbf{P}_1^t - \mathbf{P}_2^t, \\
\mathbf{b} &= -3\mathbf{P}_2 + 3\mathbf{P}_1 + \mathbf{P}_1^t + 2\mathbf{P}_2^t, \\
\mathbf{c} &= -\mathbf{P}_2^t, \\
\mathbf{d} &= \mathbf{P}_2.
\end{aligned}$$

The reversed curve is thus

$$\begin{aligned}
\mathbf{Q}(u) &= (2\mathbf{P}_2 - 2\mathbf{P}_1 - \mathbf{P}_1^t - \mathbf{P}_2^t)u^3 + (-3\mathbf{P}_2 + 3\mathbf{P}_1 + \mathbf{P}_1^t + 2\mathbf{P}_2^t)u^2 - \mathbf{P}_2^t u + \mathbf{P}_2 \\
&= (-u^3 + 2u^2)\mathbf{P}_2^t + (-u^3 + u^2 - u)\mathbf{P}_1^t
\end{aligned}$$

4 Curves

$$+ (2u^3 - 3u^2 + 1)\mathbf{P}_2 + (-2u^3 + 3u^2)\mathbf{P}_1$$

$$= (u^3, u^2, u, 1) \begin{pmatrix} -1 & -1 & 2 & -2 \\ 2 & 1 & -3 & 3 \\ -1 & 0 & 0 & 0 \\ 0 & 0 & 1 & 0 \end{pmatrix} \begin{pmatrix} \mathbf{P}_2^t \\ \mathbf{P}_1^t \\ \mathbf{P}_2 \\ \mathbf{P}_1 \end{pmatrix}. \qquad (4.204)$$

Substituting $u = 1 - t$ and reversing the geometry vector transforms this into the standard Hermite curve

$$\mathbf{Q}(t) = (t^3, t^2, t, 1) \begin{pmatrix} -1 & 0 & 0 & 0 \\ 3 & 1 & 0 & 0 \\ -3 & -2 & -1 & 0 \\ 1 & 1 & 1 & 1 \end{pmatrix} \begin{pmatrix} -1 & -1 & 2 & -2 \\ 2 & 1 & -3 & 3 \\ -1 & 0 & 0 & 0 \\ 0 & 0 & 1 & 0 \end{pmatrix}$$

$$\times \begin{pmatrix} 0 & 0 & 0 & 1 \\ 0 & 0 & 1 & 0 \\ 0 & 1 & 0 & 0 \\ 1 & 0 & 0 & 0 \end{pmatrix} \begin{pmatrix} \mathbf{P}_1 \\ \mathbf{P}_2 \\ \mathbf{P}_1^t \\ \mathbf{P}_2^t \end{pmatrix}$$

$$= (t^3, t^2, t, 1) \begin{pmatrix} 2 & -2 & 1 & 1 \\ -3 & 3 & -2 & -1 \\ 0 & 0 & 1 & 0 \\ 1 & 0 & 0 & 0 \end{pmatrix} \begin{pmatrix} \mathbf{P}_1 \\ \mathbf{P}_2 \\ \mathbf{P}_1^t \\ \mathbf{P}_2^t \end{pmatrix},$$

thereby proving that the Hermite curve is symmetric even though it is based on more than just points and even though it does not satisfy conditions (4.203).

Figure 4.91: Reversing the Tangent Vectors.

4.20 Conic Sections

The ellipse, hyperbola, and parabola (and also the circle, which is a special case of the ellipse) are called the *conic section* curves (or the *conic sections* or just plain *conics*) since they can be obtained by cutting a cone with a plane (i.e., they are the intersections of a cone and a plane).

The conics are easy to calculate and to display, so they are used a lot in applications where they can approximate the shape of other, more complex, geometric figures. Many natural motions occur along an ellipse, parabola, or hyperbola, making these curves especially useful. Planets move in ellipses; many comets move along a hyperbola (as do many colliding charged particles); objects thrown in a gravitational field follow a parabolic path.

There are several ways to define and represent these curves and this section uses a simple *geometric* definition that leads naturally to the parametric and the implicit representations of the conics.

Definition: A conic is the locus of all the points **P** that satisfy the following: The distance of **P** from a fixed point **F** (the *focus* of the conic) is proportional to its distance from a fixed line **D** (the *directrix*). Using set notation, we can write

$$\text{Conic} = \{P | PF = ePD\},$$

where e is called the *eccentricity* of the conic. It is easy to classify conics by means of their eccentricity:

$$e = \begin{cases} = 1, & \text{parabola,} \\ < 1, & \text{ellipse (the circle is the special case } e = 0\text{),} \\ > 1, & \text{hyperbola.} \end{cases}$$

In the special case where the directrix is the y axis ($x = 0$) and the focus is point $(k, 0)$, the definition results in

$$\frac{\sqrt{(x-k)^2 + y^2}}{|x|} = e, \quad \text{or} \quad (1-e^2)x^2 - 2kx + y^2 + k^2 = 0. \tag{4.205}$$

In this case, the conic is represented by a degree-2 equation. It can be shown that this is true for the general case, where the directrix and the focus can be located anywhere. It can also be shown that the inverse is also true, i.e., any degree-2 algebraic equation of the form

$$ax^2 + by^2 + 2hxy + 2fx + 2gy + c = 0 \tag{4.206}$$

represents a conic. Equation (4.206) can be used to classify the conics. If D is the determinant

$$D = \begin{vmatrix} a & h & f \\ h & b & g \\ f & g & c \end{vmatrix},$$

then Table 4.92 gives a complete classification of the conics, including degenerate cases where the conic reduces to two lines (real or imaginary) or to a point.

▶ **Exercise 4.105:** Assume that the second-degree equation

$$Ax^2 + Bxy + Cy^2 + Dx + Ey + F = 0 \tag{4.207}$$

is given. Show how to use the six parameters to determine which conic is described by this equation.

Equation (4.205) can be used to generate the familiar implicit representations of the conics. We first treat the case $e \neq 1$ by transforming $x' = x - k/(1-e^2)$. When this is substituted into Equation (4.205) (and the prime is eliminated), the result is

$$\frac{x^2}{a^2} + \frac{y^2}{b^2} = 1, \tag{4.208}$$

where $\quad a = \dfrac{ke}{1-e^2} \quad$ and $\quad b^2 = a^2(1-e^2).$

4 Curves

$ab-h^2$	Conditions			Conic
$=0$	$D \neq 0$			parabola
	$D=0$	$b \neq 0$	$g^2 - bc > 0$	2 parallel lines
			$g^2 - bc = 0$	2 parallel coincident lines
			$g^2 - bc < 0$	2 parallel imaginary lines
		$b = h = 0$	$f^2 - ac > 0$	2 parallel real lines
			$f^2 - ac = 0$	2 parallel coincident lines
			$f^2 - ac < 0$	2 parallel imaginary lines
>0	$D=0$			point (degenerate ellipse)
	$D \neq 0$		$-bD > 0$	real ellipse
			$-bD < 0$	imaginary ellipse
<0	$D=0$			2 intersecting lines
	$D \neq 0$			hyperbola

Table 4.92: Classification of Conics.

Case 1: Ellipse. The case $e < 1$ implies that both a and b are positive and $a > b$. In this case, Equation (4.208) represents the canonical ellipse. This ellipse is centered on the origin with the x and y axes being the major and minor axes of the ellipse, respectively. The major radius is a and the minor one is b. For $a = b$, this ellipse reduces to a circle. Hence, we can think of a circle as the limit of the ellipse when $e \to 0$ and $k \to \infty$.

Case 2: Hyperbola. The case $e > 1$ implies a negative a and a negative b^2 (hence an imaginary b). If we use the absolute value of the imaginary b, Equation (4.208) becomes

$$\frac{x^2}{a^2} - \frac{y^2}{b^2} = 1, \qquad a, b > 0. \tag{4.209}$$

This is a canonical hyperbola, where the x axis is the *traverse axis* and the y axis is called the *semiconjugate* or *imaginary axis*. The hyperbola consists of two distinct parts with the imaginary axis separating them. The two points $(-a, 0)$ and $(a, 0)$ are called the *vertices* of the hyperbola.

Case 3: Parabola ($e = 1$). The simple transformation $x' = x - k/2$ yields, when substituted into Equation (4.205), the canonical parabola

$$y^2 = 4ax, \qquad \text{where} \quad a = k/2 > 0, \tag{4.210}$$

with focus at $(a, 0)$ (a is thus the focal distance) and directrix $x = -a$. The origin is the *vertex* of the canonical parabola.

All the conic sections can also be described (although not in their canonical forms) by the expression

$$f(\theta) = \frac{K}{1 \pm e \cos(\theta)}.$$

For $e = 0$ this is a circle. For $0 < e < 1$ this is an ellipse. For $e = 1$ this is a parabola and for $e > 1$ it is a hyperbola.

The parametric representations of the conics are simple. We start with the ellipse. In order to show that the expression

$$\left(a\frac{1-t^2}{1+t^2}, b\frac{2t}{1+t^2}\right), \quad -\infty < t < \infty, \qquad (4.211)$$

traces out an ellipse we show that it satisfies Equation (4.208):

$$\frac{a^2\left(\frac{1-t^2}{1+t^2}\right)^2}{a^2} + \frac{b^2\left(\frac{2t}{1+t^2}\right)^2}{b^2} = \frac{1-2t^2+t^4+4t^2}{1+2t^2+t^4} = 1.$$

The first quadrant is obtained for $0 \leq t \leq 1$. To get the second quadrant, however, t has to vary from 1 to ∞. Quadrants 4 and 3 are obtained for $-\infty \leq t \leq 0$.

The canonical hyperbola is represented parametrically by

$$\left(a\frac{1+t^2}{1-t^2}, b\frac{2t}{1-t^2}\right), \quad -\infty < t < \infty. \qquad (4.212)$$

The right branch is traced out when $-1 \leq t \leq 1$, and the left branch when $-\infty \leq t \leq -1$ and $1 \leq t \leq \infty$. The two values $t = \pm 1$ thus represents hyperbola points at infinity.

The simple expression

$$(at^2, 2at), \quad -\infty < t < \infty, \qquad (4.213)$$

traces out the canonical parabola.

Equations (4.211) and (4.212) are called *rational parametrics* since they contain the parameter t in the denominator. Rational parametric curves are generally complex but can represent more shapes and are therefore more general than the nonrational ones. One disadvantage of the rational parametrics is variable velocity. Varying t in equal increments generally results in traveling along the curve in unequal steps.

In practical work, it is sometimes necessary to have conics placed anywhere in three-dimensional space, not just on the xy plane. This is done by taking a general two-dimensional conic $\mathbf{P}(t)$ (one of Equations (4.211), (4.212), or (4.213)), adding a third coordinate $z = 0$ and transforming it with the general 4×4 transformation matrix $\mathbf{T}$, Equation (3.23). Normally, such a curve is translated and rotated. It may also be scaled and sheared. The result is a three-dimensional curve of the form

$$\mathbf{P}^*(t) = \left(\frac{a_0 + a_1 t + a_2 t^2}{w_0 + w_1 t + w_2 t^2}, \frac{b_0 + b_1 t + b_2 t^2}{w_0 + w_1 t + w_2 t^2}, \frac{c_0 + c_1 t + c_2 t^2}{w_0 + w_1 t + w_2 t^2}\right)$$

$$= \left(\frac{\sum_{i=0}^{2} a_i t^i}{\sum_{i=0}^{2} w_i t^i}, \frac{\sum_{i=0}^{2} b_i t^i}{\sum_{i=0}^{2} w_i t^i}, \frac{\sum_{i=0}^{2} c_i t^i}{\sum_{i=0}^{2} w_i t^i}\right).$$

4 Curves

Denoting $x_i = a_i/w_i$, $y_i = b_i/w_i$, $z_i = c_i/w_i$, and $\mathbf{a}_i = (x_i, y_i, z_i)$, we can write this as

$$\mathbf{P}^*(t) = \frac{w_0\mathbf{a}_0 + w_1\mathbf{a}_1 t + w_2\mathbf{a}_2 t^2}{w_0 + w_1 t + w_2 t^2} = \frac{\sum_{i=0}^{2} w_i \mathbf{a}_i t^i}{\sum_{i=0}^{2} w_i t^i}. \tag{4.214}$$

This is the general rational form of the conic sections. It can also be shown that any rational parametric expression of the form (4.214) represents a conic.

4.21 Parametric Space of a Curve

A parametric space curve has the form $\mathbf{P}(t) = (x(t), y(t), z(t))$. For each value of t, it produces values for (x, y, z) that are plotted as a point in the *object space* of the curve. Since a space curve may have a complex shape, its projection on the two-dimensional output device may be ambiguous and hard to visualize. The curve has to be rotated and seen from several points of view in order to get a good idea of its shape. An alternative is to plot the three components $x(t)$, $y(t)$, and $z(t)$ separately, each as a plane parametric curve that's a function of t. Such plots are said to exist in the *parametric space* of the curve. With a little practice, it is possible to draw conclusions from the three components and their slopes about the shape of the original space curve.

An example is the parametric curve of Equation (4.30), $\mathbf{P}(t) = (-t^3 + t^2 + t, -t^3 + 2t^2, -2t^3 + 3t^2)$; its three components are

$$x(t) = -t^3 + t^2 + t, \quad y(t) = -t^3 + 2t^2, \quad z(t) = -2t^3 + 3t^2.$$

Their values at the endpoints are

$$x(0) = 0,\ y(0) = 0,\ z(0) = 0, \quad x(1) = 1,\ y(1) = 1,\ z(1) = 1.$$

We conclude that the curve starts at $(0, 0, 0)$ and ends at $(1, 1, 1)$.

The slopes of the three components are

$$x'(t) = -3t^2 + 2t + 1, \quad y'(t) = -3t^2 + 4t, \quad z'(t) = -6t^2 + 6t.$$

At the two endpoints, the slopes are

$$x'(0) = 1,\ y'(0) = 0,\ z'(0) = 0, \quad x'(1) = 0,\ y'(1) = 1,\ z'(1) = 0.$$

This means that the start tangent vector of the curve is $(1, 0, 0)$; the end tangent vector is $(0, 1, 0)$.

4.22 Curvature and Torsion

The first derivative $\mathbf{P}^t(t)$ of a parametric curve $\mathbf{P}(t)$ has been used many times, since it has a simple geometric meaning; it is the *tangent vector* of the curve. In this section, we denote the unit tangent vector at point $\mathbf{P}(i)$ by $\mathbf{T}(i)$. Thus,

$$\mathbf{T}(i) = \frac{\mathbf{P}^t(i)}{|\mathbf{P}^t(i)|}.$$

The tangent vector is an example of an *intrinsic property* of a curve. An intrinsic property of a geometric figure depends only on the figure and not on the particular choice of the coordinate axes. Any geometric figure may have intrinsic (and also extrinsic) properties. A triangle has three angles and a quadrilateral has four edges, regardless of the choice of coordinates. The tangent vector of a curve, as well as its curvature, does not depend on the particular coordinate system used. The slope of a curve, in contrast, does depend on the particular coordinates used, so it is an extrinsic property of the curve.

▶ **Exercise 4.106:** Guess a few more intrinsic and extrinsic properties of geometric figures.

This section discusses the important intrinsic properties of parametric curves. They include the principal vectors (the tangent, normal, and binormal vectors), the principal planes (the osculating, rectifying, and normal planes), and the concepts of curvature and torsion. These properties are all local and they vary from point to point on the curve. They are, thus, functions of the parameter t. Notice that these properties exist for all curves, but the present discussion assumes that the curve is represented parametrically.

4.22.1 Normal Plane

The normal plane to a curve $\mathbf{P}(t)$ at point $\mathbf{P}(i)$ is the plane that's perpendicular to the tangent $\mathbf{P}^t(i)$ and contains point $\mathbf{P}(i)$. If $\mathbf{Q}$ is a general point on the normal plane, then Figure 4.93 shows that $(\mathbf{Q} - \mathbf{P}(i)) \bullet \mathbf{P}^t(i) = 0$ (see also Sections 5.3.2 and 5.10.5). This can be written $\mathbf{Q} \bullet \mathbf{P}^t(i) - \mathbf{P}(i) \bullet \mathbf{P}^t(i) = 0$ or

$$x \cdot x_i^t + y \cdot y_i^t + z \cdot z_i^t - (x_i \cdot x_i^t + y_i \cdot y_i^t + z_i \cdot z_i^t) = 0, \quad (4.215)$$

an expression that has the familiar form $Ax + By + Cz + D = 0$ (Section 5.3.2).

4.22.2 Principal Normal Vector

The second important vector associated with a curve is the *principal normal vector* $\mathbf{N}(t)$. This unit vector is normal to the curve (and is, thus, contained in the normal plane and is also perpendicular to the tangent vector), but it is called the principal normal since it points in a special direction, the direction in which the curve is turning. The principal normal vector points toward a point called the *center of curvature* of the curve. To express $\mathbf{N}(t)$ in terms of the curve and its derivatives, we select two nearby points, t and $t + \Delta t$, on the curve. The tangent vectors at the two points are $\mathbf{a} = \mathbf{P}^t(t)$ and $\mathbf{b} = \mathbf{P}^t(t + \Delta t)$, respectively. If we subtract them

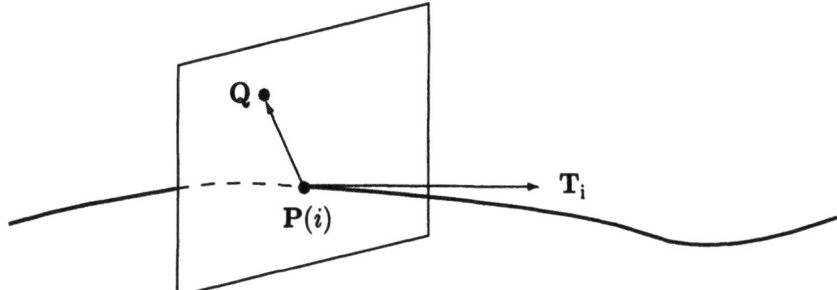

Figure 4.93: The Normal Plane.

as in Figure 4.94a, we get $c = b - a$. The difference vector c can be interpreted in two ways. On one hand, we can say that it is a small change in the tangent vector $\mathbf{P}^t(t)$, so we can denote it $\Delta\mathbf{P}^t(t)$. On the other hand, since the tangent vector can be interpreted as the velocity of the curve, any changes in it can be interpreted as acceleration, that is, the second derivative $\mathbf{P}^{tt}(t)$. We can thus write $c = \Delta\mathbf{P}^t(t) = \mathbf{P}^{tt}(t)$. The two vectors $a = \mathbf{P}^t(t)$ and $b = \mathbf{P}^t(t + \Delta t)$ define a plane and the principal normal vector lies at the intersection of this plane and the normal plane. Our task is therefore to compute a vector perpendicular to the tangent $a = \mathbf{P}^t(t)$ and contained in the plane defined by a and b.

Figure 4.94b shows vector $\vec{nl}$, which is the projection of $\mathbf{P}^{tt}(t)$ (vector $\vec{nm}$) onto $\mathbf{P}^t(t)$. Equation (A.14) tells us that the length of $\vec{nl}$ is

$$\frac{\mathbf{P}^{tt}(t) \bullet \mathbf{P}^t(t)}{|\mathbf{P}^t(t)|}.$$

Since $\vec{nl}$ is in the direction of $\mathbf{P}^t(t)$, we can write the vector $\vec{nl}$ as

$$\vec{nl} = \frac{\mathbf{P}^{tt}(t) \bullet \mathbf{P}^t(t)}{|\mathbf{P}^t(t)|} \cdot \frac{\mathbf{P}^t(t)}{|\mathbf{P}^t(t)|} = \frac{\mathbf{P}^{tt}(t) \bullet \mathbf{P}^t(t)}{|\mathbf{P}^t(t)|^2}\mathbf{P}^t(t).$$

We denote the vector $\vec{lm}$ by $\mathbf{K}(t)$ and compute it from the relation $\vec{nl} + \vec{lm} = \vec{nm} = \mathbf{P}^{tt}(t)$:

$$\mathbf{K}(t) = \mathbf{P}^{tt}(t) - \vec{nl} = \mathbf{P}^{tt}(t) - \frac{\mathbf{P}^{tt}(t) \bullet \mathbf{P}^t(t)}{|\mathbf{P}^t(t)|^2}\mathbf{P}^t(t). \qquad (4.216)$$

The principal normal vector $\mathbf{N}(t)$ is a unit vector in the direction of $\mathbf{K}(t)$, so it is given by

$$\mathbf{N}(t) = \frac{\mathbf{K}(t)}{|\mathbf{K}(t)|}.$$

▶ **Exercise 4.107:** What can we say about the nature of principal normal vector of a straight line?

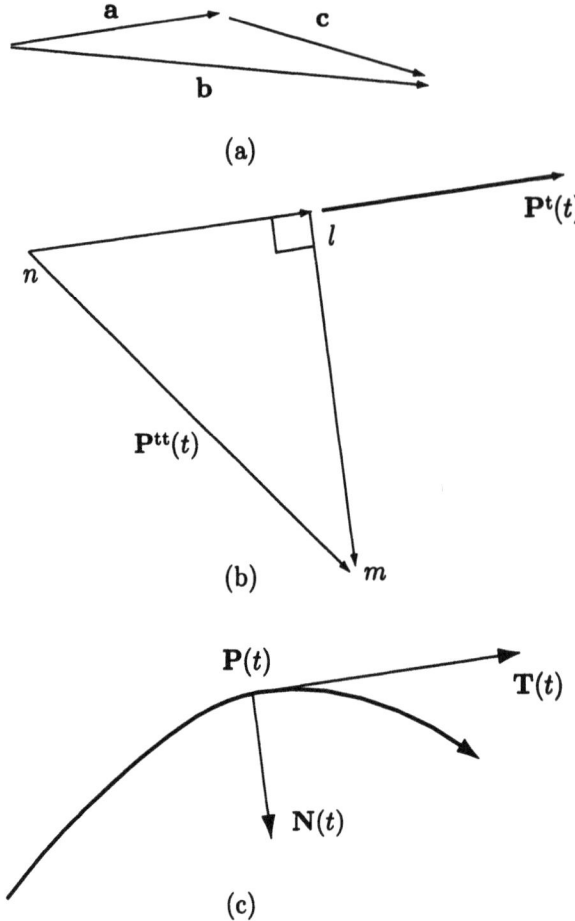

Figure 4.94: The Principal Normal Vector.

▶ **Exercise 4.108:** Calculate the principal normal vector of the PC curve $\mathbf{P}(t) = (-1,0)t^3 + (1,-1)t^2 + (1,1)t$. Notice that this curve is Equation (4.29), so we know that it goes from $(0,0)$ to $(1,0)$ with start and end tangents $(1,1)$, $(0,-1)$, respectively. Use this to check your results.

4.22.3 Binormal Vector

The third important vector associated with a curve is the *binormal vector* $\mathbf{B}(t)$. It is defined as the vector perpendicular to both the tangent and principal normal, so its definition is simply $\mathbf{B}(t) = \mathbf{T}(t) \times \mathbf{N}(t)$. Notice that it is a unit vector. Since the binormal is perpendicular to the tangent, it is contained in the normal plane. The three vectors $\mathbf{T}(t)$, $\mathbf{N}(t)$, and $\mathbf{B}(t)$ thus constitute an orthogonal coordinate system that moves along the curve as t varies, except at cusps, where they are undefined.

4.22.4 The Osculating Plane

Imagine three points h, i, and j, located close to each other on a curve. If they are not collinear, they define a plane. Now, move h and j independently closer and closer

4 Curves

to i. As these points move, the plane may change. The plane obtained at the limit is called the *osculating plane* at point i (Figure 4.95). It contains the tangent vector $\mathbf{T}(i)$ and the principal normal $\mathbf{N}(i)$. If $\mathbf{Q}$ is a general point on the osculating plane, then the plane equation is given by the determinant $|(\mathbf{Q} - \mathbf{P}(i))\, \mathbf{P}^t(i)\, \mathbf{P}^{tt}(i)| = 0$, which can be written explicitly as

$$(x - x_i)(y_i^t z_i^{tt} - y_i^{tt} z_i^t) - (y - y_i)(x_i^t z_i^{tt} - x_i^{tt} z_i^t) + (z - z_i)(x_i^t y_i^{tt} - x_i^{tt} y_i^t) = 0.$$

Another way to obtain the plane equation is to use the fact that point $\mathbf{P}(i)$ and vectors $\mathbf{T}(i)$ and $\mathbf{N}(i)$ are contained in it. Any general point $\mathbf{Q}$ in the osculating plane can, therefore, be expressed as $\mathbf{Q} = \mathbf{P}(i) + \alpha\mathbf{T}(i) + \beta\mathbf{N}(i)$, where α and β are real parameters. The osculating plane of a plane curve is, of course, the plane of the curve. The osculating plane of a straight line is undefined.

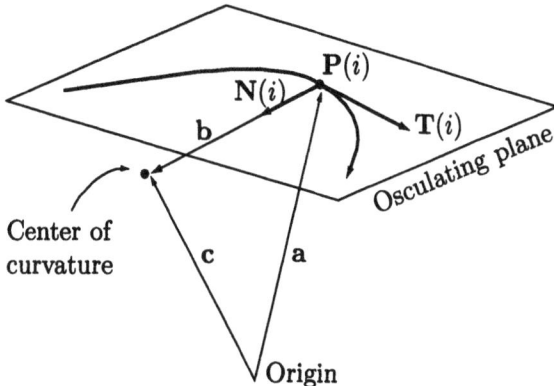

Figure 4.95: The Osculating Plane.

Incidentally, two curves joined at a point have C^2 continuity (Section 4.3.6) at the point if they have the same osculating planes and the same curvature vectors at the point.

▶ **Exercise 4.109:** (1) Calculate the Bézier curve for the four points $\mathbf{P}_0 = (0, 0, 0)$, $\mathbf{P}_1 = (1, 0, 0)$, $\mathbf{P}_2 = (2, 1, 0)$, and $\mathbf{P}_3 = (3, 0, 1)$. Notice that this is a space curve since the first three points are in the $z = 0$ plane, while the fourth one is outside that plane. (2) Calculate the (unnormalized) principal normal vector of the curve and find its values for $t = 0, 0.5$, and 1. (3) Calculate the osculating plane of the curve and find its equations for $t = 0, 0.5$, and 1 as above.

4.22.5 Rectifying Plane

The plane perpendicular to the principal normal vector of a curve is called the rectifying plane of the curve. If the curve is $\mathbf{P}(t)$, $\mathbf{N}(t)$ is its principal normal, and $\mathbf{Q}$ is a general point on the rectifying plane, then the equation of the rectifying plane at point $\mathbf{P}(i)$ is $(\mathbf{Q} - \mathbf{P}(i)) \bullet \mathbf{N}(i) = 0$. Another equation is obtained when we realize that both the tangent and binormal vectors are contained in the rectifying plane. A general point on this plane can, thus, be expressed as $\mathbf{Q} = \mathbf{P}(i) + \alpha\mathbf{T}(i) + \beta\mathbf{B}(i)$.

Figure 4.96 shows the three unit vectors and three planes associated with a particular point $\mathbf{P}(i)$ on a curve. They constitute intrinsic properties of the curve and together they form the *moving trihedron* of the curve, which can be considered a local coordinate system for the curve. The three vectors constitute the local coordinate axes and the three planes divide the space around point $\mathbf{P}(i)$ into eight octants. The curve passes through the normal plane and is tangent to both the osculating and rectifying planes.

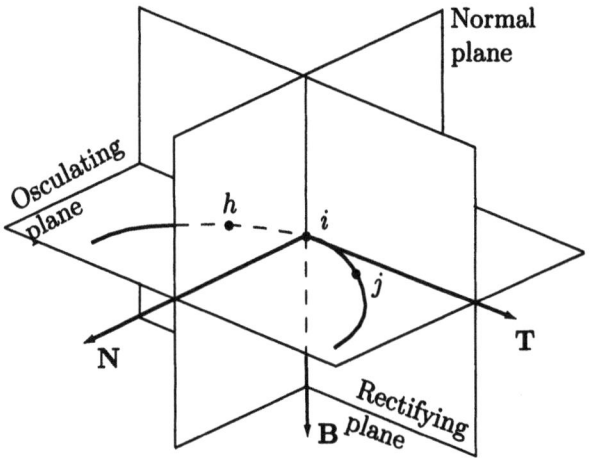

Figure 4.96: The Moving Trihedron.

4.22.6 Curvature

It is useful to define the curvature of a curve. Intuitively, the curvature should be a number that measures how much the curve deviates from a straight line. It should be large in areas where the curve wiggles, oscillates, or makes a sudden direction change; it should be small in areas where the curve is close to a straight line. It is also useful to associate a direction with the curvature, i.e., to make it a vector.

Given a parametric curve $\mathbf{P}(t)$ and a point $\mathbf{P}(i)$ on it, we calculate the first two derivatives $\mathbf{P}^t(i)$ and $\mathbf{P}^{tt}(i)$ of the curve at the point. We then construct a circle that has these same first and second derivatives and move it so it touches the point. This is called the *osculating circle* of the curve at the point. The curvature is now defined as the vector $\kappa(i)$ whose direction is from point $\mathbf{P}(i)$ to the center of this circle and whose magnitude is the reciprocal of the radius of the circle.

Using differential geometry, it can be shown that the vector

$$\frac{\mathbf{P}^t(t) \times \mathbf{P}^{tt}(t)}{|\mathbf{P}^t(t)|^3}$$

has the right magnitude. However, this vector is perpendicular to both $\mathbf{P}^t(t)$ and $\mathbf{P}^{tt}(t)$, so it is perpendicular to the osculating plane. To bring it into the plane, we

4 Curves

need to cross-product it with $\mathbf{P}^t(t)/|\mathbf{P}^t(t)|$, so the result is

$$\kappa(t) = \frac{\mathbf{P}^t(t) \times \mathbf{P}^{tt}(t) \times \mathbf{P}^t(t)}{|\mathbf{P}^t(t)|^4}. \tag{4.217}$$

Figure 4.95 shows that the curvature (vector **b**) is in the direction of the binormal $\mathbf{N}(t)$, so it can be expressed as $\kappa(t) = \rho(t)\mathbf{N}(t)$ where $\rho(t)$ is the *radius of curvature* at point $\mathbf{P}(t)$.

Given a curve $\mathbf{P}(t)$ with an arc length $s(t)$, we assume that $d\mathbf{P}/ds$ is a *unit tangent vector*:

$$\frac{d\mathbf{P}(t)}{ds} = \frac{d\mathbf{P}(t)}{dt}\frac{ds(t)}{dt} = \frac{\mathbf{P}^t(t)}{s^t(t)}. \tag{4.218}$$

Equation (4.218) shows the following:

1. $d\mathbf{P}(t)/ds$ and $\mathbf{P}^t(t)$ point in the same direction. Therefore, since $d\mathbf{P}(t)/ds$ is a unit vector, we get

$$\frac{d\mathbf{P}(t)}{ds} = \frac{\mathbf{P}^t(t)}{|\mathbf{P}^t(t)|}.$$

2. $s^t(t) = |\mathbf{P}^t(t)|$.

We now derive the expression for curvature from a different point of view. The curvature k is defined by $d^2\mathbf{P}(t)/ds^2 = k\mathbf{N}$, where $\mathbf{N}$ is the unit principal normal vector (Section 4.22.2). The problem is to express k in terms of the curve $\mathbf{P}(t)$ and its derivatives, not involving the (normally unknown) function $s(t)$. We start with

$$\frac{d^2\mathbf{P}(t)}{ds^2} = \frac{d}{ds}\left(\frac{\mathbf{P}^t(t)}{|\mathbf{P}^t(t)|}\right) = \frac{\frac{d}{dt}\left(\frac{\mathbf{P}^t(t)}{|\mathbf{P}^t(t)|}\right)}{s^t(t)}$$
$$= \frac{\frac{\mathbf{P}^{tt}(t)}{|\mathbf{P}^t(t)|} - \frac{\mathbf{P}^t(t)}{|\mathbf{P}^t(t)|^2} \cdot \frac{d|\mathbf{P}^t(t)|}{dt}}{|\mathbf{P}^t(t)|}. \tag{4.219}$$

The identity $\mathbf{A} \bullet \mathbf{A} = |\mathbf{A}|^2$ is true for any vector $\mathbf{A}(t)$ and it implies

$$\mathbf{A}(t) \bullet \mathbf{A}^t(t) = |\mathbf{A}(t)|\frac{d|\mathbf{A}(t)|}{dt}.$$

When we apply this to the vector $\mathbf{P}^t(t)$, we get

$$\frac{d^2\mathbf{P}(t)}{ds^2} = \frac{\mathbf{P}^{tt}(t)}{\mathbf{P}^t(t) \bullet \mathbf{P}^t(t)} - \frac{\mathbf{P}^t(t) \bullet \mathbf{P}^{tt}(t)}{\left(\mathbf{P}^t(t) \bullet \mathbf{P}^t(t)\right)^2}\mathbf{P}^t(t), \tag{4.220}$$

which can also be written

$$k\mathbf{N} = \frac{d^2\mathbf{P}(t)}{ds^2} = \frac{\mathbf{P}^t(t) \times \left(\mathbf{P}^{tt}(t) \times \mathbf{P}^t(t)\right)}{\left(\mathbf{P}^t(t) \bullet \mathbf{P}^t(t)\right)^2}. \tag{4.221}$$

4.22.7 Torsion

Torsion is a measure of how much a given curve deviates from a plane curve. The torsion $\tau(i)$ of a curve at a point $\mathbf{P}(i)$ is defined by means of the following two quantities:

1. Imagine a point h close to i. The curve has rectifying planes at points h and i (Figure 4.97). Denote the angle between them by θ.

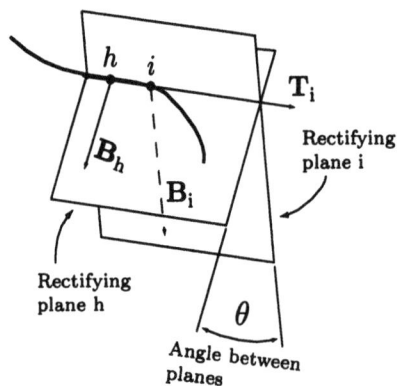

Figure 4.97: Torsion.

2. Denote by s the arc length from point h to point i.

The torsion of the curve at point i is defined as the limit of the ratio θ/s when h approaches i. Figure 4.97 shows how the rectifying plane rotates about the tangent as we move on the curve from h to i. The torsion can be expressed by means of the derivatives of the curve and by means of the curvature

$$\tau(t) = \frac{|\mathbf{P}^t(t)\,\mathbf{P}^{tt}(t)\,\mathbf{P}^{ttt}(t)|}{|\mathbf{P}^t(t) \times \mathbf{P}^t(t)|^2} = \frac{|\mathbf{P}^t(t)\,\mathbf{P}^{tt}(t)\,\mathbf{P}^{ttt}(t)|}{|\mathbf{P}^t(t)|^6}\rho(t)^2.$$

(The numerator is a determinant and the denominator is an absolute value. This expression is meaningful only when $\rho(t) < \infty$.) The torsion of a plane curve is zero.

It is interesting to note that a curve can be fully defined by specifying its curvature and torsion as functions of its arc length s. The functions $\kappa = f(s)$ and $\tau = g(s)$ uniquely define the shape of a curve (although not its location in space). An alternative is the single (implicit) function $F(\kappa, \tau, s) = 0$.

An alternative representation can be derived for a plane curve. Assume that $\mathbf{P}(t) = (x(t), y(t))$ is a curve in the xy plane. Figure 4.98 shows that its shape can be determined if its start point $\mathbf{P}(0)$ and its slope (or, equivalently, angle θ) are known as functions of the arc length s. Since θ is the angle between the tangent and the x axis, functions $x(s)$ and $y(s)$ must satisfy

$$\frac{dx}{ds} = \cos\theta, \qquad \frac{dy}{ds} = \sin\theta.$$

4 Curves

Differentiating produces

$$\frac{d^2x}{ds^2} = -\sin\theta \frac{d\theta}{ds} = -\frac{dy}{ds}\frac{d\theta}{ds}, \quad \frac{d^2y}{ds^2} = \cos\theta \frac{d\theta}{ds} = \frac{dx}{ds}\frac{d\theta}{ds}. \quad (4.222)$$

Figure 4.98 also shows that $d\theta/ds$ is the magnitude of the curvature κ, so the conclusion is that, given the curvature $\kappa(s)$ of a curve as a function of its arc length, the two functions $x(s)$, $y(s)$ can be calculated, either analytically, or point by point numerically, from the differential equations (4.222).

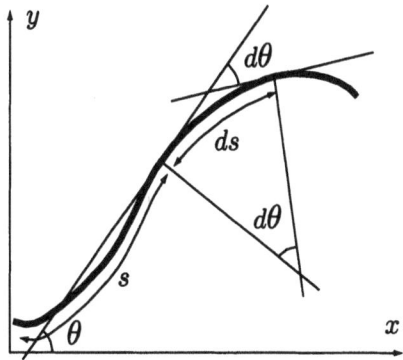

Figure 4.98: A Plane Curve.

▸ **Exercise 4.110:** Given $\kappa(s) = R$ (a constant), solve Equation (4.222) for $x(s)$ and $y(s)$. What kind of a curve is this?

4.22.8 Inflection Points

An inflection point is a point on a curve where the curvature is zero. On a straight line, every point is an inflection point. On a typical curve, an inflection point is created when the curve reverses its direction of turning (e.g., from a clockwise direction to a counterclockwise direction). From the definition of the curvature (Equation (4.217)) it follows that an inflection point satisfies

$$0 = |\mathbf{P}^t(t) \times \mathbf{P}^{tt}(t)| = \sqrt{(\mathbf{P}^t(t) \times \mathbf{P}^{tt}(t)) \bullet (\mathbf{P}^t(t) \times \mathbf{P}^{tt}(t))}.$$

Therefore,
$$(\mathbf{P}^t(t) \times \mathbf{P}^{tt}(t)) \bullet (\mathbf{P}^t(t) \times \mathbf{P}^{tt}(t)) = 0,$$

which is equivalent to

$$(\mathbf{P}^t(t) \times \mathbf{P}^{tt}(t))_x^2 + (\mathbf{P}^t(t) \times \mathbf{P}^{tt}(t))_y^2 + (\mathbf{P}^t(t) \times \mathbf{P}^{tt}(t))_z^2 = 0,$$
$$\text{or} \quad (y^t z^{tt} - z^t y^{tt})^2 + (z^t x^{tt} - x^t z^{tt})^2 + (x^t y^{tt} - y^t x^{tt})^2 = 0. \quad (4.223)$$

This is the sum of three nonnegative quantities, so each must be zero. Since

$$\frac{dy}{dx} = \frac{dy}{dt}\bigg/\frac{dx}{dt} = \frac{y^t}{x^t},$$

we get
$$\frac{d^2y}{dx^2} = \frac{d}{dt}\left(\frac{y^t}{x^t}\right)\frac{dt}{dx} = \frac{x^t y^{tt} - x^{tt} y^t}{(x^t)^3}.$$

Therefore, saying that the three quantities above are zero is the same as saying that
$$\frac{d^2y}{dx^2} = \frac{d^2x}{dz^2} = \frac{d^2z}{dy^2} = 0.$$

Equation (4.223) can be used to show that a two-dimensional parametric cubic can have at most two inflection points. We denote a general PC
$$\mathbf{P}(t) = \mathbf{a}t^3 + \mathbf{b}t^2 + \mathbf{c}t + \mathbf{d} = (a_x, a_y)t^3 + (b_x, b_y)t^2 + (c_x, c_y)t + (d_x, d_y),$$

which implies $x^t = 3a_x t^2 + 2b_x t + c_x$ and $x^{tt} = 6a_x t + b_x$, and similarly for y^t and y^{tt}. Using this notation, we write Equation (4.223) explicitly (notice that for a two-dimensional PC, only the third part is nonzero) as

$$\begin{aligned}0 &= x^t y^{tt} - y^t x^{tt} \\ &= (3a_x t^2 + 2b_x t + c_x)(6a_y t + b_y) - (3a_y t^2 + 2b_y t + c_y)(6a_x t + b_x) \\ &= 6(a_y b_x - a_x b_y)t^2 + 6(a_y c_x - a_x c_y)t + 2(b_y c_x - b_x c_y).\end{aligned}$$

This is a quadratic equation in t, so there can be at most two solutions.

4.23 The Hough Transform

This is a method for detecting patterns in bitmap images. We first show in detail how the Hough transform is used to detect straight lines, then discuss how to extend it to detect arbitrary parametric curves.

To understand the problem, let's consider the two straight segments $y = x$ and $y = x/3 + 4$ displayed on an 11×11 screen, where $0 \leq x, y \leq 10$. Looking at the screen, it is easy for humans to figure out that it features two lines. Identifying the two lines by software, however, is much harder since all that the software can "see" is the 121-bit bitmap

00000000001 00000000010 00000000100 00000001111 00000111000 00111100000
11001000000 00010000000 00100000000 01000000000 10000000000

(without the spaces), which corresponds to the pixels shown in Figure 4.99. How can a program find the equations of the lines, or even identify the fact that there are two lines, from the pixels in the bitmap? In practical cases, there may be many lines in the bitmap, with missing pixels and with some pixels at slightly wrong positions, complicating the problem even further.

The explicit equation of a straight line is $y = ax + b$, where a is the slope and b is the y intercept. For a given line, the parameters a and b are fixed and the coordinates x and y vary. The idea of the Hough transform is to reverse the roles of the parameters and the coordinates. Given a pair of coordinates (x_0, y_0), the transform calculates all the possible values of (a, b) that satisfy $y_0 = ax_0 + b$. In other words, given a point (x_0, y_0), the transform calculates all the pairs (a, b) of

4 Curves

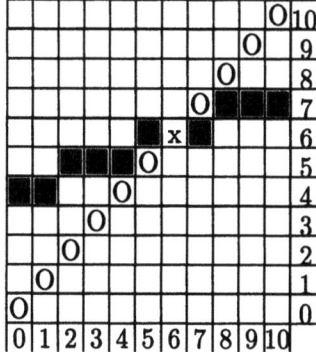

Figure 4.99: An 11×11 Bitmap.

straight lines that pass through the point. When these pairs are plotted in the ab plane, they define a line, since the values of each pair are linearly related. The ab plane is called the *parameter space*, to distinguish it from the *image space* (the xy plane) where the pixels exist.

Example: Given the point $(x_0, y_0) = (10, 10)$, we calculate the 11 pairs (a, b), where $b = 0, 1, \ldots, 9, 10$. From $10 = 10a + b$, we get $a = (10 - b)/10$. The 11 pairs are $(1, 0), (0.9, 1), (0.8, 2), (0.7, 3), \ldots, (0, 10)$.

Since there are infinitely many straight lines through any point, we have to limit the calculation to a subset of these lines (i.e., the values of a and b have to be "quantized"). We may decide, for instance, to calculate only (a, b) values that are integers in a certain range or that have just one digit to the right of the decimal point. For example, we may limit the calculation to $a = 0, 0.1, \ldots, 0.9, 1$ and $b = 0, 1, \ldots, 9, 10$ (lines with 11 slopes between 0° and 45° and 11 intercept values between 0 and 10). Each parameter can take on 11 values.

The pairs (a, b) being calculated are accumulated in a two-dimensional array ab of integers in memory, whose rows and columns correspond to values of a and b. The array should be large enough for all the possible values of pairs (a, b). In the above example, we need an array of 11×11 integers. All array elements are initially cleared. Each time a pair (a, b) is calculated, the corresponding array element is incremented by 1.

The algorithm works by scanning the bitmap, looking for bits of 1. For each 1-bit found, the following three steps are performed:

1. Calculate the screen coordinates (x, y) of the pixel.
2. Calculate all quantized pairs (a, b) for point (x, y).
3. For each pair (a, b) calculated, increment array location ab[a,b] by 1.

▶ **Exercise 4.111:** Assuming that the 121 bits of the bitmap are indexed 0 through 120 and given the row and column numbering of the pixels in Figure 4.99, figure out the row and column numbers of the pixel at bitmap index i.

This algorithm is a *transform* since it transforms points (x, y) in the image space (i.e., pixels) to points (a, b) in the parameter space (i.e., lines). It has the following properties:

1. A single point in image space is transformed to many points in the parameter space. Those points are on a line, so each image point is transformed to a line in the parameter space.

2. Any pair (a, b) defines the unique line $y = ax + b$, so any point in parameter space corresponds to a line in image space.

3. Imagine two points (x_1, y_1) and (x_2, y_2) on the line $y = mx + n$ in image space. Many lines go through each point, so each will cause many array elements to be incremented. Specifically, line $y = mx + n$ goes through both points, so each will cause array element ab[m,n] to be incremented. This element will, therefore, be incremented twice. In our example, there are 11 pixels on the line $y = x$ (for which $a = 1$, $b = 0$), so array element ab[1,0] will be incremented 11 times. Similarly, line $y = x/3 + 4$ will cause element ab[1/3,4] to be incremented 11 times. Any other elements of ab will be incremented fewer than 11 times.

4. Imagine two points (a_1, b_1) and (a_2, b_2) in parameter space. Each of them defines a line in image space. Imagine that these two lines intersect at point (x_0, y_0). All points (a, b) between (a_1, b_1) and (a_2, b_2) also define lines that intersect at (x_0, y_0).

Property 3 is the important one. It means that the two array elements ab[1,0] and ab[1/3,4] will have the largest counts. Identifying the two lines is therefore reduced to scanning array ab and finding the elements with the largest counts. Each such element ab[m,n] tells us that the line $y = mx + n$ exists in the image.

The first 1-bit is found in the bitmap at index 10. Therefore, it corresponds to the pixel at location $(10, 10)$. The parameter pairs for this point satisfy $10 = 10a + b$ or $a = (10 - b)/10$. The 11 pairs are therefore $(1, 0)$, $(0.9, 1)$, $(0.8, 2)$, $(0.7, 3),\ldots,(0, 10)$.

> No. No kidding, Cosmo. Did you ever see anything as ridiculous as me on that screen tonight?
> — Gene Kelly (as Don Lockwood) in *Singin' in the Rain* (1952).

▶ **Exercise 4.112:** The next 1-bit is found in the bitmap at index 20, so it corresponds to the pixel at location $(9, 9)$. Calculate the parameter pairs for this point.

Using the slope and intercept as parameters has the disadvantage that both can grow without a limit. A better set of parameters, the *normal* parameters, is shown in Figure 4.100. Parameter α is the angle between the x axis and the normal to the line $(0° \leq \alpha < 180°)$; β is the distance of the line from the origin $(0 \leq \beta \leq \infty)$. The straight line itself has the equation $x \cos\alpha + y \sin\alpha = \beta$, but if we consider α and β to be the variables, instead of x and y, we get a sinusoidal. Each point (x, y) in the image space is therefore transformed into a sinusoidal in the parameter space. However, when several points on a straight line are transformed into sinusoidals, all the sinusoidals intersect at one point. The four properties above also hold but have to be rephrased as follows:

1. A point (x, y) in the image space is transformed to a sinusoidal curve in the parameter space.

4 Curves

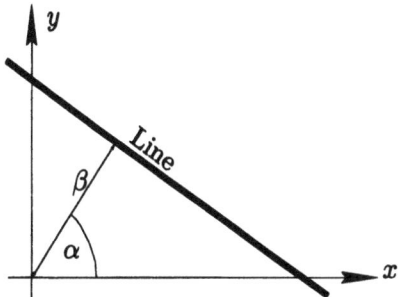

Figure 4.100: The Normal Parameters for a Line.

2. A point (α, β) in the parameter space corresponds to a straight line in the image space.

3. Points lying on a straight line in the image space correspond to sinusoidals that intersect at one point in the parameter space.

4. Points lying on a sinusoidal curve in the parameter space correspond to lines intersecting at one point in the image space.

The last feature of the Hough transform is its extension to arbitrary parametric curves. Suppose that $\mathbf{P}(t_1, t_2, \ldots, t_n)$ is a parametric curve defined by n parameters. For each n-tuple $(t_1, t_2, \ldots, t_n)$, the curve passes through a point (x, y). The Hough transform maps each point (x, y) in the image space to many points $(t_1, t_2, \ldots, t_n)$ in the n-dimensional parameter space. If $n = 2$, each point is transformed into a two-dimensional curve. For $n = 3$, each point is transformed into a three-dimensional surface. For larger values of n, each point is transformed into an n-dimensional *hypersurface*.

> "Not too good on curves," she exclaimed, and put her foot down on the gas. "Shall I take you home, Harold?"
>
> Colin Higgins, *Harold and Maude*

5
Surfaces

A surface is the next step beyond a single curve. This is illustrated by the following two features, which are demonstrated many times throughout this chapter:

1. Many curve methods can easily be extended to surfaces.
2. A surface can be drawn as a wire frame by drawing two perpendicular sets of curves (Figure 5.2b).

Surfaces are important in computer graphics since a solid object can be displayed by constructing and displaying its surface. There is no need to worry about the "inside" of the object. This chapter introduces the following types of surfaces: polygonal, bilinear, lofted, Coons, bicubic, Catmull-Rom, Bézier, Gordon, B-spline, sweep, and surfaces of revolution. It also shows how the tangent and normal vectors are calculated for the different types of surfaces.

Almost all the surfaces discussed here are represented parametrically and are based on data points, control points, or tangent vectors. However, a surface can also be represented explicitly by $z = f(x, y)$, or implicitly by $F(x, y, z) = 0$. An example of an explicit surface is $z = Ax + By + C$, representing a flat plane. A simple implicit surface is the sphere $x^2 + y^2 + z^2 - R^2 = 0$. A close relative is the ellipsoid $x^2/a^2 + y^2/b^2 + z^2/c^2 - 1 = 0$.

In practice, the implicit and explicit representations are used for polygonal surfaces or for regular solids. Curved surfaces are normally represented parametrically and are based on data points or control points.

> According to positivism, the world is all surface.
> — Karl Raimund Popper.

5.1 Input Three-Dimensional Points

When writing a program to deal with three-dimensional objects, one of the first questions is how to enter the coordinates of three-dimensional points. The simplest way is for the user to enter three numbers per point from the keyboard. The most sophisticated way is to use a three-dimensional digitizer and to digitize a model of the object. Users who like to avoid excessive typing but don't have access to a three-dimensional digitizer can use one of the following methods:

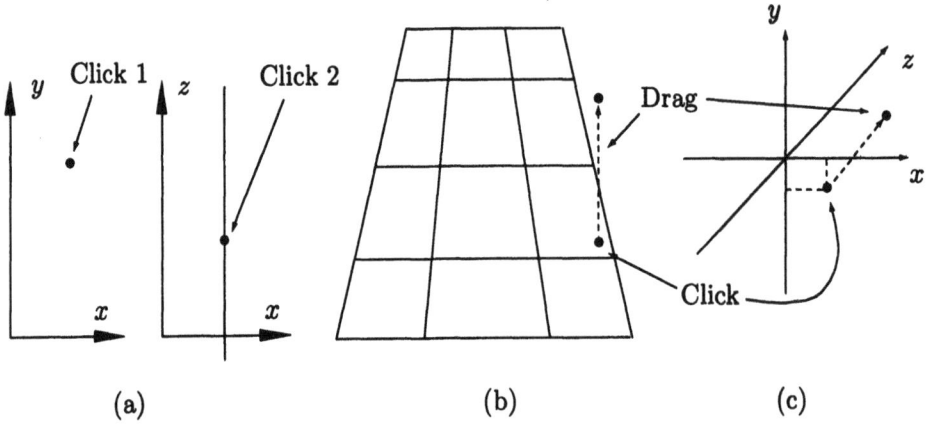

Figure 5.1: Entering Three-Dimensional Points.

1. Click the mouse to define a point on the screen. This determines the x and y coordinates of the point, and the z coordinate is then entered from the keyboard.

2. Display two coordinate systems on the screen, one for the xy coordinates and the other for xz (Figure 5.1a). The mouse is clicked in the first system, determining the x and y coordinates. The program then draws a line in the second system, corresponding to the x coordinate, and the mouse is clicked again, on that line (or close to it), to determine the z coordinate. In this case, it is useful to have a third window on the screen, showing a perspective projection of the entire coordinate system.

3. Display the xy plane, as a wire frame, on the screen (Figure 5.1b). The mouse is used to select a point on the plane (i.e., a point with $z = 0$) and is used again to drag it out of the plane in order to assign it a z coordinate. This method is handy when the points have to be equally spaced in xy, since they can be predrawn on the plane.

4. Display something similar to Figure 5.1c. The first mouse click determines the x and y coordinates. The second click should take place on or near the diagonal line representing the z axis. This click determines the z coordinate. Any number of points can therefore be entered, with two clicks per point.

Other methods can easily be developed.

5 Surfaces

> Klik hîr'
> — (Commonly seen in web pages.)

5.2 Basic Concepts

The general form of a parametric surface is $\mathbf{P}(u,w) = (f_1(u,w), f_2(u,w), f_3(u,w))$. The surface depends on two parameters, u and w, which vary independently in some range $[a,b]$ (normally, but not always, $[0,1]$). For each pair (u,w), the expression above produces the three coordinates of a point on the surface.

▶ **Exercise 5.1:** A curve can be either two-dimensional or three-dimensional. A surface, however, exists only in three dimensions, and each surface point has three coordinates. Why is it that the expression for the surface depends on two, and not on three, parameters? We would expect the surface to be of the form $\mathbf{P}(u,v,w)$, a function of three parameters. What's the explanation?

A simple example of a parametric surface is

$$\mathbf{P}(u,w) = [0.5(1-u)w + u, w, (1-u)(1-w)] \tag{5.1}$$

(this is Equation (5.9)). Such a surface is called *bilinear* since it is linear in both parameters. We use this example to discuss the concept of a surface patch and to show how a wire-frame surface can be displayed.

5.2.1 A Surface Patch

The expression $\mathbf{P}(u, 0.2)$ (where w is held fixed and u varies) depends on just one parameter and is therefore a curve on the surface. The four curves $\mathbf{P}(u,0)$, $\mathbf{P}(u,1)$, $\mathbf{P}(0,w)$, and $\mathbf{P}(1,w)$ are especially interesting. They are the *boundary curves* of the surface (Figure 5.2a). Since there are four such curves, our surface is a *patch* that has a (roughly) rectangular shape. Of special interest are the four quantities $\mathbf{P}(0,0)$, $\mathbf{P}(0,1)$, $\mathbf{P}(1,0)$, and $\mathbf{P}(1,1)$. They are the corner points of the surface patch.

We now say that the curve $\mathbf{P}(u, 0.2)$ goes on the surface in the u direction. Similarly, any curve $\mathbf{P}(u_0, w)$ where u_0 is fixed goes in the w direction. These are the two main directions on a rectangular surface patch.

A large surface is obtained by constructing a number of patches and connecting them. The method used to construct the patch should allow for smooth connection of patches.

▶ **Exercise 5.2:** Calculate the corner points and boundary curves of the bilinear surface patch of Equation (5.1).

▶ **Exercise 5.3:** Calculate the corner points and boundary curves of the surface patch

$$\mathbf{P}(u,w) = \big((c-a)u + a, (d-b)w + b, 0\big),$$

where a, b, c, and d are given constants and the parameters u and w vary independently in the range $[0,1]$. What kind of a surface is it?

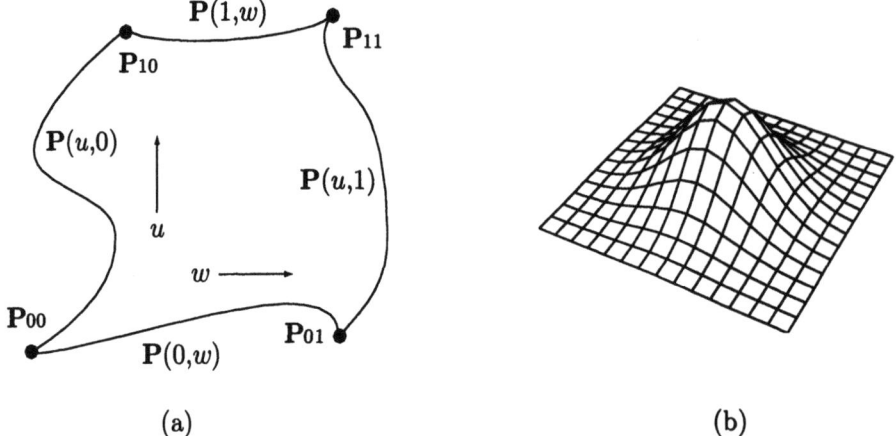

Figure 5.2: (a) A Surface Patch. (b) A Wire Frame.

5.2.2 Displaying a Surface Patch

A surface patch can be displayed either as a wire frame or as a solid surface. The pseudo-code of Figure 5.3 shows how to display a surface patch as a wire frame (Figure 5.2b). The code consists of two similar loops—one drawing the curves in the w direction, and the other drawing them in the u direction. The first loop varies u from 0 to 1 in steps of 0.2, so six curves are drawn. Each of the six is drawn by varying w in small steps (0.01 in the example). The second loop is similar and draws six curves in the u direction.

Procedure SurfacePoint receives the current values of u and w, and calculates the coordinates (x,y,z) of one surface point. Procedure PersProj uses these coordinates to calculate the screen coordinates (xs, ys) of a pixel. Finally, procedure Pixel actually displays the pixel in the desired color.

```
for u:=0 to 1 step 0.2 do            for w:=0 to 1 step 0.2 do
  begin                                begin
  for w:=0 to 1 step 0.01 do           for u:=0 to 1 step 0.01 do
    begin                                begin
    SurfacePoint(u,w,x,y,z);             SurfacePoint(u,w,x,y,z);
    PersProj(x,y,z,xs,ys);               PersProj(x,y,z,xs,ys);
    Pixel(xs,ys,color)                   Pixel(xs,ys,color)
    end                                  end
  end;                                 end;
```

Figure 5.3: Procedure for a Wire-Frame Surface.

To display a solid surface, the *normal vector* of the surface has to be calculated and a shading method applied to calculate the amount of light reflected from a surface point (Chapter 6). This must be repeated for every point on the surface.

5.3 Polygonal Surfaces

Such a surface is made of a number of flat faces, each a polygon. A polygon in such a surface is typically a triangle, since the three points of a triangle are always on the same plane. With higher-order polygons, the designer should make sure that all the points defining the polygon are on the same plane.

Each polygon is a collection of vertices (the points defining it) and edges (the lines connecting the points). Such a surface is easy to display, either as a wire frame or as a solid surface. In the former case, the edges should be displayed. In the latter case, all the points in a polygon are assigned the same color and brightness. They are all assumed to reflect the same amount of light, since the polygon is flat and has only one normal vector. A shaded polygonal surface thus appears angular and unnatural, but we will see (in Section 6.3) that it is easy to smooth out the reflection and make the polygonal surface look curved.

Three methods are described for representing such a surface in memory:

1. Explicit polygons. Each polygon is represented as a list

$$\big((x_1,y_1,z_1),(x_2,y_2,z_2),\ldots,(x_n,y_n,z_n)\big)$$

of its vertices, and it is assumed that there is an edge from point 1 to point 2, from 2 to 3, etc., and also from point n to point 1.

This representation is simple but has two disadvantages:

i. A point may be shared by several polygons, so several copies have to be stored. If the user decides to modify the point, all the copies have to be located and updated.

ii. An edge may also be shared by several polygons. When displaying the surface, such an edge will be displayed several times, slowing down the entire process.

2. Polygon definition by pointers. There is one list

$$\mathbf{V} = \big((x_1,y_1,z_1),(x_2,y_2,z_2),\ldots,(x_n,y_n,z_n)\big)$$

of all the vertices in the surface. A polygon is represented as a list of pointers, each pointing to a vertex in $\mathbf{V}$. Hence, $\mathbf{P} = (3,5,7,10)$ means that polygon $\mathbf{P}$ consists of vertices 3, 5, 7, and 10 in $\mathbf{V}$. Problem *ii* still exists.

3. Explicit edges. List $\mathbf{V}$ is as before, and there is also an edge list

$$\mathbf{E} = \big((v_1,v_6,p_3),(v_5,v_7,p_1,p_3,p_6,p_8),\ldots\big).$$

Each element of $\mathbf{E}$ represents an edge. It contains two pointers to the vertices of the edge followed by pointers to all the polygons that share the edge. Each polygon is represented by a list of pointers to $\mathbf{E}$, e.g., $\mathbf{P}_1 = (e_1, e_4, e_5)$. Problem *ii* still exists.

5.3.1 Polygon Planarity

Given a polygon defined by points $\mathbf{P}_1, \mathbf{P}_2, \ldots, \mathbf{P}_n$, we use the scalar triple product (Equation (A.13)) to test if the polygon is planar (i.e., if all its vertices $\mathbf{P}_i$ are on the same plane). Such a test is necessary only if $n > 3$. We select $\mathbf{P}_1$ as the "pivot" point and calculate the $n-1$ pivot vectors $\mathbf{v}_i = \mathbf{P}_i - \mathbf{P}_1$ for $i = 2, \ldots, n$. Next, we calculate the $n-3$ scalar triple products $\mathbf{v}_i \bullet (\mathbf{v}_2 \times \mathbf{v}_3)$ for $i = 4,\ldots,n$. If any of these products is nonzero, the polygon is not planar. Note that limited accuracy on some computers may cause an otherwise null triple product to come out as a small floating-point number.

▸ **Exercise 5.4:** Consider the polygon defined by the four points $\mathbf{P}_1 = (1,0,0)$, $\mathbf{P}_2 = (0,1,0)$, $\mathbf{P}_3 = (1,a,1)$, and $\mathbf{P}_4 = (0,-a,0)$. For what values of a will it be planar?

5.3.2 Plane Equations

A polygonal surface consists of flat polygons (normally triangles). To calculate the normal to a polygon, we first need to know the polygon's equation. The equation of a flat plane is $Ax+By+Cz+D = 0$. It seems that we need four equations in order to calculate the four unknown coefficients A, B, C, and D, but it turns out that three equations are enough. Assuming that the three points $\mathbf{P}_i = (x_i, y_i, z_i)$, $i = 1, 2, 3$, are given, we can write the four equations

$$Ax + By + Cz + D = 0,$$
$$Ax_1 + By_1 + Cz_1 + D = 0,$$
$$Ax_2 + By_2 + Cz_2 + D = 0,$$
$$Ax_3 + By_3 + Cz_3 + D = 0.$$

The first equation is true for any point (x, y, z) on the plane. We cannot solve this system of four equations in four unknowns, but we know that it has a solution if and only if its determinant is zero. The expression below assumes this and also expands the determinant by its top row:

$$0 = \begin{vmatrix} x & y & z & 1 \\ x_1 & y_1 & z_1 & 1 \\ x_2 & y_2 & z_2 & 1 \\ x_3 & y_3 & z_3 & 1 \end{vmatrix}$$

$$= x \begin{vmatrix} y_1 & z_1 & 1 \\ y_2 & z_2 & 1 \\ y_3 & z_3 & 1 \end{vmatrix} - y \begin{vmatrix} x_1 & z_1 & 1 \\ x_2 & z_2 & 1 \\ x_3 & z_3 & 1 \end{vmatrix} + z \begin{vmatrix} x_1 & y_1 & 1 \\ x_2 & y_2 & 1 \\ x_3 & y_3 & 1 \end{vmatrix} - \begin{vmatrix} x_1 & y_1 & z_1 \\ x_2 & y_2 & z_2 \\ x_3 & y_3 & z_3 \end{vmatrix}.$$

This is of the form $Ax + By + Cz + D = 0$, so we conclude that

$$A = \begin{vmatrix} y_1 & z_1 & 1 \\ y_2 & z_2 & 1 \\ y_3 & z_3 & 1 \end{vmatrix} \quad B = -\begin{vmatrix} x_1 & z_1 & 1 \\ x_2 & z_2 & 1 \\ x_3 & z_3 & 1 \end{vmatrix} \quad C = \begin{vmatrix} x_1 & y_1 & 1 \\ x_2 & y_2 & 1 \\ x_3 & y_3 & 1 \end{vmatrix} \quad D = -\begin{vmatrix} x_1 & y_1 & z_1 \\ x_2 & y_2 & z_2 \\ x_3 & y_3 & z_3 \end{vmatrix}.$$
(5.2)

▸ **Exercise 5.5:** Calculate the expression of the plane containing the z axis and passing through the point $(1,1,0)$.

5 Surfaces

> We operate with nothing but things which do not exist, with lines, planes, bodies, atoms, divisible time, divisible space—how should explanation even be possible when we first make everything into an image, into our own image!
> — Friedrich Nietzsche.

▶ **Exercise 5.6:** In the plane equation $Ax + By + Cz + D = 0$ if $D = 0$, then the plane passes through the origin. Assuming $D \neq 0$, we can write the same equation as $x/a + y/b + z/c = 1$, where $a = -D/A$, $b = -D/B$, and $c = -D/C$. What is the geometrical interpretation of a, b, and c?

In some practical situations, the normal to the plane as well as one point on the plane, are known. It is easy to derive the plane equation in such a case.

We assume that $\mathbf{N}$ is the (known) normal vector to the plane, $\mathbf{P}_1$ is a known point, and $\mathbf{P}$ is any point in the plane. The vector $\mathbf{P} - \mathbf{P}_1$ is perpendicular to $\mathbf{N}$, so their dot product $\mathbf{N} \bullet (\mathbf{P} - \mathbf{P}_1)$ equals zero. Since the dot product is associative, we can write $\mathbf{N} \bullet \mathbf{P} = \mathbf{N} \bullet \mathbf{P}_1$. The dot product $\mathbf{N} \bullet \mathbf{P}_1$ is just a number, to be denoted by s, so we get

$$\mathbf{N} \bullet \mathbf{P} = s \quad \text{or} \quad N_x x + N_y y + N_z z - s = 0. \tag{5.3}$$

Equation (5.3) can now be written as $Ax + By + Cz + D = 0$, where $A = N_x$, $B = N_y$, $C = N_z$, and $D = -s = -\mathbf{N} \bullet \mathbf{P}_1$. The three unknowns A, B, and C are, thus, simply the components of the normal vector and D can be calculated from any known point $\mathbf{P}_1$ on the plane. The expression $\mathbf{N} \bullet \mathbf{P} = s$ is a useful equation of the plane and is used later.

▶ **Exercise 5.7:** Given $\mathbf{N}(u, w) = (1, 1, 1)$ and $\mathbf{P}_1 = (1, 1, 1)$, calculate the plane equation.

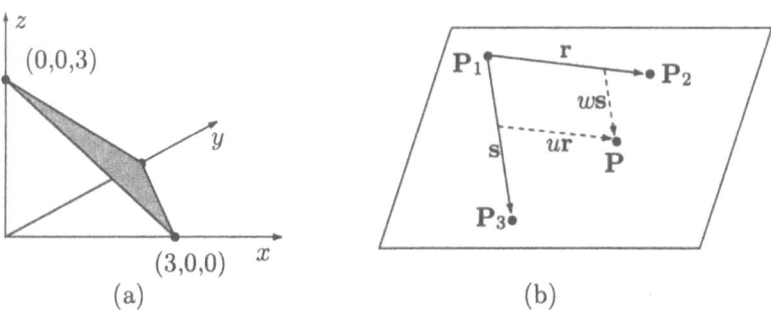

Figure 5.4: (a). A Plane. (b) Three Points on a Plane.

Note that the direction of the normal in this case is unimportant. Substituting $(-A, -B, -C)$ for (A, B, C) would also change the sign of D, resulting in the same equation. However, the direction of the normal is important when the surface is to be shaded. We want the normal, in such a case, to point *outside* the surface.

This has to be done by the user, since the computer has no idea of the shape of the surface and the meaning of "inside" and "outside." In the case where a plane is defined by three points, the direction of the normal can be specified by arranging the three points (in the data structure in memory) in a certain order.

It is also easy to get the equation of a plane when three points on the plane, P_1, P_2, P_3, are known. In order for the points to define a plane, they should not be collinear. We consider the vectors $r = P_2 - P_1$ and $s = P_3 - P_1$ a local coordinate axes on the plane. Any point P on the plane can be expressed as a linear combination $P = ur + ws$, where u and w are real numbers. Since r and s are local coordinates on the plane, the position of point P relative to the origin is expressed as (Figure 5.4b)

$$P(u, w) = P_1 + ur + ws, \quad -\infty < u, w < \infty. \quad (5.4)$$

▶ **Exercise 5.8:** Given the three points $P_1 = (3,0,0)$, $P_2 = (0,3,0)$, and $P_3 = (0,0,3)$, write the equation of the plane defined by them.

5.3.3 Space Division

An infinite plane divides the entire three-dimensional space into two parts. We can call them "outside" and "inside" (or "above" and "below"), and define the outside direction as the direction pointed by the normal. Using the plane equation, $N \bullet P = s$, it is possible to tell if a given point P_i lies inside, outside, or on the plane. All that's necessary is to examine the sign of the dot product $N \bullet (P_i - P)$, where P is any point on the plane different from P_i.

This dot product can also be written $|N| |P_i - P| \cos \theta$, where θ is the angle between the normal N and the vector $P_i - P$. The sign of the dot product equals the sign of $\cos \theta$, and Figure 5.5a shows that for $-90° < \theta < 90°$, point P_i lies outside the plane, for $\theta = 90°$, point P_i lies on the plane, and for $\theta > 90°$, P_i lies inside the plane.

> The regular division of the plane into congruent figures evoking an association in the observer with a familiar natural object is one of these hobbies or problems... I have embarked on this geometric problem again and again over the years, trying to throw light on different aspects each time. I cannot imagine what my life would be like if this problem had never occurred to me; one might say that I am head over heels in love with it, and I still don't know why.
> — M. C. Escher.

5.3.4 Turning Around on a Polygon

When moving along the edges of a polygon from vertex to vertex, we make a turn at each vertex. Sometimes, the "sense" of the turn (left or right) is important. The trouble is that the terms "left" and "right" are relative, depending on the location of the observer, and are therefore ambiguous. Consider Figure 5.5b. It shows two edges, **a** and **b**, of a "thick" polygon, with two arrows pointing from **a** to **b**. Imagine each arrow to be a bug crawling on the polygon. The bug on the top considers the

5 Surfaces

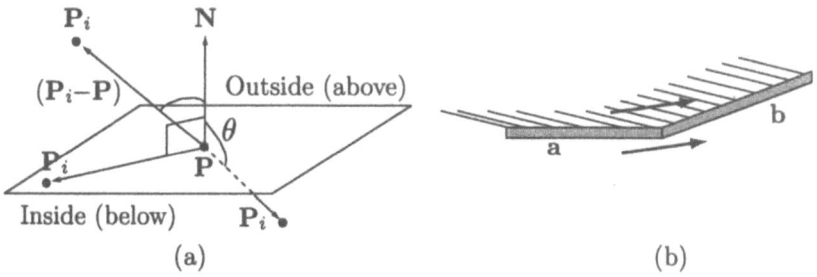

Figure 5.5: (a) Space Division. (b) Turning On A Polygon

turn from **a** to **b** a left turn, while the bug crawling on the bottom considers the same turn to be a "right" one.

We, therefore, prefer to define terms such as "positive turn" and "negative turn," that depend on the polygon and on the coordinate axes, but not on the position of any observer. To define these terms, we consider the plane defined by the vectors **a** and **b** (if they are parallel, they don't define any plane, but then there is no sense talking about turning from **a** to **b**). The cross product **a** × **b** is a vector perpendicular to the plane. It can point in the direction of the normal **N** to the plane, or in the opposite direction. In the former case, we say that the turn from **a** to **b** is positive; in the latter case, the turn is said to be negative.

To calculate the sense of the turn, we simply check the sign of the triple scalar product **N** • (**a** × **b**). A positive sign implies a positive turn.

▶ **Exercise 5.9:** Why?

5.3.5 Convex Polygons

Given a polygon, we select two arbitrary points on the edges and connect them with a straight line. If for any two such points the line is fully contained in the polygon, then the polygon is called *convex*. Another way to define a convex polygon is to say that a line can intersect such a polygon at only two points (unless the line is identical to one of the edges).

The sense of a turn (positive or negative) can also serve to define a convex polygon. When traveling from vertex to vertex in such a polygon all turns should have the same sense. They should all be positive or all negative. In contrast, when traveling along a concave polygon, both positive and negative turns must be made (Figure 5.6).

We can think of a polygon as a set of points in two dimensions. The concept of a set of points, however, exists in any number of dimensions. A set of points is convex if it satisfies the definition regardless of the number of dimensions. One important concept associated with a set of points is the *convex hull* of the set. This is the set of "extreme" points that satisfies the following: The set obtained by connecting the points of the convex hull contains all the points of the set. (A simple, two-dimensional analogy is to consider the points nails driven into a board. A rubber band placed around all the nails and stretched will identify the points that constitute the convex hull.) A simple algorithm for finding the convex hull is outlined in Section 5.4.2.

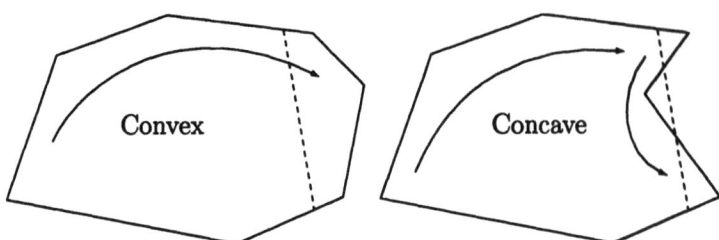

Figure 5.6: Convex and Concave Polygons.

5.3.6 Line and Plane Intersection

Given a plane $\mathbf{N} \bullet \mathbf{P} = s$ and a line $\mathbf{P} = \mathbf{P}_1 + \mu \mathbf{d}$ (Equation (A.12), page 704), it is easy to calculate their intersection point. We simply substitute the value of $\mathbf{P}$ in the plane equation to get $\mathbf{N} \bullet (\mathbf{P}_1 + \mu \mathbf{d}) = s$. This results in $\mu = (s - \mathbf{N} \bullet \mathbf{P}_1)/(\mathbf{N} \bullet \mathbf{d})$. Thus, we compute the value of μ and substitute it in the line equation, to get the point of intersection. Such a process is important in ray tracing (Section 6.5), for which the intersections of light rays and polygons are calculated all the time.

▶ **Exercise 5.10:** The intersection of a line parallel to a plane is either the entire line (if the line happens to be in the plane) or is empty. How do we distinguish these cases from the equation above?

5.3.7 Triangles

A polygonal surface is often constructed of triangles. Each triangle is flat, but it is not an infinite plane. We therefore need to modify the plane equation to describe just the area inside the triangle

Given any three noncollinear points $\mathbf{P}_1$, $\mathbf{P}_2$, and $\mathbf{P}_3$ in three dimensions, we first derive the equation of the (infinite) plane defined by them. Following that, we limit ourselves to just that part of the plane that's inside the triangle. We start with the two vectors $(\mathbf{P}_2 - \mathbf{P}_1)$ and $(\mathbf{P}_3 - \mathbf{P}_1)$. They can serve as local coordinate axes on the plane (even though they are not normally perpendicular), with point $\mathbf{P}_1$ as the local origin. The linear combination $u(\mathbf{P}_2 - \mathbf{P}_1) + w(\mathbf{P}_3 - \mathbf{P}_1)$, where both u and w can take any real values, expresses every point on the plane in this local coordinate system. To get the same points relative to the origin, we simply add the local origin $\mathbf{P}_1$. Hence, the plane equation is

$$\mathbf{P}_1 + u(\mathbf{P}_2 - \mathbf{P}_1) + w(\mathbf{P}_3 - \mathbf{P}_1) = \mathbf{P}_1(1 - u - w) + \mathbf{P}_2 u + \mathbf{P}_3 w. \quad (5.5)$$

To limit the area covered to just the triangle, we note that Equation (5.5) yields

$$\mathbf{P}_1, \text{ when } u = 0 \text{ and } w = 0,$$
$$\mathbf{P}_2, \text{ when } u = 1 \text{ and } w = 0,$$
$$\mathbf{P}_3, \text{ when } u = 0 \text{ and } w = 1.$$

The entire triangle can therefore be obtained by varying u and w under the conditions $u \geq 0, w \geq 0$, and $u + w \leq 1$.

5 Surfaces

▸ **Exercise 5.11:** Given the three points $P_1 = (10, -5, 4)$, $P_2 = (8, -4, 3.2)$, and $P_3 = (8, 4, 3.2)$, calculate the triangle defined by them.

▸ **Exercise 5.12:** Given the three points $P_1 = (10, -5, 4)$, $P_2 = (8, -4, 3.2)$, and $P_3 = (12, -6, 4.8)$, calculate the triangle defined by them.

> If triangles had a God, He'd have three sides.
> — Yiddish proverb.

5.3.8 Triangle Centers

Even though the triangle is the simplest polygon, it is possible to define its "center point" in many different ways. The five most common definitions are the following:

Centroid: Given a triangle ABC, let D, E, and F be the midpoints of sides BC, CA, and AB, respectively (Figure 5.7a). They meet at one point P, called the centroid of the triangle (Section 4.1). The centroid is the only point P for which the areas of the three triangles BCP, CAP, and ABP are equal.

An interesting property of the centroid is that it can be defined for arbitrary shapes, not just triangles. It is possible, for example, to calculate the centroid of a country. Just draw the map of the country on a piece of cardboard and cut it out. Hang it on a string from an arbitrary point X, and draw a vertical line from X. Repeat for another point Y. The two lines meet at the centroid of the map (Figure 5.7b).

Incenter: Let D, E, and F be the points at which the bisectors of angles A, B, and C meet sides BC, CA, and AB, respectively (Figure 5.7c). The points where the bisectors meet is the incenter of the triangle ABC. From its definition, it is clear that the incenter is equidistant from the three triangle edges, so we can consider this distance the radius of a circle. This is the *incircle* of the triangle.

Circumcenter: Let D, E, and F be the perpendicular bisectors of sides BC, CA, and AB, respectively (Figure 5.7d). They meet at the circumcenter of triangle ABC. Since the circumcenter is, by definition, equidistant from the three triangle corners, it becomes the center of the circumcircle of the triangle, part of which is shown in the figure. The radius of the circumcircle is the distance of the circumcenter from any of the corners.

Orthocenter: Let D, E, and F be the points where the altitudes through A, B, and C intersect sides BC, CA, and AB, respectively. The point where the three altitudes meet is the orthocenter of triangle ABC (Figure 5.7e).

The above four points were known to the ancients. The next one took centuries to discover.

Fermat Point: Let A'BC be the equilateral triangle constructed on side BC. Let AB'C and ABC' be the equilateral triangles constructed on sides CA and AB, respectively. The point where lines AA', BB', and CC' meet (Figure 5.7f) is the Fermat point of the triangle.

Trilinear Coordinates: This is a way to measure the position of an arbitrary point P inside a triangle. The trilinear coordinates (sometimes called the trilinears)

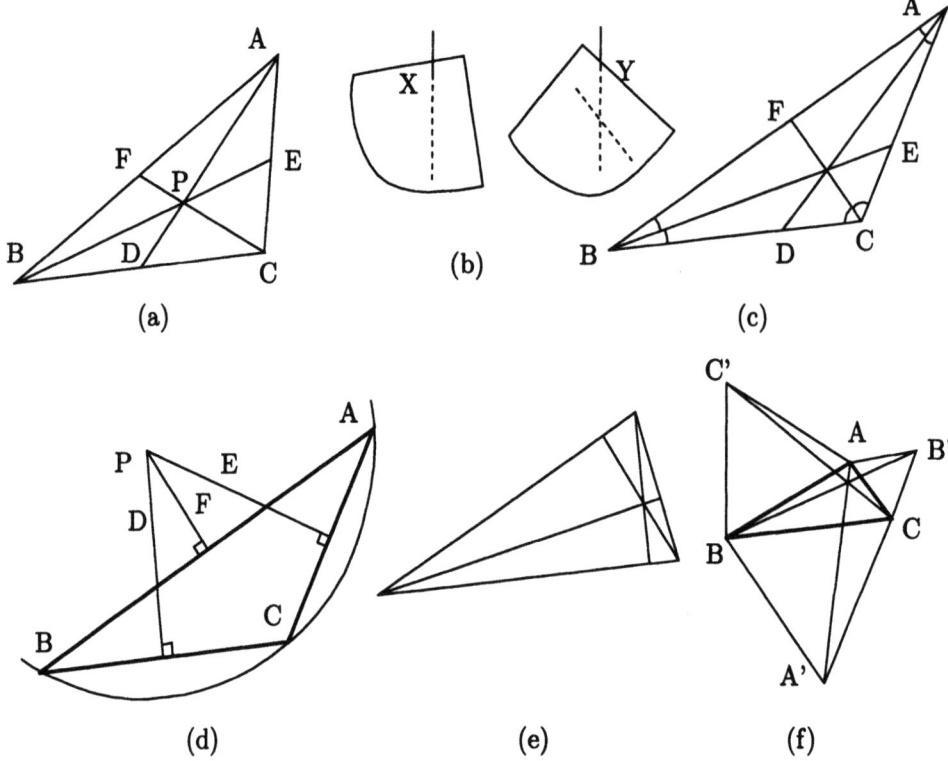

Figure 5.7: Triangle Centers.

of P are its distances from the edges BC, CA, and AB (notice that the order of the edges is important).

The incenter of triangle ABC, for example, has trilinears $(1,1,1)$ since it is equidistant from the three edges. The three corners A, B, and C have trilinears $(1,0,0)$, $(0,1,0)$, and $(0,0,1)$, respectively. The centroid has trilinears $(1/a, 1/b, 1/c)$ where a, b, and c are the sizes of the three triangle edges.

▶ **Exercise 5.13:** What are the trilinears of the circumcenter, orthocenter, and Fermat point of a triangle ABC?

For more information, see [Triangles 98] or the [Kimberling 94].

We conclude with the words of Blaise Pascal on triangles (see [Knuth 84] page 315).

> I turn, in the following treatises, to various uses of those triangles whose generator is unity. But I leave out many more than I include; it is extraordinary how fertile in properties this triangle is. Everyone can try his hand.

5.4 Delaunay Triangulation

A polygonal surface is defined by means of a set of points and a set of straight lines connecting them and creating the polygons. In some cases, the user has to specify which points should be connected by lines, but in the general case, it is possible to develop a method to automatically connect points that are near neighbors by straight lines, such that the entire surface becomes a set of triangles. The concept of triangulating a surface is called *Delaunay triangulation*, after the Russian mathematician B. Delaunay (or Deloné), who described it early in the century [Delaunay 34].

Definition: A Delaunay two-dimensional triangular network consists of a set of nonoverlapping triangles where no point in the network is enclosed by the circumcircle of any triangle.

When we want to "place" a solid object in the computer, the object should first be digitized. This is done by measuring the coordinates of points either manually or with a three-dimensional digitizer. Such a device has a stylus that is used to touch points on the object. The coordinates of these points are determined by the digitizer and are input into the computer. The following are examples:

1. When a museum decides to place its sculpture collection in a computer, it digitizes each piece of sculpture, varying the resolution in proportion to the value of a sculpture. Valuable pieces are digitized at high resolution, resulting in a large file, but also making it possible to store (and later watch) every small detail in the computer.

2. When a new housing subdivision is about to be developed, a survey is done where, among other things, the coordinates of many points are measured. These are used to create a three-dimensional terrain model in a computer that can later be used to determine, for example, what the view would be from any point.

When all the points have been input, a rough display of the surface (car, terrain, etc.) is created by connecting points with straight segments to form a set of triangles. This process is called triangulation.

It is possible to develop different algorithms for triangulation, but the most common algorithms are incremental. Such an algorithm consists of a loop, where in each iteration, one point is added to the partly completed polygonal surface, until all the points have been added. The algorithm described here assumes two-dimensional points, but its extension to three-dimensional points is straightforward.

Figure 5.8 shows eight points already connected in a seven triangle, partly-finished polygonal surface. The figure shows a circle that passes through the three corners of each triangle (i.e., it passes though three of the eight data points). Such a circle is called the *circumcircle* of the triangle. We assume that the next iteration of the loop should add point P_9. Figures 5.9 and 5.10 show the steps involved.

1. Determine the circumcircles that contain the new point (Section 5.4.1). In our example, they correspond to the three triangles $P_1P_2P_5$, $P_1P_5P_6$, and $P_1P_6P_3$.

2. Make a list of all edges of those triangles: P_1P_2, P_1P_5, P_2P_5, P_1P_5, P_1P_6, P_5P_6, P_1P_6, P_1P_3, and P_6P_3.

5.4 Delaunay Triangulation

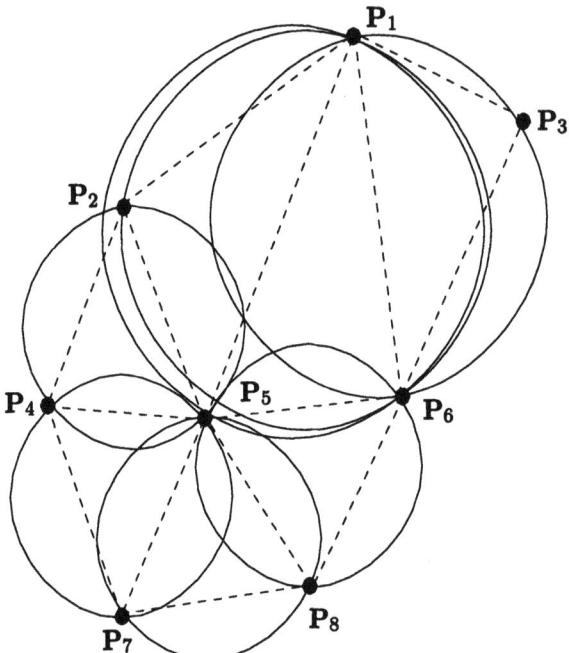

Figure 5.8: Triangulating Eight Points.

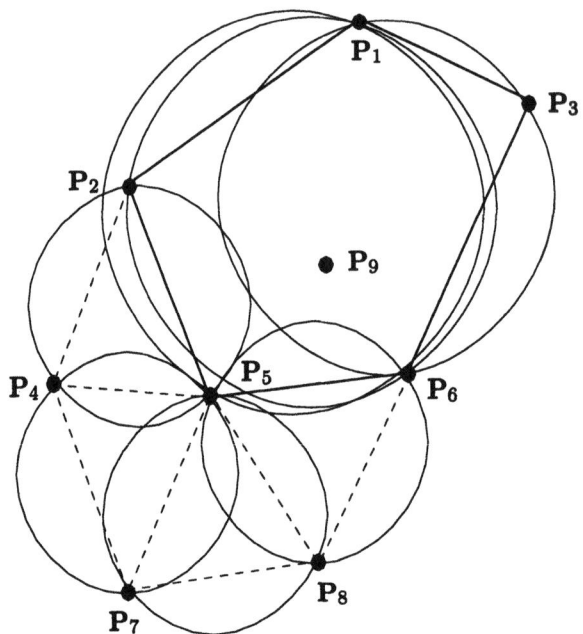

Figure 5.9: Adding Point Nine (Part I).

5 Surfaces

3. Delete any edges that appear more than once. Edges P_1P_5 and P_1P_6 are deleted in our example. This eliminates two triangles and creates the five-sided polygon shown in Figure 5.9.

4. Connect the new point P_9 to the vertices of the new polygon. The five edges P_9P_1, P_9P_2, P_9P_3, P_9P_5, and P_9P_6 are constructed, which creates the five new triangles shown in Figure 5.10.

The only remaining question is how to start the loop. This is done by computing three "imaginary" points to become the corners of an "imaginary" triangle containing all the data points (Section 5.4.2). In this way, there is already one triangle to add to when the loop starts. When the loop terminates, this triangle is deleted, together with its three points and all their edges (edges that have been added during the loop). This implies that every new point added during the loop must be contained in the large "imaginary" triangle. Thus, each iteration adds two triangles to the existing mesh, so the total number of triangles is twice the number of points. At the end of the loop, some triangles are deleted, so their final number is less than that.

It is intuitively clear, just by looking at the figures, that a set of points can be triangulated in a number of different ways. The Delaunay triangulation has the following desirable properties:

1. If three points are joined by the algorithm in a triangle, no other point will be located inside the circumcircle of this triangle.

2. Given four neighboring points that define a quadrilateral, this quadrilateral is split, in one of the loop iterations, in a way that maximizes the smaller of the internal angles. This property implies that the triangles produced by the algorithm are as close to equilateral as possible. The algorithm does not create thin, wedge-shaped triangles if at all possible.

3. The triangulation is unique, i.e., it is independent of the order in which the points are considered, except for the degenerate case where four points lie at the corners of a rectangle. Such points end up being triangulated in one of the two possible ways, depending on which of them is considered first. If unique triangulation is important, one of these points should be moved a little.

4. In cases where there are regions of high-density points and regions of low-density points, the Delaunay triangulation produces a large number of small triangles in the former case and a small number of large triangles in the latter case.

5. Suppose that n points are given and are triangulated. If the result is not satisfactory, more points can be digitized and added, one by one, to the existing triangular mesh; there is no need to start from the beginning. This is a result of the incremental nature of the algorithm.

The algorithm may fail in the following "degenerate" cases:

1. Two identical points. One of them should either be slightly moved or completely removed.

2. Three collinear points. Again, one should either be slightly moved or completely removed.

3. Four points lie on a circle and their Voronoi vertices (see discussion in Section 5.4.4) are identical.

430   5.4 Delaunay Triangulation

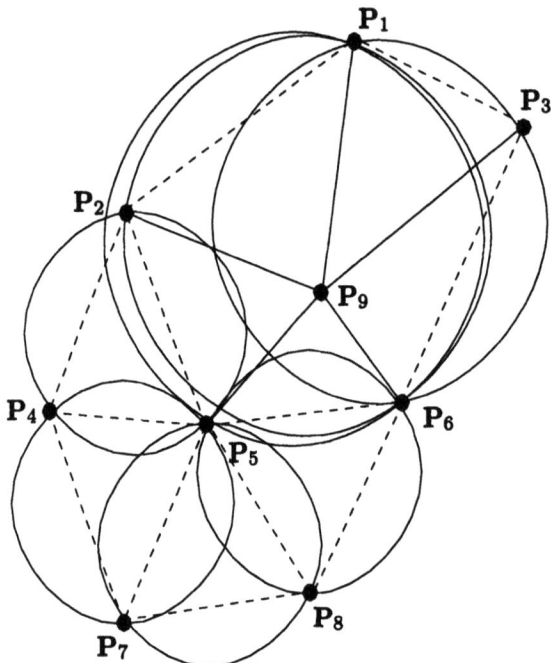

Figure 5.10: Adding Point Nine (Part II).

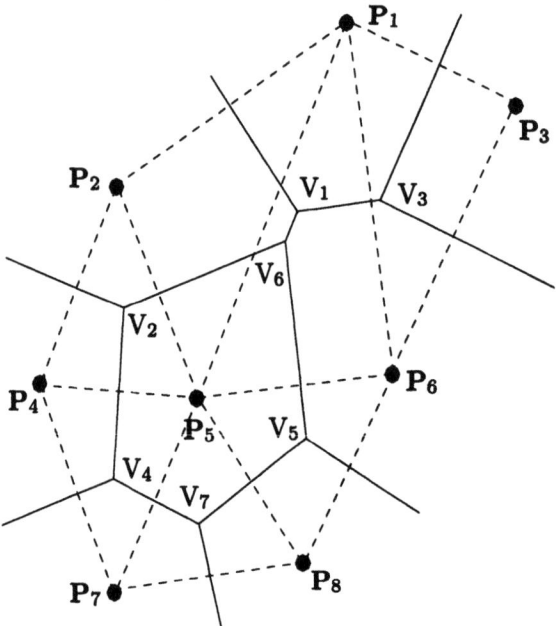

Figure 5.11: Voronoi Diagram of Eight Points.

5 Surfaces

The complexity of the algorithm is $O(n^2)$, but it can be optimized to $O(n \log n)$. The algorithm can be extended to three dimensions by dealing with spheres instead of circles. The main difference is that points should not be positioned directly above each other. Assuming that the points are displayed on a screen that's the xy plane and the z axis is perpendicular to the screen, any point A directly above another point B should be moved a little since otherwise the edge AB would degenerate to a single point.

Two references for algorithm complexity are [Skiena 1998] and [Sedgewick 1997].

5.4.1 Circumcircles

Given three points $\mathbf{P}_1$, $\mathbf{P}_2$, and $\mathbf{P}_3$ that are not collinear, what is the circle defined by them? (Since any three noncollinear points define a triangle, fitting a circle through them amounts to circumscribing a triangle.)

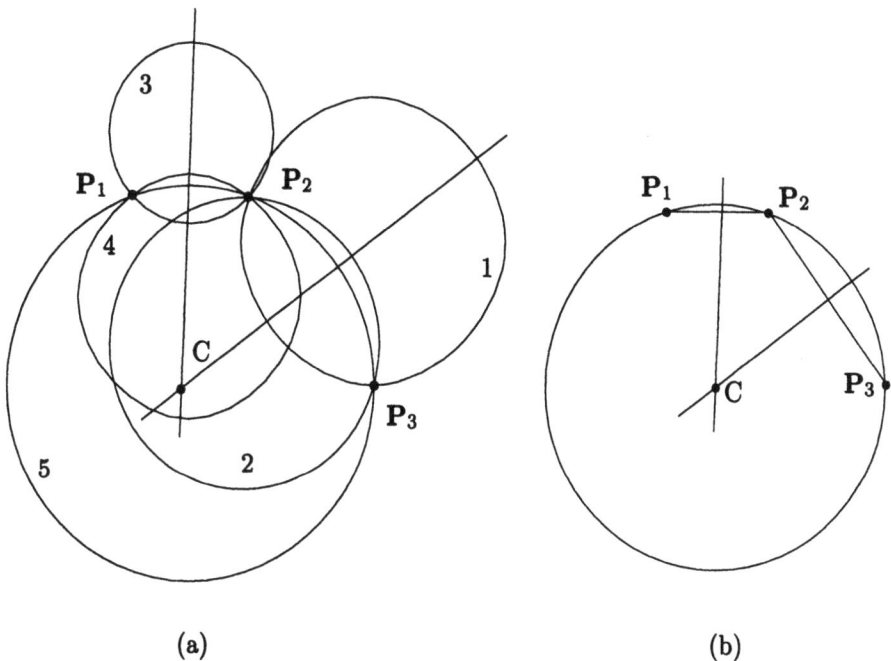

(a) (b)

Figure 5.12: A Bounding Circle for Three Points.

Figure 5.12a shows two circles (numbered 1 and 2) that pass through the two points $\mathbf{P}_2$ and $\mathbf{P}_3$. Clearly, there are infinitely many circles that pass through the two points, and the centers of all these circles are located on the perpendicular bisector of the points (Equation (10.2)). Two more circles, numbered 3 and 4, are shown, that pass through points $\mathbf{P}_1$ and $\mathbf{P}_2$. It is not hard to be convinced that the bounding circle of points $\mathbf{P}_1$, $\mathbf{P}_2$, and $\mathbf{P}_3$ is centered at the intersection C of the two perpendicular bisectors (circle 5). The radius of this circle is the distance between C and either of the three points (Figure 5.12b).

Following are two explicit approaches to this problem:

1. If **a**, **b**, and **c** are three two-dimensional points, then an expression for the circle $(\mathbf{P}, R)$ through them is

$$P_x = (b_y a_x^2 - c_y a_x^2 - b_y^2 a_y + c_y^2 a_y + b_x^2 c_y + a_y^2 b_y$$
$$+ c_x^2 a_y - c_y^2 b_y - c_x^2 b_y - b_x^2 a_y + b_y^2 c_y - a_y^2 c_y)/D,$$
$$P_y = (a_x^2 c_x + a_y^2 c_x + b_x^2 a_x - b_x^2 c_x + b_y^2 a_x - b_y^2 c_x$$
$$- a_x^2 b_x - a_y^2 b_x - c_x^2 a_x + c_x^2 b_x - c_y^2 a_x + c_y^2 b_x)/D,$$

where $D = 2(a_y c_x + b_y a_x - b_y c_x - a_y b_x - c_y a_x + c_y b_x)$ and the radius of the circle is

$$R^2 = (a_x - P_x)^2 + (a_y - P_y)^2.$$

When $D = 0$, the circle degenerates to a point.

▶ **Exercise 5.14:** Try to write this expression using fewer multiplications/divisions.

2. Expand the expression $(x-a)^2 + (y-b)^2 = R^2$ to yield $x^2 - 2ax + a^2 + y^2 - 2by + b^2 = R^2$. This is nonlinear in a, b, and R, so it seems like the wrong way to calculate these quantities. However, if we denote $q = R^2 - a^2 - b^2$ and rearrange, we get the equation

$$(2x)a + (2y)b + q = x^2 + y^2, \tag{5.6}$$

which is linear in a, b, and q. Applying Equation (5.6) to the three points gives

$$P_x = \frac{\det\begin{pmatrix}(a_x^2 + a_y^2) & 2a_y & 1\\ (b_x^2 + b_y^2) & 2b_y & 1\\ (c_x^2 + c_y^2) & 2c_y & 1\end{pmatrix}}{\det\begin{pmatrix}2a_x & 2a_y & 1\\ 2b_x & 2b_y & 1\\ 2c_x & 2c_y & 1\end{pmatrix}} = \frac{\det\begin{pmatrix}(b_x^2 + b_y^2 - a_x^2 - a_y^2) & (b_y - a_y)\\ (c_x^2 + c_y^2 - a_x^2 - a_y^2) & (c_y - a_y)\end{pmatrix}}{2\det\begin{pmatrix}(b_x - a_x) & (b_y - a_y)\\ (c_x - a_x) & (c_y - a_y)\end{pmatrix}},$$

$$P_y = \frac{\det\begin{pmatrix}(b_x - a_x) & (b_x^2 + b_y^2 - a_x^2 - a_y^2)\\ (b_y - a_y) & (c_x^2 + c_y^2 - a_x^2 - a_y^2)\end{pmatrix}}{2\det\begin{pmatrix}(b_x - a_x) & (b_y - a_y)\\ (c_x - a_x) & (c_y - a_y)\end{pmatrix}}.$$

5.4.2 Initial Triangle

The only remaining operation to be discussed regarding Delaunay triangulation is the following: Given a set of points, how can we find a triangle (not necessarily the smallest one) that contains all of them. The method described here proceeds in steps (but see also the next subsection).

1. Find the *convex hull* of the set of points. The convex hull is defined in Section 5.3.5. Finding it is a classical problem in computational geometry. The approach illustrated here is simple and slow. Faster algorithms are described, for example, in [O'Rourke 94]. Our approach is to identify the interior points of the

5 Surfaces

set. Any point not identified is an extreme point, i.e., part of the convex hull of the set. The idea is that a point $\mathbf{P}_i$ is interior if (and only if) it is inside (or on the boundary of) some triangle whose corners are three points of the set and are not identical to $\mathbf{P}_i$. The following pseudo-code illustrates this process. It consists of four nested loops, so its complexity is $O(n^4)$, very slow.

```
for each i do
  for each j ≠ i do
    for each k ≠ j ≠ i do
      for each l ≠ k ≠ j ≠ i do
        if P_l is on or in the triangle P_iP_jP_k
        then P_l is interior
        endif;
endallfor;
```

2. Compute the distance between every pair of points in the convex hull. Select the pair with the longest distance and construct the circle that passes through its two points, with the center midway between them. This is the *bounding circle* of the original set of points (the smallest circle that contains all the points).

3. Assume that the bounding circle has radius R and is centered at (x_0, y_0). The required triangle has corners at $(x_0 - R(1+\sqrt{2}), y_0 - R)$, $(x_0 + R(1+\sqrt{2}), y_0 - R)$, and $(x_0, y_0 + R\sqrt{2})$.

▶ **Exercise 5.15:** Given a circle of radius R about the origin, show that the triangle with corners at $\mathbf{P}_1 = (-R(1+\sqrt{2}), -R)$, $\mathbf{P}_2 = (R(1+\sqrt{2}), -R)$, and $\mathbf{P}_3 = (0, R\sqrt{2})$ bounds the circle from the outside (Figure Ans.26).

5.4.3 Alternative Methods

Here is an alternative approach to finding an enclosing triangle. We test all the points in the set and find four points, L, R, T, and B, with extreme left, right, top, and bottom coordinates, respectively. Notice that there may be more than one of each (there are two L points in Figure 5.13a), but we can select any of them. Thus, the rectangle bounding the set of points is easy to determine. Its height is $h = y_T - y_B$ and half its width is $w = (x_R - x_L)/2$. The center of its bottom edge is at point $(x_0, y_0) = ((x_L + x_R)/2, y_B)$. It's easy to show that one triangle that encloses this rectangle (Figure 5.13b) has corners at $\mathbf{P}_1 = (x_0 - w - h, y_0)$, $\mathbf{P}_2 = (x_0 + w + h, y_0)$, and $\mathbf{P}_3 = (x_0, y_0 + w + h)$.

▶ **Exercise 5.16:** Prove this!

Other methods are also possible. One example is the following:

1. Compute the convex hull of the set of points.

2. Connect each point in this hull to its successor. This creates a polygon bounding the entire set.

3. Select one point on the convex hull at random, and connect it to every other point. This divides the bounding polygon into triangles.

4. Now that we have some triangles, continue as before, adding the remaining points one by one.

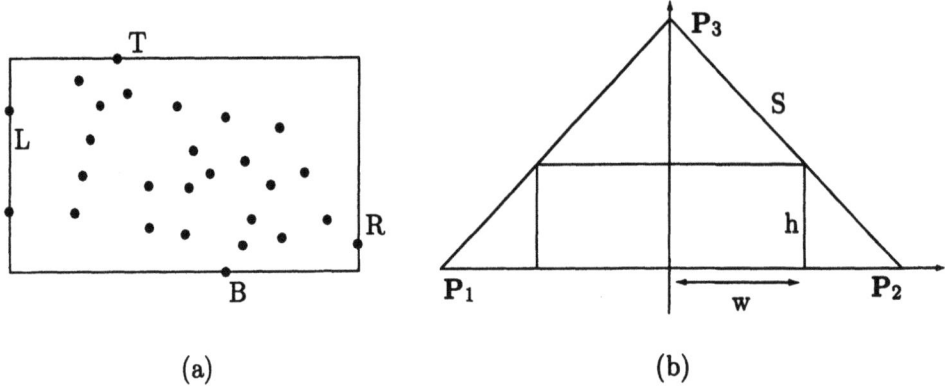

Figure 5.13: A Bounding Rectangle for a Set of Points.

5.4.4 Voronoi Diagrams

Imagine a Petri dish ready for growing bacteria. Eight bacteria of different types are simultaneously placed in it at different points and immediately start multiplying. We assume that their colonies grow at the same rate. Initially, each colony consists of a growing circle around one of the starting points. After a while, some of them meet and stop growing in the meeting area due to lack of food. The final result is that the entire dish gets divided into eight areas, one around each of the eight starting points, such that all the points within area i are closer to starting point i than to any other start point. Such areas are called *Voronoi regions* or *Dirichlet tessellations*. Figure 5.11 shows the Voronoi regions (the solid lines) for our eight points superimposed on the Delaunay triangles (the dashed lines). It turns out that Delaunay triangles and Voronoi diagrams are *complementary* (or *dual*) geometrical constructs. Each is easy to generate from the other. A close look shows that each solid line is the perpendicular bisector of one of the dashed lines.

At the time of writing there are several Java applets that demonstrate the concepts discussed here. A typical example is [Zhao 98].

 Charlie Chan: Truth, like oil, will in time rise to surface.

— Sidney Toller (as Charlie Chan) in *Charlie Chan's Murder Cruise* (1940).

5.5 Bilinear Surfaces

A polygon is the simplest type of surface since it is flat. The bilinear surface is the simple nonflat surface since it is fully defined by means of its four corner points. The four boundary curves of this surface patch are straight lines. Linear interpolation is used to calculate the coordinates of points on the surface. Since this patch is completely defined by its four corner points, it cannot have a very complex shape. Nevertheless it may be highly curved. If the four corners are coplanar, the bilinear patch defined by them is flat.

5 Surfaces

Let the corner points be P_{00}, P_{01}, P_{10}, and P_{11}. The top and bottom boundary curves are straight lines and are easy to calculate (Figure 5.14). They are $P(u,0) = (P_{10} - P_{00})u + P_{00}$ and $P(u,1) = (P_{11} - P_{01})u + P_{01}$.

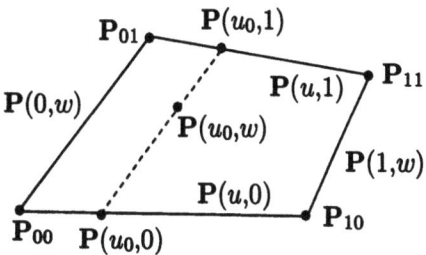

Figure 5.14: A Bilinear Surface.

To linearly interpolate between these boundary curves, we first calculate two corresponding points $P(u_0, 0)$ and $P(u_0, 1)$, one on each curve, then connect them with a straight line $P(u_0, w)$. The two points are

$$P(u_0, 0) = (P_{10} - P_{00})u_0 + P_{00} \quad \text{and} \quad P(u_0, 1) = (P_{11} - P_{01})u_0 + P_{01},$$

and the line is

$$\begin{aligned} P(u_0, w) &= (P(u_0, 1) - P(u_0, 0))w + P(u_0, 0) \\ &= \Big[(P_{11} - P_{01})u_0 + P_{01} - ((P_{10} - P_{00})u_0 + P_{00})\Big]w \\ &\quad + (P_{10} - P_{00})u_0 + P_{00}. \end{aligned}$$

The expression for the entire surface is obtained when we release the parameter u from its fixed value u_0 and let it vary:

$$\begin{aligned} P(u, w) &= P_{00}(1-u)(1-w) + P_{01}(1-u)w + P_{10}u(1-w) + P_{11}uw \\ &= \sum_{i=0}^{1}\sum_{j=0}^{1} P_{ij} B_{1i}(u) B_{1j}(w), \end{aligned} \qquad (5.7)$$

where the B functions are the Bernstein polynomials of degree 1 (they crop up in unexpected places). This is a simple expression that's linear in both u and w.

Figure 5.15 shows a bilinear surface together with the *Mathematica* code that produced it (plus the coordinates of the four corner points and the final, simplified expression of the surface). It is possible to see that even though every line in the u or in the w directions on this surface is straight, the surface itself is curved.

```
Clear[bilinear,pnts,u,w];
<<:Graphics:ParametricPlot3D.m;
pnts=ReadList["Points",{Number,Number,Number}, RecordLists->True]
bilinear[u_,w_]:=pnts[[1,1]](1-u)(1-w)+pnts[[1,2]]u(1-w) \
 +pnts[[2,1]]w(1-u)+pnts[[2,2]]u w;
Simplify[bilinear[u,w]]
g1=Graphics3D[{AbsolutePointSize[5], Table[Point[pnts[[i,j]]],
 {i,1,2},{j,1,2}]}]
g2=ParametricPlot3D[bilinear[u,w],{u,0,1,.05},{w,0,1,.05},
 Compiled->False, DisplayFunction->Identity]
Show[g1,g2, ViewPoint->{0.063, -1.734, 2.905}]
{{0, 0, 1}, {1, 1, 1}, {1, 0, 0}, {0, 1, 0}}
{u + w - 2 u w, u, 1 - w}
```

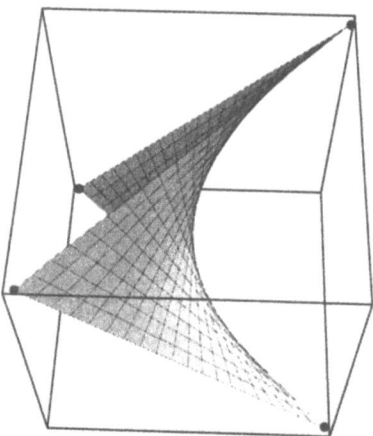

Figure 5.15: A Bilinear Surface.

5.5.1 Example 1

We select the four points $\mathbf{P}_{00} = (0,0,1)$, $\mathbf{P}_{10} = (1,0,0)$, $\mathbf{P}_{01} = (1,1,1)$, and $\mathbf{P}_{11} = (0,1,0)$ (Figure 5.15) and apply Equation (5.7). The surface patch is

$$P(u,w) = (0,0,1)(1-u)(1-w) + (1,1,1)(1-u)w + (1,0,0)u(1-w) + (0,1,0)uw$$
$$= (u + w - 2uw, w, 1 - u). \qquad (5.8)$$

It is easy to check the expression by substituting $u = 0, 1$ and $w = 0, 1$, which reduces the expression to the four corner points. The tangent vectors can easily be calculated. They are

$$\frac{\partial \mathbf{P}(u,w)}{\partial u} = (1-2w, 0, -1), \quad \frac{\partial \mathbf{P}(u,w)}{\partial w} = (1-2u, 1, 0).$$

The first vector lies in the xz plane, and the second lies in the xy plane.

5 Surfaces

5.5.2 Example 2

We select the four points $\mathbf{P}_{00} = (0,0,1)$, $\mathbf{P}_{10} = (1,0,0)$, $\mathbf{P}_{01} = (0.5,1,0)$, and $\mathbf{P}_{11} = (1,1,0)$ and apply Equation (5.7). The resulting surface patch is

$$P(u,w) = (0,0,1)(1-u)(1-w) + (0.5,1,0)(1-u)w + (1,0,0)u(1-w) + (1,1,0)uw$$
$$= \bigl(0.5(1-u)w + u, w, (1-u)(1-w)\bigr). \tag{5.9}$$

Note that the y coordinate is simply w. This means that points with the same w value, such as $\mathbf{P}(0.1, w)$ and $\mathbf{P}(0.5, w)$ will have the same y coordinate and will, therefore, be on the same horizontal line. Also, the z coordinate is a simple function of u and w, varying from 1 (when $u = w = 0$) to 0 as we move toward $u = 1$ or $w = 1$.

The boundary curves are very easy to calculate from Equation (5.9). Here are two of them:

$$\mathbf{P}(0, w) = (0.5w, w, 1-w), \qquad \mathbf{P}(u,1) = (0.5(1-u) + u, 1, 0).$$

The tangent vectors can also be obtained from Equation (5.9):

$$\frac{\partial \mathbf{P}(u,w)}{\partial u} = (-0.5w + 1, 0, w - 1), \quad \frac{\partial \mathbf{P}(u,w)}{\partial w} = (0.5(1-u), 1, u-1). \tag{5.10}$$

The first is a vector in the xz plane, while the second is a vector in the $y = 1$ plane. The following two tangent values are especially simple: $\frac{\partial \mathbf{P}(u,1)}{\partial u} = (0.5, 0, 0)$ and $\frac{\partial \mathbf{P}(1,w)}{\partial w} = (0, 1, 0)$. The first is a vector in the x direction and the second is a vector in the y direction.

Finally, let's calculate the *normal vector* to the surface. This vector is normal to the surface at any point, so it is perpendicular to the two tangent vectors $\partial \mathbf{P}(u,w)/\partial u$ and $\partial \mathbf{P}(u,w)/\partial w$ and is, therefore, the *cross-product* (Equation (A.10), page 703) of these vectors. The calculation is straightforward:

$$\mathbf{N}(u,w) = \frac{\partial \mathbf{P}}{\partial u} \times \frac{\partial \mathbf{P}}{\partial w} = (1-w, 0.5(1-u), 1 - 0.5w). \tag{5.11}$$

There are two ways of satisfying ourselves that Equation (5.11) is the correct expression for the normal:

1. It is easy to prove, by directly calculating the *dot products*, that the normal vector of Equation (5.11) is perpendicular to both tangents of Equation (5.10).

2. A closer look at the coordinates of our points shows that three of them have a z coordinate of zero and only one, namely $\mathbf{P}_{00}$, has $z = 1$. This means that the surface approaches a flat xy surface as one moves away from point $\mathbf{P}_{00}$. It also means that the normal should approach the z direction when u and w move away from zero, and it should move away from that direction when u and w approach zero. It is, in fact, easy to confirm the following limits:

$$\lim_{u,w \to 1} \mathbf{N}(u,w) = (0, 0, 0.5), \qquad \lim_{u,w \to 0} \mathbf{N}(u,w) = (1, 0.5, 1).$$

▶ **Exercise 5.17:** (1) Calculate the bilinear surface for the points $(0,0,0)$, $(1,0,0)$, $(0,1,0)$, and $(1,1,1)$. (2) Guess the implicit representation $z = F(x,y)$ of this surface. (3) What curve results from the intersection of this surface with the plane $z = k$ (parallel to the xy plane). (4) What curve results from the intersection of this surface with a plane containing the z axis?

> The scale, properly speaking, does not permit the measure of the intelligence, because intellectual qualities are not superposable, and therefore cannot be measured as linear surfaces are measured.
> —Alfred Binet

5.5.3 Example 3

The four points $\mathbf{P}_{00} = (0,0,1)$, $\mathbf{P}_{10} = (1,0,0)$, $\mathbf{P}_{01} = (0,1,0)$, and $\mathbf{P}_{11} = (0,1,0)$ create a *triangular surface patch* since two of them are identical:

$$P(u,w) = (0,0,1)(1-u)(1-w) + (0,1,0)(1-u)w + (1,0,0)u(1-w) + (0,1,0)uw$$
$$= \bigl(u(1-w), w, (1-u)(1-w)\bigr).$$

Notice that the boundary curve $\mathbf{P}(u,1)$ degenerates to the single point $(0,1,0)$, i.e., it does not depend on u.

▶ **Exercise 5.18:** Calculate the tangent vectors and the normal vector of this surface.

▶ **Exercise 5.19:** Given the two points $\mathbf{P}_{00} = (-1,-1,0)$ and $\mathbf{P}_{10} = (1,-1,0)$, consider them the endpoints of a straight segment $\mathbf{L}_1$.

Task 1: Calculate the endpoints of the three straight segments $\mathbf{L}_2$, $\mathbf{L}_3$, and $\mathbf{L}_4$. Each should be translated one unit above its predecessor on the y axis and should be rotated $60°$ about the y axis, as shown in Figure 5.16. Denote these endpoints

$$\mathbf{P}_{00} \to \mathbf{P}_{01} \to \mathbf{P}_{02} \to \mathbf{P}_{03} \quad \text{and} \quad \mathbf{P}_{10} \to \mathbf{P}_{11} \to \mathbf{P}_{12} \to \mathbf{P}_{13}.$$

Task 2: Calculate the three bilinear surface patches

$$\mathbf{P}_1(u,w) = \mathbf{P}_{00}(1-u)(1-w) + \mathbf{P}_{01}(1-u)w + \mathbf{P}_{10}u(1-w) + \mathbf{P}_{11}uw,$$
$$\mathbf{P}_2(u,w) = \mathbf{P}_{01}(1-u)(1-w) + \mathbf{P}_{02}(1-u)w + \mathbf{P}_{11}u(1-w) + \mathbf{P}_{12}uw,$$
$$\mathbf{P}_3(u,w) = \mathbf{P}_{02}(1-u)(1-w) + \mathbf{P}_{03}(1-u)w + \mathbf{P}_{12}u(1-w) + \mathbf{P}_{13}uw.$$

5 Surfaces

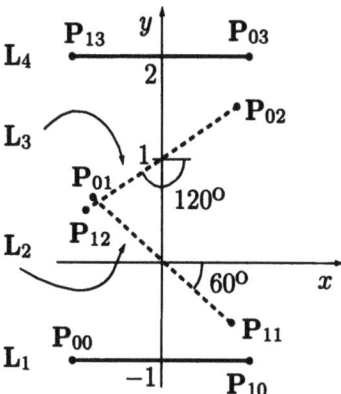

Figure 5.16: Four Straight Segments for Exercise 5.19.

5.6 Lofted Surfaces

This kind of surface patch is bounded by two straight lines and two arbitrary curves. We assume that the two opposite boundary curves $\mathbf{P}(0,w)$ and $\mathbf{P}(1,w)$ are straight lines and the other two, namely $\mathbf{P}(u,0)$ and $\mathbf{P}(u,1)$, are arbitrary given curves. Surface lines in the w direction should, therefore, be straight, while lines in the u direction should be a blend of $\mathbf{P}(u,0)$ and $\mathbf{P}(u,1)$. The blend of the two curves is simply $(1-w)\mathbf{P}(u,0) + w\mathbf{P}(u,1)$, and this blend, which is linear in w, becomes the expression of the surface

$$\mathbf{P}(u,w) = (1-w)\mathbf{P}(u,0) + w\mathbf{P}(u,1).$$

It is linear in w, guaranteeing straight lines in the w direction. Going in the u direction, we travel on a curve whose shape depends on the value of w. For $w_0 \approx 0$, the curve $\mathbf{P}(u, w_0)$ is close to the boundary curve $\mathbf{P}(u,0)$. For $w_0 \approx 1$, it is close to the boundary curve $\mathbf{P}(u,1)$. For $w_0 = 0.5$, it is $0.5\mathbf{P}(u,0) + 0.5\mathbf{P}(u,1)$, an equal mixture of the two.

Note that this kind of surface is fully defined by specifying the two boundary curves. The four corner points are implicit in these curves. These surfaces are sometimes called *ruled*, because straight lines are an important part of their description.

Example 1: We start with the six points $\mathbf{P}_1 = (-1, 0, 0)$, $\mathbf{P}_2 = (0, -1, 0)$, $\mathbf{P}_3 = (1, 0, 0)$, $\mathbf{P}_4 = (-1, 0, 1)$, $\mathbf{P}_5 = (0, -1, 1)$, and $\mathbf{P}_6 = (1, 0, 1)$. Because of the special coordinates of the points, the surface is easy to visualize (Figure 5.17). This helps to intuitively make sense of the expressions for the tangent vectors and the normal. Note especially that the left and right edges of the surface are in the xz plane, whereas we will see that all the other lines in the w direction have a small negative y component.

We proceed in six steps.

1. As the top boundary curve, $\mathbf{P}(u,1)$, we select the quadratic polynomial passing through the top three points $\mathbf{P}_4$, $\mathbf{P}_5$, and $\mathbf{P}_6$. There is only one such curve and it has the form $\mathbf{P}(u,1) = \mathbf{A} + \mathbf{B}u + \mathbf{C}u^2$, where the coefficients $\mathbf{A}$, $\mathbf{B}$, and

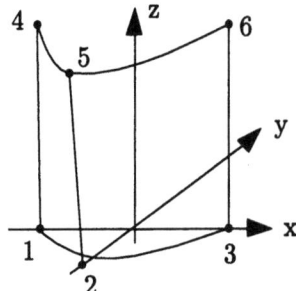

Figure 5.17: A Lofted Surface.

C have to be calculated. We use the fact that the curve passes through the three points to set up the three equations $\mathbf{P}(0,1) = \mathbf{P}_4$, $\mathbf{P}(0.5,1) = \mathbf{P}_5$, and $\mathbf{P}(1,1) = \mathbf{P}_6$, which are written explicitly as

$$\mathbf{A} + \mathbf{B} \times 0 + \mathbf{C} \times 0^2 = (-1,0,1),$$
$$\mathbf{A} + \mathbf{B} \times 0.5 + \mathbf{C} \times 0.5^2 = (0,-1,1),$$
$$\mathbf{A} + \mathbf{B} \times 1 + \mathbf{C} \times 1^2 = (1,0,1).$$

These are easy to solve and result in $\mathbf{A} = (-1,0,1)$, $\mathbf{B} = (2,-4,0)$, and $\mathbf{C} = (0,4,0)$. The top boundary curve is thus $\mathbf{P}(u,1) = \big(2u-1, 4u(u-1), 1\big)$.

2. As the bottom boundary curve, we select the Bézier curve defined by the three points $\mathbf{P}_1$, $\mathbf{P}_2$, and $\mathbf{P}_3$. The curve is

$$\mathbf{P}(u,0) = \sum_{i=0}^{2} B_{2i}(u)\mathbf{P}_{i+1}$$
$$= (1-u)^2(-1,0,0) + 2u(1-u)(0,-1,0) + u^2(1,0,0)$$
$$= \big(2u-1, -2u(1-u), 0\big).$$

3. The surface can now easily be calculated:

$$\mathbf{P}(u,w) = \mathbf{P}(u,0)(1-w) + \mathbf{P}(u,1)w$$
$$= \big(2u-1, 2u(u-1)(1+w), w\big).$$

(Notice that it does not pass through $\mathbf{P}_2$.)

4. The two tangent vectors are immediately obtained

$$\frac{\partial \mathbf{P}}{\partial u} = \big(2, 2(2u-1)(1+w), 0\big), \quad \frac{\partial \mathbf{P}}{\partial w} = \big(0, 2u(u-1), 1\big).$$

5. The normal, as usual, is the cross-product of the tangents $\mathbf{N}(u,w) = \big(2(2u-1)(1+w), -2, 4u(u-1)\big)$.

5 Surfaces

6. The most important feature of this example is the ease with which we can interpret the expressions for the tangents and the normal. This is possible because of the simple shape and orientation of the surface (again, see Figure 5.17). Look at the expressions and make sure you understand the following points:

- The two boundary curves are very similar. One difference between them is, of course, the x and z coordinates. However, the *only* important difference is in the y coordinate. Both curves are quadratic polynomials in u, but although $\mathbf{P}(u,1)$ passes through the three top points, $\mathbf{P}(u,0)$ passes only through the first and last ones.

- The tangent in the u direction, $\partial \mathbf{P}/\partial u$, features $z = 0$; it is a vector in the xy plane. At the bottom of the surface, where $w = 0$, it changes direction from $(2,-2,0)$ (when $u = 0$) to $(2,2,0)$ (when $u = 1$), both 45° directions in the xy plane. However, at the top, where $w = 1$, the tangent changes direction from $(2,-4,0)$ to $(2,4,0)$, both 63° directions. This is because the top boundary curve goes deeper in the y direction.

- The tangent in the w direction, $\partial \mathbf{P}/\partial w$ features $x = 0$; it is a vector in the yz plane. Its z coordinate is a constant 1, and its y coordinate varies from 0 (on the left, where $u = 0$), to -0.5 (in the middle, where $u = 0.5$), and back to 0 (on the right, where $u = 1$). On the left and right edges of the surface, this vector is, therefore a vertical $(0,0,1)$. In the middle, it is $(0,-0.5,1)$, making a negative half-step in y for each step in z.

- The normal vector features $y = -2$ with a small z component. It, therefore, points mostly in the negative y direction, and a little in x. At the bottom ($w = 0$), it varies from $(-2,-2,0)$, to $(0,-2,-1)$,* and ends in $(2,-2,0)$. At the top ($w = 1$), it varies from $(-4,-2,0)$, to $(0,-2,-1)$, and ends in $(4,-2,0)$. The top boundary curve is deeper, causing the tangent to be more in the y direction and the normal to be more in the x direction, than on the bottom boundary curve.

▶ **Exercise 5.20:** (a) Given the two three-dimensional points $\mathbf{P}_1 = (-1,-1,0)$ and $\mathbf{P}_2 = (1,-1,0)$, calculate the straight line from $\mathbf{P}_1$ to $\mathbf{P}_2$. This will become the bottom boundary curve of a lofted surface.

(b) Given the three three-dimensional points $\mathbf{P}_4 = (-1,1,0)$, $\mathbf{P}_5 = (0,1,1)$, and $\mathbf{P}_6 = (1,1,0)$, calculate the quadratic polynomial $\mathbf{P}(t) = \mathbf{A}t^2 + \mathbf{B}t + \mathbf{C}$ that passes through them. This will become the top boundary curve of the surface.

(c) Calculate the expression of the lofted surface patch and the coordinates of the center point $\mathbf{P}(0.5,0.5)$.

5.6.1 Example 2: A Double Helix

The two-dimensional parametric curve $(\cos t, \sin t)$ is, of course, a circle (of radius one unit, centered on the origin). As a result, the three-dimensional curve $(\cos t, \sin t, t)$ is a helix spiraling around the z axis upward from the origin. The similar curve $(\cos(t+\pi), \sin(t+\pi), t)$ is another helix, at a 180° phase difference

---

* it has a small z component, reflecting the fact that the surface is not completely vertical at $u = 0.5$.

with the first. We consider these the two boundary curves of a lofted surface and create the entire surface as a linear interpolation of the two curves. Hence,

$$\mathbf{P}(u, w) = (\cos u, \sin u, u)(1 - w) + (\cos(t + \pi), \sin(t + \pi), u)w,$$

where $0 \leq w \leq 1$, and u can vary in any range. The two curves form a double helix, so the surface looks like a twisted ribbon. Figure 5.18 shows such a surface, together with the *Mathematica* code that generated it.

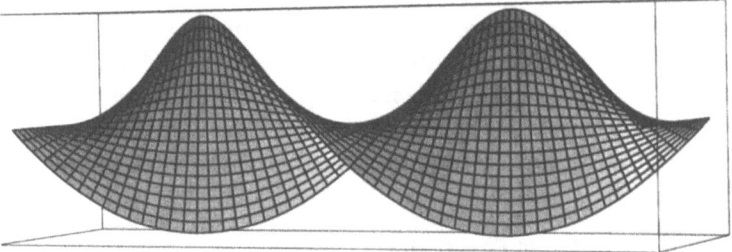

Figure 5.18: The Double Helix as a Lofted Surface.

```
Clear[loftedSurf]; (* double helix as a lofted surface *)
<<:Graphics:ParametricPlot3D.m;
loftedSurf:={Cos[u],Sin[u],u}(1-w)+{Cos[u+Pi],Sin[u+Pi],u}w;
ParametricPlot3D[loftedSurf, {u,0,2Pi,.1},{w,0,1},
ViewPoint->{-3.977, 0.143, 1.732}]
```

Code for Figure 5.18.

▶ **Exercise 5.21:** Calculate the expression of a cone as a lofted surface. Assume that the vertex of the cone is located at the origin, and the base is a circle of radius R, centered on the z axis and located on the plane $z = H$.

▶ **Exercise 5.22:** Derive the expression for a square pyramid where each face is a lofted surface. Assume that the base is a square, $2a$ units on a side, centered about the origin on the xy plane. The top is point $(0, 0, H)$.

5.6.2 Example 3: A Cusp

Given the two curves $\mathbf{P}_1(u) = (8, 4, 0)u^3 - (12, 9, 0)u^2 + (6, 6, 0)u + (-1, 0, 0)$ and $\mathbf{P}_2(u) = (2u - 1, 4u(u - 1), 1)$, the lofted surface defined by them is easy to calculate. Notice that the curves pass through the points $\mathbf{P}_1(0) = (-1, 0, 0)$, $\mathbf{P}_1(0.5) = (0, 5/4, 0)$, $\mathbf{P}_1(1) = (1, 1, 0)$, $\mathbf{P}_2(0) = (-1, 0, 1)$, $\mathbf{P}_2(0.5) = (0, -1, 1)$, and $\mathbf{P}_2(1) = (1, 0, 1)$, which makes it easy to visualize the surface (Figure 5.19). The tangent vectors of the two curves are

$$\mathbf{P}_1^u(u) = (24, 12, 0)u^2 - (24, 18, 0)u + (6, 6, 0), \quad \mathbf{P}_2^u(u) = (2, 8u - 4, 0).$$

5 Surfaces

Notice that $\mathbf{P}_1^u(0.5) = (0,0,0)$ implying that $\mathbf{P}_1(u)$ has a cusp at $u = 0.5$. The lofted surface defined by the two curves is

$$\mathbf{P}(u,w) = \big(4u^2(2u-3)(1-w)-4uw+6u-1, u^2(4u-9)(1-w)+4u^2w-10uw+6u, w\big).$$

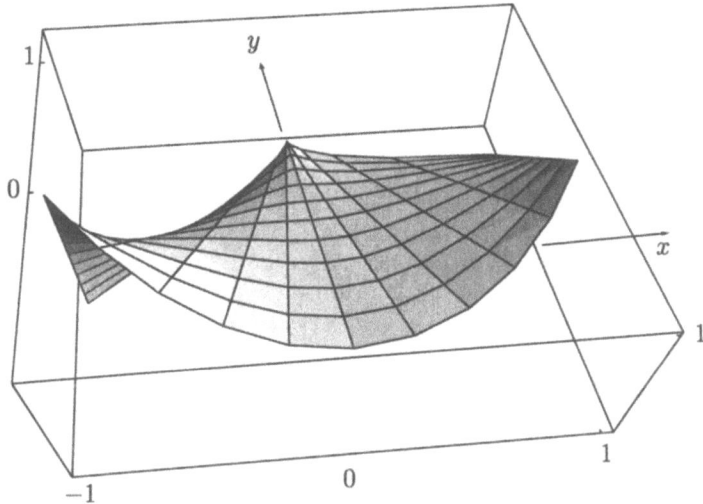

Figure 5.19: A Lofted Surface Patch.

> Now, look Gwen, y'know if we're gonna keep living together in this loft, we're gonna have to have some rules.
> — Leah Remini (as Terri Reynolds) in *Fired Up* (1997)

▶ **Exercise 5.23:** Calculate the tangent vector of this surface in the u direction, and compute its value at the cusp.

5.7 Coons Surfaces

This type of surface is based on the pioneering work on Steven Anson Coons at MIT in the 1960s. His efforts are summarized in [Coons 64] and [Coons 67].

We start with the linear Coons surface, which is a generalization of lofted surfaces. This type of surface patch is defined by its four boundary curves. All four boundary curves are given, and none has to be a straight line. The final expression $\mathbf{P}(u,w)$ of the surface should satisfy the following: (1) it should be symmetric in u and w, (2) it should be an interpolation of $\mathbf{P}(u,0)$ and $\mathbf{P}(u,1)$ in one direction, and of $\mathbf{P}(0,w)$ and $\mathbf{P}(1,w)$ in the other direction, and (3) the four curves should intersect at the corner points, so we can assume that these points are known.

The first step is to construct two lofted surfaces from the two sets of opposite boundary curves. They are $\mathbf{P}_a(u,w) = \mathbf{P}(0,w)(1-u) + \mathbf{P}(1,w)u$ and $\mathbf{P}_b(u,w) = \mathbf{P}(u,0)(1-w) + \mathbf{P}(u,1)w$.

Next, we tentatively attempt to create the final surface $\mathbf{P}(u,w)$ as the sum $\mathbf{P}_a(u,w) + \mathbf{P}_b(u,w)$. It is clear that this is not the expression we are looking for because it does not converge to the right curves at the boundaries. For $u = 0$, for example, we want $\mathbf{P}(u,w)$ to converge to the boundary curve $\mathbf{P}(0,w)$. The sum above, however, converges to $\mathbf{P}(0,w) + \mathbf{P}(0,0)(1-w) + \mathbf{P}(0,1)w$. We, thus, have to subtract $\mathbf{P}(0,0)(1-w) + \mathbf{P}(0,1)w$. Similarly, for $u = 1$, the sum converges to $\mathbf{P}(1,w) + \mathbf{P}(1,0)(1-w) + \mathbf{P}(1,1)w$, so we have to subtract $\mathbf{P}(1,0)(1-w) + \mathbf{P}(1,1)w$. For $w = 0$, we have to subtract $\mathbf{P}(0,0)(1-u) + \mathbf{P}(1,0)u$, and for $w = 1$, we should subtract $\mathbf{P}(0,1)(1-u) + \mathbf{P}(1,1)u$.

Note that the expressions $\mathbf{P}(0,0)$, $\mathbf{P}(0,1)$, $\mathbf{P}(1,0)$, and $\mathbf{P}(1,1)$ are simply the four corner points. A better notation for them may be $\mathbf{P}_{00}$, $\mathbf{P}_{01}$, $\mathbf{P}_{10}$, and $\mathbf{P}_{11}$.

Coons' idea was to define the surface as the sum above, with the right quantities subtracted. He, therefore, ended up with the linear Coons surface $\mathbf{P}(u,w) = \mathbf{P}_a(u,w) + \mathbf{P}_b(u,w) - \mathbf{P}_{ab}(u,w)$, where

$$\mathbf{P}_{ab}(u,w) = \mathbf{P}_{00}(1-u)(1-w) + \mathbf{P}_{01}(1-u)w + \mathbf{P}_{10}u(1-w) + \mathbf{P}_{11}uw.$$

Note that $\mathbf{P}_a$ and $\mathbf{P}_b$ are lofted surfaces, whereas $\mathbf{P}_{ab}$ is a bilinear surface. The final expression is

$$\mathbf{P}(u,w) = \mathbf{P}_a(u,w) + \mathbf{P}_b(u,w) - \mathbf{P}_{ab}(u,w)$$
$$= (1-u, u) \begin{pmatrix} \mathbf{P}(0,w) \\ \mathbf{P}(1,w) \end{pmatrix} + (1-w, w) \begin{pmatrix} \mathbf{P}(u,0) \\ \mathbf{P}(u,1) \end{pmatrix}$$
$$- (1-u, u) \begin{pmatrix} \mathbf{P}_{00} & \mathbf{P}_{01} \\ \mathbf{P}_{10} & \mathbf{P}_{11} \end{pmatrix} \begin{pmatrix} 1-w \\ w \end{pmatrix} \quad (5.12)$$
$$= (1-u, u, 1) \begin{pmatrix} -\mathbf{P}_{00} & -\mathbf{P}_{01} & \mathbf{P}(0,w) \\ -\mathbf{P}_{10} & -\mathbf{P}_{11} & \mathbf{P}(1,w) \\ \mathbf{P}(u,0) & \mathbf{P}(u,1) & (0,0,0) \end{pmatrix} \begin{pmatrix} 1-w \\ w \\ 1 \end{pmatrix}. \quad (5.13)$$

Equation (5.12) is more useful than Equation (5.13) since it shows how the surface is defined in terms of the two barycentric pairs $(1-u, u)$ and $(1-w, w)$. They are the *blending functions* of the linear Coons surface. It turns out that many pairs of barycentric functions $(f_1(u), f_2(u))$ and $(g_1(w), g_2(w))$ can serve as blending functions, out of which more general Coons surfaces can be constructed. All that the blending functions have to satisfy is

$$\begin{array}{l} f_1(0) = 1, \quad f_1(1) = 0, \quad f_2(0) = 0, \quad f_2(1) = 1, \quad f_1(u) + f_2(u) = 1, \\ g_1(0) = 1, \quad g_1(1) = 0, \quad g_2(0) = 0, \quad g_2(1) = 1, \quad g_1(w) + g_2(w) = 1. \end{array} \quad (5.14)$$

Example: We select the following four boundary curves

$$\mathbf{P}_{u0} = (u, 0, \sin(\pi u)), \quad \mathbf{P}_{u1} = (u, 1, \sin(\pi u)),$$
$$\mathbf{P}_{0w} = (0, w, \sin(\pi w)), \quad \mathbf{P}_{1w} = (1, w, \sin(\pi w)).$$

Each is one-half of a sine wave. The first two go along the x axis, and the last two go along the y axis. They meet at the four corner points $\mathbf{P}_{00} = (0,0,0)$, $\mathbf{P}_{01} = (0,1,0)$,

5 Surfaces

$\mathbf{P}_{10} = (1,0,0)$, and $\mathbf{P}_{11} = (1,1,0)$. The surface and the *Mathematica* code that produced it are shown in Figure 5.20. Note the Simplify command, which displays the final, simplified expression of the surface {u, w, Sin[Pi u] + Sin[Pi w]}.

```
<<:Graphics:ParametricPlot3D.m;
Clear[p00,p01,p10,p11,pu0,pu1,p0w,p1w];
p00:={0,0,0}; p01:={0,1,0};
p10:={1,0,0}; p11:={1,1,0};
pu0:={u,0,Sin[Pi u]};
pu1:={u,1,Sin[Pi u]};
p0w:={0,w,Sin[Pi w]};
p1w:={1,w,Sin[Pi w]};
Simplify[
{1-u,u}.{p0w,p1w}+{1-w,w}.{pu0,pu1}
-p00(1-u)(1-w)-p01(1-u)w
-p10(1-w)u-p11 u w]
ParametricPlot3D[%,
{u,0,1,.2},{w,0,1,.2},
PlotRange->All,
AspectRatio->Automatic,
RenderAll->False,
Ticks->{{1},{0,1},{0,1}},
Prolog->AbsoluteThickness[.4]]
```

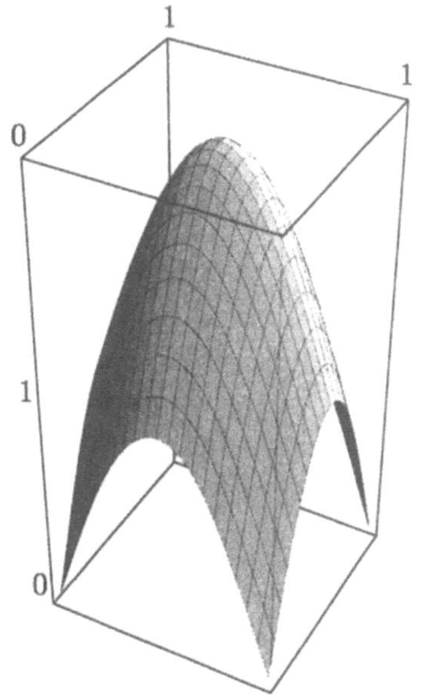

Figure 5.20: A Coons Surface.

Example: Given the four corner points $\mathbf{P}_{00} = (-1,-1,0)$, $\mathbf{P}_{01} = (-1,1,0)$, $\mathbf{P}_{10} = (1,-1,0)$, and $\mathbf{P}_{11} = (1,1,0)$ (notice that they are on the xy plane), we calculate the four boundary curves of a linear Coons surface patch as follows:

1. We select boundary curve $\mathbf{P}(0,w)$ as the straight line from $\mathbf{P}_{00}$ to $\mathbf{P}_{01}$:

$$\mathbf{P}(0,w) = \mathbf{P}_{00}(1-w) + \mathbf{P}_{01}w = (-1, 2w-1, 0).$$

2. We place the two points $(1,-0.5,0.5)$ and $(1,0.5,-0.5)$ between $\mathbf{P}_{10}$ and $\mathbf{P}_{11}$ and calculate boundary curve $\mathbf{P}(1,w)$ as the cubic Lagrange polynomial (Equation (4.19)) determined by these four points

$$\mathbf{P}(1,w) = \frac{1}{2}(w^3, w^2, w, 1) \begin{pmatrix} -9 & -27 & 27 & 9 \\ 18 & -45 & 36 & -9 \\ -11 & 18 & -9 & 2 \\ 2 & 0 & 0 & 0 \end{pmatrix} \begin{pmatrix} (1,-1,0) \\ (1,-0.5,0.5) \\ (1,0.5,-0.5) \\ (1,1,0) \end{pmatrix}$$
$$= (1, (-4 - w + 27w^2 - 18w^3)/4, 27(w - 3w^2 + 2w^3)/4).$$

3. The single point $(0,-1,-0.5)$ is placed between points $\mathbf{P}_{00}$ and $\mathbf{P}_{10}$ and boundary curve $\mathbf{P}(u,0)$ is calculated as the quadratic Lagrange polynomial (Equa-

tion (4.16)) determined by these three points:

$$\mathbf{P}(u,0) = (u^2, u, 1) \begin{pmatrix} 2 & -4 & 2 \\ -3 & 4 & -1 \\ 1 & 0 & 0 \end{pmatrix} \begin{pmatrix} (-1,-1,0) \\ (0,-1,-.5) \\ (1,-1,0) \end{pmatrix}$$
$$= (2u - 1, -1, 2u^2 - 2u).$$

4. Similarly, a new point $(0, 1, .5)$ is placed between points $\mathbf{P}_{01}$ and $\mathbf{P}_{11}$, and boundary curve $\mathbf{P}(u, 1)$ is calculated as the quadratic Lagrange polynomial determined by these three points:

$$\mathbf{P}(u,1) = (u^2, u, 1) \begin{pmatrix} 2 & -4 & 2 \\ -3 & 4 & -1 \\ 1 & 0 & 0 \end{pmatrix} \begin{pmatrix} (-1,1,0) \\ (0,1,.5) \\ (1,1,0) \end{pmatrix}$$
$$= (2u - 1, 1, -2u^2 + 2u).$$

The four boundary curves together with the four corner points are used to express the surface patch as given by Equation (5.12):

$$\mathbf{P}(u,w) = (1-u, u, 1) \begin{pmatrix} -(-1,-1,0) & -(-1,1,0) \\ -(1,-1,0) & -(1,1,0) \\ (2u-1,-1,2u^2-2u) & (2u-1,1,-2u^2+2u) \end{pmatrix}$$
$$\begin{matrix} (-1, 2w-1, 0) \\ (1, (-4-w+27w^2-18w^3)/4, 27(w-3w^2+2w^3)/4) \\ 0 \end{matrix} \Bigg) \begin{pmatrix} 1-w \\ w \\ 1 \end{pmatrix}.$$

This is simplified with the help of appropriate software and becomes

$$\mathbf{P}(u,w) = (-1 + 2u + (1-u)(1-w) - u(1-w) + (-1+2u)(1-w)$$
$$+ (1-u)w - uw + (-1+2u)w,$$
$$- 1 + (1-u)(1-w) + u(1-w) + 2w - (1-u)w$$
$$- uw + (1-u)(-1+2w) + u(-4-w+27w^2-18w^3)/4,$$
$$(-2u+2u^2)(1-w) + (2u-2u^2)w + 27u(w-3w^2+2w^3)/4).$$

The surface patch and the eight points involved are shown in Figure 5.21.

5.7.1 Higher-Degree Coons Surfaces

One possible pair of blending functions is the cubic Hermite polynomials

$$H_{3,0}(t) = B_{3,0}(t) + B_{3,1}(t) = (1-t)^3 + 3t(1-t)^2 = 1 + 2t^3 - 3t^2,$$
$$H_{3,3}(t) = B_{3,2}(t) + B_{3,3}(t) = 3t^2(1-t) + t^3 = 3t^2 - 2t^3,$$

where $B_{n,i}(t)$ are the Bernstein polynomials, Equation (4.104). The sum $H_{3,0}(t) + H_{3,3}(t)$ is identically 1 (because the Bernstein polynomials are barycentric), so these

5 Surfaces

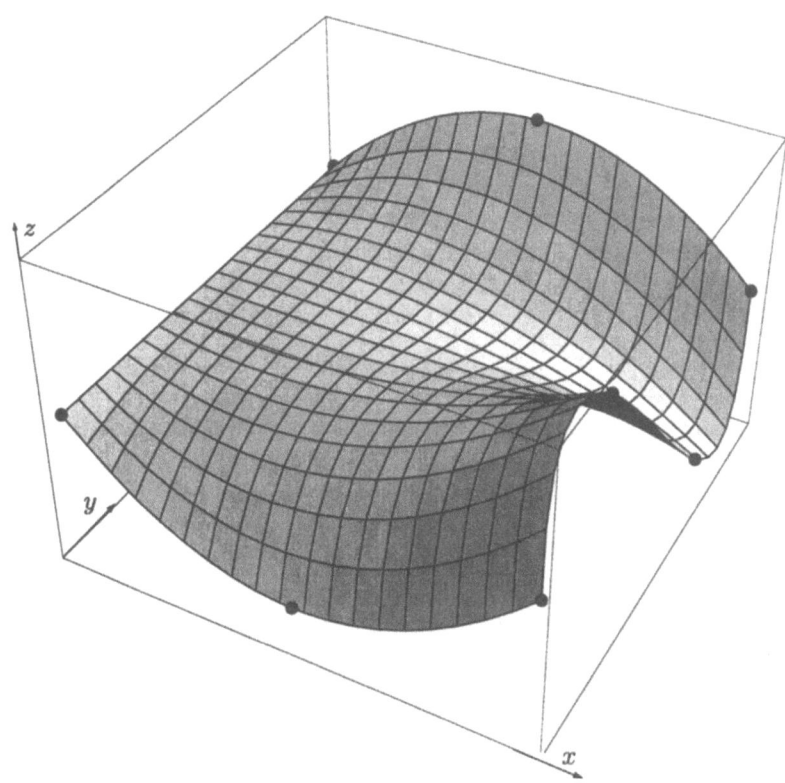

Figure 5.21: A Coons Surface Patch.

```
p00={-1,-1,0}; p01={-1,1,0}; p10={1,-1,0}; p11={1,1,0};
pnts={p00,p01,p10,p11,{1,-1/2,1/2},{1,1/2,-1/2},
  {0,-1,-1/2},{0,1,1/2}};
p0w[w_]:={-1,2w-1,0};
p1w[w_]:={1,(-4-w+27w^2-18w^3)/4,27(w-3w^2+2w^3)/4};
pu0[u_]:={2u-1,-1,2u^2-2u};
pu1[u_]:={2u-1,1,-2u^2+2u};
p[u_,w_]:=(1-u)p0w[w]+u p1w[w]+(1-w)pu0[u]+w pu1[u] \
  -p00(1-u)(1-w)-p01(1-u)
w-p10 u(1-w)-p11 u w;
g1=Graphics3D[{AbsolutePointSize[5], Table[Point[pnts[[i]]],
{i,1,8}]}];
g2=ParametricPlot3D[p[u,w], {u,0,1},{w,0,1}, Compiled->False,
Ticks->{{-1,1},{-1,1},{-1,1}}, DisplayFunction->Identity];
Show[g1,g2]
```

Code for Figure 5.21.

functions can be used to construct the *bicubic Coons surface*. Its expression is

$$\mathbf{P}(u,w) = (H_{3,0}(u), H_{3,3}(u), 1) \begin{pmatrix} -\mathbf{P}_{00} & -\mathbf{P}_{01} & \mathbf{P}(0,w) \\ -\mathbf{P}_{10} & -\mathbf{P}_{11} & \mathbf{P}(1,w) \\ \mathbf{P}(u,0) & \mathbf{P}(u,1) & 0 \end{pmatrix} \begin{pmatrix} H_{3,0}(w) \\ H_{3,3}(w) \\ 1 \end{pmatrix}$$

$$= (1 + 2u^3 - 3u^2, 3u^2 - 2u^3, 1) \begin{pmatrix} -\mathbf{P}_{00} & -\mathbf{P}_{01} & \mathbf{P}(0,w) \\ -\mathbf{P}_{10} & -\mathbf{P}_{11} & \mathbf{P}(1,w) \\ \mathbf{P}(u,0) & \mathbf{P}(u,1) & (0,0,0) \end{pmatrix}$$

$$\times \begin{pmatrix} 1 + 2w^3 - 3w^2 \\ 3w^2 - 2w^3 \\ 1 \end{pmatrix}. \tag{5.15}$$

One advantage of the bicubic Coons surface patch is that it is especially easy to connect smoothly to other patches of the same type. This is because its blending functions satisfy

$$\left.\frac{dH_{3,0}(t)}{dt}\right|_{t=0} = 0, \quad \left.\frac{dH_{3,0}(t)}{dt}\right|_{t=1} = 0, \quad \left.\frac{dH_{3,3}(t)}{dt}\right|_{t=0} = 0, \quad \left.\frac{dH_{3,3}(t)}{dt}\right|_{t=1} = 0. \tag{5.16}$$

Figure 5.22 shows two bicubic Coons surface patches, $\mathbf{P}(u,w)$ and $\mathbf{Q}(u,w)$, connected along their boundary curves $\mathbf{P}(u,1)$ and $\mathbf{Q}(u,0)$, respectively. The condition for patch connection is, of course, $\mathbf{P}(u,1) = \mathbf{Q}(u,0)$. The condition for smooth connection is

$$\left.\frac{\partial \mathbf{P}(u,w)}{\partial w}\right|_{w=1} = \left.\frac{\partial \mathbf{Q}(u,w)}{\partial w}\right|_{w=0}. \tag{5.17}$$

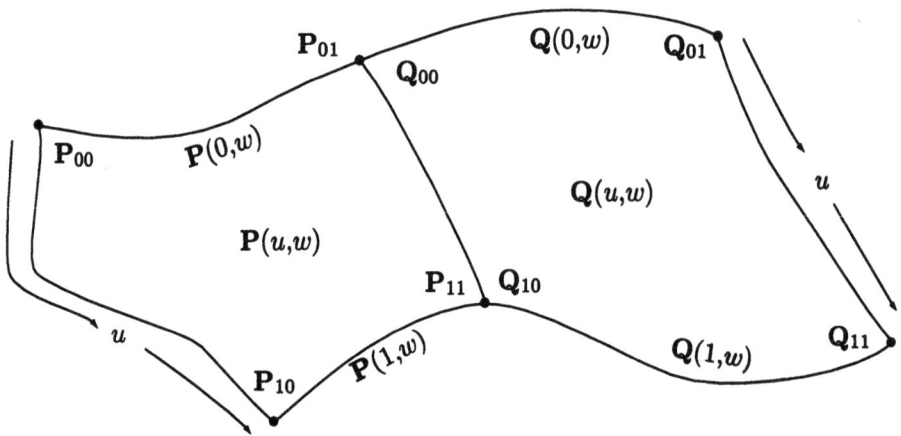

Figure 5.22: Smooth Connection of Bicubic Coons Surface Patches.

5 Surfaces

The partial derivatives of $\mathbf{P}(u,w)$ are easy to calculate from Equation (5.15):

$$\left.\frac{\partial \mathbf{P}(u,w)}{\partial w}\right|_{w=1} = H_{3,0}(u)\left.\frac{d\mathbf{P}(0,w)}{dw}\right|_{w=1} + H_{3,3}(u)\left.\frac{d\mathbf{P}(1,w)}{dw}\right|_{w=1},$$
$$\left.\frac{\partial \mathbf{Q}(u,w)}{\partial w}\right|_{w=0} = H_{3,0}(u)\left.\frac{d\mathbf{Q}(0,w)}{dw}\right|_{w=0} + H_{3,3}(u)\left.\frac{d\mathbf{Q}(1,w)}{dw}\right|_{w=0}. \tag{5.18}$$

(All other terms vanish because the blending functions satisfy Equation (5.16).) The condition for smooth connection, Equation (5.17), is therefore satisfied if

$$\left.\frac{d\mathbf{P}(0,w)}{dw}\right|_{w=1} = \left.\frac{d\mathbf{Q}(0,w)}{dw}\right|_{w=0} \quad \text{and} \quad \left.\frac{d\mathbf{P}(1,w)}{dw}\right|_{w=1} = \left.\frac{d\mathbf{Q}(1,w)}{dw}\right|_{w=0},$$

or, expressed in words, if the two boundary curves $\mathbf{P}(0,w)$ and $\mathbf{Q}(0,w)$ on the $u=0$ side of the patch connect smoothly, and the same for the two boundary curves $\mathbf{P}(1,w)$ and $\mathbf{Q}(1,w)$ on the $u=1$ side of the patch.

The reader should now find it easy to appreciate the advantage of the degree-5 Hermite blending functions

$$H_{5,0}(t) = B_{5,0}(t) + B_{5,1}(t) + B_{5,2}(t) = 1 - 10t^3 + 15t^4 - 6t^5,$$
$$H_{5,5}(t) = B_{5,3}(t) + B_{5,4}(t) + B_{5,5}(t) = 10t^3 - 15t^4 + 6t^5.$$

They are based on the Bernstein polynomials $B_{5,i}(t)$ and, thus, satisfy the conditions of Equation (5.14) and have the additional property that their first *and* second derivatives are zero for $t=0$ and for $t=1$. The degree-5 Coons surface constructed by them is

$$\mathbf{P}_5(u,w) = (H_{5,0}(u), H_{5,5}(u), 1) \begin{pmatrix} -\mathbf{P}_{00} & -\mathbf{P}_{01} & \mathbf{P}(0,w) \\ -\mathbf{P}_{10} & -\mathbf{P}_{11} & \mathbf{P}(1,w) \\ \mathbf{P}(u,0) & \mathbf{P}(u,1) & 0 \end{pmatrix} \begin{pmatrix} H_{5,0}(w) \\ H_{5,5}(w) \\ 1 \end{pmatrix}. \tag{5.19}$$

Adjacent patches of this type of surface are easy to connect with G^2 continuity. All that's necessary is to have two pairs of boundary curves $\mathbf{P}(0,w)$, $\mathbf{Q}(0,w)$ and $\mathbf{P}(1,w)$, $\mathbf{Q}(1,w)$, where the two curves of each pair connect with G^2 continuity.

5.7.2 The Tangent Matching Coons Surface

The original aim of Coons was to construct a surface patch where all four boundary curves are specified by the user. Such patches are easy to compute and the conditions for connecting them smoothly are simple. It is possible to extend the original ideas of Coons to a surface patch where the user specifies the four boundary curves and also four functions that describe how (in what direction) this surface approaches its boundaries. Figure 5.23 should make this clear. It shows a rectangular surface patch with some curves of the form $\mathbf{P}(u, w_i)$. Each of these curves goes from boundary curve $\mathbf{P}(0,w)$ to the opposite boundary curve $\mathbf{P}(1,w)$ by varying its parameter u from 0 to 1. Each has a different value of w_i. When such a curve reaches its end, it is moving in a certain, well-defined direction shown in the diagram. The end tangent vectors of these curves are different and we can imagine a function that yields these tangents as we move along the boundary curve $\mathbf{P}(1,w)$, varying w from 0 to 1. A good name for such a function is $\mathbf{P}_u(1,w)$, where the subscript u indicates that this tangent of the surface is in the u direction, the index 1 indicates the tangent at the end ($u = 1$), and the w indicates that this tangent vector is a function of w.

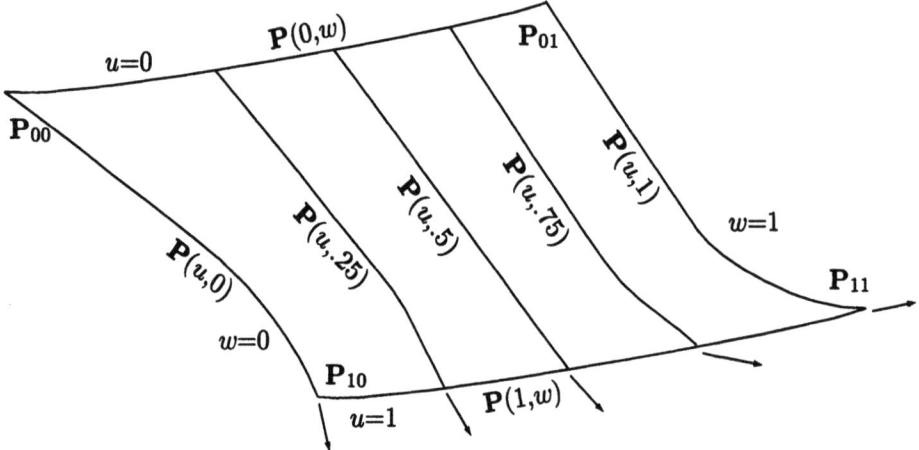

Figure 5.23: Tangent Matching in a Coons Surface.

There are four such functions, namely

$$\mathbf{P}_u(0,w), \quad \mathbf{P}_u(1,w), \quad \mathbf{P}_w(u,0), \text{ and } \mathbf{P}_w(u,1).$$

Assuming that the user provides these functions, as well as the four boundary curves, our task is to obtain an expression $\mathbf{P}(u,w)$ for the surface that will satisfy the following:

1. When we substitute 0 or 1 for u and w in $\mathbf{P}(u,w)$, we get the four given corner points and the four given boundary curves. This condition can be expressed as the eight constraints

$$\mathbf{P}(0,0) = \mathbf{P}_{00}, \quad \mathbf{P}(0,1) = \mathbf{P}_{01}, \quad \mathbf{P}(1,0) = \mathbf{P}_{10}, \quad \mathbf{P}(1,1) = \mathbf{P}_{11},$$
$\mathbf{P}(0,w), \quad \mathbf{P}(1,w), \quad \mathbf{P}(u,0), \text{ and } \mathbf{P}(u,1)$ are the given boundary curves.

2. When we substitute 0 or 1 for u and w in the partial first derivatives of $\mathbf{P}(u,w)$, we get the four given tangent functions and their values at the four corner points. This condition can be expressed as the 12 constraints

$$\left.\frac{\partial \mathbf{P}(u,w)}{\partial u}\right|_{u=0} = \mathbf{P}_u(0,w), \quad \left.\frac{\partial \mathbf{P}(u,w)}{\partial u}\right|_{u=1} = \mathbf{P}_u(1,w),$$

5 Surfaces

$$\left.\frac{\partial \mathbf{P}(u,w)}{\partial w}\right|_{w=0} = \mathbf{P}_w(u,0), \quad \left.\frac{\partial \mathbf{P}(u,w)}{\partial w}\right|_{w=1} = \mathbf{P}_w(u,1),$$

$$\left.\frac{\partial \mathbf{P}(u,w)}{\partial u}\right|_{u=0,w=0} = \mathbf{P}_u(0,0), \quad \left.\frac{\partial \mathbf{P}(u,w)}{\partial u}\right|_{u=0,w=1} = \mathbf{P}_u(0,1),$$

$$\left.\frac{\partial \mathbf{P}(u,w)}{\partial u}\right|_{u=1,w=0} = \mathbf{P}_u(1,0), \quad \left.\frac{\partial \mathbf{P}(u,w)}{\partial u}\right|_{u=1,w=1} = \mathbf{P}_u(1,1),$$

$$\left.\frac{\partial \mathbf{P}(u,w)}{\partial w}\right|_{u=0,w=0} = \mathbf{P}_w(0,0), \quad \left.\frac{\partial \mathbf{P}(u,w)}{\partial w}\right|_{u=0,w=1} = \mathbf{P}_w(0,1).$$

$$\left.\frac{\partial \mathbf{P}(u,w)}{\partial w}\right|_{u=1,w=0} = \mathbf{P}_w(1,0), \quad \left.\frac{\partial \mathbf{P}(u,w)}{\partial w}\right|_{u=1,w=1} = \mathbf{P}_w(1,1).$$

3. When we substitute 0 or 1 for u and w in the partial second derivatives of $\mathbf{P}(u,w)$, we get the four first derivatives of the given tangent functions at the four corner points. This condition can be expressed as the four constraints

$$\left.\frac{\partial^2 \mathbf{P}(u,w)}{\partial u \partial w}\right|_{u=0,w=0} = \left.\frac{d\mathbf{P}_u(0,w)}{dw}\right|_{w=0} = \left.\frac{d\mathbf{P}_u(u,0)}{du}\right|_{u=0} \stackrel{\text{def}}{=} \mathbf{P}_{uw}(0,0),$$

$$\left.\frac{\partial^2 \mathbf{P}(u,w)}{\partial u \partial w}\right|_{u=0,w=1} = \left.\frac{d\mathbf{P}_u(0,w)}{dw}\right|_{w=1} = \left.\frac{d\mathbf{P}_u(u,1)}{du}\right|_{u=0} \stackrel{\text{def}}{=} \mathbf{P}_{uw}(0,1),$$

$$\left.\frac{\partial^2 \mathbf{P}(u,w)}{\partial u \partial w}\right|_{u=1,w=0} = \left.\frac{d\mathbf{P}_u(1,w)}{dw}\right|_{w=0} = \left.\frac{d\mathbf{P}_u(u,0)}{du}\right|_{u=1} \stackrel{\text{def}}{=} \mathbf{P}_{uw}(1,0),$$

$$\left.\frac{\partial^2 \mathbf{P}(u,w)}{\partial u \partial w}\right|_{u=1,w=1} = \left.\frac{d\mathbf{P}_u(1,w)}{dw}\right|_{w=1} = \left.\frac{d\mathbf{P}_u(u,1)}{du}\right|_{u=1} \stackrel{\text{def}}{=} \mathbf{P}_{uw}(1,1).$$

This is a total of 24 constraints. A derivation of this type of surface can be found in [Beach 91]. Here, we only quote the final result

$$\mathbf{P}(u,w) = \bigl(B_0(u), B_1(u), C_0(u), C_1(u), 1\bigr)\mathbf{M}\begin{pmatrix} B_0(w) \\ B_1(w) \\ C_0(w) \\ C_1(w) \\ 1 \end{pmatrix}, \qquad (5.20)$$

where $\mathbf{M}$ is the 5×5 matrix

$$\mathbf{M} = \begin{pmatrix} -\mathbf{P}_{00} & -\mathbf{P}_{01} & -\mathbf{P}_w(0,0) & -\mathbf{P}_w(0,1) & \mathbf{P}(0,w) \\ -\mathbf{P}_{10} & -\mathbf{P}_{11} & -\mathbf{P}_w(1,0) & -\mathbf{P}_w(1,1) & \mathbf{P}(1,w) \\ -\mathbf{P}_u(0,0) & -\mathbf{P}_u(0,1) & -\mathbf{P}_{uw}(0,0) & -\mathbf{P}_{uw}(0,1) & \mathbf{P}_u(0,w) \\ -\mathbf{P}_u(1,0) & -\mathbf{P}_u(1,1) & -\mathbf{P}_{uw}(1,0) & -\mathbf{P}_{uw}(1,1) & \mathbf{P}_u(1,w) \\ \mathbf{P}(u,0) & \mathbf{P}(u,1) & \mathbf{P}_w(u,0) & \mathbf{P}_w(u,1) & (0,0,0) \end{pmatrix}. \qquad (5.21)$$

The two blending functions $B_0(t)$, $B_1(t)$ can be any functions satisfying conditions (5.14) and (5.16). Examples are the pairs $H_{3,0}(t)$, $H_{3,3}(t)$ and $H_{5,0}(t)$, $H_{5,5}(t)$

defined earlier. The two blending functions $C_0(t)$ and $C_1(t)$ should satisfy

$$C_0(0) = 0, \quad C_0(1) = 0, \quad C_0'(0) = 1, \quad C_0'(1) = 0,$$
$$C_1(0) = 0, \quad C_1(1) = 0, \quad C_1'(0) = 0, \quad C_1'(1) = 1.$$

One choice is the pair $C_0(t) = t - 2t^2 + t^3$ and $C_1(t) = -t^2 + t^3$.

Such a surface patch is hard to specify. The user has to input the four boundary curves and four tangent functions, a total of eight functions. He then has to calculate the coordinates of the 4 corner points and the other 12 quantities required by the matrix of Equation (5.21). The advantage of this type of surface is that once fully specified, such a surface patch is easy to connect smoothly to other patches of the same type since the tangents along the boundaries are fully specified by the user.

5.7.3 The Triangular Coons Surface

A triangular surface patch is bounded by three boundary curves and has three corner points. Such surface patches are handy in situations such as the one depicted in Figure 5.25, where a triangular Coons patch is used to smoothly connect two perpendicular lofted surface patches. Section 5.13 discusses the triangular Bézier surface patch which is commonly used in practice. Our approach to constructing the triangular Coons surface is to merge two of the four corner points and explore the behavior of the resulting surface patch. We arbitrarily decide to set $\mathbf{P}_{01} = \mathbf{P}_{11}$, which reduces the boundary curve $\mathbf{P}(u, 1)$ to a single point (Figure 5.24). The expression for this triangular surface patch is

$$\mathbf{P}(u,w) = \bigl(B_0(u), B_1(u), 1\bigr) \begin{pmatrix} -\mathbf{P}_{00} & -\mathbf{P}_{11} & \mathbf{P}(0,w) \\ -\mathbf{P}_{10} & -\mathbf{P}_{11} & \mathbf{P}(1,w) \\ \mathbf{P}(u,0) & \mathbf{P}_{11} & (0,0,0) \end{pmatrix} \begin{pmatrix} B_0(w) \\ B_1(w) \\ 1 \end{pmatrix}, \quad (5.22)$$

where the blending functions $B_0(t)$, $B_1(t)$ can be the pair $H_{3,0}$ and $H_{3,3}$, or the pair $H_{5,0}$ and $H_{5,5}$, or any other pair of blending functions satisfying Equations (5.14) and (5.16).

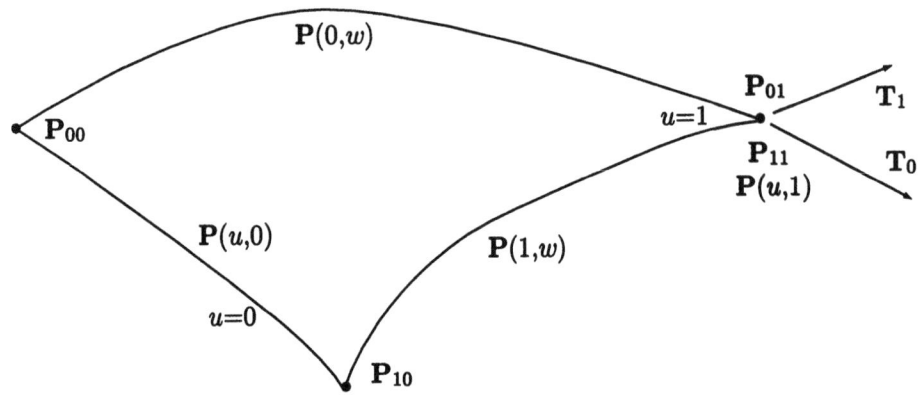

Figure 5.24: A Triangular Coons Surface Patch.

5 Surfaces

The tangent vector of the surface along the degenerate boundary curve $\mathbf{P}(u,1)$ is given by Equation (5.18):

$$\left.\frac{\partial \mathbf{P}(u,w)}{\partial w}\right|_{w=1} = B_0(u) \left.\frac{d\mathbf{P}(0,w)}{dw}\right|_{w=1} + B_1(u) \left.\frac{d\mathbf{P}(1,w)}{dw}\right|_{w=1}. \quad (5.23)$$

This tangent vector is thus a linear combination of the two tangents

$$\mathbf{T}_0 \stackrel{\text{def}}{=} \left.\frac{d\mathbf{P}(0,w)}{dw}\right|_{w=1} \quad \text{and} \quad \mathbf{T}_1 \stackrel{\text{def}}{=} \left.\frac{d\mathbf{P}(1,w)}{dw}\right|_{w=1},$$

and thus lies in the plane defined by them. As u varies from 0 to 1, this tangent vector swings from $\mathbf{T}_0$ to $\mathbf{T}_1$ while the curve $\mathbf{P}(u,1)$ stays at the common point $\mathbf{P}_{01} = \mathbf{P}_{11}$. Once this behavior is grasped, the reader should be able to accept the following statement: The triangular patch will be well behaved in the vicinity of the common point if this tangent vector does not reverse its movement while swinging from $\mathbf{T}_0$ to $\mathbf{T}_1$. If it starts moving toward $\mathbf{T}_1$, then reverses and goes back toward $\mathbf{T}_0$, then reverses again, the surface may have a *fold* close to the common point. To guarantee this smooth behavior of the tangent vector, the blending functions $B_0(t)$ and $B_1(t)$ must satisfy one more condition, namely $B_0(t)$ should be monotonically decreasing in t and $B_1(t)$ should be monotonically increasing in t. Both sets of blending functions $H_{3,0}, H_{3,3}$ and $H_{5,0}, H_{5,5}$ satisfy this condition, and can, thus, be used to construct triangular Coons surface patches.

▸ **Exercise 5.24:** What happens if the blending functions of the triangular Coons surface patch do not satisfy the condition of Equation (5.16)?

> By happy chance we saw
> A twofold image: on a grassy bank
> A snow-white ram, and in the crystal flood
> Another and the same!
>
> — William Wordsworth.

▸ **Exercise 5.25:** Given the four points $\mathbf{P}_{00} = (0,0,1)$, $\mathbf{P}_{10} = (1,0,0)$, $\mathbf{P}_{01} = (0.5,1,0)$, and $\mathbf{P}_{11} = (1,1,0)$, calculate the Coons surface defined by them, assuming straight lines as boundary curves. What type of a surface is this?

5.7.4 Summarizing Example

The surface shown in Figure 5.25 consists of four (intentionally separated) patches. A flat bilinear patch **B** at the top, two lofted patches **L** and **F** on both sides, and a triangular Coons patch **C** filling up the corner.

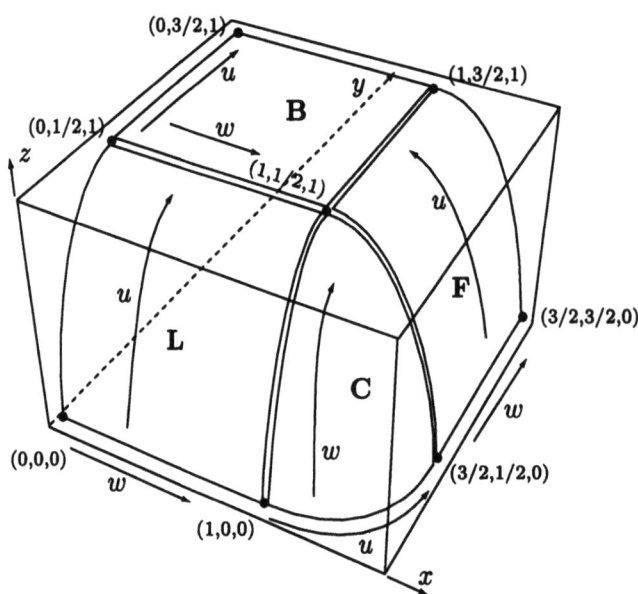

Figure 5.25: Bilinear, Lofted, and Coons Surface Patches.

```
b[u_,w_]:={0,1/2,1}(1-u)(1-w)+{1,1/2,1}(1-u)w
  +{0,3/2,1}(1-w)u+{1,3/2,1}u w;
H={{2,-2,1,1},{-3,3,-2,-1},{0,0,1,0},{1,0,0,0}};
lu0={u^3,u^2,u,1}.H.{{0,0,0},{0,1/2,1},{0,0,1},{0,1,0}};
lu1={u^3,u^2,u,1}.H.{{1,0,0},{1,1/2,1},{0,0,1},{0,1,0}};
l[u_,w_]:=lu0(1-w)+lu1 w;
fu0={u^3,u^2,u,1}.H.{{3/2,1/2,0},{1,1/2,1},{0,0,1},{-1,0,0}};
fu1={u^3,u^2,u,1}.H.{{3/2,3/2,0},{1,3/2,1},{0,0,1},{-1,0,0}};
f[u_,w_]:=fu0(1-w)+fu1 w;
cu0={u^3,u^2,u,1}.H.{{1,0,0},{3/2,1/2,0},{1,0,0},{0,1,0}};
cu1={1,1/2,1};
c0w={w^3,w^2,w,1}.H.{{1,0,0},{1,1/2,1},{0,0,1},{0,1,0}};
c1w={w^3,w^2,w,1}.H.{{3/2,1/2,0},{1,1/2,1},{0,0,1},{-1,0,0}};
c[u_,w_]:=(1-u)c0w+u c1w+(1-w)cu0+w cu1 \
  -(1-u)(1-w){1,0,0}-u(1-w){3/2,1/2,0}-w(1-u)cu1- u w cu1;
g1=ParametricPlot3D[b[u,w], {u,0,1},{w,0,1}]
g2=ParametricPlot3D[l[u,w], {u,0,1},{w,0,1}]
g3=ParametricPlot3D[f[u,w], {u,0,1},{w,0,1}]
g4=ParametricPlot3D[c[u,w], {u,0,1},{w,0,1}]
Show[g1,g2,g3,g4]
```

Code for Figure 5.25.

5 Surfaces

The bilinear patch is especially simple since it is defined by its four corner points. Its expression is

$$\mathbf{B}(u,w) = (0,1/2,1)(1-u)(1-w) + (1,1/2,1)(1-u)w$$
$$+ (0,3/2,1)(1-w)u + (1,3/2,1)uw$$
$$= (w, 1/2 + u, 1).$$

The calculation of lofted patch **L** starts with the two boundary curves $\mathbf{L}(u,0)$ and $\mathbf{L}(u,1)$. Each is calculated using Hermite interpolation since its extreme tangents, as well as its endpoints, are easy to figure out from the diagram. The boundary curves are

$$\mathbf{L}(u,0) = (u^3, u^2, u, 1)\mathbf{H}\big((0,0,0),(0,1/2,1),(0,0,1),(0,1,0)\big)^T,$$
$$\mathbf{L}(u,1) = (u^3, u^2, u, 1)\mathbf{H}\big((1,0,0),(1,1/2,1),(0,0,1),(0,1,0)\big)^T,$$

where **H** is the Hermite basis matrix, Equation (4.26). Surface patch L is thus

$$\mathbf{L}(u,w) = \mathbf{L}(u,0)(1-w) + \mathbf{L}(u,1)w = (w, u^2/2, u + u^2 - u^3).$$

Lofted patch **F** is calculated similarly. Its boundary curves are

$$\mathbf{F}(u,0) = (u^3, u^2, u, 1)\mathbf{H}\big((3/2,1/2,0),(1,1/2,1),(0,0,1),(-1,0,0)\big)^T,$$
$$\mathbf{F}(u,1) = (u^3, u^2, u, 1)\mathbf{H}\big((3/2,3/2,0),(1,3/2,1),(0,0,1),(-1,0,0)\big)^T,$$

and the patch itself is

$$\mathbf{F}(u,w) = \mathbf{F}(u,0)(1-w) + \mathbf{F}(u,1)w = ((3-u^2)/2, 1/2 + w, u + u^2 - u^3).$$

The triangular Coons surface **C** has corner points $\mathbf{C}_{00} = (1,0,0)$, $\mathbf{C}_{10} = (3/2, 1/2, 0)$, and $\mathbf{C}_{01} = \mathbf{C}_{11} = (1, 1/2, 1)$. Its bottom boundary curve is

$$\mathbf{C}(u,0) = (u^3, u^2, u, 1)\mathbf{H}\big((1,0,0),(3/2,1/2,0),(1,0,0),(0,1,0)\big)^T,$$

and its top boundary curve $\mathbf{C}(u,1)$ is the multiple point $\mathbf{C}_{01} = \mathbf{C}_{11}$. The two boundary curves in the w direction are

$$\mathbf{C}(0,w) = (w^3, w^2, w, 1)\mathbf{H}\big((1,0,0),(3/1,1/2,1),(0,0,1),(0,1,0)\big)^T,$$
$$\mathbf{C}(1,w) = (w^3, w^2, w, 1)\mathbf{H}\big((3/1,1/2,0),(1,1/2,1),(0,0,1),(-1,0,0)\big)^T,$$

and the surface patch itself equals

$$\mathbf{C}(u,w) = (1-u)\mathbf{C}(0,w) + u\mathbf{C}(1,w) + (1-w)\mathbf{C}(u,0) + w\mathbf{C}(u,1)$$
$$- (1-u)(1-w)1,0,0 - u(1-w)3/2,1/2,0 - w(1-u)\mathbf{C}_{11} - uw\mathbf{C}_{11}$$
$$= ((2 + u^2(-1+w) - u(-2+w+w^2))/2,$$
$$(-u^2(-1+w) - u(-1+w)w + w^2)/2, w + w^2 - w^3).$$

5.8 The Cartesian Product

The concept of *blending* has been and will be mentioned many times in connection with curve design. Section 4.2 introduces this important concept, Section 4.14.6 discusses the use of Bézier curves for blending, and Section 8.5 discusses blending as an interpolation technique for computer animation. This section shows how blending can be used in surface design. We start with two parametric curves $\mathbf{Q}(u) = \sum_{i=1}^{n} f_i(u)\mathbf{Q}_i$ and $\mathbf{R}(w) = \sum_{i=1}^{m} g_i(w)\mathbf{R}_i$ where $\mathbf{Q}_i$ and $\mathbf{R}_i$ can be points or vectors. Now examine the function

$$\mathbf{P}(u,w) = \sum_{i=1}^{n}\sum_{j=1}^{m} f_i(u)g_j(w)\mathbf{P}_{ij} = \sum_{i=1}^{n}\sum_{j=1}^{m} h_{ij}(u,w)\mathbf{P}_{ij}, \qquad (5.24)$$

where $h_{ij}(u,w) = f_i(u)g_j(w)$. The function $\mathbf{P}(u,w)$ describes a surface since it is a function of the two independent parameters u and w. For any value of the pair (u,w), the function computes a weighted sum of the quantities $\mathbf{P}_{ij}$. If these quantities are triplets, then $\mathbf{P}(u,w)$ returns a triplet (x,y,z) that can be interpreted as the three-dimensional coordinates of a point. When u and w vary over their ranges independently, $\mathbf{P}(u,w)$ describes all the three-dimensional points of a surface patch.

> I don't blend in at a family picnic.
> — Batman in *Batman Forever* (1995)

This technique of blending quantities $\mathbf{P}_{ij}$ into a surface by means of weights taken from two curves is called *Cartesian product*, although the names *tensor product* and *cross-product* are also sometimes used. The quantities $\mathbf{P}_{ij}$ can be points, tangent vectors, or second derivatives. Equation (5.24) can also be written

$$\mathbf{P}(u,w) = \bigl(f_1(u),\ldots,f_n(u)\bigr) \begin{pmatrix} \mathbf{P}_{11} & \mathbf{P}_{12} & \cdots & \mathbf{P}_{1m} \\ \vdots & \vdots & & \vdots \\ \mathbf{P}_{n1} & \mathbf{P}_{n2} & \cdots & \mathbf{P}_{nm} \end{pmatrix} \begin{pmatrix} g_1(w) \\ \vdots \\ g_m(w) \end{pmatrix}. \qquad (5.25)$$

Notice that it uses a matrix composed of nonscalar quantities (triplets). Even more important, Equation (5.24), combined with the isotropic principle (Section 4.18), tells us that if all $\mathbf{P}_{ij}$ are points, then the surface $\mathbf{P}(u,w)$ is independent of the particular coordinate axes used if $\sum_{ij} h_{ij}(u,w) = 1$. If the two original curves $\mathbf{Q}(u)$ and $\mathbf{R}(w)$ are isotropic, then it's easy to see that the surface is also isotropic since

$$\sum_{ij} h_{ij}(u,w) = \sum_i \sum_j f_i g_j = \Bigl(\sum_j g_j\Bigr)\Bigl(\sum_i f_i\Bigr) = 1.$$

5 Surfaces

5.9 The Biquadratic Surface Patch

This is a simple example of a surface constructed as a Cartesian product. We start with two quadratic (degree 2) polynomials

$$\mathbf{Q}(u) = \sum_{i=0}^{2} f_i(u)\mathbf{Q}_i \quad \text{and} \quad \mathbf{R}(w) = \sum_{i=0}^{2} g_i(w)\mathbf{R}_i$$

and construct the *biquadratic* surface

$$\mathbf{P}(u,w) = \sum_{i=0}^{2}\sum_{j=0}^{2} f_i(u)g_j(w)\mathbf{P}_{ij}. \tag{5.26}$$

Different constructions are possible depending on the geometric meaning of the nine quantities $\mathbf{P}_{ij}$. The following sections discuss two such constructions.

5.9.1 Nine Points

Equation (4.16) (duplicated below) gives the quadratic standard Lagrange polynomial that interpolates three given points:

$$\mathbf{P}_{2std}(t) = (t^2, t, 1) \begin{pmatrix} 2 & -4 & 2 \\ -3 & 4 & -1 \\ 1 & 0 & 0 \end{pmatrix} \begin{pmatrix} \mathbf{P}_0 \\ \mathbf{P}_1 \\ \mathbf{P}_2 \end{pmatrix}. \tag{4.16}$$

The biquadratic surface in this case is given by

$$\begin{aligned}\mathbf{P}(u,w) = (u^2, u, 1) &\begin{pmatrix} 2 & -4 & 2 \\ -3 & 4 & -1 \\ 1 & 0 & 0 \end{pmatrix} \begin{pmatrix} \mathbf{P}_{22} & \mathbf{P}_{21} & \mathbf{P}_{20} \\ \mathbf{P}_{12} & \mathbf{P}_{11} & \mathbf{P}_{10} \\ \mathbf{P}_{02} & \mathbf{P}_{01} & \mathbf{P}_{00} \end{pmatrix} \\ &\times \begin{pmatrix} 2 & -4 & 2 \\ -3 & 4 & -1 \\ 1 & 0 & 0 \end{pmatrix}^T \begin{pmatrix} w^2 \\ w \\ 1 \end{pmatrix},\end{aligned} \tag{5.27}$$

where the nine quantities $\mathbf{P}_{ij}$ are points defining this surface patch. They should be roughly equally spaced over the surface.

It is also possible to construct similar biquadratic surfaces from the expressions for the uniform and nonuniform quadratic Lagrange polynomials, Equations (4.14) and (4.15).

5.9.2 Points and Tangents

Section 4.7.9 discusses a variation on the Hermite segment where two points $\mathbf{P}_1$ and $\mathbf{P}_2$ and just one tangent vector $\mathbf{P}_1^t$ are known. The curve segment is given by

Equation (4.47), which is duplicated here:

$$\begin{aligned}\mathbf{P}(t) &= (\mathbf{P}_2 - \mathbf{P}_1 - \mathbf{P}_1^t)t^2 + \mathbf{P}_1^t t + \mathbf{P}_1 \\ &= (-t^2 + 1)\mathbf{P}_1 + t^2\mathbf{P}_2 + (-t^2 + t)\mathbf{P}_1^t \\ &= (t^2, t, 1)\begin{pmatrix} -1 & 1 & -1 \\ 0 & 0 & 1 \\ 1 & 0 & 0 \end{pmatrix}\begin{pmatrix} \mathbf{P}_1 \\ \mathbf{P}_2 \\ \mathbf{P}_1^t \end{pmatrix}.\end{aligned} \quad (4.47)$$

The tangent vector is $\mathbf{P}^t(t) = 2at + b = 2(\mathbf{P}_2 - \mathbf{P}_1 - \mathbf{P}_1^t)t + \mathbf{P}_1^t$, which implies that the end tangent is $\mathbf{P}^t(1) = 2(\mathbf{P}_2 - \mathbf{P}_1) - \mathbf{P}_1^t$. The biquadratic surface constructed as the Cartesian product of two such curves is given by

$$\mathbf{P}(u,w) = (u^2, u, 1)\begin{pmatrix} -1 & 1 & -1 \\ 0 & 0 & 1 \\ 1 & 0 & 0 \end{pmatrix}\begin{pmatrix} \mathbf{Q}_{22} & \mathbf{Q}_{21} & \mathbf{Q}_{20} \\ \mathbf{Q}_{12} & \mathbf{Q}_{11} & \mathbf{Q}_{10} \\ \mathbf{Q}_{02} & \mathbf{Q}_{01} & \mathbf{Q}_{00} \end{pmatrix}\begin{pmatrix} -1 & 0 & 1 \\ 1 & 0 & 0 \\ -1 & 1 & 0 \end{pmatrix}\begin{pmatrix} w^2 \\ w \\ 1 \end{pmatrix}, \quad (5.28)$$

where the nine quantities $\mathbf{Q}_{ij}$ still have to be assigned geometric meaning. This is done by computing $\mathbf{P}(u, w)$ and its partial derivatives for certain values of the parameters. Simple experimentation yields

$$\begin{aligned}\mathbf{P}(0,0) &= \mathbf{Q}_{22}, \quad \mathbf{P}(0,1) = \mathbf{Q}_{21}, \quad \mathbf{P}(1,0) = \mathbf{Q}_{12}, \quad \mathbf{P}(1,1) = \mathbf{Q}_{11}, \\ \mathbf{P}^u(0,0) &= \mathbf{Q}_{02}, \quad \mathbf{P}^u(0,1) = \mathbf{Q}_{01}, \quad \mathbf{P}^w(0,0) = \mathbf{Q}_{20}, \quad \mathbf{P}^w(1,0) = \mathbf{Q}_{10}, \\ &\qquad\qquad \mathbf{P}^{uw}(0,0) = \mathbf{Q}_{00}.\end{aligned}$$

This shows that the surface can be written

$$\begin{aligned}\mathbf{P}(u,w) &= (u^2, u, 1)\begin{pmatrix} -1 & 1 & -1 \\ 0 & 0 & 1 \\ 1 & 0 & 0 \end{pmatrix}\begin{pmatrix} \mathbf{P}(0,0) & \mathbf{P}(0,1) & \mathbf{P}^w(0,0) \\ \mathbf{P}(1,0) & \mathbf{P}(1,1) & \mathbf{P}^w(1,0) \\ \mathbf{P}^u(0,0) & \mathbf{P}^u(0,1) & \mathbf{P}^{uw}(0,0) \end{pmatrix} \\ &\quad \times \begin{pmatrix} -1 & 0 & 1 \\ 1 & 0 & 0 \\ -1 & 1 & 0 \end{pmatrix}\begin{pmatrix} w^2 \\ w \\ 1 \end{pmatrix} \\ &= (u^2, u, 1)\begin{pmatrix} -1 & 1 & -1 \\ 0 & 0 & 1 \\ 1 & 0 & 0 \end{pmatrix}\begin{pmatrix} \mathbf{P}_{00} & \mathbf{P}_{01} & \mathbf{P}_{00}^w \\ \mathbf{P}_{10} & \mathbf{P}_{11} & \mathbf{P}_{10}^w \\ \mathbf{P}_{00}^u & \mathbf{P}_{01}^u & \mathbf{P}_{00}^{uw} \end{pmatrix}\begin{pmatrix} -1 & 0 & 1 \\ 1 & 0 & 0 \\ -1 & 1 & 0 \end{pmatrix}\begin{pmatrix} w^2 \\ w \\ 1 \end{pmatrix}.\end{aligned} \quad (5.29)$$

This type of surface is thus defined by the following nine quantities:
  The four corner points $\mathbf{P}_{00}$, $\mathbf{P}_{01}$, $\mathbf{P}_{10}$, and $\mathbf{P}_{11}$.
  The two tangents in the u direction at points $\mathbf{P}_{00}$ and $\mathbf{P}_{01}$.
  The two tangents in the w direction at points $\mathbf{P}_{00}$ and $\mathbf{P}_{10}$.
  The second derivative at point $\mathbf{P}_{00}$.
  The first eight quantities have simple geometric meaning, but the second derivative, that is also called a *twist vector*, has no simple geometrical interpretation. It can simply be set to zero. Several methods exist to help estimate the twist vectors of biquadratic and bicubic surface patches. The simple method described here is

5 Surfaces

useful when a surface is constructed out of several such patches. We start by looking at the twist vector of a bilinear surface. Differentiating Equation (5.7) twice, with respect to u and w, produces the simple, fixed expression

$$\mathbf{P}^{uw}(u,w) = \mathbf{P}_{00} - \mathbf{P}_{01} - \mathbf{P}_{10} + \mathbf{P}_{11}, \tag{5.30}$$

which is independent of both parameters. This expression is used to estimate the twist vectors of all the patches that constitute a biquadratic or a bicubic surface. Figure 5.26 is an idealized diagram of such a surface showing some individual patches. The first step is to use Equation (5.30) to calculate a vector $\mathbf{T}_i$ for patch i using the four corner points of the patch. Vectors $\mathbf{T}_i$ are then averaged to provide estimates for the four twist vectors of every patch.

The principle is: A corner point $\mathbf{P}_i$ with one index i belongs to just one patch (patch i) and is one of the four corner points of the entire surface ($\mathbf{P}_1$, $\mathbf{P}_4$, $\mathbf{P}_9$, and $\mathbf{P}_c$ of Figure 5.26). The twist vector estimated for such a point is $\mathbf{T}_i$, the vector previously calculated for patch i. A point $\mathbf{P}_{ij}$ with two indexes ij is common to two patches i and j and is located on the boundary of the entire surface (examples are $\mathbf{P}_{15}$ and $\mathbf{P}_{59}$). The twist vector estimated for such a point is the average $(\mathbf{T}_i + \mathbf{T}_j)/2$. A point $\mathbf{P}_{ijkl}$ with four indexes is common to four patches. The twist vector estimated for such a point is the average $(\mathbf{T}_i + \mathbf{T}_j + \mathbf{T}_k + \mathbf{T}_l)/4$.

This method works well as a first estimate. After the surface is drawn, the twist vectors calculated by this method may have to be modified to bring the surface closer to its required shape.

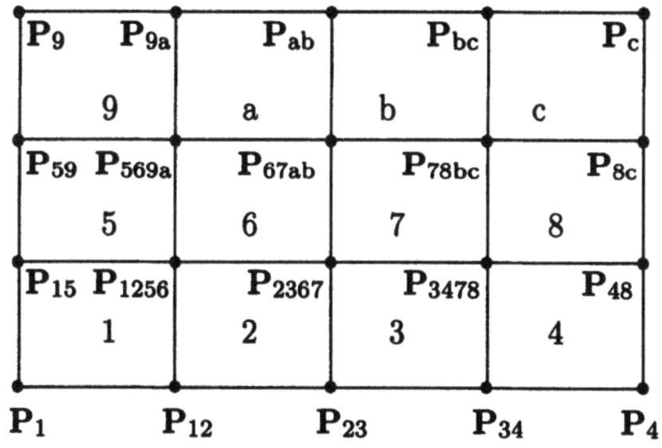

Figure 5.26: Estimating Twist Vectors.

5.10 The Bicubic Surface Patch

The parametric cubic (PC) curve (Equation (4.8)) is useful, since it can be used when either four points, or two points and two tangent vectors, are known. The PC curve can easily be extended to a bicubic surface patch by means of the *Cartesian product*.

A PC curve has the form $\mathbf{P}(t) = \sum_{i=0}^{3} \mathbf{a}_i t^i$. Two such curves, $\mathbf{P}(u)$ and $\mathbf{P}(w)$, can be combined to form the Cartesian product surface patch

$$\mathbf{P}(u,w)$$
$$= \sum_{i=0}^{3} \sum_{j=0}^{3} \mathbf{a}_{ij} u^i w^j$$
$$= \mathbf{a}_{33} u^3 w^3 + \mathbf{a}_{32} u^3 w^2 + \mathbf{a}_{31} u^3 w + \mathbf{a}_{30} u^3 + \mathbf{a}_{23} u^2 w^3 + \mathbf{a}_{22} u^2 w^2 + \mathbf{a}_{21} u^2 w + \mathbf{a}_{20} u^2$$
$$+ \mathbf{a}_{13} u w^3 + \mathbf{a}_{12} u w^2 + \mathbf{a}_{11} u w + \mathbf{a}_{10} u + \mathbf{a}_{03} w^3 + \mathbf{a}_{02} w^2 + \mathbf{a}_{01} w + \mathbf{a}_{00}$$
$$= (u^3, u^2, u, 1) \begin{pmatrix} \mathbf{a}_{33} & \mathbf{a}_{32} & \mathbf{a}_{31} & \mathbf{a}_{30} \\ \mathbf{a}_{23} & \mathbf{a}_{22} & \mathbf{a}_{21} & \mathbf{a}_{20} \\ \mathbf{a}_{13} & \mathbf{a}_{12} & \mathbf{a}_{11} & \mathbf{a}_{10} \\ \mathbf{a}_{03} & \mathbf{a}_{02} & \mathbf{a}_{01} & \mathbf{a}_{00} \end{pmatrix} \begin{pmatrix} w^3 \\ w^2 \\ w \\ 1 \end{pmatrix}, \quad \text{where } 0 \le u, w \le 1. \quad (5.31)$$

This is a double cubic polynomial (hence the name *bicubic*) with 16 terms, where each of the 16 coefficients $\mathbf{a}_{ij}$ is a triplet. When w is set to a fixed value w_0, Equation (5.31) becomes $\mathbf{P}(u, w_0)$, which is a PC curve. The same is true for $\mathbf{P}(u_0, w)$. The conclusion is that curves that go on this surface in the u or in the w directions are PC curves. The four boundary curves are therefore also PC curves.

Notice that the shape and location of the surface depend on all 16 coefficients. Any change in any of them produces a different surface patch. Equation (5.31) is the algebraic representation of the bicubic patch. In order to use it in practice, the 16 unknown coefficients have to be expressed in terms of known geometrical quantities, such as points, tangent vectors, or second derivatives.

Four types of bicubic surfaces are discussed below. The first is based on 16 data points. The second is constructed from four known curves. The third is defined by four data points, eight tangent vectors, and four twist vectors. The fourth is a flat plane.

> Milo... glanced curiously at the strange circular room, where sixteen tiny arched windows corresponded exactly to the sixteen points of the compass. Around the entire circumference were numbers from zero to three hundred and sixty, marking the degrees of the circle, and on the floor, walls, tables, chairs, desks, cabinets, and ceiling were labels showing their heights, widths, depths, and distances to and from each other.
>
> —Norton Juster, *The Phantom Tollbooth*.

5.10.1 Sixteen Points

We start with the sixteen given points

$$\begin{matrix} \mathbf{P}_{03} & \mathbf{P}_{13} & \mathbf{P}_{23} & \mathbf{P}_{33} \\ \mathbf{P}_{02} & \mathbf{P}_{12} & \mathbf{P}_{22} & \mathbf{P}_{32} \\ \mathbf{P}_{01} & \mathbf{P}_{11} & \mathbf{P}_{21} & \mathbf{P}_{31} \\ \mathbf{P}_{00} & \mathbf{P}_{10} & \mathbf{P}_{20} & \mathbf{P}_{30}. \end{matrix}$$

5 Surfaces

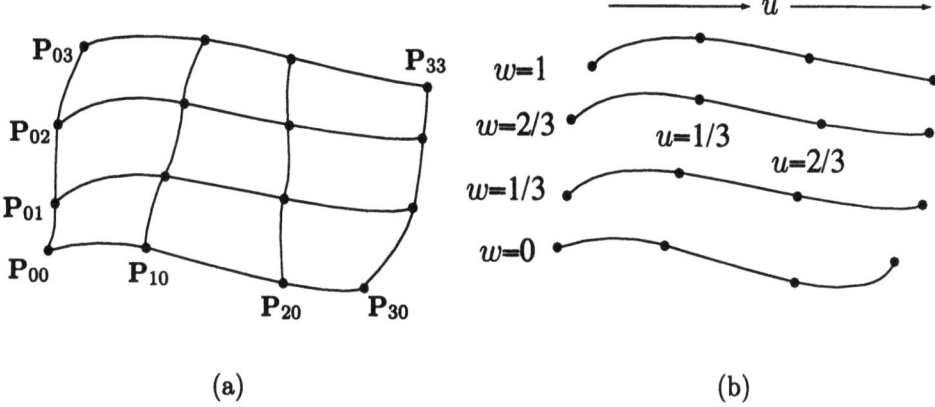

(a)   (b)

Figure 5.27: (a) Sixteen Points. (b) Four Curves.

We assume that the points are (roughly) equally spaced on the rectangular surface patch as shown in Figure 5.27a. We know that the bicubic surface has the form

$$\mathbf{P}(u,w) = \sum_{i=0}^{3}\sum_{j=0}^{3} \mathbf{a}_{ij} u^i w^j, \tag{5.32}$$

where each of the 16 coefficients $\mathbf{a}_{ij}$ is a triplet. To calculate the 16 unknown coefficients, we write 16 equations, each based on one of the given points

$$\begin{array}{llll}
\mathbf{P}(0,0) = \mathbf{P}_{00}, & \mathbf{P}(0,1/3) = \mathbf{P}_{01}, & \mathbf{P}(0,2/3) = \mathbf{P}_{02}, & \mathbf{P}(0,1) = \mathbf{P}_{03}, \\
\mathbf{P}(1/3,0) = \mathbf{P}_{10}, & \mathbf{P}(1/3,1/3) = \mathbf{P}_{11}, & \mathbf{P}(1/3,2/3) = \mathbf{P}_{12}, & \mathbf{P}(1/3,1) = \mathbf{P}_{13}, \\
\mathbf{P}(2/3,0) = \mathbf{P}_{20}, & \mathbf{P}(2/3,1/3) = \mathbf{P}_{21}, & \mathbf{P}(2/3,2/3) = \mathbf{P}_{22}, & \mathbf{P}(2/3,1) = \mathbf{P}_{23}, \\
\mathbf{P}(1,0) = \mathbf{P}_{30}, & \mathbf{P}(1,1/3) = \mathbf{P}_{31}, & \mathbf{P}(1,2/3) = \mathbf{P}_{32}, & \mathbf{P}(1,1) = \mathbf{P}_{33}.
\end{array}$$

After solving, the final expression for the surface patch can be written

$$\mathbf{P}(u,w) = (u^3, u^2, u, 1)\mathbf{N} \begin{pmatrix} \mathbf{P}_{33} & \mathbf{P}_{32} & \mathbf{P}_{31} & \mathbf{P}_{30} \\ \mathbf{P}_{23} & \mathbf{P}_{22} & \mathbf{P}_{21} & \mathbf{P}_{20} \\ \mathbf{P}_{13} & \mathbf{P}_{12} & \mathbf{P}_{11} & \mathbf{P}_{10} \\ \mathbf{P}_{03} & \mathbf{P}_{02} & \mathbf{P}_{01} & \mathbf{P}_{00} \end{pmatrix} \mathbf{N}^T \begin{pmatrix} w^3 \\ w^2 \\ w \\ 1 \end{pmatrix}, \tag{5.33}$$

where

$$\mathbf{N} = \begin{pmatrix} -4.5 & 13.5 & -13.5 & 4.5 \\ 9.0 & -22.5 & 18 & -4.5 \\ -5.5 & 9.0 & -4.5 & 1.0 \\ 1.0 & 0 & 0 & 0 \end{pmatrix}.$$

is the basis matrix used to blend four points in a PC (Equation (4.11)). As mentioned, this type of surface patch has only limited use because it cannot have a very complex shape. A larger surface, made up of a number of such patches, can be created, but it is hard to connect the individual patches smoothly.

Even though this type of surface has limited use in graphics, it can be used for two-dimensional *polynomial interpolation*. Imagine three-dimensional points arranged in a two-dimensional grid. We want to use their values in a weighted sum to predict the value of a new point at the center of the grid. It makes sense to assign more weights to points that are closer to the center, and a natural way to achieve this is to calculate the surface patch $\mathbf{P}(u,w)$ that passes through all the points in the grid and to use the value $\mathbf{P}(0.5, 0.5)$ as the interpolated value at the center of the grid.

Section 9.11 is an example of the use of this approach. The problem is to interpolate the values of a group of 4×4 pixels in an image in order to predict the value of a pixel at the center of this group. The simple solution is to calculate the surface patch defined by the 16 pixels and to use the surface point $\mathbf{P}(0.5, 0.5)$ as the interpolated value of the pixel at the center of the group. Substituting $u = 0.5$ and $w = 0.5$ in Equation (5.33) produces

$$\mathbf{P}(0.5, 0.5)$$
$$= 0.00390625\mathbf{P}_{00} - 0.0351563\mathbf{P}_{01} - 0.0351563\mathbf{P}_{02} + 0.00390625\mathbf{P}_{03}$$
$$- 0.0351563\mathbf{P}_{10} + 0.316406\mathbf{P}_{11} + 0.316406\mathbf{P}_{12} - 0.0351563\mathbf{P}_{13}$$
$$- 0.0351563\mathbf{P}_{20} + 0.316406\mathbf{P}_{21} + 0.316406\mathbf{P}_{22} - 0.0351563\mathbf{P}_{23}$$
$$+ 0.00390625\mathbf{P}_{30} - 0.0351563\mathbf{P}_{31} - 0.0351563\mathbf{P}_{32} + 0.00390625\mathbf{P}_{33}.$$

The 16 coefficients are the ones used in Table 9.29.

▶ **Exercise 5.26:** The center point of the surface is calculated as a weighted sum of the 16 equally spaced data points. It makes sense to assign small weights to points located away from the center, but our result assigns *negative* weights to eight of the 16 points. Explain the meaning of negative weights and show what role they play in interpolating the center of the surface.

Readers who find it hard to follow the above details should compare the way two-dimensional polynomial interpolation is presented here to the way it is discussed by [Press and Flannery 88]. The following quotation is from their page 125: "... The formulas that obtain the c's from the function and derivative values are just a complicated linear transformation, with coefficients which, having been determined once, in the mists of numerical history, can be tabulated and forgotten."

> Seated at his disorderly desk, caressed by a counterpane of drifting tobacco haze, he would pore over the manuscript, crossing out, interpolating, re-arguing, and then referring to volumes on his shelves.
> —Christopher Morley, *The Haunted Bookshop*.

5.10.2 Four Curves

A variant of the previous method starts with four curves (any curves, not just PCs), $\mathbf{P}_0(u)$, $\mathbf{P}_1(u)$, $\mathbf{P}_2(u)$, and $\mathbf{P}_3(u)$, roughly parallel, all going in the u direction (Figure 5.27b). It is possible to select 4 points on each curve $\mathbf{P}_i(0)$, $\mathbf{P}_i(1/3)$,

5 Surfaces

$\mathbf{P}_i(2/3)$, and $\mathbf{P}_i(1)$, for a total of 16 points. The surface patch can then easily be constructed from Equation (5.33).

5.10.3 An Example

The surface of Figure 5.28 is defined by the following four curves (shown in the diagram in an inset). All go along the x axis, at different y values, and are sine curves (with different phases) along the z axis.

$$\mathbf{P}_0(u) = (u, 0, \sin(\pi u)), \quad \mathbf{P}_1(u) = (u, 1 + u/10, \sin(\pi(u + 0.1))),$$
$$\mathbf{P}_2(u) = (u, 2, \sin(\pi(u + 0.2))), \quad \mathbf{P}_3(u) = (u, 3 + u/10, \sin(\pi(u + 0.3))),$$

The *Mathematica* code of Figure 5.28 shows how matrix `basis` is created with the 16 points

$$\begin{pmatrix} \mathbf{P}_0(0) & \mathbf{P}_0(.33) & \mathbf{P}_0(.67) & \mathbf{P}_0(1) \\ \mathbf{P}_1(0) & \mathbf{P}_1(.33) & \mathbf{P}_1(.67) & \mathbf{P}_1(1) \\ \mathbf{P}_2(0) & \mathbf{P}_2(.33) & \mathbf{P}_2(.67) & \mathbf{P}_2(1) \\ \mathbf{P}_3(0) & \mathbf{P}_3(.33) & \mathbf{P}_3(.67) & \mathbf{P}_3(1) \end{pmatrix}.$$

5.10.4 Bicubic Hermite Patch

The four corner points of the patch are given, in this type of surface, together with the two end tangents of the boundary curves at each point (a total of eight tangents). This makes for 12 known quantities, but 4 more are needed in order to calculate the 16 unknowns of Equation (5.32). The other four quantities are usually selected as the second derivatives of the surface at the corner points. They are called *twist vectors*.

To calculate the surface, 16 equations are written, expressing the way we require the surface to behave. For example, we want $\mathbf{P}(u, w)$ to approach the corner point $\mathbf{P}_{01}$ when $u \to 0$ and $w \to 1$. We also want $\mathbf{P}(0, w)$ to equal the PC between points $\mathbf{P}_{00}$ and $\mathbf{P}_{01}$. The equations are

$$\mathbf{P}_{00} = \mathbf{a}_{00},$$
$$\mathbf{P}_{10} = \mathbf{a}_{30} + \mathbf{a}_{20} + \mathbf{a}_{10} + \mathbf{a}_{00},$$
$$\mathbf{P}_{01} = \mathbf{a}_{03} + \mathbf{a}_{02} + \mathbf{a}_{01} + \mathbf{a}_{00},$$
$$\mathbf{P}_{11} = \mathbf{a}_{33} + \mathbf{a}_{32} + \mathbf{a}_{31} + \mathbf{a}_{30} + \mathbf{a}_{23} + \mathbf{a}_{22} + \mathbf{a}_{21} + \mathbf{a}_{20}$$
$$+ \mathbf{a}_{13} + \mathbf{a}_{12} + \mathbf{a}_{11} + \mathbf{a}_{10} + \mathbf{a}_{03} + \mathbf{a}_{02} + \mathbf{a}_{01} + \mathbf{a}_{00},$$
$$\mathbf{P}_{00}^u = \mathbf{a}_{10},$$
$$\mathbf{P}_{00}^w = \mathbf{a}_{01},$$
$$\mathbf{P}_{10}^u = 3\mathbf{a}_{30} + 2\mathbf{a}_{20} + \mathbf{a}_{10},$$
$$\mathbf{P}_{10}^w = \mathbf{a}_{31} + \mathbf{a}_{21} + \mathbf{a}_{11} + \mathbf{a}_{01},$$
$$\mathbf{P}_{01}^u = \mathbf{a}_{13} + \mathbf{a}_{12} + \mathbf{a}_{11} + \mathbf{a}_{10},$$
$$\mathbf{P}_{01}^w = 3\mathbf{a}_{03} + 2\mathbf{a}_{02} + \mathbf{a}_{01},$$
$$\mathbf{P}_{11}^u = 3\mathbf{a}_{33} + 3\mathbf{a}_{32} + 3\mathbf{a}_{31} + 3\mathbf{a}_{30} + 2\mathbf{a}_{23} + 2\mathbf{a}_{22} + 2\mathbf{a}_{21}$$
$$+ 2\mathbf{a}_{20} + \mathbf{a}_{13} + \mathbf{a}_{12} + \mathbf{a}_{11} + \mathbf{a}_{10},$$

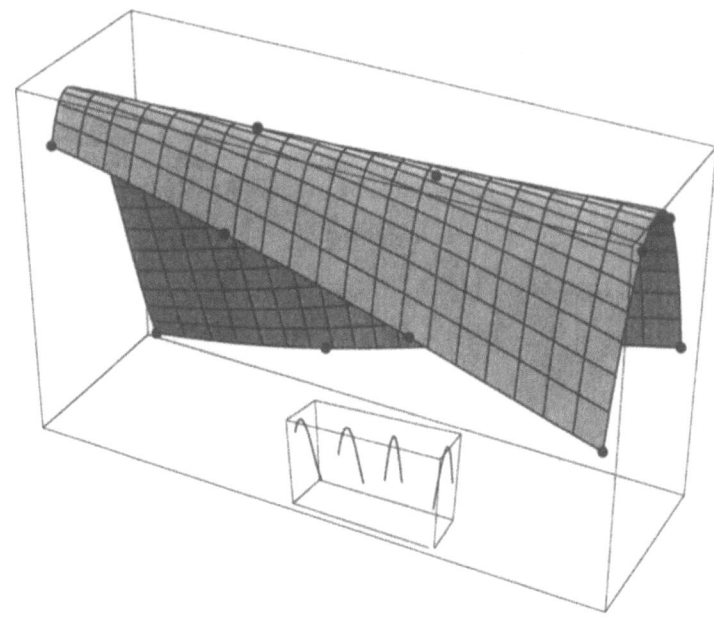

```
Clear[p0,p1,p2,p3,basis,fourP,g0,g1,g2,g3,g4,g5];
p0[u_]:={u,0,Sin[Pi u]}; p1[u_]:={u,1+u/10,Sin[Pi(u+.1)]};
p2[u_]:={u,2,Sin[Pi(u+.2)]}; p3[u_]:={u,3+u/10,Sin[Pi(u+.3)]};
(* matrix 'basis' has dimensions 4x4x3 *)
basis:={{p0[0],p0[.33],p0[.67],p0[1]},{p1[0],p1[.33],p1[.67],p1[1]},
{p2[0],p2[.33],p2[.67],p2[1]},{p3[0],p3[.33],p3[.67],p3[1]}};
fourP:= (* basis matrix for a 4-point curve *)
{{-4.5,13.5,-13.5,4.5},{9,-22.5,18,-4.5},{-5.5,9,-4.5,1},{1,0,0,0}};
prt[i_]:= (* extracts component i from the 3rd dimen of 'basis' *)
basis[[Range[1,4],Range[1,4],i]];
coord[i_]:= (* calc. the 3 parametric components of the surface *)
{u^3,u^2,u,1}.fourP.prt[i].Transpose[fourP].{w^3,w^2,w,1};
g0=ParametricPlot3D[p0[u], {u,0,1}]
g1=ParametricPlot3D[p1[u], {u,0,1}]
g2=ParametricPlot3D[p2[u], {u,0,1}]
g3=ParametricPlot3D[p3[u], {u,0,1}]
g4=Graphics3D[{AbsolutePointSize[4],
Table[Point[basis[[i,j]]],{i,1,4},{j,1,4}]}];
g5=ParametricPlot3D[{coord[1],coord[2],coord[3]},
{u,0,1,.05},{w,0,1,.05}, DisplayFunction->Identity];
Show[g0,g1,g2,g3, ViewPoint->{-2.576, -1.365, 1.718},
 Ticks->False, DisplayFunction->$DisplayFunction]
Show[g4,g5, ViewPoint->{-2.576, -1.365, 1.718},
 DisplayFunction->$DisplayFunction]
```

Figure 5.28: A Four-Curve Surface.

5 Surfaces

$$\mathbf{P}_{11}^w = 3a_{33} + 2a_{32} + a_{31} + 3a_{23} + 2a_{22} + a_{21} + 3a_{13}$$
$$+ 2a_{12} + a_{11} + 3a_{03} + 2a_{02} + a_{01},$$
$$\mathbf{P}_{00}^{uw} = a_{11},$$
$$\mathbf{P}_{10}^{uw} = 3a_{31} + 2a_{21} + a_{11},$$
$$\mathbf{P}_{01}^{uw} = 3a_{13} + 2a_{12} + a_{11},$$
$$\mathbf{P}_{11}^{uw} = 9a_{33} + 6a_{32} + 3a_{31} + 6a_{23} + 4a_{22}$$
$$+ 2a_{21} + 3a_{13} + 2a_{12} + a_{11}.$$

The solutions express the 16 coefficients a_{ij} in terms of the 4 corner points, 8 tangent vectors, and 4 twist vectors:

$$a_{01} = \mathbf{P}_{00}^w,$$
$$a_{02} = -2\mathbf{P}_{00}^w - \mathbf{P}_{01}^w - 3\mathbf{P}_{00} + 3\mathbf{P}_{01},$$
$$a_{03} = \mathbf{P}_{00}^w + \mathbf{P}_{01}^w + 2\mathbf{P}_{00} - 2\mathbf{P}_{01},$$
$$a_{10} = \mathbf{P}_{00}^u,$$
$$a_{11} = \mathbf{P}_{00}^{uw},$$
$$a_{12} = -2\mathbf{P}_{00}^{uw} - \mathbf{P}_{01}^{uw} - 3\mathbf{P}_{00}^u + 3\mathbf{P}_{01}^u,$$
$$a_{13} = \mathbf{P}_{00}^{uw} + \mathbf{P}_{01}^{uw} + 2\mathbf{P}_{00}^u - 2\mathbf{P}_{01}^u,$$
$$a_{20} = -2\mathbf{P}_{00}^u - \mathbf{P}_{10}^u - 3\mathbf{P}_{00} + 3\mathbf{P}_{10},$$
$$a_{21} = -2\mathbf{P}_{00}^{uw} - \mathbf{P}_{10}^{uw} - 3\mathbf{P}_{00}^w + 3\mathbf{P}_{10}^w,$$
$$a_{22} = 4\mathbf{P}_{00}^{uw} + 2\mathbf{P}_{01}^{uw} + 2\mathbf{P}_{10}^{uw} + \mathbf{P}_{11}^{uw} + 6\mathbf{P}_{00}^u - 6\mathbf{P}_{01}^u + 3\mathbf{P}_{10}^u - 3\mathbf{P}_{11}^u + 6\mathbf{P}_{00}^w$$
$$+ 3\mathbf{P}_{01}^w - 6\mathbf{P}_{10}^w - 3\mathbf{P}_{11}^w + 9\mathbf{P}_{00} - 9\mathbf{P}_{01} - 9\mathbf{P}_{10} + 9\mathbf{P}_{11},$$
$$a_{23} = -2\mathbf{P}_{00}^{uw} - 2\mathbf{P}_{01}^{uw} - \mathbf{P}_{10}^{uw} - \mathbf{P}_{11}^{uw} - 4\mathbf{P}_{00}^u + 4\mathbf{P}_{01}^u - 2\mathbf{P}_{10}^u + 2\mathbf{P}_{11}^u - 3\mathbf{P}_{00}^w$$
$$- 3\mathbf{P}_{01}^w + 3\mathbf{P}_{10}^w + 3\mathbf{P}_{11}^w - 6\mathbf{P}_{00} + 6\mathbf{P}_{01} + 6\mathbf{P}_{10} - 6\mathbf{P}_{11},$$
$$a_{30} = \mathbf{P}_{00}^u + \mathbf{P}_{10}^u + 2\mathbf{P}_{00} - 2\mathbf{P}_{10},$$
$$a_{31} = \mathbf{P}_{00}^{uw} + \mathbf{P}_{10}^{uw} + 2\mathbf{P}_{00}^w - 2\mathbf{P}_{10}^w,$$
$$a_{32} = -2\mathbf{P}_{00}^{uw} - \mathbf{P}_{01}^{uw} - 2\mathbf{P}_{10}^{uw} - \mathbf{P}_{11}^{uw} - 3\mathbf{P}_{00}^u + 3\mathbf{P}_{01}^u - 3\mathbf{P}_{10}^u + 3\mathbf{P}_{11}^u - 4\mathbf{P}_{00}^w$$
$$- 2\mathbf{P}_{01}^w + 4\mathbf{P}_{10}^w + 2\mathbf{P}_{11}^w - 6\mathbf{P}_{00} + 6\mathbf{P}_{01} + 6\mathbf{P}_{10} - 6\mathbf{P}_{11},$$
$$a_{33} = \mathbf{P}_{00}^{uw} + \mathbf{P}_{01}^{uw} + \mathbf{P}_{10}^{uw} + \mathbf{P}_{11}^{uw} + 2\mathbf{P}_{00}^u - 2\mathbf{P}_{01}^u + 2\mathbf{P}_{10}^u - 2\mathbf{P}_{11}^u + 2\mathbf{P}_{00}^w + 2\mathbf{P}_{01}^w$$
$$- 2\mathbf{P}_{10}^w - 2\mathbf{P}_{11}^w + 4\mathbf{P}_{00} - 4\mathbf{P}_{01} - 4\mathbf{P}_{10} + 4\mathbf{P}_{11}.$$

When Equation (5.32) is written in terms of these values, it becomes

$$\mathbf{P}(u,w) = (u^3, u^2, u, 1)\mathbf{H} \begin{pmatrix} \mathbf{P}_{00} & \mathbf{P}_{01} & \mathbf{P}_{00}^w & \mathbf{P}_{01}^w \\ \mathbf{P}_{10} & \mathbf{P}_{11} & \mathbf{P}_{10}^w & \mathbf{P}_{11}^w \\ \mathbf{P}_{00}^u & \mathbf{P}_{01}^u & \mathbf{P}_{00}^{uw} & \mathbf{P}_{01}^{uw} \\ \mathbf{P}_{10}^u & \mathbf{P}_{11}^u & \mathbf{P}_{10}^{uw} & \mathbf{P}_{11}^{uw} \end{pmatrix} \mathbf{H}^T \begin{pmatrix} w^3 \\ w^2 \\ w \\ 1 \end{pmatrix} \quad (5.34)$$
$$= \mathbf{UHBH}^T\mathbf{W}^T,$$

where H is the Hermite matrix, Equation (4.26). The quantities $\mathbf{P}_{ij}^{uw}$ are the twist

5.10.5 The Plane as a Bicubic Patch

The flat plane is a special case of the bicubic Hermite patch. This section discusses several approaches to constructing a plane.

Approach 1: We assume that the three corner points $\mathbf{P}_{00}$, $\mathbf{P}_{01}$, and $\mathbf{P}_{10}$ of the plane are given. Equation (5.4) expresses the plane as $\mathbf{P}(u,w) = \mathbf{P}_1 + u\mathbf{r} + w\mathbf{s}$. If we assume that both u and w vary in the range $[0,1]$, then Figure 5.29 shows that

$$\begin{aligned}
\mathbf{P}_{00} &= \mathbf{P}_1, & \mathbf{P}_{11} &= \mathbf{P}_1 + \mathbf{r} + \mathbf{s}, \\
\mathbf{P}_{01} &= \mathbf{P}_1 + \mathbf{s}, & \mathbf{P}_{10} &= \mathbf{P}_1 + \mathbf{r}, \\
\mathbf{P}^u_{00} &= \mathbf{P}_{10} - \mathbf{P}_{00} = \mathbf{r}, & \mathbf{P}^u_{11} &= \mathbf{P}_{11} - \mathbf{P}_{01} = \mathbf{r}, \\
\mathbf{P}^u_{01} &= \mathbf{P}_{11} - \mathbf{P}_{01} = \mathbf{r}, & \mathbf{P}^u_{10} &= \mathbf{P}_{10} - \mathbf{P}_{00} = \mathbf{r}, \\
\mathbf{P}^w_{00} &= \mathbf{P}_{01} - \mathbf{P}_{00} = \mathbf{s}, & \mathbf{P}^w_{11} &= \mathbf{P}_{11} - \mathbf{P}_{10} = \mathbf{s}, \\
\mathbf{P}^w_{01} &= \mathbf{P}_{01} - \mathbf{P}_{00} = \mathbf{s}, & \mathbf{P}^w_{10} &= \mathbf{P}_{11} - \mathbf{P}_{10} = \mathbf{s}.
\end{aligned}$$

All twist vectors are zero since they are the second derivatives of a linear function. When these quantities are substituted in Equation (5.34), it reduces to

$$\mathbf{P}(u,w) = (u^3, u^2, u, 1)\mathbf{H} \begin{pmatrix} \mathbf{P}_1 & \mathbf{P}_1+\mathbf{s} & \mathbf{s} & \mathbf{s} \\ \mathbf{P}_1+\mathbf{r} & \mathbf{P}_1+\mathbf{r}+\mathbf{s} & \mathbf{s} & \mathbf{s} \\ \mathbf{r} & \mathbf{r} & 0 & 0 \\ \mathbf{r} & \mathbf{r} & 0 & 0 \end{pmatrix} \mathbf{H}^T \begin{pmatrix} w^3 \\ w^2 \\ w \\ 1 \end{pmatrix}. \quad (5.35)$$

(Notice that corner point $\mathbf{P}_{11}$ doesn't have to be given in advance, since both $\mathbf{r}$ and $\mathbf{s}$ can be calculated without it.)

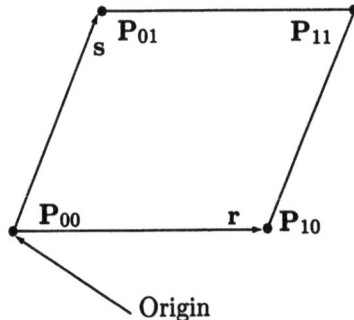

Figure 5.29: Bicubic Plane (Approach 1).

Approach 2: We again assume that the three corner points $\mathbf{P}_{00}$, $\mathbf{P}_{01}$, and $\mathbf{P}_{10}$ of the plane are given. The two tangent vectors at point $\mathbf{P}_{00}$ are simply the

5 Surfaces

differences $\mathbf{P}_{00}^u = \mathbf{P}_{01} - \mathbf{P}_{00}$, $\mathbf{P}_{00}^w = \mathbf{P}_{10} - \mathbf{P}_{00}$ (Figure 5.30). Since this plane is a parallelogram, the tangent vectors at the other three corners are the same:

$$\mathbf{P}_{01}^u = \mathbf{P}_{10}^u = \mathbf{P}_{11}^u = \mathbf{P}_{00}^u, \quad \mathbf{P}_{01}^w = \mathbf{P}_{10}^w = \mathbf{P}_{11}^w = \mathbf{P}_{00}^w.$$

The fourth corner point $\mathbf{P}_{11}$ is obtained by starting at $\mathbf{P}_{00}$, moving to $\mathbf{P}_{01}$ by adding vector $\mathbf{P}_{01} - \mathbf{P}_{00}$, then moving to $\mathbf{P}_{11}$ by adding vector $\mathbf{P}_{10} - \mathbf{P}_{00}$. The result is

$$\mathbf{P}_{11} = \mathbf{P}_{00} + (\mathbf{P}_{01} - \mathbf{P}_{00}) + (\mathbf{P}_{10} - \mathbf{P}_{00}) = \mathbf{P}_{01} + \mathbf{P}_{10} - \mathbf{P}_{00}.$$

When these quantities are substituted in Equation (5.34), it becomes

$$\mathbf{P}(u,w) = (u^3, u^2, u, 1)\mathbf{HMH}^T \begin{pmatrix} w^3 \\ w^2 \\ w \\ 1 \end{pmatrix}, \qquad (5.36)$$

where $\mathbf{M}$ is the matrix

$$\mathbf{M} = \begin{pmatrix} \mathbf{P}_{00} & \mathbf{P}_{01} & \mathbf{P}_{01} - \mathbf{P}_{00} & \mathbf{P}_{01} - \mathbf{P}_{00} \\ \mathbf{P}_{10} & \mathbf{P}_{01} + \mathbf{P}_{10} - \mathbf{P}_{00} & \mathbf{P}_{01} - \mathbf{P}_{00} & \mathbf{P}_{01} - \mathbf{P}_{00} \\ \mathbf{P}_{10} - \mathbf{P}_{00} & \mathbf{P}_{10} - \mathbf{P}_{00} & 0 & 0 \\ \mathbf{P}_{10} - \mathbf{P}_{00} & \mathbf{P}_{10} - \mathbf{P}_{00} & 0 & 0 \end{pmatrix}.$$

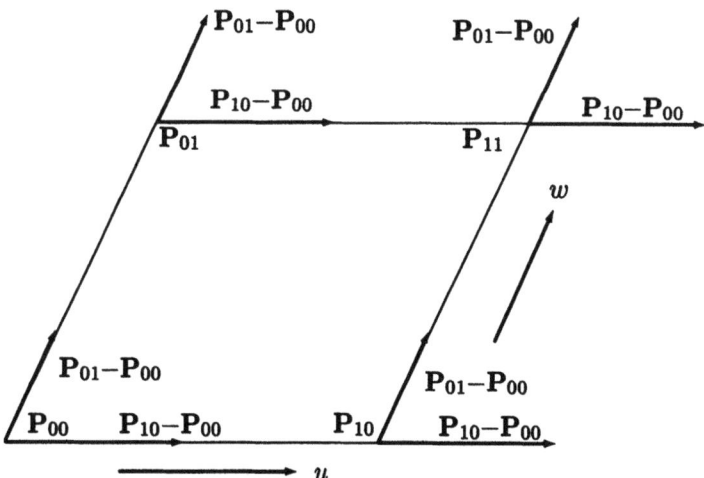

Figure 5.30: Bicubic Plane (Approach 2).

Approach 3: We assume that one corner point $\mathbf{P}_{00}$ and its two tangents $\mathbf{P}_{00}^w$ and $\mathbf{P}_{00}^u$ are given. Figure 5.31 shows how the other three corners and six tangents can be expressed in a general way in terms of these three quantities together with

18 more numbers, a_{ij}, b_{ij}, c_{ij}, d_{ij}, e_{ij}, and f_{ij}. The resulting plane is still flat but is no longer a rectangle or a parallelogram. Each of its four boundary lines can be a (plane) curve. The geometry matrix in this case is

$$\begin{pmatrix} \mathbf{P}_{00} & \mathbf{P}_{00} + a_{01}\mathbf{P}^u_{00} + b_{01}\mathbf{P}^w_{00} \\ \mathbf{P}_{00} + a_{10}\mathbf{P}^u_{00} + b_{10}\mathbf{P}^w_{00} & \mathbf{P}_{00} + a_{11}\mathbf{P}^u_{00} + b_{11}\mathbf{P}^w_{00} \\ \mathbf{P}^u_{00} & c_{01}\mathbf{P}^u_{00} + d_{01}\mathbf{P}^w_{00} \\ c_{10}\mathbf{P}^u_{00} + d_{10}\mathbf{P}^w_{00} & c_{11}\mathbf{P}^u_{00} + d_{11}\mathbf{P}^w_{00} \\ & \mathbf{P}^w_{00} & e_{01}\mathbf{P}^u_{00} + f_{01}\mathbf{P}^w_{00} \\ & e_{10}\mathbf{P}^u_{00} + f_{10}\mathbf{P}^w_{00} & e_{11}\mathbf{P}^u_{00} + f_{11}\mathbf{P}^w_{00} \\ & 0 & 0 \\ & 0 & 0 \end{pmatrix}. \tag{5.37}$$

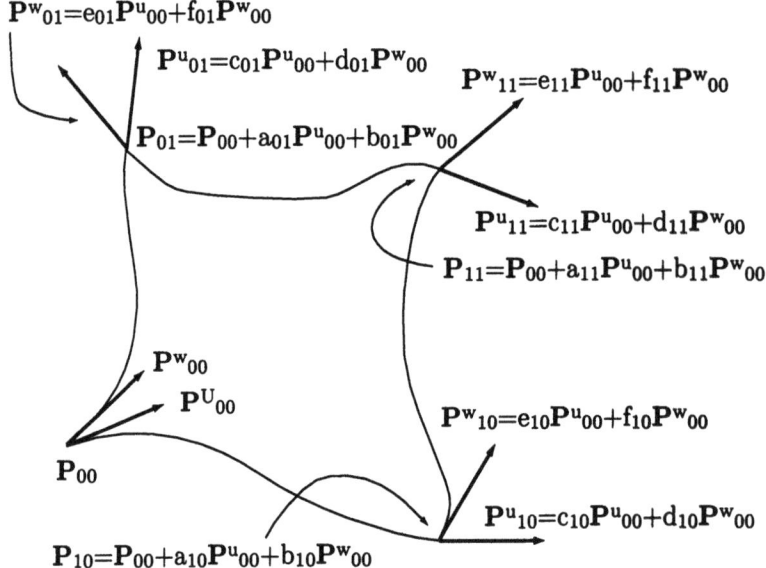

Figure 5.31: Bicubic Plane (Approach 3).

5.11 Catmull-Rom Surfaces

The Catmull-Rom curve can easily be extended to a surface that's fully defined by 16 points. In analogy to the Catmull-Rom curve segment—which involves 4 points but only passes through the 2 interior ones—a single Catmull-Rom surface patch involves 16 points, but passes only through the 4 middle points and spans the area delimited by them.

We start with a group of $m \times n$ data points roughly arranged in a rectangle. We look at all the overlapping groups that consist of 16 adjacent points, and we calculate a surface patch for each group. Some of the groups are shown in Figure 5.32.

5 Surfaces

$$
\begin{array}{lllll}
P_{40}P_{41}P_{42}P_{43} & P_{41}P_{42}P_{43}P_{44} & P_{42}P_{43}P_{44}P_{45} & \cdots & P_{4,n-3}P_{4,n-2}P_{4,n-1}P_{4n} \\
P_{30}P_{31}P_{32}P_{33} & P_{31}P_{32}P_{33}P_{34} & P_{32}P_{33}P_{34}P_{35} & \cdots & P_{3,n-3}P_{3,n-2}P_{3,n-1}P_{3n} \\
P_{20}P_{21}P_{22}P_{23} & P_{21}P_{22}P_{23}P_{24} & P_{22}P_{23}P_{24}P_{25} & \cdots & P_{2,n-3}P_{2,n-2}P_{2,n-1}P_{2n} \\
P_{10}P_{11}P_{12}P_{13} & P_{11}P_{12}P_{13}P_{14} & P_{12}P_{13}P_{14}P_{15} & \cdots & P_{1,n-3}P_{1,n-2}P_{1,n-1}P_{1n} \\
\\
P_{30}P_{31}P_{32}P_{33} & P_{31}P_{32}P_{33}P_{34} & P_{32}P_{33}P_{34}P_{35} & \cdots & P_{3,n-3}P_{3,n-2}P_{3,n-1}P_{3n} \\
P_{20}P_{21}P_{22}P_{23} & P_{21}P_{22}P_{23}P_{24} & P_{22}P_{23}P_{24}P_{25} & \cdots & P_{2,n-3}P_{2,n-2}P_{2,n-1}P_{2n} \\
P_{10}P_{11}P_{12}P_{13} & P_{11}P_{12}P_{13}P_{14} & P_{12}P_{13}P_{14}P_{15} & \cdots & P_{1,n-3}P_{1,n-2}P_{1,n-1}P_{1n} \\
P_{00}P_{01}P_{02}P_{03} & P_{01}P_{02}P_{03}P_{04} & P_{02}P_{03}P_{04}P_{05} & \cdots & P_{0,n-3}P_{0,n-2}P_{0,n-1}P_{0n}
\end{array}
$$

Figure 5.32: Points for a Catmull-Rom Surface Patch.

The expression of the surface is

$$\mathbf{P}(u,w) = (u^3, u^2, u, 1)\mathbf{BPB}^T \begin{pmatrix} w^3 \\ w^2 \\ w \\ 1 \end{pmatrix},$$

where **B** is the parabolic blending matrix of Equation (4.92):

$$\mathbf{B} = \begin{pmatrix} -0.5 & 1.5 & -1.5 & 0.5 \\ 1 & -2.5 & 2 & -0.5 \\ -0.5 & 0 & 0.5 & 0 \\ 0 & 1 & 0 & 0 \end{pmatrix}$$

and **P** is a matrix consisting of the 16 points participating in the patch

$$\mathbf{P} = \begin{pmatrix} P_{i+3,j} & P_{i+3,j+1} & P_{i+3,j+2} & P_{i+3,j+3} \\ P_{i+2,j} & P_{i+2,j+1} & P_{i+2,j+2} & P_{i+2,j+3} \\ P_{i+1,j} & P_{i+1,j+1} & P_{i+1,j+2} & P_{i+1,j+3} \\ P_{i,j} & P_{i,j+1} & P_{i,j+2} & P_{i,j+3} \end{pmatrix}.$$

Notice that the surface patches span the area bounded by the four points P_{11}, $P_{1,n-1}$, $P_{m-1,1}$, and $P_{m-1,n-1}$. If we want the surface to span the area bounded by the four corner points P_{00}, P_{0n}, P_{m0}, and P_{mn}, we need to duplicate the two extreme rows and columns of points, by analogy with the Catmull-Rom curve.

Example: Given the following coordinates for 16 points in file CRpoints

```
0 0 0     1   0 0    2     0 0    3 0 0
0 1 0    .5  .5 1   2.5   .5 0    3 1 0
0 2 0    .5 2.5 0   2.5  2.5 1    3 2 0
0 3 0     1   3 0    2     3 0    3 3 0
```

the *Mathematica* code of Figure 5.33 reads the file and generates the Catmull-Rom patch. Note how the patch spans only the four center points and how the z coordinates of 0 and 1 create the particular shape of the patch.

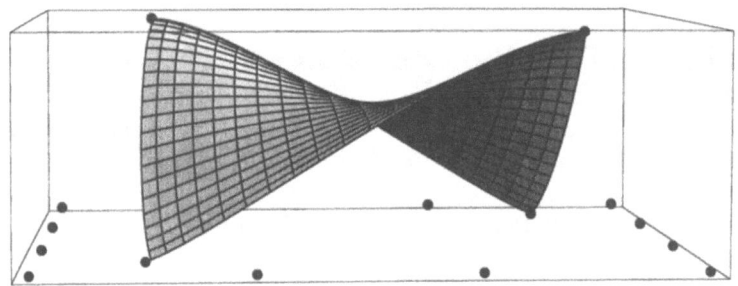

Figure 5.33: A Catmull-Rom Patch.

```
0 0 0      1  0 0    2     0 0   3 0 0
0 1 0     .5 .5 1   2.5   .5 0   3 1 0
0 2 0     .5 2.5 0  2.5 2.5 1    3 2 0
0 3 0      1 3 0    2    3 0     3 3 0
```

```
<<:Graphics:ParametricPlot3D.m;
Clear[Pt,Bm,CRpatch,g1,g2,g3];
Pt=ReadList["CRpoints",{Number,Number,Number},
 RecordLists->True];
Bm:={{-.5,1.5,-1.5,.5},{1,-2.5,2,-.5},
 {-.5,0,.5,0},{0,1,0,0}};
CRpatch[i_]:= (* 1st patch, rows 1-4 *)
{u^3,u^2,u,1}.Bm.Pt[[{1,2,3,4},{1,2,3,4},i]].
 Transpose[Bm].{w^3,w^2,w,1};
g1=Graphics3D[{AbsolutePointSize[4],
Table[Point[Pt[[i,j]]],{i,1,4},{j,1,4}]}];
g2=ParametricPlot3D[{CRpatch[1],CRpatch[2],CRpatch[3]},
{u,0,.98,.1},{w,0,1,.1}, DisplayFunction->Identity];
Show[g1,g2, ViewPoint->{-4.322, 0.242, 0.306},
 DisplayFunction->$DisplayFunction]
```

Points and Code for Figure 5.33.

5 Surfaces

Extended Example: We now add four more points to file CRpoints, and use rows 2–5 to calculate and draw another patch. Notice the five values of y compared to the four values of x. The code of Figure 5.34 reads the extended file and generates and displays both patches. Each patch spans four points, but they share the two points $(0.5, 2.5, 0)$ and $(2.5, 2.5, 1)$. Note how they connect smoothly.

5.12 Rectangular Bézier Surfaces

The Bézier surface patch, like its relative the Bézier curve, is an important type of surface and is used much in practice. We discuss the rectangular and the triangular Bézier surface methods, and this section covers the former.

We start with an $(m+1) \times (n+1)$ grid of control points arranged in a roughly rectangular shape:

$$
\begin{array}{cccc}
\mathbf{P}_{m,0} & \mathbf{P}_{m,1} & \cdots & \mathbf{P}_{m,n} \\
\vdots & \vdots & & \vdots \\
\mathbf{P}_{1,0} & \mathbf{P}_{1,1} & \cdots & \mathbf{P}_{1,n} \\
\mathbf{P}_{0,0} & \mathbf{P}_{0,1} & \cdots & \mathbf{P}_{0,n}
\end{array}
$$

and define the surface patch by requiring it to satisfy the following two properties:

1. The four boundary curves. They are to be Bézier curves, and each of them is defined by just one row or column of control points. The bottom boundary curve, for example, is defined only by the bottom row $\mathbf{P}_{0,j}$ of points.

2. The interpolation between boundary curves. Any point on the surface that's not located on any of the boundary curves is on a Bézier curve. This curve starts and ends on two opposite boundary curves and is defined by all the control points.

The top ($w = 1$) and bottom ($w = 0$) boundary curves (Figure 5.35) are now easy to define, since each depends on just one row of points

$$\mathbf{P}(u,0) = \sum_{j=0}^{n} \mathbf{P}_{0,j} B_{n,j}(u) \quad \text{and} \quad \mathbf{P}(u,1) = \sum_{j=0}^{n} \mathbf{P}_{m,j} B_{n,j}(u).$$

In a similar way, we can use the other rows of control points to define more curves. These auxiliary curves are useful in defining the surface but are not themselves located on the surface. The second row, $\mathbf{P}_{1,j}$, can be used to define a Bézier curve that we will call $\mathbf{P}(u, w_0)$:

$$\mathbf{P}(u, w_0) = \sum_{j=0}^{n} \mathbf{P}_{1,j} B_{n,j}(u).$$

The third row $\mathbf{P}_{2,j}$ can be used to define

$$\mathbf{P}(u, w_1) = \sum_{j=0}^{n} \mathbf{P}_{2,j} B_{n,j}(u).$$

5.12 Rectangular Bézier Surfaces

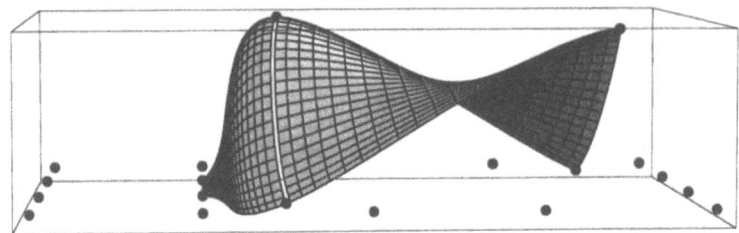

Figure 5.34: Two Catmull-Rom Patches.

```
0 0 0    1  0 0   2    0 0   3 0 0
0 1 0   .5 .5 1   2.5 .5 0   3 1 0
0 2 0   .5 2.5 0  2.5 2.5 1  3 2 0
0 3 0   1  3  0   2   3   0  3 3 0
0 4 0   1  4  0   2   4   0  3 4 0
```

```
<<:Graphics:ParametricPlot3D.m;
Clear[Pt,Bm,CRpatch,CRpatchM,g1,g2,g3];
Pt=ReadList["CRpoints",{Number,Number,Number},
 RecordLists->True];
Bm:={{-.5,1.5,-1.5,.5},{1,-2.5,2,-.5},
 {-.5,0,.5,0},{0,1,0,0}};
CRpatch[i_]:= (* 1st patch, rows 1-4 *)
{u^3,u^2,u,1}.Bm.Pt[[{1,2,3,4},{1,2,3,4},i]].
 Transpose[Bm].{w^3,w^2,w,1};
CRpatchM[i_]:= (* 2nd patch, rows 2-5 *)
{u^3,u^2,u,1}.Bm.Pt[[{2,3,4,5},{1,2,3,4},i]].
 Transpose[Bm].{w^3,w^2,w,1};
g1=Graphics3D[{AbsolutePointSize[4],
Table[Point[Pt[[i,j]]],{i,1,5},{j,1,4}]}];
g2=ParametricPlot3D[{CRpatch[1],CRpatch[2],CRpatch[3]},
{u,0,.98,.1},{w,0,1,.1}, DisplayFunction->Identity];
g3=ParametricPlot3D[{CRpatchM[1],CRpatchM[2],CRpatchM[3]},
{u,0,1,.1},{w,0,1,.1}, DisplayFunction->Identity];
Show[g1,g2,g3, ViewPoint->{-4.322, 0.242, 0.306},
DisplayFunction->$DisplayFunction]
```

Points and Code for Figure 5.34.

5 Surfaces

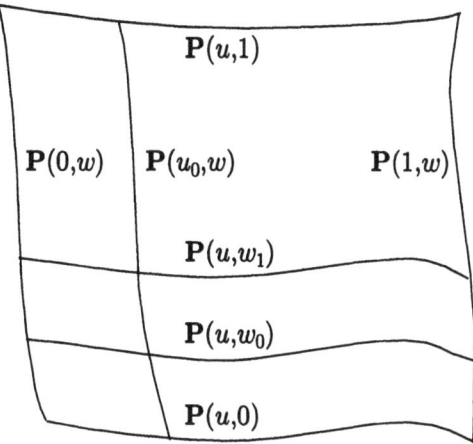

Figure 5.35: A Rectangular Surface Patch.

We now consider a general Bézier curve on the surface, in the w direction. Since it is a Bézier curve, it must have the form

$$\mathbf{P}(u_0, w) = \sum_{i=0}^{m} \mathbf{M}_i B_{m,i}(w),$$

where the $\mathbf{M}_i$ are some hypothetical points along $\mathbf{P}(u_0, w)$.

The most important step in defining the surface is to define point M_i. It is defined as a point on the auxiliary curve $\mathbf{P}(u, w_i)$. Specifically, $\mathbf{M}_i$ is defined as the point

$$\mathbf{P}(u_0, w_i) = \sum_{j=0}^{n} \mathbf{P}_{i,j} B_{n,j}(u_0).$$

This is a point on the Bézier curve defined by the row of control points $\mathbf{P}_{i0}, \mathbf{P}_{i1}, \ldots, \mathbf{P}_{in}$. The expression for curve $\mathbf{P}(u_0, w)$ can now be written as

$$\mathbf{P}(u_0, w) = \sum_{i=0}^{m} \mathbf{M}_i B_{m,i}(w) = \sum_{i=0}^{m} \left[\sum_{j=0}^{n} \mathbf{P}_{i,j} B_{n,j}(u_0) \right] B_{m,i}(w),$$

and the entire surface can now be obtained by releasing u_0 (Equation (5.39)).

This type of surface patch can, therefore, be viewed as two continuous sets of Bézier curves. One set of curves is obtained by varying w and these curves go in the u direction. The other set contains curves going in the w direction, where each curve is obtained for a different value of u.

An alternative derivation of the expression for the surface is to start with the top and bottom boundary curves and interpolate between them as follows: For any other value $w = w_0$, we want a curve $\mathbf{P}(u, w_0)$ on the surface, satisfying:

1. $\mathbf{P}(u, w_0)$ is a Bézier curve.
2. $\mathbf{P}(u, w_0)$ approaches $\mathbf{P}(u, 0)$ when $w \to 0$, and $\mathbf{P}(u, 1)$ when $w \to 1$.

Since $\mathbf{P}(u, w_0)$ is a Bézier curve, its form must be $\mathbf{P}(u, w_0) = \sum_{j=0}^{n} \mathbf{M}_j B_{n,j}(u)$, where the $\mathbf{M}_j$ are certain control points. For small values of w, points $\mathbf{M}_j$ should be close to $\mathbf{P}_{0,j}$; for large values of w (close to 1), they should be close to points $\mathbf{P}_{m,j}$. A good way of defining points $\mathbf{M}_j$ is as a weighted sum of points $\mathbf{P}_{i,j}$. Thus, $\mathbf{M}_j = \sum_{i=0}^{m} \mathbf{P}_{i,j} Weight_i(w_0)$. The best choice for the weights is the same Bernstein polynomials $B_{m,i}(w_0)$. This choice not only provides the right weights but guarantees that the final expression for the surface is symmetric with respect to u and w. We thus select

$$\mathbf{M}_0 = \sum_{i=0}^{m} \mathbf{P}_{i,0} B_{m,i}(w_0),\ \mathbf{M}_1 = \sum_{i=0}^{m} \mathbf{P}_{i,1} B_{m,i}(w_0),\ \ldots,\ \mathbf{M}_n = \sum_{i=0}^{m} \mathbf{P}_{i,n} B_{m,i}(w_0),$$

and the expression for a general curve on the surface becomes

$$\mathbf{P}(u, w_0) = \sum_{j=0}^{n} \mathbf{M}_j B_{n,j}(u) = \sum_{j=0}^{n} \left[\sum_{i=0}^{m} \mathbf{P}_{i,j} B_{m,i}(w_0) \right] B_{n,j}(u). \tag{5.38}$$

It is easy to verify that when $w \to 0$, point $\mathbf{M}_j$ approaches

$$\sum_{i=0}^{m} \mathbf{P}_{i,j} B_{m,i}(0) = \mathbf{P}_{0,j} B_{m,0}(0) = \mathbf{P}_{0,j},$$

and the curve $\mathbf{P}(u, w_0)$ approaches the bottom boundary curve $\mathbf{P}(u, 0)$. Also, when $w \to 1$, each point $\mathbf{M}_j$ approaches $\mathbf{P}_{m,j}$ and the curve $\mathbf{P}(u, w_0)$ approaches the top boundary curve $\mathbf{P}(u, 1)$.

The general expression for the surface is obtained by substituting w for w_0:

$$\mathbf{P}(u, w) = \sum_{i=0}^{m} \sum_{j=0}^{n} B_{m,i}(w) \mathbf{P}_{i,j} B_{n,j}(u)$$

$$= (B_{m,0}(w), B_{m,1}(w), \ldots, B_{m,m}(w))\, \mathbf{P} \begin{pmatrix} B_{n,0}(u) \\ B_{n,1}(u) \\ \vdots \\ B_{n,n}(u) \end{pmatrix}$$

$$= \mathbf{B}_m(w)\, \mathbf{P}\, \mathbf{B}_n(u), \tag{5.39}$$

where
$$\mathbf{P} = \begin{pmatrix} \mathbf{P}_{0,0} & \mathbf{P}_{0,1} & \cdots & \mathbf{P}_{0,n} \\ \mathbf{P}_{1,0} & \mathbf{P}_{1,1} & \cdots & \mathbf{P}_{1,n} \\ \vdots & \vdots & \ddots & \vdots \\ \mathbf{P}_{m,0} & \mathbf{P}_{m,1} & \cdots & \mathbf{P}_{m,n} \end{pmatrix}.$$

Notice that it is a Cartesian product. The surface can also be expressed, by analogy with Equation (4.108), as

$$\mathbf{P}(u, w) = \mathbf{U} \mathbf{N} \mathbf{P} \mathbf{N}^T \mathbf{W}^T, \tag{5.40}$$

5 Surfaces

where $\mathbf{U} = (u^3, u^2, u, 1)$, $\mathbf{W} = (w^3, w^2, w, 1)$, and $\mathbf{N}$ is defined by Equation (4.108).

Figure 5.36 is an example of a biquadratic Bézier surface patch with the *Mathematica* code that generated it. Notice how the surface is anchored at the four corner points and how the other control points pull the surface toward them.

Example: Given the six three-dimensional points

$$\begin{array}{ccc} \mathbf{P}_{10} & \mathbf{P}_{11} & \mathbf{P}_{12} \\ \mathbf{P}_{00} & \mathbf{P}_{01} & \mathbf{P}_{02} \end{array}$$

the corresponding Bézier surface is generated in the following three steps:

1. Find the orders m and n of the surface. Since the points are numbered starting from 0, the two orders of the surface are $m = 1$ and $n = 2$.

2. Calculate the weight functions $B_{1i}(w)$ and $B_{2j}(u)$. For $m = 1$, we get

$$B_{1i}(w) = \binom{1}{i} w^i (1-w)^{1-i},$$

which yields the two functions

$$B_{10}(w) = \binom{1}{0} w^0 (1-w)^{1-0} = 1 - w, \quad B_{11}(w) = \binom{1}{1} w^1 (1-w)^{1-1} = w.$$

For $n = 2$, we get

$$B_{2j}(u) = \binom{2}{j} u^j (1-u)^{2-j},$$

which yields the three functions

$$B_{20}(u) = \binom{2}{0} u^0 (1-u)^{2-0} = (1-u)^2,$$

$$B_{21}(u) = \binom{2}{1} u^1 (1-u)^{2-1} = 2u(1-u),$$

$$B_{22}(u) = \binom{2}{2} u^2 (1-u)^{2-2} = u^2.$$

3. Substitute the weight functions in the general expression for the surface (Equation (5.39)):

$$\begin{aligned} \mathbf{P}(u, w) &= \sum_{i=0}^{1} \sum_{j=0}^{2} B_{1i}(w) \mathbf{P}_{ij} B_{2j}(u) \\ &= B_{10}(w) \sum_{j=0}^{2} \mathbf{P}_{0j} B_{2j}(u) + B_{11}(w) \sum_{j=0}^{2} \mathbf{P}_{1j} B_{2j}(u) \\ &= (1-w) \left[\mathbf{P}_{00} B_{20}(u) + \mathbf{P}_{01} B_{21}(u) + \mathbf{P}_{02} B_{22}(u) \right] \\ &\quad + w \left[\mathbf{P}_{10} B_{20}(u) + \mathbf{P}_{11} B_{21}(u) + \mathbf{P}_{12} B_{22}(u) \right] \end{aligned}$$

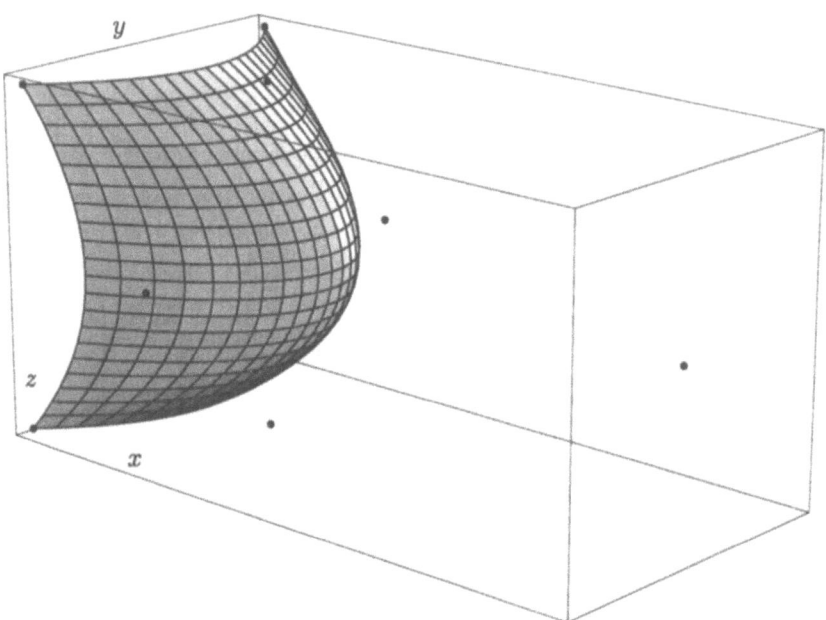

Figure 5.36: A Biquadratic Bézier Surface Patch.

```
n=2;
Clear[pwr,bern,spnts,n,bzSurf,g1,g2];
<<:Graphics:ParametricPlot3D.m
spnts={{{0,0,0},{1,0,1},{0,0,2}},
 {{1,1,0},{4,1,1},{1,1,2}}, {{0,2,0},{1,2,1},{0,2,2}}};
(* Handle Indeterminate condition *)
pwr[x_,y_]:=If[x==0 && y==0, 1, x^y];
bern[n_,i_,u_]:=Binomial[n,i]pwr[u,i]pwr[1-u,n-i]
bzSurf[u_,w_]:=Sum[spnts[[i+1,j+1]]bern[n,i,u]bern[n,j,w],
 {i,0,n}, {j,0,n}]
g2=ParametricPlot3D[bzSurf[u,w],{u,0,1,.05}, {w,0,1,.05}]
g1=Graphics3D[{AbsolutePointSize[3],
 Table[Point[spnts[[i,j]]],{i,1,n+1},{j,1,n+1}]}]
Show[g1,g2, ViewPoint->{2.783, -3.090, 1.243}]
```

Code for Figure 5.36.

5 Surfaces

$$
\begin{aligned}
&= (1-w)\left[\mathbf{P}_{00}(1-u)^2 + \mathbf{P}_{01}2u(1-u) + \mathbf{P}_{02}u^2\right] \\
&\quad + w\left[\mathbf{P}_{10}(1-u)^2 + \mathbf{P}_{11}2u(1-u) + \mathbf{P}_{12}u^2\right] \\
&= \mathbf{P}_{00}(1-w)(1-u)^2 + \mathbf{P}_{01}(1-w)2u(1-u) + \mathbf{P}_{02}(1-w)u^2 \\
&\quad + \mathbf{P}_{10}w(1-u)^2 + \mathbf{P}_{11}w2u(1-u) + \mathbf{P}_{12}wu^2. \quad (5.41)
\end{aligned}
$$

The final expression is linear in w since the surface is defined by just two points in the w direction. Surface lines in this direction are straight. In the u direction, where the surface is defined by three points, each line is a polynomial of degree 2 in u.

A good way to check the final expression is to calculate it for the four values $u = 0$, $u = 1$ and $w = 0$, $w = 1$. This should yield the coordinates of the four corner points.

The entire surface can now be easily displayed, as a wire frame, by performing two loops. One draws curves in the u direction and the other draws them in the w direction. Notice that the expression for the patch is the same regardless of the particular points used. The user may change the points to modify the surface, and the new surface can be displayed by calculating Equation (5.41).

▸ **Exercise 5.27:** Given the 3×4 array of control points

$$
\begin{aligned}
&\mathbf{P}_{20} = (0,2,0) \quad \mathbf{P}_{21} = (1,2,1) \quad \mathbf{P}_{22} = (2,2,1) \quad \mathbf{P}_{23} = (3,2,0) \\
&\mathbf{P}_{10} = (0,1,0) \quad \mathbf{P}_{11} = (1,1,1) \quad \mathbf{P}_{12} = (2,1,1) \quad \mathbf{P}_{13} = (3,1,0) \\
&\mathbf{P}_{00} = (0,0,0) \quad \mathbf{P}_{01} = (1,0,1) \quad \mathbf{P}_{02} = (2,0,1) \quad \mathbf{P}_{03} = (3,0,0),
\end{aligned}
$$

calculate the corresponding order-2×3 Bézier surface patch.

5.12.1 Joining Rectangular Bézier Patches

It is easy, although tedious, to explore the conditions for the smooth joining of two Bézier surface patches. Figure 5.37 shows a typical example of this problem. We see parts of two patches $\mathbf{P}$ and $\mathbf{Q}$. It is not hard to see that the former is based on 4×5 control points and the latter on $4 \times n$ points, where $n \geq 2$. It is also easy to see that they are joined such that the eight control points along the joint satisfy $\mathbf{P}_{i4} = \mathbf{Q}_{i0}$ for $i = 0, 1, 2, 3$.

The condition for smooth joining of the two surface patches is that the two tangent vectors at the common boundary point in the same direction, although they may have different magnitudes (see Figure 5.37). The condition is thus easy to express as

$$
\left.\frac{\partial \mathbf{P}(u,w)}{\partial w}\right|_{w=1} = \alpha \left.\frac{\partial \mathbf{Q}(u,w)}{\partial w}\right|_{w=0}.
$$

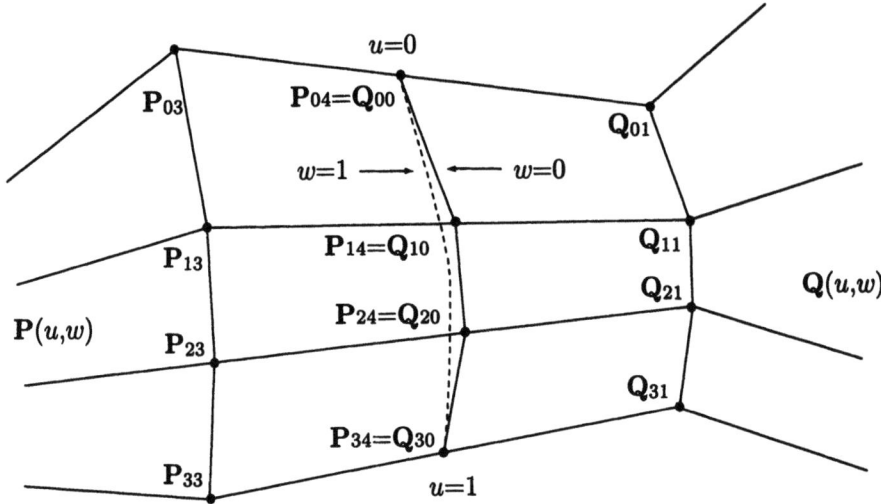

Figure 5.37: Smoothly Joining Rectangular Bézier Patches.

The two tangents are calculated from Equation (5.40) (and the $\mathbf{B}_3$ and $\mathbf{B}_4$ matrices given by Figure 4.36). For the first patch, we have

$$\left.\frac{\partial \mathbf{P}(u,w)}{\partial w}\right|_{w=1} = (u^3, u^2, u, 1)\mathbf{B}_3 \begin{pmatrix} \mathbf{P}_{00} & \mathbf{P}_{01} & \mathbf{P}_{02} & \mathbf{P}_{03} & \mathbf{P}_{04} \\ \mathbf{P}_{10} & \mathbf{P}_{11} & \mathbf{P}_{12} & \mathbf{P}_{13} & \mathbf{P}_{14} \\ \mathbf{P}_{20} & \mathbf{P}_{21} & \mathbf{P}_{22} & \mathbf{P}_{23} & \mathbf{P}_{24} \\ \mathbf{P}_{30} & \mathbf{P}_{31} & \mathbf{P}_{32} & \mathbf{P}_{33} & \mathbf{P}_{34} \end{pmatrix} \mathbf{B}_4^T \left.\begin{pmatrix} 4w^3 \\ 3w^2 \\ 2w \\ 1 \\ 0 \end{pmatrix}\right|_{w=1}$$

$$= 4(u^3, u^2, u, 1)\mathbf{B}_3^T \begin{pmatrix} \mathbf{P}_{04} - \mathbf{P}_{03} \\ \mathbf{P}_{14} - \mathbf{P}_{13} \\ \mathbf{P}_{24} - \mathbf{P}_{23} \\ \mathbf{P}_{34} - \mathbf{P}_{33} \end{pmatrix}.$$

Similarly, for the second patch,

$$\left.\frac{\partial \mathbf{Q}(u,w)}{\partial w}\right|_{w=0} = 4(u^3, u^2, u, 1)\mathbf{B}_3^T \begin{pmatrix} \mathbf{Q}_{01} - \mathbf{Q}_{00} \\ \mathbf{Q}_{11} - \mathbf{Q}_{10} \\ \mathbf{Q}_{21} - \mathbf{Q}_{20} \\ \mathbf{Q}_{31} - \mathbf{Q}_{30} \end{pmatrix}.$$

The conditions for a smooth join are therefore

$$\begin{pmatrix} \mathbf{P}_{04} - \mathbf{P}_{03} \\ \mathbf{P}_{14} - \mathbf{P}_{13} \\ \mathbf{P}_{24} - \mathbf{P}_{23} \\ \mathbf{P}_{34} - \mathbf{P}_{33} \end{pmatrix} = \alpha \begin{pmatrix} \mathbf{Q}_{01} - \mathbf{Q}_{00} \\ \mathbf{Q}_{11} - \mathbf{Q}_{10} \\ \mathbf{Q}_{21} - \mathbf{Q}_{20} \\ \mathbf{Q}_{31} - \mathbf{Q}_{30} \end{pmatrix},$$

or $\mathbf{P}_{i4} - \mathbf{P}_{i3} = \alpha(\mathbf{Q}_{i1} - \mathbf{Q}_{i0})$ for $i = 0, 1, 2, 3$. This can also be expressed by saying

5 Surfaces

that the three points $\mathbf{P}_{i3}$, $\mathbf{P}_{i4} = \mathbf{Q}_{i0}$, and $\mathbf{Q}_{i1}$ should be on a straight line, although not necessarily equally spaced.

Example: Each of the two patches in Figure 5.38 is based on 3×3 points ($n = 2$). The patches are smoothly connected along the line defined by common points $(0,2,0)$, $(0,0,0)$, and $(0,-2,0)$. Note that in the diagram they are slightly separated, but this was done deliberately. The smooth connection is obtained by making sure that the points $(-2,2,0)$, $(0,2,0)$, and $(2,2,0)$ are collinear (find the other two collinear triplets). The coordinates of the points are

$$
\begin{array}{cccccc}
-2,2,2 & -2,2,0 & 0,2,0 & 0,2,0 & 2,2,0 & 2,2,-2 \\
-4,0,2 & -4,0,0 & 0,0,0 & 0,0,0 & 4,0,0 & 4,0,-2 \\
-2,-2,2 & -2,-2,0 & 0,-2,0 & 0,-2,0 & 2,-2,0 & 2,-2,-2
\end{array}
$$

The famous Utah teapot was designed in the 1960s at the University of Utah by digitizing a real teapot (now at the computer museum in Boston) and creating 32 smoothly connected Bézier patches defined by a total of 306 control points. [Crow 87] has a detailed description. The coordinates of the points are publicly available, as is a program to display the entire surface. The program is part of a public-domain general three-dimensional graphics package called SIPP (SImple Polygon Processor). SIPP was originally written in Sweden and is distributed by the Free Software Foundation [Free 98]. It can be downloaded anonymously from several sources and for different platforms.

> She finished pouring the tea and put down the pot.
> "That's an old teapot," remarked Harold.
> "Sterling silver," said Maude wistfully. "It was my dear mother-in-law's, part of a dinner set of fifty pieces. It was sent to me, one of the few things that survived." Her voice trailed off and she absently sipped her tea.
> — Colin Higgins, *Harold and Maude*.

5.12.2 Bicubic Bézier and Hermite Patches

The order-3×3 rectangular Bézier surface patch based on 16 control points $\mathbf{P}_{ij}$ is expressed as

$$\mathbf{P}(u,w) = (u^3, u^2, u, 1)\mathbf{NPN}^T \begin{pmatrix} w^3 \\ w^2 \\ w \\ 1 \end{pmatrix} = \mathbf{UNPN}^T\mathbf{W}^T,$$

where $\mathbf{N}$ is given by Equation (4.107),

$$\mathbf{N} = \begin{pmatrix} -1 & 3 & -3 & 1 \\ 3 & -6 & 3 & 0 \\ -3 & 3 & 0 & 0 \\ 1 & 0 & 0 & 0 \end{pmatrix}.$$

5.12 Rectangular Bézier Surfaces

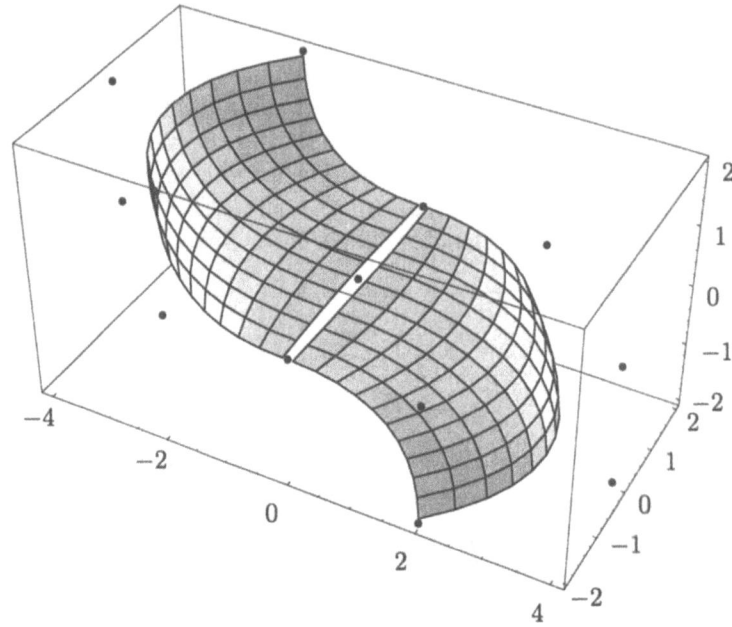

```
n=2; Clear[n,bern,p1,p2,g3,bzSurf,patch];
<<:Graphics:ParametricPlot3D.m
p1={{{-2,2,2},{-2,2,0},{0,2,0}},
 {{-4,0,2},{-4,0,0},{0,0,0}},
 {{-2,-2,2},{-2,-2,0},{0,-2,0}}};
p2={{{0,2,0},{2,2,0},{2,2,-2}},
 {{0,0,0},{4,0,0},{4,0,-2}},
 {{0,-2,0},{2,-2,0},{2,-2,-2}}};
pwr[x_,y_]:=If[x==0 && y==0, 1, x^y];
bern[n_,i_,u_]:=Binomial[n,i]pwr[u,i]pwr[1-u,n-i]
bzSurf[p_]:={Sum[p[[i+1,j+1,1]]bern[n,i,u]bern[n,j,w],
 {i,0,n,1}, {j,0,n,1}],
 Sum[p[[i+1,j+1,2]]bern[n,i,u]bern[n,j,w],
 {i,0,n,1}, {j,0,n,1}],
 Sum[p[[i+1,j+1,3]]bern[n,i,u]bern[n,j,w],
 {i,0,n,1}, {j,0,n,1}]};
patch[s_]:=
ParametricPlot3D[bzSurf[s],{u,0,1,.1}, {w,0.02,.98,.1}];
g3=Graphics3D[{AbsolutePointSize[3],
 Table[Point[p1[[i,j]]],{i,1,n+1},{j,1,n+1}]}]
g4=Graphics3D[{AbsolutePointSize[3],
 Table[Point[p2[[i,j]]],{i,1,n+1},{j,1,n+1}]}]
Show[patch[p1],patch[p2],g3,g4,
 DisplayFunction->$DisplayFunction]
```

Figure 5.38: Two Bézier Surface Patches.

5 Surfaces

This expression is similar to the one for the bicubic Hermite patch $\mathbf{UHBH}^T\mathbf{W}^T$, where $\mathbf{B}$ is defined by Equation (5.34). Comparing these expressions by setting $\mathbf{NPN}^T = \mathbf{HBH}^T$ gives us the specific matrix $\mathbf{B}$ needed to express the 3×3 Bézier patch in bicubic form

$$\mathbf{B} = \begin{pmatrix} \mathbf{P}_{00} & \mathbf{P}_{03} \\ \mathbf{P}_{30} & \mathbf{P}_{33} \\ 3(\mathbf{P}_{10} - \mathbf{P}_{00}) & 3(\mathbf{P}_{13} - \mathbf{P}_{03}) \\ 3(\mathbf{P}_{30} - \mathbf{P}_{20}) & 3(\mathbf{P}_{33} - \mathbf{P}_{23}) \end{pmatrix}$$

$$\begin{pmatrix} 3(\mathbf{P}_{01} - \mathbf{P}_{00}) & 3(\mathbf{P}_{03} - \mathbf{P}_{02}) \\ 3(\mathbf{P}_{31} - \mathbf{P}_{30}) & 3(\mathbf{P}_{33} - \mathbf{P}_{32}) \\ 9(\mathbf{P}_{00} - \mathbf{P}_{10} - \mathbf{P}_{01} + \mathbf{P}_{11}) & 9(\mathbf{P}_{02} - \mathbf{P}_{12} - \mathbf{P}_{03} + \mathbf{P}_{13}) \\ 9(\mathbf{P}_{20} - \mathbf{P}_{30} - \mathbf{P}_{21} + \mathbf{P}_{31}) & 9(\mathbf{P}_{22} - \mathbf{P}_{32} - \mathbf{P}_{33} + \mathbf{P}_{33}) \end{pmatrix}.$$

The 8 tangent vectors and 4 twist vectors are expressed in terms of the 16 control points. Each tangent is the difference of 2 control points located on the boundary of the control polyhedron (i.e., of the form $\mathbf{P}_{0j}$ or $\mathbf{P}_{i0}$), while the twists are computed by means of all 16 points.

5.12.3 A Bézier Sphere

Section 4.14.23 shows how the Bézier curve can approximate a circle to a high precision. Specifically, Exercise 4.82 shows how to place the four points $(1,0)$, $(1,k)$, $(k,1)$, and $(0,1)$, where $k \approx 0.5523$, in order to construct a curve (Equation (Ans.32)) whose maximum deviation from a true quarter circle is just 0.027%.

This section shows how to construct an approximate sphere out of eight identical Bézier surface patches, one of which is shown in Figure 5.39a. The idea is to define a degenerate surface patch where one boundary curve degenerates to a single point and each of the other three boundary curves is an approximate quarter circle. Each of the eight patches is defined by 4×4 control points arranged as in Figure 5.39b. The four points $\mathbf{P}_{00}$, $\mathbf{P}_{10}$, $\mathbf{P}_{20}$, and $\mathbf{P}_{30}$ are identical and equal $(0,0,1)$. The four points $\mathbf{P}_{03} = (1,0,0)$, $\mathbf{P}_{13} = (1,k,0)$, $\mathbf{P}_{23} = (k,1,0)$, and $\mathbf{P}_{33} = (0,1,0)$ are located on the xy plane. The group $\mathbf{P}_{10}$, $\mathbf{P}_{11}$, $\mathbf{P}_{12}$, and $\mathbf{P}_{12}$ have the same relative positions and are located on a plane rotated $30°$ from positive x to positive y. The *Mathematica* code of Figure 5.40 calculates all the points and has produced the following expression of the surface:

$$\mathbf{P}(u,w) = \big(0.211374u(7.83872 - 1.48457u - 1.62319u^2 - 3.15057w + 0.596685uw$$
$$+ 2.55388u^2w - 5.45694w^2 + 1.03349uw^2 - 1.93069u^2w^2$$
$$+ 0.768792w^3 - 0.145601uw^3 + 1.u^2w^3),$$
$$- 0.211374u(-11.7581w + 2.22686uw + 1.6925u^2w + 3.15057w^2$$
$$- 0.596685uw^2 - 1.06931u^2w^2 + 0.768792w^3 - 0.145601uw^3 + 1u^2w^3),$$
$$0.3431(2.9146 - 3.9146u^2 + 1.u^3)\big).$$

5.12 Rectangular Bézier Surfaces

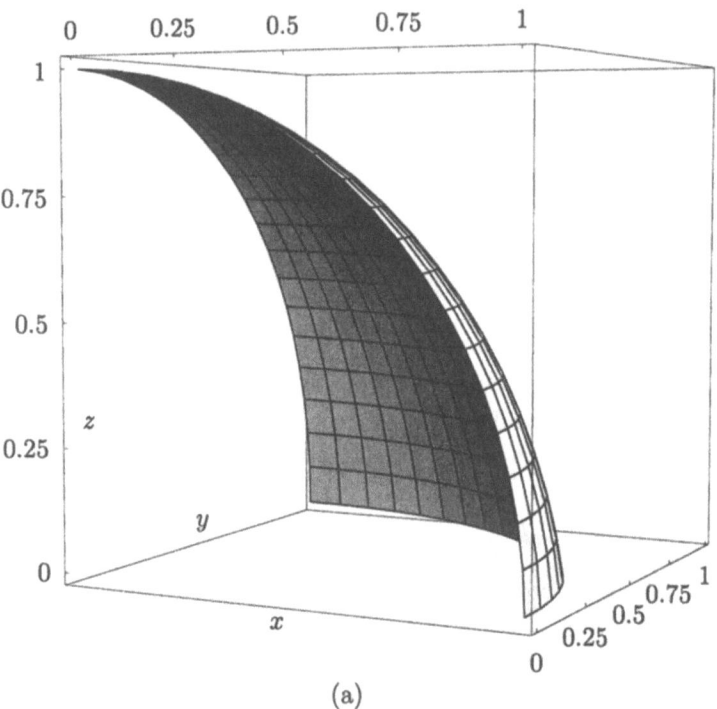

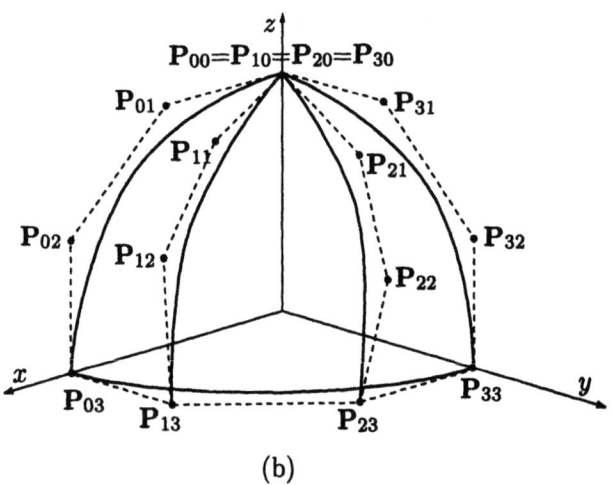

Figure 5.39: A Bézier Patch for a Sphere.

5 Surfaces

```
(* Sphere made of 8 Bezier patches *)
Clear[u,w,patch];
al3=Sin[30. Degree]; be3=Cos[30. Degree];
p00=p10=p20=p30={0,0,1}; p03={1,0,0}; p33={0,1,0};
t3={{be3,al3,0},{-al3,be3,0},{0,0,1}};
k=0.5523;
p13={1,k,0}; p23={k,1,0};
p02={1,0,k}; p01={k,0,1};
p32={0,1,k}; p31={0,k,1};
p11=p01.t3; p12=p02.t3;
t6={{al3,be3,0},{-be3,al3,0},{0,0,1}};
p21=p01.t6; p22=p02.t6;
b30[t_]:=(1-t)^3; b31[t_]:=3t(1-t)^2;
b32[t_]:=3t^2(1-t); b33[t_]:=t^3;
patch[u_,w_]:=b30[w](b30[u]p00+b31[u]p01+b32[u]p02+b33[u]p03) \
 +b31[w](b30[u]p10+b31[u]p11+b32[u]p12+b33[u]p13) \
 +b32[w](b30[u]p20+b31[u]p21+b32[u]p22+b33[u]p23) \
 +b33[w](b30[u]p30+b31[u]p31+b32[u]p32+b33[u]p33);
Factor[patch[u,w]]
ParametricPlot3D[patch[u,w],{u,0,1}, {w,0,1},
 DefaultFont->{"cmr10", 10}, Prolog->AbsoluteThickness[.5],
 Compiled->False, ViewPoint->{1.908, -3.886, 0.306}];
```

Figure 5.40: Code for a Bézier Patch.

5.13 Triangular Bézier Surfaces

Historically, the triangular, not the rectangular, was the first surface patch derived with Bézier methods. It was developed in 1959 by de Casteljau at Citroën.

The triangular Bézier patch is based on control points $\mathbf{P}_{ijk}$ arranged in a roughly triangular shape. Each control point is three-dimensional and is assigned three indexes ijk such that $0 \le i,j,k \le n$ and $i+j+k = n$. The value of n is selected by the user depending on how complex the patch should be and how fast the computer is. Generally, a large n allows for a finer control of surface details but involves more computations. In the convention used here, the first index, i, corresponds to the left side of the triangle, the second index, j, to the base, and the third one, k, corresponds to the right side. The indexing convention for the case $n = 4$ is shown below. There are $n + 1$ points on each side of the triangle, and because of the way the points are arranged, there is a total of $\frac{1}{2}(n+1)(n+2)$ control points:

$$\mathbf{P}_{040}$$
$$\mathbf{P}_{031}\ \mathbf{P}_{130}$$
$$\mathbf{P}_{022}\ \mathbf{P}_{121}\ \mathbf{P}_{220}$$
$$\mathbf{P}_{013}\ \mathbf{P}_{112}\ \mathbf{P}_{211}\ \mathbf{P}_{310}$$
$$\mathbf{P}_{004}\ \mathbf{P}_{103}\ \mathbf{P}_{202}\ \mathbf{P}_{301}\ \mathbf{P}_{400}$$

5.13 Triangular Bézier Surfaces

The surface patch itself is defined by the trinomial theorem (Equation (4.102)) as

$$\mathbf{P}(u,v,w) = \sum_{i+j+k=n} \mathbf{P}_{ijk}\frac{n!}{i!\,j!\,k!}u^i v^j w^k, \quad \text{where} \quad u+v+w=1. \tag{5.42}$$

Note that even though $\mathbf{P}(u,v,w)$ seems to depend on three parameters, it only depends on two since their sum is constant.

The three boundary curves are obtained from Equation (5.42) by setting each of the three parameters in turn to zero. Setting, for example, $u = 0$ causes all terms of Equation (5.42) except those with $i = 0$ to vanish. The result is

$$\mathbf{P}(0,v,w) = \sum_{j+k=n} \mathbf{P}_{0jk}\frac{n!}{j!\,k!}v^j w^k. \tag{5.43}$$

Since $v + w = 1$, Equation (5.43) can be written

$$\mathbf{P}(v) = \sum_{j+k=n} \mathbf{P}_{0jk}\frac{n!}{j!\,k!}v^j(1-v)^k = \sum_{j=0}^{n} \mathbf{P}_{0j,n-j}\frac{n!}{j!\,(n-j)!}v^j(1-v)^{n-j}, \tag{5.44}$$

and this is a Bézier curve.

Example: We illustrate the case $n = 2$. There should be three control points on each side of the triangle, for a total of $\frac{1}{2}(2+1)(2+2) = 6$ points. We select simple coordinates:

$$(1,3,1)$$
$$(0.5,1,0)\ (1.5,1,0)$$
$$(0,0,0)\ (1,0,-1)\ (2,0,0)$$

Note that four points have $z = 0$ and are, therefore, on the same plane. It is only the other two points, with $z = \pm 1$, that cause this surface to be nonflat.

The expression of the surface is

$$\mathbf{P}(u,v,w) = \sum_{i+j+k=2} \mathbf{P}_{ijk}\frac{n!}{i!\,j!\,k!}u^i v^j w^k$$
$$= \mathbf{P}_{002}\frac{2!}{0!\,0!\,2!}w^2 + \mathbf{P}_{101}\frac{2!}{1!\,0!\,1!}uw + \mathbf{P}_{200}\frac{2!}{2!\,0!\,0!}u^2$$
$$+ \mathbf{P}_{011}\frac{2!}{0!\,1!\,1!}vw + \mathbf{P}_{110}\frac{2!}{1!\,1!\,0!}uv + \mathbf{P}_{020}\frac{2!}{0!\,2!\,0!}v^2$$
$$= (0,0,0)w^2 + (1,0,-1)2uw + (2,0,0)u^2$$
$$+ (0.5,1,0)2vw + (1.5,1,0)2uv + (1,3,1)v^2$$
$$= (2uw + 2u^2 + vw + 3uv + v^2,\, 2vw + 2uv + 3v^2,\, -2uw + v^2).$$

It is now easy to verify that the following special values of u, v, and w produce the

5 Surfaces

three corner points:

u	v	w	point
0	0	1	(0,0,0)
0	1	0	(1,3,1)
1	0	0	(2,0,0)

But how is the surface displayed? It turns out that such a surface patch is displayed as a mesh of *three* families of curves (compare this to the two families in the case of a rectangular surface patch). Each family consists of curves that are roughly parallel to one side of the triangle (Figure 5.41a,b).

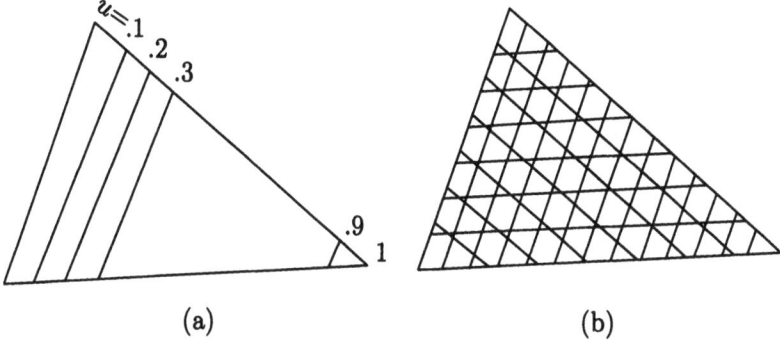

Figure 5.41: (a) Lines in the u Direction. (b) The Complete Surface Patch.

▶ **Exercise 5.28:** Write pseudo-code to draw the three families of curves.

5.13.1 Joining Triangular Bézier Patches

The triangular Bézier surface patch is commonly used in practice since a large surface is many times easier to break up into triangular patches rather than rectangular ones. It is therefore important to discover the conditions for smooth joining of these surface patches. The conditions should be expressed in terms of constraints on the control points.

These constraints are now developed for cubic surface patches, but the principles are the same for higher-degree patches. The idea is to calculate three vectors that are tangent to the surface at the common boundary curve. Intuitively, the condition for a smooth join is that these vectors be coplanar (they can, of course, have different magnitudes). We proceed in three steps:

Step 1. Figure 5.42 shows two triangular Bézier cubic patches, $\mathbf{P}(u,v,w)$ and $\mathbf{Q}(u,v,w)$, joined at the common boundary curve $\mathbf{P}(0,v,w) = \mathbf{Q}(0,v,w)$. Equation (5.44) shows how the boundary curves can be expressed as Bézier curves. Our common boundary curve can thus be written

$$\mathbf{P}(v) = \sum_{j+k=3} \frac{3!}{j!\,k!} v^j (1-v)^{3-j} \mathbf{P}_{0jk}.$$

This is easy to differentiate with respect to v and the result is

$$\frac{d\mathbf{P}(v)}{dv} = 3v^2(\mathbf{P}_{030} - \mathbf{P}_{021}) + 6v(1-v)(\mathbf{P}_{021} - \mathbf{P}_{012}) + 3(1-v)^2(\mathbf{P}_{012} - \mathbf{P}_{003})$$
$$= 3v^2\mathbf{B}_3 + 6v(1-v)\mathbf{B}_2 + (1-v)^2\mathbf{B}_1, \qquad (5.45)$$

where each of the $\mathbf{B}_i$ vectors is defined as the difference of two control points. They can be seen in the figure as thick arrows going from $\mathbf{P}_{003}$ to $\mathbf{P}_{030}$.

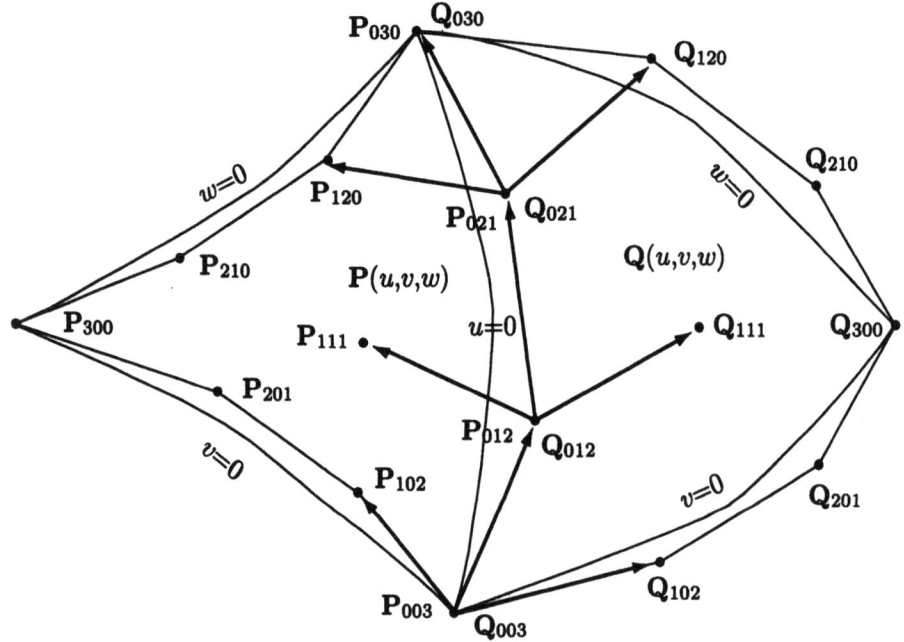

Figure 5.42: Joining Triangular Bézier Patches Smoothly.

Step 2. Another vector is computed that's tangent to the patch $\mathbf{P}(u,v,w)$ along the common boundary. This is done by calculating the tangent vector to the surface in the u direction and substituting $u = 0$. We first write the expression for the surface patch without the parameter w (it can be eliminated because $w = 1-u-v$):

$$\mathbf{P}(u,v) = \sum_{i+j+k=3} \frac{3!}{i!\,j!\,k!} u^i v^j (1-u-v)^k \mathbf{P}_{ijk}.$$

This is easy to differentiate with respect to u and it yields

$$\left.\frac{\partial \mathbf{P}(u,v)}{\partial u}\right|_{u=0} = 3v^2(\mathbf{P}_{120} - \mathbf{P}_{021}) + 6v(1-v)(\mathbf{P}_{111} - \mathbf{P}_{012})$$
$$+ 3(1-v)^2(\mathbf{P}_{102} - \mathbf{P}_{003}) \qquad (5.46)$$
$$= 3v^2\mathbf{A}_3 + 6v(1-v)\mathbf{A}_2 + 3(1-v)^2\mathbf{A}_1,$$

5 Surfaces

where each of the $\mathbf{A}_i$ vectors is again defined as the difference of two control points. They can be seen in the figure as thick arrows going, for example, from $\mathbf{P}_{003}$ to $\mathbf{P}_{102}$.

Step 3. The third vector is the tangent to the other surface patch $\mathbf{Q}(u,v,w)$ along the common boundary. It is expressed as

$$\left.\frac{\partial \mathbf{Q}(u,v)}{\partial u}\right|_{u=0} = 3v^2(\mathbf{Q}_{120} - \mathbf{Q}_{021}) + 6v(1-v)(\mathbf{Q}_{111} - \mathbf{Q}_{012}) \\ + 3(1-v)^2(\mathbf{Q}_{102} - \mathbf{Q}_{003}) \\ = 3v^2\mathbf{C}_3 + 6v(1-v)\mathbf{C}_2 + 3(1-v)^2\mathbf{C}_1, \quad (5.47)$$

where each of the $\mathbf{C}_i$ vectors is again defined as the difference of two control points. They can be seen in the figure as thick arrows going, for example, from $\mathbf{Q}_{003}$ to $\mathbf{Q}_{102}$.

The condition for smooth joining is that the vectors defined by Equations (5.45) through (5.47) be coplanar for any value of v. This can be expressed as

$$3v^2\mathbf{B}_3 + 6v(1-v)\mathbf{B}_2 + (1-v)^2\mathbf{B}_1 \\ = \alpha(3v^2\mathbf{A}_3 + 6v(1-v)\mathbf{A}_2 + 3(1-v)^2\mathbf{A}_1) \\ + \beta(3v^2\mathbf{C}_3 + 6v(1-v)\mathbf{C}_2 + 3(1-v)^2\mathbf{C}_1), \quad (5.48)$$

or, equivalently,

$$v^2(\mathbf{B}_3 - \alpha\mathbf{A}_3 - \beta\mathbf{C}_3) + 2v(1-v)(\mathbf{B}_2 - \alpha\mathbf{A}_2 - \beta\mathbf{C}_2) + (1-v)^2(\mathbf{B}_1 - \alpha\mathbf{A}_1 - \beta\mathbf{C}_1) = 0.$$

Since this should hold for any value of v, it can be written as the set of three equations:

$$\begin{aligned} \mathbf{B}_1 &= \alpha\mathbf{A}_1 + \beta\mathbf{C}_1, \\ \mathbf{B}_2 &= \alpha\mathbf{A}_2 + \beta\mathbf{C}_2, \\ \mathbf{B}_3 &= \alpha\mathbf{A}_3 + \beta\mathbf{C}_3. \end{aligned} \quad (5.49)$$

Each of the three sets of vectors $\mathbf{B}_i$, $\mathbf{A}_i$, and $\mathbf{C}_i$ ($i = 1, 2, 3$) should thus be coplanar. This condition can be expressed for the control points by saying that each of the three quadrilaterals given by

$$\begin{aligned} &\mathbf{P}_{003} = \mathbf{Q}_{003}, \quad \mathbf{P}_{102}, \quad \mathbf{P}_{012} = \mathbf{Q}_{012}, \quad \mathbf{Q}_{102}, \\ &\mathbf{P}_{012} = \mathbf{Q}_{012}, \quad \mathbf{P}_{111}, \quad \mathbf{P}_{021} = \mathbf{Q}_{021}, \quad \mathbf{Q}_{111}, \\ &\mathbf{P}_{021} = \mathbf{Q}_{021}, \quad \mathbf{P}_{120}, \quad \mathbf{P}_{030} = \mathbf{Q}_{030}, \quad \mathbf{Q}_{120}, \end{aligned}$$

should be planar. In the special case $\alpha = \beta = 1$, each quadrilateral should be a square. Otherwise, each should have the same ratio of height to width.

5.14 Converting Bézier Patches

Creating surface patches and connecting them to form a large, complex surface is sometimes a demanding job, requiring time, experience, and concentration. It is therefore no wonder that some designers prefer to work with rectangular patches, while others use mostly triangular ones. This section shows examples of conversion between the two. A rectangular Bézier patch based on control points $\mathbf{R}_{ij}$ can be converted to two triangular patches, identical in shape to the original one, by dividing the rectangle into two triangles and calculating new control points $\mathbf{T}_{ijk}$ based on the original ones, $\mathbf{R}_{ij}$. Similarly, two triangular patches sharing a boundary curve can be converted to a single rectangular patch by calculating new control points $\mathbf{R}_{ij}$ from the known ones, $\mathbf{T}_{ijk}$. An alternative is to convert a single triangular patch into a degenerate rectangular patch where two corner points are identical (and the boundary curve connecting them therefore shrinks to a point).

Imagine a rectangular grid of $m \times n$ control points. The corresponding rectangular Bézier patch $\mathbf{R}(u, w)$ is a polynomial of degree m in u and degree n in w. It is, thus, a *bivariate* polynomial of total degree $m + n$. When dividing the rectangle into two triangles that have the same shape as the original rectangle, each triangular patch $\mathbf{T}(u, v, w)$ should also be a polynomial of total degree $m + n$. Converting from a rectangular patch to a triangular one is therefore more useful since it results in two patches each having the same degree, but both are defined by more control points. This makes it easier to fine-tune the shape of the surface.

This section is based on [Lischinski 92] and [Lischinski 94] and it illustrates the subject by presenting three examples.

Example 1: Converting quadratic triangular patches into biquadratic rectangular ones. Given the triangular patch

$$T(u,v,w) = w^2\mathbf{T}_{002} + 2uw\mathbf{T}_{101} + u^2\mathbf{T}_{200} + 2vw\mathbf{T}_{011} + 2uv\mathbf{T}_{110} + v^2\mathbf{T}_{020} \quad (5.50)$$

based on the six control points $\mathbf{T}_{ijk}$, we want to convert it to a rectangular patch of the form

$$\begin{aligned}\mathbf{R}(s,t) = &\big((1-s)^2\mathbf{R}_{00} + 2s(1-s)\mathbf{R}_{01} + s^2\mathbf{R}_{02}\big)(1-t)^2 \\ &+ \big((1-s)^2\mathbf{R}_{10} + 2s(1-s)\mathbf{R}_{11} + s^2\mathbf{R}_{12}\big)2t(1-t) \\ &+ \big((1-s)^2\mathbf{R}_{20} + 2s(1-s)\mathbf{R}_{21} + s^2\mathbf{R}_{22}\big)t^2,\end{aligned} \quad (5.51)$$

based on the nine control points $\mathbf{R}_{ij}$. The idea is to create a degenerate rectangular patch where one boundary curve is shrunk to a single point. Figure 5.43b shows the two patches where the three points $\mathbf{R}_{20}$, $\mathbf{R}_{21}$, and $\mathbf{R}_{22}$ are set equal to $\mathbf{T}_{020}$. The relation between the two patches can be established (Figure 5.43a) by the parameter substitution $u = s(1-t)$ and $v = t$. It is easy to see the correspondence between the points:

$$\begin{aligned}&\mathbf{R}_{00} = \mathbf{T}_{002}, \quad \mathbf{R}_{01} = \mathbf{T}_{101}, \quad \mathbf{R}_{02} = \mathbf{T}_{200}, \quad \mathbf{R}_{10} = \mathbf{T}_{011}, \\ &\mathbf{R}_{12} = \mathbf{T}_{110}, \quad \mathbf{R}_{20} = \mathbf{R}_{21} = \mathbf{R}_{22} = \mathbf{T}_{020}.\end{aligned} \quad (5.52)$$

5 Surfaces

The only nontrivial question is how to define point $\mathbf{R}_{11}$ in terms of the $\mathbf{T}_{ijk}$'s. Since this point is in the middle of the rectangular patch, where $s = t = 1/2$, it must correspond to $\mathbf{T}(1/4, 1/2)$, a constraint that produces the equation

$$\frac{1}{16}\mathbf{T}_{002} + \frac{1}{8}\mathbf{T}_{101} + \frac{1}{16}\mathbf{T}_{200} + \frac{1}{4}\mathbf{T}_{011} + \frac{1}{4}\mathbf{T}_{110} + \frac{1}{4}\mathbf{T}_{020}$$
$$= \frac{1}{16}\mathbf{R}_{00} + \frac{1}{8}\mathbf{R}_{01} + \frac{1}{16}\mathbf{R}_{02} + \frac{1}{8}\mathbf{R}_{10} \qquad (5.53)$$
$$+ \frac{1}{4}\mathbf{R}_{11} + \frac{1}{8}\mathbf{R}_{12} + \frac{1}{16}\mathbf{R}_{20} + \frac{1}{8}\mathbf{R}_{21} + \frac{1}{16}\mathbf{R}_{22}.$$

Using the correspondences of Equation (5.52), the solution is $\mathbf{R}_{11} = (\mathbf{T}_{011} + \mathbf{T}_{110})/2$.

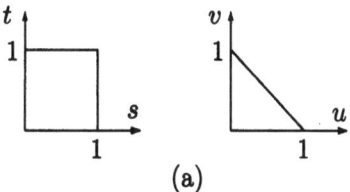

(a)

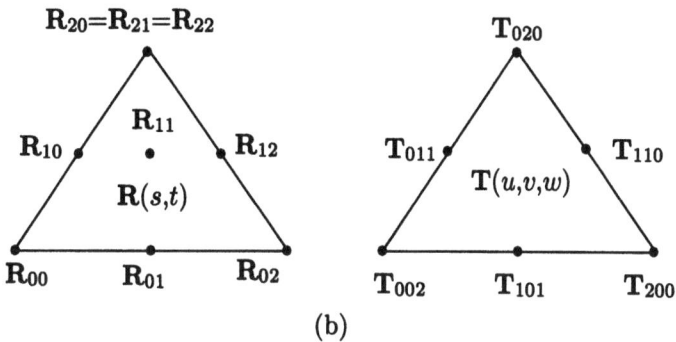

(b)

Figure 5.43: Rectangular and Triangular Patches.

This solution is easy to verify by substituting it in Equation (5.51). Using the relations $u = s(1-t)$, $v = t$, and $w = 1 - u - v$, this should yield the triangular patch of Equation (5.50):

$$\begin{aligned}
\mathbf{R}(s,t) &= \big((1-s)^2\mathbf{R}_{00} + 2s(1-s)\mathbf{R}_{01} + s^2\mathbf{R}_{02}\big)(1-t)^2 \\
&\quad + \big((1-s)^2\mathbf{R}_{10} + 2s(1-s)\mathbf{R}_{11} + s^2\mathbf{R}_{12}\big)2t(1-t) \\
&\quad + \big((1-s)^2\mathbf{R}_{20} + 2s(1-s)\mathbf{R}_{21} + s^2\mathbf{R}_{22}\big)t^2, \\
&= (1-s)^2(1-t)^2\mathbf{T}_{002} + 2s(1-s)(1-t)^2\mathbf{T}_{101} + s^2(1-t)^2\mathbf{T}_{200} \\
&\quad + 2t(1-s)^2(1-t)\mathbf{T}_{011} + 4ts(1-t)(1-s)\frac{1}{2}(\mathbf{T}_{011}+\mathbf{T}_{110}) \\
&\quad + 2ts^2(1-t)\mathbf{T}_{110} + t^2(1-s)^2\mathbf{T}_{020} + 2st^2(1-s)\mathbf{T}_{020} + s^2t^2\mathbf{T}_{020} \\
&= w^2\mathbf{T}_{002} + 2uw\mathbf{T}_{101} + u^2\mathbf{T}_{200} + 2vw\mathbf{T}_{011} + 2uv\mathbf{T}_{110} + v^2\mathbf{T}_{020} \\
&= \mathbf{T}(u,v,w).
\end{aligned}$$

The last lines of the equation above were obtained from the three identities

$$t^2(1-s)^2 + 2st^2(1-s) + s^2t^2 = t^2 = v^2,$$
$$2t(1-t)(1-s)^2 + 2ts(1-t)(1-s) = 2t(1-t)(1-s) = 2vw,$$
$$2ts^2(1-t) + 2ts(1-t)(1-s) = 2ts(1-t) = 2uv.$$

Example 2: Converting cubic triangular patches into bicubic rectangular ones. This is done similarly. Figure 5.44 suggests the relations

$$\begin{aligned}
\mathbf{R}_{00} &= \mathbf{T}_{003}, \quad \mathbf{R}_{01} = \mathbf{T}_{102}, \quad \mathbf{R}_{02} = \mathbf{T}_{201}, \quad \mathbf{R}_{03} = \mathbf{T}_{300}, \\
\mathbf{R}_{10} &= \mathbf{T}_{012}, \quad \mathbf{R}_{13} = \mathbf{T}_{210}, \\
\mathbf{R}_{20} &= \mathbf{T}_{021}, \quad \mathbf{R}_{23} = \mathbf{T}_{120}, \\
\mathbf{R}_{30} &= \mathbf{R}_{31} = \mathbf{R}_{32} = \mathbf{R}_{33} = \mathbf{T}_{030}.
\end{aligned} \quad (5.54)$$

The remaining four points, $\mathbf{R}_{11}$, $\mathbf{R}_{12}$, $\mathbf{R}_{21}$, and $\mathbf{R}_{22}$, are calculated using the relations $u = s(1-t)$ and $v = t$. They produce the four equations

$$\mathbf{T}(\tfrac{2}{9},\tfrac{1}{3}) = \mathbf{R}(\tfrac{1}{3},\tfrac{1}{3}), \quad \mathbf{T}(\tfrac{4}{9},\tfrac{1}{3}) = \mathbf{R}(\tfrac{2}{3},\tfrac{1}{3}),$$
$$\mathbf{T}(\tfrac{1}{9},\tfrac{2}{3}) = \mathbf{R}(\tfrac{1}{3},\tfrac{2}{3}), \quad \mathbf{T}(\tfrac{2}{9},\tfrac{2}{3}) = \mathbf{R}(\tfrac{2}{3},\tfrac{2}{3}),$$

whose solutions are

$$\mathbf{R}_{11} = (\mathbf{T}_{012} + 2\mathbf{T}_{111})/3, \quad \mathbf{R}_{12} = (2\mathbf{T}_{111} + \mathbf{T}_{210})/3,$$
$$\mathbf{R}_{21} = (2\mathbf{T}_{021} + \mathbf{T}_{120})/3, \quad \mathbf{R}_{22} = (\mathbf{T}_{021} + 2\mathbf{T}_{120})/3.$$

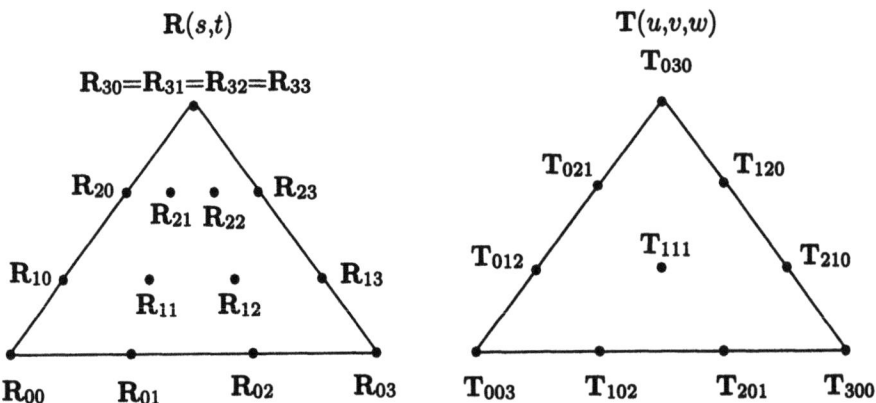

Figure 5.44: Rectangular and Triangular Patches.

Example 3: Converting biquadratic rectangular patches into quartic triangular patches. This case is more complicated than the previous ones since a triangle can be converted into a degenerate rectangle, but a general rectangle cannot be converted

5 Surfaces

into a triangle. The simplest approach is to convert one rectangular patch into two smoothly joined triangles, as shown in Figure 5.45. A biquadratic rectangular Bézier patch $\mathbf{R}(s,t)$ is given by a biquadratic polynomial, one that's quadratic in both s and t and, thus, has a total degree of 4. This is why each of the triangular patches must be given by a quartic (degree 4) polynomial. Each triangular patch thus requires 15 control points $\mathbf{T}_{ijk}$ and is expressed by

$$\begin{aligned}\mathbf{T}(u,v,w) = &\, w^4\mathbf{T}_{004} + 4uw^3\mathbf{T}_{103} + 6u^2w^2\mathbf{T}_{202} + 4u^3w\mathbf{T}_{301}\\ &+ u^4\mathbf{T}_{400} + 4vw^3\mathbf{T}_{013} + 12uvw^2\mathbf{T}_{112} + 12u^2vw\mathbf{T}_{211}\\ &+ 4u^3v\mathbf{T}_{310} + 6v^2w^2\mathbf{T}_{121} + 6u^2v^2\mathbf{T}_{220}\\ &+ 4v^3w\mathbf{T}_{031} + 4uv^3\mathbf{T}_{130} + v^4\mathbf{T}_{040}\end{aligned} \quad (5.55).$$

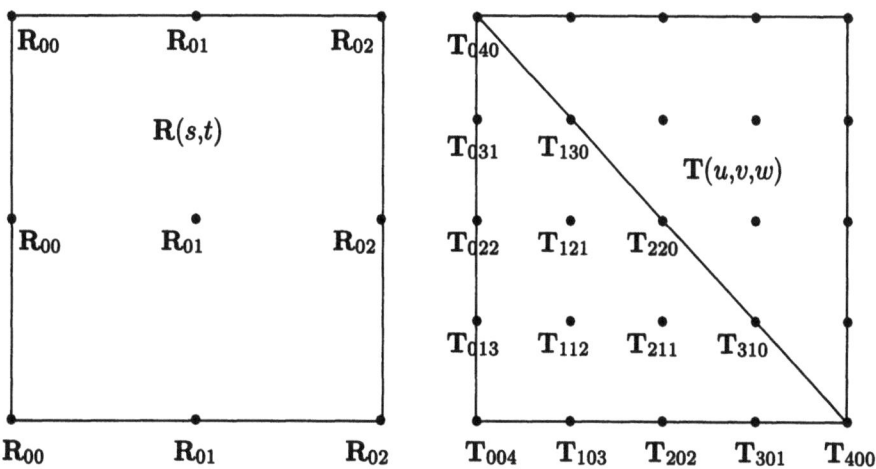

Figure 5.45: One Rectangular and Two Triangular Patches.

The three control points $\mathbf{R}_{0j}$, $j = 0,1,2$, define the boundary curve $\mathbf{R}(0,t)$. This curve is identical to the boundary curve $\mathbf{T}(u,0,w)$ that's defined by the five control points $\mathbf{T}_{i0k}$. We, therefore, need to express these five points in terms of the $\mathbf{R}_{0j}$ points. This is done by elevating the degree of boundary curve $\mathbf{R}(0,t)$ from 2 to 4. Exercise 4.70 shows that the result is

$$\mathbf{T}_{004} = \mathbf{R}_{00}, \quad \mathbf{T}_{103} = (\mathbf{R}_{00} + \mathbf{R}_{01})/2,$$
$$\mathbf{T}_{202} = (\mathbf{R}_{00} + 4\mathbf{R}_{01} + \mathbf{R}_{02})/6,$$
$$\mathbf{T}_{301} = (\mathbf{R}_{01} + \mathbf{R}_{02})/2, \quad \mathbf{T}_{400} = \mathbf{R}_{02}.$$

The boundary curve $\mathbf{R}(s,0)$ is similarly defined by the three points $\mathbf{R}_{i0}$, $i = 0,1,2$. This curve is identical to boundary curve $\mathbf{T}(0,v,w)$ that's defined by the five points $\mathbf{T}_{0jk}$. These five points are thus given by

$$\mathbf{T}_{004} = \mathbf{R}_{00}, \quad \mathbf{T}_{013} = (\mathbf{R}_{00} + \mathbf{R}_{10})/2,$$

$$T_{022} = (R_{00} + 4R_{10} + R_{20})/6,$$
$$T_{031} = (R_{10} + R_{20})/2, \quad T_{040} = R_{20}.$$

The five points T_{ij0} on the common boundary of the two triangles can easily be computed when we notice that they define the "diagonal" curve $R(1-t,t)$ on the biquadratic patch. This curve is given by

$$\begin{aligned}
R(1-t,t) &= t^2(1-t)^2 R_{00} + 2t(1-t)^3 R_{01} + (1-t)^4 R_{02} \\
&\quad + 2t^3(1-t) R_{10} + 4t^2(1-t)^2 R_{11} + 2t(1-t)^3 R_{12} \\
&\quad + t^4 R_{20} + 2t^3(1-t) R_{21} + t^2(1-t)^2 R_{22} \\
&= (1-t)^4 R_{02} + 4t(1-t)^3[(R_{01} + R_{12})/2] \\
&\quad + 6t^2(1-t)^2[(R_{00} + 4R_{11} + R_{22})/6] + 4t^3(1-t)[(R_{10} + R_{21})/2] + t^4 R_{20} \\
&= (1-t)^4 T_{400} + 4t(1-t)^3[(R_{01} + R_{12})/2] \\
&\quad + 6t^2(1-t)^2[(R_{00} + 4R_{11} + R_{22})/6] + 4t^3(1-t)[(R_{10} + R_{21})/2] + t^4 T_{040}.
\end{aligned}$$

Comparing the above expression to the boundary curve $T(1-t,t)$ on the triangle given by Equation (5.55) shows that

$$T_{310} = (R_{01} + R_{12})/2, \quad T_{220} = (R_{00} + 4R_{11} + R_{22})/6, \quad T_{130} = (R_{10} + R_{21})/2.$$

All that remains is the three interior points T_{112}, T_{121}, and T_{211}. They are calculated by solving the three simultaneous equations

$$T(\tfrac{1}{4},\tfrac{1}{4}) = R(\tfrac{1}{4},\tfrac{1}{4}), \quad T(\tfrac{1}{2},\tfrac{1}{4}) = R(\tfrac{1}{2},\tfrac{1}{4}), \quad T(\tfrac{1}{4},\tfrac{1}{2}) = R(\tfrac{1}{4},\tfrac{1}{2}).$$

The solutions are

$$\begin{aligned}
T_{112} &= (2R_{00} + R_{01} + R_{10} + 2R_{11})/6, \\
T_{121} &= (R_{00} + 2R_{10} + 2R_{11} + R_{21})/6, \\
T_{211} &= (R_{00} + 2R_{01} + 2R_{11} + R_{12})/6.
\end{aligned}$$

The 15 control points of the other triangle are symmetric to the ones calculated here.

5.14.1 Reparametrizing the Bézier Surface

We illustrate the method described here by applying it to the bicubic Bézier surface patch. The expression for this patch is given by Equations (5.40) and (5.39):

$$\begin{aligned}
P(u,w) &= \sum_{i=0}^{3}\sum_{j=0}^{3} B_{3,i}(u) P_{i,j} B_{3,j}(w) \\
&= \sum_{i=0}^{3}\sum_{j=0}^{3} (u^3, u^2, u, 1) M P M^{-1} (w^3, w^2, w, 1)^T,
\end{aligned}$$

5 Surfaces

where $\mathbf{M}$ is the basis matrix

$$\mathbf{M} = \begin{pmatrix} -1 & 3 & -3 & 1 \\ 3 & -6 & 3 & 0 \\ -3 & 3 & 0 & 0 \\ 1 & 0 & 0 & 0 \end{pmatrix}$$

and $\mathbf{P}$ is the 4×4 matrix of control points

$$\begin{pmatrix} \mathbf{P}_{3,0} & \mathbf{P}_{3,1} & \mathbf{P}_{3,2} & \mathbf{P}_{3,3} \\ \mathbf{P}_{2,0} & \mathbf{P}_{2,1} & \mathbf{P}_{2,2} & \mathbf{P}_{2,3} \\ \mathbf{P}_{1,0} & \mathbf{P}_{1,1} & \mathbf{P}_{1,2} & \mathbf{P}_{1,3} \\ \mathbf{P}_{0,0} & \mathbf{P}_{0,1} & \mathbf{P}_{0,2} & \mathbf{P}_{0,3} \end{pmatrix}.$$

This surface patch can be reparametrized with the method of Section 4.14.12. We select part of patch $\mathbf{P}(u,w)$, e.g., the part where u varies from a to b, and define it as a new patch $\mathbf{Q}(u,w)$ where both u and w vary in the range $[0,1]$. The method discussed here shows how to obtain the control points $\mathbf{Q}_{ij}$ of patch $\mathbf{Q}(u,w)$ as functions of a, b and points $\mathbf{P}_{ij}$.

> B-splines are the defacto standard that drives today's sophisticated computer graphics applications. This method is also responsible for the developments that have transformed computer-aided geometric design from the era of hand-built models and manual measurements to fast computations and three-dimensional renderings.

Suppose we want to reparametrize the "left" part of $\mathbf{P}(u,w)$, i.e., the part where $0 \leq u \leq 0.5$. Applying the methods of Section 4.14.12, we select $a = 0$, $b = 0.5$ and can write

$$\mathbf{P}(u/2, w) = (u^3, u^2, u, 1)\mathbf{MBPM}^{-1}(w^3, w^2, w, 1)^T,$$

where $\mathbf{B}$ is given by Equation (4.121):

$$\mathbf{B} = \begin{pmatrix} (1-a)^3 & 3(a-1)^2 a & 3(1-a)a^2 & a^3 \\ (a-1)^2(1-b) & (a-1)(-2a-b+3ab) & a(a+2b-3ab) & a^2 b \\ (1-a)(-1+b)^2 & (b-1)(-a-2b+3ab) & b(2a+b-3ab) & ab^2 \\ (1-b)^3 & 3(b-1)^2 b & 3(1-b)b^2 & b^3 \end{pmatrix}.$$

Exercise 4.74 shows that selecting $a = 0$ and $b = 0.5$ reduces matrix $\mathbf{B}$ to

$$\mathbf{B} = \begin{pmatrix} 1 & 0 & 0 & 0 \\ \frac{1}{2} & \frac{1}{2} & 0 & 0 \\ \frac{1}{4} & \frac{1}{2} & \frac{1}{4} & 0 \\ \frac{1}{8} & \frac{3}{8} & \frac{3}{8} & \frac{1}{8} \end{pmatrix}.$$

The new control points for our surface patch are thus

$$\begin{pmatrix} Q_{3,0} & Q_{3,1} & Q_{3,2} & Q_{3,3} \\ Q_{2,0} & Q_{2,1} & Q_{2,2} & Q_{2,3} \\ Q_{1,0} & Q_{1,1} & Q_{1,2} & Q_{1,3} \\ Q_{0,0} & Q_{0,1} & Q_{0,2} & Q_{0,3} \end{pmatrix} = \begin{pmatrix} 1 & 0 & 0 & 0 \\ \frac{1}{2} & \frac{1}{2} & 0 & 0 \\ \frac{1}{4} & \frac{1}{2} & \frac{1}{4} & 0 \\ \frac{1}{8} & \frac{3}{8} & \frac{3}{8} & \frac{1}{8} \end{pmatrix} \begin{pmatrix} P_{3,0} & P_{3,1} & P_{3,2} & P_{3,3} \\ P_{2,0} & P_{2,1} & P_{2,2} & P_{2,3} \\ P_{1,0} & P_{1,1} & P_{1,2} & P_{1,3} \\ P_{0,0} & P_{0,1} & P_{0,2} & P_{0,3} \end{pmatrix}$$

$$= \begin{pmatrix} P_{3,0} & P_{3,1} \\ \frac{1}{2}P_{3,0} + \frac{1}{2}P_{2,0} & \frac{1}{2}P_{3,1} + \frac{1}{2}P_{2,1} \\ \frac{1}{4}P_{3,0} + \frac{1}{2}P_{2,0} + \frac{1}{4}P_{1,0} & \frac{1}{4}P_{3,1} + \frac{1}{2}P_{2,1} + \frac{1}{4}P_{1,1} \\ \frac{1}{8}P_{3,0} + \frac{3}{8}P_{2,0} + \frac{3}{8}P_{1,0} + \frac{1}{8}P_{0,0} & \frac{1}{8}P_{3,1} + \frac{3}{8}P_{2,1} + \frac{3}{8}P_{1,1} + \frac{1}{8}P_{1,0} \\ P_{3,2} & P_{3,3} \\ \frac{1}{2}P_{3,2} + \frac{1}{2}P_{2,2} & \frac{1}{2}P_{3,3} + \frac{1}{2}P_{2,3} \\ \frac{1}{4}P_{3,2} + \frac{1}{2}P_{2,2} + \frac{1}{4}P_{1,2} & \frac{1}{4}P_{3,3} + \frac{1}{2}P_{2,3} + \frac{1}{4}P_{1,3} \\ \frac{1}{8}P_{3,2} + \frac{3}{8}P_{2,2} + \frac{3}{8}P_{1,2} + \frac{1}{8}P_{2,0} & \frac{1}{8}P_{3,3} + \frac{3}{8}P_{2,3} + \frac{3}{8}P_{1,3} + \frac{1}{8}P_{3,0} \end{pmatrix}.$$

In general, suppose we want to reparametrize that portion of patch $\mathbf{P}(u,w)$ where $a \le u \le b$ and $c \le w \le d$. We can write

$\mathbf{Q}(u,w)$
$= \mathbf{P}([b-a]u + a, [d-c]w + c)$

$= \left(([b-a]u+a)^3, ([b-a]u+a)^2, ([b-a]u+a), 1 \right) \mathbf{M} \cdot \mathbf{P} \cdot \mathbf{M}^{-1} \begin{pmatrix} ([d-c]w+c)^3 \\ ([d-c]w+c)^2 \\ [d-c]w+c \\ 1 \end{pmatrix}$

$= (u^3, u^2, u, 1) \mathbf{A}_{ab} \mathbf{M} \cdot \mathbf{P} \cdot \mathbf{M}^T \cdot \mathbf{A}_{cd}^T (w^3, w^2, w, 1)^T$
$= (u^3, u^2, u, 1) \mathbf{M} (\mathbf{M}^{-1} \cdot \mathbf{A}_{ab} \cdot \mathbf{M}) \mathbf{P} (\mathbf{M}^T \cdot \mathbf{A}_{cd}^T \cdot (\mathbf{M}^T)^{-1}) \mathbf{M}^T (w^3, w^2, w, 1)^T$
$= (u^3, u^2, u, 1) \mathbf{M} \cdot \mathbf{B}_{ab} \cdot \mathbf{P} \cdot \mathbf{B}_{cd}^T \cdot \mathbf{M}^T (w^3, w^2, w, 1)^T$
$= (u^3, u^2, u, 1) \mathbf{M} \cdot \mathbf{Q} \cdot \mathbf{M}^T (w^3, w^2, w, 1)^T, \qquad (5.56)$

where $\mathbf{B}_{ab} = \mathbf{M}^{-1} \cdot \mathbf{A}_{ab} \cdot \mathbf{M}$, $\mathbf{B}_{cd}^T = \mathbf{M}^T \cdot \mathbf{A}_{cd}^T \cdot (\mathbf{M}^T)^{-1}$, $\mathbf{Q} = \mathbf{B}_{ab} \cdot \mathbf{P} \cdot \mathbf{B}_{cd}^T$, and

$$\mathbf{A}_{ab} = \begin{pmatrix} (b-a)^3 & 0 & 0 & 0 \\ 3a(b-a)^2 & (b-a)^2 & 0 & 0 \\ 3a^2(b-a) & 2a(b-a) & b-a & 0 \\ a^3 & a^2 & a & 1 \end{pmatrix}.$$

The elements of $\mathbf{Q}$ depend on a, b, c, and d, and the $\mathbf{P}_{ij}$'s and are quite complex. They can be produced by the following *Mathematica* code:

```
B={{(1 - a)^3, 3*(-1 + a)^2*a, 3*(1 - a)*a^2, a^3},
   {(-1 + a)^2*(1 - b), (-1 + a)*(-2*a - b + 3*a*b),
    a*(a + 2*b - 3*a*b),
    a^2*b}, {(1 - a)*(-1 + b)^2, (-1 + b)*(-a - 2*b + 3*a*b),
    b*(2*a + b - 3*a*b), a*b^2},
```

5 Surfaces

```
     {(1 - b)^3, 3*(-1 + b)^2*b, 3*(1 - b)*b^2, b^3}};
  TB={{(1 - c)^3, (-1 + c)^2*(1 - d), (1 - c)*(-1 + d)^2,
     (1 - d)^3},
    {3*(-1 + c)^2*c, (-1 + c)*(-2*c - d + 3*c*d),
     (-1 + d)*(-c - 2*d + 3*c*d), 3*(-1 + d)^2*d},
    {3*(1 - c)*c^2, c*(c + 2*d - 3*c*d), d*(2*c + d - 3*c*d),
     3*(1 - d)*d^2},
    {c^3, c^2*d, c*d^2, d^3}};
  P={{P30,P31,P32,P33},{P20,P21,P22,P23},
    {P10,P11,P12,P13},{P00,P01,P02,P03}};
  Q=Simplify[B.P.TB]
```

5.15 The Gregory Patch

Gregory has originally developed this method to extend the Coons surface patch. The Gregory method, however, becomes very practical when used to extend the bicubic Bézier patch. Recall that such a patch is based on $4 \times 4 = 16$ control points (Figure 5.46a). We can divide the 16 points into 2 groups: the interior points, consisting of the 4 points $\mathbf{P}_{11}$, $\mathbf{P}_{12}$, $\mathbf{P}_{21}$, and $\mathbf{P}_{22}$, and the boundary points, consisting of the remaining 12 points. Experience shows that there are too few interior points to fine-tune the shape of the patch. Moving point $\mathbf{P}_{11}$, for example, affects both the direction from $\mathbf{P}_{01}$ to $\mathbf{P}_{11}$, and the direction from $\mathbf{P}_{10}$ to $\mathbf{P}_{11}$.

The idea in the Gregory patch is to split each of the four interior points into two points. Hence, instead of point $\mathbf{P}_{11}$, e.g., there should be two points $\mathbf{P}_{110}$ and $\mathbf{P}_{111}$, both in the vicinity of the original $\mathbf{P}_{11}$. Moving $\mathbf{P}_{110}$ affects the shape of the patch only in the direction from $\mathbf{P}_{10}$ to $\mathbf{P}_{110}$. The shape of the patch around point $\mathbf{P}_{01}$ is not changed (at least, not significantly). The Gregory patch is thus defined by 20 points (Figure 5.46b): 8 interior points and 12 boundary points. Points $\mathbf{P}_{110}$ and $\mathbf{P}_{111}$ can initially be set equal to $\mathbf{P}_{11}$, then moved interactively in different directions to get the right shape of the surface.

To calculate the surface, we first define 16 new points $\mathbf{Q}_{ij}$, then use Equation (5.39) with the new points as control points and with $n = m = 3$. Twelve of the $\mathbf{Q}$ points are boundary and are identical to the boundary $\mathbf{P}$ points. The remaining four $\mathbf{Q}$ points are interior and each is calculated from a pair of interior $\mathbf{P}$ points. The definitions are

$$\mathbf{Q}_{11}(u,w) = \frac{u\mathbf{P}_{110} + w\mathbf{P}_{111}}{u+w}, \qquad \mathbf{Q}_{21}(u,w) = \frac{(1-u)\mathbf{P}_{210} + w\mathbf{P}_{211}}{1-u+w},$$

$$\mathbf{Q}_{12}(u,w) = \frac{u\mathbf{P}_{120} + (1-w)\mathbf{P}_{121}}{u+1-w}, \qquad \mathbf{Q}_{22}(u,w) = \frac{(1-u)\mathbf{P}_{220} + (1-w)\mathbf{P}_{221}}{1-u+1-w}.$$

Note that $\mathbf{Q}_{11}(u,w)$ is a barycentric sum of two $\mathbf{P}$ points, so it is well defined. Even though u and w are independent and each is varied from 0 to 1 independently of the other, the sum is always a point on the straight line connecting $\mathbf{P}_{110}$ to $\mathbf{P}_{111}$. The same is true for the other three interior $\mathbf{Q}$ points.

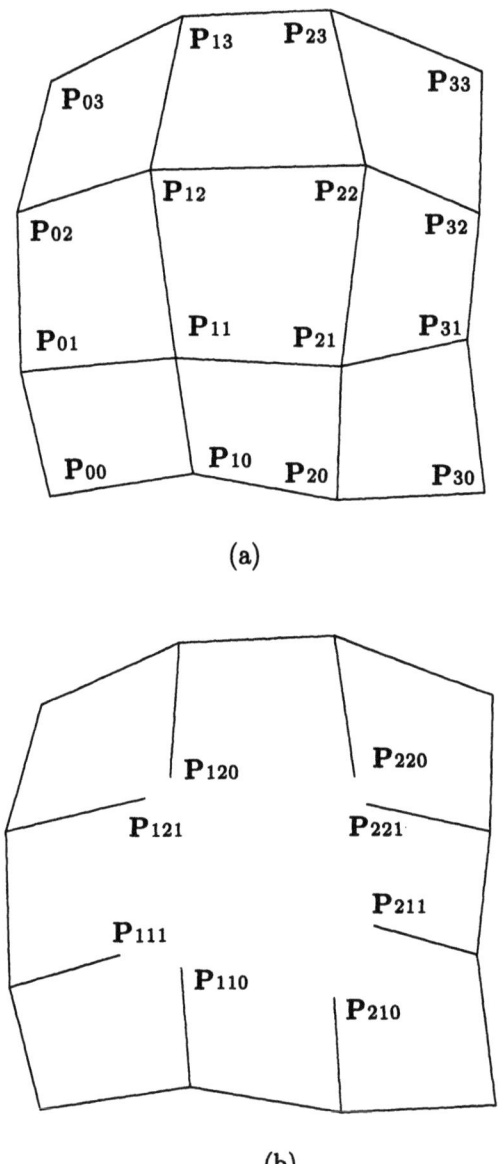

Figure 5.46: (a) A Bicubic Bézier Patch. (b) A Gregory Patch.

5 Surfaces

After calculating the new points, the Gregory patch is defined as the bicubic Bézier patch

$$\mathbf{P}(u,w) = \sum_{i=0}^{3}\sum_{j=0}^{3} B_{3,i}(w)\mathbf{Q}_{i,j}B_{3,j}(u).$$

5.15.1 The Gregory Tangent Vectors

The first derivatives of the Gregory patch are more complex than those of the bicubic Bézier patch since four of the control points depend on the parameters u and w. The derivatives are

$$\frac{\partial \mathbf{P}(u,w)}{\partial u}$$
$$= \sum_{i=0}^{3}\sum_{j=0}^{3} \frac{dB_{3,i}(u)}{du} B_{3,j}(w)\mathbf{Q}_{i,j}(u,w) + \sum_{i=0}^{3}\sum_{j=0}^{3} B_{3,i}(u)B_{3,j}(w)\frac{\partial \mathbf{Q}_{i,j}(u,w)}{\partial u},$$
$$\frac{\partial \mathbf{P}(u,w)}{\partial w}$$
$$= \sum_{i=0}^{3}\sum_{j=0}^{3} B_{3,i}(u)\frac{dB_{3,j}(w)}{dw}\mathbf{Q}_{i,j}(u,w) + \sum_{i=0}^{3}\sum_{j=0}^{3} B_{3,i}(u)B_{3,j}(w)\frac{\partial \mathbf{Q}_{i,j}(u,w)}{\partial w}.$$

Each derivative is the sum of two similar terms, each of which has the same format as a derivative of the bicubic Bézier patch. Therefore, only one procedure is needed to calculate the derivatives numerically. This procedure is called twice for each partial derivative. The second call involves the derivatives of the control points, which are shown below.

The 12 boundary $\mathbf{Q}$ points don't depend on u or w, so their derivatives are zero. The derivatives of the eight interior points are

$$\frac{\partial \mathbf{Q}_{11}(u,w)}{\partial u} = \frac{w(\mathbf{P}_{110} - \mathbf{P}_{111})}{(u+w)^2}, \quad \frac{\partial \mathbf{Q}_{11}(u,w)}{\partial w} = \frac{u(\mathbf{P}_{110} - \mathbf{P}_{111})}{(u+w)^2},$$
$$\frac{\partial \mathbf{Q}_{21}(u,w)}{\partial u} = \frac{w(\mathbf{P}_{210} - \mathbf{P}_{211})}{(1-u+w)^2}, \quad \frac{\partial \mathbf{Q}_{21}(u,w)}{\partial w} = \frac{(1-u)(\mathbf{P}_{210} - \mathbf{P}_{211})}{(1-u+w)^2},$$
$$\frac{\partial \mathbf{Q}_{12}(u,w)}{\partial u} = \frac{(1-w)(\mathbf{P}_{120} - \mathbf{P}_{121})}{(u+1-w)^2}, \quad \frac{\partial \mathbf{Q}_{12}(u,w)}{\partial w} = \frac{u(\mathbf{P}_{120} - \mathbf{P}_{121})}{(u+1-w)^2},$$
$$\frac{\partial \mathbf{Q}_{22}(u,w)}{\partial u} = \frac{(1-w)(\mathbf{P}_{220} - \mathbf{P}_{221})}{(1-u+1-w)^2}, \quad \frac{\partial \mathbf{Q}_{22}(u,w)}{\partial w} = \frac{(1-u)(\mathbf{P}_{220} - \mathbf{P}_{221})}{(1-u+1-w)^2}.$$

After the first derivatives (the tangent vectors) are calculated numerically at a point, they are used to numerically calculate the normal vector at the point.

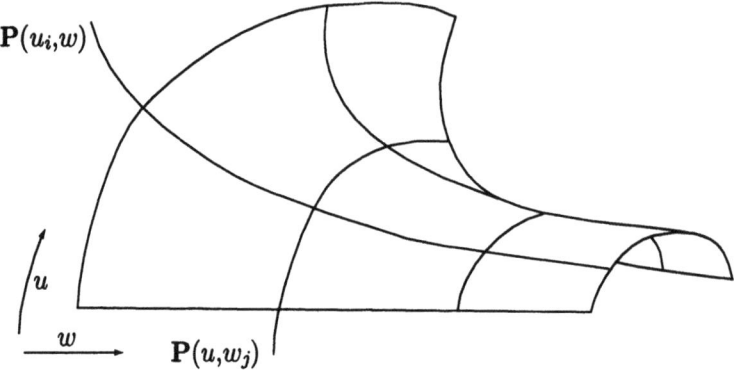

Figure 5.47: A Gordon Surface.

5.16 Gordon Surfaces

These are a generalization of Coons surfaces. A linear Coons surface is fully defined by means of four boundary curves, so its shape cannot be too complex. A Gordon surface (Figure 5.47) is defined by means of two families of curves, one in each of the u and w directions. It can have very complex shapes and is a good candidate for use in applications where realism is important.

We denote the curves by $\mathbf{P}(u_i,w)$, where $i = 0,\ldots,m$, and $\mathbf{P}(u,w_j)$, $j = 0,\ldots,n$. The main idea is to find an expression for a surface $\mathbf{P}_a(u,w)$ that interpolates the first family of curves, add it to a similar expression for a surface $\mathbf{P}_b(u,w)$ that interpolates the second family of curves, and subtract a surface $\mathbf{P}_{ab}(u,w)$ that represents multiple contributions from $\mathbf{P}_a$ and $\mathbf{P}_b$.

The first surface, $\mathbf{P}_a(u,w)$, should interpolate the family of $m+1$ curves $\mathbf{P}(u_i,w)$. When moving on this surface in the u direction (fixed w), we want to intersect all $m+1$ curves. For a given, fixed w, we, therefore, need to find a curve that will pass through the $m+1$ **points** $\mathbf{P}(u_i,w)$. Such a curve is, of course, our old friend the Lagrange polynomial (Section 4.4). We write it as $\mathbf{P}_a(u,w) = \sum_{i=0}^{m} \mathbf{P}(u_i,w)L_i^m(u)$, and it is valid for any value of w. Similarly, we can write the second surface as the Lagrange polynomial $\mathbf{P}_b(u,w) = \sum_{j=0}^{n} \mathbf{P}(u,w_j)L_j^n(w)$.

The surface representing multiple contributions is similar to the bilinear part of Equation (5.12). It is

$$\mathbf{P}_{ab}(u,w) = \sum_{i=0}^{m}\sum_{j=0}^{n} \mathbf{P}(u_i,w_j)L_i^m(u)L_j^n(w),$$

and the final expression of the Gordon surface is $\mathbf{P}(u,w) = \mathbf{P}_a(u,w) + \mathbf{P}_b(u,w) - \mathbf{P}_{ab}(u,w)$. Note that the $(m+1) \times (n+1)$ points $\mathbf{P}(u_i,w_j)$ should be located on *both* curves. For such a surface to make sense, the curves have to intersect.

5 Surfaces

5.17 Uniform B-Spline Surfaces

This type of surface is constructed as a Cartesian product of two uniform B-spline curves. The biquadratic B-spline surface patch, for example, is fully defined by nine control points and is given by

$$\mathbf{P}(u,w) = \left(\frac{1}{2}\right)^2 (u^2, u, 1) \begin{pmatrix} 1 & -2 & 1 \\ -2 & 2 & 0 \\ 1 & 1 & 0 \end{pmatrix} \begin{pmatrix} \mathbf{P}_{00} & \mathbf{P}_{01} & \mathbf{P}_{02} \\ \mathbf{P}_{10} & \mathbf{P}_{11} & \mathbf{P}_{12} \\ \mathbf{P}_{20} & \mathbf{P}_{21} & \mathbf{P}_{22} \end{pmatrix}$$
$$\times \begin{pmatrix} 1 & -2 & 1 \\ -2 & 2 & 0 \\ 1 & 1 & 0 \end{pmatrix}^T \begin{pmatrix} w^2 \\ w \\ 1 \end{pmatrix}. \tag{5.57}$$

Its four corner points are not the four extreme control points, but

$$\mathbf{K}_{00} = \mathbf{P}(0,0) = \frac{1}{4}(\mathbf{P}_{00} + \mathbf{P}_{01} + \mathbf{P}_{10} + \mathbf{P}_{11}),$$
$$\mathbf{K}_{01} = \mathbf{P}(0,1) = \frac{1}{4}(\mathbf{P}_{01} + \mathbf{P}_{02} + \mathbf{P}_{11} + \mathbf{P}_{12}),$$
$$\mathbf{K}_{10} = \mathbf{P}(1,0) = \frac{1}{4}(\mathbf{P}_{10} + \mathbf{P}_{11} + \mathbf{P}_{20} + \mathbf{P}_{21}),$$
$$\mathbf{K}_{11} = \mathbf{P}(1,1) = \frac{1}{4}(\mathbf{P}_{11} + \mathbf{P}_{12} + \mathbf{P}_{21} + \mathbf{P}_{22}). \tag{5.58}$$

Notice that corner point $\mathbf{K}_{00}$ can be written

$$\mathbf{K}_{00} = \frac{1}{2}\left(\frac{\mathbf{P}_{00} + \mathbf{P}_{01}}{2} + \frac{\mathbf{P}_{10} + \mathbf{P}_{11}}{2}\right).$$

This point is thus located midway between points $(\mathbf{P}_{00}+\mathbf{P}_{01})/2$ and $(\mathbf{P}_{10}+\mathbf{P}_{11})/2$. Figure 5.48a shows this location, as well as the locations of the other three corner points, for the case where the control points are equally spaced.

Example: Given the nine points

$$\mathbf{P}_{00} = (0,0,0), \quad \mathbf{P}_{01} = (0,1,0), \quad \mathbf{P}_{02} = (0,2,0),$$
$$\mathbf{P}_{10} = (1,0,0), \quad \mathbf{P}_{11} = (1,1,1), \quad \mathbf{P}_{12} = (1,2,0),$$
$$\mathbf{P}_{20} = (2,0,0), \quad \mathbf{P}_{21} = (2,1,0), \quad \mathbf{P}_{22} = (2,2,0),$$

the biquadratic B-spline surface patch defined by them is given by the simple expression

$$\mathbf{P}(u,w) = (u+1/2, w+1/2, (-1-2u+2u^2)(-1-2w+2w^2)/4).$$

Its four corner points are

$$\mathbf{K}_{00} = \mathbf{P}(0,0) = \left(\frac{1}{2}, \frac{1}{2}, \frac{1}{4}\right), \quad \mathbf{K}_{01} = \mathbf{P}(0,1) = \left(\frac{1}{2}, \frac{3}{2}, \frac{1}{4}\right),$$

5.17 Uniform B-Spline Surfaces

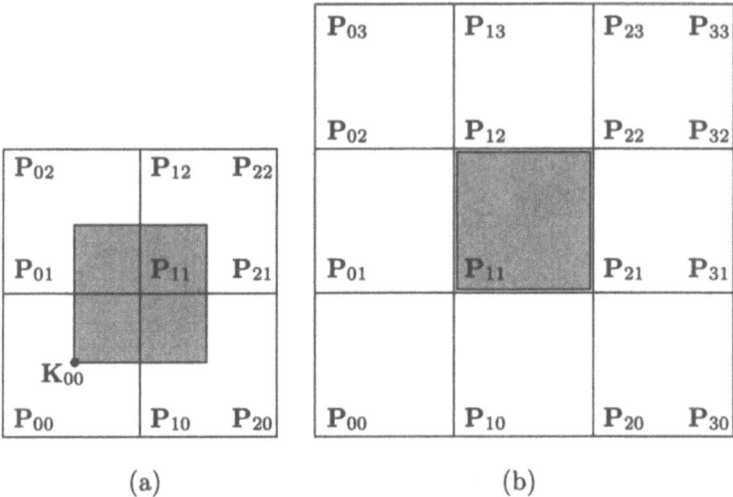

Figure 5.48: Idealized B-Spline Surface Patches.

$$\mathbf{K}_{10} = \mathbf{P}(1,0) = \left(\frac{3}{2}, \frac{1}{2}, \frac{1}{4}\right), \quad \mathbf{K}_{11} = \mathbf{P}(1,1) = \left(\frac{3}{2}, \frac{3}{2}, \frac{1}{4}\right).$$

Figure 5.49 shows the relation between this surface and its control points.

▸ **Exercise 5.29:** Calculate the midpoint $\mathbf{P}(1/2, 1/2)$ of this patch.

The bicubic B-spline patch is defined by 16 control points and is given by

$$\mathbf{P}(u,w) = \left(\frac{1}{6}\right)^2 (u^3, u^2, u, 1) \begin{pmatrix} -1 & 3 & -3 & 1 \\ 3 & -6 & 3 & 0 \\ -3 & 0 & 3 & 0 \\ 1 & 4 & 1 & 0 \end{pmatrix}$$
$$\times \begin{pmatrix} \mathbf{P}_{00} & \mathbf{P}_{01} & \mathbf{P}_{02} & \mathbf{P}_{03} \\ \mathbf{P}_{10} & \mathbf{P}_{11} & \mathbf{P}_{12} & \mathbf{P}_{13} \\ \mathbf{P}_{20} & \mathbf{P}_{21} & \mathbf{P}_{22} & \mathbf{P}_{23} \\ \mathbf{P}_{30} & \mathbf{P}_{31} & \mathbf{P}_{32} & \mathbf{P}_{33} \end{pmatrix} \begin{pmatrix} -1 & 3 & -3 & 1 \\ 3 & -6 & 3 & 0 \\ -3 & 0 & 3 & 0 \\ 1 & 4 & 1 & 0 \end{pmatrix}^T \begin{pmatrix} w^3 \\ w^2 \\ w \\ 1 \end{pmatrix}. \quad (5.59)$$

Its four corner points are

$$\mathbf{K}_{00} = \mathbf{P}(0,0)$$
$$= \frac{1}{36}(\mathbf{P}_{00} + \mathbf{P}_{02} + 4\mathbf{P}_{10} + 4\mathbf{P}_{12} + \mathbf{P}_{20} + 4\mathbf{P}_{01} + 16\mathbf{P}_{11} + 4\mathbf{P}_{21} + \mathbf{P}_{22}),$$
$$\mathbf{K}_{01} = \mathbf{P}(0,1)$$
$$= \frac{1}{36}(\mathbf{P}_{01} + 4\mathbf{P}_{02} + \mathbf{P}_{03} + 4\mathbf{P}_{11} + 16\mathbf{P}_{12} + 4\mathbf{P}_{13} + \mathbf{P}_{21} + 4\mathbf{P}_{22} + \mathbf{P}_{23}),$$
$$\mathbf{K}_{10} = \mathbf{P}(1,0) \quad (5.60)$$
$$= \frac{1}{36}(\mathbf{P}_{10} + \mathbf{P}_{12} + 4\mathbf{P}_{20} + 4\mathbf{P}_{22} + \mathbf{P}_{30} + 4\mathbf{P}_{11} + 16\mathbf{P}_{21} + 4\mathbf{P}_{31} + \mathbf{P}_{32}),$$

5 Surfaces

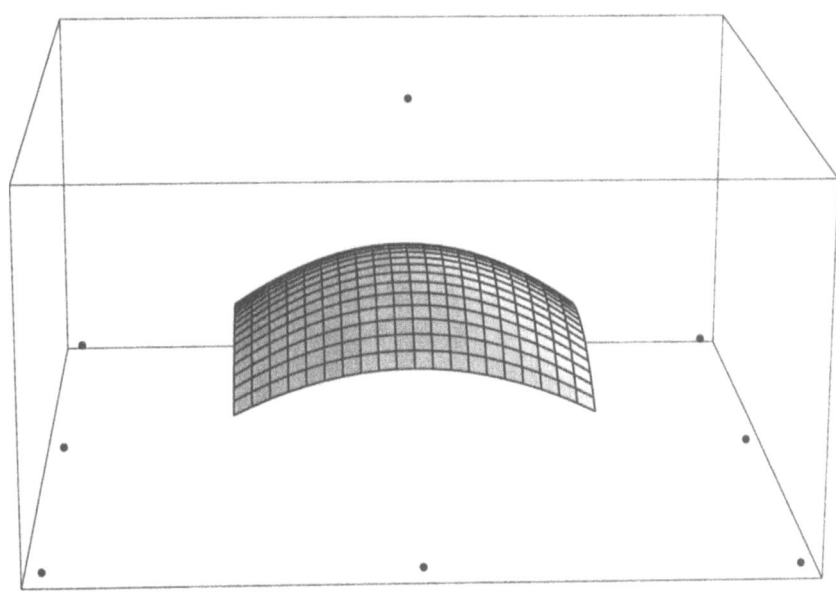

Figure 5.49: A Biquadratic B-Spline Surface Patch.

```
<<:Graphics:ParametricPlot3D.m
Clear[Pnts,Quadr,QuSr,g1,g2];
p00={0,0,0}; p01={0,1,0}; p02={0,2,0};
p10={1,0,0}; p11={1,1,1}; p12={1,2,0};
p20={2,0,0}; p21={2,1,0}; p22={2,2,0};
Quadr={{1,-2,1},{-2,2,0},{1,1,0}};
Pnts={{p00,p01,p02},{p10,p11,p12},{p20,p21,p22}};
g1=Graphics3D[{AbsolutePointSize[3],
 Table[Point[Pnts[[i,j]]],{i,1,3},{j,1,3}]}];
prt[i_]:=Pnts[[Range[1,3],Range[1,3],i]];
(* extracts component i from the 3rd dimen of 'Pnts' *)
QuSr[u_,w_,i_]:= \
 (1/4) {u^2,u,1}.Quadr.prt[i].Transpose[Quadr].{w^2,w,1}
g2=ParametricPlot3D[{QuSr[u,w,1],QuSr[u,w,2],QuSr[u,w,3]},
 {u,0,1,.05},{w,0,1,.05}];
Show[g1,g2, ViewPoint->{-0.196, -4.177,1.160}]
```

Control Points and Code for Figure 5.49.

$$\mathbf{K}_{11} = \mathbf{P}(1,1)$$
$$= \frac{1}{36}(\mathbf{P}_{11} + 4\mathbf{P}_{12} + \mathbf{P}_{13} + 4\mathbf{P}_{21} + 16\mathbf{P}_{22} + 4\mathbf{P}_{23} + \mathbf{P}_{31} + 4\mathbf{P}_{32} + \mathbf{P}_{33}).$$

Each is a barycentric sum of nine control points. Notice that the first corner point can be rewritten in the form

$$\mathbf{K}_{00} = \frac{1}{6}\left[\frac{1}{6}(\mathbf{P}_{00}+4\mathbf{P}_{10}+\mathbf{P}_{20}) + \frac{4}{6}(\mathbf{P}_{01}+4\mathbf{P}_{11}+\mathbf{P}_{21}) + \frac{1}{6}(\mathbf{P}_{02}+4\mathbf{P}_{12}+\mathbf{P}_{22})\right]. \quad (5.61)$$

This point is, therefore, the weighted sum of three points, each the weighted sum of three control points. Its precise location depends on the positions of the nine points involved.

▶ **Exercise 5.30:** What is the value of $\mathbf{K}_{00}$ for the special case where the control points are equally spaced?

The other three corner points can be expressed similarly. If all 16 points are equally spaced, the bicubic surface patch has its corners at the 4 control points $\mathbf{P}_{11}$, $\mathbf{P}_{21}$, $\mathbf{P}_{12}$, and $\mathbf{P}_{22}$ (Figure 5.48b shows an idealized diagram).

Large B-spline surfaces can be constructed from these bicubic patches by starting with a mesh of $(m+1) \times (n+1)$ control points $\mathbf{P}_{00}$ through $\mathbf{P}_{mn}$, dividing it into $(m-2) \times (n-2)$ overlapping groups of 4×4 points each, and using Equation (5.59) to calculate a cubic patch for each group. The individual patches will not only connect at their joint points but will have C^2 continuity along their boundaries.

To show that the bicubic patches connect at the joints, we note how joint point $\mathbf{K}_{01}$ can be obtained from joint $\mathbf{K}_{00}$ by incrementing the second indices of the nine control points involved in their expressions (Equation (5.60)). The same thing is true for joints $\mathbf{K}_{10}$ and $\mathbf{K}_{11}$. Similarly, joint point $\mathbf{K}_{10}$ can be obtained from $\mathbf{K}_{00}$ by incrementing the first index of each control point, and the same is true for joints $\mathbf{K}_{01}$ and $\mathbf{K}_{11}$.

To show first-order continuity we calculate, for example, the two tangent vectors $\mathbf{P}^t(u,0)$ and $\mathbf{P}^t(u,1)$:

$$\begin{aligned}\mathbf{P}^t(u,0) = (&\mathbf{P}_{00} - \mathbf{P}_{02} - \mathbf{P}_{20} + \mathbf{P}_{22} - 2\mathbf{P}_{00}u + 2\mathbf{P}_{02}u + 4\mathbf{P}_{10}u \\ &- 4\mathbf{P}_{12}u - 2\mathbf{P}_{20}u + 2\mathbf{P}_{22}u + \mathbf{P}_{00}u^2 - \mathbf{P}_{02}u^2 - 3\mathbf{P}_{10}u^2 \\ &+ 3\mathbf{P}_{12}u^2 + 3\mathbf{P}_{20}u^2 - 3\mathbf{P}_{22}u^2 - \mathbf{P}_{30}u^2 + \mathbf{P}_{32}u^2)/4, \\ \mathbf{P}^t(u,1) = (&\mathbf{P}_{01} - \mathbf{P}_{03} - \mathbf{P}_{21} + \mathbf{P}_{23} - 2\mathbf{P}_{01}u + 2\mathbf{P}_{03}u + 4\mathbf{P}_{11}u \\ &- 4\mathbf{P}_{13}u - 2\mathbf{P}_{21}u + 2\mathbf{P}_{23}u + \mathbf{P}_{01}u^2 - \mathbf{P}_{03}u^2 - 3\mathbf{P}_{11}u^2 \\ &+ 3\mathbf{P}_{13}u^2 + 3\mathbf{P}_{21}u^2 - 3\mathbf{P}_{23}u^2 - \mathbf{P}_{31}u^2 + \mathbf{P}_{33}u^2)/4.\end{aligned} \quad (5.62)$$

Equation (5.62) shows that $\mathbf{P}^t(u,1)$ can be obtained from $\mathbf{P}^t(u,0)$ by incrementing the second index of every control point involved. Equation (5.63) illustrates the

5 Surfaces

same property for the second derivatives, thereby showing second-order continuity:

$$\begin{aligned}
\mathbf{P}^{tt}(u,0) = &\mathbf{P}_{00} - 2\mathbf{P}_{01} + \mathbf{P}_{02} - 2\mathbf{P}_{10} + 4\mathbf{P}_{11} - 2\mathbf{P}_{12} + \mathbf{P}_{20} \\
&- 2\mathbf{P}_{21} + \mathbf{P}_{22} - \mathbf{P}_{00}u + 2\mathbf{P}_{01}u - \mathbf{P}_{02}u + 3\mathbf{P}_{10}u - 6\mathbf{P}_{11}u \\
&+ 3\mathbf{P}_{12}u - 3\mathbf{P}_{20}u + 6\mathbf{P}_{21}u - 3\mathbf{P}_{22}u + \mathbf{P}_{30}u - 2\mathbf{P}_{31}u + \mathbf{P}_{32}u, \\
\mathbf{P}^{tt}(u,1) = &\mathbf{P}_{01} - 2\mathbf{P}_{02} + \mathbf{P}_{03} - 2\mathbf{P}_{11} + 4\mathbf{P}_{12} - 2\mathbf{P}_{13} + \mathbf{P}_{21} \\
&- 2\mathbf{P}_{22} + \mathbf{P}_{23} - \mathbf{P}_{01}u + 2\mathbf{P}_{02}u - \mathbf{P}_{03}u + 3\mathbf{P}_{11}u - 6\mathbf{P}_{12}u \\
&+ 3\mathbf{P}_{13}u - 3\mathbf{P}_{21}u + 6\mathbf{P}_{22}u - 3\mathbf{P}_{23}u + \mathbf{P}_{31}u - 2\mathbf{P}_{32}u + \mathbf{P}_{33}u.
\end{aligned} \tag{5.63}$$

5.17.1 Relation to Other Surfaces

The short discussion that follows shows how the uniform bicubic B-spline surface patch can be expressed as either a bicubic Coons or a bicubic Bézier patch.

Bicubic Coons and B-Spline Patches: A bicubic B-spline surface patch can be written in the form of a bicubic Coons patch. That patch (Equation (5.34), duplicated below) is defined in terms of four corner points, eight tangent vectors, and four twist vectors. These 16 quantities (the elements of matrix $\mathbf{C}$ below) can be expressed in terms of the 16 control points $\mathbf{P}_{ij}$ defining the B-spline patch. The idea is to equate the expression for the Coons surface

$$\mathbf{Q}(u,w) = (u^3, u^2, u, 1)\mathbf{H} \begin{pmatrix} \mathbf{Q}_{00} & \mathbf{Q}_{01} & \mathbf{Q}_{00}^w & \mathbf{Q}_{01}^w \\ \mathbf{Q}_{10} & \mathbf{Q}_{11} & \mathbf{Q}_{10}^w & \mathbf{Q}_{11}^w \\ \mathbf{Q}_{00}^u & \mathbf{Q}_{01}^u & \mathbf{Q}_{00}^{uw} & \mathbf{Q}_{01}^{uw} \\ \mathbf{Q}_{10}^u & \mathbf{Q}_{11}^u & \mathbf{Q}_{10}^{uw} & \mathbf{Q}_{11}^{uw} \end{pmatrix} \mathbf{H}^T \begin{pmatrix} w^3 \\ w^2 \\ w \\ 1 \end{pmatrix} = \mathbf{UHCH}^T\mathbf{W}^T, \tag{5.34}$$

with that of the B-spline surface, Equation (5.59), and solve for the 16 elements of matrix $\mathbf{C}$. This process is straightforward and the solutions are

$$\mathbf{Q}_{00} = \frac{1}{6}\left(\frac{\mathbf{P}_{00}}{6} + \frac{4\mathbf{P}_{10}}{6} + \frac{\mathbf{P}_{20}}{6}\right) + \frac{4}{6}\left(\frac{\mathbf{P}_{01}}{6} + \frac{4\mathbf{P}_{11}}{6} + \frac{\mathbf{P}_{21}}{6}\right) + \frac{1}{6}\left(\frac{\mathbf{P}_{02}}{6} + \frac{4\mathbf{P}_{12}}{6} + \frac{\mathbf{P}_{22}}{6}\right),$$

$$\mathbf{Q}_{01} = \frac{1}{6}\left(\frac{\mathbf{P}_{01}}{6} + \frac{4\mathbf{P}_{11}}{6} + \frac{\mathbf{P}_{21}}{6}\right) + \frac{4}{6}\left(\frac{\mathbf{P}_{02}}{6} + \frac{4\mathbf{P}_{12}}{6} + \frac{\mathbf{P}_{22}}{6}\right) + \frac{1}{6}\left(\frac{\mathbf{P}_{03}}{6} + \frac{4\mathbf{P}_{13}}{6} + \frac{\mathbf{P}_{23}}{6}\right),$$

$$\mathbf{Q}_{10} = \frac{1}{6}\left(\frac{\mathbf{P}_{10}}{6} + \frac{4\mathbf{P}_{20}}{6} + \frac{\mathbf{P}_{30}}{6}\right) + \frac{4}{6}\left(\frac{\mathbf{P}_{11}}{6} + \frac{4\mathbf{P}_{21}}{6} + \frac{\mathbf{P}_{31}}{6}\right) + \frac{1}{6}\left(\frac{\mathbf{P}_{12}}{6} + \frac{4\mathbf{P}_{22}}{6} + \frac{\mathbf{P}_{32}}{6}\right),$$

$$\mathbf{Q}_{11} = \frac{1}{6}\left(\frac{\mathbf{P}_{11}}{6} + \frac{4\mathbf{P}_{21}}{6} + \frac{\mathbf{P}_{31}}{6}\right) + \frac{4}{6}\left(\frac{\mathbf{P}_{12}}{6} + \frac{4\mathbf{P}_{22}}{6} + \frac{\mathbf{P}_{32}}{6}\right) + \frac{1}{6}\left(\frac{\mathbf{P}_{13}}{6} + \frac{4\mathbf{P}_{23}}{6} + \frac{\mathbf{P}_{33}}{6}\right),$$

$$\mathbf{Q}_{00}^u = \frac{1}{6}\left(\frac{\mathbf{P}_{20} - \mathbf{P}_{00}}{2}\right) + \frac{4}{6}\left(\frac{\mathbf{P}_{21} - \mathbf{P}_{01}}{2}\right) + \frac{1}{6}\left(\frac{\mathbf{P}_{22} - \mathbf{P}_{02}}{2}\right),$$

$$\mathbf{Q}_{01}^u = \frac{1}{6}\left(\frac{\mathbf{P}_{21} - \mathbf{P}_{01}}{2}\right) + \frac{4}{6}\left(\frac{\mathbf{P}_{22} - \mathbf{P}_{02}}{2}\right) + \frac{1}{6}\left(\frac{\mathbf{P}_{23} - \mathbf{P}_{03}}{2}\right),$$

$$\mathbf{Q}_{10}^u = \frac{1}{6}\left(\frac{\mathbf{P}_{30} - \mathbf{P}_{10}}{2}\right) + \frac{4}{6}\left(\frac{\mathbf{P}_{31} - \mathbf{P}_{11}}{2}\right) + \frac{1}{6}\left(\frac{\mathbf{P}_{32} - \mathbf{P}_{12}}{2}\right),$$

$$\mathbf{Q}_{11}^u = \frac{1}{6}\left(\frac{\mathbf{P}_{31} - \mathbf{P}_{11}}{2}\right) + \frac{4}{6}\left(\frac{\mathbf{P}_{32} - \mathbf{P}_{12}}{2}\right) + \frac{1}{6}\left(\frac{\mathbf{P}_{33} - \mathbf{P}_{13}}{2}\right),$$

$$\mathbf{Q}_{00}^{w} = \frac{1}{2}\left(\frac{\mathbf{P}_{02}}{6} + \frac{4\mathbf{P}_{12}}{6} + \frac{\mathbf{P}_{22}}{6}\right) - \frac{1}{2}\left(\frac{\mathbf{P}_{00}}{6} + \frac{4\mathbf{P}_{10}}{6} + \frac{\mathbf{P}_{20}}{6}\right),$$

$$\mathbf{Q}_{01}^{w} = \frac{1}{2}\left(\frac{\mathbf{P}_{03}}{6} + \frac{4\mathbf{P}_{13}}{6} + \frac{\mathbf{P}_{23}}{6}\right) - \frac{1}{2}\left(\frac{\mathbf{P}_{01}}{6} + \frac{4\mathbf{P}_{11}}{6} + \frac{\mathbf{P}_{21}}{6}\right),$$

$$\mathbf{Q}_{10}^{w} = \frac{1}{2}\left(\frac{\mathbf{P}_{12}}{6} + \frac{4\mathbf{P}_{22}}{6} + \frac{\mathbf{P}_{32}}{6}\right) - \frac{1}{2}\left(\frac{\mathbf{P}_{10}}{6} + \frac{4\mathbf{P}_{20}}{6} + \frac{\mathbf{P}_{30}}{6}\right),$$

$$\mathbf{Q}_{11}^{w} = \frac{1}{2}\left(\frac{\mathbf{P}_{13}}{6} + \frac{4\mathbf{P}_{23}}{6} + \frac{\mathbf{P}_{33}}{6}\right) - \frac{1}{2}\left(\frac{\mathbf{P}_{11}}{6} + \frac{4\mathbf{P}_{21}}{6} + \frac{\mathbf{P}_{31}}{6}\right),$$

$$\mathbf{Q}_{00}^{uw} = \frac{1}{2}\left(\frac{\mathbf{P}_{22} - \mathbf{P}_{02}}{2}\right) - \frac{1}{2}\left(\frac{\mathbf{P}_{20} - \mathbf{P}_{00}}{2}\right),$$

$$\mathbf{Q}_{01}^{uw} = \frac{1}{2}\left(\frac{\mathbf{P}_{23} - \mathbf{P}_{03}}{2}\right) - \frac{1}{2}\left(\frac{\mathbf{P}_{21} - \mathbf{P}_{01}}{2}\right),$$

$$\mathbf{Q}_{10}^{uw} = \frac{1}{2}\left(\frac{\mathbf{P}_{32} - \mathbf{P}_{12}}{2}\right) - \frac{1}{2}\left(\frac{\mathbf{P}_{30} - \mathbf{P}_{10}}{2}\right),$$

$$\mathbf{Q}_{11}^{uw} = \frac{1}{2}\left(\frac{\mathbf{P}_{33} - \mathbf{P}_{13}}{2}\right) - \frac{1}{2}\left(\frac{\mathbf{P}_{31} - \mathbf{P}_{11}}{2}\right).$$

Bézier and B-Spline Bicubic Patches: A bicubic B-spline surface patch can also be written in the form of a bicubic Bézier patch. The bicubic Bézier patch is fully defined by 16 control points $\mathbf{Q}_{ij}$ (the elements of matrix $\mathbf{P}$ of Equation (5.40)). They can be expressed in terms of the 16 control points $\mathbf{P}_{ij}$ defining the B-spline patch. The idea is to equate the expressions for the bicubic Bézier and B-spline surface patches and solve for the elements of matrix $\mathbf{P}$. The solutions are

$$\mathbf{Q}_{00} = \frac{1}{6}\left(\frac{\mathbf{P}_{00}}{6} + \frac{4\mathbf{P}_{10}}{6} + \frac{\mathbf{P}_{20}}{6}\right) + \frac{4}{6}\left(\frac{\mathbf{P}_{01}}{6} + \frac{4\mathbf{P}_{11}}{6} + \frac{\mathbf{P}_{21}}{6}\right) + \frac{1}{6}\left(\frac{\mathbf{P}_{02}}{6} + \frac{4\mathbf{P}_{12}}{6} + \frac{\mathbf{P}_{22}}{6}\right),$$

$$\mathbf{Q}_{01} = \frac{4}{6}\left(\frac{\mathbf{P}_{01}}{6} + \frac{4\mathbf{P}_{11}}{6} + \frac{\mathbf{P}_{21}}{6}\right) + \frac{2}{6}\left(\frac{\mathbf{P}_{02}}{6} + \frac{4\mathbf{P}_{12}}{6} + \frac{\mathbf{P}_{32}}{6}\right),$$

$$\mathbf{Q}_{02} = \frac{2}{6}\left(\frac{\mathbf{P}_{01}}{6} + \frac{4\mathbf{P}_{11}}{6} + \frac{\mathbf{P}_{21}}{6}\right) + \frac{4}{6}\left(\frac{\mathbf{P}_{02}}{6} + \frac{4\mathbf{P}_{12}}{6} + \frac{\mathbf{P}_{32}}{6}\right),$$

$$\mathbf{Q}_{03} = \frac{1}{6}\left(\frac{\mathbf{P}_{01}}{6} + \frac{4\mathbf{P}_{11}}{6} + \frac{\mathbf{P}_{21}}{6}\right) + \frac{4}{6}\left(\frac{\mathbf{P}_{02}}{6} + \frac{4\mathbf{P}_{12}}{6} + \frac{\mathbf{P}_{22}}{6}\right) + \frac{1}{6}\left(\frac{\mathbf{P}_{03}}{6} + \frac{4\mathbf{P}_{13}}{6} + \frac{\mathbf{P}_{23}}{6}\right),$$

$$\mathbf{Q}_{10} = \frac{1}{6}\left(\frac{4\mathbf{P}_{10} + 2\mathbf{P}_{20}}{6}\right) + \frac{4}{6}\left(\frac{4\mathbf{P}_{11} + 2\mathbf{P}_{21}}{6}\right) + \frac{1}{6}\left(\frac{4\mathbf{P}_{12} + 2\mathbf{P}_{22}}{6}\right),$$

$$\mathbf{Q}_{11} = \frac{4}{6}\left(\frac{4\mathbf{P}_{11} + 2\mathbf{P}_{21}}{6}\right) + \frac{2}{6}\left(\frac{4\mathbf{P}_{12} + 2\mathbf{P}_{22}}{6}\right),$$

$$\mathbf{Q}_{12} = \frac{2}{6}\left(\frac{4\mathbf{P}_{11} + 2\mathbf{P}_{21}}{6}\right) + \frac{4}{6}\left(\frac{4\mathbf{P}_{12} + 2\mathbf{P}_{22}}{6}\right),$$

$$\mathbf{Q}_{13} = \frac{1}{6}\left(\frac{4\mathbf{P}_{11} + 2\mathbf{P}_{21}}{6}\right) + \frac{4}{6}\left(\frac{4\mathbf{P}_{12} + 2\mathbf{P}_{22}}{6}\right) + \frac{1}{6}\left(\frac{4\mathbf{P}_{13} + 2\mathbf{P}_{23}}{6}\right),$$

5 Surfaces

$$Q_{20} = \frac{1}{6}\left(\frac{4\mathbf{P}_{10} + 2\mathbf{P}_{20}}{6}\right) + \frac{4}{6}\left(\frac{4\mathbf{P}_{11} + 2\mathbf{P}_{21}}{6}\right) + \frac{1}{6}\left(\frac{4\mathbf{P}_{12} + 2\mathbf{P}_{22}}{6}\right),$$

$$Q_{21} = \frac{4}{6}\left(\frac{2\mathbf{P}_{11} + 4\mathbf{P}_{21}}{6}\right) + \frac{2}{6}\left(\frac{2\mathbf{P}_{12} + 4\mathbf{P}_{22}}{6}\right),$$

$$Q_{22} = \frac{2}{6}\left(\frac{2\mathbf{P}_{11} + 4\mathbf{P}_{21}}{6}\right) + \frac{4}{6}\left(\frac{2\mathbf{P}_{12} + 4\mathbf{P}_{22}}{6}\right),$$

$$Q_{23} = \frac{1}{6}\left(\frac{2\mathbf{P}_{11} + 4\mathbf{P}_{21}}{6}\right) + \frac{4}{6}\left(\frac{2\mathbf{P}_{12} + 4\mathbf{P}_{22}}{6}\right) + \frac{1}{6}\left(\frac{2\mathbf{P}_{13} + 4\mathbf{P}_{23}}{6}\right),$$

$$Q_{30} = \frac{1}{6}\left(\frac{\mathbf{P}_{10}}{6} + \frac{4\mathbf{P}_{20}}{6} + \frac{\mathbf{P}_{30}}{6}\right) + \frac{4}{6}\left(\frac{\mathbf{P}_{11}}{6} + \frac{4\mathbf{P}_{21}}{6} + \frac{\mathbf{P}_{31}}{6}\right) + \frac{1}{6}\left(\frac{\mathbf{P}_{12}}{6} + \frac{4\mathbf{P}_{22}}{6} + \frac{\mathbf{P}_{32}}{6}\right),$$

$$Q_{31} = \frac{4}{6}\left(\frac{\mathbf{P}_{11} + 4\mathbf{P}_{21} + \mathbf{P}_{31}}{6}\right) + \frac{2}{6}\left(\frac{\mathbf{P}_{12} + 4\mathbf{P}_{22} + \mathbf{P}_{32}}{6}\right),$$

$$Q_{32} = \frac{2}{6}\left(\frac{\mathbf{P}_{11} + 4\mathbf{P}_{21} + \mathbf{P}_{31}}{6}\right) + \frac{4}{6}\left(\frac{\mathbf{P}_{12} + 4\mathbf{P}_{22} + \mathbf{P}_{32}}{6}\right),$$

$$Q_{33} = \frac{1}{6}\left(\frac{\mathbf{P}_{11}}{6} + \frac{4\mathbf{P}_{21}}{6} + \frac{\mathbf{P}_{31}}{6}\right) + \frac{4}{6}\left(\frac{\mathbf{P}_{12}}{6} + \frac{4\mathbf{P}_{22}}{6} + \frac{\mathbf{P}_{32}}{6}\right) + \frac{1}{6}\left(\frac{\mathbf{P}_{13}}{6} + \frac{4\mathbf{P}_{23}}{6} + \frac{\mathbf{P}_{33}}{6}\right).$$

5.17.2 An Interpolating Bicubic Patch

The uniform bicubic B-spline surface patch is defined by 16 control points. A mesh of $(m+1) \times (n+1)$ control points can be used to calculate $(m-2) \times (n-2)$ such patches. Each patch has four corner points, but since the patches are connected, the total number of joint points is $(m-1) \times (n-1)$. This section shows how to solve the opposite problem, namely given a mesh of $(m-1) \times (n-1)$ data points $\mathbf{Q}_{1,1}$ through $\mathbf{Q}_{m-1,n-1}$, how to calculate the bicubic B-spline surface that passes through them.

The given data points $\mathbf{Q}_{ij}$ are considered the joint points of the unknown surface and Equation (5.61) shows how they are related to the (yet unknown) control points $\mathbf{P}_{00}$ through $\mathbf{P}_{mn}$:

$$\begin{aligned}\mathbf{Q}_{ij} = \frac{1}{6}\Big[&\frac{1}{6}(\mathbf{P}_{i-1,j-1} + 4\mathbf{P}_{i,j-1} + \mathbf{P}_{i+1,j-1}) \\ &+ \frac{4}{6}(\mathbf{P}_{i-1,j} + 4\mathbf{P}_{i,j} + \mathbf{P}_{i+1,j}) \\ &+ \frac{1}{6}(\mathbf{P}_{i-1,j+1} + 4\mathbf{P}_{i,j+1} + \mathbf{P}_{i+1,j+1})\Big].\end{aligned} \quad (5.64)$$

Equation (5.64) can be written $(m-1) \times (n-1)$ times, once for each given data point $\mathbf{Q}_{ij}$. The number of equations needed, however, is $(m+1) \times (n+1)$. We use the relation

$$(m+1) \times (n+1) = (m-1) \times (n-1) + 2m + 2n,$$

to figure out how many more equations are needed. The extra equations are obtained by the user specifying the vectors shown in Figure 5.50. There are $m-1$

vectors going from boundary control points $\mathbf{P}_{i,0}$ to the "bottom" data points $\mathbf{Q}_{i1}$. There are $m-1$ more such vectors going from the boundary control points $\mathbf{P}_{i,n+1}$ to the "top" data points $\mathbf{Q}_{i,n}$. In addition, there are $2(n-1)$ vectors going from the "left" and "right" boundary control points to the extreme data points $\mathbf{Q}_{1,j}$ and $\mathbf{Q}_{m-1,j}$. Finally, there are four vectors going from the four corner control points to the four corner data points. Once all $2(n-1)+2(m-1)+4$ vectors have been specified, a system of $(m+1) \times (n+1)$ linear equations can be set and solved, to yield the control points.

If the surface should be closed along one dimension, some of the vectors don't have to be specified. For example, if the surface of Figure 5.50 should be closed in the vertical direction (i.e., if it should resemble a horizontal cylinder), then the bottom row of control points $\mathbf{P}_{i,0}$ should be duplicated and renamed $\mathbf{P}_{i,4}$, and the top row $\mathbf{P}_{i,3}$ should be duplicated and renamed $\mathbf{P}_{i,-1}$. Two extra rows of surface patches should be calculated, but every patch now has control points above and below it, so the $2(m-1)$ vertical vectors need not be specified by the user.

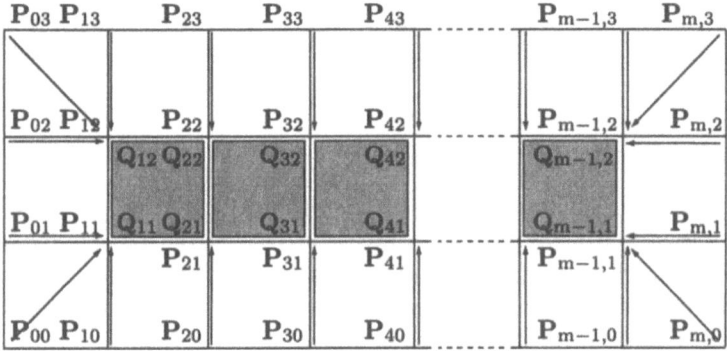

Figure 5.50: An Interpolating B-Spline Surface.

5.17.3 The Quadratic-Cubic B-Spline Surface

This type of surface patch is defined by a 3×4 mesh of control points and its expression is a Cartesian product of the quadratic and cubic B-spline curves:

$$\mathbf{P}(u,w) = \left(\frac{1}{2}\right)\left(\frac{1}{6}\right)(u^2, u, 1)\begin{pmatrix} 1 & -2 & 1 \\ -2 & 2 & 0 \\ 1 & 1 & 0 \end{pmatrix}$$

$$\times \begin{pmatrix} \mathbf{P}_{00} & \mathbf{P}_{01} & \mathbf{P}_{02} & \mathbf{P}_{03} \\ \mathbf{P}_{10} & \mathbf{P}_{11} & \mathbf{P}_{12} & \mathbf{P}_{13} \\ \mathbf{P}_{20} & \mathbf{P}_{21} & \mathbf{P}_{22} & \mathbf{P}_{23} \end{pmatrix} \begin{pmatrix} -1 & 3 & -3 & 1 \\ 3 & -6 & 3 & 0 \\ -3 & 0 & 3 & 0 \\ 1 & 4 & 1 & 0 \end{pmatrix}^T \begin{pmatrix} w^3 \\ w^2 \\ w \\ 1 \end{pmatrix}.$$

Figure 5.51 is an example.

5 Surfaces

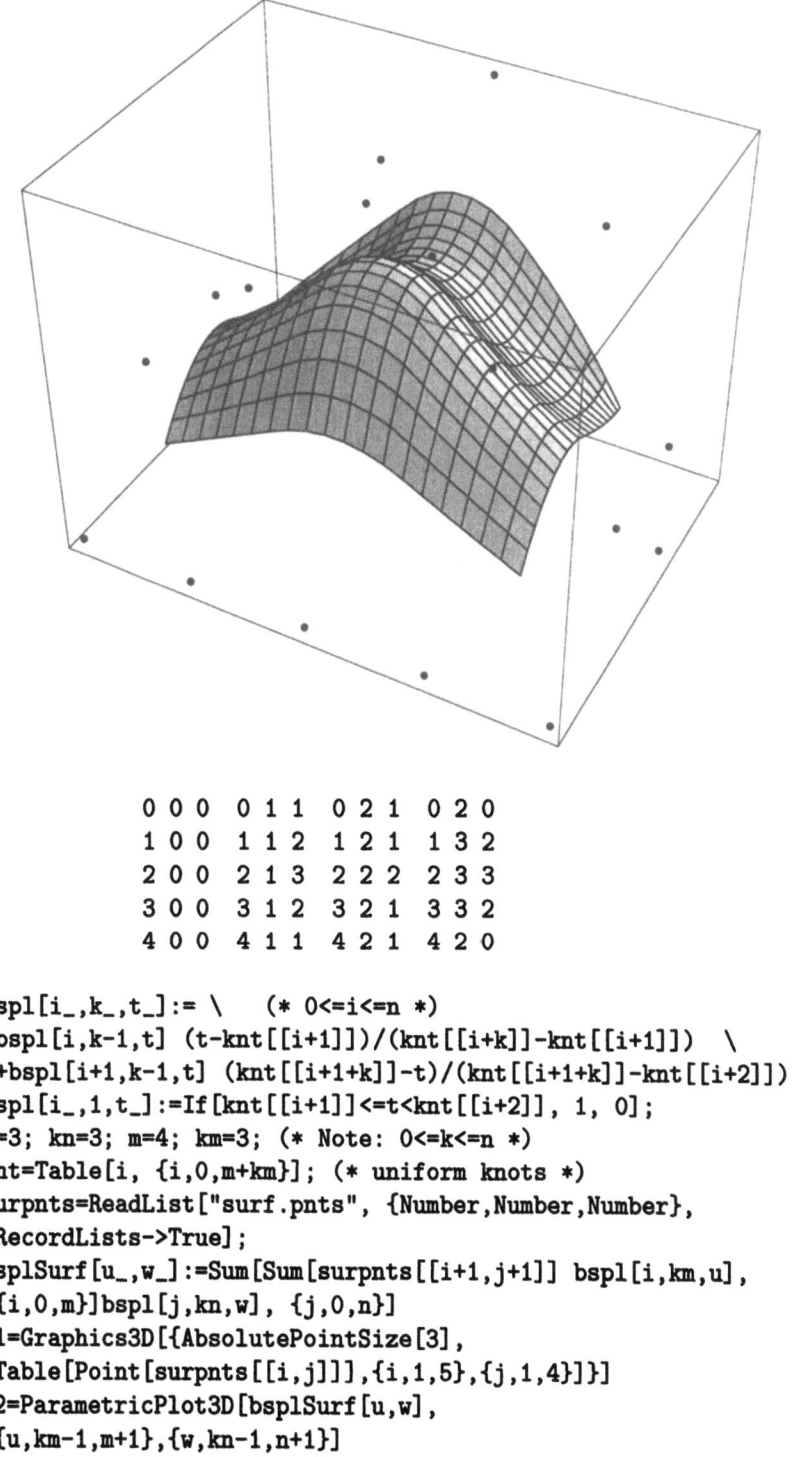

```
0 0 0    0 1 1    0 2 1    0 2 0
1 0 0    1 1 2    1 2 1    1 3 2
2 0 0    2 1 3    2 2 2    2 3 3
3 0 0    3 1 2    3 2 1    3 3 2
4 0 0    4 1 1    4 2 1    4 2 0
```

```
bspl[i_,k_,t_]:= \    (* 0<=i<=n *)
 bspl[i,k-1,t] (t-knt[[i+1]])/(knt[[i+k]]-knt[[i+1]])  \
 +bspl[i+1,k-1,t] (knt[[i+1+k]]-t)/(knt[[i+1+k]]-knt[[i+2]])
bspl[i_,1,t_]:=If[knt[[i+1]]<=t<knt[[i+2]], 1, 0];
n=3; kn=3; m=4; km=3; (* Note: 0<=k<=n *)
knt=Table[i, {i,0,m+km}]; (* uniform knots *)
surpnts=ReadList["surf.pnts", {Number,Number,Number},
 RecordLists->True];
bsplSurf[u_,w_]:=Sum[Sum[surpnts[[i+1,j+1]] bspl[i,km,u],
 {i,0,m}]bspl[j,kn,w], {j,0,n}]
g1=Graphics3D[{AbsolutePointSize[3],
 Table[Point[surpnts[[i,j]]],{i,1,5},{j,1,4}]}]
g2=ParametricPlot3D[bsplSurf[u,w],
 {u,km-1,m+1},{w,kn-1,n+1}]
Show[g1,g2, PlotRange->All]
```

Figure 5.51: A Quadratic-Cubic B-Spline Surface Patch.

5.17.4 Biquadratic B-Spline Surface by Subdivision

The general idea of subdivision has been introduced in Section 4.15, where Chaikin's algorithm for curves is discussed. Generating the quadratic B-spline curve by subdivision is described in Section 4.16.3. This material should be reviewed before reading ahead. The idea of subdivision can be extended to those surfaces that are defined by a mesh of control points. We use the biquadratic B-spline surface patch as an example. Such a patch is constructed by Equation (5.57) from a mesh of 3×3 control points $\mathbf{P}_{ij}$. We denote this patch by BSP and the original points by $\mathbf{P}^0_{ij}$. The principle of constructing BSP by subdivision is to find a way to subdivide the mesh of original points into a finer mesh with more points $\mathbf{P}^1_{ij}$ and, as a result, with more subpatches. If this is done right, the new control points $\mathbf{P}^1_{ij}$ will be closer to the ideal surface BSP than the original ones. When this process is repeated, we get more and more control points $\mathbf{P}^k_{ij}$ that get closer and closer to BSP. At the limit, we end up with infinitely many points, each of which lies on the surface. In practice, we stop the subdivision process after a finite number k of steps and generate the surface by connecting points $\mathbf{P}^k_{ij}$ with straight segments.

The rule for subdividing a biquadratic B-spline surface patch $\mathbf{P}(u,w)$ uses reparametrization to calculate four new patches $\mathbf{Q}(u,w)$. The idea of reparametrization has been introduced in Section 4.14.12 for curves and has been extended for Bézier surface patches in Section 5.14.1. It can easily be modified for the biquadratic B-spline surface by rewriting Equation (5.56) in the form

$$\mathbf{Q}(u,w) = \mathbf{P}([b-a]u+a, [d-c]w+c)$$

$$= (([b-a]u+a)^2, ([b-a]u+a), 1)\mathbf{M} \cdot \mathbf{P} \cdot \mathbf{M}^{-1} \begin{pmatrix} ([d-c]w+c)^2 \\ [d-c]w+c \\ 1 \end{pmatrix}$$

$$= (u^2, u, 1)\mathbf{A}_{ab}\mathbf{M} \cdot \mathbf{P} \cdot \mathbf{M}^T \cdot \mathbf{A}^T_{cd}(w^2, w, 1)^T$$
$$= (u^2, u, 1)\mathbf{M}(\mathbf{M}^{-1} \cdot \mathbf{A}_{ab} \cdot \mathbf{M})\mathbf{P}(\mathbf{M}^T \cdot \mathbf{A}^T_{cd} \cdot (\mathbf{M}^T)^{-1})\mathbf{M}^T(w^2, w, 1)^T$$
$$= (u^2, u, 1)\mathbf{M} \cdot \mathbf{B}_{ab} \cdot \mathbf{P} \cdot \mathbf{B}^T_{cd} \cdot \mathbf{M}^T(w^2, w, 1)^T$$
$$= (u^2, u, 1)\mathbf{M} \cdot \mathbf{Q} \cdot \mathbf{M}^T(w^2, w, 1)^T,$$

where

$$\mathbf{M} = \begin{pmatrix} 1 & -2 & 1 \\ -2 & 2 & 1 \\ 1 & 0 & 0 \end{pmatrix}, \quad \mathbf{A}_{ab} = \begin{pmatrix} (b-a)^2 & 0 & 0 \\ 2a(b-a) & b-a & 0 \\ a^2 & a & 1 \end{pmatrix}$$

($\mathbf{M}$ is the basis matrix for the biquadratic B-spline surface),

$$\mathbf{P} = \begin{pmatrix} \mathbf{P}_{00} & \mathbf{P}_{01} & \mathbf{P}_{02} \\ \mathbf{P}_{10} & \mathbf{P}_{11} & \mathbf{P}_{12} \\ \mathbf{P}_{20} & \mathbf{P}_{21} & \mathbf{P}_{22} \end{pmatrix},$$

$$\mathbf{B}_{ab} = \mathbf{M}^{-1} \cdot \mathbf{A}_{ab} \cdot \mathbf{M}$$
$$= \begin{pmatrix} ((1-a)(1-2a+b))/2 & (1+3a-4a^2-b+2ab)/2 & a^2-(ab)/2 \\ 1/2-a/2-b/2+(ab)/2 & (1+a+b-2ab)/2 & (ab)/2 \\ ((1+a-2b)(1-b))/2 & (1-a+3b+2ab-4b^2)/2 & -(ab)/2+b^2 \end{pmatrix},$$

5 Surfaces

$$\mathbf{B}_{cd}^T = \mathbf{M}^T \cdot \mathbf{A}_{cd}^T \cdot (\mathbf{M}^T)^{-1}$$

$$= \begin{pmatrix} ((1-c)(1-2c+d))/2 & 1/2-c/2-d/2+(cd)/2 & ((1+c-2d)(1-d))/2 \\ (1+3c-4c^2-d+2cd)/2 & (1+c+d-2cd)/2 & (1-c+3d+2cd-4d^2)/2 \\ c^2-(cd)/2 & (cd)/2 & -(cd)/2+d^2 \end{pmatrix},$$

and

$$\mathbf{Q} = \mathbf{B}_{ab} \cdot \mathbf{P} \cdot \mathbf{B}_{cd}^T. \tag{5.65}$$

The elements of $\mathbf{Q}$ depend on the four parameters a, b, c, and d, and on the $\mathbf{P}_{ij}$'s. Once the four parameters are known, matrix $\mathbf{Q}$ is easy to calculate symbolically by appropriate mathematical software.

The rule for subdividing a biquadratic B-spline surface patch $\mathbf{P}(u,w)$ is as follows: Use reparametrization to calculate the four surface patches defined by the following sets of parameters:

$$a=0, \ b=0.5, \ c=0, \ d=0.5, \quad a=0.5, \ b=1, \ c=0, \ d=0.5,$$
$$a=0, \ b=0.5, \ c=0.5, \ d=1, \quad a=0.5, \ b=1, \ c=0.5, \ d=1.$$

The basic idea is shown in idealized form in Figure 5.53a. Each of the new patches is defined by 9 points, but some of the new points are identical, so the 4 patches are fully defined by 16 points. The first of the four patches is constructed by setting $a=0$, $b=0.5$, $c=0$, $d=0.5$ (this is, thus, a reparametrization of the "upper left" quarter of the original surface patch) and then applying Equation (5.65). The resulting nine control points $\mathbf{P}_{ij}^1$ are

$$\begin{aligned}
\mathbf{P}_{00}^1 &= \frac{1}{16}(9\mathbf{P}_{00}^0 + 3\mathbf{P}_{10}^0 + 3\mathbf{P}_{01}^0 + \mathbf{P}_{11}^0), \\
\mathbf{P}_{01}^1 &= \frac{1}{16}(3\mathbf{P}_{00}^0 + \mathbf{P}_{10}^0 + 9\mathbf{P}_{01}^0 + 3\mathbf{P}_{11}^0), \\
\mathbf{P}_{02}^1 &= \frac{1}{16}(9\mathbf{P}_{01}^0 + 3\mathbf{P}_{11}^0 + 3\mathbf{P}_{02}^0 + \mathbf{P}_{12}^0), \\
\mathbf{P}_{10}^1 &= \frac{1}{16}(3\mathbf{P}_{00}^0 + 9\mathbf{P}_{10}^0 + \mathbf{P}_{01}^0 + 3\mathbf{P}_{11}^0), \\
\mathbf{P}_{11}^1 &= \frac{1}{16}(\mathbf{P}_{00}^0 + 3\mathbf{P}_{10}^0 + 3\mathbf{P}_{01}^0 + 9\mathbf{P}_{11}^0), \qquad (5.66) \\
\mathbf{P}_{12}^1 &= \frac{1}{16}(3\mathbf{P}_{01}^0 + 9\mathbf{P}_{11}^0 + \mathbf{P}_{02}^0 + 3\mathbf{P}_{12}^0), \\
\mathbf{P}_{20}^1 &= \frac{1}{16}(9\mathbf{P}_{10}^0 + 3\mathbf{P}_{20}^0 + 3\mathbf{P}_{11}^0 + \mathbf{P}_{21}^0), \\
\mathbf{P}_{21}^1 &= \frac{1}{16}(3\mathbf{P}_{10}^0 + \mathbf{P}_{20}^0 + 9\mathbf{P}_{11}^0 + 3\mathbf{P}_{21}^0), \\
\mathbf{P}_{22}^1 &= \frac{1}{16}(9\mathbf{P}_{11}^0 + 3\mathbf{P}_{21}^0 + 3\mathbf{P}_{12}^0 + \mathbf{P}_{22}^0).
\end{aligned}$$

These points can be interpreted geometrically in two ways:

1. The original surface patch has four faces and each new control point is located on one of these faces. Such a point is a weighted sum of the four points on

its face where the weights are 9/16, 3/16, 3/16, and 1/16. There are four weight patterns shown in Figure 5.52.

2. Take the two points $\mathbf{P}_{00}^0$ and $\mathbf{P}_{10}^0$ and add them with weights 3/4 and 1/4. Do the same thing with $\mathbf{P}_{01}^0$ and $\mathbf{P}_{11}^0$. Add the two results also with weights 3/4 and 1/4, and this produces point $\mathbf{P}_{00}^1$. Each new point is, thus, the sum of two quantities, each a sum of two points on the same edge, where all the sums use weights of 3/4 and 1/4. Recall that these are the weights used by the original Chaikin's algorithm.

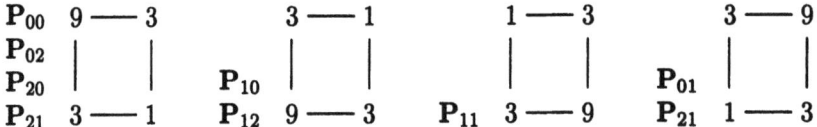

Figure 5.52: The Four Weight Patterns.

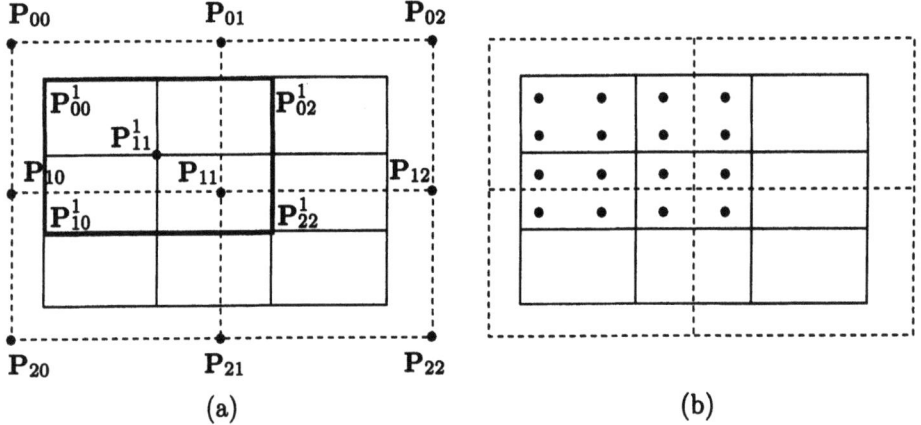

Figure 5.53: The First Two Subdivision Steps.

Using each of the other three sets of parameters to reparametrize the surface results in nine more points, only some of which are new. The total number of points $\mathbf{P}_{ij}^1$ is $9+3+3+1 = 16$, four new points for each of the four faces of the original surface. The new mesh of 16 points is used to calculate $4 \times 9 = 36$ points $\mathbf{P}_{ij}^2$ (4 points for each of the 9 faces, Figure 5.53b) and so on. When enough points have been obtained, the surface can be generated by connecting the points with straight lines. It becomes a polygonal surface made of four-sided polygons (quadrilaterals).

An examination of both parts of Figure 5.53 suggests that the subdivision process produces smaller and smaller meshes of control points, thus generating smaller and smaller surface patches. It is easy to show that this is not so. Let $\mathbf{Q}(u, w)$ denote the reparametrization of the "upper left" quarter of the original surface patch $\mathbf{P}(u, w)$. These two patches are based on different meshes of control points, but we show that they have the same "upper left" corner point, i.e., $\mathbf{Q}(0,0) = \mathbf{P}(0,0)$. The corner point $\mathbf{P}(0,0)$ of a general biquadratic B-spline surface patch

5 Surfaces

$\mathbf{P}(u,w)$ is shown by Equation (5.58) to be

$$\mathbf{P}(0,0) = \frac{1}{4}(\mathbf{P}_{00} + \mathbf{P}_{01} + \mathbf{P}_{10} + \mathbf{P}_{11}).$$

The corner point $\mathbf{Q}(0,0)$ is therefore

$$\begin{aligned}
\mathbf{Q}(0,0) &= \frac{1}{4}(\mathbf{P}_{00}^1 + \mathbf{P}_{01}^1 + \mathbf{P}_{10}^1 + \mathbf{P}_{11}^1) \\
&= \frac{1}{4 \cdot 16}[(9\mathbf{P}_{00}^0 + 3\mathbf{P}_{10}^0 + 3\mathbf{P}_{01}^0 + \mathbf{P}_{11}^0) + (3\mathbf{P}_{00}^0 + \mathbf{P}_{10}^0 + 9\mathbf{P}_{01}^0 + 3\mathbf{P}_{11}^0) \\
&\quad + (3\mathbf{P}_{00}^0 + 9\mathbf{P}_{10}^0 + \mathbf{P}_{01}^0 + 3\mathbf{P}_{11}^0) + (\mathbf{P}_{00}^0 + 3\mathbf{P}_{10}^0 + 3\mathbf{P}_{01}^0 + 9\mathbf{P}_{11}^0) \\
&= \frac{1}{4}(\mathbf{P}_{00}^0 + \mathbf{P}_{01}^0 + \mathbf{P}_{10}^0 + \mathbf{P}_{11}^0)] \\
&= \mathbf{P}(0,0).
\end{aligned}$$

It turns out that even though consecutive steps of the subdivision process result in smaller meshes, those meshes converge to a limit and don't shrink indefinitely.

5.17.5 Bicubic B-Spline Surface by Subdivision

The technique used in this section to subdivide a bicubic B-spline surface patch is similar to the one used in Section 5.17.4 to subdivide the biquadratic B-spline surface.

The rule for subdividing a bicubic B-spline surface patch $\mathbf{P}(u,w)$ uses reparametrization to calculate four new patches $\mathbf{Q}(u,w)$. This is done by rewriting Equation (5.56) in the form

$$\begin{aligned}
\mathbf{Q}(u,w) &= \mathbf{P}([b-a]u + a, [d-c]w + c) \\
&= (([b-a]u+a)^3, ([b-a]u+a)^2, ([b-a]u+a), 1)\mathbf{M} \cdot \mathbf{P} \cdot \mathbf{M}^{-1} \begin{pmatrix} ([d-c]w+c)^3 \\ ([d-c]w+c)^2 \\ [d-c]w+c \\ 1 \end{pmatrix} \\
&= (u^3, u^2, u, 1)\mathbf{A}_{ab}\mathbf{M} \cdot \mathbf{P} \cdot \mathbf{M}^T \cdot \mathbf{A}_{cd}^T (w^3, w^2, w, 1)^T \\
&= (u^3, u^2, u, 1)\mathbf{M}(\mathbf{M}^{-1} \cdot \mathbf{A}_{ab} \cdot \mathbf{M})\mathbf{P}(\mathbf{M}^T \cdot \mathbf{A}_{cd}^T \cdot (\mathbf{M}^T)^{-1})\mathbf{M}^T(w^3, w^2, w, 1)^T \\
&= (u^3, u^2, u, 1)\mathbf{M} \cdot \mathbf{B}_{ab} \cdot \mathbf{P} \cdot \mathbf{B}_{cd}^T \cdot \mathbf{M}^T(w^3, w^2, w, 1)^T \\
&= (u^3, u^2, u, 1)\mathbf{M} \cdot \mathbf{Q} \cdot \mathbf{M}^T(w^3, w^2, w, 1)^T, \quad (5.67)
\end{aligned}$$

where

$$\mathbf{M} = \begin{pmatrix} -1 & 3 & -3 & 1 \\ 3 & -6 & 3 & 0 \\ -3 & 0 & 3 & 0 \\ 1 & 4 & 1 & 0 \end{pmatrix}, \quad \mathbf{A}_{ab} = \begin{pmatrix} (b-a)^2 & 0 & 0 \\ 2a(b-a) & b-a & 0 \\ a^2 & a & 1 \end{pmatrix}$$

5.17 Uniform B-Spline Surfaces

(**M** is the basis matrix for the bicubic B-spline surface),

$$\mathbf{P} = \begin{pmatrix} \mathbf{P}_{00} & \mathbf{P}_{01} & \mathbf{P}_{02} & \mathbf{P}_{03} \\ \mathbf{P}_{10} & \mathbf{P}_{11} & \mathbf{P}_{12} & \mathbf{P}_{13} \\ \mathbf{P}_{20} & \mathbf{P}_{21} & \mathbf{P}_{22} & \mathbf{P}_{23} \\ \mathbf{P}_{30} & \mathbf{P}_{31} & \mathbf{P}_{32} & \mathbf{P}_{33} \end{pmatrix},$$

$\mathbf{B}_{ab} = \mathbf{M}^{-1} \cdot \mathbf{A}_{ab} \cdot \mathbf{M}$

$$= \begin{pmatrix}
((1-a)(1-5a+6a^2+3b-7ab+2b^2))/6 & (4-22a^2+18a^3+20ab-21a^2b-4b^2+6ab^2)/6 & 1/6+a+(11a^2)/6-3a^3-b/2-(5ab)/3+(7a^2b)/2+b^2/3-ab^2 & a^3-(7a^2b)/6+(ab^2)/3 \\
((a-1)(-1+2a-2ab+b^2))/6 & (4-4a^2-4ab+6a^2b+2b^2-3ab^2)/6 & 1/6+a/2+a^2/3+(ab)/3-a^2b-b^2/6+(ab^2)/2 & (a(2a-b)b)/6 \\
((a-1)(1+a-2b)(b-1))/6 & (4+2a^2-4ab-3a^2b-4b^2+6ab^2)/6 & 1/6-a^2/6+b/2+(ab)/3+(a^2b)/2+b^2/3-ab^2 & (ab(-a+2b))/6 \\
((1-b)(1+3a+2a^2-5b-7ab+6b^2))/6 & (4-4a^2+20ab+6a^2b-22b^2-21ab^2+18b^3)/6 & 1/6-a/2+a^2/3+b-(5ab)/3-a^2b+(11b^2)/6+(7ab^2)/2-3b^3 & (a^2b)/3-(7ab^2)/6+b^3
\end{pmatrix},$$

$\mathbf{B}_{cd}^T = \mathbf{M}^T \cdot \mathbf{A}_{cd}^T \cdot (\mathbf{M}^T)^{-1}$

$$= \begin{pmatrix}
((1-c)(1-5c+6c^2+3d-7cd+2d^2))/6 & ((-1+c)(-1+2c-2cd+d^2))/6 & ((-1+c)(1+c-2d)(-1+d))/6 & ((1-d)(1+3c+2c^2-5d-7cd+6d^2))/6 \\
(4-22c^2+18c^3+20cd-21c^2d-4d^2+6cd^2)/6 & (4-4c^2-4cd+6c^2d+2d^2-3cd^2)/6 & (4+2c^2-4cd-3c^2d-4d^2+6cd^2)/6 & (4-4c^2+20cd+6c^2d-22d^2-21cd^2+18d^3)/6 \\
1/6+c+(11c^2)/6-3c^3-d/2-(5cd)/3+(7c^2d)/2+d^2/3-cd^2 & 1/6+c/2+c^2/3+(cd)/3-c^2d-d^2/6+(cd^2)/2 & 1/6-c^2/6+d/2+(cd)/3+(c^2d)/2+d^2/3-cd^2 & 1/6-c/2+c^2/3+d-(5cd)/3-c^2d+(11d^2)/6+(7cd^2)/2-3d^3 \\
c^3-(7c^2d)/6+(cd^2)/3 & (c(2c-d)d)/6 & (cd(-c+2d))/6 & (c^2d)/3-(7cd^2)/6+d^3
\end{pmatrix},$$

5 Surfaces

and
$$\mathbf{Q} = \mathbf{B}_{ab} \cdot \mathbf{P} \cdot \mathbf{B}_{cd}^T. \tag{5.68}$$

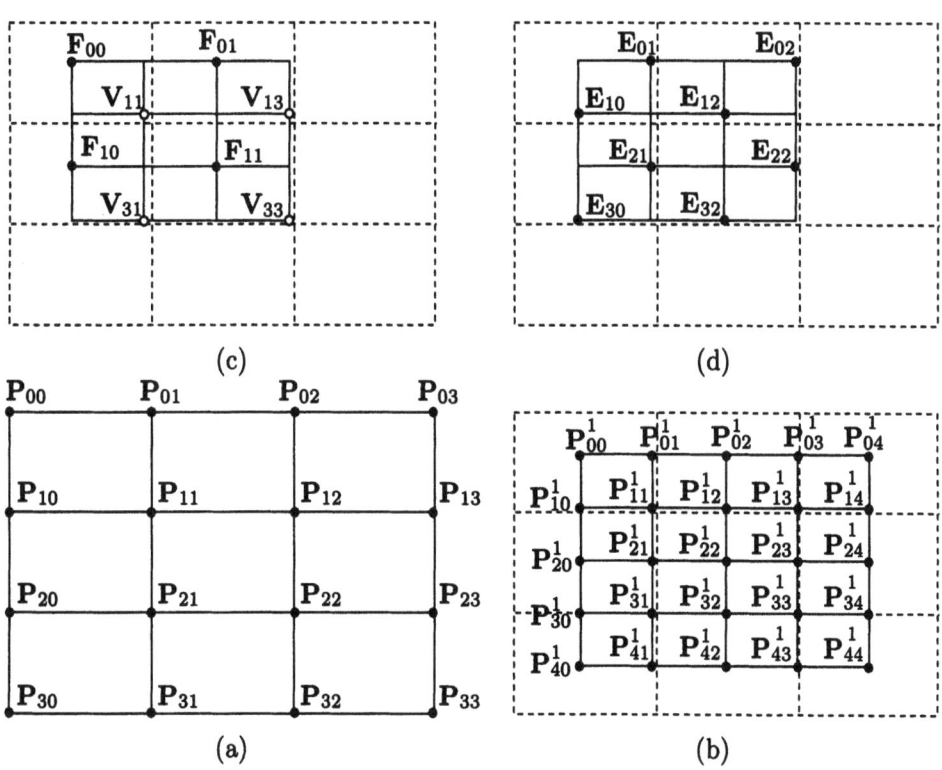

Figure 5.54: The First Subdivision Step.

The rule for subdividing a bicubic B-spline surface patch $\mathbf{P}(u, w)$ is: Use reparametrization to calculate the four surface patches defined by the following sets of parameters:

$$a = 0, \ b = 0.5, \ c = 0, d = 0.5, \quad a = 0.5, \ b = 1, \ c = 0, \ d = 0.5,$$
$$a = 0, \ b = 0.5, \ c = 0.5, \ d = 1, \quad a = 0.5, \ b = 1, \ c = 0.5, \ d = 1.$$

The basic idea is shown in idealized form in Figure 5.54a,b. Each of the new patches is defined by 16 points, but some of the new points are identical, so the 4 patches are fully defined by 25 points. The first of the four patches is constructed by setting $a = 0$, $b = 0.5$, $c = 04$, and $d = .5$ (this is thus a reparametrization of the "upper left" quarter of the original surface patch) and applying Equation (5.68). The resulting 16 control points $\mathbf{P}_{ij}^1$ are

$$\mathbf{P}_{00}^1 = \frac{1}{4}(\mathbf{P}_{00}^0 + \mathbf{P}_{10}^0 + \mathbf{P}_{01}^0 + \mathbf{P}_{11}^0),$$

$$\mathbf{P}^1_{01} = \frac{1}{16}(\mathbf{P}^0_{00} + \mathbf{P}^0_{10} + 6(\mathbf{P}^0_{01} + \mathbf{P}^0_{11}) + \mathbf{P}^0_{02} + \mathbf{P}^0_{12}),$$

$$\mathbf{P}^1_{02} = \frac{1}{4}(\mathbf{P}^0_{01} + \mathbf{P}^0_{11} + \mathbf{P}^0_{02} + \mathbf{P}^0_{12}),$$

$$\mathbf{P}^1_{03} = \frac{1}{16}(\mathbf{P}^0_{01} + \mathbf{P}^0_{11} + 6(\mathbf{P}^0_{02} + \mathbf{P}^0_{12}) + \mathbf{P}^0_{03} + \mathbf{P}^0_{13}),$$

$$\mathbf{P}^1_{01} = \frac{1}{16}(\mathbf{P}^0_{00} + \mathbf{P}^0_{01} + 6(\mathbf{P}^0_{10} + \mathbf{P}^0_{11}) + \mathbf{P}^0_{20} + \mathbf{P}^0_{21}),$$

$$\mathbf{P}^1_{11} = \frac{1}{64}(\mathbf{P}^0_{00} + 6\mathbf{P}^0_{10} + \mathbf{P}^0_{20} + 6(\mathbf{P}^0_{01} + 6\mathbf{P}^0_{11} + \mathbf{P}^0_{21}) + \mathbf{P}^0_{02} + 6\mathbf{P}^0_{12} + \mathbf{P}^0_{22}),$$

$$\mathbf{P}^1_{12} = \frac{1}{16}(\mathbf{P}^0_{01} + \mathbf{P}^0_{12} + 6(\mathbf{P}^0_{11} + \mathbf{P}^0_{12}) + \mathbf{P}^0_{21} + \mathbf{P}^0_{22}),$$

$$\mathbf{P}^1_{13} = \frac{1}{64}(\mathbf{P}^0_{01} + 6\mathbf{P}^0_{11} + \mathbf{P}^0_{21} + 6(\mathbf{P}^0_{02} + 6\mathbf{P}^0_{12} + \mathbf{P}^0_{22}) + \mathbf{P}^0_{03} + 6\mathbf{P}^0_{13} + \mathbf{P}^0_{23}),$$

$$\mathbf{P}^1_{20} = \frac{1}{4}(\mathbf{P}^0_{10} + \mathbf{P}^0_{20} + \mathbf{P}^0_{11} + \mathbf{P}^0_{21}), \tag{5.69}$$

$$\mathbf{P}^1_{21} = \frac{1}{16}(\mathbf{P}^0_{10} + \mathbf{P}^0_{20} + 6(\mathbf{P}^0_{11} + \mathbf{P}^0_{21}) + \mathbf{P}^0_{12} + \mathbf{P}^0_{22}),$$

$$\mathbf{P}^1_{22} = \frac{1}{4}(\mathbf{P}^0_{11} + \mathbf{P}^0_{21} + \mathbf{P}^0_{12} + \mathbf{P}^0_{22}),$$

$$\mathbf{P}^1_{23} = \frac{1}{16}(\mathbf{P}^0_{11} + \mathbf{P}^0_{21} + 6(\mathbf{P}^0_{12} + \mathbf{P}^0_{22}) + \mathbf{P}^0_{13} + \mathbf{P}^0_{23}),$$

$$\mathbf{P}^1_{30} = \frac{1}{16}(\mathbf{P}^0_{10} + \mathbf{P}^0_{11} + 6(\mathbf{P}^0_{20} + \mathbf{P}^0_{21}) + \mathbf{P}^0_{30} + \mathbf{P}^0_{31}),$$

$$\mathbf{P}^1_{31} = \frac{1}{64}(\mathbf{P}^0_{10} + 6\mathbf{P}^0_{20} + \mathbf{P}^0_{30} + 6(\mathbf{P}^0_{11} + 6\mathbf{P}^0_{21} + \mathbf{P}^0_{31}) + \mathbf{P}^0_{12} + 6\mathbf{P}^0_{22} + \mathbf{P}^0_{32}),$$

$$\mathbf{P}^1_{32} = \frac{1}{16}(\mathbf{P}^0_{11} + \mathbf{P}^0_{12} + 6(\mathbf{P}^0_{21} + \mathbf{P}^0_{22}) + \mathbf{P}^0_{31} + \mathbf{P}^0_{32}),$$

$$\mathbf{P}^1_{33} = \frac{1}{64}(\mathbf{P}^0_{11} + 6\mathbf{P}^0_{21} + \mathbf{P}^0_{31} + 6(\mathbf{P}^0_{12} + 6\mathbf{P}^0_{22} + \mathbf{P}^0_{32}) + \mathbf{P}^0_{13} + 6\mathbf{P}^0_{23} + \mathbf{P}^0_{33}).$$

These points can be classified into face points, edge points, and vertex points. The four face points are (Figure 5.54c) $\mathbf{F}_{00} = \mathbf{P}^1_{00}$, $\mathbf{F}_{01} = \mathbf{P}^1_{02}$, $\mathbf{F}_{10} = \mathbf{P}^1_{20}$, and $\mathbf{F}_{11} = \mathbf{P}^1_{22}$. Each is the average of four corner points of one face of the original patch. The eight edge points are (Figure 5.54d)

$$\mathbf{E}_{01} = \mathbf{P}^1_{01}, \quad \mathbf{E}_{02} = \mathbf{P}^1_{03}, \quad \mathbf{E}_{10} = \mathbf{P}^1_{10}, \quad \mathbf{E}_{12} = \mathbf{P}^1_{12},$$
$$\mathbf{E}_{21} = \mathbf{P}^1_{21}, \quad \mathbf{E}_{22} = \mathbf{P}^1_{23}, \quad \mathbf{E}_{30} = \mathbf{P}^1_{30}, \quad \mathbf{E}_{32} = \mathbf{P}^1_{32}.$$

Each is the average of two face points and the two points $\mathbf{P}^0_{ij}$ that are closest to it. The remaining four points are called vertex points. There is one vertex point for each interior vertex of the original mesh. The vertex points are shown in Figure 5.54c and they have the form $\mathbf{V} = (\mathbf{Q} + 2\mathbf{R} + \mathbf{S})/4$, where $\mathbf{S}$ is an interior vertex, $\mathbf{Q}$ is the average of the four face points located on the faces adjacent to $\mathbf{S}$, and $\mathbf{R}$ is the average of the midpoints of the four edges that meet at $\mathbf{S}$. As an example, consider the interior vertex $\mathbf{P}_{11}$ (Figure 5.55). If we denote $\mathbf{S} = \mathbf{P}_{11}$, then $\mathbf{Q}$ is the average

5 Surfaces

of the four face points $\mathbf{F}_{00}$, $\mathbf{F}_{01}$, $\mathbf{F}_{10}$, and $\mathbf{F}_{11}$, and $\mathbf{R}$ is the average of the midpoints of the four edges $\mathbf{P}_{01}\mathbf{P}_{11}$, $\mathbf{P}_{10}\mathbf{P}_{11}$, $\mathbf{P}_{12}\mathbf{P}_{11}$, and $\mathbf{P}_{21}\mathbf{P}_{11}$ (the points labeled × in the figure). This interior vertex therefore corresponds to vertex point

$$\frac{1}{4}(\mathbf{Q} + 2\mathbf{R} + \mathbf{S})$$
$$= \frac{1}{4}(\mathbf{F}_{00} + \mathbf{F}_{01} + \mathbf{F}_{10} + \mathbf{F}_{11})$$
$$+ \frac{2}{4}\left(\frac{\mathbf{P}_{01} + \mathbf{P}_{11}}{2} + \frac{\mathbf{P}_{10} + \mathbf{P}_{11}}{2} + \frac{\mathbf{P}_{12} + \mathbf{P}_{11}}{2} + \frac{\mathbf{P}_{21} + \mathbf{P}_{11}}{2}\right) + \mathbf{P}_{11}$$
$$= \frac{1}{16}((\mathbf{P}_{00} + \mathbf{P}_{10} + \mathbf{P}_{01} + \mathbf{P}_{11}) + (\mathbf{P}_{10} + \mathbf{P}_{20} + \mathbf{P}_{11} + \mathbf{P}_{21})$$
$$+ (\mathbf{P}_{01} + \mathbf{P}_{11} + \mathbf{P}_{02} + \mathbf{P}_{12}) + (\mathbf{P}_{11} + \mathbf{P}_{21} + \mathbf{P}_{12} + \mathbf{P}_{22}))$$
$$+ \frac{1}{4}(\mathbf{P}_{01} + \mathbf{P}_{10} + \mathbf{P}_{21} + \mathbf{P}_{12} + 4\mathbf{P}_{11}) + \mathbf{P}_{11}$$
$$= \frac{1}{16}(\mathbf{P}_{00} + 6\mathbf{P}_{10} + 6\mathbf{P}_{01} + 36\mathbf{P}_{11} + \mathbf{P}_{20} + 6\mathbf{P}_{21} + \mathbf{P}_{02} + 6\mathbf{P}_{12} + \mathbf{P}_{22})$$
$$= \mathbf{P}_{11}^{1}.$$

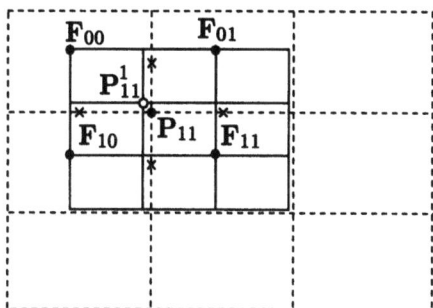

Figure 5.55: Constructing Vertex Point $\mathbf{P}_{11}^{1}$.

Here are the rules for calculating all 25 points $\mathbf{P}_{ij}^{1}$:

1. Construct one face point for each face of the original mesh. This point is the average of all the points defining the face.

2. Construct one edge point for each interior edge of the original mesh. This point is the average of the midpoint of the edge and the two face points of the faces adjacent to the edge.

3. Construct one vertex point for each interior vertex of the original mesh. This point is the average of (1) four face points, (2) four midpoints of edges, and (3) one interior vertex.

Since the original bicubic mesh consists of 9 faces, 12 interior edges, and 4 interior vertices, the first subdivision step results in 9 face points, 12 edge points, and 4 vertex points, a total of 25 points.

It should be noted that even though the mesh resulting from each subdivision is smaller than its predecessor, they don't shrink to a point but converge to a limit. All these meshes define the same bicubic B-spline surface.

▶ **Exercise 5.31:** Equation (5.60) shows that the "top left" corner of a bicubic B-spline patch is given by

$$P(0,0) = \frac{1}{36}(P_{00} + P_{02} + 4P_{10} + 4P_{12} + P_{20} + 4P_{01} + 16P_{11} + 4P_{21} + P_{22}).$$

Show that this is still the same corner of the patch after one subdivision.

5.18 Surfaces of Revolution

Such a surface is obtained when a space curve is rotated a full 360° about some axis in space. The curve is, thus, the *profile* of the surface. If the rotation axis is known, then the rotation matrix $T(\theta)$ about it is given by Equation (3.27). If the space curve is expressed by $P(u)$, where $0 \leq u \leq 1$, then the complete surface has the form $P(u, w) = P(u)T(w)$, where $0 \leq u \leq 1$ and $0 \leq w \leq 2\pi$. Varying u moves us along the curve, whereas changing w moves us in a circle about the rotation axis.

Example: Given the parametric curve $P(u) = (f(u), 0, g(u))$ in the xz plane, we can revolve it around the z axis using the rotation matrix

$$T_z(w) = \begin{pmatrix} \cos w & \sin w & 0 \\ -\sin w & \cos w & 0 \\ 0 & 0 & 1 \end{pmatrix} \quad (5.70)$$

to get the surface

$$P(u)T_z(w) = (f(u)\cos w, f(u)\sin w, g(u)), \quad \text{where } 0 \leq u \leq 1 \text{ and } 0 \leq w \leq 2\pi.$$

Example: Given the five points $P_1 = (0, 1, 0)$, $P_2 = (1, 1, 0)$, $P_3 = (2, 2, 0)$, $P_4 = (1.5, 3, 0)$, and $P_5 = (1.5, 5, 0)$, we calculate $P(u)$ as their Bézier curve

$$P(u) = \left(4t - 6t^3 + 2t^4 + (3/2)t^4, -4(t-1)^3 t + 12(t-1)^2 t^2 - 12(t-1)t^3 + 5t^4 + (t-1)^4, 0\right).$$

Since all the z coordinates are zero, the curve is in the xy plane. We arbitrarily decide to rotate it about the y axis, so the rotation matrix is

$$T_y(w) = \begin{pmatrix} \cos w & 0 & \sin w \\ 0 & 1 & 0 \\ -\sin w & 0 & \cos w \end{pmatrix}. \quad (5.71)$$

The surface expression is

$$P(u)T_y(w) = \big(4t - 6t^3 + 7t^4/2\big)\cos w,$$
$$(t-1)^4 - 4(-1+t)^3 t + 12(t-1)^2 t^2 - 12(t-1)t^3 + 5t^4,$$
$$(4t - 6t^3 + 7t^4/2)\sin w\big).$$

5 Surfaces

Such a surface is easy to display. To display it as a wire frame, just write a double loop in which u is varied from 0 to 1, and w is varied from 0 to 2π, in any desired steps. To display it as a solid surface, a similar double loop should cover every pixel (i.e., should iterate in very small steps) and should calculate the normal to the surface at the pixel and, from it, the intensity of light reflected from the pixel (Chapter 6).

Following are other examples of surfaces of revolution:

1. A sphere of radius R, is generated by rotating a half-circle 360° about some axis. Figure 5.56a shows the half-circle $\mathbf{P}(u) = (R\cos u, R\sin u, 0)$ in the xy plane. A sphere $\mathbf{P}(u,w)$ is obtained when this half-circle is rotated about the y axis:

$$\mathbf{P}(u,w) = \mathbf{P}(u)\mathbf{T}_y(w) = (R\cos u \cos w, R\sin u, R\cos u \sin w), \tag{5.72}$$

where $-\pi/2 \leq u \leq \pi/2$ and $0 \leq w \leq 2\pi$. It is not hard to see (Figure 5.56b) that curves of constant w are meridians of longitude. As u varies from $-\pi/2$ to $\pi/2$, we travel on a semicircle (the profile of the surface) on the sphere. Similarly, varying w for a constant u takes us along a latitude. The north pole is obtained for $u = \pi/2$ (and any w). The equator is the curve obtained when varying w for $u = 0$.

▶ **Exercise 5.32:** Derive the expression for the same sphere centered at (x_0, y_0, z_0).

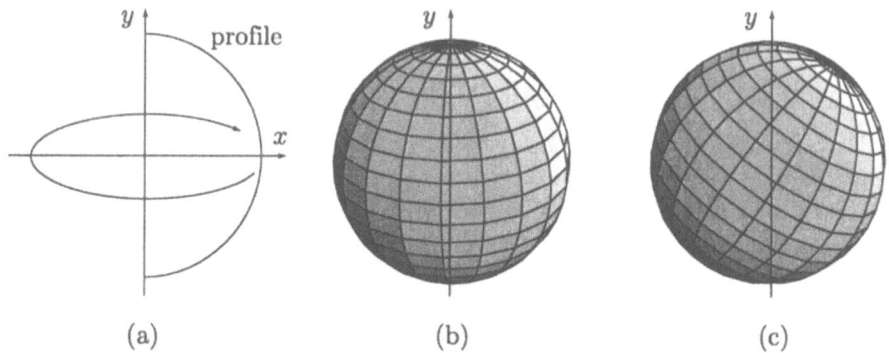

Figure 5.56: The Sphere as a Surface of Revolution.

▶ **Exercise 5.33:** Tilt the sphere of Equation (5.72) θ degrees about the z axis (Figure 5.56c).

▶ **Exercise 5.34:** Derive the expression of the sphere that's obtained when the half-circle in the xz plane is rotated 360° about the z axis.

2. An ellipsoid with radii a and b is obtained by rotating, for example, the ellipse $\mathbf{P}(u) = (a\cos u, b\sin u, 0)$ about the y axis. After translating by (x_0, y_0, z_0), the result is

$$(x_0 + a\cos u \cos w, y_0 + b\sin u, z_0 + a\cos u \sin w),$$

where $-\pi/2 \leq u \leq \pi/2$ and $0 \leq w \leq 2\pi$.

▸ **Exercise 5.35:** Derive the equation of a torus as a surface of revolution. Assume that the torus is centered at the origin, and its two radii are R and r (Figure 5.57). The surface is created by drawing the circle of radius r centered at $(R, 0, 0)$, and rotating it 360° about the z axis.

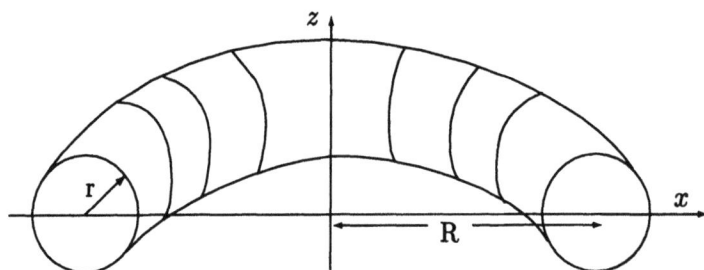

Figure 5.57: The Torus as a Surface of Revolution.

3. Figure 5.58a,b shows a chalice as a surface of revolution and its profile.

Generating surfaces of revolution with a rotation matrix is simple but slow since it requires the use of trigonometric functions. An alternative method is described below.

Two given curves

$$\mathbf{P}(u) = (P_x(u), P_y(u), P_z(u)) \text{ and } \mathbf{C}(w) = (C_x(w), C_y(w), C_z(w))$$

can be combined as follows:

$$\mathbf{S}(u, w) = \big(P_x(u)C_x(w), P_y(u)C_y(w), P_z(u)C_z(w)\big), \qquad (5.73)$$

and it's easy to show that $\mathbf{S}(u, w)$ is a surface. When u is fixed at a value u_0, expression (5.73) becomes

$$\begin{aligned}\mathbf{S}(u_0, w) &= \big(P_x(u_0)C_x(w), P_y(u_0)C_y(w), P_z(u_0)C_z(w)\big) \\ &= \big(\alpha C_x(w), \beta C_y(w), \gamma C_z(w)\big),\end{aligned}$$

which is a curve in the w direction. For each u_0, we thus have a curve in the w direction and, similarly, for each value w_0, we have a curve going in the u direction. The only condition is that none of the components of the curves be identical to zero. If, for example, $C_x(w) = 0$, then the x component of $\mathbf{S}(u_0, w)$ is always zero, so it degenerates from a surface to a curve in the yz plane.

Equation (5.73) can be used to construct a surface of revolution if $\mathbf{C}(w)$ is a circle or an arc. To explain our approach, let's first restrict the discussion to curves that are cubic polynomial segments. Such a curve has the form $\mathbf{P}(u) = (u^3, u^2, u, 1)\mathbf{MP}$, where $\mathbf{M}$ is a 4×4 basis matrix and $\mathbf{P}$ is a geometry vector, a

5 Surfaces

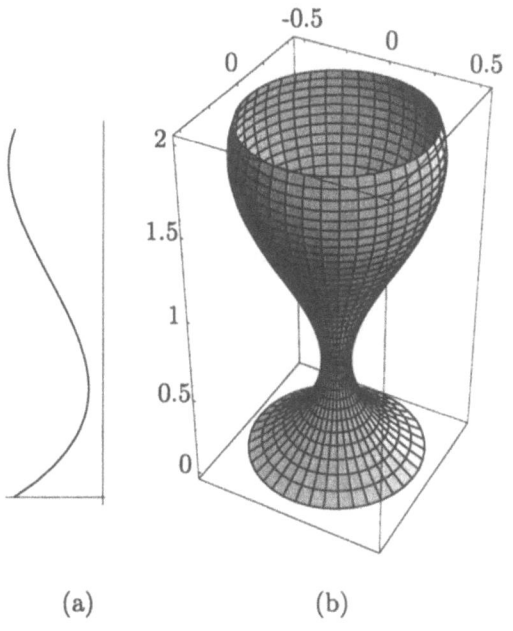

(a) (b)

Figure 5.58: A Chalice as a Surface of Revolution.

```
(* A Chalice *)
<<:Graphics:SurfaceOfRevolution.m
(* the profile *)
ParametricPlot[{.5u^3-.3u^2-.5u-.2,u+1},{u,-1,1},
 AspectRatio->Automatic]
(* the surface *)
SurfaceOfRevolution[{.5u^3-.3u^2-.5u-.2,u+1},{u,-1,1},
 PlotPoints->40]
```

Code for Figure 5.58.

4-tuple of points (although tangent vectors or other vectors can also be used). We can write such a curve in the form

$$\begin{aligned}\mathbf{P}(u) &= \bigl(F_0(u), F_1(u), F_2(u), F_3(u)\bigr) \begin{pmatrix} \mathbf{P}_0 \\ \mathbf{P}_1 \\ \mathbf{P}_2 \\ \mathbf{P}_3 \end{pmatrix} \\ &= F_0(u)\mathbf{P}_0 + F_1(u)\mathbf{P}_1 + F_2(u)\mathbf{P}_2 + F_3(u)\mathbf{P}_3 \\ &= \sum_{i=0}^{3} F_i(u)\mathbf{P}_i.\end{aligned}$$

(See, for example, Equations (4.24), (4.106), and (4.159).) Similarly, curve $\mathbf{C}(w)$

can be expressed as

$$\mathbf{C}(w) = (w^3, w^2, w, 1)\mathbf{NC}$$

$$= \big(G_0(w), G_1(w), G_2(w), G_3(w)\big) \begin{pmatrix} \mathbf{C}_0 \\ \mathbf{C}_1 \\ \mathbf{C}_2 \\ \mathbf{C}_3 \end{pmatrix}$$

$$= G_0(w)\mathbf{C}_0 + G_1(w)\mathbf{C}_1 + G_2(w)\mathbf{C}_2 + G_3(w)\mathbf{C}_3$$

$$= \sum_{i=0}^{3} G_i(w)\mathbf{C}_i.$$

Now, consider the x component of the surface resulting from the product of two such curves:

$$\mathbf{S}_x(u,w) = \left[\sum_{i=0}^{3} F_i(u) P_{xi}\right]\left[\sum_{j=0}^{3} G_j(w) C_{xj}\right]$$

$$= \sum_{i,j=0}^{3} F_i(u) P_{xi} C_{xj} G_j(w)$$

$$= \sum_{i,j=0}^{3} F_i(u) Q_{xij} G_j(w)$$

$$= \big(F_0(u), F_1(u), F_2(u), F_3(u)\big)\mathbf{Q}_x \begin{pmatrix} G_0(w) \\ G_1(w) \\ G_2(w) \\ G_3(w) \end{pmatrix},$$

where Q_{xij} is defined as the product $P_{xi}C_{xj}$ and similarly for the y and z components. The elements $\mathbf{Q}_{ij}$ of matrix $\mathbf{Q}$ are, therefore, triplets of the form

$$\mathbf{Q}_{ij} = (Q_{xij}, Q_{yij}, Q_{zij}) = \big(P_{xi}C_{xj}, P_{yi}C_{yj}, P_{zi}C_{zj}\big) \tag{5.74}$$

and the entire surface can be expressed as a typical bicubic patch

$$\mathbf{S}(u,w) = \big(F_0(u), F_1(u), F_2(u), F_3(u)\big)\mathbf{Q}\begin{pmatrix} G_0(w) \\ G_1(w) \\ G_2(w) \\ G_3(w) \end{pmatrix}$$

$$= (u^3, u^2, u, 1)\mathbf{MQN}^T \begin{pmatrix} w^3 \\ w^2 \\ w \\ 1 \end{pmatrix}. \tag{5.75}$$

Equation (5.75) can be generalized to cases where the constructing curves $\mathbf{C}(w)$ and $\mathbf{P}(u)$ are not cubic polynomials.

5 Surfaces

Once the designer has an idea of the shape of the surface, it may not be too hard to select two curves that will produce this shape. The problem is to place the surface at the right location in space. The location of the surface depends both on the types and the locations of the curves used. Imagine, for example, that two cubic Bézier curves are used to construct such a surface. One curve starts and ends at control points $\mathbf{P}_0$ and $\mathbf{P}_3$, and the other goes from $\mathbf{C}_0$ to $\mathbf{C}_3$. The resulting surface will be a bicubic Bézier patch anchored at the four corner points:

$$\mathbf{Q}_{00} = (P_{x0}C_{x0}, P_{y0}C_{y0}, P_{z0}C_{z0}), \quad \mathbf{Q}_{01} = (P_{x0}C_{x1}, P_{y0}C_{y1}, P_{z0}C_{z1}),$$
$$\mathbf{Q}_{10} = (P_{x1}C_{x0}, P_{y1}C_{y0}, P_{z1}C_{z0}), \quad \mathbf{Q}_{11} = (P_{x1}C_{x1}, P_{y1}C_{y1}, P_{z1}C_{z1}).$$

There is no reason why these points will happen to be in the right locations and it may take some effort to change the coordinates of all the control points to move the curves to other locations without changing their shape, in order to move points $\mathbf{Q}_{ij}$ to the right locations. The use of this surface method may therefore be limited, but it is useful for surfaces of revolution. Imagine the problem of designing a machine part with circular symmetry. If the part is to be manufactured under computer control, the location of the part in three-dimensional space may be unimportant since the machine making it is only interested in its shape.

In order to apply Equation (5.75) to create a surface of revolution we need one curve $\mathbf{P}(u)$ to serve as a "profile" and another curve $\mathbf{C}(w)$ that's a circle, an ellipse, or an arc. As an example, consider the approximate circles obtained by cubic uniform B-splines of Section 4.16.6. We place four points $\mathbf{C}_i$ in the way explained in that section to make curve $\mathbf{C}(w)$ an approximate circle or circular arc. If curve $\mathbf{P}(u)$ is also expressed as a cubic B-spline, then Equation (5.75) becomes the bicubic B-spline patch:

$$\mathbf{S}(u,w) = \left(\frac{1}{6}\right)^2 (u^3, u^2, u, 1) \begin{pmatrix} -1 & 3 & -3 & 1 \\ 3 & -6 & 3 & 0 \\ -3 & 0 & 3 & 0 \\ 1 & 4 & 1 & 0 \end{pmatrix} \mathbf{Q} \begin{pmatrix} -1 & 3 & -3 & 1 \\ 3 & -6 & 3 & 0 \\ -3 & 0 & 3 & 0 \\ 1 & 4 & 1 & 0 \end{pmatrix}^T \begin{pmatrix} w^3 \\ w^2 \\ w \\ 1 \end{pmatrix}$$
(5.76)

(compare with Equation (5.59)). The surface is created in two steps. In step 1, the surface control points $\mathbf{Q}_{ij}$ are calculated. If $\mathbf{P}(u)$ is based on the $n+1$ points $\mathbf{P}_0$ through $\mathbf{P}_n$ and $\mathbf{C}(w)$ is based on the $m+1$ control points $\mathbf{C}_0$ through $\mathbf{C}_m$, then matrix $\mathbf{Q}$ is of order $(n+1) \times (m+1)$. In step 2, Equation (5.76) is applied $(n-1) \times (m-1)$ times to calculate all the surface patches. If the surface should make a complete revolution, then curve $\mathbf{C}(w)$ should be closed. The number of control points in this case is the same, but the number of patches is $(n-1) \times (m+1)$. If curve $\mathbf{P}(u)$ is also closed (as in a torus), then $(n+1) \times (m+1)$ surface patches are needed.

If $\mathbf{C}(w)$ should be a full circle, at least four control points $\mathbf{C}_i$ are needed and the (closed) curve consists of four segments. If curve $\mathbf{P}(u)$ (the "profile" of the surface) is open and is defined by $n+1$ points, it consists of $n-1$ segments. The total number of surface control points $\mathbf{Q}_{ij}$ is, in such a case, $4 \times (n+1)$ and the entire surface of revolution consists of $4 \times (n-1)$ patches.

5.18 Surfaces of Revolution

Example: We select the third quarter-circle segment $\mathbf{P}_4(t)$ of Equation (Ans.33) and denote it by $\mathbf{C}(w)$:

$$\mathbf{C}(w) = \frac{1}{4}(2t^3 - 6t^2 + 4, -2t^3 + 6t, 1).$$

It is defined by the four control points $\mathbf{C}_0 = (0, -3/2, 1)$, $\mathbf{C}_1 = (3/2, 0, 1)$, $\mathbf{C}_2 = (0, 3/2, 1)$, and $\mathbf{C}_3 = (-3/2, 0, 1)$ and it goes from $(1, 0, 1)$ to $(0, 1, 1)$. Notice that we have located $\mathbf{C}(w)$ on the $z = 1$ plane, so none of its components is identical to zero. For the curve profile $\mathbf{P}(u)$, we select the cubic B-spline segment defined by the four control points $\mathbf{P}_0 = (0, 0, 0)$, $\mathbf{P}_1 = (-1, 1, 0)$, $\mathbf{P}_2 = (-1, 1, 3)$, and $\mathbf{P}_3 = (0, 0, 3)$. These points are located on the $x = -y$ plane and go from $z = 0$ to $z = 3$, so none of the three components of $\mathbf{P}(u)$ is zero. Matrix $\mathbf{Q}$ is shown in Table 5.59. Figure 5.60 shows the surface itself and the *Mathematica* code that generated it.

The location of this surface in space may sometimes be a problem and should therefore be discussed. Since our quarter circle goes from $(1, 0, 1)$ to $(0, 1, 1)$, we intuitively expect the profile $\mathbf{P}(u)$ to be rotated from direction $(1, 0)$ (the positive x axis) to direction $(0, 1)$ (the positive y axis). A direct check, however, shows that the four corners of this patch are $\mathbf{S}(0, 0) = (-0.833, 0, 0.5)$, $\mathbf{S}(0, 1) = (-0.833, 0, 2.5)$, $\mathbf{S}(1, 0) = (0, 0.833, 0.5)$, and $\mathbf{S}(1, 1) = (0, 0.833, 2.5)$. The profile has thus been rotated from direction $(-0.833, 0)$ to direction $(0, 0.833)$ because of its particular original location (as defined, the profile is located on the $x = -y$ plane).

Because of the high symmetry of surfaces of revolution, especially those that go through a complete revolution, their precise position in space may not be important, so our method may be useful for this type of surface.

The method developed here can be used with any type of parametric curves, not just B-splines and not just PCs. Equation (5.77) shows how a standard quadratic Lagrange polynomial (Equation (4.16)) can be combined with a degree-4 Bézier curve to form a surface patch based on 3×5 points

$$\mathbf{Q}_{ij} = (Q_{xij}, Q_{yij}, Q_{zij}) = (P_{xi}C_{xj}, P_{yi}C_{yj}, P_{zi}C_{zj}).$$

The surface expression is

$$\mathbf{S}(u, w) = (u^2, u, 1) \begin{pmatrix} 2 & -4 & 2 \\ -3 & 4 & -1 \\ 1 & 0 & 0 \end{pmatrix} \mathbf{Q} \begin{pmatrix} 1 & -4 & 6 & -4 & 1 \\ -4 & 12 & -12 & 4 & 0 \\ 6 & -12 & 6 & 0 & 0 \\ -4 & 4 & 0 & 0 & 0 \\ 1 & 0 & 0 & 0 & 0 \end{pmatrix}^T \begin{pmatrix} w^4 \\ w^3 \\ w^2 \\ w \\ 1 \end{pmatrix},$$
(5.77)

where

$$\mathbf{Q} = \begin{pmatrix} Q_{00} & Q_{01} & Q_{02} & Q_{03} & Q_{04} \\ Q_{10} & Q_{11} & Q_{12} & Q_{13} & Q_{14} \\ Q_{20} & Q_{21} & Q_{22} & Q_{23} & Q_{24} \end{pmatrix}.$$

5 Surfaces

	(0,0,0)	(−1,1,0)	(−1,1,3)	(0,0,3)
(0,−3/2,1)	(0,0,0)	(0,−3/2,0)	(0,−3/2,3)	(0,0,3)
(3/2,0,1)	(0,0,0)	(−3/2,0,0)	(−3/2,0,3)	(0,0,3)
(0,3/2,1)	(0,0,0)	(0,3/2,0)	(0,3/2,3)	(0,0,3)
(3/2,0,1)	(0,0,0)	(3/2,0,0)	(3/2,0,3)	(0,0,3)

Table 5.59: Matrix **Q** for Surface of Revolution Example.

```
<<:Graphics:ParametricPlot3D.m;    (* Surface of revolution *)
Clear[basis,Cubi];  (* as a combination of 2 cubic B-splines *)
(* matrix 'basis' has dimensions 4x4x3 *)
basis={{{0,0,0},{0,-3/2,0},{0,-3/2,3},{0,0,3}}
 ,{{0,0,0},{-3/2,0,0},{-3/2,0,3},{0,0,3}}
 ,{{0,0,0},{0,3/2,0},{0,3/2,3},{0,0,3}},{{0,0,0}
 ,{3/2,0,0},{3/2,0,3},{0,0,3}}};
Cubi={{-1,3,-3,1},{3,-6,3,0},{-3,0,3,0},{1,4,1,0}};
prt[i_]:=basis[[Range[1,4],Range[1,4],i]];
(* 'prt' extracts component i from the 3rd dimen of 'basis' *)
coord[i_]:={u^3,u^2,u,1}.Cubi.prt[i].Transpose[Cubi].{w^3,w^2,w,1};
ParametricPlot3D[{coord[1],coord[2],coord[3]}/36,
{u,0,1,.1},{w,0,1,.1},
Prolog->AbsoluteThickness[.5],ViewPoint->{1.736, -0.751, -0.089}]
```

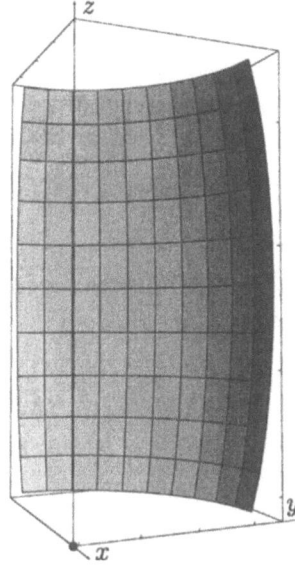

Figure 5.60: A Quarter-Circle Surface of Revolution Made of B-Splines.

5.18.1 Matrix Equations of Quadric Surfaces

Spheres, cones, and cylinders are quadric surfaces, and it is well known that such surfaces can be represented implicitly by $\mathbf{k}\mathbf{Q}\mathbf{k}^T = 0$, where $\mathbf{k} = (x, y, z, 1)$ and $\mathbf{Q}$ is a symmetric, 4×4 matrix. Here are three examples:

1. A canonical sphere (centered on the origin) of radius r is represented by the implicit expression $x^2 + y^2 + z^2 - r^2 = 0$. The same sphere centered at point $\mathbf{C} = (c_1, c_2, c_3)$ is represented by the similar implicit expression $(x - c_1)^2 + (y - c_2)^2 + (z - c_3)^2 - r^2 = 0$. This can also be written

$$(x, y, z, 1) \begin{pmatrix} 1 & 0 & 0 & -c_1 \\ & 1 & 0 & -c_2 \\ & & 1 & -c_3 \\ & & & |\mathbf{C}|^2 - r^2 \end{pmatrix} \begin{pmatrix} x \\ y \\ z \\ 1 \end{pmatrix} = 0.$$

(Only the upper right half of the symmetric matrix is shown.)

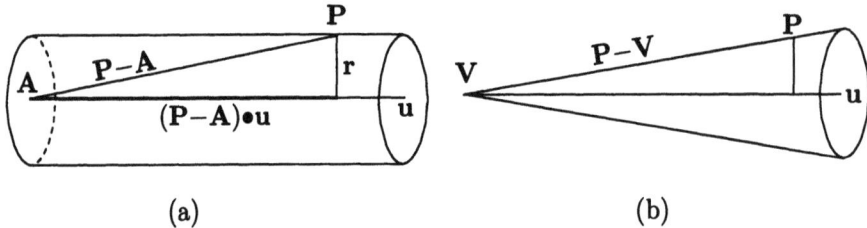

(a)   (b)

Figure 5.61: (a) A Cylinder. (b) A Cone.

2. Consider the cylinder of radius r whose axis is the straight segment $\mathbf{A} + t\mathbf{u}$, where $\mathbf{u} = (u_1, u_2, u_3)$ is a unit vector and $\mathbf{A}$ is the vector (a_1, a_2, a_3). If $\mathbf{P} = (x, y, z)$ is any point on the cylinder, then a simple application of Pythagoras' theorem gives

$$|\mathbf{P} - \mathbf{A}|^2 = |(\mathbf{P} - \mathbf{A}) \bullet \mathbf{u}| + r^2 \tag{5.78}$$

(Figure 5.61a).

The first term, $|\mathbf{P} - \mathbf{A}|^2$, gives rise to the symmetric matrix $\mathbf{M}$ and the second term, $|(\mathbf{P} - \mathbf{A}) \bullet \mathbf{u}|$, is represented by matrix $\mathbf{N}$:

$$\mathbf{M} = \begin{pmatrix} 1 & 0 & 0 & -a_1 \\ & 1 & 0 & -a_2 \\ & & 1 & -a_3 \\ & & & |\mathbf{A}|^2 \end{pmatrix}, \quad \mathbf{N} = \begin{pmatrix} u_1^2 & u_1 u_2 & u_1 u_3 & -(\mathbf{A} \bullet \mathbf{u})u_1 \\ & u_2^2 & u_2 u_3 & -(\mathbf{A} \bullet \mathbf{u})u_2 \\ & & u_3^2 & -(\mathbf{A} \bullet \mathbf{u})u_3 \\ & & & (\mathbf{A} \bullet \mathbf{u})^2 \end{pmatrix}.$$

Combining these two with the third term (r^2) yields the implicit matrix expression for the cylinder:

$$(x, y, z, 1) \begin{pmatrix} 1 - u_1^2 & -u_1 u_2 & -u_1 u_3 & -a_1 + (\mathbf{A} \bullet \mathbf{u})u_1 \\ & 1 - u_2^2 & -u_2 u_3 & -a_2 + (\mathbf{A} \bullet \mathbf{u})u_2 \\ & & 1 - u_3^2 & -a_3 + (\mathbf{A} \bullet \mathbf{u})u_3 \\ & & & |\mathbf{A}|^2 - (\mathbf{A} \bullet \mathbf{u})^2 - r^2 \end{pmatrix} \begin{pmatrix} x \\ y \\ z \\ 1 \end{pmatrix} = 0.$$

5 Surfaces

The equation $(x - x_0)^2 + (y - y_0)^2 = R^2$ describes a cylinder parallel to the z axis, with radius R and a center at (x_0, y_0). To generalize this to a cylinder about an arbitrary axis, we proceed in two steps.

Step 1. We first consider a cylinder of radius R centered on the line L that passes through the point $(x_0, y_0, 0)$ and is parallel to the z axis. A point $\mathbf{P} = (x, y, z)$ is on that cylinder when the distance from $\mathbf{P}$ to the axis equals R (Figure 5.62a). The square of the distance from $\mathbf{P}$ to the axis is $(x - x_0)^2 + (y - y_0)^2$ and this must equal R^2.

Step 2. We now assume a cylinder about an arbitrary line, and derive its expression by using the same idea. We assume a line $\mathbf{L}$ through point $\mathbf{P}_0 = (x_0, y_0, z_0)$ in the direction of the unit vector $\mathbf{u} = (u_1, u_2, u_3)$. We further assume that a cylinder is centered on $\mathbf{L}$ and that $\mathbf{P} = (x, y, z)$ is any point on the cylinder. The point $\mathbf{Q}$ on $\mathbf{L}$ nearest $\mathbf{P}$ is (Figure 5.62b)

$$\mathbf{Q} = \mathbf{P}_0 + \big((x - x_0)u_1 + (y - y_0)u_2 + (z - z_0)u_3\big)\mathbf{u}.$$

To get the equation of the cylinder, we calculate the square of the distance from $\mathbf{P}$ to $\mathbf{Q}$ and set it equal to R^2.

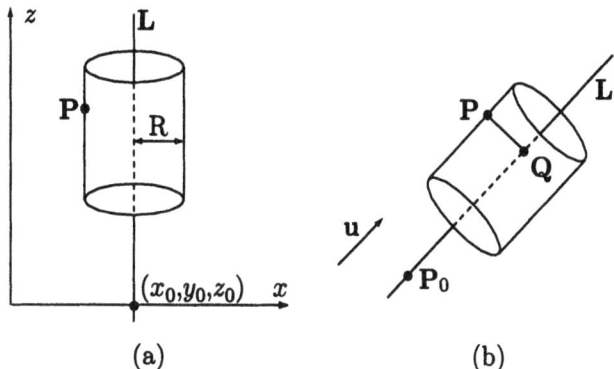

Figure 5.62: A Cylinder.

An alternative approach uses parametric lines. Suppose that line $\mathbf{L}$ goes through point $\mathbf{P}_0$ and lies in the direction of the unit vector $\mathbf{w}$. Suppose also that $\mathbf{u}$ and $\mathbf{v}$ are unit vectors such that $\mathbf{u}$, $\mathbf{v}$, and $\mathbf{w}$ are mutually perpendicular. The parametric equation of the cylinder in this case is

$$\mathbf{P}(t, \theta) = \mathbf{P}_0 + \mathbf{u}\cos\theta + \mathbf{v}\sin\theta + t\mathbf{w}, \text{ where } 0 \leq \theta \leq 2\pi \text{ and } t \text{ is real.}$$

3. The third example is the cone with axis direction $\mathbf{u} = (u_1, u_2, u_3)$, a vertex at $\mathbf{V} = (v_1, v_2, v_3)$, and cone angle α (where $\mathbf{u}$ is a unit vector). For any point $\mathbf{P}$ on the cone that is not the vertex, the vector $\mathbf{P} - \mathbf{V}$ makes an angle of either α or $\pi - \alpha$ with the axis direction $\mathbf{u}$ (Figure 5.62b). The property of the dot product therefore suggests that

$$\frac{\mathbf{p} - \mathbf{V}}{|\mathbf{p} - \mathbf{V}|} \bullet \mathbf{u} = \pm\cos\alpha. \tag{5.79}$$

After squaring and rearranging, this becomes

$$(\mathbf{P} \bullet \mathbf{u} - \mathbf{V} \bullet \mathbf{u})^2 - \cos^2 \alpha (\mathbf{P} - \mathbf{V}) \bullet (\mathbf{P} - \mathbf{V}) = 0.$$

The implicit matrix representation can now be written using considerations similar to the ones for case 2 (the cylinder). The result is

$$\begin{pmatrix} u_1^2 - \cos^2 \alpha & u_1 u_2 & u_1 u_3 & v_1 \cos^2 \alpha - (\mathbf{u} \bullet \mathbf{V}) u_1 \\ & u_2^2 - \cos^2 \alpha & u_2 u_3 & v_2 \cos^2 \alpha - (\mathbf{u} \bullet \mathbf{V}) u_2 \\ & & u_3^2 - \cos^2 \alpha & v_3 \cos^2 \alpha - (\mathbf{u} \bullet \mathbf{V}) u_3 \\ & & & (\mathbf{u} \bullet \mathbf{V})^2 - |\mathbf{V}|^2 \cos^2 \alpha \end{pmatrix}.$$

Equations (5.78) and (5.79) make it easy to classify points as being inside, on, or outside the surface. For the cylinder, we define $f(\mathbf{P}, \mathbf{A}, \mathbf{u}) = |\mathbf{P} - \mathbf{A}|^2 - |(\mathbf{P} - \mathbf{A}) \bullet \mathbf{u}|$. It turns out that

$$f(\mathbf{P}, \mathbf{A}, \mathbf{u}) \begin{cases} > r^2, & \mathbf{P} \text{ is outside the cylinder,} \\ = r^2, & \mathbf{P} \text{ is on the cylinder,} \\ < r^2, & \mathbf{P} \text{ is inside the cylinder.} \end{cases}$$

For the cone, we consider the angle between vectors $\mathbf{P} - \mathbf{V}$ and $\mathbf{u}$. If this angle is less than α or greater than $\pi - \alpha$, then $\mathbf{P}$ lies inside the cone. We, thus, end up with the three cases

$$\left(\frac{\mathbf{P} - \mathbf{V}}{|\mathbf{P} - \mathbf{V}|} \bullet \mathbf{u} \right)^2 \begin{cases} > \cos^2 \alpha, & \mathbf{P} \text{ is inside the cone,} \\ = \cos^2 \alpha, & \mathbf{P} \text{ is on the cone,} \\ < \cos^2 \alpha, & \mathbf{P} \text{ is outside the cone.} \end{cases}$$

5.19 Sweep Surfaces

This type of surface is obtained when a space curve $\mathbf{C}(u)$ is transformed according to a transformation rule $\mathbf{T}(w)$. The transformation must include translation and/or rotation, and may also include scaling, shearing, etc. We say that the surface is *swept* by the curve when it is transformed. The expression for the surface is simply the product $\mathbf{P}(u, w) = \mathbf{C}(u) \cdot \mathbf{T}(w)$. The transformation $\mathbf{T}$ is a 4×4 matrix, so vector $\mathbf{C}$ should be written in homogeneous coordinates, as the 4-tuple $\mathbf{C}(u) = (x(u), y(u), z(u), 1)$.

The simplest example is the translation of a straight line. Imagine the line $\mathbf{C}(u) = (u, 0, 0, 1)$ where $0 \leq u \leq 1$, translated along the y axis by the transformation matrix

$$\mathbf{T}(w) = \begin{pmatrix} 1 & 0 & 0 & 0 \\ 0 & 1 & 0 & 0 \\ 0 & 0 & 1 & 0 \\ 0 & w & 0 & 1 \end{pmatrix},$$

where $0 \leq w \leq 1$. The surface $\mathbf{P}(u, w) = \mathbf{C}(u) \cdot \mathbf{T}(w) = (u, w, 0, 1)$ swept by this line equals (after dividing by the fourth element) $\mathbf{P}(u, w) = (u, w, 0)$. This surface

The Quadric Surfaces

Ellipsoid, real	$\left(\dfrac{x}{a}\right)^2 + \left(\dfrac{y}{b}\right)^2 + \left(\dfrac{z}{c}\right)^2 = 1.$
Ellipsoid, imaginary	$\left(\dfrac{x}{a}\right)^2 + \left(\dfrac{y}{b}\right)^2 + \left(\dfrac{z}{c}\right)^2 = -1.$
Cone, real	$\left(\dfrac{x}{a}\right)^2 + \left(\dfrac{y}{b}\right)^2 - \left(\dfrac{z}{c}\right)^2 = 0.$
Cone, imaginary	$\left(\dfrac{x}{a}\right)^2 + \left(\dfrac{y}{b}\right)^2 + \left(\dfrac{z}{c}\right)^2 = 0.$
Cylinder, real (Elliptic)	$\left(\dfrac{x}{a}\right)^2 + \left(\dfrac{y}{b}\right)^2 = 1.$
Cylinder, imaginary (Elliptic)	$\left(\dfrac{x}{a}\right)^2 + \left(\dfrac{y}{b}\right)^2 = -1.$
Cylinder (Hyperbolic)	$\left(\dfrac{x}{a}\right)^2 - \left(\dfrac{y}{b}\right)^2 = 1.$
Cylinder (Parabolic)	$x^2 + y^2 = 0.$
Elliptic paraboloid	$\left(\dfrac{x}{a}\right)^2 + \left(\dfrac{y}{b}\right)^2 + 2z = 0.$
Hyperbolic paraboloid	$\left(\dfrac{x}{a}\right)^2 - \left(\dfrac{y}{b}\right)^2 + 2z = 0.$
Single-sheet hyperboloid	$\left(\dfrac{x}{a}\right)^2 - \left(\dfrac{y}{b}\right)^2 - \left(\dfrac{z}{c}\right)^2 = 1.$
Double-sheet hyperboloid	$\left(\dfrac{x}{a}\right)^2 - \left(\dfrac{y}{b}\right)^2 - \left(\dfrac{z}{c}\right)^2 = -1.$

is simply the square, on the xy plane, whose opposite corners are the origin and point $(1,1,0)$.

A slightly more complex example is the line $\mathbf{C}(u) = (u,0,0,1)$, where $0 \le u \le 1$ (which goes from the origin to point $(1,0,0)$) translated a distance α along the z axis while being rotated 360° about that axis. The transformation matrix is

$$\mathbf{T}(w) = \begin{pmatrix} \cos(2\pi w) & \sin(2\pi w) & 0 & 0 \\ -\sin(2\pi w) & \cos(2\pi w) & 0 & 0 \\ 0 & 0 & 1 & 0 \\ 0 & 0 & \alpha w & 1 \end{pmatrix}.$$

The surface is $\mathbf{P}(u,w) = (u\cos(2\pi w), u\sin(2\pi w), \alpha w, 1)$. For $w = 0.5$, it reduces to the line $(0, u, 0.5\alpha)$ (in the y direction), and for $w = 1$, it becomes the line $(u, 0, \alpha)$ (in the original x direction, but at a distance α along z).

▶ **Exercise 5.36:** Calculate the sweep surface obtained when line $\mathbf{C}(u) = (3u, 0, 0, 1)$ is translated along the z axis and at the same time translated in the y direction on a sine curve.

▶ **Exercise 5.37:** Calculate the half-sphere produced when the quarter circle

$$\mathbf{C}(u) = \left(\frac{1-u^2}{1+u^2}, \frac{2u}{1+u^2}, 0, 1\right), \quad \text{where} \quad 0 \le u \le 1,$$

is rotated 360° about the y axis.

▶ **Exercise 5.38:** Calculate the expression of a cone as a sweep surface. Assume that the cone is created by constructing the line from the origin to point $(R, 0, H)$, and rotating it 360° about the z axis.

> ...treat Nature by the sphere, the cylinder and the cone...
> — Paul Cézanne

Note: The basic sweep surface $\mathbf{C}(u)\mathbf{T}(w)$ can be extended to the product $\mathbf{C}(u,w)\mathbf{T}(w)$ of a surface and a transformation. This product is still a sweep surface since $\mathbf{C}(u,w)$ reduces to a curve for any value of w. We can think of $\mathbf{C}(u,w)$ as a curve that's a function of the parameter u but also depends on w. As w is varied, $\mathbf{C}(u,w)$ yields different curves and each is transformed differently.

Example: The parametric equation of the Möbius strip is

$$\mathbf{P}(s,t) = \bigl(\cos(s) + t\cos(s/2) \times \cos(s), \sin(s) + t\cos(s/2) \times \sin(s), t\sin(s/2)\bigr),$$

where $0 \le s \le 2\pi$ and $-0.3 \le t \le 0.3$ (typically). This can be viewed as the product of the surface

$$\mathbf{C}(s,t) = (1 + t\cos(s/2), 1 + t\cos(s/2), t)$$

and the scaling transformation

$$\mathbf{T}(s) = \begin{pmatrix} \cos(s) & 0 & 0 \\ 0 & \sin(s) & 0 \\ 0 & 0 & \sin(s/2) \end{pmatrix}.$$

See Figure 5.63 for an illustration. Notice that the surface $\mathbf{C}(s,t)$ (also shown in the figure) depends on s, so it changes shape when s is varied during the sweep.

Example: A sweep surface that's a product of the surface $\mathbf{C}(u,w) = (u, 1, u+2)w + (-u, 1, u-2)(1-w)$ and a rotation about the z axis. Note that $\mathbf{C}(u,w)$ varies from the curve $\mathbf{C}(u,0) = (-u, 1, u-2)$ to the straight line $\mathbf{C}(u,1) = (u, 1, u+2)$ while being rotated.

5 Surfaces

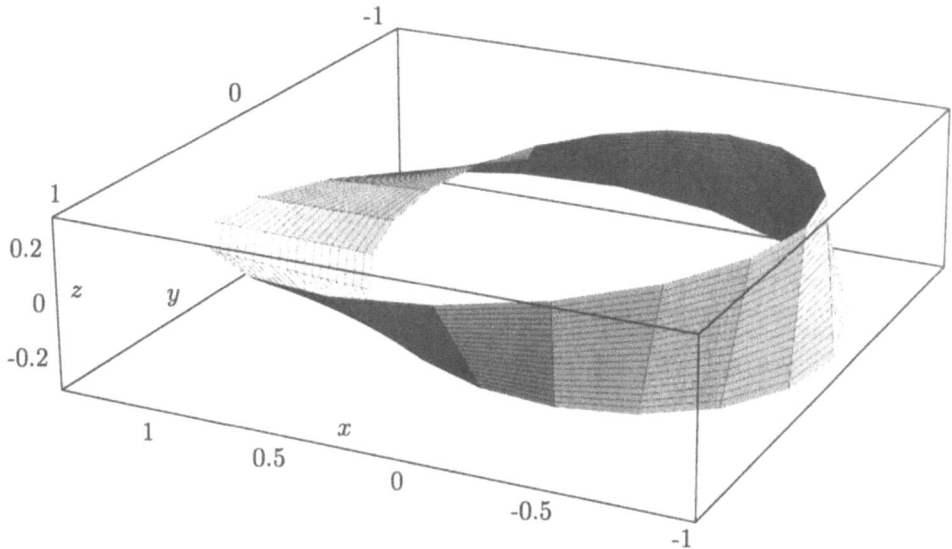

Figure 5.63: The Möbius Strip as a Sweep Surface.

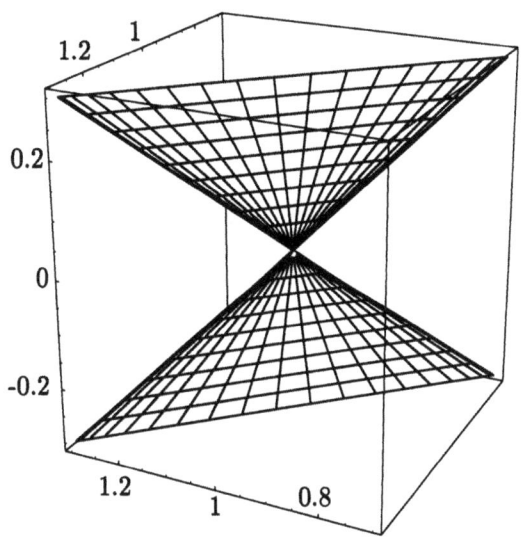

Untransformed Surface $C(s,t)$ for Figure 5.63.

```
Clear[mob,c,rot,s,t];
(* mob[s_,t_]:=
 {Cos[s]+t Cos[s/2]Cos[s],Sin[s]+t Cos[s/2]Sin[s],t Sin[s/2]}; *)
c[s_,t_]:={1+t Cos[s/2],1+t Cos[s/2],t};
rot[s_]:={{Cos[s],0,0},{0,Sin[s],0},{0,0,Sin[s/2]}};
ParametricPlot3D[c[s,t].rot[s], {s,0,2Pi,0.05}, {t,-0.3,+0.3},
 ViewPoint->{-1.790, 3.662, 1.490}]
```

Code for Figure 5.63.

(* A Sweep Surface.
 Curve Cu[u,w] times matrix Trn[w] *)
<<:Graphics:ParametricPlot3D.m;
Clear[Cu,Trn];
Cu[u_,w_]:={u,1,u+2}w+{-u,1,u-2}(1-w);
Trn[w_]:={
 {Cos[2Pi w],Sin[2Pi w],0},
 {-Sin[2Pi w],Cos[2Pi w],0},
 {0,0,1}};
ParametricPlot3D[
 {Cu[u,w].Trn[w][[1]],Cu[u,w].Trn[w][[2]],
 Cu[u,w].Trn[w][[3]]},
 {u,0,1,.2},{w,0,1,.2}, Ticks->None,
 PlotRange->All, AspectRatio->Automatic,
 RenderAll->False, Prolog->AbsoluteThickness[.4],
 ViewPoint->{-0.510, -1.365, 1.210}]

5.20 Polygonal Surfaces by Subdivision

Polygonal surfaces have been discussed in Section 5.3. Such a surface is normally obtained by measuring the coordinates of points on an object, either manually or with a three-dimensional digitizer. The designer then selects a set of points and the software connects those points with straight segments, obtaining a polygon. This is how the original mesh of points is converted to a set of polygons. The only condition is that the polygons be flat. The entire polygonal surface can then be shaded using Gouraud or Phong shading (Section 6.3). If the result is not smooth enough, it can be improved by subdividing the original mesh of points, which is why the subdivision of polygonal surfaces is important.

5.20.1 Doo Sabin Surfaces

The method described here is due to Donald Doo and Malcolm Sabin [Doo and Sabin 78]. They observed that the method used in Section 5.17.4 to subdivide a biquadratic B-spline surface patch generates each new point $\mathbf{P}^1_{ij}$ as a weighted sum of four points: a vertex point, two edge points, and a face point. For example, Equation (5.66) gives point $\mathbf{P}^1_{00}$ as

$$\mathbf{P}^1_{00} = \frac{1}{16}(9\mathbf{P}^0_{00} + 3\mathbf{P}^0_{10} + 3\mathbf{P}^0_{01} + \mathbf{P}^0_{11}),$$

so we write it in the form

$$\begin{aligned}\mathbf{P}^1_{00} &= \frac{1}{16}(9\mathbf{P}^0_{00} + 3\mathbf{P}^0_{10} + 3\mathbf{P}^0_{01} + \mathbf{P}^0_{11}) \\ &= \frac{1}{16}\big(4\mathbf{P}^0_{00} + 2(\mathbf{P}^0_{00} + \mathbf{P}^0_{01}) + 2(\mathbf{P}^0_{00} + \mathbf{P}^0_{10}) + (\mathbf{P}^0_{00} + \mathbf{P}^0_{01} + \mathbf{P}^0_{10} + \mathbf{P}^0_{11})\big) \\ &= \frac{1}{4}\big(4\mathbf{P}^0_{00} + (\mathbf{P}^0_{00} + \mathbf{P}^0_{01})/2 + (\mathbf{P}^0_{00} + \mathbf{P}^0_{10})/2 + (\mathbf{P}^0_{00} + \mathbf{P}^0_{01} + \mathbf{P}^0_{10} + \mathbf{P}^0_{11})/4\big)\end{aligned}$$

5 Surfaces

$$= \frac{1}{4}(4\mathbf{V} + \mathbf{E}_1 + \mathbf{E}_2 + \mathbf{F}),$$

where $\mathbf{V}$ is the vertex point $\mathbf{P}^0_{00}$, $\mathbf{E}_1$ is the average of $\mathbf{P}^0_{00}$ and $\mathbf{P}^0_{01}$ (i.e., it is located midway between them), $\mathbf{E}_2$ is the average of $\mathbf{P}^0_{00}$ and $\mathbf{P}^0_{10}$, and $\mathbf{F}$ is the average of the four corners of the polygon being subdivided.

The idea of Doo and Sabin is to subdivide a mesh of points that consists of any polygons, not just quadrilaterals, by performing the following steps:

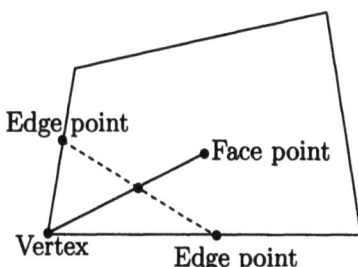

Figure 5.64: Edge and Face Points.

1. Consider a vertex $\mathbf{P}^0_i$ on the original mesh (Figure 5.64). It is located on a certain face F (and perhaps on other faces as well) and is at the intersection of two edges, $E1$ and $E2$ (two of the edges that form F). Create a new point $\mathbf{P}^1_i$ as a weighted average of $\mathbf{P}^0_i$, the two edge points adjacent to $\mathbf{P}^0_i$ (i.e., the center points of $E1$ and $E2$), and the face point that's the average of all the vertices forming F. Repeat this for every vertex $\mathbf{P}^0_i$. See Figure 5.65a for an example.

2. Consider face F again. It now contains some new points $\mathbf{P}^1_i$. Connect them so that they form a new polygon. This polygon will become a face in the new, subdivided surface. Repeat for all faces F (Figure 5.65b).

3. Consider again a vertex $\mathbf{P}^0_i$ on the original mesh. Such a vertex is normally common to several faces. For each of those faces, find the new point that's nearest $\mathbf{P}^0_i$. Connect those points to each other to form a new polygon. This polygon will also become a face in the new, subdivided surface. Repeat for all vertices $\mathbf{P}^0_i$ (Figure 5.65c).

4. Consider an edge of the original mesh of points. There will normally be two faces adjacent to this edge and they will have new points $\mathbf{P}^1_i$. Connect the new points around the edge to form a new polygon. This polygon will also become a face in the new, subdivided surface. Repeat this step for all edges (Figure 5.65d).

Notice that the new mesh may contain all kinds of polygons, not just triangles or quadrilaterals.

5.20.2 Catmull-Clark Surfaces

The method described here is due to Edwin Catmull and Jim Clark [Catmull and Clark 78] and is an extension of the method of Section 5.17.5 to arbitrary polygonal surfaces. We have seen that subdividing a bicubic B-spline surface patch generates each new point $\mathbf{P}^1_{ij}$ as either a face point, an edge point, or a vertex point. A

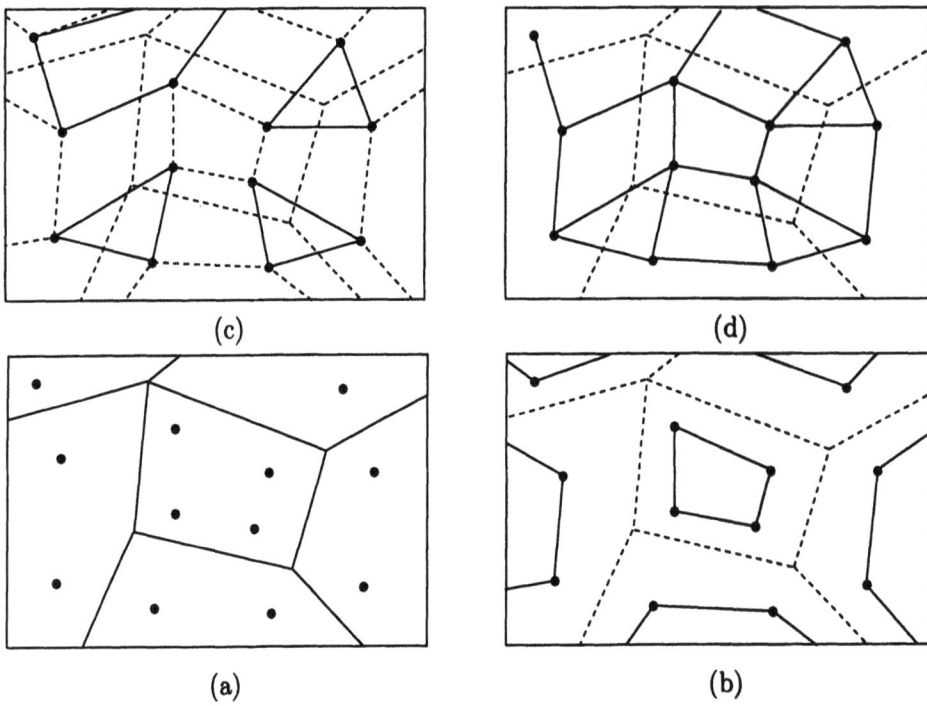

Figure 5.65: The First Doo-Sabin Subdivision Step.

Catmull-Clark surface patch starts with an arbitrary polygonal surface and subdivides it by generating new face, edge, and vertex points and connecting them in a simple way. The rules for generating the points are the following:

1. A face point is calculated for each face of the original mesh. The point is simply the average of all the points that bound the face.

2. An edge point is created for each interior edge of the polygonal surface. The point is the average of the midpoint of the edge and of the two face points on both sides of the edge.

3. A vertex point is generated for each interior vertex **P** of the original mesh. The point is the average of **Q**, 2**R**, and **S**$(n-3)/4$, where **Q** is the average of the face points on all the faces adjacent to **P**, **R** is the average of the midpoints of all the edges incident on **P**, and **S** is simply **P** itself.

Figure 5.66 shows an example. We start with a mesh of eight vertices defining six polygons, two rectangles, and two triangles (notice that the polygons may have any number of sides, not just three or four). This surface has six faces, seven interior edges, and two interior vertices. The six new face points are shown in Figure 5.66a as small circles. Each is the average of the points bounding its face. Figure 5.66b shows the midpoints of the edges as small ×'s and the seven new edge points as diamonds. In Figure 5.66c, we select one of the two interior vertices as **S**, temporarily connect the four face points surrounding it (just to show who they are), and calculate **Q** (shown as a small "+") as their average. In Figure 5.66d, we

5 Surfaces

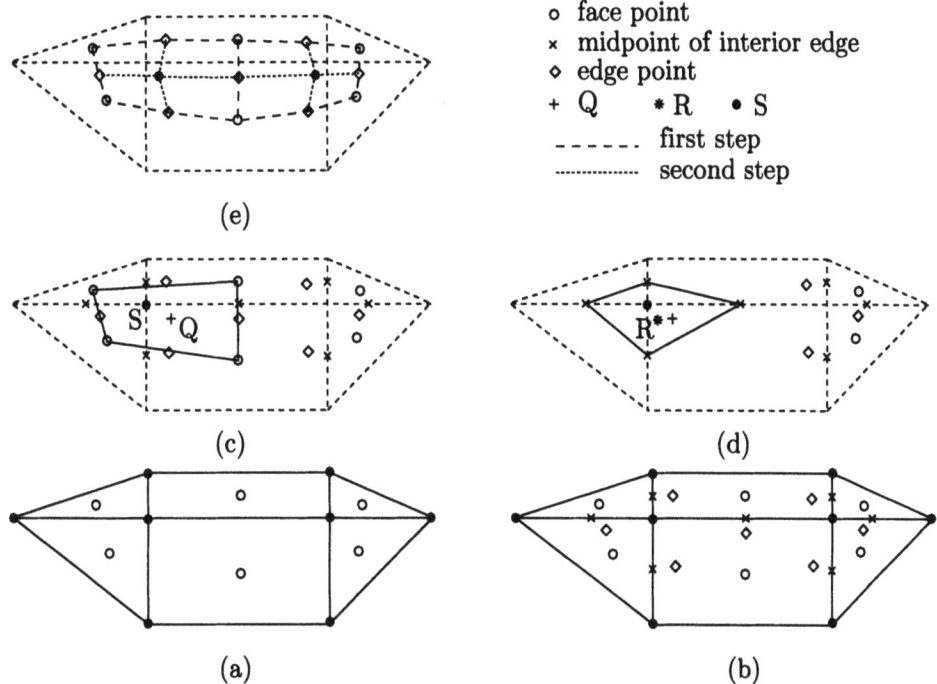

Figure 5.66: The First Catmull-Clark Subdivision Step.

show how **R** (shown as an asterisk) is calculated as the average of four midpoints of edges (temporarily connected).

After the new points have been generated, they are connected according to the following rules:

1. Each face point is connected to all the edge points of the interior edges bounding its face. These are shown as large dashes in Figure 5.66e.

2. Each new vertex point is connected to all the edge points that were used in calculating it. These lines are shown as small dashes in Figure 5.66e.

Notice that even though the original polygonal mesh may have polygons with any number of sides, the new, subdivided mesh will consist of quadrilaterals (four-sided polygons) only.

5.21 Curves on Surfaces

Up to now, we have considered curves going on a surface patch either in the u or in the w directions. It is also possible to calculate and draw curves that go in other directions on a surface. The idea is to vary u and w simultaneously in a controlled way. If the surface is expressed by $\mathbf{P}(u,w)$ and we want a curve $\mathbf{C}(t)$ on it, we need functions $u(t)$ and $w(t)$ that will produce a pair of parameters (u,w), each in the range $[0,1]$, for any value of $0 \le t \le 1$. The curve will then be given by $\mathbf{C}(t) = \mathbf{P}(u(t), w(t))$.

As an example, consider the bilinear surface patch of Equation (5.8), Section 5.5.1: $\mathbf{P}(u,w) = (u + w - 2uw, w, 1 - u)$ defined by the four corner points

$\mathbf{P}_{00} = (0,0,1)$, $\mathbf{P}_{10} = (1,0,0)$, $\mathbf{P}_{01} = (1,1,1)$, and $\mathbf{P}_{11} = (0,1,0)$. Since this is a bilinear patch, curves in the u or in the w directions are straight lines (Figures 5.15 and 5.67a). However, since the patch is curved, not flat, we can expect curves in other directions to be curved as well. It is especially easy to derive expressions for curves that go in a "45°" direction on this surface patch (i.e., from one corner to the diagonally opposite corner). Figure 5.67b shows the curve from $\mathbf{P}_{01}$ to $\mathbf{P}_{10}$. It is obtained when u varies from 0 to 1 and, at the same time, w varies from 1 to 0. The two functions in this case are $u(t) = t$ and $w(t) = 1 - t$. When substituted in Equation (5.8), they produce

$$\mathbf{C}(t) = (t + 1 - t - 2t(1-t), 1-t, 1-t) = (2t^2 - 2t + 1, 1-t, 1-t).$$

A simple check verifies that this curve really goes from $\mathbf{C}(0) = (1,1,1) = \mathbf{P}_{01}$ to $\mathbf{C}(1) = (1,0,0) = \mathbf{P}_{10}$. Notice that it is not a straight line.

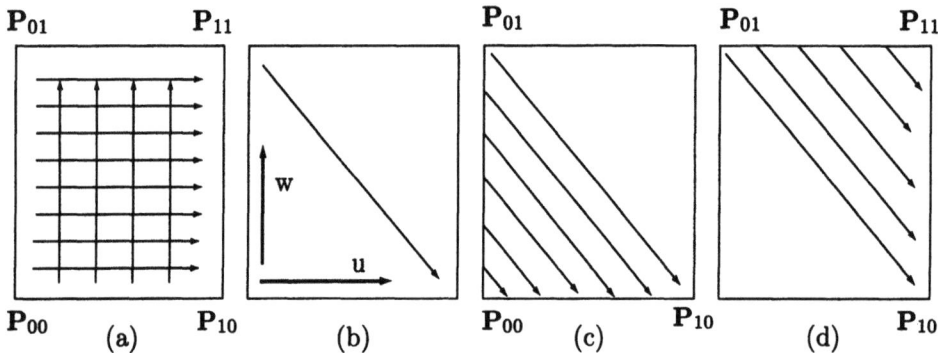

Figure 5.67: Diagonal Curves on a Bilinear Patch.

Figure 5.67c shows the family of curves located "below the main diagonal" of our surface patch. These curves are obtained with the pair $u(t) = t$, $w(t) = a - t$, where a is held constant for each curve. For each curve, the parameter t varies in the range $[0, a]$ and the entire family is obtained when a varies from 0 to 1. A general curve in this family is thus

$$\mathbf{C}(t) = (t + a - t - 2t(a-t), a - t, 1 - t) = (2t^2 - 2ta + a, a - t, 1 - t).$$

A simple check verifies that for $a = 1$ (where t goes from 0 to 1), the curve really goes from $\mathbf{C}(0) = (1,1,1) = \mathbf{P}_{01}$ to $\mathbf{C}(1) = (1,0,0) = \mathbf{P}_{10}$. However, when $a = 0$ (t goes from 0 to 0), the curve reduces to the single point $\mathbf{C}(0) = (0,0,1) = \mathbf{P}_{00}$.

▶ **Exercise 5.39:** Derive the expression for the family of curves "above the main diagonal" (Figure 5.67d).

▶ **Exercise 5.40:** Consider the order-2 × 3 Bézier surface patch of Exercise 5.27. Calculate the diagonal curve that goes on this surface from $\mathbf{P}_{03} = (3,0,0)$ to $\mathbf{P}_{20} = (0,2,0)$. What kind of a curve is this?

5.22 Surface Normals

Our aim is to display realistically looking, solid surfaces. This is done by means of shading (Chapter 6), and requires the calculation of the normal to the surface at every point. The normal is the vector that's perpendicular to the surface at the point. It can be defined in two ways.

1. We imagine a flat plane touching the surface at the point (this is called the *osculating plane*). The normal is the vector that's perpendicular to this plane.
2. We calculate two tangent vectors to the surface at the point. The normal is the vector that's perpendicular to both tangents.

The following shows how to calculate the normal vectors for various types of surfaces.

- The normal to the implicit surface $F(x,y,z) = 0$ at point (x_0, y_0, z_0) is the vector
$$\left(\frac{\partial F(x_0, y_0)}{\partial x}, \frac{\partial F(x_0, y_0)}{\partial y}, \frac{\partial F(x_0, y_0)}{\partial z} \right).$$

Example: The ellipsoid $x^2/a^2 + y^2/b^2 + z^2/c^2 - 1 = 0$. A partial derivative would be, for example, $\partial f/\partial x = 2x/a^2$, so the normal is

$$\left(\frac{2x}{a^2}, \frac{2y}{b^2}, \frac{2z}{c^2} \right) \quad \text{which is in the same direction as} \quad \left(\frac{x}{a^2}, \frac{y}{b^2}, \frac{z}{c^2} \right).$$

The normal at point $(0, 0, -c)$ would, therefore, be $(0, 0, -c/c^2) = (0, 0, -1/c)$. This is a vector in the direction $(0, 0, -1)$.

▶ **Exercise 5.41:** What is the normal to the explicit surface $z = f(x,y)$ at point (x_0, y_0)?

> No money, no job, no rent. Hey, I'm back to normal.
> — Mickey Rourke (as Henry Chinaski) in *Barfly* (1987).

- The normal to the parametric surface $\mathbf{P}(u, w)$ is calculated in two steps. In step 1, the two tangent vectors $\mathbf{U} = \partial \mathbf{P}(u, w)/\partial u$ and $\mathbf{V} = \partial \mathbf{P}(u, w)/\partial w$ are calculated. In step 2, the normal is calculated as their cross-product $\mathbf{U} \times \mathbf{V}$ (Equation (A.10), page 703).

- The normal to a polygon in a polygonal surface can be calculated as shown for an implicit surface. The (implicit) plane equation is $F(x, y, z) = Ax + By + Cz + D = 0$, so the normal is $\left(\frac{\partial F}{\partial x}, \frac{\partial F}{\partial y}, \frac{\partial F}{\partial z} \right)$, which is simply (A, B, C). Another way of calculating the normal, especially suited to triangles, is to find two vectors on the surface and calculate their cross-product. Two suitable vectors are $\mathbf{U} = \mathbf{P}_1 - \mathbf{P}_2$ and $\mathbf{V} = \mathbf{P}_1 - \mathbf{P}_3$. Their cross product is

$$\mathbf{U} \times \mathbf{V} = (U_2 V_3 - U_3 V_2, U_3 V_1 - U_1 V_3, U_1 V_2 - U_2 V_1).$$

Example: A polygon with the four vertices $(1,1,-1)$, $(1,1,1)$ $(1,-1,1)$, and $(1,-1,-1)$. All vertices have $x=1$, so they are on the $x=1$ plane, which means that the normal should be a vector in the x direction. The calculation is straightforward:

$$\mathbf{U} = (1,1,1) - (1,1,-1) = (0,0,2),$$
$$\mathbf{V} = (1,-1,1) - (1,1,-1) = (0,-2,2),$$
$$\mathbf{U} \times \mathbf{V} = (0-(-4), 0-0, 0-0) = (4,0,0).$$

This is a vector in the right direction.

▶ **Exercise 5.42:** What will happen if we calculate $\mathbf{U}$ as $(1,1,-1) - (1,1,1)$?

▶ **Exercise 5.43:** Find the normal to the pyramid face of Equation (Ans.39).

▶ **Exercise 5.44:** Find the normal to the cone of Equation (Ans.38).

▶ **Exercise 5.45:** Construct a cylinder as a sweep surface and find its normal vector. Assume that the cylinder is swept when the line from $(-a, 0, R)$ to $(a, 0, R)$ is rotated $360°$ about the x axis.

> You know how on a flat surface, which has only two dimensions,
> we can represent a figure of a three-dimensional solid.
>
> H. G. Wells, *The Time Machine*.

6
Rendering

Rendering is a general name for methods that display a realistic-looking three-dimensional solid object on a two-dimensional output device (normally screen or paper). Displaying a wire-frame surface, for example, is one way of rendering it. Rendering includes texture mapping, the calculation and display of shadows, and the use of colors. The most common method of rendering, however, is shading.

6.1 Introduction

Producing a computer-generated image is a multistep process:

1. The designer/user has to specify the objects in the image, their shapes, positions, orientations, and surface color/texture.

2. He should select the viewer's position and direction of view. The computer then transforms the points in the image to create a perspective projection.

3. An algorithm should now be executed, to determine what parts of the image are visible to the viewer. This is the hidden-surface problem. All parts of all objects in the image must be checked, and only those that are supposed to be visible to the viewer are actually displayed.

4. The objects in the image are made to look real by simulating lighting. The designer/user has to define the light source (or sources), their positions, shapes, intensities, and colors. The light emanating from any surface in the picture is a combination of (1) light coming from light sources (direct lighting) and reflected by the surface, (2) light coming from other surfaces (indirect) and reflected by our surface, (3) light generated by the surface (if it happens to be one of the light sources), and (4) light transmitted by the surface (if it happens to be transparent or translucent).

5. The image is now displayed by rendering software that computes the amount and color of light reaching the viewer's eye from any point in the image, and then displays that point.

Modern graphics workstations may have special hardware to implement perspective projections, hidden-surface elimination, and direct illumination. Everything else requires software. The most important rendering task done by software is illumination, both direct and indirect. The latter is important when the image contains shiny surfaces, each reflecting some of the others. Two methods for indirect illumination are currently popular, *ray tracing* (Section 6.5) and *radiosity*. These methods use opposite approaches to compute the light reflection.

Ray tracing was introduced by Turner Whitted of Bell Laboratories in 1979. The main idea is to trace the path of a ray from the eye through each pixel on the screen into the image. If the ray strikes a surface, the algorithm spawns reflected or refracted rays, which, in turn, are traced to see if they intersect any other surfaces. The final color and intensity of each pixel are determined by adding up the light contributed by each spawned ray.

Ray tracing produces realistic images but is view dependent. This means that the entire computation must be repeated when the viewer's position is changed. Ray tracing is also too slow to generate real-time sequences of pictures, since each image may take minutes or more to compute.

The radiosity method, developed at Cornell in 1984, is view independent: Given an image, the calculations need be made only once. Once the global illumination has been determined, it is easy to create a series of images by moving the viewer to different viewpoints. Indeed, the method can be used to generate real-time sequences of images, which makes it useful for applications such as flight simulation and architecture (walking through a newly designed structure).

Radiosity uses conservation of energy to compute the light intensity for each surface in a picture made of ideal diffuse surfaces (either light sources or reflectors). An equation is written for the radiosity of each surface in the image—the intensity of light emanating from the surface—as a function of the radiosity of all the other surfaces.

6.2 A Simple Shading Model

The simplest shading technique simulates light reflection from the surface to be rendered. It assumes a light source (which itself may not have to be displayed) at a certain point. It assumes that the viewer is located at the center of perspective and that there is an environment around the object—such as walls and furniture—that can reflect more light on the object and can cast shadows. It then uses a mathematical model to calculate the intensity (and color) of the light reflected from every pixel on the surface of the object. Such a calculation may be very complex, depending on the model used. Figure 6.1 shows how a flat drawing can be made to look real (i.e., three-dimensional) by simulating reflection. It is easy to tell which of the four buttons are convex and which is concave. It is also easy to tell that the bevels around the buttons in Figure 6.1c,d point outside.

The resulting shaded image depends on the following:

- The light source. Its intensity, color, shape, direction, and distance. It can be a point source, or a large one, such as a window.

- The surface of the object. It can vary from shiny to dull, from smooth to

6 Rendering

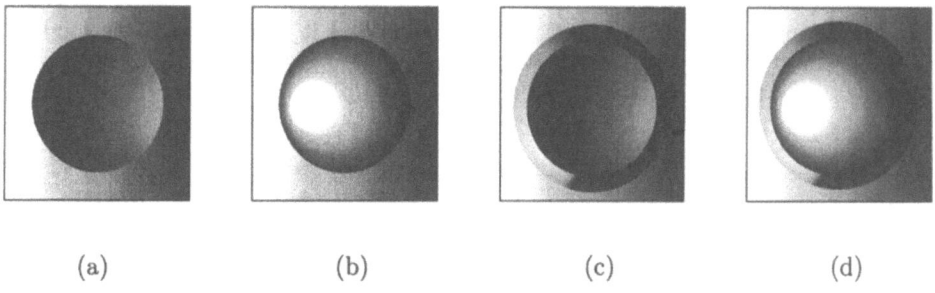

(a)   (b)   (c)   (d)

Figure 6.1: Light Reflection from Buttons.

rough, and from bright to dark. It can have several colors, can be opaque, transparent or translucent (diffusing light so that objects beyond it cannot be distinguished).

■ The environment. Objects seen in empty space, without any background to reflect light on them, look harsh. Imagine a spaceship in deep space, away from any reflecting planets. Those parts of the ship illuminated by direct starlight are very bright, while parts that are in the shade are completely dark. The result is that we see the ship mostly in black and white, with few grays or colors. A realistic shading model should, therefore, consider light reflection from other objects and from nearby walls.

In general, a ray of light striking a surface is partly absorbed, partly transmitted (and also refracted), and partly reflected. The following three sections briefly discuss these phenomena. Following these, we describe the simple shading model.

6.2.1 Absorption

This phenomenon depends on the material and on the wavelength of the light. Material that absorbs all wavelengths except blue (which it reflects or transmits) looks blue. Absorption of light, therefore, determines the color and brightness of the object.

6.2.2 Refraction

When light moves from one medium to another, such as from air to glass to water, its speed changes. The denser the medium, the slower the light travels. When we talk about the speed of light, we implicitly mean its speed in vacuum (where it is fastest). The result of the speed change is that a beam of light changes its direction of motion (it *bends*) when it enters a different medium. This phenomenon is called *refraction*. Notice that it affects all electromagnetic radiation (X-rays, radio waves, microwaves, etc.), not just light. Figure 6.2a shows why refraction is important in computer graphics. When a ray of light moves from air to glass and again to air, it bends twice, in opposite directions, so it comes out of the glass in its original direction but with its position shifted. An observer looking at an object through the glass will, thus, see the object shifted away from its original position. Realistic-looking computer-generated images should, therefore, simulate the details of refraction.

Figure 6.2e shows a ray of light entering a slab of glass from air at an angle α to the normal. Inside the glass, the ray bends and it now moves at an angle β with respect to the normal. The rule of refraction, discovered experimentally by the Dutch mathematician Willebrord Snell in 1621, is

$$\frac{\sin\alpha}{\sin\beta} = \frac{C_1}{C_2} = C,$$

where C_1 and C_2 are the speeds of light in air and glass, respectively, and C, their ratio, is called the *refraction coefficient* of air and glass. This is Snell's law. (The *index of refraction* of a medium M is defined as the ratio of the speed of light in a vacuum and the speed of light in M. Hence, the refraction coefficient of two media is the ratio of their refraction indexes.)

How does the change of speed cause the light to change its direction? This can be explained by means of a general physical principle called *the principle of least time*. It was proposed by Pierre Fermat around 1650, so it is sometimes called *Fermat's principle*. It says that light chooses the particular path in air and glass that takes the *shortest time* to traverse. Using this principle, it is easy to prove that the path of least time is the one obeying Snell's law. Figure 6.2b shows an analogous situation. A lifeguard is stationed on a beach and there is a swimmer in the water. The swimmer starts drowning and the lifeguard starts running toward him. The best path for the lifeguard (from the point of view of the swimmer) is that of least time. Path *b* is a straight line. This may be the intuitive choice of many lifeguards, but it may not be the path of least time since swimming is slower than running. Path *d* minimizes the swimming time, but there is no guarantee that it is the right path. Intuitively, it seems that the right path is somewhere between paths *b* and *d* since it is clear that paths such as *a* and *e* require longer times.

We now show that the path of least time is the one that satisfies Snell's law. The proof is short and elegant and its geometry is shown in Figure 6.2c. We assume that the best path is the one that hits the water at point **P**, and we then try another path that hits at point **Q**, close to **P**. Figure 6.2d shows how the curve of travel time versus point of hit has a minimum at **P**. Since point **Q** is close to **P** and since the curve is continuous, we expect only a very small difference in the travel times of the rays that hit the water at points **P** and **Q**. Another way to express this is to say that we expect the travel times of the two rays to be essentially the same in the first approximation, because the curve of Figure 6.2d is close to flat (horizontal) at point **P**.

The proof should, therefore, figure out the difference between the travel times along the two paths and set that difference to zero. This will generate an equation whose solution should produce Snell's law. The first step is to draw a perpendicular to LP that goes through point **Q**. This shows that path LPS has to travel on the beach a distance of a units longer than path LQS. It takes a/C_1 time units to travel distance a. The second step is to draw a perpendicular to line QS that goes through point **P**. This shows that path LQS has to travel in the water a distance b longer than path LPS. It takes b/C_2 time units to travel distance b. Denoting by d the distance PQ, we get $\sin\alpha = a/d$ and $\sin\beta = b/d$ from elementary trigonometry. We

6 Rendering

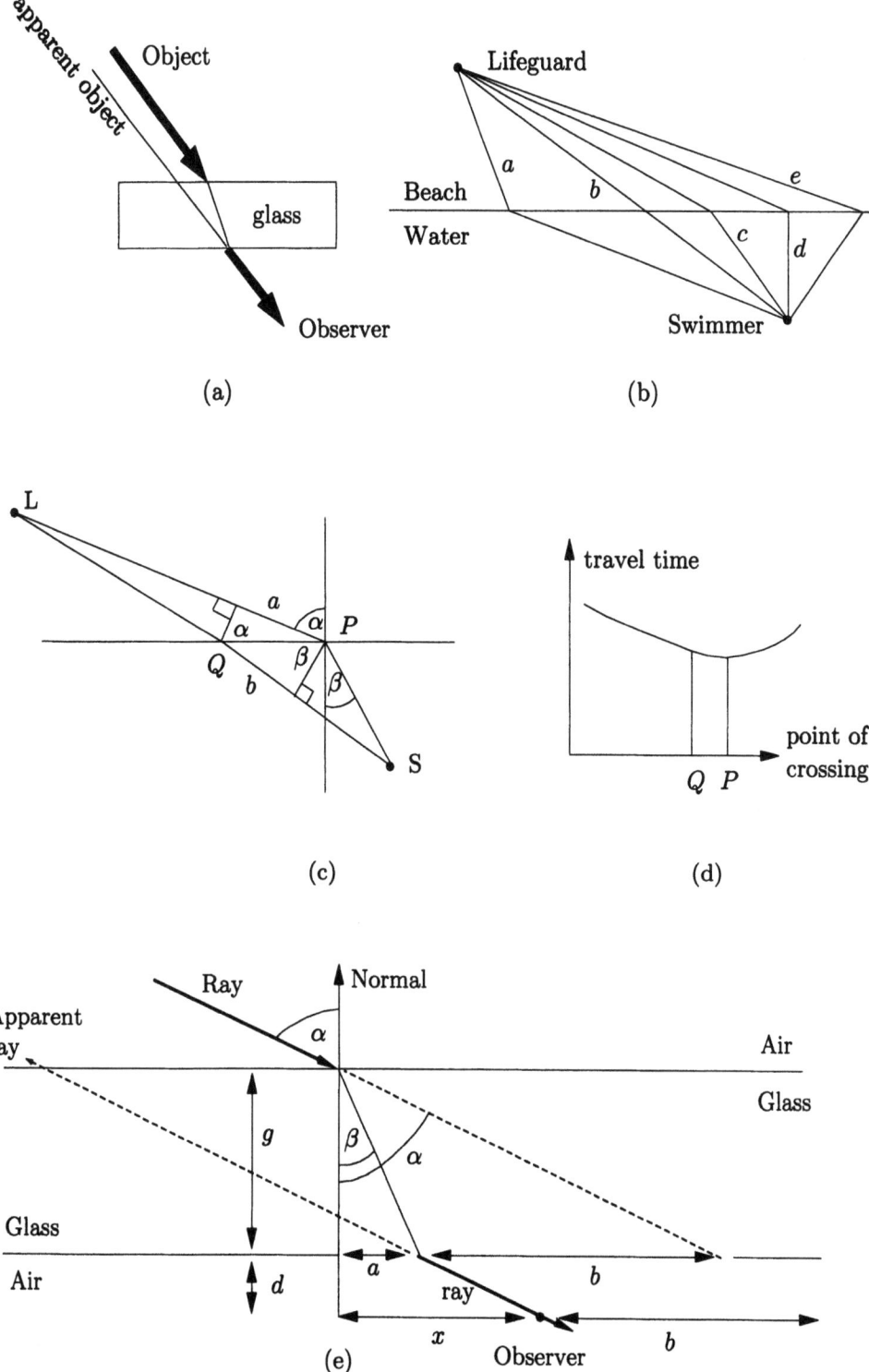

Figure 6.2: Refraction.

can now write

$$\text{time}(a) = \text{time}(b)$$
$$\Rightarrow \frac{a}{C_1} = \frac{b}{C_2}$$
$$\Rightarrow \frac{d\sin\alpha}{C_1} = \frac{d\sin\beta}{C_2}$$
$$\Rightarrow \frac{\sin\alpha}{C_1} = \frac{\sin\beta}{C_2}$$
$$\Rightarrow \frac{\sin\alpha}{\sin\beta} = \frac{C_1}{C_2}.$$

▶ **Exercise 6.1:** Prove Snell's law using just elementary calculus and trigonometry.

Snell's law is mathematically simple, so people generally like to think of it as an explanation of refraction. A ray of light hits a glass surface and bends by the right amount, depending on the angle of incidence. It is easy for us to imagine that the light "knows" at what angle it hits and what medium it is entering, so it changes its direction of motion accordingly. The least-time principle, on the other hand, even though more elegant, is harder to accept as an explanation. The problem is: How does light know in advance what the least time path is? When we humans are faced with such a problem, we have to try different paths, we hesitate, we need to perform calculations, but light does not hesitate, does not seem to try different paths, and always selects the right path confidently.

Quantum electrodynamics provides a completely different explanation to refraction. Advanced readers are referred to pages 49–52 of [Feynman 85]. Another book by Feynman, *The Feynman Lectures on Physics*, (volume 1, chapter 26, page 5) describes a few common phenomena, such as a mirage, that are caused by refraction.

Figure 6.2e illustrates the refraction problem as it typically occurs in practice (i.e., in computer graphics applications). A ray of light goes through a thick slab of glass and is observed on the other side of the glass. The known quantities are the angle of incidence α, the thickness g of the glass, the vertical distance d between the observer and the glass, and the refraction coefficient $C = C_1/C_2$. The unknown quantity is x, the horizontal distance between the observer and the point of incidence. Once x is known, we know where the observer should be positioned in order to see the (refracted) ray. The derivation is elementary and uses similar triangles:

$$a = g\tan\beta, \quad a + b = g\tan\alpha,$$
$$b = g\tan\alpha - a = g\tan\alpha - g\tan\beta, \quad x + b = (g + d)\tan\alpha,$$
$$x = g\tan\alpha + d\tan\alpha - b$$
$$= g\tan\alpha + d\tan\alpha - g\tan\alpha + g\tan\beta$$
$$= d\tan\alpha + g\tan\beta$$
$$= d\tan\alpha + \frac{g\sin\alpha}{\sqrt{C^2 - \sin^2\alpha}}. \qquad (6.1)$$

6 Rendering

The last equality is true because

$$\tan\beta = \frac{\sin\beta}{\cos\beta} = \frac{\sin\alpha\sin\beta}{\sin\alpha\sqrt{1-\sin^2\beta}}$$
$$= \frac{\sin\alpha\sin\beta}{\sqrt{\sin^2\alpha - \sin^2\alpha\sin^2\beta}}$$
$$= \frac{\sin\alpha}{\sqrt{\left(\frac{\sin\alpha}{\sin\beta}\right)^2 - \sin^2\alpha}} = \frac{\sin\alpha}{\sqrt{C^2-\sin^2\alpha}}.$$

It is easy to test Equation (6.1) for the case $C_1 = C_2$, where the light goes from a medium M to the same medium M. In this case, there should be no refraction, so the equation should yield $\alpha = \beta$. Substituting $C = 1$ in Equation (6.1) yields

$$x = d\tan\alpha + \frac{g\sin\alpha}{\sqrt{1-\sin^2\alpha}} = d\tan\alpha + \frac{g\sin\alpha}{\cos\alpha} = (d+g)\tan\alpha.$$

Figure 6.2e shows that $x = (d+g)\tan\alpha$ implies $b = 0$ or $\alpha = \beta$.

▶ **Exercise 6.2:** The SI (Système International) definition of the meter was adopted at the 1983 Conference Générale des Poids et Mesures. It says "the meter is the length of the path traveled by light in vacuum during a time interval of 1/299,792,458 of a second." This defines the speed of light in vacuum to be exactly 299,792,458 m/s. The speed of light in typical glass fiber is roughly 33% less, or about 200,000 km/s. The refraction coefficient C from vacuum to glass is, therefore, approximately 1.5. Using Equation (6.1), calculate and plot the distance x as a function of the angle of incidence α for the case $d = 0$ and $g = 1$.

6.2.3 Reflection

When a ray of light hits a mirror, it gets reflected. The direction of reflection is determined by the following simple rule: The angle of reflection equals the angle of incidence (the angles are measured between the rays of light and the normal to the surface). This rule can also be elegantly deduced from Fermat's principle. Figure 6.3 shows a ray going from point **A** to a mirror M, getting reflected, and arriving at point **B**. What path takes the ray from **A** to the mirror and to **B** in the least time?

Consider path ADB. The travel time from **A** to **D** is minimal, but the travel time from **D** to **B** is much longer. If we move a bit to the right and let the ray hit the mirror at, say, point **E**, we slightly increase the travel time AE but greatly decrease the travel time EB. To find the best point, we use an elegant "trick." We construct an imaginary point **B'** on the other side of the mirror, at the same distance as **B**. The total travel time AEB equals the travel time AEB', since the speed of light on both sides of the mirror is the same. It is now clear that the minimal-time path AB' is also the minimal distance path AB', i.e., a straight line. We denote by **C** the point where this line intercepts the mirror. Since line ACB' is straight, and since BF=FB', we conclude that angle BCF equals angle B'CF which,

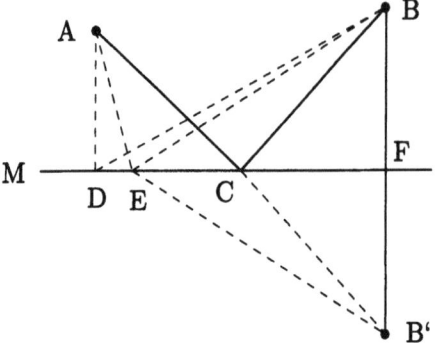

Figure 6.3: Least Time in Reflection.

in turn, equals angle ACM. This implies that the angle between direction AC and the normal equals the angle between direction CB and the normal.

In the case of reflection, the light speed is always the same, so times are proportional to distances. In the case of refraction, however, the path of minimal time is different from that of minimal distance.

An ideal mirror reflects all the light that hits it, and each ray hitting a point on the surface is reflected in one direction, such that the angle of reflection equals that of incidence. If a viewer happens to be in that direction, looking at the point, he will see a reflection of the light source at the point (in the color of the light source, not that of the surface). Such an ideal reflection is called *specular*. Calculating the specular reflection from a point requires the knowledge of the normal to the surface at the point, and the position of the viewer.

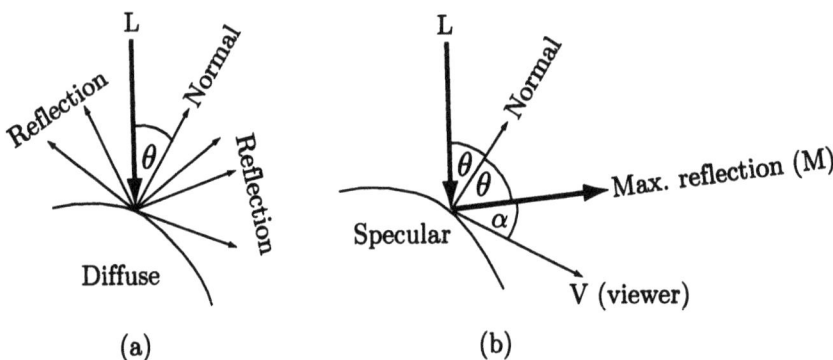

Figure 6.4: Diffuse and Specular Reflection.

An ideal dull surface reflects each ray of light in all directions, since every point on the surface has many microfacets pointing in different directions. A viewer always sees the same intensity reflected from a given point, regardless of his position. He still sees different reflections from different points, since some points may be farther

6 Rendering

away from the light source, or may be pointing away from it. This type of reflection is called *diffuse* (Figure 6.4).

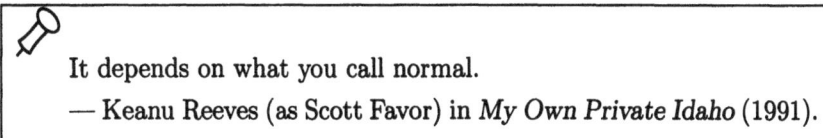

It depends on what you call normal.
— Keanu Reeves (as Scott Favor) in *My Own Private Idaho* (1991).

6.2.4 A Reflection Model

We now realize that every point on a surface emits three types of light: diffuse and specular reflected light, and transmitted (refracted) light. Each light ray leaving the surface is a sum of these three contributions. We are now ready to discuss a simple shading model that simulates only reflection (no refraction) but produces acceptable results.

Diffuse reflection: The intensity of diffuse reflection from a point depends on the intensity I_p and direction $\mathbf{L}$ of the light source, on the direction $\mathbf{N}$ of the normal to the surface at the point, and on the coefficient of diffuse reflection k_d (a number between 0 and 1). According to Lambert's law, the intensity is $I_p k_d \cos\theta$.

Figure 6.5 illustrates this law. In Figure 6.5a, a wide light beam hits a surface at a right angle. In Figure 6.5b, the same beam hits the surface at an angle θ to the normal. Elementary trigonometry shows that the surface area covered by the beam is now greater by a factor of $1/\cos\theta$. Since the same amount of light is now spread over a larger area, the reflection is weaker, by the same factor.

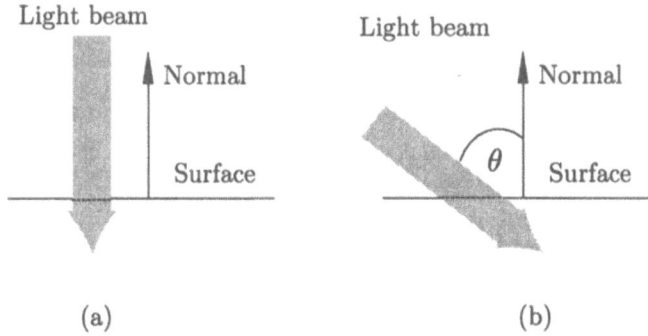

Figure 6.5: Diffuse Reflection at an Angle.

If $\mathbf{L}$ and $\mathbf{N}$ are unit vectors, then $\cos\theta = \mathbf{L} \bullet \mathbf{N}$. To simplify the calculation, we can sometimes assume that the light source is at infinity, i.e., all the light rays arriving at the surface are parallel, so $\mathbf{L}$ is the same for all points on the surface.

The intensity that reaches the viewer depends on their distance R from the point, so our model of diffuse reflection should be modified to give $I_p k_d (\mathbf{L} \bullet \mathbf{N})/R^2$ as the intensity of light reaching the viewer from a given point. Notice that R varies as we move on the surface from point to point.

Specular reflection: Looking at a shiny surface, we may see a highlight at a certain point. The reflection at that point is strong and it has the color of the light source, rather than that of the surface. It also has the shape of the light source. As we move around the surface, the highlight moves with us on the surface. This is specular reflection.

The simple specular reflection model assumes **L** and **N** as before, and a viewer at **V**. The reflection is mostly in direction **M**, and the intensity seen by the viewer depends on the material and on the angle α (Figure 6.4b). The smaller α, the stronger the intensity seen by the viewer from this specific point on the surface. The intensity is, thus, proportional to $\cos\alpha$, and to include the properties of the material, we use $\cos^n \alpha$, where n is an integer that depends on the material. For a perfect mirror, $n = \infty$ (i.e., the specular reflection is strictly in the **M** direction). For a rougher surface, n is normally in the range 1–100.

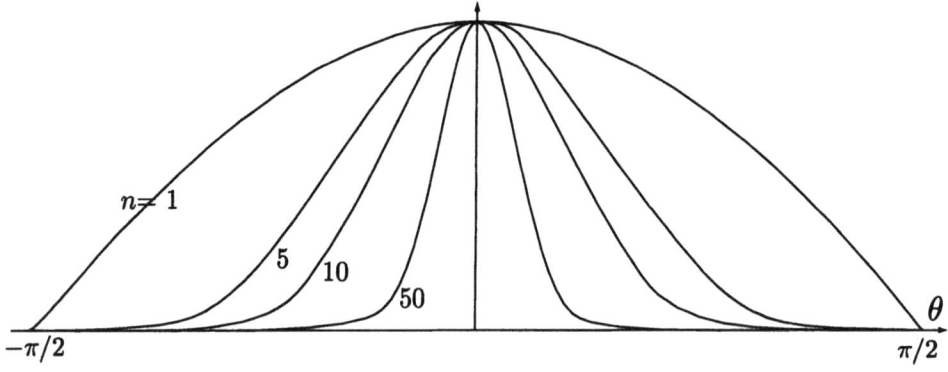

Figure 6.6: The Behavior of $\cos^n \theta$.

Figure 6.6 illustrates the behavior of $\cos^n \theta$ for $n = 1, 5, 10, 50$. It is clear that for large values of n, the function is almost always zero.

The intensity of specular reflection that reaches the viewer is (assuming that **M** and **V** are unit vectors) $I_p k_s \cos^n \alpha / R^2 = I_p k_s (\mathbf{M} \bullet \mathbf{V})^n / R^2$. The quantity k_s is the coefficient of specular reflection, a parameter that depends on the material.

Transparent objects transmit light, but also reflect some of it. We know from everyday experience that, looking through a sheet of glass, the angle between the line of sight and the glass surface determines how clearly we see through the glass. When a ray of light strikes the glass at a 90° angle, almost all of it is transmitted and refracted; very little is absorbed or reflected. The opposite is true when the ray hits the glass at a grazing angle. Such a ray is mostly reflected. Thus, a transparent object reflects light in a special way. The amount of reflection depends in a complex way on the angle of incidence and also on the wavelength. We say that such a surface has a coefficient of specular reflection that's a function of both the angle and the wavelength.

In practice, it is hard to calculate vector **M** because it has to point in a certain direction, and also be in the same plane as **L** and **N**, so the dot product **M** • **V**

6 Rendering

should preferably be replaced by something that involves just the vectors **L**, **N**, and **V**. We note that

$$\mathbf{V} \bullet \mathbf{L} = \cos(2\theta + \alpha)$$
$$= \cos(2\theta)\cos(\alpha) - \sin(2\theta)\sin\alpha$$
$$= \cos\alpha[\cos^2(\theta) - \sin^2\theta] - 2\sin\theta\cos\theta\sin\alpha,$$

and

$$\mathbf{N} \bullet \mathbf{V} = \cos(\theta + \alpha)$$
$$= \cos\theta\cos\alpha - \sin\theta\sin\alpha.$$

Combining these expressions yields $\mathbf{M}\bullet\mathbf{V} = 2(\mathbf{N}\bullet\mathbf{L})(\mathbf{N}\bullet\mathbf{V}) - \mathbf{V}\bullet\mathbf{L}$. The intensity of specular reflection can now be expressed as

$$\frac{I_p k_s[2(\mathbf{N}\bullet\mathbf{L})(\mathbf{N}\bullet\mathbf{V}) - \mathbf{V}\bullet\mathbf{L}]^n}{R^2}.$$

The calculation is more intensive than in the case of diffuse reflection, since it involves determining the vectors **N** and **V** for every pixel (if the light source cannot be assumed to be at infinity, then **L** also has to be calculated).

Example: Assuming a surface with a normal in the y direction, $\mathbf{N} = (0,1,0)$, a light source at 45° in the xy plane $\mathbf{L} = (-0.7071, 0.7071, 0)$, and a viewer at 30° from the x axis in the same plane, $\mathbf{V} = (0.866, 0.5, 0)$ (Figure 6.7). We get $\mathbf{N}\bullet\mathbf{L} = 0.7071$, $\mathbf{N}\bullet\mathbf{V} = 0.5$, and $\mathbf{V}\bullet\mathbf{L} = -0.2588$, which yields $2(\mathbf{N}\bullet\mathbf{L})(\mathbf{N}\bullet\mathbf{V}) - \mathbf{V}\bullet\mathbf{L} = 2\times 0.7071\times 0.5 + 0.2588 = 0.966$. A relatively high reflection (96.6% of the maximum value), since the viewer is only 15° away from the direction of maximum reflection. Including the effects of $I_p k_s$ and R^2 normally reduces this number. The effect of n can now easily be illustrated.

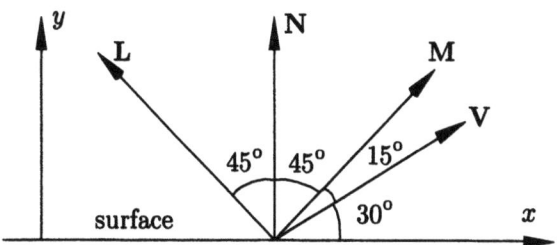

Figure 6.7: Specular Reflection Example.

For $n = 10$, the value above goes down to 70.7%, and for $n = 100$ (a very shiny surface), it drops all the way down to 3.14%! With a shiny surface, even an offset of 15° is enough to reduce the reflection highlight to almost nothing.

Ambient Reflection: In most cases, we can account for multiple reflections from nearby objects with the simple model $I_a k_a$, where I_a is the intensity of *ambient reflection* and k_a is the coefficient of this reflection (a parameter that depends on the material).

In summary, our simple shading model thus assigns to each pixel an intensity I of reflected light reaching the viewer of

$$I = I_a k_a + \frac{I_p[k_d(\mathbf{L} \bullet \mathbf{N}) + k_s(\mathbf{M} \bullet \mathbf{V})^n]}{R^2}.$$

Color Shading: The reflection coefficients k_d and k_a depend on the color of the incident light. A shading model for a color output device should therefore calculate three intensities I_R, I_G, and I_B for each pixel, and use them to determine the color of the pixel. For example, $I_G = I_{aG}k_{aG} + I_{pG}[k_{dG}(\mathbf{L} \bullet \mathbf{N}) + k_s(\mathbf{M} \bullet \mathbf{V})^n]/R^2$.

More sophisticated rendering models can calculate shadows and take into account multiple reflections, transparent objects, and shadows. They can also deal with complex objects such as waves and clouds. They are the reason why rendering is considered a computationally intensive application.

> Teraflop club: /te'r*-flop kluhb/ [FLOP = Floating Point Operation] n. A mythical association of people who consume outrageous amounts of computer time in order to produce a few simple pictures of glass balls with intricate ray-tracing techniques. Caltech professor James Kajiya is said to have been the founder.
>
> —Eric Raymond, *The Hacker's Dictionary*.

6.3 Gouraud and Phong Shading

A polygonal surface is especially easy to shade, since we can assume that all the pixels of a polygon reflect the same amount of light. The shading model being used needs, therefore, to be applied just once to each polygon. The resulting surface, however, looks angular and unnatural. Fortunately, there are two simple methods that result in a better looking surface by smoothing out the shading. These are the Gouraud and Phong methods. The former interpolates intensities and is discussed below. The latter interpolates normal vectors and its implementation details are similar.

Gouraud's method [Gouraud 71] smooths out the shading of a polygonal surface by calculating reflection intensities at the corner points of each polygon and interpolating these intensities at every pixel on the polygon. It proceeds in four steps:

1. The normal vectors $\mathbf{N}_i$ are calculated for all polygons i (Figure 6.8a).

2. Vertex normals $\mathbf{N}_v$ are calculated for each vertex v by averaging the surface normals of all the polygons sharing the vertex. Figure 6.8a shows one such normal, calculated at the intersection of four triangles. Its value is $\mathbf{N}_v = (\mathbf{N}_1 + \mathbf{N}_2 + \mathbf{N}_3 + \mathbf{N}_4)/4$.

3. Vertex intensities I_v are calculated for all the vertices of the surface by using the normal vectors $\mathbf{N}_v$ of step 2 and any desired shading model.

6 Rendering

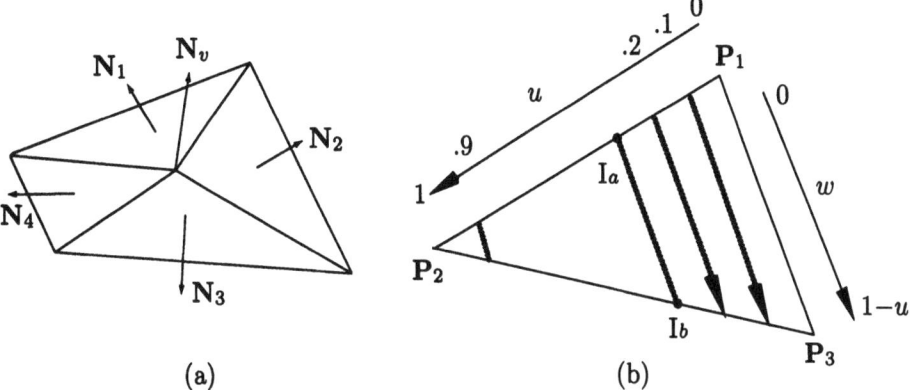

Figure 6.8: Gouraud Shading.

4. Each polygon is shaded, scan line by scan line, by interpolating the reflection intensities at its vertices. The scanning is done using Equation (5.5), which is duplicated here:

$$\mathbf{P}_1 + u(\mathbf{P}_2 - \mathbf{P}_1) + w(\mathbf{P}_3 - \mathbf{P}_1) = \mathbf{P}_1(1 - u - w) + \mathbf{P}_2 u + \mathbf{P}_3 w. \qquad (5.5)$$

For each scan line, two intensities, I_a and I_b, are interpolated from the vertex intensities I_1, I_2, and I_3 (Figure 6.8b) and are then used to interpolate an intensity I_p for every pixel on the line. Figure 6.9a shows a procedure for scanning a triangle. Figure 6.9b shows the start and end points of each scan line.

▶ **Exercise 6.3:** Modify the above triangle scanning procedure to handle a four-sided polygon.

Gouraud's shading is a simple method that usually produces good results. Its main disadvantage is the case where there should be a small shiny reflection (a highlight) inside a polygon. The method only calculates the reflection intensities at the corners, so it never finds out about the highlight.

Phong shading is similar, but it overcomes this problem by interpolating the normal vectors rather than the intensities. It ends up with an interpolated normal $\mathbf{N}_p$ at every pixel p and uses any shading method to calculate the reflection intensity from the point. This enhances mostly specular reflection.

6.4 Palette Optimization

The color lookup table was discussed on page 8. In this section, we assume a lookup table of 256 entries. Before displaying an image, the table has to be loaded with a palette of 256 colors, and **only** these colors can later be displayed on the screen.

The general problem of rendering consists of the separate tasks of calculating the surfaces and of displaying them. Displaying a surface is also a two-part problem. First, a shading algorithm is executed to determine the ideal color of each pixel, then another algorithm is needed, to pick up the best palette color, the one that's nearest the ideal color. It is, therefore, crucial to load the lookup table with the

```
procedure Gouraud(P1,P2,P3,I1,I2,I3);
real I; point P;
for u:=0 to 1 step 0.1 do
 for w:=0 to 1-u step 0.001 do
  I:=I1*(1-u-w)+I2*u+I3*w;
  P:=P1*(1-u-w)+P2*u+P3*w;
  Pixel(P,I);
end;
```

(a)

Value of u	Range of w	Scan Line from	to
0	0 − 1	P_1	P_3
.1	0 − .9	$.9P_1 + .1P_2$	$.1P_2 + .9P_3$
.2	0 − .8	$.8P_1 + .8P_2$	$.2P_2 + .8P_3$
⋮			⋮
.9	0 − .1	$.1P_1 + .9P_2$	$.9P_2 + .1P_3$
1	0 − 0	P_2	P_2

(b)

Figure 6.9: Scanning a Triangle.

right 256 colors, the ones that best "represent" the image to be displayed. This is the problem of *palette optimization*.

A simple solution is to count the number of times each color occurs in the original image and to load into the lookup table the 256 colors most common in the image. This is a simple solution that has one drawback—a color that occurs in just a few pixels in the original image may be crucial to our understanding (or our enjoyment) of the image. A better method should load the lookup table with the most common colors of the image, but also with a representative of every other color that happens to appear in the image.

A good example is an image of a beach scene. The dominant colors are blue (water and sky), yellow (sand), and white (clouds). However, there may be a person in the image, wearing a green swimsuit. Clearly, the lookup table needs to have at least one shade of green in it, even though it is going to be used in only a few pixels.

> Examining an image, one would be at a loss to know whether it owed its shape to the original source of light or to the details of the intervening gravitational field. The only difference between the appearance of the surface of the lake and that of the night sky is that the former depends on reflection and the latter on refraction.
>
> —H. C. von Baeyer, *The Fermi Solution*, 1993.

6 Rendering

A good, although not fast, palette optimization method is *median-cut color quantization*. It starts with the RGB cube (Section 7.4.1) where each axis is labeled from 0 to 255, and it ends up cutting the cube into 256 rectangular blocks, each containing about the same number of picture colors. In the beach example, there will be many blocks in the blue, yellow, and white regions of the RGB cube, but there will be at least one block in the green region. The last step is to select the color in the middle of each block and to load the lookup table with those 256 colors. Here are the steps:

1. Determine the extreme values of red in the picture colors. If no color in the picture has red below, say, 18 and above 240, then those parts of the RGB cube corresponding to red below 18 and above 240 are ignored (we can imagine them being cut off the cube and thrown away, since the picture has no colors in those parts). The same thing is done for the green and blue dimensions of the RGB cube.

2. The longest side of the remaining cube is now determined. Let's say that it is the red side. All the colors in the picture are sorted by their red values and the median color is picked up. The median is that shade of red that has equal numbers of shades of red above and below it. If it is, for example, $(56, x, y)$, then there are equal numbers of colors in the picture with red < 56 and with red > 56. The RGB cube is now cut at red $= 56$, producing two rectangular blocks.

3. The above process is now applied to the two blocks created in step 2. Each is cut at a median, which produces four blocks.

4. The process of cutting blocks is repeated 4 more times, for a total of 8 times, producing $2^8 = 256$ blocks. Because the cuts are always done at the median, each of the final 256 blocks has the same number of picture colors.

5. The center of each of the blocks is calculated, using the corner coordinates of the block, and is added to the color lookup table. An even better result is obtained by averaging, in each block, all the picture colors included in the block, but this is even more time-consuming.

6.5 Ray Tracing

This is a sophisticated shading method. It can calculate shading for an image that consists of several objects, several light sources, and a complex environment. It produces striking results, especially when the picture involves shiny objects reflecting each other in complex ways.

The principle is to start at the observer and trace rays from the eye to every pixel on the screen. The scene is assumed to be behind the screen and each ray is continued behind the screen until it hits part of the scene. The ray is then considered a reflection from that point, and the original ray that created that reflection is calculated and traced.

Figure 6.10 shows a scene consisting of a room, a light source, a mirror, and a crystal ball. The first ray is traced backward from the observer's eye to point A on the ball. Tracing it back brings us to the mirror and to the light source. This means that point A should show strong reflection. The ray traced back to point B behaves differently. It also originates at the light bulb but is reflected from a wall and from the floor before hitting the ball. Point B should, therefore, show weak reflection.

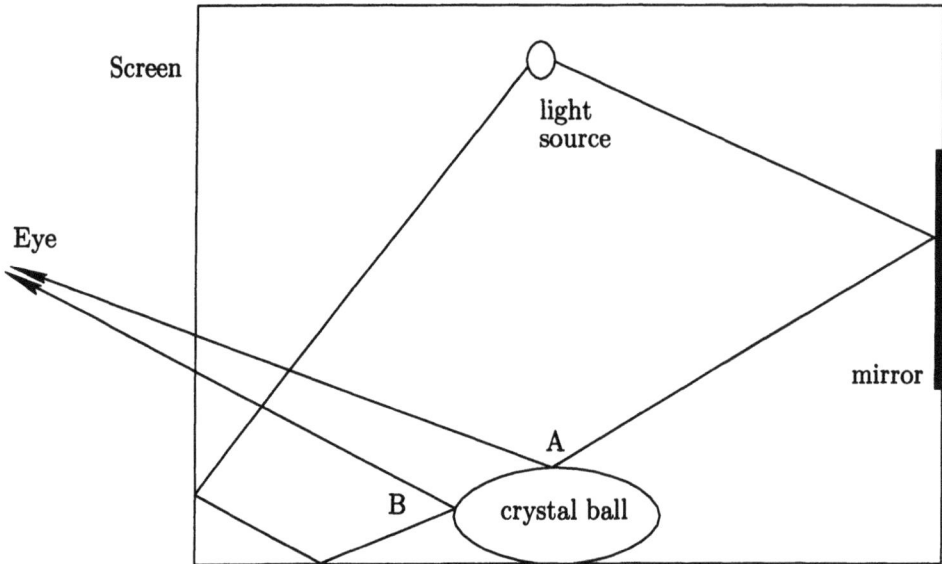

Figure 6.10: The Ray Tracing Principle.

Recall that a ray emitted from a given point is the sum of three contributions: light generated at the point, light reflected by it, and light transmitted through the object and leaving through the point. Imagine a light source to the right of the crystal ball. Rays of light will travel through the ball from right to left and some will go through point B. The ray from B to the eye, thus, consists of two parts, and the ray tracing software must be able to identify them and trace each of them individually. The tracing path from the eye to B *splits* at B, and two rays have to be traced from that point. In principle, each of the two may split again, which is one reason why ray tracing is so complex to implement.

A good, detailed reference for ray tracing is [Glassner 89], but see also Chapter 8 of [Hill 90] for a simplified, two-dimensional example. Many other references are available in the standard computer graphics literature.

6.6 Texturing

Texturing is a method commonly used to add realism to an image. The idea is to create a table with texture values (black and white, or colors) and map it onto the surface. The texture table is just a small bitmap, where each entry describes the color of a pixel. In practice, the table is an array $T[m,n]$ of values. Texturing is done in one of two ways:

1. A surface $\mathbf{P}(u,w)$ is given and has to be textured. The entire surface has to be scanned and each pixel should be assigned a value from the texture table. Normally, the surface is much bigger than the table. The table covers only a small part of the surface, and several copies of the table have to be used to texture the entire surface. The surface is scanned by varying u and w independently from 0 to 1, and a function is needed that maps each pair (u,w) to a pair of indexes (i,j) in the texture table (where $1 \leq i \leq m$ and $1 \leq j \leq n$).

6 Rendering

2. Several surfaces are given and the entire scene is shaded by ray tracing. Instead of shading each surface independently, we follow light rays from the eye of the observer through the screen and into the scene. Such a ray may hit a surface $\mathbf{P}(u,w)$ at point $\mathbf{P}=(x,y,z)$. We have to find the pair (u,w) that corresponds to this point and map this pair to a pair of indexes (i,j) in the texture table.

Regardless of the method used, texturing is affected by the topology of the surface. The texture table can be considered a flat rectangle. If the surface is anything other than a flat plane, then mapping the table to the surface may introduce distortions (in the same way that mapping the spherical Earth to a flat sheet of paper always involves distortions). This is why there is no single mapping that is good for all surfaces. Mapping functions have to be derived for common, regular surfaces, such as a sphere, a cone, a cylinder, or a torus. When given a general surface, the user should decide which of the known mapping functions to use, based on the shape of the surface given.

Example: Mapping a cylinder. This is a simple example because it is easy to wrap a rectangle on a cylinder without distortions. We start with the parametric equation of the cylinder (Equation (Ans.41)):

$$\mathbf{P}(u,w) = \bigl(a(2u-1), R\sin w, R\cos w\bigr),$$

where $0 \le u \le 1$ and $0 \le w \le 2\pi$. This describes a cylinder that's $2a$ pixels long, with a radius R, centered on the origin and pointing in the x direction. We assume a texture table $T(i,j)$, where $0 \le i \le m-1$ and $0 \le j \le n-1$.

The quantity $\text{round}(u(2a-1))$ has integer values that vary from 0 (when $u=0$) to $2a-1$ (when $u=1$). We, therefore, define $i = \text{round}(u(2a-1)) \bmod m$. This assigns index i values between 0 and $m-1$ and causes several copies of the texture table to be laid side by side along the cylinder.

For j, we similarly start with $\text{round}(w(2\pi R-1))$. This quantity gets integer values in the range 0 (when $w=0$) to $2\pi R - 1$ (when $w=1$). The index j itself is now defined as $j = \text{round}(w(2\pi R-1)) \bmod n$.

The surface can now be displayed by a double loop, on u and on w, where for each point, we perform two steps:

1. The normal vector at the point is calculated and a shading model is used to calculate the reflection intensity from the point.

2. A pair of indexes (i,j) is calculated, and the value $T(i,j)$ is used to modify the reflected intensity of step 1.

When ray tracing is used, we start with a ray that intersects the cylinder at a point (x,y,z). We need to find the values of (u,w) that correspond to that point. We consider the single equation

$$\bigl(a(2u-1), R\sin w, R\cos w\bigr) = (x,y,z),$$

a system of three equations in the two unknowns u and w. The solutions are

$$u = \left(\frac{x}{a}+1\right)/2, \qquad w = \arcsin(y/R) = \arccos(z/R).$$

Once u and w are known, a pair of indexes (i,j) can be calculated as above.

Example: Texturing a Sphere. A sphere of radius R, centered at the origin is expressed by:

$$(R\cos u \cos w, R\cos u \sin w, R\sin u),$$

where $-\pi/2 \le u \le \pi/2$ and $0 \le w \le 2\pi$. A large circle on this sphere has length $2\pi R$, so a meridian (constant w) from the south to the north pole covers πR pixels. The quantity $u + \pi/2$ varies in the range $[0, \pi]$, so $(u + \pi/2)R$ varies in the range $[0, \pi R]$. We, therefore, define the first mapping index as $i = \text{round}((u + \pi/2)R) \bmod m$.

Varying w for a constant u takes us along a latitude of radius $R\cos u$. This radius varies from R at the equator (where $u = 0$) to zero at the poles (where $u = \pi/2$ or $-\pi/2$). The circumference of this circle is, therefore, $2\pi R\cos u$, and half the circumference is $\pi R\cos u$.

Since w varies in the range $[0, 2\pi]$, the quantity $(w/2)R\cos u$ varies in the range $[0, \pi R\cos u]$. The second mapping index, j, is, therefore, defined as $j = \text{round}((w/2)R\cos u) \bmod n$.

The distortion can be observed by noticing that when $u = \pi/2$, the general expression of the sphere reduces to point $(0, 0, R)$, the north pole, regardless of w. The first texture index for this point is $i = \text{round}(\pi R) \bmod m$. This is a number in the range $[0, m-1]$. However, the second texture index for this point is $j = \text{round}((w/2)R\cos(\pi/2)) \bmod n = 0$.

When ray tracing is used, a ray may intersect the sphere at point (x, y, z). We need to calculate the corresponding values of (u, w). The equation

$$(R\cos u \cos w, R\cos u \sin w, R\sin u) = (x, y, z)$$

is used as a system of three equations with the two unknowns u and w. The solutions are $u = \arcsin(z/R)$ and $\tan w = y/x \Rightarrow w = \arctan(y/x)$.

6.7 Bump Mapping

Shading models are supposed to create realistically looking surfaces, but they result in surfaces that are ideally smooth and, hence, look artificial. Real surfaces are never completely smooth. They may have stains, holes, or cracks in them. At the very least, a real surface features small irregularities. Bump mapping is a method that simulates such irregularities.

The principle is to perturb the normal to the surface at every point before the point is shaded and use the new normal to shade the point. The perturbation can be random, or it may be based on a table. A good example is the surface of a strawberry. It is shiny but not smooth. Anyone familiar with this fruit knows that its surface varies in a complex but regular way. It is possible to prepare a small table that describes the variation in surface height over a small region of the strawberry and use the table repeatedly to create a "strawberry" effect on any surface.

If the bumps are to be random, no special algorithm is needed. A normal vector $\mathbf{N} = (N_x, N_y, N_z)$ is first normalized, then perturbed by adding random numbers to each of its three components. A component of a unit vector is always in the range $[-1, 1]$, so the random numbers should be drawn from a smaller interval, say, $[-0.001, 0.001]$.

> Bump: A small area raised above the level of the surrounding surface; protuberance.
>
> — A dictionary definition.

If the bumps are not random, we assume that a bump table $B(u, w)$, that gives the bump size for every surface point, is available. Note that B is a table of numbers, not of points or vectors. The value of each table entry $B(u, w)$ indicates by how much the corresponding surface point $\mathbf{P}(u, w)$ is to be raised (or lowered, if $B(u, w)$ is negative). Such a bump table can be generated by a drawing/painting program, or by scanning a picture. The surface is denoted, as usual, by $\mathbf{P}(u, w)$, the two partial derivatives are denoted $\mathbf{P}_u(u, w)$ and $\mathbf{P}_w(u, w)$, the normal is the cross-product $\mathbf{N}(u, w) = \mathbf{P}_u(u, w) \times \mathbf{P}_w(u, w)$, and the unit normal vector is denoted by $\mathbf{n}(u, w)$. The mapping rule is as follows: Each surface point should be raised by $B(u, w)\mathbf{n}$ (note that this defines both the magnitude and direction of the raise). We use the notation $\mathbf{P}^*(u, w) = \mathbf{P}(u, w) + B(u, w) \cdot \mathbf{n}$, where $\mathbf{P}^*(u, w)$ is the new surface point. The new tangent vectors are

$$\mathbf{P}^*_u(u, w) = \mathbf{P}_u(u, w) + B_u(u, w) \cdot \mathbf{n} + B(u, w) \cdot \mathbf{n}_u,$$
$$\mathbf{P}^*_w(u, w) = \mathbf{P}_w(u, w) + B_w(u, w) \cdot \mathbf{n} + B(u, w) \cdot \mathbf{n}_w.$$

Next, we note that the derivatives $\mathbf{n}_u$ and $\mathbf{n}_w$ of the unit normal depend on the curvature of the surface. If the surface is not highly curved, the magnitudes of those derivatives are small and they can be ignored. We can, therefore, write $\mathbf{P}^*_u \approx \mathbf{P}_u + B_u \cdot \mathbf{n}$ and $\mathbf{P}^*_w \approx \mathbf{P}_w + B_w \cdot \mathbf{n}$.

The normal $\mathbf{N}^*(u, w)$ to the new surface $\mathbf{P}^*(u, w)$ is the cross-product

$$\begin{aligned}\mathbf{N}^* &= \mathbf{P}^*_u \times \mathbf{P}^*_w \\ &= (\mathbf{P}_u + B_u \cdot \mathbf{n}) \times (\mathbf{P}_w + B_w \cdot \mathbf{n}) \\ &= \mathbf{P}_u \times \mathbf{P}_w + B_u \cdot \mathbf{n} \times \mathbf{P}_w + B_w \cdot \mathbf{P}_u \times \mathbf{n} \\ &= \mathbf{N} + B_u \cdot \mathbf{n} \times \mathbf{P}_w + B_w \cdot \mathbf{P}_u \times \mathbf{n} \\ &= \mathbf{N} + \alpha \mathbf{P}_u + \beta \mathbf{P}_w.\end{aligned}$$

It is the sum of the original normal $\mathbf{N}$ and of two cross-products, each scaled by a derivative of B. The first cross-product is perpendicular to both $\mathbf{P}_w$ and $\mathbf{n}$, so it points in the direction of $\mathbf{P}_u$. The second cross product is perpendicular to both $\mathbf{P}_u$ and $\mathbf{n}$, so it points in the direction of $\mathbf{P}_w$.

The new normal is now used to shade the surface point. Note that we don't actually move or "bump" the surface point $\mathbf{P}(u, w)$ to $\mathbf{P}^*(u, w)$. We just calculate the new normal $\mathbf{N}^*$ and use it to shade the point.

It is also important to note that the bump map B itself is not used by the algorithm, only its derivatives. In practice, good values for the derivatives can be obtained by subtracting

$$B_u(u,w) = B(u+s,w) - B(u-s,w),$$
$$B_w(u,w) = B(u,w+s) - B(u,w-s).$$

where s is a convenient step size, and the indexes $u \pm s$, $w \pm s$ should be calculated modulo the size of the bump table.

> He began to think that the source and secret of this ghostly light might be in the adjoining room, from whence, on further tracing it, it seemed to shine.
>
> Charles Dickens, *A Christmas Carol*

MIX
Papier aus verantwortungsvollen Quellen
Paper from responsible sources
FSC® C105338

If you have any concerns about our products,
you can contact us on
ProductSafety@springernature.com

In case Publisher is established outside the EU,
the EU authorized representative is:
**Springer Nature Customer Service Center GmbH
Europaplatz 3, 69115 Heidelberg, Germany**

Printed by Libri Plureos GmbH
in Hamburg, Germany

Computer Graphics
and
Geometric Modeling

Springer Science+Business Media, LLC

David Salomon

Computer Graphics and Geometric Modeling

With 335 Illustrations

 Springer

David Salomon
Department of Computer Science
California State University
Northridge, CA 91330-8281
USA

Library of Congress Cataloging-in-Publication Data
Salomon, D. (David), 1938–
    Computer graphics and geometric modeling / David Salomon.
       p.   cm.
    Includes bibliographical references and index.
    ISBN 978-1-4612-7170-3   ISBN 978-1-4612-1504-2 (eBook)
    DOI 10.1007/978-1-4612-1504-2
    1. Computer graphics.   2. Mathematical models.   I. Title.
T385.S243   1999
006.6—dc21   98-33424

Printed on acid-free paper.

© 1999 Springer Science+Business Media New York
Originally published by Springer-Verlag New York, Inc. in 1999
Softcover reprint of the hardcover 1st edition 1999

All rights reserved. This work may not be translated or copied in whole or in part without the written permission of the publisher Springer Science+Business Media, LLC, except for brief excerpts in connection with reviews or scholarly analysis. Use in connection with any form of information storage and retrieval, electronic adaptation, computer software, or by similar or dissimilar methodology now known or hereafter developed is forbidden.
The use of general descriptive names, trade names, trademarks, etc., in this publication, even if the former are not especially identified, is not to be taken as a sign that such names, as understood by the Trade Marks and Merchandise Marks Act, may accordingly be used freely by anyone.

Production managed by Jenny Wolkowicki; manufacturing supervised by Jeffrey Taub.
Camera-ready copy provided by the author.

9 8 7 6 5 4 3 2 1

ISBN 978-1-4612-7170-3

To my family, without whose help, patience, and support this book would not have been written.

Preface

Joseph-Louis Lagrange (1736–1813), one of the greatest mathematicians of the 18th century, made important contributions to the theory of numbers and to analytical and celestial mechanics. His most important work is *Mécanique Analytique* (1788), the textbook on which all subsequent work in this field is based. A contemporary reader is surprised to find no diagrams or figures of any kind in this book on mechanics. This reflects one extreme approach to graphics, namely considering it unimportant or even detracting as a teaching tool and not using it. Today, of course, this approach is unthinkable. Graphics, especially computer graphics, is commonly used in texts, advertisements, and movies to illustrate concepts, to emphasize points being discussed, and to entertain.

Our approach to graphics has been completely reversed since the days of Lagrange, and it seems that much of this change is due to the use of computers. Computer graphics today is a mature, successful, and growing field. It is used by many people for many purposes and it is enjoyed by even more people. One criterion for the maturity of a field of study is its size. When a certain discipline becomes so big that no one person can keep all of it in their head, we say that that discipline has matured (or has come of age). This is what happened to computer graphics in the last decade or so. It is now a large field consisting of many subfields such as curve and surface design, rendering methods, and computer animation. Even a person who has written a book covering the entire field cannot claim that they keep all that material in their head all the time, which is precisely the reason why textbooks are being written.

> Lagrange was born in Turin, Italy and his original name was Giuseppe Luigi LaGrangia. Here is a quote from the preface to his book:
> "The reader will find no figures in this work. The methods which I set forth do not require either constructions or geometrical or mechanical reasonings, but only algebraic operations, subject to a regular and uniform rule of procedure."

The material presented here has been developed during many years of teaching computer graphics. It has been revised and distilled many times, with many examples and exercises added. The text emphasizes the mathematics behind computer graphics and is intended for readers who want to understand how graphics programs work and how present-day computer graphics can generate and display realistic-looking curves, surfaces, and solid objects.

Most of the necessary mathematical background (such as vectors and matrices) is covered in the Appendix. However, some math concepts that are used only once (such as the mediation operator and points vs. vectors) are discussed right where they are introduced.

The many exercises sprinkled in the text are not a cosmetic feature. They deal with important topics, and should be worked out. Answers are provided but they should be consulted only as a last resort.

The main topics covered in this book are the following:

1. Scan conversion methods. These are used to select the best pixels for generating lines, circles, and other geometrical figures.

2. Geometric transformations and projections. This topic starts with simple concepts of translating and rotating objects, and ends with a complete picture of how an observer (or a camera) can be made to move around a three-dimensional scene, step by step, and project it on a two-dimensional screen. Perspective projections (a traditionally confusing topic) are discussed in two ways, first using the traditional approach, found in most texts, then using a novel, coordinate-free, vector-based approach. As a bonus, transformations and projections in many dimensions are discussed, as well as several interesting types of nonlinear projections.

3. Curves and surfaces. A realistic-looking object is calculated and displayed on the screen by simulating the light reflected from its surface, which is why surfaces are important in computer graphics. The key to constructing a surface, however, is to know how to draw a single curve. This is the main reason why curves are also important. The discussion is mathematical and it proceeds from curves to wireframe surfaces, to the calculation of the normal to a surface, to the rendering of a solid surface. Quite a few methods are discussed, each illustrated by examples.

4. Several other topics are presented, such as CRT operation, antialiasing, computer animation, color and color perception, halftoning and dithering, polygonal surfaces, and compression of graphics files.

An important feature of this text is the attention to "orphans." Those are topics that most texts on computer graphics either mention briefly or completely ignore. Examples are perspective projections, curves, surfaces, quaternions, and image compression. The reader will find that this text discusses orphans in great detail, including numerous examples and exercises.

The book is intended as a textbook for a two-course sequence on computer graphics and geometric modeling for graduate and advanced undergraduate students. However, it is the author's belief that the book can also serve as a professional book, widening the horizons of professionals in other fields who are interested in a thorough, mathematical exposition of the principles, methods, and techniques used in computer graphics. The mathematical background required for a complete understanding of this material includes polynomials, vectors, matrices, determinants, and differentiation.

Preface

Historical Notes

The history of computer graphics started in the early 1950s. This is very early, considering that the history of the modern digital electronic computer itself started in the late 1940s. However, due to high hardware prices, the field was originally the domain of a few lucky individuals, and it was only in the 1970s that it started growing fast and eventually became the wide discipline that we know today. Here is a short chronology.

By 1951 the Whirlwind computer installed at MIT had two 16-inch graphics displays (actually, modified oscilloscopes). Surprisingly, there were no immediate users.

Plotters came into use as graphics output devices in 1953.

In 1955, the SAGE air defense system started its operations. It used vector-scanned monitors as its main output and light pens as its input devices.

Digital Equipment Corporation (DEC) was founded in 1957. It started making minicomputers that were later used in the early development of computer graphics.

Light pens came into wide use in 1958, the same year as the first microfilm recorder.

In 1959, a partnership of General Motors and IBM produced the first piece of drawing software, the DAC-1 (Design Augmented by Computers). Users could input the three-dimensional description of a car, view the car in perspective, and rotate it.

It was in the 1960s that the field got its first big push. In 1961, Ivan Sutherland developed Sketchpad, a drawing program, as his Ph.D. thesis at MIT. Sketchpad used a light pen as its main input device and an oscilloscope (modified to do vector scan) as its output device. The first version handled two-dimensional figures only, and was later extended to draw, transform, and project three-dimensional objects, and also to perform engineering calculations such as stress analysis. One important feature of Sketchpad was its ability to recognize constraints. The user could draw, e.g., a rough square, then instruct the software to convert it to an exact square. Another feature was the ability to deal with objects, not just individual curve segments. The user could build an object out of segments, then ask the software to scale it. For information and images related to Sketchpad, see http://www.sun.com/960710/feature3/sketchpad.html#sketch.

At about the same time, Steven Russell, another MIT student, developed the first video game, *Spacewar*. This program was written for the PDP-1 and was later used by DEC salesmen to attract customers for that minicomputer.

In 1963, the first computer-generated film, titled *Simulation of a two-giro gravity attitude control system*, was created by E. E. Zajac at Bell laboratories. Other researchers at Bell, Boeing, and Lawrence Radiation Laboratory followed soon with more films.

The first digitizer, the RAND tablet, appeared in 1964.

In the mid-1960s, interest in computer graphics was picking up. More and more companies started projects involving graphics, which gave IBM the idea of developing the first graphics terminal, the IBM 2250. At about the same time, David Evans and Ivan Sutherland cofounded their company which made, among other things, vector scan displays. Those displays are historically important since they gave a tremendous boost to computer graphics throughout the sixties.

In 1966, Sutherland developed the first three-dimensional head-mounted display (HMD). It displayed a stereoscopic pair of wire-frame images. This device was rediscovered in the 1980s and is commonly used today in virtual-reality applications.

In the late 1960s, both Sutherland and Evans were invited to develop a program in computer science at the University of Utah in Salt Lake City. Computer graphics quickly became the specialty of their department, and for years maintained its position as the primary world center for this field. Many important methods and techniques were developed at the UU computer graphics lab, among them illumination models, hidden-surface algorithms, and basic rendering techniques for polygonal surfaces. Names of UU students such as Phong Bui-Tuong, Henri Gouraud, James Blinn, and Ed Catmull are associated with many basic algorithms still in use today. A short history of computer graphics at UU can be found at URL
http://www.cs.utah.edu/~riloff/cs-history.html.

Computer graphics in the 1960s was expensive since hardware was expensive. There were no personal computers or workstations. Users had to pay for mainframe time by the second or buy expensive minicomputers. Display monitors used vector scan and were black and white. The result was that only computer professionals could do computer graphics and the software was noninteractive and nonportable.

The advent of the microprocessor, in the mid-1970s, was another factor in the rapid advance of computer graphics. Personal computers appeared on the market and suddenly anyone could afford to own a computer. This encouraged the formation of small companies that developed computer animation, mostly to be used in TV commercials. Names such as Abel and Associates, Information International Inc., Digital Effects, and Systems Simulation Ltd. became well known and produced short pieces that demonstrated dazzling effects.

SIGGRAPH, the Special Interest Group on Computer Graphics (part of the ACM), held its first conference in 1973. It attracted 1200 attendees and later conferences boasted as many as 30,000 participants and hundreds of exhibitors.

The famous Utah teapot (see page 479) was developed in 1975. This is perhaps the best known three-dimensional model in computer graphics. The original teapot this model is based on can be seen at the Computer Museum in Boston.

It was during the 1970s that activity in basic computer graphics research started moving from UU first to NYIT, the New York Institute of Technology, then to Lucasfilm. Computer animation and computer painting were two topics seriously developed at those places.

The technique of (and hardware for) raster scan was developed in the 1970s by Richard Shoupe at Xerox Palo Alto Research Center (PARC). Workers in the field soon realized the advantages of raster scan and the word "pixel" entered the field of computer graphics.

Like any other mature discipline, computer graphics eventually got its first periodic publication. *Computer Graphics World* started carrying news and reviews in late 1977.

Fractals, developed by Benoit Mandelbrot in the 1960s and 1970s, were applied to computer graphics in the late 1970s by Loren Carpenter and others.

Preface

Ray tracing, a sophisticated rendering method, was developed by Turner Whitted of Bell labs and published in 1980.

Silicon Graphics Inc. (SGI) was founded in 1982 and has been building high-performance graphics computers since.

The technique of particle systems was developed in the early eighties at Lucasfilm. Morphing was developed at the same time at NYIT.

The data glove, very popular today for virtual-reality applications, was developed at Atari in 1983.

Radiosity came out of Cornell University in 1984. This is a sophisticated rendering method that simulates light reflection between surfaces by determining the exchange of energy between them.

GUI, graphical user interfaces, appeared in 1984 with the release of the first Macintosh computer.

In 1985 came the first ISO standard, the High Sierra, for CD-ROMs. The Commodore Amiga personal computer was also introduced the same year. It immediately became popular for what today are called multimedia applications.

The 1980s saw the emergence of raster-scan display monitors as the main graphics output device. This technology has benefited from experience gained with television and has resulted in the cheap, reliable color monitors of today.

The late 1980s and early 1990s also saw the developments of graphics standards such as GKS and PHIGS.

MS Windows 3.0 was first shipped in 1990 and, of course, gave a tremendous boost to the concept of GUI. More and more applications were developed to run under MS Windows.

Released in 1997, *Toy Story* is the first full-length (79 minutes, which translates to more than 114,000 animation frames at 24 frames per second) feature film that's completely computer-animated. It represents a milestone in computer graphics and marks the beginning of an era when computer graphics rendering techniques have become so sophisticated that viewers may find it impossible to tell if an image is real or if it is a mathematical model being rendered.

Shaw's plays are the price we pay for Shaw's prefaces.
— James Agate

Resources for Computer Graphics

As is natural to expect, the World Wide Web has many resources for computer graphics. There are also many periodicals on various aspects of graphics. Here is a list of some of the most important resources, current as of 1998.

- http://www.siggraph.org/ is the official home page of SIGGRAPH, the special interest group for graphics, one of many SIGs that are part of the ACM.

- http://www.siggraph.org/conferences/fundamentals has course notes from SIGGRAPH conferences.

- `http://www.primenet.com/~grieggs/cg_faq.html` by John Grieggs contains answers to frequently asked questions on graphics, as well as pointers to other resources.

- The most recent version of Richard Parent's book on computer animation is at `http://www.cis.ohio-state.edu/~parent/book/outline.html`.

- `http://mambo.ucsc.edu/psl/cg.html` is a jumping point to many sites that deal with computer graphics.

- A similar site is `http://www.cs.rit.edu/~ncs/graphics.html` that also has many links to CG sites.

- A very extensive site of computer-graphics-related pointers is `http://ls7-www.informatik.uni-dortmund.de/html/englisch/servers.html`.

- Search the Internet under "history of computer graphics" for many sites.

- It is also a good idea to search the Internet for subjects such as computer graphics, computer animation, image processing, computer vision, and computer-assisted design (CAD).

- *IEEE Computer Graphics and Applications* is a technical journal carrying research papers and news. See `http://computer.org/cga`.

- *Animation Magazine* is a monthly publication covering the entire animation field, computer and otherwise. Check either at `http://animag.com/` or URL `http://www.bcdonline.com/animag/`.

- *Computer Graphics World* is a monthly publication concentrating on news (`http://www.cgw.com/`).

- *Digital Imaging* is a bimonthly reporting on the digital imaging industry.

A Word on Notation

It is common to represent nonscalar quantities such as points, vectors, and matrices with boldface. Below are example of the notation used here:

x, y, z, t, u, v	Italics are used for scalar quantities such as coordinates and parameters.
$\mathbf{P}, \mathbf{Q}_i, \mathbf{v}, \mathbf{M}$	Boldface is used for points, vectors, and matrices.
$\vec{CP}$	An alternative notation for vectors, used when the two endpoints of the vector are known.
$\mathbf{P}(t), \mathbf{P}(u,v)$	Boldface with arguments is used for nonscalar functions such as curves and surfaces.
$\begin{pmatrix} a_{11} & a_{12} \\ a_{21} & a_{22} \end{pmatrix}$	Parentheses are used for matrices.

Preface

$\begin{vmatrix} a_{11} & a_{12} \\ a_{21} & a_{22} \end{vmatrix}$	Vertical bars are used for determinants.
$\|\mathbf{v}\|$	The absolute value (length) of vector $\mathbf{v}$.
$\mathbf{A}^T$	The transpose of matrix $\mathbf{A}$.
$x^*, \mathbf{P}^*$	The transformed values of scalars and points.
$f^u(u), \mathbf{P}^t(t), \mathbf{P}^{tt}(t)$	The derivatives (first, second,...) of scalar and vector functions.
$\dfrac{df(u)}{du}, \dfrac{d\mathbf{P}(t)}{dt}$	Alternative notation for derivatives.
$\dfrac{df^2(u)}{du^2}, \dfrac{d\mathbf{P}^2(t)}{dt^2}$	Alternative notation for higher-order derivatives.
$\dfrac{\partial f(u,v)}{\partial u}, \dfrac{\partial \mathbf{P}(u,v)}{\partial v}$	Partial derivatives.
$f(x)\|_{x_0}$ or $f(x_0)$	Value of function $f(x)$ at point x_0.
$\sum_{i=1}^{n} x_i$	The sum $x_1 + x_2 + \cdots + x_n$.
$\prod_{i=1}^{n} x_i$	The product $x_1 x_2 \ldots x_n$.

▶ **Exercise 1:** What is the meaning of $(\mathbf{P}_1, \mathbf{P}_2, \mathbf{P}_3, \mathbf{P}_4)$?

⏵ The attention symbol, shown on the left, is used to attract the reader's attention when important concepts are introduced.

The left triangle ◂ is the QED symbol, indicating the end of a proof (of which there are just a few in this book).

A book of this magnitude cannot be written without the help and encouragement of many people, but this book is an exception. The author would like to thank Nelson H. F. Beebe who went over the manuscript, made important suggestions, and pointed out many errors. J. Robert Henderson rendered valuable help with mathematical topics. Apart from this, the entire text is the product of the author who alone is responsible for any remaining mistakes, errors, and omissions.

The author welcomes any comments, suggestions and corrections. They should be e-mailed to david.salomon@csun.edu. An errata list, as well as other information, will be kept on the author's web page http://www.ecs.csun.edu/~dxs.

Fasten your seatbelts, it's going to be a bumpy night!
— Bette Davis (as Margo Channing) *All About Eve (1950)*

Contents

Preface vii

1. **First Principles** 1
   1. Graphics Output — 2
   2. Bitmap Scaling — 16
   3. Bitmap Rotation — 20
   4. A Practical Drawing Program — 23
2. **Scan-Converting Methods** 27
   1. Scan-Converting Lines — 27
   2. Midpoint Subdivision — 28
   3. DDA Methods — 28
   4. Double-Step DDA — 34
   5. Best-Fit DDA — 38
   6. Scan-Converting in Parallel — 39
   7. Scan-Converting Circles — 42
   8. Thick Curves — 49
   9. Antialiasing — 50
3. **Transformations and Projections** 59
   1. Introduction — 59
   2. Two-Dimensional Transformations — 61
   3. Windowing — 88
   4. Clipping — 90
   5. Three-Dimensional Transformations — 91
   6. Transforming the Coordinate System — 102
   7. Projections — 103
   8. Parallel Projections — 103
   9. Perspective Projections — 112
   10. Application: Stereo Image — 137
   11. The Viewing Volume — 141
   12. Going Beyond the Third Dimension — 144
   13. Nonlinear Projections — 153

4. Curves — 173

1	Points and Vectors	174
2	Parametric Blending	180
3	Curve Representations	181
4	The Lagrange Polynomial	198
5	The Newton Polynomial	205
6	Spline Methods for Curves	206
7	Hermite Interpolation	207
8	The Cubic Spline Curve	225
9	The Quadratic Spline	247
10	Cardinal Splines	248
11	Parabolic Blending: Catmull-Rom Curves	251
12	Kochanek-Bartels Splines	258
13	Fitting a PC to Experimental Points	262
14	The Bézier Curve	266
15	Subdivision Curves	321
16	The B-Spline	328
17	The Beta Spline	389
18	Barycentric Sums Revisited	393
19	Symmetry in Curves	394
20	Conic Sections	397
21	Parametric Space of a Curve	401
22	Curvature and Torsion	402
23	The Hough Transform	410

5. Surfaces — 415

1	Input Three-Dimensional Points	416
2	Basic Concepts	417
3	Polygonal Surfaces	419
4	Delaunay Triangulation	427
5	Bilinear Surfaces	434
6	Lofted Surfaces	439
7	Coons Surfaces	443
8	The Cartesian Product	456
9	The Biquadratic Surface Patch	457
10	The Bicubic Surface Patch	459
11	Catmull-Rom Surfaces	468
12	Rectangular Bézier Surfaces	471
13	Triangular Bézier Surfaces	483
14	Converting Bézier Patches	488
15	The Gregory Patch	495
16	Gordon Surfaces	498
17	Uniform B-Spline Surfaces	499
18	Surfaces of Revolution	516
19	Sweep Surfaces	526
20	Polygonal Surfaces by Subdivision	530
21	Curves on Surfaces	533
22	Surface Normals	535

Contents

6. Rendering ——————————————————————— **537**

1	Introduction	537
2	A Simple Shading Model	538
3	Gouraud and Phong Shading	548
4	Palette Optimization	549
5	Ray Tracing	551
6	Texturing	552
7	Bump Mapping	554

7. Color ——————————————————————— **557**

1	Color and the Eye	557
2	The HLS Color Model	559
3	The HSV Color Model	559
4	The RGB Color Model	561
5	Additive and Subtractive Colors	563
6	Complementary Colors	567
7	Spectral Density	567
8	The CIE Standard	571

8. Computer Animation ——————————————————————— **575**

1	Background	575
2	Interpolating Positions	578
3	Interpolating Orientations: I	583
4	Interpolating Orientations: II	593
5	Nonuniform Interpolation	600
6	Morphing	606
7	Free-Form Deformations	607

9. Image Compression ——————————————————————— **609**

1	Introduction	610
2	Variable-Size Codes	611
3	Run-Length Encoding	612
4	Fax Compression	615
5	Cell Encoding	622
6	Quadtrees	624
7	Progressive Image Compression	630
8	FELICS	634
9	The Golomb Code	642
10	Progressive FELICS	643
11	MLP	646
12	Differential Lossless Image Compression	654
13	Wavelets	656

10. Short Topics — 661

1	Graphics Standards	661
2	Boundary Fill	668
3	Halftoning	669
4	Dithering	671
5	Fractals	680
6	A Fractal Line	681
7	Branching Rules	684
8	Iterated Function Systems (IFS)	685
9	Image Processing	688

A. Mathematical Topics — 693

1	Fourier Transforms	693
2	Forward Differences	698
3	Coordinate Systems	700
4	Vector Algebra	702
5	Matrices	709
6	Trigonometric Identities	711
7	The Greek Alphabet	715
8	Complex Numbers	716
9	Quaternions	717
10	Groups	719
11	Fields	720

References — 723

Answers to Exercises — 733

Index — 833

> To me style is just the outside of content, and content the inside of style, like the outside and the inside of the human body both go together, they can't be separated.
> — Jean-Luc Godard

Computer Graphics
and
Geometric Modeling

7
Color

7.1 Color and the Eye

We rarely find ourselves in complete darkness. In fact, most of the time we are flooded with light. This is why people get used to having light around and, as a result, they only rarely ask themselves *what is light*?

The best current explanation is that light can be understood (or interpreted) in two different ways, as a wave, or as a stream of particles. The latter interpretation sees light as a stream of massless particles, called *photons*, moving at a constant speed. The most important attribute of a photon is its frequency, since the photon's energy is proportional to it. Photons are useful in physics to explain a multitude of phenomena (such as the interaction of matter and light). In computer graphics, the most important property of light is its color, which is why we use the former interpretation and we consider light to be a wave.

What "waves" (or undulates) in light is the electric and magnetic fields. When a region of space is flooded with light, those fields change periodically as we move from point to point in space. If we stay at one point, the fields also change periodically with time. Visible light is thus a (small) part of the electromagnetic spectrum (Figure 7.1) that includes radio waves, ultraviolet, infrared, X-rays, and other types of radiation.

The most important properties of a wave are its frequency f, its wavelength λ, and its speed. Light moves, obviously, at the speed of light (in vacuum, it is $c \approx 3 \times 10^{10}$ cm/s). The three quantities are related by $f\lambda = c$. It is important to realize that the speed of light depends on the medium in which it moves. As light moves from vacuum to air to glass, it slows down (in glass, the speed of light is about $0.65c$). Its wavelength also decreases, but its frequency remains constant. Nevertheless, it is customary to relate colors to the wavelength and not to the frequency. Since visible light has very short wavelengths, a convenient unit is the nanometer (1 nm=10^{-9} m).

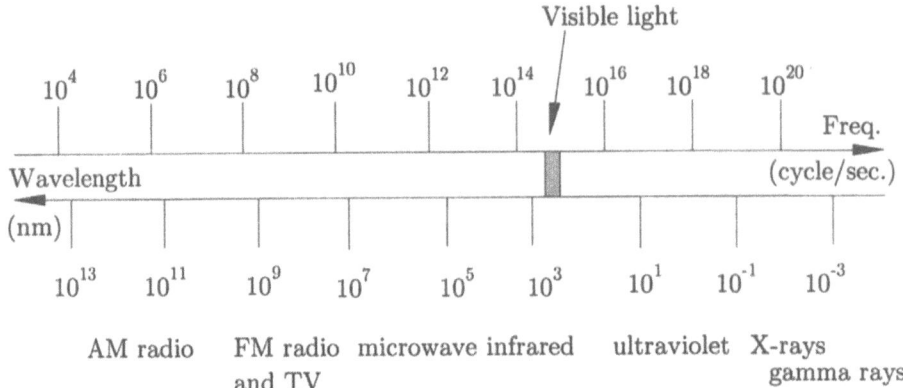

Figure 7.1: The Electromagnetic Spectrum.

Visible light ranges from about 400 nm to about 700 nm and the color is determined by the wavelength. A wavelength of 420 nm, for example, corresponds to pure violet, while 620 nm is perceived by the eye as pure red. Using special lasers, it is possible to create pure (monochromatic) light consisting of one wavelength (Figure 7.2a). Most light sources, however, output light that's a mixture of several (or even many) wavelengths, normally with one wavelength dominating (Figures 7.2b and 7.8).

The colors of the spectrum that are most visible to the human eye are (Figure 7.7) violet (390–430), blue-violet (460–480), cyan, green (490–530), yellow (550–580), orange (590–640), and red (650–800).

White light is a mixture of all wavelengths, but what is gray light? It turns out that the wavelength of light is not its only important attribute. The *intensity* is another attribute that should be considered. Gray light is a mixture of all wavelengths, but at a low intensity. When doing computer graphics, the main problem is how to specify the precise color of each pixel to the hardware. In many real-life situations, it is sufficient to say "I think I would like a navy blue suit," but computer hardware requires, of course, a much more precise specification. It, therefore, comes as a surprise to discover that color can be completely specified by just three parameters. Their meanings depend on the particular *color model* used. The RGB model is popular in computer graphics. In the printing industry, the CMYK is normally used. Many artists use the HLS model. These models are discussed below.

Figure 7.2b shows a simplified diagram of light smeared over the entire range of visible wavelengths, with a spike at about 620 nm (red), where it has a much higher intensity. This light can be described by specifying its *hue*, *saturation*, and *luminance*. The hue of the color is its dominant wavelength—620 nm in our example. The luminance is related to the intensity of the light. It is defined as the total power included in the spectrum and it is proportional to the area under the curve, which is $L = (700 - 400)A + B(D - A)$. The saturation is defined as the percentage of the luminance that resides in the dominant wavelength. In our case, it is $B(D - A)/L$. When there is no dominant wavelength (i.e., when $D = A$ or $B = 0$), the saturation

7 Color

is zero and the light is white. Large saturation means either large $D - A$ or small L. In either case, there is less white in the color and we see more of the red hue. Large saturation therefore corresponds to *pure color*.

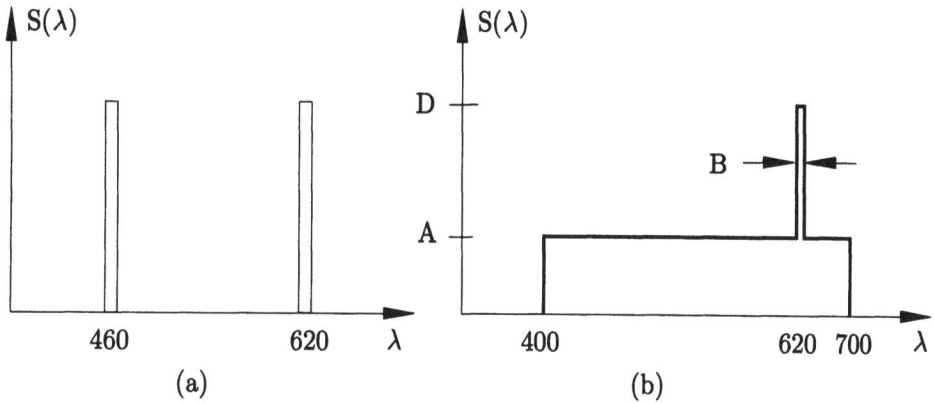

Figure 7.2: (a) Pure Colors. (b) A Dominant Wavelength.

7.2 The HLS Color Model

This model was introduced in 1978 by Tektronics, aiming for an intuitive way to specify colors. The name stands for hue, lightness, and saturation. Lightness (or value) refers to the amount of black in the color. It controls the brightness of the color. Maximum lightness always creates white, regardless of the hue. Minimum lightness results in black. Saturation (or chroma) refers to the amount of white in the color. It controls the purity or vividness of the color. Low saturation means more white in the color, resulting in a pastel color. Very low saturation results in a washed-out color. For a pure, vivid color, the saturation should be maximum. The achromatic colors black, white, and gray have zero saturation and differ in their values (Plate 1).

The HLS model is summarized in the double cone of Figure 7.3. The vertical axis corresponds to L (lightness). It starts at zero (black) at the bottom and ends at one (white) at the top. The distance from the central axis corresponds to S (saturation). All points on the axis have zero saturation, so they correspond to shades of gray. Points farther away from the axis have more saturation; they correspond to more vivid colors. The H parameter (hue) corresponds to the hue of the color. This parameter is measured as an angle of rotation around the hexagon.

7.3 The HSV Color Model

The HSV model also uses hue, saturation, and value (lightness). It is summarized in the cone of Figure 7.4. This is a single cone where the value V, which corresponds to lightness, goes from 0 (black) at the bottom, to 1 (white) at the flat top. The S and H parameters have the same meanings as in the double HLS cone.

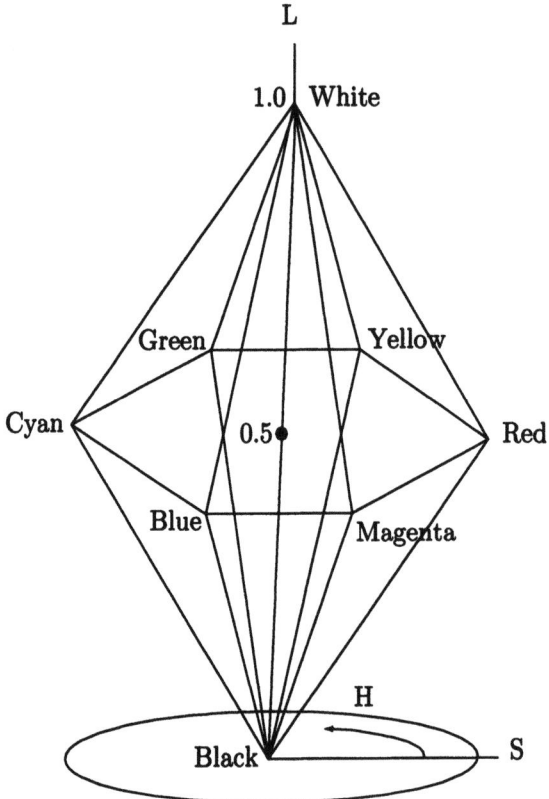

Figure 7.3: The HLS Double Hexcone.

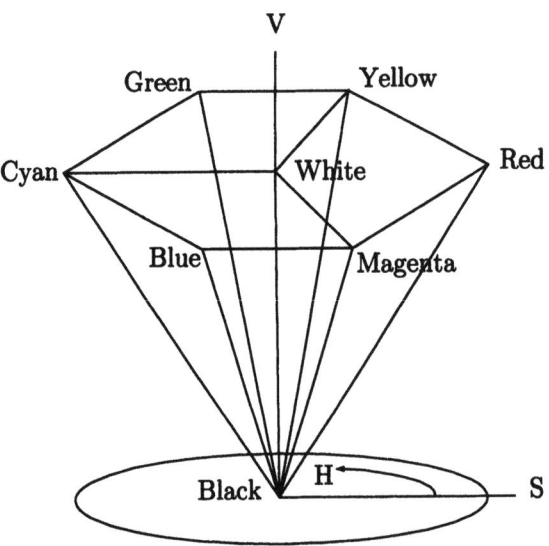

Figure 7.4: The HSV Hexcone.

7.4 The RGB Color Model

A *primary hue* is a color in a color model that cannot be made from the other colors used in that model. Primary hues serve as a basis for mixing and creating all other colors in the color model. Any color created by mixing *two* primary hues in a color model is a *secondary hue* in that model.

In the RGB color model, the three primaries are Red, Green, and Blue. They can be combined, two at a time (Plate 1) to create the secondary hues. Magenta (pinkish hue) is (R+B), cyan (bluish hue) is (B+G), and yellow is (R+G). There are two reasons for using the red, green, and blue colors as primaries: (1) the cones in the eye are very sensitive to these colors (Figure 7.7) and (2) adding red, green, and blue can produce many colors (although not all colors, see discussion of RGB color gamut on page 573).

The RGB color model is useful in computer graphics because of the way color CRTs work. They creates different colors by light emitted from phosphors of different types. The colors are then mixed in the eye of the observer, creating the impression of a perfect mixture. Assuming a range of $[0, 255]$ for each RGB color component, here are some examples of mixed colors:

$$\text{red} = (255, 0, 0),\ \text{magenta} = (255, 0, 255),\ \text{white} = (255, 255, 255),$$
$$50\%\ \text{gray} = (127, 127, 127),\ \text{light gray} = (25, 25, 25).$$

> It's the weird color-scheme that freaks me. Every time you try to operate one of these weird black controls, which are labeled in black on a black background, a small black light lights up black to let you know you've done it!
>
> — Mark Wing-Davey (as Zaphod Beeblebrox) in *The Hitchhiker's Guide to the Galaxy* (1981).

7.4.1 The RGB Cube

The *color gamut* of a color model is the entire range of colors that can be produced by the model. The color gamut of the RGB model can be summarized in a diagram shaped like a cube (Figure 7.5). Every point in the cube has three coordinates (r, g, b)—each in the range $[0, 1]$—which give the intensities of red, green, and blue of the point. Small values, close to $(0, 0, 0)$, mean a dark shade, whereas anything close to $(1, 1, 1)$ is very bright. Point $(1, 1, 1)$ itself corresponds to pure white. Point $(1, 0, 0)$ corresponds to red and point $(0, 1, 0)$ corresponds to green. Therefore, point $(1, 1, 0)$ describes a mixture of red and green, that is, yellow.

The RGB cube is useful because coordinates of points in it can readily be translated into values stored in the color lookup table of the computer.

▶ **Exercise 7.1:** (Easy.) What colors correspond to the diagonal line connecting the black and white corners of the RGB cube?

7.4 The RGB Color Model

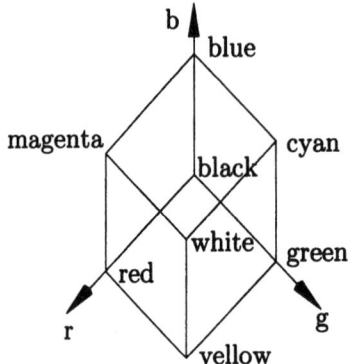

Figure 7.5: The RGB Cube.

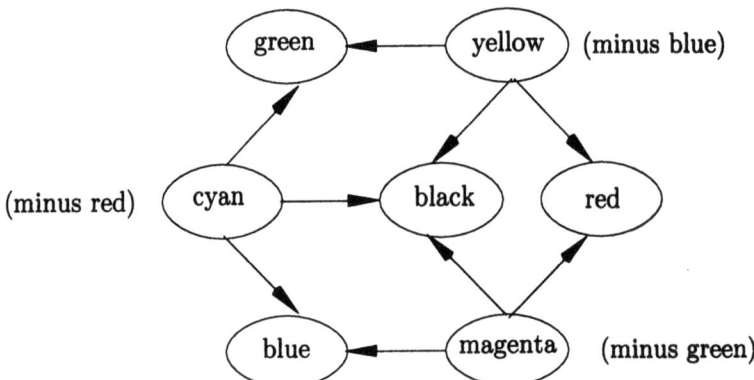

Figure 7.6: RGB and CMYK Relationships.

> For example, the human eye and its controlling software implicitly embody the false theory that yellow light consists of a mixture of red and green light (in the sense that yellow light gives us the same sensation as a mixture of red light and green light does). In reality, all three types of light have different frequencies and cannot be created by mixing light of other frequencies. The fact that a mixture of red and green light appears to us to be yellow light has nothing whatever to do with the properties of light, but is a property of our eyes. It is a result of a design compromise that occurred at some time during our distant ancestors' evolution.
>
> —David Deutsch, *The Fabric Of Reality.*

7 Color

7.5 Additive and Subtractive Colors

To create a mixture of several colors, we sometimes have to add them and sometimes have to subtract them. Imagine a white wall in a dark room. There is no light for the wall to reflect, so it looks black. We now shine red light on it. Since the wall is white (reflects all colors), it will reflect the red light and will look red. The same thing is true for green light. If we now shine both red and green colors on the wall, it will reflect both, which our brain interprets as yellow. We say that in this case the colors are added.

To understand the concept of subtracting colors, imagine a white sheet of paper in a bright environment. The paper reflects all colors, so it looks white. If we want to paint a certain spot red, we have to cover it with a chemical (red paint) that absorbs all colors except red. We say that the red paint *subtracts* the green and blue from the original white reflection, so the spot now reflects just red light. Similarly, if we want a yellow spot, we have to use yellow paint, which is a substance that absorbs blue and reflects red and green.

We conclude that in the case where we shine white light on a reflecting surface, we have to subtract colors in order to get the precise color that we want. In the case where we shine light of several colors on such a surface, we have to add colors to get any desired mixture (Plate 1).

> The various colours that may be obtained by the mixture of other colours, are innumerable. I only propose here to give the best and simplest modes of preparing those which are required for use. Compound colours, formed by the union of only two colours, are called by painters virgin tints. The smaller the number of colours of which any compound colour is composed, the purer and the richer it will be. They are prepared as follows:
> —Daniel Young, 1861, *Young's Translation of Scientific Secrets.*

7.5.1 Subtractive Color Models

There are two such models: painter's pigments and printing pigments.

Painter's Pigments: The primary colors of this color model are red, yellow, and blue. They were chosen because past artists believed that they were pure colors, containing no traces of any other colors. These three primaries can be mixed, two at a time, to produce the secondaries purple (R+B), green (B+Y), and orange (Y+R). Mixing equal amounts of all three primaries subtracts all colors and, hence, yields black.

Printing Pigments: This color model is also known as *process color* and is the result of development in color ink and printing processes. The three primaries are magenta, yellow, and cyan. The three secondaries are blue (M+C), red (M+Y), and green (C+Y). Mixing equal amounts of all three primaries should yield black, but, because of the properties of real inks, this black is normally not dark enough. In practice, true black is included in this model as an artificial fourth primary (also because black ink is cheaper). It is used when grayscale or black printing is required. The model is therefore sometimes called CMYK (K for black, to avoid confusion with blue) and color printing is known as the four-color process. Figure 7.6 shows the relationships between the three CMY primaries and their secondaries. Plate 2 shows examples of CMYK colors.

Because of the particular primaries and secondaries of the CMY model there is a simple relationship between it and the RGB model. The relation is

$$(r,g,b) = (1,1,1) - (c,m,y).$$

This relationship shows that, for example, increasing the amount of cyan in a color, reduces the amount of red.

Traditional color printing uses color separation. The first step is to photograph the original image through different color filters. Each filter separates a primary color from the multicolored original. A blue filter separates the yellow parts of the original and creates a transparency with those parts printed in black and white. A red filter separates the cyan parts and a green filter separates the magenta parts. Another transparency is prepared, with the black parts. Each of the four transparencies is then converted to a halftone image (Section 10.3) and the four images become masters for the final printing. They are placed in different stages of the printing machine and, as the paper moves through the machine, each stage adds halftone dots of colored ink to the paper. The result is a picture made of four halftone grids, each in one of the CMYK colors. The grids are not superimposed on the paper but are printed offset. The eye sees dots colored in the four primaries, and the brain creates a mixed color that depends on the number of halftone dots of each primary.

When such a color print is held close to the eye, the individual dots in the four colors can be seen. Today, there are dye sublimation printers that mix wax of different dyes inside the printer to create a drop of wax of the right color which is then deposited on the paper. No halftoning is used. The result is a picture in vivid colors, but these printers are currently too expensive for home use. For more information on color printing and the CMYK model, see [Stone et al. 88].

To simplify the task of editors and graphics designers, several color standards have been developed. Instead of figuring out the ratios of CMYK, the graphics designer looks at a table that has many color samples, selects one, and uses its name to specify it to the printer. One such standard in common use today is the PANTONE matching system. It is described in [Pantone 91].

▶ **Exercise 7.2:** A surface has a certain color because of its ability to absorb and reflect light. A surface that absorbs most of the light frequencies appears dark; a surface that reflects most frequencies appears bright. What colors are absorbed and what are reflected by a yellow surface?

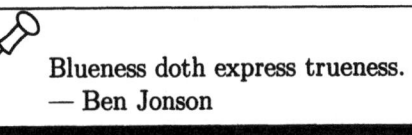

Blueness doth express trueness.
— Ben Jonson

7 Color

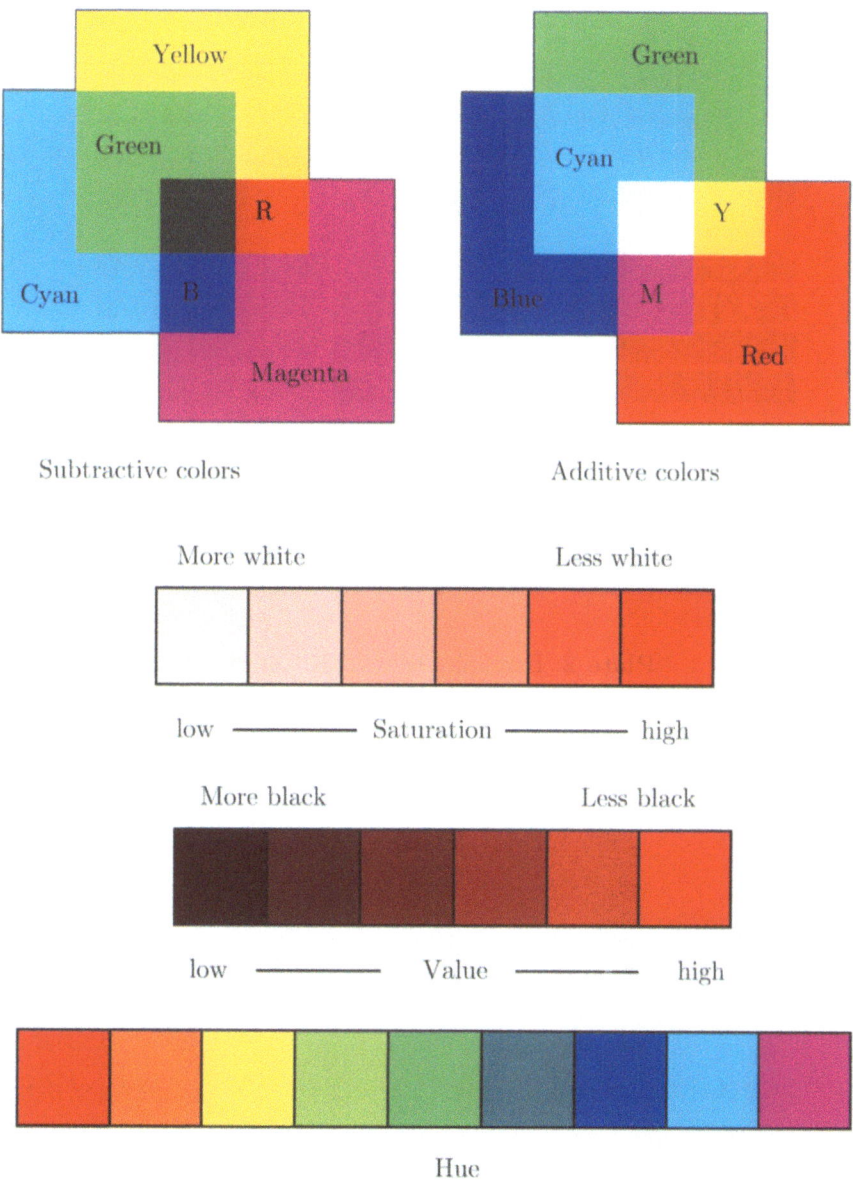

Plate 1: Examples in Color

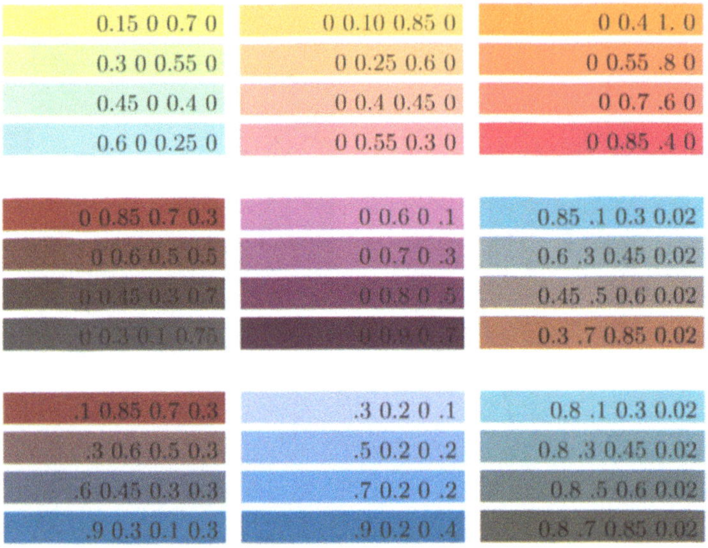

Plate 2: Examples of CMYK Colors.

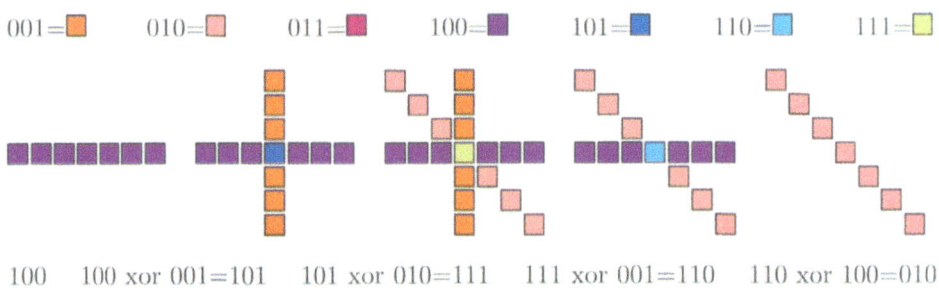

Plate 3: Drawing and Erasing Color Lines in XOR Mode.

7 Color

7.6 Complementary Colors

The concept of complementary colors is based on the idea that two colors appear psychologically harmonious if their mixture produces white. Imagine the entire color spectrum. The sum of all the colors produces white. If we subtract one color, say, blue, the sum of the remaining colors produces the complementary color, yellow. Blue and yellow are thus complementary colors (a dyad) in an additive color model. Other dyads are green and magenta, red and cyan, yellow-orange and cyan-blue, cyan-green and red-magenta, and yellow-green and blue-violet.

Subtractive Complementary Colors: These are based on the idea that two colors look harmonious if their mixture yields a shade of gray. The subtractive dyads are yellow and violet, red and green, blue and orange, yellow-orange and blue-violet, blue-green and red-orange, and yellow-green and red-violet.

Complementary colors produce strong visual contrast, which creates a feeling of color vibrations or activity.

▸ **Exercise 7.3:** Is there such a thing as additive color triads?

> **Color**
>
> Colors, like features, follow the changes of the emotions.
> — Pablo Picasso
> There is no blue without yellow and without orange.
> — Vincent Van Gogh

7.7 Spectral Density

A laser is capable of emitting "pure" light, i.e., just one wavelength. Most light sources, however, emit "dirty" light that's a mixture of many wavelengths, normally with one dominating. For each light source, the graph of light intensity as a function of the wavelength λ is called the *spectral density* of the light source. Figure 7.8 shows the spectral densities of several typical light sources.

These simple diagrams illustrate one problem in attempting to specify color systematically and unambiguously. Several different spectral densities may be perceived by us as identical. When the colors created by these spectral densities are placed side by side, we find it impossible to distinguish between them. The first step in solving the problem is color matching. Suppose that we use a color model defined by the three primaries $A(\lambda)$, $B(\lambda)$, and $C(\lambda)$ and we have a color described by the spectral density $S(\lambda)$. How can we express $S(\lambda)$ in terms of our three primaries? One way to do this is to shine a spot of $S(\lambda)$ on a white screen and, right next to it, a spot of light $P(\lambda) = \alpha A(\lambda) + \beta B(\lambda) + \gamma C(\lambda)$ created by mixing the three primaries (where $0 \leq \alpha, \beta, \gamma \leq 1$). Now change the amounts of α, β, and γ until a trained observer agrees that the spots are indistinguishable. We can now say that, in some sense, $S(\lambda)$ and $P(\lambda)$ are identical, and write $S = P$.

In what sense is the preceding true? It turns out that the above statement is meaningful because of a remarkable property of colors. Suppose that two spectral

Human Vision

We see light that enters the eye and falls on the retina, where there are two types of photosensitive cells. They contain pigments that absorb visible light and hence give us the sense of vision. One type is the *rods*, which are numerous, are spread all over the retina, and respond only to light and dark. They are very sensitive and can respond to a single photon of light. There are about 110,000,000 to 125,000,000 rods in the eye [Osterberg 35]. The other type is the *cones*, located in one small area of the retina (the fovea). They number about 6,400,000, are sensitive to color, but require more intense light, in the order of hundreds of photons. Incidentally, the cones are very sensitive to red, green, and blue (Figure 7.7), which is one reason why CRTs use these colors as primaries.

Each of the light sensors in the eye, rods, and cones sends a light sensation to the brain that's essentially a pixel, and the brain combines these pixels to a continuous image. The human eye is, thus, similar to a digital camera. Once this is realized, we naturally want to compare the resolution of the eye to that of a modern digital camera. Current digital cameras have from 300,000 sensors (for a cheap camera) to about six million sensors (for a high-quality one). Thus, the eye features a much higher resolution, but its effective resolution is even higher if we consider that the eye can move and refocus itself about three to four times a second. This means that in a single second, the eye can sense and send to the brain about half a billion pixels. Assuming that our camera takes a snapshot once a second, the ratio of the resolutions is about 100. (See also Section 1.1.7.)

Certain colors—such as red, orange, and yellow—are psychologically associated with heat. They are considered *warm* and cause a picture to appear larger and closer than it really is. Other colors—such as blue, violet, and green—are associated with cool things (air, sky, water, ice) and are therefore called *cool* colors. They cause a picture to look smaller and farther away.

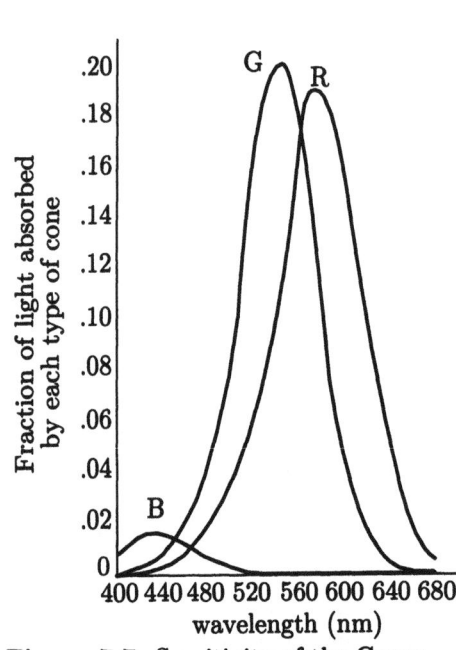

Figure 7.7: Sensitivity of the Cones

7 Color

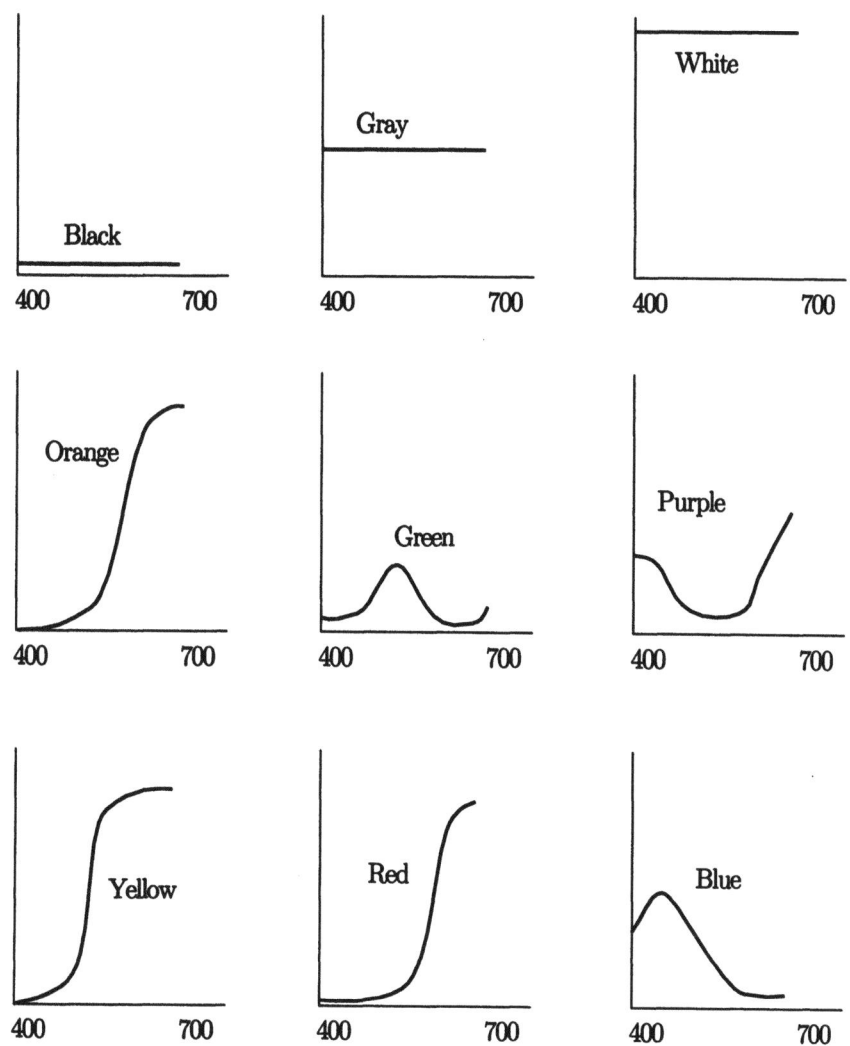

Figure 7.8: Some Spectral Densities.

densities $S(\lambda)$ and $P(\lambda)$ have the same perceived color, so we write $S = P$. We now select another color Q and shine it on both spots S and P. We know from experience that the two new spots would also be indistinguishable. This means that we can use the symbol $+$ for adding lights and we can describe the two spots by $S(\lambda)+Q(\lambda)$ and $P(\lambda)+Q(\lambda)$. In short, we can say "if $S = P$, then $S+Q = P+Q$." The same is true for changing intensities. If $S = P$, then $\alpha S = \alpha P$ for any intensity α. We, therefore, end up with a vector algebra for colors, where a color can be treated as a three-dimensional vector, with the usual vector operations.

Given the above, we can select a color model based on three primaries, A, B, and C, and can represent any color S as a linear combination of the primaries $S = \alpha A + \beta B + \gamma C$. We can say that the vector (α, β, γ) is the representation of S in the basis (A, B, C). Equivalently, we can say that S is represented as the point (α, β, γ) in the three-dimensional space defined by the vectors $A = (1,0,0)$, $B = (0,1,0)$, and $C = (0,0,1)$.

Since three-dimensional graphs are hard to draw on paper, we would like to artificially reduce the representation from three to two dimensions. This is done by realizing that the vector $(2\alpha, 2\beta, 2\gamma)$ represents the same color as (α, β, γ), only twice as bright. We, therefore, restrict ourselves to vectors (α, β, γ), where $\alpha + \beta + \gamma = 1$. These are vectors normalized to unit brightness. All vectors of unit brightness lie in the $\alpha + \beta + \gamma = 1$ plane (see Section 5.3.2 for the equation of a plane) and it is this plane (which, of course, is two-dimensional) that we plot on paper. We can specify a color by using two numbers, say, α and β and calculate γ as $1 - \alpha - \beta$.

For the RGB color model, we now select the pure spectral colors using trained human experts. The idea is to shine a spot of a pure color, say, 500 nm, on a screen and, right next to it, a spot that's a mixture $(r, g, b = 1-r-g)$ of the three primaries of the RGB model. The values of r and g are varied until the observer judges the two spots to be indistinguishable. The point (r, g, b) is then plotted in the three-dimensional RGB color space. When all the pure colors have been plotted in this way, the points are connected to form a smooth curve, $\mathbf{P}(\lambda) = (r(\lambda), g(\lambda), b(\lambda))$. This is the *pure spectral color curve* of the RGB model (Figure 7.9).

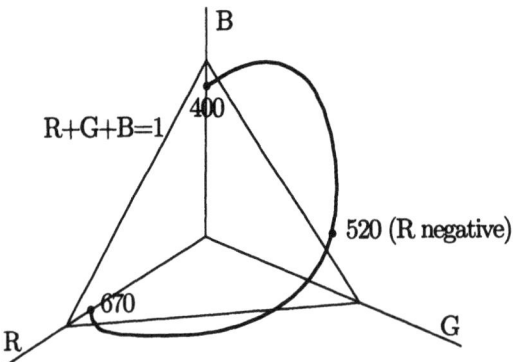

Figure 7.9: Pure RGB Spectral Color Curve.

An important property of the curve is that some of r, g, and b may sometimes have to be negative. An example is $\lambda \approx 520$ nm, where r turns out to be negative. What is the meaning of adding a negative quantity of green in a color defined by, for example, $S = 0.8R - 0.1G + 0.3B$? The way to understand this is to write the equation as $S + 0.1G = 0.8R + 0.3B$. It now means that color S cannot be constructed from the RGB primaries, but color $S + 0.1G$ can. The important conclusion is that not every color can be created in the RGB model! This is illustrated in Figure 7.10. For some colors we can only create an approximation. This is true for all color models that can be created in practice.

> Did I ever tell you my favorite color is blue?
> — Jürgen Prochnow (as Sutter Cane) in *In the Mouth of Madness*.

7 Color

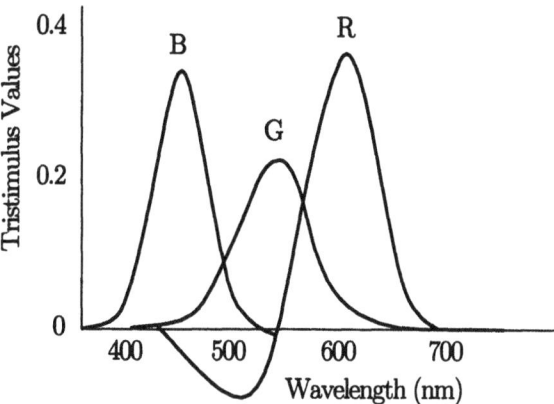

Figure 7.10: RGB Color Combinations.

7.8 The CIE Standard

This standard was created in 1931 by the International Committee on Illumination (Commission Internationale de l'Éclairage). It is based on three carefully chosen artificial color primaries X, Y, and Z. They don't correspond to any real colors, but they have the important property that any real color can be represented as a linear combination $xX + yY + zZ$, where $x + y + z = 1$ and none of x, y, and z are negative (Figure 7.11).

The plane $x + y + z = 1$ in the XYZ space is called the CIE chromaticity diagram (Figure 7.12). The curve of pure spectral color in the CIE diagram covers all the pure colors, from 420 nm to 660 nm. It is shaped like a horseshoe.

Point $w = (0.310, 0.316)$ in the CIE diagram is special and is called "illuminant white." It is assumed to be the fully unsaturated white and is used in practice to match to colors that should be pure white.

▶ **Exercise 7.4:** If illuminant white is pure white, why isn't it on the curve of pure spectral color in the CIE diagram?

The CIE diagram provides a standard for describing colors. There are instruments that, given a color sample, calculate the (x, y) coordinates of the color in the diagram. Also, given the CIE coordinates of a color, those instruments can generate a sample of the color. The diagram can also be used for useful color calculations. Here are some examples:

1. Given two points a and b in the diagram (Figure 7.12), the line connecting them has the form $(1 - \alpha)a + \alpha b$ for $0 \le \alpha \le 1$. This line shows all the colors that can be created by adding varying amounts of colors a and b.

2. Imagine two points, such as c and d in Figure 7.12. They are on the opposite sides of illuminant white and, therefore, correspond to complementary colors.

▶ **Exercise 7.5:** Why is that true?

3. The dominant wavelength of any color, such as e in Figure 7.12 can be measured in the diagram. Just draw a straight line from illuminant white w to e and continue it until it intercepts the curve of pure spectral color. Then read the

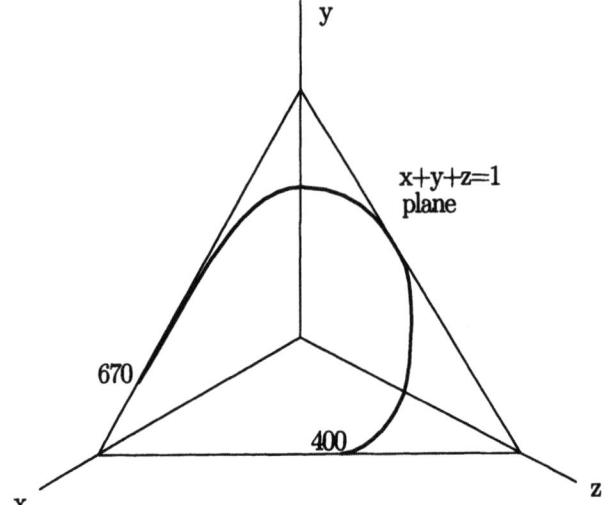

Figure 7.11: Pure Spectral Color Curve.

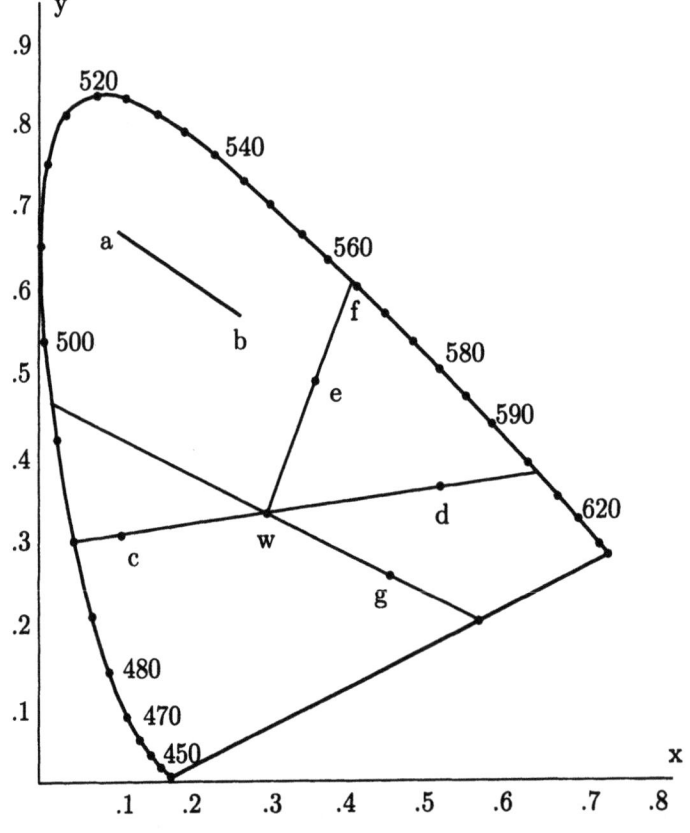

Figure 7.12: The CIE Chromaticity Diagram.

7 Color

wavelength at the interception point f (564 nm in our example).

▶ **Exercise 7.6:** How can the saturation of color e be calculated from the diagram?

▶ **Exercise 7.7:** What is the dominant wavelength of point g in the CIE diagram?

> Have you lost your mind? What color is this bill?
> — Lori Petty (as Georgia "George" Sanders) in *Lush Life* (1996).

4. The *color gamut* of a device is the range of colors that can be displayed by the device. This can also be calculated with the CIE diagram. An RGB monitor, for example, can display combinations of red, green, and blue, but what colors are included in those combinations? To find the color gamut of an RGB monitor, we first have to find the locations of pure red, green, and blue in the diagram [these are points $(0.628, 0.330)$, $(0.285, 0.590)$, and $(0.1507, 0.060)$], then connect them with straight lines. The color gamut consists of all the colors within the resulting triangle. Each can be expressed as a linear combination of red, green, and blue, with non-negative coefficients (a convex combination).

Interestingly, because of the shape of the horseshoe, no three colors on or inside it can serve as ideal primaries. No matter what three points we select, some colors will be outside the triangle defined by them. This means that no set of three primaries can be used to create all the colors. The RGB set has the advantage that the triangle created by it is large, and thus contains many colors. The CMY triangle, for example, is much smaller by comparison. This is another reason for using red, green, and blue as the RGB primaries.

> While wondering what he should do in this emergency he came upon a girl sitting by the roadside. She wore a costume that struck the boy as being remarkably brilliant: her silken waist being of emerald green and her skirt of four distinct colors – blue in front, yellow at the left side, red at the back and purple at the right side. Fastening the waist in front were four buttons—the top one blue, the next yellow, a third red and the last purple.
>
> L. Frank Baum, *The Marvelous Land of Oz*

8
Computer Animation

Webster defines "animate" as "to give life to; to make alive." This is precisely what we feel when we watch a well-made piece of animation, and this is the reason why traditional animation has always been popular and why computer animation is such a successful field. In fact, computer graphics has developed over the years in three stages. The first stage was to display a single image consisting of smooth, curved, realistic-looking surfaces. The second stage was to create and display an entire animation made of many *frames*, where each frame is an image. The third stage is *virtual reality*, where the user can interact with the animation.

> There is increasing interaction between images and language. One might say that living in society today is almost like living in a vast comic strip.
> — Jean-Luc Godard (as Narrator) in *Deux ou trois choses que je sais d'elle* (1966).

8.1 Background

Animation is based on the way our eye and brain work. If the eye is presented with a slow sequence of images, the brain interprets them as separate. If the images are speeded up more and more, the brain starts interpreting them first as motion with flickers, then as continuous motion, and, finally, as a blur. The physiological property that allows our eye and brain to turn a sequence of individual images into a continuous stream is called *persistence of vision*.

The rate of animation should be fast enough to create the perception of continuous motion but slow enough so as not to waste resources. In practice, *playback rates* of 24 or 30 frames per second are normally used. With cheap animation, however, each frame may be displayed several consecutive times, producing a *sampling rate* (or *update rate*) that's much lower than the playback rate.

The computer is used to automate parts of the overall task of animation, letting the animator work on an abstract level, concentrating on scene design and specifying the important information. In practice, this means that the animator enters information about the state of the animation at certain *key frames* and the software uses this to create the images for all the frames by interpolating between consecutive key frames. We say that the computer does the *in-betweening*.

Different pieces of animation can have different characteristics. Artistic animation, cheap cartoons, and flight simulation are all animations, but they are very different in their approach, attention to detail, the use of color, and the amount of information displayed. As a result, different tools, techniques, and algorithms have to be used, depending on the type of animation at hand.

Computer animation is divided into *computer-assisted* (or two-dimensional) animation and *computer-generated* (or three-dimensional) animation. The former uses the computer to interpolate between two-dimensional shapes, whereas the latter uses it to build three-dimensional objects, to move both camera and objects along their paths, and to stop and take a snapshot at each frame. The term two-and-a-half-dimensional animation is also sometimes used. It refers to two-dimensional animation where each frame consists of several shapes drawn on separate slides. They represent objects at different distances from the viewer (for example, a nearby dog and trees in the background) and are moved different distances between frames (the background trees are shifted to the right while the dog is running to the left) to simulate parallax.

A complete piece of animation is sometimes called a *presentation*. It consists of a number of *acts*, where each act is broken down into several *scenes*. A scene is made of several *shots* or *sequences* of animation, each a succession of *animation frames*, where there is a small change in scene and camera position between consecutive frames. The hierarchy is thus

$$\text{piece} \rightarrow \text{act} \rightarrow \text{scene} \rightarrow \text{sequence} \rightarrow \text{frame}.$$

Each sequence is tested before it is actually produced, by displaying it with low-quality rendering and a small number of frames. Objects may be displayed as wire frames, or without removal of hidden parts, or in low resolution. The camera may be moved large distances between frames. The test is played back and watched by animators, which may lead them to change features such as timing, the camera path, the arrangement of objects, or the background color.

Time is an important parameter in animation and should be discussed further. Time is used in animation as a discrete quantity and can be changed by changing the number of frames. Speeding up an action is traditionally done by deleting certain frames, while slowing down the animation requires adding new frames. In traditional animation, the new frames added are identical to existing ones. For example, if every other frame is duplicated, the same sequence takes 50% longer to run. In computer animation, time can be controlled in a sophisticated way and there is no need to add or delete frames. If the animator decides, based on a test, to slow down an n-frame sequence by, say, 50%, the software is simply told to recreate the entire sequence from scratch using 50% more frames ($1.5n$ frames instead of n). The entire action is interpolated between the frames and the result is that no two frames are identical.

8 Computer Animation

[ToyStory 98] has a history of computer graphics, including the history of computer animation.

Early computer animation used film as the output medium. Either 16-, 35-, or 70-mm film was used at 24 frames per second (fps). Very high quality can be achieved with 70-mm film, but high resolution (at least $2K \times 2K$) is needed. The advantage of film is high resolution (Section 1.1.7), a large number of colors, and insensitivity of the medium to magnetic fields. The disadvantages are the need for developing (the film cannot be watched immediately) and non-reusability of the medium (the same film cannot be used twice).

It is possible to place a camera in front of the computer screen and shoot frames, but the resulting photos are distorted because of the curvature of the CRT screen. Another drawback is the fact that the screen is refreshed all the time. Taking a quick shot may produce a picture that's partly bright (from parts of the screen that have just been refreshed) and partly dark. The shot should thus be slow, covering several screen refreshes, or it should be synchronized with the refresh so that the camera shutter remains open during an entire screen refresh. This is why better results are obtained with a film recorder. Such a device has a special flat screen and can be used with different cameras to take high-quality pictures.

If the animation is produced for television or for home entertainment, where it is going to be played back from a VCR, it makes sense to record it on video tape. Such a tape can be viewed immediately, is easy to copy, lasts a long time, and can be reused. Its main disadvantage is its low resolution. The NTSC standard calls for 525 scan lines per image, of which only 480 actually contain the image. The NTSC aspect ratio is 4:3, leading to 640 pixels per scan line. A resolution of 480×640 can at most be considered intermediate, not high. The proposed HDTV standard (Section 1.1.3) doubles both the horizontal and vertical resolutions and uses a 16:9 aspect ratio. This will presumably result in a high-resolution video image.

Placing a video camera in front of the computer screen involves the same problems as with film. It is, therefore, better to output the bitmap from memory directly to the camera, which is done by means of a special interface card plugged into the computer.

The main problems in computer animation are as follows:

1. How to display, on the screen, only those parts of the scene that would be seen by an actual camera located at a certain point. This involves general perspective projection and clipping. In computer animation, the term *camera* replaces the word *observer*. This term refers to what is displayed on the screen (what we want the camera to see). In the computer, the camera is represented by several numbers describing its position, direction of view, an "up" direction, the distance k between it and the projection screen, and the two viewing half-angles h and v (Section 3.9.8).

2. How to move the camera along any desired path and rotate it during movement so it always points to the *center of interest* (generally a different direction in each frame). Its "up" direction may also have to be rotated to achieve the desired animation effects.

3. How to move the scene along another path (mathematically this is the same as problem 2) and move parts of the scene in different ways (imagine a person walking, moving hands and feet in a complex pattern).

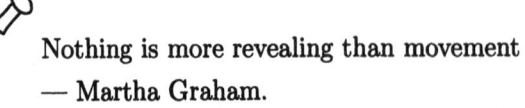

Nothing is more revealing than movement
— Martha Graham.

A typical *sequence* in a computer-generated piece of animation involves a camera moving smoothly along a curved path around a scene composed of objects. The objects may also move at the same time. Creating such a sequence involves the following tasks:

1. Defining the camera path. This may be a long, complex curve but the software should be able to follow it and to stop at many points (frames) for a snapshot. The frames should be equally spaced if uniform camera speed and smooth animation are important. Special effects may require the camera to accelerate or to slow down. The problem is that a typical parametric curve $\mathbf{P}(t)$ has variable velocity; varying t in equal increments advances unequal segments on the curve.

2. At each point, the camera may have to be rotated so that it points in the right direction (normally directly at the scene, but sometimes off it). This is where *spherical interpolation* is used (Section 8.3).

3. When the camera is properly positioned, a snapshot is taken. This is done by projecting the scene (or part of it) on the screen, which is assumed to be perpendicular to the line of sight of the camera, at a distance of k units from it. The y axis of the screen should be in the "up" direction of the camera. Perspective projection is normally used, since an image generated by other types of projection may look unnatural.

4. The objects constituting the scene may also have to be moved and rotated (imagine a camera flying over a moving train). This task can use the techniques and tools developed for tasks 1 and 2. In fact, the case where only the camera moves and the objects of the scene are stationary is special and is referred to as a "walk-through" or a "flyby."

Perspective projection has been discussed in detail in Section 3.9. Here we show simple ways to approach the first 2 tasks.

8.2 Interpolating Positions

Task 1, defining a curve and moving along it at a constant speed, can be done by an interpolating Bézier curve. Section 4.14.18 shows how such a curve can be constructed. Given a set of $n+1$ points $\mathbf{P}_0, \mathbf{P}_1, \ldots, \mathbf{P}_n$, the curve goes from $\mathbf{P}_1$ to $\mathbf{P}_{n-1}$ (not from $\mathbf{P}_0$ to $\mathbf{P}_n$) and is constructed of $n-2$ segments $\mathbf{P}_i(t)$, each connecting one pair of points. The pairs are $(\mathbf{P}_1,\mathbf{P}_2)$, $(\mathbf{P}_2,\mathbf{P}_3)$, up to $(\mathbf{P}_{n-2},\mathbf{P}_{n-1})$. Each segment $\mathbf{P}_i(t)$ is based on four points, the two exterior points are $\mathbf{P}_i$ and $\mathbf{P}_{i+1}$ and the two interior ones, $\mathbf{X}_i$ and $\mathbf{Y}_i$, are automatically calculated by Equation (4.128), duplicated here

$$\mathbf{X}_i = \mathbf{P}_i + \frac{1}{6}(\mathbf{P}_{i+1} - \mathbf{P}_{i-1}); \quad \mathbf{Y}_i = \mathbf{P}_{i+1} - \frac{1}{6}(\mathbf{P}_{i+2} - \mathbf{P}_i). \qquad (4.128)$$

No segments connect $\mathbf{P}_0$ to $\mathbf{P}_1$ or $\mathbf{P}_{n-1}$ to $\mathbf{P}_n$. The two extreme points, $\mathbf{P}_0$ and $\mathbf{P}_n$, are used as guide points, to control the initial and final directions of the curve. Point

8 Computer Animation

X_1 is obtained by adding vector $(P_2 - P_0)/6$ to point P_1. Point Y_{n-2} is similarly obtained by subtracting vector $(P_n - P_{n-2})/6$ from point P_{n-1}. Figure 8.1 shows such a curve. The method of Section 4.14.19 is similar and can also be used.

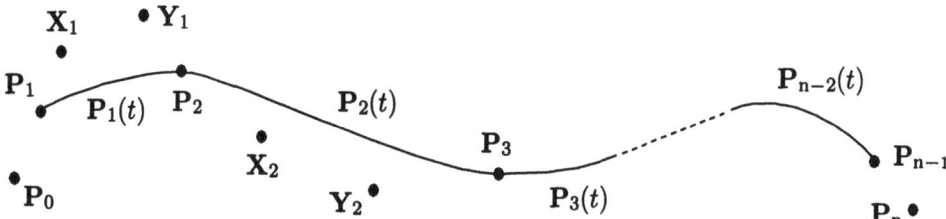

Figure 8.1: An Interpolating Bézier Curve for $n + 1$ Points.

In practical animation work, the animator should have a rough idea of the shape of the path along which the camera should move. The animator inputs the coordinates of $n-1$ key points P_i on the path (they should be fairly close to each other and roughly equally spaced) followed by the two extreme guide points P_0 and P_n to control the start and end directions of the path. The $n-1$ points are called the animation *key frames*. The software calculates and displays all the interior points X_i and Y_i, and the $n-2$ individual Bézier segments $P_i(t)$. The path (which is normally three-dimensional) is then examined by rotating it and watching it from different directions. If the path is not satisfactory, it can be edited either by deleting some key frames, moving them, or adding new ones. The interior points can also be manually repositioned at this time.

When the right path is finally obtained, the software moves the camera along the path, segment by segment, varying t from 0 to 1 in F steps, called *frames*, (where F is a parameter) for **each** of the $n-2$ segments. The value of t for frame f is, thus, given by $t = (f-1)/(F-1)$. The original $n+1$ points are converted in this way to $n-2$ Bézier segments that produce the final $(n-2)F$ equally spaced animation frames. At each frame, the camera is rotated to point in the right direction and a snapshot taken. In professional jargon, this process is called *in-betweening*. The computer stops the camera and generates F frames between each pair of key frames supplied by the animator.

> And when a damp Fell round the path of Milton, in his hand The thing became a trumpet; whence he blew Soul-animating strains,–alas! too few.
> —William Wordsworth, *Scorn Not The Sonnet.*

8.2.1 Constant Speed: I

To obtain smooth animation, the camera should move along its path in equal steps, covering equal arc lengths in each step. In principle, this can be achieved by a parameter substitution. Suppose that we substitute some parameter $s(t)$ for t, such that our curve becomes $P(s)$ instead of $P(t)$. Clearly, the best choice for s is the arc length. If $s(0.2)$ is the length of the curve from its start $P(0)$ to point $P(0.2)$, then incrementing s in equal steps will advance equal arc lengths on the curve. Section A.4.5 shows that the arc length of the entire curve $P(t)$ is given by the integral

$$\int |dP(t)| = \int_0^1 |P^t(t)|\, dt.$$

The arc length $s(u)$ from $t = 0$ to $t = u$ is, thus, given by

$$s(u) = \int_0^u |P^t(t)|\, dt.$$

The trouble is that such integrals are normally impossible to calculate analytically; they must be done numerically (they belong to the family of *elliptic integrals*). A simple alternative that's sometimes satisfactory is to calculate a large number of points on the curve, to replace the curve with the polyline made by these points, and to calculate approximate arc lengths by calculating the lengths of the polyline segments. The steps are as follows:

1. Vary t from 0 to 1 in $n+1$ small, equal steps and calculate $n+1$ points $P(t)$ on the curve.

2. Calculate the n straight line distances between the points.

3. Accumulate the distances of step 2, such that accumulated distance i will give the total (approximate) distance from the start of the curve to point $t = i$.

4. Divide all the accumulated distances by the last one, resulting in a table T of normalized accumulated distances (whose values are between 0 and 1).

5. Find the entries in T that are closest to the required arc lengths. Suppose, for example, that we want to select the six points on the curve where the normalized accumulated arc lengths are 0, 0.2, 0.4, 0.6, 0.8, and 1. We find the entries of table T that are the closest to these values. Assume that these are entries 1, 43, 61, 78, 95, and 100. The parameter t should be set to the normalized values of these six entries.

Figure 8.2 is a listing of *Mathematica* code that illustrates this method for a four-point Bézier curve. A similar approach is presented in Section 4.14.15.

> Speed is scarcely the noblest virtue of graphic composition, but it has its curious rewards. There is a sense of getting somewhere fast, which satisfies a native American urge.
> — James Thurber.

8.2.2 Constant Speed: II

Given a space curve $P(t) = (x(t), y(t), z(t))$, we denote by $\text{Len}(t_1, t_2)$ the arc length from $P(t_1)$ to $P(t_2)$. This section discusses a numerical approach—proposed by [Guenther and Parent 90] and called *adaptive subdivision*—to two problems:

8 Computer Animation

```
p0={0,1}; p1={5,1}; p2={5,0}; p3={4,.5};
Bez[t_]:=(1-t)^3p0+3t(1-t)^2p1+3t^2(1-t)p2+t^3p3;
tbl=Table[Bez[t], {t,0,1,.01}];
(* tab1 is a list of lengths of straight segments *)
tab1=Table[Sqrt[(tbl[[i+1,1]]-tbl[[i,1]])^2
 +(tbl[[i+1,2]]-tbl[[i,2]])^2], {i,1,100}];
(* tab2 is a list of accumulated lengths *)
tab2={tab1[[1]]};
Do[tab2=Append[tab2,tab1[[i]]+tab2[[i-1]]],{i,2,100}];
tab2=tab2/tab2[[100]]; (* normalize tab2 *)
tab3={0}; d=.1;
(* tab3 is a list of non-equally-spaced parameter values *)
Do[If[tab2[[i]]>d, {tab3=Append[tab3,i/100], d=d+.1}], {i,1,100}];
tab3=Append[tab3,1];
len=Length[tab3];
tab4=Table[Bez[tab3[[i]]], {i,1,len}];
 (* use tab3 as the parameter values *)
ListPlot[tab4] (* display equally-spaced points *)
ListPlot[tbl]  (* display 101 non-equally-spaced points *)
```

Figure 8.2: Normalized Accumulated Arc Lengths.

Problem 1. Given values t_1 and t_2 of the time parameter, calculate $\text{Len}(t_1, t_2)$ numerically.

Problem 2. Given a value t_1 and an arc length s, find a value t_2 such that $\text{Len}(t_1, t_2) = s$.

We first show why these problems are important. If we want to move the animation camera along $\mathbf{P}(t)$ at a constant speed, we can use problem 1 to find the length $S = \text{Len}(0, 1)$ of the entire curve, then divide it into $n - 1$ equal parts $s = S/(n-1)$ and use problem 2 to find values $t_1 = 0 < t_2 < t_3 < \cdots < t_n = 1$ such that

$$\text{Len}(t_1, t_2) = \text{Len}(t_2, t_3) = \cdots = \text{Len}(t_{n-1}, t_n) = s.$$

The t_i values should then be used to specify n equally spaced frames along the curve. A similar method can be used for more complex cases where we want to move the camera at a nonuniform speed along the curve. We divide S into parts s_i of different sizes and use problem 2 to find values t_i such that $\text{Len}(t_i, t_{i+1}) = s_i$. Acceleration will result if $s_i < s_{i+1}$, but any nonuniform motion can be generated by carefully selecting the values of s_i. Here is how to approach the two problems.

Problem 1: Section A.4.5 shows that the arc length of a curve $\mathbf{P}(t)$ is given by

$$\int_0^1 |\mathbf{P}^t(t)| \, dt.$$

Since $\mathbf{P}(t) = (x(t), y(t), z(t))$, we get

$$\mathbf{P}^t(t) = \frac{d\mathbf{P}(t)}{dt} = \left(\frac{dx(t)}{dt}, \frac{dy(t)}{dt}, \frac{dz(t)}{dt}\right),$$

and from this,

$$|\mathbf{P}^t(t)| = \sqrt{\left(\frac{dx(t)}{dt}\right)^2 + \left(\frac{dy(t)}{dt}\right)^2 + \left(\frac{dz(t)}{dt}\right)^2}. \qquad (8.1)$$

Gaussian quadrature is used to numerically integrate Equation (8.1) from 0 to t_i for certain values of i. Each such integral results in an arc length s_i from the start of the curve to point $\mathbf{P}(t_i)$. The pairs of values (t_i, s_i) are stored in a table and are later used to solve problem 1 in the following way. Given two values t_1 and t_2, the arc length $\text{Len}(t_1, t_2)$ is calculated as follows:

1. Find entry i in the table such that $t_i \le t_1 < t_{i+1}$.
2. Using Gaussian quadrature, integrate Equation (8.1) from t_i to t_1 to obtain arc length s_1 (if $t_i = t_1$, skip the integration and set $s_1 = 0$).
3. Set $\text{Len}(0, t_1) = s_i + s_1$.
4. Do the same thing for t_2. Find entry j in the table such that $t_j \le t_2 < t_{j+1}$, integrate from t_j to t_2 to obtain s_2 (or set $s_2 = 0$, if $t_j = t_2$), and set $\text{Len}(0, t_2) = s_j + s_2$.
5. Subtract $\text{Len}(0, t_2) - \text{Len}(0, t_1) = s_j + s_2 - (s_i + s_1)$ to obtain $\text{Len}(t_1, t_2)$.

The question is what values of i to select for the table, and the answer should now be obvious. Since we integrate from t_i to t_1, we can relate the distance between two consecutive values t_i and t_{i+1} to the curvature of $\mathbf{P}(t)$ in that region. If the curvature is low (the curve between points $\mathbf{P}(t_i)$ and $\mathbf{P}(t_{i+1})$ is close to a straight line), we can place t_{i+1} well away from t_i. The integral from t_i to t_1 would be done over a region of the curve that may be long but is close to a straight line. The result would therefore be quick and accurate. If the curvature is high, the two values t_i and t_{i+1} have to be close by. The integral from t_i to t_1 would be done in this case over a curvy but *short* region of the curve, so, again, it would be accurate.

Instead of calculating the curvature, the method uses a recursive procedure Subdivide, and a threshold parameter **eps**. The procedure is given a range $[t_l, t_r]$, it uses Gaussian integration to find the arc length s_{lr} of this range, then divides the range in the middle $t_m = (t_l + t_r)/2$, integrates each part to get arc lengths s_{lm} and s_{mr}, and calculates the difference $|s_{lr} - (s_{lm} + s_{mr})|$. If this difference is less than **eps**, the procedure assumes that the curvature of $\mathbf{P}(t)$ in the region $[t_l, t_r]$ is small enough and it stores the pair $(t_m, s_{0l} + s_{lm})$ in the table. Otherwise, it calls itself recursively for the two ranges $[t_l, t_m]$ and $[t_m, t_r]$. Figure 8.3 lists C++ code for this procedure.

Problem 2: Given a value t_1 and an arc length s, find a value t_2 such that $\text{Len}(t_1, t_2) = s$. We define a function $f(t) = \text{Len}(t_1, t) - s$ that reduces problem 2 to that of finding a zero of $f(t)$ in the range $[t_1, 1]$. Perhaps the simplest method for finding a zero of a function is binary subdivision. The range $[t_1, 1]$ is divided in

8 Computer Animation

```c
#include <stdio.h>
#include <math.h> // for function fabs
float totl_arc; // global variable
void Add_tabl(float, float);
float Gauss(float, float);
float Subdivide(float left, float right, float full_intr, float eps){
float mid, left_arc, right_arc, left_sub;
 mid=(left+right)/2;
 left_arc=Gauss(left,mid);
 right_arc=Gauss(mid,right);
 if(fabs(full_intr-left_arc-right_arc)<eps)
  {left_sub=Subdivide(left,mid,left_arc,eps/2.0);
  totl_arc=totl_arc+left_sub;
  Add_tabl(mid,totl_arc);
  return(Subdivide(mid,right,right_arc,eps/2.0)+left_sub);}
 else
  return(left_arc+right_arc);
}
int main(){
float left, right, full_intr, eps;
left=0; right=1.0; totl_arc=0; eps=0.001;
 full_intr=Gauss(left,right);
 Subdivide(left,right,full_intr,eps);
 }
```

Figure 8.3: Procedure Subdivide.

the middle, $t_m = (t_1 + 1)/2$. If $f(t_m) = 0$ (or if it is very close to 0), we are done. Otherwise, if $f(t_1)$ and $f(t_m)$ have the same sign, we divide the range $[t_1, t_m]$ in the middle and perform the same tests. Otherwise, we divide the range $[t_m, 1]$.

Finding the zero of a function can also be done by using the well-known Newton-Raphson method (Section 4.13). This is a fast method, but it has two disadvantages.

1. It requires the derivative of the function. In our case, the derivative depends on the particular curve used, so it has to be implemented by the user for each curve separately.

2. The derivative may be zero, or very close to zero. Since this method divides by the derivative, the division may result in overflow.

8.3 Interpolating Orientations: I

Now comes the second task. The animator should supply the animation software with the data needed for orienting the camera. This process is based on the following fact, proved by Leonhard Euler in 1752. Imagine a rigid object positioned at point **P** and having a certain orientation. We now send the object flying through space. It may roll and tumble in a complicated way, but its position and orientation at any moment can be completely described by two transformations, a translation from **P** to its present position and a rotation through an angle θ about some axis **v**.

Our imaginary camera may be considered such an object. It may have to move around the scene along a complicated path and change its orientation all the time, so that it always points in the right direction. The animator should have a rough idea of how to move the camera and in what direction it should look. The animator should, therefore, input $n-2$ direction vectors (the directions in which the camera should look when positioned at the *key frames*) and the animation software should use these vectors to interpolate between key frames and determine the orientation of the camera at any point.

Before we discuss how to interpolate the direction vectors, we have to introduce one more complication, namely the "up" vector. Imagine a camera placed in the xy plane, looking in the positive x direction $(1,0,0)$ with its top pointing in the positive z direction. We now rotate the camera in small steps until it points in the positive y direction $(0,1,0)$ with its top still pointing in the positive z direction (Figure 8.4a). At any time during this rotation the camera points in a direction $(a, b, 0)$, i.e., somewhere in the xy plane. Now, imagine that while rotating the camera from x to y, the animator also wants to rotate it about its direction of view $(a, b, 0)$ such that when it reaches its final direction, its top will be pointing in the *negative* z direction (Figure 8.4b). If such an effect is called for, then the animator also has to specify an "up" direction in each key frame. These "up" vectors should be interpolated between key frames and should be used to indicate the top of the screen each time a snapshot is taken. (When the software calls a procedure to project the scene on the screen, it should transfer to the procedure, as parameters, the position of the camera, its direction of view, the direction of the "top" of the screen, and any other necessary data.) The interpolation method discussed below should therefore be applied to the direction of view of the camera, as well as to its "up" direction, if this direction is explicitly defined.

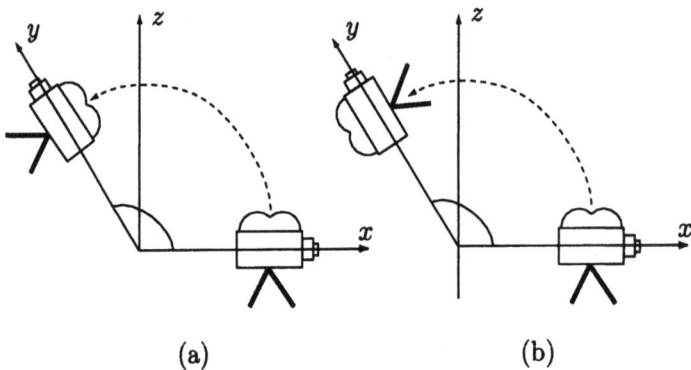

(a)   (b)

Figure 8.4: Illustrating the "Up" Direction.

(One special, important case of camera orientation, namely the case where the camera follows a moving object along its path, should be mentioned. Imagine a camera following an airplane, repeating all its maneuvers while staying the same distance behind it all the time. This case is easy to implement. When the camera is located at point $\mathbf{P}_i(t)$, it should look at point $\mathbf{P}_i(t+f)$, where f is a constant.

8 Computer Animation

If f is negative, then the camera is located in front of the object, flying backward and constantly looking at the object.)

For each of the $n-1$ key frames, the animator has to input the direction $\mathbf{D}_i$ in which the camera should be looking. The software interpolates these vectors to orient the camera between successive key frames. Figure 8.5 shows several vectors $\mathbf{D}_i$. To interpolate $\mathbf{D}_i$ and $\mathbf{D}_{i+1}$ we need to compute the angle θ_i between them. This is done by first normalizing the two direction vectors (dividing each by its length to obtain unit vectors), then computing their dot product $\mathbf{D}_i \bullet \mathbf{D}_{i+1}$ which equals $\cos \theta_i$. We will see that $\sin \theta_i$ is also necessary, but it can always be calculated as $\sin \theta_i = \pm\sqrt{1 - \cos^2 \theta_i}$.

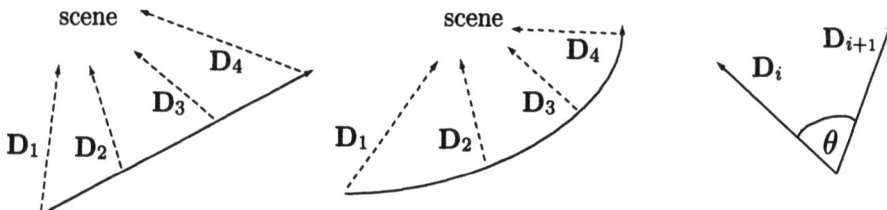

Figure 8.5: Rotating the Camera at Key Frames.

▶ **Exercise 8.1:** How does the software decide what sign to use for $\sqrt{1 - \cos^2 \theta}$?

We now need a function to correctly interpolate direction vectors. For each of the $n-2$ segments $\mathbf{P}_i(t)$ that constitute the camera path, we start with a direction $\mathbf{D}_i$ in animation frame 1 (where $t=0$) and end with a direction $\mathbf{D}_{i+1}$ in animation frame F ($t=1$). A linear interpolation $(1-t)\mathbf{D}_i + t\mathbf{D}_{i+1}$ is simple but produces nonuniform moves. When t is varied from 0 to 1 in equal steps, the interpolation steps are not the same; they start large and get smaller. The reason is that rotation has to do with spherical symmetry, whereas linear interpolation has to do with straight lines.

To derive a proper interpolation function, we have to think in terms of moving along a circular arc. We start with a two-dimensional example. Imagine two unit vectors $\mathbf{D}_1$ and $\mathbf{D}_2$ in two-dimensional space. The vector $\mathbf{D}_l$ that's defined as the combination $\mathbf{D}_l(t) = (1-t)\mathbf{D}_1 + t\mathbf{D}_2$ rotates from $\mathbf{D}_1$ to $\mathbf{D}_2$ such that its tip moves along a straight line (i.e., the magnitude of $\mathbf{D}_l$ keeps changing, Figure 8.6a). To move from $\mathbf{D}_1$ to $\mathbf{D}_2$ along a circular arc, we should use *spherical interpolation* (Figure 8.6b). We use the expression

$$\mathbf{D}_s(t) = \frac{\sin((1-t)\theta)}{\sin \theta}\mathbf{D}_1 + \frac{\sin(t\theta)}{\sin \theta}\mathbf{D}_2, \qquad (8.2)$$

where θ is the angle between vectors $\mathbf{D}_1$ and $\mathbf{D}_2$ (note that $\mathbf{D}_1 \bullet \mathbf{D}_2 = \cos \theta$, since these are unit vectors). $\mathbf{D}_s(t)$ is a unit vector that changes direction from $\mathbf{D}_1$ (when $t=0$) to $\mathbf{D}_2$ (when $t=1$) in equal increments. Its tip describes a circular arc (Figure 8.6b).

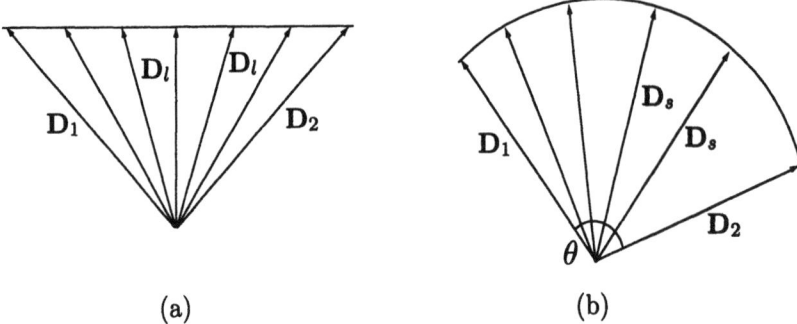

Figure 8.6: Linear and Spherical Interpolations.

▸ **Exercise 8.2:** Prove the above claim.

Another way to look at spherical interpolation is to consider all the two-dimensional unit vectors. They have the same size, but they point in all possible directions. Placing them with their tails at the origin creates a unit circle about the origin. Spherically interpolating two such vectors, D_i and D_{i+1}, is equivalent to moving along an arc on this circle.

The same is true for two unit vectors in three-dimensional space. We can imagine all the three-dimensional unit vectors to form a unit sphere. Spherically interpolating two such vectors is equivalent to moving along a great arc on this sphere. A good spherical interpolation function for direction vectors D_i and D_{i+1} would therefore be

$$D_{i+t} = \frac{\sin((1-t)\theta)}{\sin\theta} D_i + \frac{\sin(t\theta)}{\sin\theta} D_{i+1}. \tag{8.3}$$

When t is incremented in equal steps, this function produces smooth movements of the animation camera from the orientation specified by D_i to that specified by D_{i+1}. A two-dimensional vector $D = (x, y)$ can be considered a complex number. When we use this interpretation, the spherical interpolation of two unit vectors can also be written $D_i(D_i^{-1}D_{i+1})^t$ (complex numbers can be multiplied, they have an inverse, and they can be raised to a power). When t varies from 0 to 1, this expression varies from D_i to D_{i+1}. Since both D_i and D_{i+1} are unit vectors, the absolute value of this triple product is 1.

Spherical interpolation involves a division by $\sin\theta$, so the case $\sin\theta = 0$ should be discussed. This case occurs when $\theta = 0°$ or $\theta = 180°$, but the latter case can be excluded since it does not make sense to use two direction vectors going in opposite directions in two consecutive key frames (see Exercise 8.3). The case $\theta = 0°$ means two parallel direction vectors in two consecutive key frames. This case is common (it means that the camera's orientation should not change between two consecutive key frames), so the interpolation software should check for it and perform the trivial interpolation $D_{i+t} = D_i$.

▸ **Exercise 8.3:** Explain why the case $\theta = 180°$ can be excluded.

8 Computer Animation

Table 8.7 illustrates the difference between linear and spherical interpolations. Given the two unit vectors $\mathbf{D}_1 = (1, 0)$ and $\mathbf{D}_2 = (0, 1)$ with a 90° angle between them, the table shows the results of the linear interpolation $(1 - t)\mathbf{D}_1 + t\mathbf{D}_2$ and the spherical interpolation

$$\frac{\sin(90(1-t))}{\sin 90°}\mathbf{D}_1 + \frac{\sin(90t)}{\sin 90°}\mathbf{D}_2$$

for 11 values of t. The spherical interpolation results in equal increments of 9°, while the linear interpolation results in angle increments (row "Diff" in the table) that initially get bigger, then get smaller.

t:	.1	.2	.3	.4	.5	.6	.7	.8	.9	1
Linear:	6.34	14.04	23.20	33.69	45.00	56.31	66.80	75.96	83.66	90.00
Diff:	6.34	7.70	9.16	10.49	11.31	11.31	10.49	9.16	7.70	6.34
Spherical:	9	18	27	36	45	54	62	72	81	90

Table 8.7: Linear and Spherical Interpolations.

```
(* Two interpolations of vectors with 90 deg *)
d1={1,0}; d2={0,1};
(* Generate 11 linearly interpolated vectors in 'vec' *)
vec=Table[(1-t)d1+t d2,{t,0,1,.1}];
(* Normalize these vectors *)
Do[vec[[i]]=vec[[i]]/Sqrt[vec[[i,1]]^2+vec[[i,2]]^2], {i,1,11}];
(* Show them *)
Table[ArcCos[vec[[1]].vec[[i+1]]]/Degree, {i,1,10}]
Table[ArcCos[vec[[i]].vec[[i+1]]]/Degree, {i,1,10}]
(* Generate 11 spherically interpolated vectors in 'vec' *)
vec=Table[(Sin[90(1-t)Degree]d1+Sin[90t Degree]d2),{t,0,1,.1}];
(* Normalize these vectors *)
Do[vec[[i]]=vec[[i]]/Sqrt[vec[[i,1]]^2+vec[[i,2]]^2], {i,1,11}];
(* Show them *)
Table[ArcCos[vec[[1]].vec[[i+1]]]/Degree, {i,1,10}]
Table[ArcCos[vec[[i]].vec[[i+1]]]/Degree, {i,1,10}]
```

Mathematica Code for Table 8.7.

8.3.1 Summary

Animating the camera starts with the animator specifying the key frames. For key frame i, the animator should specify the camera position $\mathbf{P}_i$, its direction of view $\mathbf{D}_i$, and if necessary, also its "up" direction (the values of k, h, and v may also vary from one key frame to the next). The software prepares the entire path as described in Section 8.2. It then varies the time parameter t in F steps for each path segment, to obtain all the $(n-2)F$ frames. For each frame, the two direction vectors $\mathbf{D}_i$ and $\mathbf{D}_{i+1}$ are spherically interpolated, the camera is pointed in the new direction, and the entire scene is projected on the projection plane (the screen), a process that may require clipping. Each path segment should be short and should not deviate much from a straight line, so varying t in equal steps would cover roughly equal distances on the segment even though the velocity of the segment is normally variable.

> I'm appropriately animate for a human being in the context in which I exist.
> — Woody Allen in *Wild Man Blues* (1998).

8.3.2 Example 1

This example is in three dimensions, but to make it easier to visualize the way the camera moves, we restrict the scene and the camera path to the xy plane. The scene is assumed to be located about the origin and the camera path, Figure 8.8a, is assumed to be in the xy plane. The camera should thus start at point $\mathbf{P}_1$, pointing toward the origin (i.e., in the positive y direction) and should rotate about the z axis as it moves, in order to always point toward the origin. We define the camera path by means of the seven points

$$\mathbf{P}_0 = (1.5, -2, 0), \quad \mathbf{P}_1 = (0, -2, 0), \quad \mathbf{P}_2 = (-2, 0, 0), \quad \mathbf{P}_3 = (1.5, 2, 0),$$
$$\mathbf{P}_4 = (5, 0, 0), \quad \mathbf{P}_5 = (3, -2, 0), \quad \mathbf{P}_6 = \mathbf{P}_0.$$

The two extreme points $\mathbf{P}_0$ and $\mathbf{P}_6$ control the start and end directions of the path, respectively. The path itself is made of four segments defined by means of the four overlapping groups $\mathbf{P}_0\mathbf{P}_1\mathbf{P}_2\mathbf{P}_3$, $\mathbf{P}_1\mathbf{P}_2\mathbf{P}_3\mathbf{P}_4$, $\mathbf{P}_2\mathbf{P}_3\mathbf{P}_4\mathbf{P}_5$, and $\mathbf{P}_3\mathbf{P}_4\mathbf{P}_5\mathbf{P}_6$.

Equation (4.128) is used to calculate the four sets of $\mathbf{X}$ and $\mathbf{Y}$ points:

$$\mathbf{X}_1 = \mathbf{P}_1 + \tfrac{1}{6}(\mathbf{P}_2 - \mathbf{P}_0) = \left(-\tfrac{7}{12}, -\tfrac{5}{3}, 0\right), \quad \mathbf{Y}_1 = \mathbf{P}_2 - \tfrac{1}{6}(\mathbf{P}_3 - \mathbf{P}_1) = \left(-\tfrac{9}{4}, -\tfrac{2}{3}, 0\right),$$
$$\mathbf{X}_2 = \mathbf{P}_2 + \tfrac{1}{6}(\mathbf{P}_3 - \mathbf{P}_1) = \left(-\tfrac{21}{12}, \tfrac{2}{3}, 0\right), \quad \mathbf{Y}_2 = \mathbf{P}_3 - \tfrac{1}{6}(\mathbf{P}_4 - \mathbf{P}_2) = \left(\tfrac{2}{6}, 2, 0\right),$$
$$\mathbf{X}_3 = \mathbf{P}_3 + \tfrac{1}{6}(\mathbf{P}_4 - \mathbf{P}_2) = \left(\tfrac{8}{3}, 2, 0\right), \quad \mathbf{Y}_3 = \mathbf{P}_4 - \tfrac{1}{6}(\mathbf{P}_5 - \mathbf{P}_3) = \left(\tfrac{19}{4}, \tfrac{2}{3}, 0\right),$$
$$\mathbf{X}_4 = \mathbf{P}_4 + \tfrac{1}{6}(\mathbf{P}_5 - \mathbf{P}_3) = \left(\tfrac{21}{4}, -\tfrac{2}{3}, 0\right), \quad \mathbf{Y}_4 = \mathbf{P}_5 - \tfrac{1}{6}(\mathbf{P}_6 - \mathbf{P}_4) = \left(\tfrac{43}{12}, -\tfrac{5}{3}, 0\right).$$

The path is, thus, made of the four Bézier segments:

$$\mathbf{P}_1(t) = (1-t)^3 \mathbf{P}_1 + 3t(1-t)^2 \mathbf{X}_1 + 3t^2(1-t)\mathbf{Y}_1 + t^3 \mathbf{P}_2,$$
$$\mathbf{P}_2(t) = (1-t)^3 \mathbf{P}_2 + 3t(1-t)^2 \mathbf{X}_2 + 3t^2(1-t)\mathbf{Y}_2 + t^3 \mathbf{P}_3,$$
$$\mathbf{P}_3(t) = (1-t)^3 \mathbf{P}_3 + 3t(1-t)^2 \mathbf{X}_3 + 3t^2(1-t)\mathbf{Y}_3 + t^3 \mathbf{P}_4,$$
$$\mathbf{P}_4(t) = (1-t)^3 \mathbf{P}_4 + 3t(1-t)^2 \mathbf{X}_4 + 3t^2(1-t)\mathbf{Y}_4 + t^3 \mathbf{P}_5.$$

We assume that the scene is located at the origin and the camera should always be positioned to look at it. The five direction vectors $\mathbf{D}_1$ through $\mathbf{D}_5$ are thus the vectors from each of the points $\mathbf{P}_i$ to the origin (Figure 8.8b). They are shown as unit vectors:

$$\mathbf{D}_1 = -\mathbf{P}_1 = (0, 1, 0), \quad \mathbf{D}_2 = -\mathbf{P}_2 = (1, 0, 0), \quad \mathbf{D}_3 = -\mathbf{P}_3 = (-3/5, -4/5, 0),$$
$$\mathbf{D}_4 = -\mathbf{P}_4 = (-1, 0, 0), \quad \mathbf{D}_5 = -\mathbf{P}_5 = (-0.83, 0.55, 0).$$

8 Computer Animation

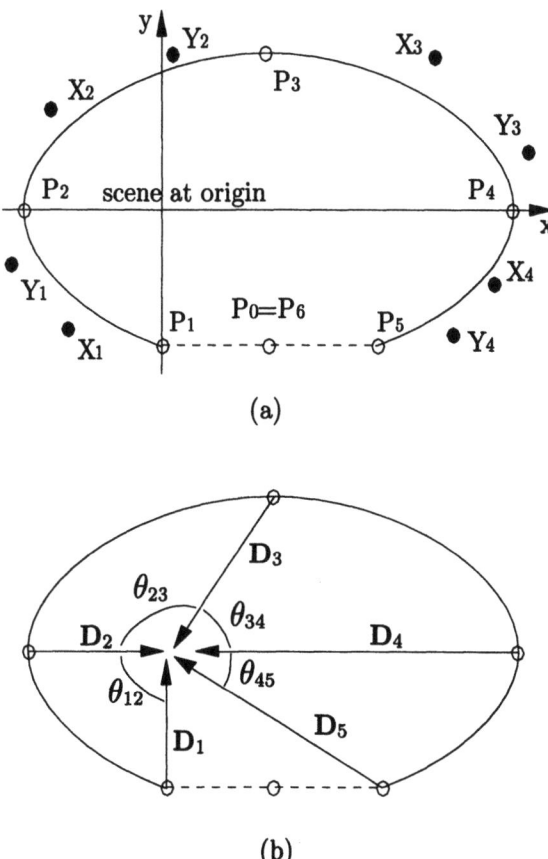

Figure 8.8: (a) A Seven-Point Animation Path. (b) Five Direction Vectors.

The angles between them are calculated by means of dot products:

$$(\mathbf{D}_1 \bullet \mathbf{D}_2) = 0 \rightarrow \theta_{12} = 90°, \quad (\mathbf{D}_2 \bullet \mathbf{D}_3) = -3/5 \rightarrow \theta_{23} = 126.87°,$$
$$(\mathbf{D}_3 \bullet \mathbf{D}_4) = 3/5 \rightarrow \theta_{34} = 53.13°, \quad (\mathbf{D}_4 \bullet \mathbf{D}_5) = .83 \rightarrow \theta_{45} = 33.9°.$$

To produce the first animation frame, we assume that the camera is located at $\mathbf{P}_1$, looking in direction $\mathbf{D}_1$ and we take a snapshot, i.e., we assume a projection plane perpendicular to $\mathbf{D}_1$, located at a distance k from $\mathbf{P}_1$, on which the scene can now be projected.

To produce more animation frames, we start with the first Bézier segment, $\mathbf{P}_1(t)$, vary t in steps, and for each value t_i calculate a position $\mathbf{P}_1(t_i)$ on the curve and a direction $\mathbf{D}_{1+t_i}$ that's a spherical interpolation of $\mathbf{D}_1$ and $\mathbf{D}_2$:

$$\mathbf{D}_{1+t_i} = \frac{\sin(\theta_{12} - t_i\theta_{12})}{\sin\theta_{12}}\mathbf{D}_1 + \frac{\sin(t_i\theta_{12})}{\sin\theta_{12}}\mathbf{D}_2.$$

Once $\mathbf{D}_{1+t_i}$ is obtained for frame i, we use it to take a snapshot.

As an example, for $t = 0.5$ the camera should be moved to point

$$\begin{aligned}\mathbf{P}_1(0.5) &= 0.5^3(0,-2,0) + 3 \cdot 0.5 \cdot 0.5^2(-7/12,-5/3,0) \\ &\quad + 3 \cdot 0.5^2 \cdot 0.5(-9/4,-2/3,0) + 0.5^3(-2,0,0) \\ &= 0.5^3[(0,-2,0) + 3(-7/12,-5/3,0) + 3(-9/4,-2/3,0) + (-2,0,0)] \\ &= (-1.3125, -1.125, 0)\end{aligned}$$

and should look in direction $\mathbf{D}_{1+0.5}$,

$$\begin{aligned}\mathbf{D}_{1+0.5} &= \frac{\sin((1-0.5)\theta_{12})}{\sin\theta_{12}}\mathbf{D}_1 + \frac{\sin(0.5\theta_{12})}{\sin\theta_{12}}\mathbf{D}_2 \\ &= \frac{\sin 45°}{\sin 90°}(\mathbf{D}_1 + \mathbf{D}_2), \\ &= 0.7071[(0,1,0) + (1,0,0)] \\ &= (0.7071, 0.7071, 0).\end{aligned}$$

Notice that $\mathbf{D}_{1+0.5}$ is a unit vector pointing in a 45° direction in the xy plane.

We now have the new position $(-1.3125, -1.125, 0)$ and new direction of view $(0.7071, 0.7071, 0)$ of the camera and we can use a perspective projection technique, such as the one described in Section 3.9.8, to calculate the projections of all the points in the scene. We assume that the projection plane is perpendicular to $\mathbf{D}_{1+0.5} = (0.7071, 0.7071, 0)$ and is located at a distance k from the camera (where k is a user-controlled parameter that may vary from frame to frame). Notice that the technique of Section 3.9.8 assumes that two viewing half-angles h and v are given. They correspond to the size of the projection plane. Any image point that would be projected outside that size should be ignored.

▶ **Exercise 8.4:** The new camera direction is $(0.7071, 0.7071, 0)$. In order to point at the origin, the camera should be located at a point with coordinates $(-c, -c, 0)$, i.e., the x and y coordinates should be identical. We, however, got $\mathbf{P}_1(0.5) = (-1.3125, -1.125, 0)$. What's the explanation?

▶ **Exercise 8.5:** If at $\mathbf{P}_1$ the camera should look at the positive y direction, and at $\mathbf{P}_2$, at the positive x direction, then midway it should look between these directions, i.e., at 45° or in the $(1,1,0)$ direction. In fact, if we want the camera to stop, for example, at $\mathbf{P}_1$, $\mathbf{P}_2$, and at three other points equally spaced in between, we know we should point the camera at angles of 0°, 22.5°, 45°, 67.5°, and 90° to the positive y direction at the five points and there seems to be no need for the direction vectors $\mathbf{D}_i$. What's the explanation?

▶ **Exercise 8.6:** Perform the same calculation for the second Bézier segment. Find the coordinates of point $\mathbf{P}_2(0.5)$ and compute the new camera direction $\mathbf{D}_{2+0.5}$ as a spherical interpolation of $\mathbf{D}_2$ and $\mathbf{D}_3$.

▶ **Exercise 8.7:** Calculate $\mathbf{D}_{3+0.5}$ for the third path segment.

8 Computer Animation

8.3.3 Example 2

This is the same as Example 1, except that point $\mathbf{P}_2$ is moved to location $(-2, 0, 1)$. The camera path in this example is, thus, not completely contained in the $z = 0$ plane. The only direction vector that changes is $\mathbf{D}_2$, which becomes $(2, 0, -1)$ or, after normalization, $(2, 0, -1)/\sqrt{5}$. Only two angles are affected:

$$(\mathbf{D}_1 \bullet \mathbf{D}_2) = \frac{1}{\sqrt{5}}(0, 1, 0) \bullet (2, 0, -1) = 0 \to \theta_{12} = 90°,$$

$$(\mathbf{D}_2 \bullet \mathbf{D}_3) = \frac{1}{\sqrt{5}}(2, 0, -1) \bullet (-3/5, -4/5, 0) = -\frac{6}{5\sqrt{5}} \approx -0.5367 \to \theta_{23} = 122.46°.$$

The two interpolations $\mathbf{D}_{1+0.5}$ and $\mathbf{D}_{2+0.5}$ are shown

$$\mathbf{D}_{1+0.5} = \frac{\sin 45°}{\sin 90°}(\mathbf{D}_1 + \mathbf{D}_2) = 0.7071 \left[(0, 1, 0) + \frac{1}{\sqrt{5}}(2, 0, -1)\right]$$
$$= (0.6324, 0.7071, -0.3162),$$

$$\mathbf{D}_{2+0.5} = \frac{\sin 61.23°}{\sin 122.46°}(\mathbf{D}_2 + \mathbf{D}_3) = \frac{0.8766}{0.8438}\left[\frac{1}{\sqrt{5}}(2, 0, -1) + (-3/5, -4/5, 0)\right]$$
$$= (0.3059, -0.8311, -0.4646).$$

The new camera directions have a negative z component, since the camera itself is now located at points with positive z and should be looking at the origin. However, it is impossible to tell just by examining the interpolated directions whether they are the right ones. The best test is to actually implement the example in software.

8.3.4 Example 3

This is still a simple example, but this time the camera is aimed at different points in the scene while moving along its path. We assume a camera path that's a straight line from $\mathbf{P}_1 = (2, 2, 0)$ to $\mathbf{P}_2 = (1, 1, 0)$ (Figure 8.9a). The equation of this line is, of course, $\mathbf{P}_1(t) = (1-t)\mathbf{P}_1 + t\mathbf{P}_2 = (2-t, 2-t, 0)$, but notice that this equation is also easy to obtain with an interpolating Bézier curve, which is our standard method. All that's necessary is two more points, $\mathbf{P}_0 = (3, 3, 0)$ and $\mathbf{P}_3 = (0, 0, 0)$ which will define the start and end directions, respectively, of the curve, and will make it a straight line. We first calculate the two new interior points

$$\mathbf{X}_1 = \mathbf{P}_1 + \frac{1}{6}(\mathbf{P}_2 - \mathbf{P}_0) = \left(\frac{5}{3}, \frac{5}{3}, 0\right), \quad \mathbf{Y}_1 = \mathbf{P}_2 - \frac{1}{6}(\mathbf{P}_3 - \mathbf{P}_1) = \left(\frac{4}{3}, \frac{4}{3}, 0\right).$$

The curve is, as usual,

$$\mathbf{P}_1(t) = (1-t)^3 \mathbf{P}_1 + 3t(1-t)^2 \mathbf{X}_1 + 3t^2(1-t)\mathbf{Y}_1 + t^3 \mathbf{P}_2$$
$$= (1-t)^3 (2, 2, 0) + 3t(1-t)^2 \left(\frac{5}{3}, \frac{5}{3}, 0\right) + 3t^2(1-t)\left(\frac{4}{3}, \frac{4}{3}, 0\right) + t^3(1, 1, 0)$$
$$= (2-t, 2-t, 0).$$

We arbitrarily decide that at point $\mathbf{P}_1$, the camera should look at point $\mathbf{S}_1 = (1.75, 1.75, -1)$, while at $\mathbf{P}_2$, it should look at $\mathbf{S}_2 = (1.25, 1.25, -1)$. The idea is to slide the camera along its simple path while panning it, so it covers the area between $\mathbf{S}_1$ and $\mathbf{S}_2$.

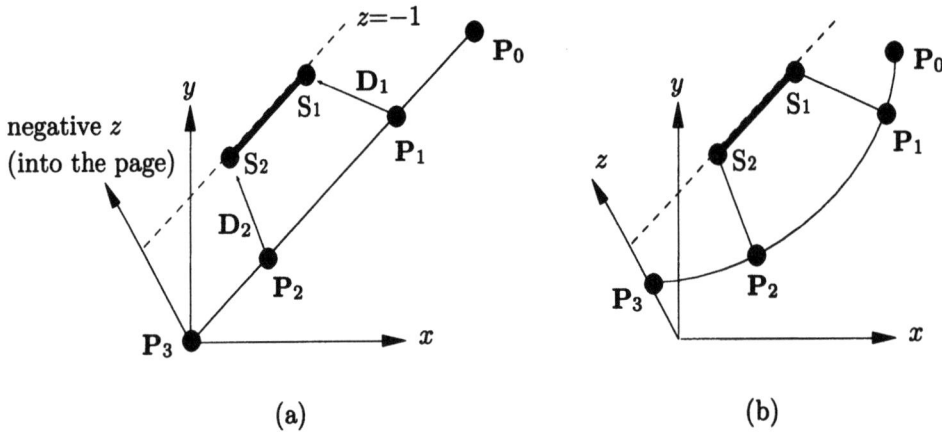

Figure 8.9: An Animation Path with Panning.

The two normalized direction vectors are

$$\mathbf{D}_1 = \mathbf{S}_1 - \mathbf{P}_1 = (1.75, 1.75, -1) - (2, 2, 0)$$
$$= (-0.25, -0.25, -1) \text{ normalized to } (-0.2357, -0.2357, -0.9428),$$
$$\mathbf{D}_2 = \mathbf{S}_2 - \mathbf{P}_2 = (1.25, 1.25, -1) - (1, 1, 0)$$
$$= (0.25, 0.25, -1) \text{ normalized to } (0.2357, 0.2357, -0.9428).$$

The angle between them is

$$\cos \theta_{12} = (\mathbf{D}_1 \bullet \mathbf{D}_2) = -0.049 - 0.049 + 0.79 = 0.7777,$$

implying $\theta_{12} = 38.94°$. For $t = 0.5$, the position of the camera midway between $\mathbf{P}_1$ and $\mathbf{P}_2$ is given by the linear interpolation $\mathbf{P}_1(0.5) = (2 - 0.5, 2 - 0.5, 0) = (1.5, 1.5, 0)$. Its direction of view is calculated by the spherical interpolation

$$\mathbf{D}_{1+0.5} = \frac{\sin 19.47°}{\sin 38.94°}(\mathbf{D}_1 + \mathbf{D}_2) = 0.5303(0, 0, -1.8856) = (0, 0, -1).$$

Both values, the position and direction of view, are easy to verify visually because of the simple geometry of the problem.

▶ **Exercise 8.8:** Change the camera path from a straight line to an arc (Figure 8.9b) by moving the two extreme guide points $\mathbf{P}_0$ and $\mathbf{P}_3$ to positions $(3, 3, -0.25)$ and $(0, 0, -0.25)$, respectively. Notice that this will not change the interpolated directions of the camera.

8 Computer Animation

8.4 Interpolating Orientations: II

The discussion so far has employed only two direction vectors, $\mathbf{D}_i$ and $\mathbf{D}_{i+1}$, to compute the new camera orientation at each frame by spherical interpolation. This, however, may lead to a sudden change in camera direction at a key frame and, hence, to nonsmooth, jerky animation. As a simple example, imagine a two-segment camera path with direction vectors at three consecutive key frames pointing, respectively, in the positive x, y, and z directions. When the camera is moved along the first segment, it will change directions from the x to the y axis, so it will always point somewhere in the xy plane. When the camera switches to the second segment, it will start pointing somewhere in the yz plane. Switching directions between the two perpendicular xy and yz planes (Figure 8.10) may cause a jerk in the animation. The usual solution is to define key frames with direction vectors that don't differ by much. An alternative may be to derive a new spherical interpolation function that interpolates several consecutive direction vectors. When the camera moves in segment i, such a function should interpolate $\mathbf{D}_i$ and $\mathbf{D}_{i+1}$ but should also assign weights to $\mathbf{D}_{i-1}$ (mostly at the start of segment i) and to $\mathbf{D}_{i+2}$ (mostly at the end of the segment).

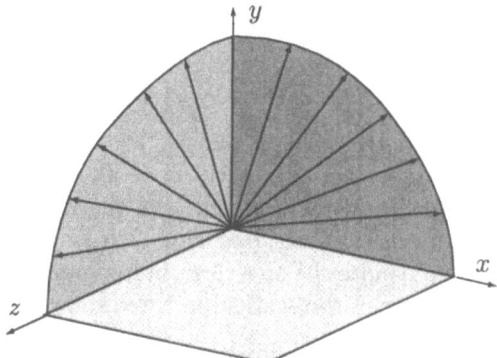

Figure 8.10: Abrupt Change of Direction.

We now show how such interpolation can be achieved. At a certain point, the animator has to input the $n-1$ direction vectors $\mathbf{D}_i$ at the key frames (in practice, the animator may input the coordinates of the points the camera should be looking at in every key frame and the software uses these points to calculate the direction vectors). In addition to these data, the animator should input two more direction vectors $\mathbf{X}_1$ and $\mathbf{Y}_{n-1}$, to provide the software with more direction information at the start and at the end of the path. The animation software then calculates "intermediate" direction vectors $\mathbf{X}_i$ and $\mathbf{Y}_i$ for each segment, in the same way the interior points are defined for the segment, i.e.,

$$\mathbf{X}_i = \mathbf{D}_i + \frac{1}{6}(\mathbf{D}_{i+1} - \mathbf{D}_{i-1}), \quad \mathbf{Y}_i = \mathbf{D}_{i+1} - \frac{1}{6}(\mathbf{D}_{i+2} - \mathbf{D}_i).$$

Once this is done, the software should do the following for each segment i of the camera path. Vary the time parameter t from 0 to 1, calculate spatial camera positions $\mathbf{P}_i(t)$, and for each value t, use spherical interpolation to interpolate the four direction vectors $\mathbf{D}_i$, $\mathbf{X}_i$, $\mathbf{Y}_i$, and $\mathbf{D}_{i+1}$ of the segment. Initially, when t is close to 0, the interpolation should give more weight to $\mathbf{X}_i$ (and thus include a contribution from $\mathbf{D}_{i-1}$). Toward the end, when t gets close to 1, the same spherical interpolation should assign more weight to $\mathbf{Y}_i$ (and, thus, to $\mathbf{D}_{i+2}$).

Our problem is to find the right way to do this kind of spherical interpolation. The first thing that comes to mind is to define the interpolated camera direction $\mathbf{D}_{i+t}$ as the standard Bézier weighted sum,

$$\mathbf{D}_{i+t} = (1-t)^3 \mathbf{D}_i + 3t(1-t)^2 \mathbf{X}_i + 3t^2(1-t) \mathbf{Y}_i + t^3 \mathbf{D}_{i+1}.$$

This certainly favors $\mathbf{X}_i$ in the early parts of the segment and favors $\mathbf{Y}_i$ in the later parts. However, since the Bézier curve has variable velocity, this kind of interpolation will produce direction vectors that are spread nonuniformly between $\mathbf{D}_i$ and $\mathbf{D}_{i+1}$. What we need in this case is to extend spherical interpolation to four vectors, and we do this by means of the de Casteljau construction of Section 4.14.7, but with spherical instead of linear mediation. As a reminder, the mediation operator $t[\![\mathbf{P}_0, \mathbf{P}_1]\!]$ between two points $\mathbf{P}_0$ and $\mathbf{P}_1$ is defined as

$$t[\![\mathbf{P}_0, \mathbf{P}_1]\!] = t\mathbf{P}_1 + (1-t)\mathbf{P}_0 = t(\mathbf{P}_1 - \mathbf{P}_0) + \mathbf{P}_0, \quad \text{where} \quad 0 \le t \le 1.$$

We now use our spherical interpolation, Equation (8.2), as a *spherical mediation operator* and apply it to construct the scaffolding of the four direction vectors in the same way it is done for four points (see page 285 and Figure Ans.20). We use the notation $[\mathbf{A}; \mathbf{B}; t]$ to denote the spherical interpolation of vectors $\mathbf{A}$ and $\mathbf{B}$ (Equation (8.3)) and we construct the scaffold in three steps.

1. Calculate the three interpolated direction vectors

$$\mathbf{P}_{01} = [\mathbf{D}_i; \mathbf{X}_i; t], \quad \mathbf{P}_{02} = [\mathbf{X}_i; \mathbf{Y}_i; t], \quad \text{and} \quad \mathbf{P}_{03} = [\mathbf{Y}_i; \mathbf{D}_{i+1}; t].$$

2. Calculate the two interpolated direction vectors $\mathbf{P}_{11} = [\mathbf{P}_{01}; \mathbf{P}_{02}; t]$ and $\mathbf{P}_{12} = [\mathbf{P}_{02}; \mathbf{P}_{03}; t]$.

3. Calculate the final interpolated direction vector $\mathbf{D}_{i+t} = [\mathbf{P}_{11}; \mathbf{P}_{12}; t]$. This becomes the direction of the camera at point $\mathbf{P}_i(t)$.

When the camera is moved to point $\mathbf{P}_i(t)$ and is oriented there, pointing in direction $\mathbf{D}_{i+t}$, we can expect smooth animation since direction $\mathbf{D}_{i+t}$ takes into account not just $\mathbf{D}_i$ and $\mathbf{D}_{i+1}$ but also "remembers" the past direction $\mathbf{D}_{i-1}$ and "anticipates" the future direction $\mathbf{D}_{i+2}$. Notice that four such direction vectors are available at every key frame, including the first and last ones, since the animator inputs the two extra direction vectors $\mathbf{X}_1$ and $\mathbf{Y}_{n-1}$ explicitly.

8 Computer Animation

8.4.1 Example 4

The same points and direction vectors of Example 1 of Section 8.3.2 are used. The normalized direction vectors are

$$\mathbf{D}_1 = -\mathbf{P}_1 = (0,1,0), \quad \mathbf{D}_2 = -\mathbf{P}_2 = (1,0,0), \quad \mathbf{D}_3 = -\mathbf{P}_3 = (-3/5, -4/5, 0),$$
$$\mathbf{D}_4 = -\mathbf{P}_4 = (-1,0,0), \quad \mathbf{D}_5 = -\mathbf{P}_5 = (-0.83, 0.55, 0).$$

We assume that the animator inputs the two extra directions, $\mathbf{X}_1 = (1,3,0)$ and $\mathbf{Y}_4 = (-1, 0.5, 0)$. All other "interior" directions are now calculated and normalized:

$$\mathbf{X}_1 = (1,3,0) \to (0.3162, 0.9487, 0),$$
$$\mathbf{Y}_1 = \mathbf{D}_2 - \frac{1}{6}(\mathbf{D}_3 - \mathbf{D}_1) = (1.1, 0.3, 0) \to (0.964764, 0.263117, 0),$$
$$\mathbf{X}_2 = \mathbf{D}_2 + \frac{1}{6}(\mathbf{D}_3 - \mathbf{D}_1) = (0.9, -0.3, 0) \to (0.948683, -0.316228, 0),$$
$$\mathbf{Y}_2 = \mathbf{D}_3 - \frac{1}{6}(\mathbf{D}_4 - \mathbf{D}_2) = (-0.2667, -0.8, 0) \to (-0.316228, -0.948683, 0),$$
$$\mathbf{X}_3 = \mathbf{D}_3 + \frac{1}{6}(\mathbf{D}_4 - \mathbf{D}_2) = (-0.9333, -0.8, 0) \to (-0.759257, -0.650791, 0),$$
$$\mathbf{Y}_3 = \mathbf{D}_4 - \frac{1}{6}(\mathbf{D}_5 - \mathbf{D}_3) = (-0.9617, -0.225, 0) \to (-0.973704, -0.227816, 0),$$
$$\mathbf{X}_4 = \mathbf{D}_4 + \frac{1}{6}(\mathbf{D}_5 - \mathbf{D}_3) = (-1.0383, 0.225, 0) \to (-0.977318, 0.211778, 0),$$
$$\mathbf{Y}_4 = (-1, 0.5, 0) \to (-0.8944, 0.4472, 0).$$

We now calculate $\mathbf{D}_{1+0.5}$ in three steps.

Step 1: Calculate the three interpolated direction vectors

$$\mathbf{P}_{01} = [\mathbf{D}_1; \mathbf{X}_1; .5] = \frac{\sin 18.43°}{\sin 9.215°}(\mathbf{D}_1 + \mathbf{X}_1) = (0.160167, 0.987089, 0),$$
$$\mathbf{P}_{02} = [\mathbf{X}_1; \mathbf{Y}_1; .5] = \frac{\sin 56.31°}{\sin 28.155°}(\mathbf{X}_1 + \mathbf{Y}_1) = (0.726438, 0.687225, 0),$$
$$\mathbf{P}_{03} = [\mathbf{Y}_1; \mathbf{D}_2; .5] = \frac{\sin 15.255°}{\sin 7.628°}(\mathbf{Y}_1 + \mathbf{D}_2) = (0.991152, 0.132733, 0).$$

Step 2: Calculate the two interpolated direction vectors

$$\mathbf{P}_{11} = [\mathbf{P}_{01}; \mathbf{P}_{02}; .5] = \frac{\sin 37.37°}{\sin 18.69°}(\mathbf{P}_{01} + \mathbf{P}_{02}) = (0.46797, 0.883741, 0),$$
$$\mathbf{P}_{12} = [\mathbf{P}_{02}; \mathbf{P}_{03}; .5] = \frac{\sin 35.78°}{\sin 17.89°}(\mathbf{P}_{02} + \mathbf{P}_{03}) = (0.902439, 0.430814, 0).$$

Step 3: Calculate the final interpolated direction vector

$$\mathbf{D}_{1+0.5} = [\mathbf{P}_{11}; \mathbf{P}_{12}; 0.5] = \frac{\sin 36.58°}{\sin 18.29°}(\mathbf{P}_{11} + \mathbf{P}_{12}) = (0.721658, 0.692245, 0).$$

This becomes the direction of the camera at point $\mathbf{P}_1(0.5) = (-1.3125, -1.125, 0)$. Notice that it differs from the 45° direction calculated in Example 1, since it depends on the choice of the "exterior" direction $\mathbf{X}_1$ that was input by the animator.

8.4.2 Interpolating Orientations: III

This approach to the problem of interpolating orientations uses *quaternions*. These mathematical entities are introduced in Section A.9. Their application to general rotations is discussed in Section 3.5.4. The reader should review these sections prior to reading this section. Quaternions can be used in computer animation to interpolate orientations between key frames because of two facts:

1. A general rotation of θ degrees about an axis $\mathbf{u}$ can be expressed by the unit quaternion $\mathbf{q} = [\cos(\theta/2), \sin(\theta/2)\mathbf{u}]$.

2. When a rigid object is sent flying through space, it may roll and tumble in a complicated way, but, at any moment, its position and orientation can be completely described by two transformations—a translation from its initial position to its present position and a rotation of θ degrees about some axis $\mathbf{u}$.

We can imagine all the unit quaternions to form a unit four-dimensional sphere. Spherically interpolating two unit quaternions is thus equivalent to moving along a great arc on this sphere. The technique is identical to the one discussed in Section 8.3 for vectors. Interpolating camera orientation between two key frames by using quaternions is done in the following steps:

1. The animator inputs the data for all the key frames. The software uses this to calculate the directions of view $\mathbf{D}_i$ for each key frame i.

2. The software "positions" the camera at the preferred point $(0, 0, -k)$, looking in the positive z direction [i.e., in direction $(0, 0, 1)$].

3. A quaternion $\mathbf{q}_i$ is calculated for each key frame i, describing the rotation that would bring the camera *from its initial orientation* $(0, 0, 1)$ to its orientation in key frame i.

4. The software goes into a loop where it moves the camera along its path, segment by segment. In segment i (the segment between key frames i and $i+1$), the time parameter t is incremented from 0 to 1 in F steps. In each step t_m, the camera is translated to position $\mathbf{P}_i(t_m)$ and is reoriented by rotating it. The main point is that the translation is done *from the initial position* $(0, 0, -k)$, and the rotation is done *from the initial orientation* $(0, 0, 1)$. The software uses t_m to spherically interpolate the two quaternions $\mathbf{q}_i$ and $\mathbf{q}_{i+1}$ to a quaternion $\mathbf{q}_{i+t_m}$. Quaternion $\mathbf{q}_i$ describes a rotation from $(0, 0, 1)$ to key frame i. Similarly, $\mathbf{q}_{i+1}$ describes a rotation from $(0, 0, 1)$ to key frame $i+1$. Thus, their interpolation describes a rotation that will bring the camera from its initial orientation $(0, 0, 1)$ to the orientation it should have at point $\mathbf{P}_i(t_m)$.

(The discussion of Section 8.4 suggests that four quaternions, instead of two, should participate in any interpolation. The software should, thus, calculate two auxiliary quaternions $\mathbf{X}_i$ and $\mathbf{Y}_i$ for each segment and use the scaffolding construction on $\mathbf{q}_i$, $\mathbf{X}_i$, $\mathbf{Y}_i$, and $\mathbf{q}_{i+1}$ to calculate $\mathbf{q}_{i+t_m}$.)

5. Once $\mathbf{q}_{i+t_m}$ is obtained, the software uses it to generate a rotation matrix $\mathbf{M}$ according to Equation (3.28). This matrix is then used to take a snapshot. The

8 Computer Animation

snapshot can be taken using the methods of Section 3.9.5 or 3.9.8. An alternative is to use the technique of Section 3.9.2, which is the one used here. The principle is the following: We know that the camera had to be translated from its initial position $(0,0,-k)$ to its present position $\mathbf{P}_i(t_m)$ and rotated according to $\mathbf{q}_{i+t_m}$. The software simply applies the two *reverse* transformations (and in reverse order) to every point of the scene, thus "bringing the scene to the camera" (which remains in its preferred position) instead of bringing the camera to the scene. Once the scene is brought to the camera, any point in the scene can be projected using the standard projection matrix $\mathbf{T}_p$, Equation (3.39).

8.4.3 Example 5

The camera is located at the preferred point $(0,0,-k)$ (here, we assume that $0 < k < 1$), looking in the preferred direction $\mathbf{D} = (0,0,1)$. Three key frames are defined, at points $\mathbf{P}_1 = (-1,0,2)$, $\mathbf{P}_2 = (0,0,1)$, and $\mathbf{P}_3 = (1,0,2.5)$ (Figure 8.11a, where a right-handed coordinate system implies that the y axis should come out of the page). The center of interest (the direction the camera should be looking at) is arbitrarily selected as point $(0,0,2)$. The three direction vectors are, thus,

$$\mathbf{D}_1 = (0,0,2) - (-1,0,2) = (1,0,0),$$
$$\mathbf{D}_2 = (0,0,2) - (0,0,1) = (0,0,1),$$
$$\mathbf{D}_3 = (0,0,2) - (1,0,2.5) = (-1,0,-0.5), \quad \text{normalized to } (-0.8944, 0, -0.4472).$$

The angles between each direction vector and the original direction $\mathbf{D}$ are (Figure 8.11b)

$$\cos\theta_1 = \mathbf{D} \bullet \mathbf{D}_1 = (0,0,1) \bullet (1,0,0) = 0 \to \theta_1 = 90°,$$
$$\cos\theta_2 = \mathbf{D} \bullet \mathbf{D}_2 = (0,0,1) \bullet (0,0,1) = 1 \to \theta_2 = 0°,$$
$$\cos\theta_3 = \mathbf{D} \bullet \mathbf{D}_3 = (0,0,1) \bullet (-.8944, 0, -.4472) = -.4472 \to \theta_3 = 116.56°.$$

The quaternions for the three key frames can now be calculated:

$$\mathbf{q}_1 = [\cos(\theta_1/2), \sin(\theta_1/2)(0,1,0)] = (0.7071, 0, 0.7071, 0),$$
$$\mathbf{q}_2 = [\cos(\theta_2/2), \sin(\theta_2/2)(0,1,0)] = (1, 0, 0, 0),$$
$$\mathbf{q}_3 = [\cos(\theta_3/2), \sin(\theta_3/2)(0,-1,0)] = (0.5258, 0, -0.8506, 0).$$

Notice that $\mathbf{q}_1$ corresponds to a clockwise rotation about the positive y axis, whereas $\mathbf{q}_3$ corresponds to a clockwise rotation about the *negative* y axis (Figure 8.11b). This is the reason for using direction $(0,-1,0)$ as the rotation axis for the latter. The axis of rotation for quaternion $\mathbf{q}_i$ is simply the cross-product $\mathbf{D} \times \mathbf{D}_i$, where the unnormalized form of $\mathbf{D}_i$ is used.

The quaternions are now used to calculate, as an example, the two interpolations $\mathbf{q}_{1+0.5}$ and $\mathbf{q}_{2+0.5}$. From $\mathbf{q}_1 \bullet \mathbf{q}_2 = 0.7071$, we find that the angle between $\mathbf{q}_1$ and $\mathbf{q}_2$ is $45°$. Similarly, $\mathbf{q}_2 \bullet \mathbf{q}_3 = 0.5258$ implies that the angle between them is $58.28°$. (Notice that these are angles between quaternions, not between the direction

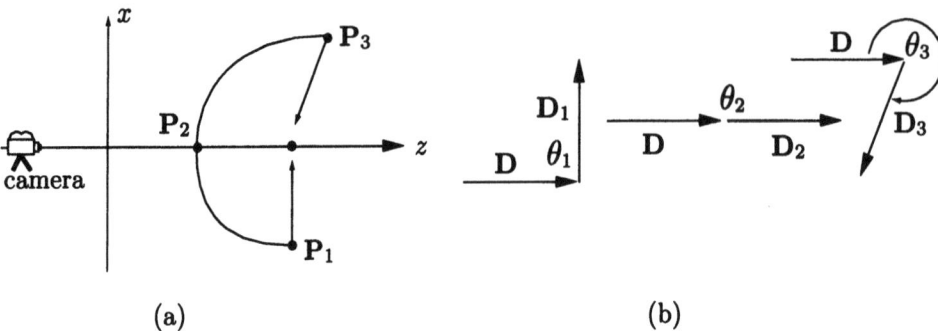

Figure 8.11: A Three-Point Animation Path.

vectors.) Using these angles, we get

$$q_{1+0.5} = \frac{\sin(45°/2)}{\sin 45°}[(0.7071, 0, 0.7071, 0) + (1, 0, 0, 0)]$$

$$= \frac{0.3829}{0.7071}(1.7071, 0, 0.7071, 0)$$

$$= (0.9239, 0, 0.3829, 0) = (\cos 22.5°, 0, \sin 22.5°, 0),$$

$$q_{2+0.5} = \frac{\sin(58.28°/2)}{\sin 58.28°}[(1, 0, 0, 0) + (0.5258, 0, -0.8506, 0)]$$

$$= \frac{0.4869}{0.85}(1.5258, 0, 0.8506, 0)$$

$$= (0.8734, 0, 0.4869, 0) = (\cos -29.14°, 0, \sin -29.14°, 0).$$

Quaternion $q_{1+0.5}$ thus generates a rotation of $22.5 \times 2 = 45°$ from the initial direction $(0,0,1)$ about the y axis. Quaternion $q_{2+0.5}$ corresponds to a rotation of $-29.14 \times 2 = -58.28°$ *from the same initial direction* about the same axis. (Notice how the camera has to be rotated in opposite directions for $q_{1+0.5}$ and $q_{2+0.5}$.)

Here are the details of the snapshots for the first two key frames. At P_1, the camera has to go (Figure 8.12) through the three transformations (1) translate to the origin (k units in the positive z direction), (2) rotate 90° clockwise about the positive y axis, and (3) translate one unit in the negative x and two units in the positive z directions. Since we leave the camera in place, we have to apply the *reverse transformations in reverse order* to the scene: (4) translate one unit in the positive x and two units in the negative z directions, (5) rotate 90° about the origin, counterclockwise about the positive y axis, and (6) translate k units in the negative z direction. Notice how the relative positions of the camera and scene are the same in parts (3) and (6) of the figure.

At P_2, the camera has to go through the two transformations (Figure 8.13): (1) translate to the origin (k units in the positive z direction), and (2) translate one unit in the positive z direction. We again apply the reverse transformations in reverse order to the scene, (3) translate one unit in the negative z direction, and (4) translate k units in the negative z direction.

8 Computer Animation

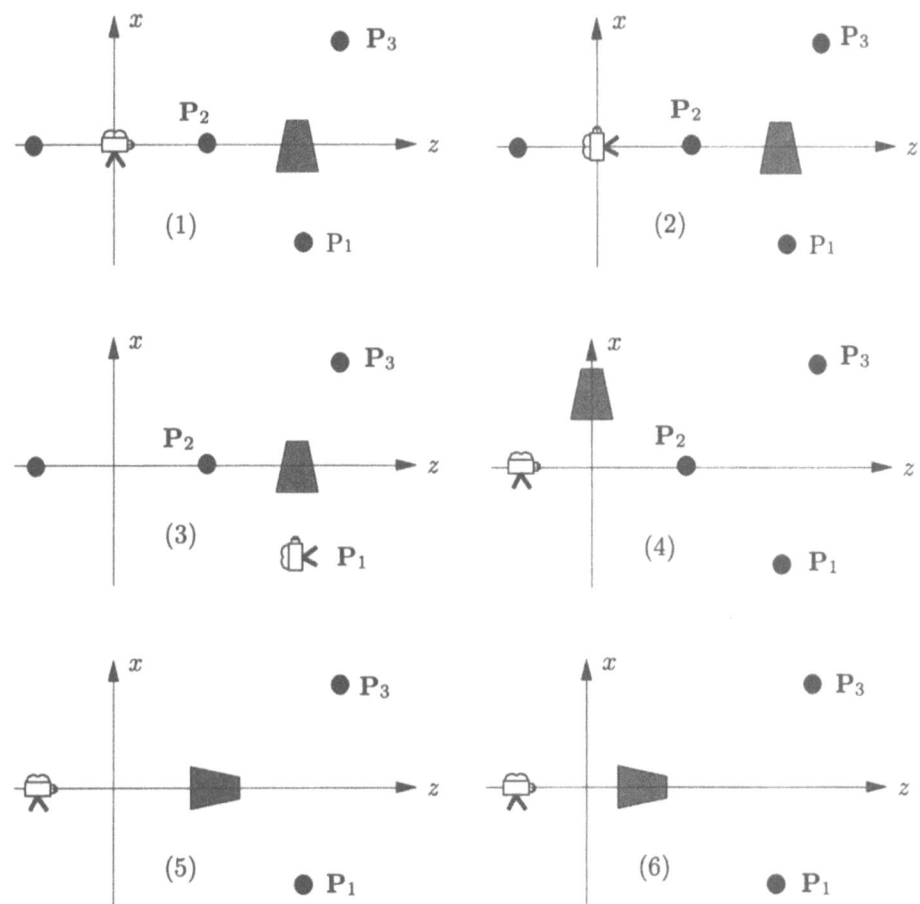

Figure 8.12: Camera (1–3) and Scene (4–6) Transformations for P_1.

▶ **Exercise 8.9:** Describe the transformations for key frame 3.

A general snapshot is, thus, taken as follows:

1. Determine the camera position $P_i(t_m)$. Assume that this is point (a, b, c).
2. Calculate q_{i+t_m} by interpolating q_i and q_{i+1}. Assume that this is quaternion (w, x, y, z).
3. The camera is translated to the origin. This is done by matrix T_1 below.
4. The camera is rotated by matrix M, Equation (3.28).
5. The camera is translated the rest of the way to point (a, b, c), i.e., by an amount $(a, b, c - k)$ using matrix T_2:

$$T_1 = \begin{pmatrix} 1 & 0 & 0 & 0 \\ 0 & 1 & 0 & 0 \\ 0 & 0 & 1 & 0 \\ 0 & 0 & k & 1 \end{pmatrix}, \quad T_2 = \begin{pmatrix} 1 & 0 & 0 & 0 \\ 0 & 1 & 0 & 0 \\ 0 & 0 & 1 & 0 \\ a & b & c-k & 1 \end{pmatrix},$$

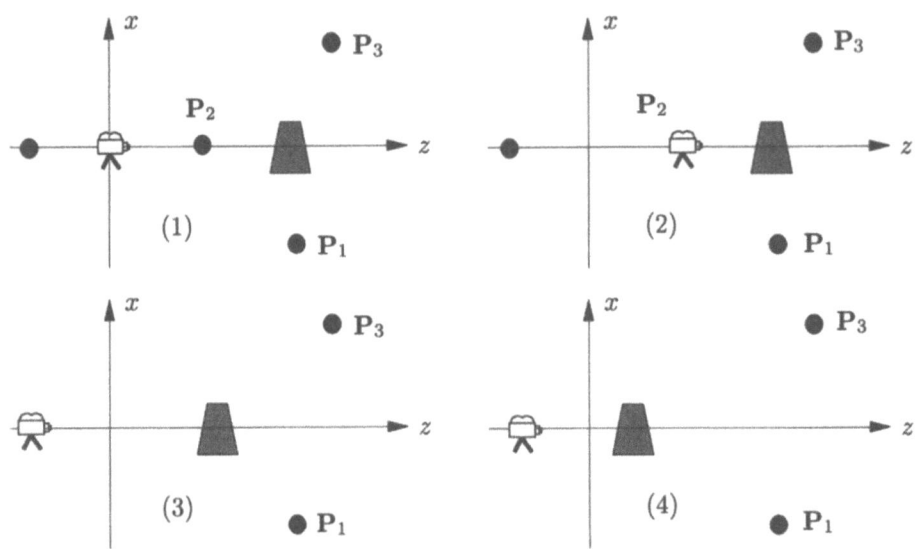

Figure 8.13: Camera (1–2) and Scene (3–4) Transformations for $\mathbf{P}_2$.

$$\mathbf{M} = \begin{pmatrix} 1 - 2y^2 - 2z^2 & 2xy + 2wz & 2xz - 2wy & 0 \\ 2xy - 2wz & 1 - 2x^2 - 2z^2 & 2yz - 2wx & 0 \\ 2xz + 2wy & 2yz - 2wx & 1 - 2x^2 - 2y^2 & 0 \\ 0 & 0 & 0 & 1 \end{pmatrix}.$$

6. Since we want to apply the reverse transformations to the scene (and in reverse order), we multiply each point of the scene by

$$\begin{pmatrix} 1 & 0 & 0 & 0 \\ 0 & 1 & 0 & 0 \\ 0 & 0 & 1 & 0 \\ -a & -b & -c+k & 1 \end{pmatrix} \cdot \mathbf{M}^{-1} \cdot \begin{pmatrix} 1 & 0 & 0 & 0 \\ 0 & 1 & 0 & 0 \\ 0 & 0 & 1 & 0 \\ 0 & 0 & -k & 1 \end{pmatrix} \cdot \mathbf{T}_p,$$

where $\mathbf{T}_p$ is the standard projection matrix, Equation (3.39). (The inverse of matrix $\mathbf{M}$ can be calculated in general by appropriate software, but it is too big and complex to show here.)

8.5 Nonuniform Interpolation

Spherical interpolation has been specifically developed for uniform change of orientation. Varying t in equal steps produces direction vectors that are uniformly distributed between directions $\mathbf{D}_i$ and $\mathbf{D}_{i+1}$. Sometimes, however, nonuniform changes of orientation and/or position are required. In such cases, a function $T(t)$ is needed, such that varying t in equal steps will vary $T(t)$ from 0 to 1 in unequal steps. If T is used to position the camera, this will simulate acceleration or deceleration of the animation. If T is used to interpolate camera orientation, this will simulate rotating the camera at nonuniform rates. The methods presented here for nonuniform interpolation are based on the concept of blending (Section 4.2).

8.5.1 Quadratic and Cubic Blending

The well-known expression $P(t) = (1-t)P_0 + tP_1$ (Equation (4.4)) can be considered a blending of the two values P_0 and P_1. It blends a $(1-t)$ fraction of P_0 with a t fraction of P_1. The weights (or fractions) should add up to 1. Since this expression is linear in t, we can call it *linear blending*.

It is possible to blend values (numbers, points, vectors, etc.) in nonlinear ways. Section 4.14.6 is a short discussion of the concepts involved. Nonlinear blending seems the best approach for nonuniform interpolation and we start by exploring quadratic blending, i.e., ways to blend two quantities by using weights that employ t^2. The simplest approach is to generalize the linear expression above by squaring t and $(1-t)$. This results in $(1-t)^2 P_1 + t^2 P_2$, which varies from P_1 (for $t = 0$) to P_2 (for $t = 1$). However, this expression is clearly wrong since the two weights $(1-t)^2$ and t^2 do not add up to 1 and thus cannot serve as fractions. It is possible to correct this by artificially adding the missing term $2t(1-t)$ [recall that $(1-t)^2 + 2t(1-t) + t^2 = 1$]. We now notice that the term $2t(1-t)$ is zero when $t = 0$ and also when $t = 1$. It, therefore, does not affect the blending at the extreme values, but it must have an effect on the blending in between. We can therefore multiply this term by any quantity P_w and find out how various values of P_w affect the blending. Figure 8.14a shows a blending of the form $(1-t)^2 P_1 + 2t(1-t)P_w + t^2 P_2$ created by the *Mathematica* code of Figure 8.14c. The figure shows the results of blending $P_0 = 0$ and $P_1 = 1$ for five values of P_w ranging from 0 to 1. Notice that certain values of P_w create blendings outside the range $[P_0, P_1]$, but, in general, quadratic blending can give satisfactory results in many, perhaps most, practical cases.

Similarly, if we try a cubic blend by simply writing $P(t) = (1-t)^3 P_1 + t^3 P_2$, we get the same problem. Cubic blending can be achieved by adding four terms with weights t^3, $3t^2(1-t)$, $3t(1-t)^2$, and $(1-t)^3$. Figure 8.14b shows the results of cubically blending the two values $P_0 = 0$ and $P_3 = 1$. It calculates

$$t^3 P_0 + 3t^2(1-t)P_1 + 3t(1-t)^2 P_2 + (1-t)^3 P_3,$$

for the five sets of "interior" weights (P_1, P_2) set to $(0, 0.1)$, $(0.2, 0.3)$, $(0.333, 0.667)$, $(0.7, 0.8)$, and $(0.9, 1)$.

We next notice that the expressions for the linear, quadratic, and cubic blends are identical to the parametric sums used to construct the Bézier curve. This suggests a way to define parametric blends for cases where complex behavior of $T(t)$ is required. In general, a parametric blend $T(t)$ that uses the $n-1$ parameters $P_1, P_2, \ldots, P_{n-1}$ to blend the two quantities P_0 and P_n should have the form

$$T(t) = \sum_{i=0}^{n} P_i B_{ni}(t),$$

where $B_{ni}(t)$ are the Bernstein polynomials of degree n The fact that the Bézier curve is an ideal tool for blending numbers also suggests how to get smooth blending across key frames. We know from Section 4.14.5 how to connect individual Bézier

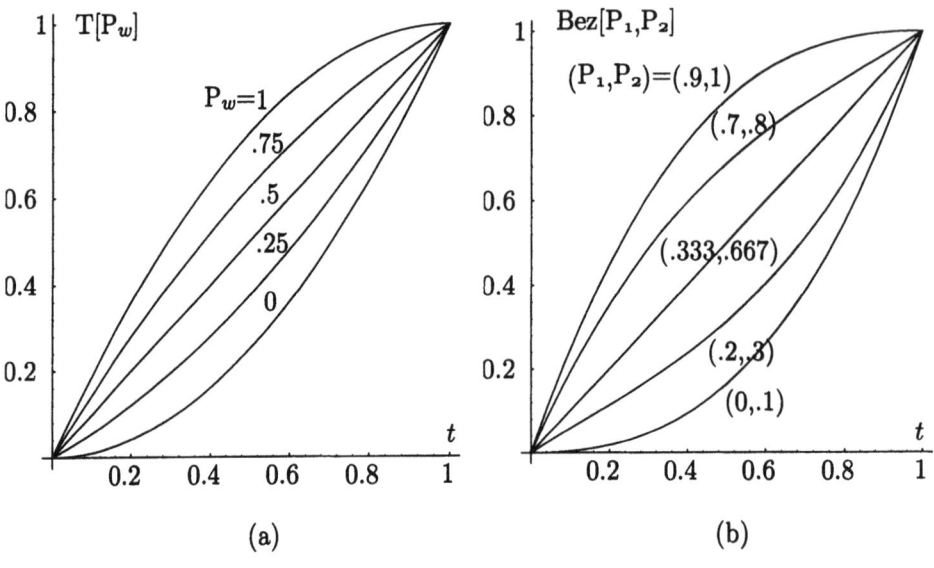

Figure 8.14: (a) Quadratic Blending. (b) Cubic Blending. (c) Code.

segments smoothly. The same idea can be used when numbers are blended. Suppose, for example, that we use a five-point (i.e., $n = 4$) Bézier blending to advance $T(t)$ nonuniformly from 0 to 1 in each of our key frames. For each key frame, we therefore have to select $P_0 = 0$, $P_4 = 1$, plus three parameters P_1, P_2, and P_3 to control the precise way T varies. If we want T to have the same speed on both sides of a key frame, we have to make sure that the difference $P_4 - P_3 = 1 - P_3$ in the segment to the left of the key frame equals the difference $P_1 - P_0 = P_1 - 0$ in the segment to the right of the same key frame. If we select, for example, $P_3 = 0.9$ in one key frame, then we should select $P_1 = 0.1$ in the next key frame.

It is also possible to use the Hermite interpolation (Section 4.7) to blend two values v_1 and v_2 in different proportions by using two parameters s and e.

Hermite interpolation has been developed to create a curve by blending two points and two tangent vectors. It can also be applied to blend any two numbers

8 Computer Animation

v_1 and v_2, by using two user-defined "rates of change" s and e. Equation (4.26) can be modified by substituting v_1 for $\mathbf{P}_1$, v_2 for $\mathbf{P}_2$, and s and e for $\mathbf{P}_1^t$ and $\mathbf{P}_2^t$, respectively. The result is

$$T(t) = (t^3, t^2, t, 1) \begin{pmatrix} 2 & -2 & 1 & 1 \\ -3 & 3 & -2 & -1 \\ 0 & 0 & 1 & 0 \\ 1 & 0 & 0 & 0 \end{pmatrix} \begin{pmatrix} v_1 \\ v_2 \\ s \\ e \end{pmatrix}.$$

The quantities s and e can be considered the start and end "slopes" or rates of change of the blending. Figure 8.15a illustrates the results of the five Hermite interpolations of $v_1 = 0$ and $v_2 = 1$ for $e = 0$ and s values ranging from 0 to 4. It is easy to see how s affects the start slope of the interpolation. Figure 8.15b shows the results of similar interpolations for identical s and e values ranging from 0 to 4. Here both the start and end slopes are affected.

Ease-in/Ease-out: This term refers to nonuniform velocity that starts with acceleration, gradually changes to constant speed, then decelerates. Figure 8.16 shows a typical example. One way to achieve this effect is to set parameters $0 \le a \le b \le 1$ and use the sine function to interpolate and define a parameter $T(t)$ that accelerates when t varies from 0 to a, decelerates when t varies from b to 1, and is linear in the range $[a, b]$. Mathematically, this is expressed by Equation (8.4) (where the second line scales $T(t)$ to the range $[0, 1]$). Figure 8.16 illustrates the result. Notice that the precise shape of $T(t)$ depends on the values of a and b.

$$T_0(t) = \begin{cases} \dfrac{2a}{\pi} \sin\left(\dfrac{\pi}{2} \cdot \dfrac{t-a}{a}\right), & t < a, \\ \dfrac{2(1-b)}{\pi} \sin\left(\dfrac{\pi}{2} \cdot \dfrac{t-b}{1-b}\right) + \dfrac{2a}{\pi} + b - a, & t > b, \\ \dfrac{2a}{\pi} + t - a, & a \le t \le b. \end{cases}$$

$$T(t) = T_0(t) / \left(\dfrac{2a}{\pi} + \dfrac{2(1-b)}{\pi} + b - a\right). \tag{8.4}$$

▶ **Exercise 8.10:** Calculate the acceleration of $T(t)$ in the initial interval $[0, a]$ and its deceleration in the final interval $[b, 1]$.

The same effect of ease-in/ease-out can be obtained from physical considerations without the use of the sine function, by integrating speed to obtain position. We first decide what speed $v(t)$ we want in each subrange of $[0, 1]$, then integrate $v(t)$ to obtain the position $T(t)$ as a function of t in each subrange. Equation (8.5) describes a speed $v(t)$ that increases from zero to a certain value V in subrange $[0, a)$, decreases from V to zero in subrange $(b, 1]$, and is constant in between:

$$v(t) = \begin{cases} V \cdot \dfrac{t}{a}, & t < a, \\ V, & a \le t \le b, \\ V - V \cdot \dfrac{t-b}{1-b} = V \cdot \dfrac{1-t}{1-b}, & t > b. \end{cases} \tag{8.5}$$

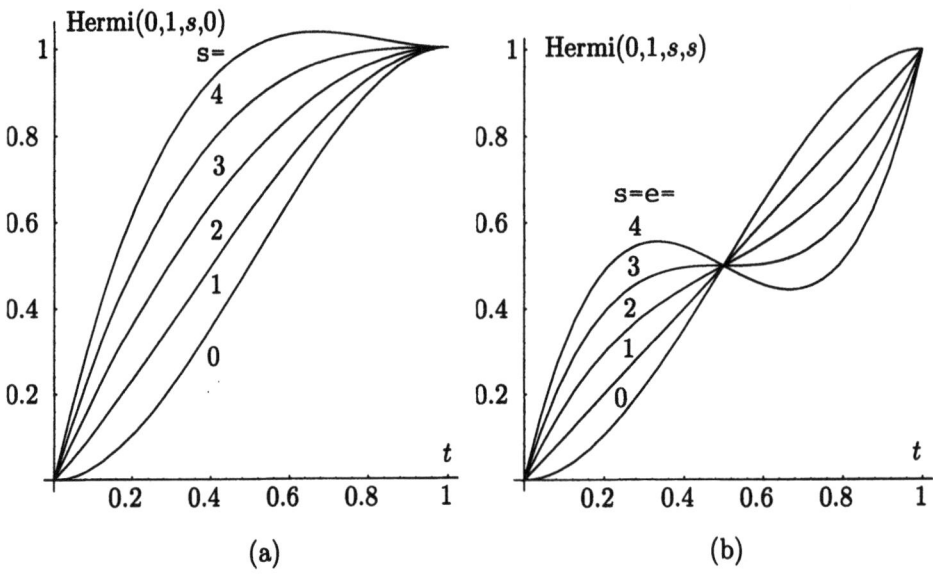

```
Clear[T,H,Hermi]; (* Hermite Interpolation *)
T={t^3,t^2,t,1};
H={{2,-2,1,1},{-3,3,-2,-1},{0,0,1,0},{1,0,0,0}};
(*B={0,1,0,0};*)
Hermi[v1_,v2_,s_,e_]:=Plot[T.H.{v1,v2,s,e},{t,0,1},
 AspectRatio->Automatic, Prolog->AbsoluteThickness[.4]];
Show[Hermi[0,1,0,0], Hermi[0,1,1,1], Hermi[0,1,2,2],
 Hermi[0,1,3,3], Hermi[0,1,4,4]]
```

Figure 8.15: Hermite Interpolation.

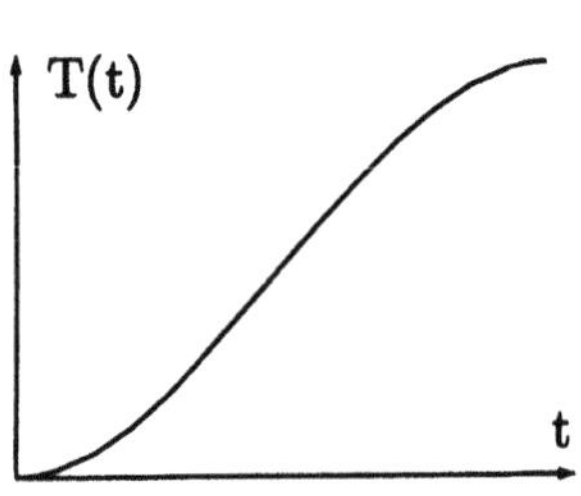

```
Clear[fa,fb,fm,den,a,b];
a=.1; b=.3;
fa:=2a(Sin[Pi(t-a)/(2a)]+1)/Pi;
fb:=Sin[Pi(t-b)/(2(1-b))]2(1-b)/Pi+2a/Pi+b-a;
fm:=2a/Pi+t-a;
den=2a/Pi+2(1-b)/Pi+b-a;
T:=If[t<a,fa/den,If[t>b,fb/den,fm/den]];
Plot[T, {t,0,1}, AspectRatio->1]
```

Figure 8.16: Ease-in/Ease-out with a Sine Function.

8 Computer Animation

We can find the total distance traveled in each subrange by integrating $v(t)$:

$$\int_0^a \frac{Vt}{a} dt = \frac{1}{2}\frac{V}{a}(a^2 - 0),$$

$$\int_a^b V \, dt = V(b-a),$$

$$\int_b^1 V\frac{1-t}{1-b} dt = \frac{V(1-b)^2}{2(1-b)} = \frac{1}{2}V(1-b).$$

The total distance for the three subranges is thus

$$\frac{1}{2}Va + V(b-a) + \frac{1}{2}V(1-b) = \frac{b-a+1}{2}.$$

If we want this distance to equal one unit, we should select $V = 2/(b-a+1)$. The distance $T(t)$ traveled in the first subrange is

$$\int_0^t \frac{Vt}{a} dt = \frac{Vt^2}{2a}.$$

For $a \leq t \leq b$, the total distance traveled from the start of the curve is

$$\frac{1}{2}Va + \int_a^t V \, dt = \frac{1}{2}Va + V(t-a).$$

(Notice that for $t = a$, this equals the distance traveled in the first subrange.) For $b \leq t \leq 1$, the total distance traveled from the start of the curve is

$$\frac{V}{2}a + V(b-a) + \int_b^t V\frac{1-t}{1-b} dt = \frac{V}{2}a + V(b-a) + \left[-\frac{V(1-t)^2}{2(1-b)} + \frac{V(1-b)^2}{2(1-b)}\right]$$

$$= \frac{V}{2}a + V(b-a) + \frac{V}{2(1-b)}[-2b + b^2 + 2t - t^2].$$

These methods can be generalized to obtain other types of nonuniform speeds.

> When graphing a function, the width of the line should be inversely proportional to the precision of the data.
>
> — Marvin J. Albinak

8.6 Morphing

The technique of in-betweening is one of the main advantages of computer animation. This technique has been mentioned before, but it can also be implemented by means of *morphing*. The idea in morphing is for an artist or designer to prepare two key frames of animation and use software to generate all the in-between frames automatically.

The *metamorphosis* of an object, a topic that came to be known as *morphing*, is the case where two pictures are painted by an artist and are designated as the first and last frames of a scene. The artist specifies what points on the first and last frames correspond to each other and the computer then creates several intermediate frames by interpolating each point.

The word Morphing is derived from the Greek $\mu o \rho \phi \epsilon$, meaning form or shape.

Let's assume that point $\mathbf{P}_1$ in the first frame corresponds to point $\mathbf{P}_2$ in the last frame and that four intermediate frames are needed. The coordinates of the point in the four frames are simply $t[\mathbf{P}_1, \mathbf{P}_2]$, where $t = 0.2, 0.4, 0.6, 0.8$. Figure 8.17 is a simple example of morphing. To us, it seems that only two objects are involved, a start face and an end face. To the computer, however, each component, such as eye, nose, and mouth, has to be transformed separately.

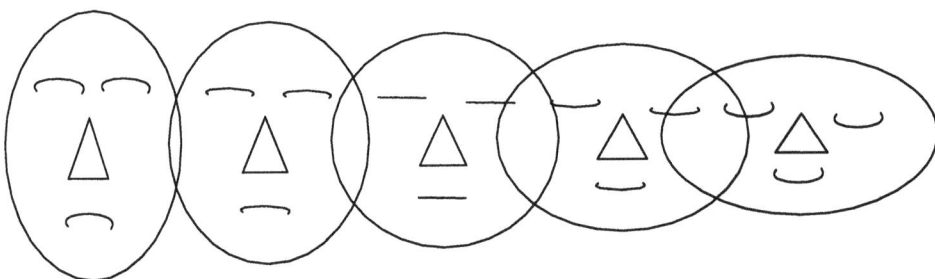

Figure 8.17: An Example of Morphing.

You know the funny thing about morphin? You don't appreciate it till you can't do it anymore!
— David Yost (as Billy) in *Mighty Morphin Power Rangers: The Movie* (1995).

8 Computer Animation

8.7 Free-Form Deformations

This is a modern technique based on old concepts. It uses a grid of control points to distort an image in a systematic way, in order to achieve special effects.

Figure 8.18 shows a fish in a grid being distorted in two different ways. Such effects can be useful in computer animation. The principle is to build a regular grid around the original image and to place control points at strategic locations on the grid (points P_{11} through P_{33} in Figure 8.18). The points are then moved in any desired way to deform the control grid. The image is then displayed, point by point, where each point is first transformed, based on the new coordinates of the control points.

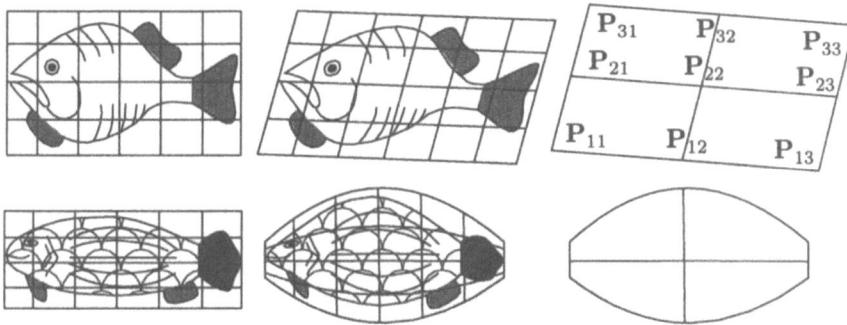

Figure 8.18: Free-Form Deformations.

Before displaying the image, we first have to determine the extreme coordinates of the control points. Let's denote them x_{min}, y_{min}, x_{max}, and y_{max}. A point $\mathbf{P} = (x, y)$ on the object being displayed is transformed to $\mathbf{P}^* = (x^*, y^*)$ in two steps:

1. Its position relative to the extreme coordinates is first calculated by

$$u = \frac{(x - x_{min})}{(x_{max} - x_{min})} \quad \text{and} \quad w = \frac{(y - y_{min})}{(y_{max} - y_{min})}.$$

2. Its new coordinates are calculated using quadratic interpolation:

$$\begin{aligned}(x^*, y^*) &= \left((1-u)^2, 2u(1-u), u^2\right) \begin{pmatrix} P_{11} & P_{12} & P_{13} \\ P_{21} & P_{22} & P_{23} \\ P_{31} & P_{32} & P_{33} \end{pmatrix} \begin{pmatrix} (1-w)^2 \\ 2w(1-w) \\ w^2 \end{pmatrix} \\ &= (B_{20}(u), B_{21}(u), B_{22}(u)) \begin{pmatrix} P_{11} & P_{12} & P_{13} \\ P_{21} & P_{22} & P_{23} \\ P_{31} & P_{32} & P_{33} \end{pmatrix} \begin{pmatrix} B_{20}(w) \\ B_{21}(w) \\ B_{22}(w) \end{pmatrix}.\end{aligned}$$

This is repeated for every point in the image. If the image is made of straight lines, only the two endpoints of each line have to be transformed.

If the image is complex, or if a complicated deformation is needed, the grid can be made bigger and more control points used. The only difference is that higher-order Bernstein polynomials need be used.

This technique can also be extended to three-dimensional images. The control points must be arranged in a three-dimensional grid and the process is similar. In each step, three parameters, u, v, and w, are calculated and used to transform a point $\mathbf{P} = (x, y, z)$ to a point $\mathbf{P}^* = (x^*, y^*, z^*)$.

To produce an animation sequence of F frames, we start with two sets of control points: an initial and a final. We then calculate F in-between sets and display the image by using each set.

> The walks and courts appeared as lines and squares of white, while the soldiers and servants moved about like tiny animated dots.
>
> Harold Bell Wright, *The Uncrowned King*

9
Image Compression

Modern computers employ graphics extensively. Window-based operating systems display the disk's file directory graphically. The progress of many system operations, such as downloading a file, may also be displayed graphically. Many applications provide a graphical user interface (GUI), which makes it easier to use the program and to interpret displayed results. Computer graphics is used in many areas in everyday life to convert many types of complex information to images. Images are thus important, but they tend to be large! Since modern computer hardware can display many colors, it is common to have a pixel represented internally as a 24-bit number, where the percentages of red, green, and blue occupy 8 bits each. Such a 24-bit pixel can specify one of $2^{24} \approx 16.78$ million colors. An image at a resolution of 512×512 made of such pixels thus occupies 786,432 bytes. At a resolution of 1024×1024, it gets four times as large, requiring 3,145,728 bytes. Movies are also commonly used with computers, making for even larger images. This is why image compression is so important. An important feature of image compression is that it can be lossy. An image, after all, exists for people to look at, so when it is compressed, it is acceptable to lose image features for which the human eye is not sensitive. This is one of the main ideas behind JPEG and other *lossy* image compression methods.

The idea of losing image information becomes more palatable when we consider how digital images are created. Here are three examples: (1) a real-life image may be scanned from a photograph or a painting and digitized (converted to pixels), (2) an image may be recorded by a video camera that creates pixels and stores them directly in memory, and (3) an image may be painted on the screen by means of a paint program. In all these cases, the image loses some information when it is digitized. The fact that the viewer is willing to put up with this loss suggests that further loss of information might be tolerable if done carefully.

9.1 Introduction

How should an image be compressed? Text is compressed with either statistical methods or dictionary-based methods. These are unsatisfactory for color or grayscale images, and here is why.

Statistical methods work best when the symbols being compressed have different probabilities (Section 9.2). An input stream where all symbols have the same probability will not compress, even if it is not random. It turns out that, in a continuous-tone color or grayscale image, the different colors or shades of gray have roughly the same probabilities. This is why statistical methods are not a good choice for compressing such images, and why new approaches are necessary. Images with color discontinuities, where adjacent pixels have widely different colors, compress better with statistical methods.

> That's what you've done with your force field. You've compressed the energy of years into a moment.
> — Robert Lansing (as Scott Nelson) in *4D Man* (1959).

Dictionary-based compression methods also tend to be unsuccessful in dealing with continuous-tone images. Such an image typically contains adjacent pixels with similar colors, but does not contain repeating patterns. Even an image that contains repeated patterns such as vertical lines may lose them when digitized. A vertical line in the original image may become slightly slanted when the image is digitized (Figure 9.1), so the pixels in a scan row may end up having slightly different colors from those in the preceding and following rows, resulting in a dictionary with short strings.

An ideal vertical rule is shown in (a). In (b), the rule is assumed to be perfectly digitized into 10 pixels, laid vertically. However, if the image is placed in the digitizer slightly slanted, the digitizing process may be imperfect, and the resulting pixels might look as in (c).

(a) (b) (c)

Figure 9.1: Perfect and Imperfect Digitizing.

Traditional methods are thus unsatisfactory for image compression (lossy or lossless), so novel approaches to image compression are discussed here. Even though the methods described in this chapter are different, they are all based on the same principle. The principle of image compression states that if we select a pixel in an image at random, there is a good chance that its neighbors will have the same color. All the different image compression methods use this principle, although in different ways.

▶ **Exercise 9.1:** Consider a random image. If we select a pixel in such an image, chances are its neighbors will be different. How can such an image be compressed?

9.2 Variable-Size Codes

These codes are used both by statistical methods for text compression and by some image compression methods, which is why they are introduced here.

Text files are commonly used with computers. Such a file is a stream of ASCII codes, so it is made of 128 symbols. A general (binary) file is normally divided into 8-bit bytes so it consists of 256 different symbols. To simplify our dscussion of variable-size codes, let's assume that our data are made of just the four different symbols, a_1, a_2, a_3, and a_4. If they appear in our data with equal probability (= 0.25 each), then we can simply assign them the four 2-bit codes 00, 01, 10, and 11.

In practice, however, the symbols occur in data files with different probabilities. Table 9.2 illustrates such a case where a_1 appears in the data (on average) about half the time, a_2 and a_3 have equal probabilities, and a_4 is rare. In this case, we can benefit by assigning our symbols codes of different sizes such as Code1 and Code2 shown in the table.

Symbol	Prob.	Code1	Code2
a_1	0.49	1	1
a_2	0.25	01	01
a_3	0.25	010	000
a_4	0.01	001	001

Table 9.2: Variable-Size Codes.

Code1 is designed such that the most common symbol, a_1, is assigned the shortest code. When long data strings are transmitted using Code1, the average size (the number of bits per symbol) is $1\times0.49 + 2\times0.25 + 3\times0.25 + 3\times0.01 = 1.77$, which is very close to the theoretical minimum. An interesting example is the 20-symbol string $a_1a_3a_2a_1a_3a_3a_4a_2a_1a_1a_2a_2a_1a_1a_3a_1a_1a_2a_3a_1$, where the 4 symbols occur with (approximately) the right frequencies. Encoding this string with Code1 yields the 37 bits:

1|010|01|1|010|010|001|01|1|1|01|01|1|1|010|1|1|01|010|1

(without the vertical bars). Using 37 bits to encode 20 symbols yields an average size of 1.85 bits/symbol, not far from the calculated average size. (The reader should bear in mind that our examples are short. To get results close to the best that's theoretically possible, an input stream with at least thousands of symbols is needed.)

However, when we attempt to *decode* the above binary string, it becomes obvious that Code1 is bad. The first bit is 1 and, since only a_1 has this code, it must be the first symbol. The next bit is 0, but the codes of a_2, a_3, and a_4 all start with a 0, so the decoder has to read the next bit. It is 1, but the codes of both a_2

and a_3 start with 01. The decoder does not know whether to decode the string as $1|010|01\ldots$, which is $a_1a_3a_2\ldots$, or as $1|01|001\ldots$, which is $a_1a_2a_4\ldots$. Thus, Code1 is *ambiguous*. In contrast, Code2, which has the same average size as Code1, can be decoded unambiguously.

The property of Code2 that makes it so much better than Code1 is called the *prefix property*. This property requires that once a certain bit pattern has been assigned as the code of a symbol, no other codes should start with that pattern (the pattern cannot be the *prefix* of any other code). Once the string "1" was assigned as the code of a_1, no other codes could start with 1 (i.e., they all had to start with 0). Once "01" was assigned as the code of a_2, no other codes could start with 01 (they all had to start with 00). This is why the codes of a_3 and a_4 had to start with 00. Naturally, they became 000 and 001.

Designing variable-size codes is done by following two principles: (1) assign short codes to the more frequent symbols and (2) obey the prefix property. Following these principles produces short, unambiguous codes, but not necessarily the best (i.e., shortest) ones. In addition to these principles, an algorithm is needed that always produces the shortest code. The only input to such an algorithm is the frequencies (or the probabilities) of the symbols in the alphabet. Two such algorithms, arithmetic coding and the Huffman method, are commonly used and are widely discussed in the literature (see, for example, [Salomon 97]).

A prefix code is a variable-size code that satisfies the prefix property. The binary representation of the integers does not satisfy the prefix property. Another disadvantage of this representation is that the size n of the set of integers being coded has to be known in advance since it determines the code size, which is $\lfloor 1 + \log_2 n \rfloor$. In some applications, a prefix code is required to code a set of integers whose size is not known in advance. Several such codes, most of which are due to P. Elias [Elias 75], are presented in Sections 9.8.1, 9.8.2, and 9.9.

9.3 Run-Length Encoding

This method (also called RLE), scans the bitmap, row by row, looking for *runs* of pixels of the same color. If the bitmap starts, for example, with 17 white pixels, followed by 1 black one, followed by 55 white ones, etc., then only the numbers 17, 1, 55,... have to be saved.

The method assumes that the bitmap starts with a white pixel. If this is not true, then the bitmap starts with zero white pixels and the sequence saved should start with 0. The resolution of the bitmap should also be saved, at the start of the file.

The size of the compressed file depends on the complexity of the image. The more detail, the worse the compression.

▶ **Exercise 9.2:** We want to express the quality of a file compression by a single number, the compression ratio. What is a good definition for it?

▶ **Exercise 9.3:** What would be the compressed file in the case of the 6×8 bitmap of Figure 9.3?

Run-length encoding can also be used with colors or shades of gray. Each run of pixels of the same color is encoded as a pair (run-length, pixel value). The run

9 Image Compression

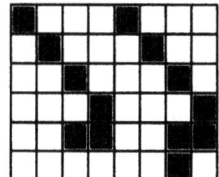

Figure 9.3: Bitmap for Exercise 9.3.

length usually occupies 1 byte, allowing for runs of up to 255 pixels. The pixel value occupies as many bytes as necessary, depending on the number of colors (typically between 1 and 3 bytes).

Example: An 8-bit deep grayscale bitmap that starts with

$$12, 12, 12, 12, 12, 12, 12, 12, 12, 35, 76, 112, 67, 87, 87, 87, 5, 5, 5, 5, 5, 5, 1, \ldots$$

is compressed into $\boxed{9}$,12,35,76,112,67,$\boxed{3}$,87, $\boxed{6}$,5,1..., where the boxed numbers indicate counts. The problem is to distinguish between a byte containing a color value (such as 12) and one containing a count (such as 9). Here are some solutions:

1. If the image is limited to just 128 colors, we can devote one bit in each byte to indicate whether the byte contains a color value or a count.

2. We can reduce the number of colors from 256 to 255 and reserve the unused value as a flag to precede every byte with count. If the reserved value is, say, 255, then the sequence above becomes

$$255, \boxed{9}, 12, 35, 76, 112, 67, 255, \boxed{3}, 87, 255, \boxed{6}, 5, 1 \ldots$$

3. Again, one bit is devoted to each byte to indicate whether the byte contains a color value or a count. This time, however, these extra bits are written, as a compact array, at the start of the compressed file. The compact array corresponding to the compressed sequence $\boxed{9}$,12,35,76,112,67,$\boxed{3}$,87, $\boxed{6}$,5,1... is 10000010100.... The size of this compact array is, of course, one-eighth the size of the compressed file (it contains 1 bit for each byte of the compressed file) so it increases the size of that file by 12.5%.

4. A group of m pixels that are all different is preceded by a byte with the negative value $-m$. The above sequence is encoded by

$$9, 12, -4, 35, 76, 112, 67, 3, 87, 6, 5, ?, 1, \ldots$$

(the value of the byte with ? is positive or negative depending on what follows the pixel of 1). The worst case is a sequence of pixels (p_1, p_2, p_2) repeated n times throughout the bitmap. It is encoded as $(-1, p_1, 2, p_2)$, four numbers instead of the original three! If each pixel requires 1 byte, then the original 3 bytes are expanded into 4 bytes instead of being compressed. If each pixel requires three bytes, then the original 3 pixels (which constitute nine bytes) are compressed into eight (1+3+1+3) bytes, a factor of 9/8.

Three more points should be mentioned:

1. Since the run length cannot be zero, it makes sense to write the [run length minus 1] on the compressed file. Hence the pair (3, 87) means a run of *four* pixels of color 87. This way, a run can be up to 256 pixels long.

2. In color images, it is common to have each pixel stored as 3 bytes, representing the intensities of the red, green, and blue components of the pixel. In such a case, runs of each color should be encoded separately. Thus, the pixels (171, 85, 34), (172, 85, 35) (172, 85, 30), and (173, 85, 33) should be separated into the three sequences (171, 172, 172, 173,...), (85, 85, 85, 85,...), and (34, 35, 30, 33,...). Each sequence should be run-length encoded separately. This means that any method for compressing grayscale images can be applied to color images as well.

3. It is preferable to encode each row of the bitmap individually. Hence, if a row ends with four pixels of color 87 and the following row starts with nine such pixels, it is better to write 4, 87, 9, 87 on the compressed file rather than 13, 87. The reason is that sometimes the user may decide to accept or reject an image just by seeing its general shape, without any details. If each line is encoded individually, the decoding algorithm can start by decoding and displaying lines 1, 6, 11, ..., continue with lines 2, 7, 12, ..., etc. The image is built on the screen gradually, in several steps, and it is possible to get an idea of what's in the image at an early stage.

Another advantage of encoding each bitmap row individually is to make it possible to extract just part of an encoded image (rows k through l, say). Yet another application is to merge two compressed images without having to decompress them first.

▶ **Exercise 9.4:** There is another, obvious, reason why each bitmap row should be coded individually. What is it?

If this idea is adopted, then the compressed file must contain information on where each bitmap row starts in the file. This can be done by writing a header at the start of the file that contains a group of 4 bytes (32 bits) for each bitmap row. The kth such group contains the offset (in bytes) from the start of the file to the start of the information for row k. This increases the size of the compressed file but may still offer a good trade-off between space (size of compressed file) and time (time to decide whether to accept or reject the image).

Disadvantages of run-length encoding are as follows:

1. When the image is modified, the run lengths change and the compression has to be completely redone.

2. If the image is complex, the compressed file may turn out to be bigger than a raw dump of the bitmap, so run-length encoding may sometimes expand an image instead of compressing it. Imagine an image with many vertical lines. When it is scanned horizontally, it produces very short runs, resulting in very bad compression, or even no compression at all.

3. Most run lengths are short and require fewer than 8 bits. Some are longer than 256 and will not fit in a byte. Writing the run lengths as bytes on the compressed file is simple but wasteful. A better idea is to further encode the run lengths and to write the codes on the compressed file. Codes of different sizes should be assigned to the run lengths such that common run lengths would be assigned short

9 Image Compression

codes and such that the codes could be read unambiguously from the compressed file by the decoder. The fax compression methods of Section 9.4 do just that.

4. Imagine an image with many vertical lines. When it is scanned horizontally, it produces very short runs, resulting in very bad compression or even expansion. A good, practical RLE image compressor should be able to scan the bitmap by rows, columns, or in zigzag and it may automatically try all three ways on every bitmap compressed to achieve the best compression.

▶ **Exercise 9.5:** Given the following 8 × 8 bitmap, use RLE to compress it, first row by row, then column by column. Describe the results in detail.

> Hey, I went shopping, Buster, to furnish your perfect house, to build your perfect image, to be your perfect Beverly Hills wife!
> — Shelley Long (as Phyllis Nefler) in *Troop Beverly Hills* (1989).

9.3.1 Lossy Compression

It is possible to get even better compression ratios if short runs are ignored. Such a method loses information when compressing an image, but, sometimes, this is acceptable to the user. (Notable examples where no loss is acceptable are X-ray images and astronomical images taken by large telescopes, where the price of an image is astronomical.)

A lossy run-length encoding algorithm should start by asking the user for the longest run that should still be ignored. If the user says, for example, 3, then the program ignores all runs of 1, 2, or 3 identical pixels. The compressed data "6,8,1,2,4,3,11,2" would be saved, in this case, as "6,8,7,16" where 7 is the sum $1 + 2 + 4$ and 16 is the sum $3 + 11 + 2$.

9.4 Fax Compression

Data compression is especially important when images are transmitted, because the user is typically waiting at the receiver, eager to get the result fast. Documents transferred between fax machines are sent as bitmaps, so a standard data compression method was needed when those machine became popular. Several methods were developed and proposed by the ITU-T.

The ITU-T is one of four permanent parts of the International Telecommunications Union (ITU), based in Geneva, Switzerland (http://www.itu.ch/). It issues recommendations for standards applying to modems, packet-switched interfaces, V.24 connectors, etc. Although it has no power of enforcement, the standards it recommends are generally accepted and adopted by industry. Until March 1993, the ITU-T was known as the Consultative Committee for International Telephone and Telegraph (CCITT).

The first data compression standards developed by the ITU-T were T2 (also known as group 1) and T3 (group 2). They are now obsolete and have been replaced by T4 (group 3) and T6 (group 4). Group 3 is currently used by all fax machines designed to operate with the Public Switched Telephone Network (PSTN). These are the machines we have at home and they operate at maximum speeds of 9600 baud. Group 4 is used by fax machines designed to operate on a digital network, such as ISDN. They have typical speeds of 64K baud. Both methods can produce compression ratios of 10:1 or better, reducing the transmission time of a typical page to about a minute with the former and a few seconds with the latter.

> You've reached the Szalinskis. At the sound of the beep, please leave your message, your fax, or your binary file.
>
> — Rick Moranis (as Wayne Szalinski) in *Honey, We Shrunk Ourselves* (1997).

9.4.1 One-Dimensional Coding

A fax machine scans a document line by line, converting each line to small black and white dots called *pels* (from Picture ELement). The horizontal resolution is always 8.05 pels per millimeter (about 205 pels per inch). An 8.5-in. wide scan line is thus converted to 1728 pels. The T4 standard, though, recommends scanning only about 8.2 in., thus producing 1664 pels per scan line (these numbers, as well as the ones in the next paragraph, are all to within $\pm 1\%$ accuracy).

The vertical resolution is either 3.85 scan lines per millimeter (standard mode) or 7.7 lines/mm (fine mode). Many fax machines have also a very fine mode, where they scan 15.4 lines/mm. Table 9.4 assumes a 10-in. high page (254 mm) and shows the total number of pels per page and typical transmission times for the three modes without compression. The times are long, which shows how important data compression is in fax transmissions.

To develop the group 3 code, the ITU-T counted all the run lengths of white and black pels on a set of eight "training" documents (that they felt represented typical text and images sent by fax) and used the Huffman algorithm to assign a variable-size code to each run length. The most common run lengths were found to be two, three, and four black pixels, so they were assigned the shortest codes (Table 9.5). Next come run lengths of two to seven white pixels. They were assigned slightly longer codes. Most run lengths were rare and were assigned long, 12-bit codes.

▶ **Exercise 9.6:** A run length of 1664 white pels was assigned the short code 011000. Why is this length so common?

Since run lengths can be long, the Huffman algorithm was modified. Codes were assigned to run lengths of 1 to 63 pels (they are the termination codes of Table 9.5a) and to run lengths that are multiples of 64 pels (the make-up codes of Table 9.5b). Group 3 is, thus, a *modified Huffman code* (also called MH). The code

Scan lines	Pels per line	Pels per page	Time (s)	Time (min)
978	1664	1.670M	170	2.82
1956	1664	3.255M	339	5.65
3912	1664	6.510M	678	11.3

Ten inches equal 254 mm. The number of pels is in the millions and the transmission times, at 9600 baud without compression, are between 3 and 11 min., depending on the mode. However, if the page is shorter than 10 in., or if most of it is white, the compression ratio can be 10:1 or better, resulting in transmission times of between 17 and 68 s.

Table 9.4: Fax Transmission Times.

of a run length is either a single termination code (if the run length is short) or one or more make-up codes, followed by one termination code (if it is long). Here are some examples:

1. A run length of 12 white pels is coded as 001000.
2. A run length of 76 white pels (=64+12) is coded as 11011|001000 (without the vertical bar).
3. A run length of 140 white pels (=128+12) is coded as 10010|001000.
4. A run length of 64 black pels (=64+0) is coded as 0000001111|0000110111.
5. A run length of 2561 black pels (2560+1) is coded as 000000011111|010.

▶ **Exercise 9.7:** There are no runs of length zero. Why then were codes assigned to runs of zero black and white pels?

▶ **Exercise 9.8:** An 8.5-in.-wide scan line results in 1728 pels, so how can there be a run of 2561 consecutive pels?

Each scan line is coded separately and its code is terminated with the special end-of-line (EOL) code 00000000001. Each line also gets one white pel appended to it on the left when it is scanned. This is done to remove any ambiguity when the line is decoded on the receiving side. After reading the EOL for the previous line, the receiver assumes that the new line starts with a run of white pels and it ignores the first of them. Examples:

1. The 14-pel line 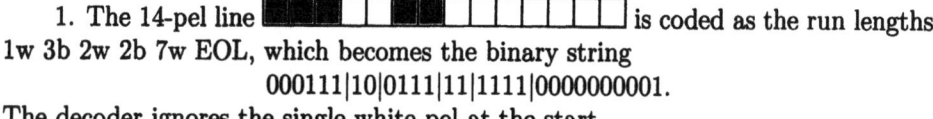 is coded as the run lengths 1w 3b 2w 2b 7w EOL, which becomes the binary string
000111|10|0111|11|1111|0000000001.
The decoder ignores the single white pel at the start.

2. The line is coded as the run lengths 3w 5b 5w 2b EOL, which becomes the binary string 1000|0011|1100|11|0000000001. The decoder starts the line with two white pels.

The group 3 code has no error correction, but many errors can be detected. Because of the nature of the Huffman code, even one bad bit in the transmission

(a)

Run length	White code-word	Black code-word	Run length	White code-word	Black code-word
0	00110101	0000110111	32	00011011	000001101010
1	000111	010	33	00010010	000001101011
2	0111	11	34	00010011	000011010010
3	1000	10	35	00010100	000011010011
4	1011	011	36	00010101	000011010100
5	1100	0011	37	00010110	000011010101
6	1110	0010	38	00010111	000011010110
7	1111	00011	39	00101000	000011010111
8	10011	000101	40	00101001	000001101100
9	10100	000100	41	00101010	000001101101
10	00111	0000100	42	00101011	000011011010
11	01000	0000101	43	00101100	000011011011
12	001000	0000111	44	00101101	000001010100
13	000011	00000100	45	00000100	000001010101
14	110100	00000111	46	00000101	000001010110
15	110101	000011000	47	00001010	000001010111
16	101010	0000010111	48	00001011	000001100100
17	101011	0000011000	49	01010010	000001100101
18	0100111	0000001000	50	01010011	000001010010
19	0001100	00001100111	51	01010100	000001010011
20	0001000	00001101000	52	01010101	000000100100
21	0010111	00001101100	53	00100100	000000110111
22	0000011	00000110111	54	00100101	000000111000
23	0000100	00000101000	55	01011000	000000100111
24	0101000	00000010111	56	01011001	000000101000
25	0101011	00000011000	57	01011010	000001011000
26	0010011	000011001010	58	01011011	000001011001
27	0100100	000011001011	59	01001010	000000101011
28	0011000	000011001100	60	01001011	000000101100
29	00000010	000011001101	61	00110010	000001011010
30	00000011	000001101000	62	00110011	000001100110
31	00011010	000001101001	63	00110100	000001100111

(b)

Run length	White code-word	Black code-word	Run length	White code-word	Black code-word
64	11011	0000001111	1344	011011010	0000001010011
128	10010	000011001000	1408	011011011	0000001010100
192	010111	000011001001	1472	010011000	0000001010101
256	0110111	000001011011	1536	010011001	0000001011010
320	00110110	000000110011	1600	010011010	0000001011011
384	00110111	000000110100	1664	011000	0000001100100
448	01100100	000000110101	1728	010011011	0000001100101
512	01100101	0000001101100	1792	00000001000	same as
576	01101000	0000001101101	1856	00000001100	white
640	01100111	0000001001010	1920	00000001101	from this
704	011001100	0000001001011	1984	000000010010	point
768	011001101	0000001001100	2048	000000010011	
832	011010010	0000001001101	2112	000000010100	
896	011010011	0000001110010	2176	000000010101	
960	011010100	0000001110011	2240	000000010110	
1024	011010101	0000001110100	2304	000000010111	
1088	011010110	0000001110101	2368	000000011100	
1152	011010111	0000001110110	2432	000000011101	
1216	011011000	0000001110111	2496	000000011110	
1280	011011001	0000001010010	2560	000000011111	

Table 9.5: Group 3 and 4 Fax Codes: (a) Termination Codes. (b) Make-Up Codes.

9 Image Compression

can cause the receiver to get out of synchronization and to produce a string of wrong pels. This is why each scan line is encoded separately. If the receiver detects an error, it skips bits, looking for an EOL. In this way, at most one scan line can be received incorrectly. If the receiver does not see an EOL after a certain number of lines, it assumes a high error rate and it aborts the process, notifying the transmitter. Since the codes are between 2 and 12 bits long, the receiver can detect an error if it cannot decode a valid code after reading 12 bits.

Each page of the coded document is preceded by one EOL and is followed by six EOL codes. Because each line is coded separately, this method is a *one-dimensional coding* scheme. The compression ratio depends on the image. Images with large contiguous black or white areas (text or black-and-white diagrams) can be highly compressed. Images with many short runs can sometimes produce negative compression. This is especially true in the case of images with shades of gray (mostly photographs). Such shades are produced by halftoning, which covers areas with many alternating black and white pels (runs of length 1).

▶ **Exercise 9.9:** What is the compression factor for runs of length 1 (strictly alternating pels)?

The T4 standard also allows for fill bits to be inserted between the data bits and the EOL. This is done in cases where a pause is necessary, or where the total number of bits transmitted for a scan line must be a multiple of 8. The fill bits are zeros.

Example: The binary string 000111|10|0111|11|1111|0000000001 becomes

$$000111|10|0111|11|1111|0000|0000000001$$

after four zeros are added as fill bits, bringing the total length of the string to 32 bits (= 8×4). The decoder sees the four zeros of the fill, followed by the nine zeros of the EOL, followed by the single 1, so it knows that it has encountered a fill followed by an EOL.

See http://www.cis.ohio-state.edu/htbin/rfc/rfc804.html for a description of group 3.

At the time of writing, the T.4 and T.6 recommendations can also be found at src.doc.ic.ac.uk in directory /computing/ccitt/ccitt-standards/1988/, as files 7_3_01.ps.gz and 7_3_02.ps.gz.

9.4.2 Two-Dimensional Coding

This method was developed because one-dimensional coding does not produce good results for images with gray areas. Two-dimensional coding is optional on fax machines that use group 3, but it is the only method used by machines intended to work on a digital network. When a fax machine using group 3 supports two-dimensional coding as an option, each EOL is followed by one extra bit, to indicate the compression method used for the next scan line. That bit is 1 if the next line is encoded with one-dimensional coding, and it is 0 if the next line is encoded with two-dimensional coding.

The two-dimensional coding method is also called MMR, for *modified-modified READ*, where READ stands for *relative element address designate*. The term modified-modified is used since this is a modification of one-dimensional coding, which itself is a modification of the original Huffman method. The *two-dimensional coding* method works by comparing the current scan line (called the *coding line*) to its predecessor (which is called the *reference line*) and recording the differences between them, the assumption being that two consecutive lines in a document will normally differ by just a few pels. The method assumes that there is an all-white line above the page, which is used as the reference line for the first scan line of the page. After coding the first line, it becomes the reference line and the second scan line is coded. Similarly to one-dimensional coding, each line is assumed to start with a white pel, which is ignored by the receiver.

The two-dimensional coding method is less reliable than one-dimensional coding since an error in decoding a line will cause errors in decoding all its successors and will propagate through the entire document. This is why the T.4 (group 3) standard includes a requirement that says that after a line is encoded with the one-dimensional method, at most $K-1$ lines will be encoded with the two-dimensional coding method. For standard resolution, $K=2$, while for a fine resolution, $K=4$. The T.6 standard (group 4) does not have this requirement and uses two-dimensional coding exclusively.

Scanning the coding line and comparing it to the reference lines results in three cases or modes. The mode is identified by comparing the next run length on the reference line [$(b_1 b_2)$ in Figure 9.6] with both the current run length $(a_0 a_1)$ and the next one $(a_1 a_2)$ on the coding line. Each of these three runs can be black or white. The three modes are (see also flowchart in Figure 9.9) as follows:

1. **Pass mode.** This is the case when $(b_1 b_2)$ is to the left of $(a_1 a_2)$. b_2 is to the left of a_1 (Figure 9.6a). This mode does not include the case where b_2 is on top of a_1. When this mode is identified, the length of run $(b_1 b_2)$ is coded using the codes of Table 9.5 and is transmitted. Pointer a_0 is moved below b_2, and the four values b_1, b_2, a_1, and a_2 are updated.

2. **Vertical mode.** $(b_1 b_2)$ overlaps $(a_1 a_2)$ by not more than three pels (Figure 9.6b1,b2). Assuming that consecutive lines do not differ by much, this is the most common case. When this mode is identified, one of seven codes is produced (Table 9.7) and is transmitted. Pointers are updated as in case 1 above.

3. **Horizontal mode.** $(b_1 b_2)$ overlaps $(a_1 a_2)$ by more than three pels (Figure 9.6c1,c2). When this mode is identified, the lengths of runs $(a_0 a_1)$ and $(a_1 a_2)$ are coded using the codes of Table 9.5 and are transmitted. Pointers are updated as in cases 1 and 2 above.

When scanning starts, pointer a_0 is set to an imaginary white pel on the left of the coding line. a_1 is set to point to the first black pel on the coding line. (Since a_0 corresponds to an imaginary pel, the first run length is $|a_0 a_1| - 1$.) Pointer a_2 is set to the first white pel following that. Pointers b_1 and b_2 are set to point to the start of the first and second runs on the reference line, respectively.

After identifying the current mode and transmitting codes according to Table 9.7, a_0 is updated as shown in the flow chart and the other four pointers are updated relative to the new a_0. The process continues until the end of the coding line is reached. The encoder assumes an extra pel on the right of the line, with a color opposite that of the last pel.

9 Image Compression

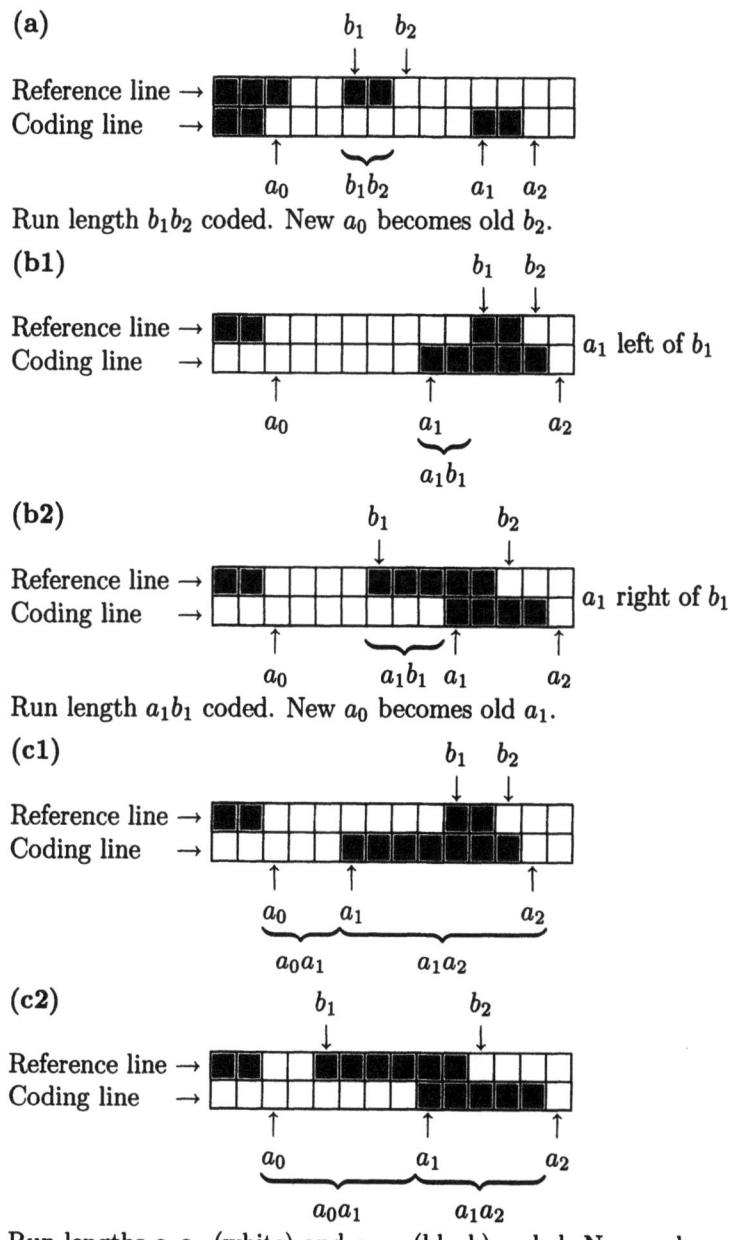

Figure 9.6: Five Run-Length Configurations: (a) Pass Mode. (b) Vertical Mode. (c) Horizontal Mode.

Notes:
1. a_0 is the first pel of a new codeword and can be black or white.
2. a_1 is the first pel to the right of a_0 with a different color.
3. b_1 is the first pel on the reference line to the right of a_0 with a different color.
4. b_2 is the first pel on the reference line to the right of b_1 with a different color.

Mode	Run length to be encoded	Abbreviation	Codeword
Pass	b_1b_2	P	0001+coded length of b_1b_2
Horizontal	a_0a_1, a_1a_2	H	001+coded length of a_0a_1 and a_1a_2
Vertical	$a_1b_1 = 0$	V(0)	1
	$a_1b_1 = -1$	VR(1)	011
	$a_1b_1 = -2$	VR(2)	000011
	$a_1b_1 = -3$	VR(3)	0000011
	$a_1b_1 = +1$	VL(1)	010
	$a_1b_1 = +2$	VL(2)	000010
	$a_1b_1 = +3$	VL(3)	0000010
Extension			0000001000

Table 9.7: Codes for the Group 4 Method.

The extension code in Table 9.7 is used to abort the encoding process prematurely, before reaching the end of the page. This is necessary if the rest of the page is transmitted in a different code or even in uncompressed form.

9.5 Cell Encoding

That's all for fax compression. We now consider an image stored in a bitmap and displayed on a screen. Let's start with the case where the image consists of just text, with each character occupying the same area, say, 8×8 pixels (large enough for a 5×7 or a 6×7 character and some spacing). Assuming a set of 256 characters (but see insert on Unicode on page 625), each cell can be encoded as an 8-bit pointer pointing to a 256-entry table, where each entry contains the description of an 8×8 character as a 64-bit string. The compression factor is thus 64/8, or 8. Figure 9.8 shows the letter "H" both as a bitmap and as a 64-bit string.

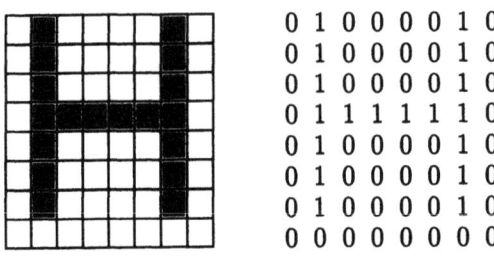

Figure 9.8: An 8×8 Letter "H".

> I am a stone cutter. The Pharaoh likes his images cut deep.
> — John Derek (as Joshua) in *The Ten Commandments* (1956).

9 Image Compression

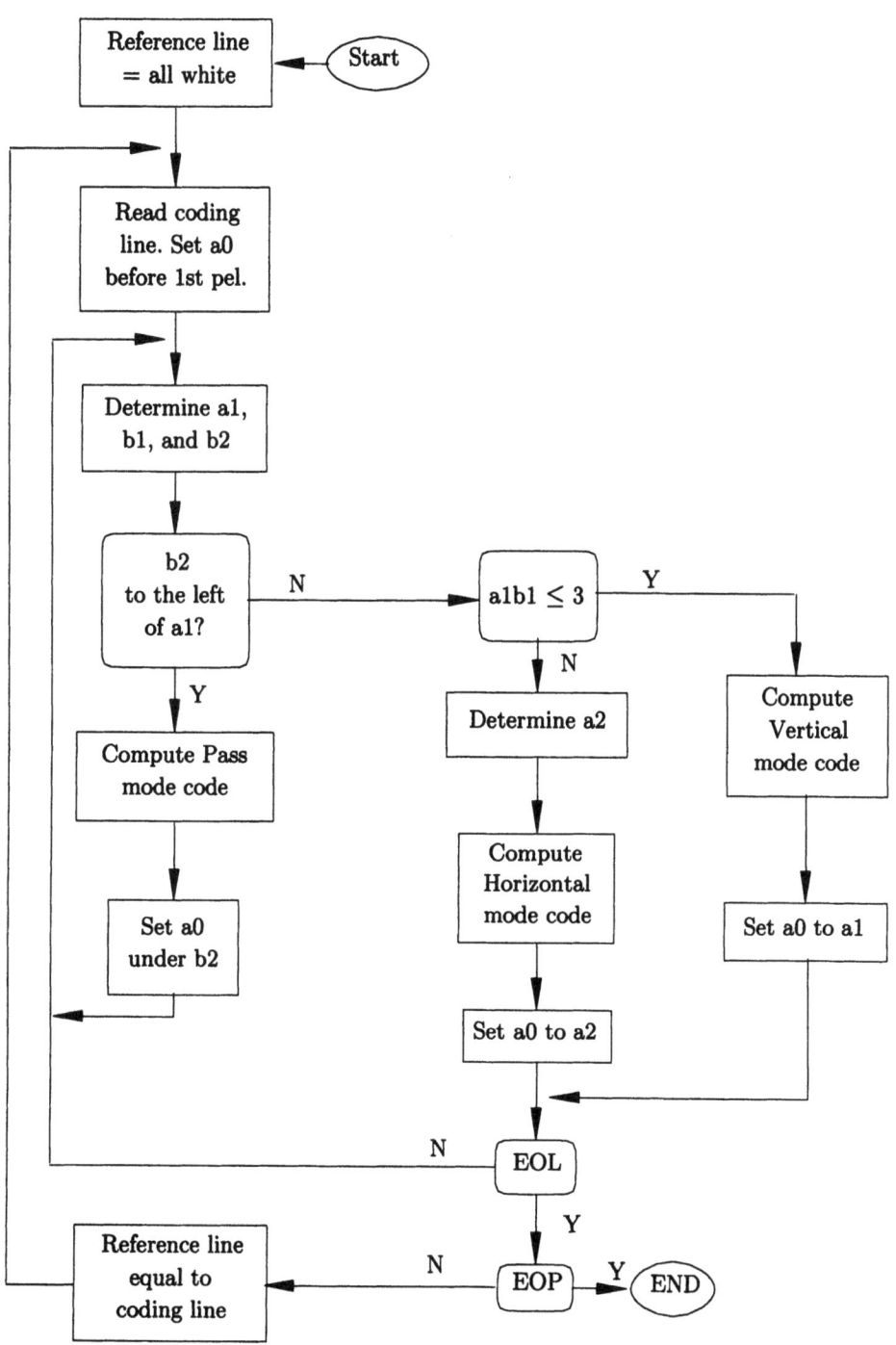

Figure 9.9: MMR Flowchart.

Cell encoding is not very useful for text, which can always be represented with 8 bits per character (but see page 625 for Unicode), but this method can also be extended to an image that consists of straight lines only. The entire bitmap is divided into cells of, say, 8×8 pixels, and is scanned cell by cell. The first cell is stored in entry 0 of a table and is encoded (i.e., written on the compressed file) as the pointer 0. Each subsequent cell is searched in the table. If found, its index in the table becomes its code and is written on the compressed file. Otherwise, it is added to the table. With 8×8 cells, each of the 64 pixels can be black or white, so the total number of different cells is $2^{64} \approx 1.8\times 10^{19}$, a huge number. However, some patterns never appear, as they don't represent any possible combination of line segments. Also, many cells are translated or reflected versions of other cells (Figure 9.10). All this brings the total number of distinct cells to just 108 [Jordan and Barrett 74]. These 108 cells can be stored in ROM and used many times to compress images.

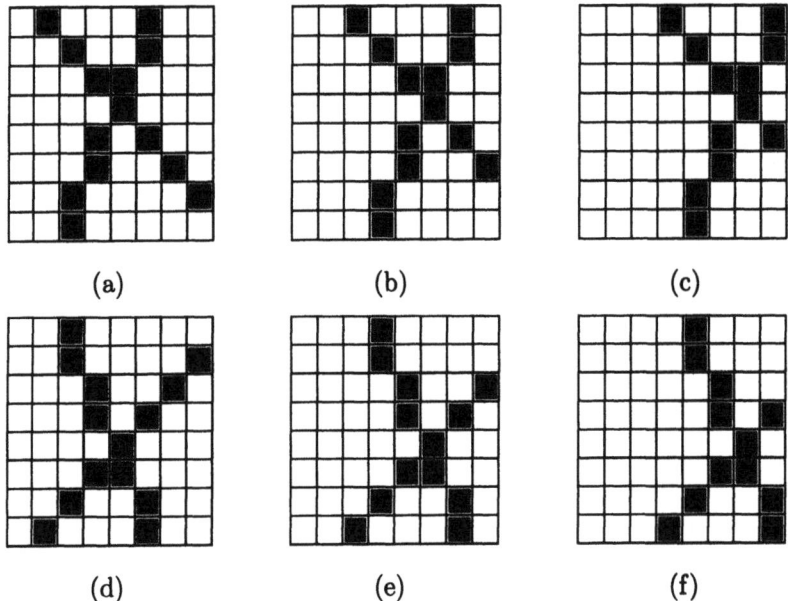

Figure 9.10: Six 8×8 Bitmaps Translated (a–c) and Reflected (d–f).

9.6 Quadtrees

This is a data structure suitable for saving an image in compressed form. It is useful for paintings or any image whose elements are not simple geometric figures. The bitmap is divided into four quadrants. Each homogeneous quadrant is saved as one node of the quadtree. Any nonhomogeneous quadrant is recursively divided into four smaller ones. Figure 9.11 shows a simple example:

The 8×8 bitmap of Figure 9.11a results in a quadtree (Figure 9.11b) with 21 nodes. Sixteen of them are leaves (each containing the color of 1 quadrant, 0 for white, 1 for black) and the other 5 (the small circles) are interior nodes containing

Unicode

Unicode is an international standard, started in 1990, for character coding designed to support the interchange, processing, and display of the written texts of the diverse languages of the modern world. In addition, it supports classical and historical texts of many written languages.

Each character or symbol is assigned a unique 16-bit code in Unicode, allowing for $2^{16} = 65,536$ codes, of which only 38,885 have (in version 2) been assigned so far. These codes cover the characters used by the principal written languages of the Americas, Europe, the Middle East, Africa, India, Asia, and Pacifica. To keep Unicode simple, no complex modes or escape symbols are used. Currently, there are about 18,000 unused code values for future expansion. Unicode also reserves over 6000 codes for private use, which software and hardware developers can assign internally for their own characters and symbols.

ASCII, with its 128, 7-bit codes is a subset of Unicode, while Unicode itself is a subset of a much bigger code, the International Standard ISO/IEC 10646-1 code of 1993, whose 32-bit codes can represent 4,294,967,296 characters. Unicode's design is based on ASCII, but goes far beyond ASCII's limited ability to encode only the Latin alphabet. The Unicode Standard provides the capacity to encode all of the characters and symbols used by the major written languages of the world, as well as by mathematics and various engineering disciplines.

Unicode provides a consistent coding format for international character sets and brings order to a chaotic state of affairs that has made it difficult to exchange text files across language borders. Computer users who deal with multilingual text—international business people, linguists, researchers, scientists, and others—will find that the Unicode Standard greatly simplifies their work. Mathematicians and technicians, who regularly use mathematical symbols and other technical characters, will also find the Unicode Standard valuable.

The Unicode Standard defines codes for characters used in every major language written today. Scripts include Latin, Greek, Cyrillic, Armenian, Hebrew, Arabic, Devanagari, Bengali, Gurmukhi, Gujarati, Oriya, Tamil, Telugu, Kannada, Malayalam, Thai, Lao, Georgian, Tibetan, Japanese kana, the complete set of modern Korean hangul, and a unified set of Chinese/Japanese/Korean (CJK) ideographs.

Also included are punctuation marks, diacritics, mathematical symbols, technical symbols, arrows, dingbats, etc. Diacritics are modifying character marks such as the tilde (~) that are used in conjunction with base characters to encode accented or vocalized letters (ñ, for example). Some modern written languages are not yet supported or are only partially supported due to a need for further research into the encoding needs of certain scripts.

Unicode is currently used in the operating systems Inferno and Plan 9 from Lucent Technologies, BeOS, Java OS, Microsoft Windows NT and CE, and NeXT OS. It is also used in the Limbo (from Lucent Technologies) and Java programming languages.

For more information, see [Unicode 96] and [Unicode 98].

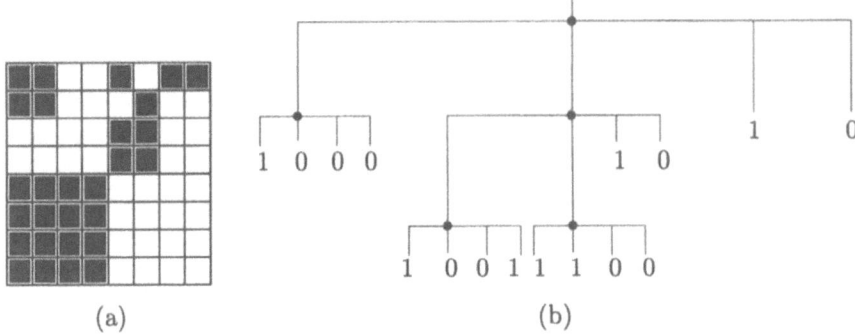

Figure 9.11: A Quadtree.

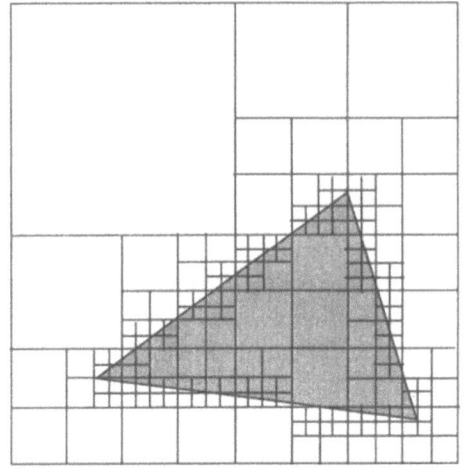

Figure 9.12: Another Quadtree.

only pointers. The quadrant numbering used is $\binom{0\ 1}{2\ 3}$. Figure 9.12 is another example of a quadtree.

The size of a quadtree depends on the complexity of the picture. Assuming a bitmap size of $2^N \times 2^N$, one extreme case is where all the pixels are the same. The quadtree in this case has one node. At the other extreme is the case where each quadrant, even the smallest one, is heterogeneous. The lowest level of the quadtree has, in such a case, $2^N \times 2^N = 4^N$ nodes. The level directly above it has a quarter of that number (4^{N-1}) and the level above that one has 4^{N-2} nodes. Hence, the total number of nodes is $4^0 + 4^1 + \cdots + 4^{N-1} + 4^N = (4^{N+1} - 1)/3 \approx 4^N(4/3) \approx 1.33 \times 4^N = 1.33 \times (2^N \times 2^N)$. In the worst case, the quadtree, thus, has $\approx 33\%$ more nodes than the number of pixels (the bitmap size) causing considerable expansion.

How is a quadtree generated? We outline two approaches. The first approach is recursive. It constructs a quadtree with leaves that have data fields with values B and W. All interior nodes have data fields=Gray. The procedure makes four recursive calls for the four quadrants of the bitmap. Each call returns a root for a subtree.

9 Image Compression

If the root is B or W, then it is a single leaf. If it is G, then it is more than a leaf. The procedure then checks the four roots (see below).

The last recursive calls get a quadrant of size 2×2 (the case L=2). Such a call generates four nodes with the colors (B, W) of the four pixels.

Each call then checks the four nodes. If all four have the same color (but not gray), they are replaced by one node T with that same color. If they are different (or if they are all gray), they become the children of a single node T and the color of T is set to Gray.

The procedure is recursive, a maximum recursion depth of N (where the bitmap is of size $2^N \times 2^N$) where N is typically in the range 8–12 (for a resolution range of 256×256 to $4K \times 4K$).

An advantage of this method is that it tests each pixel only once. Some unnecessary nodes are generated but are disposed of very quickly. The running time is the total number of the recursive calls. Originally, there is a single call (for the full bitmap of size $2^N \times 2^N$), which generates four recursive ones [for size $(2^N - 1) \times (2^N - 1)$], each generating four more. Hence, in the second level of recursion, there are $16 = 4^2$ calls, each handling a part of size $(2^N - 2) \times (2^N - 2)$. The lowest recursion level makes 4^{N-1} recursive calls, each for a 2×2 part of the bitmap. The total number of calls is thus $(1/3) \times 4^N$, which is a quarter of the size of a complete quadtree.

The second approach is nonrecursive. It starts by building the complete quadtree (assuming that all quadrants are heterogeneous) and then checking the assumption. Every time a quadrant is tested and found to be homogeneous, the four nodes corresponding to its four quarters are deleted from the quadtree. This process proceeds from the bottom (the leaves) up toward the root.

Three steps are involved:

Step 1. A complete quadtree of height N is constructed. It contains levels $0, 1, \ldots, N$, where level k has 4^k nodes.

Step 2. All $2^N \times 2^N$ pixels are copied from the bitmap into the leaves (the lowest level) of the quadtree.

Step 3. The tree is scanned level by level, from the bottom (level N) to the root (level 0). When level k is scanned, its 4^k nodes are examined in groups of four, the four nodes in each group having a common parent. If the four nodes in a group are leaves and have the same color (i.e., if they represent a homogeneous quadrant), they are deleted and their parent is changed from an interior node to a leaf having the same color.

Analysis of the time complexity: A complete quadtree has $\approx 1.33 \times 4^N$ nodes and, since each is tested once, the number of operations (in step 3) is 1.33×4^N. Step 1 also requires 1.33×4^N operations and step 2 requires 4^N operations. The total number of operations is thus $(1.33 + 1.33 + 1) \times 4^N = 3.66 \times 4^N$. We are faced with comparing the first method, which requires $(1/3) \times 4^N$ steps, with the second method, which needs 3.66×4^N operations. Since N usually varies in the narrow range 8–12, the difference is not very significant. Similarly, an analysis of storage requirements shows that the first method uses just the amount of memory required by the final quadtree, whereas the second one uses all the storage needed for a complete quadtree.

9.6.1 Octrees

This is a generalization of quadtrees to three-dimensional objects. Each node in an octree is either a leaf (containing the color of a homogeneous octant of the object) or is an interior node with eight children. Whereas quadtrees are used as a data structure for saving and compressing an image, the use of octrees is mostly for manipulating three-dimensional objects. An octree representing an object can be used to determine if a certain point is interior or exterior to the object. It can be used to determine what parts of the object are hidden from view. (Depending on the location of the viewer, there exists an easily determined traversal sequence that, when applied recursively, will visit octants in such a way that no octant later in the sequence can obscure an octant visited earlier.) Set operations can easily be performed on objects represented by octrees. It is possible, by simultaneously traversing two octrees, to determine which octants are common to both objects and, hence, to generate the logical **and** of the two. Geometric transformations can also be performed on objects represented by octrees.

The two texts by H. Samet [Samet 90] are a detailed reference to quadtrees (and also to octrees).

9.6.2 Prefix Compression

This is a variant of quadtrees proposed by [Anedda and Felician 88]. We start with a $2^n \times 2^n$ image. Each quadrant in the quadtree of this image is numbered 0, 1, 2, or 3, a 2-bit number. Each subquadrant has a two-digit (i.e., a 4-bit) number and each subsubquadrant, a three-digit number. As the quadrants get smaller, their numbers get longer. When this numbering scheme is carried down to individual pixels, the number of a pixel turns out to be n digits or $2n$ bits long. Prefix compression is designed for bilevel images with text or diagrams where the number of black pixels is relatively small. It is not suitable for grayscale, color, or any image that contains many black pixels, such as a painting. The method is best explained by an example. Figure 9.13 shows the pixel numbering in an 8×8 image (i.e., $n = 3$) and also a simple 8×8 image containing of 18 black pixels. Each pixel number is three digits long and range from 000 to 333.

```
000 001 010 011   100 101 110 111
002 003 012 013   102 103 112 113
020 021 030 031   120 121 130 131
022 023 032 033   122 123 132 133

200 201 210 211   300 301 310 311
202 203 212 213   302 303 312 313
220 221 230 231   320 321 330 331
222 223 232 233   322 323 332 333
```

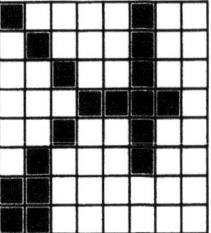

Figure 9.13: Example of Prefix Compression.

The first step is to use quadtree methods to figure the three-digit id numbers of the 18 black pixels. They are 000 101 003 103 030 121 033 122 123 132 210 301 203 303 220 221 222 223.

9 Image Compression

The next step is to select a *prefix* value P. We select $P = 2$, a choice that's justified below. The code of a pixel is now divided into P prefix digits followed by $3 - P$ suffix digits. The last step goes over the sequence of black pixels and selects all the pixels with the same prefix. The first prefix is 00, so all the pixels that start with 00 are selected. They are 000 and 003. They are removed from the original sequence and are compressed by writing the token 00|1|0|3 on the output stream. The first part of this token is a prefix (00), the second part is a count (1), and the rest are the suffixes of the two pixels having prefix 00. Notice that a count of one implies two pixels: the count is always 1 less than the number of pixels being counted. Sixteen pixels now remain in the original sequence and the first of them has prefix 10. The two pixels with this prefix are removed and compressed by writing the token 10|1|1|3 on the output stream. This continues until the original sequence is empty. The final result is the nine-token string

$$00|1|0|3, 10|1|1|3, 03|1|0|3, 12|2|1|2|3, 13|0|2, 21|0|0, 30|1|1|3, 20|0|3, 22|3|0|1|2|3$$

(without the commas). Such a string can be decoded uniquely since each segment starts with a two-digit prefix, followed by a one-digit count c, followed by $c + 1$ one-digit suffixes.

In general the prefix is P digits long and the count and each suffix are $n - P$ digits each. The maximum number of suffixes in a segment is, therefore, 4^{n-P}. The maximum size of a segment is, thus, $P + (n - P) + 4^{n-P}(n - P)$ digits. Each segment corresponds to a different prefix. A prefix has P digits, each between 0 and 3, so the maximum number of segments is 4^P. The entire compressed string thus occupies at most

$$4^P \left[P + (n - P) + 4^{n-P}(n - P) \right] = n \cdot 4^P + 4^n(n - P)$$

digits. To find the optimum value of P, we differentiate the above expression with respect to P:

$$\frac{d}{dP} \left[n \cdot 4^P + 4^n(n - P) \right] = n \cdot 4^P \ln 4 - 4^n,$$

and set the derivative to 0. The solution is

$$4^P = \frac{4^n}{n \cdot \ln 4} \quad \text{or} \quad P = \log_4 \left[\frac{4^n}{n \cdot \ln 4} \right] = \frac{1}{2} \log_2 \left[\frac{4^n}{n \cdot \ln 4} \right].$$

For $n = 3$, this yields

$$P \approx \frac{1}{2} \log_2 \left[\frac{4^3}{3 \times 1.386} \right] = \frac{\log_2 15.388}{2} = 3.944/2 = 1.97.$$

This is why $P = 2$ was selected earlier.

A downside of this method is that some pixels may be assigned numbers with different prefixes, even though they are near neighbors. This happens when they are located in different quadrants. An example is pixels 123 and 301 of Figure 9.13.

Improvement: The count field was arbitrarily set to one digit (2 bits). The maximum count is, thus, $3(=11_2)$, i.e., four pixels. It is possible to have a variable-size count field containing a variable-size code, such as the unary code of Section 9.8.1. In this way, a single token could compress any number of pixels.

9.7 Progressive Image Compression

Most modern image compression methods are either progressive or optionally so. Progressive compression is an attractive choice when compressed images are transmitted over a communications line and are decompressed and viewed in real time. When such an image is received and decompressed, the decoder can very quickly display the entire image in a low-quality format, then improve the display quality as more and more of the image is being received and decompressed. A user watching the image develop on the screen can normally recognize most of the image features after only 5–10% of it have been decompressed.

This should be compared to raster-scan image compression. When an image is raster scanned and compressed, a user normally cannot tell much about the image when only 5–10% of it has been decompressed and displayed. Since images are supposed to be viewed by humans, progressive compression makes sense even in cases where it is slower or less efficient than nonprogressive.

Perhaps a good way to think of progressive image compression is to imagine that the encoder compresses the most important image information first, then compresses less important information and appends it to the compressed stream, and so on. This explains why all progressive image compression methods have a natural lossy option: simply stop compressing at a certain point. The user can control the amount of loss by means of a parameter that tells the encoder how soon to stop the progressive encoding process. The sooner encoding is stopped, the better the compression ratio and the higher the data loss.

It is useful to think of progressive decoding as the process of improving image features over time, and this can be done in three ways:

1. Encode spatial frequency data progressively. An observer watching such an image being decoded sees the image changing from blurred to sharp. Methods that work this way typically feature medium-speed encoding and slow decoding.

2. Start with a gray image and add colors or shades of gray to it. An observer watching such an image being decoded will see all the image details from the start and will see them improve as more color is continuously added to them. Vector quantization methods use this kind of progressive compression. Such a method normally features slow encoding and fast decoding.

3. Encode the image in layers, where early layers consist of a few large, low-resolution pixels, followed by later layers with smaller, higher-resolution pixels. When an observer watches such an image being decoded, he will see more details added to the image over time. Such a method thus adds details (or resolution) to the image as it is being decompressed. This way of progressively encoding an image is called *pyramid coding*. Most progressive methods use this principle, so this section discusses general ideas for implementing pyramid coding.

Assuming that the image size is $2^n \times 2^n = 4^n$ pixels, the simplest method that comes to mind when attempting to do progressive compression is to calculate each

9 Image Compression

pixel of layer $i-1$ as the average of a group of 2×2 pixels of layer i. Thus, layer n is the entire image, layer $n-1$ contains $2^{n-1}\times 2^{n-1} = 4^{n-1}$ large pixels of size 2×2, and so on, down to layer 1, with $4^{n-n}=1$ large pixel, representing the entire image. If the image isn't too large, all the layers can be saved in memory. The pixels are then written on the compressed stream in reverse order, starting with layer 1. The single pixel of layer 1 is the "parent" of the four pixels of layer 2, each of which is the parent of four pixels in layer 3, and so on. The total number of pixels in the pyramid is

$$4^0 + 4^1 + \cdots + 4^{n-1} + 4^n = (4^{n+1}-1)/3 \approx 4^n(4/3) \approx 1.33 \times 4^n = 1.33(2^n \times 2^n),$$

33% more than the original number!

A simple way to bring the total number of pixels in the pyramid down to 4^n is to include only three of the four pixels of a group in layer i and to compute the value of the fourth pixel by using the parent of the group (from the preceding layer, $i-1$) and its three siblings.

Example: Figure 9.14c shows a 4×4 image that becomes the third layer in its progressive compression. Layer 2 is shown in Figure 9.14b where, for example, pixel 81.25 is the average of the four pixels 90, 72, 140, and 23 of layer 3. The single pixel of layer 1 is shown in Figure 9.14a.

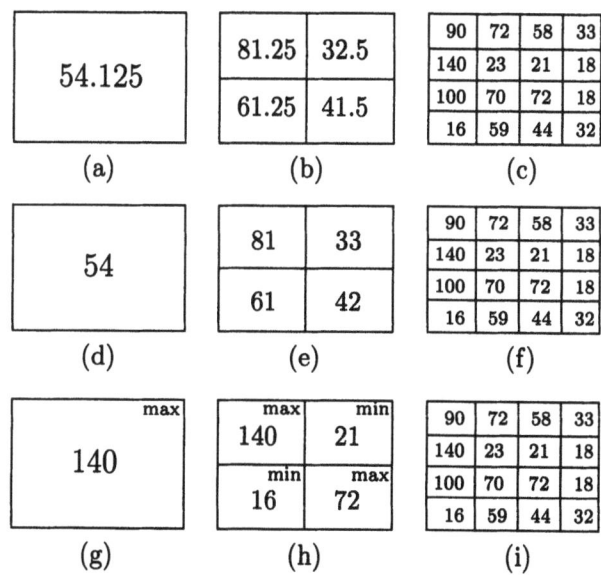

Figure 9.14: Progressive Image Compression.

The compressed file should contain just the numbers

$$54.125,\ 32.5, 41.5, 61.25,\ 72, 23, 140,\ 33, 18, 21,\ 18, 32, 44,\ 70, 59, 16$$

(properly encoded, of course), from which all the missing pixel values can easily be calculated. The missing pixel 81.25, for example, can be calculated from $(x + 32.5 + 41.5 + 61.25)/4 = 54.125$.

A small complication with this method is that averages of integers may be nonintegers. If we want our pixel values to remain integers, we either have to lose precision or to keep using longer and longer integers. Assuming that pixels are represented by 8 bits, adding four 8-bit integers produces a 10-bit integer. Dividing it by 4, to create the average, reduces the sum back to an 8-bit integer but some precision may be lost. If we don't want to lose precision, we should represent our second layer pixels as 10-bit numbers and our first layer (single) pixel as a 12-bit number. Figure 9.14d,e,f shows the results of rounding off our pixel values and thus losing some image information. The contents of the compressed file in this case should be

$$54, 33, 42, 61, 72, 23, 140, 33, 18, 21, 18, 32, 44, 70, 59, 16.$$

The first missing pixel 81 of layer 3 can be calculated from the equation $(x + 33 + 42 + 61)/4 = 54$, which yields the (slightly wrong) value 80.

▶ **Exercise 9.10:** Show that the sum of four n-bit numbers is an $n + 2$-bit number.

A better method is to let the parent of a group help in calculating the values of its four children. This can be done by calculating the differences between the parent and children and writing the differences (suitably coded) in layer i of the compressed stream. The decoder decodes the differences, then uses the parent from layer $i - 1$ to compute the values of the four pixels. Either Huffman or arithmetic coding can be used to encode the differences. If all the layers are calculated and saved in memory, then the distribution of difference values can be found and used to achieve the best statistical compression.

If there is no room in memory for all the layers, a simple adaptive model can be implemented. It starts by assigning a count of 1 to every difference value (to avoid the zero-probability problem [Salomon 97]). When a particular difference is calculated, it is assigned a probability and is encoded according to its count and its count is then updated. It is a good idea to update the counts by incrementing them by a value > 1, since this way the original counts of 1 become insignificant very quickly.

Some improvement can be achieved if the parent is used to help calculate the values of three child pixels and then these three plus the parent are used to calculate the value of the fourth pixel of the group. If the four pixels of a group are a, b, c, and d, then their average is $v = (a + b + c + d)/4$. The average becomes part of layer $i - 1$, and layer i need only contain the three differences $k = a - b$, $l = b - c$, and $m = c - d$. Once the decoder has read and decoded the three differences, it can use their values, together with the value of v from the previous layer, to compute the values of the four pixels of the group. Calculating v by a division by 4 still causes the loss of 2 bits, but this 2-bit quantity can be isolated before the division and retained by encoding it separately, following the three differences.

The parent pixel of a group does not have to be its average. One alternative is to select the maximum (or the minimum) pixel of a group as the parent. This has the advantage that the parent is identical to one of the pixels in the group. The encoder has to encode just three pixels in each group and the decoder decodes three pixels (or differences) and uses the parent as the fourth pixel, to complete the group. When encoding consecutive groups in a layer, the encoder should alternate between selecting the maximum and the minimum as parents, since always selecting the same creates progressive layers that are either too dark or too bright. Figure 9.14g,h,i shows the three layers in this case.

The compressed file should contain the numbers

$$140, (0), 21, 72, 16, (3), 90, 72, 23, (3), 58, 33, 18, (0), 18, 32, 44, (3), 100, 70, 59,$$

where the numbers in parentheses are 2 bits each. They tell where (in what quadrant) the parent from the previous layer should go. Quadrant numbering is $\begin{pmatrix} 0 & 1 \\ 3 & 2 \end{pmatrix}$

Selecting the median of a group is a little slower than selecting the maximum or the minimum, but it improves the appearance of the layers during progressive decompression. In general, the median of a sequence $(a_1, a_2, \ldots, a_n)$ is an element a_i such that half the elements (or very close to half) are no larger than a_i and the other half are no smaller. If the four pixels of a group satisfy $a < b < c < d$, then either b or c can be considered the median pixel of the group. The main advantage of selecting the median as the group's parent is that it tends to smooth large differences in pixel values that may occur because of one, extreme pixel. In the group 1, 2, 3, 100, for example, selecting 2 or 3 as the parent is much more representative than selecting the average. Finding the median of four pixels requires a few comparisons, but calculating the average requires a division by 4 (or, alternatively, a right shift).

> Median: The middle number in a given sequence of numbers, or the average of the two middle numbers when the sequence has an even number of numbers.

Once the median has been selected and encoded as part of layer $i - 1$, the remaining three pixels can be encoded in layer i by encoding their (three) differences, preceded by a 2-bit code telling which of the four is the parent. Another small advantage of using the median is that once the decoder reads this 2-bit code, it knows how many of the three pixels are smaller and how many are greater than the median. If the code says, for example, that one pixel is smaller and the other two are greater than the median and the decoder reads a pixel that's smaller than the median, it knows that the next two pixels decoded will be greater than the median. This knowledge changes the distribution of the differences and it can be taken advantage of by using three count tables to estimate probabilities when the differences are encoded. One table is used when a pixel is encoded that the decoder will know is greater than the median. Another table is used to encode pixels that the decoder will know are smaller than the median, and the third table is used for pixels where the decoder will not know their relations to the median in advance. This improves compression by a few percent and is another example of how adding more features to a compression method brings diminishing returns.

9.8 FELICS

FELICS is an acronym for Fast, Efficient, Lossless Image Compression System [Howard and Vitter 93]. It is a special-purpose compression method designed for grayscale images. It is fast and it generally produces good compression. However, it cannot compress an image to below 1 bit per pixel, so it is not a good choice for highly redundant images.

The principle of FELICS is to code each pixel with a variable-size code based on the values of two of its previously seen neighbor pixels. Figure 9.15a shows the two known neighbors A and B of some pixels P. For a general pixel, these are the neighbors above it and to its left. For a pixel in the top row, these are its two left neighbors (except for the first two pixels of the image). For a pixel in the leftmost column, these are the first two pixels of the line above it. Notice that the first two pixels of the image don't have any previously seen neighbors but since there are only two of them, they can be output without any encoding, causing just a slight degradation in the overall compression.

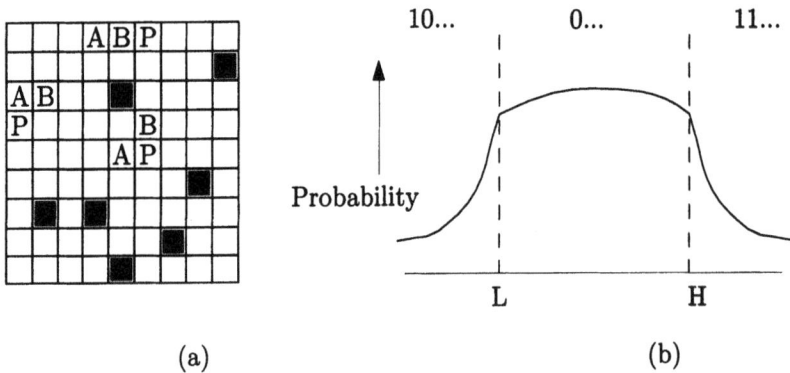

Figure 9.15: (a) The Two Neighbors. (b) The Three Regions.

Consider the two neighbors A and B of a pixel P. We use A, B, and P to denote both the three pixels and their intensities (grayscale values). We denote by L and H the neighbors with the smaller and the larger intensities, respectively. Pixel P should be assigned a variable-size code depending on where the intensity P is located relative to L and H. There are three cases:

1. The intensity of pixel P is between L and H (it is located in the central region of Figure 9.15b). This case is known experimentally to occur in about half the pixels and P is assigned, in this case, a code that starts with 0. (A special case occurs when L=H. In such a case, the range [L, H] consists of one value only and the chance that P will have that value is small.) The probability that P will be in this central region is almost, but not completely, flat, so P should be assigned a binary code that has about the same size in the entire region but is slightly shorter at the center of the region.

2. The intensity of P is lower than L (P is in the left region). The code assigned to P in this case starts with 10.

9 Image Compression

3. P's intensity is greater than H. P is assigned a code that starts with 11.

When pixel P is in one of the outer regions, the probability that its intensity will differ from L or H by much is small, so P can be assigned a long code in these cases.

The code assigned to P should, thus, depend heavily on whether P is in the central region or in one of the outer regions. Here is how the code is assigned when P is in the central region. We need H–L+1 variable-size codes that will not differ much in size and will, of course, satisfy the prefix property (Section 9.2). We let $k = \lfloor \log_2(H - L + 1) \rfloor$ and compute integers a and b by

$$a = 2^{k+1} - (H - L + 1), \quad b = 2(H - L + 1 - 2^k).$$

(Example: If H–L=9, then $k = 3$, $a = 2^{3+1} - (9+1) = 6$, and $b = 2(9+1-2^3) = 4$.) We now select the a codes $2^k - 1, 2^k - 2, \ldots$ expressed as k-bit numbers and the b codes $0, 1, 2, \ldots$ expressed as $k + 1$-bit numbers. In the example, the a codes are $8 - 1 = 111$, $8 - 2 = 110$ through $8 - 6 = 010$, and the b codes are 0000, 0001, 0010, and 0011. The a short codes are assigned to values of P in the middle of the central region and the b long codes are assigned to values of P closer to L or H. Notice that b is even, so the b codes can always be divided evenly. Table 9.16 shows how 10 such codes can be assigned in the case L=15, H=24.

Pixel P	Region code	Pixel code
L=15	0	0000
16	0	0010
17	0	010
18	0	011
19	0	100
20	0	101
21	0	110
22	0	111
23	0	0001
H=24	0	0011

Table 9.16: The Codes for the Central Region.

When P is in one of the outer regions, say the upper one, the value P–H should be assigned a variable-size code whose size can grow quickly as P–H gets larger (since P–H will rarely be large). One way to do this is to select a small non-negative integer m (typically 0, 1, 2, or 3) and to assign the integer n a two-part code. The second part is the lower m bits of n and the first part, the unary code of [n without its lower m bits] (see Exercise 9.12 for the unary code). Example: If $m = 2$, then the integer $n = 1101_2$ is assigned the code 110|01 since 110 is the unary code of 11. This code is a special case of the Golomb code, (Section 9.9), where the parameter b is a power of 2 (it equals 2^m). Table 9.17 shows some examples of this

Pixel		Region		$m =$		
P	P–H	code	0	1	2	3
H+1=25	1	11	0	00	000	0000
26	2	11	10	01	001	0001
27	3	11	110	100	010	0010
28	4	11	1110	101	011	0011
29	5	11	11110	1100	1000	0100
30	6	11	111110	1101	1001	0101
31	7	11	1111110	11100	1010	0110
32	8	11	11111110	11101	1011	0111
33	9	11	111111110	111100	11000	10000
...	...	...	...		...	

Table 9.17: The Codes for an Outer Region.

code for $m = 0, 1, 2, 3$ and for $n = 1, 2, \ldots, 9$. The value of m used in any particular compression job can be selected as a parameter by the user.

▶ **Exercise 9.11:** Given the 4×4 bitmap of Figure 9.18, calculate the FELICS codes for the three pixels with values 8, 7, and 0.

2	5	7	12
3	0	11	10
2	1	8	15
4	13	11	9

Figure 9.18: A 4×4 Bitmap.

9.8.1 The Unary Code

The *unary code* of the non-negative integer n is defined as $n - 1$ ones followed by one zero (Table 9.19) or, alternatively, as $n - 1$ zeros followed by a single one. The length of the unary code for the integer n is n bits.

n	Code	Alt. Code
1	0	1
2	10	01
3	110	001
4	1110	0001
5	11110	00001

Table 9.19: Some Unary Codes.

9 Image Compression

▶ **Exercise 9.12:** Discuss the use of the unary code as a variable-size code.

It is also possible to define general unary codes, also known as start-step-stop codes. Such a code depends on a triplet (start, step, stop) of integer parameters, and codewords are created to code symbols used in the data, such that the nth codeword consists of n ones, followed by one zero, followed by all the combinations of a bits where $a = \text{start} + n \times \text{step}$. If $a = \text{stop}$, then the single zero preceding the a bits is dropped. The number of codes for a given triplet is finite and depends on the choice of parameters. Tables 9.20 and 9.21 show the 680 codes of (3,2,9) and the 2044 codes of (2,1,10).

n	$a = 3 + n \cdot 2$	nth Codeword	Number of codewords	Range of integers
0	3	$0xxx$	$2^3 = 8$	0–7
1	5	$10xxxxx$	$2^5 = 32$	8–39
2	7	$110xxxxxxx$	$2^7 = 128$	40–167
3	9	$111xxxxxxxxx$	$2^9 = 512$	168–679
		Total	680	

Table 9.20: The General Unary Code (3,2,9).

n	$a = 2 + n \cdot 1$	nth Codeword	Number of codewords	Range of integers
0	2	$0xx$	4	0–3
1	3	$10xxx$	8	4–11
2	4	$110xxxx$	16	12–27
3	5	$1110xxxxx$	32	28–59
...		...		...
8	10	$\underbrace{11...1}_{8}\underbrace{xx...x}_{10}$	1024	1020–2043
		Total	2044	

Table 9.21: The General Unary Code (2,1,10).

The number of different general unary codes is

$$\frac{2^{\text{stop}+\text{step}} - 2^{\text{start}}}{2^{\text{step}} - 1}.$$

Notice that this expression increases exponentially with parameter "stop," so large sets of these codes can be generated with small values of the three parameters.

▶ **Exercise 9.13:** What codes are defined by the parameters $(n, 1, n)$ and what are defined by $(0, 0, \infty)$?

▶ **Exercise 9.14:** How many codes are produced by the triplet $(1, 1, 30)$?

▶ **Exercise 9.15:** Derive the general unary code for $(10, 2, 14)$.

9.8.2 Other Prefix Codes

Four more prefix codes, due to P. Elias [Elias 75], are described in this section. We use B(n) to denote the binary representation of integer n. Thus, |B(n)| is the length, in bits, of this representation. We also use $\bar{B}(n)$ to denote B(n) without its most significant bit (which is always 1).

Code C_1 is made of two parts. To code the positive integer n, we first generate the unary code of |B(n)| (the size of the binary representation of n), then append $\bar{B}(n)$ to it. An example is $n = 16 = 10000_2$. The size of B(16) is 5, so we start with the unary code 11110 (or 00001) and append $\bar{B}(16)$=0000. The complete code is thus 11110|0000 (or 00001|0000). Another example is $n = 5 = 101_2$, whose code is 110|01. The length of $C_1(n)$ is $2\lfloor \log_2 n \rfloor + 1$ bits. Notice that this code is identical to the general unary code $(0, 1, \infty)$.

Code C_2 is a rearrangement of C_1, where each of the $\lfloor \log_2 n \rfloor$ bits of the first part (the unary code) is followed by one of the bits of the second part. Thus, code $C_2(16) = 101010100$ and $C_2(5) = 10110$.

Code C_3 starts with |B(n)| coded in C_2, followed by $\bar{B}(n)$. Thus, 16 is coded as $C_2(5) = 11101$ followed by $\bar{B}(16)$=0000, and 5 is coded as code $C_2(3) = 110$ followed by $\bar{B}(5)$=01. The size of $C_3(n)$ is $1 + \lfloor \log_2 n \rfloor + 2\lfloor \log_2(1 + \lfloor \log_2 n \rfloor) \rfloor$.

Code C_4 is multipart. We start with B(n). To the left of this, we write the binary representation of |B(n)| − 1 (the length of n, minus 1). This continues recursively, until a 2-bit number is written. A zero is then added to the right of the entire number, to make it decodable. To encode 16, we start with 10000, add $|B(16)| - 1 = 4 = 100_2$ to the left, then $|B(4)| - 1 = 2 = 10_2$ to the left of that, and, finally, a zero on the right. The result is 10|100|10000|0. For example, to encode 5, we start with 101, and add $|B(5)| - 1 = 2 = 10_2$ to the left and a zero on the right. The result is 10|101|0.

▶ **Exercise 9.16:** How does the zero on the right make the code decodable?

Table 9.22 shows examples of the four codes above as well as of B(n) and $\bar{B}(n)$.

The lengths of the four codes shown here increases as $\log_2 n$, in contrast to the length of the unary code, which increases as n. These codes are, therefore, a good choice in cases where the data consist of integers n satisfying $P(n) \approx 1/n$. Specifically, the length L of the unary code of n is $L = n$, so it is ideal for the case where $P(n) = 2^{-L} = 2^{-n}$. The length of code $C_1(n)$ is $L = 1 + 2\lfloor \log_2 n \rfloor = \log_2 2 + \log_2 n^2 = \log_2(2n^2)$, so it is ideal for the case where

$$P(n) = 2^{-L} = \frac{1}{2n^2}.$$

The length of code $C_3(n)$ is

$$L = 1 + \lfloor \log_2 n \rfloor + 2\lfloor \log_2(1 + \lfloor \log_2 n \rfloor) \rfloor = \log_2 2 + 2\lfloor \log \log_2 2n \rfloor + \lfloor \log_2 n \rfloor,$$

so it is ideal for the case where

$$P(n) = 2^{-L} = \frac{1}{2n(\log_2 n)^2}.$$

9 Image Compression

n	Unary	$B(n)$	$\overline{B}(n)$	C_1	C_2	C_3	C_4
1	0	1		1|	1	1|	0
2	10	10	0	10|0	100	100|0	10|0
3	110	11	1	10|1	110	100|1	11|0
4	1110	100	00	110|00	10100	110|00	10|100|0
5	11110	101	01	110|01	10110	110|01	10|101|0
6	111110	110	10	110|10	11100	110|10	10|110|0
7	...	111	11	110|11	11110	110|11	10|111|0
8		1000	000	1110|000	1010100	10100|000	11|1000|0
9		1001	001	1110|001	1010110	10100|001	11|1001|0
10		1010	010	1110|010	1011100	10100|010	11|1010|0
11		1011	011	1110|011	1011110	10100|011	11|1011|0
12		1100	100	1110|100	1110100	10100|100	11|1100|0
13		1101	101	1110|101	1110110	10100|101	11|1101|0
14		1110	110	1110|110	1111100	10100|110	11|1110|0
15		1111	111	1110|111	1111110	10100|111	11|1111|0
16		10000	0000	11110|0000	101010100	10110|0000	10|100|10000|0
31		11111	1111	11110|1111	111111110	10110|1111	10|100|11111|0
32		100000	00000	111110|00000	10101010100	11100|00000	10|101|100000|0
63		111111	11111	111110|11111	11111111110	11100|11111	10|101|111111|0
64		1000000	000000	1111110|000000	1010101010100	11110|000000	10|110|1000000|0
127		1111111	111111	1111110|111111	1111111111110	11110|111111	10|110|1111111|0
128		10000000	0000000	11111110|0000000	101010101010100	1010100|0000000	10|111|10000000|0
255		11111111	1111111	11111110|1111111	111111111111110	1010100|1111111	10|111|11111111|0

Table 9.22: Some Prefix Codes.

n	Unary	C_1	C_3
1	0.5	0.5000000	
2	0.25	0.1250000	0.2500000
3	0.125	0.0555556	0.0663454
4	0.0625	0.0312500	0.0312500
5	0.03125	0.0200000	0.0185482
6	0.015625	0.0138889	0.0124713
7	0.0078125	0.0102041	0.0090631
8	0.00390625	0.0078125	0.0069444

Table 9.23: Ideal Probabilities of Eight Integers for Three Codes.

Table 9.23 shows the ideal probabilities that the first eight positive integers should have for the above three codes to be used.

More prefix codes for the positive integers, appropriate for special applications, may be designed by the following general approach. Select positive integers v_i and combine them in a list V (which may be finite or infinite according to needs). The code of the positive integer n is prepared in the following three steps:

1. Find k such that

$$\sum_{i=1}^{k-1} v_i < n \leq \sum_{i=1}^{k} v_i.$$

2. Compute the difference

$$d = n - \sum_{i=1}^{k-1} v_i - 1.$$

The largest value of n is $\sum_1^k v_i$, so the largest value of d is $\sum_i^k v_i - \sum_1^{k-1} v_i - 1 = v_k - 1$, a number that can be written in $\lceil \log_2 v_k \rceil$ bits. d is encoded, using the standard binary code, either in this number of bits, or if $d < 2^{\lceil \log_2 v_k \rceil} - v_k$, it is encoded in $\lfloor \log_2 v_k \rfloor$ bits.

3. Encode n in two parts. Start with k encoded in some prefix code and concatenate the binary code of d. Since k is coded in a prefix code, any decoder would know how many bits to read for k. After reading and decoding k, the decoder can compute the value $2^{\lceil \log_2 v_k \rceil} - v_k$ and, thus, know how many bits to read for d.

A simple example is the infinite sequence $V = (1, 2, 4, 8, \ldots, 2^{i-1}, \ldots)$ with k coded in unary. The integer $n = 10$ satisfies

$$\sum_{i=1}^{3} v_i < 10 \leq \sum_{i=1}^{4} v_i,$$

so $k = 4$ (with unary code 1110) and $d = 10 - \sum_{i=1}^{3} v_i - 1 = 2$. Thus, the code of 10 is 1110|010.

See also the Golomb code, Section 9.9 and the subexponential code of Section 9.10.1.

> "Ah, Roy," said Evry, shaking his head. "Roy Biv! Tell me. Did he ever sign himself Roy G. Biv... He changed his name to that. If you were American you'd understand. It's a mnemonic. The rainbow. Red, orange, yellow, green. Blue, indigo, violet. He wanted to please everyone. That's Roy. Poor Roy."
>
> — Martin Amis, *The Information*

Number Bases

Decimal numbers use base 10. The number 2037_{10}, for example, is

$$2\times10^3 + 0\times10^2 + 3\times10^1 + 7\times10^0.$$

We can say that 2037 is the sum of the digits 2, 0, 3, and 7, each weighted by a power of 10. Fractions are represented in the same way, using negative powers of 10. Thus, $0.82 = 8\times10^{-1} + 2\times10^{-2}$ and $300.7 = 3\times10^2 + 7\times10^{-1}$.

Binary numbers use base 2. Such a number is represented as a sum of its digits, each weighted by a power of 2. Thus,

$$101.11_2 = 1\times2^2 + 0\times2^1 + 1\times2^0 + 1\times2^{-1} + 1\times2^{-2}.$$

Since there is nothing special about 10 or 2,* it should be easy to convince yourself that any positive integer $n > 1$ can serve as the basis for representing numbers. Such a representation requires n "digits" (if $n > 10$, we use the 10 digits and the letters A, B, C...) and it represents the number $d_3 d_2 d_1 d_0 . d_{-1}$ as the sum of the "digits" d_i, each multiplied by a power of n; thus,

$$d_3 d_2 d_1 d_0 . d_{-1} = d_3 n^3 + d_2 n^2 + d_1 n^1 + d_0 n^0 + d_{-1} n^{-1}.$$

The base for a number system does not have to consist of powers of an integer, but can be any *superadditive* sequence that starts with 1.

Definition: A superadditive sequence $a_0, a_1, a_2, \ldots$ is one where any element a_i is greater than the sum of all its predecessors. An example is 1, 2, 4, 8, 16, 32, 64,..., where each element equals 1 plus the sum of all its predecessors. This sequence consists of the familiar powers of 2, so we know that any integer can be expressed by it using just the digits 0 and 1 (the two bits). Another example is 1, 3, 6, 12, 24, 50,... where each element equals 2 plus the sum of all its predecessors. It is easy to see that any integer can be expressed by it using just the digits 0, 1, and 2 (the three base-3 digits, also called trits).

Given a positive integer k, the sequence $1, 1+k, 2+2k, 4+4k, \ldots, 2^i(1+k)$ is superadditive since each element equals the sum of all its predecessors plus k. Any non-negative integer can be *uniquely* represented in such a system as a number $x \ldots xxy$, where x are bits and y is in the range $[0, k]$.

In contrast, a general superadditive sequence, such as 1, 8, 50, 3102, can be used to represent integers, but not uniquely. The number 50, for example, equals $8\times6 + 1 + 1$, so it can be represented as $0062 = 0\times3102 + 0\times50 + 6\times8 + 2\times1$, but also as $0100 = 0\times3102 + 1\times50 + 0\times8 + 0\times1$.

It can be shown that the sum $1 + r + r^2 + \cdots + r^k$ is less than r^{k+1} for any real number $r > 1$, which implies that the powers of any real number $r > 1$ can serve as the base of a number system that uses the digits $0, 1, 2, \ldots, d$ for some d.

---

* Actually, there is. Two is the smallest integer that can be a base for a number system. Ten is the number of our fingers.

The number $\phi = \frac{1}{2}(1+\sqrt{5}) \approx 1.618$ is the well-known Golden Ratio. It can serve as the base of a number system that uses the two binary digits. Thus, for example, $100.1_\phi = \phi^2 + \phi^{-1} \approx 3.23_{10}$.

Some real bases have special properties. For example, any positive integer R can be expressed as $R = b_1 F_1 + b_2 F_2 + b_3 F_3 + b_4 F_5 + \cdots$ (that's $b_4 F_5$, not $b_4 F_4$), where b_i are either 0 or 1 and the F_i are the Fibonacci numbers $1, 2, 3, 5, 8, 13, \ldots$. This representation has the interesting property that the string $b_1 b_2 \ldots$ does not contain any adjacent 1's. As an example, the integer 33 equals the sum $1 + 3 + 8 + 21$, so it is expressed in the Fibonacci base as the 7-bit number 1010101.

A non-negative integer can be represented as a finite sum of binomials

$$n = \binom{a}{1} + \binom{b}{2} + \binom{c}{3} + \binom{d}{4} + \cdots, \quad \text{where } 0 \leq a < b < c < d \ldots$$

are integers and $\binom{i}{n}$ is the binomial $\frac{i!}{n!(i-n)!}$. This is the *binomial number system*.

9.9 The Golomb Code

The *Golomb code* for non-negative integers n [Golomb 66] can be an effective Huffman code. The code depends on the choice of a parameter b. The first step is to compute the two quantities

$$q = \left\lfloor \frac{n-1}{b} \right\rfloor \quad \text{and} \quad r = n - qb - 1,$$

following which, the code is constructed of two parts; the first is the value of $q+1$, coded in unary (Exercise 9.12) and the second, the binary value of r coded in either $\lfloor \log_2 b \rfloor$ bits (for the small remainders) or in $\lceil \log_2 b \rceil$ bits (for the large ones). Choosing, for example, $b = 3$ produces three possible remainders, 0, 1, and 2. They are coded 0, 10, and 11, respectively. Choosing $b = 5$ produces the five remainders 0 through 4 that are coded 00, 01, 100, 101, and 110. Table 9.24 shows some examples of the Golomb code for $b = 3$ and $b = 5$.

n:	1	2	3	4	5	6	7	8	9	10
$b = 3$:	0|0	0|10	0|11	10|0	10|10	10|11	110|0	110|10	110|11	1110|0
$b = 5$:	0|00	0|01	0|100	0|101	10|110	10|00	10|01	10|100	10|101	110|110

Table 9.24: Some Golomb Codes for $b = 3$ and $b = 5$.

Imagine an input data stream consisting of positive integers where the probability of integer n appearing in the data is $P(n) = (1-p)^{n-1} p$, for some $0 \leq p \leq 1$. It can be shown that the Golomb code is an optimal code for these data if b is chosen such that

$$(1-p)^b + (1-p)^{b+1} \leq 1 < (1-p)^{b-1} + (1-p)^b.$$

9 Image Compression

Given the right data, it is easy to generate the best variable-size codes without going through the Huffman algorithm.

9.10 Progressive FELICS

The original FELICS method can easily be extended to progressive compression of images because of its main principle. FELICS scans the image row by row (raster scan) and encodes a pixel based on the values of two of its (previously seen and encoded) neighbors. Progressive FELICS [Howard and Vitter 94] works similarly, but it scans the pixels in levels. Each level uses the k pixels encoded in all previous levels to encode k more pixels, so the number of encoded pixels doubles after each level. Assuming that the image consists of $n \times n$ pixels and the first level starts with just four pixels, consecutive levels result in

$$4, 8, \ldots, \frac{n^2}{8}, \frac{n^2}{4}, \frac{n^2}{2}, n^2$$

pixels. The number of levels is thus the number of terms, $2\log_2 n - 1$, in this sequence.

▶ **Exercise 9.17:** Prove this!

Figure 9.25 shows the pixels encoded in most of the levels of a 16×16-pixel image. Figure 9.26 shows how the pixels of each level are selected. In Figure 9.26a there are $8 \times 8 = 64$ pixels, one-quarter of the final number, arranged in a square grid. Each group of four pixels is used to encode a new pixel, so Figure 9.26b has 128 pixels, half the final number. The image of Figure 9.26b is then rotated 45° and scaled by factors of $\sqrt{2} \approx 1.414$ in both directions, to produce Figure 9.26c, which is a square grid that looks exactly like Figure 9.26a. The next step (not shown in the figure) is to use every group of 4×4 pixels in Figure 9.26c to encode a pixel, thereby encoding the remaining 128 pixels. In practice, there is no need to actually rotate and scale the image; the program simply alternates between xy and diagonal coordinates.

Each group of four pixels is used to encode the pixel at its center. Notice that in early levels, the four pixels of a group are far from each other and are, thus, not correlated, resulting in poor compression. However, the last two levels encode three-quarters of the total number of pixels and these levels contain compact groups. Two of the four pixels of a group are selected to encode the center pixel and are designated L and H. Experience shows that the best choice for L and H is the two median pixels (page 633), the ones with the middle values (i.e., not the maximum or the minimum pixels of the group). Ties can be resolved in any way, as long as it is done consistently. If the two medians in a group are the same, then the median and the minimum (or the median and the maximum) pixels can be selected. The two selected pixels, L and H, are used to encode the center pixel in the same way FELICS uses two neighbors to encode a pixel. The only difference is that a new prefix code (Section 9.10.1) is used, instead of the Golomb code.

▶ **Exercise 9.18:** Why is it important to resolve ties in a consistent way?

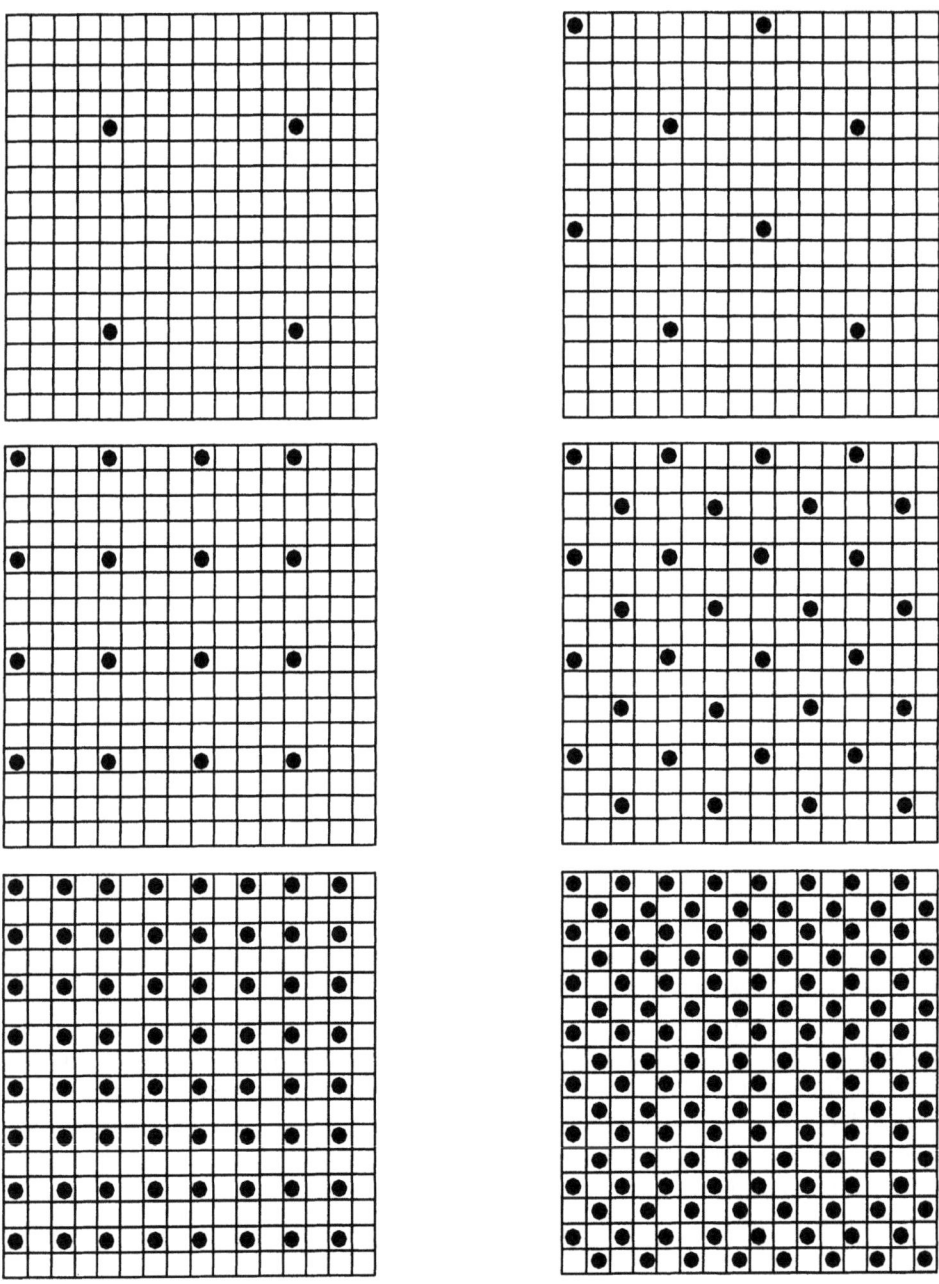

Figure 9.25: Some Levels of a 16×16 Image.

9 Image Compression

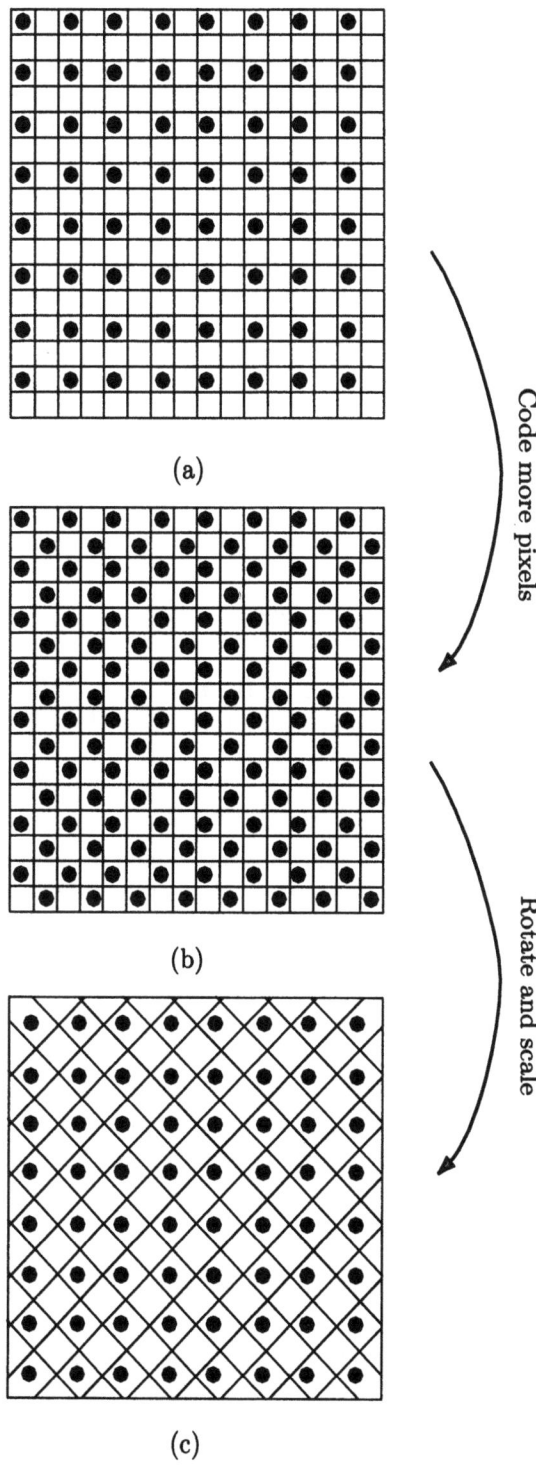

Figure 9.26: Rotation and Scaling.

9.10.1 Subexponential Code

In early levels, the four pixels used to encode a pixel are far from each other. As more levels are progressively encoded, the groups get more compact, so their pixels get closer. The encoder uses the absolute difference between the L and H pixels in a group (the *context* of the group) to encode the pixel at the center of the group, but a given absolute difference means more for late levels than for early ones, because the groups of late levels are smaller, so their pixels are more correlated. The encoder should, therefore, scale the difference by a weight parameter s that gets heavier from level to level. The precise value of s is not critical and experiments recommend the value 12.

The prefix code used by progressive FELICS is called *subexponential*. Like the Golomb code (Section 9.9), this new code depends on a parameter $k \geq 0$. The main feature of the subexponential code is its length. For integers $n < 2^{k+1}$, the code length increases linearly with nm, but for larger n, it increases logarithmically. The subexponential code of the non-negative integer n is computed in two steps. In the first step, values b and u are calculated by

$$b = \begin{cases} k & \text{if } n < 2^k, \\ \lfloor \log_2 n \rfloor & \text{if } n \geq 2^k \end{cases} \quad \text{and} \quad u = \begin{cases} 0 & \text{if } n < 2^k, \\ b - k + 1 & \text{if } n \geq 2^k. \end{cases}$$

In the second step, the unary code of u (in $u + 1$ bits) followed by the b least significant bits of n becomes the subexponential code of n. The total size of the code is thus

$$u + 1 + b = \begin{cases} k + 1, & \text{if } n < 2^k, \\ 2\lfloor \log_2 n \rfloor - k + 2, & \text{if } n \geq 2^k. \end{cases}$$

Table 9.27 shows examples of the subexponential code for various values of n and k. It can be shown that for a given n, the code lengths for consecutive values of k differ by at most 1.

If the value of the pixel to be encoded lies between those of L and H, the pixel is encoded as in FELICS. If it lies outside the range [L, H], the pixel is encoded by using the subexponential code where the value of k is selected by the following rule.

Suppose that the current pixel P to be encoded has context C. The encoder maintains a cumulative total, for some reasonable values of k, of the code length the encoder would have if it had used that value of k to encode all pixels encountered so far in context C. The encoder then uses the k value with the smallest cumulative code length to encode P.

9.11 MLP

Text compression methods can use context to predict (i.e., to estimate the probability of) the next character of text. Using context to predict the intensity of the next pixel in image compression is more complex because: (1) An image is two-dimensional, allowing for many possible contexts and (2) a digital image is normally the result of digitizing an analog image. The intensity of any individual pixel is thus determined by the details of digitization and may differ from the "ideal" intensity.

The multilevel progressive method (MLP) described here [Howard and Vitter 92a] is a computationally intensive, lossless method for compressing grayscale

n	k = 0	k = 1	k = 2	k = 3	k = 4	k = 5
0	0|	0|0	0|00	0|000	0|0000	0|00000
1	10|	0|1	0|01	0|001	0|0001	0|00001
2	110|0	10|0	0|10	0|010	0|0010	0|00010
3	110|1	10|1	0|11	0|011	0|0011	0|00011
4	1110|00	110|00	10|00	0|100	0|0100	0|00100
5	1110|01	110|01	10|01	0|101	0|0101	0|00101
6	1110|10	110|10	10|10	0|110	0|0110	0|00110
7	1110|11	110|11	10|11	0|111	0|0111	0|00111
8	11110|000	1110|000	110|000	10|000	0|1000	0|01000
9	11110|001	1110|001	110|001	10|001	0|1001	0|01001
10	11110|010	1110|010	110|010	10|010	0|1010	0|01010
11	11110|011	1110|011	110|011	10|011	0|1011	0|01011
12	11110|100	1110|100	110|100	10|100	0|1100	0|01100
13	11110|101	1110|101	110|101	10|101	0|1101	0|01101
14	11110|110	1110|110	110|110	10|110	0|1110	0|01110
15	11110|111	1110|111	110|111	10|111	0|1111	0|01111
16	111110|0000	11110|0000	1110|0000	110|0000	10|0000	0|10000

Table 9.27: Some Subexponential Codes.

images. It uses context to predict the intensities of pixels and then uses arithmetic coding to encode the difference between the prediction and the actual value of a pixel (the error). The Laplace distribution is used to estimate the probability of the error. The method combines four separate steps: (1) pixel sequencing, (2) prediction (image modeling), (3) error modeling (by means of the Laplace distribution), and (4) arithmetically encoding the errors.

MLP is also progressive, encoding the image in levels, where the pixels of each level are selected as in progressive FELICS. When the image is decoded, each level adds details to the entire image, not just to certain parts, so a user can view the image as it is being decoded and decide in real time whether to accept or reject it. This feature is useful when an image has to be selected from a large archive of compressed images. The user can browse through images very fast, without having to wait for any image to be completely decoded. Another advantage of progressive compression is that it provides a natural lossy option. The encoder may be told to stop encoding before it reaches the last level (thereby encoding only half the total number of pixels) or before it reaches the next to last level (encoding only a quarter of the total number of pixels). Such an option results in an excellent compression ratio but a loss of image data. The decoder may be told to use interpolation to determine the intensities of any missing pixels.

Like any compression method for grayscale images, MLP can be used to compress color images. The original color image should be separated into three color components and each component compressed individually as a grayscale image. Following is a detailed description of the individual MLP encoding steps.

Pixel Sequencing: Pixels are selected in levels, as in progressive FELICS, where each level encodes the same number of pixels as all the preceding levels combined, thereby doubling the number of encoded pixels. This means that the last level encodes half the number of pixels, the level preceding it encodes a quarter of the total number, and so on. The first level should start with at least 4 pixels, but may also start with 8, 16, or any desired power of 2.

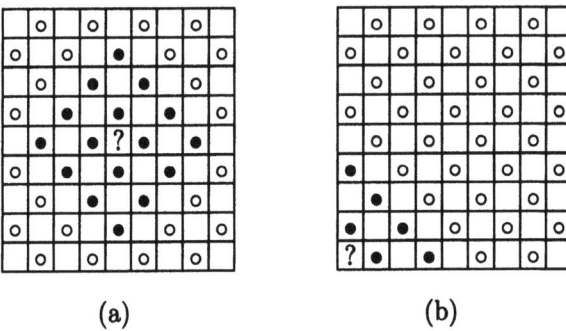

Figure 9.28: (a) Sixteen Neighbors. (b) Six Neighbors.

Prediction: A pixel is predicted by calculating a weighted average of 16 of its known neighbors. Keep in mind that pixels are not raster scanned but are encoded (and thus also decoded) in levels. When decoding the pixels of level L, the MLP decoder has already decoded all the pixels of all the preceding levels and it can use their values (gray levels) to predict values of pixels of L. Figure 9.28a shows the situation when the MLP encoder processes the last level. Half the pixels have already been encoded in previous levels, so they will be known to the decoder when the last level is decoded. The encoder can, thus, use a diamond-shaped group of 4×4 pixels (shown in black) from previous levels to predict the pixel at the center of the group. This group is the *context* of the pixel. Compression methods that scan the image row by row (raster scan) can use only pixels above and to the left of pixel P to predict P. Because of the progressive nature of MLP, it can use a symmetric context, which produces more accurate predictions. On the other hand, the pixels of the context are not near neighbors and may even be (in early levels) far from the predicted pixel.

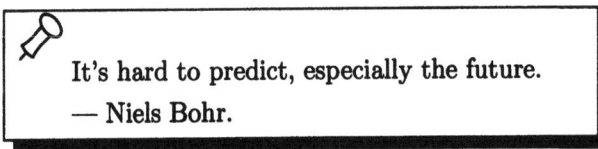

It's hard to predict, especially the future.
— Niels Bohr.

Table 9.29 shows the 16 weights used for a group. They are calculated by polynomial interpolation (Section 5.10.1) and are normalized such that they add up to 1. (Notice that in Table 9.29a the weights are not normalized; they add up to 256. When these integer weights are used, the weighted sum should be divided by 256.) To predict a pixel near an edge, where some of the 16 neighbors are missing

9 Image Compression

			1				
		−9		−9			
	−9	81			−9		
1	81		81		1		
	−9	81			−9		
		−9		−9			
			1				

(a)

		0.0039		
	−0.0351		−0.0351	
	−0.0351	0.3164	−0.0351	
0.0039	0.3164	0.3164	0.0039	
	−0.0351	0.3164	−0.0351	
	−0.0351		−0.0351	
		0.0039		

(b)

Table 9.29: Sixteen Weights. (a) Integers. (b) Normalized.

(as in Figure 9.28b), only those neighbors that exist are used, and their weights are renormalized, to bring their sum to 1.

▶ **Exercise 9.19:** Why do the weights have to add up to 1?

▶ **Exercise 9.20:** Show how to renormalize the six weights needed to predict the pixel at the bottom left corner of Figure 9.28b.

The encoder predicts all the pixels of a level by using the diamond-shaped group of 4×4 (or fewer) "older" pixels around each pixel of the level. This is the *image model* used by MLP.

9.11.1 MLP Error Modeling

Assume that the weighted sum of the 16 near neighbors of pixel P is R. R is, thus, the value predicted for P. The prediction error, E, is simply the difference R–P. Assuming an image with 16 gray levels (4 bits per pixel), the largest value of E is 15 (when R=15 and P=0) and the smallest is −15. Depending on the image, we can expect most of the errors to be small integers, either zero or close to it. Few errors should be ±15 or close to that. Experiments with a large number of images have produced the error distribution shown in Figure 9.30a. This is a symmetric, narrow curve, with a sharp peak, indicating that most errors are small and are, thus, concentrated at the top. Such a curve has the shape of the well-known Laplace distribution with mean 0. This is a statistical distribution, similar to the normal (Gaussian) distribution (Section 10.6.1) but narrower and sharply peaked. The general Laplace distribution with variance V and mean m is given by

$$L(V, x) = \frac{1}{\sqrt{2V}} \exp\left(-\sqrt{\frac{2}{V}}|x - m|\right).$$

Table 9.31 shows some values for the Laplace distributions with $m = 0$ and $V = 3, 4, 5$, and 1000. Figure 9.30b shows the graphs of the first three of those. It is clear that as V grows, the graph becomes lower and wider, with a less-pronounced peak.

The reason for the factor $1/\sqrt{2V}$ in the definition of the Laplace distribution is to scale the area under the curve of the distribution to 1. Because of this, it is easy to use the curve of the distribution to calculate the probability of any error value.

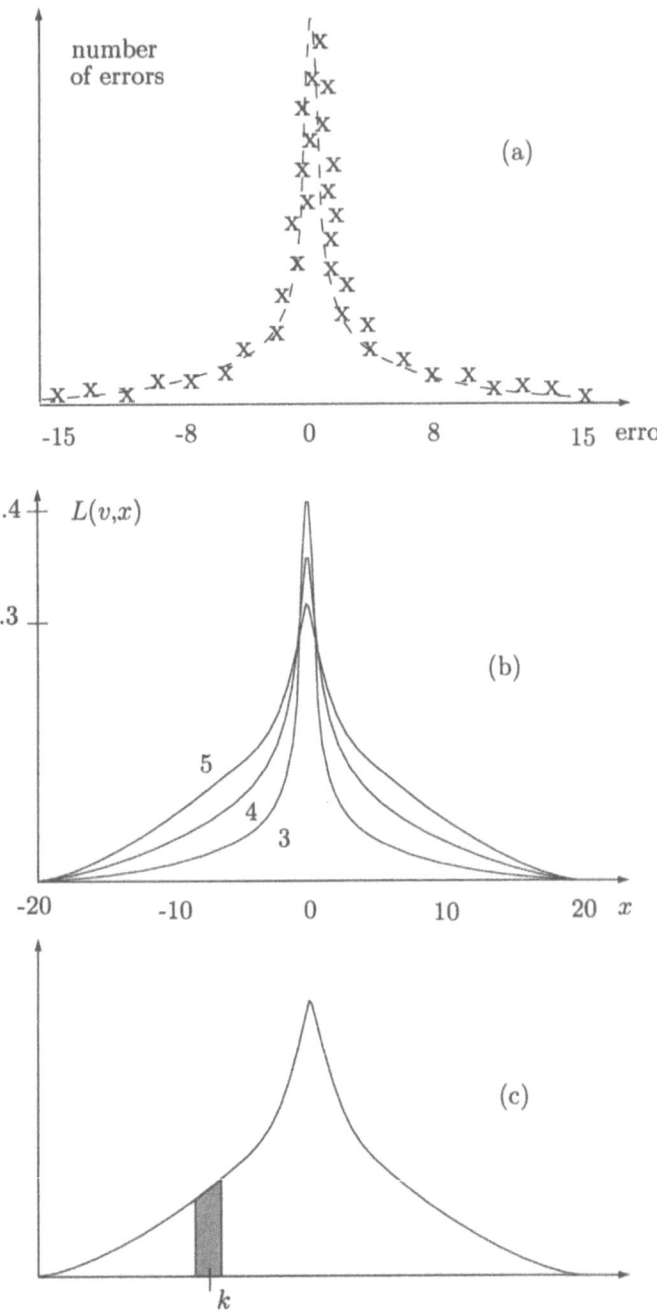

Figure 9.30: (a) Error Distribution. (b) Laplace Distributions. (c) Probability of k.

9 Image Compression

	x					
V	0	2	4	6	8	10
3:	0.408248	0.0797489	0.015578	0.00304316	0.00059446	0.000116125
4:	0.353553	0.0859547	0.020897	0.00508042	0.00123513	0.000300282
5:	0.316228	0.0892598	0.025194	0.00711162	0.00200736	0.000566605
1,000:	0.022360	0.0204475	0.018698	0.0170982	0.0156353	0.0142976

Table 9.31: Some Values of the Laplace Distribution with $V = 3, 4, 5$, and 1000.

Figure 9.30c shows a gray strip one unit wide, under the curve of the distribution, centered about an error value of k. The area of this strip equals the probability of any error E having the value k. Mathematically, the area is the integral

$$P_V(k) = \int_{k-.5}^{k+.5} \frac{1}{\sqrt{2V}} \exp\left(-\sqrt{\frac{2}{V}}|x|\right) dx, \tag{9.1}$$

▶ and this is the key to encoding the errors. With 4-bit pixels, error values are in the range $[-15, +15]$. When an error k is obtained, the MLP encoder encodes it arithmetically with a probability computed by Equation (9.1). In practice, both encoder and decoder should have a table with all the possible probabilities precomputed.

▸ **Exercise 9.21:** What is the indefinite integral of the Laplace distribution?

The only remaining point to discuss is what value of the variance V should be used in Equation (9.1). Both the encoder and decoder need to know this value. It is clear, from Figure 9.30b, that using a large variance (which corresponds to a low, flat distribution) results in too low a probability estimate for small error values k. The arithmetic encoder would produce an unnecessarily long code in such a case. On the other hand, using a small variance (which corresponds to a high, narrow distribution) would allocate too low probabilities to large error values k. The choice of variance is, therefore, important. An ideal method to estimate the variance should assign the best variance value to each error and should involve no overhead (i.e., no extra data should be written on the compressed stream to help the decoder estimate the variances). Here are some approaches to variance selection:

1. The Laplace distribution was adopted as the MLP error distribution after many experiments with real images. The distribution obtained by all those images has a certain value of V and this value should be used by MLP. This is a simple approach that can be used in fast versions of MLP. However, it is not adaptive (since it always uses the same variance) and, thus, does not result in best compression performance for all images.

2. (A two-pass compression job.) Each compression should start with a pass where all the error values are computed and their variance calculated. This variance should be used, after the first pass, to calculate a table of probabilities from Equation (9.1). The table should be used in the second pass, where the error values are encoded and written on the compressed stream. Compression performance would be excellent, but any two-pass job is slow. Notice that the entire table can

be written at the start of the compressed stream, thereby greatly simplifying the task of the decoder.

▶ **Exercise 9.22:** Show an example where this approach is practical (i.e., when slow encoding is unimportant, but a fast decoder and excellent compression are).

3. Every time an error is obtained and is about to be arithmetically coded, use some method to estimate the variance associated with that error. Quantize the estimate and use a number of precomputed probability tables, one for each quantized variance value, to compute the probability of the error. This is computationally intensive but may be a good compromise between approaches 1 and 2 above.

We now need to discuss how the variance of an error can be estimated and we start with an explanation of the concept of variance. Variance is a statistical concept defined for a sequence of values $a_1, a_2, \ldots, a_n$. It measures how elements a_i vary by calculating the differences between them and the average A of the sequence, which is given, of course, by $A = (1/n)\sum a_i$. This is why the curves of Figure 9.30b that correspond to smaller variances are narrower; their values are concentrated near the average, which in this case is zero. The sequence $(5,5,5)$, for example, has average 5 and variance 0, since every element of the sequence equals the average. The sequence $(0,5,10)$ also has average 5 but should have a nonzero variance since two of its elements differ from the average. In general, the variance of the sequence a_i is defined as the non-negative quantity

$$V = \sum_1^n (a_i - A)^2,$$

so the variance of $(0, 5, 10)$ is $(0-5)^2 + (5-5)^2 + (10-5)^2 = 50$. Statisticians also use a quantity called *standard deviation* (denoted σ) that is defined as the square root of the variance.

We now discuss several ways to estimate the variance of a prediction error E.

3.1. Equation (9.1) gives the probability of an error E with value k, but this probability depends on V. We can consider $P_V(k)$ a function of the two variables V and k, and find the optimal value of V by solving the equation $\partial P_V(k)/\partial V = 0$. The solution is $V = 2k^2$, but this method is not practical since the decoder does not know k (it is trying to decode k so it can find the value of pixel P) and, thus, cannot mirror the operations of the encoder. It is possible to write the values of all the variances on the compressed stream, but this would significantly reduce the compression ratio. This method can be used to encode a particular image in order to find the best compression ratio of the image and compare it to what is achieved in practice.

3.2. While the pixels of a level are being encoded, consider their errors E a sequence of numbers and find its variance V. Use V to encode the pixels of the next level. The number of levels is never very large, so all the variance values can be written (arithmetically encoded) on the compressed stream, resulting in fast decoding. The variance used to encode the first level should be a user-controlled parameter whose value is not critical since that level contains just a few pixels. Since MLP quantizes a variance to 1 of 37 values (see below) each variance written on

9 Image Compression

the compressed stream is encoded in just $\log_2 37 \approx 5.21$ bits, a negligible overhead. The obvious disadvantage of this method is that it disregards local concentrations of identical or very similar pixels in the same level.

3.3. (Similar to 3.2.) While the pixels of a level are being encoded, collect the prediction errors of each block of $b \times b$ pixels and use them to calculate a variance which will be used to encode the pixels inside this block in the next level. The variance values for a level can also be written on the compressed stream following all the encoded errors for that level, so the decoder could use them without having to compute them. Parameter b should be adjusted by experiments and the authors recommend the value $b = 16$. This method entails significant overhead and, thus, may degrade compression performance.

3.4. (This is a later addition to MLP, see [Howard and Vitter 92b].) A *variability index* is computed, by both the encoder and decoder, for each pixel. This index should depend on the amount by which the pixel differs from its near neighbors. The variability indexes of all the pixels in a level are then used to adaptively estimate the variances for the pixels, based on the assumption that pixels with similar variability index should use Laplace distributions with similar variances. The method proceeds in the following steps:

1. Variability indexes are calculated for all the pixels of the current level, based on values of pixels in the preceding levels. This is done by the encoder and is later mirrored by the decoder. After several tries, the developers of MLP have settled on a simple way to calculate the variability index. It is calculated as the variance of the four nearest neighbors of the current pixel (the neighbors are from preceding levels, so the decoder can mirror this operation).

2. The variance estimate V is set to some initial value. The choice of this value is not critical, as V is going to be updated later many times. The decoder chooses this value in the same way.

3. The pixels of the current level are sorted in variability index order. The decoder can mirror this even though it still does not have the values of these pixels (the decoder has already calculated the values of the variability index in step 1, since they depend on pixels of previous levels).

4. The encoder loops over the sorted pixels in decreasing order (from large variability indexes to small ones). For each pixel:

4.1. The encoder calculates the error E of the pixel and sends E and V to the arithmetic encoder. The decoder mirrors this step. It knows V, so it can decode E.

4.2. The encoder updates V by

$$V \leftarrow f \times V + (1 - f)E^2,$$

where f is a smoothing parameter (experience suggests a large value, such as 0.99, for f). This is how V is adapted from pixel to pixel, using the errors E. Because of the large value of f, V is decreased in small steps. This means that later pixels (those with small variability indexes) will get small variances assigned. The idea is that compressing pixels with large variability indexes is less sensitive to accurate values of V.

As the loop progresses, V gets assigned more accurate values and these are used to compress pixels with small variability indexes, which are more sensitive to variance values. Notice that the decoder can mirror this step since it has already decoded E in step 4.1. Notice also that the arithmetic encoder writes the encoded error values on the compressed stream in decreasing variability index order, not row by row. The decoder can mirror this too, since it has already sorted the pixels in this order in step 3.

This method gives excellent results but is even more computationally intensive than the original MLP. (End of method 3.4.)

Using one of the above four methods, the variance V is estimated. Before using V to encode error E, V is quantized to 1 of 37 values as shown in Table 9.32. For example, if the estimated variance value is 0.31, it is quantized to 7. The quantized value is then used to select 1 of 37 precomputed probability tables (in our example, table 7, precomputed for variance value 0.290, is selected) prepared using Equation (9.1) and the value of error E is used to index that table. The value retrieved from the table is the probability that's sent to the arithmetic encoder, together with the error value, to arithmetically encode error E.

Variance range	Var. used	Variance range	Var. used	Variance range	Var. used
0.005–0.023	0.016	2.882–4.053	3.422	165.814–232.441	195.569
0.023–0.043	0.033	4.053–5.693	4.809	232.441–326.578	273.929
0.043–0.070	0.056	5.693–7.973	6.747	326.578–459.143	384.722
0.070–0.108	0.088	7.973–11.170	9.443	459.143–645.989	540.225
0.108–0.162	0.133	11.170–15.627	13.219	645.989–910.442	759.147
0.162–0.239	0.198	15.627–21.874	18.488	910.442–1285.348	1068.752
0.239–0.348	0.290	21.874–30.635	25.875	1285.348–1816.634	1506.524
0.348–0.502	0.419	30.635–42.911	36.235	1816.634–2574.021	2125.419
0.502–0.718	0.602	42.911–60.123	50.715	2574.021–3663.589	3007.133
0.718–1.023	0.859	60.123–84.237	71.021	3663.589–5224.801	4267.734
1.023–1.450	1.221	84.237–118.157	99.506	5224.801–7247.452	6070.918
1.450–2.046	1.726	118.157–165.814	139.489	7247.452–10195.990	8550.934
2.046–2.882	2.433				

Table 9.32: Thirty-Seven Quantized Variance Values.

MLP is, thus, one of the many compression methods that implement a model to estimate probabilities and use arithmetic coding to do the actual compression.

Table 9.33 is a pseudo-code summary of MLP encoding.

9.12 Differential Lossless Image Compression

There is always a trade-off between speed and performance, so there is always a demand for fast compression methods as well as for methods that are slow but very efficient. The differential method of this section, due to Sayood and Anderson [Sayood and Anderson 92], belongs to the former class. It is fast and simple to implement, while offering good, albeit not spectacular, performance.

```
for each level L do
 for every pixel P in level L do
  Compute a prediction R for P using a group from level L-1;
  Compute E=R-P;
  Estimate the variance V to be used in encoding E;
  Quantize V and use it as an index to select a Laplace table LV;
  Use E as an index to table LV and retrieve LV[E];
  Use LV[E] as the probability to arithmetically encode E;
 endfor;
 Determine the pixels of the next level (rotate & scale);
endfor;
```

Table 9.33: MLP Encoding.

The principle is to compare each pixel p to a certain *reference pixel*, which is one of its immediate neighbors, and encode p in two parts—a prefix, which is the number of most significant bits of p that are identical to those of the reference pixel, and a suffix, which is the remaining least significant bits of p. For example, if the reference pixel is 10110010 and p is 10110100, then the prefix is 5, since the five most significant bits of p are identical to those of the reference pixel, and the suffix is 00. Notice that the remaining three least significant bits in this case are 100, but the suffix does not have to include the 1 since the decoder can easily deduce its value.

▶ **Exercise 9.23:** How can the decoder do this?

The prefix is thus an integer in the range [0, 8] and compression can be improved by encoding the prefix further. Huffman coding is a good choice for this purpose, with either a fixed set of nine Huffman codes or with adaptive codes. The suffix can be any number of bits between 0 and 8, so there are 256 possible suffixes. Since this number is relatively large and since we expect most suffixes to be small, it makes sense to write the suffix on the output stream unencoded.

This method encodes each pixel by using a different number of bits. The encoder generates bits until it has eight or more of them, then outputs a byte. The decoder can easily mimic this. All that it has to know is the location of the reference pixel and the Huffman codes. In the example above, if the Huffman code of 6 is, say, 010, the code of p will be the 5 bits 010|00.

The only remaining point to be discussed is the selection of the reference pixel. It should be a near neighbor of the current pixel p and it should be known to the decoder when p is decoded. The rules for selecting the reference pixel are simple. The very first pixel of an image is written on the output stream unencoded. For every other pixel in the first scan line, the reference pixel is its immediate left neighbor. For the first pixel on subsequent scan lines, the reference pixel is the one above it. For every other pixel, it is possible to select the reference pixel in one of three ways: (1) the pixel immediately to its left, (2) the pixel above it, or (3) the pixel on the left, but, if the resulting prefix is less than a predetermined threshold, the pixel above it.

An example of case 3 is a threshold value of 3. Initially, the reference pixel for p is its left neighbor, but if this results in a prefix value of 0, 1, or 2, the reference pixel is changed to the one above p, regardless of the prefix value which is then produced.

This method assumes 1 byte per pixel (256 colors or grayscale values). If a pixel is defined by 3 bytes, the image should be separated into three parts and the method applied individually to each part.

▸ **Exercise 9.24:** Can this method be used for images with 16 grayscale values (where each pixel is 4 bits and a byte contains two pixels)?

9.13 Wavelets

Wavelets have been developed in the 1980s as an alternative to the windowed Fourier transform for digital signal processing. They have since found many applications in signal processing and computer graphics, in addition to their use in data compression. This section presents a simplified approach to the use of wavelets for image compression. We first discuss the wavelet compression of grayscale images, then show how it can be extended to color images.

An image is a two-dimensional array of pixel values. To illustrate the main concept of wavelet compression, we start with a single row of pixel values, i.e., a one-dimensional array of n values. For simplicity, we assume that n is a power of 2. Consider the array of eight values $(1, 2, 3, 4, 5, 6, 7, 8)$. We first compute the four averages $(1+2)/2 = 3/2$, $(3+4)/2 = 7/2$, $(5+6)/2 = 11/2$, and $(7+8)/2 = 15/2$. It is impossible to reconstruct the original eight values from these four averages, so we also compute the four differences $(1-2)/2 = -1/2$, $(3-4)/2 = -1/2$, $(5-6)/2 = -1/2$, and $(7-8)/2 = -1/2$. These differences are also called *detail coefficients*.

It is easy to see that the array $(3/2, 7/2, 11/2, 15/2, -1/2, -1/2, -1/2, -1/2)$, which is made of the four averages and four differences, can be used to reconstruct the original eight values. This array has eight values, but its last four components, the differences, tend to be small numbers, which helps in compression. Encouraged by this, we repeat the process on the four averages, the large components of our array. They are transformed into two averages and two differences, yielding the array

$$(10/4, 26/4, -4/4, -4/4, -1/2, -1/2, -1/2, -1/2).$$

The next, and last, iteration of this process transforms the first two components of the new array into one average (the average of all eight components of the original array) and one difference

$$(36/8, -16/8, -4/4, -4/4, -1/2, -1/2, -1/2, -1/2).$$

The last array is called the *wavelet transform* of the original array.

Because of the differences, the wavelet transform tends to have numbers smaller than the original pixel values, so it is easier to compress by using RLE, perhaps combined with move-to-front and Huffman coding (see [Salomon 97] for these methods). Lossy compression can be obtained if some of the smaller differences are changed to zero.

9 Image Compression

It is useful to associate with each iteration a quantity called *resolution*, which is defined as the number of remaining averages at the end of the iteration. The resolutions after each of the three iterations above are $4 (= 2^2)$, $2 (= 2^1)$, and $1 (= 2^0)$. Theory shows that it is better to normalize each component of the wavelet transform by dividing it by the square root of the resolution. (This has to do with representing the components of the wavelet transform by means of the *orthonormal Haar transform* [Stollnitz et al. 96] and will not be discussed here.) Our example wavelet transform thus becomes

$$\left(\frac{36/8}{\sqrt{2^0}}, \frac{-16/8}{\sqrt{2^0}}, \frac{-4/4}{\sqrt{2^1}}, \frac{-4/4}{\sqrt{2^1}}, \frac{-1/2}{\sqrt{2^2}}, \frac{-1/2}{\sqrt{2^2}}, \frac{-1/2}{\sqrt{2^2}}, \frac{-1/2}{\sqrt{2^2}}\right).$$

If the normalized wavelet transform is used, it can be formally proved that ignoring the smallest differences is the best choice for lossy wavelet compression since it creates the smallest loss of image information.

The two procedures of Figure 9.34 illustrate how the normalized wavelet transform of an array of n components (where n is a power of 2) can be computed. Reconstructing the original array from the normalized wavelet transform is illustrated by the pair of procedures of Figure 9.35.

These procedures seem at first to be different from the averages and differences discussed earlier. They don't compute averages since they divide by $\sqrt{2}$ instead of by 2; the first one starts by dividing the entire array by $\sqrt{n}$ and the second one ends by doing the reverse. The final result, however, is the same as that shown above. Starting with array $(1, 2, 3, 4, 5, 6, 7, 8)$, the three iterations of procedure NWTcalc result in

$$\left(\frac{3}{\sqrt{2^4}}, \frac{7}{\sqrt{2^4}}, \frac{11}{\sqrt{2^4}}, \frac{15}{\sqrt{2^4}}, \frac{-1}{\sqrt{2^4}}, \frac{-1}{\sqrt{2^4}}, \frac{-1}{\sqrt{2^4}}, \frac{-1}{\sqrt{2^4}}\right)$$
$$= \left(\frac{10}{\sqrt{2^5}}, \frac{26}{\sqrt{2^5}}, \frac{-4}{\sqrt{2^5}}, \frac{-4}{\sqrt{2^5}}, \frac{-1}{\sqrt{2^4}}, \frac{-1}{\sqrt{2^4}}, \frac{-1}{\sqrt{2^4}}, \frac{-1}{\sqrt{2^4}}\right),$$
$$= \left(\frac{36}{\sqrt{2^6}}, \frac{-16}{\sqrt{2^6}}, \frac{-4}{\sqrt{2^5}}, \frac{-4}{\sqrt{2^5}}, \frac{-1}{\sqrt{2^4}}, \frac{-1}{\sqrt{2^4}}, \frac{-1}{\sqrt{2^4}}, \frac{-1}{\sqrt{2^4}}\right),$$
$$= \left(\frac{36/8}{\sqrt{2^0}}, \frac{-16/8}{\sqrt{2^0}}, \frac{-4/4}{\sqrt{2^1}}, \frac{-4/4}{\sqrt{2^1}}, \frac{-1/2}{\sqrt{2^2}}, \frac{-1/2}{\sqrt{2^2}}, \frac{-1/2}{\sqrt{2^2}}, \frac{-1/2}{\sqrt{2^2}}\right).$$

9.13.1 Two-Dimensional Wavelet Transform

Once the concept of a wavelet transform is grasped, it's easy to generalize it to a complete, two-dimensional image. This can be done in two ways, called the *standard decomposition* and the *nonstandard decomposition*.

The former (Figure 9.36) starts by computing the wavelet transform of every row of the image. This results in a transformed image where the first column contains averages and all the other columns contain differences. The standard algorithm then computes the wavelet transform of every column. This results in one average value at the top-left corner, with the rest of the top row containing averages of differences and with all other pixel values transformed into differences.

```
procedure NWTcalc(a:array of real, n:int);
 comment n is the array size (a power of 2)
 a:=a/√n comment divide entire array
 j:=n;
 while j≥ 2 do
  NWTstep(a, j);
  j:=j/2;
 endwhile;
end;

procedure NWTstep(a:array of real, j:int);
 for i=1 to j/2 do
  b[i]:=(a[2i-1]+a[2i])/√2;
  b[j/2+i]:=(a[2i-1]-a[2i])/√2;
 endfor;
 a:=b; comment move entire array
end;
```

Figure 9.34: Computing the Normalized Wavelet Transform.

```
procedure NWTreconst(a:array of real, n:int);
 j:=2;
 while j≤n do
  NWTRstep(a, j);
  j:=2j;
 endwhile
 a:=a√n; comment multiply entire array
end;

procedure NWTRstep(a:array of real, j:int);
 for i=1 to j/2 do
  b[2i-1]:=(a[i]+a[j/2+i])/√2;
  b[2i]:=(a[i]-a[j/2+i])/√2;
 endfor;
 a:=b; comment move entire array
end;
```

Figure 9.35: Restoring from a Normalized Wavelet Transform.

```
procedure StdCalc(a:array of real, n:int);
 comment array size is nxn (n = power of 2)
 for r=1 to n do
  NWTcalc(row r of a, n);
 endfor;
 for c=n to 1 do comment loop backwards
  NWTcalc(col c of a, n);
 endfor;
end;

procedure StdReconst(a:array of real, n:int);
 for c=n to 1 do comment loop backwards
  NWTreconst(col c of a, n);
 endfor;
 for r=1 to n do
  NWTreconst(row r of a, n);
 endfor;
end;
```

Figure 9.36: The Standard Image Wavelet Transform.

```
procedure NStdCalc(a:array of real, n:int);
 a:=a/√n comment divide entire array
 j:=n;
 while j≥ 2 do
  for r=1 to j do
   NWTstep(row r of a, j);
  endfor;
  for c=j to 1 do comment loop backwards
   NWTstep(col c of a, j);
  endfor;
  j:=j/2;
 endwhile;
end;

procedure NStdReconst(a:array of real, n:int);
 j:=2;
 while j≤n do
  for c=j to 1 do comment loop backwards
   NWTRstep(col c of a, j);
  endfor;
  for r=1 to j do
   NWTRstep(row r of a, j);
  endfor;
  j:=2j;
 endwhile
 a:=a√n; comment multiply entire array
end;
```

Figure 9.37: The Nonstandard Image Wavelet Transform.

The latter method computes the wavelet transform of the image by alternating between rows and columns. The first step is to calculate averages and differences for all the rows (just one iteration, not the entire wavelet transform). This creates averages in the left half of the image, and differences in the right half. The second step is to calculate averages and differences for all the columns, which results in averages in the top-left quadrant of the image and differences elsewhere. Steps 3 and 4 operate on the rows and columns of that quadrant, resulting in averages concentrated in the top-left octant. Pairs of steps continue recursively on smaller and smaller quadrants, until only one average is left, at the top-left corner of the image, and all other pixel values have been reduced to differences. This process is summarized in Figure 9.37.

Either method, standard and nonstandard, results in a transformed—although not yet compressed—image that has one average at the top-left corner and smaller numbers, differences or averages of differences, everywhere else. This can now be compressed by using a combination of methods, such as RLE, move-to-front, and Huffman coding. If lossy compression is acceptable, some of the smallest differences should be replaced by zeros, which may create run lengths of zeros, making the use of RLE more desirable.

Color Images: So far, we have assumed that each pixel is a single number (i.e., we have a single-component image, in which all pixels are shades of the same color, normally gray). Any compression method for single-component images can be extended to color (three-component) images by separating the three components and compressing each individually. If the compression method is lossy, it makes sense to convert the three image components from their original color representation, that is normally RGB, to the YIQ color representation. The Y component of this representation is called *luminance* and the I and Q (the chrominance) components are responsible for the color information. The advantage of this color representation is that the human eye is most sensitive to Y and least sensitive to Q. A lossy method should thus leave the Y component alone and delete some data from the I and more data from the Q components, resulting in good compression and in loss for which the eye is not very sensitive.

It is interesting to note that U.S. color television transmission also takes advantage of the YIQ representation. Signals are broadcast with bandwidths of 4 MHz for Y, 1.5 MHz for I, and only 0.6 MHz for Q.

> As soon as we use words like "image," we are already thinking of how one shape corresponds to the other—of how you might move one shape to bring it into coincidence with the other. Bilateral symmetry means that if you reflect the left half in a mirror, then you obtain the right half. Reflection is a mathematical concept, but it is not a shape, a number, or a formula. It is a *transformation*—that is, a rule for moving things around.
>
> Ian Stewart, *Nature's Numbers*, 1995

10
Short Topics

Several topics are grouped together in this chapter because they are short and because they don't fit in any of the other chapters. It is the author's belief, however, that "short" does not imply "unimportant," so it is recommended that the reader give these topics the same attention given to the "long" ones.

10.1 Graphics Standards

Standards are useful in many aspects of everyday life. One notable example is the automobile. The "user interface" of a car, including the pedals and shift lever, is standard so a person entering a new car finds the same familiar controls. Any car manufacturer that deviates from this standard runs the risk of not being able to sell its products. Standards can also be useful in computer graphics, where so many algorithms and so much useful software is developed and implemented all the time. A graphics standard is a special-purpose programming language that includes commands and procedures for graphics, that is independent of the particular graphics output device used and that is implemented on all common platforms. A piece of software written in such a language is useful since it can easily be executed on different computers without any modifications.

An early attempt to create such a standard is GKS, the *Graphical Kernel System*. It was an international effort, published in 1985 by the ISO (the International Standards Organization) as standard #7942. It has also been adopted by ANSI (the American National Standards Institute) as their standard X3.124–1985 [ANSI 85]. Two detailed references are [Enderle et al. 87] and [Hopgood et al. 86]. Following is a short description of GKS, attempting to convey its "flavor" to the reader. Before we start, it should be mentioned that there are other, more modern standards, the most important of which are PHIGS and POSTSCRIPT (the latter is also described in this section).

GKS is used inside a program and its operation depends on the programming language used. The idea is to write a program to calculate the individual components of an image, then use GKS commands inside the program to actually draw these components. The original GKS definition was developed with Fortran 77 in mind, which is why the GKS commands have a Fortran "flavor."

GKS provides machine-independent notation to specify and manipulate images. An image is constructed from simple building blocks (primitives). GKS offers five primitives plus commands that specify additional attributes for the primitives. A primitive is similar to a procedure call in a typical programming language in the sense that when a primitive is used, the user has to specify the values of *parameters*. In addition to the parameters, the primitives have *attributes*, which are features that don't change very often. They are specified by additional GKS commands. An example of a primitive is text. Each time text is to be displayed, the user has to specify the text itself (the string of characters to be displayed) and its location as parameters, but the font and the text size are attributes since we can expect several consecutive strings of text to have the same font and size. The five primitives are as follows:

1. *polyline*: to draw a sequence of straight, connected line segments.
2. *polymarker*: to draw a sequence of points, all with the same symbol.
3. *fill area*: to fill a polygon with a pattern and draw it.
4. *text*: to display text in different fonts, sizes, and orientations.
5. *cell array*: to display an image in grayscale or color.

Polyline: The general format of this primitive is POLYLINE(N, XPTS, YPTS), where N is the number of points to be connected with straight segments, and XPTS and YPTS are two arrays with the x and y coordinates of the points. Notice that the polyline consists of $N-1$ segments. The polyline is an important primitive since it can be used to draw curves by drawing a large number of short straight segments. Each polyline has attributes that are specified by means of the command

SET POLYLINE REPRESENTATION(WS,PLI,LT,LW,PLCI)

where WS is the platform id (WS stands for WorkStation) and PLI is the polyline index. The idea is to predefine several types of polylines, to be used on different platforms, and to assign each an identifying number (an index). For example, after executing the commands

SET POLYLINE REPRESENTATION(1,3,...solid....)
SET POLYLINE REPRESENTATION(2,3,...dashed...)
SET POLYLINE REPRESENTATION(3,3,...dotted...)

the program can draw a polyline of index 3, and this polyline will be drawn in either solid, dashed, or dotted style depending on whether the current platform has id 1, 2, or 3.

The LT parameter specifies the *line type*, which can be solid (LT=1), dashed (2), dotted (3), or dashed-dotted (4). Other values may be added, but they are nonstandard.

Parameter LW specifies the *linewidth scale factor*. This is a real quantity, giving the width of the polyline segments relative to the width of a standard line on the particular platform used (the width of a standard line is typically one pixel).

PLCI is the color parameter. Instead of specifying the color itself, this parameter is a pointer to an RGB color table.

10 Short Topics

After setting the different polyline representations at the start of the program, a particular representation is selected by the command SET POLYLINE INDEX(N). This command is normally followed by several polylines, following which, another SET POLYLINE INDEX may be used, to select another representation.

Polymarker: The general format of this primitive is POLYMARKER(N, XPTS, YPTS), where N is the number of points to be drawn, and XPTS and YPTS are two arrays with the x and y coordinates of the points. The symbol actually drawn at each points is selected by the command
$$\text{SET POLYMARKER REPRESENTATION(WS,INDX,MT,MS,PMCI)}$$
where WS and INDX are the workstation and index numbers, and PMCI is the color specification. Parameter MS is the marker-size scale factor, a real number specifying the size of the marker relative to the standard marker size on the platform being used. The MT parameter specifies the actual marker symbol to be used. The five standard values of this parameter are
$$1\ \text{``.''},\quad 2\ \text{``+''},\quad 3\ \text{``*''},\quad 4\ \text{``0''},\ \text{and}\ 5\ \text{``×''},$$
and any platform may have its own nonstandard values.

After setting the different marker styles at the start of the program, a particular style is selected by the command SET POLYMARKER INDEX(N).

Fill Area: The general format of this primitive is FILL AREA(N, XPTS, YPTS). There is a SET FILL AREA INDEX(N) command as well as a
$$\text{SET FILL AREA REPRESENTATION(WS,INDX,IS,SI,FACI)}$$
command. The WS and INDX parameters should be familiar by now. FACI is a pointer to a color table. Parameter IS specifies the interior style of the fill area. It can be one of HOLLOW, SOLID, PATTERN, or HATCH, where the last three require more specifications.

Text: The general format of this primitive is TEXT(X,Y,STRING). It draws the string of text with its bottom-left corner at point (X,Y). There are additional commands to specify the font, size, and orientation of the text.

GKS is a large system, including hundreds of commands and specifications, but the above gives an idea of what it is like to use GKS. A three-dimensional GKS standard, called GKS-3D, was also developed. However, developments in computer graphics in the late 1980s and during the 1990s have made GKS obsolete, and today it is rarely used.

PHIGS (Programmer's Hierarchical Interactive Graphics Standard) is a more sophisticated standard [Hopgood and Duce 91], that offers commands for modeling as well as for drawing. It also features hierarchical structure of images and makes it possible to edit individual parts of an image. PHIGS also has capabilities for color specifications and surface rendering. An extension, called PHIGS+, was later developed [Howard et al. 91] to include shading of surfaces.

The *computer graphics interface* (CGI) is a standard for interface methods. The *computer graphics metafile* (CGM) is a standard for image files.

The *Initial Graphics Exchange Specification* (IGES) is the standard for the interchange of design information between CAD/CAM software systems. Its official definition is in [IGES 86].

POSTSCRIPT is a page-description language designed by Adobe Inc. in the mid-1980s to serve as a device-independent language where any desired page can be described independently of any output device. The page can then be displayed or printed on any graphics output device that has a POSTSCRIPT interpreter. The main reference is [Adobe Systems Inc. 90] but the excellent tutorial [Adobe Systems Inc. 85] is perhaps a better place to start.

The interpreter is device dependent (i.e., any POSTSCRIPT output device has to have its own interpreter). It reads a plain text file that includes the POSTSCRIPT description of the page, and converts the POSTSCRIPT commands (which are also called operators) to printer-specific commands. This is how the page can be printed on different printers. The results produced by the printers are not identical since they depend on the printer's resolution, quality, and number of colors. A high-resolution image printed on a low resolution printer will come out in low resolution. Similarly, a color image printed on a black-and-white printer will come out in black and white.

Many current laser printers have a built-in POSTSCRIPT interpreter. The StyleScript software package, from Infowave, Inc. [Infowave Inc. 98], is a POSTSCRIPT interpreter for some inkjet printers. GhostScript, by L. Peter Deutsch [Ghostscript 98] is a free POSTSCRIPT interpreter that can display an image on the computer screen and on dozens of other devices. It has been ported to various platforms.

POSTSCRIPT includes commands that make it possible to specify graphics elements and place them at precise locations on the page. There are three major types of graphics elements: text, geometric figures, and digitized images.

Text: Any string of characters, from any font, can be specified and placed on the page at any location and in any orientation. The text may also follow a curve.

Geometric Figures: Straight segments, curves, and areas can be defined and placed on the page. The areas can be filled with any pattern or color and an area can be clipped to the boundary of any other area.

Sampled Images: These can be obtained by digitizing any drawing, painting, photograph, or any other image. The sampled image can be placed on the page in any orientation and can also be scaled (which normally reduces its quality).

The POSTSCRIPT language uses the following three important terms:

Current Page: This starts blank and gets filled with graphics elements as more and more POSTSCRIPT *painting operators* are being executed. However, the page is not printed until the **showpage** command has been executed. Any graphics object placed on the page obscures anything behind it, since POSTSCRIPT assumes that all colors are opaque. Nothing is transparent or translucent.

Current Path: The **newpath** operator starts a new current path. A POSTSCRIPT path is a set of graphics elements such as points, lines, curves, and areas. It is constructed by POSTSCRIPT's *path operators*. A path does not have to be contiguous on the paper (i.e., it may consist of disconnected parts) and it does not automatically get drawn on the paper. To actually place marks on the page, the user has to stroke the path and fill it. The term *stroke* refers to the edge of the path. It can be thick or thin, it can be in any shade of gray or in any color, and it can be a pattern. The path can be filled with white, black, gray, any color, or any pattern. Figure 10.1 shows examples of paths with different strokes (a)–(c) and fills (d)–(f). The current path is terminated when the POSTSCRIPT interpreter encounters the next **newpath** or a **showpage**.

10 Short Topics

Clipping Path: The initial clipping path is the entire current page. At any time, the user can specify any path as the current clipping path. Following that, when a mark is placed on the page, only those parts of the mark that are inside the clipping path will be drawn and the rest will be clipped.

> I was staring through the cage of those meticulous ink strokes at an absolute beauty.
> — F. Murray Abraham (as Antonio Salieri) in *Amadeus* (1984).

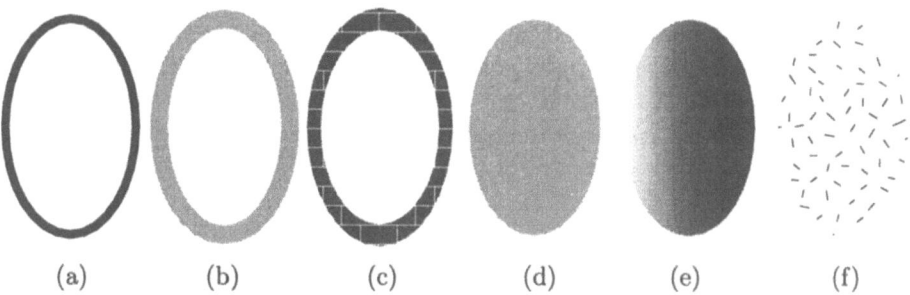

(a)   (b)   (c)   (d)   (e)   (f)

Figure 10.1: Examples of Stroke and Fill.

The reason for the name POSTSCRIPT is that the language uses a stack and the *postfix* notation, where an operator follows its operands. The language is also extensible in the sense that once a new procedure is defined, it can be used as any built-in operator. The following example illustrates the use of the POSTSCRIPT stack.

```
-6
18
add
```

When the POSTSCRIPT interpreter reads the first number −6, it simply pushes it into the stack since it has nothing else to do with it. The same is true for the second number. When the **add** operator is found and is executed, it pops the top two stack elements, adds them, and places the sum at the top of the stack. In general, the rule is that any numbers being input are placed on the stack and that operators look for their operands in the stack, remove them, and place their results on the stack.

Notice, however, that a stack element does not have to be a number. It can be a string of characters, for example.

▶ **Exercise 10.1**: Guess how the following is executed by the POSTSCRIPT interpreter "3.2 11.6 sub"?

The next example defines a path, strokes it, and prints the page:

```
1 newpath
2 72 144 moveto
3 216 72 lineto
4 stroke
5 showpage
```

Line 1 starts a new path that's terminated by line 5. Line 2 pushes the numbers 72 and 144 into the stack and executes moveto. This command pops the top two stack elements, uses them as the (x, y) coordinates of a point on the page, and moves an imaginary pen to that point (without drawing anything). Line 3 is similar, but the lineto operator draws a line as it moves the pen. Line 4 strokes the path, which makes it visible. The precise stroke used depends on the values of several POSTSCRIPT parameters. Finally, line 5 causes the page to be printed (or displayed, if GhostScript or something similar is used). The result is a straight line from $(72, 144)$ to $(216, 72)$. POSTSCRIPT uses a default coordinate system with an origin at the bottom-left corner of the page and with 72 coordinate units per inch. Expressed in inches, the coordinates of the endpoints of our line will therefore be $(1, 2)$ and $(3, 1)$. The origin, orientation, and units of the coordinate system can, of course, be changed by the user.

There are also rmoveto and rlineto operators that consider their operands as relative coordinates. A square, for example, is drawn by the (relatively long) POSTSCRIPT program

```
1 newpath
2 288 288 moveto
3 0 72 rlineto
4 72 0 rlineto
5 0 -72 rlineto
6 -72 0 rlineto
7 4 setlinewidth
8 stroke showpage
```

where lines 3–6 draw four 1-in. segments and line 7 sets the stroke width to four coordinate units (i.e., 4/72 inch when the default is used). Since a square is a closed path, it is better to close it automatically. This is done by replacing the last line segment "-72 0 rlineto" in line 6 with the operator closepath. The square can be filled by saying fill instead of (or in addition to) stroke. The default fill is black, but a fill of 50% gray can be specified by the commands .5 setgray fill.

If a certain image calls for many squares, it is best to define the above program as a *procedure*. Before showing how this is done, we need to discuss the POSTSCRIPT dictionaries. A common language dictionary is a set of pairs where each pair consists of a word and its definition. Similarly, a POSTSCRIPT dictionary is a set of pairs, where each pair consists of a *key* and its *value*. At any time, there are at least two dictionaries: the system dictionary and the user dictionary. The former contains the predefined POSTSCRIPT operators and the latter contains the user-defined procedures and variables.

When the interpreter reads an item from the input file, it searches the user dictionary, then the system dictionary for the item. If it finds the item as a key in one of the dictionaries, it uses the associated value to decide what to do next. If the item is not found, the PostScript interpreter generates an error.

The user may create new user dictionaries and PostScript maintains a dictionary stack. Initially, this stack contains the user dictionary at the top and the system dictionary below it. As more dictionaries are created, they are added at the top of the dictionary stack. The topmost dictionary is called the current dictionary.

If a sequence of PostScript operators is used a lot in a program, it can be defined as a procedure (i.e., it can become a user-defined operator). It is assigned a name and both its name and its definition are stored as the key and value, respectively, of a pair in the current dictionary. Defining a new operator is done by the **def** operator. As an example, we show how to define the sequence of commands for a square as a procedure called **square**. The code is

```
/square
{newpath
moveto
0 72 rlineto
72 0 rlineto
0 -72 rlineto
closepath}
def
```

where the slash "/" in the first line indicates that the string **square** (that is going to be the procedure's name) is to be pushed into the stack. The curled braces are also pushed, with their contents, into the stack. The stack thus contains two items, but notice that neither is a number. The **def** operator on the last line pops two items from the stack and enters them into the current dictionary. The top item becomes the value and the item below it (the string **square**) becomes the key. Once this is done, the string **square** can be used, since it can be found in the current dictionary (i.e., its value is known). An example is

```
72 144 square stroke
288 288 square 4 setlinewidth stroke
0 288 square .5 setgray fill
showpage
```

Notice how each use of **square** is preceded by pushing two numbers into the stack. They can be considered the parameters of procedure **square** and they specify the location of the square on the page.

Curves can be drawn by the operator **x1 y1 x2 y2 x3 y3 curveto**. This draws a cubic Bézier curve segment from the current position of the pen to point (x_3, y_3), using (x_1, y_1) and (x_2, y_2) as the two intermediate control points.

Adobe Illustrator is an example of a graphics program, available for several platforms, that produces its output in PostScript. Another example is **dvips**, by Tomas Rokicki. It translates a **dvi** file, which is the main output produced by the typesetting program TEX, to PostScript.

10.2 Boundary Fill

The problem of boundary fill (also known as polygon fill) is to fill a given area, bounded by pixels of a certain color, with pixels of another color. We denote the boundary pixels by ● and the fill pixels by ⊗. The area can be of any shape, but it must be completely bounded. If the boundary is not complete, the fill algorithm may not know where to stop and it may spill out of the area. To specify the area to the fill program, the user has to point to and select one of the interior pixels. This pixel is called the *seed*.

A straightforward algorithm is the following:

1. Set the seed pixel to the fill color ⊗.
2. Push the four nearest neighbors of the seed into a stack, unless any of them is a boundary pixel ●, or has already been colored.
3. Pop the stack. Set that pixel to the fill color, and push its four near neighbors, as indicated in step 2, into the stack.
4. Repeat step 3 until the stack is empty.

This method is slow because of the excessive use of the stack. Also, it may easily push thousands of pixels into the stack and overflow it. A better algorithm is outlined below. It still uses a stack, but it works with rows of pixels rather than individual ones. The stack is used to indicate future rows to be filled, instead of future pixels, so its use is not excessive.

Consider the area in Figure 10.2a. The seed pixel is shown as ⊙. The algorithm is the following:

1. Fill the line where the seed pixel is located with the fill color ⊗. Scanning left and right of the seed as far as necessary until a boundary pixel ● is reached (Figure 10.2b).
2. Examine the two rows right above and below the current row (if there are none, go to step 3). Scan each of the two rows from right to left looking for pixels that lie immediately to the left of a boundary pixel and that haven't been colored already. Those pixels are pushed into the stack. Figure 10.2c shows three such pixels, labeled in the order in which they were found.
3. The pixel at the top of the stack is popped out (if the stack is empty, the algorithm terminates) and steps 1 and 2 are repeated on that pixel. The result is shown in Figure 10.2d, where two new pixels, numbered 3 and 4, have been pushed into the stack. Repeating steps 2 and 3 results in Figure 10.2e. After a few more repetitions, the situation (which is now much advanced) is as shown in Figure 10.2f.

It should be an easy exercise to apply this algorithm to the rest of the example and fill up the entire figure. Note that the region to be filled may include holes and may also be concave, but it should be connected. For the purposes of boundary fill, a region such as in Figure 10.2g is considered two separate, disconnected areas, consisting of five pixels each.

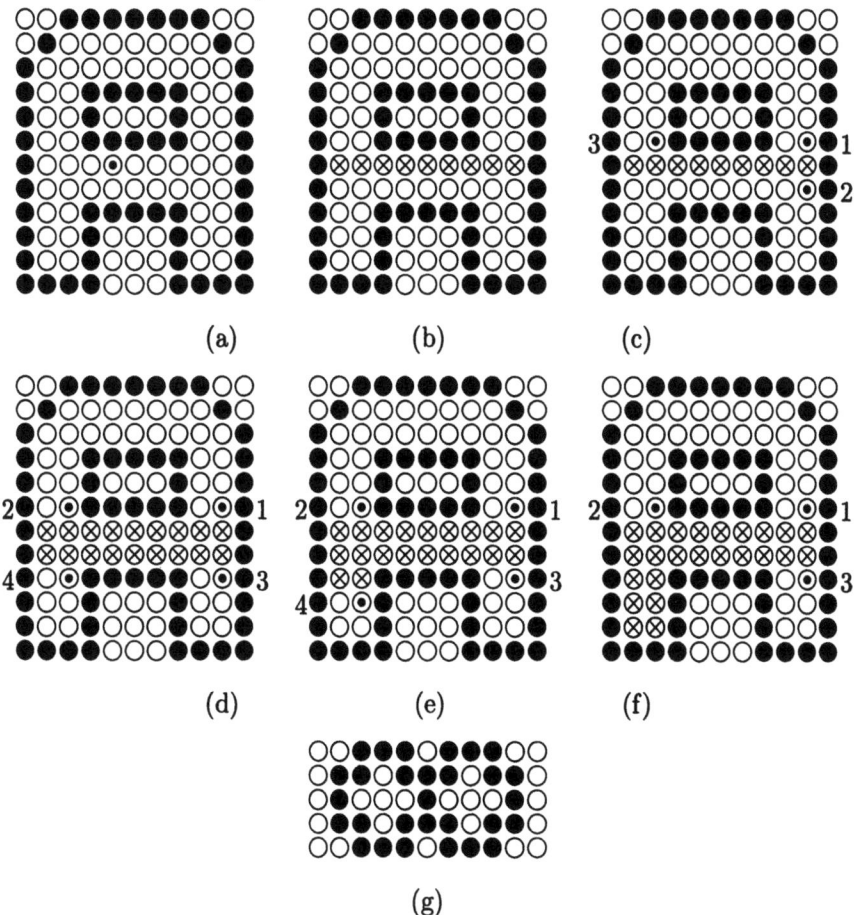

Figure 10.2: Boundary Fill.

10.3 Halftoning

Color CRTs are common nowadays, but the really high-resolution CRTs are normally monochromatic. Also, most high-resolution, high-quality printers are black and white. Halftoning is a method that makes it possible to display images with shades of gray on a black and white (i.e., bilevel) output device. The trade-off is loss of resolution. Instead of small, individual pixels, halftoning uses groups of pixels where only some of the pixels in a group are black. Halftoning is important since it makes it possible to print pictures consisting of more than just black and white on a black and white printer. It is commonly used in newspapers and books. A classic reference is [Ulichney 87].

The human eye can resolve details as small as 1 minute of arc under normal light conditions. This is called the *visual acuity*. If we view a very small area from a normal viewing distance, our eyes cannot see the details in the area and end up integrating them, such that we only see an average intensity coming from the area. This property is called *spatial integration* and is very nicely demonstrated by

Figure 10.3. The figure consists of black circles and dots on a white background, but spatial integration creates the effect of a gray background.

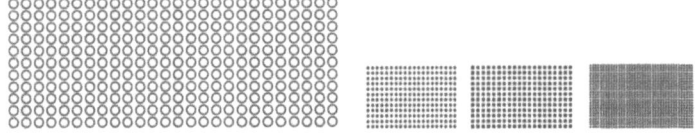

Figure 10.3: Gray Backgrounds Obtained by Spatial Integration.

The principle of halftoning is to use groups of $n \times n$ pixels (with n usually in the range 2–4) and to set some of the pixels in a group to black. Depending on the black-to-white percentage of pixels in a group, the group appears to have a certain shade of gray. An $n \times n$ group of pixels contains n^2 pixels and can therefore provide $n^2 + 1$ levels of gray. The only practical problem is to find the best pattern for each of those levels. The $n^2 + 1$ pixel patterns selected must satisfy the following conditions:

1. Areas covered by copies of the same pattern should not show any textures.

2. Any pixel set to black for pattern k must also be black in all patterns of intensity levels $> k$. This is considered a good *growth sequence* and it minimizes the differences between patterns of successive intensities.

3. The patterns should grow from the center of the $n \times n$ area, to create the effect of a growing dot.

4. All the black pixels of a pattern must be adjacent to each other. This property is called *clustered dot halftoning* and is important if the output is intended for a printer (laser printers cannot always fully reproduce small isolated dots). If the output is intended for CRT only, then *dispersed dot halftoning* can be used, where the black pixels of a pattern are not adjacent.

As a simple example of condition 1, a pattern such as

must be avoided, since large areas with level-3 groups would produce long horizontal lines. Other patterns may result in similar, annoying textures. With a 2×2 group, such effects may be impossible to avoid. The best that can be done is to use the patterns ▢▢ ▨▢ ▨▨ ▨▨ ▨▨ .

A 3×3 group provides for more possibilities. The 10 patterns below ($= 3^2 + 1$) are the best ones possible (reflections and rotations of these patterns are considered identical) and usually avoid the problem above. They were produced by the matrix

$$\begin{bmatrix} 7 & 9 & 5 \\ 2 & 1 & 4 \\ 6 & 3 & 8 \end{bmatrix}$$

using the following rule: To create a group with intensity n, only cells with values $\leq n$ in the above matrix should be black.

The halftone method is not limited to a monochromatic display. Imagine a display with four levels of gray per pixel (2-bit pixels). Each pixel is either black or can be in one of three other levels. A 2×2 group consists of four pixels, each of which can be in one of three levels of gray or in black. The total number of levels is therefore $4 \times 3 + 1 = 13$. One possible set of the 13 patterns is shown below.

10.4 Dithering

The downside of halftoning is loss of resolution. It is also possible to display continuous-tone images (i.e., images with different shades of gray) on a bilevel device *without* losing resolution. Such methods are sometimes called *dithering* and their trade-off is loss of image detail. If the device resolution is high enough and if the image is watched from a suitable distance, then our eyes perceive an image in grayscale, but with fewer details than in the original.

The dithering problem can be phrased as follows: given an $m \times n$ array A of pixels in grayscale, calculate an array B of the same size with zeros and ones (corresponding to white and black pixels, respectively) such that for every pixel $B[i,j]$ the average value of the pixel and a group of its near neighbors will approximately equal the normalized value of $A[i,j]$. (Assume that pixel $A[i,j]$ has an integer value I in the interval $[0,a]$; then its normalized value is the fraction I/a. It is in the interval $[0,1]$.)

The simplest dithering method uses a threshold and the single test: Set $B[i,j]$ to white (0) if $A[i,j]$ is bright enough (i.e., less than the value of the threshold); otherwise, set $B[i,j]$ to black (1). This method is fast and simple but generates very poor results, as the next example shows, so it is never used in practice. As an example, imagine a human head. The hair is generally darker than the face below it, so the simple threshold method may quantize the entire hair area to black and the entire face area to white, a very poor, unrecognizable, and unacceptable result. (It should be noted, however, that some images are instantly recognizable even in just black and white, as Figure 10.4 aptly demonstrates.) This method can be improved a little by using a different, random threshold for each pixel, but even this produces low-quality results.

Four approaches to dithering, namely ordered dither, constrained average dithering, diffusion dither, and dot diffusion, are presented in this section. Another approach, called ARIES, is discussed in [Roetling 76] and [Roetling 77].

10.4.1 Ordered Dither

The principle of this method is to paint a pixel $B[i,j]$ black or leave it white, depending on the intensity of pixel $A[i,j]$ **and** on its position in the picture [i.e., on its coordinates (i,j)]. If $A[i,j]$ is a dark shade of gray, then $B[i,j]$ should ideally be dark thus, it is painted black most of the time, but sometimes it is left white. The decision whether to paint it black or white depends on its coordinates i and j. The opposite is true for a bright pixel. This method is described in [Jarvis et al. 76].

Figure 10.4: A Familiar Black-and-White Image.

> The giant panda resembles a bear, although anatomically it is more like a raccoon. It lives in the high bamboo forests of central China. Its body is mostly white, with black limbs, ears, and eye patches. Adults weigh 200 to 300 lb (90 to140 kg).

The method starts with an $m\times n$ dithering matrix D_{mn} which is used to determine the color (black = 1 or white = 0) of all the B pixels. In the example below we assume that the A pixels have 16 gray levels, with 0 as white and 15 as black. The dithering matrices for $n = 2$ and $n = 4$ are shown below. The idea in these matrices is to minimize the amount of texture in areas with a uniform gray level.

$$D_{22} = \begin{bmatrix} 0 & 2 \\ 3 & 1 \end{bmatrix}, \qquad D_{44} = \begin{bmatrix} 0 & 8 & 2 & 10 \\ 12 & 4 & 14 & 6 \\ 3 & 11 & 1 & 9 \\ 15 & 7 & 13 & 5 \end{bmatrix}.$$

The rule is: Given a pixel $A[x,y]$ calculate $i = x \bmod m$, $j = y \bmod n$, then select black (i.e., set $B[x,y]$ to 1) if $A[x,y] \geq D_{mn}[i,j]$ and select white otherwise.

To see how the dithering matrix is used, imagine a large, uniform area in the image A where all the pixels have a gray level of 4. Since 4 is the fifth of 16 levels, we would like to end up with 5/16 of the pixels in the area set to black (ideally they should be randomly distributed in this area). When a row of pixels is scanned in this area, y is incremented, but x does not change. Since i depends on x, and j depends on y, the pixels scanned are compared to one of the rows of matrix D_{44}. If this happens to be the first row, then we end up with the sequence 10101010... in bitmap B.

10 Short Topics

When the next line of pixels is scanned, x and, as a result, i have been incremented, so we look at the next row of D_{44}, that produces the pattern 01000100... in B. The final result is an area in B that looks like

$$10101010...$$
$$01000100...$$
$$10101010...$$
$$00000000...$$

Ten out of the 32 pixels are black, but $10/32 = 5/16$. The black pixels are not randomly distributed in the area, but their distribution does not create annoying patterns either.

▶ **Exercise 10.2:** Assume that image A has three large uniform areas with gray levels 0, 1, and 15 and calculate the pixels that go into bitmap B for these areas.

Ordered dither is easy to understand if we visualize copies of the dither matrix laid next to each other on top of the bitmap. Figure 10.5 shows a 6×12 bitmap with six copies of a 4×4 dither matrix laid on top of it. The threshold for dithering a pixel $A[i,j]$ is that element of the dither matrix that happens to lie on top of $A[i,j]$.

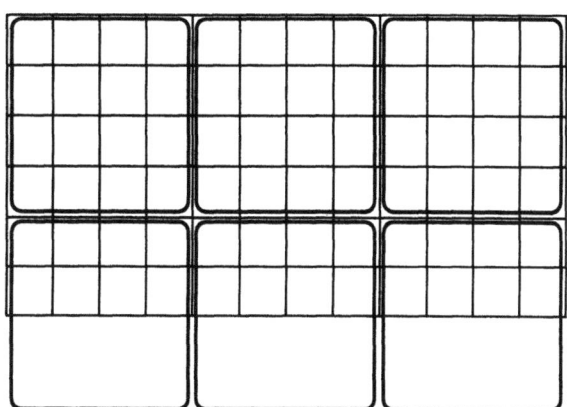

Figure 10.5: Ordered Dither.

Matrix D_{44} above was created from D_{22} by the recursive rule

$$D_{nn} = \begin{pmatrix} 4D_{n/2,n/2} & 4D_{n/2,n/2} + 2U_{n/2,n/2} \\ 4D_{n/2,n/2} + 3U_{n/2,n/2} & 4D_{n/2,n/2} + U_{n/2,n/2} \end{pmatrix}, \qquad (10.1)$$

where U_{nn} is an $n \times n$ matrix with all ones. Other matrices are easy to generate with this rule.

▶ **Exercise 10.3:** Use the rule of Equation (10.1) to construct D_{88}.

The basic rule of ordered dither can be generalized as follows: Given a pixel $A[x,y]$, calculate $i = x \bmod m$ and $j = y \bmod n$, then select black (i.e., assign

$B[x,y] \leftarrow 1)$ if $Ave[x,y] \geq D_{mn}[i,j]$, where $Ave[x,y]$ is the average of the 3×3 group of pixels centered on $A[x,y]$. This is computationally more intensive but tends to produce better results in most cases since it considers the average brightness of a group of pixels.

Ordered dither is a simple, fast method, but it tends to create images that have been described by various people as "computerized," "cold," or "artificial." The reason for that is probably the recursive nature of the dithering matrix.

10.4.2 Constrained Average Dithering

In cases where high speed is not important, this method [Jarvis and Roberts 76] gives good results, although it involves more computations than ordered dither. The idea is to compute, for each pixel $A[i,j]$, the average $\bar{A}[i,j]$ of the pixel and its eight near neighbors. The pixel is then compared to a threshold of the form

$$\gamma + \left(1 - \frac{2\gamma}{M}\right) \bar{A}[i,j],$$

where γ is a user-selected parameter and M is the maximum value of $A[i,j]$. Notice that the threshold can have values in the range $[\gamma, M - \gamma]$. The final step is to compare $A[i,j]$ to the threshold and set $B[i,j]$ to 1 if $A[i,j] \geq$ threshold and to 0 otherwise.

The main advantage of this method is edge enhancement. An example is Figure 10.6 that shows a 6×8 bitmap A, where pixels have 4-bit values. Most pixels have a value of 0, but the bitmap also contains a thick slanted line indicated in the figure. It separates the 0 (white) pixels in the upper-left and bottom-right corners from the darker pixels in the middle.

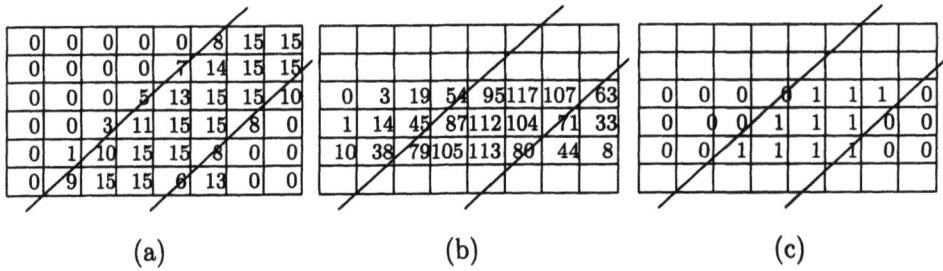

Figure 10.6: Constrained Average Dithering.

Figure 10.6a shows how the pixels around the line have values approaching the maximum (which is 15). Figure 10.6b shows (for some pixels) the average of the pixel and its eight near neighbors (the averages are shown as integers, so an average of 54 really indicates 54/9). The result of comparing these averages to the threshold (which in this example is 75) is shown in Figure 10.6c. It is easy to see how the thick line is sharply defined in the bilevel image.

10 Short Topics

> So I'm a ditherer? Well, I'm jolly well going to dither, then!
> — Roland Young (as Cosmo Topper) in *Topper* (1937).

10.4.3 Diffusion Dither

Imagine a photograph, rich in color, being digitized by a scanner that can distinguish millions of colors. The result may be an image file where each pixel $A[i,j]$ is represented by, say, 24 bits. The pixel may have one of 2^{24} colors. Now, imagine that we want to display this file on a computer that can only display 256 colors simultaneously on the screen. A good example is a computer using a color lookup table whose size is 3×256 bytes (page 8).

We begin by loading a palette of 256 colors into the lookup table (see Section 6.4 for a good method to select such a palette). Each pixel $A[i,j]$ of the original image will now have to be displayed on the screen as a pixel $B[i,j]$ in 1 of the 256 palette colors. The diagram below shows a pixel $A[i,j]$ with original color $(255, 52, 80)$. If we decide to assign pixel $B[i,j]$ the palette color $(207, 62, 86)$, then we are left with a difference of $A[i,j] - B[i,j] = (48, -10, -6)$. This difference is called the *color error* of the pixel.

R=255		R=207		R=48
G=52	−	G=62	=	G=−10
B=80		B=86		B=−6

Large color errors degrade the quality of the displayed image, so an algorithm is needed to minimize the total color error of the image. Diffusion dither does this by distributing the color errors among all the pixels such that the total color error for the entire image is zero (or very close to zero).

The algorithm is very simple. Pixels $A[i,j]$ are scanned line by line from top to bottom. In each line, they are scanned from left to right. For each pixel, the algorithm performs the following:

1. Pick up the palette color that's nearest the original pixel's color. This palette color is stored in the destination bitmap $B[i,j]$.
2. Calculate the color error $A[i,j] - B[i,j]$ for the pixel.
3. Distribute this error to four of $A[i,j]$'s nearest neighbors that haven't been scanned yet (the one on the right and the three ones centered below) according to the *Floyd-Steinberg filter* [Floyd and Steinberg 75] (where the X represents the current pixel):

	X	7/16
3/16	5/16	1/16

Consider the example of Figure 10.7a. The current pixel is $(255, 52, 80)$ and we (arbitrarily) assume that the nearest palette color is $(207, 62, 86)$. The color error is $(48, -10, -6)$ and is distributed as shown in Figure 10.7c. The five nearest neighbors are assigned new colors as shown in Figure 10.7b. The algorithm is shown in Figure 10.8a where the weights $p1$, $p2$, $p3$, and $p4$ can be assigned either the Floyd-Steinberg values 7/16, 3/16, 5/16, and 1/16 or any other values.

Before	R=255 G=52 B=80	R=178 G=20 B=60
R=192 G=45 B=75	R=250 G=49 B=83	R=191 G=31 B=72

(a)

After	R=207 G=62 B=86	R=199 G=16 B=57
R=201 G=43 B=74	R=265 G=46 B=81	R=194 G=30 B=72

(b)

$\frac{7}{16} \times 48 = 21, \quad \frac{1}{16} \times 48 = 3, \quad \frac{5}{16} \times 48 = 15, \quad \frac{3}{16} \times 48 = 9,$

$\frac{7}{16} \times (-10) = -4, \quad \frac{1}{16} \times (-10) = -1, \quad \frac{5}{16} \times (-10) = -3, \quad \frac{3}{16} \times (-10) = -2,$

$\frac{7}{16} \times (-6) = -3, \quad \frac{1}{16} \times (-6) = 0, \quad \frac{5}{16} \times (-6) = -2, \quad \frac{3}{16} \times (-6) = -1.$

(c)

Figure 10.7: Diffusion Dither.

```
for i := 1 to m do
  for j := 1 to n do
  begin
    B[i,j]:=SearchPalette(A[i,j]);

    err := A[i,j] − B[i,j];
    A[i, j + 1] := A[i, j + 1] + err ∗ p1;
    A[i + 1, j − 1] := A[i + 1, j − 1] + err ∗ p2;
    A[i + 1, j] := A[i + 1, j] + err ∗ p3;
    A[i + 1, j + 1] := A[i + 1, j + 1] + err ∗ p4;
  end.
```

(a)

```
for i := 1 to m do
  for j := 1 to n do
  begin
    if A[i,j] < 0.5 then B[i,j] := 0
      else B[i,j] := 1;
    err := A[i,j] − B[i,j];
    A[i, j + 1] := A[i, j + 1] + err ∗ p1;
    A[i + 1, j − 1] := A[i + 1, j − 1] + err ∗ p2;
    A[i + 1, j] := A[i + 1, j] + err ∗ p3;
    A[i + 1, j + 1] := A[i + 1, j + 1] + err ∗ p4;
  end.
```

(b)

Figure 10.8: Diffusion Dither Algorithm. (a) For Color. (b) For Bilevel.

The total color error may not be exactly zero because the method does not work well for the leftmost column and for the bottom row of pixels. However, the results can be quite good if the palette colors are carefully selected.

This method can easily be applied to the case of a monochromatic display (or any *bilevel* output device), as shown by the pseudo-code of Figure 10.8b, where $p1$, $p2$, $p3$, and $p4$ are the four error diffusion parameters. They can be the ones already given (i.e., 7/16, 3/16, 5/16, and 1/16) or different ones, but their sum should be 1.

▶ **Exercise 10.4:** Consider an all-gray image, where $A[i, j]$ is a real number in the range $[0, 1]$ and it equals 0.5 for all pixels. What image B would be generated by diffusion dither in this case?

▶ **Exercise 10.5:** Imagine a grayscale image consisting of a single row of pixels where pixels have real values in the range $[0, 1]$. The value of each pixel p is compared to the threshold value of 0.5 and the error is propagated to the neighbor on the right. Show the result of dithering a row of pixels all with values 0.5.

Error diffusion can also be used for color printing. A typical low-end ink-jet color printer has four ink cartridges for cyan, magenta, yellow, and black ink. The printer places dots of ink on the page such that each dot has one of the four colors. If a certain area on the page should have color L, where L isn't any of CMYK, then L can be simulated by dithering. This is done by printing adjacent dots in the area with CMYK colors such that the eye (which integrates colors in a small area) will perceive color L. This can be done with error diffusion where the palette consists of the four colors cyan $(255, 0, 0)$, magenta $(0, 255, 0)$, yellow $(0, 0, 255)$, and black $(255, 255, 255)$ and the error for a pixel is the difference between the pixel color and the nearest palette color.

A slightly simpler version of error diffusion is the *minimized average error* method. The errors are not propagated but rather calculated and saved in a separate table. When a pixel $A[x, y]$ is examined, the error table is used to look up the errors $E[x + i, y + j]$ already computed for some previously seen neighbors $A[i, j]$ of the pixel. Pixel $B[x, y]$ is assigned a value of 0 or 1 depending on the corrected intensity:

$$A[x,y] + \frac{1}{\sum_{ij} \alpha_{ij}} \sum_{ij} \alpha_{ij} E[x+i, y+j].$$

The new error, $E[x, y] = A[x, y] - B[x, y]$, is then added to the error table to be used for future pixels. The quantities α_{ij} are weights assigned to the near neighbors of $A[x, y]$. They can be assigned in many different ways, but they should assign more weight to nearby neighbors, so the following is a typical example:

$$\alpha = \begin{pmatrix} 1 & 3 & 5 & 3 & 1 \\ 3 & 5 & 7 & 5 & 3 \\ 5 & 7 & x & - & - \end{pmatrix},$$

where x is the current pixel $A[x, y]$ and the weights are defined for some previously seen neighbors above and to the left of $A[x, y]$. If the weights add up to 1, then the corrected intensity above is simplified and becomes

$$A[x,y] + \sum_{ij} \alpha_{ij} E[x+i, y+j].$$

Floyd-Steinberg error diffusion generally produces better results than ordered dither but has two drawbacks, namely it is serial in nature and it sometimes produces annoying "ghosts." Diffusion dither is serial since the near neighbors of $B[i,j]$ cannot be calculated until the calculation of $B[i,j]$ is complete and the error $A[i,j] - B[i,j]$ has been distributed to the four near neighbors of $A[i,j]$. To understand why ghosts are created, imagine a dark area positioned above a bright area (for example, a dark sky above a bright sea). When the algorithm works on the last dark A pixels, a lot of error is distributed below, to the first bright A pixels (and also to the right). When the algorithm gets to the first bright pixels, they have collected so much error from above that they may no longer be bright, creating perhaps several rows of dark B pixels. It has been found experimentally that ghosts can be "exorcised" by scaling the A pixels before the algorithm starts. For example, each $A[i,j]$ pixel can be replaced by $0.1 + 0.8A[i,j]$, which "softens" the differences in brightness (the contrast) between the dark and bright pixels, thereby reducing the ghosts. This solution also changes all the pixel intensities, but the eye is less sensitive to absolute intensities than to changes in contrast, so changing intensities may be acceptable in many practical situations.

10.4.4 Dot Diffusion

This section is based on [Knuth 87], a very detailed article containing thorough analysis and actual images dithered using several different methods. The dot diffusion algorithm is somewhat similar to diffusion dither, it also produces good quality, sharp bilevel images, but it is not serial in nature and may be easier to implement on a parallel computer.

We start with the 8×8 *class matrix* of Figure 10.9a. The way this matrix was constructed will be discussed later. For now, we simply consider it a permutation of the integers $(0, 1, \ldots, 63)$, which we call *classes*. The class number k of a pixel $A[i,j]$ is found at position (i,j) of the class matrix. The main algorithm is shown in Figure 10.10.

The algorithm computes all the pixels of class 0 first, then those of class 1, and so on. Procedure Distribute is called for every class k and diffuses the error *err* to those near neighbors of $A[i,j]$ whose class numbers exceed k. The algorithm distinguishes between the four orthogonal neighbors and the four diagonal neighbors of $A[i,j]$. If a neighbor is $A[u,v]$, then the former type satisfies $(u-i)^2 + (v-j)^2 = 1$, while the latter type is identified by $(u-i)^2 + (v-j)^2 = 2$. It is reasonable to distribute more of the error to the orthogonal neighbors than to the diagonal ones, so a possible weight function is weight$(x,y) = 3 - x^2 - y^2$. For an orthogonal neighbor, either $(u-i)$ or $(v-j)$ equals 1, so weight$(u-i, v-j) = 2$, while for a diagonal neighbor, both $(u-i)$ and $(v-j)$ equal 1, so weight$(u-i, v-j) = 1$. Procedure Distribute is listed in pseudo-code in Figure 10.10.

Once the coordinates (i,j) of a pixel $A[i,j]$ are known, the class matrix gives the pixel's class number k that is independent of the color (or brightness) of the

10 Short Topics

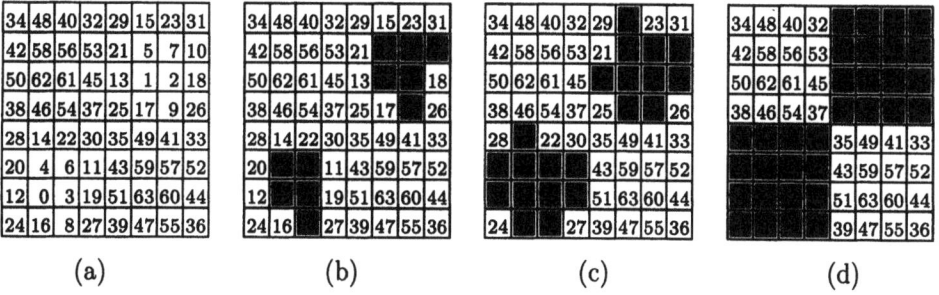

Figure 10.9: 8×8 Matrices for Dot Diffusion.

```
for k := 0 to 63 do
 for all (i,j) of class k do
  begin
  if A[i,j] < .5 then B[i,j] := 0 else B[i,j] := 1;
  err := A[i,j] − B[i,j];
  Distribute(err,i,j,k);
  end.

procedure Distribute(err,i,j,k);
 w := 0;
 for all neighbors A[u,v] of A[i,j] do
  if class(u,v) > k then w := w+weight(u − i, v − j);
 if w > 0 then for all neighbors A[u,v] of A[i,j] do
  if class(u,v) > k then A[u,v] := A[u,v] + err×weight(u − i, v − j)/w;
end;
```

Figure 10.10: The Dot Diffusion Algorithm.

pixel. The class matrix also gives the classes of the eight near neighbors of $A[i,j]$, so those neighbors whose classes exceed k can be selected and linked in a list. It is a good idea to construct those lists once and for all, since this speeds up the algorithm considerably.

It remains to show how the class matrix, Figure 10.9a, was constructed. The main consideration is the relative positioning of small and large classes. Imagine a large class surrounded, in the class matrix, by smaller classes. An example is class 63, which is surrounded by the "lower classes" 43, 59, 57, 51, 60, 39, 47, and 55. A little thinking shows that as the algorithm iterates toward 63, more and more error is absorbed into pixels that belong to this class, regardless of their brightness. A large class surrounded by lower classes is therefore undesirable and may be called a "baron." The class matrix of Figure 10.9a has just two barons. Similarly, "near-baron" positions, which have only one higher-class neighbor, are undesirable and should be avoided. Our class matrix has just two of them.

▶ **Exercise 10.6:** What are the barons and near-barons of our class matrix?

▶ **Exercise 10.7:** Consider an all-gray image where $A[i,j] = 0.5$ for all pixels. What image B would be generated by dot diffusion in this case?

Another important consideration is the positions of consecutive classes in the class matrix. Figure 10.9b,c,d show the class matrix after 10, 21, and 32 of its lowest classes have been blackened. It is easy to see how the black areas form 45° grids that grow and eventually form a 2×2 checkerboard. This helps create diagonal, rather than rectilinear dot patterns in the bilevel array B, and we know from experience that such patterns are less noticeable to the eye. Figure 10.11a shows a class matrix with just one baron and one near-baron, but it is easy to see how the lower classes are mostly concentrated at the bottom-left corner of the matrix.

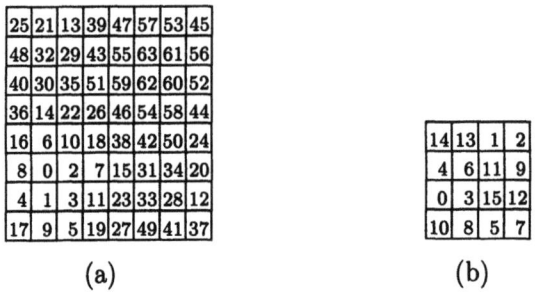

Figure 10.11: Two Class Matrices for Dot Diffusion.

A close examination of the class matrix shows that the class numbers in positions (i,j) and $(i,j+4)$ always add up to 63. This means that the grid pattern of $63-k$ white pixels after k steps is identical to the grid pattern of $63-k$ black pixels after $63-k$ steps, shifted right four positions. This relation between the dot pattern and the diffusion pattern is the reason for the name *dot diffusion*.

▶ **Exercise 10.8:** Figure 10.11b shows a 4 × 4 class matrix. Identify the barons, near-barons, and grid patterns.

Experiments with the four methods described in this section seem to indicate that the dot diffusion method produces best results for printing because it tends to generate contiguous areas of black pixels, rather than "checkerboard" areas of alternating black and white. Modern laser and ink-jet printers have resolutions of 600 dpi or more, but they generally cannot produce a high-quality checkerboard of 300 black and 300 white alternating pixels per inch.

10.5 Fractals

A fractal is a mathematical object that has detailed structure no matter how closely we look at it, or how much we magnify it. Many different types of fractals are known, but all are defined by means of infinite recursion, infinite sums, or other operations that are performed infinitely many times. Certain types of fractals are useful in computer graphics since they can be used to create complex patterns such as terrain, clouds, and dust.

[Barnsley 93] is a good introduction to fractals. The following sections discuss several types of fractals.

10.6 A Fractal Line

This is perhaps the simplest example of a fractal. A straight-line segment (a, b) is converted to a "rough" path by the following process:

1. Find the perpendicular bisector of the segment (Equation (10.2)).
2. Draw a random number t and locate the point m on the bisector at distance t from the segment.
3. Replace the original segment with the two segments (a, m) and (m, b).
4. Repeat the process recursively on the two segments, to obtain four segments. The recursion stops when a segment is created whose length is smaller than a certain threshold parameter (or after a predetermined number of steps).

Note that Equation (10.2) allows for positive and negative values of t. Our random numbers t should, therefore, be signed. Also, most should be small, with only a few large ones. This suggests using random numbers that have a *Gaussian distribution* (Section 10.6.1) instead of a uniform distribution. Traditionally, a Gaussian distribution with zero mean and a standard deviation of 1 is used. The program multiplies each random number by a factor f (which amounts to scaling the standard deviation by f). Small values of f result in a jagged line close to the original one. The *persistence H* of the curve is defined as

$$f = 2^{(0.5-H)}.$$

Figure 10.12 shows three steps in the line fractalization process. In each step, the original line is shown as solid, and the two new ones are shown as dashed segments. After three steps, there are eight segments randomly placed around the original line. After just 10 steps, 1024 segments are created. After 20 steps, there are more than a million segments. The result is a rough, meandering path, similar to the one created by Brownian motion.

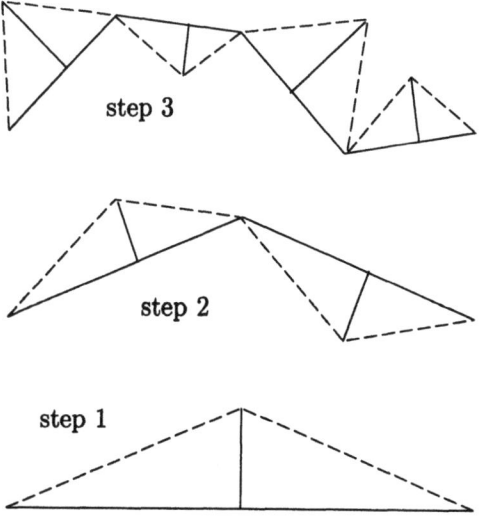

Figure 10.12: Fractalizing a Line.

This fractalization process can easily be extended to two dimensions. When applied to a flat plane, it results in a polygonal surface resembling rough terrain. It is also possible to create a mountain-shaped polygonal surface this way.

10.6.1 Gaussian Distribution

The Gaussian (also known as the Normal) distribution is an important statistical tool used in many branches of science. It provides a good model for continuous distributions that occur in many everyday situations. Examples are the following:

1. The distribution of peoples' heights. Most people are of medium height. Few are tall or short. Even fewer are very tall or very short. Practically no one is a giant or a dwarf. Imagine a sample of people whose height is known. If the sample is large enough and is not biased, the graph describing the number of people of height h as a function of h will look very similar to Figure 10.13.

2. The speed of gas molecules. The molecules of a gas are in constant motion. They move randomly, collide with each other and with objects around them, and change their velocities all the time. However, most molecules in a given volume of gas move at about the same speed, and only a few move much faster or much slower than this speed. This speed is related to the temperature of the gas. The higher this average speed, the hotter the gas feels to us.

3. The results of throwing two dice are distributed normally. Each die yields a number in the range $[1, 6]$, so throwing two dice yields a number in the range $[2, 12]$. A result of 2 can only be obtained if both dice happen to fall on 1. A result of 12, similarly, is only obtained if both dice fall on 6. A result of 6, however, is much more common since it is obtained when the two dice yield (1,5), (5,1), (2,4), (4,2), or (3,3).

The Gaussian distribution with mean m and standard deviation s is defined as

$$f(x) = \frac{1}{s\sqrt{2\pi}} \exp\left\{-\frac{1}{2}\left(\frac{x-m}{s}\right)^2\right\}.$$

This function has a maximum for $x = m$ (at the mean), where its value is $f(m) = 1/(s\sqrt{2\pi})$. It is also symmetric about $x = m$ since it depends on x according to $(x-m)^2$. It has the general "bell" shape of Figure 10.13. At $x = m + s$ and $x = m - s$, its value is

$$f(m \pm s) = \frac{1}{s\sqrt{2\pi}} e^{-\frac{1}{2}} \approx \frac{0.6065}{s\sqrt{2\pi}},$$

which means that at one standard deviation from the mean, it has dropped to about 60% of its maximum. At two standard deviations, it drops to about 0.1353 of its maximum value.

The total area under the Normal curve is one unit. The area one standard deviation from the mean equals 0.682. At two standard deviations, it equals 0.9545, and at three standard deviations, it is approximately 0.9973.

The precise shape of the curve depends on s. As s increases, the curve gets wider. For small values of s, the curve approaches a narrow spike of height $1/(s\sqrt{2\pi})$ placed at $x = m$.

10 Short Topics

When we talk about random numbers, we usually mean numbers that are uniformly distributed over a certain interval $[a, b]$. If we divide interval $[a, b]$ into equal-size subintervals, any of them would contain the same amount of random numbers. It is possible to draw random numbers that have different distributions, such as Gaussian. When we compute many random numbers that are normally (i.e., Gaussian) distributed with mean m and standard deviation s, then count the amount y of these numbers in a small subinterval $[x, x + \epsilon]$, plot (x, y) as a point, and repeat for many subintervals, we get the Normal distribution with mean m and standard deviation s.

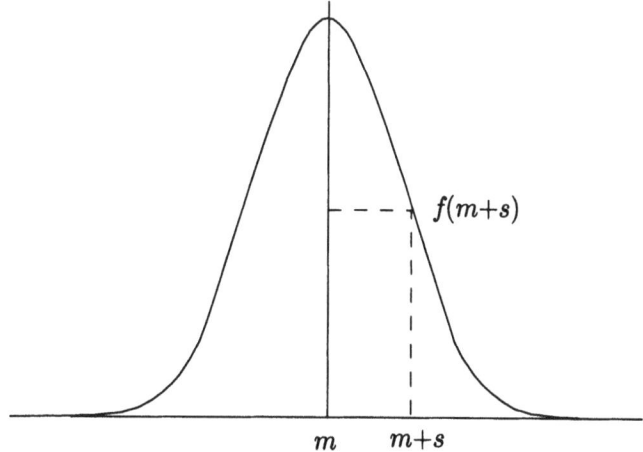

Figure 10.13: Gaussian (Normal) Distribution.

Here are two ways to compute random numbers that are normally distributed with mean 0 and standard deviation 1.

1. Draw n uniformly distributed random numbers R_i in the range $[-a, +a]$ for any real a and calculate their average $(1/n) \sum R_i$. This is the first of the normally distributed random numbers N_i. Repeat this process to get N_2, N_3, and so on. The larger n, the closer to Normal will be the distribution of these numbers. The reason that the N_i are normally distributed is that it is rare for the average of R_i to be $-a$ or $+a$ or close to these values, but it is common for it to be around 0. This is an aspect of the *law of large numbers* that says: If R_i are random numbers of any distribution, then the averages $(1/n) \sum R_i$ are normally distributed.

2. Method 1 is simple but slow since n should be large. The *Polar method* (see [Knuth 81] Vol. 2, Sec. 3.4.1) is more efficient. Let U_1 and U_2 be two uniformly distributed random numbers in the range $[0, 1]$. We calculate two normally-distributed random numbers X_1, X_2 by the following two simple steps:

Step 1. Compute $V_1 := 2U_1 - 1$, $V_2 := 2U_2 - 1$, and $S := V_1^2 + V_2^2$.

Step 2. If $S \geq 1$, go to step 1; else compute $X_1 := V_1 \sqrt{\frac{-2 \ln S}{S}}$, $X_2 := V_2 \sqrt{\frac{-2 \ln S}{S}}$.

Once a sequence N_i is obtained of normally distributed random numbers with mean 0 and standard deviation 1, it is easy to convert them to normally distributed random numbers with mean m and standard deviation s. Just transform each N_i to $m + N_i \times s$.

Assuming that a function Rnd(), which returns uniformly distributed random numbers in the range $[0, 1]$, is given. Gaussian random numbers with 0 mean and a standard deviation of 1 can be obtained by the following:

```
x:=0.0;
for i:=1 to 12 do x:=x+Rnd();
Gauss:=x-6.0;
```

> Because of the very nature of the tables, it did not seem necessary to proofread every page of the final manuscript in order to catch random errors.
> — *A Million Random Digits With 100,000 Normal Deviates*, RAND Corp., 1955.

10.6.2 The Perpendicular Bisector

Given a straight segment from **P** to **Q**, the problem is to calculate a line perpendicular to the segment and passing through its middle. The slope vector of the segment is $(Q_x - P_x, Q_y - P_y)$ and the slope of the perpendicular line can be obtained by the negate and exchange rule (Equation (3.6)). It is $(-(Q_y - P_y), Q_x - P_x)$. The midpoint of the segment is $((P_x+Q_x)/2, (P_y+Q_y)/2)$, so the perpendicular bisector can be expressed as

$$((P_x+Q_x)/2 - (Q_y-P_y)t, (P_y+Q_y)/2 + (Q_x-P_x)t), \text{ where } -\infty \le t \le \infty. \quad (10.2)$$

▶ **Exercise 10.9:** Calculate the perpendicular bisector of the segment from $(1, 1)$ to $(3, 3)$.

10.7 Branching Rules

These rules produce fractals that look like a tree with branches growing from the trunk and splitting into smaller branches. The rules are very simple and don't require the use of random numbers. A vertical bar is drawn, representing the trunk. At its top, two lines are drawn, at predetermined angles, with sizes that are certain fractions of the length L of the trunk. These are the main branches. Out of each branch, two more branches are grown. The process continues recursively.

Figure 10.14a shows two branches, at 10° and 60° to the trunk, with lengths of $0.9L$ and $0.6L$, respectively. A complete tree, after several recursive iterations, is shown in Figure 10.14b.

Impressive results are obtained when the calculations are done in three dimensions and each iteration rotates the plane of the two new branches 90° relative to the preceding branches. Perspective projection should be used, of course, when displaying or printing the results. More realism can be added by making each branch thinner than its parent or by drawing them as real branches, perhaps with leaves added, instead of just straight segments.

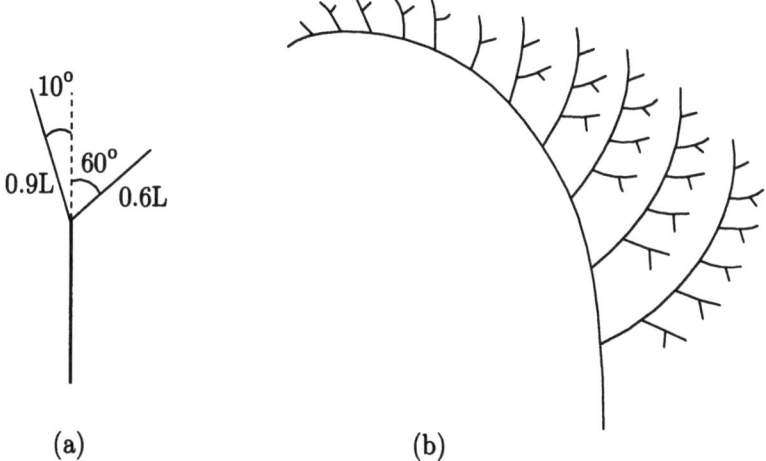

Figure 10.14: (a) Simple Branching. (b) A Two-Dimensional Tree.

10.8 Iterated Function Systems (IFS)

This type of fractal can be used to create beautiful complex two-dimensional shapes (see Figure 10.15) and save them in extremely small files. Once such a pattern is deemed useful it can be saved as just a few numbers. IFS can thus be considered an efficient graphics compression method (with compression factors of 10,000 or greater; see, for example, [Salomon 97]) but is described here as a method that creates nice patterns. A graphics shape created by IFS is uniquely defined by a set of affine transformations. The only rule is that the scale factors must be smaller than 1 (contraction). To save the pattern on a file, only the transformations need be saved. (As a reminder, a two-dimensional affine transformation is completely defined by a 3×3 matrix with $(0,0,1)$ in the last column. Therefore, only the remaining six numbers need be saved.)

A simple example is the set of three transformations

$$\mathbf{T}_1 = \begin{pmatrix} 0.5 & 0 & 0 \\ 0 & 0.5 & 0 \\ 8 & 8 & 1 \end{pmatrix}, \quad \mathbf{T}_2 = \begin{pmatrix} 0.5 & 0 & 0 \\ 0 & 0.5 & 0 \\ 96 & 16 & 1 \end{pmatrix}, \quad \mathbf{T}_3 = \begin{pmatrix} 0.5 & 0 & 0 \\ 0 & 0.5 & 0 \\ 120 & 60 & 1 \end{pmatrix}.$$

We first discuss the concept of the *fixed point*. Imagine the sequence $\mathbf{P}_1 = \mathbf{P}_0 \mathbf{T}_1$, $\mathbf{P}_2 = \mathbf{P}_1 \mathbf{T}_1 \ldots$, where $\mathbf{T}_1$ is applied repeatedly to create a sequence of points. It is easy to prove that $\lim_{k \to \infty} \mathbf{P}_k = (2m, 2n)$, where m and n are integers. This point is called the *fixed point* of $\mathbf{T}_1$ and it does not depend on the particular starting point $\mathbf{P}_0$ selected.

Proof: We denote $\mathbf{P}_0 = (x_0, y_0)$. The first two iterations yield

$$\mathbf{P}_1 = \mathbf{P}_0 \mathbf{T}_1 = (0.5 x_0 + 8, 0.5 y_0 + 8)$$
$$\mathbf{P}_2 = \mathbf{P}_1 \mathbf{T}_1 = (0.5(0.5 x_0 + 8) + 8, 0.5(0.5 y_0 + 8) + 8).$$

It is easy to see (and also to prove by induction) that $x_n = 0.5^n x_0 + 0.5^{n-1} 8 +$

$0.5^{n-2}8 + \cdots + 0.5^1 8 + 8$. In the limit $x_n = 0.5^n x_0 + 8\sum_{i=0}^{\infty} 0.5^i$, which adds up to $8 \times 2 = 16$ regardless of x_0. ◄

Now it is easy to show that for the above transformations, with scale factors of 0.5 and no shearing, each new point in the sequence moves half the remaining distance toward the fixed point. Given a point $\mathbf{P}_i = (x_i, y_i)$, the point midway between $\mathbf{P}_i$ and the fixed point $(16, 16)$ is

$$\big((x_i + 16)/2, (y_i + 16)/2\big) = (0.5x_i + 8, 0.5y_i + 8) = (x_{i+1}, y_{i+1}) = \mathbf{P}_{i+1}.$$

Consequently, for the particular above transformations, there is no need to use the transformation matrix. At each step of the iteration, point $\mathbf{P}_{i+1}$ is obtained by $(\mathbf{P}_i + (2m, 2n))/2$. For other transformations, matrix multiplication is necessary to compute point $\mathbf{P}_{i+1}$.

In general, every affine transformation where the scale and shear factors are less than 1 has a fixed point, but it may not be easy to find it.

The principle of the IFS method is now easy to describe. A set of transformations (an IFS code) is selected. A sequence of points is calculated and plotted by starting with an arbitrary point $\mathbf{P}_0$, selecting a transformation at random, and applying it to $\mathbf{P}_0$, transforming it into a point $\mathbf{P}_1$, applying another transformation at random, and so on. Every point is plotted and, gradually, the object begins to take shape on the screen. The shape of the object is called the *attractor* of the IFS code and it depends on the IFS code selected. The shape also depends slightly on the particular selection of $\mathbf{P}_0$. It is best to choose $\mathbf{P}_0$ as one of the fixed points of the IFS code (if they are known in advance). In such a case, all the points in the sequence will lie inside the attractor. For any other choice of $\mathbf{P}_0$, a finite number of points will lie outside the attractor, but eventually they will move into the attractor and stay there.

It is surprising that the attractor does not depend on the precise order of the transformations used. This result has been proved by the mathematician John Elton.

Another surprising property is that the random numbers used to select the next transformation don't have to be uniformly distributed; they can be weighted. Transformation $\mathbf{T}_1$, for example, may be selected at random 50% of the time, transformation $\mathbf{T}_2$, 30%, and transformation $\mathbf{T}_3$, 20%. The shape being generated does not depend on the weights, but the computation time does. The weights, of course, have to add up to 1 and cannot be 0.

The three transformations of Table 10.16 create an attractor in the form of a Sierpinski triangle (see box). The translation factors determine the coordinates of the three triangle corners. Figure 10.17a,b shows transformations that create attractors in the form of a fern and a coastline (the notation used is $\begin{pmatrix} a & b \\ c & d \end{pmatrix} + \begin{pmatrix} m \\ n \end{pmatrix}$).

The program in Figure 10.18 calculates and displays IFS attractors for any given set of transformations.

10 Short Topics

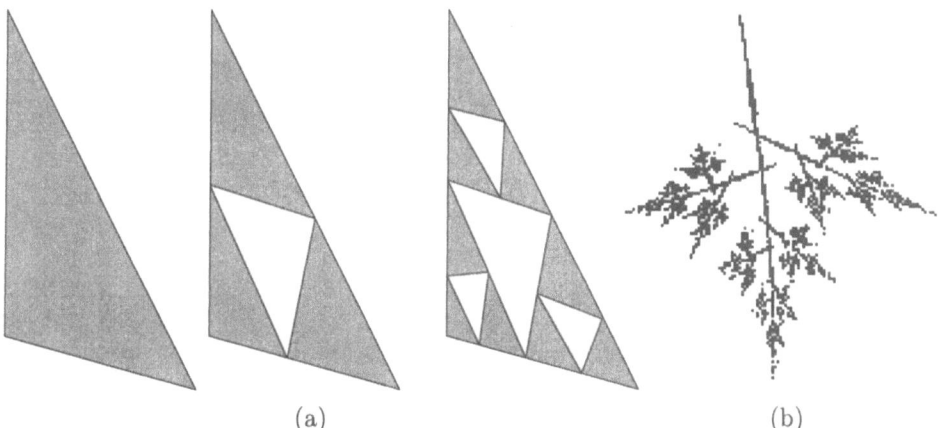

Figure 10.15: (a) Sierpinski Triangle. (b) A Leaf.

a	b	c	d	m	n	a	b	c	d	m	n
0	−28	0	29	151	92	64	0	0	64	82	6
−2	37	−31	29	85	103	0	−80	−22	1	243	151
−1	18	−18	−1	88	147	2	−48	0	50	160	80

(All the numbers are shown as integers, but the first four numbers should be divided by 100, to make them < 1. The last two numbers are the translation factors.)

Table 10.16: Attractor for The Sierpinski Triangle.

The Sierpinski triangle (Figure 10.15a) is defined recursively. Start with any triangle, find the midpoint of each edge, and connect the three midpoints to obtain a new triangle, fully contained in the original one. Cut the new triangle out. The newly created hole now divides the original triangle into three smaller ones. Repeat the process on each of the smaller triangles. At the limit, there is no area left in the triangle. It resembles Swiss cheese without any cheese, just holes.

```
5
    0   -28    0   29  151   92        4
   64     0    0   64   82    6      -17  -26   34  -12   84   53
   -2    37  -31   29   85  103       25  -20   29   17  192   57
   17   -51  -22    3  183  148       35    0    0   35   49    3
   -1    18  -18   -1   88  147       25   -6    6   25  128   28
```
                (a)   (b)

Figure 10.17: (a) A Fern. (b) A Coastline.

```
PROGRAM IFS;
USES ScreenIO, Graphics, MathLib;

CONST LB = 5; Width = 490; Height = 285;
(* LB=left bottom corner of window *)

VAR i,k,x0,y0,x1,y1,NumTransf: INTEGER;
Transf: ARRAY[1..6,1..10] OF INTEGER;
Params:TEXT;
filename:STRING;

BEGIN (* main *)
Write('params file='); Readln(filename);
Assign(Params,filename); Reset(Params);
Readln(Params,NumTransf);
FOR i:=1 TO NumTransf DO
Readln(Params,Transf[1,i],Transf[2,i],Transf[3,i],
 Transf[4,i],Transf[5,i],Transf[6,i]);
OpenGraphicWindow(LB,LB,Width,Height,'IFS shape');
SetMode(paint);

x0:=100; y0:=100;

REPEAT
k:=RandomInt(1,NumTransf+1);
x1:=Round((x0*Transf[1,k]+y0*Transf[2,k])/100)+Transf[5,k];
y1:=Round((x0*Transf[3,k]+y0*Transf[4,k])/100)+Transf[6,k];
Dot(x1,y1); x0:=x1; y0:=y1;
UNTIL Button()=TRUE;
  ScBOL; ScWriteStr('Hit a key & close this window to quit');
  ScFreeze;
END.
```

Figure 10.18: Calculate and Display IFS Attractors.

10.9 Image Processing

This is the name of a very large group of techniques that modify a given image in ways that make it more useful or more interesting. Many times, an image taken by a satellite needs to be sharpened or painted with false colors. On the other hand, an artist may want to take a sharply focused photograph and intentionally blur it, or make it look as if it was originally painted by watercolors, or make it resemble an image embossed on paper—in order to achieve interesting effects.

The input image to be processed is always in the form of a bitmap, with one or more bits per pixel. The processing software must be given the three dimensions of the bitmap (number of rows, number of columns, and number of bits per pixel)

10 Short Topics

and it normally creates the new image in another bitmap, pixel by pixel. A typical image processing algorithm consists of a loop that iterates over all the pixels of the image, processing each in the same way. The original pixel stays in the original bitmap (since its value may be needed to process neighboring pixels) and the newly calculated pixel is stored in the new bitmap.

The techniques described here produce very different results, but most are based on the same principle. The principle is to define a small matrix of values, called a *convolution kernel*, to place it centered on the current pixel, to multiply the value of the pixel and the values of its neighbors by the values of the kernel, to add the results, and to store the sum in the new bitmap, as the value of the newly calculated pixel. If the color of a pixel is specified by means of three numbers (normally the red, green, and blue components), then this process is applied separately to each of the three components of the current pixel.

Blurring: This is achieved by the convolution kernel of Figure 10.19b. The values of the current pixel and eight of its nearest neighbors are multiplied by the weights shown and the products added. The result is a color value that still has 20% of the old pixel value but has contributions of 8% and 12% from neighboring pixels. Note that the weights add up to 1. As an example, suppose that the current pixel is the center one of Figure 10.19a. The calculation is

$$5\times 0.8 + 5\times 0.12 + 8\times 0.8 + 19\times 0.12 + 5\times 0.20 + 8\times 0.12$$
$$+ 11\times 0.8 + 1\times 0.12 + 8\times 0.8 = 30.56,$$

yielding a value of 31 for the new pixel.

Note that more blurring can be achieved by having a 5×5 convolution kernel (the weights should add up to 1) which will spread the color of a pixel to 24 of its neighbors.

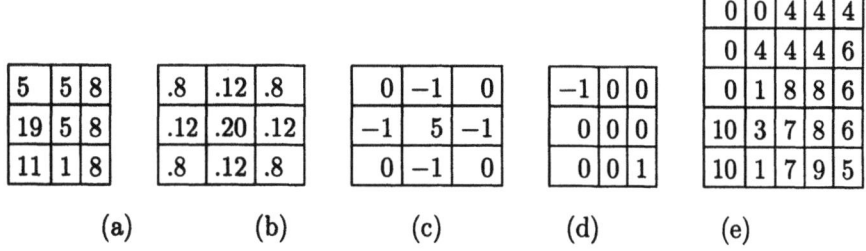

Figure 10.19: Image Processing Techniques.

Sharpening: This is achieved by the convolution kernel of Figure 10.19c. The weights again add up to 1, but the negative weights magnify any contrasts between the original pixels. More sharpening may be achieved by repeating the process on the new bitmap.

Embossing: This is achieved by the convolution kernel of Figure 10.19d. Note that the weights here add up to 0. To understand how this works, we should think of pixels along an edge, as opposed to pixels away from an edge. Pixels located away from an edge tend to be similar and we can call them "background" pixels. The convolution kernel of Figure 10.19d sets such pixels to 0 or close to 0. In contrast, if the current pixel is part of an edge (if it is a nonbackground pixel), then its two diagonal neighbors should have different colors and our kernel will create a new nonzero pixel.

Our kernel creates a white background since it sets all background pixels to zero. Visually, it is better to have a medium gray background, and this is easily achieved by adding 128 to each pixel generated. Background pixels will now have a value of 128 (or close to 128) and nonbackground ones will have any values. If they have a value greater than 255, it can be truncated to 8 bits (calculated modulo 256).

Note that the embossing kernel can be written in a number of ways. All that is needed are the numbers 1 and -1 in opposite corners. Different kernels create the effect of the light hitting the embossed picture from different directions.

Watercoloring: This is achieved by examining a group of neighbors centered on the current pixel and replacing the original value of the pixel by the median of the group. Assuming that the current pixel is the center one in Figure 10.19e, we sort the values in the group of 5×5 neighbor pixels to get

$$0,0,0,0,1,1,3,4,4,4,4,4,4,5,6,6,6,7,7,8,8,8,9,10,10.$$

The median value is 4 (since there are 12 smaller values and 12 greater ones). Our center pixel of 8 is thus replaced, in the new bitmap, by 4. If the result is too soft, it can later be sharpened.

10.9.1 An Alternative Approach

It is possible to process images by means of the important relation $(1-t)\mathbf{P}_0 + t\mathbf{P}_1$ (this is the well-known Equation (4.4)). The idea is to blend two images (Section 4.2) by interpolating (i.e., using $0 \leq t \leq 1$) or by extrapolating (using $t < 0$ or $t > 1$) them. Values of $t > 1$ subtract part of $\mathbf{P}_0$ while scaling $\mathbf{P}_1$. Negative values of t do the reverse.

The examples below show how a general image $\mathbf{P}_0$ can be blended with a special image $\mathbf{P}_1$ (a mask) to obtain the following useful results:

Brightness: We select a bitmap of all black as the mask $\mathbf{P}_1$. Interpolation darkens the image while extrapolation brightens it. The original image is obtained for $t = 0$.

Contrast: We compute the average intensity I of all the pixels in the original image. We build the mask as a gray bitmap where every pixel has value I. Interpolation ($0 \leq t \leq 1$) reduces contrast, while extrapolation increases it. Negative values of t generate inverted images. The average intensity of the final image is always I.

Saturation: We first compute the luminance of every pixel in $\mathbf{P}_0$ and set the corresponding pixel in the mask $\mathbf{P}_1$ to a shade of gray with that luminance. The mask is then used to change the luminance of every pixel in the original image $\mathbf{P}_0$. Interpolation decreases saturation, while extrapolation increases it. Negative t also inverts the hue of image $\mathbf{P}_0$.

10 Short Topics

Sharpening: Section 2.9.4 discusses convolution. Sharpening and blurring are examples of convolutions. If the mask is a blurred version of the original, then interpolation blurs the original and extrapolation sharpens it.

For more information, see [Haeberli and Voorhies 94].

> Most processes used for halftoning consist of linear and nonlinear elements. Neural networks offer the possibility of combining these elements in a general and flexible structure. Image binarization methods can be analysed and transferred to neural structures and typical neural learning algorithms offer new ways to treat the halftoning problem.
>
> —T. Tuttass and O. Bryngdahl

A
Mathematical Topics

This chapter discusses most of the mathematical background needed for a thorough understanding of the material presented in the book. It has been mentioned in the Preface, however, that math concepts which are only used once (such as the mediation operator and points vs. vectors) are discussed right where they are introduced.

> Do not worry too much about your difficulties in mathematics, I can assure you that mine are still greater.
> — Albert Einstein.

A.1 Fourier Transforms

Our curves are functions of an arbitrary parameter t. For functions used in science and engineering, *time* is often the parameter (or the independent variable). We, therefore, say that a function $g(t)$ is represented in the *time domain*. Since a typical function oscillates, we can think of it as being similar to a wave and we may try to represent it as a wave (or as a combination of waves). When this is done, we have the function $G(f)$, where f stands for the frequency of the wave, and we say that the function is represented in the *frequency domain*. This turns out to be a useful concept, since many operations on functions are easy to carry out in the frequency domain.

Transforming a function between the time and frequency domains is easy when the function is *periodic*, but it can also be done for certain nonperiodic functions. The present discussion is restricted to periodic functions.

Definition: A function $g(t)$ is periodic if (and only if) there exists a constant P such that $g(t+P) = g(t)$ for all values of t (Figure A.1a). P is called the *period* of the function. If several such constants exist, only the smallest of them is considered the period.

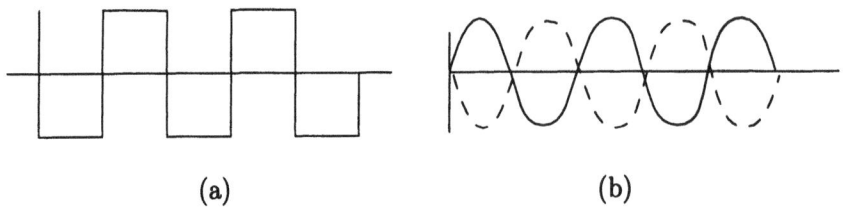

Figure A.1: Periodic Functions.

A periodic function has four important attributes: its amplitude, period, frequency, and phase. The amplitude of the function is the maximum value it has in any period. The frequency f is the inverse of the period ($f = 1/P$). It is expressed in cycles per second, or Hertz (Hz). The phase is the least understood of the four attributes. It measures the position of the function within a period and it is easy to visualize when a function is compared to its own copy. Examine the two sinusoids in Figure A.1b: They are identical, but out of phase. One follows the other at a fixed interval called the *phase difference*. We can write them as $g_1(t) = A\sin(2\pi ft)$ and $g_2(t) = A\sin(2\pi ft + \theta)$. The phase difference between them is θ, but we can also say that the first one has no phase, while the second one has a phase of θ. (By the way, this example also shows that cosine is a sine function with a phase of $\theta = \pi/2$.)

To understand the concept of frequency domain, let's look at two simple examples. The function $g(t) = \sin(2\pi ft) + (1/3)\sin(2\pi(3f)t)$ is a combination of two sine waves with amplitudes 1 and 1/3 and with frequencies f and $3f$, respectively. They are shown in Figure A.2a,b. The sum (Figure A.2c) is also periodic, with frequency f (the smaller of the two frequencies). The frequency domain of $g(t)$ is a function consisting of just the two points $(f, 1)$ and $(3f, 1/3)$ (Figure A.2h). It indicates that the original (time domain) function is made up of frequency f with amplitude 1 and frequency $3f$ with amplitude $1/3$.

This example is extremely simple, since it involves just two frequencies. When a function involves several frequencies that are integer multiples of some lowest frequency, the latter is called the *fundamental frequency* of the function.

Not every function has a simple frequency domain representation. Consider the single square pulse in Figure A.2d. Its time domain is

$$g(t) = \begin{cases} 1, & -a/2 \leq t \leq a/2, \\ 0, & \text{elsewhere}, \end{cases}$$

but its frequency domain is as in Figure A.2e. It consists of all the frequencies from 0 to ∞, with amplitudes that drop continuously. This means that the time domain representation, even though simple, consists of all possible frequencies, with lower frequencies contributing more and higher ones contributing less and less.

In general, a periodic function can be represented in the frequency domain as the sum of (phase shifted) sine waves with frequencies that are integer multiples (harmonics) of some fundamental frequency. However, the square pulse of Figure A.2d is not periodic. It turns out that frequency domain concepts can be

A Mathematical Topics

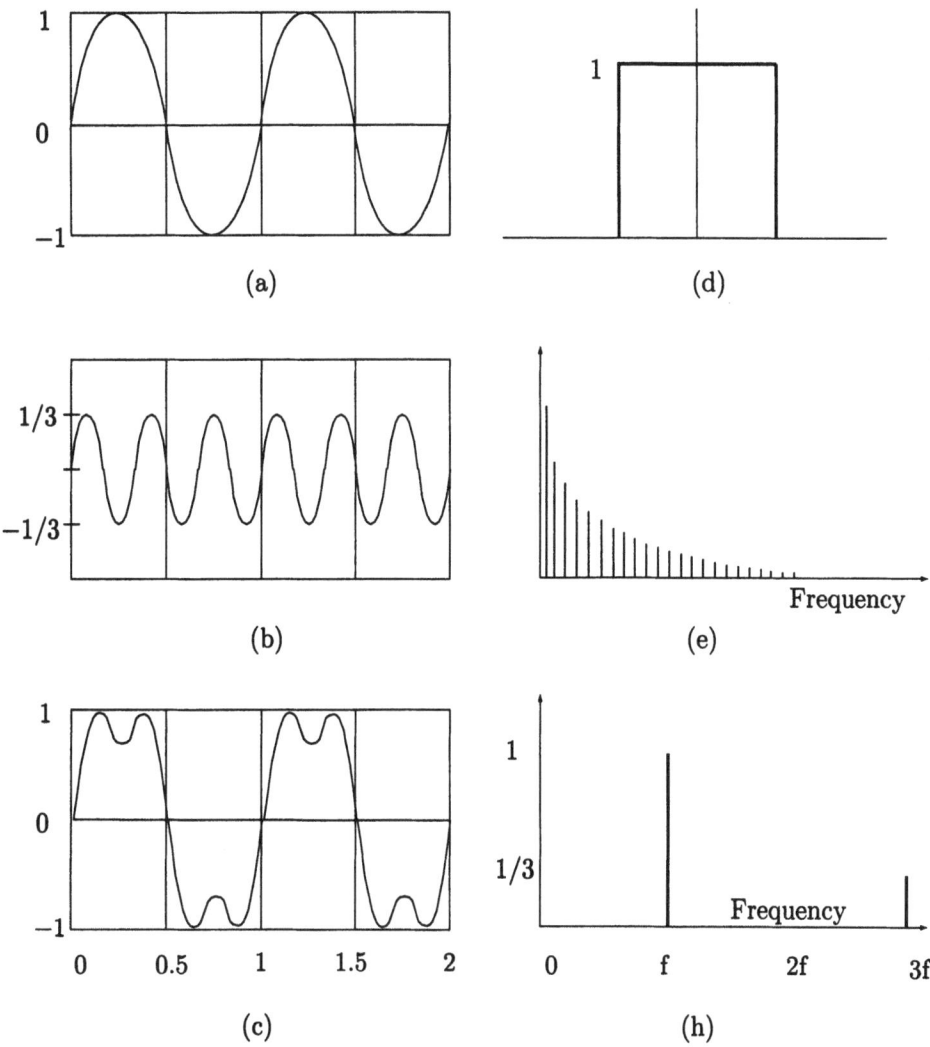

Figure A.2: Time and Frequency Domains.

applied to a nonperiodic function, but only if it is nonzero over a finite range (like our square pulse). Such a function is represented as the sum of (phase shifted) sine waves with all kinds of frequencies, not just harmonics.

The *spectrum* of the frequency domain is the range of frequencies it contains. For the function of Figure A.2c,h, the spectrum is the two frequencies f and $3f$. For the function of Figure A.2d,e, it is the entire range $[0, \infty]$. The *bandwidth* of the frequency domain is the width of the spectrum. It is $2f$ in our first example and ∞ in the second one.

Another important concept that should be mentioned is the *dc component* of the function. The time domain of a function may include a component of zero frequency. Engineers call this component the *direct current*, so the rest of us have adopted the term "dc component." Figure A.3a is identical to Figure A.2c except that it oscillates from 0 to 2, instead of from −1 to +1. The frequency domain (Figure A.3b) now has an added point at (0, 1), representing the dc component.

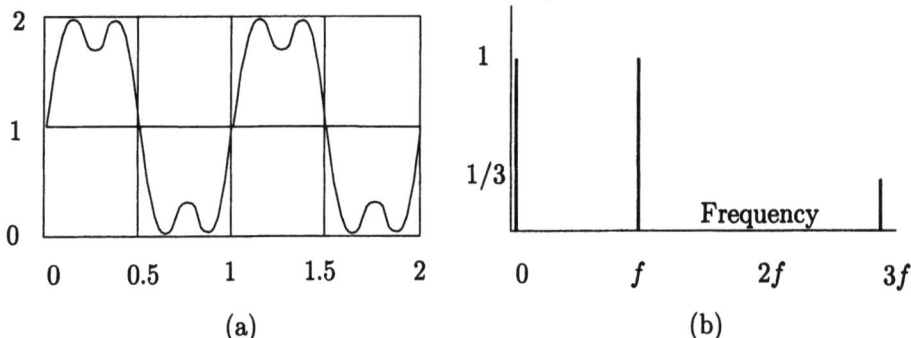

Figure A.3: Time and Frequency Domains with a dc Component.

The entire concept of the two domains is due to the French mathematician Joseph Fourier. He proved a fundamental theorem that says that every periodic function can be represented as the sum of sine and cosine functions. He also showed how to transform a function between the time and frequency domains. If the shape of the function is far from a regular wave, its Fourier expansion will include an infinite number of frequencies. For a continuous function $g(t)$, the Fourier transform and its inverse are given by

$$G(f) = \int_{-\infty}^{\infty} g(t)[\cos(2\pi ft) - i\sin(2\pi ft)]\, dt,$$

$$g(t) = \int_{-\infty}^{\infty} G(f)[\cos(2\pi ft) + i\sin(2\pi ft)]\, df.$$

In computer graphics, we normally have discrete functions that take just n (equally spaced) values. In such a case, the discrete transform is

$$G(f) = \sum_{t=0}^{n-1} g(t)[\cos(2\pi ft/n) - i\sin(2\pi ft/n)], \quad 0 \le f \le n-1.$$

Its inverse is

$$g(t) = \frac{1}{n}\sum_{f=0}^{n-1} G(f)[\cos(2\pi ft/n) + i\sin(2\pi ft/n)], \quad 0 \le t \le n-1.$$

Note that $G(f)$ is complex, so it can be written $G(f) = R(f) + iI(f)$. For any value of f, the amplitude (or magnitude) of G is given by $|G(f)| = \sqrt{R^2(f) + I^2(f)}$.

Note how the function of Figure A.2c that's obtained by adding the simple functions of Figure A.2a,b starts to resemble a square pulse. It turns out that

A Mathematical Topics

we can bring it closer to a square pulse (like the one of Figure A.1a) by adding $(1/5)\sin(2\pi(5f)t)$, $(1/7)\sin(2\pi(7f)t)$, and so on. We say that the Fourier series of a square wave with amplitude A and frequency f is the infinite sum

$$A\sum_{k=1}^{\infty} \frac{1}{k}\sin(2\pi kft),$$

where successive terms have smaller and smaller amplitudes.

We now apply these concepts to computer graphics. Imagine that we have a black-and-white photograph and we want to store it in the computer so it can be edited and displayed. One way to do this is to scan the photograph line by line. For all practical purposes, we can assume that the photograph has infinite resolution (its shades of gray vary continuously, but see also Section 1.1.7). An ideal scan would, therefore, result in an infinite sequence of numbers and they can be considered the values of an (continuous) intensity function $i(t)$. In practice, we can only store a finite sequence in memory, so we have to select a finite number of values (a sample) $i(1), i(2), \ldots, i(n)$. This process is known as *sampling*.

Intuitively, sampling seems a trade-off between quality and price. The bigger the sample, the better the quality of the final image, but more hardware (more memory and higher screen resolution) is required, resulting in higher costs. This intuitive conclusion, however, is not completely true. Sampling theory tells us that we can sample an image and reconstruct it in memory without loss of quality if we can do the following:

1. Transform the intensity function from the time domain $i(t)$ to the frequency domain $I(f)$.

2. Find the maximum frequency f_m.

3. Sample $i(t)$ at a rate *greater than or equal* $2f_m$ (for example, if $f_m = 22,000$ Hz, generate samples at the rate of 44,000 Hz or higher).

4. Store the sampled values in the bitmap. The resulting picture should be equal in quality to the original photograph.

There are two points to consider. The first is that f_m could be infinite. In such a case, a value f_m should be selected such that frequencies greater than f_m do not contribute much (have low amplitudes). There is some loss of image quality in such a case. The second point is that the bitmap (and, consequently, the resolution) may be too small for the sample generated in step 3. In such a case, a smaller sample has to be taken, again resulting in a loss of image quality.

The result above was proved by H. Nyquist, and the quantity $2f_m$ is called the Nyquist rate. It is used in many practical situations. The normal range of a human hearing, for instance, is between 16 Hz and 22,000 Hz. When sound is digitized, it should, therefore, be sampled at the Nyquist rate of 44,000 Hz (or higher). Anything lower than that would result in distortions. This is why music recorded on a CD has such high quality.

A.2 Forward Differences

This is an incremental method for evaluating polynomials. Only the initial steps require multiplications, and all subsequent evaluations can be done with additions only. The method can be applied to polynomials of any degree, but, to simplify the discussion, it is described here for cubic polynomials.

We start with a general cubic polynomial $P(x) = ax^3 + bx^2 + cx + d$. Given a value h to be denoted the *step*, we use multiplications to calculate this polynomial at a point x_0, and also to calculate the first three differences of the polynomial for h. Following this, it is possible to compute the values $P(x_0 + h)$, $P(x_0 + 2h)$,... using just additions. The first step is to construct the forward differences:

1. The first difference is

$$\Delta P(x_0) = P(x_0 + h) - P(x_0)$$
$$= a(x_0 + h)^3 + b(x_0 + h)^2 + c(x_0 + h) + d - \left(ax_0^3 + bx_0^2 + cx_0 + d\right)$$
$$= (3ah)x_0^2 + (3ah^2 + 2bh)x_0 + (ah^3 + bh^2 + ch). \qquad (A.1)$$

This is a quadratic polynomial in x_0.

2. We consider the function $f(x_0) = \Delta P(x_0)$. Its first difference is the second difference of $P(x)$.

$$\Delta^2 P(x_0) = \Delta f(x_0) = f(x_0 + h) - f(x_0)$$
$$= (3ah)(x_0 + h)^2 + (3ah^2 + 2bh)(x_0 + h) + (ah^3 + bh^2 + ch)$$
$$\quad - \left((3ah)x_0^2 + (3ah^2 + 2bh)x_0 + (ah^3 + bh^2 + ch)\right)$$
$$= (6ah^2)x_0 + (6ah^3 + 2bh^2). \qquad (A.2)$$

This is a linear polynomial in x_0.

3. Similarly, we define $g(x_0) = \Delta f(x_0) = \Delta^2 P(x_0)$ and calculate its first forward difference, that's the third difference of the original polynomial:

$$\Delta^3 P(x_0) = \Delta g(x_0) = g(x_0 + h) - g(x_0)$$
$$= (6ah^2)(x_0 + h) + (6ah^3 + 3dh^2) - \left((6ah^2)x_0 + (6ah^3 + 3dh^2)\right)$$
$$= 6ah^3. \qquad (A.3)$$

This is a constant.

The next step is to write Equations (A.1) through (A.3) as

$$P(x_0 + h) = \Delta P(x_0) + P(x_0), \qquad (A.4)$$
$$\Delta P(x_0 + h) = \Delta^2 P(x_0) + \Delta P(x_0), \qquad (A.5)$$
$$\Delta^2 P(x_0 + h) = \Delta^3 P(x_0) + \Delta^2 P(x_0), \qquad (A.6)$$
$$\Delta^3 P(x_0) = 6ah^3, \qquad (A.7)$$

and the last step is to use Equations (A.4) through (A.7) to calculate the values of $P(x_0)$, $\Delta P(x_0)$, $\Delta^2 P(x_0)$, and $\Delta^3 P(x_0)$.

A Mathematical Topics

Given a step h, the value of $P(x_0 + h)$ can now be calculated from Equation (A.4) using just one addition. If we also use Equations (A.5) and (A.6) to calculate $\Delta P(x_0 + h)$ and $\Delta^2 P(x_0 + h)$, then we can use point $x_0 + h$ as our new starting point x_1 and calculate $P(x_1 + h)$ [which equals $P(x_0 + 2h)$].

Example 1: It is common to select $x_0 = 0$, which reduces Equations (A.1)–(A.3) to

$$P(0) = d, \quad \Delta P(0) = ah^3 + bh^2 + ch, \quad \Delta^2 P(0) = 6ah^3 + 2bh^2, \quad \Delta^3 P(0) = 6ah^3. \tag{A.8}$$

We select the simple polynomial $P(x) = x^3 + 1$ and select $h = 0.5$ to calculate $P(0.5)$, $P(1)$, and $P(1.5)$. The first step uses Equation (A.8) to calculate $P(0) = 1$, $\Delta P(0) = 1/8$, $\Delta^2 P(0) = 6/8$, and $\Delta^3 P(0) = 6/8$. Equation (A.4) gives us $P(0.5) = 1/8 + 1 = 9/8$. Equation (A.5) results in $\Delta P(0.5) = 7/8$, which can be used in Equation (A.4) again to give $P(1) = 7/8 + 9/8 = 2$.

From Equation (A.6) we get $\Delta^2 P(0.5) = 12/8$, which can immediately be used to produce $\Delta P(1) = \Delta P(0.5) + \Delta^2 P(0.5) = 7/8 + 12/8 = 19/8$. This, in turn, can be used to yield $P(1.5) = 19/8 + 2 = 35/8$.

It is easy to write a program that evaluates Equations (A.4) through (A.6) repeatedly to calculate the values $P(x_0 + h)$, $P(x_0 + 2h)\ldots$. The main problem is round-off errors that may accumulate. This is a problem with any incremental method and it may be serious in a computer with low-precision integers (many modern microprocessors support high-precision 80-bit floating-point numbers, but only 16-bit integers).

Example 2: Forward differences are applied to the problem of fast drawing of a circle. We use the approximate circle given by the parametric cubic polynomial of Equation (Ans.32)

$$\mathbf{P}(t) = (0.3431, -0.3431)t^3 - (1.3431, 0.3138)t^2 + (0, 1.6569)t + (1, 0),$$

and select $t_0 = 0$ and $h = 0.01$. The first three differences are

$$\begin{aligned}\Delta \mathbf{P}(0) &= (0.3431, -0.3431) \times 0.01^3 - (1.3431, 0.3138) \times 0.01^2 + (0, 1.6569) \times 0.01 \\ &= (-0.000133967, 0.0165372769), \\ \Delta^2 \mathbf{P}(0) &= 6(0.3431, -0.3431) \times 0.01^3 - 2(1.3431, 0.3138) \times 0.01^2 \\ &= (-0.0002665614, -0.0000648186), \\ \Delta^3 \mathbf{P}(0) &= 6(0.3431, -0.3431) \times 0.01^3 = (0.0000020586, -0.0000020586),\end{aligned}$$

which is all that's needed to write the algorithm below.

```
x:=1; y:=0;
dx:=-0.000133967; dy:=0.0165372769;
ddx:=-0.0002665614; ddy:=-0.0000648186;
dddx:=0.0000020586; dddy:=-0.0000020586;
pixel4(x,y);
for i:=1 to numpixel do
  x:=x+dx; y:=y+dy;
```

```
dx:=dx+ddx; dy:=dy+ddy;
ddx:=ddx+dddx; ddy:=ddy+dddy;
pixel4(x,y);
endfor;
```

See also Section 4.14.3 where this method is used for a fast calculation of the Bézier curve.

A.3 Coordinate Systems

The two-dimensional Cartesian coordinate system is well known. The position of a point in such a system is specified by its distances from the coordinate axes. If the coordinates of a point **P** are (x, y), then x is the distance of **P** from the y axis and y is its distance from the x axis. Figure A.4a shows two ways to position such a system on a screen, where the origin can be either at the bottom-left or at the top-left corner.

The Cartesian coordinate systems used in practice have straight, perpendicular axes. Such coordinate systems are called *rectangular*. In principle, the angle between the axes can be any (except zero) and the axes themselves don't have to be straight. Figure A.4b shows an *oblique* and a *curvilinear* Cartesian coordinate systems. It is easy to see how the position of a point is measured in such systems.

The *polar* coordinate system is based on one axis where an origin O has been chosen. The coordinates (r, θ) of a point **P** are determined (Figure A.5) by its distance r from the origin, and the angle θ between the axis and the straight segment OP. To convert from polar to Cartesian coordinates, we designate the existing axis the x axis and add the y axis at the origin. Simple trigonometry then shows that $x = r \cos\theta$ and $y = r \sin\theta$, which makes it easy to convert from Cartesian to polar coordinates:

$$r = \sqrt{x^2 + y^2}, \quad \theta = \tan^{-1}\left(\frac{y}{x}\right).$$

A curve is specified in polar coordinates by a relation of the form $r = f(\theta)$, so it is clear that any such curve can be expressed in Cartesian coordinates as $x = f(\theta)\cos\theta$, $y = f(\theta)\sin\theta$. This is a parametric representation where θ is the parameter.

The polar coordinate system is based on the circle, so circular curves may have simple representations in this system. The circle itself becomes $r = $ constant, the Archimedean spiral is $r = a\theta$, and the logarithmic spiral is $a\, r\, e^{b\theta}$, where a and b are constants.

A deeper look at polar coordinates suggests that other geometric figures, such as the ellipse, parabola, and hyperbola, can be used as the basis of a coordinate system.

The right-hand and the left-hand three-dimensional Cartesian coordinate systems are shown in Figure A.6 together with their relation to the screen. These systems are discussed in Section 3.5.1.

The *cylindrical* coordinate system (Figure A.7a) specifies the position of a three-dimensional point by the triplet (r, θ, z). A surface of constant r is a cylinder and a surface of constant θ is a plane containing the z axis.

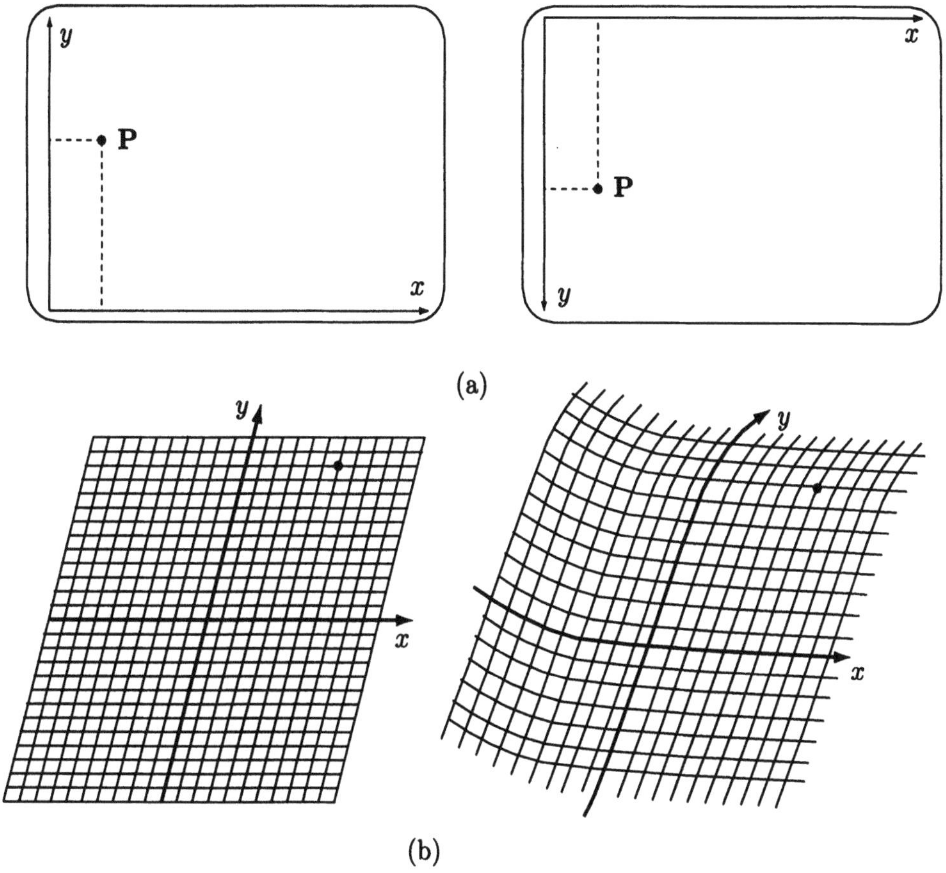

A Mathematical Topics

(a)

(b)

Figure A.4: Two-Dimensional Cartesian Coordinate Systems.

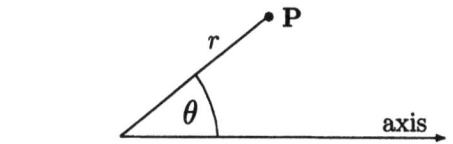

Figure A.5: Polar Coordinate System.

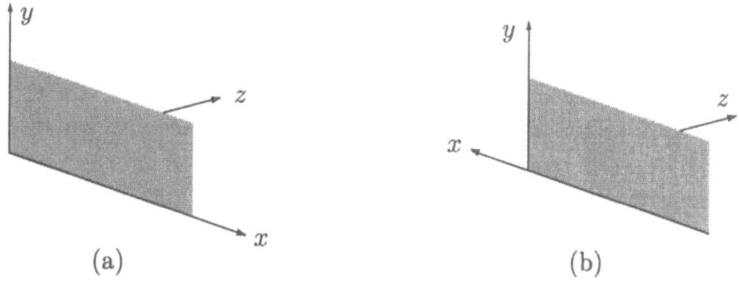

(a) (b)

Figure A.6: Three-Dimensional Cartesian Coordinate Systems.

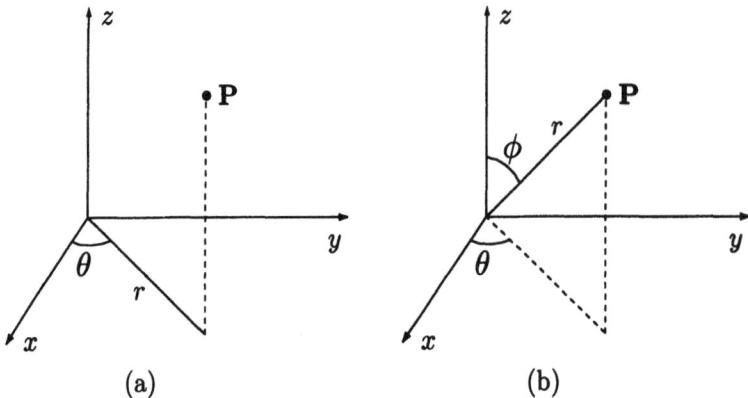

Figure A.7: Cylindrical and Spherical Coordinate Systems.

▶ **Exercise A.1:** What is a surface of constant z in cylindrical coordinates?

The conversion formulas $x = r\cos\theta$ and $y = r\sin\theta$ are trivial to figure out.

The *spherical* coordinate system (Figure A.7b) uses the similar triplet (r, θ, ϕ). It is easy to see why a surface of constant r is a sphere and why a surface of constant θ is a plane containing the z axis.

▶ **Exercise A.2:** What is the surface of constant ϕ in spherical coordinates?

Converting between Cartesian and spherical coordinates is done by

$$x = r\sin\phi\cos\theta, \quad y = r\sin\phi\sin\theta, \quad z = r\cos\phi,$$

$$r = \sqrt{x^2 + y^2 + z^2}, \quad \phi = \cos^{-1}\left(\frac{z}{r}\right), \quad \theta = \arctan(y, x),$$

where the two-argument arctan function is defined by

$$\arctan(y, x) = \begin{cases} \tan^{-1}(y/x) & \text{if } x > 0, \\ \pi + \tan^{-1}(y/x) & \text{if } x < 0, \\ \pi/2 & \text{if } x = 0 \text{ and } y > 0, \\ -\pi/2 & \text{if } x = 0 \text{ and } y < 0. \end{cases}$$

A.4 Vector Algebra

A vector is a mathematical entity with two attributes: direction and magnitude. The magnitude of vector $\mathbf{P} = (x, y, z)$ (also called its *absolute value*) is $|\mathbf{P}| = \sqrt{x^2 + y^2 + z^2}$. The direction of a vector can be expressed by the cosines of the angles it makes with the coordinate axes $x/|\mathbf{P}|$, $y/|\mathbf{P}|$, and $z/|\mathbf{P}|$. Note that the vector $(x/|\mathbf{P}|, y/|\mathbf{P}|, z/|\mathbf{P}|)$ has a magnitude (or *absolute value*) of 1 (it is a *unit vector*).

The three unit vectors in the directions of the coordinate axes are traditionally denoted $\mathbf{i} = (1, 0, 0)$, $\mathbf{j} = (0, 1, 0)$, and $\mathbf{k} = (0, 0, 1)$.

A Mathematical Topics

A.4.1 Operations on Vectors

Vector addition is defined by adding the individual elements of the vectors being added: $\mathbf{P}+\mathbf{Q} = (P_x, P_y, P_z) + (Q_x, Q_y, Q_z) = (P_x+Q_x, P_y+Q_y, P_z+Q_z)$. This operation is both commutative $(\mathbf{P}+\mathbf{Q} = \mathbf{Q}+\mathbf{P})$ and associative $(\mathbf{P}+(\mathbf{Q}+\mathbf{T}) = (\mathbf{P}+\mathbf{Q})+\mathbf{T})$. Subtraction of vectors $(\mathbf{P}-\mathbf{Q})$ is done similarly and results in the vector from $\mathbf{Q}$ to $\mathbf{P}$.

Vectors can be multiplied in three different ways:

1. The multiplication of a scalar by a vector is defined by $\alpha \mathbf{P} = (\alpha x, \alpha y, \alpha z)$. It changes the magnitude of the vector (by a factor α), but not its direction. This operation is distributive with respect to vector addition or subtraction, $\alpha(\mathbf{P} \pm \mathbf{Q}) = \alpha \mathbf{P} \pm \alpha \mathbf{Q}$.

2. The dot product of two vectors is denoted by $\mathbf{P} \bullet \mathbf{Q}$ and is defined as the scalar

$$(P_x, P_y, P_z)(Q_x, Q_y, Q_z)^T = \mathbf{P}\mathbf{Q}^T = P_x Q_x + P_y Q_y + P_z Q_z.$$

This also equals $|\mathbf{P}||\mathbf{Q}|\cos\theta$, where θ is the angle between the vectors. The dot product of perpendicular vectors (also called *orthogonal vectors*) is thus zero. The dot product is commutative, $\mathbf{P} \bullet \mathbf{Q} = \mathbf{Q} \bullet \mathbf{P}$, and is also distributive with respect to vector addition or subtraction, $\mathbf{P} \bullet (\mathbf{Q} \pm \mathbf{T}) = \mathbf{P} \bullet \mathbf{Q} \pm \mathbf{P} \bullet \mathbf{T}$.

The triple product $(\mathbf{P} \bullet \mathbf{Q})\mathbf{R}$ is sometimes useful. It can be represented as

$$\begin{aligned}(\mathbf{P} \bullet \mathbf{Q})\mathbf{R} &= (P_x Q_x + P_y Q_y + P_z Q_z)(R_x, R_y, R_z) \\ &= \big((P_x Q_x + P_y Q_y + P_z Q_z)R_x, (P_x Q_x + P_y Q_y + P_z Q_z)R_y, \\ &\quad (P_x Q_x + P_y Q_y + P_z Q_z)\big)R_z \\ &= (Q_x, Q_y, Q_z) \begin{pmatrix} P_x R_x & P_y R_x & P_z R_x \\ P_x R_y & P_y R_y & P_z R_y \\ P_x R_z & P_y R_z & P_z R_z \end{pmatrix} \\ &= \mathbf{Q}(\mathbf{P}\mathbf{R}), \end{aligned} \qquad (A.9)$$

where the notation $(\mathbf{PR})$ stands for the 3×3 matrix.

3. The cross product of two vectors (also called the *vector product*) is denoted by $\mathbf{P} \times \mathbf{Q}$ and is defined as the vector

$$(P_2 Q_3 - P_3 Q_2, -P_1 Q_3 + P_3 Q_1, P_1 Q_2 - P_2 Q_1). \qquad (A.10)$$

It is easy to show that $\mathbf{P} \times \mathbf{Q}$ is perpendicular to both $\mathbf{P}$ and $\mathbf{Q}$.

▶ **Exercise A.3:** Prove it!

The following expressions show how $\mathbf{P} \times \mathbf{Q}$ can be expressed by means of a determinant:

$$\begin{aligned}\mathbf{P} \times \mathbf{Q} &= \begin{vmatrix} \mathbf{i} & \mathbf{j} & \mathbf{k} \\ P_1 & P_2 & P_3 \\ Q_1 & Q_2 & Q_3 \end{vmatrix} = \mathbf{i}\begin{vmatrix} P_2 & P_3 \\ Q_2 & Q_3 \end{vmatrix} - \mathbf{j}\begin{vmatrix} P_1 & P_3 \\ Q_1 & Q_3 \end{vmatrix} + \mathbf{k}\begin{vmatrix} P_1 & P_2 \\ Q_1 & Q_2 \end{vmatrix} \\ &= (P_2 Q_3 - P_3 Q_2, -P_1 Q_3 + P_3 Q_1, P_1 Q_2 - P_2 Q_1),\end{aligned}$$

or, alternatively, by means of a matrix

$$= (Q_1, Q_2, Q_3) \begin{pmatrix} 0 & P_3 & -P_2 \\ -P_3 & 0 & P_1 \\ P_2 & -P_1 & 0 \end{pmatrix}. \tag{A.11}$$

▸ **Exercise A.4:** The cross-product $\mathbf{P} \times \mathbf{Q}$ is perpendicular to both $\mathbf{P}$ and $\mathbf{Q}$. In what direction does it point?

The cross-product is not commutative and is not associative. It is, however, distributive with respect to addition or subtraction of vectors. Hence, $\mathbf{P} \times (\mathbf{Q} \pm \mathbf{T}) = \mathbf{P} \times \mathbf{Q} \pm \mathbf{P} \times \mathbf{T}$.

The magnitude of $\mathbf{P} \times \mathbf{Q}$ equals $|\mathbf{P}||\mathbf{Q}|\sin\theta$, where θ is the angle between the two vectors. The cross-product, therefore, has a simple geometric interpretation. Its magnitude equals the area of the parallelogram defined by the two vectors.

▸ **Exercise A.5:** Given that $\mathbf{P} \times \mathbf{Q} = 0$, what does it tell us about the vectors involved?

As an example, the vector equation of a straight line is shown below for the case where the direction of the line and one point on the line are known. Assume that $\mathbf{d}$ is a unit vector in the direction of the line and $\mathbf{P}_1$ is a given point on the line. The equation of the entire line is

$$\mathbf{P}(t) = \mathbf{P}_1 + t\mathbf{d}, \tag{A.12}$$

where t can take any real value.

▸ **Exercise A.6:** Derive the vector line equation for the straight segment between two given points $\mathbf{P}_1$ and $\mathbf{P}_2$.

A.4.2 The Scalar Triple Product

The scalar triple product of three vectors, $\mathbf{P}$, $\mathbf{Q}$, and $\mathbf{R}$, is defined as

$$\begin{aligned} S = \mathbf{P} \bullet (\mathbf{Q} \times \mathbf{R}) &= P_1(Q_2 R_3 - Q_3 R_2) + P_2(Q_3 R_1 - Q_1 R_3) + P_3(Q_1 R_2 - Q_2 R_1) \\ &= \begin{vmatrix} P_1 & P_2 & P_3 \\ Q_1 & Q_2 & Q_3 \\ R_1 & R_2 & R_3 \end{vmatrix}. \end{aligned} \tag{A.13}$$

Interchanging two rows in a determinant changes its sign, so interchanging rows twice leaves the determinant unchanged. This is why the triple product is not affected by a cyclic permutation of its three components. We can therefore write

$$S = \mathbf{P} \bullet (\mathbf{Q} \times \mathbf{R}) = \mathbf{Q} \bullet (\mathbf{R} \times \mathbf{P}) = \mathbf{R} \bullet (\mathbf{P} \times \mathbf{Q}).$$

The triple product has a simple geometric interpretation. It equals the volume of the parallelepiped defined by the three vectors. An important corollary is: If the three vectors are coplanar, then the parallelepiped defined by them has volume zero, implying that their scalar triple product is zero. This property is used to determine whether or not a given polygon is planar (Section 5.3.1).

A Mathematical Topics

A.4.3 Projecting a Vector

A common and useful operation on vectors is projecting a vector **a** on another vector **b**. The idea is to break vector **a** up into two perpendicular components **c** and **d**, such that **c** is in the direction of **b**.

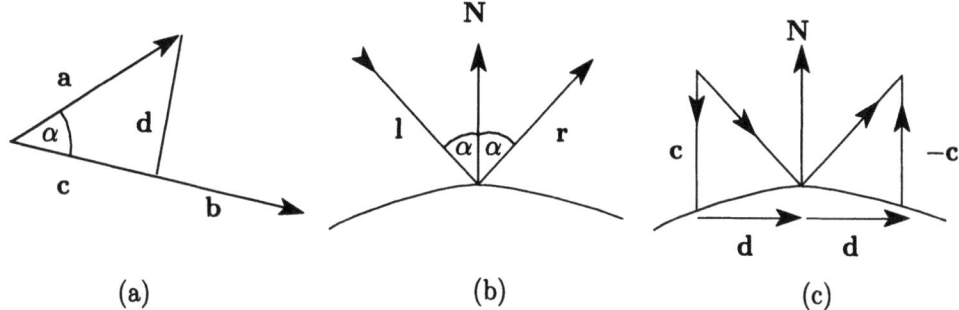

Figure A.8: Projecting a Vector.

Figure A.8a shows that $\mathbf{a} = \mathbf{c} + \mathbf{d}$ and $|\mathbf{c}| = |\mathbf{a}|\cos\alpha$. On the other hand, $\mathbf{a} \bullet \mathbf{b} = |\mathbf{a}||\mathbf{b}|\cos\alpha$, yielding the magnitude of **c**:

$$|\mathbf{c}| = |\mathbf{a}|\frac{(\mathbf{a} \bullet \mathbf{b})}{|\mathbf{a}||\mathbf{b}|} = \frac{(\mathbf{a} \bullet \mathbf{b})}{|\mathbf{b}|}. \tag{A.14}$$

The direction of **c** is identical to the direction of **b**, so we can write vector **c** as

$$\mathbf{c} = |\mathbf{c}|\frac{\mathbf{b}}{|\mathbf{b}|} = \frac{(\mathbf{a} \bullet \mathbf{b})}{|\mathbf{b}|^2}\mathbf{b}. \tag{A.15}$$

Example: Given vectors $\mathbf{a} = (2,1)$ and $\mathbf{b} = (1,0)$, it is easy to calculate

$$\mathbf{c} = \frac{(\mathbf{a} \bullet \mathbf{b})}{|\mathbf{b}|^2}\mathbf{b} = \frac{2\times 1 + 1\times 0}{1^2 + 0^2}(2,0) = (4,0), \qquad \mathbf{d} = \mathbf{a} - \mathbf{c} = (-2,1).$$

▶ **Exercise A.7:** The above projection method works also for three-dimensional vectors. Given vectors $\mathbf{a} = (2,1,3)$ and $\mathbf{b} = (1,0,-1)$, calculate the projection of **a** on **b**.

A.4.4 An Application

We apply vector projections to the calculation of the direction of reflection. Figure A.8b shows a ray of light **l** reflecting from a surface at a point with a normal vector **N**. The ray is reflected in a direction **r** such that the angle of incidence equals the angle of reflection. Assuming that **l** and **N** are given and that **N** is a unit vector, we calculate **r**. The idea is to project **l** in the direction of **N**, yielding

$$\mathbf{c} = \frac{\mathbf{l} \bullet \mathbf{N}}{|\mathbf{N}|^2}\mathbf{N} = (\mathbf{l} \bullet \mathbf{N})\mathbf{N}, \qquad \mathbf{d} = \mathbf{l} - \mathbf{c}. \tag{A.16}$$

Vector **r** is then given as the difference (Figure A.8c) $\mathbf{r} = \mathbf{d} - \mathbf{c} = \mathbf{l} - 2\mathbf{c} = \mathbf{l} - 2(\mathbf{l} \bullet \mathbf{N})\mathbf{N}$.

Equation (A.16) implies $|\mathbf{r}| = |\mathbf{l}|$. In practice, the intensity of the reflected light is lower than that of the incident beam and is determined by a shading model.

A.4.5 Length and Area of Curves

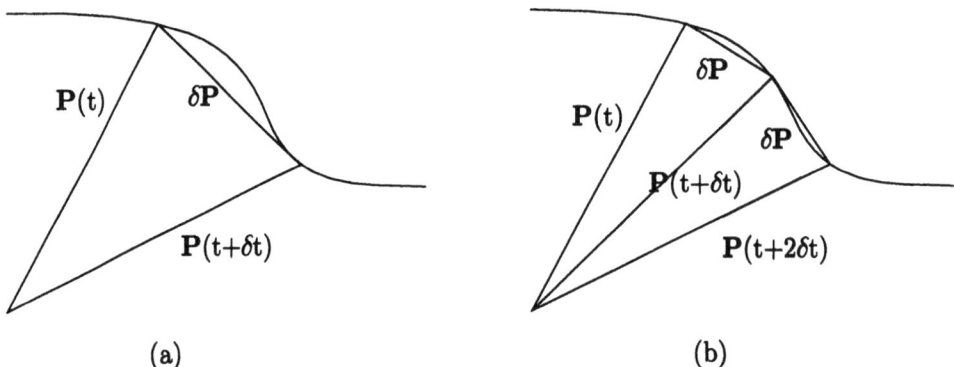

Figure A.9: Arc Length and Area of a Curve.

To calculate the arc length of the curve $\mathbf{P}(t)$, $0 \le t \le 1$, we first divide the arc into a large number of short, straight segments of length $\delta\mathbf{P}$. The length of the arc equals approximately the sum $\sum \delta\mathbf{P}$. Figure A.9a shows that $\delta\mathbf{P} = \mathbf{P}(t+\delta t) - \mathbf{P}(t)$. On the other hand, we can write $\delta\mathbf{P}(t)/\delta t \approx \mathbf{P}^t(t)$. In the limit, when $\delta\mathbf{P} \to 0$, we can write $d\mathbf{P}(t)/dt = \mathbf{P}^t(t)$ or $d\mathbf{P}(t) = \mathbf{P}^t(t)dt$ and get the exact arc length by replacing the sum $\sum \delta\mathbf{P}$ with the integral

$$\int |d\mathbf{P}(t)| = \int_0^1 |\mathbf{P}^t(t)|\, dt.$$

To find the area subtended at the origin by the vectors $\mathbf{P}(0)$ and $\mathbf{P}(1)$ and the curve, we again divide the curve into many straight segments of length $\delta\mathbf{P}$ and create the narrow triangles shown in Figure A.9b. The area of each triangular slice is $(1/2)\mathbf{P}(t) \times \delta\mathbf{P}$ or $(1/2)\mathbf{P}(t) \times \mathbf{P}^t(t)\delta t$, so, in the limit, the total area is the integral

$$1/2 \int_0^1 \mathbf{P}(t) \times \mathbf{P}^t(t)\, dt.$$

Note that the above expression is a vector. Its magnitude is the area and its direction is perpendicular to the plane defined by $\mathbf{P}(0)$ and $\mathbf{P}(1)$.

A.4.6 Example: Area of Planar Polygons

Vectors can be used to calculate areas of polygons (see Section 5.3.7 for the area of a triangle). Given a planar polygon with consecutive vertices $P_0, \ldots, P_n$, we denote

A Mathematical Topics

by $\mathbf{P}_i$ the vector from the origin to point P_i. The area of the polygon is then given by one of the following expressions:

$$\frac{1}{2}\left|\sum_{i=0}^{n}\mathbf{P}_i \times \mathbf{P}_{i+1}\right|, \qquad \frac{1}{2}\left|\mathbf{N}\bullet\left\{\sum_{i=0}^{n}\mathbf{P}_i \times \mathbf{P}_{i+1}\right\}\right|,$$

where $\mathbf{P}_{n+1} = \mathbf{P}_0$. The expression on the left applies to polygons that lie in the xy plane. The one on the right applies to polygons that lie in some plane perpendicular to a given unit vector $\mathbf{N}$. These expressions hold even for nonconvex polygons.

The first expression can be summarized in an easy-to-remember way. Assuming that the vertices $\mathbf{P}_i = (x_i, y_i)$ are enumerated counterclockwise, the area can be expressed by

$$\frac{1}{2}\left\|\begin{matrix}x_1 & x_2 & x_3 & \cdots & x_n & x_1 \\ y_1 & y_2 & y_3 & & y_n & y_1\end{matrix}\right\| \stackrel{\text{def}}{=} (x_1y_2 + x_2y_3 + \cdots + x_ny_1) - (x_2y_1 + x_3y_2 + \cdots + x_1y_n).$$

The operator pair $\| \cdots \|$ is defined as the sum of the products of the "downward" diagonals, minus the sum of the products of the "upward" diagonals.

Pick's theorem: The coordinates of a pixel are integers, so we can think of pixels as points of a grid. A polygon made of pixels is, thus, a *lattice polygon*. There is an elegant formula, called Pick's theorem, for the area of a lattice polygon. Given a polygon whose boundary consists of connected nonintersecting straight segments, its area is $I + B/2 - 1$, where I is the number of interior lattice points and B is the number of boundary lattice points. For example, the area of the simple lattice polygon of Figure A.10 is $31 + 15/2 - 1 = 37.5$.

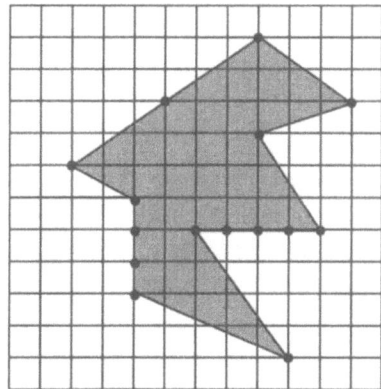

Figure A.10: Illustration of Pick's Theorem.

A.4.7 Example: Volume of Polyhedra

Assume that a polyhedron with planar polygonal faces $S_0, \ldots, S_n$ is given. We denote by $\mathbf{Q}_i$ any point on face S_i, and by $\mathbf{N}_i$ the outward pointing unit normal vector to face S_i. The volume of the polygon is then given by

$$\frac{1}{3}\left|\sum_{i=0}^{n}(\mathbf{Q}_i \bullet \mathbf{N}_i)\text{area}(S_i)\right|,$$

where $\mathbf{Q}_i$ is the vector from the origin to point Q_i.

We can apply the previous example to the areas of the faces. Denote by $\mathbf{P}_{0j},\ldots,\mathbf{P}_{mj}$ the vertices of face S_j enumerated counterclockwise with respect to $\mathbf{N}_j$; then, the area of S_j is

$$\frac{1}{2}\left|\mathbf{N}_j \bullet \left\{\sum_{j=0}^{m}\mathbf{P}_{kj} \times \mathbf{P}_{k+1,j}\right\}\right|.$$

We simplify the result above in two ways: (1) Instead of a general point on face S_j, we choose vertex $\mathbf{P}_{0j}$, and (2) we express the unit normal to face S_j in terms of the three vertices $\mathbf{P}_{0j}$, $\mathbf{P}_{1j}$, and $\mathbf{P}_{2j}$.

$$\mathbf{N}_j = \frac{(\mathbf{P}_{1j} - \mathbf{P}_{0j}) \times (\mathbf{P}_{2j} - \mathbf{P}_{0j})}{|(\mathbf{P}_{1j} - \mathbf{P}_{0j}) \times (\mathbf{P}_{2j} - \mathbf{P}_{0j})|}.$$

Combining all this results in the following expression for the volume:

$$\frac{1}{6}\left|\left[\sum_j(\mathbf{P}_{0j} \bullet \mathbf{N}_j)\left(\left|\mathbf{N}_j \bullet \left\{\sum_k \mathbf{P}_{kj} \times \mathbf{P}_{k+1,j}\right\}\right|\right)\right]\right|.$$

Note, again, that this expression holds even for nonconvex polyhedra.

A.4.8 Tensors

Let $\mathbf{v} = (v_x, v_y, v_z)$ be a unit vector. Its three components depend on its direction. They will change when the vector changes direction or when the coordinate system is transformed. However, they always satisfy $v_x^2 + v_y^2 + v_z^2 = 1$. Let $\mathbf{u}$ be another unit vector. We can create the nine products $v_i u_j$ and arrange them in a 3×3 matrix:

$$\mathbf{T} = \begin{pmatrix} v_x u_x & v_x u_y & v_x u_z \\ v_y u_x & v_y u_y & v_y u_z \\ v_z u_x & v_z u_y & v_z u_z \end{pmatrix}.$$

When the coordinate system is translated and/or rotated, the nine components of $\mathbf{T}$ change, but not independently. They reflect the fact that the lengths of the vectors and the angle between them are invariant under these transformations. Matrix $\mathbf{T}$ thus obeys special transformation rules and is called a *tensor*. A tensor is any matrix that obeys these rules, not just matrix $\mathbf{T}$ above.

A Mathematical Topics

> My dear sir! In this wireless age any owl rooster can peck up bostoons. But whoewaxed he so anquished? Was he vector victored of victim vexed?
>
> —James Joyce, *Finnegans Wake*.

A.5 Matrices

A matrix **T** is a rectangular array of numbers, where each element a_{ij} is identified by its row and column. Matrix $\mathbf{T}_1$ below is "generic," with m rows and n columns. Matrix $\mathbf{T}_2$ is diagonal, matrix $\mathbf{T}_3$ is symmetric, and $\mathbf{T}_4$ is an identity matrix.

$$\mathbf{T}_1 = \begin{pmatrix} a_{11} & a_{12} & \cdots & a_{1n} \\ a_{21} & a_{22} & \cdots & a_{2n} \\ \vdots & \vdots & \ddots & \vdots \\ a_{m1} & a_{m2} & \cdots & a_{mn} \end{pmatrix}, \quad \mathbf{T}_2 = \begin{pmatrix} a_{11} & 0 & 0 & 0 \\ 0 & a_{22} & 0 & 0 \\ 0 & 0 & a_{33} & 0 \\ 0 & 0 & 0 & a_{44} \end{pmatrix},$$

$$\mathbf{T}_3 = \begin{pmatrix} 33 & -17 & 201 & -5 \\ -17 & 66 & 26 & -68 \\ 201 & 26 & 21 & -9 \\ -5 & -68 & -9 & 0 \end{pmatrix}, \quad \mathbf{T}_4 = \begin{pmatrix} 1 & 0 & 0 & 0 \\ 0 & 1 & 0 & 0 \\ 0 & 0 & 1 & 0 \\ 0 & 0 & 0 & 1 \end{pmatrix}.$$

The transpose of matrix **A** (denoted $\mathbf{A}^T$) is obtained from **A** by reflecting all the elements with respect to the main diagonal. A symmetric matrix equals its transpose.

The rule for matrix addition/subtraction is $c_{ij} = a_{ij} \pm b_{ij}$, where $\mathbf{C} = \mathbf{A} \pm \mathbf{B}$. The rule for matrix multiplication is more complex. It is $c_{ij} = \sum_{k=1}^{n} a_{ik} b_{kj}$. Each element of **C** is, thus, the dot product of a row of **A** and a column of **B**. In the dot product, corresponding elements are multiplied and the results are summed. In order for the multiplication to make sense, each row of **A** must have the same size as a column of **B**. Matrices **A** and **B** can, therefore, be multiplied only if the number of columns of **A** equals the number of rows of **B**. Note that matrix multiplication is not commutative (i.e., $\mathbf{AB} \neq \mathbf{BA}$ in general).

Tensor Products. This is a special case of a matrix multiplication. If **A** is a column vector and **B** is a row vector (each with n elements), then their tensor product **C** is defined by $c_{ij} = a_i b_j$. For example,

$$\begin{pmatrix} 4 \\ -2 \\ 3 \end{pmatrix} (1, -1, 5) = \begin{pmatrix} 4 & -4 & 20 \\ -2 & 2 & -10 \\ 3 & -3 & 15 \end{pmatrix}.$$

A square matrix has a determinant, denoted either $\det \mathbf{A}$ or $|\mathbf{A}|$, that is a function. Its argument is the square matrix and its result is a single number. The determinant of the 2×2 matrix $\begin{pmatrix} a & b \\ c & d \end{pmatrix}$ is defined as $ad - bc$. The determinant of a larger matrix can be calculated by the rule (note the alternating signs)

$$\begin{vmatrix} a_{11} & a_{12} & a_{13} \\ a_{21} & a_{22} & a_{23} \\ a_{31} & a_{32} & a_{33} \end{vmatrix} = a_{11} \begin{vmatrix} a_{22} & a_{23} \\ a_{32} & a_{33} \end{vmatrix} - a_{12} \begin{vmatrix} a_{21} & a_{23} \\ a_{31} & a_{33} \end{vmatrix} + a_{13} \begin{vmatrix} a_{21} & a_{22} \\ a_{31} & a_{32} \end{vmatrix}.$$

Matrix division is not defined, but certain matrices have an *inverse*. The inverse of **A** is denoted $\mathbf{A}^{-1}$ and has the property that $\mathbf{A}\mathbf{A}^{-1} = \mathbf{A}^{-1}\mathbf{A} = I$, where **I** is the *identity matrix* (with ones on the diagonal and zeros elsewhere).

The inverse of the 2×2 matrix

$$\mathbf{T} = \begin{pmatrix} a & b \\ c & d \end{pmatrix} \quad \text{is} \quad \mathbf{T}^{-1} = \frac{1}{ad - bc} \begin{pmatrix} d & -b \\ -c & a \end{pmatrix}. \tag{A.17}$$

It exists only if $ad - bc \neq 0$ or, equivalently, if $\det \mathbf{T} \neq 0$. A matrix whose determinant is zero is called *singular*. The inverse of the 3×3 transformation matrix

$$\mathbf{T} = \begin{pmatrix} a & b & 0 \\ c & d & 0 \\ m & n & 1 \end{pmatrix} \quad \text{is} \quad \mathbf{T}^{-1} = \frac{1}{ad - bc} \begin{pmatrix} d & -b & 0 \\ -c & a & 0 \\ cn - dm & bm - an & 1 \end{pmatrix}.$$

(see Equation (3.22)). In general, however, the calculation of the inverse is not trivial and can be found in any text on linear algebra and in [Press and Flannery 88]. Not every matrix has an inverse. A matrix that has an inverse is called nonsingular. It is easy to prove that if matrix **A** equals the product $\mathbf{T}_1 \cdot \mathbf{T}_2 \cdot \mathbf{T}_3$, then its inverse $\mathbf{A}^{-1}$ equals the product $\mathbf{T}_3^{-1} \cdot \mathbf{T}_2^{-1} \cdot \mathbf{T}_1^{-1}$. This fact is used in Section 3.9.2.

A matrix where all the off-diagonal elements are 0 is called *diagonal*. A matrix with nonzero elements only on the main diagonal and the 1's immediately above and below it is called *tridiagonal*.

An $m \times n$ matrix where the absolute value of every diagonal element a_{ii} exceeds the sum of the absolute values of the off-diagonal elements in row i is called *rowwise diagonally dominant*. Similarly, a matrix where the absolute value of every a_{jj} exceeds the sum of the absolute values of the off-diagonal elements in column j is called *columnwise diagonally dominant*. Either type is called a *diagonally dominant* matrix. Such a matrix satisfies one of the conditions

$$|a_{ii}| > \sum_{j \neq i} |a_{ij}|, \quad (i = 1, \ldots, m), \quad \text{or} \quad |a_{jj}| > \sum_{i \neq j} |a_{ij}|, \quad (j = 1, \ldots, n). \tag{A.18}$$

▶ **Exercise A.8:** Assume that we change the relational ">" in the above conditions to the weaker relational "≥". Would that change the intuitive meaning of the word "dominant"?

The 3×3 matrix of Equation (A.19) is an example of a rowwise diagonally dominant matrix:

$$\mathbf{A} = \begin{pmatrix} 1 & 0 & 0 \\ 0 & 1 & 0 \\ 0.6 & 0.6 & 1 \end{pmatrix}. \tag{A.19}$$

A Mathematical Topics

▶ **Exercise A.9:** Show an example of a columnwise diagonally dominant matrix.

> For I was not then aware that amongst the fossils from Bahia Blanca there was a horse's tooth hidden in the matrix.
> — Charles Darwin, *Voyage of the Beagle.*

It can be shown that a diagonally dominant matrix is *nonsingular*. This is what makes these matrices important in computer graphics and in other practical applications. If a system of linear algebraic equations has a diagonally dominant coefficient matrix, then this system is guaranteed to have a (unique) solution. For example, the cubic spline basis matrix, Equation (4.57), is tridiagonal and thus diagonally dominant, guaranteeing the existence of a unique cubic spline.

The following is a summary of the properties of matrix operations:

$$k(\mathbf{A}+\mathbf{B}) = k\mathbf{A}+k\mathbf{B}, \quad (k+m)\mathbf{A} = k\mathbf{A}+m\mathbf{A}, \quad k(m\mathbf{A}) = (km)\mathbf{A} = m(k\mathbf{A}),$$
$$\mathbf{A}(\mathbf{BC}) = (\mathbf{AB})\mathbf{C}, \quad \mathbf{A}(\mathbf{B}+\mathbf{C}) = \mathbf{AB}+\mathbf{AC}, \quad (\mathbf{A}+\mathbf{B})\mathbf{C} = \mathbf{AB}+\mathbf{AC},$$
$$(\mathbf{A}+\mathbf{B})^T = \mathbf{A}^T+\mathbf{B}^T, \quad (k\mathbf{A})^T = k\mathbf{A}^T, \quad (\mathbf{AB})^T = \mathbf{B}^T\mathbf{A}^T,$$
$$\mathbf{A}+\mathbf{B} = \mathbf{B}+\mathbf{A}, \quad \mathbf{A}+(\mathbf{B}+\mathbf{C}) = (\mathbf{A}+\mathbf{B})+\mathbf{C},$$
$$\mathbf{A} = \mathbf{T}_1\mathbf{T}_2\mathbf{T}_3 \to \mathbf{A}^{-1} = \mathbf{T}_3^{-1}\mathbf{T}_2^{-1}\mathbf{T}_1^{-1},$$
$$\mathbf{A}(k\mathbf{B}) = k(\mathbf{AB}) = (k\mathbf{A})\mathbf{B}.$$

[Turnbull 98] has a short history of determinants and matrices.

> All problems in computer graphics can be solved with a matrix inversion.
> — James F. Blinn, 1993.

A.6 Trigonometric Identities

Basic Identities

$$\tan\alpha = \frac{\sin\alpha}{\cos\alpha}, \quad \cot\alpha = \frac{\cos\alpha}{\sin\alpha} = \frac{1}{\tan\alpha}, \quad \csc\alpha = \frac{1}{\sin\alpha}, \quad \sec\alpha = \frac{1}{\cos\alpha};$$
$$\sin(-\alpha) = -\sin\alpha, \quad \cos(-\alpha) = \cos\alpha, \quad \tan(-\alpha) = -\tan\alpha;$$
$$\sin^2\alpha + \cos^2\alpha = 1, \quad \tan^2\alpha + 1 = \sec^2\alpha, \quad \cot^2\alpha + 1 = \csc^2\alpha.$$

Sum and Difference Identities

$$\cos(\alpha \pm \beta) = \cos\alpha\cos\beta \mp \sin\alpha\sin\beta, \quad \sin(\alpha \pm \beta) = \sin\alpha\cos\beta \pm \cos\alpha\sin\beta,$$
$$\tan(\alpha \pm \beta) = \frac{\tan\alpha \pm \tan\beta}{1 \mp \tan\alpha\tan\beta}.$$

Cofunction Identities

$$\sin(\pi/2 - \alpha) = \cos\alpha, \quad \cos(\pi/2 - \alpha) = \sin\alpha, \quad \tan(\pi/2 - \alpha) = \cot\alpha.$$

Multiple-Angle and Half-Angle Identities

$$\cos 2\alpha = \cos^2 \alpha - \sin^2 \alpha = 1 - 2\sin^2 \alpha = 2\cos^2 \alpha - 1, \quad \sin 2\alpha = 2\sin\alpha\cos\alpha,$$

$$\tan 2\alpha = \frac{2\tan\alpha}{1-\tan^2\alpha};$$

$$\cos(\alpha/2) = \pm\sqrt{(1+\cos\alpha)/2}, \quad \sin(\alpha/2) = \pm\sqrt{(1+\cos\alpha)/2};$$

$$\tan(\alpha/2) = \pm\sqrt{\frac{1-\cos\alpha}{1+\cos\alpha}} = \frac{\sin\alpha}{1+\cos\alpha} = \frac{1-\cos\alpha}{\sin\alpha}.$$

Sum and Product Identities

$$\sin\alpha + \sin\beta = 2\sin\left(\frac{\alpha+\beta}{2}\right)\cos\left(\frac{\alpha-\beta}{2}\right),$$

$$\sin\alpha - \sin\beta = 2\cos\left(\frac{\alpha+\beta}{2}\right)\sin\left(\frac{\alpha-\beta}{2}\right).$$

$$\cos\alpha + \cos\beta = 2\cos\left(\frac{\alpha+\beta}{2}\right)\cos\left(\frac{\alpha-\beta}{2}\right),$$

$$\cos\alpha - \cos\beta = -2\sin\left(\frac{\alpha+\beta}{2}\right)\sin\left(\frac{\alpha-\beta}{2}\right).$$

$$\sin\alpha\cos\beta = \frac{1}{2}[\sin(\alpha+\beta)+\sin(\alpha-\beta)],$$

$$\cos\alpha\sin\beta = \frac{1}{2}[\sin(\alpha+\beta)-\sin(\alpha-\beta)].$$

$$\cos\alpha\cos\beta = \frac{1}{2}[\cos(\alpha+\beta)+\cos(\alpha-\beta)],$$

$$\sin\alpha\sin\beta = -\frac{1}{2}[\cos(\alpha+\beta)-\cos(\alpha-\beta)].$$

Note that the last identity also implies

$$\cos^2\alpha = \frac{1}{2}(\cos(2\alpha)+1), \quad \sin^2\alpha = \frac{1}{2}(1-\cos(2\alpha)).$$

Laws of Sines and Cosines: Any triangle with sides a, b, and c and angles α, β, and γ satisfies the law of sines $a/\sin\alpha = b/\sin\beta = c/\sin\gamma$ and the law of cosines

$$a^2 = b^2 + c^2 - 2bc\cos\alpha, \quad b^2 = a^2 + c^2 - 2ac\cos\beta, \quad c^2 = a^2 + b^2 - 2ab\cos\gamma.$$

A.6.1 Example: Great Circle Distances

This example applies trigonometry and vector analysis to the practical problem of finding the distance between two points on the surface of the Earth.

We initially assume that Earth is a sphere of radius R. The distance between two points P and Q on a sphere is the length of the arc connecting them, an arc

A Mathematical Topics

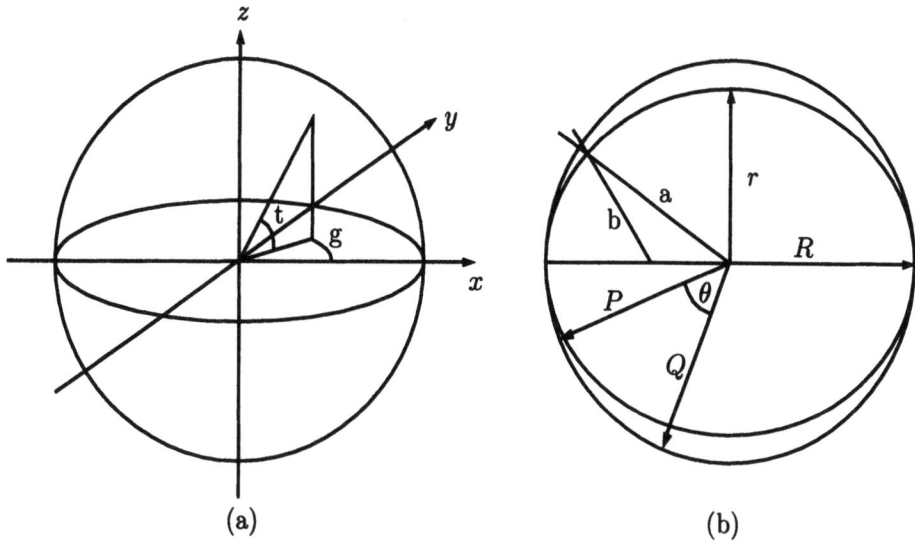

Figure A.11: (a) Spherical to Cartesian. (b) Sphere and Spheroid.

that is part of a great circle on the sphere. Figure A.11b shows that the length of the circular arc between points P and Q is $R\theta$. We, therefore, have to express θ in terms of the coordinates of the two points. We assume that point P has spherical coordinates (latitude and longitude) t_1 and g_1. The spherical coordinates of Q are similarly (t_2, g_2). The first step is to convert the spherical coordinates to Cartesian ones. From Figure A.11a, it is easy to derive the relations

$$x = R\cos(t)\cos(g), \quad y = R\cos(t)\sin(g), \quad z = R\sin(t).$$

The next step is to use the dot product of vectors $\mathbf{P}$ and $\mathbf{Q}$. Recall that the dot product is defined as $\mathbf{P} \bullet \mathbf{Q} = P_x Q_x + P_y Q_y + P_z Q_z$ but also equals $|\mathbf{P}||\mathbf{Q}|\cos\theta$. In our case, $|\mathbf{P}| = |\mathbf{Q}| = R$. We, therefore, write

$$\begin{aligned} R \times R \times \cos\theta &= R \times R[\cos(t_1)\cos(g_1)\cos(t_2)\cos(g_2) \\ &\quad + \cos(t_1)\sin(g_1)\cos(t_2)\sin(g_2) + \sin(t_1)\sin(t_2)] \\ &= R^2[\cos(t_1)\cos(t_2)\cos(g_2 - g_1) + \sin(t_1)\sin(t_2)]. \end{aligned}$$

The angle between the two points is thus

$$\theta = \arccos[\cos(t_1)\cos(t_2)\cos(g_2 - g_1) + \sin(t_1)\sin(t_2)],$$

and the length of the arc between them is $R\theta$ (in radians) or $(2\pi/360)R\theta$ (in degrees).

Note that $\cos(a - b) = \cos(b - a)$, so reversing the two points leaves $\mathbf{P} \bullet \mathbf{Q}$ unchanged.

In reality, Earth is not a sphere but an oblate spheroid, squashed at the poles. It has an equatorial radius

$$R = 6{,}378{,}137 \pm 2 \text{ meters} = 20{,}925{,}646.3254 \text{ feet}$$
$$= 3443 \text{ nautical miles} = 3963 \text{ statute miles},$$

and a polar radius

$$r = 6{,}356{,}752 \text{ meters} = 20{,}855{,}486 \text{ feet}$$
$$= 3432 \text{ nautical miles} = 3950 \text{ statute miles}.$$

The intersection of Earth with a plane that goes through the poles is thus an ellipse with radii R and r. Unfortunately, arc lengths in an ellipse are given by the elliptic integral

$$R \int_0^\theta \sqrt{1 - e^2 \cos^2 \theta}\, d\theta$$

[this is the arc length from point $(R, 0)$ to point $(R\cos\theta, r\sin\theta)$], which is impossible to calculate analytically, only numerically. We, therefore, calculate the arc length approximately by selecting the radius at the point with latitude $T = (t_1 + t_2)/2$ and calculating the quantity $R(T)\theta$. The radius $R(T)$ at any latitude T is given by

$$\frac{R \times r}{\sqrt{R^2 - (R^2 - r^2)\cos^2 T}}.$$

▶ **Exercise A.10:** Prove the above relation.

> When people thought the Earth was flat, they were wrong. When people thought the Earth was spherical they were wrong. But if you think that thinking the Earth is spherical is just as wrong as thinking the Earth is flat, then your view is wronger than both of them put together.
> —Isaac Asimov, *The Relativity of Wrong*.

Another point that needs to be mentioned is the difference between geocentric and geodetic latitudes. A geocentric latitude is defined as the angle between the equatorial plane and a line that goes from the center to the surface. This latitude was used in the derivation above (line "a" in Figure A.11b). However, the latitudes given in maps and globes for points on Earth are normally geodetic ones. The geodetic latitude of a point on the surface of a spheroid is the angle between the equatorial plane and a line through the point that is normal to the surface at the point (line "b" in Figure A.11b). Converting from geodetic to geocentric latitudes is done by

$$\text{Geocentric latitude} = \arctan\left[\left(\frac{r}{R}\right)^2 \tan(\text{geodetic latitude})\right].$$

A Mathematical Topics

The last point has to do with the calculation of the arccos function. Most calculators support just arctan, so here are two useful conversion formulas.

$$\arcsin\theta = \begin{cases} \theta = 1, & \pi/2, \\ \theta = -1, & -\pi/2, \\ \text{else}, & \arctan\frac{\theta}{\sqrt{1-\theta^2}}, \end{cases} \qquad \arccos\theta = \begin{cases} \theta = 1, & 0, \\ \theta = -1, & \pi, \\ \text{else}, & \frac{\pi}{2} - \arctan\frac{\theta}{\sqrt{1-\theta^2}}. \end{cases}$$

Tests: The following *Mathematica* code implements the great circle distance calculation. It is tested for the distance between Chicago and Honolulu (approximately 4235 miles), but notice the test for the distance between the poles, where the latitude of the south pole is negative! Notice, also, that longitudes can vary either from 0 to 180 or from −90 to 90.

```
Clear[geodLat1,long1,geodLat2,long2,rEquat,rPolar,rAverge,teta];
rEquat=3949.90276422; rPolar=3963.19059194; (* in miles *)
(* 'Degree' converts degrees to radians below *)
geodLat1=Degree 42; long1=Degree 88;    (* Chicago *)
geodLat2=Degree 21.3; long2=Degree 158; (* Honolulu *)
(* geodLat1=Degree 90; long1=Degree 88;   North pole *)
(* geodLat2=-Degree 90; long2=Degree 88;  South pole *)
lat1=ArcTan[(rPolar/rEquat)^2 Tan[geodLat1]];
lat2=ArcTan[(rPolar/rEquat)^2 Tan[geodLat2]];
rAverge=(rEquat*rPolar)/
  Sqrt[rEquat^2-(rEquat^2-rPolar^2)*Cos[(lat1+lat2)/2]^2];
teta=ArcCos[Cos[lat1]Cos[lat2]Cos[long2-long1]+Sin[lat1]Sin[lat2]];
N[teta*rAverge]
```

> This was something new and very interesting, and he felt that it was of greater import than trigonometry, which he never could understand. It was like a window on life that he had a chance of peeping through, and he looked with a wildly beating heart.
> —W. Somerset Maugham, *Of Human Bondage*.

A.7 The Greek Alphabet

A	α	alpha	I	ι	iota	P	ρ ϱ	rho
B	β	beta	K	κ	kappa	Σ	σ ς	sigma
Γ	γ	gamma	Λ	λ	lambda	T	τ	tau
Δ	δ	delta	M	μ	mu	Y	υ	upsilon
E	ϵ	epsilon	N	ν	nu	Φ	ϕ φ	phi
Z	ζ	zeta	Ξ	ξ	xi	X	ξ	xi
H	η	eta	O	o	omicron	Ψ	ψ	psi
Θ	θ	theta	Π	π ϖ	pi	Ω	ω	omega

A.8 Complex Numbers

Complex numbers are expressed in terms of the special number i that is defined as $\sqrt{-1}$ and, hence, satisfies $i \times i = i^2 = -1$. Any complex number z can be represented either as the sum $a+bi$ or as the pair (a,b), where a and b are real. The *conjugate* of z is denoted z^* and is defined as $a - bi$. Complex conjugates roughly correspond to negative real numbers. The sum of the real numbers a and $-a$ is zero and the sum $z + z^*$ is $2a$, which is real. The *magnitude* or *absolute value* of a complex number is denoted $|z|$ and is defined as $\sqrt{z \cdot z^*} = \sqrt{a^2 + b^2}$. The sum and the difference of the complex numbers $a + bi$, $c + di$ are the obvious $(a+b) \pm (c+d)i$. The product makes use of the relation $i^2 = -1$ and is $(a+bi)(c+di) = (ac - bd) + (ad + bc)i$. The *inverse*, z^{-1}, of z is defined as $z^*/|z|$. It corresponds to the reciprocal $1/a$ of a real number a. The division z_1/z_2 is easy to perform for $|z_2| \neq 0$:

$$\frac{z_1}{z_2} = \frac{z_1 z_2^*}{z_2 z_2^*} = \frac{(a+bi)(c-di)}{c^2 + d^2} = \left(\frac{ac+bd}{c^2+d^2}, \frac{bc-ad}{c^2+d^2}\right).$$

The multiplication rule of complex numbers can be interpreted as a rotation in two dimensions. This is easy to see if we consider the product of the two complex numbers (x, y) and $(\cos \theta, \sin \theta)$.

$$\begin{aligned}(x, y) \cdot (\cos \theta, \sin \theta) &= (x \cos \theta - y \sin \theta, x \sin \theta + y \cos \theta) \\ &= (x, y) \begin{pmatrix} \cos \theta & \sin \theta \\ -\sin \theta & \cos \theta \end{pmatrix}.\end{aligned} \quad (A.20)$$

This product rotates the two-dimensional point (x, y) through an angle θ about the origin. It should be noted that quaternions, which are introduced in Section A.9, have a multiplication rule that can be interpreted as a rotation in three dimensions (this is discussed in Section 3.5.4).

We normally use the *Cartesian coordinates* (a, b) of a point **P**. The *polar coordinates* of **P** are (r, θ), where $r = \sqrt{a^2 + b^2}$ is the distance of the point from the origin and $\theta = \arctan(b/a)$ is the angle between the x axis and vector **r**. The quantities r and θ are also called the *absolute value* and the *argument* (**arg** for short) of the complex number $z = (a, b) = a + ib$. The polar coordinates can be obtained from the Cartesian ones by $(a, b) = (r \cos \theta, r \sin \theta)$. Since the complex number $z = (a, b)$ can be interpreted as a two-dimensional point, it has the polar representation $z = (r \cos \theta, r \sin \theta) = r(\cos \theta + i \sin \theta)$. This representation is useful in many applications.

The famous Euler formula

$$e^{i\theta} = \cos \theta + i \sin \theta$$

allows us to write $z = re^{i\theta}$, a representation that makes it easy to multiply/divide complex numbers

$$z_1 z_2 = r_1 r_2 e^{i(\theta_1 + \theta_2)}, \quad \frac{z_1}{z_2} = r_1 r_2 e^{i(\theta_1 - \theta_2)},$$

A Mathematical Topics

and even extract roots

$$\sqrt[n]{z} = \sqrt[n]{r}\left[\cos\left(\frac{\theta + 2k\pi}{n}\right) + i\sin\left(\frac{\theta + 2k\pi}{n}\right)\right], \quad k = 0, 1, \ldots, n-1.$$

The n roots of z can be visualized as equally spaced points lying on the circumference of a circle of radius $\sqrt[n]{r}$ whose center is at the origin. Connecting them produces an n-sided regular polygon.

▶ **Exercise A.11:** (Mathematical.) We know that $i = \sqrt{-1}$. What is $\sqrt{i}$?

▶ **Exercise A.12:** While we are at it, what are i^i and $\ln i$?

> The shortest path between two truths in the real domain passes through the complex domain.
> —Jacques Hadamard.

A.9 Quaternions

Complex numbers can be interpreted as points in the xy plane. The complex number (a, b) can be interpreted as the point with coordinates (a, b). Is it possible to define *hypercomplex* numbers of the form (a, b, c) that could be interpreted as three-dimensional points? This question bothered the Irish mathematician William Rowan Hamilton (1805–1865) for a long time. The problem was that multiplying complex numbers can be interpreted as a rotation in two dimensions (Section 3.5.4) so it made sense to require that multiplying the new hypercomplex numbers would be equivalent to a rotation in three dimensions. Readers of this book know (from Section 3.5.3) that a general rotation in three dimensions is fully defined by four numbers: one for the rotation angle and three for the rotation axis. Three numbers are not enough to fully specify such a rotation.

Hamilton could not come up with a reasonable rule to multiply hypercomplex numbers that are triplets, and he eventually discovered, in October 1843, that he needed to add a fourth component to his triplets, i.e., turn them into 4-tuples, in order to multiply them in a way that made sense. He called these new entities *quaternions*. Using modern notation, a quaternion $\mathbf{q}$ can be represented using 2×2 matrices of complex numbers:

$$\mathbf{q} = \begin{pmatrix} z & w \\ -w^* & z^* \end{pmatrix} = \begin{pmatrix} a + ib & c + id \\ -c + id & a - ib \end{pmatrix},$$

where z and w are complex numbers and a, b, c, and d are real. This can also be written (by analogy with the complex numbers $a \cdot 1 + b \cdot i$) as $\mathbf{q} = a\mathbf{U} + b\mathbf{I} + c\mathbf{J} + d\mathbf{K}$, where

$$\mathbf{U} = \begin{pmatrix} 1 & 0 \\ 0 & 1 \end{pmatrix}, \quad \mathbf{I} = \begin{pmatrix} i & 0 \\ 0 & -i \end{pmatrix}, \quad \mathbf{J} = \begin{pmatrix} 0 & 1 \\ -1 & 0 \end{pmatrix}, \quad \text{and } \mathbf{K} = \begin{pmatrix} 0 & i \\ i & 0 \end{pmatrix}.$$

(Note that **U**, not **I**, is used here to denote the identity matrix. These matrices are closely related to the Pauli spin matrices used in particle physics.) From the above definitions, it follows that $\mathbf{I}^2 = -\mathbf{U}$, $\mathbf{J}^2 = -\mathbf{U}$, and $\mathbf{K}^2 = -\mathbf{U}$. We, therefore, conclude that **I**, **J**, and **K** are three different solutions of the matrix equation $\mathbf{X}^2 = -\mathbf{U}$. They should, thus, be considered the square roots of minus the identity matrix.

Quaternions can also be considered elements of a four-dimensional *vector space*, one of whose bases is given by

$$\mathbf{i} = \begin{pmatrix} 0 & 1 & 0 & 0 \\ -1 & 0 & 0 & 0 \\ 0 & 0 & 0 & 1 \\ 0 & 0 & -1 & 0 \end{pmatrix}, \quad \mathbf{j} = \begin{pmatrix} 0 & 0 & 0 & -1 \\ 0 & 0 & -1 & 0 \\ 0 & 1 & 0 & 0 \\ 1 & 0 & 0 & 0 \end{pmatrix},$$

$$\mathbf{k} = \begin{pmatrix} 0 & 0 & -1 & 0 \\ 0 & 0 & 0 & 1 \\ 1 & 0 & 0 & 0 \\ 0 & -1 & 0 & 0 \end{pmatrix}, \quad \mathbf{1} = \begin{pmatrix} 1 & 0 & 0 & 0 \\ 0 & 1 & 0 & 0 \\ 0 & 0 & 1 & 0 \\ 0 & 0 & 0 & 1 \end{pmatrix}.$$

Quaternions satisfy the following identities, also known as Hamilton's Rules,

$$\mathbf{i}^2 = \mathbf{j}^2 = \mathbf{k}^2 = -1, \quad \mathbf{ij} = -\mathbf{ji} = \mathbf{k}, \quad \mathbf{jk} = -\mathbf{kj} = \mathbf{i}, \quad \mathbf{ki} = -\mathbf{ik} = \mathbf{j}.$$

They have the following multiplication table:

	1	i	j	k
1	1	i	j	k
i	i	-1	k	-j
j	j	-k	-1	i
k	k	j	-i	-1

The eight quaternions ± 1, $\pm \mathbf{i}$, $\pm \mathbf{j}$, and $\pm \mathbf{k}$ form a group of order 8 with multiplication as the group operation.

Quaternions can also be interpreted as a combination of a scalar and a vector. They are thus closely related to 4-vectors. Using this interpretation, a quaternion q can be represented as the sum $\mathbf{q} = w + x\mathbf{i} + y\mathbf{j} + z\mathbf{k}$, the 4-tuple (x, y, z, w), or the pair $[s, \mathbf{v}]$, where $s = w$ and $\mathbf{v} = (x, y, z)$.

The conjugate quaternion is given by $\mathbf{q}^* = w - x\mathbf{i} - y\mathbf{j} - z\mathbf{k}$. The sum/difference of two quaternions is the obvious

$$\mathbf{q}_1 \pm \mathbf{q}_2 = (w_1 + w_2) \pm (x_1 + x_2)\mathbf{i} \pm (y_1 + y_2)\mathbf{j} \pm (z_1 + z_2)\mathbf{k} = [s_1 \pm s_2, (\mathbf{v}_1 \pm \mathbf{v}_2)],$$

and the product is the nonobvious

$$\begin{aligned}\mathbf{q}_1 \cdot \mathbf{q}_2 &= (w_1 w_2 - x_1 x_2 - y_1 y_2 - z_1 z_2) + (w_1 x_2 + x_1 w_2 + y_1 z_2 - z_1 y_2)\mathbf{i} \\ &\quad + (w_1 y_2 - x_1 z_2 + y_1 w_2 + z_1 x_2)\mathbf{j} + (w_1 z_2 + x_1 y_2 - y_1 x_2 + z_1 w_2)\mathbf{k} \\ &= [(s_1 s_2 - \mathbf{v}_1 \bullet \mathbf{v}_2), (s_1 \mathbf{v}_2 + s_2 \mathbf{v}_1 + \mathbf{v}_1 \times \mathbf{v}_2)].\end{aligned}$$

A Mathematical Topics

Quaternion product is associative, i.e., $(\mathbf{q}_1\mathbf{q}_2)\mathbf{q}_3 = \mathbf{q}_1(\mathbf{q}_2\mathbf{q}_3)$, but it is not commutative.

An appropriate measure of the size of a quaternion is its *norm*, defined as

$$|\mathbf{q}| = \mathbf{q}\cdot\mathbf{q}^* = \mathbf{q}^*\cdot\mathbf{q} = \sqrt{w^2 + x^2 + y^2 + z^2} = \sqrt{s^2 + x^2 + y^2 + z^2}.$$

It is easy to verify that the norm is multiplicative, $|\mathbf{q}_1\mathbf{q}_2| = |\mathbf{q}_1|\,|\mathbf{q}_2|$ (i.e., the norm of a product equals the product of the two individual norms). A *unit quaternion* is one for which $|\mathbf{q}| = 1$.

The inverse of a quaternion is given by

$$\mathbf{q}^{-1} = \frac{\mathbf{q}^*}{(\mathbf{q}\mathbf{q}^*)} = \frac{\mathbf{q}^*}{|\mathbf{q}|^2} = \frac{\mathbf{q}^*}{w^2 + x^2 + y^2 + z^2},$$

so quaternion division $\mathbf{q}_1/\mathbf{q}_2$ (except by zero) is performed by multiplying $\mathbf{q}_1$ by the inverse $\mathbf{q}_2^{-1}$. It's easy to verify that $\mathbf{q}\mathbf{q}^{-1} = [1, (0,0,0)] = [0, \mathbf{0}]$.

[Baker 98] is a large quaternion bibliography, including a collection of articles. [Bailey 98] discusses many properties of quaternions.

> Every morning in the early part of the above-cited month [Oct. 1843] on my coming down to breakfast, your brother William Edwin and yourself used to ask me, "Well, Papa, can you multiply triplets?" Whereto I was always obliged to reply, with a sad shake of the head, "No, I can only add and subtract them."
> —William Rowan Hamilton.

A.10 Groups

A mathematical group is a (finite or infinite) collection of "elements" that has the following properties.

1. There is an operation defined on the elements of the group such that the group is closed under this operation. For convenience, this operation is called "multiplication" and is denoted "·" although the group's elements do not have to be numbers. The term "closed under multiplication" means that the product of two group elements is also an element of the group; that is, if A and B are group elements, then $A \cdot B$ is also a group element.

2. One of the group elements, denoted I, satisfies $I \cdot A = A \cdot I = A$ for any element A of the group. This a "do nothing" element, also called the *group identity*.

3. Each group element, except the identity, must have an "inverse." An inverse of an element A is another group element, denoted A^{-1}, that satisfies $A \cdot A^{-1} = A^{-1} \cdot A = I$. The inverse undoes whatever A does. For example, the inverse of a translation upward is a translation downward. The inverse of a rotation 90° clockwise is a rotation 90° counterclockwise.

▶ **Exercise A.13:** What is the inverse of a reflection?

4. The group operation must be associative, i.e., $(A \cdot B) \cdot C = A \cdot (B \cdot C)$ for any group elements A, B, and C. The product of three elements can thus be written without using parentheses.

Note that the group operation does not have to be commutative. There are, therefore, commutative and noncommutative groups. The former are called *Abelian* after the mathematician Abel.

An example of a group is all the transformations of the plane that leave certain patterns invariant. Figure A.12 illustrates the symmetry of the so-called p1 plane symmetry group. It is easy to see how this pattern remains invariant after any translations along two axes (any two nonparallel axes would do for a p1 group). The group operation of a plane symmetry group is combining two transformations. There are 17 plane symmetry groups, as described in [Joyce 98]

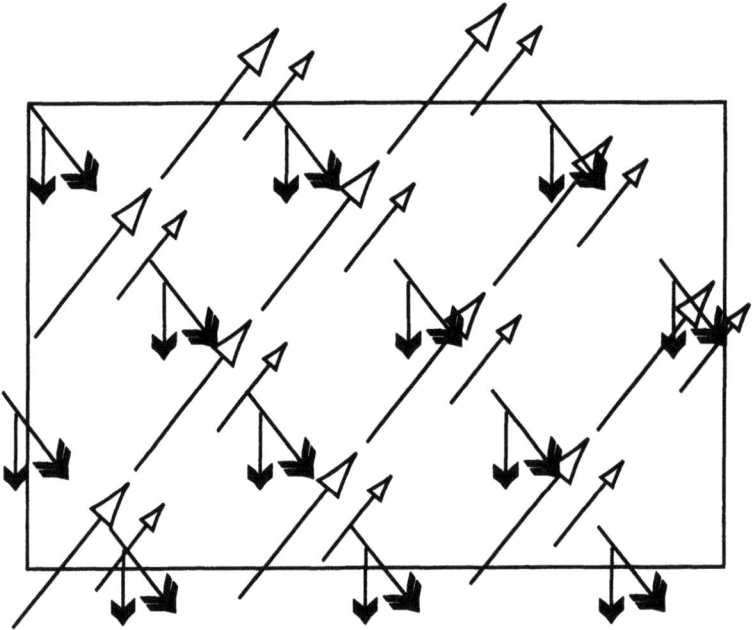

Figure A.12: A p1 Plane Symmetry Example.

A.11 Fields

A *mathematical field* is defined as a set of mathematical elements that satisfy the following:

1. Two operations, called addition (denoted "+") and multiplication (denoted ".") are defined on the field elements. The multiplication sign can be omitted if this does not cause an ambiguity.

2. Both operations are commutative, associative, and distributive, i.e.,

$$z1 + z2 = z2 + z1 \quad \text{and} \quad (z1 + z2) + z3 = z1 + (z2 + z3),$$
$$z1 \cdot z2 = z2 \cdot z1 \quad \text{and} \quad (z1 \cdot z2)z3 = z1(z2 \cdot z3),$$
$$(z1 + z2)z3 = z1 \cdot z3 + z2 \cdot z3 \quad \text{and} \quad z3(z1 + z2) = z3 \cdot z1 + z3 \cdot z1.$$

A Mathematical Topics

3. The "+" operation has an identity, denoted 0 and also an inverse, denoted $-z$, such that $z + (-z) = 0$.

4. The multiplication operation has an identity, denoted 1, and also an inverse, denoted z^{-1} (for all field elements except 0) satisfying $z \cdot z^{-1} = z^{-1} \cdot z = 1$.

The complex numbers constitute a field. They behave like the real numbers. We can add, multiply, subtract, and divide them and we can use any of the standard algebraic manipulations. Other examples of fields are the real numbers and the rational numbers.

The quaternions satisfy all the field requirements except that quaternion multiplication is not commutative. In general, $\mathbf{q_1 q_2} \neq \mathbf{q_2 q_1}$. As a result, the quaternions do not form a field but something called a *division ring*. We can perform arithmetic with quaternions, but we should be careful to note the order of multiplication.

The set of all $n \times n$ matrices with standard matrix addition and multiplication as operations constitutes a ring but not a division ring, because not every matrix has an inverse. The set of all three-dimensional vectors with vector addition and cross-product as operations also constitutes a ring, but not a division ring, since vectors don't have a multiplicative identity.

> Life is good for only two things, discovering mathematics and teaching mathematics
>
> Siméon Poisson

References

> **The Professor's Guilt List**
>
> Talking about a bibliography, there's an English professor at San Jose State University in California who has prepared a wish list consisting of literary works that English majors are supposed to have read before receiving their Bachelor degree. He complains that most English majors don't read even a small fraction of this list, a complaint that few would object to. Compare your literacy level to that of his students by checking the list at URL
>   `http://www.legendinc.com/FolderLinc/CyberFour/GuiltList`
> (since most URLs are short lived, try to search for "guiltlist").

Abbott, Edwin A. (1880) *Flatland: A Romance of Many Dimensions*, New York, Dover Thrift Editions (reprinted 1992).

Abelson, H., and A. A. diSessa (1982) *Turtle Geometry*, Cambridge, MA, MIT Press.

Adams, J. Alan (1988) *Descriptive Geometry and Geometric Modeling: A Basis for Design*, New York, Holt, Rinehart and Winston.

Adobe Systems Inc. (1985) *PostScript Language Tutorial and Cookbook*, Reading, MA, Addison-Wesley.

Adobe Systems Inc. (1990) *PostScript Language Reference Manual*, 2nd edition, Reading, MA, Addison-Wesley.

Alfeld, Peter, Marian Neamtu, and Larry L. Schumaker (1995) "Circular Bernstein-Bézier Polynomials," in *Mathematical Methods for Curves and Surfaces*, M. Daehlen et al., eds. Nashville, Vanderbilt University Press, pp. 11–20.

Ammeraal, L. (1988) *Interactive Three-Dimensional Computer Graphics*, New York, John Wiley.

Ammeraal, L. (1986) *Programming Principles in Computer Graphics*, New York, John Wiley.

Anedda, C. and L. Felician (1988) "P-Compressed Quadtrees for Image Storing," *The Computer Journal*, **31**(4):353–357.

Angell, Ian O. (1985) *Computer Geometric Art*, New York, Dover.

ANSI (1985), Standard X3.124-1985, available from ANSI, 1430 Broadway, New York, NY 10018.

Artwick, Bruce A. (1984) *Applied Concepts in Microcomputer Graphics*, Englewood Cliffs, NJ, Prentice-Hall.

Arvo, James (ed.) (1991) *Graphics Gems II*, Boston, MA, Academic Press.

Bailey, Kim R. (1998) http://www.rust.net/~kgeoinfo/quat.htm.

Baker, Henry G., (1998) ftp://ftp.netcom.com/pub/hb/hbaker/quaternion/.

Banchoff, Thomas (1996) *Beyond the Third Dimension: Geometry, Computer Graphics, and Higher Dimensions*, New York, W. H. Freeman & Co. (Scientific American Library Series).

Barnsley, M. (1993) *Fractals Everywhere*, 2nd edition, Boston, MA, Academic Press.

Barsky, Brian A. (1988) *Computer Graphics and Geometric Modeling Using Beta-splines*, New York, Springer-Verlag.

Bartels, Richard H., et al. (1987) *An Introduction to Splines For Use in Computer Graphics and Geometric Modeling*, Los Altos, CA, Morgan Kaufmann Publications.

Battisti, Eugenio (1981) *Filippo Brunelleschi*, translated by Robert Erich Wolf, New York, Rizzoli.

Beach, Robert C. (1991) *An Introduction to the Curves and Surfaces of Computer-Aided Design*, New York, Van Nostrand Reinhold.

Blinn, J. (1987) "How Many Ways Can You Draw a Circle?" *IEEE Computer Graphics and Applications*, **7**(8):39–44, August.

Burger, Peter (1989) *Interactive Computer Graphics: Functional, Procedural, and Device-Level Methods*, Reading, MA, Addison-Wesley.

Catmull, E., and J. Clark (1978) "Recursively Generated B-Spline Surfaces on Arbitrary Topological Meshes," *Computer-Aided Design* **10**(6):350–355, Sept.

Chaikin, G. (1974) "An Algorithm for High-Speed Curve Generation," *Computer Graphics and Image Processing*, **3**:346–349.

Cohen, E., et al., (1980) "Discrete B-Splines and Subdivision Techniques in Computer Aided Geometric Design and Computer Graphics," *Computer Graphics and Image Processing*, **14**:87–111.

Cohen, E., et al. (1985) "Algorithms For Degree Raising of Splines," *ACM Transactions on Graphics*, **4**:171–181.

Coons, Steven A. (1964) "Surfaces for Computer-Aided Design of Space Figures," Cambridge, MA, MIT Project MAC, report MAC-M-253, January.

Coons, Steven A. (1967) "Surfaces for Computer-Aided Design of Space Forms," Cambridge, MA, MIT Project MAC TR-41, June.

Bibliography

Crow, Frank (1987) "The Origins of the Teapot," *IEEE Computer Graphics and Applications*, **7**(1):8–19, January.

Daehlen, Morten, T. Lyche, and L. L. Schumaker (eds.) (1995) *Mathematical Methods for Curves and Surfaces*, Nashville, TN, Vanderbilt University Press.

Davis, Philip J. (1963) *Interpolation and Approximation*, Waltham, MA, Blaisdell Publishing, and New York, Dover Publications, 1975.

DeBoor, Carl, (1972) "On Calculating With B-Splines," *Journal of Approximation Theory*, **6**:50–62.

Delaunay B. (1934) "Sur La Sphere Vide," *Bulletin of Academy of Sciences of the USSR*, pp. 793–800.

DeVore R. et al. (1992) "Image Compression Through Wavelet Transform Coding," *IEEE Transactions on Information Theory* **38**(2):719–746, March.

Doo, Donald, and M. Sabin (1978) "Behavior of Recursive Division Surfaces Near Extraordinary Points," *Computer-Aided Design*, **10**(6):356–360, Sept.

Elias, P. (1975) "Universal Codeword Sets and Representations of the Integers," *IEEE Transactions on Information Theory*, IT-21(2):194–203, March.

Encarnacao, Jose L. (1992) *Fractal Geometry and Computer Graphics*, Berlin, Springer-Verlag.

Enderle, Gunter, K. Kansy, and G. Pfaff (1987) *Computer Graphics Programming: GKS, The Graphics Standard*, Berlin, Springer-Verlag.

Ernst, Bruno (1976) *The Magic Mirror of M. C. Escher*, New York, Random House.

Falcidieno, B. (ed.) (1992) *Computer Graphics and Mathematics*, Berlin, Springer-Verlag.

Farin G. (1992) *Curves and Surfaces for CAGD*, 3rd edition, Boston, MA, Academic Press.

Farin G. (1998) *NURBS Curves and Surfaces*, Wellesley, MA, AK Peters.

Feynman, R. (1985) *QED, The Strange Theory of Light and Matter*, Princeton, NJ, Princeton University Press.

Fiume, Eugene L. (1989) *The Mathematical Structure of Raster Graphics*, London, Academic Press.

Flat Earth Sociey (1998), see URL http://www.flat-earth.org/ for information.

Floyd, R., and L. Steinberg (1975) "An Adaptive Algorithm for Spatial Gray Scale," in *Society for Information Display 1975 Symposium Digest of Technical Papers*, p. 36.

Foley J., and A. Van Dam (1994) *Fundamentals of Interactive Computer Graphics*, 2nd edition, Reading, MA, Addison-Wesley.

Free Software Foundation (1998), 675 Mass. Ave., Cambridge, MA 02139, USA.

Freeman, H. (ed.) (1980) *Tutorial and Selected Readings in Interactive Computer Graphics*, Silver Springs, MD, IEEE Computer Society Press.

Ghostscript (1998) http://www.cs.wisc.edu/~ghost/index.html.

Glassner, Andrew S. (1989) *An Introduction to Ray Tracing*, London, Academic Press.

Glassner, Andrew S. (ed.) (1990) *Graphics Gems*, London, Academic Press.

Golomb, S. W. (1966) "Run-Length Encodings," *IEEE Transactions on Information Theory*, IT-12(3):399–401.

Gouraud, Henri (1971) "Continuous-Shading of Curved Surfaces," *IEEE Transactions on Computers*, C-20(6):623–629, June. (Reprinted in [Freeman 80].)

Gravesen, J. (1993) "Adaptive Subdivision and the Length of Bézier Curves," Technical Report 472, The Danish Center for Applied Mathematics and Mechanics, Technical University of Denmark.

Green, Stuart (1991) *Parallel Processing for Computer Graphics*, Cambridge, MA, MIT Press.

Guenter B., and R. Parent (1990) "Computing the Arc Lengths of Parametric Curves," *IEEE Computer Graphics and Applications* **10**(3):72–78, May.

Haeberli, P., and D. Voorhies (1994) "Image Processing by Linear Interpolation and Extrapolation," *IRIS Universe Magazine* (28), August.

Hearn D., and J. P. Baker (1997) *Computer Graphics*, 2nd edition, Englewood Cliffs, NJ, Prentice-Hall.

Heckbert, Paul S., ed., (1994) *Graphics Gems IV*, Boston, MA, AP Professional.

Herman, I. (1992) *The Use of Projective Geometry in Computer Graphics*, New York, Springer Verlag.

Hill, F. S. (1998) *Computer Graphics*, 2nd edition New York, MacMillan.

Hobby, John (1986) "Smooth, Easy to Compute Interpolating Splines," *Discrete and Computational Geometry* **1**:123–140.

Hoffmann, Christoph M. (1989) *Geometric and Solid Modeling: An Introduction*, San Mateo, CA, Morgan Kaufmann.

Hosaka, M. (1992) *Modeling of Curves and Surfaces in CAD/CAM*, New York, Springer-Verlag.

Hopgood, F. R. A. et al. (1986) *Introduction to the Graphical Kernel System (GKS)*, London, Academic Press.

Hopgood, F. R. A., and D. A. Duce (1991) *A Primer for PHIGS*, Chichester, John Wiley.

Howard, Paul G., and J. S. Vitter, (1993) "Fast and Efficient Lossless Image Compression," in *Proceedings of the 1993 Data Compression Conference*, J. Storer, ed., Los Alamitos, CA, IEEE Computer Society Press, pp. 351–360.

Howard, Paul G., and J. S. Vitter (1992a) "New Methods for Lossless Image Compression Using Arithmetic Coding," *Information Processing and Management* **28**(6):765–779.

Bibliography

Howard, Paul G., and J. S. Vitter (1992b) "Error Modeling for Hierarchical Lossless Image Compression," in *Proceedings of the 1992 Data Compression Conference*, J. Storer, ed., Los Alamitos, CA, IEEE Computer Society Press, pp. 269–278.

Howard, Paul G., and J. S. Vitter, (1993) "Fast and Efficient Lossless Image Compression," in *Proceedings of the 1993 Data Compression Conference*, J. Storer, ed., Los Alamitos, CA, IEEE Computer Society Press, pp. 351–360.

Howard, P. G., and J. S. Vitter (1994) "Fast Progressive Lossless Image Compression," Proceedings of the Image and Video Compression Conference, *IS&T/SPIE 1994 Symposium on Electronic Imaging: Science & Technology*, Vol. 2186, San Jose, CA, p. 98–109. February.

Howard, T. L. J., et al. (1991) *A Practical Introduction to PHIGS and PHIGS Plus*, Wokingham, Addison-Wesley.

Huntley, H. E. (1970) *The Divine Proportion: A Study in Mathematical Beauty*, New York, Dover Publications.

IGES (1986) *Initial Graphics Exchange Specifications*, version 3.0, Doc. #NB-SIR 86-3359, National Bureau of Standards, Gaithersburg, MD.

Infowave Inc. (1998) see URL http://www.infowave.net/.

Jarvis, J. F., C. N. Judice, and W. H. Ninke (1976) "A Survey of Techniques for the Image Display of Continuous Tone Pictures on Bilevel Displays" *Computer Graphics and Image Processing* **5**(1):13–40.

Jarvis, J. F. and C. S. Roberts (1976) "A New Technique for Displaying Continuous Tone Images on a Bilevel Display" *IEEE Transactions on Communications* **24**(8):891–898, August.

Jarvis, P. (1990) "Implementing CORDIC Algorithms," *Dr. Dobb's Journal*, 152–158, October,

Jordan, B. W., and R. C. Barrett (1974) "A Cell Organized Raster Display for Line Drawings," *Communications of the ACM*, **17**(2):70–77.

Joyce, David E., (1998) http://aleph0.clarku.edu/~djoyce/wallpaper/.

Kalay, Yehuda E. (1987) *Graphic Introduction to Programming*, New York, John Wiley.

Kalay, Yehuda E. (1989) *Modeling Objects and Environments*, New York, John Wiley.

Kay, David C. (1992) *Graphics File Formats*, New York, Windcrest/McGraw-Hill.

Kimberling, C., (1994) "Central Points and Central Lines in the Plane of a Triangle," *Mathematical Magazine* **67**:163–187.

Kirk, David, ed., (1992) *Graphics Gems III*, San Diego, CA, Harcourt Brace Jovanovich.

Knuth, Donald E., (1981) *The Art of Computer Programming*, Reading, MA, Addison-Wesley.

Knuth, Donald E., (1984) *The TEXBook*, Reading, MA, Addison-Wesley.

Knuth, Donald E., (1986) *The Metafont Book*, Reading, MA, Addison-Wesley.

Knuth, Donald E., (1987) "Digital Halftones by Dot Diffusion," *ACM Transactions on Graphics* **6**(4):245–273.

Kochanek, D. H. U., and R. H. Bartels (1984) "Interpolating Splines with Local Tension, Continuity, and Bias Control," *Computer Graphics* **18**(3):33–41 (Proceedings SIGGRAPH '84).

Lamé (1998) is a Java applet at URL http://www-groups.dcs.st-andrews.ac.uk/ file ~history/Java/Lame.html.

Lee, E. (1986) "Rational Bézier Representations for Conics," in *Geometric Modeling*, Farin, G., editor, Philadelphia, SIAM Publications, pp. 3–27.

L'Engle, Madeleine (1962) *A Wrinkle in Time*, New York, Farrar, Straus, and Giroux.

Lewell, John (1985) *Computer Graphics: A Survey of Current Techniques and Applications*, New York, Van Nostrand Reinhold.

Lindenmayer, A. (1968) "Mathematical Models for Cellular Interaction in Development," *Journal of Theoretical Biology* **18**:280–315.

Lischinski D. (1992) "Converting Bézier Triangles Into Rectangular Patches," in *Graphics Gems III*, Kirk, David, ed., San Diego, CA, Harcourt Brace Jovanovich, pp. 256–261.

Lischinski D. (1994) "Converting Rectangular Patches Into Bézier Triangles," in *Graphics Gems IV*, Heckbert, Paul S., ed., Boston, MA, AP Professional, pp. 278–282.

Manning, J. R. (1974) "Continuity Conditions for Spline Curves," *Computer Journal* **17**(4):181–186, May.

Mantyla, Martti (1988) *An Introduction to Solid Modeling*, Gaithersburg, MD, Computer Science Press.

May, C. P. (1962) *James Clerk Maxwell and Electromagnetism*, New York, Franklin Watts.

Mesdag Documentation Society (1998) is at http://www.mesdag.com/index.html.

Mielke, Bruce (1991) *Integrated Computer Graphics*, St. Paul, MN, West Publishers.

Mortenson, Michael E. (1985) *Geometric Modeling*, New York, John Wiley.

Mortenson, Michael E. (1989) *Computer Graphics: An Introduction to the Mathematics and Geometry*, New York, Industrial Press.

Mortenson, Michael E. (1995) *Geometric Transformations*, New York, Industrial Press.

O'Rourke, Joseph (1994) *Computational Geometry in C*, Cambridge, Cambridge University Press.

Osterberg, G. (1935) "Topography of the Layer of Rods and Cones in the Human Retina," *Acta Ophthalmologica*, suppl. 6:1–103.

Bibliography

Paeth, Alan W. (1986) "A Fast Algorithm for General Raster Rotation," in *Proceedings Graphics Interface '86*, Canadian Information Processing Society, pp. 77–81.

Paeth, Alan W. (editor) (1995) *Graphics Gems V*, Boston, MA, AP Professional.

Pantone Inc., (1991) *PANTONE Color Formula Guide 1000*, Pantone Inc.

Penna, Michael A. (1986) *Projective Geometry and its Applications to Computer Graphics*, Englewood Cliffs, NJ, Prentice-Hall.

Piegl L., and W. Tiller (1996) *The NURBS Book*, Berlin, Springer-Verlag.

Plass, Michael, and Maureen Stone (1983) "Curve-Fitting with Piecewise Parametric Cubics," *ACM Transactions on Computer Graphics*, **17**(3):229–239, July.

Pokorny C. K. (1994) *Computer Graphics*, 2nd edition, Wilsonville, OR, Franklin Beedle.

Prautzsch, H., (1984) "A Short Proof of the Oslo Algorithm," *Computer Aided Geometric Design*, **1**:95–96.

Press, W. H., B. P. Flannery, et al. (1988) *Numerical Recipes in C: The Art of Scientific Computing*, Cambridge University Press. (Also available online from http://www.nr.com/.)

Pritchard, D. H. (1977) "US Color Television Fundamentals—a Review," *IEEE Transactions on Consumer Electronics* CE-23(4):467–478, November.

Prusinkiewicz, Przemyslaw (1989) *Lindenmayer Systems, Fractals, and Plants*, New York, Springer-Verlag.

Richardson, Malcolm (1989) *Modern Computer Graphics*, Boston, Blackwell Scientific Publications.

Roetling, P. G. (1976) "Halftone Method with Edge Enhancement and Moiré Suppression," *Journal of the Optical Society of America*, **66**:985–989.

Roetling, P. G. (1977) "Binary Approximation of Continuous Tone Images," *Photography Science and Engineering*, **21**:60–65.

Rogers D., and C. Adams (1992) *Mathematical Elements for Computer Graphics*, 2nd edition, New York, McGraw-Hill.

Rogers, David (1997) *Procedural Elements for Computer Graphics*, 2nd edition, New York, McGraw-Hill.

Salmon, Rod (1987) *Computer Graphics: Systems and Concepts*, Reading, MA, Addison-Wesley.

Salomon, David (1997) *Data Compression: The Complete Reference*, New York, Springer-Verlag.

Samet, Hanan (1990a) *Applications of Spatial Data Structures: Computer Graphics, Image Processing, and GIS*, Reading, MA, Addison-Wesley.

Samet, Hanan (1990b) *The Design and Analysis of Spatial Data Structures*, Reading, MA, Addison-Wesley.

Sayood K., and K. Robinson (1992) "A Differential Lossless Image Compression Scheme," *IEEE Transactions on Signal Processing* **40**(1):236–241, January.

Sedgewick, Robert (1997) *Algorithms in C: Parts 1–4: Fundamentals, Data Structures, Sorting, Searching*, Reading, MA, Addison-Wesley.

Skiena, Steven S. (1998), *The Algorithm Design Manual*, Berlin, Springer-Verlag.

Skov, Anders (1992) URL http://www.mi.aau.dk/~askov/index.html (also private communication).

Snyder, John M. (1992) *Generative Modeling for Computer Graphics and CAD: Symbolic Shape Design Using Interval Analysis*, Boston, MA, Academic Press.

Stollnitz E. J., T. D. DeRose, and D. H. Salesin (1996) *Wavelets for Computer Graphics*, San Francisco, CA, Morgan Kaufmann.

Stone, M. W. et al., (1988) "Color Gamut Mapping and the Printing of Digital Color Images," *ACM Transactions on Graphics* **7**(3):249–292, October.

Swartzlander, Earl E. (1990) *Computer Arithmetic*, Silver Spring, MD, IEEE Computer Society Press.

Taylor, W. (1992) *The Geometry of Computer Graphics*, Pacific Grove, CA, Wadsworth and Brooks/Cole.

Theoharis, T. (1989) *Algorithms for Parallel Polygon Rendering*, New York, NY, Springer-Verlag.

Thomas, Frank, and Ollie Johnstone (1981) *Disney Animation: The Illusion of Life*, New York, Abbeville Press.

ToyStory (1998), http://www.toystory.com/about/history/history.html.

Triangles (1998), Many triangles' centers at http://www.evansville.edu/ directory ~ck6/tcenters/index.html.

Turnbull, Herbert Westren (1998) is URL http://www-groups.dcs.st-and.ac.uk/ file ~history/HistTopics/Matrices_and_determinants.html.

Ulichney, Robert (1987) *Digital Halftoning*, Cambridge, MA, MIT Press.

Unicode (1998) http://www.unicode.org/.

Unicode Consortium (1996) *The Unicode Standard, Version 2.0*, Reading, MA, Addison-Wesley.

Vachss, Raymond (1987) "The CORDIC Magnification Function," *IEEE Micro*, **7**(5)83–84, October.

Volder, Jack E. (1959) "The CORDIC Trigonometric Computing Technique," *IRE Transactions on Electronic Computers*, **EC-8**:330–334.

Walther, John S. (1971) "A Unified Algorithm for Elementary Functions," *Proceedings of Spring Joint Computer Conference*, **38**:379–385.

Watt, Alan H. (1989) *Fundamentals of Three-Dimensional Computer Graphics*, Reading, MA, Addison-Wesley.

Watt, Alan, and Mark Watt (1992) *Advanced Animation and Rendering: Theory and Practice*, Reading, MA, Addison-Wesley.

Whitman, Scott (1992) *Multiprocessor Methods for Computer Graphics Rendering*, Boston, MA, Jones and Bartlett.

Bibliography

Yamaguchi, F. (1988) *Curves and Surfaces in Computer Aided Geometric Design*, Berlin, Springer-Verlag.

Wu, X., and J. G. Rokne (1987) "Double-Step Incremental Generation of Lines and Circles," *Computer Vision, Graphics, and Image Processing* **37**:331–344.

Zhao, Zhiyuan (1998) is an applet at
http://ra.cfm.ohio-state.edu/~zhao/algorithms/algorithms.html.

Grip a holy bib
Anagram of *bibliography*

Answers to Exercises

1: This is a row vector whose four elements are points and, thus, are themselves vectors (pairs in two dimensions and triplets in three dimensions).

1.1: At present, virtual reality renders images and sounds and allows for some user interaction with the simulated environment. Extrapolating this, we predict that the next step will be to compute a simulation that covers the entire human sensory range, including the visual, kinesthetic (tactile and emotional feelings), olfactory (smell), gustatory (taste), and auditory senses. Such a simulation would create a perfect illusion, completely overriding the normal functioning of the senses and fooling the user into believing that they really are experiencing the simulated, virtual environment. To understand how such a simulation can be done let's consider, for example, the sense of smell. It can be simulated by preparing chemicals that have the needed smells and using them during a virtual-reality session. However, a better approach is to find out how the sense of smell works. Once this is understood, it may be possible to send the brain the same signals normally sent by the nerves from the nose, and this way directly stimulate the brain and create the sensation of any desired smells.

1.2: The Golden Ratio $\phi \approx 1.618$ has traditionally been considered the aspect ratio that's most pleasing to the eye. With this in mind, it is easy to see that 1.77 is the better aspect ratio.

1.3: This is easy. The index is $(r-1-y)c+x$.

1.4: Each millimeter has 100 pixels, so there are 10^4 pixels in each millimeter squared. The area of the film is $24 \times 36 = 864\,\text{mm}^2$, yielding a total of 8,640,000 pixels. Incidentally, the aspect ratio of this film is $36/24 = 1.5$, fairly close to the Golden Ratio.

1.5: The same factors that determine picture quality on film. Among them are the film temperature when the picture is taken, the developing time, and the resolving power of the camera lens. The latter is a major factor which also depends on the aperture. A low-quality lens may also resolve the center part of the chart better than it does the off-center parts. As a result of the many factors involved, the measured resolution of a particular film may change over time and can change even on the same roll!

1.6: Since each truth table has four, 1-bit entries, there can be 16 combinations of the entries, leading to 16 possible logical operations. Most, however, are not useful and are never used in practice. The last example in Table 1.9, for example, creates a zero if its first input is zero. If its first input is one, it creates the opposite of the second input. This is rarely, if ever, useful.

1.7: Yes! When dragging an object or rubber banding it, the program has to draw an outline, erase it, draw a slightly different outline, etc., very quickly.

1.8: Yes! Imagine 4-bit pixels (16 colors). If a pixel is drawn in color 1010 and we have a new source of color 1110, then the xor pixel being drawn will be 0100. If we now erase, say, the 1010 pixel, the result will be the xor of 0100 and 1010, which is 1110. It's easy to demonstrate that this method works even for three or more objects intersecting at a point (see Plate 3).

1.9: Figure Ans.1 illustrates the results of the 4/10 shrinking.

Figure Ans.1: 4/10 Bitmap Shrinking by Copying.

1.10: Direct calculation produces

$$(-1,-1) \to (-1,0), \quad (-1,0) \to (-0.8, 0.8), \quad (-1,1) \to (0,1),$$
$$(0,-1) \to (-0.8, -0.8), \quad (0,0) \to (0,0), \quad (0,1) \to (0.8, 0.8),$$
$$(1,-1) \to (0,-1), \quad (1,0) \to (0.8, -0.8), \quad (1,1) \to (1,0).$$

Answers to Exercises

If we round 0.8 to 1, we get

$$(-1,-1) \to (-1,0), \quad (-1,0) \to (-1,1), \quad (-1,1) \to (0,1),$$
$$(0,-1) \to (-1,-1), \quad (0,0) \to (0,0), \quad (0,1) \to (1,1),$$
$$(1,-1) \to (0,-1), \quad (1,0) \to (1,-1), \quad (1,1) \to (1,0).$$

A direct check verifies that in this case, every destination pixel is the mapping of some source pixel and no two source pixels map to the same destination.

1.11: Once the program detects a click, it inputs the cursor coordinates and checks the corresponding location of the codemap. In our example, it finds serial number 2. The program examines location 2 of the geometric data structure, finds that graphics object with serial number 2 is a circle, and finds the pointer to its specific data. That data consists of the radius and the coordinates of the center point. The program then calculates all the pixels of the circle (using the same scan-converting method that was originally used to draw it) and highlights them. Certain points may be highlighted with a different color or made larger (Figure 1.20f), making them more obvious to the user. These are called *anchors* and can later be used by the user to drag or reshape the circle.

1.12: Another array, the size of the codemap, may be declared, whose elements are boolean. Alternatively, the codemap may be constructed as an array of structures, each consisting of a code field and a flag (boolean) field.

2.1: It is the small step. If the line has a small slope (i.e., it is close to horizontal), small changes in x cause only small changes in y. If $x = 4.32$ corresponds to $y = 6.15$, then $x = 4.33$ may correspond to, say, $y = 6.27$. Both values are rounded to the pixel $(4,6)$. A good algorithm should create a new pixel in each iteration.

2.2: The slope a of the line equals $\Delta y/\Delta x = (6-2)/(4-1) = 4/3$. Since $\Delta y > \Delta x$ we set $G = 1/a = 3/4$ and $H = 1$. The loop iterates from $L = 1$ to $L = max(\Delta x, \Delta y) + 1 = 5$. The five pixels generated are

x:	1	1.75	2.5	3.25	4
round(x):	1	2	3	3	4
y:	2	3	4	5	6

The length of the line equals $\sqrt{\Delta^2 x + \Delta^2 y} = \sqrt{3^2 + 4^2} = 5$. It is identical to the number of pixels.

2.3: For simple DDA, we get a slope of $(5-1)/(5-1) = 1$. The x coordinate is incremented from 1 to 5 in steps of 1. The y coordinate is incremented from the initial y value of 1 to the final value in steps of the slope, which is also 1. The points are thus $(1,1), (2,2)$ through $(5,5)$.

For the quadrantal DDA, we start with $\Delta x = 4$, $\Delta y = 4$ and Err = 0. Table Ans.2 summarizes the nine steps of the loop. The results are

736 Answers to Exercises

Step	Plot	Err > 0?	Update	New Err
1	(1,1)	No	$y \leftarrow 2$	4
2	(1,2)	Yes	$x \leftarrow 2$	$4-4=0$
3	(2,2)	No	$y \leftarrow 3$	4
4	(2,3)	Yes	$x \leftarrow 3$	$4-4=0$
5	(3,3)	No	$y \leftarrow 4$	4
6	(3,4)	Yes	$x \leftarrow 4$	$4-4=0$
7	(4,4)	No	$y \leftarrow 5$	4
8	(4,5)	Yes	$x \leftarrow 5$	$4-4=0$
9	(5,5)	No	$y \leftarrow 6$	4

Table Ans.2: A Quadrantal DDA Example.

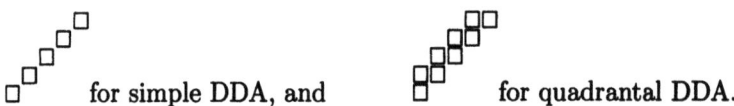

for simple DDA, and   for quadrantal DDA.

2.4: We select an initial value Err $= \Delta x/2 = 0.5$ for a better looking line. The seven steps of the algorithm are summarized in Table Ans.3.

Step	Plot	Err < 0?	Update	New Err
1	(1,1)	No	$y \leftarrow 2$	$1.5 - 1 = .5$
2	(1,2)	No	$y \leftarrow 3$	$.5 - 1 = -.5$
3	(1,3)	Yes	$x \leftarrow 2, y \leftarrow 4$	$-.5 + 6 - 1 = 4.5$
4	(2,4)	No	$y \leftarrow 5$	$4.5 - 1 = 3.5$
5	(2,5)	No	$y \leftarrow 6$	$3.5 - 1 = 2.5$
6	(2,6)	No	$y \leftarrow 7$	$2.5 - 1 = 1.5$
7	(2,7)			

Table Ans.3: An Octantal DDA Example.

After reversing the y coordinates, we get the seven pixels $(1,-1)$, $(1,-2)$, $(1,-3)$, $(2,-4)$, $(2,-5)$, $(2,-6)$, and $(2,-7)$.

2.5: It is true that such a line has an ideal shape, but it is dimmer than a horizontal or a vertical line. Compare the following two lines:

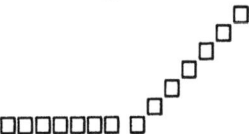

The first goes from (1,1) to (7,1). It is horizontal, it consists of seven pixels, and its length is also 7. The second line goes from (1,1) to (7,7). It is slanted at 45°, is made of seven pixels, but its length is $\sqrt{2} \times 7 \approx 10$. The two lines will not have the same brightness! To correct this, we can add three pixels to the second line, distributing them as evenly as possible:

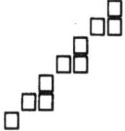

The two lines will now have the same brightness, but the 45° line will not look as precise as before.

We, thus, end up with a trade-off. The octantal DDA method produces 45° lines that look great but are dim compared to other lines. Other methods may produce 45° lines that are bright but don't look that great.

2.6: For the first line, the steps are listed in Table Ans.4a and the result is 01000100 or 00100010. For the second line, the steps are listed in Table Ans.4b and the final result is either 0100010010 or 0100100010.

x	y	str1	str2		x	y	str1	str2
8	2	0	1		10	3	0	1
6	2	0	1		7	3	0	1
4	2	0	10		4	3	0	10
2	2	0	010		1	3	0	010
					1	2	0010	010
					1	1	0100010	010

(a)   (b)

Table Ans.4: Examples of Best-Fit DDA.

2.7: The explicit equation of a straight line is $y = ax+b = (\Delta y/\Delta x)x+b$. If point (x_0, y_0) lies on the line, then $y_0 = (\Delta y/\Delta x)x_0 + b$. This yields $x_0 \Delta y - y_0 \Delta x = b\Delta x$ and $b\Delta x$ does not depend on x_0 or y_0.

2.8: The reason for selecting $n = 18$ is that $90/5 = 18$. Figure Ans.5 shows the 18 points. The coordinates of the points are also shown, as well as the *Mathematica* code that did the calculations.

2.9: Consider point **T** in Figure 2.13a. Its coordinates are $(0, b)$, so its distance $d/2$ from any of the foci satisfies $(d/2)^2 = b^2 + c^2$. Now consider point **L**. Its distances from the two foci are $a-c$ and $a+c$, so it tells us that $(a-c)+(a+c) = d$ or $a = d/2$. The result is $a^2 = (d/2)^2 = b^2 + c^2$.

2.10: Consider point **L** of Figure 2.13b. Both the maximum value of d' and the minimum value of d occur at this point. We can thus write $d'_{max} = d_{min} + F$. In general, the oval satisfies $d + 2d' = S$. From these two relations, we get

$$S = d + 2d' = d_{min} + 2d'_{max} = 3d_{min} + 2F$$

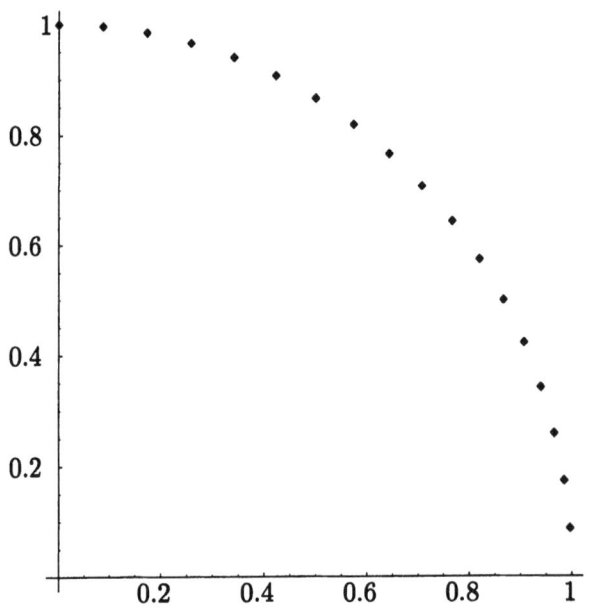

Figure Ans.5: Circle in Polar Coordinates ($\Delta\theta = 5°$).

```
Clear[L];
n=18; delta=5 Degree; R=1;
xk=R; yk=0;
dcos=Cos[delta]//N; dsin=Sin[delta]//N;
L={};
Do[xn=xk dcos-yk dsin; yn=xk dsin+yk dcos;
xk=xn; yk=yn; L=Append[L,{xn,yn}], {k,0,n-1}];
L
ListPlot[L, Prolog->AbsolutePointSize[3],
 AspectRatio->Automatic]
```

Mathematica Code for Figure Ans.5

0.996,0.087	0.985,0.174	0.966,0.259	0.940,0.342	0.906,0.423	0.866,0.500
0.819,0.574	0.766,0.643	0.707,0.707	0.643,0.766	0.574,0.819	0.500,0.866
0.423,0.906	0.342,0.940	0.259,0.966	0.174,0.985	0.087,0.996	0.000,1.000

Coordinates of 18 Points of Figure Ans.5

> **Curves.**
>
> Head down against the wind,
> surf pounding to my right,
> I notice the pattern the sand makes
> as it blows along the beach,
> filling in footprints,
> covering chevron streaks left by the falling tide.
> The sand moves like smoke from a chimney,
> or water weed in a smoothly-flowing stream,
> or the curve—I forget its name—
> drawn by tying a pencil to a thread
> unwinding from a spool.
> There are connections here.
> My mind struggles clumsily, glimpsing
> an elegance I long to comprehend.
> —Maureen Eppstein: *Poems*.
> First published in: Quantum Tao (1996)

or $d_{\min} = (S - 2F)/3$. Similarly, both the minimum value of d' and the maximum value of d occur at point **R** of Figure 2.13b, which enables us to write

$$S = d + 2d' = d_{\max} + 2d'_{\min} = 3d_{\max} - 2F$$

or $d_{\max} = (S + 2F)/3$.

2.11: No. Table Ans.2 shows an example where applying quadrantal DDA from $(1,1)$ to $(5,5)$ produces nine pixels, the last two of which are $(4,5)$ and $(5,5)$. However, when applying the same algorithm from $(5,5)$ to $(1,1)$ the first two pixels are $(5,5)$ and $(5,4)$.

2.12: No. Table 2.2 shows an example where a pixel is drawn twice.

2.13: The simplest way is to first calculate all the pixels in octant 2 and store them in a table, then use them to calculate and draw the pixels of octant 1, octant 2, and so on up to octant 8. This is a good method because the table size is reasonably small even for large circles.

2.14: The two given points imply $\Delta x = 1$ and $\Delta y = 6$. Equal intensities imply $I_p = n - I_p$ or $I_p = 7.5$. If Err is positive, this means $15 \times \text{Err}/6 = 7.5$ or $\text{Err} = 3$. If Err is negative, this means $-15 \times \text{Err}/1 = 7.5$ or $\text{Err} = -0.5$.

3.1: Function f_1 is not onto since point $(-1, 0)$ is not the mapping of any real point. This function is also not one-to-one since the two different points (a, b) and $(-a, b)$ map to (a^2, b). Function f_2, however, is a valid geometric transformation.

3.2: No. It is easy to come up with examples of two transformations f and g such that $f \circ g \neq g \circ f$. One example is a 90° counterclockwise rotation about the origin and a reflection about the x axis. When the point $(1,0)$ is first rotated 90° about the origin and then reflected about the x axis, it is first moved to $(0,1)$, then ends up at $(0,-1)$. If the same point is first reflected and then rotated, it first moves to itself and then to $(0,1)$.

3.3: This is a direct application of Equation (3.3). The result is

$$A(b_{11}x^* + b_{12}y^*)^2 + B(b_{11}x^* + b_{12}y^*)(b_{21}x^* + b_{22}y^*) + C(b_{21}x^* + b_{22}y^*)^2$$
$$+ D(b_{11}x^* + b_{12}y^*) + E(b_{21}x^* + b_{22}y^*) + F = 0,$$

which is a second-degree curve.

3.4: A point (x,y) on a circle with radius R satisfies $x^2 + y^2 = R^2$ or $(x/R)^2 + (y/R)^2 = 1$. The transformed point (x^*, y^*) on an ellipse should satisfy $(x/a)^2 + (y/b)^2 = 1$. It is easy to guess that the transformation rule is $x^* = ax/R$, $y^* = by/R$, but this can also be proved as follows: The general scaling transformation is $x^* = k_1 x$, $y^* = k_2 y$. For the transformed point to be on an ellipse, it should satisfy $(k_1 x/a)^2 + (k_2 y/b)^2 = 1$, which can be simplified to $k_1^2 b^2 x^2 + k_2^2 a^2 y^2 = a^2 b^2$. Substituting $y^2 = R^2 - x^2$ yields

$$(k_1^2 b^2 - k_2^2 a^2) x^2 = a^2 b^2 - k_2^2 a^2 R^2.$$

This equation must hold for every value of x and this is possible only if $k_1^2 b^2 - k_2^2 a^2 = 0$ and $a^2 b^2 - k_2^2 a^2 R^2 = 0$. Solving these equations yields $k_1 = a/R$ and $k_2 = b/R$.

3.5: The transformation can be written $(x,y) \to (x, -x+y)$, so $(1,0) \to (1,-1)$, $(3,0) \to (3,-3)$, $(1,1) \to (1,0)$, and $(3,1) \to (3,-2)$. The original rectangle is, thus, transformed into a parallelogram.

3.6: From $\cos 45° = 0.7071$ and $\tan 45° = 1$, we get the 45°-rotation matrix as the product:

$$\begin{pmatrix} 0.7071 & 0 \\ 0 & 0.7071 \end{pmatrix} \begin{pmatrix} 1 & -1 \\ 1 & 1 \end{pmatrix}.$$

Figure Ans.6 shows how a 2×2 square centered on the origin (Figure Ans.6a) is first shrunk to about 70% of its original size (Figure Ans.6b), then sheared by the second matrix according to $(x^*, y^*) = (x+y, -x+y)$, and becomes the rotated diamond shape of Figure Ans.6c. Direct calculations show how the two original corners $(-1,1)$ and $(1,1)$ are transformed to $(0, 1.4142)$ and $(1.4142, 0)$, respectively.

3.7: Figure 3.3 gives the polar coordinates $\mathbf{P} = (r, \alpha)$ and $\mathbf{P}^* = (r, \phi) = (r, \alpha - \theta)$. We are looking for a matrix $\mathbf{T} = \begin{pmatrix} a & b \\ c & d \end{pmatrix}$ such that $\mathbf{P}^* = \mathbf{PT}$ and it does not take much to figure out that

$$\mathbf{T} = \begin{pmatrix} 1 & -\theta/r \\ 0 & 1 \end{pmatrix}.$$

Answers to Exercises

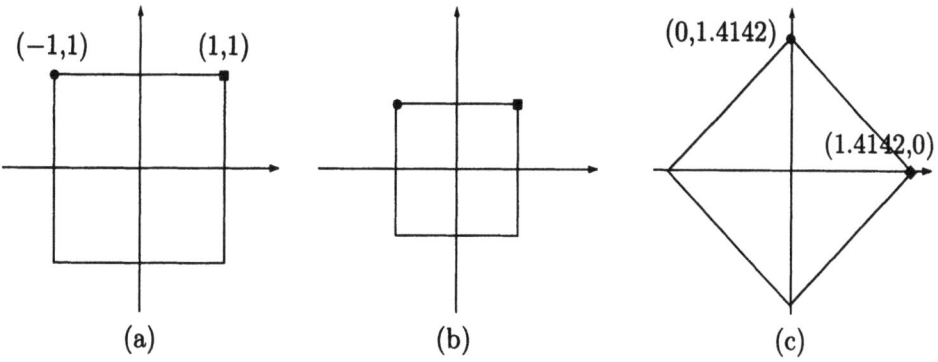

Figure Ans.6: A 45° Rotation as Scaling and Shearing.

3.8: A reflection about the x axis transforms a point (x,y) to a point $(x,-y)$. A reflection about $y=-x$ similarly transforms a point (x,y) to a point $(-y,-x)$ (this is matrix $\mathbf{T}_3$ of Equation (3.5)). The combination of these two transformations thus transforms (x,y) to $(y,-x)$, which is another form of the negate and exchange rule, corresponding to a 90° clockwise rotation about the origin. This rotation can also be expressed by the matrix (compare to Equation (3.6))

$$\begin{pmatrix} \cos 90° & \sin 90° \\ -\sin 90° & \cos 90° \end{pmatrix} = \begin{pmatrix} 0 & 1 \\ -1 & 0 \end{pmatrix}.$$

3.9: The determinant of this matrix equals

$$\left(\frac{1-t^2}{1+t^2}\right)^2 - \frac{-4t^2}{(1+t^2)^2} = \frac{(1-t^2)^2 + 4t^2}{(1+t^2)^2} = +1,$$

which shows that it generates pure rotation. Also, if we denote this matrix by

$$\begin{pmatrix} a_{11} & a_{12} \\ a_{21} & a_{22} \end{pmatrix},$$

it is easy to see that $a_{11} = a_{22}$, $a_{12} = -a_{21}$, $a_{11}^2 + a_{12}^2 = 1$, and $a_{21}^2 + a_{22}^2 = 1$. These properties are all satisfied by a rotation matrix.

3.10: The determinant of this matrix is

$$\left(\frac{a}{A}\right)^2 - \frac{b}{A}\left(-\frac{b}{A}\right) = \frac{a^2+b^2}{A^2}.$$

It equals 1 for $A = \pm\sqrt{a^2+b^2}$ but cannot equal -1 since it is the quotient of the two non-negative numbers a^2+b^2 and A^2. We, consequently, conclude that this matrix represents pure rotation. An example is $a = b = 1$, which produces $A = \pm\sqrt{2} \approx \pm 1.414$. The rotation matrices for this case are

$$\begin{pmatrix} 1/\sqrt{2} & 1/\sqrt{2} & 0 \\ -1/\sqrt{2} & 1/\sqrt{2} & 0 \\ 0 & 0 & 1 \end{pmatrix} = \begin{pmatrix} 0.7071 & 0.7071 & 0 \\ -0.7071 & 0.7071 & 0 \\ 0 & 0 & 1 \end{pmatrix},$$

$$\begin{pmatrix} -1/\sqrt{2} & -1/\sqrt{2} & 0 \\ 1/\sqrt{2} & -1/\sqrt{2} & 0 \\ 0 & 0 & 1 \end{pmatrix} = \begin{pmatrix} -0.7071 & -0.7071 & 0 \\ 0.7071 & -0.7071 & 0 \\ 0 & 0 & 1 \end{pmatrix},$$

and they correspond to 45° rotations about the origin.

3.11: The combined transformation matrix is the product

$$\begin{pmatrix} 1 & 0 & 0 \\ 0 & -1 & 0 \\ 0 & 0 & 1 \end{pmatrix} \begin{pmatrix} 1 & 0 & 0 \\ 0 & 1 & 0 \\ -1 & -1 & 1 \end{pmatrix} \begin{pmatrix} \cos 180° & -\sin 180° & 0 \\ \sin 180° & \cos 180° & 0 \\ 0 & 0 & 1 \end{pmatrix} = \begin{pmatrix} -1 & 0 & 0 \\ 0 & 1 & 0 \\ 1 & 1 & 1 \end{pmatrix}.$$

This matrix combines a reflection of the x coordinates with a one-unit translation in the x and y directions. Applying it to the four points yields $(0, 2)$, $(0, 0)$, $(2, 2)$, and $(2, 0)$. This is the same square but is now located in the first quadrant (Figure Ans.7).

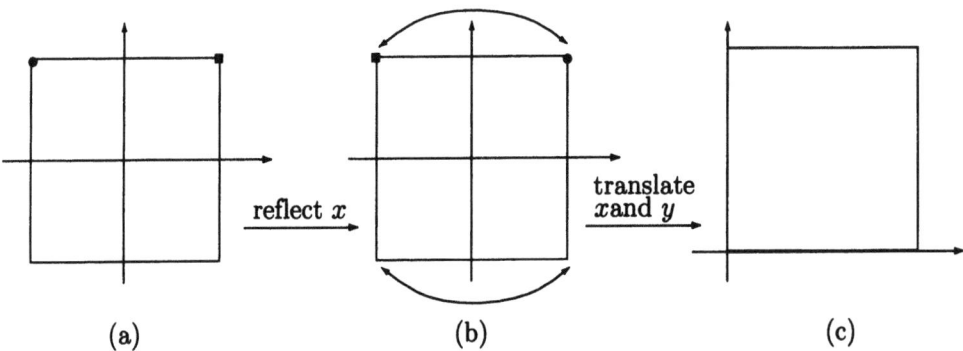

Figure Ans.7: An x Reflection and Translation.

3.12: Using angles ϕ and θ from Figure 3.3 but assuming that the rotation is counterclockwise about (x_0, y_0), we get

$$x^* = x_0 + (x - x_0)\cos\theta - (y - y_0)\sin\theta,$$
$$y^* = y_0 + (x - x_0)\sin\theta + (y - y_0)\cos\theta.$$

Answers to Exercises

We are looking for a matrix that satisfies

$$(x^*, y^*, 1) = (x, y, 1) \begin{pmatrix} a & b & 0 \\ c & d & 0 \\ m & n & 1 \end{pmatrix}.$$

The simple solution is

$$\mathbf{T} = \begin{pmatrix} \cos\theta & \sin\theta & 0 \\ -\sin\theta & \cos\theta & 0 \\ x_0(1-\cos\theta) + y_0\sin\theta & y_0(1-\cos\theta) - x_0\sin\theta & 1 \end{pmatrix}.$$

In a similar way, it can be shown that a clockwise rotation about (x_0, y_0) is produced by

$$\mathbf{T} = \begin{pmatrix} \cos\theta & -\sin\theta & 0 \\ \sin\theta & \cos\theta & 0 \\ x_0(1-\cos\theta) - y_0\sin\theta & y_0(1-\cos\theta) + x_0\sin\theta & 1 \end{pmatrix}.$$

3.13: If a point $\mathbf{P} = (x, y, 1)$ is reflected to a point $\mathbf{P}^* = (x^*, y^*, 1) = (y-1, x+1, 1)$ about the line $y = x + 1$, then the midpoint between them [which is $(\mathbf{P} + \mathbf{P}^*)/2 = (x+y-1, y+x+1)/2$] should be on the line. It's easy to see that it is because its y coordinate equals 1 plus its x coordinate.

3.14: This is easily done with the help of the right software, and the result is

$$\begin{pmatrix} 0.5 & 0.866 & 0 \\ 0.866 & -0.5 & 0 \\ -0.866 & 1.5 & 1 \end{pmatrix}.$$

3.15: Such a thing is possible, but would not improve the algorithm. Transforming a point from octant 1 to octant 2 is done by reflecting it about the 45° line $y = x$. A point (x, y) is thus transformed to the point (y, x). The similar transformation between half-octants amounts to reflection about the 22.5° line $y = ax$ (where $a = \tan 22.5° \approx 0.414$). This transforms point (x, y) to $(0.7071x + 0.7071y, 0.7071x - 0.7071y)$ (see the following proof) and would slow down the algorithm, since it involves real-number arithmetic.

Proof: Let's denote $\alpha = \sin 22.5°$, $\beta = \cos 22.5°$. To reflect about the 22.5° line, we rotate clockwise by 22.5°, reflect about the x axis, and rotate back. The combined transformation matrix is

$$\begin{pmatrix} \beta & -\alpha & 0 \\ \alpha & \beta & 0 \\ 0 & 0 & 1 \end{pmatrix} \begin{pmatrix} 1 & 0 & 0 \\ 0 & -1 & 0 \\ 0 & 0 & 1 \end{pmatrix} \begin{pmatrix} \beta & \alpha & 0 \\ -\alpha & \beta & 0 \\ 0 & 0 & 1 \end{pmatrix}$$

$$= \begin{pmatrix} \beta^2 - \alpha^2 & 2\alpha\beta & 0 \\ 2\alpha\beta & \alpha^2 - \beta^2 & 0 \\ 0 & 0 & 1 \end{pmatrix} \approx \begin{pmatrix} .7071 & .7071 & 0 \\ .7071 & -.7071 & 0 \\ 0 & 0 & 1 \end{pmatrix}.$$

The last equality is true because

$$0.7071 \approx \sin 45° = \sin 22.5° \cos 22.5° + \cos 22.5° \sin 22.5° = 2\alpha\beta,$$
$$0.7071 \approx \cos 45° = \cos 22.5° \cos 22.5° - \sin 22.5° \sin 22.5° = \beta^2 - \alpha^2.$$

3.16: In order for the general line $ax + by + c = 0$ to pass through the origin, it must satisfy $c = 0$. This implies $y = -(a/b)x$, so $-a/b$ is the slope (i.e., $\tan\theta$) and a and b equal $\sin\theta$ and $\cos\theta$, respectively, up to a sign. This also implies $a^2 + b^2 = 1$ and $ab = \sin\theta\cos\theta$. When this is substituted in Equation (3.12), it reduces to

$$\begin{aligned} x^* &= x - 2a(ax+by) = x(1-2a^2) - 2aby \\ &= x\cos(2\theta) + y\sin(2\theta), \\ y^* &= y - 2b(ax+by) = -2abx + y(1-2b^2) \\ &= x\sin(2\theta) - y\cos(2\theta). \end{aligned} \qquad \text{(Ans.1)}$$

3.17: Reflecting a point (x, y) about the line $y = c$ moves it to $(x, 2c - y)$. Reflecting this about line $y = 0$ simply reverses the y coordinate. The two reflections thus move (x, y) to $(x, y - 2c)$. This is a translation of $-2c$ units in the y direction.

3.18: Starting with $\sin 90° = 1$, $\cos 90° = 0$, we multiply the matrices to get

$$\begin{pmatrix} 0 & 1 & 0 \\ 2 & 0 & 0 \\ 0 & 0 & 1 \end{pmatrix} \begin{pmatrix} 0 & -1 & 0 \\ 1 & 0 & 0 \\ 0 & 0 & 1 \end{pmatrix} = \begin{pmatrix} 1 & 0 & 0 \\ 0 & -2 & 0 \\ 0 & 0 & 1 \end{pmatrix},$$

which is a reflection and scaling in the y dimension.

3.19: Direct multiplication yields

$$\begin{pmatrix} \cos\theta_1\cos\theta_2 - \sin\theta_1\sin\theta_2 & -\cos\theta_1\sin\theta_2 - \cos\theta_2\sin\theta_1 & 0 \\ \sin\theta_1\cos\theta_2 + \cos\theta_1\sin\theta_2 & -\sin\theta_1\sin\theta_2 + \cos\theta_1\cos\theta_2 & 0 \\ 0 & 0 & 1 \end{pmatrix}$$
$$= \begin{pmatrix} \cos(\theta_1+\theta_2) & -\sin(\theta_1+\theta_2) & 0 \\ \sin(\theta_1+\theta_2) & \cos(\theta_1+\theta_2) & 0 \\ 0 & 0 & 1 \end{pmatrix},$$

proving that two-dimensional rotations are additive.

3.20: Direct multiplication yields

$$\mathbf{T}_1\mathbf{T}_2 = \begin{pmatrix} 1+bc & b & 0 \\ c & 1 & 0 \\ 0 & 0 & 1 \end{pmatrix}.$$

This is a combination of shearing and scaling in the x direction. It is pure shearing only if $bc = 0$. This shows that shearing is not an additive transformation.

3.21: The product of the three shears is

$$\begin{pmatrix} 1 & a \\ 0 & 1 \end{pmatrix}\begin{pmatrix} 1 & 0 \\ b & 1 \end{pmatrix}\begin{pmatrix} 1 & c \\ 0 & 1 \end{pmatrix} = \begin{pmatrix} ab+1 & a+abc+c \\ b & bc+1 \end{pmatrix}.$$

When we equate this to the standard rotation matrix

$$\begin{pmatrix} \cos\theta & -\sin\theta \\ \sin\theta & \cos\theta \end{pmatrix},$$

we end up with

$$a = c = \frac{\cos\theta - 1}{\sin\theta} = -\tan\frac{\theta}{2}, \qquad b = \sin\theta,$$

which shows how to calculate a, b, and c from θ. Notice that both $(\cos\theta - 1)$ and $\sin\theta$ approach zero for small angles and the ratio of small numbers is hard to calculate with any precision, which is why it is preferable to use $\tan(\theta/2)$ instead. This particular combination of transformations does not save any time because we still have to calculate $\sin\theta$ and $\cos\theta$ in order to obtain a, b, and c. Still, it is an interesting, unexpected result that's illustrated in Figure Ans.8 for $\theta = 45°$.

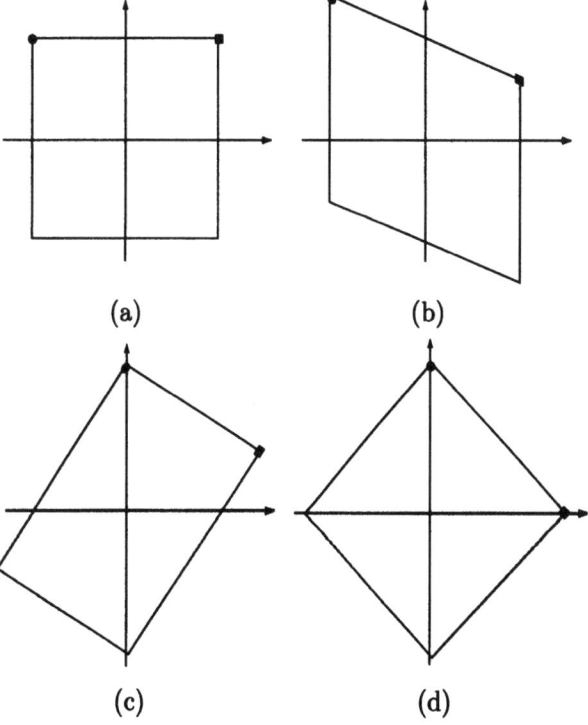

Figure Ans.8: A 45° Rotation as Three Successive Shearings.

3.22: The transformation matrices are

$$\begin{pmatrix} \cos\theta & -\sin\theta & 0 \\ \sin\theta & \cos\theta & 0 \\ 0 & 0 & 1 \end{pmatrix} \begin{pmatrix} a & 0 & 0 \\ 0 & d & 0 \\ 0 & 0 & 1 \end{pmatrix} \begin{pmatrix} \cos\theta & \sin\theta & 0 \\ -\sin\theta & \cos\theta & 0 \\ 0 & 0 & 1 \end{pmatrix}$$

$$= \begin{pmatrix} a\cos^2\theta + d\sin^2\theta & (d-a)\cos\theta\sin\theta & 0 \\ (d-a)\cos\theta\sin\theta & a\sin^2\theta + d\cos^2\theta & 0 \\ 0 & 0 & 1 \end{pmatrix}.$$

When $a = d$, this reduces to

$$\begin{pmatrix} a & 0 & 0 \\ 0 & a & 0 \\ 0 & 0 & 1 \end{pmatrix},$$

which does not depend on θ! This proves that uniform scaling produces identical results regardless of the particular axes used.

3.23: We simply multiply

$$\begin{pmatrix} \cos\theta & -\sin\theta & 0 \\ \sin\theta & \cos\theta & 0 \\ 0 & 0 & 1 \end{pmatrix} \begin{pmatrix} 1 & b & 0 \\ c & 1 & 0 \\ 0 & 0 & 1 \end{pmatrix} \begin{pmatrix} \cos\theta & \sin\theta & 0 \\ -\sin\theta & \cos\theta & 0 \\ 0 & 0 & 1 \end{pmatrix}$$

$$= \begin{pmatrix} \cos^2\theta - c\sin\theta\cos\theta - & \sin\theta\cos\theta - c\sin^2\theta + & 0 \\ -b\sin\theta\cos\theta + \sin^2\theta & +b\cos^2\theta - \sin\theta\cos\theta & \\ \sin\theta\cos\theta + c\cos^2\theta - & \sin^2\theta + c\sin\theta\cos\theta + & 0 \\ -b\sin^2\theta - \sin\theta\cos\theta & +b\sin\theta\cos\theta + \cos^2\theta & \\ 0 & 0 & 1 \end{pmatrix}$$

$$= \begin{pmatrix} 1 - (b+c)\sin\theta\cos\theta & b\cos^2\theta - c\sin^2\theta & 0 \\ c\cos^2\theta - b\sin^2\theta & 1 + (b+c)\sin\theta\cos\theta & 0 \\ 0 & 0 & 1 \end{pmatrix}.$$

This expression does depend on θ! When $b = c = 0$, the expression reduces to the identity matrix. However, when $b = c \neq 0$, this does not reduce to anything simple or elegant.

3.24: A direct scaling of point $\mathbf{P} = (x, y)$ relative to (x_0, y_0) is done by

$$x^* = x_0 + (x - x_0)s_x = x \cdot s_x + x_0(1 - s_x),$$
$$y^* = y_0 + (y - y_0)s_y = y \cdot s_y + y_0(1 - s_y).$$

Using matrix notation, this is written as

$$(x^*, y^*, 1) = (x, y, 1) \begin{pmatrix} s_x & 0 & 0 \\ 0 & s_y & 0 \\ x_0(1 - s_x) & y_0(1 - s_y) & 1 \end{pmatrix}. \qquad (\text{Ans.2})$$

Answers to Exercises

Performing the same transformation by means of translation, scaling, and reverse translation is done by the matrix product

$$\begin{pmatrix} 1 & 0 & 0 \\ 0 & 1 & 0 \\ -x_0 & -y_0 & 1 \end{pmatrix} \begin{pmatrix} s_x & 0 & 0 \\ 0 & s_y & 0 \\ 0 & 0 & 1 \end{pmatrix} \begin{pmatrix} 1 & 0 & 0 \\ 0 & 1 & 0 \\ x_0 & y_0 & 1 \end{pmatrix},$$

which produces the same result.

3.25: Substituting $k_1 = k_2 = k$ in Equation (3.16) yields

$$\begin{pmatrix} k^2 & 0 & 0 \\ 0 & k^2 & 0 \\ k(1-k)x_1 + (1-k)x_2 & k(1-k)y_1 + (1-k)y_2 & 1 \end{pmatrix}.$$

This is equivalent to a single scaling by a factor k^2 about point

$$\mathbf{P}_c = \frac{k(1-k)}{1-k^2}\mathbf{P}_1 + \frac{1-k}{1-k^2}\mathbf{P}_2 = \frac{k}{1+k}\mathbf{P}_1 + \frac{1}{1+k}\mathbf{P}_2.$$

3.26: Using homogeneous coordinates we transform

$$(t^2, t, 1) \begin{pmatrix} -1 & 0 & 1 \\ 0 & 2 & 0 \\ 1 & 0 & 1 \end{pmatrix} = (1 - t^2, 2t, 1 + t^2),$$

which, after dividing by the third component, becomes the point

$$\left(\frac{1-t^2}{1+t^2}, \frac{2t}{1+t^2} \right).$$

This point satisfies the relation $x^2 + y^2 = 1$, so it is located on the unit circle (see also the various circle representations on page 182).

3.27: The following *Mathematica* code

```
t14=2^14;
Print["(x*=",(8192-(2 14189.))/t14,",y*=",(14189.+(2 8192))/t14,")"]
Print["(x*=",Cos[60 Degree]-2. Sin[60 Degree],
  ",y*=",Sin[60 Degree]+2. Cos[60 Degree], ")"]
```

calculates the rotated point twice, first using integers, then using the *Mathematica* built-in sine and cosine functions. The results are identical: $(x^* = -1.23206, y^* = 1.86603)$.

For an 80° rotation, the code

```
t14=2^14;
```

```
Print["(x*=",(2845.-(2 16135.))/t14,",y*=",(16135.+(2 2845.))/t14,
  ")"]
Print["(x*=",Cos[80 Degree]-2. Sin[80 Degree],
  ",y*=",Sin[80 Degree]+2. Cos[80 Degree], ")"]
```

produces the slightly different results: $(x^* = -1.79596, y^* = 1.33209)$ and $(x^* = -1.79597, y^* = 1.3321)$.

3.28: From the definition of θ_i we know that the ratio $\tan\theta_{i+1}/\tan\theta_i$ is $1/2$. Small angles satisfy $\tan\theta \approx \theta$, so we conclude that the ratio θ_{i+1}/θ_i equals approximately $1/2$, except for the first few θ_i's. This can also be confirmed by manually checking the ratios from Table 3.10. Given an infinite sequence of numbers $t, t/2, t/4, \ldots, t/2^i$, we can express every number from 0 (which is obtained by subtracting all the numbers in the sequence from the first one) to $2t$ (which is obtained by adding all the numbers in the sequence). Our sequence of θ_i is finite and the ratios of consecutive elements isn't always precisely $1/2$, but [Walther 71] proves that every number in the range $[0, 90°)$ can be reached, up to a certain precision, by adding and subtracting a number of consecutive θ_i's.

3.29: The method proposed here is based on the fact that the magnitude of the rotated vector (x^*, y^*) should be identical to that of the original vector (x, y). This can be achieved by first normalizing (x^*, y^*), then multiplying it by the magnitude of (x, y):

$$(x^*, y^*) \leftarrow (x^*, y^*) \frac{\sqrt{x^2+y^2}}{\sqrt{x^{*2}+y^{*2}}} = (x^*, y^*)\sqrt{\frac{x^2+y^2}{x^{*2}+y^{*2}}},$$

a calculation involving four exponentiations, one division, one multiplication, and one square root.

3.30: The traditional way of calculating a sine function is by its power series

$$\sin(\theta) = \frac{\theta}{1!} - \frac{\theta^3}{3!} + \frac{\theta^5}{5!} - \frac{\theta^7}{7!} + \cdots,$$

and similarly for cosine. These series, however, converge very slowly, so many multiplications and divisions are needed. If a graphics application needs just rotations, the method of Section 3.2.3 may be simpler and faster than CORDIC. The advantage of using CORDIC is that it can be adapted to the calculation of many different functions. A general software package that is concerned not just with rotations may benefit from using CORDIC.

3.31: From the definition $k = \sqrt{a^2+c^2}$, it follows that $k = 0$ implies $a = c = 0$. In this case, the similarity becomes $x^* = m$, $y^* = n$, and this is not a transformation because it is not one-to-one.

Answers to Exercises

3.32: Transforming point $(x-2P_x+2Q_x, y-2P_y+2Q_y)$ through another halfturn yields

$$(x-2P_x+2Q_x, y-2P_y+2Q_y, 1)\begin{pmatrix} -1 & 0 & 0 \\ 0 & -1 & 0 \\ 2R_x & 2R_y & 1 \end{pmatrix}$$

$$= (-x+2P_x-2Q_x+2R_x, -y+2P_y-2Q_y+2R_y, 1).$$

Comparing this with Equation (3.19) shows that the result of three halfturns is a halfturn about the point $S = P - Q + R$. Writing this as $S - P = R - Q$ shows that **PQRS** is a parallelogram (Figure 3.13c). Thus, point **S** completes the original three points to a parallelogram.

3.33: The first part results in

$$(x^*, y^*) = (x, y)\begin{pmatrix} 3 & 4 & 0 \\ -2 & 5 & 0 \\ 1 & -6 & 1 \end{pmatrix}.$$

The decomposition is simple since $A = \sqrt{9+16} = 5$:

$$\begin{pmatrix} 1 & 0 & 0 \\ 14/25 & 1 & 0 \\ 0 & 0 & 1 \end{pmatrix}\begin{pmatrix} 5 & 0 & 0 \\ 0 & 23/5 & 0 \\ 0 & 0 & 1 \end{pmatrix}\begin{pmatrix} 3/5 & 4/5 & 0 \\ -4/5 & 3/5 & 0 \\ 0 & 0 & 1 \end{pmatrix}\begin{pmatrix} 1 & 0 & 0 \\ 0 & 1 & 0 \\ 1 & -6 & 1 \end{pmatrix}.$$

3.34: From Equation (3.22), we get the following,

1. For scaling: The inverse of

$$\begin{pmatrix} a & 0 & 0 \\ 0 & d & 0 \\ 0 & 0 & 1 \end{pmatrix} \text{ is } \frac{1}{ad}\begin{pmatrix} d & 0 & 0 \\ 0 & a & 0 \\ 0 & 0 & 1 \end{pmatrix},$$

which is also scaling, by factors $1/a$ and $1/d$.

2. For shearing: The inverse of

$$\begin{pmatrix} 1 & b & 0 \\ c & 1 & 0 \\ 0 & 0 & 1 \end{pmatrix} \text{ is } \frac{1}{-bc}\begin{pmatrix} 1 & -b & 0 \\ -c & 1 & 0 \\ 0 & 0 & 1 \end{pmatrix},$$

which is a combination of shearing and scaling.

3. For rotation: The inverse of

$$\begin{pmatrix} \cos\theta & -\sin\theta & 0 \\ \sin\theta & \cos\theta & 0 \\ 0 & 0 & 1 \end{pmatrix}$$

is

$$\frac{1}{\cos^2\theta + \sin^2\theta}\begin{pmatrix} \cos\theta & \sin\theta & 0 \\ -\sin\theta & \cos\theta & 0 \\ 0 & 0 & 1 \end{pmatrix}.$$

This is a rotation in the opposite direction.

4. For translation: The inverse of

$$\begin{pmatrix} 1 & 0 & 0 \\ 0 & 1 & 0 \\ m & n & 1 \end{pmatrix} \text{ is } \begin{pmatrix} 1 & 0 & 0 \\ 0 & 1 & 0 \\ -m & -n & 1 \end{pmatrix}.$$

This is a reverse of the original translation.

3.35: We denote the transformation matrix $\begin{pmatrix} a & b \\ c & d \end{pmatrix}$ and write the four equations

$$\mathbf{P}_i \begin{pmatrix} a & b \\ c & d \end{pmatrix} = \mathbf{P}_i^*, \quad \text{for } 1 \leq i \leq 4.$$

These are easy to solve and yield $a = 6$, $b = 1$, $c = 2$, and $d = 3$.

3.36: They should be executed either by using a negative angle or by moving the minus sign in the rotation matrices to the other sine function.

3.37: The product $\mathbf{T}_r \mathbf{R}_x \mathbf{T}_{rr}$ yields

$$\begin{pmatrix} 1 & 0 & 0 & 0 \\ 0 & \cos\theta & \sin\theta & 0 \\ 0 & -\sin\theta & \cos\theta & 0 \\ 0 & m(\cos\theta - 1) - n\sin\theta & n(\cos\theta - 1) + m\sin\theta & 1 \end{pmatrix}.$$

Substituting $\theta = 30°$ produces the matrix

$$\begin{pmatrix} 1 & 0 & 0 & 0 \\ 0 & 0.866 & 0.5 & 0 \\ 0 & -0.5 & 0.866 & 0 \\ 0 & 0.634 & -0.366 & 1 \end{pmatrix},$$

which transforms point $(1, 2, 3, 1)$ to $(1, 0.866, 3.232, 1)$.

3.38: Using the rule for quaternion multiplication and the three trigonometric identities

$$\cos\theta = \cos^2\tfrac{\theta}{2} - \sin^2\tfrac{\theta}{2}, \quad \sin\theta = 2\sin\tfrac{\theta}{2}\cos\tfrac{\theta}{2}, \quad \cos\theta = 1 - 2\sin^2\tfrac{\theta}{2},$$

Answers to Exercises

we can write

$$\begin{aligned}
\mathbf{q} \cdot [0, \mathbf{P}] \cdot \mathbf{q}^{-1} &= [\cos\tfrac{\theta}{2}, \mathbf{u}\sin\tfrac{\theta}{2}] \cdot [0, \mathbf{P}] \cdot [\cos\tfrac{\theta}{2}, -\mathbf{u}\sin\tfrac{\theta}{2}] \\
&= \{[\cos\tfrac{\theta}{2}, \mathbf{u}\sin\tfrac{\theta}{2}] \cdot [0, \mathbf{P}]\} \cdot [\cos\tfrac{\theta}{2}, -\mathbf{u}\sin\tfrac{\theta}{2}] \\
&= [-\sin\tfrac{\theta}{2}(\mathbf{u} \bullet \mathbf{P}), \cos\tfrac{\theta}{2}\mathbf{P} + \sin\tfrac{\theta}{2}(\mathbf{u} \times \mathbf{P})] \cdot [\cos\tfrac{\theta}{2}, -\mathbf{u}\sin\tfrac{\theta}{2}] \\
&= [-\sin\tfrac{\theta}{2}\cos\tfrac{\theta}{2}(\mathbf{u} \bullet \mathbf{P}) + \sin\tfrac{\theta}{2}\cos\tfrac{\theta}{2}(\mathbf{P} \bullet \mathbf{u}) - \sin^2\tfrac{\theta}{2}(\mathbf{u} \times \mathbf{P}) \bullet \mathbf{u}, \\
&\quad \sin^2\tfrac{\theta}{2}(\mathbf{u} \bullet \mathbf{P})\mathbf{u} + \cos^2\tfrac{\theta}{2}\mathbf{P} + \sin\tfrac{\theta}{2}\cos\tfrac{\theta}{2}(\mathbf{u} \times \mathbf{P}) \\
&\quad - \sin\tfrac{\theta}{2}\cos\tfrac{\theta}{2}(\mathbf{P} \times \mathbf{u}) - \sin^2\tfrac{\theta}{2}(\mathbf{u} \times \mathbf{P}) \times \mathbf{u}] \\
&= [0, \sin^2\tfrac{\theta}{2}(\mathbf{u} \bullet \mathbf{P})\mathbf{u} + \cos^2\tfrac{\theta}{2}\mathbf{P} + 2\sin\tfrac{\theta}{2}\cos\tfrac{\theta}{2}(\mathbf{u} \times \mathbf{P}) \\
&\quad - \sin^2\tfrac{\theta}{2}(\mathbf{P} - (\mathbf{u} \bullet \mathbf{P})\mathbf{u})] \\
&= [0, 2\sin^2\tfrac{\theta}{2}(\mathbf{u} \bullet \mathbf{P})\mathbf{u} + (\cos^2\tfrac{\theta}{2} - \sin^2\tfrac{\theta}{2})\mathbf{P} + 2\sin\tfrac{\theta}{2}\cos\tfrac{\theta}{2}(\mathbf{u} \times \mathbf{P})] \\
&= [0, (1 - \cos\theta)(\mathbf{u} \bullet \mathbf{P})\mathbf{u} + \cos\theta\mathbf{P} + \sin\theta(\mathbf{u} \times \mathbf{P})] \\
&= [0, (\mathbf{u} \bullet \mathbf{P})\mathbf{u} + \cos\theta(\mathbf{P} - (\mathbf{u} \bullet \mathbf{P})\mathbf{u}) + \sin\theta(\mathbf{u} \times \mathbf{P})],
\end{aligned}$$

that is Equation (3.26).

3.39: They could be (a) a cube, (b) the same cube seen edge on, and (c) the same cube seen rotated through 30° with one front edge and one back edge.

3.40: Given $s_z = 0.625$, we calculate θ and ϕ

$$\theta = \sin^{-1}\left(\pm\frac{0.625}{\sqrt{2}}\right) = \sin^{-1}(\pm 0.44194) = \pm 26.23°,$$

$$\phi = \sin^{-1}\left(\pm\frac{0.625}{\sqrt{2 - 0.625^2}}\right) = \sin^{-1}(\pm 0.49266) = \pm 29.52°.$$

3.41: Equation (3.32) shows that $s_x^2 = s_z^2$ is equivalent to

$$\cos^2\phi + \sin^2\phi\sin^2\theta = \sin^2\phi + \cos^2\phi\sin^2\theta.$$

This can be simplified to $(\sin^2\phi - \cos^2\phi)\cos^2\theta = 0$, with the two solutions $\cos^2\theta = 0 \to \theta = \pm 90°$ and $\sin^2\phi - \cos^2\phi = 0$, which implies $\sin\phi = \pm\cos\phi$ and results in $\phi = 90° \pm 45°$ and $270° \pm 45°$.

3.42: Lines on a plane perpendicular to the observer (i.e., a plane parallel to the xy plane).

3.43: The terms *clockwise* and *counterclockwise* fully describe rotations in two dimensions. Our example, however, is in three dimensions, where rotations are more complex and can have more directions. Instead of saying "counterclockwise," we should say that our rotation is from the negative x direction to the negative z direction. The rule on page 96 says that this is precisely the rotation produced by matrix (3.40).

3.44: Because of the special orientation of the projection plane. This equation says that any point satisfying $\alpha x = -\beta z$ is on the projection plane, no matter what its y coordinate is.

3.45: The case $\theta = 0$ means $\alpha = 0$ and $\beta = 1$. Matrix (3.41) reduces to

$$\begin{pmatrix} k & 0 & 0 & 0 \\ 0 & k & 0 & 0 \\ 0 & 0 & 0 & 1 \\ 0 & 0 & 0 & k \end{pmatrix} = k \begin{pmatrix} 1 & 0 & 0 & 0 \\ 0 & 1 & 0 & 0 \\ 0 & 0 & 0 & r \\ 0 & 0 & 0 & 1 \end{pmatrix}.$$

The case $\theta = 45°$ implies $\alpha = \beta = 1/\sqrt{2}$. Matrix (3.41) reduces to

$$\begin{pmatrix} k/2 & 0 & -k/2 & 1/\sqrt{2} \\ 0 & k & 0 & 0 \\ -k/2 & 0 & k/2 & 1/\sqrt{2} \\ 0 & 0 & 0 & k \end{pmatrix}.$$

The case $\theta = 90°$ means $\alpha = 1$ and $\beta = 0$. Matrix (3.41) reduces to

$$\begin{pmatrix} 0 & 0 & 0 & 1 \\ 0 & k & 0 & 0 \\ 0 & 0 & k & 0 \\ 0 & 0 & 0 & k \end{pmatrix} = k \begin{pmatrix} 0 & 0 & 0 & r \\ 0 & 1 & 0 & 0 \\ 0 & 0 & 1 & 0 \\ 0 & 0 & 0 & 1 \end{pmatrix}.$$

3.46: Direct multiplication yields

$$(\beta l, m, -\alpha l, 1) \begin{pmatrix} k\beta^2 & 0 & -k\alpha\beta & \alpha \\ 0 & k & 0 & 0 \\ -k\alpha\beta & 0 & k\alpha^2 & \beta \\ 0 & 0 & 0 & k \end{pmatrix}$$
$$= (kl\beta^3 + kl\alpha^2\beta, mk, -kl\alpha\beta^2 - kl\alpha^3, l\alpha\beta - l\alpha\beta + k)$$
$$= (kl\beta, km, -kl\alpha, k).$$

The transformed point is $\mathbf{P}^* = (l\beta, m, -l\alpha) = \mathbf{P}$. Point $\mathbf{P}$ is thus transformed to itself! This is because $\mathbf{P}$ resides on the projection plane. The equation of the plane is $\alpha x = -\beta z$ and a simple check verifies that the coordinates of point $\mathbf{P}$ satisfy this.

3.47: The steps are similar to the ones used to derive matrix (3.41):

- Calculate the line from the observer to a general point $\mathbf{P} = (l, m, n)$:

$$(l + k\alpha, m + k\beta\gamma, n + k\beta\delta)t + (-k\alpha, -k\beta\gamma, -k\beta\delta).$$

- Use the relation $(-k\alpha, -k\beta\gamma, -k\beta\delta) \bullet (x, y, z) = 0$ to calculate the equation of the projection plane. This is trivial and the equation is $-xk\alpha - yk\beta\gamma - zk\beta\delta = 0$.

Answers to Exercises

- Calculate the value of t_0 at the intersection point. From

$$[(l+k\alpha)t_0 - k\alpha]k\alpha + [(m+k\beta\gamma)t_0 - k\beta\gamma]k\beta\gamma + [(n+k\beta\delta)t_0 - k\beta\delta]k\beta\delta = 0,$$

we get

$$t_0 = \frac{k(\alpha^2 + \beta^2\gamma^2 + \beta^2\delta^2)}{(l+k\alpha)\alpha + (m+k\beta\gamma)\beta\gamma + (n+k\beta\delta)\beta\delta}$$

$$= \frac{k(\alpha^2 + \beta^2\gamma^2 + \beta^2\delta^2)}{l\alpha + m\beta\gamma + n\beta\delta + k(\alpha^2 + \beta^2\gamma^2 + \beta^2\delta^2)}.$$

- The coordinates of the projected point can now be calculated. The x^* coordinate is

$$x^* = (l+k\alpha)t_0 - k\alpha = (l+k\alpha)\frac{k(\alpha^2 + \beta^2\gamma^2 + \beta^2\delta^2)}{l\alpha + m\beta\gamma + n\beta\delta + k(\alpha^2 + \beta^2\gamma^2 + \beta^2\delta^2)} - k\alpha$$

$$= \frac{lk\beta^2(\gamma^2 + \delta^2) - mk\alpha\beta\gamma - nk\alpha\beta\delta}{l\alpha + m\beta\gamma + n\beta\delta + k(\alpha^2 + \beta^2\gamma^2 + \beta^2\delta^2)}.$$

- The y^* coordinate is

$$y^* = (m+k\beta\gamma)t_0 - k\beta\gamma$$

$$= (m+k\beta\gamma)\frac{k(\alpha^2 + \beta^2\gamma^2 + \beta^2\delta^2)}{l\alpha + m\beta\gamma + n\beta\delta + k(\alpha^2 + \beta^2\gamma^2 + \beta^2\delta^2)} - k\beta\gamma$$

$$= \frac{-lk\alpha\beta\gamma + mk(\alpha^2 + \beta^2\delta^2) - nk\beta^2\gamma\delta}{l\alpha + m\beta\gamma + n\beta\delta + k(\alpha^2 + \beta^2\gamma^2 + \beta^2\delta^2)}.$$

- The z^* coordinate is

$$z^* = (n+k\beta\delta)t_0 - k\beta\delta$$

$$= (n+k\beta\delta)\frac{k(\alpha^2 + \beta^2\gamma^2 + \beta^2\delta^2)}{l\alpha + m\beta\gamma + n\beta\delta + k(\alpha^2 + \beta^2\gamma^2 + \beta^2\delta^2)} - k\beta\delta$$

$$= \frac{-lk\alpha\beta\delta - mk\beta^2\gamma\delta + nk(\alpha^2 + \beta^2\gamma^2)}{l\alpha + m\beta\gamma + n\beta\delta + k(\alpha^2 + \beta^2\gamma^2 + \beta^2\delta^2)}.$$

- The projection matrix is now easy to calculate. It is

$$\begin{pmatrix} k\beta^2(\gamma^2+\delta^2) & -k\alpha\beta\gamma & -k\alpha\beta\delta & \alpha \\ -k\alpha\beta\gamma & k(\alpha^2+\beta^2\delta^2) & -k\beta^2\gamma\delta & \beta\gamma \\ -k\alpha\beta\delta & -k\beta^2\gamma\delta & k(\alpha^2+\beta^2\gamma^2) & \beta\delta \\ 0 & 0 & 0 & k(\alpha^2+\beta^2\gamma^2+\beta^2\delta^2) \end{pmatrix}. \quad \text{(Ans.3)}$$

To check our result, we consider the special case of no rotation about the x axis. In this case, $\phi = 0$, $\gamma = 0$, and $\delta = 1$. It is easy to see that matrix (Ans.3) reduces to matrix (3.41).

3.48: After two rotations, the observer may end up at any point in space, but the projection plane still passes through the origin. This is why our case is not completely general.

3.49: Recall that the basic rule of perspective projection is to connect an image point to the observer with a line that intercepts the projection plane. The observer and the image points should, therefore, be on different sides of the projection plane. In our case point $(0,0,0)$ is behind the observer, so it is on the same side of the projection plane as the observer and, consequently, it does not make sense to project it.

3.50: Direct multiplication yields

$$(\beta l, m, -\alpha l, 1) \begin{pmatrix} \beta & 0 & 0 & \alpha r \\ 0 & 1 & 0 & 0 \\ -\alpha & 0 & 0 & \beta r \\ 0 & 0 & 0 & 1 \end{pmatrix} = (l\beta^2 + l\alpha^2, m, 0, lr\alpha\beta - lr\alpha\beta + 1) = (l, m, 0, 1),$$

so the transformed point is $\mathbf{P}^* = (l, m, 0)$. Figure Ans.9 shows that point $\mathbf{P} = (\beta l, m, -\alpha l)$ resides on the projection plane. After the transformations, it is still located on the projection plane, only now this is the xy plane.

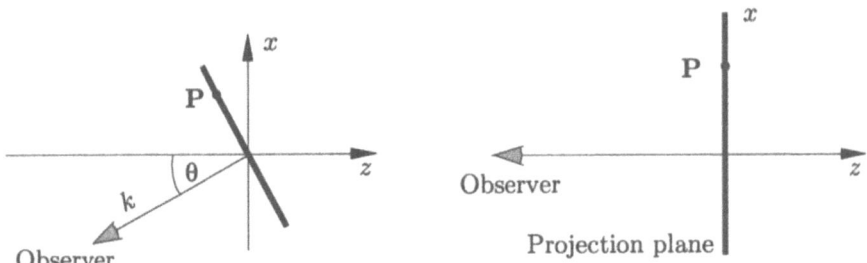

Figure Ans.9: Transforming and Projecting.

3.51: Because they project on different projection planes. Matrix (3.41) projects on plane $\alpha x = -\beta z$, where the z coordinate is proportional to the x coordinate, whereas matrix (3.43) projects on the xy plane, where the z coordinate is zero.

3.52: This is easily done with the help of appropriate software. The results for the two cases are

$$\begin{pmatrix} 1 & 0 & 0 & 0 \\ 0 & 1 & 0 & 0 \\ 0 & 0 & 0 & r \\ -a & -b & 0 & -cr \end{pmatrix}, \quad \begin{pmatrix} 1 & 0 & 0 & 0 \\ 0 & 1 & 0 & 0 \\ 0 & 0 & 0 & r \\ 0 & 0 & 0 & 1 \end{pmatrix}. \quad \text{(Ans.4)}$$

Notice how the second matrix of (Ans.4) is the standard perspective projection matrix $\mathbf{T}_p$ of Equation (3.39).

Answers to Exercises

3.53: We substitute $(a, b, c) = (0, 1, 0)$ and $(d, e, f) = (0, 1/\sqrt{2}, 1/\sqrt{2})$ in matrix (3.44). The transformation is, therefore,

$$(0, 1, 10, 1) \begin{pmatrix} 1 & 0 & 0 & 0 \\ 0 & \frac{1}{\sqrt{2}} & 0 & \frac{r}{\sqrt{2}} \\ 0 & \frac{-1}{\sqrt{2}} & 0 & \frac{r}{\sqrt{2}} \\ 0 & \frac{-1}{\sqrt{2}} & 0 & \frac{-r}{\sqrt{2}} \end{pmatrix} = \left(0, \frac{1 - 10 - 1}{\sqrt{2}}, 0, \frac{r + 10r - r}{\sqrt{2}}\right),$$

so $\mathbf{P}^* = (0, -1/r, 0) = (0, -k, 0) = (0, -\sqrt{8}, 0)$.

3.54: One vanishing point per stair—created by the two parallel lines at the top and bottom of the riser—except for the stair that happens to be perpendicular to the direction of observation.

3.55: When T_1 or T_2 gets large, the object is magnified. However, when T_3 gets large, the object is scaled in the z direction *relative to the origin*. All the z coordinates become big, effectively moving the object away from the observer. When all three scale factors get large, the magnification in the x and y directions is canceled out by the effect of moving away in the z direction, so the object does not seem to change in size.

3.56: We first calculate α:

$$\alpha = \frac{|\mathbf{a}|^2}{\mathbf{a} \bullet (\mathbf{p} - \mathbf{b})} = \frac{8}{(0, 2, 2) \bullet (x - 0, y - 1, z - 0)} = \frac{4}{y + z - 1}.$$

(Note that $\mathbf{P} = (0, 1, 10)$, implying $\alpha = 4/(1 + 10 - 1) = 2/5$.)

Next, we calculate vector $\mathbf{d}$:

$$\begin{aligned}
\mathbf{d} &= \mathbf{b} + \alpha(\mathbf{p} - \mathbf{b}) \\
&= (0, 1, 0) + \frac{4}{y + z - 1}(x, y - 1, z) \\
&= \frac{4}{y + z - 1}(x, (5y - z - 5)/4, z).
\end{aligned}$$

[A check verifies that $\mathbf{P} = (0, 1, 10) \Rightarrow \mathbf{d} = (0, 1, 4)$.]

Vector $\mathbf{c}$ can now be calculated:

$$\begin{aligned}
\mathbf{c} &= \alpha(\mathbf{p} - \mathbf{b}) - \mathbf{a} \\
&= \frac{4}{y + z - 1}(x, y - 1, z) - (0, 2, 2) \\
&= \frac{4}{y + z - 1}(x, (y - z - 1)/2, -(y - z - 1)/2).
\end{aligned}$$

The screen coordinates are, thus,

$$\mathbf{u}\bullet\mathbf{c} = (1,0,0)\bullet\frac{4}{y+z-1}(x,(y-z-1)/2,-(y-z-1)/2)$$
$$=\frac{4x}{y+z-1}.$$
$$\mathbf{w}\bullet\mathbf{c} = (0,1/\sqrt{2},-1/\sqrt{2})\bullet\frac{4}{y+z-1}(x,(y-z-1)/2,-(y-z-1)/2)$$
$$=\frac{4(y-z-1)}{\sqrt{2}(y+z-1)}.$$

Again, a direct check verifies that $\mathbf{P} = (0,1,10)$ results in

$$\mathbf{u}\bullet\mathbf{c} = 0, \quad\text{and}\quad \mathbf{w}\bullet\mathbf{c} = \frac{4(1-10-1)}{\sqrt{2}(1+10-1)} = \frac{-4}{\sqrt{2}} = -\sqrt{8}.$$

Also, the screen coordinates of point $\mathbf{P} = (0,5,4)$ are

$$\mathbf{u}\bullet\mathbf{c} = 0, \quad\text{and}\quad \mathbf{w}\bullet\mathbf{c} = \frac{4(5-4-1)}{\sqrt{2}(5+4-1)} = 0,$$

as should be expected (why?).

3.57: Figure Ans.10 shows that in a right-handed coordinate system, the positive y axis is in the direction of vector $\mathbf{w}$ and vector $\mathbf{u}$ is in the direction of *negative* x.

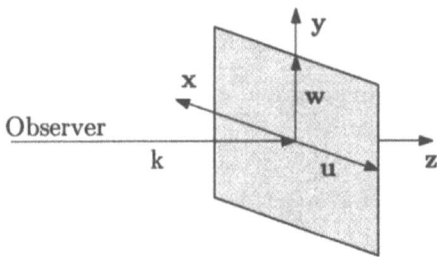

Figure Ans.10: A Right-Handed Coordinate System.

3.58: The proof is straightforward but a little messy. We start with two three-dimensional points $\mathbf{P}_1 = (x_1,y_1,z_1)$ and $\mathbf{P}_2 = (x_2,y_2,z_2)$. Their projections are

$$\mathbf{P}_1^* = \left(\frac{x_1 k}{k+z_1}, \frac{y_1 k}{k+z_1}, \frac{z_1}{k+z_1}\right) \quad\text{and}\quad \mathbf{P}_2^* = \left(\frac{x_2 k}{k+z_2}, \frac{y_2 k}{k+z_2}, \frac{z_2}{k+z_2}\right).$$

Now consider the two lines $\mathbf{P}(t) = \mathbf{P}_1 + (\mathbf{P}_2 - \mathbf{P}_1)t$ and $\mathbf{P}^*(u) = \mathbf{P}_1^* + (\mathbf{P}_2^* - \mathbf{P}_1^*)u$. We need to prove that every point on $\mathbf{P}(t)$ is transformed to a point on $\mathbf{P}^*(u)$, where u depends on t, k, $\mathbf{P}_1$, and $\mathbf{P}_2$ only.

Answers to Exercises

The coordinates of a general point on $\mathbf{P}(t)$ are

$$\left(\frac{(x_1 + (x_2 - x_1)t)k}{k + z_1 + (z_2 - z_1)t}, \frac{(y_1 + (y_2 - y_1)t)k}{k + z_1 + (z_2 - z_1)t}, \frac{(z_1 + (z_2 - z_1)t)k}{k + z_1 + (z_2 - z_1)t}\right).$$

The coordinates of a general point on $\mathbf{P}^*(u)$ are

$$\left(\frac{x_1 k}{k + z_1} + \left(\frac{x_2 k}{k + z_2} - \frac{x_1 k}{k + z_1}\right)u, \frac{y_1 k}{k + z_1} + \left(\frac{y_2 k}{k + z_2} - \frac{y_1 k}{k + z_1}\right)u,\right.$$

$$\left.\frac{z_1}{k + z_1} + \left(\frac{z_2}{k + z_2} - \frac{z_1}{k + z_1}\right)u\right).$$

In order for the points to be equal, the following two equations have to hold:

$$\frac{(x_1 + (x_2 - x_1)t)k}{k + z_1 + (z_2 - z_1)t} = \frac{x_1 k}{k + z_1} + \left(\frac{x_2 k}{k + z_2} - \frac{x_1 k}{k + z_1}\right)u,$$

$$\frac{(z_1 + (z_2 - z_1)t)k}{k + z_1 + (z_2 - z_1)t} = \frac{z_1}{k + z_1} + \left(\frac{z_2}{k + z_2} - \frac{z_1}{k + z_1}\right)u.$$

(There are actually three equations but the second one, for y, is equivalent to the first one, so it is not included here.) Because of the way the depth transformation is defined, both equations are satisfied if u is defined by

$$u = t \frac{k + z_2}{k + z_1 + (z_2 - z_1)t}.$$

Note that $t = 0 \Rightarrow u = 0$ and $t = 1 \Rightarrow u = 1$.

3.59: Since k is scaled by the same factor of 5, we should scale e by this factor, bringing it down from 2.5 to 0.5.

3.60: The tangent of half the angle is $(W/2)/k = 1/2$. Therefore, half the angle equals 26.5° and the entire field of view is twice that, or 53° wide.

3.61: Time is certainly *a* fourth dimension but not *the* fourth dimension.

3.62: The rule for constructing a multidimensional pyramid is similar to that for the tetrahedron. If we imagine a one-dimensional pyramid to be a straight segment and a two-dimensional pyramid to be an equilateral triangle, then the rule is as follows: An n-dimensional pyramid is constructed by selecting the midpoint of the $(n - 1)$-dimensional **cube** and dragging it into the nth dimension. Figure Ans.11 illustrates the construction of the two-dimensional and three-dimensional pyramids.

The rule for an octahedron is different. We imagine the one-dimensional octahedron to be a straight segment and the two-dimensional octahedron to be a diamond-shaped square. The rule for constructing multidimensional octahedrons is as follows: An n-dimensional octahedron is constructed by selecting the midpoint of the $(n-1)$-dimensional octahedron and dragging it into the nth dimension twice, in the positive and the negative directions. Figure Ans.12 illustrates the construction of the two-dimensional and three-dimensional octahedrons.

> A child['s]... first geometrical discoveries are topological... If you ask him to copy a square or a triangle, he draws a closed circle.
> — Jean Piaget.

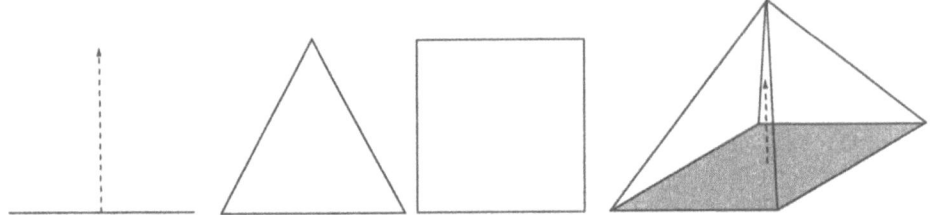

Figure Ans.11: Pyramids of Various Dimensions.

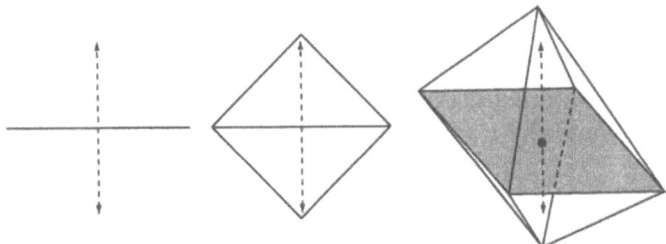

Figure Ans.12: Octahedrons of Various Dimensions.

3.63: A segment of size m can be considered as "made of" m unit segments. A square of side length m can be subdivided into m^2 unit squares, so its area is m^2. A cube of side length m can be filled with m^3 unit cubes, so its volume is m^3. By extension, a hypercube of side length m can be "hyperfilled" with m^4 unit hypercubes, so its hypervolume is m^4.

3.64: They are the xy, xz, xw, yz, yw, and zw planes.

3.65: In two dimensions, we can translate a point $\mathbf{P} = (x, y)$ by $\mathbf{P} + (t_x, t_y)$ and this can be extended to n dimensions in an obvious way.

3.66: The three-dimensional cube is defined by eight corner points (x, y, z) where the coordinates are either zero or one. After deleting any two of the three coordinates, we are left with eight points, four of which are (0) and the other four are (1). The result is the straight segment from zero to one. Trying to visualize the shape of

Answers to Exercises

a four-dimensional cube by looking at its two-dimensional projections is comparable to trying to visualize a three-dimensional cube by looking at this straight segment.

3.67: The z^* coordinate depends on Z in the sense that point **P** should be projected on the cylinder only if $|z^*| \leq Z$.

3.68: Figure Ans.13a shows a cylinder of radius R centered on the origin with its axis in the z direction. We start with a circle in the xy plane. The circle's equation is $(R\cos(2\pi t), R\sin(2\pi t), 0)$. The circle is now rotated θ degrees about the y axis, as shown in Figure Ans.13b. The new circle is given by

$$(R\cos(2\pi t), R\sin(2\pi t), 0) \begin{pmatrix} \cos\theta & 0 & \sin\theta \\ 0 & 1 & 0 \\ -\sin\theta & 0 & \cos\theta \end{pmatrix}$$
$$= (R\cos(2\pi t)\cos\theta, R\sin(2\pi t), R\cos(2\pi t)\sin\theta).$$

Figure Ans.13c shows that in order to convert this tilted circle into an ellipse, its x and z coordinates should be scaled by a factor of $1/\cos\theta$. The equation of this ellipse is thus

$$(R\cos(2\pi t), R\sin(2\pi t), R\cos(2\pi t)\tan\theta). \tag{Ans.5}$$

In order to prove that this is an ellipse, we can rotate it back to the xy plane. The result is

$$(R\cos(2\pi t), R\sin(2\pi t), R\cos(2\pi t)\tan\theta) \begin{pmatrix} \cos\theta & 0 & -\sin\theta \\ 0 & 1 & 0 \\ \sin\theta & 0 & \cos\theta \end{pmatrix}$$
$$= (R(\cos\theta + \sin^2\theta)\cos(2\pi t), R\sin(2\pi t), 0),$$

an expression that satisfies $x^2/a^2 + y^2/b^2 = 1$ for $a = R(\cos\theta + \sin^2\theta)$ and $b = R$. Figure Ans.13d shows the unrolled cylinder, cut along the $y = 0$ line, with the origin at its center.

The behavior of the resulting flat curve can be figured out when we notice that the x and y coordinates of the ellipse of Equation (Ans.5) form a circle, which is a curve with constant speed. This means that when the curve is flattened, it has constant speed in the horizontal direction (i.e., incrementing t in equal steps moves us equal horizontal increments on the unrolled cylinder). The vertical behavior of the flattened curve is determined by the z coordinate of Equation (Ans.5), and this coordinate behaves like a sine curve with an amplitude $R\tan\theta$. The result is the parametric curve

$$\Big((2t-1)\pi R, R\tan\theta\cos\big[(2t-1)\pi\big]\Big), \quad 0 \leq t \leq 1.$$

As t varies from zero to one, the horizontal coordinate varies from $-\pi R$ to $+\pi R$ and the vertical coordinate varies as a sine curve from -1 to zero to one, back to zero, and ends up at -1.

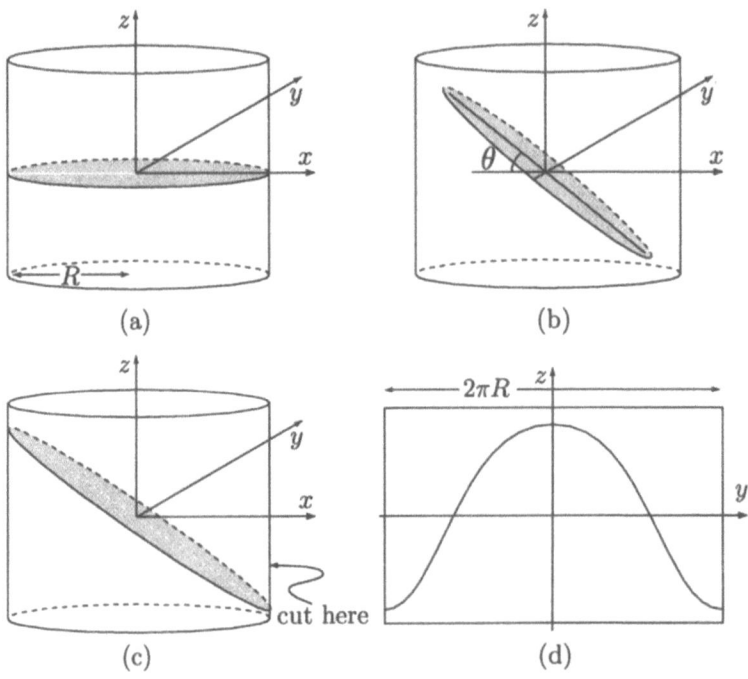

Figure Ans.13: Ellipse and Sinusoidal.

It is also interesting to consider the curvature of this sine curve. The curvature is essentially given by the second derivative, which, in the case of $\sin(t)$, equals $-\sin(t)$. Being interested only in the absolute magnitude of the curvature, we can disregard the minus sign. The result is that for $t = 0$ and $t = \pi$ the curvature is zero, while for $t = \pi/2$, it is maximum. The conclusion is that when a straight line is projected by curved perspective into a sinusoidal, those parts of the line that are close to the observer become highly curved while the distant parts remain straight or close to straight.

> You're wasting that panorama on me, Nan. Save it for Dave Slade.
> — Robert McWade (as District Attorney) in *Ladies They Talk About* (1933)

3.69: The image is circular because the main mirror is circular. It has a hole in the middle because the main mirror has a hole in it (more accurately, because light hitting the top of the main mirror, around its hole, cannot reach the secondary mirror).

3.70: It is easy to see from Equation (3.54) that $z = k \rightarrow z^* = k/2$.

Answers to Exercises

3.71: Figure Ans.14 shows a circle C that does not pass through the origin (notice that the circle of inversion itself is not shown). We construct the line L from O through the center of C and examine the intersection points **P** and **Q**. Their projections **P*** and **Q*** must be on L. We select an arbitrary point **R** on C and denote its projection **R***. From $OP \cdot OP^* = OQ \cdot OQ^* = OR \cdot OR^*$, we get $OR^*/OP = OP^*/OR$, indicating that the two triangles ORP and OR^*P^* are similar. This implies that angles OP^*R^* and ORP are equal and also angles OQ^*R^* and ORQ are equal. We subtract angles and find that $OP^*R^* - OQ^*R^* = ORP - ORQ = 90°$, which implies that angle $P^*R^*Q^* = 90°$. Since this is true for a general point **R**, we conclude that all the points R^* (that together constitute the projection of C) are located on the circle C^* centered on L with diameter P^*Q^*.

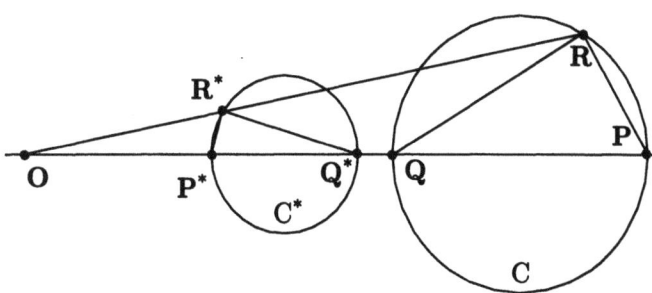

Figure Ans.14: Circular Inversion of a Circle.

3.72: The rule $\mathbf{P}^* = 1/\mathbf{P}$ is generalized to $\mathbf{P}^* = R^2/\mathbf{P}$. This projection retains all the features mentioned in the text with regard to the unit circle.

4.1: Yes, since the relation $\mathbf{P}_1 = \alpha \mathbf{P}_0$ equals $(24, -6) = \alpha(12, -3)$ with the solution $\alpha = 0.5$ and $\mathbf{P}_2 = \beta \mathbf{P}_0$ equals $(18, -4.5) = \beta(12, -3)$ which solves to $\beta = 1.5$.

4.2: Clearly, we can write $\mathbf{P}_3 = \gamma \mathbf{P}_0$ for the fourth point. We only need to know the single number γ instead of the two coordinates of $\mathbf{P}_3$. It is now clear that n collinear points can be represented by one point plus $n - 1$ numbers.

4.3: It is a point. Figure 4.2b shows the two sums $\mathbf{P}_1^* = \mathbf{P}_1 + \mathbf{v}$ and $\mathbf{P}_2^* = \mathbf{P}_2 + \mathbf{v}$. It is easy to see that the relative positions of $\mathbf{P}_1^*$ and $\mathbf{P}_2^*$ are the same as those of $\mathbf{P}_1$ and $\mathbf{P}_2$. Another way to look at the sum $\mathbf{P} + \mathbf{v}$ is to observe that it moves us away from **P**, which is a point, in a certain direction and by a certain distance. We thus conclude that we get to another point.

4.4: Three two-dimensional points are independent if they are not collinear. The three corners of a triangle cannot, of course, be on the same line and are, therefore, independent. As a result, the three components of Equation (4.3), which are based on the coordinates of the corner points, are independent.

4.5: It is always true that $P_0 = 1 \cdot P_0 + 0 \cdot P_1 + 0 \cdot P_2$, so the barycentric coordinates of P_0 are $(1, 0, 0)$. Points outside the triangle have barycentric coordinates, some of which are negative and others are greater than 1.

4.6: This is easy. The centroid is given by $(1/3)P_0 + (1/3)P_1 + (1/3)P_2$.

4.7: We can look at this sum in two ways:

1. As the sum $(P + v) + (-Q + w)$. we know that $P + v$ is a point and so is $-Q + w$. This is, therefore, the sum of points and it equals the vector from point $-Q + w$ to point $P + v$.

2. As the sum $(P - Q) + v + w$. This is the sum of three vectors, so it is a vector.

4.8: This expression is an attempt to find a parametric cubic polynomial that's close to a circle in the first quadrant. The general form of such a polynomial is $P(t) = at^3 + bt^2 + ct + d$ where the coefficients a, b, c, and d are pairs of numbers and t is a parameter varying in the interval $[0, 1]$. To determine the four coefficient pairs, we need four equations, so we require that the polynomial and the circle be identical at four points. For $t = 0$, we require that $P(0) = (1, 0)$ and, for $t = 1$, that $P(1) = (0, 1)$. In addition, we select the two equally spaced values $t = 1/3$ and $2/3$ and require that $P(1/3) = (\cos 30°, \sin 30°)$ and $P(1/3) = (\cos 60°, \sin 60°)$. The four equations are thus

$$P(0) = at^3 + bt^2 + ct + d|_{t=0} = (1, 0),$$
$$P(1/3) = at^3 + bt^2 + ct + d|_{t=1/3} = (\cos 30°, \sin 30°),$$
$$P(2/3) = at^3 + bt^2 + ct + d|_{t=2/3} = (\cos 60°, \sin 60°),$$
$$P(1) = at^3 + bt^2 + ct + d|_{t=1} = (0, 1),$$

and the solutions are $a = (0.441, -0.441)$, $b = (-1.485, -0.162)$, $c = (0.044, 1.603)$, and $d = (1, 0)$. The cubic polynomial is therefore

$$P(t) = (0.441, -0.441)t^3 + (-1.485, -0.162)t^2 + (0.044, 1.603)t + (1, 0).$$

This polynomial is just an approximation. At other values of t, it passes close to the circle but not on it.

4.9: It is easy to see from Figure 4.5a that $d = R \cos(\pi/n)$, so $R - d = R(1 - \cos(\pi/n))$. This expression approaches zero for large n.

4.10: This velocity is variable since it slows down from $P^t(0) = (0, 2)$ (a speed of $\sqrt{0^2 + 2^2} = 2$) to $P^t(1) = (-1, 0)$ (a speed of $\sqrt{(-1)^2 + 0^2} = 1$). Notice that the term "speed" refers to a scalar, whereas "velocity" is a vector, having both direction and magnitude.

4.11: We know that the two-dimensional parametric curve $(\cos t, \sin t)$ is a circle of radius 1, centered on the origin. As a result, the three-dimensional curve $(\cos t, \sin t, t)$ is a helix spiraling around the z axis upward from the origin.

Answers to Exercises

4.12: Three approaches are discussed.

Approach 1: The general two-dimensional line $y = ax + b$ goes through point $(0, b)$ and its direction is the vector $(1, a)$ (for each step in the x direction, make a steps in the y direction). We can therefore express it as

$$\mathbf{P}(t) = \mathbf{P}_0 + t\mathbf{v} = (0, b) + t(1, a).$$

Applying Equation (4.5), we get point $\mathbf{Q}$:

$$\begin{aligned}
\mathbf{Q} &= \mathbf{P}_0 + \frac{(\mathbf{P} - \mathbf{P}_0) \bullet \mathbf{v}}{\mathbf{v} \bullet \mathbf{v}} \mathbf{v} \\
&= (0, b) + \frac{(P_x, P_y - b) \bullet (1, a)}{(1, a) \bullet (1, a)} (1, a) \qquad \text{(Ans.6)} \\
&= \left(\frac{aP_y + P_x - ab}{a^2 + 1}, \frac{a^2 P_y + aP_x + b}{a^2 + 1} \right).
\end{aligned}$$

Hence, the distance between $\mathbf{P}$ and $\mathbf{Q}$ is

$$\begin{aligned}
D &= \sqrt{(P_x - Q_x)^2 + (P_y - Q_y)^2} \\
&= \sqrt{\left(P_x - \frac{aP_y + P_x - ab}{a^2 + 1} \right)^2 + \left(P_y - \frac{a^2 P_y + aP_x + b}{a^2 + 1} \right)^2} \\
&= \sqrt{\left(\frac{a^2 P_x - aP_y + ab}{a^2 + 1} \right)^2 + \left(\frac{P_y - aP_x - b}{a^2 + 1} \right)^2} \\
&= \sqrt{(P_y - aP_x - b)^2} = \frac{|P_y - aP_x - b|}{\sqrt{1 + a^2}};
\end{aligned}$$

same as Equation (4.6)

Approach 2: We denote the line $y = ax + b$ by L_1. We find $\mathbf{Q}$, the point on L_1 closest to $\mathbf{P}$, by calculating the equation of the line that's (1) perpendicular to L_1 and (2) goes through $\mathbf{P}$. Denote this line (L_2) by $y = Ax + B$. Since L_2 is perpendicular to L_1, its slope is $-1/a$. The requirement that it goes through $\mathbf{P}$ gives us an equation for B:

$$P_y = -\frac{1}{a} P_x + B, \quad \text{whose solution is} \quad B = P_y + \frac{P_x}{a}.$$

Line L_2 is thus

$$y = -\frac{1}{a} x + \left(P_y + \frac{P_x}{a} \right).$$

The intersection of the two lines yields point $\mathbf{Q}$:

$$ax + b = -\frac{1}{a} x + \left(P_y + \frac{P_x}{a} \right) \quad \text{yields} \quad x = \frac{aP_y + P_x - ab}{a^2 + 1},$$

and
$$y = a\left(\frac{aP_y + P_x - ab}{a^2 + 1}\right) + b = \frac{a^2 P_y + aP_x + b}{a^2 + 1}.$$

Point **Q** is thus
$$\mathbf{Q} = \left(\frac{aP_y + P_x - ab}{a^2 + 1}, \frac{a^2 P_y + aP_x + b}{a^2 + 1}\right),$$

which is the same as that given by Equation (Ans.6).

Approach 3: Any point **Q** on the line has coordinates $(x, ax + b)$. The distance D between **P** and a general point **Q** on the line is therefore
$$D(x) = \sqrt{(P_x - x)^2 + (P_y - ax - b)^2}.$$

This distance is a function of x and we are looking for that x value for which $D(x)$ has a minimum. Instead of differentiating $D(x)$ (tedious because of the square root), we differentiate $D^2(x)$, noting that both functions $D(x)$ and $D^2(x)$ have a minimum at the same value of x. The result is
$$\frac{d}{dx} D^2(x) = -2(P_x - x) - 2a(P_y - ax - b)$$
$$= -2P_x - 2aP_y + 2ab + 2x + 2a^2 x.$$

When this is equated to zero, we find that $D(x)$ has a minimum for
$$x = \frac{aP_y + P_x - ab}{a^2 + 1}.$$

Point **Q** is, thus,
$$\mathbf{Q} = (x, ax + b) = \left(\frac{aP_y + P_x - ab}{a^2 + 1}, \frac{a^2 P_y + aP_x + b}{a^2 + 1}\right),$$

same as Equation (Ans.6). The distance is, thus, given by Equation (4.6)

4.13: Direct calculation shows that in the former case, both α and β have the indefinite value 0/0. In the latter case, they are both of the form $x/0$, where x is nonzero.

See also Section 4.23 for an interesting approach to the problem of detecting lines and other patterns using just the bitmap pixels.

4.14: The two simple curves $x(t)$ and $y(t)$ defined below are identical. When drawn in the xt or yt plane, each is a horizontal line followed by a 45° line:
$$x(t) = y(t) = \begin{cases} 0.5, & 0 \le t \le 0.5, \\ t, & 0.5 \le t \le 1. \end{cases}$$

Answers to Exercises

The curve itself is now defined parametrically:

$$\mathbf{P}(t) = (x(t), y(t)) = \begin{cases} (0.5, 0.5), & 0 \leq t \leq 0.5, \\ (t, t), & 0.5 \leq t \leq 1. \end{cases}$$

In the range $0 \leq t \leq 0.5$ the curve stays at point $(0.5, 0.5)$. Then, when $0.5 \leq t \leq 1$, the curve moves smoothly from $(0.5, 0.5)$ to $(1, 1)$.

4.15: The following functions are degree-3 polynomials in t and are not straight lines:

$$x(t) = 2t^3 - 3t^2 + 2, \quad y(t) = -4t^3 + 6t^2 + 1, \quad z(t) = -2t^3 + 3t^2 + 3.$$

When combined to form a parametric space curve, the result is

$$\mathbf{P}(t) = \bigl(x(t), y(t), z(t)\bigr) = (-1, 2, 1)(-2t^3 + 3t^2) + (2, 1, 3).$$

A simple change of parameter $T = -2t^3 + 3t^2$ yields $\mathbf{P}(T) = (-1, 2, 1)T + (2, 1, 3)$, a straight line from point $(2, 1, 3)$ to point $(-1, 2, 1) + (2, 1, 3) = (1, 3, 4)$. Notice that $t = 0 \to T = 0$ and $t = 1 \to T = 1$. The expression $(-2t^3 + 3t^2)$ also happens to be function $F_2(t)$ of Equation (4.25).

4.16: The curve can be written $\mathbf{P}(t) = \mathbf{P} + (\mathbf{Q} - \mathbf{P})[2\alpha t + (1 - 2\alpha)t^2]$. We define $T = 2\alpha t + (1 - 2\alpha)t^2$ and substitute T for t as the parameter. Note that $t = 0 \Rightarrow T = 0$ and $t = 1 \Rightarrow T = 1$. The curve can now be written $\mathbf{P}(T) = \mathbf{P} + (\mathbf{Q} - \mathbf{P})T$ (where $0 \leq T \leq 1$), which is linear in T and is, therefore, a straight line. This is a (sometimes baffling) property of parametric curves. A substitution of the parameter does not change the shape of the curve and can be used to shed light on its behavior. Intuitively, the reason our curve is a straight line is that the same vector $(\mathbf{Q} - \mathbf{P})$ is used in the coefficients of both t and t^2.

4.17: Such a polynomial is fully defined by three coefficients $\mathbf{A}$, $\mathbf{B}$, and $\mathbf{C}$ that can be considered three-dimensional points and any three points are on the same plane.

4.18: We can gain an insight to the shape of the n-degree polynomial $P(x) = \sum_{i=0}^{n} A_i x^i$ by writing the equation $P(x) = 0$. This is an nth-degree equation in the unknown x and, therefore, has n solutions (some may be identical or complex). Each solution is an x value for which the polynomial becomes zero. As x is varied, the polynomial crosses the x axis n times, so it oscillates between positive and negative values.

4.19: This is straightforward:

$$\begin{aligned}
\mathbf{P}(2/3) &= (0, -9)(2/3)^3 + (-4.5, 13.5)(2/3)^2 + (4.5, -3.5)(2/3) \\
&= (0, -8/3) + (-2, 6) + (3, -7/3) \\
&= (1, 1) = \mathbf{P}_3.
\end{aligned}$$

4.20: We use the relations $\sin 30° = \cos 60° = 0.5$ and the approximation $\cos 30° = \sin 60° \approx 0.866$. The four points are $\mathbf{P}_1 = (1,0)$, $\mathbf{P}_2 = (\cos 30°, \sin 30°) = (0.866, 0.5)$, $\mathbf{P}_3 = (0.5, 0.866)$, and $\mathbf{P}_4 = (0,1)$. The relation $\mathbf{A} = \mathbf{NP}$ becomes

$$\begin{pmatrix} a \\ b \\ c \\ d \end{pmatrix} = \mathbf{A} = \mathbf{NP} = \begin{pmatrix} -4.5 & 13.5 & -13.5 & 4.5 \\ 9.0 & -22.5 & 18 & -4.5 \\ -5.5 & 9.0 & -4.5 & 1.0 \\ 1.0 & 0 & 0 & 0 \end{pmatrix} \begin{pmatrix} (1,0) \\ (0.866, 0.5) \\ (0.5, 0.866) \\ (0,1) \end{pmatrix}.$$

The solutions are

$\mathbf{a} = -4.5(1,0) + 13.5(0.866, 0.5) - 13.5(0.5, 0.866) + 4.5(0,1) = (0.441, -0.441),$
$\mathbf{b} = 19(1,0) - 22.5(0.866, 0.5) + 18(0.5, 0.866) - 4.5(0,1) = (-1.485, -0.162),$
$\mathbf{c} = -5.5(1,0) + 9(0.866, 0.5) - 4.5(0.5, 0.866) + 1(0,1) = (0.044, 1.603),$
$\mathbf{d} = 1(1,0) - 0(0.866, 0.5) + 0(0.5, 0.866) - 0(0,1) = (1,0).$

The PC is thus $\mathbf{P}(t) = (0.441, -0.441)t^3 + (-1.485, -0.162)t^2 + (0.044, 1.603)t + (1,0)$. The midpoint is $\mathbf{P}(0.5) = (0.7058, 0.7058)$, only 0.2% away from the midpoint of the arc, which is at $(\cos 45°, \sin 45°) \approx (0.7071, 0.7071)$.

(See also Exercise 4.8.)

4.21: From the definitions of the relative coordinates, we get $\mathbf{P}_2 = \Delta_1 + \mathbf{P}_1$, $\mathbf{P}_3 = \Delta_2 + \mathbf{P}_2 = \Delta_1 + \Delta_2 + \mathbf{P}_1$, and $\mathbf{P}_4 = \Delta_3 + \mathbf{P}_3 = \Delta_1 + \Delta_2 + \Delta_3 + \mathbf{P}_1$. When this is substituted in Equations (4.9) and (4.11), they become

$$\mathbf{P}(t) = \mathbf{G}(t)\mathbf{P} = \mathbf{T}(t)\mathbf{NP}$$

$$= (t^3, t^2, t, 1) \begin{pmatrix} -4.5 & 13.5 & -13.5 & 4.5 \\ 9.0 & -22.5 & 18 & -4.5 \\ -5.5 & 9.0 & -4.5 & 1.0 \\ 1.0 & 0 & 0 & 0 \end{pmatrix} \begin{pmatrix} \mathbf{P}_1 \\ \mathbf{P}_2 \\ \mathbf{P}_3 \\ \mathbf{P}_4 \end{pmatrix}$$

$$= (t^3, t^2, t, 1) \begin{pmatrix} -4.5 & 13.5 & -13.5 & 4.5 \\ 9.0 & -22.5 & 18 & -4.5 \\ -5.5 & 9.0 & -4.5 & 1.0 \\ 1.0 & 0 & 0 & 0 \end{pmatrix} \begin{pmatrix} \mathbf{P}_1 \\ \mathbf{P}_1 + \Delta_1 \\ \mathbf{P}_1 + \Delta_1 + \Delta_2 \\ \mathbf{P}_1 + \Delta_1 + \Delta_2 + \Delta_3 \end{pmatrix}.$$

Selecting, for example, $\Delta_1 = (2,0)$, $\Delta_2 = (0,2)$, and $\Delta_3 = (1,1)$ produces

$$\mathbf{P}(t) = (t^3, t^2, t, 1) \begin{pmatrix} -4.5 & 13.5 & -13.5 & 4.5 \\ 9.0 & -22.5 & 18 & -4.5 \\ -5.5 & 9.0 & -4.5 & 1.0 \\ 1.0 & 0 & 0 & 0 \end{pmatrix} \begin{pmatrix} \mathbf{P}_1 \\ \mathbf{P}_1 + (2,0) \\ \mathbf{P}_1 + (2,2) \\ \mathbf{P}_1 + (3,3) \end{pmatrix}$$

$$= \mathbf{P}_1 + (12t - 22.5t^2 + 13.5t^3, -6t + 22.5t^2 - 13.5t^3).$$

It is now clear that the three relative coordinates fully determine the shape of the curve but do not fix its position in space. The value of $\mathbf{P}_1$ is needed for that.

Answers to Exercises

4.22: The new equations are easy enough to set up. With the help of *Mathematica*, they are also easy to solve. The code

```
Solve[{d==p1,
    a al^3+b al^2+c al+d==p2,
    a be^3+b be^2+c be+d==p3,
    a+b+c+d==p4},{a,b,c,d}];
ExpandAll[Simplify[%]]
```

(where al and be stand for α and β, respectively) produces the (messy) solutions

$$a = -\frac{\mathbf{P}_1}{\alpha\beta} + \frac{\mathbf{P}_2}{-\alpha^2 + \alpha^3 + \alpha\beta - \alpha^2\beta}$$
$$+ \frac{\mathbf{P}_3}{\alpha\beta - \beta^2 - \alpha\beta^2 + \beta^3} + \frac{\mathbf{P}_4}{1 - \alpha - \beta + \alpha\beta},$$
$$b = \mathbf{P}_1\left(-\alpha + \alpha^3 + \beta - \alpha^3\beta - \beta^3 + \alpha\beta^3\right)/\gamma + \mathbf{P}_2\left(-\beta + \beta^3\right)/\gamma$$
$$+ \mathbf{P}_3\left(\alpha - \alpha^3\right)/\gamma + \mathbf{P}_4\left(\alpha^3\beta - \alpha\beta^3\right)/\gamma,$$
$$c = -\mathbf{P}_1\left(1 + \frac{1}{\alpha} + \frac{1}{\beta}\right) + \frac{\beta\mathbf{P}_2}{-\alpha^2 + \alpha^3 + \alpha\beta - \alpha^2\beta}$$
$$+ \frac{\alpha\mathbf{P}_3}{\alpha\beta - \beta^2 - \alpha\beta^2 + \beta^3} + \frac{\alpha\beta\mathbf{P}_4}{1 - \alpha - \beta + \alpha\beta},$$
$$d = \mathbf{P}_1,$$

where

$$\gamma = (-1 + \alpha)\alpha(-1 + \beta)\beta(-\alpha + \beta).$$

From here, the basis matrix immediately follows:

$$\begin{pmatrix} -\frac{1}{\alpha\beta} & \frac{1}{-\alpha^2+\alpha^3\alpha\beta-\alpha^2\beta} & \frac{1}{\alpha\beta-\beta^2-\alpha\beta^2+\beta^3} & \frac{1}{1-\alpha-\beta+\alpha\beta} \\ \frac{-\alpha+\alpha^3+\beta-\alpha^3\beta-\beta^3+\alpha\beta^3}{\gamma} & \frac{-\beta+\beta^3}{\gamma} & \frac{\alpha-\alpha^3}{\gamma} & \frac{\alpha^3\beta-\alpha\beta^3}{\gamma} \\ -\left(1+\frac{1}{\alpha}+\frac{1}{\beta}\right) & \frac{\beta}{-\alpha^2+\alpha^3+\alpha\beta-\alpha^2\beta} & \frac{\alpha}{\alpha\beta-\beta^2-\alpha\beta^2+\beta^3} & \frac{\alpha\beta}{1-\alpha-\beta+\alpha\beta} \\ 1 & 0 & 0 & 0 \end{pmatrix}.$$

A direct check, again using *Mathematica*, for $\alpha = 1/3$ and $\beta = 2/3$ produces the basis matrix of Equation (4.11).

4.23: This is the case $n = 1$. The general form of the LP is, therefore, $y = \sum_{i=0}^{1} y_i L_i^1$. The weight functions are easy to calculate:

$$L_0^1 = \frac{x - x_1}{x_0 - x_1}, \quad L_1^1 = \frac{x - x_0}{x_1 - x_0},$$

and the curve is, therefore,

$$y = y_0 L_0^1 + y_1 L_1^1$$
$$= y_0 \frac{x - x_1}{x_0 - x_1} + y_1 \frac{x - x_0}{x_1 - x_0}$$
$$= x \frac{y_0 - y_1}{x_0 - x_1} + \frac{y_1 x_0 - y_0 x_1}{x_0 - x_1}$$
$$= ax + b.$$

This is a straight line.

4.24: Since there are just two points, the only knots are $t_0 = 0$ and $t_1 = 1$. The weight functions are

$$L_0^1 = \frac{t - t_1}{t_0 - t_1} = 1 - t, \qquad L_1^1 = \frac{t - t_0}{t_1 - t_0} = t,$$

and the curve is, therefore,

$$\mathbf{P}(t) = \mathbf{P}_0 L_0^1 + \mathbf{P}_1 L_1^1 = (1 - t)\mathbf{P}_0 + t\mathbf{P}_1.$$

This is a straight line expressed parametrically.

4.25: Since the three points are approximately equally spaced, it makes sense to use knot values $t_0 = 0$, $t_1 = 1/2$, and $t_2 = 1$. The first step is to calculate the three basis functions $L_i^2(t)$:

$$L_0^2 = \frac{\Pi_{j \neq 0}}{\Pi_{j \neq 0}} = \frac{(t - t_1)(t - t_2)}{(t_0 - t_1)(t_0 - t_2)} = 2(t - 1/2)(t - 1),$$
$$L_1^2 = \frac{\Pi_{j \neq 1}}{\Pi_{j \neq 1}} = \frac{(t - t_0)(t - t_2)}{(t_1 - t_0)(t_1 - t_2)} = -4t(t - 1),$$
$$L_2^2 = \frac{\Pi_{j \neq 2}}{\Pi_{j \neq 2}} = \frac{(t - t_0)(t - t_1)}{(t_2 - t_0)(t_2 - t_1)} = 2t(t - 1/2).$$

The LP is now easy to calculate:

$$\mathbf{P}(t) = (0,0)2(t - 1/2)(t - 1) - (0,1)4t(t - 1) + (1,1)2t(t - 1/2)$$
$$= (2t^2 - t, -2t^2 + 3t). \tag{Ans.7}$$

This is a quadratic (degree-2) parametric polynomial and a simple test verifies that it passes through the three given points.

Answers to Exercises

4.26: We set the knots to $t_0 = 0$, $t_1 = 1/3$, $t_2 = 2/3$, and $t_3 = 1$. The first step is to calculate the four basis functions $L_i^3(t)$:

$$L_0^3 = \frac{\Pi_{j \neq 0}}{\Pi_{j \neq 0}} = \frac{(t-t_1)(t-t_2)(t-t_3)}{(t_0-t_1)(t_0-t_2)(t_0-t_3)} = -4.5t^3 + 9t^2 - 5.5t + 1,$$

$$L_1^3 = \frac{\Pi_{j \neq 1}}{\Pi_{j \neq 1}} = \frac{(t-t_0)(t-t_2)(t-t_3)}{(t_1-t_0)(t_1-t_2)(t_1-t_3)} = 13.5t^3 - 22.5t^2 + 9t,$$

$$L_2^3 = \frac{\Pi_{j \neq 2}}{\Pi_{j \neq 2}} = \frac{(t-t_0)(t-t_1)(t-t_3)}{(t_2-t_0)(t_2-t_1)(t_2-t_3)} = -13.5t^3 + 18t^2 - 4.5t,$$

$$L_3^3 = \frac{\Pi_{j \neq 2}}{\Pi_{j \neq 2}} = \frac{(t-t_0)(t-t_1)(t-t_2)}{(t_3-t_0)(t_3-t_1)(t_3-t_2)} = 4.5t^3 - 4.5t^2 + t.$$

The LP is now easy to calculate:

$$\mathbf{P}(t) = (-4.5t^3 + 9t^2 - 5.5t + 1)\mathbf{P}_1 + (13.5t^3 - 22.5t^2 + 9t)\mathbf{P}_2$$
$$+ (-13.5t^3 + 18t^2 - 4.5t)\mathbf{P}_3 + (4.5t^3 - 4.5t^2 + t)\mathbf{P}_4.$$

This is identical to Equation (4.9).

4.27: The first step is to calculate the basis functions

$$N_0(t) = 1, \quad N_1(t) = t - t_0 = t, \quad N_2(t) = (t-t_0)(t-t_1) = t(t-1/2).$$

The next step is to calculate the three coefficients

$$\mathbf{A}_0 = \mathbf{P}_0 = (0,0),$$

$$\mathbf{A}_1 = \frac{\mathbf{P}_1 - \mathbf{P}_0}{t_1 - t_0} = \frac{(0,1) - (0,0)}{1/2} = (0,2),$$

$$\mathbf{A}_2 = \frac{\frac{(1,1)-(0,1)}{1-1/2} - \frac{(0,1)-(0,0)}{1/2-0}}{1-0} = (2,-2).$$

The polynomial can now be calculated:

$$\mathbf{P}(t) = 1 \times (0,0) + t(0,2) + t(t-1/2)(2,-2) = (2t^2 - t, -2t^2 + 3t).$$

It is, of course, identical to the LP calculated in Exercise 4.25.

4.28: The curve is given by $\mathbf{P}(t) = (2t^2 - t, -2t^2 + 3t)$, so its derivative is $\mathbf{P}^t(t) = (4t-1, -4t+3)$. The three tangent vectors are $\mathbf{P}^t(t_0 = 0) = (-1, 3)$, $\mathbf{P}^t(t_1 = 1/2) = (1,1)$, and $\mathbf{P}^t(t_2 = 1) = (3,-1)$. The direction of tangent vector $(-1,3)$ is described by saying "for every three steps in the y direction, the curve moves one step in the negative x direction."

The slopes are calculated by dividing the y coordinate of a tangent vector by its x coordinate. The slopes at the three points are, therefore, $-3/1$, $1/1$, and $-1/3$. They correspond to angles of $288.44°$, $45°$, and $-18.43°$, respectively.

4.29: a. Because a quadratic polynomial is a solution to a quadratic equation and, therefore, cannot intercept the x axis more than twice. b. An inflection point (Section 4.22.8) is a point where the curvature (which is proportional to the second derivative) is zero. The general form of a parametric quadratic curve is $\mathbf{P}(t) = \mathbf{a}t^2 + \mathbf{b}t + \mathbf{c}$. Thus, its first derivative has the form $\mathbf{P}^t(t) = 2\mathbf{a}t + \mathbf{b}$ and its second derivative is $\mathbf{P}^{tt}(t) = 2\mathbf{a}$. Since $\mathbf{a}$ cannot be zero (why?), the curve has no inflection points.

4.30: We start with the PC of Equation (4.8), $\mathbf{P}(t) = (x(t), y(t)) = \mathbf{a}t^3 + \mathbf{b}t^2 + \mathbf{c}t + \mathbf{d}$, where each of the four coefficients $\mathbf{a}$, $\mathbf{b}$, $\mathbf{c}$, and $\mathbf{d}$ is a pair denoted by, for example, $\mathbf{a} = (a_x, a_y)$. The distance $s(t)$ between the PC and a point $\mathbf{P_0} = (x_0, y_0)$ is given by

$$s(t) = \sqrt{(x(t) - x_0)^2 + (y(t) - y_0)^2}.$$

Squaring both sides gives $s^2(t) = (x(t) - x_0)^2 + (y(t) - y_0)^2$. Differentiating with respect to t yields

$$2s(t)\frac{ds(t)}{dt} = 2(x(t) - x_0)\frac{dx(t)}{dt} + 2(y(t) - y_0)\frac{dy(t)}{dt}.$$

The minimum distance is thus found by solving the equation

$$2(x(t) - x_0)\frac{dx(t)}{dt} + 2(y(t) - y_0)\frac{dy(t)}{dt} = 0. \quad \text{(Ans.8)}$$

Equation (4.8) can be used to calculate

$$\frac{dx(t)}{dt} = 3a_x t^2 + 2b_x t + c_x, \quad \frac{dy(t)}{dt} = 3a_y t^2 + 2b_y t + c_y,$$

and this shows that Equation (Ans.8) is of degree 5. Such an equation can normally be solved numerically but not analytically.

4.31: When the user specifies four points, the curve should pass through the original points. By changing a point, the curve will no longer pass through the original point. When only the two endpoints are specified, the user is normally willing to consider different curves that pass through them, with different start and end directions.

4.32: Take one of these vectors, say, $(2, 1, .6)$ and divide it by its magnitude. The result is

$$\frac{(2, 1, 0.6)}{\sqrt{2^2 + 1^2 + 0.6^2}} \approx \frac{(2, 1, 0.6)}{2.93} = (0.7272, 0.3636, 0.2045).$$

The new vector points in the same direction but its magnitude is 1. Its components thus satisfy

$$\sqrt{0.7272^2 + 0.3636^2 + 0.2045^2} = 1, \quad \text{(Ans.9)}$$

so they are dependent. Any of them can be calculated from the other two using Equation (Ans.9).

4.33: Substituting $t = 0.5$ in Equation (4.23) yields

$$\mathbf{P}(0.5) = (2\mathbf{P}_1 - 2\mathbf{P}_2 + \mathbf{P}_1^t + \mathbf{P}_2^t)/8 + (-3\mathbf{P}_1 + 3\mathbf{P}_2 - 2\mathbf{P}_1^t - \mathbf{P}_2^t)/4 + \mathbf{P}_1^t/2 + \mathbf{P}_1$$
$$= \frac{1}{2}(\mathbf{P}_1 + \mathbf{P}_2) + \frac{1}{8}(\mathbf{P}_1^t - \mathbf{P}_2^t). \qquad \text{(Ans.10)}$$

The first part of this expression is the midpoint of the segment $\mathbf{P}_1 \to \mathbf{P}_2$ and the second part is the difference of the two tangents, divided by 8. Figure Ans.15 illustrates how adding $(\mathbf{P}_1^t - \mathbf{P}_2^t)/8$ to the midpoint of $\mathbf{P}_1 \to \mathbf{P}_2$ brings us to the midpoint of the curve.

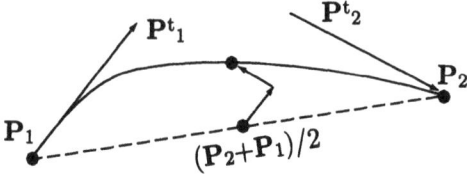

Figure Ans.15: The Midpoint $\mathbf{P}(0.5)$ of a Hermite Segment.

4.34: The Hermite segment is a cubic polynomial in t, so its third derivative is constant. It is easy to see, from Equation (4.25), that the third derivatives of the Hermite blending functions $F_i(t)$ are

$$F_1^{ttt}(t) = 12, \quad F_2^{ttt}(t) = -12, \quad F_3^{ttt}(t) = 6, \quad F_4^{ttt}(t) = 6.$$

The third derivative of the segment is, therefore,

$$\mathbf{P}^{ttt}(t) = (12\mathbf{P}_1 - 12\mathbf{P}_2 + 6\mathbf{P}_1^t + 6\mathbf{P}_2^t)$$
$$= (t^3, t^2, t, 1) \begin{pmatrix} 0 & 0 & 0 & 0 \\ 0 & 0 & 0 & 0 \\ 0 & 0 & 0 & 0 \\ 12 & -12 & 6 & 6 \end{pmatrix} \begin{pmatrix} \mathbf{P}_1 \\ \mathbf{P}_2 \\ \mathbf{P}_1^t \\ \mathbf{P}_2^t \end{pmatrix}$$
$$= \mathbf{T}(t)\mathbf{H}_{ttt}\mathbf{B}.$$

It is independent of t, so it is a vector. It is the fixed vector $12(\mathbf{P}_1 - \mathbf{P}_2) + 6(\mathbf{P}_1^t + \mathbf{P}_2^t)$.

Here are the Hermite matrix and its derivatives side by side. Use your experience to explain how each is derived from its predecessor.

$$\mathbf{H} = \begin{pmatrix} 2 & -2 & 1 & 1 \\ -3 & 3 & -2 & -1 \\ 0 & 0 & 1 & 0 \\ 1 & 0 & 0 & 0 \end{pmatrix}, \quad \mathbf{H}_t = \begin{pmatrix} 0 & 0 & 0 & 0 \\ 6 & -6 & 3 & 3 \\ -6 & 6 & -4 & -2 \\ 0 & 0 & 1 & 0 \end{pmatrix},$$

$$\mathbf{H}_{tt} = \begin{pmatrix} 0 & 0 & 0 & 0 \\ 0 & 0 & 0 & 0 \\ 12 & -12 & 6 & 6 \\ -6 & 6 & -4 & -2 \end{pmatrix}, \quad \mathbf{H}_{ttt} = \begin{pmatrix} 0 & 0 & 0 & 0 \\ 0 & 0 & 0 & 0 \\ 0 & 0 & 0 & 0 \\ 12 & -12 & 6 & 6 \end{pmatrix}.$$

4.35: The expression for $\mathbf{P}(t)$ is used to show that $\mathbf{P}(0) = (-1,0)0^3 + (1,-1)0^2 + (1,1)0 = (0,0)$ and $\mathbf{P}(1) = (-1,0)1^3 + (1,-1)1^2 + (1,1)1 = (1,0)$. The tangent vector of $\mathbf{P}(t)$ is

$$\frac{d\mathbf{P}(t)}{dt} = 3(-1,0)t^2 + 2(1,-1)t + (1,1),$$

so the two extreme tangent vectors are

$$\frac{d\mathbf{P}(0)}{dt} = 3(-1,0)0^2 + 2(1,-1)0 + (1,1) = (1,1),$$

$$\frac{d\mathbf{P}(1)}{dt} = 3(-1,0)1^2 + 2(1,-1) + (1,1) = (0,-1),$$

as should be.

4.36: Similar to the original example, we get

$$\mathbf{P}(t) = (t^3, t^2, t, 1)\mathbf{H}\,((0,0),(1,0),(2,2),(0,-1))^T$$
$$= (0,1)t^3 - (1,3)t^2 + (2,2)t.$$

It's a different polynomial and it has a different shape; yet a simple check shows that it passes through the same endpoints and has the same start and end directions.

4.37: Equation (4.26) becomes

$$\mathbf{P}(t) = (t^3, t^2, t, 1)\begin{pmatrix} 2 & -2 & 1 & 1 \\ -3 & 3 & -2 & -1 \\ 0 & 0 & 1 & 0 \\ 1 & 0 & 0 & 0 \end{pmatrix}\begin{pmatrix} \mathbf{P}_1 \\ \mathbf{P}_2 \\ (0,0) \\ (0,0) \end{pmatrix}$$
$$= (3t^2 - 2t^3)(\mathbf{P}_2 - \mathbf{P}_1) + \mathbf{P}_1. \qquad \text{(Ans.11)}$$

To find the type of the curve, we substitute $j = 3t^2 - 2t^3$ (note that $t = 0 \Rightarrow j = 0$ and $t = 1 \Rightarrow j = 1$). This results in the familiar expression $\mathbf{P}(t) = j(\mathbf{P}_2 - \mathbf{P}_1) + \mathbf{P}_1 = (1 - j)\mathbf{P}_1 + j\mathbf{P}_2$. The curve is, therefore, the straight line from $\mathbf{P}_1$ to $\mathbf{P}_2$. The (important) conclusion is as follows: If the initial and final directions of the Hermite segment are not specified, the curve will "choose" the shortest path from $\mathbf{P}_1$ to $\mathbf{P}_2$.

4.38: For case 1, we use the notation $\mathbf{P}^t(0) = \mathbf{P}_1^t$, $\mathbf{P}^t(1/2) = \mathbf{P}_2^t$, and $\mathbf{P}^t(1) = \mathbf{P}_3^t$. From $\mathbf{P}(t) = \mathbf{a}t^3 + \mathbf{b}t^2 + \mathbf{c}t + \mathbf{d}$, we get $\mathbf{P}^t(t) = 3\mathbf{a}t^2 + 2\mathbf{b}t + \mathbf{c}$, resulting in the three equations

$$3\mathbf{a}\cdot 0^2 + 2\mathbf{b}\cdot 0 + \mathbf{c} = \mathbf{P}_1^t,$$
$$3\mathbf{a}\cdot(1/2)^2 + 2\mathbf{b}\cdot(1/2) + \mathbf{c} = \mathbf{P}_2^t,$$
$$3\mathbf{a}\cdot 1^2 + 2\mathbf{b}\cdot 1 + \mathbf{c} = \mathbf{P}_3^t,$$

where the unknowns are $\mathbf{a}$, $\mathbf{b}$, $\mathbf{c}$, and $\mathbf{d}$ (notice that $\mathbf{d}$ does not participate in our equations). It is clear that $\mathbf{c} = \mathbf{P}_1^t$. The other two unknowns are solved by the simple *Mathematica* code `Solve[{3a/4+2b/2+p1==p2, 3a+2b+p1==p3}, {a,b}]`,

Answers to Exercises

which yields $\mathbf{a} = \frac{2}{3}(\mathbf{P}_1^t - 2\mathbf{P}_2^t + \mathbf{P}_3^t)$ and $\mathbf{b} = \frac{1}{2}(-3\mathbf{P}_1^t + 4\mathbf{P}_2^t - \mathbf{P}_3^t)$. The curve is thus given by

$$\mathbf{P}(t) = \mathbf{a}t^3 + \mathbf{b}t^2 + \mathbf{c}t + \mathbf{d}$$
$$= \frac{2}{3}(\mathbf{P}_1^t - 2\mathbf{P}_2^t + \mathbf{P}_3^t)t^3 + \frac{1}{2}(-3\mathbf{P}_1^t + 4\mathbf{P}_2^t - \mathbf{P}_3^t)t^2 + \mathbf{P}_1^t t + \mathbf{d},$$

which shows that the three given tangents fully determine the shape of the curve but not its position in space. The latter requires the value of $\mathbf{d}$.

For case 2, we denote $\mathbf{P}(1/3) = \mathbf{P}_1$, $\mathbf{P}(2/3) = \mathbf{P}_2$, $\mathbf{P}^t(0) = \mathbf{P}_1^t$, and $\mathbf{P}^t(1) = \mathbf{P}_2^t$. This results in the four equations

$$\mathbf{a}(1/3)^3 + \mathbf{b}(1/3)^2 + \mathbf{c}(1/3) + \mathbf{d} = \mathbf{P}_1,$$
$$\mathbf{a}(2/3)^3 + \mathbf{b}(2/3)^2 + \mathbf{c}(2/3) + \mathbf{d} = \mathbf{P}_2,$$
$$3\mathbf{a} \cdot 0^2 + 2\mathbf{b} \cdot 0 + \mathbf{c} = \mathbf{P}_1^t,$$
$$3\mathbf{a} \cdot 1^2 + 2\mathbf{b} \cdot 1 + \mathbf{c} = \mathbf{P}_2^t,$$

where the unknowns are again $\mathbf{a}$, $\mathbf{b}$, $\mathbf{c}$, and $\mathbf{d}$. It is again clear that $\mathbf{c} = \mathbf{P}_1^t$ and the other three unknowns are easily solved by the code

```
Solve[{a (1/3)^3+b (1/3)^2+p1t (1/3)+d==p1,
a (2/3)^3+b (2/3)^2+p1t (1/3)+d==p2, 3a+2b+p1t==p2t}, {a,b,d}],
```

which yields the solutions

$$\mathbf{a} = -\frac{9}{13}(-6\mathbf{P}_1 + \mathbf{P}_1^t + 6\mathbf{P}_2 - \mathbf{P}_2^t),$$
$$\mathbf{b} = \frac{1}{13}(-81\mathbf{P}_1 + 7\mathbf{P}_1^t + 81\mathbf{P}_2 - 7\mathbf{P}_2^t),$$
$$\mathbf{d} = \frac{1}{117}(180\mathbf{P}_1 - 43\mathbf{P}_1^t - 63\mathbf{P}_2 + 4\mathbf{P}_2^t).$$

The PC segment is thus

$$\mathbf{P}(t) = \mathbf{a}t^3 + \mathbf{b}t^2 + \mathbf{c}t + \mathbf{d}$$
$$= -\frac{9}{13}(-6\mathbf{P}_1 + \mathbf{P}_1^t + 6\mathbf{P}_2 - \mathbf{P}_2^t)t^3 + \frac{1}{13}(-81\mathbf{P}_1 + 7\mathbf{P}_1^t + 81\mathbf{P}_2 - 7\mathbf{P}_2^t)t^2$$
$$+ \mathbf{P}_1^t t + \frac{1}{117}(180\mathbf{P}_1 - 43\mathbf{P}_1^t - 63\mathbf{P}_2 + 4\mathbf{P}_2^t).$$

Case 3 is similar to case 2 and is not shown here.

4.39: We are looking for a parametric curve $\mathbf{P}(t)$ that's a quadratic polynomial satisfying

$$\mathbf{P}(t) = \mathbf{a}t^2 + \mathbf{b}t + \mathbf{c}, \qquad \text{(Ans.12)}$$
$$\mathbf{P}^t(t) = 2\mathbf{a}t + \mathbf{b}, \qquad \text{(Ans.13)}$$

$$\mathbf{P}(0) = \mathbf{c} = \mathbf{P}_0, \qquad \text{(Ans.14)}$$
$$\mathbf{P}(1) = \mathbf{a} + \mathbf{b} + \mathbf{c} = \mathbf{P}_2, \qquad \text{(Ans.15)}$$
$$\mathbf{P}^t(0) = \mathbf{b} = 4\alpha(\mathbf{P}_1 - \mathbf{P}_0), \qquad \text{(Ans.16)}$$
$$\mathbf{P}^t(1) = 2\mathbf{a} + \mathbf{b} = 4\alpha(\mathbf{P}_2 - \mathbf{P}_1). \qquad \text{(Ans.17)}$$

Subtracting Equation (Ans.15) from Equation (Ans.17) yields

$$\mathbf{a} = 4\alpha(\mathbf{P}_2 - \mathbf{P}_1) + \mathbf{P}_2 - \mathbf{P}_0. \qquad \text{(Ans.18)}$$

Substituting Eqs. (Ans.14), (Ans.16), and (Ans.18) in Equation (Ans.15) yields

$$\mathbf{a} + \mathbf{b} + \mathbf{c} = 4\alpha(\mathbf{P}_2 - \mathbf{P}_1) + \mathbf{P}_2 - \mathbf{P}_0 + 4\alpha(\mathbf{P}_1 - \mathbf{P}_0) + \mathbf{P}_0 = \mathbf{P}_2,$$

or $4\alpha(\mathbf{P}_2 - \mathbf{P}_0) = 2(\mathbf{P}_2 - \mathbf{P}_0)$, implying $\alpha = 0.5$. Once α is known, the curve is obtained from Equation (Ans.12) as

$$\mathbf{P}(t) = (1-t)^2 \mathbf{P}_0 + 2t(1-t)\mathbf{P}_1 + t^2 \mathbf{P}_2.$$

4.40: For $\theta = 90°$, we have $\sin\theta = 1$, $\cos\theta = 0$ and $a = 4$. Equation (4.32) becomes
$$\mathbf{P}(t) = (2t^3 - 3t^2 + 1)(0,-1) + (-2t^3 + 3t^2)(0,1)$$
$$+ (t^3 - 2t^2 + t)4(1,0) + (t^3 - t^2)4(-1,0)$$
$$= (-4t^2 + 4t, -4t^3 + 6t^2 - 1).$$

It is easy to see that $\mathbf{P}(0) = (0,-1)$, $\mathbf{P}(1) = (0,1)$, and $\mathbf{P}(0.5) = (1,0)$. At $t = 0.25$, the curve passes through point $\mathbf{P}(0.25) = (12/16, -11/16)$, whose distance from the origin is

$$\sqrt{\left(\frac{12}{16}\right)^2 + \left(\frac{11}{16}\right)^2} \approx 1.0174.$$

The deviation from a true circle at this point is therefore about 1.74%, an excellent approximation for such a large arc.

4.41: Yes. Equation (4.34) was derived for any real values of a and b, not just positive. However, when a and b become negative, the tangent vectors reverse directions, and the curve completely changes its shape. Figure 4.14 shows the curve for $\alpha = -0.4$. Another example of negative a and b is $\alpha = -1/4$, which yields $a = b = -1$, changing Equation (4.34) to $\mathbf{Q}(t) = -(6,3)t^3 + (9,5)t^2 - (1,1)t$. It is easy to verify that $\mathbf{Q}(0) = (0,0)$, $\mathbf{Q}(1) = (2,1)$, and $\mathbf{Q}(0.5) = (1, 3/8)$.

4.42: The midpoint of our curve is always $(1, 5/8 + \alpha)$. The condition $\mathbf{Q}(0.5) = (1,0)$ implies $5/8 + \alpha = 0$ or $\alpha = -5/8$. Since $a = b = 1 + 8\alpha$, we get $a = b = -4$, resulting in $\mathbf{Q}(t) = -(12,6)t^3 + (18,11)t^2 - (4,4)t$.

Answers to Exercises

4.43: Equation (4.27) yields the first derivative of the Hermite segment

$$\mathbf{P}^t(t) = (t^3, t^2, t, 1) \begin{pmatrix} 0 & 0 & 0 & 0 \\ 6 & -6 & 3 & 3 \\ -6 & 6 & -4 & -2 \\ 0 & 0 & 1 & 0 \end{pmatrix} \begin{pmatrix} (0,0) \\ (1,0) \\ \alpha(\cos\theta, \sin\theta) \\ \alpha(\cos\theta, -\sin\theta) \end{pmatrix}$$

$$= 3[(-2,0) + \alpha(2\cos\theta, 0)]t^2 + 2[(3,0) - \alpha(3\cos\theta, \sin\theta)]t + (\cos\theta, \sin\theta).$$

Because of the symmetry of the endpoints and vectors, a cusp can only occur in the middle of this curve. A cusp is the case where the tangent vector of the curve becomes indefinite, so we are looking for the value of α that's a solution of $\mathbf{P}^t(0.5) = (0,0)$.

$$\mathbf{P}^t(0.5) = \frac{3}{4}(-2,0) + \frac{3\alpha}{4}(2\cos\theta, 0) + (3,0) - \alpha(3\cos\theta, \sin\theta) + \alpha(\cos\theta, \sin\theta)$$

$$= (3/2, 0) + (-\cos\theta/2, 0).$$

It is easy to figure out that $\mathbf{P}^t(0.5) = (0,0)$ yields $\alpha = 3/\cos\theta$.

4.44: The two endpoints of $\mathbf{Q}(T)$ are $\mathbf{P}(0.25) = (0.3, 0.19)$, $\mathbf{P}(0.75) = (0.89, 0.19)$. The two extreme tangents are $0.5\mathbf{P}^t(0.25) = (0.66, 0.25)$, $.5\mathbf{P}^t(0.75) = (0.41, -0.25)$ [notice the 0.5 factor that equals $(t_j - t_i)$]. The new PC and its derivative are thus

$$\mathbf{Q}(T) = (T^3, T^2, T, 1) \begin{pmatrix} 2 & -2 & 1 & 1 \\ -3 & 3 & -2 & -1 \\ 0 & 0 & 1 & 0 \\ 1 & 0 & 0 & 0 \end{pmatrix} \begin{pmatrix} (0.3, 0.19) \\ (0.89, 0.19) \\ (0.66, 0.25) \\ (0.41, -0.25) \end{pmatrix}$$

$$= (-0.125, 0)t^3 + (0.0625, -0.25)t^2 + (0.65625, 0.25)t + (0.296875, 0.1875).$$

$$\mathbf{Q}^T(T) = (-0.375, 0)t^2 + (0.125, -0.5)t + (0.65625, 0.25).$$

Direct checks verify that $\mathbf{Q}(0) = \mathbf{P}(0.25)$, $\mathbf{Q}(1) = \mathbf{P}(0.75)$, $\mathbf{Q}^T(0) = \mathbf{P}^t(0.25)$, and $\mathbf{Q}^T(1) = \mathbf{P}^t(0.75)$.

4.45: The tangent vector of Equation (4.42) is $\mathbf{P}^t(t) = (-6t^2 + 6t)(\mathbf{P}_2 - \mathbf{P}_1)$. Its absolute value (that's the speed of the curve) is thus proportional to the function $-6t^2 + 6t$. When t varies from 0 to 1, this function goes up from 0 to a maximum of 1.5 at $t = 0.5$, then down to 0.

4.46: In this case, Equation (4.46) becomes

$$\begin{aligned} \mathbf{a} \cdot 0^2 + \mathbf{b} \cdot 0 + \mathbf{c} &= \mathbf{P}_1, \\ \mathbf{a}\Delta^2 + \mathbf{b}\Delta + \mathbf{c} &= \mathbf{P}_2, \\ 2\mathbf{a} \cdot 0 + \mathbf{b} &= \mathbf{P}_1^t. \end{aligned} \qquad \text{(Ans.19)}$$

The solutions are

$$c = \mathbf{P}_1, \quad b = \mathbf{P}_1^t, \quad \text{and} \quad a = \frac{\mathbf{P}_2}{\Delta^2} - \frac{\mathbf{P}_1}{\Delta^2} - \frac{\mathbf{P}_1^t}{\Delta}$$

and the polynomial is thus

$$\mathbf{P}(t) = (t^2, t, 1) \begin{pmatrix} -\frac{1}{\Delta^2} & \frac{1}{\Delta^2} & \frac{-1}{\Delta} \\ 0 & 0 & 1 \\ 1 & 0 & 0 \end{pmatrix} \begin{pmatrix} \mathbf{P}_1 \\ \mathbf{P}_2 \\ \mathbf{P}_1^t \end{pmatrix}. \qquad \text{(Ans.20)}$$

It is easy to see that Equation (Ans.20) reduces to Equation (4.47) for $\Delta = 1$.

4.47: We are looking for a curve of the form $\mathbf{P}(t) = at^2 + bt + c$. Its tangent vector is the derivative $\mathbf{P}^t(t) = 2at + b$. We denote the two known quantities by $\mathbf{P}^t(0) = \mathbf{P}_1^t$ and $\mathbf{P}^t(1) = \mathbf{P}_2^t$. The two equations $2a \cdot 0 + b = \mathbf{P}_1^t$ and $2a \cdot 1 + b = \mathbf{P}_2^t$ are easily solved to yield $b = \mathbf{P}_1^t$ and $a = (\mathbf{P}_2^t - \mathbf{P}_1^t)/2$. The curve is thus expressed by $\mathbf{P}(t) = \frac{1}{2}(\mathbf{P}_2^t - \mathbf{P}_1^t)t^2 + \mathbf{P}_1^t t + c$ and its derivative is $\mathbf{P}^t(t) = (\mathbf{P}_2^t - \mathbf{P}_1^t)t + \mathbf{P}_1^t$ (the straight line from $\mathbf{P}_1^t$ to $\mathbf{P}_2^t$). Notice that the two extreme tangents fully define the shape of this curve but do not fix its position in space. To place such a curve in space, we have to know the value of c. The two endpoints of this curve are $\mathbf{P}(0) = c$ and $\mathbf{P}(1) = c + \frac{1}{2}(\mathbf{P}_1^t + \mathbf{P}_2^t)$. The reader is encouraged to draw a diagram that shows the geometric meaning of adding the vector sum $\frac{1}{2}(\mathbf{P}_1^t + \mathbf{P}_2^t)$ to point c.

4.48: The curve and its first two derivatives can be expressed as usual:

$$\mathbf{P}(t) = at^2 + bt^2 + ct + d,$$
$$\mathbf{P}^t(t) = 3at^2 + 2bt + c,$$
$$\mathbf{P}^{tt}(t) = 6at + 2b.$$

For the standard case where $0 \le t \le 1$, the three conditions are expressed as $\mathbf{P}(0) = \mathbf{P}_1$, $\mathbf{P}(1) = \mathbf{P}_2$, $\mathbf{P}^t(1) = \mathbf{P}_2^t$, and $\mathbf{P}^{tt}(0) = 0$. The explicit equations are

$$\begin{aligned} a \cdot 0^3 + b \cdot 0^2 + c \cdot 0 + d &= \mathbf{P}_1, \\ a \cdot 1^3 + b \cdot 1^2 + c \cdot 1 + d &= \mathbf{P}_2, \\ 3a \cdot 1^2 + 2b \cdot 1 + c &= \mathbf{P}_2^t, \\ 6a \cdot 0 + 2b &= 0. \end{aligned} \qquad \text{(Ans.21)}$$

They are easy to solve and yield $a = \frac{1}{2}\mathbf{P}_2^t - \frac{1}{2}(\mathbf{P}_2 - \mathbf{P}_1)$, $b = 0$, $c = \frac{3}{2}(\mathbf{P}_2 - \mathbf{P}_1) - \frac{1}{2}\mathbf{P}_2^t$, and $d = \mathbf{P}_1$. The polynomial is thus

$$\begin{aligned} \mathbf{P}_{std}(t) &= \left(\frac{1}{2}\mathbf{P}_2^t - \frac{1}{2}(\mathbf{P}_2 - \mathbf{P}_1)\right)t^3 + \left(\frac{3}{2}(\mathbf{P}_2 - \mathbf{P}_1) - \frac{1}{2}\mathbf{P}_2^t\right)t + \mathbf{P}_1 \\ &= \left(\frac{1}{2}t^3 - \frac{3}{2}t + 1\right)\mathbf{P}_1 + \left(-\frac{1}{2}t^3 + \frac{3}{2}t\right)\mathbf{P}_2 + \left(\frac{1}{2}t^3 - \frac{1}{2}t\right)\mathbf{P}_2^t \end{aligned}$$

Answers to Exercises

$$= (t^3, t^2, t, 1) \begin{pmatrix} 1/2 & -1/2 & 1/2 \\ 0 & 0 & 0 \\ -3/2 & 3/2 & -1/2 \\ 1 & 0 & 0 \end{pmatrix} \begin{pmatrix} \mathbf{P}_1 \\ \mathbf{P}_2 \\ \mathbf{P}_2^t \end{pmatrix}. \qquad \text{(Ans.22)}$$

For the nonstandard case where $0 \le t \le \Delta$, Equation (Ans.22) is extended to

$$\mathbf{P}_{nstd}(t) = (t^3, t^2, t, 1) \begin{pmatrix} \frac{1}{2\Delta^3} & -\frac{1}{2\Delta^3} & \frac{1}{2\Delta^2} \\ 0 & 0 & 0 \\ -\frac{3}{2\Delta} & \frac{3}{2\Delta} & -\frac{1}{2} \\ 1 & 0 & 0 \end{pmatrix} \begin{pmatrix} \mathbf{P}_1 \\ \mathbf{P}_2 \\ \mathbf{P}_2^t \end{pmatrix}. \qquad \text{(Ans.23)}$$

4.49: By using the *same* symbol, $\mathbf{P}_{k+1}^t$, for the end tangent of $\mathbf{P}_k(t)$ and the start tangent of $\mathbf{P}_{k+1}(t)$.

4.50: The three segments are

$$\mathbf{P}_1(t) = (-\tfrac{1}{3}, -\tfrac{1}{5})t^3 + (\tfrac{1}{3}, \tfrac{6}{5})t^2 + (1, -1)t,$$
$$\mathbf{P}_2(t) = (0, -\tfrac{2}{5})t^3 + (-\tfrac{2}{3}, \tfrac{3}{5})t^2 + (\tfrac{2}{3}, \tfrac{4}{5})t + (1, 0),$$
$$\mathbf{P}_3(t) = (\tfrac{1}{3}, -\tfrac{1}{5})t^3 - (\tfrac{2}{3}, \tfrac{3}{5})t^2 + (-\tfrac{2}{3}, \tfrac{4}{5})t + (1, 1).$$

The first intermediate point should be $\mathbf{P}_1(1)$ and also $\mathbf{P}_2(0)$. A simple calculation yields

$$\mathbf{P}_1(1) = (-\tfrac{1}{3}, -\tfrac{1}{5})1^3 + (\tfrac{1}{3}, \tfrac{6}{5})1^2 + (1, -1) = (1, 0),$$
$$\mathbf{P}_2(0) = (0, -\tfrac{2}{5})0^3 + (-\tfrac{2}{3}, \tfrac{3}{5})0^2 + (\tfrac{2}{3}, \tfrac{4}{5})0 + (1, 0) = (1, 0).$$

The second intermediate point should be $\mathbf{P}_2(1)$ and also $\mathbf{P}_3(0)$. A similar calculation gives

$$\mathbf{P}_2(1) = (0, -\tfrac{2}{5})1^3 + (-\tfrac{2}{3}, \tfrac{3}{5})1^2 + (\tfrac{2}{3}, \tfrac{4}{5})1 + (1, 0) = (1, 1),$$
$$\mathbf{P}_3(0) = (\tfrac{1}{3}, -\tfrac{1}{5})0^3 - (\tfrac{2}{3}, \tfrac{3}{5})0^2 + (-\tfrac{2}{3}, \tfrac{4}{5})0 + (1, 1) = (1, 1).$$

Both tangent vectors can be obtained from the second segment. Its derivative is

$$\mathbf{P}_2^t(t) = \frac{d\mathbf{P}_2(t)}{dt} = 3(0, -\tfrac{2}{5})t^2 + 2(-\tfrac{2}{3}, \tfrac{3}{5})t + (\tfrac{2}{3}, \tfrac{4}{5}).$$

So the two vectors are

$$\mathbf{P}_2^t(0) = 3(0, -\tfrac{2}{5})0^2 + 2(-\tfrac{2}{3}, \tfrac{3}{5})0 + (\tfrac{2}{3}, \tfrac{4}{5}) = (\tfrac{2}{3}, \tfrac{4}{5}),$$
$$\mathbf{P}_2^t(1) = 3(0, -\tfrac{2}{5})1^2 + 2(-\tfrac{2}{3}, \tfrac{3}{5})1 + (\tfrac{2}{3}, \tfrac{4}{5})(-\tfrac{2}{3}, \tfrac{10}{5}).$$

The first tangent thus points in the direction $(5, 6)$ and the second one, in the direction $(-1, 3)$.

4.51: Equation (4.57) becomes

$$\begin{pmatrix} 1 & 4 & 1 & 0 \\ 0 & 1 & 4 & 1 \end{pmatrix} \begin{pmatrix} (0,0) \\ \mathbf{P}_2^t \\ \mathbf{P}_3^t \\ (-1,-1) \end{pmatrix} = \begin{pmatrix} 3[(1,1)-(0,0)] \\ 3[(0,1)-(1,0)] \end{pmatrix} = \begin{pmatrix} (3,3) \\ (-3,3) \end{pmatrix},$$

$$(0,0) + 4\mathbf{P}_2^t + \mathbf{P}_3^t = (3,3),$$

or

$$\mathbf{P}_2^t + 4\mathbf{P}_3^t + (-1,-1) = (-3,3).$$

The solutions are $\mathbf{P}_2^t = (\frac{8}{15}, \frac{8}{15})$ and $\mathbf{P}_3^t = (\frac{37}{15}, \frac{37}{15})$. The first segment, from Equation (4.26), is thus

$$\mathbf{P}_1(t) = (t^3, t^2, t, 1) \begin{pmatrix} 2 & -2 & 1 & 1 \\ -3 & 3 & -2 & -1 \\ 0 & 0 & 1 & 0 \\ 1 & 0 & 0 & 0 \end{pmatrix} \begin{pmatrix} (0,0) \\ (1,0) \\ (1,-1) \\ (\frac{8}{15}, \frac{8}{15}) \end{pmatrix}$$

$$= (-\tfrac{22}{15}, \tfrac{8}{15})t^3 + (\tfrac{37}{15}, -\tfrac{8}{15})t^2.$$

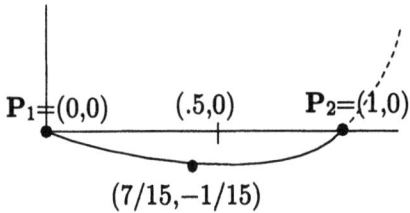

(7/15,−1/15)

Figure Ans.16

An initial direction of $(0,0)$ means that the curve will be the shortest possible (Figure Ans.16). It also means that the curve will start slowly and will speed up as it goes along. It is easy to see that

$$\mathbf{P}_1(0.5) = (-\tfrac{22}{15}, \tfrac{8}{15})\tfrac{1}{8} + (\tfrac{37}{15}, -\tfrac{8}{15})\tfrac{1}{4} = (\tfrac{7}{15}, -\tfrac{1}{15}).$$

At $t = 0.5$, the curve hasn't reached the midpoint between $\mathbf{P}_1$ and $\mathbf{P}_2$.

4.52: For the third segment, Equation (4.26) becomes

$$\mathbf{P}_3(t) = (t^3, t^2, t, 1) \begin{pmatrix} 2 & -2 & 1 & 1 \\ -3 & 3 & -2 & -1 \\ 0 & 0 & 1 & 0 \\ 1 & 0 & 0 & 0 \end{pmatrix} \begin{pmatrix} (1,1) \\ (0,1) \\ (-\frac{3}{5}, \frac{2}{3}) \\ (-\frac{6}{5}, -\frac{1}{3}) \end{pmatrix}$$

$$= (\tfrac{1}{5}, \tfrac{1}{3})t^3 - (\tfrac{3}{5}, 1)t^2 + (-\tfrac{3}{5}, \tfrac{2}{3})t + (1,1).$$

Answers to Exercises

4.53: For the third segment, Equation (4.26) becomes

$$\mathbf{P}_3(t) = (t^3, t^2, t, 1) \begin{pmatrix} 2 & -2 & 1 & 1 \\ -3 & 3 & -2 & -1 \\ 0 & 0 & 1 & 0 \\ 1 & 0 & 0 & 0 \end{pmatrix} \begin{pmatrix} (0,1) \\ (-1,0) \\ (-\frac{3}{2},0) \\ (0,-\frac{3}{2}) \end{pmatrix}$$

$$= (\tfrac{1}{2}, \tfrac{1}{2})t^3 + (0, -\tfrac{3}{2})t^2 + (-\tfrac{3}{2}, 0)t + (0, 1).$$

For the fourth segment, Equation (4.26) becomes

$$\mathbf{P}_4(t) = (t^3, t^2, t, 1) \begin{pmatrix} 2 & -2 & 1 & 1 \\ -3 & 3 & -2 & -1 \\ 0 & 0 & 1 & 0 \\ 1 & 0 & 0 & 0 \end{pmatrix} \begin{pmatrix} (-1,0) \\ (0,-1) \\ (0,-\frac{3}{2}) \\ (\frac{3}{2},0) \end{pmatrix}$$

$$= (-\tfrac{1}{2}, \tfrac{1}{2})t^3 + (\tfrac{3}{2}, 0)t^2 + (0, -\tfrac{3}{2})t + (-1, 0).$$

4.54: Equation (4.65) gives

$$\mathbf{P}_1^t = -\mathbf{P}_3^t = \frac{3}{4}(\mathbf{P}_2 - \mathbf{P}_1 - \mathbf{P}_3 + \mathbf{P}_2) - \frac{1}{4}(\mathbf{P}_2^t - \mathbf{P}_2^t)$$

$$= \frac{3}{4}(2\mathbf{P}_2 - \mathbf{P}_1 - \mathbf{P}_3)$$

$$= (0, 3/2).$$

We next substitute the anticyclic end condition in Equation (4.64), which becomes

$$(1, 4, 1) \begin{pmatrix} (0, 3/2) \\ \mathbf{P}_2^t \\ (0, -3/2) \end{pmatrix} = 3(\mathbf{P}_3 - \mathbf{P}_1) = (6, 0). \qquad \text{(Ans.24)}$$

The solution is $\mathbf{P}_2^t = (3/2, 0)$.

The first spline segment can now be calculated from Equation (4.26):

$$\mathbf{P}_1(t) = (t^3, t^2, t, 1) \begin{pmatrix} 2 & -2 & 1 & 1 \\ -3 & 3 & -2 & -1 \\ 0 & 0 & 1 & 0 \\ 1 & 0 & 0 & 0 \end{pmatrix} \begin{pmatrix} (-1,0) \\ (0,1) \\ (0,3/2) \\ (3/2,0) \end{pmatrix}$$

$$= (-\tfrac{1}{2}, -\tfrac{1}{2})t^3 + (\tfrac{3}{2}, 0)t^2 + (0, \tfrac{3}{2})t + (-1, 0).$$

Its derivative is

$$\mathbf{P}_1^t(t) = (-3/2, -3/2)t^2 + (3, 0)t + (0, 3/2),$$

so $\mathbf{P}_1^t(0) = (0, 3/2)$ and $\mathbf{P}_1^t(1) = (3/2, 0)$.

The second spline segment is similarly calculated:

$$\mathbf{P}_2(t) = (t^3, t^2, t, 1) \begin{pmatrix} 2 & -2 & 1 & 1 \\ -3 & 3 & -2 & -1 \\ 0 & 0 & 1 & 0 \\ 1 & 0 & 0 & 0 \end{pmatrix} \begin{pmatrix} (0,1) \\ (1,0) \\ (3/2,0) \\ (0,-3/2) \end{pmatrix}$$

$$= (-\tfrac{1}{2}, \tfrac{1}{2})t^3 + (0, -\tfrac{3}{2})t^2 + (\tfrac{3}{2}, 0)t + (0, 1).$$

Its derivative is

$$\mathbf{P}_2^t(t) = (-3/2, 3/2)t^2 + (0, -3)t + (3/2, 0),$$

so $\mathbf{P}_2^t(0) = (3/2, 0)$ and $\mathbf{P}_2^t(1) = (0, -3/2)$.

To compare this anticyclic cubic spline to the clamped cubic spline for the same points, we have to select the same start and end tangents, namely $\mathbf{P}_1^t = (0, 3/2)$ and $\mathbf{P}_3^t = (0, -3/2)$. When these tangents are substituted in Equation (4.57), it becomes identical to Equation (Ans.24), showing that for this particular choice of points, the clamped and anticyclic cubic splines are identical.

4.55: We denote $W = |\mathbf{P}_2 - \mathbf{P}_1|$. Figure Ans.17 shows the geometry of the problem. We start with triangle $ab\mathbf{P}_1$, from which we get $W/2 = R\sin\theta$ or

$$R = \frac{W/2}{\sin\theta}. \qquad (\text{Ans.25})$$

Exercise 4.33 tells us that the midpoint of our segment satisfies

$$\mathbf{P}(0.5) = (2\mathbf{P}_1 - 2\mathbf{P}_2 + \mathbf{P}_1^t + \mathbf{P}_2^t)/8 + (-3\mathbf{P}_1 + 3\mathbf{P}_2 - 2\mathbf{P}_1^t - \mathbf{P}_2^t)/4 + \mathbf{P}_1^t/2 + \mathbf{P}_1$$

$$= \frac{1}{2}(\mathbf{P}_1 + \mathbf{P}_2) + \frac{1}{8}(\mathbf{P}_1^t - \mathbf{P}_2^t).$$

In the figure, vectors f and g correspond to $(1/8)\mathbf{P}_1$ and $(1/8)\mathbf{P}_2$, respectively. Since $ac = a\mathbf{P}_1 = R$, we get $ab = R\cos\theta$, so $bc = R(1 - \cos\theta)$ and the distance e, which equals half bc (since the two tangents have the same magnitude) is $e = (R/2)(1 - \cos\theta)$. Noticing that the magnitude of vector f is $e/\sin\theta$, we get the final result

$$f = \frac{1}{8}|\mathbf{P}_1^t| = \frac{R(1-\cos\theta)}{2\sin\theta},$$

which implies

$$|\mathbf{P}_1^t| = \frac{4R(1-\cos\theta)}{\sin\theta} \underset{(1)}{=} \frac{4(W/2)(1-\cos\theta)}{\sin^2\theta}$$

$$= \frac{2W(1-\cos\theta)}{1-\cos^2\theta} = \frac{2W}{1+\cos\theta},$$

where Equation (Ans.25) is used at point (1).

Answers to Exercises

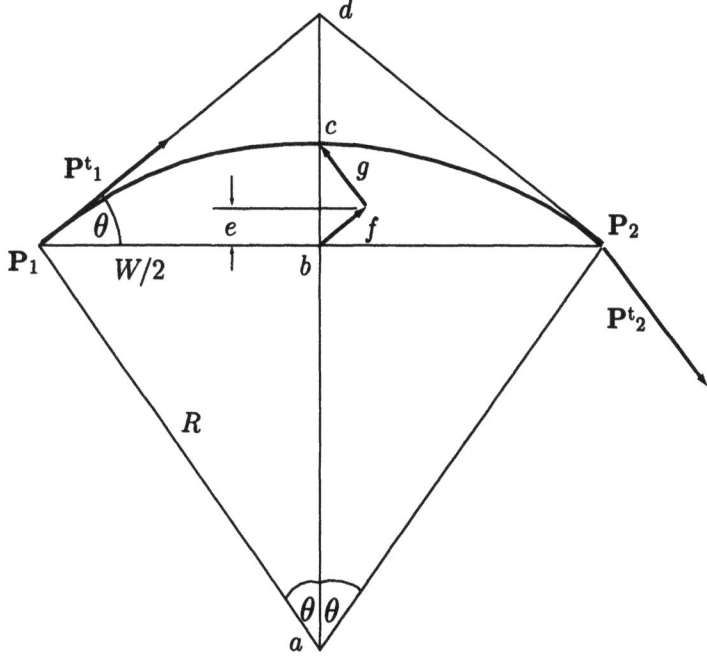

Figure Ans.17: The $2W/(1+\cos\theta)$ Rule.

4.56: The two tangent vectors reverse directions, since s becomes negative. This changes the shape of the curve completely. However, large negative values of s still produce a loose curve.

4.57: The tangent vector of the curve is easily calculated from Equation (4.91):

$$\mathbf{P}_1^t(t) = (-1.5t^2+2t-0.5)\mathbf{P}_1+(4.5t^2-5t)\mathbf{P}_2+(-4.5t^2+4t+0.5)\mathbf{P}_3+(1.5t^2-1)\mathbf{P}_4.$$

At the end ($t=1$), the tangent is $-0.5\mathbf{P}_2+0.5\mathbf{P}_4$.

The tangent vector of the next segment has the same coefficients, so its form is

$$\mathbf{P}_2^t(t) = (-1.5t^2+2t-0.5)\mathbf{P}_2+(4.5t^2-5t)\mathbf{P}_3+(-4.5t^2+4t+0.5)\mathbf{P}_4+(1.5t^2-1)\mathbf{P}_5.$$

At the start ($t=0$), this tangent also has the value $-0.5\mathbf{P}_2+0.5\mathbf{P}_4$, so the two tangents are equal at the connection points between curve segments.

4.58: The quadratic interpolation polynomial $\mathbf{Q}_e(t)$ for the three points $\mathbf{P}_1$, $\mathbf{P}_2$, and $\mathbf{P}_3$ is obtained from Equation (4.93) by incrementing all the indices. The result is

$$\mathbf{Q}_e(t) = (t^2, t, 1) \begin{pmatrix} \frac{1}{\Delta_1(\Delta_1+\Delta_2)} & -\frac{1}{\Delta_1\Delta_2} & \frac{1}{(\Delta_1+\Delta_2)\Delta_2} \\ \frac{-1}{\Delta_1+\Delta_2} - \frac{1}{\Delta_1} & \frac{1}{\Delta_1}+\frac{1}{\Delta_2} & -\frac{1}{\Delta_2}+\frac{1}{\Delta_1+\Delta_2} \\ 1 & 0 & 0 \end{pmatrix} \begin{pmatrix} \mathbf{P}_1 \\ \mathbf{P}_2 \\ \mathbf{P}_3 \end{pmatrix}.$$

(Ans.26)

Similarly, the tangent vector at $\mathbf{P}_2$ is obtained from Equation (4.94) by incrementing all the indices:

$$\begin{aligned} \mathbf{Q}_e^t(\Delta_1) &= -\frac{\Delta_2}{\Delta_1(\Delta_1+\Delta_2)}\mathbf{P}_1 + \frac{\Delta_2-\Delta_1}{\Delta_1\Delta_2}\mathbf{P}_2 + \frac{\Delta_1}{(\Delta_1+\Delta_2)\Delta_2}\mathbf{P}_3 \\ &= \left(\frac{\Delta_2}{\Delta_1+\Delta_2}\right)\left(\frac{\mathbf{P}_2-\mathbf{P}_1}{\Delta_1}\right) + \left(\frac{\Delta_1}{\Delta_1+\Delta_2}\right)\left(\frac{\mathbf{P}_3-\mathbf{P}_2}{\Delta_2}\right). \end{aligned}$$

(Ans.27)

4.59: The tangent vector leaving the start point $\mathbf{P}_3$ is given by

$$\begin{aligned} \mathbf{P}_2^t(0) &= \tfrac{1}{2}(1-T)(1+b)(1-c)(\mathbf{P}_3-\mathbf{P}_2) + \tfrac{1}{2}(1-T)(1-b)(1+c)(\mathbf{P}_4-\mathbf{P}_3) \\ &= \tfrac{1}{2}(1+0.3)(1+1)(1-0)[(4,2)-(3,1)] \\ &= (1.3, 1.3). \end{aligned}$$

The tangent vector arriving at $\mathbf{P}_2$ is given by

$$\begin{aligned} \mathbf{P}_2^t(1) &= \frac{1}{2}(1-T)(1+b)(1+c)(\mathbf{P}_4-\mathbf{P}_3) + \frac{1}{2}(1-T)(1-b)(1-c)(\mathbf{P}_5-\mathbf{P}_4) \\ &= \frac{1}{2}(1+0.3)(1+1)(1+0)[(5,1)-(4,2)] \\ &= (1.3, -1.3). \end{aligned}$$

The segment is, thus,

$$\mathbf{P}_2(t) = (t^3, t^2, t, 1) \begin{pmatrix} 2 & -2 & 1 & 1 \\ -3 & 3 & -2 & -1 \\ 0 & 0 & 1 & 0 \\ 1 & 0 & 0 & 0 \end{pmatrix} \begin{pmatrix} (4,2) \\ (5,1) \\ (1.3,1.3) \\ (1.3,-1.3) \end{pmatrix}$$
$$= (0.6, 2)t^3 + (-0.9, -4.3)t^2 + (1.3, 1.3)t + (4, 2).$$

Its tangent vector is

$$\mathbf{P}_2^t(t) = (1.8, 6)t^2 + (-1.8, -8.6)t + (1.3, 1.3).$$

In order for the curve to be horizontal, the two equations $1.8t^2 - 1.8t + 1.3 = x$ and $6t^2 - 8.6t + 1.3 = 0$ must be satisfied for some positive x. The second equation is easily solved:

$$t = \frac{8.6 \pm \sqrt{8.6^2 - 4\times 6\times 1.3}}{12} = \frac{8.6 \pm 6.54}{12} \approx 1.262, 0.172.$$

Answers to Exercises

The first solution is ignored since it is greater than 1. The second solution shows that the curve becomes horizontal at a point about 17% of the way from the start point $\mathbf{P}_3$.

4.60: Figure Ans.18 shows the points, their polygon and the *Mathematica* code that produced the three lists below. The first list contains the distances between consecutive epoints. The second is the seven values s_2 through s_8 (the cumulative distances), and the third consists of the seven values t_2 through t_7 (the first one, $t_1 = 0$, is not included).

$$(3, 4.243, 4.243, 5, 5, 1, 2.828)$$
$$(3, 7.24264, 11.4853, 16.4853, 21.4853, 22.4853, 25.3137)$$
$$(0.118513, 0.286115, 0.453718, 0.651239, 0.848761, 0.888265, 1).$$

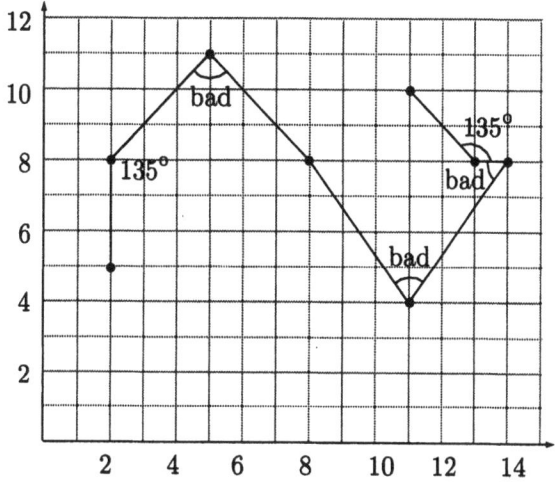

```
pnts={{2,5},{2,8},{5,11},{8,8},{11,4},{14,8},{13,8},{11,10}};
t=Table[N[Sqrt[(pnts[[i+1,1]]-pnts[[i,1]])^2
  +(pnts[[i+1,2]]-pnts[[i,2]])^2],4],{i,1,7}]
Do[t[[i+1]]=t[[i+1]]+t[[i]], {i,1,6}];
t
t=t/t[[7]]
```

Figure Ans.18: Eight Experimental Points and Their Polygon.

4.61: The *Mathematica* code

```
t={-0.1,0.1,0.2,0.3,0.4,0.6,0.8,1.2};
al=0.1; be=0.2; n=8;
scale[t_]:=t+al (n-i)/(n-1)-be (i-1)/(n-1);
```

```
t=Table[scale[t[[i]]], {i,1,8}]
```

produces the eight scaled values

0, 0.157143, 0.214286, 0.271429, 0.328571, 0.485714, 0.642857, 1.

4.62: We simply calculate the quadratic Bézier curve for the three points, and write it in the form $at^2 + bt + c$ to show that it is a parabola. Since this is a Bézier curve, its extreme tangents point in the desired directions:

$$\mathbf{P}(t) = \mathbf{P}_1(1-t)^2 + 2\mathbf{P}_2(1-t)t + \mathbf{P}_3 t^2 = (\mathbf{P}_1 - 2\mathbf{P}_2 + \mathbf{P}_3)t^2 + 2(\mathbf{P}_2 - \mathbf{P}_1)t + \mathbf{P}_1.$$

4.63: The substitution is $u = 2t - 1$, from which we get $t = (1+u)/2$ and $1 - t = (1-u)/2$. The curve of Equation (4.107) can now be written

$$\mathbf{P}(t) = \frac{1}{8}(1-u)^3 \mathbf{P}_0 + \frac{1}{4}(1+u)(1-u)^2 \mathbf{P}_1 + 2\left(\frac{1+u}{2}\right)^2 \left(\frac{1-u}{2}\right)\mathbf{P}_2 + \frac{1}{8}(1+u)^3 \mathbf{P}_3$$

$$= \frac{1}{8}(u^3, u^2, u, 1) \begin{pmatrix} -1 & 2 & -2 & 1 \\ 3 & -2 & -2 & 3 \\ -3 & -2 & 2 & 3 \\ 1 & 2 & 2 & 1 \end{pmatrix} \begin{pmatrix} \mathbf{P}_0 \\ \mathbf{P}_1 \\ \mathbf{P}_2 \\ \mathbf{P}_3 \end{pmatrix}.$$

The only difference is the basis matrix.

4.64: Direct calculation of $B_{4,i}(t)$ for $0 \le i \le 4$ yields the five functions

$B_{4,0} = (1-t)^4$, $B_{4,1} = 4t(1-t)^3$, $B_{4,2} = 6t^2(1-t)^2$, $B_{4,3} = 4t^3(1-t)$, and $B_{4,4} = t^4$.

4.65: The weights are

$$B_{1,0}(t) = \binom{1}{0} t^0 (1-t)^{1-0} = 1 \cdot 1 \cdot (1-t) = (1-t), \qquad B_{1,1}(t) = \binom{1}{1} t^1 (1-t)^{1-1} = t,$$

and the curve is $\mathbf{P}(t) = \mathbf{P}_0(1-t) + \mathbf{P}_1 t$. The straight line from $\mathbf{P}_0$ to $\mathbf{P}_1$.

4.66: Three collinear points are dependent, which means that any one of the three can be expressed as a linear combination (a weighted sum) of the other two, with barycentric weights. We therefore assume that $\mathbf{P}_1 = (1-\alpha)\mathbf{P}_0 + \alpha \mathbf{P}_2$ for some real α. The general Bézier curve for three points,

$$\mathbf{P}(t) = \mathbf{P}_0(1-t)^2 + \mathbf{P}_1 2t(1-t) + \mathbf{P}_2 t^2,$$

now becomes

$$\mathbf{P}(t) = \mathbf{P}_0(1-t)^2 + [(1-\alpha)\mathbf{P}_0 + \alpha \mathbf{P}_2] 2t(1-t) + \mathbf{P}_2 t^2,$$

Answers to Exercises

which is easily simplified to

$$\begin{aligned}\mathbf{P}(t) &= \mathbf{P}_0 + 2\alpha(\mathbf{P}_2 - \mathbf{P}_0)t + (1 - 2\alpha)(\mathbf{P}_2 - \mathbf{P}_0)t^2 \\ &= \mathbf{P}_0 + (\mathbf{P}_2 - \mathbf{P}_0)[2\alpha t + (1 - 2\alpha)t^2] \\ &= \mathbf{P}_0 + (\mathbf{P}_2 - \mathbf{P}_0)T.\end{aligned}$$ (Ans.28)

This is identical to Equation (4.7), which represents a straight line.

This case does not contradict the fact that the Bézier curve does not pass through the intermediate points. We have considered three *collinear* points, which really are only two points. The Bézier curve for two points is a straight line. Note that even with four collinear points, only two are really independent.

Note also that $\mathbf{P}_1$ does not have to be between $\mathbf{P}_0$ and $\mathbf{P}_2$. It can be one of the endpoints. The two cases $\alpha = 0$ and 1 imply that point $\mathbf{P}_1$ is identical to $\mathbf{P}_0$ or $\mathbf{P}_2$, respectively. The case $\alpha = 0.5$ means that $\mathbf{P}_1$ is midway between $\mathbf{P}_0$ and $\mathbf{P}_2$. The cases $\alpha < 0$ and $\alpha > 1$ are special. The former means that $\mathbf{P}_1$ "precedes" $\mathbf{P}_0$. The latter means that $\mathbf{P}_1$ "follows" $\mathbf{P}_2$. In these cases, the curve is no longer a straight line but goes from $\mathbf{P}_0$ toward $\mathbf{P}_1$, reverses direction without reaching $\mathbf{P}_1$, and continues to $\mathbf{P}_2$. The point where it reverses direction becomes a cusp (a "sharp corner"), where the curve has an indefinite tangent vector (Figure Ans.19).

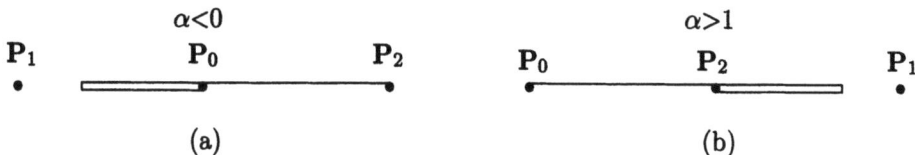

Figure Ans.19: Bézier Straight Segments.

Proof: We first show that in these cases the curve does not go through point $\mathbf{P}_1$. Equation (Ans.28) can be written

$$\mathbf{P}(t) = \mathbf{P}_0\left(1 - 2\alpha t - t^2 + 2\alpha t^2\right) + \mathbf{P}_2\left(2\alpha t + t^2 - 2\alpha t^2\right).$$

Let's see for what value of t the curve passes through point $\mathbf{P}_1 = (1-\alpha)\mathbf{P}_0 + \alpha\mathbf{P}_2$. The conditions are

$$1 - 2\alpha t - t^2 + 2\alpha t^2 = 1 - \alpha \quad \text{and} \quad 2\alpha t + t^2 - 2\alpha t^2 = \alpha.$$

These conditions yield the following quadratic equations for t:

$$\alpha - 2\alpha t + (2\alpha - 1)t^2 = 0 \quad \text{and} \quad -\alpha + 2\alpha t - (2\alpha - 1)t^2 = 0.$$

(Note that the equations are really identical.) Their solutions are

$$t = \frac{\alpha \pm \sqrt{\alpha(\alpha - 1)}}{\alpha} \quad \text{and} \quad t = \frac{-\alpha \pm \sqrt{\alpha(1 - \alpha)}}{-\alpha}.$$

The first equation has no real solution for negative α and the second one has no real solution for $\alpha > 1$. For these values of α, the curve does not pass through control point $\mathbf{P}_1$.

We now calculate the value of t for which the curve has a cusp (a sharp corner). The tangent vector of the curve is

$$\mathbf{P}^t(t) = \mathbf{P}_0\left(-2\alpha - 2t + 4\alpha t\right) + \mathbf{P}_2\left(2\alpha + 2t - 4\alpha t\right) = (2\alpha + 2t - 4\alpha t)(\mathbf{P}_2 - \mathbf{P}_0).$$

The condition for an indefinite tangent vector is, thus, $2\alpha + 2t - 4\alpha t = 0$, which happens for $t = \alpha/(2\alpha - 1)$.

The following three special cases are particularly interesting:

1. $\alpha \ll 0$. This is the case where $\mathbf{P}_1$ is far away from both $\mathbf{P}_0$ and $\mathbf{P}_2$. The limit of $\alpha/(2\alpha - 1)$ in this case is $1/2$, which means that the curve changes direction at its midpoint.

2. $\alpha = -1$. In this case point $\mathbf{P}_0$ is exactly between $\mathbf{P}_1$ and $\mathbf{P}_2$. The value of $\alpha/(2\alpha - 1)$ in this case is $1/3$ (Figure Ans.19a illustrates why this makes sense).

3. $\alpha \gg 1$. Here, $\mathbf{P}_1$ is, again, far from both $\mathbf{P}_0$ and $\mathbf{P}_2$, but in the other direction. The limit of $\alpha/(2\alpha - 1)$ in this case is, again, $1/2$. ◄

4.67: The curve is

$$\mathbf{P}(t) = (1-t)^3(0,0) + 3t(1-t)^2(1,1) + 3t^2(1-t)(0,1) + t^3(1,0)$$
$$= (4t^3 - 6t^2 + 3t, -3t^2 + 3t).$$

Its tangent vector is $\mathbf{P}^t(t) = (12t^2 - 12t + 3, -6t + 3)$, so $\mathbf{P}^t(0.5) = (0,0)$.

4.68: Figure Ans.20 shows the new points. For $\alpha = 1/2$, their values are

$$\mathbf{P}_{01} = 1/2(\mathbf{P}_0 + \mathbf{P}_1), \quad \mathbf{P}_{12} = 1/2(\mathbf{P}_1 + \mathbf{P}_2), \quad \mathbf{P}_{23} = 1/2(\mathbf{P}_2 + \mathbf{P}_3),$$
$$\mathbf{P}_{012} = 1/4(\mathbf{P}_0 + 2\mathbf{P}_1 + \mathbf{P}_2), \quad \mathbf{P}_{123} = 1/4(\mathbf{P}_1 + 2\mathbf{P}_2 + \mathbf{P}_3),$$
$$\mathbf{P}_{0123} = 1/8(\mathbf{P}_0 + 3\mathbf{P}_1 + 3\mathbf{P}_2 + \mathbf{P}_3).$$

Using matrix notation, this can be expressed as,

$$\begin{pmatrix} \mathbf{P}_0 \\ \mathbf{P}_{01} \\ \mathbf{P}_{012} \\ \mathbf{P}_{0123} \end{pmatrix} = \frac{1}{8} \begin{pmatrix} 8 & 0 & 0 & 0 \\ 4 & 4 & 0 & 0 \\ 2 & 4 & 2 & 0 \\ 1 & 3 & 3 & 1 \end{pmatrix} \begin{pmatrix} \mathbf{P}_0 \\ \mathbf{P}_1 \\ \mathbf{P}_2 \\ \mathbf{P}_3 \end{pmatrix} = M_L \mathbf{G},$$

$$\begin{pmatrix} \mathbf{P}_{0123} \\ \mathbf{P}_{123} \\ \mathbf{P}_{23} \\ \mathbf{P}_3 \end{pmatrix} = \frac{1}{8} \begin{pmatrix} 1 & 3 & 3 & 1 \\ 0 & 2 & 4 & 2 \\ 0 & 0 & 4 & 4 \\ 0 & 0 & 0 & 8 \end{pmatrix} \begin{pmatrix} \mathbf{P}_0 \\ \mathbf{P}_1 \\ \mathbf{P}_2 \\ \mathbf{P}_3 \end{pmatrix} = M_R \mathbf{G},$$

where $\mathbf{G}$ is the column consisting of the four original control points of the segment.

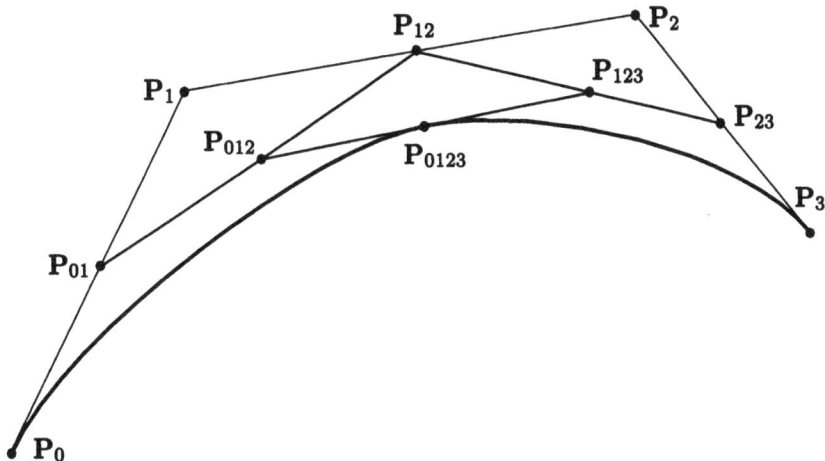

Figure Ans.20: Scaffolding for $k = 3$.

4.69: Because it is defined as

$$\mathbf{P}_{n+1}(t) = t\mathbf{P}_n(t) + (1-t)\mathbf{P}_n(t) = (t+1-t)\mathbf{P}_n(t) = \mathbf{P}_n(t).$$

The two polynomials $\mathbf{P}_{n+1}(t)$ and $\mathbf{P}_n(t)$ are different. They have different degrees and different coefficients, but for any t in the range $[0, 1]$, they have the same value.

4.70: Applying Equation (4.120) to the original three points yields the four points

$$\mathbf{P}_0, \quad (\mathbf{P}_0 + 2\mathbf{P}_1)/3, \quad (2\mathbf{P}_1 + \mathbf{P}_2)/3, \quad \text{and} \quad \mathbf{P}_2.$$

Applying the same equation to these points results in the five points

$$\mathbf{P}_0, \quad \big(\mathbf{P}_0 + 3(\mathbf{P}_0 + 2\mathbf{P}_1)/3\big)/4 = (\mathbf{P}_0 + \mathbf{P}_1)/2,$$
$$\big(2(\mathbf{P}_0 + 2\mathbf{P}_1)/3 + 2(2\mathbf{P}_1 + \mathbf{P}_2)/3\big)/4 = (\mathbf{P}_0 + 4\mathbf{P}_1 + \mathbf{P}_2)/6,$$
$$\big(2(2\mathbf{P}_1 + \mathbf{P}_2)/3 + \mathbf{P}_2\big)/4 = (\mathbf{P}_1 + \mathbf{P}_2)/2, \quad \text{and} \quad \mathbf{P}_2.$$

4.71: Equation (4.120) gives the five new control points

$$\mathbf{Q}_0 = \mathbf{P}_0 = (0,0), \quad \mathbf{Q}_1 = \frac{\mathbf{P}_0 + 3\mathbf{P}_1}{4} = \left(\frac{3}{4}, \frac{3}{2}\right), \quad \mathbf{Q}_2 = \frac{2\mathbf{P}_1 + 2\mathbf{P}_2}{4} = (2,2),$$
$$\mathbf{Q}_3 = \frac{3\mathbf{P}_2 + \mathbf{P}_3}{4} = \left(\frac{11}{4}, \frac{3}{2}\right), \quad \text{and} \quad \mathbf{Q}_4 = \mathbf{P}_3 = (2,0).$$

The original curve is

$$\mathbf{P}_3(t) = (1-t)^3(0,0) + 3t(1-t)^2(1,2) + 3t^2(1-t)(3,2) + t^3(2,0),$$

and the new one is

$$\mathbf{P}_4(t) = (1-t)^4(0,0) + 4t(1-t)^3(3/4,3/2) + 6t^2(1-t)^2(2,2)$$
$$+ 4t^3(1-t)(11/4,3/2) + t^4(2,0).$$

These polynomials seem different, but a closer look reveals that both equal the polynomial $t^3(-4,0) + t^2(3,-6) + t(3,6)$. The two curves $\mathbf{P}_3(t)$, $\mathbf{P}_4(t)$ thus, have the same shape, but $\mathbf{P}_4(t)$ is easier to reshape because it depends on five control points.

4.72: A direct check shows that the elements of every row of matrix $\mathbf{B}$ add up to 1, regardless of the values of a and b. This guarantees that each new control point $\mathbf{Q}_i$ will be a barycentric sum of the $\mathbf{P}_i$'s.

4.73: (1) For $a = 1$ and $b = a + x$, matrix $\mathbf{B}$ is

$$\mathbf{B} = \begin{pmatrix} 0 & 0 & 0 & 1 \\ 0 & 0 & -x & 1+x \\ 0 & x^2 & -2x(1+x) & (1+x)^2 \\ -x^3 & 3x^2(1+x) & -3x(1+x)^2 & (1+x)^3 \end{pmatrix}.$$

(2) The new control points are

$$\mathbf{Q}_0 = \mathbf{P}_3,$$
$$\mathbf{Q}_1 = -(\mathbf{P}_2 x) + \mathbf{P}_3(1+x),$$
$$\mathbf{Q}_2 = \mathbf{P}_1 x^2 + \mathbf{P}_3(1+x)^2 + \mathbf{P}_2(1+x)(3+x-3(1+x)),$$
$$\mathbf{Q}_3 = -(\mathbf{P}_0 x^3) + 3\mathbf{P}_1 x^2(1+x) - 3\mathbf{P}_2 x(1+x)^2 + \mathbf{P}_3(1+x)^3.$$

(3) For $x = 0.75$, they become

$$\mathbf{Q}_0 = \mathbf{P}_3,$$
$$\mathbf{Q}_1 = -0.75\mathbf{P}_2 + 1.75\mathbf{P}_3,$$
$$\mathbf{Q}_2 = 0.5625\mathbf{P}_1 - 2.625\mathbf{P}_2 + 3.0625\mathbf{P}_3,$$
$$\mathbf{Q}_3 = -0.421875\mathbf{P}_0 + 2.953125\mathbf{P}_1 - 6.890625\mathbf{P}_2 + 5.359375\mathbf{P}_3.$$

Notice how each $\mathbf{Q}_i$ is a barycentric combination of the $\mathbf{P}_i$'s.

4.74: For $a = 0$ and $b = 0.5$, matrix $\mathbf{B}$ becomes

$$\mathbf{B} = \begin{pmatrix} 1 & 0 & 0 & 0 \\ 0.5 & 0.5 & 0 & 0 \\ 0.25 & 0.5 & 0.25 & 0 \\ 0.125 & 0.375 & 0.375 & 0.125 \end{pmatrix},$$

Answers to Exercises

and the new control points are

$$Q_0 = P_0,$$
$$Q_1 = \frac{1}{2}P_0 + \frac{1}{2}P_1,$$
$$Q_2 = \frac{1}{4}P_0 + \frac{1}{2}P_1 + \frac{1}{4}P_2,$$
$$Q_3 = \frac{1}{8}P_0 + \frac{3}{8}P_1 + \frac{3}{8}P_2 + \frac{1}{8}P_3.$$

4.75: We denote by A the point midway between P_{01} and P_{012} in Figure 4.44. Similarly, let B denote the point midway between P_{123} and P_{23}. The chord $L_0^{(2)}(P)$ consists of the four segments $P_0 A$, AP_{0123}, $P_{0123}B$, and BP_3.

4.76: This is easy to verify directly. We substitute the two points P_0 and P_3 and the two tangents $3(P_1 - P_0)$ and $3(P_3 - P_2)$ in Equation (4.24)

$$P(t) = (2t^3 - 3t^2 + 1)P_0 + (-2t^3 + 3t^2)P_3 + (t^3 - 2t^2 + t)3(P_1 - P_0) + (t^3 - t^2)3(P_3 - P_2).$$

After rearranging, we get

$$P(t) = (1-t)^3 P_0 + 3t(1-t)^2 P_1 + 3t^2(1-t)P_2 + t^3 P_3,$$

which is the cubic Bézier curve defined by the four points.

4.77: This is obvious. Equation (4.127) describes a vector whose direction is from P_0 to P_3. Varying t changes the magnitude of this vector but not its direction.

4.78: Figure Ans.21 shows the curve, the points, and the code that produced them.

4.79: They are

$$C_0 = c_0(\cos 0, \sin 0) = (0,0), \quad C_1 = c_1(\cos(\pi/6), \sin(\pi/6)) \approx (0.0866, 0.05),$$
$$C_2 = c_2(\cos(\pi/3), \sin(\pi/3)) \approx (0.1, 0.1732), \quad C_3 = c_3(\cos(\pi/2), \sin(\pi/2)) = (0, 2).$$

4.80: The *Mathematica* code

```
Clear[BCtab,CBtab,Bern,den,b1,b2,t1,t2,c0,c1,c2,c3];
t1=0; t2=Pi/2; c0=2; c1=2.2; c2=1.6; c3=1;
den=Sin[t2-t1]; b1=Sin[t2-t]/den; b2=Sin[t-t1]/den;
Bern[t_]:=c0 b1^3+3 c1 b1^2 b2^1 Sin[t]+3 c2 b1^1 b2^2+c3 b2^3;
CBtab=Table[{Cos[t] Bern[t], Sin[t] Bern[t]}, {t,0,Pi/2,0.1}];
v={c0{Cos[0],Sin[0]}, c1{Cos[Pi/6],Sin[Pi/6]},
```

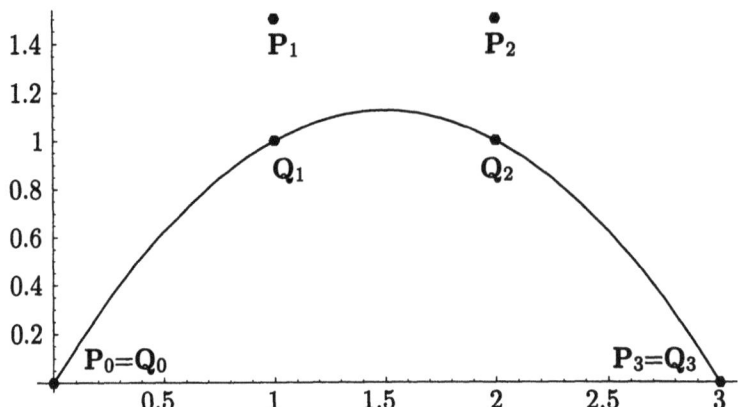

Figure Ans.21: An Interpolating Bézier Curve: III.

```
q0={0,0}; q1={1,1}; q2={2,1}; q3={3,0};
p0=q0; p1={1,3/2}; p2={2,3/2}; p3=q3;
c[t_]:=(1-t)^3 p0+3t(1-t)^2 p1+3t^2(1-t) p2+t^3 p3
g1=ListPlot[{p0,p1,p2,p3,q1,q2},
  Prolog->AbsolutePointSize[4], PlotRange->All,
  AspectRatio->Automatic, DisplayFunction->Identity]
g2=ParametricPlot[c[t], {t,0,1}, DisplayFunction->Identity]
Show[g1,g2, DisplayFunction->$DisplayFunction]
```

Code for Figure Ans.21.

```
c2{Cos[Pi/3],Sin[Pi/3]}, c3{Cos[Pi/2],Sin[Pi/2]}};
v//N
c2=1.3;
BCtab=Table[{Cos[t] Bern[t], Sin[t] Bern[t]}, {t,0,Pi/2,0.1}];
Show[ListPlot[CBtab],ListPlot[BCtab], PlotRange->All,
  AspectRatio->Automatic]
```

shows that $C_0 = (2,0)$, $C_1 = (1.9, 1.1)$, $C_2 = (0.8, 1.39)$, and $C_3 = (0,1)$. It also produces Figure Ans.22, showing the effect of changing weight c_2 from 1.6 to 1.3.

4.81: The Bézier curve can be written $\mathbf{P}(t) = (f(t), g(t))$, where f and g are the (x, y) coordinates of points on the curve. Both f and g are polynomials in t, so they can be written

$$f(t) = a_0 + a_1 t + \cdots + a_k t^k, \qquad g(t) = b_0 + b_1 t + \cdots + b_k t^k.$$

We now suppose that the curve $\mathbf{P}(t) = (f(t), g(t))$ produces the circle $(x-p)^2 + (y-q)^2 = R^2$ and will prove that this implies $a_i = b_i = 0$ for $i = 1, \ldots, k$.

The assumption implies that

$$\bigl(a_0 + a_1 t + \cdots + a_k t^k - p\bigr)^2 + \bigl(b_0 + b_1 t + \cdots + b_k t^k - q\bigr)^2 = R^2 \qquad \text{(Ans.29)}$$

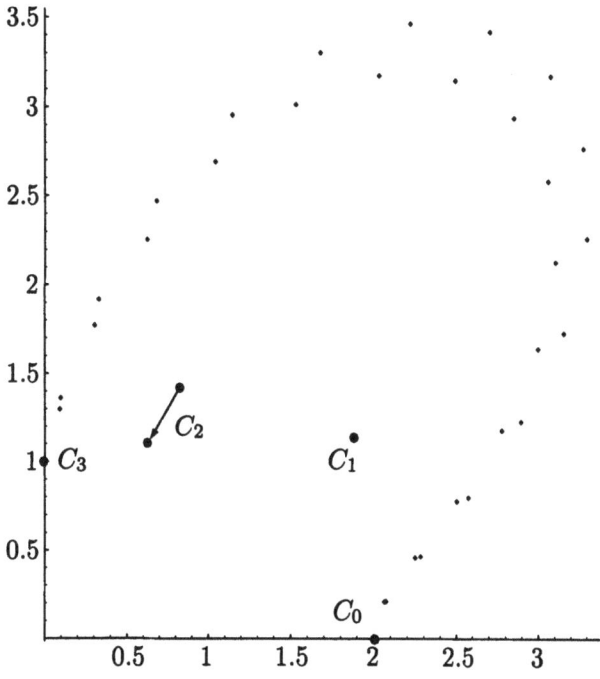

Figure Ans.22: Two Circular Bézier Curves and Their Control Points.

for all values of t. For $t = 0$, we get $(a_0 - p)^2 + (b_0 - q)^2 = R^2$, so now we can write Equation (Ans.29) as

$$\left(a_1 t + \cdots + a_k t^k\right)^2 + \left(b_1 t + \cdots + b_k t^k\right)^2 = 0. \qquad (Ans.30)$$

Carrying out the multiplications produces an expression of the type

$$\left(\cdots + (a_k^2 + b_k^2) t^{2k}\right) = 0.$$

This implies that the sum $(a_k^2 + b_k^2)$ is zero, and since it is the sum of squares, each must be zero. We can now write Equation (Ans.30) as

$$\left(a_1 t + \cdots + a_{k-1} t^{k-1}\right)^2 + \left(b_1 t + \cdots + b_{k-1} t^{k-1}\right)^2 = 0,$$

and use similar reasoning to prove that $a_{k-1} = b_{k-1} = 0$. In this way, we can show that all the coefficients of $f(t)$ and $g(t)$ are zero (except a_0 and b_0, which may create a circle consisting of one point if they satisfy $(a_0 - p)^2 + (b_0 - q)^2 = R^2$).

4.82: Because of the symmetry of a circle, the two interior points must have coordinates $\mathbf{P}_1 = (1, k)$ and $\mathbf{P}_2 = (k, 1)$. To set up an equation that will allow us to solve for k, we arbitrarily require that the midpoint of the curve $\mathbf{P}(.5)$ coincide with the midpoint of the quarter circle $(1/\sqrt{2}, 1/\sqrt{2})$ (Figure Ans.23a). The equation becomes

$$\mathbf{P}(0.5) = \left(\frac{1}{\sqrt{2}}, \frac{1}{\sqrt{2}}\right),$$

or

$$\mathbf{P}(0.5) = \sum_{i=0}^{3} \mathbf{P}_i B_{3,i}(0.5)$$
$$= \frac{1}{8}\mathbf{P}_0 + \frac{3}{8}\mathbf{P}_1 + \frac{3}{8}\mathbf{P}_2 + \frac{1}{8}\mathbf{P}_3$$
$$= \frac{1}{8}(1,0) + \frac{3}{8}(1,k) + \frac{3}{8}(k,1) + \frac{1}{8}(0,1)$$
$$= \left(\frac{3k+4}{8}, \frac{3k+4}{8}\right),$$

and the solution is $k = 4(\sqrt{2} - 1)/3 \approx 0.5523$.

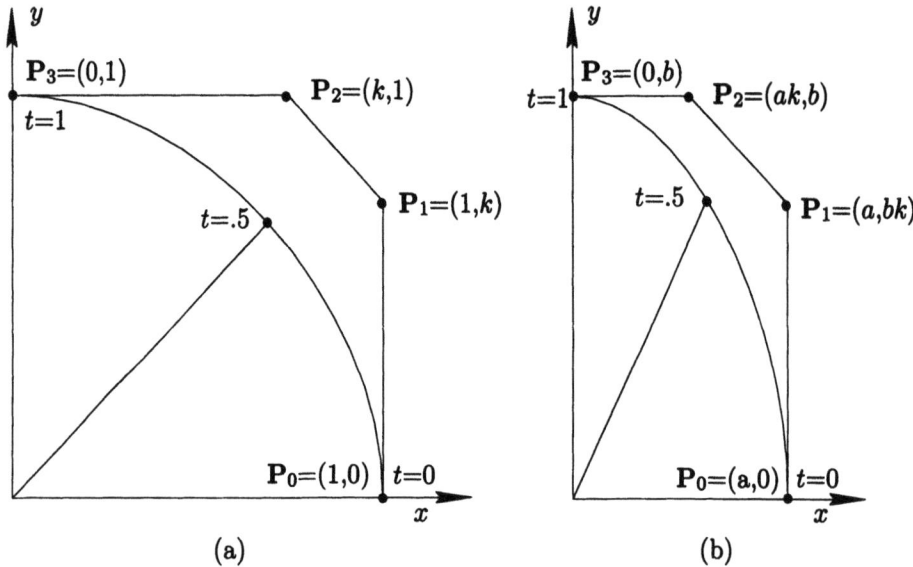

Figure Ans.23: An Almost Circular Bézier Curve.

The final expression of the curve is

$$\mathbf{P}(t) = \big((1-t)^3 + 3t(1-t)^2 + 3t^2(1-t)k, \qquad \qquad$$
$$3t(1-t)^2 k + 3t^2(1-t) + t^3\big) \qquad \text{(Ans.31)}$$
$$\approx \big(0.3431t^3 - 1.3431t^2 + 1, -0.3431t^3 - 0.3138t^2 + 1.6569t\big) .\text{(Ans.32)}$$

(See Sections A.2 and 5.12.3 for applications of this expression.) For a circle of radius R, the expression above is simply multiplied by R.

Answers to Exercises

The maximum deviation of this cubic curve from a true circle can be calculated similar to the quadratic case. The tangent vector of Equation (Ans.31) is

$$\mathbf{P}^t(t) = \bigl(6(k-1)t + 3(2-3k)t^2, 3k + 6(1-2k)t + 3(3k-2)t^2\bigr),$$

so the condition $\mathbf{P}(t) \cdot \mathbf{P}^t(t) = 0$ becomes

$$(9k^2 + 6k - 6)t + (-54k^2 + 18k + 6)t^2 + (126k^2 - 132k + 36)t^3$$
$$- 15(3k-2)^2 t^4 + 6(3k-2)t^5 = 0.$$

Numerical solution gives the five roots $t_1 = 0$, $t_2 = 0.211$, $t_3 = 0.5$, $t_4 = 0.789$, and $t_5 = 1$. The maximum distance between the origin and a point $\mathbf{P}(t)$ on this curve is thus $D(t_2) = D(t_4) = 1.00027$. The maximum deviation of this cubic curve from a true circle is this just 0.027%, much better than the quadratic approximation, although the latter is preferable in practice since it is simpler.

4.83: The implicit expression for the arc is simply the equation of the ellipse

$$\frac{x^2}{a^2} + \frac{y^2}{b^2} = 1, \quad \text{where} \quad 0 \le x \le a.$$

The two endpoints are $\mathbf{P}_0 = (a, 0)$ and $\mathbf{P}_3 = (0, b)$. Based on the symmetry shown in Figure Ans.23b, we select the other two control points with coordinates $\mathbf{P}_1 = (a, bk)$ and $\mathbf{P}_2 = (ak, b)$. To set up an equation that will allow us to solve for k, we require that the midpoint of the curve, $\mathbf{P}(0.5)$, coincide with the midpoint of the quarter arc $(a/\sqrt{2}, b/\sqrt{2})$ (Figure Ans.23b). The equation becomes

$$\mathbf{P}(.5) = \left(\frac{a}{\sqrt{2}}, \frac{b}{\sqrt{2}}\right),$$

or

$$= \sum_{i=0}^{3} \mathbf{P}_i B_{3,i}(.5)$$
$$= \frac{1}{8}\mathbf{P}_0 + \frac{3}{8}\mathbf{P}_1 + \frac{3}{8}\mathbf{P}_2 + \frac{1}{8}\mathbf{P}_3$$
$$= \frac{1}{8}(a,0) + \frac{3}{8}(a,bk) + \frac{3}{8}(ak,b) + \frac{1}{8}(0,b)$$
$$= \left(\frac{a(3k+4)}{8}, \frac{b(3k+4)}{8}\right),$$

and the solution is, again, $k = 4(\sqrt{2}-1)/3 \approx 0.5523$.

The final expression of the curve is

$$\mathbf{P}(t) = (1-t)^3(a,0) + 3t(1-t)^2(a,bk) + 3t^2(1-t)(ak,b) + t^3(0,b).$$

4.84: The parametric equation of the circular arc is $\mathbf{a}(u) = (\cos u, \sin u)$ for $-\theta \leq u \leq \theta$. The expression of the curve is

$$\mathbf{P}(t) = (1-t)^3 \mathbf{P}_0 + 3t(1-t)^2 \mathbf{P}_1 + 3t^2(1-t)\mathbf{P}_2 + t^3 \mathbf{P}_3,$$

where the four control points have to be calculated. To calculate $\mathbf{P}_0$ and $\mathbf{P}_3$, we require that the curve passes through the first and last points of the arc. This implies $\mathbf{P}_0 = (\cos\theta, -\sin\theta)$ and $\mathbf{P}_3 = (\cos\theta, \sin\theta)$. To calculate $\mathbf{P}_1$ and $\mathbf{P}_2$, we require the curve and the arc to have the same tangent vectors at their start and end points, i.e.,

$$\frac{d\mathbf{a}(-\theta)}{du} = \frac{d\mathbf{P}(0)}{dt} \quad \text{and} \quad \frac{d\mathbf{a}(\theta)}{du} = \frac{d\mathbf{P}(1)}{dt}.$$

The tangent vectors are

$$\frac{d\mathbf{a}(u)}{du} = (-\sin u, \cos u),$$

$$\frac{d\mathbf{P}(t)}{dt} = -3(1-t)^2 \mathbf{P}_0 + (3-9t)(1-t)\mathbf{P}_1 + 3t(2-3t)\mathbf{P}_2 + 3t^2 \mathbf{P}_3.$$

Equating them at the start point yields

$$(-\sin(-\theta), \cos(-\theta)) = -3\mathbf{P}_0 + 3\mathbf{P}_1 = -3(\cos\theta, -\sin\theta) + 3\mathbf{P}_1,$$

so

$$\mathbf{P}_1 = (\sin\theta + 3\cos\theta, -3\sin\theta - \cos\theta)/3,$$

and, by symmetry,

$$\mathbf{P}_2 = (\sin\theta + 3\cos\theta, 3\sin\theta + \cos\theta)/3.$$

The Bézier curve is thus

$$\mathbf{P}(t) = (1-t)^3(\cos\theta, -\sin\theta) + 3t(1-t)^2(\sin\theta + 3\cos\theta, -3\sin\theta - \cos\theta)/3$$
$$+ 3t^2(1-t)(\sin\theta + 3\cos\theta, 3\sin\theta + \cos\theta)/3 + t^3(\cos\theta, \sin\theta).$$

The midpoint of the arc is $\mathbf{a}(0.5) = (1,0)$ and that of the curve is

$$\mathbf{P}(.5) = \frac{1}{8}(\cos\theta, -\sin\theta) + \frac{3}{8}(\sin\theta + 3\cos\theta, -3\sin\theta - \cos\theta)/3$$
$$+ \frac{3}{8}(\sin\theta + 3\cos\theta, 3\sin\theta + \cos\theta)/3 + \frac{1}{8}(\cos\theta, \sin\theta)$$
$$= (\cos\theta + \frac{1}{4}\sin\theta, 0).$$

For small θ, the deviation of this curve from the true arc is small (for angles up to 45° the deviation is less than 12%). For larger angles, the curve deviates much from the arc (for $\theta = 90°$, the midpoint of the curve is $(0.25, 0)$, so it is very different from the arc).

Answers to Exercises

4.85: No, since it does not pass through the first and last control points. (However, the text shows that this is a Bézier curve, only it's defined by different control points.)

4.86: The first step of Chaikin's algorithm selects points

$$\frac{3}{4}\mathbf{P}_1 + \frac{1}{4}\mathbf{P}_0 = \frac{1}{2}\left(\frac{\mathbf{P}_0 + \mathbf{P}_1}{2} + \mathbf{P}_1\right) = \frac{\mathbf{A}+\mathbf{B}}{2} = \mathbf{M}_{ab},$$

$$\frac{3}{4}\mathbf{P}_1 + \frac{1}{4}\mathbf{P}_2 = \frac{1}{2}\left(\mathbf{P}_1 + \frac{\mathbf{P}_1 + \mathbf{P}_2}{2}\right) = \frac{\mathbf{B}+\mathbf{C}}{2} = \mathbf{M}_{bc}.$$

4.87: Often, the coordinates of pixels are integers and we know that integer division truncates the result to the nearest integer. Comparing the distance between $\mathbf{P}_1$ and $\mathbf{P}_4$ to two pixels may result in a situation in which they become identical. If this happens, then the assignment $\mathbf{P}_1 \leftarrow \mathbf{P}_4$ does not do anything, and the flow chart of Figure 4.65d shows that this results in a loop that empties the stack without doing anything useful.

4.88: Given three control points $\mathbf{P}_0$, $\mathbf{P}_1$, and $\mathbf{P}_2$, the quadratic B-spline segment defined by them is identical to the quadratic Bézier segment that goes from the midpoint of $\mathbf{P}_0\mathbf{P}_1$ to the midpoint of $\mathbf{P}_1\mathbf{P}_2$. Both segments round out the corner created by the control points with a *parabolic fillet*.

4.89: The second quadratic spline segment is also obtained from Equation (4.154):

$$\mathbf{P}_2(t) = \frac{1}{2}(t^2, t, 1)\begin{pmatrix} 1 & -2 & 1 \\ -2 & 2 & 0 \\ 1 & 1 & 0 \end{pmatrix}\begin{pmatrix} \mathbf{P}_1 \\ \mathbf{P}_2 \\ \mathbf{P}_3 \end{pmatrix}$$

$$= \frac{1}{2}(t^2 - 2t + 1)(1,1) + \frac{1}{2}(-2t^2 + 2t + 1)(2,1) + \frac{t^2}{2}(2,0)$$

$$= (-t^2/2 + t + 3/2, -t^2/2 + 1).$$

It starts at joint $\mathbf{K}_2 = \mathbf{P}_2(0) = (\frac{3}{2}, 1)$ and ends at joint $\mathbf{K}_3 = \mathbf{P}_2(1) = (2, \frac{1}{2})$. The tangent vector is $\mathbf{P}_2^t(t) = (-t+1, -t)$, showing that this segment starts going in direction $\mathbf{P}_2^t(0) = (1,0)$ and ends going in direction $\mathbf{P}_2^t(1) = (0,-1)$ (down).

4.90: We write

$$\mathbf{P}_i(0) = \frac{1}{6}(\mathbf{P}_{i-1} + 4\mathbf{P}_i + \mathbf{P}_{i+1}) = \frac{1}{3}\frac{\mathbf{P}_{i-1} + \mathbf{P}_{i+1}}{2} + \frac{2}{3}\mathbf{P}_i = \frac{1}{3}\mathbf{M} + \frac{2}{3}\mathbf{P}_i,$$

where $\mathbf{M}$ is the midpoint between $\mathbf{P}_{i-1}$ and $\mathbf{P}_{i+1}$. This shows that $\mathbf{P}_i(0)$ is located on the straight segment connecting $\mathbf{M}$ to $\mathbf{P}_i$, two-thirds of the way from $\mathbf{M}$ (Figure Ans.24). Similarly,

$$\mathbf{P}_i(1) = \frac{1}{3}\frac{\mathbf{P}_i + \mathbf{P}_{i+2}}{2} + \frac{2}{3}\mathbf{P}_{i+1} = \frac{1}{3}\mathbf{M} + \frac{2}{3}\mathbf{P}_{i+1}.$$

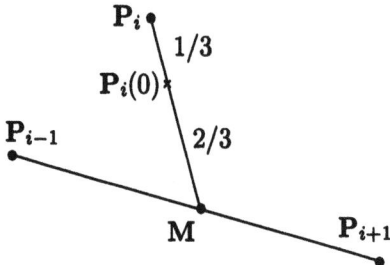

Figure Ans.24: The 2/3 Rule.

This is called the 2/3 rule.

4.91: The second cubic segment is given by Equation (4.159):

$$\mathbf{P}_1(t) = \frac{1}{6}(-t^3 + 3t^2 - 3t + 1)(0,1) + \frac{1}{6}(3t^3 - 6t^2 + 4)(1,1)$$
$$+ \frac{1}{6}(-3t^3 + 3t^2 + 3t + 1)(2,1) + \frac{t^3}{6}(2,0)$$
$$= (-t^3/6 + t + 1, -t^3/6 + 1).$$

It goes from joint $\mathbf{K}_2 = \mathbf{P}_2(0) = (1,1)$ to joint $\mathbf{K}_3 = \mathbf{P}_2(1) = (11/6, 5/6)$. The tangent vector is

$$\mathbf{P}_2^t(t) = \frac{1}{6}(-3t^2 + 6t - 3)(0,1) + \frac{1}{6}(9t^2 - 12t)(1,1)$$
$$+ \frac{1}{6}(-9t^2 + 6t + 3)(2,1) + \frac{t^2}{2}(2,0)$$
$$= (-t^2/2 + 1, -t^2/2).$$

The two extreme tangents are $\mathbf{P}_2^t(0) = (1,0)$ and $\mathbf{P}_2^t(1) = (1/2, -1/2)$.

4.92: Each of the three quadratic segments is given by Equation (4.154). The first segment is

$$\mathbf{P}_1(t) = \frac{1}{2}(t^2 - 2t + 1)(0,0) + \frac{1}{2}(-2t^2 + 2t + 1)(0,1) + \frac{t^2}{2}(1,1)$$
$$= (t^2/2, -t^2/2 + t + 1/2).$$

It goes from $\mathbf{K}_1 = \mathbf{P}_1(0) = (0, 1/2)$ to $\mathbf{K}_2 = \mathbf{P}_1(1) = (1/2, 1)$.

The second segment is

$$\mathbf{P}_2(t) = \frac{1}{2}(t^2 - 2t + 1)(0,1) + \frac{1}{2}(-2t^2 + 2t + 1)(1,1) + \frac{t^2}{2}(2,1)$$
$$= (t + 1/2, 1).$$

Answers to Exercises 797

It goes from $\mathbf{K}_2 = \mathbf{P}_2(0) = (1/2, 1)$ to $\mathbf{K}_3 = \mathbf{P}_2(1) = (3/2, 1)$. Notice that this segment is horizontal.

The third segment is

$$\mathbf{P}_3(t) = \frac{1}{2}(t^2 - 2t + 1)(1, 1) + \frac{1}{2}(-2t^2 + 2t + 1)(2, 1) + \frac{t^2}{2}(2, 0)$$
$$= (-t^2/2 + t + 3/2, -t^2/2 + 1).$$

It goes from $\mathbf{K}_3 = \mathbf{P}_3(0) = (3/2, 1)$ to $\mathbf{K}_4 = \mathbf{P}_3(1) = (2, 1/2)$.

4.93: The two segments are easy to calculate from Equation (4.159). They are

$$\mathbf{P}_3(t) = \frac{1}{6}(2t^3 - 3t^2 - 3t + 5)\mathbf{P}_2 + \frac{1}{6}(-2t^3 + 3t^2 + 3t + 1)\mathbf{P}_4,$$
$$\mathbf{P}_4(t) = \frac{1}{6}(-t^3 + 3t^2 - 3t + 1)\mathbf{P}_3 + \frac{1}{6}(t^3 - 3t^2 + 3t + 5)\mathbf{P}_4.$$

Their extreme points are thus

$$\mathbf{P}_3(0) = \frac{5}{6}\mathbf{P}_2 + \frac{1}{6}\mathbf{P}_4, \quad \mathbf{P}_3(1) = \frac{1}{6}\mathbf{P}_2 + \frac{5}{6}\mathbf{P}_4,$$
$$\mathbf{P}_4(0) = \frac{1}{6}\mathbf{P}_3 + \frac{5}{6}\mathbf{P}_4, \quad \mathbf{P}_4(1) = \mathbf{P}_4.$$

They are indicated by small crosses in Figure 4.70.

4.94: Given the four control points $\mathbf{P}_0 = \mathbf{P}_1 = \mathbf{P}_2 \neq \mathbf{P}_3$, we use Equation (4.159) to write such a segment:

$$\mathbf{P}_1(t) = \frac{1}{6}(-t^3 + 6)\mathbf{P}_0 + \frac{t^3}{6}\mathbf{P}_3 = (1 - u)\mathbf{P}_0 + u\mathbf{P}_3, \quad \text{for } u = t^3/6,$$

which shows the segment to be straight and the start point to be $\mathbf{P}_1(0) = \mathbf{P}_0$.

4.95: There are five segments. The first is defined by points $\mathbf{P}_0$ through $\mathbf{P}_3$. The second is defined by $\mathbf{P}_1$ through $\mathbf{P}_4$, and the last, by $\mathbf{P}_4$ through $\mathbf{P}_7$.

$$\mathbf{P}_1(t) = \frac{1}{6}(t^3 + 6, t^3), \quad \text{a straight line,}$$
$$\mathbf{P}_2(t) = \frac{1}{6}(3t^2 + 3t + 7, -3t^3 + 3t^2 + 3t + 1),$$
$$\mathbf{P}_3(t) = \frac{1}{6}(-3t^3 + 3t^2 + 9t + 13, 4t^3 - 6t^2 + 4),$$
$$\mathbf{P}_4(t) = \frac{1}{6}(2t^3 - 6t^2 + 6t + 22, -3t^3 + 6t^2 + 2),$$
$$\mathbf{P}_5(t) = \frac{1}{6}(24, t^3 - 3t^2 + 3t + 5), \quad \text{a vertical straight line.}$$

They meet at the four joints $(7/6, 1/6)$, $(13/6, 4/6)$, $(22/6, 2/6)$, and $(24/6, 5/6)$. Notice that the fifth segment is vertical. Figure Ans.25 shows these curves (slightly separated to indicate the joint points).

The cubic Bézier curve defined by the same points (where only four points are used) is

$$\mathbf{P}(t) = (1-t)^3(1,0) + 3t(1-t)^2(2,1) + 3t^2(1-t)(4,0) + t^3(4,1)$$
$$= (-3t^3 + 3t^2 + 3t + 1, 4t^3 - 6t^2 + 3t).$$

It goes from $(1,0)$ to $(4,1)$ but is different from the B-spline because, for example, it is never vertical. It is shown dashed in Figure Ans.25. The degree-7 Bézier curve defined by the same points is also shown (dot-dashed) in the same figure for comparison. It is clear that it is tight because of its strong attraction to the multiple points.

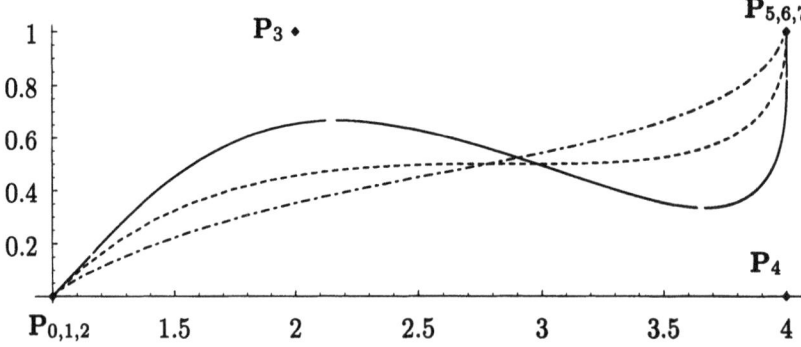

Figure Ans.25: Comparing a Uniform B-spline and a Bézier Curve for Eight Points.

4.96: They are all done with Equation (4.159) by rotating the four control points to the left in each segment. The result is

$$\mathbf{P}_2(t) = \frac{1}{6}[(-t^3 + 3t^2 - 3t + 1)(0, 3/2) + (3t^3 - 6t^2 + 4)(-3/2, 0)$$
$$+ (-3t^3 + 3t^2 + 3t + 1)(0, -3/2) + t^3(3/2, 0)]$$
$$= \frac{1}{4}(-2t^3 + 6t^2 - 4, 2t^3 - 6t),$$

$$\mathbf{P}_3(t) = \frac{1}{6}[(-t^3 + 3t^2 - 3t + 1)(-3/2, 0) + (3t^3 - 6t^2 + 4)(0, -3/2)$$
$$+ (-3t^3 + 3t^2 + 3t + 1)(3/2, 0) + t^3(0, 3/2)]$$
$$= \frac{1}{4}(-2t^3 + 6t, -2t^3 + 6t^2 - 4),$$

$$\mathbf{P}_4(t) = \frac{1}{6}[(-t^3 + 3t^2 - 3t + 1)(0, -3/2) + (3t^3 - 6t^2 + 4)(3/2, 0)$$

Answers to Exercises

$$+ (-3t^3 + 3t^2 + 3t + 1)(0, 3/2) + t^3(-3/2, 0)]$$
$$= \frac{1}{4}(2t^3 - 6t^2 + 4, -2t^3 + 6t). \tag{Ans.33}$$

4.97: Compute the midpoint $(\mathbf{S}+\mathbf{E})/2$ and normalize its coordinates.

4.98: Equation (4.166) can be written $\mathbf{P}^t(t) = (t^2 - t)[(\mathbf{P}_0 - \mathbf{P}_3) + 3(\mathbf{P}_1 - \mathbf{P}_2)]$. This is the sum of two differences of points. The first difference is the vector from $\mathbf{P}_3$ to $\mathbf{P}_0$ and the second is the vector from $\mathbf{P}_2$ to $\mathbf{P}_1$ (multiplied by 3). The tangent vector of Equation (4.166) therefore points in the direction of the sum of these vectors, and this direction does not depend on t. The size of the tangent vector depends on t, but the size affects just the speed of the spline segment, not its shape.

4.99: By substituting, for example, $t + 1$ for t in the expression for $N_{13}(t)$.

4.100: The tangent vectors of the three segments are

$$\mathbf{P}_1^t(t) = 2(t - 1)\mathbf{P}_0 + (2 - 3t)\mathbf{P}_1 + t\mathbf{P}_2,$$
$$\mathbf{P}_2^t(t) = (t - 2)\mathbf{P}_1 + (3 - 2t)\mathbf{P}_2 + (t - 1)\mathbf{P}_3,$$
$$\mathbf{P}_3^t(t) = (t - 3)\mathbf{P}_2 + (7 - 3t)\mathbf{P}_3 + 2(t - 2)\mathbf{P}_4.$$

They satisfy $\mathbf{P}_1^t(1) = \mathbf{P}_2^t(1) = \mathbf{P}_2 - \mathbf{P}_1$, and $\mathbf{P}_2^t(2) = \mathbf{P}_3^t(2) = \mathbf{P}_3 - \mathbf{P}_2$.

4.101: For the curve of Figure 4.84c, the knot vector is

$$(-3, -2, -1, 0, 1, 1, 1, 2, 3, 4, 5, 6).$$

The range of the parameter t is from $t_3 = 0$ to $t_8 = 3$ and we obtain the blending functions by direct calculations (only the last group N_{i4} of blending functions is shown):

$$N_{04}(t) = \frac{t - t_0}{t_3 - t_0} N_{03} + \frac{t_4 - t}{t_4 - t_1} N_{13} = \frac{1}{6}(1 - t)^3 \qquad \text{for } t \in [0, 1),$$

$$N_{14}(t) = \frac{t - t_1}{t_4 - t_1} N_{13} + \frac{t_5 - t}{t_5 - t_2} N_{23} = \frac{1}{12}(11t^3 - 15t^2 - 3t + 7) \qquad \text{for } t \in [0, 1),$$

$$N_{24}(t) = \frac{t - t_2}{t_5 - t_2} N_{23} + \frac{t_6 - t}{t_6 - t_3} N_{33} = \frac{1}{4}(-7t^3 + 3t^2 + 3t + 1) \qquad \text{for } t \in [0, 1),$$

$$N_{34}(t) = \frac{t - t_3}{t_6 - t_3} N_{33} + \frac{t_7 - t}{t_7 - t_4} N_{43} = \begin{cases} t^3 & \text{for } t \in [0, 1), \\ (2 - t)^3 & \text{for } t \in [1, 2), \end{cases}$$

$$N_{44}(t) = \frac{t - t_4}{t_7 - t_4} N_{43} + \frac{t_8 - t}{t_8 - t_5} N_{53} = \frac{1}{4} \begin{cases} (7t^3 - 39t^2 + 69t - 37) & \text{for } t \in [1, 2), \\ (3 - t)^3 & \text{for } t \in [2, 3), \end{cases}$$

$$N_{54}(t) = \frac{t - t_5}{t_8 - t_5} N_{53} + \frac{t_9 - t}{t_9 - t_6} N_{63} = \frac{1}{12} \begin{cases} (-11t^3 + 51t^2 - 69t + 29) & \text{for } t \in [1, 2), \\ (7t^3 - 57t^2 + 147t - 115) & \text{for } t \in [2, 3), \end{cases}$$

$$N_{64}(t) = \frac{t-t_6}{t_9-t_6}N_{63} + \frac{t_{10}-t}{t_{10}-t_7}N_{73} = \frac{1}{6}\begin{cases}(t-1)^3 & \text{for } t \in [1,2),\\ (-3t^3+21t^2-45t+31) & \text{for } t \in [2,3),\end{cases}$$

$$N_{74}(t) = \frac{t-t_7}{t_{10}-t_7}N_{73} + \frac{t_{11}-t}{t_{11}-t_8}N_{83} = \frac{1}{6}(t-2)^3 \qquad \text{for } t \in [2,3).$$

This group of blending functions can now be used to construct the five spline segments

$$\begin{aligned}\mathbf{P}_3(t) &= N_{04}(t)\mathbf{P}_0 + N_{14}(t)\mathbf{P}_1 + N_{24}(t)\mathbf{P}_2 + N_{34}(t)\mathbf{P}_3\\ &= \frac{1}{6}(1-t)^3\mathbf{P}_0 + \frac{1}{12}(11t^3-15t^2-3t+7)\mathbf{P}_1\\ &\quad + \frac{1}{4}(-7t^3+3t^2+3t+1)\mathbf{P}_2 + t^3\mathbf{P}_3, & t \in [0,1),\end{aligned}$$

$$\begin{aligned}\mathbf{P}_4(t) &= N_{14}(1)\mathbf{P}_1 + N_{24}(1)\mathbf{P}_2 + N_{34}(1)\mathbf{P}_3 + N_{44}(1)\mathbf{P}_4\\ &= \mathbf{P}_3 \quad \text{(a point)}, & t \in [1,1),\end{aligned}$$

$$\begin{aligned}\mathbf{P}_5(t) &= N_{24}(1)\mathbf{P}_2 + N_{34}(1)\mathbf{P}_3 + N_{44}(1)\mathbf{P}_4 + N_{54}(1)\mathbf{P}_5\\ &= \mathbf{P}_3 \quad \text{(a point)}, & t \in [1,1),\end{aligned}$$

$$\begin{aligned}\mathbf{P}_6(t) &= N_{34}(t)\mathbf{P}_3 + N_{44}(t)\mathbf{P}_4 + N_{54}(t)\mathbf{P}_5 + N_{64}(t)\mathbf{P}_6\\ &= (2-t)^3\mathbf{P}_3 + \frac{1}{4}(7t^3-39t^2+69t-37)\mathbf{P}_4\\ &\quad + \frac{1}{12}(-11t^3+51t^2-69t+29)\mathbf{P}_5 + \frac{1}{6}(t-1)^3\mathbf{P}_6, & t \in [1,2),\end{aligned}$$

$$\begin{aligned}\mathbf{P}_7(t) &= N_{44}(t)\mathbf{P}_4 + N_{54}(t)\mathbf{P}_5 + N_{64}(t)\mathbf{P}_6 + N_{74}(t)\mathbf{P}_7\\ &= (3-t)^3\mathbf{P}_4 + \frac{1}{12}(7t^3-57t^2+147t-115)\mathbf{P}_5\\ &\quad + (-3t^3+21t^2-45t+31)\mathbf{P}_6 + \frac{1}{6}(t-2)^3\mathbf{P}_7, & t \in [2,3).\end{aligned}$$

A direct check verifies that each segment has barycentric weights. The entire curve starts at $\mathbf{P}_3(0) = \mathbf{P}_0/6 + 7\mathbf{P}_1/12 + \mathbf{P}_2/4$ and ends at $\mathbf{P}_7(3) = (\mathbf{P}_5 + 4\mathbf{P}_6 + \mathbf{P}_7)/6$. The two join points between the segments are

$$\mathbf{P}_3(1) = \mathbf{P}_6(1) = \mathbf{P}_3, \quad \mathbf{P}_6(2) = \mathbf{P}_7(2) = \mathbf{P}_4/4 + 7\mathbf{P}_5/12 + \mathbf{P}_6/6.$$

Both segments $\mathbf{P}_4(t)$ and $\mathbf{P}_5(t)$ reduce to the single control point $\mathbf{P}_3$.

For the curve of Figure 4.84d, the knot vector is

$$(-3,-2,-1,0,1,1,1,1,2,3,4,5).$$

The range of the parameter t is from $t_3 = 0$ to $t_8 = 2$ and we get by direct calculations (again only the last group N_{i4} of blending functions is shown)

$$N_{04}(t) = \frac{t-t_0}{t_3-t_0}N_{03} + \frac{t_4-t}{t_4-t_1}N_{13} = \frac{1}{6}(1-t)^3 \qquad \text{for } t \in [0,1),$$

$$N_{14}(t) = \frac{t-t_1}{t_4-t_1}N_{13} + \frac{t_5-t}{t_5-t_2}N_{23} = \frac{1}{12}(11t^3-15t^2-3t+7) \qquad \text{for } t \in [0,1),$$

Answers to Exercises

$$N_{24}(t) = \frac{t-t_2}{t_5-t_2}N_{23} + \frac{t_6-t}{t_6-t_3}N_{33} = \frac{1}{4}(-7t^3+3t^2+3t+1) \quad \text{for } t \in [0,1),$$

$$N_{34}(t) = \frac{t-t_3}{t_6-t_3}N_{33} + \frac{t_7-t}{t_7-t_4}N_{43} = t^3 \quad \text{for } t \in [0,1),$$

$$N_{44}(t) = \frac{t-t_4}{t_7-t_4}N_{43} + \frac{t_8-t}{t_8-t_5}N_{53} = (2-t)^3 \quad \text{for } t \in [1,2),$$

$$N_{54}(t) = \frac{t-t_5}{t_8-t_5}N_{53} + \frac{t_9-t}{t_9-t_6}N_{63} = \frac{1}{4}(7t^3-39t^2+69t-37) \quad \text{for } t \in [1,2),$$

$$N_{64}(t) = \frac{t-t_6}{t_9-t_6}N_{63} + \frac{t_{10}-t}{t_{10}-t_7}N_{73} = \frac{1}{12}(-11t^3+51t^2-69t+29) \quad \text{for } t \in [1,2),$$

$$N_{74}(t) = \frac{t-t_7}{t_{10}-t_7}N_{73} + \frac{t_{11}-t}{t_{11}-t_8}N_{83} = \frac{1}{6}(t-1)^3 \quad \text{for } t \in [1,2).$$

This group of blending functions can now be used to construct the five spline segments

$$\begin{aligned}
\mathbf{P}_3(t) &= N_{04}(t)\mathbf{P}_0 + N_{14}(t)\mathbf{P}_1 + N_{24}(t)\mathbf{P}_2 + N_{34}(t)\mathbf{P}_3 \\
&= \frac{1}{6}(1-t)^3\mathbf{P}_0 + \frac{1}{12}(11t^3-15t^2-3t+7)\mathbf{P}_1 \\
&\quad + \frac{1}{4}(-7t^3+3t^2+3t+1)\mathbf{P}_2 + t^3\mathbf{P}_3, & t \in [0,1), \\
\mathbf{P}_4(t) &= N_{14}(1)\mathbf{P}_1 + N_{24}(1)\mathbf{P}_2 + N_{34}(1)\mathbf{P}_3 + N_{44}(1)\mathbf{P}_4 \\
&= \mathbf{P}_3 + \mathbf{P}_4 \quad \text{(undefined)}, & t \in [1,1), \\
\mathbf{P}_5(t) &= N_{24}(1)\mathbf{P}_2 + N_{34}(1)\mathbf{P}_3 + N_{44}(1)\mathbf{P}_4 + N_{54}(1)\mathbf{P}_5 \\
&= \mathbf{P}_3 + \mathbf{P}_4 \quad \text{(undefined)}, & t \in [1,1), \\
\mathbf{P}_6(t) &= N_{34}(t)\mathbf{P}_3 + N_{44}(t)\mathbf{P}_4 + N_{54}(t)\mathbf{P}_5 + N_{64}(t)\mathbf{P}_6 \\
&= \mathbf{P}_3 + \mathbf{P}_4 \quad \text{(undefined)}, & t \in [1,1), \\
\mathbf{P}_7(t) &= N_{44}(t)\mathbf{P}_4 + N_{54}(t)\mathbf{P}_5 + N_{64}(t)\mathbf{P}_6 + N_{74}(t)\mathbf{P}_7 \\
&= (2-t)^3\mathbf{P}_4 + \frac{1}{4}(7t^3-39t^2+69t-37)\mathbf{P}_5 \\
&\quad + \frac{1}{12}(-11t^3+51t^2-69t+29)\mathbf{P}_6 + \frac{1}{6}(t-1)^3\mathbf{P}_7, & t \in [1,2).
\end{aligned}$$

A direct check verifies that each segment has barycentric weights. The curve consists of the two separate segments $\mathbf{P}_3(t)$ and $\mathbf{P}_7(t)$. The former goes from $\mathbf{P}_0/6 + 7\mathbf{P}_1/12 + \mathbf{P}_2/4$ to $\mathbf{P}_3$ and the latter from $\mathbf{P}_4$ to $\mathbf{P}_5/4 + 7\mathbf{P}_6/12 + \mathbf{P}_7/6$. The three segments $\mathbf{P}_4(t)$, $\mathbf{P}_5(t)$, and $\mathbf{P}_6(t)$ get the undefined value $\mathbf{P}_3 + \mathbf{P}_4$ at $t=1$.

4.102: We use the parameter substitution $T = t^2/((1-t)^2+t^2)$ to write this curve in the form $(1-T)\mathbf{P}_0 + T\mathbf{P}_2$. It is now clear that this is the required line. It is also easy to see that $t = 0 \to T = 0$ and $t = 1 \to T = 1$.

4.103: It is obvious from the figure that $\mathbf{P}_0 = (0,-1)R$. To figure out the coordinates of $\mathbf{P}_2$ we notice the following:

1. The point is on the circle $x^2 + y^2 = R^2$, so it satisfies

$$x_2^2 + y_2^2 = R^2. \qquad \text{(Ans.34)}$$

2. The point is on line L. The equation of this line can be written $y = ax + b$, where the slope a equals $\tan 60° = \sqrt{3} \approx 1.732$, so we have

$$y_2 = ax_2 + b, \qquad \text{(Ans.35)}$$

where b still has to be determined.

3. $\mathbf{P}_2$ is located on the circle at a point where the tangent has a slope of $60°$. We differentiate the equation of the circle $x^2 + y^2 = R^2$ with respect to x to obtain $2x + 2y(dy/dx) = 0$ or $x = -y \cdot y'$. A slope of $60°$ means that $y' = \tan 60 = a$, so $\mathbf{P}_2$ also satisfies

$$x_2 = -y_2 a. \qquad \text{(Ans.36)}$$

Equations (Ans.34) through (Ans.36) are easy to solve. The three solutions are $y_2 = R/\sqrt{a^2+1} = 0.5R$, $x_2 = -ay_2 = -0.866R$, and $b = y_2 - ax_2 = R(1+a^2)/\sqrt{a^2+1} = 2R$.

To figure out the coordinates of $\mathbf{P}_1$, we notice that it is located on line L and its y coordinate equals $-R$. It therefore satisfies $-R = ax_1 + b$, so $x_1 = -aR = -1.732R$.

4.104: The base angle of the triangle defined by the three points is $\theta = 45°$, so a circular arc is obtained when we set $w_1 = \cos \theta = 0.7071$. Substituting the points in Equation (4.194) and setting $w_1 = 0.7071$ yields the $90°$ circular arc that goes from $\mathbf{P}_0$ to $\mathbf{P}_2$:

$$\begin{aligned}\mathbf{P}(t) &= \frac{(1-t)^2 \mathbf{P}_0 + 2w_1 t(1-t) \mathbf{P}_1 + t^2 \mathbf{P}_2}{(1-t)^2 + 2w_1 t(1-t) + t^2} \\ &= \frac{(1-t)^2(1,0) + 1.414t(1-t)(0,0) + t^2(0,1)}{(1-t)^2 + 1.414t(1-t) + t^2} R \\ &= \frac{((1-t)^2, t^2)R}{(1-t)^2 + 1.414t(1-t) + t^2}. \end{aligned} \qquad \text{(Ans.37)}$$

4.105: The particular conic generated by Equation (4.207) is determined by the sign of the discriminant $B^2 - 4AC$. The exact shape of the curve is determined by the values of all six parameters. The general rule is

$$B^2 - 4AC \begin{cases} < 0, & \text{ellipse (or circle)}, \\ = 0, & \text{parabola}, \\ > 0, & \text{hyperbola}. \end{cases}$$

Examples:

Answers to Exercises

1. The canonical circle is obtained when $A = C = 1$, $F = -R^2$, and $B = D = E = 0$.

2. A straight line is the result of $A = B = C = 0$.

3. The canonical parabola $y^2 = 2ax$ is the result of $A = B = E = F = 0$, $C = 1$, and $D = -2a$.

4.106: The attributes "vertical" and "horizontal" are extrinsic. "Cusp" and "smooth," however, are intrinsic. The length of a curve and area of a polygon or a closed curve are extrinsic since they can be changed by scaling the coordinate system. If a certain point on a surface has a tangent plane in one coordinate system (i.e., the surface is smooth in the vicinity of the point), it will have such a plane (although perhaps a different one) in any other coordinate system. This property of a surface is, thus, intrinsic.

4.107: The principal normal vector at point i points, by definition, in the direction the curve turns at the point. Since a straight line does not make any turns, its principal normal vector is undefined.

4.108: The first two derivatives are $\mathbf{P}^t(t) = (-3,0)t^2 + (2,-2)t + (1,1)$ and $\mathbf{P}^{tt}(t) = (-6,0)t + (2,-2)$. The principal normal vector (still unnormalized) is thus

$$\mathbf{N}(t) = \mathbf{P}^{tt}(t) - \frac{\mathbf{P}^{tt}(t) \bullet \mathbf{P}^t(t)}{|\mathbf{P}^t(t)^2|}\mathbf{P}^t(t) = \mathbf{P}^{tt}(t) - \left[\frac{18t^3 - 18t^2 + 2t}{9t^4 - 12t^3 + 2t^2 + 2}\right]\mathbf{P}^t(t).$$

Simple tests result in $\mathbf{N}(0) = (2,-2)$, $\mathbf{N}(.5) = (0,-2)$, and $\mathbf{N}(1) = (-4,0)$. Vector $\mathbf{N}(t)$ thus starts in direction $(1,-1)$, changes to $(0,-1)$ (down) when $t = 0.5$ (this makes sense since $\mathbf{P}^t(0.5)$ is horizontal), and finishes in direction $(-1,0)$ (i.e., in the negative x direction). It is always perpendicular to the direction of the curve.

4.109: The curve and its first two derivatives are given by

$$\mathbf{P}(t) = (1-t)^3(0,0,0) + 3t(1-t)^2(1,0,0) + 3t^2(1-t)(2,1,0) + t^3(3,0,1)$$
$$= (3t, 3t^2(1-t), t^3),$$
$$\mathbf{P}^t(t) = (3, 3t(2-3t), 3t^2),$$
$$\mathbf{P}^{tt}(t) = (0, 6-18t, 6t).$$

The unnormalized principal normal vector is given by

$$\mathbf{N}(t) = \mathbf{P}^{tt}(t) - \left[\frac{18t(2-9t+10t^2)}{9+9t^2(2-3t)^2}\right]\mathbf{P}^t(t),$$

from which we get $\mathbf{N}(0) = (0,6,0)$, $\mathbf{N}(0.5) = (0,-3,3)$, and $\mathbf{N}(1) = (-9,-3,-3)$.

The osculating plane is the solution of $\det[((x,y,z) - \mathbf{P}(t))\,\mathbf{P}^t(t)\,\mathbf{P}^{tt}(t)] = 0$. The explicit determinant is

$$\det\begin{pmatrix} x - 3t & y - 3t^2(1-t) & z - t^3 \\ 3 & 3t(2-3t) & 3t^2 \\ 0 & 6 - 18t & 6t \end{pmatrix}.$$

Thus, the osculating plane is given by $t^2x - ty + (1 - 3t)z - t^3 = 0$. At $t = 0, 0.5$ and 1 this plane has the equations $z = 0$, $0.25x - 0.5y - 0.5z - 0.125 = 0$, and $x - y - 2z - 1 = 0$, respectively.

4.110: Equation (4.222) becomes

$$\frac{d^2x}{ds^2} = -R\frac{dy}{ds}, \quad \frac{d^2y}{ds^2} = R\frac{dx}{ds}.$$

It is easy to guess that the solutions are $x(s) = R\cos(R \cdot s) + A$ and $y(s) = R\sin(R \cdot s) + B$. The curve is a circle of radius R with center at (A, B).

4.111: The row number is $(i \bmod 11)$ and the column number is $10 - (i \div 11)$, where $\div$ denotes integer-by-integer division (the quotient is truncated to the nearest integer).

4.112: The parameter pairs for pixel $(9, 9)$ satisfy $9 = 9a + b$ or $a = (9-b)/9$. The 11 pairs are, therefore, $(1, 0)$, $(8/9 \approx 0.9, 1)$, $(7/9 \approx 0.8, 2)$, $(6/9 \approx 0.7, 3),\ldots,(0, 9)$, and $(-1/9 \approx -0.1, 10)$.

5.1: Because a surface has no depth. Imagine a flat surface, such as the xy plane. Each point on this surface has just two coordinates (the third one, z, is zero) and can be located by means of these two numbers. Now, crumple this flat surface and leave it in the same three-dimensional space. Each surface point now has three coordinates (the z coordinate is no longer zero), but the same two numbers are still sufficient to locate the point on the crumpled surface.

A surface is a two-dimensional structure embedded in three-dimensional space. Each point on the surface has three coordinates, but only two numbers are needed to specify the position of the point on the surface. In contrast, a solid object requires three parameters to be expressed. The surface function $\mathbf{P}(u, w)$ evaluates to a triplet (the three coordinates of a point on the surface) for every pair (u, w) of parameters.

5.2: It is easy to show that the corner points are $\mathbf{P}_{00} = (0, 0, 1)$, $\mathbf{P}_{10} = (1, 0, 0)$, $\mathbf{P}_{01} = (0.5, 1, 0)$, and $\mathbf{P}_{11} = (1, 1, 0)$. The boundary curves are also not hard to calculate. They are

$$\mathbf{P}(0, w) = (0.5w, w, 1-w), \quad \mathbf{P}(u, 1) = (0.5(1-u) + u, 1, 0),$$
$$\mathbf{P}(1, w) = (1, w, 0), \quad \mathbf{P}(u, 0) = (u, 0, 1-u).$$

5.3: The four boundary curves are

$$\mathbf{P}(u, 0) = ((c-a)u + a, b, 0), \quad \mathbf{P}(u, 1) = ((c-a)u + a, d, 0),$$
$$\mathbf{P}(0, w) = (a, (d-b)w + b, 0), \quad \mathbf{P}(1, w) = (c, (d-b)w + b, 0).$$

Answers to Exercises

Obviously, they are straight lines. The four corner points can be obtained from the boundary curves

$$\mathbf{P}_{00} = (a,b,0), \quad \mathbf{P}_{01} = (a,d,0), \quad \mathbf{P}_{10} = (c,b,0); \quad \mathbf{P}_{11} = (c,d,0).$$

The surface patch is the flat rectangle defined by these points.

5.4: We select $(1,0,0)$ as our pivot point and calculate the three vectors $\mathbf{v}_1 = (0,1,0) - (1,0,0) = (-1,1,0)$, $\mathbf{v}_2 = (1,a,1) - (1,0,0) = (0,a,1)$, and $\mathbf{v}_3 = (0,-a,0) - (1,0,0) = (-1,-a,0)$. Next, we calculate the only scalar triple product

$$\mathbf{v}_4 \bullet (\mathbf{v}_2 \times \mathbf{v}_3) = \begin{vmatrix} -1 & -a & 0 \\ -1 & 1 & 0 \\ 0 & a & 1 \end{vmatrix} = 0.$$

It equals zero for any value of a, so the polygon is planar for any a.

5.5: The plane should pass through the three points $(0,0,0)$, $(0,0,1)$, and $(1,1,0)$. Equation (5.2) gives

$$A = \begin{vmatrix} 0 & 0 & 1 \\ 0 & 1 & 1 \\ 1 & 0 & 1 \end{vmatrix} = -1, \quad B = -\begin{vmatrix} 0 & 0 & 1 \\ 0 & 1 & 1 \\ 1 & 0 & 1 \end{vmatrix} = 1,$$

$$C = \begin{vmatrix} 0 & 0 & 1 \\ 0 & 0 & 1 \\ 1 & 1 & 1 \end{vmatrix} = 0, \quad D = -\begin{vmatrix} 0 & 0 & 0 \\ 0 & 0 & 1 \\ 1 & 1 & 0 \end{vmatrix} = 0.$$

The expression of the plane is, therefore, $-x + y = 0$

5.6: They are the points where the plane $x/a + y/b + z/c = 1$ intercepts the three coordinate axes.

5.7: $s = \mathbf{N} \bullet \mathbf{P}_1 = (1,1,1) \bullet (1,1,1) = 3$, so the plane is given by $x + y + z - 3 = 0$. It intercepts the three coordinate axes at points $(3,0,0)$, $(0,3,0)$, and $(0,0,3)$ (Figure 5.4a).

5.8: The expression is

$$\mathbf{P}(u,w) = \mathbf{P}_1 + u(\mathbf{P}_2 - \mathbf{P}_1) + w(\mathbf{P}_3 - \mathbf{P}_1) = (3,0,0) + u(-3,3,0) + w(-3,0,3).$$

5.9: If the cross-product $\mathbf{a} \times \mathbf{b}$ points in the direction of $\mathbf{N}$, the angle between them is zero. Its cosine thus equals 1, causing the dot product $\mathbf{N} \bullet (\mathbf{a} \times \mathbf{b})$ to be positive (since it is the product of the magnitudes of the vectors and the cosine of the angle between them).

5.10: If the line is parallel to the plane, then its direction vector $\mathbf{d}$ is parallel to the plane (i.e., perpendicular to the normal), resulting in $\mathbf{N} \bullet \mathbf{d} = 0$ (infinite μ). If the line is also in the plane, then $\mathbf{P}_1$ is in the plane, resulting in $s = \mathbf{N} \bullet \mathbf{P}_1$ or $s - \mathbf{N} \bullet \mathbf{P}_1 = 0$ (in this case, μ is of the form 0/0, indefinite).

5.11: We first subtract $\mathbf{P}_2 - \mathbf{P}_1 = (-2, 1, -0.8)$ and $\mathbf{P}_3 - \mathbf{P}_1 = (-2, 9, -0.8)$. The triangle is therefore given by

$$(10, -5, 4) + u(-2, 1, -0.8) + w(-2, 9, -0.8)$$
$$= \bigl(10 - 2(u+w), -5 + u + 9w, 4 - 0.8(u+w)\bigr),$$

where $u \geq 0$, $w \geq 0$, and $u + w \leq 1$.

5.12: We first subtract $\mathbf{P}_2 - \mathbf{P}_1 = (-2, 1, -0.8)$ and $\mathbf{P}_3 - \mathbf{P}_1 = (2, -1, 0.8)$. The differences are related because the points are collinear; the triangle is therefore given by $(10, -5, 4) + (-2, 1, -0.8)(u - w) = \mathbf{P}_1 + (\mathbf{P}_2 - \mathbf{P}_1)(u - w)$. It depends only on the difference $u - w$. When u and w are varied independently, the difference between them changes from -1 to 1. The triangle therefore degenerates into the straight line $\mathbf{P}(t) = \mathbf{P}_1 + (\mathbf{P}_2 - \mathbf{P}_1)t$, where $-1 \leq t \leq 1$. This line goes from $\mathbf{P}(-1) = \mathbf{P}_1 - (\mathbf{P}_2 - \mathbf{P}_1) = 2\mathbf{P}_1 - \mathbf{P}_2 = \mathbf{P}_3$ to $\mathbf{P}(1) = \mathbf{P}_1 + (\mathbf{P}_2 - \mathbf{P}_1) = \mathbf{P}_2$.

5.13: The trilinears of the circumcenter are $(\cos A, \cos B, \cos C)$. The trilinears of the orthocenter are $(\sec A, \sec B, \sec C)$. The trilinears of the Fermat point are $(\csc(A + \pi/3), \csc(B + \pi/3), \csc(C + \pi/3))$.

5.14: The following expression is more efficient, using 20 multiply/divide operations versus 57:

$$P_x = (((a_x - c_x)(a_x + c_x) + (a_1 - c_1)(a_1 + c_1))/2(b_1 - c_1)$$
$$- ((b_x - c_x)(b_x + c_x) + (b_1 - c_1)(b_1 + c_1))/2(a_1 - c_1))/D,$$
$$P_y = (((b_x - c_x)(b_x + c_x) + (b_y - c_y)(b_y + c_y))/2(a_x - c_x)$$
$$- ((a_x - c_x)(a_x + c_x) + (a_y - c_y)(a_y + c_y))/2(b_x - c_x))/D,$$

where $D = (a_x - c_x)(b_y - c_y) - (b_x - c_x)(a_y - c_y)$ and the radius is given by

$$R^2 = (c_x - p_x)^2 + (c_y - p_y)^2.$$

This approach uses Cramer's Rule to find the intersection of two perpendicular bisectors of triangle edges.

5.15: Since the circle is symmetric, it makes sense to tentatively place the three triangle corners at points $\mathbf{P}_1 = (-A, -B)$, $\mathbf{P}_2 = (A, -B)$, and $\mathbf{P}_3 = (0, C)$, where A, B, and C are parameters to be determined. Figure Ans.26 shows that $B = R$. We calculate the line L that passes through points $\mathbf{P}_1$ and $\mathbf{P}_2$. Its equation is $y = ax + b$ and we arbitrarily require $a = -1$. This makes L tangent to the circle at point $\mathbf{P} = (R/\sqrt{2}, R/\sqrt{2})$, which yields $b = R\sqrt{2}$. The equation of line L is thus

Answers to Exercises

$y = -x + R\sqrt{2}$. We use the requirement that it passes through $\mathbf{P}_2 = (A, -R)$ to get $A = R(1 + \sqrt{2})$. Similarly, passing through $\mathbf{P}_3 = (0, C)$ yields $C = R\sqrt{2}$. The three triangle points are, thus, $\mathbf{P}_1 = (-R(1+\sqrt{2}), -R)$, $\mathbf{P}_2 = (R(1+\sqrt{2}), -R)$, and $\mathbf{P}_3 = (0, R\sqrt{2})$.

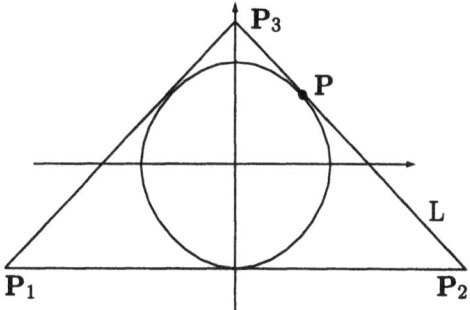

Figure Ans.26: A Triangle Bounding a Circle.

5.16: We translate the rectangle such that point (x_0, y_0) becomes the origin (Figure 5.13b). We tentatively place the three corners of a triangle at $\mathbf{P}_1 = (-A, 0)$, $\mathbf{P}_2 = (A, 0)$, and $\mathbf{P}_3 = (0, B)$, where A and B are parameters to be determined. We draw a line S with a slope of -1 from $\mathbf{P}_2$ to $\mathbf{P}_3$ such that it touches the rectangle at (w, h). The line equation is $y = -x + b$. Since it passes through (w, h), we get $b = w + h$. Since it passes through $(A, 0)$, we get $A = w + h$. Similarly, it is trivial to show that $B = w + h$.

5.17: 1. Equation (5.7) yields the expression of the surface

$$\mathbf{P}(u, w) = ((0, 0, 0)(1-u)(1-w) + (1, 0, 0)u(1-w) + (0, 1, 0)(1-u)w + (1, 1, 1)uw)$$
$$= (u, w, uw).$$

2. The implicit representation is $z = xy$. This is easy to guess because the x coordinate equals u, the y coordinate is w, and the z coordinate equals uw.

3. The two conditions $z = k$ and $z = xy$ produce $k = xy$ or $y = k/x$. This curve is a *hyperbola*.

4. The plane through the three points $(0, 0, 0)$, $(0, 0, 1)$, and $(1, 1, 0)$ contains the z axis and is especially easy to calculate. Its equation is $x - y = 0$. Intersected with $z = xy$, it yields the curve $z = x^2$, a parabola.

This is the reason why the bilinear surface is sometimes called a *hyperbolic paraboloid*.

5.18: The two tangent vectors are

$$\frac{\partial \mathbf{P}(u, w)}{\partial u} = (1 - w, 0, w - 1), \quad \frac{\partial \mathbf{P}(u, w)}{\partial w} = (-u, 1, u - 1).$$

The normal vector is

$$\mathbf{N}(u,w) = \frac{\partial \mathbf{P}(u,w)}{\partial u} \times \frac{\partial \mathbf{P}(u,w)}{\partial w} = (1-w, 1-w, 1-w) = (1-w)(1,1,1).$$

This vector does not depend on u, it always points in the $(1,1,1)$ direction, and its magnitude varies from $(1,1,1)$ for $w=0$ to the indefinite $(0,0,0)$ for $w=1$ at the multiple point $\mathbf{P}_{01} = \mathbf{P}_{11}$. Thus, the surface does not posses a normal vector at $w=1$ since the surface itself reduces to a point at this value. The reason that the normal does not depend on u is that this surface patch is flat. It is simply the triangle connecting the three points $(0,0,1)$, $(1,0,0)$, and $(0,1,0)$.

5.19: Figures 5.16 and 3.22 show that the rotation matrix should be the one given by Equation (3.24):

$$\begin{pmatrix} \cos 60° & 0 & -\sin 60° \\ 0 & 1 & 0 \\ \sin 60° & 0 & \cos 60° \end{pmatrix} = \begin{pmatrix} 0.5 & 0 & -0.866 \\ 0 & 1 & 0 \\ 0.866 & 0 & 0.5 \end{pmatrix}.$$

Applying this rotation to the two given points yields the following six points, where the translation in the y direction has already been included:

$$\mathbf{P}_{01} = (-0.5, 0, 0.866), \quad \mathbf{P}_{11} = (0.5, 0, -0.866),$$
$$\mathbf{P}_{02} = (0.5, 1, 0.866), \quad \mathbf{P}_{12} = (-0.5, 1, -0.866),$$
$$\mathbf{P}_{03} = (1, 2, 0), \quad \mathbf{P}_{13} = (-1, 2, 0).$$

The three bilinear patches are now easy to calculate:

$\mathbf{P}_1(u,w) = (-1,-1,0)(1-u)(1-w) + (-0.5, 0, .866)(1-u)w$
$\quad + (1,-1,0)u(1-w) + (0.5, 0, -0.866)uw$
$= (-1 + 2u + 0.5uw + 0.5w - 1.5uw, -1 + w, -0.866uw + 0.866w - 0.866uw),$
$\mathbf{P}_2(u,w) = (1,-1,0)(1-u)(1-w) + (0.5, 1, .866)(1-u)w$
$\quad + (0.5, 0, -0.866)u(1-w) + (-0.5, 1, -0.866)uw$
$= (1 - 0.5u - 0.5uw - 0.5w, -1 + u + uw + 2w - 2uw, -0.866u - 0.866uw + 0.866w),$
$\mathbf{P}_3(u,w) = (0.5, 1, 0.866)(1-u)(1-w) + (1, 2, 0)(1-u)w$
$\quad + (-0.5, 1, -0.866)u(1-w) + (-1, 2, 0)uw$
$= (0.5 - 1u - uw + 0.5w, 1 + 2uw + w - 2uw, 0.866 - 1.732u - 0.866w + 1.732uw).$

5.20: (a) The straight line is

$$\mathbf{P}(u,0) = \mathbf{P}_1(1-u) + \mathbf{P}_2 u = \mathbf{P}_1 + (\mathbf{P}_2 - \mathbf{P}_1)u = (-1,-1,0) + (2,0,0)u.$$

(b) For the quadratic, we set up the equations

$$(-1,1,0) = \mathbf{P}_3 = \mathbf{P}(0,1) = \mathbf{C},$$
$$(0,1,1) = \mathbf{P}_4 = \mathbf{P}(0.5,1) = 0.25\mathbf{A} + 0.5\mathbf{B} + \mathbf{C},$$
$$(1,1,0) = \mathbf{P}_5 = \mathbf{P}(1,1) = \mathbf{A} + \mathbf{B} + \mathbf{C},$$

which are solved to yield $\mathbf{A} = (0,0,-4)$, $\mathbf{B} = (2,0,4)$, and $\mathbf{C} = (-1,1,0)$. The top curve is thus

$$\mathbf{P}(u,1) = (0,0,-4)u^2 + (2,0,4)u + (-1,1,0),$$

and the surface is $\mathbf{P}(u,w) = \mathbf{P}(u,0)(1-w) + \mathbf{P}(u,1)w = (2u-1, 2w-1, 4uw(1-u))$. The center point is $\mathbf{P}(0.5, 0.5) = (0, 0, 0.5)$. Figure Ans.27 shows this surface.

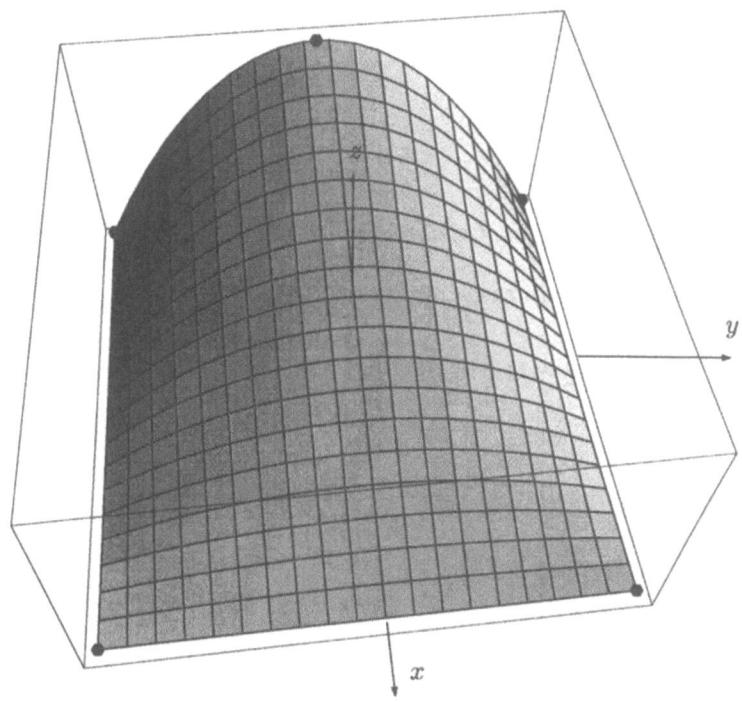

Figure Ans.27: A Lofted Surface.

```
(* A lofted surface example. Bottom boundary curve is straight *)
pnts={{-1,-1,0},{1,-1,0},{-1,1,0},{0,1,1},{1,1,0}};
g1=Graphics3D[{AbsolutePointSize[5],
        Table[Point[pnts[[i]]],{i,1,5}]}]
g2=ParametricPlot3D[{2u-1,2w-1,4u w(1-u)}, {u,0,1},{w,0,1},
DefaultFont->{"cmr10", 10}, DisplayFunction->Identity,
 AspectRatio->Automatic, Ticks->{{0,1},{0,1},{0,1}}]
Show[g1,g2, ViewPoint->{-0.139, -1.179, 1.475}]
```

Code for Figure Ans.27.

5.21: The base can be considered one boundary curve. Its equation is $\mathbf{P}_1(u) = (R\cos u, R\sin u, H)$, where $0 \le u \le 2\pi$. The other boundary curve is the vertex $\mathbf{P}_2(u) = (0,0,0)$ (it is a degenerate curve). The entire surface is obtained, as usual, by

$$\begin{aligned}\mathbf{P}(u,w) &= \mathbf{P}_1(u)w + \mathbf{P}_2(u)(1-w) \\ &= (Ru\cos u, Ru\sin u, Hw),\end{aligned} \qquad \text{(Ans.38)}$$

where $0 \le w \le 1$ and $0 \le u \le 2\pi$.

5.22: The four corner points of the base are $(-a,-a,0)$, $(-a,a,0)$, $(a,-a,0)$, and $(a,a,0)$. We select the two points $\mathbf{P}_1 = (-a,-a,0)$ and $\mathbf{P}_2 = (a,-a,0)$. The straight line between them is

$$\mathbf{P}(u) = (1-u)\mathbf{P}_1 + u\mathbf{P}_2 = (-a + 2ua, -a, 0).$$

The face defined by these points is, thus, expressed by

$$\begin{aligned}\mathbf{P}(u,w) &= \mathbf{P}(u)(1-w) + (0,0,H)w \\ &= \bigl(a(2u-1)(1-w), a(w-1), Hw\bigr).\end{aligned} \qquad \text{(Ans.39)}$$

The other three faces are calculated similarly.

5.23: The tangent vector in the u direction is

$$\frac{\partial \mathbf{P}(u,w)}{\partial u} = \Bigl(24u^2(1-w) - 24u(1-w) - 4w + 6,$$

$$12u^2(1-w) - 18u(1-w) + 8uw - 10w + 6, 0\Bigr).$$

At $u = 0.5$, this vector reduces to

$$\frac{\partial \mathbf{P}(0.5,w)}{\partial u} = \bigl(6(w-1) - 4w + 6, 6(w-1) - 6w + 6, 0\bigr) = (2w, 0, 0),$$

which implies that

$$\frac{\partial \mathbf{P}(0.5,0)}{\partial u} = (0,0,0).$$

This shows that the surface does not have a tangent at the cusp, point $(0, 5/4, 0)$.

5.24: In such, a case the tangent vector of the surface along the degenerate boundary curve $\mathbf{P}(u,1)$ is the weighted sum of the eight quantities

$$\mathbf{T}_0 = \left.\frac{d\mathbf{P}(0,w)}{dw}\right|_{w=1}, \quad \mathbf{T}_1 = \left.\frac{d\mathbf{P}(1,w)}{dw}\right|_{w=1}, \quad \mathbf{P}(u,0), \mathbf{P}(u,1), \mathbf{P}_{00}, \mathbf{P}_{01}, \mathbf{P}_{10}, \mathbf{P}_{11},$$

instead of being the simple linear combination $B_0(u)\mathbf{T}_0 + B_1(u)\mathbf{T}_1$ of Equation (5.23). As it swings from $\mathbf{T}_0$ to $\mathbf{T}_1$, this vector will not have to stay in the plane defined by $\mathbf{T}_0$ and $\mathbf{T}_1$ and may wiggle wildly in and out of this plane, causing the surface to be wrinkled in the vicinity of the common point.

Answers to Exercises

5.25: We start with the boundary curves. They are straight lines, and so are obtained from Equation (4.4):

$$\mathbf{P}(0,w) = (1-w)\mathbf{P}_{00} + w\mathbf{P}_{01}, \quad \mathbf{P}(1,w) = (1-w)\mathbf{P}_{10} + w\mathbf{P}_{11},$$
$$\mathbf{P}(u,0) = (1-u)\mathbf{P}_{00} + u\mathbf{P}_{10}, \quad \mathbf{P}(u,1) = (1-u)\mathbf{P}_{01} + u\mathbf{P}_{11}.$$

The surface expression is now obtained from Equation (5.12):

$$\begin{aligned}\mathbf{P}(u,w) &= (1-u)(1-w)\mathbf{P}_{00} + (1-u)w\mathbf{P}_{01} + u(1-w)\mathbf{P}_{10} + uw\mathbf{P}_{11} \\ &\quad + (1-w)(1-u)\mathbf{P}_{00} + (1-w)u\mathbf{P}_{10} + w(1-u)\mathbf{P}_{01} + wu\mathbf{P}_{11} \\ &\quad - (1-u)(1-w)\mathbf{P}_{00} - u(1-w)\mathbf{P}_{10} - (1-u)w\mathbf{P}_{01} - uw\mathbf{P}_{11} \\ &= (0.5(1-u)w + u, w, (1-u)(1-w)).\end{aligned}$$

Note that it is identical to the bilinear surface of Equation (5.9).

5.26: Figure Ans.28a shows a diamond-shaped grid of 16 equally spaced points. The eight points with negative weights are shown in black. Figure Ans.28b shows a cut (labeled xx in Figure Ans.28a) through four points in this surface. The cut is a curve that passes through pour data points. It is easy to see that when the two exterior (black) points are raised, the center of the curve (and, as a result, the center of the surface) gets lowered. It is now clear that points with negative weights push the center of the surface in a direction opposite that of the points.

Figure Ans.28c is a more detailed example that also shows why the four corner points should have positive weights. It shows a simple symmetric surface patch that interpolates the 16 points

$$\begin{aligned}&\mathbf{P}_{00} = (0,0,0), \quad \mathbf{P}_{10} = (1,0,1), \quad \mathbf{P}_{20} = (2,0,1), \quad \mathbf{P}_{30} = (3,0,0), \\ &\mathbf{P}_{01} = (0,1,1), \quad \mathbf{P}_{11} = (1,1,2), \quad \mathbf{P}_{21} = (2,1,2), \quad \mathbf{P}_{31} = (3,1,1), \\ &\mathbf{P}_{02} = (0,2,1), \quad \mathbf{P}_{12} = (1,2,2), \quad \mathbf{P}_{22} = (2,2,2), \quad \mathbf{P}_{32} = (3,2,1), \\ &\mathbf{P}_{03} = (0,3,0), \quad \mathbf{P}_{13} = (1,3,1), \quad \mathbf{P}_{23} = (2,3,1), \quad \mathbf{P}_{33} = (3,3,0).\end{aligned}$$

We first raise the eight boundary points from $z = 1$ to $z = 1.5$. Figure Ans.28d shows how the center point $\mathbf{P}(.5, .5)$ gets lowered from $(1.5, 1.5, 2.25)$ to $(1.5, 1.5, 2.10938)$. We next return those points to their original positions and instead raise the four corner points from $z = 0$ to $z = 1$. Figure Ans.28e shows how this raises the center point from $(1.5, 1.5, 2.25)$ to $(1.5, 1.5, 2.26563)$.

5.27: The weight functions in the u direction are

$$B_{20}(u) = (1-u)^2, \quad B_{21}(u) = 2u(1-u), \quad B_{22}(u) = u^2.$$

Those in the w direction are

$$B_{30}(w) = (1-w)^3, \quad B_{31}(w) = 3w(1-w)^2,$$
$$B_{32}(w) = 3w^2(1-w), \quad B_{33}(w) = w^3.$$

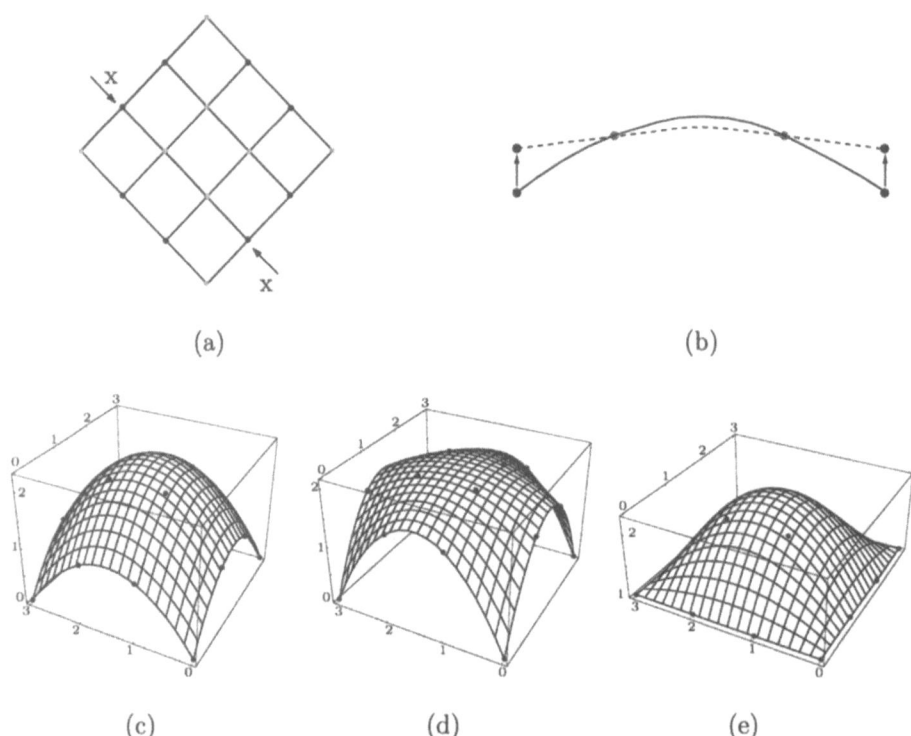

Figure Ans.28: An Interpolating Bicubic Surface Patch.

```
Clear[Nh,p,pnts,U,W];
p00={0,0,0}; p10={1,0,1}; p20={2,0,1}; p30={3,0,0};
p01={0,1,1}; p11={1,1,2}; p21={2,1,2}; p31={3,1,1};
p02={0,2,1}; p12={1,2,2}; p22={2,2,2}; p32={3,2,1};
p03={0,3,0}; p13={1,3,1}; p23={2,3,1}; p33={3,3,0};
Nh={{-4.5,13.5,-13.5,4.5},{9,-22.5,18,-4.5},
 {-5.5,9,-4.5,1},{1,0,0,0}};
pnts={{p33,p32,p31,p30},{p23,p22,p21,p20},
 {p13,p12,p11,p10},{p03,p02,p01,p00}};
U[u_]:={u^3,u^2,u,1};   W[w_]:={w^3,w^2,w,1};
(* prt [i] extracts component i from the 3rd dimen of P *)
prt[i_]:=pnts[[Range[1,4],Range[1,4],i]];
p[u_,w_]:={U[u].Nh.prt[1].Transpose[Nh].W[w],
 U[u].Nh.prt[2].Transpose[Nh].W[w], \
 U[u].Nh.prt[3].Transpose[Nh].W[w]};
 g1=ParametricPlot3D[p[u,w], {u,0,1},{w,0,1},
 Compiled->False, DisplayFunction->Identity];
 g2=Graphics3D[{AbsolutePointSize[2],
 Table[Point[pnts[[i,j]]],{i,1,4},{j,1,4}]}];
Show[g1,g2, ViewPoint->{-2.576, -1.365, 1.718}]
```

Code For Figure Ans.28.

Answers to Exercises

The surface patch is, therefore,

$$
\begin{aligned}
\mathbf{P}&(u,w) \\
&= \sum_{i=0}^{2}\sum_{j=0}^{3} B_{2i}(u)\mathbf{P}_{ij}B_{3j}(w) \\
&= B_{20}(u)[\mathbf{P}_{00}B_{30}(w) + \mathbf{P}_{01}B_{31}(w) + \mathbf{P}_{02}B_{32}(w) + \mathbf{P}_{03}B_{33}(w)] \\
&\quad + B_{21}(u)[\mathbf{P}_{00}B_{30}(w) + \mathbf{P}_{01}B_{31}(w) + \mathbf{P}_{02}B_{32}(w) + \mathbf{P}_{03}B_{33}(w)] \\
&\quad + B_{22}(u)[\mathbf{P}_{00}B_{30}(w) + \mathbf{P}_{01}B_{31}(w) + \mathbf{P}_{02}B_{32}(w) + \mathbf{P}_{03}B_{33}(w)] \\
&= (1-u)^2[(1-w)^3(0,0,0) + 3w(1-w)^2(1,0,1) \\
&\quad + 3w^2(1-w)(2,0,1) + w^3(3,0,0)] \\
&\quad + 2u(1-u)[(1-w)^3(0,1,0) + 3w(1-w)^2(1,1,1) \\
&\quad + 3w^2(1-w)(2,1,1) + w^3(3,1,0)] \\
&\quad + u^2[(1-w)^3(0,2,0) + 3w(1-w)^2(1,2,1) + 3w^2(1-w)(2,2,1) + w^3(3,2,0)] \\
&= (3w, 2u, 3w(1-w)). \hspace{5em} \text{(Ans.40)}
\end{aligned}
$$

5.28: Here is one such loop. It displays one family of (11) curves.

```
for u:=0 step 0.1 to 1 do (* 11 curves *)
  for v:=0 step 0.01 to 1-u do (* 100 pixels per curve *)
  w:=1-u-v;
  Calculate & project point P(u,v,w)
  endfor;
endfor;
```

The second family consists of the curves parallel to the base of the triangle (the main loop is on v), and the third family consists of the curves parallel to the right side.

5.29: This is point $\mathbf{P}(0.5, 0.5) = (1, 1, (-1 - 2/2 + 2/4)(-1 - 2/2 + 2/4)/4) = (1, 1, 9/16)$.

5.30: In the case of equally spaced control points, we have

$$(\mathbf{P}_{00} + \mathbf{P}_{20})/2 = \mathbf{P}_{10} \Rightarrow \frac{1}{6}(\mathbf{P}_{00} + 4\mathbf{P}_{10} + \mathbf{P}_{20}) = \mathbf{P}_{10},$$

$$(\mathbf{P}_{01} + \mathbf{P}_{21})/2 = \mathbf{P}_{11} \Rightarrow \frac{1}{6}(\mathbf{P}_{01} + 4\mathbf{P}_{11} + \mathbf{P}_{21}) = \mathbf{P}_{11},$$

$$(\mathbf{P}_{02} + \mathbf{P}_{22})/2 = \mathbf{P}_{12} \Rightarrow \frac{1}{6}(\mathbf{P}_{02} + 4\mathbf{P}_{12} + \mathbf{P}_{22}) = \mathbf{P}_{12},$$

so $\mathbf{K}_{00} = \frac{1}{6}\mathbf{P}_{10} + \frac{4}{6}\mathbf{P}_{11} + \frac{1}{6}\mathbf{P}_{12} = \mathbf{P}_{11}$.

5.31: The top left corner after one subdivision is

$$\mathbf{P}^1(0,0) = \frac{1}{36}(\mathbf{P}^1_{00} + \mathbf{P}^1_{02} + 4\mathbf{P}^1_{10} + 4\mathbf{P}^1_{12} + \mathbf{P}^1_{20} + 4\mathbf{P}^1_{01} + 16\mathbf{P}^1_{11} + 4\mathbf{P}^1_{21} + \mathbf{P}^1_{22})$$

$$= \frac{1}{36}\Big[\frac{1}{4}(\mathbf{P}^0_{00} + \mathbf{P}^0_{10} + \mathbf{P}^0_{01} + \mathbf{P}^0_{11}) + \frac{1}{4}(\mathbf{P}^0_{01} + \mathbf{P}^0_{11} + \mathbf{P}^0_{02} + \mathbf{P}^0_{12})$$

$$+ \frac{1}{4}(\mathbf{P}^0_{00} + \mathbf{P}^0_{01} + 6\mathbf{P}^0_{10} + 6\mathbf{P}^0_{11} + \mathbf{P}^0_{20} + \mathbf{P}^0_{21})$$

$$+ \frac{1}{4}(\mathbf{P}^0_{00} + 6\mathbf{P}^0_{10} + \mathbf{P}^0_{20} + 6\mathbf{P}^0_{01} + 36\mathbf{P}^0_{11} + 6\mathbf{P}^0_{21} + \mathbf{P}^0_{02} + 6\mathbf{P}^0_{12} + \mathbf{P}^0_{22})$$

$$+ \frac{1}{4}(\mathbf{P}^0_{10} + \mathbf{P}^0_{20} + 6\mathbf{P}^0_{11} + 6\mathbf{P}^0_{21} + \mathbf{P}^0_{12} + \mathbf{P}^0_{22}) + \frac{1}{4}(\mathbf{P}^0_{11} + \mathbf{P}^0_{21} + \mathbf{P}^0_{12} + \mathbf{P}^0_{22})\Big]$$

$$= \frac{1}{36}(\mathbf{P}^0_{00} + 4\mathbf{P}^0_{01} + 16\mathbf{P}^0_{11} + \mathbf{P}^0_{02} + 4\mathbf{P}^0_{12} + \mathbf{P}^0_{20} + 4\mathbf{P}^0_{21} + \mathbf{P}^0_{22})$$

$$= \mathbf{P}(0,0).$$

5.32: This is trivial. Just translate each coordinate:

$$\mathbf{P}(u,w) = \mathbf{P}(u)\mathbf{T}_y(w) = (x_0 + R\cos u \cos w, y_0 + R\sin u, z_0 + R\cos u \sin w),$$

where $-\pi/2 \le u \le \pi/2$ and $0 \le w \le 2\pi$.

5.33: This is done by multiplying the surface of Equation (5.72) by rotation matrix $\mathbf{T}_z(\theta)$. The result is

$$\mathbf{P}(u,w) = \mathbf{P}(u)\mathbf{T}_y(w)\mathbf{T}_z(\theta)$$
$$= (R\cos u \cos w \cos\theta - R\sin u \sin\theta, R\cos u \cos w \sin\theta + R\sin u \cos\theta, R\cos u \sin w),$$

where $-\pi/2 \le u \le \pi/2$ and $0 \le w \le 2\pi$. Notice how the z coordinates of this sphere don't depend on θ.

5.34: The half-circle is $\mathbf{P}(u) = (R\cos u, 0, R\sin u)$. Multiplying this by $\mathbf{T}_z(w)$ yields
$$\mathbf{P}(u,w) = (R\cos u \cos w, R\cos u \sin w, R\sin u).$$

5.35: We start with the circle $(R + r\cos u, 0, r\sin u)$, where $0 \le u \le 2\pi$. The torus is generated when this circle is rotated 360° about the z axis, by means of rotation matrix $\mathbf{T}_z(w)$:

$$\mathbf{P}(u,w) = (R + r\cos u, 0, r\sin u)\begin{pmatrix} \cos w & -\sin w & 0 \\ \sin w & \cos w & 0 \\ 0 & 0 & 1 \end{pmatrix}$$
$$= \big((R + r\cos u)\cos w, -(R + r\cos u)\sin w, r\sin u\big),$$

Answers to Exercises

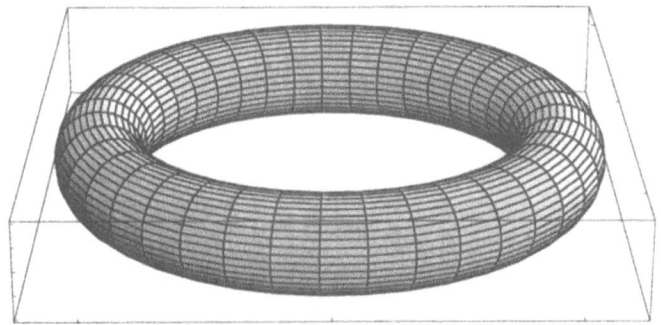

Figure Ans.29: A Torus as a Surface of Revolution.

```
R=10; r=2;    (* The Torus as a surface of revolution *)
ParametricPlot3D[
  {(R+r Cos[2Pi u])Cos[2Pi w],-(R+r Cos[2Pi u])Sin[2Pi w]
  ,r Sin[2Pi u]},{u,0,1},{w,0,1},
  ViewPoint->{-0.028, -4.034, 1.599}]
```

Code for Figure Ans.29.

where $0 \leq u, w \leq 2\pi$, or

$$\mathbf{P}(u,w) = \big((R + r\cos(2\pi u))\cos(2\pi w), -(R + r\cos(2\pi u))\sin(2\pi w), r\sin(2\pi u)\big),$$

where $0 \leq u, w \leq 1$ (Figure Ans.29).

5.36: We just need to multiply $\mathbf{C}(u)$ by the two translation matrices:

$$(u, 0, 0, 1) \begin{pmatrix} 1 & 0 & 0 & 0 \\ 0 & 1 & 0 & 0 \\ 0 & 0 & 1 & 0 \\ 0 & 0 & w & 1 \end{pmatrix} \begin{pmatrix} 1 & 0 & 0 & 0 \\ 0 & 1 & 0 & 0 \\ 0 & 0 & 1 & 0 \\ 0 & \sin w & 0 & 1 \end{pmatrix}.$$

The result is the surface $\mathbf{P}(u,w) = (u, \sin w, w)$.

5.37: This quarter circle starts at point $\mathbf{C}(0) = (1,0,0)$ on the x axis and ends at point $\mathbf{C}(1) = (0,1,0)$ on the y axis. A 360° rotation about the y axis is expressed by

$$\mathbf{T}(w) = \begin{pmatrix} \cos(2\pi w) & 0 & \sin(2\pi w) & 0 \\ 0 & 1 & 0 & 0 \\ -\sin(2\pi w) & 0 & \cos(2\pi w) & 0 \\ 0 & 0 & 0 & 1 \end{pmatrix}.$$

So the surface is

$$\frac{1}{1+u^2}\big((1-u^2)\cos(2\pi w), 2u, (1-u^2)\sin(2\pi w), 1\big).$$

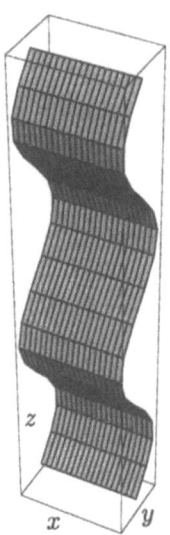

Figure Ans.30: A Sweep Surface.

```
<<:Graphics:ParametricPlot3D.m;
ParametricPlot3D[{3u,Sin[w],w}, {u,0,1},{w,0,4Pi},
Ticks->False, AspectRatio->Automatic]
```

Code for Figure Ans.30.

For $u = 0$, this reduces to $(\cos(2\pi w), 0, \sin(2\pi w))$, that's the unit circle in the xz plane. For $u = 1$, the same expression reduces to $1/2(0, 2, 0) = (0, 1, 0)$. This is the top of the half-sphere, a point that does not depend on w.

5.38: The equation of the line is

$$\mathbf{P}_1(1-u) + \mathbf{P}_2 u = (0,0,0)(1-u) + (R,0,H)u = (Ru, 0, Hu),$$

where $0 \leq u \leq 1$. Multiplying by the rotation matrix about the z axis yields

$$(Ru, 0, Hu) \begin{pmatrix} \cos w & -\sin w & 0 \\ \sin w & \cos w & 0 \\ 0 & 0 & 1 \end{pmatrix} = (Ru\cos w, -Ru\sin w, Hu),$$

where $0 \leq w \leq 2\pi$. (Compare with Equation (Ans.38).)

5.39: These curves are obtained with the pair $u(t) = t$, $w(t) = 1 + a - t$, where a is a constant. For each curve, the parameter t varies in the range $[a, 1]$ and the entire family is obtained when a varies in the range $[0, 1]$. A general curve in this family is, thus,

$$\begin{aligned}\mathbf{C}(t) &= (t + 1 + a - t - 2t(1 + a - t), 1 + a - t, 1 - t) \\ &= (2t^2 - 2t(1+a) + 1 + a, 1 + a - t, 1 - t).\end{aligned}$$

Answers to Exercises

A simple check verifies that for $a = 0$ (where t goes from 0 to 1), the curve really goes from $\mathbf{C}(0) = (1,1,1) = \mathbf{P}_{01}$ to $\mathbf{C}(1) = (1,0,0) = \mathbf{P}_{10}$. However, when $a = 1$ (t goes from 1 to 1), the curve reduces to the single point $\mathbf{C}(0) = (0,1,0) = \mathbf{P}_{11}$.

5.40: The curve is given by Equation (Ans.40) where the parameters u and w are replaced by the functions $u(t) = t$, $w(t) = 1 - t$. The result is $\mathbf{C}(t) = (3(1-t), 2t, 3t(1-t))$. This is a Bézier curve based on the three control points $\mathbf{P}_{03} = (3,0,0)$, $(0,1,1.5)$, and $\mathbf{P}_{20} = (0,2,0)$.

5.41: We can always write the explicit surface $z = f(x,y)$ as the implicit surface $f(x,y) - z = 0$. The normal is, therefore,

$$\left(\frac{\partial f(x_0, y_0)}{\partial x}, \frac{\partial f(x_0, y_0)}{\partial y}, -1\right).$$

5.42: The normal will point in the negative x direction.

5.43: The face is given by $\mathbf{P}(u,w) = (a(2u-1)(1-w), a(w-1), Hw)$. The two partial derivatives are

$$\frac{\partial \mathbf{P}}{\partial u} = (2a(1-w), 0, 0), \quad \frac{\partial \mathbf{P}}{\partial w} = (a(1-2u), a, H).$$

The normal is the cross-product (Equation (A.10), page 703) $2a(1-w)[0, -H, a]$.

To understand this result, recall that the face is the triangle defined by the three points $(-a, -a, 0)$, $\mathbf{P}_2 = (a, -a, 0)$, and $(0, 0, H)$. This explains why the x component of the normal is zero. Note that its magnitude depends on w, but its direction does not. The direction of the normal can be expressed by saying "for each H units traveled in the negative y direction, we should travel a units in the positive z direction."

5.44: The cone is defined by $\mathbf{P}(u,w) = (Ru\cos w, Ru\sin w, Hu)$. The two partial derivatives are

$$\frac{\partial \mathbf{P}}{\partial u} = (R\cos w, R\sin w, H), \quad \frac{\partial \mathbf{P}}{\partial w} = (-Ru\sin w, Ru\cos w, 0).$$

The normal is the cross-product (Equation (A.10)) $Ru(-H\cos w, -H\sin w, R)$. Note that its direction does not depend on u. When w varies, the normal rotates about the z axis, and it always has a positive component in the z direction.

5.45: The line from $(-a, 0, R)$ to $(a, 0, R)$ is given by $(a(2u-1), 0, R)$. The surface is thus given by the product of this line and the rotation matrix about the x axis:

$$\mathbf{P}(u,w) = (a(2u-1), 0, R) \begin{pmatrix} 1 & 0 & 0 \\ 0 & \cos w & -\sin w \\ 0 & \sin w & \cos w \end{pmatrix}$$

$$= (a(2u-1), R\sin w, R\cos w), \quad \text{(Ans.41)}$$

where $0 \leq u \leq 1$ and $0 \leq w \leq 2\pi$. The two partial derivatives are

$$\frac{\partial \mathbf{P}}{\partial u} = (2a, 0, 0), \qquad \frac{\partial \mathbf{P}}{\partial w} = (0, R\cos w, -R\sin w).$$

The normal is the cross-product $2aR(0, \sin w, \cos w)$. Note that it is perpendicular to the x axis. When w varies, the normal rotates about the x axis.

6.1: We use Figure Ans.31, where the vertical distance between the light source and the observer is $y_1 + y_2$ and the horizontal distance is L. The only variable is x, the horizontal distance between the light source and the point of incidence. It is easy to see that when x gets smaller, α gets smaller and β gets bigger. The problem is to find that value of x for which the travel time of the ray along the two paths D_1 and D_2 is minimal.

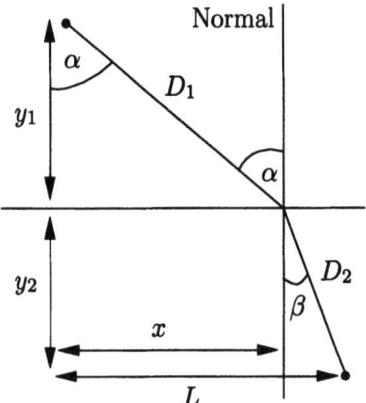

Figure Ans.31: Snell's Law.

Before we start, we prepare the auxiliary relations

$$\tan \alpha = \frac{x}{y_1} \Rightarrow \frac{d}{dx}\tan \alpha = \frac{1}{\cos^2 \alpha}\frac{d\alpha}{dx} = \frac{1}{y_1} \Rightarrow \frac{y_1}{\cos^2 \alpha}\frac{d\alpha}{dx} = 1,$$

$$\tan \beta = \frac{L-x}{y_2} \Rightarrow \frac{d}{dx}\tan \beta = \frac{1}{\cos^2 \beta}\frac{d\beta}{dx} = \frac{-1}{y_2} \Rightarrow \frac{y_2}{\cos^2 \beta}\frac{d\beta}{dx} = -1.$$

The travel times of the ray along the two paths are

$$t_1 = \frac{D_1}{C_1} = \frac{y_1}{C_1 \cos \alpha} \quad \text{and} \quad t_2 = \frac{D_2}{C_2} = \frac{y_2}{C_2 \cos \beta},$$

Answers to Exercises

so the derivative, with respect to x, of the total time is

$$\frac{d}{dx}(t_1+t_2) = \frac{-y_1(-\sin\alpha)}{C_1\cos^2\alpha}\frac{d\alpha}{dx} + \frac{-y_2(-\sin\beta)}{C_2\cos^2\beta}\frac{d\beta}{dx}$$

$$= \frac{\sin\alpha}{C_1}\frac{y_1}{\cos^2\alpha}\frac{d\alpha}{dx} + \frac{\sin\beta}{C_2}\frac{y_2}{\cos^2\beta}\frac{d\beta}{dx}$$

$$= \frac{\sin\alpha}{C_1} - \frac{\sin\beta}{C_2}.$$

To find the minimum, we equate $\frac{d}{dx}(t_1+t_2) = 0$, which yields $\sin\alpha/\sin\beta = C_1/C_2$.

6.2: The *Mathematica* code `Plot[Sin[a]/Sqrt[1.5^2-Sin[a]^2], {a,0,Pi/2}]` produces Figure Ans.32.

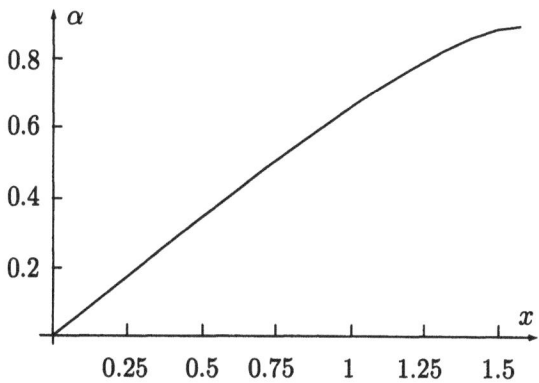

Figure Ans.32: Refraction Distance in Glass.

6.3: First, the polygon should be tested for planarity (Section 5.3.1). If it passes the test, the pseudo-code `Gouraud4` procedure of Figure Ans.33 can be used to scan it, to calculate the interpolated intensity I for every pixel P, and to display the results.

Another possibility is to divide the four-sided polygon into two triangles.

7.1: All the shades of gray.

7.2: A yellow surface absorbs blue and reflects green and red.

7.3: Yes! These are three colors that produce white when mixed. Examples are red, green, and blue; cyan, magenta, and yellow.

7.4: Because white isn't a pure color; it is a mixture of all colors.

7.5: Recall that the sum of a dyad is white. Since illuminant white is in the middle of the line connecting c and d, it is obtained by adding equal amounts of them ($0.5c + 0.5d$). This is why they are complementary.

```
procedure Gouraud4(P1,P2,P3,P4,I1,I2,I3,I4);
real I, Ia, Ib; point P, Pa, Pb;
for u:=0 to 1 step 0.1 do
Ia:=I2*(1-u)+I1*u; Ib:=I3*(1-u)+I4*u;
Pa:=P2*(1-u)+P1*u; Pb:=P3*(1-u)+P4*u;
  for w:=0 to 1-u step 0.001 do
  I:=Ia*(1-w)+Ib*w;
  P:=Pa*(1-w)+Pb*w;
  Pixel(P,I);
end;
```

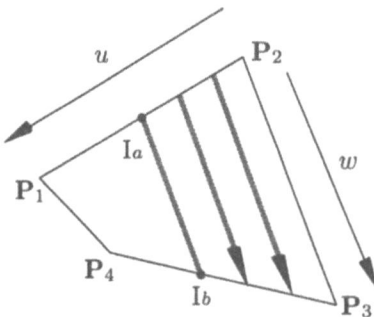

Figure Ans.33: Scan Procedure for a Four-Sided Polygon.

7.6: Saturation refers to the amount of white in a color. Point f corresponds to full saturation, whereas illuminant white corresponds to no saturation. The saturation of the color of point e is, therefore, the ratio of the distances fw/ew.

7.7: If we continue the line from w to g, it intercepts the pure spectral curve at the bottom, an area that does not correspond to any wavelength. We therefore continue the line in the opposite direction until it intercepts the pure spectral curve at h and we say that the dominant wavelength of point g is 497_c (where c stands for "complement").

8.1: Elementary trigonometry shows that $\sin\theta$ is negative in the range $3\pi/2 < \theta < 2\pi$.

8.2: Imagine two unit vectors $\mathbf{D}_1 = (x_1, y_1)$ and $\mathbf{D}_2 = (x_2, y_2)$ in two dimensions. Denote the angle between them by θ. Since they are unit vectors they satisfy $x_1^2 + y_1^2 = 1$, $x_2^2 + y_2^2 = 1$ and $\mathbf{D}_1 \bullet \mathbf{D}_2 = x_1 x_2 + y_1 y_2 = \cos\theta$. We have to prove that the weighted sum

$$\frac{\sin(\theta - t\theta)}{\sin\theta}\mathbf{D}_1 + \frac{\sin(t\theta)}{\sin\theta}\mathbf{D}_2, \quad 0 \le t \le 1,$$

is a unit vector for any t. The proof for three-dimensional vectors follows by analogy.

Answers to Exercises

Proof: We start with $\sin^2\theta + \cos^2\theta = 1$. From this, we get (using several trigonometric identities):

$\sin^2\theta = 1 + \cos\theta[-\cos\theta]$
$= 1 + \cos\theta[\cos(\theta - 2t\theta) - \cos\theta - \cos(\theta - 2t\theta)]$ $\quad$ < add & sub $\cos(\theta - 2t\theta)$ >
$= 1 + \cos\theta[2\sin(\theta - t\theta)\sin(t\theta) - \cos(\theta - 2t\theta)]$ $\quad$ < from $\sin\alpha\sin\beta = \ldots$ >
$= 1 - \cos\theta\cos(\theta - 2t\theta) + 2\sin(\theta - t\theta)\sin(t\theta)\cos\theta$
$= 1 - 0.5[\cos(2\theta - 2t\theta) + \cos(2t\theta)] + 2\sin(\theta - t\theta)\sin(t\theta)\cos\theta$
$\quad$ < previous line is from $\cos\alpha\cos\beta = \ldots$ >
$= 0.5[1 - \cos(2\theta - 2t\theta)] + 2\sin(\theta - t\theta)\sin(t\theta)\cos\theta + 0.5[1 - \cos(2t\theta)]$
$= \sin^2(\theta - t\theta) + 2\sin(\theta - t\theta)\sin(t\theta)\cos\theta + \sin^2(t\theta)$ $\quad$ < from $\sin^2\alpha = \ldots$ >
$= \sin^2(\theta - t\theta)[x_1^2 + y_1^2] + 2\sin(\theta - t\theta)\sin(t\theta)[x_1x_2 + y_1y_2] + \sin^2(t\theta)[x_2^2 + y_2^2]$
$= \sin^2(\theta - t\theta)x_1^2 + 2\sin(\theta - t\theta)\sin(t\theta)x_1x_2 + \sin^2(t\theta)x_2^2$
$\quad + \sin^2(\theta - t\theta)y_1^2 + 2\sin(\theta - t\theta)\sin(t\theta)y_1y_2 + \sin^2(t\theta)y_2^2$
$= \left[\sin^2(\theta - t\theta)x_1 + \sin^2(t\theta)x_2\right]^2 + \left[\sin^2(\theta - t\theta)y_1 + \sin^2(t\theta)y_2\right]^2,$

which implies

$$\left[\frac{\sin^2(\theta - t\theta)}{\sin\theta}x_1 + \frac{\sin^2(t\theta)}{\sin\theta}x_2\right]^2 + \left[\frac{\sin^2(\theta - t\theta)}{\sin\theta}y_1 + \frac{\sin^2(t\theta)}{\sin\theta}y_2\right]^2 = 1.$$

(End of Proof.)

8.3: This case means that the camera should point in opposite directions in two consecutive key frames. Thus, there are two equally preferred ways to swing the camera from the start direction to the end direction and the software cannot decide which of them to select.

8.4: It is important to understand that the new camera direction is interpolated spherically between $\mathbf{D}_1$ and $\mathbf{D}_2$, but the position on the path is interpolated *linearly* between $\mathbf{P}_1(0)$ and $\mathbf{P}_1(1)$. This is why we got the correct direction but a slightly wrong position for $t = 0.5$. This is also why it is important to have short-path segments. In a long segment, any errors created by the nonuniform velocity of the curve may get magnified and may place the camera a large distance away from the correct positions.

It should also be mentioned that spherical interpolation is a natural choice for rotations because rotations move points in circular arcs. At the same time, spherical interpolation is not a good choice for interpolating positions on a curve because large parts of a curve may be straight or close to straight.

8.5: Our example is simple. The entire camera path is contained in the xy plane and the camera is rotated about the z axis only. This is why it is easy to figure out the directions of the camera at various key points. In general, the axis of rotation of the camera is not fixed, which is why rotating the camera is a complex task. Imagine the (still relatively simple) case where at $\mathbf{P}_1$ the camera looks in the positive y direction and at $\mathbf{P}_2$ it should look in direction $(1,1,1)$. Now, it is much harder to tell in what direction it should look at any point in between. This is why direction vectors are necessary.

8.6: The coordinates of point $\mathbf{P}_2(0.5)$ are

$$\begin{aligned}\mathbf{P}_2(0.5) &= 0.5^3(-2,0,0) + 3 \cdot 0.5 \cdot 0.5^2(-21/12, 2/3, 0) \\ &\quad + 3 \cdot 0.5^2 \cdot 0.5(2/6, 2, 0) + 0.5^3(1.5, 2, 0) \\ &= 0.5^3[(-2,0,0) + (-21/4, 2, 0) + (1, 6, 0) + (3/2, 2, 0)] \\ &= (-19/32, 5/4, 0).\end{aligned}$$

The angle θ_{23} between $\mathbf{D}_2$ and $\mathbf{D}_3$ is $126.87°$. The spherical interpolation of $\mathbf{D}_2$, $\mathbf{D}_3$ at $t = 0.5$ is thus

$$\begin{aligned}\mathbf{D}_{2+0.5} &= \frac{\sin((1-.5)126.87)}{\sin 126.87°}\mathbf{D}_2 + \frac{\sin(0.5 \cdot 126.87)}{\sin 126.87°}\mathbf{D}_3 \\ &= \frac{\sin 63.44°}{\sin 126.87°}(\mathbf{D}_2 + \mathbf{D}_3) \\ &= \frac{0.8944}{0.8}[(1,0,0) + (-3/5, -4/5, 0)] \\ &= (0.4472, -0.8944, 0).\end{aligned}$$

8.7: The angle between $\mathbf{D}_3$ and $\mathbf{D}_4$ is $\theta_{34} = 53.13°$. The spherical interpolation of $\mathbf{D}_3$, $\mathbf{D}_4$ at $t = 0.5$ is thus

$$\begin{aligned}\mathbf{D}_{3+0.5} &= \frac{\sin((1-0.5)53.13)}{\sin 53.13°}\mathbf{D}_3 + \frac{\sin(0.5 \cdot 53.13)}{\sin 53.13°}\mathbf{D}_4 \\ &= \frac{\sin 26.57°}{\sin 53.13°}(\mathbf{D}_3 + \mathbf{D}_4) \\ &= \frac{0.4472}{0.8}[(-3/5, -4/5, 0) + (-1, 0, 0)] \\ &= 0.559(-8/5, -4/5, 0).\end{aligned}$$

8.8: We first calculate the two new interior points

$$\mathbf{X}_1 = \mathbf{P}_1 + \frac{1}{6}(\mathbf{P}_2 - \mathbf{P}_0) = (1.667, 1.667, -0.0417),$$

$$\mathbf{Y}_1 = \mathbf{P}_2 - \frac{1}{6}(\mathbf{P}_3 - \mathbf{P}_1) = (1.333, 1.333, -0.0417).$$

The curve is the usual Bézier weighted sum

$$\begin{aligned}\mathbf{P}_1(t) &= (1-t)^3\mathbf{P}_1 + 3t(1-t)^2\mathbf{X}_1 + 3t^2(1-t)\mathbf{Y}_1 + t^3\mathbf{P}_2 \\ &= (1-t)^3(2,2,0) + 3t(1-t)^2(1.667,1.667,-0.0417) \\ &\quad + 3t^2(1-t)(1.333,1.333,-0.0417) + t^3(1,1,0).\end{aligned}$$

8.9: Figure Ans.34 shows that the camera transformations are (1) translate to the origin (k units in the positive z direction), (2) rotate 58.28° counterclockwise about the positive y axis, and (3) translate one unit in the positive x and 2.5 units in the positive z directions. The reverse transformations, applied in reverse order on the scene are, (4) translate one unit in the negative x and 2.5 units in the negative z directions, (5) rotate 58.28° about the origin, clockwise about the positive y axis, and (6) translate k units in the negative z direction.

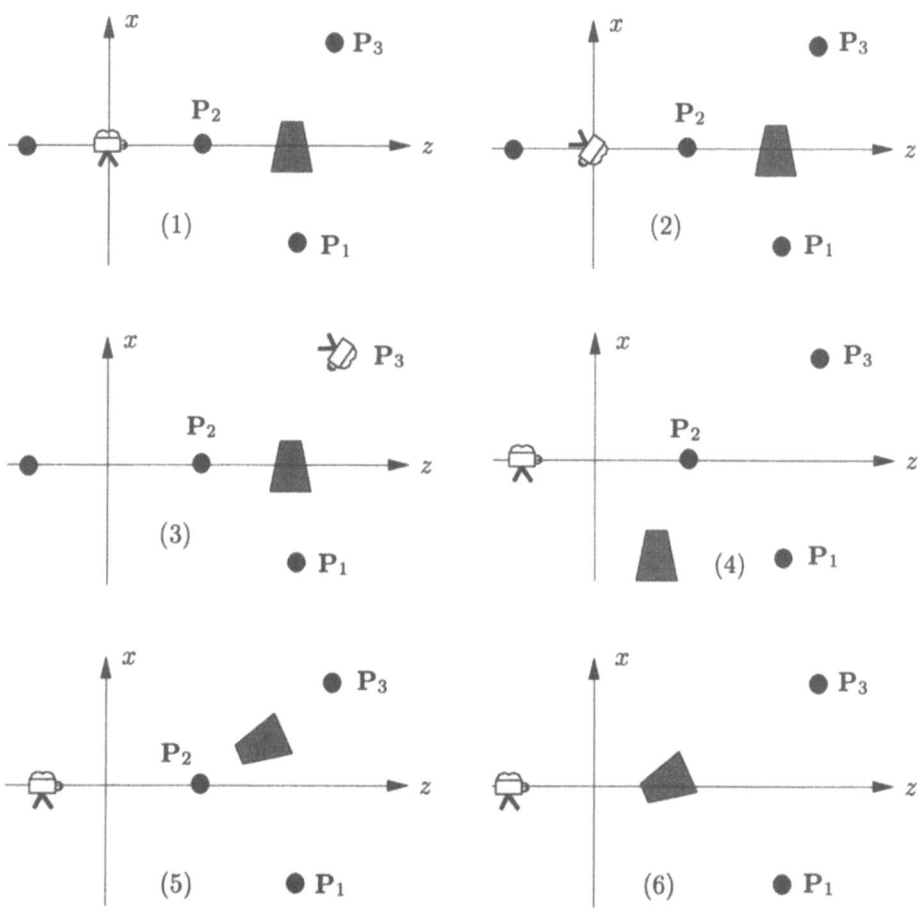

Figure Ans.34: Camera and Scene Transformations for $\mathbf{P}_3$.

8.10: The first two derivatives of
$$\frac{2a}{\pi}\sin\left(\frac{\pi}{2}\cdot\frac{t-a}{a}\right)$$
are
$$\cos\left(\frac{\pi(t-a)}{2a}\right) \quad \text{and} \quad -\frac{\pi}{2a}\sin\left(\frac{\pi(t-a)}{2a}\right).$$
They correspond to the velocity and acceleration of the curve, respectively. The accelerations at $t=0$ and $t=a$ are thus
$$-\frac{\pi}{2a}\sin(-\pi/2) = \pi/2a \quad \text{and} \quad -\frac{\pi}{2a}\sin(0) = 0, \text{ respectively.}$$

The first two derivatives of
$$\frac{2(1-b)}{\pi}\sin\left(\frac{\pi}{2}\cdot\frac{t-b}{1-b}\right) + \frac{2a}{\pi} + b - a$$
are
$$\cos\left(\frac{\pi(t-b)}{2(1-b)}\right) \quad \text{and} \quad -\frac{\pi\sin\left(\frac{\pi(t-b)}{2(1-b)}\right)}{2(1-b)}.$$
The second derivative varies from 0 (for $t=b$) to $-\pi/(2(1-b))$ (for $t=1$).

9.1: It cannot. In general, random data, be it text, images, or anything else, cannot be compressed.

9.2: The simplest definition is the ratio (size of the compressed file)/(size of the original file). A value of 0.6 means that the file occupies 60% of its original size after compression. Values > 1 mean a compressed file larger than the original one. Another, more common, definition is the simple expression (1 − compressed size/original size) × 100. A value of 60 means that the compressed file occupies 40% of its original size (or that the compression has resulted in a savings of 60%).

9.3: 6,8,0,1,3,1,4,1,3,1,4,1,3,1,4,1,3,1,2,2,2,2,6,1,1. If each number occupies a byte on the compressed file, then the size of the file is 25 bytes, compared to a bitmap size of only 6×8 bits = 6 bytes. The method does not work for small images.

9.4: Run-length encoding of images is based on the idea that adjacent pixels tend to be identical. The last pixel of a row, however, has no reason to be identical to the first pixel of the following row.

9.5: Each of the first four rows yields the eight runs 1,1,1,2,1,1,1,eol. Rows 6 and 8 yield the four runs 0,7,1,eol each. Rows 5 and 7 yield the two runs 8,eol each. The total number of runs (including the eol's) is, thus, 44.

When compressing by columns, columns 1, 3, and 6 yield the five runs 5,1,1,1,eol each. Columns 2, 4, 5, and 7 yield the six runs 0,5,1,1,1,eol each. Column 8 gives 4,4,eol, so the total number of runs is 42. This image is thus "balanced" with respect to rows and columns.

Answers to Exercises

9.6: A typical fax machine scans lines that are about 8.2 in. wide ($\approx$ 208 mm). Therefore, a blank scan line produces 1664 consecutive white pels.

9.7: Such codes are needed for cases such as Example 4, where the run length is 64,128 or any length for which a make-up code has been assigned.

9.8: There may be fax machines (now or in the future) built for wider paper, so the group 3 code was designed to accommodate them.

9.9: The code of a run length of one white pel is 000111 and that of one black pel is 010. Two consecutive pels of different colors are, thus, coded into nine bits. Since the uncoded data requires just two bits (01 or 10), the compression factor is 9/2=4.5 (the compressed file is 4.5 times larger than the uncompressed one).

9.10: This is shown by multiplying the largest four n-bit number, $\underbrace{11\ldots1}_{n}$ by 4, which is easily done by shifting it 2 positions to the left. The result is the $n+2$-bit number $\underbrace{11\ldots1}_{n}00$.

9.11: The two previously seen neighbors of P=8 are A=1 and B=11. P is, thus, in the central region, where all codes start with a zero and L=1, H=11. The computations are straightforward

$$k = \lfloor\log_2(11-1+1)\rfloor = 3, \qquad a = 2^{3+1} - 11 = 5, \qquad b = 2(11 - 2^3) = 6.$$

Table Ans.35 lists the five 3-bit codes and six 4-bit codes for the central region. The code for 8 is, thus, 0|111.

The two previously seen neighbors of P=7 are A=2 and B=5. Thus, P is in the right outer region, where all codes start with 11 and L=2, H=7. We are looking for the code of $7 - 5 = 2$. Choosing $m = 1$ yields, from Table 9.17, the code 11|01.

The two previously seen neighbors of P=0 are A=3 and B=5. Thus, P is in the left outer region, where all codes start with 10 and L=3, H=5. We are looking for the code of $3 - 0 = 3$. Choosing $m = 1$ yields, from Table 9.17, the code 10|100.

9.12: It is easy to see that the unary code satisfies the prefix property, so it definitely can be used as a variable-size code. Since its length L satisfies $L = n$, we get $2^{-L} = 2^{-n}$, so it makes sense to use it in cases were the input data consists of integers n with probabilities $P(n) \approx 2^{-n}$. If the data lend themselves to the use of the unary code, the entire Huffman algorithm can be skipped and the codes of all the symbols can easily and quickly be constructed before compression or decompression starts.

9.13: The triplet $(n, 1, n)$ defines the standard n-bit binary codes, as can be verified by direct construction. The number of such codes is easily seen to be

$$\frac{2^{n+1} - 2^n}{2^1 - 1} = 2^n.$$

Pixel P	Region code	Pixel code
1	0	0000
2	0	0010
3	0	0100
4	0	011
5	0	100
6	0	101
7	0	110
8	0	111
9	0	0001
10	0	0011
11	0	0101

Table Ans.35: The Codes for a Central Region.

The triplet $(0, 0, \infty)$ defines the codes 0, 10, 110, 1110,... which are the unary codes but are assigned to the integers 0, 1, 2,... instead of to 1, 2, 3,...

9.14: The number is $(2^{30} - 2^1)/(2^1 - 1) \approx$ one billion.

9.15: This is straightforward. Table Ans.36 shows the code. There are only three different codewords, since "start" and "stop" are so close, but there are many codes, since "start" is large.

n	$a = 10 + n \cdot 2$	nth Codeword	Number of codewords	Range of integers
0	10	$0\underbrace{x...x}_{10}$	$2^{10} = 1K$	0–1023
1	12	$10\underbrace{xx...x}_{12}$	$2^{12} = 4K$	1024–5119
2	14	$11\underbrace{xx...xx}_{14}$	$2^{14} = 16K$	5120–21503
		Total	21504	

Table Ans.36: The General Unary Code (10,2,14).

9.16: Each part of C_4 is the standard binary code of some integer, so it starts with a 1. A part that starts with a 0 signals to the decoder that this is the last bit of the code.

9.17: The first term can be written

$$4 = \frac{n^2}{2^{-2}n^2} = \frac{n^2}{2^{2\log n - 2}}.$$

Terms in the sequence thus contain powers of 2 that go from 0 to $2\log_2 n - 2$, showing that there are $2\log_2 n - 1$ terms.

Answers to Exercises 827

9.18: Because the decoder has to resolve ties in the same way as the encoder.

9.19: Because this will result in a weighted sum whose value is in the same range as the values of the pixels. If pixel values are, for example, in the range [0, 15] and the weights add up to 2, a prediction may result in values of up to 30.

9.20: Each of the three weights 0.0039, −0.0351, and 0.3164 is used twice. The sum of the weights is, thus, 0.5704 and the result of dividing each weight by this sum is 0.0068, −0.0615, and 0.5547. It is easy to verify that the sum of the renormalized weights 2(0.0068 − 0.0615 + 0.5547) equals 1.

9.21: Using appropriate mathematical software, it is easy to obtain this integral separately for negative and non-negative values of x.

$$\int L(V, x)\, dx = \begin{cases} \dfrac{-1}{V \exp\left(\sqrt{\frac{2}{V}} x\right)}, & x \geq 0, \\ \dfrac{1}{\sqrt{2V}} \exp\left(\sqrt{\frac{2}{V}} x\right), & x < 0. \end{cases}$$

9.22: One such example is an archive of static images. NASA has a large archive of images taken by various satellites. They should be kept highly compressed, but they never change, so each image has to be compressed only once. A slow encoder is therefore acceptable but a fast decoder is certainly handy. Another example is an art collection. Many museums have digitized their collections of paintings, and those are also static.

9.23: When the decoder inputs the 5, it knows that the difference between p (the pixel being decoded) and the reference pixel starts at position 6 (counting from left). Since bit 6 of the reference pixel is 0, that of p must be 1.

9.24: Yes, but compression would suffer. One way to apply this method is to separate each byte into two 4-bit pixels and encode each pixel separately. This approach is bad, since the prefix and suffix of a 4-bit pixel may often require more than 4 bits. Another approach is to ignore the fact that a byte contains two pixels and use the method as originally described. This may still compress the image, but it is not very efficient, as the following example illustrates.

Example: The two bytes 1100|1101 and 1110|1111 represent four pixels, each differing from its two immediate neighbors by its least significant bit. The four pixels thus have similar colors (or grayscales). Comparing consecutive pixels results in prefixes of 3 or 2, but comparing the two bytes produces the prefix 2.

10.1: The two numbers are placed on the stack and the sub(tract) executes 3.2 − 11.6, removes the top two stack elements, and places the result, −8.4, on the stack.

$$A[x,y] = \qquad 0 \qquad\qquad\qquad 1 \qquad\qquad\qquad 15$$

```
          10001000...     10001000...     11111111...
          00000000...     00000000...     11111111...
          00000000...     00100010...     11111111...
          00000000...     00000000...     11111111...
```

Figure Ans.37: Ordered Dither: Three Uniform Areas.

10.2: Direct calculations using matrix D_{44} produce the areas shown in Figure Ans.37. The three areas have black pixel percentages of 1/16, 2/16, and 16/16, respectively.

10.3: A direct application of Equation (10.1) yields

$$D_{88} = \begin{array}{|cccccccc|} \hline 0 & 32 & 8 & 40 & 2 & 34 & 10 & 42 \\ 48 & 16 & 56 & 24 & 50 & 18 & 58 & 26 \\ 12 & 44 & 4 & 36 & 14 & 46 & 6 & 38 \\ 60 & 28 & 52 & 20 & 62 & 30 & 54 & 22 \\ 3 & 35 & 11 & 43 & 1 & 33 & 9 & 41 \\ 51 & 19 & 59 & 27 & 49 & 17 & 57 & 25 \\ 15 & 47 & 7 & 39 & 13 & 45 & 5 & 37 \\ 63 & 31 & 55 & 23 & 61 & 29 & 53 & 21 \\ \hline \end{array}.$$

10.4: A checkerboard pattern. This can be seen by manually simulating the algorithm of Figure 10.8b for a few pixels.

10.5: We assume that the test is

if $p \geq 0.5$, then $p := 1$ else $p := 0$; add the error $0.5 - p$ to the next pixel q.

The first pixel is thus set to 1 and the error of $0.5 - 1 = -0.5$ is added to the second pixel, changing it from 0.5 to 0. The second pixel is set to 0 and the error, which is $0 - 0 = 0$, is added to the third pixel, leaving it at 0.5. The third pixel is thus set to 1 and the error of $0.5 - 1 = -0.5$ is added to the fourth pixel, changing it from 0.5 to 0. The results are

$$\boxed{.5|.5|.5|.5|.5} \to \boxed{1|0|.5|.5|.5} \to \boxed{1|0|1|0|.5} \to \boxed{1|0|1|0|1}$$

10.6: Direct examination shows that the barons are 62 and 63 and the near-barons are 60 and 61.

10.7: A checkerboard pattern, similar to the one produced by diffusion dither. This can be seen by manually executing the algorithm of Figure 10.10 for a few pixels.

10.8: Classes 14, 15, and 10 are barons. Classes 12 and 13 are near-barons. The class numbers in positions (i,j) and $(i, j+2)$ add up to 15.

Answers to Exercises

10.9: The slope vector of the segment is $(3-1, 3-1) = (2,2)$, so the slope of the perpendicular is $(-2, 2)$. The midpoint of the segment is $((1+3)/2, (1+3)/2) = (2,2)$, so the perpendicular bisector is $(2-2t, 2+2t)$.

A.1: It is, of course, the xy plane, same as in Cartesian coordinates.

A.2: It is a cone with apex at the origin.

A.3: This is easily proved by showing that both dot products $(\mathbf{P} \times \mathbf{Q}) \bullet \mathbf{P}$ and $(\mathbf{P} \times \mathbf{Q}) \bullet \mathbf{Q}$ equal zero:

$$(\mathbf{P} \times \mathbf{Q}) \bullet \mathbf{P} = P_1(P_2Q_3 - P_3Q_2) + P_2(-P_1Q_3 + P_3Q_1) + P_3(P_1Q_2 - P_2Q_1) = 0,$$

and similarly for $(\mathbf{P} \times \mathbf{Q}) \bullet \mathbf{Q}$.

A.4: In the special case where $\mathbf{i} = (1,0,0)$ and $\mathbf{j} = (0,1,0)$, it is easy to verify that the product $\mathbf{i} \times \mathbf{j}$ equals $(0,0,1) = \mathbf{k}$. The triplet $(\mathbf{i}, \mathbf{j}, \mathbf{i} \times \mathbf{j} = \mathbf{k})$, thus, has the handedness of the coordinate system (it is either right-handed or left-handed, depending on the coordinate system). In a right-handed coordinate system, the right-hand rule makes it easy to predict the direction of $\mathbf{P} \times \mathbf{Q}$. The rule is as follows: If your thumb points in the direction of $\mathbf{P}$ and your second finger, in the direction of $\mathbf{Q}$, then your middle finger will point in the direction of $\mathbf{P} \times \mathbf{Q}$. In a left-handed coordinate system, a similar left-hand rule applies.

A.5: They either have the same direction, or they point in opposite directions.

A.6: We are looking for a vector $\mathbf{P}(t)$ that's linear in t and that satisfies $\mathbf{P}(0) = \mathbf{P}_1$ and $\mathbf{P}(1) = \mathbf{P}_2$. It is easy to guess that

$$\mathbf{P}(t) = (1-t)\mathbf{P}_1 + t\mathbf{P}_2 = t(\mathbf{P}_2 - \mathbf{P}_1) + \mathbf{P}_1$$

satisfies both conditions. As a reminder, this is Equation (4.4). It is important and is used often in graphics.

A.7: This is not especially hard;

$$\mathbf{c} = \frac{2 \cdot 1 + 1 \cdot 0 + 3 \cdot (-1)}{1^2 + 0^2 + (-1)^2}(1, 0, -1) = (-1/2, 0, 1/2),$$
$$\mathbf{d} = \mathbf{a} - \mathbf{c} = (2.5, 1, 2.5).$$

A.8: Yes. The diagonal elements of matrix $\begin{pmatrix} 1 & 1 \\ 1 & 1 \end{pmatrix}$ are not dominant, yet this matrix satisfies the weaker form of the conditions of Equation (A.18).

A.9: The transpose of matrix $\mathbf{A}$ of Equation (A.19).

A.10: A little thinking shows that the radius at a point on a spheroid depends only on the latitude and not on the longitude. The problem thus becomes that of finding the radius p of an ellipse as a function of the angle θ. This is a simple application of polar coordinates. The canonical equation of an ellipse $x^2/R^2 + y^2/r^2 = 1$ is easy to translate to polar coordinates, by using the standard relations $x = p\cos\theta$, $y = p\sin\theta$. The result is

$$\frac{p^2 \cos^2\theta}{R^2} + \frac{p^2 \sin^2\theta}{r^2} = 1,$$

which can be written

$$p^2 \left[\frac{\cos^2\theta}{R^2} + \frac{1}{r^2} - \frac{\cos^2\theta}{r^2}\right] = 1$$

or

$$p^2 \left[\frac{r^2 \cos^2\theta + R^2 - R^2 \cos^2\theta}{R^2 r^2}\right] = 1,$$

from which we get

$$p = \frac{Rr}{\sqrt{R^2 + (r^2 - R^2)\cos^2\theta}}.$$

A.11: We start with $i = \cos(\pi/2) + i\sin(\pi/2)$, from which we get

$$\sqrt{i} = \left(\cos\frac{\pi}{2} + i\sin\frac{\pi}{2}\right)^{1/2}.$$

By DeMoivre's theorem, this equals

$$\cos\frac{\pi}{4} + i\sin\frac{\pi}{4} = \frac{1}{\sqrt{2}} + i\frac{1}{\sqrt{2}} = \frac{1+i}{\sqrt{2}}.$$

A simple check gives

$$\left(\frac{1+i}{\sqrt{2}}\right)^2 = \frac{1 + 2i + i^2}{2} = i.$$

(DeMoivre's theorem states that $\sin(nx) + i\cos(nx)$ is one of the values of $(\sin x + i\cos x)^n$.)

A.12: We start with the elegant formula

$$e^{it} = \cos t + i\sin t.$$

Substituting $t = \pi/2$ yields

$$e^{i\pi/2} = i\sin(\pi/2) = i.$$

Both sides are now raised to the ith power, yielding

$$i^i = (e^{i\pi/2})^i = e^{i^2 \pi/2} = e^{-\pi/2}.$$

Answers to Exercises 831

Since $a = \exp(\ln a)$ for any a, we also get

$$e^{\ln i} = i = e^{i\pi/2},$$

from which it is clear that $\ln i = i\pi/2$.

By the way, the exponential function e^z is defined for any complex number $z = x + iy$ by

$$e^z = 1 + \frac{z}{1!} + \frac{z^2}{2!} + \frac{z^3}{3!} + \cdots.$$

A.13: The inverse of a reflection, surprisingly enough, is itself.

> Keep the faculty of effort alive in you
> by a little gratuitous exercise every day
>
> William James, 1890 *The Principles of Psychology*

Index

180° fisheye projection, 153–155
180° rotation, 60, 81–82
2/3 rule, 796
360° panoramic projection, 153, 155–165
80x87 coprocessors, 77
90° rotation, 65

Abel and Associates, x
Abel, Niels Henrik (1802–1829), 720
Abraham, Frank Murray, 665
absolute value
  of a complex number, 716
  of a vector, 702
absorption (of light), 539
ACM, x
additive colors, 563–564
Adobe Illustrator (program), 305
aerial perspective (in art), 114
affine transformations, 61, 75, 102, 176, 394, 685
  definition of, 179
Agate, James, xi
Alberti, Leon Battista (Renaissance painter), 114
Albinak, Marvin J., 605
Alger, Horatio, 57
Allen, Woody, 346, 588
alphabet, Greek, 715
ambient reflection (of light), 547
Amiga personal computer, xi
Amis, Kingsley, (Colophon)
Amis, Martin, 640
amplitude (of a function), 694

Anderson, Karen, 654
animation (computer), 1, 456, 575–608
antialiasing, 50–56
  and double-step DDA, 35
  Pitteway-Watkinson method, 53
approximating curve, 181, 266
Arabic (unicode), 625
arc
  and Bézier curve, 319, 794
  circular
    and animation, 585
    and circular Bézier Curve, 315
    and fair splines, 241
    and Hermite segment, 241
    and surfaces of revolution, 521
    PC approximation, 197
    with B-splines, 341, 343
  elliptic, 319
  great (on a sphere), 596
  on a great circle on a sphere, 712
  in panoramic projection, 157
arc length
  and chord length, 190
  and constant speed of curves, 580
  calculation of, 706
  in curvature, 407
  curvature and torsion, 408, 409
  in ellipse, 714
  in smooth animation, 579, 582
Archimedean spiral, 700
area of curve, 706
area of polygon, 706–707
arg (of a complex number), 716

ARIES, 671
arithmetic coding, 612
Armenian (unicode), 625
Asimov, Isaac, 714
aspect ratio
  definition of, 3
  of HDTV, 5
  of television, 3, 577
Atari, xi
attractor (of IFS), 686
axonometric projections, 105–109

B-spline, 174, 321, 328–389
  2/3 rule, 796
  algebraic definition of, 358
  and circles, 341–345, 384
  cubic, 334–351
    related to Bézier curves, 346–348
  geometric definition of, 357
  nonuniform, 368–376
    matrix form, 376–381
  open, 364–368
  order (definition of), 337
  quadratic, 329–334
    related to Bézier curve, 326, 331, 795
  rational, 321, 383–389
  subdividing, 381–383
  tension in, 329, 345–346
  uniform, 329–363
    collinear points, 337–339
B-spline surfaces
  and Bézier, 504–505
  and Hermite, 503–504
  bicubic, 500–506
    subdivision, 511–516
  biquadratic, 499–500
    subdivision, 508–511
  interpolating, 505–506
Baeyer, H. C. von, 550
Balzac, Honoré de, 186
Banchoff, Thomas, 153
Bartels, Richard H., 258
barycentric coordinates, 177, 178
  circular, 315
  in a triangle, 177
barycentric sum, 177, 279, 283, 315, 393–394
barycentric weights, 177, 190, 209, 393–394
Batman (quoted), 456

Baudrillard, Jean, 144
Baum, L. Frank, 573
Beebe, Nelson H. F., xiii
Bell, Eric Temple, 49
Bengali (unicode), 625
BeOS (unicode), 625
Bernshteĭn, Sergeĭ Natanovich, 270
Bernstein polynomials, 174, 270, 368, 387, 435, 446, 601, 608
  in *Mathematica*, 270
Bessel's algorithm, 256–258
Bessel, Friedrich Wilhelm, 257
best fit, 13
best-fit DDA, 38–39
Beta splines, 174, 389–393
Bézier blending, 283, 601–602
Bézier curve, 174, 266–321, 336
  alternative representation, 273
  and arc, 319, 794
  and Catmull-Rom, 304
  and circle, 317–321, 384, 481, 790, 793
  and sine wave, 319, 321
  as a linear interpolation, 283–286
  as a spherical interpolation, 594
  as special case of B-spline, 365–368
  Bernstein form of, 269–283
  chord of, 296, 297, 300
  circular, 189, 314–317
  control polygon, 296, 300
  converting, 301–302
  cubic, 23, 272
  degree 4, 272
  degree elevation, 291–293
  disadvantages of, 328
  fast calculation of, 275–279, 700
  in PostScript, 667
  in *Mathematica*, 270
  interpolating, 304–314
  length of, 296–297
  linear, 273
  parameter substitution, 272, 784
  quadratic, 272
  rational, 317, 321
  related to B-spline, 321, 326, 331, 346–348, 795
  reparametrizing, 293–296
  representations of, 273–275
  speed of, 297–300
  subdividing, 290–291
  tension in, 302–304

Index

Bézier methods (history of), 267
Bézier, Pierre Etienne, 267, 270
Bézier surfaces, 471–495
  and a sphere, 481
  and B-spline, 504–505
  converting, 488–492
  in *Mathematica*, 475
  rectangular, 471–481
  reparametrizing, 492–495, 508
  triangular, 179, 452, 483–487
bias (in curves), 174, 259–261, 389, 390
bicubic
  surfaces, 459–468
bicubic Coons surface, 448
bilinear surfaces, 417, 434–438, 534
  triangular, 438
Binet, Alfred, 438
binomial coefficient, 269, 270
binomial theorem, 267–269, 275, 279
binormal vector, 404
biquadratic surfaces, 457–459
bisector (perpendicular), 425, 431, 434, 681, 684, 806, 829
BitBlt, 13–16
bitmap, 7
  and cache, 50
  and drawing, 23–26
  as an array, 7
  bitplanes, 7
  for color, 7
  erasing from, 14
  rotating, 20–23
  scaling, 16–18, 28
bitplanes (in a bitmap), 7
blending
  Bernstein polynomials, 270
  Bézier, 283, 601–602
  Catmull-Rom surface, 469
  in computer animation, 600–605
  Coons surface, 444
  cubic, 283, 601
  Hermite, 207–225, 602–603
  Hermite derivatives, 210–212
  images, 690
  linear, 601
  lofted surface, 439
  parabolic, 251–256
  parametric, 180, 456, 600–605
  in a PC, 194, 196
  quadratic, 283, 601

Blinn, James F., x, 42, 711
blurring an image, 689
Bohr, Niels, 648
boundary curves, 417, 434, 435, 437–439, 441–443, 445, 449, 455, 463, 498, 804, 811
  degenerate, 453, 488, 810
  three, 452, 484
boundary fill, 668
bounding circle, 433
bounding rectangle, 39
branching rules, 684
Bresenham-Michener circle method, 44–46, 50, 71, 264
Bridges, Jeff, 123
Brownian motion (and fractals), 681
Brunelleschi, Filippo (and perspective), 114
Bryngdahl, O., 691
buddy system, 13
Bui-Tuong, Phong, x, 548
bump mapping, 554–556

cabinet projection, 110
cameras, panoramic, 163–165
cardinal splines, 174, 248–251, 302, 345
Carpenter, Loren, x
Cartesian coordinate system, 700
Cartesian product, 456, 459, 474, 709
Cassai, Tommaso, *see* Masaccio
Casteljau, Paul de Faget de, 267, 270, 283, 378, 483, 594
Catmull, Edwin, x, 531
Catmull-Clark surfaces, 531–533
Catmull-Rom
  curves, 174, 250–258
  surfaces, 468–471
cavalier projection, 110
cell encoding (image compression), 622–624
center of curvature, 402
centers of triangle, 179, 425–426
centroid, 178, 425, 762
Cézanne, Paul (1839–1906), 528
CGI standard, 663
CGM standard, 663
Chaikin, George Merrill, 324, 328
Chaikin's algorithm, 322–328, 333, 348, 508, 510
chord of the Bézier curve, 296, 297, 300

chroma, *see* saturation
CIE color diagram, 571–573
circle, 46–49, 63
   and B-spline, 341–345, 384
   and Bézier curve, 317–321, 384, 481, 790, 793
   and Hermite curve, 212–214
   and surfaces of revolution, 521
   and torus, 183
   as a conic section, 803
   bounding n points, 433
   Bresenham-Michener method, 44–46, 50, 71, 264
   forward differences, 699
   in polar coordinates, 42–44
   parametric representation of, 46, 212–214, 317–321, 341–345, 762
   representations of, 182–183
   scaled to ellipse, 63, 740
   scan converting, 42–49
   through three points, 431
circle inversion, 153, 170–171
circular arc
   and animation, 585
   and circular Bézier Curve, 315
   and fair splines, 241
   and Hermite segment, 241
   and surfaces of revolution, 521
   with B-splines, 341, 343
   PC approximation of, 197
circular barycentric coordinates, 315
circular Bernstein polynomials, 315
circular Bézier curve, 189, 314–317
circumcenter, 425
   trilinears of, 806
circumcircle, 425, 427, 431–432
circumscribing a triangle, 431
Clark, Jim, 531
clipping, 89–91
   Cohen-Sutherland method, 90–91
   in three-dimensions, 142
CMYK color model, 558, 563, 564, 573, 677
codemap (in drawing), 23–26
codes
   Golomb, 635, 640, 642, 646
   prefix, 638–642
   start-step-stop, 637
   subexponential, 640, 646
   unary, 636–642

   variable-size, 634–642
Cohen-Sutherland clipping method, 90–91
collinear points (in B-spline), 337–339
color, 557–573
   adding, 563
   additive, 563–564
   as wavelength, 557
   complementary, 567
   cool, 568
   gamut, 561, 573
   model, 558
   primary, 561
   pure, 559
   secondary, 561
   spectral density, 567–570
   subtracting, 563
   subtractive, 563–564
   transparent, 15
   warm, 568
color image compression, 647, 656, 660
color images (and grayscale compression), 614
color lookup table, 8, 549, 561, 675
   and bitmap scaling, 17
   and color quantization, 551
   and xor drawing, 15
color printing, 564
   and dithering, 677
color shading, 548
complementary colors, 567
complementary geometrical constructs, 434
complex compression methods, 633
complex numbers, 716–717
compressing images, 624
   cell encoding, 622–624
   fax, 615–622
      group 3, 616–619
   FELICS, 634–636
   lossy methods, 615
   MLP, 646–654
   quadtrees, 624–627
      prefix compression, 628–630
   run-length encoding, 612–615
computer aided geometric design (CAGD), 267
computer animation, 1, 121, 174, 180, 195, 258–261, 308, 456, 575–608
   and bias, 260

Index

history of, 577
in-betweening, 576, 579, 606
computer arithmetic (limited accuracy), 420
computer art, 173
computer graphics (history of), ix–xi, 577
computer graphics (neglected topics in), viii
computer-aided manufacturing, 173
cone
  as a lofted surface, 442
  as a quadric surface, 524, 525
  as a sweep surface, 528
  normal vector, 536
  texturing, 553
cones (in the retina), 568
conformal projection, 170
conic sections, 174, 212, 321, 397–401
  and NURBS, 387–389
  approximated, 212–214
constrained average dithering, 671, 674
continuity (in curves), 174, 258–261
continuity parameter, 259
control points, 174, 181, 262, 266, 271, 328, 415
  and convex hull, 282
  and curve speed, 298
  and degree elevation, 291
  and scaffolding, 286
  auxiliary, 305
  collinear, 337
  multiple, 339–341
  reversing, 271
control polygon of the Bézier curve, 296, 300
convex hull, 282, 423, 432
convex hull property, 282
convex polygons, 143, 423
convex polyhedrons, 143
convolution, 56
convolution kernel, 689
cool colors, 568
Coons, Steven Anson, 267, 443
Coons surfaces
  bicubic, 446–449
  degree-5, 449
  linear, 443–455
  tangent matching, 449–452
  triangular, 452–453
coordinate systems, 700–702

Cartesian, 700
  three-dimensional, 700
curvilinear, 700
cylindrical, 700
oblique, 700
polar, 700
rectangular, 700
spherical, 702
coordinate transformation, 88
CORDIC rotations, 77–81
Cox, M., 363, 378
Cox-DeBoor formula, 363, 376–378, 381
Cramer's Rule, 806
cross-product, 703
CRT, 2–3
  and interlaced scan, 8
  and RGB, 561
  color, 3
  refreshing, 9
CRT controller, 3, 13
cube (multidimensional), 145
cubic blending, 174, 283, 601
cubic Bézier curve, 23, 272
cubic polynomials, 193–195
cubic splines, 174, 207, 225–246, 312, 711
  anticyclic, 234–235
  clamped, 228, 230
  closed, 232, 235–237
  cyclic, 232–234
  fair, 173, 241–246
  four points, 195
  nonuniform, 237–240
  normalized, 237
  periodic, 232, 235
  relaxed, 230–232
  uniform, 237
curvature (definition of), 406–407
curved perspective, 159–163
curves, viii, 173–413
  approximating, 181
  area of, 706
  B-spline, 328–389
    subdividing, 381–383
  Bessel's algorithm, 256–258
  bias in, 259–261, 389, 390
  binormal vector, 404
  by subdivision, 321–328, 333–334, 348–351
  Bézier, 266–321, 336
    and Catmull-Rom, 304

as special case of B-spline, 365–368
converting, 301–302
cubic, 23, 272
degree 4, 272
degree elevation, 291–293
disadvantages of, 328
fast calculation of, 275–279, 700
in PostScript, 667
interpolating, 304–314
length, 296–297
linear, 273
parameter substitution, 272, 784
quadratic, 272
related to B-spline, 321, 346–348
reparametrizing, 293–296
representations of, 273–275
speed, 297–300
subdividing, 290–291
tension, 302–304
Catmull-Rom, 250–258
center of curvature, 402
circular Bézier, 189, 314–317
continuity in, 258–261
converting, 301–302
curvature of, 406–407
degenerate, 221
explicit representation of, 181
extrinsic properties of, 402, 803
fair, 241
Hermite, 207–225, 302, 345
  degenerate, 220–225
  indefinite tangent vectors, 772
  nonuniform, 224–225
  quadratic, 223–224, 457
  special, 220–225
  straight, 221–223
  symmetry of, 396–397
implicit representation of, 181
independence of the axes, 188
inflection points, 409–410
interpolating, 181
intrinsic properties of, 402, 803
introduction to, 181–198
length of, 580, 706
nonpolynomial, 189
normal plane, 402
object space, 401
osculating circle, 406
osculating plane, 404–405
parametric

fitting to epoints, 262–266
nonpolynomials, 189
parametric (three-dimensional), 190
parametric nonpolynomial, 189
parametric representation of, 181
parametric space, 401
periodic, 235
pleasing to the eye, 241
principal normal vector, 402–404, 407
radius of curvature, 407
rectifying plane, 405–406
representations of, 181–198
reversing direction, 219
special, 221
speed of, 762
subdividing, 287–291
on a surface, 533–534
symmetry in, 394–397
tangent vector, 190, 195, 206, 211, 227, 231, 402
  and continuity, 191, 192
  and slope, 181
  antiequal, 234
  B-spline, 331, 332, 336, 337, 353, 796
  beta spline, 390
  Bézier, 271, 281
  cardinal spline, 248
  definition of, 182
  direction of, 769
  equal, 232
  extreme, 227–231, 238, 305, 313, 353
  Hermite, 207
  indeterminate, 211, 286, 775, 785, 786
  Kochanek-Bartels, 258, 260, 782
  magnitude, 208, 216
  phantom points, 336
  reversed, 396
  speed, 391
  tension, 249
tension in, 174, 248–251, 302–304, 312–314, 345–346, 389, 391
torsion, 408
velocity of, 195, 762
curvilinear coordinate system, 700
cusp, 188, 189, 217, 259, 266, 286, 775, 785, 786, 803
in a cubic B-spline, 340
cyclorama, 157
cylinder
as a quadric surface, 524, 525

Index

as a sweep surface, 536
  normal vector, 536
  texturing, 553–554
cylindrical (sphere projection), 167
cylindrical coordinate system, 700
Cyrillic (unicode), 625

da Vinci, Leonardo, 114
Darwin, Charles, 711
data compression
  dictionary-based methods, 610
  differential image, 654–656
  diminishing returns, 633
  FELICS, 634–636
  image
    color, 614
    differential, 654–656
    grayscale, 614
    progressive, 630–633
  images, viii, 609–660
  IFS, 685
  MLP, 646–654
  model, 632
  progressive FELICS, 643–646
  progressive image, 630–633
  run-length encoding, 612–615
  statistical methods, 610
  text, 610
  two-pass, 651
  wavelets, 656–660
data glove, xi
data points, 181, 262
Davis, Bette, xiii
Davis, Philip J., 206
dc component (of a function), 695
DDA methods, 28–49
  and bitmap scaling, 28
  best fit, 38–39
  bitmap scaling, 18
  circles, 42–47
  double-step, 34–38
  octantal, 33–34
  quadrantal, 31–33, 50, 739
  simple, 29–30
  symmetrical, 30–31, 50
DeBoor, Carl, 363, 378
decomposing transformations, 84–85
degenerate curves, 221
degenerate Hermite segments, 220–221
degree elevation, 291–293

degreees of freedom in a PC, 262
Delaunay triangulation, 427–434
DeMoivre's theorem, 830
Derek, John, 622
determinant, 221, 398, 408, 803
  and area of ellipse, 47
  and cross-product, 703
  and plane equation, 420
  definition of, 709
  history of, 711
  in plane equation, 405
  pure reflection, 64
  pure rotation, 64
  row interchange, 704
  use in scaling, 47, 63
Deutsch, David, 1, 562
Deutsch, L. Peter, 664
Devanagari (unicode), 625
diacritics (unicode), 625
diagonal matrix, 709, 710
diagonally dominant matrix, 228, 353, 710–711
Dickens, Charles, 102, 556
differential image compression, 654–656
diffuse reflection, 545
diffusion dither, 671, 675–678
digital camera, 568
Digital Effects Corp., x
Digital Equipment Corporation (DEC), ix
digitizer (three-dimensional), 416, 427, 530
dimetric projections, 108
direction cosines, 99, 702
direction of a vector, 702
Dirichlet tessellations, see Voronoi diagrams
distance of line and point, 185
dithering, 671–680
  ARIES, 671
  color printing, 677
  constrained average, 674
  diffusion dither, 675–678
  dot diffusion, 678–680
  minimized average error, 677
  ordered dither, 671–674
division of matrices, 710
Doo, Donald, 530
Doo Sabin surfaces, 530–531
dot diffusion, 671, 678–680

dot product, 703
double helix, 235, 441–442
double scaling, 74–75
double-step DDA, 34–38
drawing and illustrating, 23–26
dual geometrical constructs, 434
Dürer, Albrecht, 114
dvi file (output of TEX), 667
dvips, 667
dynamic memory allocation, 13

Earth
    as oblate spheroid, 714
    equatorial radius of, 714
    great circle distances, 712
    its orbit around the Sun, 47
    mapping of, 167, 169, 553, 712
eccentricity of the ellipse, 47
egg (shape of), 47
Einstein, Albert, 693
electromagnetic radiation (and refraction), 539
electromagnetic spectrum, 557, 558
Elias, Peter, 612, 638
Eliot, George, 121
ellipse, 46–49, 63, 212
    and surfaces of revolution, 521
    arc length, 714
    area of, 47
    canonical, 46
    definition of, 46
    eccentricity of, 47
ellipsoid
    implicit, 415
    normal of, 535
    parametric, 517
elliptic arc, 319
elliptic integrals, 580, 714
Elton, John, 686
embossing an image, 689
epoints, *see* experimental points
Eppstein, Maureen, 739
equatorial radius of Earth, 714
equiareal transformation, 90
Escher, Maurits Cornelis, 129, 154, 155, 159, 161, 162, 422
Euclid's algorithm (and best fit DDA), 38
Euler, Leonhard, 583, 716
Euler's formula, 716
Euripides, 125

Evans, David, ix, x
experimental points, 262
explicit polygons, 419
explicit representation of lines, 71, 184
explicit surfaces, 415
    normal of, 535
    written as implicit, 817
extrinsic properties, 402, 803
extrusion (for a multidimensional cube), 145
eye (resolution of), 568
eye and spatial integration, 3, 8, 53, 561, 669, 677

fair curve, 241
false perspective projection, 153–154
fax compression, 615–622
    group 3, 616–619
FELICS, 634–636
    progressive, 643–648
Fermat, Pierre, 540, 543
Fermat point, 425
    trilinears of, 806
Fermat's principle, 540
Feynman, Richard Phillips, 542
field (definition of), 720–721
fill (in PostScript), 664
filled polygons, 53
fillet (parabolic), 795
filling an area, 668
film (resolution of), 9–13
filtering, 55–56
first fit, 13
fisheye projection, 153–155
fitting a spline to epoints, 262–266
Fitzgerald, F. Scott, 313
flight simulation (example), 143
Floyd-Steinberg filter, 675
fold in a surface, 453
font (definition of), 312
font design, 241
Fortran 77 (and GKS), 662
forward differences, 199, 276, 698–700
    and circles, 699
four-color process, 564
Fourier, Jean Baptiste Joseph, 696
Fourier transform, 656, 693–697
    in computer graphics, 697
fractals, x, 680–686
    and L-systems, 94

Index

IFS, 685–686
line, 681–682
mountain, 682
terrain, 682
trees, 684
frequencies of symbols, 611, 612
frequency, 694
  domain, 693
  of a function, 694
functions
  one-to-one, 59
  onto, 59
  periodic, 693
fundamental frequency (of a function), 694

Galilei, Galileo, 183
Gaussian distribution, 649, 681–684
Gaussian random numbers, 684
geocentric latitude, 714
geodetic latitude, 714
geometric transformations, 59–102
  definition of, 60
  in n dimensions, 147–149
geometrical constructs
  complementary, 434
  dual, 434
Georgian (unicode), 625
Gerhart, Susan, 351
GhostScript (PostScript preview), 664, 666
GIF89 graphics file format, 15
GKS, xi, 661–663
glide reflection, 82–83
global control of the Bézier curve, 282
Godard, Jean-Luc, xviii, 575
Gogh, Vincent Van, 567
Golden Ratio, 66, 80, 85, 642, 733
  and 35 mm film, 733
Golomb code, 635, 640, 642, 646
Gordon surfaces, 201, 498
Gordon, William J., 267
Gould, Stephen Jay, 851
Gouraud, Henri, x, 548
Gouraud shading, 180, 530, 548–549
Graham, Martha, 578
Graphical Kernel System, see GKS
graphics standards, xi, 661–667
  GKS, 661–663
  IGES, 384, 663

PHIGS, 663
PostScript, 663–667
graphics windows, 13–16
grayscale image compression (extended to color images), 614
great arc (on sphere), 596
great circle, 183
great circle distances, 712–715
Greek (unicode), 625
Greek alphabet, 715
Gregory surfaces, 495–497
Grieggs, John, xii
groups
  definition of, 719–720
  of transformations, 60
GUI, xi, 609
guiltlist (literacy list), 723
Gujarati (unicode), 625
Gurmukhi (unicode), 625

Hadamard, Jacques, 717
half-angles (for perspective), 134, 577, 590
halftones, 564, 619, 669–671
halfturns, 81–82
Hamilton, William Rowan, 717, 719
Hartley, L. P., 192
HDTV
  aspect ratio of, 3–6
  high resolution, 577
  resolution of, 3–6
  standards used in, 3–6
Hebrew (unicode), 625
Hein, Piet, 48
helix curve, 235, 441, 762
helix curve (double), 235, 441–442
Henderson, J. Robert, xiii
Hermite blending, 207–225, 602–603
  and spline fit, 266
  fair, 241, 242
Hermite, Charles, 207
Hermite interpolation, 173, 207–225, 312, 602
  and B-spline curves, 302, 345
  and Bézier curves, 302
  and circle, 212–214
  and parabola, 212
  and spline fit, 266
  degenerate, 220–225
  fair, 241, 242

indefinite tangent vectors, 772
is symmetric, 396–397
nonuniform, 224–225
quadratic, 223–224, 457
special, 220–225
straight, 221–223
Hermite surfaces
and B-spline, 503–504
Hertz (Hz), 694
Herzog, Werner, 2
hidden-surface elimination, 537, 538
Higgins, Colin, 413, 479
High Sierra (CD-ROM standard), xi
history of computer animation, 577
history of computer graphics, ix–xi, 577
HLS color model, 558–559
Hobby, John Douglas, 312
Hockney, David, 165
Holy Trinity (fresco by Masaccio), 114
homogeneous coordinates, 65–68
and rational B-spline, 384
Horner's rule, 199
Hough transform, 410–413
HSV color model, 559
hue, 559
definition of, 558
primary, 561
secondary, 561
Huffman algorithm, 612
Huffman coding
and wavelets, 656, 660
in image compression, 655
Hulcher, Charles A. (camera inventor), 164
Hulcherama panoramic camera, 164
hyperbola, 212, 252, 807
hyperbolic paraboloid (as a bilinear surface), 807

IBM 2250 graphics terminal, ix
identity matrix, 709, 710
IFS, see iterated function systems
Illustrator (Adobe), 305, 667
image compression, viii, 609–660
differential, 654–656
FELICS, 634–636
IFS, 685
MLP, 646–654
prefix compression, 628–630
principle of, 610

progressive, 630–633
median, 633
progressive FELICS, 643–646
reasons for, 609
wavelets, 656–660
image processing, 688–691
implicit representation of lines, 71, 184
implicit surfaces, 415, 438, 524, 807
canonical sphere, 524
cone, 526
cylinder, 524
ellipsoid, 415
normal of, 535
plane, 535
sphere, 415
improper rotations, 84
in-betweening (in computer animation), 576, 579, 606
incenter, 425
incircle, 425
index of refraction, 540
Inferno (unicode), 625
inflection points, 193, 206, 262, 409–410, 770
in a PC, 410
Information International Inc., x
Infowave, Inc., 664
infrared, 557
inkjet color printing, 677
interlacing scan lines, 5
interpolating curve, 181
interpolating polynomials, 198, 461–462
intrinsic properties, 402, 803
inverse of a matrix, 710
ISO, xi
isometric projections, 108, 109, 167
isometry, 89
isotropic principle, 393, 394, 456
iterated function systems, 685–686

James, William, 831
Japanese (unicode), 625
Java OS (unicode), 625
Jonson, Ben, 564
Joyce, James, 105, 110, 169, 709
JPEG, 609
Juster, Norton, 460

Kajiya, James, 548
Kannada (unicode), 625

Index

Kelly, Gene, 412
Kepler, Johannes, 85
knots (in curve design), 200
Kochanek, Doris H., 258
Kochanek-Bartels splines, 174, 258–261, 390
Korean hangul (unicode), 625

L systems, 94–96
Laertius, Diogenes, 235
Lagrange polynomial, 173, 198–204, 498, 767
  cubic, 203–204
  quadratic, 201–203, 257, 258, 457
Lagrange, Joseph Louis, vii, 199
Lambert's law, 545
Lamé, Gabriel, 48
Lansing, Robert, 610
Lao (unicode), 625
Laplace distribution, 647, 649–654
large numbers (law of), 683
Latin (unicode), 625
latitude
  geocentric, 714
  geodetic, 714
lattice polygon, 707
law of large numbers, 683
Leahy, Joe, 304
least squares (in curve design), 262
least time principle
  and reflection, 543
  and refraction, 540
left-handed coordinate system, 91
length of curve, 580, 706
light, 557–559
  speed of, 543
  visible, 558, 568
light absorption, 539
light reflection, 173, 543–549
  ambient, 547
  diffuse, 545
  specular, 544, 546
lightness, 559
  definition of, 559
Limbo (unicode), 625
Lindenmayer, Aristid, 94
line
  as a degenerate conic section, 803
  distance from point, 185

  explicit representation of, 27, 71, 184, 410
  fractalized, 681–682
  implicit representation of, 71, 184
  intercept representation of, 185
  intersection of segments, 187
  intersection with plane, 424
  normal parametric representation of, 185, 412
  parametric representation of, 185, 768, 772, 784
  standard form of, 184
  vector equation, 185, 704
linear Bézier curve, 273, 316
linear blending, 601
linear Coons surface, 443
Lischinski, Daniel, 488
Littlewood, J. E., 198
lofted surfaces, 439–443
  cone, 442
  pyramid, 442
  twisted ribbon, 442
logarithmic spiral, 700
LOGO (programming language), 94
Long, Shelley, 615
loop in a curve, 193, 217, 262, 266
LP, *see* Lagrange polynomial
Lucasfilm, x, xi
luminance (definition of), 558

Macintosh computer, xi
magnitude of a vector, 702
Malayalam (unicode), 625
Mandelbrot, Benoit, x
Mann, Thomas, 171
Mansfield, L, 378
mapping the Earth, 167, 169, 553, 712
Masaccio (Renaissance painter), 114
matrices
  definition and operations, 709–711
  diagonal, 710
  diagonally dominant, 228, 353, 710–711
  division (undefined), 710
  history of, 711
  identity, 710
  inverse, 87, 96, 218, 600, 710, 749
  nonsingular, 228, 710
  orthogonal, 99
  orthonormal, 64
  singular, 710
  square, 709

tensor product, 709
transpose, 709
tridiagonal, 228, 710
Maugham, William Somerset, 715
Maxwell, James Clerk, 47
McLuhan, Marshall, 174
McWade, Robert, 760
median, definition of, 633
median-cut color quantization, 551
mediation operator, viii, 266, 283, 284, 290, 291, 594, 693
  spherical, 594
Mercator projection, 167–169
Mercator, Gerhardus, 167
Meriwether, Lee, 149
Mesdag Panorama, 157
Mesdag, Hendrik Willem, 157
metamorphosis, *see* morphing
meter (definition of), 543
microscopic projection, 153, 166–167
midpoint subdivision, 28
MIMD (Multi Instruction, Multi Data) computers, 39
minimized average error, 677
MLP, 646–654
Möbius strip, 528
Moivre, Abraham de (1667–1754), 830
Moore, Henry, 153
Moranis, Rick, 616
Morley, Christopher, 462
morphing, xi, 606
mountain (as a fractal), 682
move-to-front method (and wavelets), 656, 660
MS Windows, xi, 88
multimedia applications, xi

n-point perspective, 128–130
NASA, 827
negate and exchange rule, 65, 77, 684, 741
New York Institute of Technology, x
Newton, Isaac, 188, 199, 205, 269, 279
Newton polynomial, 173, 201, 205–206
Newton-Raphson method, 264, 583
NeXT OS (unicode), 625
Nietzsche, Friedrich, 421
Noblex panoramic camera, 164
nonlinear projections, viii, 153–171
nonsingular matrix, 228, 710

nonuniform parametric representation, 190–191
normal
  calculation of, 535–536
  of cone, 536
  of cylinder, 536
  of pyramid, 536
normal plane, 402
notation used in this book, xii–xiii, 188
NTSC, 3, 577
NURBS, *see* rational B-splines
Nyquist rate, 697
Nyquist, Harry, 697

object space, 401
oblate spheroid, 714
oblique coordinate system, 700
oblique projections, 109–112
octahedron (multidimensional), 757
octantal DDA, 33–34
octrees, 628
one-to-one (function), 60
onto (function), 59
ordered dither, 671–674
Oriya (unicode), 625
orphans (neglected topics in CG), viii
orthocenter, 425
  trilinears of, 806
orthographic projections, 104–105
orthonormal matrix, 64
osculating circle, 406
osculating plane, 404–405, 535
Oslo algorithm, 381–383
Ovals, 47
Overhauser splines, 174

Paeth, Alan W., 20
painter's pigments, 563
painting operators (in PostScript), 664
Palermo, David, 163
palette, 8, 549, 675
  optimization, 549–551
panda (black/white image), 672
panoramic cameras, 163–165
panoramic projection, 153, 155–165
PANTONE matching system, 564
parabola, 212, 252, 807
  and Bézier curve, 272
  and Hermite curve, 212
  and triangle, 212
  as a conic section, 803

Index

as a fillet, 795
parabolic blending, 251–256
parallax, 139
　simulation of, 576
parallel computers, 39, 55
parallel projections, 103–112, 155
　in n dimensions, 149
parametric blending, 180, 456, 600–605
parametric cubic, 193–195, 206–212
　algebraic representation of, 206
　circle approximating, 762
　distance to a point, 207, 770
　four points, 195
　geometric representation of, 208
　Hermite, 207–225
　　symmetry of, 396–397
　inflection points, 410
　PC, 193
parametric curves
　and polynomials, 193–195
　continuity of, 191–192
　cubic, 193–195
　cusp in, 189, 217
　degreees of freedom, 262
　fitting to epoints, 262–266
　intrinsic properties of, 402
　least squares, 262
　loops in, 193, 217, 262
　non-polynomials, 444
　nonpolynomials, 189
　nonuniform, 190–191
　properties of, 188–190
　rational, 400
　reparametrization, 219, 220
　substitution of parameter, 191, 192, 216, 219, 272, 339, 392, 395, 397, 765, 772, 784, 801
　three-dimensional, 190
　uniform, 190–191
　velocity of, 182
parametric representation of lines, 185
parametric space of a curve, 401
Parent, Richard, xii
particle paradigm (in parametric curves), 182
particle systems, xi
Pascal, Blaise, 426
Pascal triangle, 267–269
path operators (in PostScript), 664
Pauli, Wolfgang, 718

PC, *see* parametric cubic
pel, 616
Perec, Georges, 65
period (of a function), 694
periodic curves, 235
periodic functions, 693
perpendicular bisector, 425, 431, 434, 681, 684, 806, 829
perspective projection, viii, 112–141, 155
　and NURBS, 384
　curved, 159–163
　depth projection, 137
　in n dimensions, 151–153
　n points, 128–130
　spherical panoramic, 161–163
　two points, 129
　vanishing point, 114, 128
Petri dish, 434
Petty, Lori, 573
phase (of a function), 694
PHIGS, xi, 661, 663
Phong shading, 530, 548–549
Piaget, Jean, 758
Picasso, Pablo, 567
Pick's theorem, 707
pinhole camera, 165
pitch (in rotation), 94, 95
Pitteway-Watkinson method, 53
pixels
　and color lookup table, 8
　in color, 7, 9, 15, 734
　definition of, 7
　origin of word, x
　raster scan, 8
　transparent color, 15
planarity test for polygons, 419–420
plane
　equation of, 420–422, 466–468
　intersection with line, 424
Plan 9 (unicode), 625
points
　collinear (in B-spline), 337–339
　control, 181, 266, 271, 328, 415
　　and convex hull, 282
　　and curve speed, 298
　　and degree elevation, 291
　　and scaffolding, 286
　　auxiliary, 305
　　collinear, 337
　　multiple, 339–341

reversing, 271
data, 181
distance from line, 185
experimental, 262
how to input, 416–417
inflection, 193, 206, 262, 409–410, 770
operations on, 174–180
Poisson, Siméon, 721
polar coordinate system, 700
polar radius of Earth, 714
polygonal surfaces, 419–434
    Catmull-Clark, 531–533
    Doo Sabin, 530–531
    fractals, 682
    subdivision, 530–533
polygons
    area of, 706–707
    as space dividers, 422
    convex, 143, 423
    filled, 53
    lattice, 707
    planarity, 419–420
    plane equation, 420–422
    triangle, 424–425
    turning on, 422–423
polyhedra (volume of), 707–708
polyhedrons
    convex, 143
polyline, 195, 299, 580
    in B-spline, 340
    in GKS, 662
polynomials
    and parametric curves, 193–195
    Bernstein, 174, 270
    circular Bernstein, 315
    cubic, 193–195
    definition of, 199
    forward differences, 199, 698–700
    Horner's rule, 199
    interpolating, 198, 461–462
    not a circle, 317, 319, 384, 790–791
    parametric cubic, 206
    trigonometric, 316
Popper, Karl Raimund, 415
Portal panoramic lens system, 165
PostScript, 663–667
Pound, Ezra, 54
prefix codes, 612, 638–642
prefix compression, 628–630
prefix property (definition of), 612

primary hue, 561
principal normal vector, 402–404, 407
principle of image compression, 610
principle of least time
    and reflection, 543
    and refraction, 540
printing in color, 677
printing pigments, 563
Prior, Matthew, 146
probability model, 632
process color, 563
Prochnow, Jürgen, 570
progressive FELICS, 643–648
progressive image compression, 630–633
    FELICS, 643–646
    lossy option, 630, 647
    median, 633
    MLP, 646–654
projections
    aerial perspective, 114
    axonometric, 105–109
    circle inversion, 153, 170–171
    conformal, 170
    dimetric, 108
    false perspective, 153–154
    fisheye, 153–155
    general rule, 104
    isometric, 108, 109, 167
    Mercator, 167–169
    microscopic, 153, 166–167
    nonlinear, viii, 153–171
    oblique, 109–112
    orthographic, 104–105
    panoramic, 153, 155–165
    parallel, 103–112, 155
        in n dimensions, 149
    perspective, viii, 112–141, 155
        and NURBS, 384
        curved, 159–163
        in n dimensions, 151–153
        spherical panoramic, 161–163
    slicing, 149–151
    sphere, 153, 167–169
        cylindrical, 167
        stereographic, 167, 169
    stereographic, 169
    telescopic, 153, 165–166
    trimetric, 107
Prokofieff, Sergei, 206
pure color, 559

Index

pyramid
  as a lofted surface, 442
  normal vector of, 536
pyramid coding (in progressive compression), 630
pyramid (multidimensional), 757

quadrantal DDA, 31–33, 50, 739
quadratic
  blending, 174, 283, 601
  interpolation, 607
quadratic Bézier curve, 272
quadratic splines, 174, 247–248
quadric surfaces, 524–526
quadtrees, 624–627
  prefix compression, 628–630
quaternions, viii, 99–102, 180, 716–719
  and spin matrices, 718
Quirk, Randolph, 184

radio waves, 557
radiosity, xi, 538
radius of curvature, 407
RAND Corp., 684
random scan, 2
raster scan, x, xi, 2, 7–9
rational B-splines (NURBS), 383–389
rational Bézier curves, 317, 321
rational parametric curves, 400
ray tracing, 538, 551–552
Raymond, Eric, 7, 51, 548
reconstructing transformations, 86–87
Recorde, Robert, 112
rectangular Bézier patches, 471–481
  converting to triangular, 488–492
rectangular coordinate system, 700
rectifying plane, 405–406
Reeves, Keanu, 545
refinement methods, *see* subdivision methods
reflection, 63, 90
  direction of, 705
  glide, 82–83
  two consecutive ones, 72
reflection of light, 173, 543–549
refraction, 539–543
refraction coefficient, 540
refreshing CRT, 9
Remini, Leah, 443
reparametrization of Bézier curves, 293

reparametrization of curves, 219, 220
reparametrizing Bézier curves, 293–296
reparametrizing Bézier patches, 492–495, 508
resolution
  of film, 9–13
  of HDTV, 3–6
  of television, 3
  of the eye, 568
reverse transformations, 102
reversing a curve, 219
RGB
  color model, 558, 561
  cube, 551, 561
  reasons for using, 561, 568, 573
Riesenfeld, Richard, 328
right-hand rule, 829
right-handed coordinate system, 91, 95
rigid transformations, 393, 596
RLE, *see* run-length encoding
rods (in the retina), 568
Rokicki, Tomas, 667
roll, 94, 95
rotation, 63, 90, 394
  180°, 81–82
  90°, 65
  as three shears, 20
  bitmap, 20–23
  CORDIC, 77–81
  in n dimensions, 147–149
  matrix, 64
  pitch, 94
  roll, 94
  yaw, 94
Rourke, Mickey, 535
Rozenberg, Grzegorz, 96
rubber banding, 734
rule of projections, 104, 113
ruled surfaces, 439
run-length encoding, 612–615
  and wavelets, 656, 660
Rushdie, Salman, 118
Russell, Bertrand, 169
Russell, Steven, ix

Sabin, Malcolm, 530
Samet, Hanan, 628
sampling, 697
Santa Maria del Fiore (Cathedral), 114
saturation, 559, 573

definition of, 558
full, 820
Sayood, Khalid, 654
scaffolding method of de Casteljau, 285, 286, 291, 321–324, 378, 787
scaling, 62, 90, 155, 394
  a bitmap, 16–18, 28
  double, 74–75
scan converting, 27–47
  circles, 42–47
  lines, 27–42
  on an MIMD computer, 39–42
  thick curves, 49
secondary hue, 561
shading, 537–556
  color, 548
  Gouraud, 180, 530, 548–549
  Phong, 530, 548–549
shadow mask, 3
sharpening an image, 689
shearing, 63, 90
  three successive, 20, 74
Shelley, Percy Bysshe, 155
Shoupe, Richard, x
Sierpinski triangle, 686, 687
SIGGRAPH, x
similarities, 81
similarity, 90
simple DDA, 29–30
sine wave (and Bézier curve), 319, 321
singular matrix, 710
SIPP (SImple Polygon Processor), 479
Sketchpad, ix
Skov, Anders T. K., 159
slicing (as a projection), 149–151
small numbers (easy to compress), 656
Smith, B. Sidney, 154
Snell, Willebrord van Roijen, 540
Snell's law, 540, 542
space division by a plane, 422
special curves, 221
special Hermite segments, 220–221
spectral density, 567–570
specular reflection, 544, 546
speed (of parametric curves), 183, 222, 762, 775
speed of light, 543
sphere
  as a Bézier surface, 481
  as a quadric surface, 524

half, 528, 816
implicit, 415
parametric, 517
texturing, 553–554
sphere projections, 153, 167–169
  Mercator, 167–169
spherical coordinate system, 702
spherical interpolation, 585, 593
  not good for interp. positions, 821
  proof of, 821
spherical panoramic perspective, 161–163
spiral
  Archimedean, 700
  logarithmic, 700
spiral curve, 441, 762
splines
  as a piecewise curve, 193
  B, 321, 328–389
    and circles, 341–345, 384
    cubic, 334–351
    matrix form, 376–381
    nonuniform, 368–376
    open, 364–368
    quadratic, 329–334
    rational, 321, 383–389
    tension, 329, 345–346
    uniform, 329–363
  Beta, 389–393
  cardinal, 248–251, 302, 345
  cubic, 174, 207, 225–246, 312, 711
    anticyclic, 234–235
    clamped, 228, 230
    closed, 232, 235–237
    cyclic, 232–234
    fair, 173, 241–246
    four points, 195
    Hermite, 207–225
    nonuniform, 237–240
    normalized, 237
    periodic, 232, 235
    relaxed, 230–232
    uniform, 237
  fitting to epoints, 262–266
  Kochanek-Bartels, 258–261, 390
  quadratic, 247–248
  tension in, 174, 248–251, 302–304, 312–314, 345–346, 389, 391
square matrix, 709
standards
  GKS, 661–663

Index

IGES, 384, 663
PHIGS, 663
PostScript, 663–667
standards (graphics), 661–667
start-step-stop codes, 637
stereo images, x, 1, 137–141
stereographic (sphere projection), 167, 169
stereographic projection, 169
stereoscope, 140
Stewart, Ian, 62, 660
straight Hermite segments, 221–223
straight lines, *see* line
strawberry (and bump mapping), 554
stroke (in PostScript), 664
Strunk, William, 92
StyleScript (PostScript interpreter), 664
subdividing curves, 287–291, 333–334, 348–351
subdividing surfaces, 508–516, 530–533
subdivision methods for curves, 321–328, 333–334, 348–351
subexponential code, 640, 646
substitution of parameter (in curves), 191, 192, 216, 219, 272, 339, 392, 395, 397, 765, 772, 784, 801
subtractive colors, 563–564
summary of transformations, 89–90
superellipse, 48
  superness of, 48
superellipsoid, 48
supersampling, 54–55
surfaces, viii, 415–536
  B-spline, 499–516
    and Bézier, 504–505
    and Hermite, 503–504
  bicubic, 459–468
  bilinear, 417, 434–438, 534
    triangular, 438
  biquadratic, 457–459
  boundary curves, 417, 434, 435, 437, 439, 441–443, 445, 449, 455, 463, 498, 804, 811
    degenerate, 438, 453, 488, 810
    three, 452, 484
  Bézier
    and B-spline, 504–505
    converting, 488–492
    rectangular, 471–481
    reparametrizing, 492–495, 508

  triangular, 452, 483–487
  Cartesian product, 456
  Catmull-Clark, 531–533
  Catmull-Rom, 468–471
  Coons
    bicubic, 446–449
    degree-5, 449
    linear, 443–455
    tangent matching, 449–452
    triangular, 452–453
  curves on, 533–534
  Doo Sabin, 530–531
  explicit representation of, 415
  fold in, 453
  Gordon, 201, 498
  Gregory, 495–497
  Hermite, 463–468, 479–481
    and B-spline, 503–504
  implicit representation of, 415, 438, 524, 807
  lofted, 439–443
  normal, 535–536
  of revolution, 516–522
    torus, 518, 814
  osculating plane, 535
  patch, 417
  polygonal, 419–434
    fractals, 682
    subdivision, 530–533
  quadric, 524–526
  ruled, 439
  subdividing, 508–516, 530–533
  sweep, 526–530
  wire frame, 417
  wrinkles in, 810
Sutherland, Ivan, ix, x
sweep surfaces, 526–530
  cone, 528
  cylinder, 536
symmetric matrix, 709
symmetrical DDA, 30–31, 50
symmetry in curves, 394–397
Systems Simulation Ltd., x

Tamil (unicode), 625
tangent vector, 190, 206, 211, 227, 231, 402
  and continuity, 191, 192
  and slope, 181
  antiequal, 234

B-spline, 331, 332, 336, 337, 353, 796
beta spline, 390
Bézier, 271, 281
cardinal spline, 248
definition of, 182
direction of, 769
equal, 232
extreme, 227–231, 238, 305, 313, 353
Hermite, 207
indeterminate, 211, 286, 775, 785, 786
Kochanek-Bartels, 258, 260, 782
magnitude, 208, 216
of a PC, 195
phantom points, 336
reversed, 396
speed, 391
tension, 249
Taylor expansion, 314
Taylor series, 276–278
teapot (Utah), x, 479
Tektronics (and the HLS model), 559
Tektronix storage tube, 50
telescope (and a parabola), 252
telescopic projection, 153, 165–166
television
  aspect ratio of, 3
  resolution of, 3
  scan line interlacing, 5
  standards used in, 3, 660
Telugu (unicode), 625
tension (in curves), 174, 248–251, 302–304, 312–314, 345–346, 389, 391
tensor product, *see* Cartesian product
terrain (as a fractal), 682
tetrahedron (multidimensional), 146
TeX (typesetting program), 667
texturing, 552–554
  a cone, 553
  a cylinder, 553–554
  a sphere, 553–554
Thai (unicode), 625
thin lens equation, 166
three-dimensional transformations, 91–102
Thurber, James, 580
Tibetan (unicode), 625
time domain, 693
Toller, Sidney, 434
Tolstoy, Lev, 57
torsion (definition of), 408

torus, 183
  as a surface of revolution, 518, 814
Toy Story (as a milestone), xi
transformations, 59–102
  180° rotation, 81–82
  affine, 61, 75, 102, 394, 685
  decomposing, 84–85
  definition of, 60
  equiareal, 90
  glide reflection, 82–83
  halfturns, 81–82
  improper rotations, 84
  in n dimensions, 147–149
  reconstructing, 86–87
  reverse, 102
  rigid, 393, 596
  similarities, 81
  summary of, 89–90
  three-dimensional, 91–102
  two-dimensional, 61–90
translation, 65, 393, 394
translucent surface, 537, 539
transparent color, 15
transparent surface, 537, 539
transpose of a matrix, 709
trees (as fractals), 684
triangle, 424–425
  barycentric coordinates, 177
  centers of, 179, 425–426
  centroid, 178, 425, 762
  circumcenter, 425
  circumscribing, 431
  Fermat point, 425
  God of, 425
  incenter, 425
  trilinear coordinates, 425
triangular Bézier patches, 483–487
  converting to rectangular, 488–492
triangular surface patches
  bilinear, 438
  Bézier, 483–487
  Coons, 452–453
tridiagonal matrix, 228, 710
trigonometric identities, 711–712
trigonometric polynomial, 316
trilinear coordinates, 425
trimetric projections, 107
trinomial theorem, 269, 484
trit, 641
Turgenyev, Ivan, 208

Index

turning (on a polygon), 422–423
Tuttass, T., 691
twist vectors, 458, 460, 463, 465
two-dimensional transformations, 61–90
two-pass compression, 651
Tze, Sun, 288

Uccello, Paolo (Renaissance painter), 114
ultraviolet, 557
unary code, 630, 635–642, 646, 825
   general, 637
unicode, 622, 624, 625
uniform parametric representation, 190–191
Utah teapot, x, 479

vanishing points, 112, 114, 117
   number of, 128–130
variable-size codes, 611–612, 634–642
   designing of, 612
variance (definition of), 652
vector equation of a line, 185
vector scan, ix, x, 2, 6–7
vectors, 702–708
   absolute value, 702
   addition, 703
   compared to points, viii, 174, 693
   cross-product, 703
      direction of, 704, 829
   direction cosines, 702
   direction of, 702
   dot product, 703
   magnitude of, 702
   orthogonal, 703
   projection, 705–706
   scaling of, 394
   spherical interpolation of, 586
   translation of, 175, 394
   unit, 702
   vector product, 703
velocity of curves, 195, 762

viewing volume, 141–143
viewport, 88
Vinci, Leonardo da, 114
virtual reality, x, xi, 1, 575
   and four dimensions, 144
vision (human), 568
visual acuity, 669
Voronoi diagrams, 434

Warhol, Andy, 19
warm colors, 568
watercoloring an image, 690
wavelets, 656–660
Webster (dictionary), 575
Wells, Herbert George, 536
Wheeler, John A., (Colophon)
White, Elwyn Brooks, 92
Whitehead, Alfred North, 147
Whitted, Turner, xi, 538
window (into the image), 88
window manager, 13
windows (graphics), 13–16
Windows NT (unicode), 625
Wing-Davey, Mark, 561
Wood, John, 161
Wordsworth, William, 453, 579
Wright, Harold Bell, 608
wrinkles in a surface, 810

X-rays, 539, 557, 615
Xerox Palo Alto Research Center (PARC), x

yaw, 94, 95
Yost, David, 606
Young, Daniel, 563
Young, Roland, 675

Zajac, E. E., ix
zero-probability problem, 632

In the bad old days, the index was a list of prohibited books; may we now, in a more enlightened age, ban books without indexes?
— Stephen Jay Gould, *An Urchin in the Storm* (1987)

Colophon

This book was started in the early 1980s, as class notes. However, most of it was written during 1997–1998. It was designed by the author and was typeset by him with the TEX typesetting system developed by D. Knuth. The text and tables were done with Textures, a commercial TEX implementation for the Macintosh. The diagrams were done with Adobe Illustrator, also on the Macintosh. The following remarks illustrate the amount of work that went into it:

- The book contains about 362,500 words, consisting of about 1,986,700 characters.

- The text is typeset mainly in font cmr10, but 30 other fonts were used.

- The raw index file contained 3176 items.

- There are about 1060 cross-references in the book.

- The 335 figures occupy a total of 22,617,000 bytes on the author's hard disk.

- Just correcting the proofs took three weeks of hard labor. The following quote (by Kingsley Amis about the proofs of *Lucky Jim*) reflects the author's feelings about this work:

"Correcting the proofs is an efficient device for making you hate what you have written."

As mentioned in the Preface, the author welcomes any comments, suggestions, and corrections. They should be emailed to `david.salomon@csun.edu`. An errata list, as well as other relevant information, will be kept on the author's web page `http://www.ecs.csun.edu/~dxs`.

The best way to learn, after teaching, is writing
John A. Wheeler, 1998, *Geons, Black Holes, and Quantum Foam.*

If you have any concerns about our products,
you can contact us on
ProductSafety@springernature.com

In case Publisher is established outside the EU,
the EU authorized representative is:
Springer Nature Customer Service Center GmbH
Europaplatz 3, 69115 Heidelberg, Germany

Printed by Libri Plureos GmbH
in Hamburg, Germany